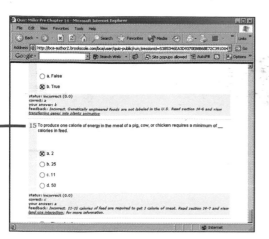

Environmental Science ⊕ Now™

A personalized learning plan focusing on the help YOU need!

Environmental ScienceNow is a powerful diagnostic tool that can help you assess how well you understand each chapter topic, starting with a pretest. The program responds to all incorrect answers, providing you with a personalized learning plan. It directs you to the precise text sections and interactive animations where you'll find the right answer. Then, to assess your progress, a quiz on the entire chapter assesses your mastery of chapter concepts.

Narrated animations tied to the text . . . nearly 100 in all

From colorful animations that show complex processes and relationships unfolding on screen to interactions that give you many opportunities to explore figures from the text by clicking on various parts of the figures, this CD-ROM offers the kind of tutorial support never before possible for this course. For a complete list of animation/interaction topics, please turn the page ▶

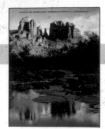

In this 14th Edition

More than 100 *Case Studies* provide a window on the world of environmental issues and problems!

Ch #	Case Study	Ch #	Case Study
1	Living in an Exponential Age	17	Should Oil and Gas Development Be Allowed in the Arctic National Wildlife Refuge? To Drill or Not to Drill
1	The Tragedy of the Commons—Degrading Free Renewable Resources	17	The Yucca Mountain Storage Site for High-Level Radioactive Wastes—Controversy over Desert Burial
2	Near Extinction of the American Bison		
2	Aldo Leopold and His Land Ethic	18	The Coming Energy-Efficiency and Renewable-Energy Revolution
3	An Environmental Lesson from Easter Island	19	The Big Killer
4	Have You Thanked the Insects Today?	19	Are Hormonally Active Agents a Human Health Threat? Serious Concern but Inconclusive Evidence
4	What Species Rule the World? Small Matters!		
5	Earth: The Just-Right, Resilient Planet	19	Are We Losing Ground in Our Struggle against Infectious Bacteria?
6	Blowing in the Wind: A Study of Connections	19	The Global Tuberculosis Epidemic—A Growing Threat
7	Why Should We Care about Coral Reefs?	19	How Serious Is the Global Threat from HIV and AIDS? A Rapidly Growing Health Threat
8	Flying Foxes: Keystone Species in Tropical Forests		
8	Why Are Amphibians Vanishing? Warnings from Frogs	19	Malaria: A Deadly Parasitic Disease That Is Making a Comeback
8	Why Are Sharks Important Species? Culling the Oceans and Helping Improve Human Health	20	When Is a Lichen Like a Canary?
		20	South Asia's Massive Brown Cloud—Choking in China and India
9	Sea Otters: Are They Back from the Brink of Extinction?	20	Are You Being Exposed to Radioactive Radon Gas? Test the Air in Your Home
10	Slowing Population Growth in Thailand: A Success Story		
10	Should the United States Encourage or Discourage Immigration?	20	Should We Use the Marketplace to Reduce Pollution? Emissions Trading
10	Family Planning in Iran: A Success Story	20	What Should We Do about Air Pollution from Older Coal-Burning Facilities? A Burning Controversy
10	Case Studies: India and China		
11	Reintroducing Wolves to Yellowstone	21	A.D. 2060: Green Times on Planet Earth
11	How Should U.S. National Forests Be Managed? An Ongoing Controversy	21	Warning Signals from the Earth's Ice and Snow: Meltdowns Are Under Way
11	National Parks in the United States	22	Learning Nature's Ways to Purify Sewage
11	What Has Costa Rica Done to Protect Some of Its Land from Degradation? A Global Conservation Leader	22	India's Ganges River: Religion, Poverty, and Health
		22	Lake Washington—A Success Story
11	How Much Wilderness Has Been Protected in the United States? Fighting for Crumbs and Losing	22	Pollution in the Great Lakes—Hopeful Progress
		22	The Chesapeake Bay: An Estuary in Trouble
11	Ecological Restoration of a Tropical Dry Forest in Costa Rica	23	Along Came a Spider—Biological Pest Control
12	The Passenger Pigeon: Gone Forever	23	How Successful Have Pesticides Been in Reducing Crop Losses in the United States? Barely Holding the Line
12	Why Should We Care about Bats?		
12	How Do Human Activities Affect Bird Species? A Disturbing Message from the Birds	23	Revisiting DDT: From Riches to Rags
		24	Love Canal: There Is No "Away"
12	Deliberate Introduction of the Kudzu Vine: Unintended Consequences	24	How Much Solid Waste Does the United States Produce? Affluenza in Action
12	The Termite from Hell	24	How Much Wastepaper Is Being Recycled? Encouraging News
12	Exploding Deer Populations in the United States: Should We Put Bambi on Birth Control?	24	Is It Feasible to Recycle Plastics? Some Problems
		24	A Black Day in Bhopal, India
12	The Rising Demand for Bushmeat in Africa: Hungry People Trying to Survive	25	The Ecocity Concept in Curitiba, Brazil
		25	What Are the Advantages of Urbanization? Concentrating People Helps
12	What Has the Endangered Species Act Accomplished?		
12	Using Reconciliation Ecology to Protect Bluebirds	25	What Are the Disadvantages of Urbanization? Concentrating People Has Some Harmful Effects
13	A Biological Roller Coaster Ride in Lake Victoria		
13	Should Commercial Whaling Be Resumed? An Ongoing Controversy	25	How Do the Urban Poor in Developing Countries Live—Living on the Edge with Ingenuity and Hope
13	Near Extinction of the Blue Whale: Big Species Are Easy to Kill		
13	Restoring the Florida Everglades: Will It Work?	25	Mass Transit in the United States—Destroying a Great System
13	Can the Great Lakes Survive Repeated Invasions by Alien Species? They Keep Coming	25	How Is New Urbanism Creating More Livable Spaces? Returning to Traditional Neighborhood Development
		25	Chattanooga, Tennessee—From Brown to Green
13	Managing the Columbia River Basin for People and Salmon: A Difficult Balancing Act	26	How Important Are Natural Resources
		26	What Is Cost-Benefit Analysis, and How Can It Be Improved? Weighing Costs and Benefits to Make Choices
14	Growing Perennial Crops on the Kansas Prairie		
14	Industrial Food Production in the United States: A Success Story	26	What Is the Role of the World Bank in Economic Development? A Controversial Organization
14	Soil Erosion in the United States Today: Some Hopeful News		
14	The Dust Bowl: An Environmental Lesson from Nature	26	How Are Germany and the Netherlands Working to Achieve More Environmentally Sustainable Economies? Leading the Way
14	Can China's Population Be Fed? A Precarious Situation		
15	Water Conflicts in the Middle East	27	Rescuing a River
15	Freshwater Resources in the United States—Unequal Distribution	27	What Is Environmental Leadership? An Option for Each of Us
15	The Colorado River Basin—An Overtapped Resource	27	How Is Environmental Policy Made in the United States? A Complicated and Thorny Process
15	China's Three Gorges Dam—A Controversial Project		
15	The Aral Sea Disaster—A Glaring Example of Unintended Consequences	27	What Are the Major Types of Environmental Laws in the United States? A Variety of Approaches
15	The California Water Transfer Project—Bringing Water to a Desert		
15	Canada's James Bay Watershed Transfer Project—Rearranging Nature	27	Environmental Action by Students in the United States—Making a Difference
15	The Shrinking Ogallala Aquifer—Using Up Natural Capital		
15	Living on the Floodplains in Bangladesh—Danger for the Poor	27	Is Encouraging Global Free Trade Environmentally Helpful or Harmful? A Difficult Issue
16	The Mining Law of 1872		
17	Bitter Lessons from Chernobyl	28	Biosphere2: A Lesson in Humility
17	How Much Oil Does the United States Have? Rapidly Dwindling Supplies		

Living in the Environment
Principles, Connections, and Solutions
FOURTEENTH EDITION

G. TYLER MILLER, JR.

President, Earth Education and Research

THOMSON

BROOKS/COLE

Australia • Canada • Mexico • Singapore • Spain • United Kingdom • United States

THOMSON

BROOKS/COLE

Publisher: *Jack Carey*
Developmental Editor: *Scott Spoolman*
Production Project Manager: *Andy Marinkovich*
Technology Project Manager: *Keli Sato Amann*
Senior Assistant Editor: *Suzannah Alexander*
Editorial Assistants: *Jana Davis, Chris Ziemba*
Permission Editor/Photo Researcher: *Linda L. Rill*
Marketing Manager: *Ann Caven*
Marketing Assistant/Associate: *Leyla Jowza*
Advertising Project Manager: *Nathaniel Bergson-Michelson*
Print/Media Buyer: *Karen Hunt*

Production Management, Copyediting, and Composition: *Thompson Steele, Inc.*
Interior Illustration: *Precision Graphics; Sarah Woodward; Darwin and Vally Hennings; Tasa Graphic Arts, Inc.; Alexander Teshin Associates; John and Judith Walker; Rachel Ciemma; Victor Royer, Electronic Publishing Services, Inc.; J/B Woosley Associates; Kerry Wong, ScEYEnce Studios.*
Cover Image: *Cathedral Rock (Sedona, Arizona), C. Frank/ImageState/ Panoramic Images*
Text and Cover Printer: *R. R. Donnelley (Willard)*
Title Page Photograph: *Crater Lake, Oregon (Jack Carey)*

For more information about our products, contact us at:
Thomson Learning Academic Resource Center
1-800-423-0563
For permission to use material from this text, contact us by:
Phone: 1-800-730-2214
Fax: 1-800-730-2215
Web: http://www.thomsonrights.com

Library of Congress Control Number: 2002106327

Student Edition: ISBN 0-534-99729-5

Annotated Instuctor's Edition: ISBN 0-534-99730-9

International Student Edition: ISBN 0-534-99728-7
(Not for sale in the United States)

Brooks/Cole-Thomson Learning
511 Forest Lodge Road
Pacific Grove, CA 93950
USA

Asia
Thomson Learning
5 Shenton Way #01-01
UIC Building
Singapore 068808

Australia
Nelson Thomson Learning
102 Dodds Street
South Melbourne, Victoria 3205
Australia

Canada
Nelson Thomson Learning
1120 Birchmount Road
Toronto, Ontario M1K 5G4
Canada

Europe/Middle East/Africa
Thomson Learning
High Holborn House
50/51 Bedford Row
London WC1R 4LR
United Kingdom

Latin America
Thomson Learning
Seneca, 53
Colonia Polanco
11560 Mexico D.F.
Mexico

Spain
Paraninfo Thomson Learning
Calle/Magallanes, 25
28015 Madrid, Spain

Brief Contents

Detailed Contents   iv

Preface   xiv

Introduction: Learning Skills   1

PART I

HUMANS AND SUSTAINABILITY: AN OVERVIEW

1   Environmental Problems, Their Causes, and Sustainability   5

2   Environmental History: Learning from the Past   20

PART II

SCIENCE AND ECOLOGICAL PRINCIPLES

3   Science, Systems, Matter, and Energy   32

4   Ecosystems: What Are They and How Do They Work?   55

5   Evolution and Biodiversity   87

6   Climate and Terrestrial Biodiversity   101

7   Aquatic Biodiversity   127

8   Community Ecology   143

9   Population Ecology   163

10   Applying Population Ecology: The Human Population   176

PART III

SUSTAINING BIODIVERSITY

11   Sustaining Terrestrial Biodiversity: Managing and Protecting Ecosystems   194

12   Sustaining Biodiversity: The Species Approach   224

13   Sustaining Aquatic Biodiversity   251

PART IV

SUSTAINING NATURAL RESOURCES

14   Food and Soil Resources   273

15   Water Resources   305

16   Geology and Nonrenewable Mineral Resources   331

17   Nonrenewable Energy Resources   350

18   Energy Efficiency and Renewable Energy   379

PART V

SUSTAINING ENVIRONMENTAL QUALITY

19   Risk, Toxicology, and Human Health   409

20   Air Pollution   433

21   Climage Change and Ozone Loss   461

22   Water Pollution   41

23   Pest Management   518

24   Solid and Hazardous Waste   532

25   Sustainable Cities   563

PART VI

SUSTAINING HUMAN SOCIETIES

26   Economics, Environment, and Sustainability   583

27   Politics, Environment, and Sustainability   605

28   Environmental Worldviews, Ethics, and Sustainability   630

Appendices   A1

Glossary   G1

Index   I1

Detailed Contents

NASA

Introduction: Learning Skills 1

PART I

HUMANS AND SUSTAINABILITY: AN OVERVIEW

1   Environmental Problems, Their Causes, and Sustainability 5

   Case Study: Living in an Exponential Age 5

1-1   Living More Sustainably 6

1-2   Population Growth, Economic Growth, Economic Development, and Globalization 7

1-3   Resources 9

   Case Study: The Tragedy of the Commons—Degrading Free Renewable Resources 9

1-4   Pollution 11

1-5   Environmental and Resource Problems: Causes and Connections 12

1-6   Is Our Present Course Sustainable? 16

2   Environmental History: Learning from the Past 20

   Case Study: Near Extinction of the American Bison 20

2-1   Cultural Changes and the Environment 21

2-2   Environmental History of the United States: The Tribal and Frontier Eras 24

2-3   Environmental History of the United States: The Early Conservation Era (1832–1960) 24

2-4   Environmental History of the United States: The Environmental Era (1960–2004) 26

   Individuals Matter: *Rachel Carson* 27

2-5   **Case Study: Aldo Leopold and His Land Ethic 30**

PART II

SCIENCE AND ECOLOGICAL PRINCIPLES

3   Science, Systems, Matter, and Energy 32

   Case Study: An Environmental Lesson from Easter Island 32

3-1   The Nature of Science 33

   Connections: *What Is Harming the Robins?* 35

3-2   Models and Behavior of Systems 36

3-3   Matter 38

3-4   Energy 44

3-5   The Law of Conservation of Matter: A Rule We Cannot Break 46

3-6   Nuclear Changes 48

3-7   Energy Laws: Two Rules We Cannot Break 51

3-8   Matter and Energy Laws and Environmental Problems 52

4 Ecosystems: What Are They and How Do They Work? 55

Case Study: Have You Thanked the Insects Today? 55

4-1 The Nature of Ecology 56

Case Study: What Species Rule the World? Small Matters! 56

4-2 The Earth's Life-Support Systems 59

4-3 Ecosystem Components 61

4-4 Energy Flow in Ecosystems 67

4-5 Primary Productivity of Ecosystems 70

4-6 Soils 72

4-7 Matter Cycling in Ecosystems 76

4-8 How Do Ecologists Learn about Ecosystems? 84

5 Evolution and Biodiversity 87

Case Study: Earth: The Just-Right, Resilient Planet 87

5-1 Origins of Life 88

5-2 Evolution and Adaptation 88

5-3 Ecological Niches and Adaptation 91

Spotlight: *Cockroaches: Nature's Ultimate Survivors* 92

5-4 Speciation, Extinction, and Biodiversity 94

5-5 What Is the Future of Evolution? 97

6 Climate and Terrestrial Biodiversity 101

Case Study: Blowing in the Wind: A Story of Connections 101

6-1 Weather: A Brief Introduction 102

6-2 Climate: A Brief Introduction 105

6-3 Biomes: Climate and Life on Land 110

6-4 Desert Biomes 113

6 5 Grassland, Tundra, and Chaparral Biomes 115

6-6 Forest Biomes 120

6-7 Mountain Biomes 125

7 Aquatic Biodiversity 127

Case Study: Why Should We Care about Coral Reefs? 127

7-1 Aquatic Environments 128

7-2 Saltwater Life Zones 130

7-3 Freshwater Life Zones 137

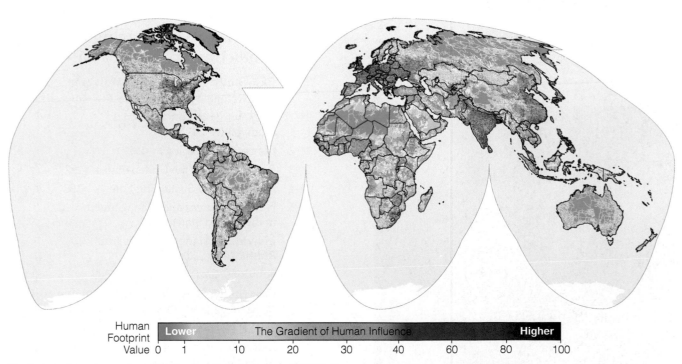

Human Footprint Value	Lower	The Gradient of Human Influence						Higher
0	1	10	20	30	40	60	80	100

The human footprint on the earth's land surface (Data from Wildlife Conservation Society and the Center for International Earth Science Information Network at Columbia University [CIESIN]. Reprinted by permission.)

8 Community Ecology 143

Case Study: Flying Foxes: Keystone Species in Tropical Forests 143

8-1 Community Structure and Species Diversity 144

8-2 Types of Species 146

Case Study: Why Are Amphibians Vanishing? Warnings from Frogs 147

Connections: *Why Should We Care about Alligators?* 149

8-3 Species Interactions: Competition and Predation 150

Case Study: Why Are Sharks Important Species? Culling the Oceans and Helping Improve Human Health 151

8-4 Species Interactions: Parasitism, Mutualism, and Commensalism 154

8-5 Ecological Succession: Communities in Transition 156

8-6 Ecological Stability, Complexity, and Sustainability 160

9 Population Ecology 163

Case Study: Sea Otters: Are They Back from the Brink of Extinction? 163

9-1 Population Dynamics and Carrying Capacity 164

9-2 Reproductive Patterns and Survival 168

9-3 Effects of Genetic Variations on Population Size 171

9-4 Human Impacts on Natural Systems: Learning from Nature 171

Connections: *Ecological Surprises* 173

10 Applying Population Ecology: The Human Population 176

Case Study: Slowing Population Growth in Thailand: A Success Story 176

10-1 Factors Affecting Human Population Size 177

Case Study: Should the United States Encourage or Discourage Immigration? 183

10-2 Population Age Structure 184

10-3 Solutions: Influencing Population Size 187

Case Study: Family Planning in Iran: A Success Story 190

10-4 **Case Studies: India and China** 191

10-5 Cutting Global Population Growth 192

PART III
SUSTAINING BIODIVERSITY

11 Sustaining Terrestrial Biodiversity: Managing and Protecting Ecosystems 194

Case Study: Reintroducing Wolves to Yellowstone 194

11-1 Human Impacts on Terrestrial Biodiversity 195

11-2 Conservation Biology 197

11-3 Public Lands in the United States 197

11-4 Managing and Sustaining Forests 199

11-5 Forest Resources and Management in the United States 205

Individuals Matter: *Butterfly in a Redwood Tree* 206

Solutions: *Goats to the Rescue* 209

Case Study: How Should U.S. National Forests Be Managed? An Ongoing Controversy 209

11-6 Tropical Deforestation 210

Individuals Matter: *Kenya's Green Belt Movement* 214

11-7 National Parks 214

Case Study: National Parks in the United States 214

11-8 Nature Reserves 215

Gene Alexander/USDA

Tree plantation

U.S. Department of Agriculture

Fast growing Kenaf for making paper, Texas

Case Study: What Has Costa Rica Done to Protect Some of Its Land from Degradation? A Global Conservation Leader   216

Case Study: How Much Wilderness Has Been Protected in the United States? Fighting for Crumbs and Losing   219

11-9   Ecological Restoration   220

Case Study: Ecological Restoration of a Tropical Dry Forest in Costa Rica   221

11-10   What Can We Do?   221

12   Sustaining Biodiversity: The Species Approach   224

Case Study: The Passenger Pigeon: Gone Forever   224

12-1   Species Extinction   225

12-2   Importance of Wild Species   229

Case Study: Why Should We Care about Bats? Ecological Allies   230

Connections: *Biophilia*   231

12-3   Extinction Threats from Habitat Loss and Degradation   231

Case Study: How Do Human Activities Affect Bird Species? A Disturbing Message from the Birds   233

12-4   Extinction Threats from Nonnative Species   234

Case Study: Deliberate Introduction of the Kudzu Vine: Unintended Consequences   234

Case Study: The Termite from Hell   237

Case Study: Exploding Deer Populations in the United States: Should We Put Bambi on Birth Control?   238

12-5   Extinction Threats from Poaching and Hunting   238

Case Study: The Rising Demand for Bushmeat in Africa: Hungry People Trying to Survive   239

12-6   Other Extinction Threats   239

12-7   Protecting Wild Species: The Research and Legal Approach   240

Case Study: What Has the Endangered Species Act Accomplished?   244

12-8   Protecting Wild Species: The Sanctuary Approach   245

12-9   Reconciliation Ecology   247

Case Study: Using Reconciliation Ecology to Protect Bluebirds?   248

13   Sustaining Aquatic Biodiversity   251

Case Study: A Biological Roller Coaster Ride in Lake Victoria   251

13-1   An Overview of Aquatic Biodiversity   252

13-2   Human Impacts on Aquatic Biodiversity   253

13-3   Protecting and Sustaining Marine Biodiversity   257

Kudzu taking over a house and a truck

NOAA Corps Collection/Photographer: Commander John Bortniak, NOAA Corps. (ret.)

Coral reef sanctuary, Tortugas Marine Ecological Reserve, Florida Keys

Individuals Matter: *Killing Invader Species and Saving Shipping Companies Money* 257

Case Study: Should Commercial Whaling Be Resumed? An Ongoing Controversy 259

Case Study: Near Extinction of the Blue Whale: Big Species are Easy to Kill 260

13-4 Managing and Sustaining the World's Marine Fisheries 263

13-5 Protecting, Sustaining, and Restoring Wetlands 265

Individuals Matter: *Restoring a Wetland* 266

Case Study: Restoring the Florida Everglades: Will It Work? 266

13-6 Protecting, Sustaining, and Restoring Lakes and Rivers 267

Case Study: Can the Great Lakes Survive Repeated Invasions by Alien Species? They Keep Coming 267

Case Study: Managing the Columbia River Basin for People and Salmon: A Difficult Balancing Act 268

Individuals Matter: *The Man Who Planted Trees to Restore a Stream* 270

PART IV

SUSTAINING NATURAL RESOURCES

14 Food and Soil Resources 273

Case Study: Growing Perennial Crops on the Kansas Prairie by Copying Nature 273

14-1 How is Food Produced? 274

14-2 Producing Food by Green Revolution and Traditional Techniques 276

Case Study: Industrial Food Production in the United States: A Success Story 276

Individuals Matter: *Low-Tech Sustainable Agriculture in Africa* 279

14-3 Soil Erosion and Degradation 279

Case Study: Soil Erosion in the United States Today: Some Hopeful News 280

Case Study: The Dust Bowl: An Environmental Lesson from Nature 281

14-4 Soil Conservation 284

14-5 Food Production, Nutrition, and Environmental Effects 286

Case Study: Can China's Population Be Fed? A Precarious Situation 289

14-6 Increasing Crop Production 291

14-7 Producing More Meat 294

Connections: *Some Environmental Consequences of Meat Production* 295

14-8 Catching and Raising More Fish and Shellfish 297

14-9 Government Agricultural Policy 301

14-10 Sustainable Agriculture 302

Don Alcorn/National Maritime Fisheries

Hawaiian monk seal's mouth caught in plastic

Severe soil salinization

Severe topsoil erosion in Arizona

15 Water Resources 305

Case Study: Water Conflicts in the Middle East 305

15-1 Water's Importance and Unique Properties 306

15-2 Supply, Renewal, and Use of Water Resources 307

Case Study: Freshwater Resources in the United States—Unequal Distribution 309

15-3 Too Little Water 311

15-4 Using Dams and Reservoirs to Supply More Water 313

Case Study: The Colorado River Basin—An Overtapped Resource 314

Case Study: China's Three Gorges Dam— A Controversial Project 315

15-5 Transferring Water from One Place to Another 316

Case Study: The Aral Sea Disaster— A Glaring Example of Unintended Consequences 316

Case Study: The California Water Transfer Project—Bringing Water to a Desert 317

Case Study: Canada's James Bay Watershed Transfer Project—Rearranging Nature 318

15-6 Tapping Groundwater, Converting Saltwater to Freshwater, Seeding Clouds, and Towing Icebergs and Big Baggies 318

Case Study: The Shrinking Ogallala Aquifer—Using Up Natural Capital 320

15-7 Reducing Water Waste 322

Spotlight: *Running Short of Water in Las Vegas, Nevada* 326

15-8 Too Much Water 327

Case Study: Living on Floodplains in Bangladesh—Danger for the Poor 327

15-9 A More Sustainable Water Future 329

16 Geology and Nonrenewable Mineral Resources 331

Case Study: The Mining Law of 1872 331

16-1 Geologic Processes 332

16-2 Internal and External Geologic Processes 332

16-3 Natural Geologic Hazards: Earthquakes and Volcanic Eruptions 336

16-4 Minerals, Rocks, and the Rock Cycle 338

16-5 Finding, Removing, and Processing Nonrenewable Mineral Resources 340

16-6 Environmental Effects of Using Mineral Resources 343

16-7 Supplies of Mineral Resources 345

17 Nonrenewable Energy Resources 350

Case Study: Bitter Lessons from Chernobyl 350

17-1 Envaluating Energy Resources 351

Oil shale rock (left) and the shale oil (right) extracted from it

17-2   Oil   355

Case Study: How Much Oil Does the United States Have? Rapidly Dwindling Supplies   356

Case Study: Should Oil and Gas Development Be Allowed in the Arctic National Wildlife Refuge? To Drill or Not to Drill   359

17-3   Natural Gas   362

17-4   Coal   364

17-5   Nuclear Energy   366

Case Study: The Yucca Mountain Storage Site for High-Level Radioactive Wastes— Controversy over Desert Burial   373

18   Energy Efficiency and Renewable Energy   379

Case Study: The Coming Energy- Efficiency and Renewable-Energy Revolution   379

18-1   The Importance of Improving Energy Efficiency   380

18-2   Ways to Improve Energy Efficiency   382

Connections: *The Real Cost of Gasoline in the United States*   384

18-3   Using Renewable Energy to Provide Heat and Electricity   391

18-4   Producing Electricity from the Water Cycle   395

18-5   Producing Electricity from Wind   396

18-6   Producing Energy from Biomas   397

18-7   Geothermal Energy   400

18-8   Hydrogen   401

Spotlight: *Producing Hydrogen from Green Algae Found in Pond Scum*   404

18-9   Entering the Age of Decentralized Micropower   404

18-10   A Sustainable Energy Strategy   405

PART V
SUSTAINING ENVIRONMENTAL QUALITY

19   Risk, Toxicology, and Human Health   409

Case Study: The Big Killer   409

19-1   Risk, Probability, and Hazards   410

19-2   Toxicology: Assessing Chemical Hazards   410

19-3   Chemical Hazards   416

Case Study: Are Hormonally Active Agents a Human Health Threat? Serious Concern but Inconclusive Evidence   416

19-4   Biological Hazards: Disease in Developed and Developing Countries   419

Case Study: Are We Losing Ground in Our Struggle against Infectious Bacteria? Growing Germ Resistance to Antibiotics   419

Case Study: The Global Tuberculosis Epidemic—A Growing Threat   421

Case Study: How Serious Is the Global Threat from HIV and AIDS? A Rapidly Growing Health Threat   423

Case Study: Malaria: A Deadly Parasitic Disease That Is Making a Comeback   424

19-5   Risk Analysis   428

20   Air Pollution   433

Case Study: When Is a Lichen Like a Canary?   433

20-1   Structure and Science of the Atmosphere   434

20-2   Outdoor Air Pollution   435

Spotlight: *Air Pollution in the Past: The Bad Old Days*   437

20-3   Photochemical and Industrial Smog   439

Case Study: South Asia's Massive Brown Cloud—Choking in China and India   442

20-4   Regional Outdoor Air Pollution from Acid Deposition   444

20-5   Indoor Air Pollution   449

Case Study: Are You Being Exposed to Radioactive Radon Gas? Test the Air in Your Home   450

20-6   Effects of Air Pollution on Living Organisms and Materials   451

20-7   Preventing and Reducing Air Pollution   453

Case Study: Should We Use the Marketplace to Reduce Pollution? Emissions Trading   454

Case Study: What Should We Do about Air Pollution from Older Coal-Burning Facilities? A Burning Controversy   456

21   Climate Change and Ozone Loss   461

Case Study: A.D. 2060: Green Times on Planet Earth   461

21-1   Past Climate Change   462

21-2   The Earth's Natural Greenhouse Effect   464

21-3   Climate Change and Human Activities   465

Case Study: Warning Signals from the Earth's Ice and Snow: Meltdowns Are Under Way   466

21-4   Projecting Future Changes in the Earth's Temperature   469

21-5   Factors Affecting the Earth's Temperature   472

21-6   Possible Effects of a Warmer World   474

21-7   Dealing with the Threat of Global Warming   477

21-8   What Is Being Done to Reduce Greenhouse Gas Emissions?   481

21-9   Ozone Depletion in the Stratosphere   484

21-10   Protecting the Ozone Layer   488

Individuals Matter: *Ray Turner and His Refrigerator*   489

22   Water Pollution   491

Case Study: Learning Nature's Ways to Purify Sewage   491

22-1   Types, Effects, and Sources of Water Pollution   492

22-2   Pollution of Freshwater Streams   495

Case Study: India's Ganges River: Religion, Poverty, and Health   497

22-3   Pollution of Freshwater Lakes   497

Case Study: Lake Washington—A Success Story   499

Case Study: Pollution in the Great Lakes— Hopeful Progress   500

22-4   Pollution of Groundwater   501

22-5   Ocean Pollution   504

Case Study: The Chesapeake Bay: An Estuary in Trouble   505

22-6   Preventing and Reducing Surface Water Pollution   509

22-7   Drinking Water Quality   514

23   Pest Management   518

Case Study: Along Came a Spider—Biological Pest Control   518

23-1   Pesticides: Types and Uses   519

23-2   The Case for Pesticides   521

23-3   The Case against Pesticides   522

Case Study: How Successful Have Pesticides Been in Reducing Crop Losses in the United States? Barely Holding the Line   524

Spotlight: *A Superbug Called the Silverleaf Whitefly*   524

Connections: *What Goes Around Can Come Around*   525

23-4   Pesticide Regulation   525

Case Study: Revisiting DDT—from Riches to Rags   526

23-5   Alternatives to Conventional Chemical Pesticides   527

Ocean Arks International

Solar-aquatic sewage treatment system

24 Solid and Hazardous Waste 532

Case Study: Love Canal: There Is No "Away" 532

24-1 Wasting Resources 533

Case Study: How Much Solid Waste Does the United States Produce? Affluenza in Action 533

24-2 Producing Less Waste 535

24-3 The Ecoindustrial Revolution and Selling Services Instead of Things 536

Individuals Matter: *Ray Anderson* 538

24-4 Reuse 538

24-5 Recycling 540

Case Study: How Much Wastepaper Is Being Recycled? Encouraging News 543

Case Study: Is It Feasible to Recycle Plastics? Some Problems 543

24-6 Burning and Burying Solid Waste 545

24-7 Hazardous Waste 548

Case Study: A Black Day in Bhopal, India 549

24-8 Case Studies: Lead, Mercury, and Dioxins 554

24-9 Hazardous Waste Regulation in the United States 558

Spotlight: *Using Honeybees to Detect Toxic Pollutants* 559

24-10 Achieving a Low-Waste Society 560

The Love Canal housing development near Niagara Falls, New York

New York State Department of Environmental Conservation

25 Sustainable Cities 563

Case Study: The Ecocity Concept in Curitiba, Brazil 563

25-1 Urbanization and Urban Growth 564

25-2 Urban Resource and Environmental Problems 568

Case Study: What Are the Advantages of Urbanization? Concentrating People Helps 568

Case Study: What Are the Disadvantages of Urbanization? Concentrating People Has Some Harmful Effects 568

Case Study: How Do the Urban Poor in Developing Countries Live—Living on the Edge with Ingenuity and Hope 570

25-3 Transportation and Urban Development 571

Connections: *How Can Reducing Crime Help the Environment?* 572

Case Study: Mass Transit in the United States—Destroying a Great System 576

25-4 Urban Land-Use Planning and Control 576

Case Study: How Is New Urbanism Creating More Livable Spaces? Returning to Traditional Neighborhood Development 579

25-5 Making Urban Areas More Livable and Sustainable 580

Case Study: Chattanooga, Tennessee—From Brown to Green 581

Part VI
SUSTAINING HUMAN SOCIETIES

26 Economics, Environment, and Sustainability 583

Case Study: How Important Are Natural Resources? 583

26-1 Economic Resources and Systems 584

26-2 Economists' Views of Pollution Control and Resource Management 586

Case Study: What Is Cost-Benefit Analysis, and How Can It Be Improved? Weighing Costs and Benefits to Make Choices 589

26-3 Monitoring Environmental Progress 590

26-4 Harmful External Costs and Full-Cost Pricing 592

26-5 Ways to Improve Environmental Quality and Shift to Full-Cost Pricing 593

Biosphere 2, Tucson, Arizona

26-6 Reducing Poverty to Improve
Environmental Quality and Human Well-
Being 598

**Case Study: What Is the Role of the World Bank
in Economic Development? Controversy over
Big Loans 598**

26-7 Making the Transition to More
Environmentally Sustainable
Economies 600

**Case Study: How Are Germany and the
Netherlands Working to Achieve More
Environmentally Sustainable Economies?
Leading the Way 600**

Solutions: *Microloans to the Poor* 601

27 Politics, Environment, and Sustainability 605

Case Study: Rescuing a River 605

27-1 Environmental and Political Challenges
for This Century 606

27-2 Dealing with Environmental Problems in
Democracies 607

27-3 Developing, Influencing, and Implementing
Environmental Policy 607

**Case Study: What Is Environmental
Leadership? An Option for Each of Us 608**

Individuals Matter: *Environmental Careers* 610

**Case Study: How Is Environmental Policy
Made in the United States? A Complicated
and Thorny Process 611**

27-4 Environmental Law 616

**Case Study: What Are the Major Types of
Environmental Laws in the United States?
A Variety of Approaches 618**

27-5 Environmental Groups and Their
Opponents 618

**Case Study: Environmental Action
by Students in the United States—Making
a Difference 620**

Solutions: *How Can We Improve U.S.
Environmental Laws and Regulations? Time
for a Checkup* 622

27-6 Global Environmental Policy 622

**Case Study: Is Encouraging Global Free
Trade Environmentally Helpful or Harmful?
A Difficult Issue 625**

Individuals Matter: *Severn Cullis-Suzuki* 627

28 Environmental Worldviews, Ethics, and Sustainability 630

**Case Study: Biosphere 2: A Lesson
in Humility 630**

28-1 Environmental Worldviews in Industrial
Societies 631

28-2 Life-Centered and Earth-Centered
Environmental Worldviews 633

28-3 Living More Sustainably 636

Spotlight: *What Are Our Basic Needs?* 639

Appendices

1 Units of Measure A1

2 Major Events in U.S. Environmental
History A2

3 Some Basic Chemistry A8

4 Classifying and Naming Species A11

5 Brief History of the Age of Oil A12

Glossary G1

Index I1

For Instructors

What Are the Major Features of This Book? Science Based, Solutions Oriented, and Flexible Use

This book is designed for introductory courses on environmental science. It is an *interdisciplinary* study of how nature works and how things in nature are interconnected. About 40% of the book is devoted to providing a scientific base needed to understand environmental problems and to evaluate possible solutions, 30% presents environmental problems, and another 30% presents and evaluates solutions to these problems.

This book is *science based, solution oriented,* and *flexible.* It is divided into six major parts (see Brief Contents, p. iii). A highly flexible format allows instructors to use almost any course outline. I suggest that instructors use Chapter 1 to provide an overview of environmental problems and solutions and Chapters 3 through 9 to provide a base of scientific principles and concepts. The remaining chapters—2 and 10 through 28—can be used in virtually any order or omitted as desired. In addition, sections within chapters can be omitted or rearranged to meet instructor needs.

Each chapter begins with a brief *case study* designed to capture interest and set the stage for the material that follows. In addition to these 28 case studies, 77 other case studies are found throughout the book; they provide a more in-depth look at specific environmental problems and their possible solutions. Twenty-one Guest Essays are available on the website for this book.

This book is an integrated study of environmental problems, connections, and solutions. The seven integrative themes are *natural capital, sustainability, pollution prevention and waste reduction, population and exponential growth, energy and energy efficiency, solutions to environmental problems,* and *the importance of individuals working together to bring about environmental change.*

To get an overview of this book I urge you to look at the brief table of contents (p. iii)

This book has 567 illustrations (111 of them new to this edition) designed to present complex ideas in understandable ways and to relate learning to the real world.

I have not cited specific sources of information within the text. This is rarely done for an introductory-level text in any field, and it would interrupt the flow of the material. Instead, on the website for each chapter you will find readings, Internet site references, and references to complete articles that can be accessed online on the InfoTrac® College Edition supplement available free to qualified users of this book. These sources back up most of the content of this book and serve as springboards to further information and ideas. Placing these references on the website also allows me to update them regularly.

Instructors wanting a book covering this material with a different emphasis, organization, and length can use one of my three other books written for various types of environmental science courses: *Environmental Science,* 10th edition (538 pages, Brooks/Cole, 2004), *Sustaining the Earth: An Integrated Approach,* 7th edition (333 pages, Brooks/Cole, 2005), and *Essentials of Ecology,* 3rd edition (272 pages, Brooks/Cole, 2005).

To help ensure that the material is accurate and up to date, I have consulted more than 10,000 research sources in the professional literature and about the same number of Internet sites. I have also benefited from the more than 250 experts and teachers (see list on pp. xviii–xx) who have provided detailed reviews of this and my other three books in this field.

How Is the Presentation of Views Balanced? Trade-Offs

There are always *trade-offs* involved in making and implementing environmental decisions. The challenge for an author is to give a fair and balanced presentation of opposing viewpoints, advantages and disadvantages of various technologies and proposed solutions to environmental problems, and good and bad news about environmental problems without injecting personal bias. This allows readers to make up their own minds about important environmental issues. To study a subject as important as environmental science and end up with no conclusions, opinions, and beliefs would mean that both the teacher and student have failed. However, students should reach conclusions only by using using critical thinking to evaluate opposing ideas and to understand the trade-offs involved.

A few examples of my efforts to provide a balanced presentation are **(1)** the advantages and disadvantages of reducing birth rates (pp. 187–188), **(2)** Case Study: Should Oil and Natural Gas Development Be

Allowed in the Arctic National Wildlife Refuge? (p. 359), **(3)** advantages and disadvantages of pesticides (pp. 521–525), and **(4)** 45 trade-offs diagrams that give the advantages and disadvantages of various environmental technologies and solutions to environmental problems.

Why Has the Chapter Order Changed? Satisfying Text Users

There are hundreds of different ways to organize the chapters in this book to fit the needs of different instructors. That is why one of the key features of this book is the *flexibility* for changing the order of chapters once the basic science has been covered.

Since the first edition of this book came out almost 30 years ago I have used different chapter orders, usually in response to feedback from instructors. I have decided to put the three chapters on biodiversity (Chapters 11, 12, and 13) after the chapters in Part II on Science and Ecological Principles for two reasons.

First, the loss and degradation of biodiversity is one of the most serious environmental problems. *Second,* the majority of instructors using this book prefer that I put all of the biology together (Chapters 4 through 13).

There is no best chapter order but this one makes sense to most text users, and I plan to stick with this order from now on. Because of the book's flexibility, anyone who wants to use a different order can easily do so, as users have done for almost three decades.

What Are the Major Changes in the Fourteenth Edition? The Most Significant Revision Ever

This is the most significant revision since the first edition. Major changes include the following:

- Chapter order changed to put all of the biology (ecology and biodiversity, Chapters 4–13) material together.

- Entire book rewritten with help from a developmental editor and three in-depth reviews

- New *How Would You Vote* feature enables students to vote online on 69 critical environmental issues and to get national and global tallies of the results

- Simpler, more learning-oriented writing style

- Improved learning system. Each subsection begins with a question. Then brief answers to the question are given by a short headline and by a one-sentence summary of the subsection Here is an example:

What Is Biodiversity? Variety Is the Spice of Life.

A vital renewable resource is the biodiversity found in the earth's variety of genes, species, ecosystems, and ecosystem processes.

Text of the subsection then follows.

- Simpler design with fewer boxes and bulleted lists

- 111 new figures and 85 improved or updated figures out of a total of 567 figures

- 45 *Trade-Offs* diagrams and 12 new *What Can You Do?* diagrams

- 105 Case Studies, including one that opens each chapter

- Reduction of basic text by 116 pages. (About half of the cuts came from editing and rewriting, and the remainder came from transferring the Guest Essays and detailed lists of chapter review questions to the book's website. This allows use of more Guest Essays and integration of the review questions with other learning materials on the website.)

- Important new topics: introduction giving a simple summary of learning skills, *natural capital* (see top half of back cover), *affluenza, future implications of genetic engineering, foundation species, four principles of sustainability* (see bottom half of back cover), *reconciliation ecology, ownership of water resources, nanotechnology, Canada's huge oil sand reserves, General Motors' fuel cell car of the future, bioterrorism, Asia's brown cloud, pollution of India's Ganges River, phytoremediation, revisiting DDT, light pollution, global environmental policies, expanded treatment of stewardship environmental worldview.*

- Extensive updating. It is essentially impossible for readers to detect the over 5,600 updates in this book based on information and data published in 2002, 2003, and 2004. There are over 420 references throughout the book that specifically cite these years and an estimated 5,600 additional updates that are not cited by these years.

What In-Text Study Aids Are Provided? Tried and True Tools Plus Some New Ones

Each chapter begins with a few general questions to reveal how it is organized and what students will be learning. When a new term is introduced and defined, it is printed in boldface type. A glossary of all key terms is located at the end of the book.

Questions are used as titles for all subsections so readers know the focus of the material that follows. In effect, this is a built-in set of learning objectives. A brief attention-grabbing headline follows each subsection question to give students a general idea of the content of the subsection. This is followed by a single sentence that summarizes the key material in each subsection. In effect, this provides readers with a built-in running summary of the material (see example at left). I believe this is a more effective learning tool than providing a summary of the entire chapter at the end of each chapter.

Each chapter ends with a set of *Critical Thinking* questions and *Projects* to encourage students to think critically and apply what they have learned to their lives. The website for the book also contains a set of *Review Questions* covering *all* of the material in the chapter, which can be used as a study guide for students. Some instructors download this from the website and give students a list of the particular questions they expect their students to answer.

What Internet and Online Study Aids Are Available?

Qualified users of this textbook have free access to the *Brooks/Cole Biology and Environmental Science Resource Center*. Access the online resource material for this book by logging on at

http://biology.brookscole.com/miller14

At this website you will find the following material for each chapter:

- Flash Cards, which allow you to test your mastery of the Terms and Concepts to Remember for each chapter

- Environmental Quizzes, which provide a multiple-choice practice quiz

- Student Guide to InfoTrac® College Edition, which will lead you to Critical Thinking Projects that use InfoTrac College Edition as a research tool

- References, which lists the major books and articles consulted in writing this chapter

- A brief What You Can Do list addressing key environmental problems

- Hypercontents, which takes you to an extensive list of websites with news, research, and images related to individual sections of the chapter

Qualified adopters of this textbook also have free access to *WebTutor Toolbox on WebCT* and *Blackboard* at:

http://e.thomsonlearning.com

It provides access to a full array of study tools, including flashcards (with audio), practice quizzes, online tutorials, and web links.

Teachers and students using *new* copies of this textbook also have free and unlimited access to *InfoTrac® College Edition*. This fully searchable online library gives users access to complete environmental articles from several hundred periodicals dating back over the past 25 years.

Other student learning tools include:

- *Audio version for study and review.* Students can listen to the chapters in this book while walking, traveling, sitting in your room, or working out. They can use the pin code enclosed in every new copy of this book to download book chapters free from the Web to any MP3 player.

- An interactive CD-ROM, *Student CD with Environmental ScienceNow,* that integrates concept summaries and almost 100 engaging animations and interactions based on figures from the text with flashcards and quizzing on the Web.

- *Essential Study Skills for Science Students* by Daniel D. Chiras. This book includes chapters on developing good study habits, sharpening memory, getting the most out of lectures, labs, and reading assignments, improving test-taking abilities, and becoming a critical thinker. Your instructor can have this book bundled FREE with your textbook.

- *Laboratory Manual* by C. Lee Rockett and Kenneth J. Van Dellen. This manual includes a variety of laboratory exercises, workbook exercises, and projects that require a minimum of sophisticated equipment.

The following supplementary materials are available to instructors adopting this book:

- *Multimedia Manager.* This CD-ROM, free to adopters, allows you to create custom lectures using over 2,000 pieces of high-resolution artwork, images, and QuickTime movies from the CD and the web, assemble database files, and create Microsoft PowerPoint lectures using text slides and figures from the textbook. This program's editing tools allow use of slides from other lectures, modification or removal of figure labels and leaders, insertion of your own slides, saving slides as JPEG images, and preparation of lectures for use on the Web.

- *Transparency Masters and Acetates.* Includes 100 color acetates of line art and nearly 450 black-and-white master sheets of key diagrams for making overhead transparencies. Free to adopters.

- *CNN™ Today Videos.* These videos, updated annually, contain short clips of news stories about environmental news. Adopters can receive 5 CDs or video tapes containing more than 120, 2- to 5-minute, video clips shot over the last few years. Each year after, adopters will receive a new free CD or video tape of new clips.

- Two videos, *In the Shadow of the Shuttle: Protecting Endangered Species,* and *Costa Rica: Science in the Rainforest,* are available to adopters.

- *Instructor's Manual with Test Bank.* Free to adopters.

- *ExamView.* Allows you to easily create and customize tests, see them on the screen exactly as they will print, and print them out.

Help Me Improve This Book

Let me know how you think this book can be improved; if you find any errors, bias, or confusing explanations, please e-mail me about them at

http://mtg89@hotmail.com

Most errors can be corrected in subsequent printings of this edition rather than waiting for a new edition.

Acknowledgments

I wish to thank the many students and teachers who have responded so favorably to the 13 previous editions of *Living in the Environment,* the 10 editions of *Environmental Science,* and the 6 editions of *Sustaining the Earth,* the 2 editions of *Essentials of Ecology,* and who have corrected errors and offered many helpful suggestions for improvement. I am also deeply indebted to the more than 250 reviewers, who pointed out errors and suggested many important improvements in the various editions of these three books. Any errors and deficiencies left are mine.

I am particularly indebted to Scott Spoolman who served as a developmental editor for this new edition and who made numerous suggestions for improving this book. I am also indebted to Michael L. Cain, John Pichtel, and Jennifer Rivers who did in-depth reviews of the 13th edition of this book and provided many helpful suggestions. I also thank Sue Holt for her helpful review of the environmental economics chapter.

My deep gratitude also goes to Paul M. Rich, who served as coauthor for the ecology, biodiversity, and urban land use chapters of the eleventh edition of this book and made many other important improvements to the book. He was immensely helpful during a time when a health problem prevented me from working full-time.

The members of the talented production team, listed on the copyright page, have made vital contributions as well. My thanks also go to production editors Andy Marinkovich at Wadsworth and Andrea Fincke at Thompson Steele; copy editor Anita Wagner; page layout artist Bonnie Van Slyke; Brooks/Cole's hardworking sales staff; and Keli Amann, Chris Evers, Cathy Miller, Steve Bolinger, Joy Westberg, and the other members of the talented team who developed the multimedia, website, and advertising materials associated with this book.

I also thank C. Lee Rockett and Kenneth J. Van Dellen for developing the *Laboratory Manual* to accompany this book; Jane Heinze-Fry for her work on concept mapping and *Environmental Articles, Critical Thinking and the Environment: A Beginner's Guide;* Paul Blanchard for his excellent work on the *Instructor's Manual,* and the people who have translated this book into seven different languages for use throughout much of the world.

My deepest thanks go to Jack Carey, biology publisher at Brooks/Cole, for his encouragement, help, 38 years of friendship, and superb reviewing system. It helps immensely to work with the best and most experienced editor in college textbook publishing.

I dedicate this book to the earth and to Kathleen Paul Miller, my wife and research associate.

G. Tyler Miller, Jr.

Guest Essayists

Guest Essays by the following authors are available online at the website for this book.

M. Kat Anderson, ethnoecologist with the National Plant Center of the USDA's Natural Resource Conservation Service; **Lester R. Brown,** president, Earth Policy Institute; **Alberto Ruz Buenfil,** environmental activist, writer, and performer; **Robert D. Bullard,** professor of sociology and director of the Environmental Justice Resource Center at Clark Atlanta University; **Michael Cain,** Bowdoin College; **Herman E. Daly,** senior research scholar at the school of Public Affairs, University of Maryland; **Lois Marie Gibbs,** director, Center for Health, Environment, and Justice; **Garrett Hardin,** professor emeritus (now deceased) of human ecology, University of California, Santa Barbara; **Paul G. Hawken,** environmental author and business leader; **Jane Heinze-Fry,** author, teacher, and consultant in environmental education; **Amory B.** **Lovins,** energy policy consultant and director of research, Rocky Mountain Institute; **Lester W. Milbrath,** director of the research program in environment and society, State University of New York, Buffalo; **Peter Montague,** director, Environmental Research Foundation; **Norman Myers,** tropical ecologist and consultant in environment and development; **David W. Orr,** professor of environmental studies, Oberlin College; **Noel Perrin,** adjunct professor of environmental studies, Dartmouth College; **John Pichtel,** Ball State University; **David Pimentel,** professor of entomology, Cornell University; **Andrew C. Revkin,** environmental author and environmental reporter for the *New York Times;* and **Nancy Wicks,** ecopioneer and director of Round Mountain Organics; **Donald Worster,** environmental historian and professor of American history, University of Kansas.

Reviewers

Barbara J. Abraham, Hampton College; Donald D. Adams, State University of New York at Plattsburgh; Larry G. Allen, California State University, Northridge; Susan Allen-Gil, Ithaca College; James R. Anderson, U.S. Geological Survey; Mark W. Anderson, University of Maine; Kenneth B. Armitage, University of Kansas; Samuel Arthur, Bowling Green State University; Gary J. Atchison, Iowa State University; Marvin W. Baker, Jr., University of Oklahoma; Virgil R. Baker, Arizona State University; Ian G. Barbour, Carleton College; Albert J. Beck, California State University, Chico; W. Behan, Northern Arizona University; Keith L. Bildstein, Winthrop College; Jeff Bland, University of Puget Sound; Roger G. Bland, Central Michigan University; Grady Blount II, Texas A&M University, Corpus Christi; Georg Borgstrom, Michigan State University; Arthur C. Borror, University of New Hampshire; John H. Bounds, Sam Houston State University; Leon F. Bouvier, Population Reference Bureau; Daniel J. Bovin, Université Laval; Michael F. Brewer, Resources for the Future, Inc.; Mark M. Brinson, East Carolina University; Dale Brown, University of Hartford; Patrick E. Brunelle, Contra Costa College; Terrence J. Burgess, Saddleback College North; David Byman, Pennsylvania State University, Worthington–Scranton; Michael L. Cain, Bowdoin College, Lynton K. Caldwell, Indiana University; Faith Thompson Campbell, Natural Resources Defense Council, Inc.; Ray Canterbery, Florida State University; Ted J. Case, University of San Diego; Ann Causey, Auburn University; Richard A. Cellarius, Evergreen State University; William U. Chandler, Worldwatch Institute; F. Christman, University of North Carolina, Chapel Hill; Lu Anne Clark, Lansing Community College; Preston Cloud, University of California, Santa Barbara; Bernard C. Cohen, University of Pittsburgh; Richard A. Cooley, University of California, Santa Cruz; Dennis J. Corrigan; George Cox, San Diego State University; John D. Cunningham, Keene State College; Herman E. Daly, University of Maryland; Raymond F. Dasmann, University of California, Santa Cruz; Kingsley Davis, Hoover Institution; Edward E. DeMartini, University of California, Santa Barbara; Charles E. DePoe, Northeast Louisiana University; Thomas R. Detwyler, University of Wisconsin; Peter H. Diage, University of California, Riverside; Lon D. Drake, University of Iowa; David DuBose, Shasta College; Dietrich Earnhart, University of Kansas; T. Edmonson, University of Washington; Thomas Eisner, Cornell University; Michael Esler, Southern Illinois University; David E. Fairbrothers, Rutgers University; Paul P. Feeny, Cornell University; Richard S. Feldman, Marist College; Nancy Field, Bellevue Community College; Allan Fitzsimmons, University of Kentucky; Andrew J. Friedland, Dartmouth College; Kenneth O. Fulgham, Humboldt State University; Lowell L. Getz, University of Illinois at Urbana–Champaign; Frederick F. Gilbert, Washington State University; Jay Glassman, Los Angeles Valley College; Harold Goetz, North Dakota State University; Jeffery J. Gordon, Bowling Green State University;

Eville Gorham, University of Minnesota; Michael Gough, Resources for the Future; Ernest M. Gould, Jr., Harvard University; Peter Green, Golden West College; Katharine B. Gregg, West Virginia Wesleyan College; Paul K. Grogger, University of Colorado at Colorado Springs; L. Guernsey, Indiana State University; Ralph Guzman, University of California, Santa Cruz; Raymond Hames, University of Nebraska, Lincoln; Raymond E. Hampton, Central Michigan University; Ted L. Hanes, California State University, Fullerton; William S. Hardenbergh, Southern Illinois University at Carbondale; John P. Harley, Eastern Kentucky University; Neil A. Harriman, University of Wisconsin, Oshkosh; Grant A. Harris, Washington State University; Harry S. Hass, San Jose City College; Arthur N. Haupt, Population Reference Bureau; Denis A. Hayes, environmental consultant; Stephen Heard, University of Iowa; Gene Heinze-Fry, Department of Utilities, Commonwealth of Massachusetts; Jane Heinze-Fry, environmental educator; John G. Hewston, Humboldt State University; David L. Hicks, Whitworth College; Kenneth M. Hinkel, University of Cincinnati; Eric Hirst, Oak Ridge National Laboratory; Doug Hix, University of Hartford; S. Holling, University of British Columbia; Sue Holt, Cabrillo College; Donald Holtgrieve, California State University, Hayward; Michael H. Horn, California State University, Fullerton; Mark A. Hornberger, Bloomsburg University; Marilyn Houck, Pennsylvania State University; Richard D. Houk, Winthrop College; Robert J. Huggett, College of William and Mary; Donald Huisingh, North Carolina State University; Marlene K. Hull, IBM; David R. Inglis, University of Massachusetts; Robert Janiskee, University of South Carolina; Hugo H. John, University of Connecticut; Brian A. Johnson, University of Pennsylvania, Bloomsburg; David I. Johnson, Michigan State University; Mark Jonasson, Crafton Hills College; Agnes Kadar, Nassau Community College; Thomas L. Keefe, Eastern Kentucky University; Nathan Keyfitz, Harvard University; David Kidd, University of New Mexico; Pamela S. Kimbrough; Jesse Klingebiel, Kent School; Edward J. Kormondy, University of Hawaii–Hilo/West Oahu College; John V. Krutilla, Resources for the Future, Inc.; Judith Kunofsky, Sierra Club; E. Kurtz; Theodore Kury, State University of New York at Buffalo; Steve Ladochy, University of Winnipeg; Mark B. Lapping, Kansas State University; Tom Leege, Idaho Department of Fish and Game; William S. Lindsay, Monterey Peninsula College; E. S. Lindstrom, Pennsylvania State University; M. Lippiman, New York University Medical Center; Valerie A. Liston, University of Minnesota; Dennis Livingston, Rensselaer Polytechnic Institute; James P. Lodge, air pollution consultant; Raymond C. Loehr, University of Texas at Austin; Ruth Logan, Santa Monica City College; Robert D. Loring, DePauw University; Paul F. Love, Angelo State University; Thomas Lovering, University of California, Santa Barbara; Amory B. Lovins, Rocky Mountain Institute; Hunter Lovins, Rocky Mountain Institute; Gene A. Lucas, Drake University; Claudia Luke; David Lynn; Timothy F. Lyon, Ball State University; Stephen Malcolm, Western Michigan University; Melvin G. Marcus, Arizona State University; Gordon E. Matzke, Oregon State University; Parker Mauldin, Rockefeller Foundation; Marie McClune, The Agnes Irwin School (Rosemont, Pennsylvania); Theodore R. McDowell, California State University; Vincent E. McKelvey, U.S. Geological Survey; Robert T. McMaster, Smith College; John G. Merriam, Bowling Green State University; A. Steven Messenger, Northern Illinois University; John Meyers, Middlesex Community College; Raymond W. Miller, Utah State University; Arthur B. Millman, University of Massachusetts, Boston; Fred Montague, University of Utah; Rolf Monteen, California Polytechnic State University; Ralph Morris, Brock University, St. Catherine's, Ontario, Canada; Angela Morrow, Auburn University; William W. Murdoch, University of California, Santa Barbara; Norman Myers, environmental consultant; Brian C. Myres, Cypress College; A. Neale, Illinois State University; Duane Nellis, Kansas State University; Jan Newhouse, University of Hawaii, Manoa; Jim Norwine, Texas A&M University, Kingsville; John E. Oliver, Indiana State University; Carol Page, copyeditor; Eric Pallant, Allegheny College; Charles F. Park, Stanford University; Richard J. Pedersen, U.S. Department of Agriculture, Forest Service; David Pelliam, Bureau of Land Management, U.S. Department of Interior; Rodney Peterson, Colorado State University; Julie Phillips, De Anza College; John Pichtel, Ball State University; William S. Pierce, Case Western Reserve University; David Pimentel, Cornell University; Peter Pizor, Northwest Community College; Mark D. Plunkett, Bellevue Community College; Grace L. Powell, University of Akron; James H. Price, Oklahoma College; Marian E. Reeve, Merritt College; Carl H. Reidel, University of Vermont; Charles C. Reith, Tulane University; Roger Revelle, California State University, San Diego; L. Reynolds, University of Central Arkansas; Ronald R. Rhein, Kutztown University of Pennsylvania; Charles Rhyne, Jackson State University; Robert A. Richardson, University of Wisconsin; Benjamin F. Richason III, St. Cloud State University; Jennifer Rivers, Northeastern University; Ronald Robberecht, University of Idaho; William Van B. Robertson, School of Medicine, Stanford University; C. Lee Rockett, Bowling Green State University; Terry D. Roelofs, Humboldt State University; Christopher Rose, California Polytechnic State University; Richard G. Rose, West Valley College; Stephen T. Ross, University of Southern Mississippi; Robert E.

Roth, Ohio State University; Arthur N. Samel, Bowling Green State University; Floyd Sanford, Coe College; David Satterthwaite, I.E.E.D., London; Stephen W. Sawyer, University of Maryland; Arnold Schecter, State University of New York; Frank Schiavo, San Jose State University; William H. Schlesinger, Ecological Society of America; Stephen H. Schneider, National Center for Atmospheric Research; Clarence A. Schoenfeld, University of Wisconsin, Madison; Henry A. Schroeder, Dartmouth Medical School; Lauren A. Schroeder, Youngstown State University; Norman B. Schwartz, University of Delaware; George Sessions, Sierra College; David J. Severn, Clement Associates; Paul Shepard, Pitzer College and Claremont Graduate School; Michael P. Shields, Southern Illinois University at Carbondale; Kenneth Shiovitz; F. Siewert, Ball State University; E. K. Silbergold, Environmental Defense Fund; Joseph L. Simon, University of South Florida; William E. Sloey, University of Wisconsin, Oshkosh; Robert L. Smith, West Virginia University; Val Smith, University of Kansas; Howard M. Smolkin, U.S. Environmental Protection Agency; Patricia M. Sparks, Glassboro State College; John E. Stanley, University of Virginia; Mel Stanley, California State Polytechnic University, Pomona; Norman R. Stewart, University of Wisconsin, Milwaukee; Frank E. Studnicka, University of Wisconsin, Platteville; Chris Tarp, Contra Costa College; Roger E. Thibault, Bowling Green State University; William L. Thomas, California State University, Hayward; Shari Turney, copyeditor; John D. Usis, Youngstown State University; Tinco E. A. van Hylckama, Texas Tech University; Robert R. Van Kirk, Humboldt State University; Donald E. Van Meter, Ball State University; Gary Varner, Texas A&M University; John D. Vitek, Oklahoma State University; Harry A. Wagner, Victoria College; Lee B. Waian, Saddleback College; Warren C. Walker, Stephen F. Austin State University; Thomas D. Warner, South Dakota State University; Kenneth E. F. Watt, University of California, Davis; Alvin M. Weinberg, Institute of Energy Analysis, Oak Ridge Associated Universities; Brian Weiss; Margery Weitkamp, James Monroe High School (Granada Hills, California); Anthony Weston, State University of New York at Stony Brook; Raymond White, San Francisco City College; Douglas Wickum, University of Wisconsin, Stout; Charles G. Wilber, Colorado State University; Nancy Lee Wilkinson, San Francisco State University; John C. Williams, College of San Mateo; Ray Williams, Rio Hondo College; Roberta Williams, University of Nevada, Las Vegas; Samuel J. Williamson, New York University; Ted L. Willrich, Oregon State University; James Winsor, Pennsylvania State University; Fred Witzig, University of Minnesota at Duluth; George M. Woodwell, Woods Hole Research Center; Robert Yoerg, Belmont Hills Hospital; Hideo Yonenaka, San Francisco State University; Malcolm J. Zwolinski, University of Arizona.

Introduction: Learning Skills

Students who can begin early in their lives to think of things as connected, even if they revise their views every year, have begun the life of learning.

MARK VAN DOREN

Why Is It Important to Study Environmental Science? Learning How the Earth Works

Environmental science may be the most important course you will ever take.

Welcome to *environmental science*—an *interdisciplinary* study of how nature works and how things in nature are interconnected. This book is an integrated and science-based study of environmental problems, connections, and solutions.

Environmental issues affect every part of your life and are an important part of the news stories presented on television and in newspapers and magazines. Thus, the concepts, information, and issues discussed in this book and the course you are taking should be useful to you now and in the future.

Understandably, I am biased. But *I strongly believe that environmental science is the single most important course in your education.* What could be more important than learning how the earth works, how we are affecting its life support system, and how we can reduce our environmental impact? In India every college student must take an introductory course in environmental science.

We live in an incredibly challenging era. There is a growing awareness that during this century we need to make a new cultural transition in which we learn how to live more sustainably by not degrading our life support system.

I hope this book and the course you are taking will help you learn about the exciting challenges we face. More important, I hope it will stimulate you to become involved in this change in the way we view and treat the earth that sustains us, other life, and all economies.

How Did I Become Involved with Environmental Problems? Individuals Matter

I became involved in environmental science and education after hearing a scientific lecture in 1966 by Dean Cowie.

In 1966, I heard Dean Cowie, a physicist with the U.S. Geological Survey, give a lecture on the problems of population growth and pollution. Afterward I went to him and said, "If even a fraction of what you have said is true, I will feel ethically obligated to give up my research on the corrosion of metals and devote the rest of my life to research and education on environmental problems and solutions. Frankly, I do not want to change my life, and I am going into the literature to try to show that your statements are either untrue or grossly distorted."

After 6 months of study I was convinced of the seriousness of these and other environmental problems. Since then, I have been studying, teaching, and writing about them. This book summarizes what I have learned in more than three decades of trying to understand environmental principles, problems, connections, and solutions.

How Can You Improve Your Study and Learning Skills? Becoming an Efficient and Effective Learner

Learning how to learn is life's most important skill.

Maximizing your ability to learn should be one of your most important lifetime educational goals. This involves continually trying to *improve your study and learning skills.*

This has a number of payoffs. You can learn more and do this more efficiently. You will also have more time to pursue other interests besides studying without feeling guilty about always being behind. It can also help you get better grades and live a more fruitful and rewarding life.

Here are some *general study and learning skills*

Get organized. Becoming more efficient at studying gives you more time for other interests.

Make daily to-do lists in writing. Put items in order of importance, focus on the most important tasks, and assign a time to work on these items. Because life is full of uncertainties, you will be lucky to accomplish half of the items on your daily list. Shift your schedule as needed to accomplish the most important items.

Set up a study routine in a distraction-free environment. Develop a written daily study schedule and stick to it. Study in a quiet, well-lighted space. Work sitting at a desk or table—not lying down on a couch or bed. Take breaks every hour or so. During each break take several deep breaths and move around to help you stay more alert and focused.

Avoid procrastination—putting work off until another time. Do not fall behind on your reading and other assignments. Accomplish this by setting aside a particular time for studying each day and making it a part of your daily routine.

Do not eat dessert first. Otherwise, you may never get to the main meal (studying). When you have accomplished your study goals then reward yourself with play (dessert).

Make hills out of mountains. It is psychologically difficult to climb a mountain such as reading an entire book, reading a chapter in a book, writing a paper, or cramming to study for a test. Instead, break such large tasks (mountains) down into a series of small tasks (hills). Each day read a few pages of a book or chapter, write a few paragraphs of a paper, and review what you have studied and learned. As Henry Ford put it, "Nothing is particularly hard if you divide it into small jobs."

Look at the big picture first. Get an overview of an assigned reading by looking at the main headings or chapter outline. In this textbook, I provide a list of the main questions that are the focus of each chapter.

Ask and answer questions as you read. For example, what is the main point of this section or paragraph? To help you do this I start each subsection with a question that the material is designed to answer. Then to help you further I follow this question with a brief headline to give you an idea of how the question will be answered. Then I provide a one-sentence summary of the key material in the subsection. This provides you with a running summary of the text. I find this type of summary to be more helpful than providing a harder to digest summary at the end of each chapter. My goal is to present the material in more manageable bites. You can also use the one-sentence summaries as a way to review what you have learned. Putting them all together gives you a summary of the chapter.

Focus on key terms. Use the glossary in your textbook to look up the meaning of terms or words you do not understand. Make flash cards for learning key terms and concepts and review them frequently. This book shows all key terms in **boldfaced** type and lesser but still important terms in *italicized* type. Flash cards for testing your mastery of key terms for each chapter are available on the website for this book.

Interact with what you read. I do this by marking key sentences and paragraphs with a highlighter or pen. I put an asterisk in the margin next to an idea I think is important and double asterisks next to an idea I think

is especially important. I write comments in the margins, such as *Beautiful, Confusing, Misleading,* or *Wrong.* I fold down the top corner of pages with highlighted passages and the top and bottom corners of especially important pages. This way, I can flip through a chapter or book and quickly review the key ideas. The website for this book has a list of review questions and a multiple-choice practice quiz.

Use the audio version of this book for study and review. You can listen to the chapters in this book while walking, traveling, sitting in your room, or working out. You can use the pin code enclosed in every new copy of this book to download book chapters free from the Web to any MP3 player.

Review to reinforce learning. Before each class review the material you learned in the previous class and read the assigned material. Review, fill in, and organize your notes as soon as possible after each class.

Become a better note taker. Do not try to take down everything your instructor says. Instead take down main points and key facts using your own shorthand system. Fill in and organize your notes after class.

Write out answers to questions to focus and reinforce learning. Answer questions at the end of each chapter or those assigned to you and the review questions on the website for each chapter. Do this in writing as if you were turning them in for a grade. Save your answers for review and preparation for tests.

Use the buddy system. Study with a friend or become a member of a study group to compare notes, review material, and prepare for tests. Explaining something to someone else is a great way to focus your thoughts and reinforce your learning. If available, attend review sessions offered by instructors or teaching assistants.

Learn your instructor's test style. Does your instructor emphasize multiple choice, fill-in-the-blank, true-or-false, factual, thought, or essay questions? How much of the test will come from the textbook and how much from lecture material? Adapt your learning and studying methods to this style. You may disagree with this style and feel that it does not adequately reflect what you know. But the reality is that your instructor is in charge.

Become a better test taker. Avoid cramming. Eat well and get plenty of sleep before a test. Arrive on time or early. Calm yourself and increase your oxygen intake by taking several deep breaths. Do this about every 10–15 minutes. Look over the test and answer the questions you know well first. Then work on the harder ones. Use the process of elimination to narrow down the choices for multiple-choice questions. Getting it down to two choices gives you a 50% chance of guessing the right answer. For essay questions, organize your thoughts before you start writing. If you have no idea what a question means, make an educated guess. You might get some partial credit and avoid a zero. Another strategy for getting some credit is to show your knowledge and reasoning by writing: "If this question means so and so, then my answer is _____."

Develop an optimistic outlook. Try to be a "glass is half-full" rather than a "glass is half-empty" person. Pessimism, fear, anxiety, and excessive worrying (especially over things you have no control over) are destructive and lead to inaction. Try to keep your energizing feelings of optimism slightly ahead of your immobilizing feelings of pessimism. Then you will always be moving forward.

Take time to enjoy life. Every day take time to laugh and enjoy nature, beauty, and friendship. Becoming an effective and efficient learner is the best way to do this without getting behind and living under a cloud of guilt and anxiety.

How Can You Improve Your Critical Thinking Skills? Detecting Baloney

Learning how to think critically is a skill you will need throughout your life.

Every day we are exposed to a sea of information, ideas, and opinions. How do we know what to believe and why? Do the claims seem reasonable or exaggerated?

Critical thinking involves developing skills to help you analyze and evaluate the validity of information and ideas you are exposed to and to make decisions. Critical thinking skills help you decide rationally what to believe or what to do. This involves examining information and conclusions or beliefs in terms of the evidence and chain of logical reasoning that supports them. Critical thinking helps you distinguish between facts and opinions, evaluate evidence and arguments, take and defend an informed position on issues, integrate information and see relationships, and apply your knowledge to dealing with new and different problems. Here are some basic skills for learning how to think more critically.

Question everything and everybody. Be skeptical, as any good scientist is. Do not believe everything you hear or read, including the content of this textbook. Evaluate all information you receive. Seek other sources and opinions. As Albert Einstein put it, "The important thing is not to stop questioning."

Do not believe everything you read on the Internet. The Internet is a wonderful and easily accessible source of information. It is also a useful way to find alternative information and opinions on almost any subject or issue—much of it not available in the mainstream media and scholarly articles. However, because the Internet is so open, anyone can write anything they want with no editorial control or peer evaluation—the method in which scientific or other experts in an area review and comment on an article before it is accepted for publication in a scholarly journal. As a result, evaluating information on the Internet is one of the best ways to put into practice the principles of critical thinking discussed here. Use and enjoy the Internet, but think critically and proceed with caution.

Identify and evaluate your personal biases and beliefs. Each of us has biases and beliefs taught to us by sources such as parents, teachers, friends, role models, and experience. What are your basic beliefs and biases? Where did they come from? What basic assumptions are they based on? How sure are you that your beliefs and assumptions are right and why? According to William James, "A great many people think they are thinking when they are merely rearranging their prejudices."

Be open-minded, flexible, and humble. Be open to considering different points of view, suspend judgment until you gather more evidence, and be capable of changing your mind. Recognize that there may be a number of useful and acceptable solutions to a problem and that very few issues are black or white. There are usually valid points on both (or many) sides of an issue. One way to evaluate divergent views is to get into another person's head or walk in his or her shoes. How do they see or view the world? What are their basic assumptions and beliefs? Is their position logically consistent with their assumptions and beliefs? And be humble about what you know. According to Will Durant, "Education is a progressive discovery of our own ignorance."

Evaluate how the information related to an issue was obtained. Are the statements made based on firsthand knowledge or research or on hearsay? Are unnamed sources used? Is the information based on reproducible and widely accepted scientific studies (*sound* or *consensus science*, p. 36) or preliminary scientific results that may be valid but need further testing (*frontier science*, p. 36)? Is the information based on a few isolated stories or experiences (*anecdotal information*) instead of carefully controlled studies? Is it based on unsubstantiated and widely doubted scientific information or beliefs (*junk science* or *pseudoscience*)? You need to know how to detect junk science, as discussed on p. 36.

Question the evidence and conclusions presented. What are the conclusions or claims? What evidence is presented to support them? Does the evidence support them? Is there a need to gather further evidence to test the conclusions? Are there other, more reasonable conclusions?

Try to identify and assess the assumptions and beliefs of those presenting evidence and drawing conclusions. What is their expertise in this area? Do they have any unstated assumptions, beliefs, biases, or values? Do they have a personal agenda? Can they benefit financially or politically from acceptance of their evidence and conclusions? Would investigators with different basic assumptions or beliefs take the same data and come to different conclusions?

Do the arguments used involve common logical fallacies or debating tricks? Here are six of many examples. *First,* attack the presenter of an argument rather than the argument itself. *Second,* appeal to emotion rather than facts and logic. *Third,* claim that if one piece of evidence or one conclusion is false, then all other pieces of evidence and conclusions are false. *Fourth,* say that a conclusion is false because it has not been scientifically proven (scientists can never prove anything absolutely but they can establish degrees of reliability, as discussed on p. 35). *Fifth,* inject irrelevant or misleading information to divert attention from important points. *Sixth,* present only either/or alternatives when there may be a number of alternatives.

Become a seeker of wisdom, not a vessel of information. Develop a written list of principles, concepts, and rules to serve as guidelines in evaluating evidence and claims and making decisions. Continually evaluate and modify this list on the basis of experience. Many people believe that the main goal of education is to learn as much as you can by concentrating on gathering more and more information—much of it useless or misleading. I believe that the primary goal is to know as little as possible. This is done by learning how to sift through mountains of facts and ideas to find the few *nuggets of wisdom* that are the most useful in understanding the world and in making decisions. This takes a firm commitment to learning how to think logically and critically and continually flushing less valuable and thought-clogging information from our minds. This book is full of facts and numbers, but they are useful only to the extent that they lead to an understanding of key and useful ideas, scientific laws, concepts, principles, and connections. A major goal of the study of environmental science is to find out how nature works and sustains itself (*environmental wisdom*) and use *principles of environmental wisdom* to help make our human societies and economies more sustainable and thus more beneficial and enjoyable. As Sandra Carey put it, "Never mistake knowledge for wisdom. One helps you make a living; the other helps you make a life."

If you want more detailed information on study habits and critical thinking, see the booklet *Essential Study Skills for Science Students* by Daniel D. Chiras that comes with most versions of this book.

How Have I Attempted to Achieve Balance? Evaluate Trade-Offs

There are no simple answers to the environmental problems we face.

There are always *trade-offs* involved in making and implementing environmental decisions. My challenge is to give a fair and balanced presentation of different viewpoints, advantages and disadvantages of various technologies and proposed solutions to environmental problems, and good and bad news about environmental problems without injecting personal bias.

Studying a subject as important as environmental science and ending up with no conclusions, opinions, and beliefs means that both the teacher and student have failed. However, such conclusions should be based on using critical thinking to evaluate different ideas and to understand the trade-offs involved.

How Can You Improve This Book? Keep Those Ideas Coming

I welcome your help in improving this book.

Researching and writing a book that covers and connects ideas in such a wide variety of disciplines is a challenging and exciting task. Almost every day I learn about some new connection in nature.

In a book this complex, there are bound to be some errors—some typographical mistakes that slip through and some statements that you might question based on your knowledge and research. My goal is to provide you with an interesting, accurate, balanced, and challenging book that furthers your understanding of this vital subject. I have also attempted to provide balance in presenting various sides of key environmental issues.

I invite you to contact me and point out any remaining bias, correct any errors you find, and suggest ways to improve this book. Over decades of teaching some of my best teachers have been students taking my classes and reading my textbooks. Please e-mail your suggestions to me at

http://mtg89@hotmail.com

Now start your journey into this fascinating and important study of how the earth works and how we can leave the planet in at least as good a shape as we found it. Have fun.

Study nature love nature, stay close to nature. It will never fail you.
FRANK LLOYD WRIGHT

1 Environmental Problems, Their Causes, and Sustainability

CASE STUDY
Living in an Exponential Age

Two ancient kings enjoyed playing chess, with the winner claiming a prize from the loser. After one match, the winner asked the loser to pay him by placing one grain of wheat on the first square of the chessboard, two on the second, four on the third, and so on, with the number doubling on each square until all 64 were filled.

The losing king, thinking he was getting off easy, agreed with delight. It was the biggest mistake he ever made. He bankrupted his kingdom because the number of grains of wheat he had promised was probably more than all the wheat that has ever been harvested!

This fictional story illustrates the concept of **exponential growth,** in which a quantity increases at a constant rate per unit of time such as 2% a year. Exponential growth is deceptive. It starts off slowly, but after only a few doublings, it grows to enormous numbers because each doubling is more than the total of all earlier growth.

Here is another example. Fold a piece of paper in half to double its thickness. If you could continue doubling the thickness of the paper 42 times, the stack would reach from the earth to the moon, 386,400 kilometers (240,000 miles) away. If you could double it 50 times, the folded paper would almost reach the sun, 149 million kilometers (93 million miles) away!

Between 1950 and 2004, the world's population increased exponentially from 2.5 billion to 6.4 billion and may increase to somewhere between 8 billion and 12 billion people by the end of this century (Figure 1-1).

Global economic output—some of it environmentally beneficial and some of it environmentally harmful—is a rough measure of the hu-

man use of the earth's resources. It has increased sevenfold since 1950 and is projected to continue increasing exponentially at a rapid rate.

Despite a 22-fold increase in economic growth since 1900, almost *one of every two people in the world try to survive on an income of less than $3 (U.S.) per day.*

Such poverty affects environmental quality because to survive many of the poor must deplete and degrade local forests, grasslands, soil, and wildlife.

Biologists estimate that human activities are causing premature extinction of the earth's species at an exponential rate of 0.1% to 1% a year—an irreversible loss of the earth's great variety of life forms, or *biodiversity.* In various parts of the world, forests, grasslands, wetlands, and coral reefs continue to disappear or become degraded as the human ecological footprint continues to spread exponentially across the globe.

There is growing concern that exponential growth in human activities such as burning fossil fuels and clearing forests will change the earth's climate during this century. This could ruin some areas for farming, shift water supplies, alter and reduce biodiversity, and disrupt economies in various parts of the world.

Exponential growth plays a key role in five important and interconnected environmental issues: *population growth, resource use and waste, poverty, loss of biological diversity,* and *global climate change. Great news.* We have solutions to these problems that we could implement within a few decades, as you will learn in this book.

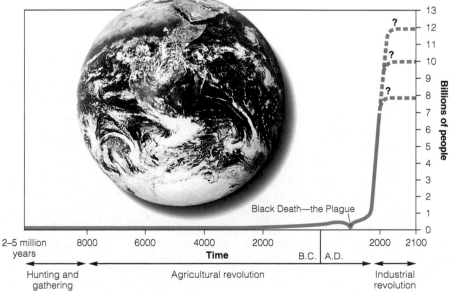

Figure 1-1 The *J*-shaped curve of past exponential world population growth, with projections to 2100. Notice that exponential growth starts off slowly, but as time passes the curve becomes increasingly steep. The current world population of 6.4 billion people is projected to reach 8–12 billion people sometime this century. (This figure is not to scale.) (Data from World Bank and United Nations; photo courtesy of NASA)

Alone in space, alone in its life-supporting systems, powered by inconceivable energies, mediating them to us through the most delicate adjustments, wayward, unlikely, unpredictable, but nourishing, enlivening, and enriching in the largest degree—is this not a precious home for all of us? Is it not worth our love?

BARBARA WARD AND RENÉ DUBOS

This chapter presents an overview of environmental problems, their causes, controversy over their seriousness, and ways we can live more sustainably. It discusses these questions:

- What keeps us alive? What is an environmentally sustainable society?

- How fast is the human population increasing? What are economic growth, economic development, and globalization?

- What are the earth's main types of resources? How can they be depleted or degraded?

- What are the principal types of pollution? What can we do about pollution?

- What are the basic causes of today's environmental problems? How are these causes connected?

- Is our current course sustainable? What is environmentally sustainable development?

1-1 LIVING MORE SUSTAINABLY

What Is the Difference between Environment, Ecology, and Environmental Science? Defining Some Basic Terms

Environmental science is a study of how the earth works, how we interact with the earth, and how to deal with environmental problems.

Environment is everything that affects a living organism (any unique form of life). **Ecology** is a biological science that studies the relationships between living organisms and their environment.

This textbook is an introduction to **environmental science,** an interdisciplinary study that uses information from the physical sciences and social sciences to learn how the earth works, how we interact with the earth, and how to deal with environmental problems. Environmental science involves integrating ideas from the *natural world* (*biosphere*) and our *cultural world* (*culturesphere*).

Environmentalism is a social movement dedicated to protecting the earth's life support systems for us and other species. Members of the environmental community include *ecologists, environmental scientists, conservation biologists, conservationists, preservationists, restorationists,* and *environmentalists.*

What Keeps Us Alive? The Sun and the Earth's Natural Capital

All life and economies depend on energy from the sun (solar capital) and the earth's resources and ecological services (natural capital).

Our existence, lifestyles, and economies depend completely on the sun and the earth, a blue and white island in the black void of space (Figure 1-1). To economists, *capital* is wealth used to sustain a business and to generate more wealth. For example, suppose you invest $100,000 of capital and get a 10% return on your money. In a year you get $10,000 in income from interest and increase your wealth to $110,000.

By analogy, we can think of energy from the sun as **solar capital. Solar energy** includes direct sunlight and indirect forms of renewable solar energy such as *wind power, hydropower* (energy from flowing water), and *biomass* (direct solar energy converted to chemical energy and stored in biological sources of energy such as wood).

Similarly, we can think of the planet's air, water, soil, wildlife, forest, rangeland, fishery, mineral, and energy resources and the processes of natural purification, recycling, and pest control as **natural resources** or **natural capital** (Figure 1-2). See the Guest Essay by Paul Hawken on the website for this chapter.

Natural capital consists of *resources* (orange in Figure 1-2) and *ecological services* (green in Figure 1-2) that support and sustain the earth's life and economies. This priceless natural capital that nature provides at no cost to us plus the natural biological income it supplies can sustain the planet and our economies indefinitely as long as we do not deplete them. Examples of *biological income* are renewable supplies of wood, fish, grassland for grazing, and underground water for drinking and irrigation.

What Is an Environmentally Sustainable Society? One That Preserves Natural Capital and Lives Off Its Income

An environmentally sustainable society meets the basic resource needs of its people indefinitely without degrading or depleting the natural capital that supplies these resources.

An **environmentally sustainable society** meets the current needs of its people for food, clean water, clean air, shelter, and other basic resources without compromising the ability of future generations to meet their needs. *Living sustainably* means living off natural income replenished by soils, plants, air, and water and not depleting or degrading the earth's natural capital that supplies this biological income.

Imagine you win $1 million in a lottery. Invest this capital at 10% interest per year, and you will have a

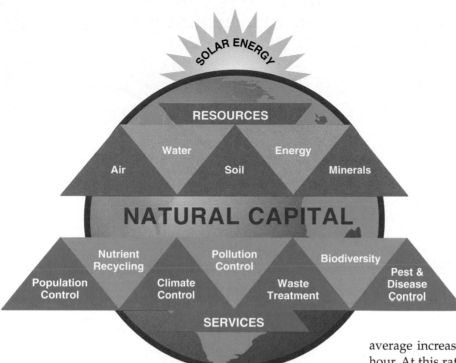

Figure 1-2 The earth's *natural capital.* Energy from the sun (solar capital) and the earth's natural capital provide resources (orange) and ecological services (green) that support and sustain the earth's life and economies. Wedges from this diagram will be used near the titles of various chapters to indicate the components of natural capital that are the primary focus of such chapters. This diagram also appears on the back cover of this book.

sustainable annual income of $100,000 without depleting your capital. If you spend $200,000 a year, your $1 million will be gone early in the 7th year and even if you spend only $110,000 a year, you will be bankrupt early in the 18th year.

The lesson here is an old one: *Protect your capital and live off the income it provides.* Deplete, waste, or squander your capital, and you move from a sustainable to an unsustainable lifestyle.

The same lesson applies to the earth's natural capital. According to many environmentalists and leading scientists, we are living unsustainably by wasting, depleting, and degrading the earth's natural capital at an accelerating rate.

Some people disagree. They contend that environmentalists have exaggerated the seriousness of population, resource, and environmental problems. They also believe we can overcome these problems by human ingenuity, economic growth, and technological advances.

1-2 POPULATION GROWTH, ECONOMIC GROWTH, ECONOMIC DEVELOPMENT, AND GLOBALIZATION

How Rapidly Is the Human Population Growing? Pretty Fast

The rate at which the world's population is growing has slowed but is still growing pretty rapidly.

Currently the world's population is growing exponentially at a rate of about 1.25% a year. This does not seem like a very fast rate. But it added about 80 million people (6.4 billion × 0.0125 = 80 million) to the world's population in 2004, an average increase of 219,000 people a day, or 9,100 an hour. At this rate it takes only about 3 days to add the 651,000 Americans killed in battle in all U.S. wars and only 1.6 years to add the 129 million people killed in all wars fought in the past 200 years!

How much is 80 million? Suppose you spend 1 second saying hello to each of the 80 million new people added this year for 24 hours a day—no sleeping, eating or anything else allowed. How long would this handshaking marathon take? Answer: 2.5 years. By then there would be about 192 million more people to shake hands with. Exponential growth is astonishing!

What Is the Difference between Economic Growth and Economic Development? More Stuff and Better Living Standards

Economic growth provides people with more goods and services and economic development uses economic growth to improve living standards.

Economic growth is an increase in the capacity of a country to provide people with goods and services. Accomplishing this increase requires population growth (more producers and consumers), more production and consumption per person, or both.

Economic growth is usually measured by the percentage change in a country's **gross domestic product (GDP):** the annual market value of all goods and services produced by all firms and organizations, foreign and domestic, operating within a country. Changes in a country's standard of living is measured by **per capita GDP:** the GDP divided by the total population at midyear.

Economic development is the improvement of living standards by economic growth. The United Nations

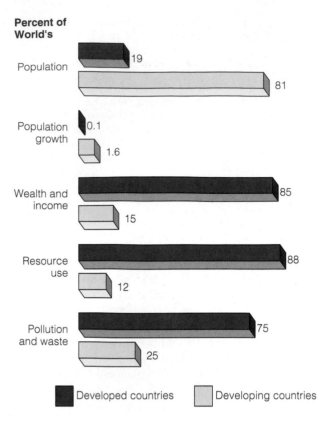

Percent of World's

Population
19
81

Population growth
0.1
1.6

Wealth and income
85
15

Resource use
88
12

Pollution and waste
75
25

◼ Developed countries ◻ Developing countries

Figure 1-3 Comparison of developed and developing countries. (Data from United Nations and the World Bank)

(UN) classifies the world's countries as economically developed or developing based primarily on their degree of industrialization and their per capita GDP.

The **developed countries** (with 1.2 billion people) include the United States, Canada, Japan, Australia, New Zealand, and the countries of Europe. Most are highly industrialized and have high average per capita GDP. All other nations (with 5.2 billion people) are classified as **developing countries,** most of them in Africa, Asia, and Latin America. Some are *middle-income, moderately developed countries* and others are *low-income countries.*

Figure 1-3 compares some key characteristics of developed and developing countries. About 97% of the projected increase in the world's population is expected to take place in developing countries (Figure 1-4).

Figure 1-5 summarizes some of the benefits (*good news*) and harm (*bad news*) caused mostly by economic development. It shows effects of the wide and increasing gap between the world's haves and have-nots.

What Is Globalization? Being Connected

We live in a world that is increasingly interconnected through economic, cultural, and environmental interdependence.

You have probably heard about **globalization:** the process of social, economic, and environmental global

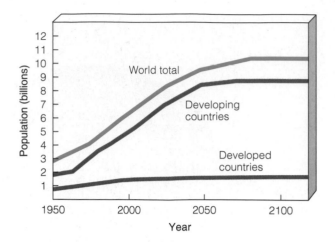

Figure 1-4 Past and projected population size for developed countries, developing countries, and the world, 1950–2100. Developing countries are expected to account for 97% of the 2.5 billion people projected to be added to the world's population between 2004 and 2050. (Data from United Nations)

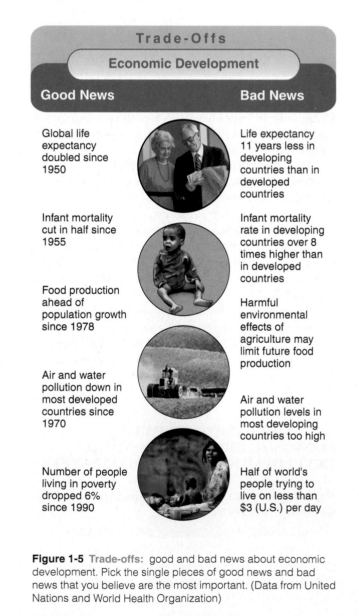

Trade-Offs

Economic Development

Good News | **Bad News**

Global life expectancy doubled since 1950 | Life expectancy 11 years less in developing countries than in developed countries

Infant mortality cut in half since 1955 | Infant mortality rate in developing countries over 8 times higher than in developed countries

Food production ahead of population growth since 1978 | Harmful environmental effects of agriculture may limit future food production

Air and water pollution down in most developed countries since 1970 | Air and water pollution levels in most developing countries too high

Number of people living in poverty dropped 6% since 1990 | Half of world's people trying to live on less than $3 (U.S.) per day

Figure 1-5 **Trade-offs:** good and bad news about economic development. Pick the single pieces of good news and bad news that you believe are the most important. (Data from United Nations and World Health Organization)

changes that lead to an increasingly interconnected world. It involves increasing exchanges of people, products, services, capital, and ideas across international borders.

Factors accelerating globalization include information and communication technologies, human mobility, and international trade and investment. Modern communication via cell phones and the Internet also allows powerless people throughout the world to share ideas and to band together to bring about change from the bottom up.

This decentralized network, where everyone has access to everyone else, represents a *democratization of learning and communication* that is unprecedented in human history. A sustainable community or country recognizes that it is part of a larger global economic and ecological system and that it cannot be sustainable unless these larger systems are also sustainable.

1-3 RESOURCES

What Is a Resource? Things We Need or Want

We obtain resources from the environment to meet our needs and wants.

From a human standpoint, a **resource** is anything obtained from the environment to meet our needs and wants. Examples include food, water, shelter, manufactured goods, transportation, communication, and recreation. On our short human time scale, we classify the material resources we get from the environment as *perpetual, renewable,* or *nonrenewable,* as shown in Figure 1-6.

Some resources, such as solar energy, fresh air, wind, fresh surface water, fertile soil, and wild edible plants, are directly available for use. Other resources, such as petroleum (oil), iron, groundwater (water found underground), and modern crops, are not directly available. They become useful to us only with some effort and technological ingenuity. For example, petroleum was a mysterious fluid until we learned how to find and extract it and refine it into gasoline, heating oil, and other products that we could sell at affordable prices.

What Are Perpetual and Renewable Resources? Resources That Can Last

Resources renewed by natural processes are sustainable if we do not use them faster than they are replenished.

Solar energy is called a **perpetual resource** because on a human time scale it is renewed continuously. It is expected to last at least 6 billion years as the sun completes its life cycle as a star.

On a human time scale, a **renewable resource** can be replenished fairly rapidly (from hours to several

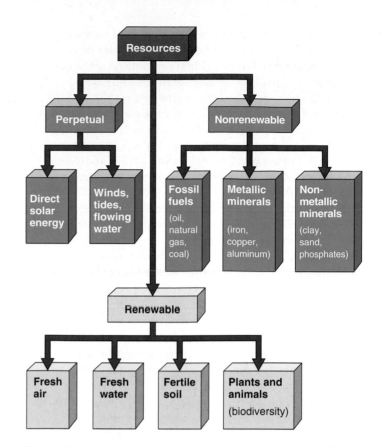

Figure 1-6 Natural capital: major types of material resources. This scheme is not fixed; renewable resources can become nonrenewable if used for a prolonged period at a faster rate than natural processes renew them.

decades) through natural processes. But this works only as long as the resource is not used up faster than it is replaced. Examples of renewable resources are forests, grasslands, wild animals, fresh water, fresh air, and fertile soil.

Renewable resources can be depleted or degraded. The highest rate at which a renewable resource can be used *indefinitely* without reducing its available supply is called its **sustainable yield.**

When we exceed a renewable resource's natural replacement rate, the available supply begins to shrink, a process known as **environmental degradation.** Examples include urbanization of productive land, excessive topsoil erosion, pollution, deforestation (temporary or permanent removal of large expanses of forest for agriculture or other uses), groundwater depletion, overgrazing of grasslands by livestock, and reduction in the earth's forms of wildlife (biodiversity) by elimination of habitats and species.

Case Study: The Tragedy of the Commons— Degrading Free Renewable Resources

Renewable resources that are freely available to everyone can be degraded.

One cause of environmental degradation of renewable resources is the overuse of **common-property** or **free-access resources.** No individual owns these resources, and they are available to users at little or no charge.

Examples include clean air, the open ocean and its fish, migratory birds, wildlife species, publicly owned lands (such as national forests and national parks), gases of the lower atmosphere, and space.

In 1968, biologist Garrett Hardin (1915–2003) called the degradation of renewable free-access resources the **tragedy of the commons.** It happens because each user reasons, "If I do not use this resource, someone else will. The little bit I use or pollute is not enough to matter, and such resources are renewable."

With only a few users, this logic works. But the cumulative effect of many people trying to exploit a free-access resource eventually exhausts or ruins it. Then no one can benefit from it, and that is the tragedy.

One solution is to use free-access resources at rates well below their estimated sustainable yields by reducing population, regulating access to the resources, or both. Some communities have established rules and traditions to regulate and share their access to common-property resources such as ocean fisheries, grazing lands, and forests. Governments have also enacted laws and international treaties to regulate access to commonly owned resources such as forests, national parks, rangelands, and fisheries in coastal waters.

Another solution is to *convert free-access resources to private ownership.* The reasoning is that if you own something, you are more likely to protect your investment.

This sounds good, but private ownership is not always the answer. One problem is private owners do not always protect natural resources they own when this conflicts with protecting their financial capital or increasing their profits. For example, some private forest owners can make more money by clear-cutting the timber, selling the degraded land, and investing their profits in other timberlands or businesses.

A second problem is that this approach is not practical for global common resources—such as the atmosphere, the open ocean, most wildlife species, and migratory birds—that cannot be divided up and converted to private property.

What Is Our Ecological Footprint? Our Growing Environmental Impact

Supplying each person with renewable resources and absorbing the wastes from such resource use creates a large ecological footprint or environmental impact.

The **per capita ecological footprint** is the amount of biologically productive land and water needed to supply each person or population with the renewable resources they use and to absorb or dispose of the wastes from such resource use. It measures the average environmental impact of individuals in different countries and areas. In other words, it is a measure of how much of the earth's natural capital and biological income each of us uses.

Bad news. Humanity's ecological footprint per person exceeds the earth's biological capacity to replenish renewable resources and absorb waste by about 15% (Figure 1-7, right). If these estimates are correct, *it will*

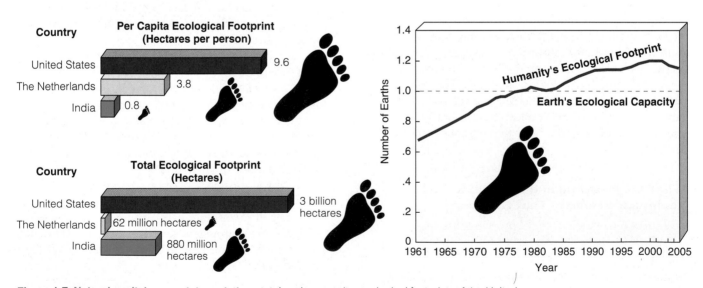

Figure 1-7 Natural capital use and degradation: total and per capita *ecological footprints* of the United States, the Netherlands, and India (left). The *ecological footprint* is a measure of the biologically productive areas of the earth required to produce the renewable resources required per person and absorb or break down the wastes produced by such resource use. Currently, humanity's average ecological footprint per person is 15% higher than the earth's biological capacity per person (right). (Data from William Rees and Mathis Wackernagel, *Redefining Progress*, 2004)

take the resources of 1.15 planet earths to indefinitely support our current use of renewable resources!

The ecological footprint of each person in developed countries is large compared to that in developing countries (Figure 1-7, left). The per capita ecological footprint of the United States is nearly double the country's biological capacity per person—explaining why the country spreads its ecological footprint by importing large quantities of renewable resources from other countries. You can estimate your ecological footprint by going to the website www.redefiningprogress.org/. Also, see the Guest Essay by Michael Cain on the website for this chapter.

This eventually unsustainable situation is expected to get worse as affluence increases in both developed and developing countries. According to William Rees and Mathis Wachernagel, developers of the ecological footprint concept, it would take the land area of about *four more planet earths* for the rest of the world to reach U.S. levels of consumption with existing technology. Clearly, such consumption patterns cannot be sustained.

A new country with a large and growing ecological footprint is emerging. China has the world's largest population and hopes to increase its total and per capita economic growth, which will increase the ecological footprints of its people. See the Guest Essay on this topic by Norman Myers on the website for this chapter.

What Are Nonrenewable Resources? Resources We Can Deplete

Nonrenewable resources can be economically depleted to the point where it costs too much to obtain what is left.

Nonrenewable resources exist in a fixed quantity or stock in the earth's crust. On a time scale of millions to billions of years, geological processes can renew such resources. But on the much shorter human time scale of hundreds to thousands of years, these resources can be depleted much faster than they are formed.

These exhaustible resources include *energy resources* (such as coal, oil, and natural gas that cannot be recycled), *metallic mineral resources* (such as iron, copper, and aluminum that can be recycled), and *nonmetallic mineral resources* (such as salt, clay, sand, and phosphates that usually are difficult or too costly to recycle).

Figure 1-8 shows the production and depletion cycle of a nonrenewable energy or mineral resource. We never completely exhaust such a resource, but it becomes *economically depleted* when the costs of extracting and using what is left exceed its economic value. At that point, we have six choices: try to find more, recycle or reuse existing supplies (except for nonrenewable energy resources, which cannot be recycled or reused), waste less, use less, try to develop a substitute, or wait millions of years for more to be produced.

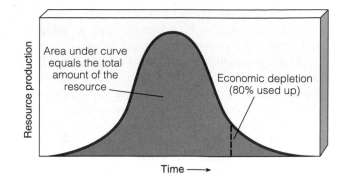

Figure 1-8 Full production and exhaustion cycle of a nonrenewable resource such as copper, iron, oil, or coal. Usually, a nonrenewable resource is considered *economically depleted* when 80% of its total supply has been extracted and used. Normally, it costs too much to extract and process the remaining 20%.

Some nonrenewable mineral resources, such as copper and aluminum, can be recycled or reused to extend supplies. **Recycling** involves collecting waste materials, processing them into new materials, and selling these new products. For example, discarded aluminum cans can be crushed and melted to make new aluminum cans or other aluminum items that consumers can buy. Recycling means nothing if we do not close the loop by buying products that are made from or contain recycled materials. **Reuse** is using a resource again in the same form. For example, glass bottles can be collected, washed, and refilled many times.

Recycling nonrenewable metallic resources takes much less energy, water, and other resources and produces much less pollution and environmental degradation than exploiting virgin metallic resources. Reusing such resources takes even less energy and other resources and produces less pollution and environmental degradation than recycling.

1-4 POLLUTION

Where Do Pollutants Come From, and What Are Their Harmful Effects? Threats to Health and Survival

Pollutants are chemicals found at high enough levels in the environment to cause harm to people or other organisms.

Pollution is the presence of substances at high enough levels in air, water, soil, or food to threaten the health, survival, or activities of humans or other organisms. Pollutants can enter the environment naturally (for example, from volcanic eruptions) or through human or anthropogenic activities (for example, from burning coal). Most pollution from human activities occurs in or near urban and industrial areas, where pollution sources such as cars and factories are concentrated. Industrialized agriculture is also a major source of pollution. Most pollutants are unintended by products of useful activities such as burning coal to generate electricity, driving cars, and growing crops.

Some pollutants contaminate the areas where they are produced and some are carried by wind or flowing water to other areas. Pollution does not respect the neat territorial political lines we draw on maps.

The pollutants we produce come from two types of sources. **Point sources** of pollutants are single, identifiable sources. Examples are the smokestack of a coal-burning power plant, the drainpipe of a factory, and the exhaust pipe of an automobile. **Nonpoint sources** of pollutants are dispersed and often difficult to identify. Examples are pesticides sprayed into the air or blown by the wind into the atmosphere and runoff of fertilizers and pesticides from farmlands, golf courses, and suburban lawns and gardens into streams and lakes. It is much easier and cheaper to identify and control pollution from point sources than from widely dispersed nonpoint sources.

Pollutants can have three types of unwanted effects. *First,* they can disrupt or degrade life-support systems for humans and other species. *Second,* they can damage wildlife, human health, and property. *Third,* they can be nuisances such as noise and unpleasant smells, tastes, and sights.

Solutions: What Can We Do about Pollution? Prevention Pays

We can try to clean up pollutants in the environment or prevent them from entering the environment.

We use two basic approaches to deal with pollution. One is **pollution prevention,** or **input pollution control,** which reduces or eliminates the production of pollutants. The other is **pollution cleanup,** or **output pollution control,** which involves cleaning up or diluting pollutants after they have been produced.

Environmentalists have identified three problems with relying primarily on pollution cleanup. *First,* it is only a temporary bandage as long as population and consumption levels grow without corresponding improvements in pollution control technology. For example, adding catalytic converters to car exhaust systems has reduced some forms of air pollution. But increases in the number of cars and in the distance each travels have reduced the effectiveness of this approach.

Second, cleanup often removes a pollutant from one part of the environment only to cause pollution in another. For example, we can collect garbage, but the garbage is then *burned* (perhaps causing air pollution and leaving toxic ash that must be put somewhere), *dumped* into streams, lakes, and oceans (perhaps causing water pollution), or *buried* (perhaps causing soil and groundwater pollution).

Third, once pollutants have entered and become dispersed into the environment at harmful levels, it usually costs too much to reduce them to acceptable levels.

Both pollution prevention (front-of-the-pipe) and pollution cleanup (end-of-the-pipe) solutions are needed. But environmentalists and some economists urge us to put more emphasis on prevention because it works better and is cheaper than cleanup. As Benjamin Franklin observed long ago, "An ounce of prevention is worth a pound of cure."

1-5 ENVIRONMENTAL AND RESOURCE PROBLEMS: CAUSES AND CONNECTIONS

What Are Key Environmental Problems and Their Basic Causes? The Big Five

The major causes of environmental problems are population growth, wasteful resource use, poverty, poor environmental accounting, and ecological ignorance.

We face a number of interconnected environmental and resource problems, as listed in Figure 1-9. The first step in dealing with these problems is to identify their

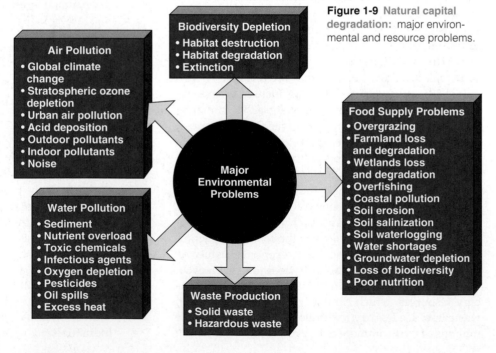

Figure 1-9 Natural capital degradation: major environmental and resource problems.

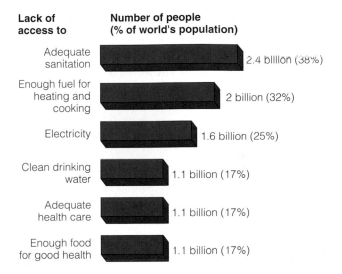

Causes of Environmental Problems

- Rapid population growth
- Unsustainable resource use
- Poverty
- Not including the environmental costs of economic goods and services in their market prices
- Trying to manage and simplify nature with too little knowledge about how it works

Figure 1-10 Environmentalists have identified five basic causes of the environmental problems we face.

underlying causes, listed in Figure 1-10 and sometimes known as the *big five*.

Four of these causes are rapid population growth (p. 7), poverty (discussed below), and excessive and wasteful use of resources (discussed on p. 14) A fourth is failure to include the harmful environmental costs of items in their market prices, discussed in Chapter 26. This in turn is a policy and political failure to address this issue. The fifth, inadequate understanding of how the earth works, is discussed throughout this book.

What Is the Relationship between Poverty and Environmental Problems? Being Poor Is Bad for People and the Earth

Poverty is a major threat to human health and the environment.

Many of the world's poor do not have access to the basic necessities for a healthy, productive, and decent life, as listed in Figure 1-11. Their daily lives are focused on getting enough food, water, and fuel (for cooking and heat) to survive. Desperate for land to grow enough food, many of the world's poor people deplete and degrade forests, soil, grasslands, and wildlife for short-term survival. They do not have the luxury of worrying about long-term environmental quality or sustainability.

Another problem for the poor is living in areas with high levels of air and water pollution and with a great risk of natural disasters such as floods, earthquakes, hurricanes, and volcanic eruptions. And they usually must take jobs—if they can find them—with unhealthy and unsafe working conditions at very low pay.

Poverty also affects population growth. Poor people often have many children as a form of economic security. Their children help them grow food, gather fuel (mostly wood and dung), haul drinking water, tend livestock, work, and beg in the streets. The children also help their parents survive in their old age before they die, typically in their 50s in the poorest countries. The poor do not have retirement plans, social security, or government-sponsored health plans.

Many of the world's desperately poor die prematurely from four preventable health problems. One is *malnutrition* from a lack of protein and other nutrients needed for good health (Figure 1-12). The second is increased susceptibility to normally nonfatal infectious diseases, such as diarrhea and measles, because of their weakened condition from malnutrition. A third factor is lack of access to clean drinking water. A fourth factor is severe respiratory disease and premature death from

Lack of access to	Number of people (% of world's population)
Adequate sanitation	2.4 billion (38%)
Enough fuel for heating and cooking	2 billion (32%)
Electricity	1.6 billion (25%)
Clean drinking water	1.1 billion (17%)
Adequate health care	1.1 billion (17%)
Enough food for good health	1.1 billion (17%)

Figure 1-11 Natural capital degradation: some harmful effects of poverty. Which two of these effects do you believe is the most harmful? (Data from United Nations, World Bank, and World Health Organization)

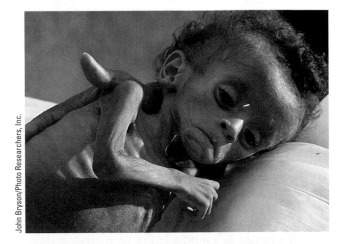

John Bryson/Photo Researchers, Inc.

Figure 1-12 One in every three children under age 5, such as this Brazilian child, suffers from malnutrition. According to the World Health Organization, each day at least 13,700 children under age 5 die prematurely from malnutrition and infectious diseases from drinking contaminated water and other causes.

inhaling indoor air pollutants produced by burning wood or coal for heat and cooking in open fires or in poorly vented stoves. According to the World Health Organization, these four factors cause premature death for at least 7 million of the poor a year.

This premature death of about 19,200 human beings per day is equivalent to 48 fully loaded 400-passenger jumbo jet planes crashing every day with no survivors! Two-thirds of those dying are children under age 5.

What Is the Relationship between Resource Consumption and Environmental Problems? Affluenza

Many consumers in developed countries have become addicted to buying more and more stuff in their search for fulfillment and happiness.

Affluenza ("af-loo-EN-zuh") is a term used to describe the unsustainable addiction to overconsumption and materialism exhibited in the lifestyles of affluent consumers in the United States and other developed countries. It is based on the assumption that buying more and more things can, should, and does buy happiness.

Most people infected with this contagious *shop-till-you-drop virus* have some telltale symptoms. They feel overworked, have high levels of debt and bankruptcy, suffer from increasing stress and anxiety, have declining health, and feel unfulfilled in their quest to accumulate more and more stuff. As humorist Will Rogers said, "Too many people spend money they haven't earned to buy things they don't want, to impress people they don't like." For some, shopping until you drop means shopping until you go bankrupt. Between 1998 and 2001, more Americans declared bankruptcy than graduated from college.

Globalization and global advertising are now spreading the virus throughout much of the world. Affluenza has an enormous environmental impact. It takes about 27 tractor-trailer loads of resources per year to support one American. This amounts to 7.9 billion truckloads of resources a year to support the U.S. population. Stretched end-to-end, these trucks would more than reach the sun!

What can we do about affluenza? The first step for addicts is to admit they have a problem. Then they begin steps to kick their addiction by going on a stuff diet. For example, before buying anything a person with the affluenza addiction should ask: Do I really need this or merely want it? Can I buy it secondhand (reuse)? Can I borrow it from a friend or relative? Another withdrawal strategy: Do not hang out with other addicts. Shopaholics should avoid malls as much as they can.

After a lifetime of studying the growth and decline of the world's human civilizations, historian Arnold Toynbee summarized the true measure of a civilization's growth in what he called the *law of progressive sim-*

plification: "True growth occurs as civilizations transfer an increasing proportion of energy and attention from the material side of life to the nonmaterial side and thereby develop their culture, capacity for compassion, sense of community, and strength of democracy."

How Can Affluence Help Increase Environmental Quality? Another Side of the Story

Affluent countries have more money for improving environmental quality.

Some analysts point out that affluence need not lead to environmental degradation. Instead, it can lead people to become more concerned about environmental quality, and it provides money for developing technologies to reduce pollution, environmental degradation, and resource waste. This explains why most of the important environmental progress made since 1970 has taken place in developed countries.

In the United States, the air is cleaner, drinking water is purer, most rivers and lakes are cleaner, and the food supply is more abundant and safer than in 1970. Also, the country's total forested area is larger than it was in 1900 and most energy and material resources are used more efficiently. Similar advances have been made in most other affluent countries. Affluence financed these improvements in environmental quality.

How Are Environmental Problems and Their Causes Connected? Exploring Connections

Environmental quality is affected by interactions between population size, resource consumption, and technology.

Once we have identified environmental problems and their root causes, the next step is to understand how they are connected to one another. The three-factor model in Figure 1-13 is a starting point.

According to this simple model, the environmental impact (**I**) of a population on a given area depends on three factors: the number of people (**P**), the average resource use per person (affluence, **A**), and the beneficial and harmful environmental effects of the technologies (**T**) used to provide and consume each unit of resource and to control or prevent the resulting pollution and environmental degradation.

In developing countries, population size and the resulting degradation of renewable resources as the poor struggle to stay alive tend to be the key factors in total environmental impact (Figure 1-13, top). In such countries per capita resource use is low.

In developed countries, high rates of per capita resource use and the resulting high levels of pollution and environmental degradation per person usually are the key factors determining overall environmental impact (Figure 1-13, bottom) and a country's ecological

Developing Countries

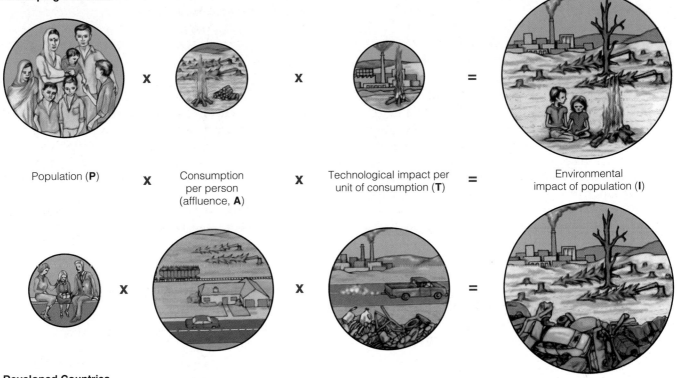

Population (**P**)   **X**   Consumption per person (affluence, **A**)   **X**   Technological impact per unit of consumption (**T**)   **=**   Environmental impact of population (**I**)

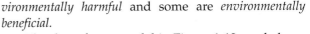

Developed Countries

Figure 1-13 Connections: simplified model of how three factors—number of people, affluence, and technology—affect the environmental impact of the population in developing countries (top) and developed countries (bottom).

footprint per person (Figure 1-7). For example, the average U.S. citizen consumes about 35 times as much as the average citizen of India and 100 times as much as the average person in the world's poorest countries. *Thus poor parents in a developing country would need 70 to 200 children to have the same lifetime resource consumption as 2 children in a typical U.S. family.*

Some forms of technology, such as polluting factories and motor vehicles and energy-wasting devices, increase environmental impact by raising the T factor in the equation. But other technologies, such as pollution control and prevention, solar cells, and energy-saving devices, lower environmental impact by decreasing the T factor. In other words, some forms of technology are *en-vironmentally harmful* and some are *environmentally beneficial.*

The three-factor model in Figure 1-13 can help us understand how key environmental problems and some of their causes are connected. It can also guide us in seeking solutions. However, these problems involve a number of poorly understood interactions between many more factors than those in this simplified model, as outlined in Figure 1-14. Look at the interactions shown in this figure.

Figure 1-14 Connections: major components and interactions within and between the earth's life-support system and the human sociocultural system (culturesphere). The goal of environmental science is to learn as much as possible about these complex interactions.

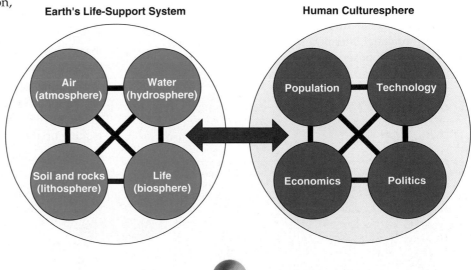

1-6 IS OUR PRESENT COURSE SUSTAINABLE?

Are Things Getting Better or Worse? The Answer Is Both

There is good and bad environmental news.

Experts disagree about how serious our environmental problems are and what we should do about them. Some analysts believe human ingenuity, technological advances, and economic growth and development will allow us to clean up pollution to acceptable levels, find substitutes for resources that become scarce, and keep expanding the earth's ability to support more humans, as we have done in the past. They accuse many scientists and environmentalists of exaggerating the seriousness of the problems we face and failing to appreciate the progress we have made in improving quality of life and protecting the environment.

Environmentalists and many leading scientists disagree with this view. They cite evidence that we are degrading and disrupting many of the world's life-support systems for us and other species at an accelerating rate. They are greatly encouraged by the progress we have made in increasing average life expectancy, reducing infant mortality, increasing food supplies, and reducing many forms of pollution—especially in developed countries. But they point out that we need to use the earth in a way that is more sustainable for present and future human generations and other species that support us and other forms of life.

The most useful answer to the question of whether things are getting better or worse is *both*. Some things are getting better, some worse.

Our challenge is not to get trapped into confusion and inaction by listening primarily to either of two groups of people. One group consists of *technological optimists*. They tend to overstate the situation by telling us to be happy and not worry, because technological innovations and conventional economic growth and development will lead to a wonderworld for everyone. Leave the driving to us because we know best.

The second group consists of *environmental pessimists* who overstate the problems to the point where our environmental situation seems hopeless. According to the noted conservationist Aldo Leopold, "I have no hope for a conservation based on fear."

X *HOW WOULD YOU VOTE?** * Is the society you live in on an unsustainable path? Cast your vote online at http://biology .brookscole.com/miller14.

---

*To cast your vote, go to the website for the book listed above and then go to the appropriate chapter (in this case Chapter 1). In most cases you will be able to compare how you voted with others using this book throughout the United States and the world.

How Should We Live? A Clash of Environmental Worldviews

The way we view the seriousness of environmental problems and how to solve them depends on our environmental worldview.

The differing views about how serious our environmental problems are and what we should do about them arise mostly out of differing environmental worldviews. Your **environmental worldview** is how you think the world works, what you think your role in the world should be, and what you believe is right and wrong environmental behavior (**environmental ethics**).

People who have widely differing environmental worldviews can take the same data, be logically consistent, and arrive at quite different conclusions because they start with different assumptions and values.

Some people in today's industrial consumer societies have a **planetary management worldview**. Here are the basic environmental beliefs of this worldview:

- As the planet's most important species, we are in charge of nature.

- We will not run out of resources because of our ability to develop and find new ones.

- The potential for global economic growth is essentially unlimited.

- Our success depends on how well we manage the earth's life-support systems, mostly for our own benefit.

A second environmental worldview, known as the **stewardship worldview,** consists of the following major beliefs:

- We are the planet's most important species but we have an ethical responsibility to care for the rest of nature.

- We will probably not run out of resources but they should not be wasted.

- We should encourage environmentally beneficial forms of economic growth and discourage environmentally harmful forms of economic growth.

- Our success depends on how well we can manage the earth's life-support systems for our benefit and for the rest of nature.

Another environmental worldview, known as the **environmental wisdom worldview,** is based on the following major beliefs, which are the opposite of those making up the planetary management worldview:

- Nature exists for all species, not just for us and we are not in charge of the earth.

- The earth's resources are limited, should not be wasted, and are not all for us.

- We should encourage earth-sustaining forms of economic growth and discourage earth-degrading forms.

- Our success depends on learning how the earth sustains itself and integrating such lessons from nature (environmental wisdom) into the ways we think and act.

What Are the Greatest Environmental Problems We Face Now and in the Future? The Big Picture

Poverty and malnutrition, smoking, infectious diseases, water shortages, biodiversity loss, and climate changes are the most serious environmental problems we face.

Figure 1-15 ranks major environmental problems on a time scale in terms of the estimated number of people prematurely killed annually today and over the next hundred years.

From this diagram you can see that we should focus our money, minds, and hearts on reducing the environmental risks from *poverty, malnutrition, unsafe drinking water, smoking, air pollution, infectious diseases (AIDS, TB, malaria, and hepatitis B), water shortages, climate changes, and loss and degradation of biodiversity.* The poor in developing countries bear the brunt of most of these serious problems.

X *How Would You Vote?* What do you think is our most serious environmental problem? Cast your vote online at http://biology.brookscole.com/miller14.

What Is Environmentally Sustainable Economic Development? Rewarding Environmentally Beneficial Activities

Environmentally sustainable economic development rewards environmentally beneficial and sustainable activities and discourages environ-mentally harmful and unsustainable activities.

During this century, many analysts call for us to put much greater emphasis on **environmentally sustainable economic development.** Figure 1-16 (p. 18) lists some of the shifts involved in implementing such an *environmental*, or *sustainability*, *revolution* during this century based on this concept. Study this figure carefully.

This type of development uses economic rewards (government subsidies and tax breaks) to *encourage* environmentally beneficial and more sustainable forms of economic growth and economic penalties (government taxes and regulations) to *discourage* environmentally harmful and unsustainable forms of economic growth.

Throughout this book I try to give you a balanced view of good and bad environmental news. Try not to be overwhelmed or immobilized by the bad environmental news, because there is also some *great environmental news.* We have made immense progress in improving the human condition and dealing with many environmental problems. We are learning a great deal about how nature works and sustains itself. And we have numerous scientific, technological, and economic solutions available to deal with the environmental problems we face.

The challenge is to make creative use of our economic and political systems to implement such solutions. One key is to recognize that most economic and political change comes about as a result of individual actions and individuals acting together to bring about change by grassroots action from the bottom up. *Good news.* Social scientists suggest it takes only about 5–10% of the population of a country or of the world to bring about major social change. Anthropologist Margaret Mead summarized our potential for change: "Never doubt that a small group

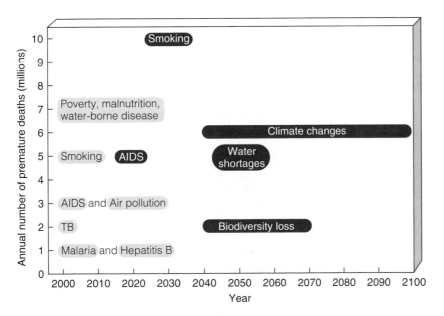

Figure 1-15 Priorities: ranking of major environmental risks in terms of the estimated number of people prematurely killed annually now (yellow) and over the next hundred years (red). Some scientists consider biodiversity loss and climate change the two most serious ecological risks to humans and other species. Estimates of deaths from biodiversity loss and climate change 50 or more years into the future are difficult to make and could be higher or lower than those shown here. (Data from UN Food and Agriculture Organization, World Health Organization, United Nations Environment Program, U.S. Centers for Disease Control and Prevention, and the World Bank)

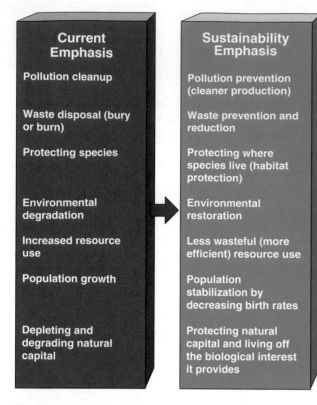

Current Emphasis	Sustainability Emphasis
Pollution cleanup	Pollution prevention (cleaner production)
Waste disposal (bury or burn)	Waste prevention and reduction
Protecting species	Protecting where species live (habitat protection)
Environmental degradation	Environmental restoration
Increased resource use	Less wasteful (more efficient) resource use
Population growth	Population stabilization by decreasing birth rates
Depleting and degrading natural capital	Protecting natural capital and living off the biological interest it provides

Figure 1-16 Solutions: some shifts involved in the *environmental* or *sustainability revolution.*

of thoughtful, committed citizens can change the world. Indeed, it is the only thing that ever has."

We live in exciting times during what might be called a *hinge of cultural history.* Indeed, if I had to pick a time to live, it would be the next 50 years as we face the challenge of developing more environmentally sustainable societies.

What's the use of a house if you don't have a decent planet to put it on?

HENRY DAVID THOREAU

CRITICAL THINKING

1. Do you favor instituting policies designed to reduce population growth and stabilize **(a)** the size of the world's population as soon as possible and **(b)** the size of the U.S. population (or the population of the country where you live) as soon as possible? Explain. If you agree that population stabilization is desirable, what three major policies would you implement to accomplish this goal?

2. List **(a)** three forms of economic growth you believe are environmentally unsustainable and **(b)** three forms you believe are environmentally sustainable.

3. Give three examples of how you cause environmental degradation as a result of the tragedy of the commons.

4. When you read that about 19,200 human beings die prematurely each day (13 per minute) from preventable malnutrition and infectious disease, do you **(a)** doubt whether it is true, **(b)** not want to think about it, **(c)** feel hopeless, **(d)** feel sad, **(e)** feel guilty, or **(f)** want to do something about this problem?

5. How do you feel when you read that **(1)** the average American consumes about 35 times more resources than the average Indian citizen, **(2)** human activities lead to the premature extinction of at least 10 species per day, and **(3)** human activities are projected to make the earth's climate warmer: **(a)** skeptical about their accuracy, **(b)** indifferent, **(c)** sad, **(d)** helpless, **(e)** guilty, **(f)** concerned, or **(g)** outraged? Which of these feelings help perpetuate such problems, and which can help alleviate them?

6. See if you are infected by the affluenza bug by indicating whether you agree or disagree with the following statements.
  a. I am willing to work at a job I despise so I can buy lots of stuff.
  b. When I am feeling down, I like to go shopping to make myself feel better.
  c. I would rather be shopping right now.
  d. I owe more than $1,000 on my credit cards.
  e. I usually make only the minimum payment on my monthly credit card bills.
  f. I am running out of room to store my stuff.
If you agree with three of these statements, you are infected with affluenza. If you agree with more than three, you have a serious case of affluenza. Compare your answers with those of your classmates and discuss the effects of the results on the environment and your feelings of happiness.

7. Explain why you agree or disagree with each of the following statements: **(a)** humans are superior to other forms of life, **(b)** humans are in charge of the earth, **(c)** all economic growth is good, **(d)** the value of other species depends only on whether they are useful to us, **(e)** because all species eventually become extinct we should not worry about whether our activities cause the premature extinction of a species, **(f)** all species have an inherent right to exist, **(g)** nature has an almost unlimited storehouse of resources for human use, **(h)** technology can solve our environmental problems, **(i)** I do not believe I have any obligation to future generations, and **(j)** I do not believe I have any obligation to other species.

8. What are the basic beliefs of your environmental worldview? Are the beliefs of your environmental worldview consistent with your answers to question 7? Are your environmental actions consistent with your environmental worldview?

PROJECTS

1. What are the major resource and environmental problems where you live? Which of these problems affect you directly? Have these problems gotten better or worse during the last 10 years?

2. Write two-page scenarios describing what your life and that of any children you may have might be like 50 years from now if **(a)** we continue on our present path; **(b)** we shift to more sustainable societies throughout most of the world.

3. Make a list of the resources you truly need. Then make another list of the resources you use each day only because you want them. Finally, make a third list of resources you want and hope to use in the future. Compare your lists with those compiled by other members of your class, and relate the overall result to the tragedy of the commons (p. 9).

4. Use the library or the Internet to find out bibliographic information about *Barbara Ward, René Dubos,* and *Henry David Thoreau,* whose quotes appear at the beginning and end of this chapter.

5. Make a concept map of this chapter's major ideas using the section heads, subheads, and key terms (in boldface type). Look on the website for this book for information about making concept maps.

LEARNING ONLINE

The website for this book contains study aids and many ideas for further reading and research. They include a chapter summary, review questions for the entire chapter, flash cards for key terms and concepts, a multiple-choice practice quiz, interesting Internet sites, references, and a guide for accessing thousands of InfoTrac® College Edition articles. Log on to

http://biology.brookscole.com/miller14

Then click on the Chapter-by-Chapter area, choose Chapter 1, and select a learning resource.

2 Environmental History: Learning from the Past

Biodiversity

CASE STUDY
Near Extinction of the American Bison

In 1500, before Europeans settled North America, 30–60 million North American bison—commonly known as the buffalo—grazed the plains, prairies, and woodlands over much of the continent.

These animals were once so numerous that in 1832 a traveler wrote, "As far as my eye could reach the country seemed absolutely blackened by innumerable herds." A single herd on the move might thunder past for hours.

For centuries, several Native American tribes depended heavily on bison. Typically they killed only enough animals to meet their needs for food, clothing, and shelter. They also burned dried feces of these animals, known as "buffalo chips," to cook food and provide heat.

By 1906, the once vast range of the bison had shrunk to a tiny area, and the species had been driven nearly to extinction (Figure 2-1). How did this happen? It began when settlers moving west after the Civil War upset the sustainable balance between Native Americans and bison. Several Plains tribes traded bison skins to settlers for steel knives and firearms, which allowed them to kill more bison.

But it was the new settlers who caused the most relentless slaughter. As railroads spread westward in the late 1860s, railroad companies hired professional bison hunters—including Buffalo Bill Cody—to supply construction crews with meat. Passengers also gunned down bison from train windows for sport, leaving the carcasses to rot.

Commercial hunters shot millions of bison for their hides and tongues (considered a delicacy), leaving most of the meat to rot. "Bone pickers" collected the bleached bones that whitened the prairies and shipped them east to be ground up as fertilizer.

Farmers shot bison because they damaged crops, fences, telegraph poles, and sod houses. Ranchers killed them because they competed with cattle and sheep for pasture. The U.S. Army killed at least 12 million bison as part of its campaign to subdue

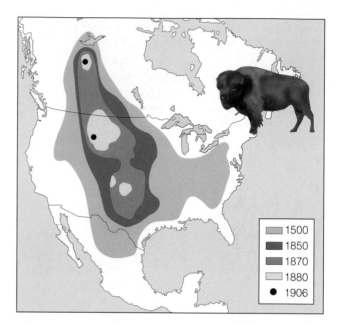

Figure 2-1 The historical range of the bison shrank severely between 1500 and 1906, mostly because of unregulated and deliberate overhunting.

Legend: 1500, 1850, 1870, 1880, ● 1906

the Plains tribes by killing off their primary source of food.

Between 1870 and 1875, at least 2.5 million bison were slaughtered each year. Only 85 bison were left by 1892. They were given refuge in Yellowstone National Park and protected by an 1893 law that forbids the killing of wild animals in national parks.

In 1905, 16 people formed the American Bison Society to protect and rebuild the captive population. Soon thereafter, the federal government established the National Bison Range near Missoula, Montana. Today an estimated 350,000 bison are alive, about 97% of them on privately owned ranches.

Some wildlife conservationists have suggested restoring large herds of bison on public lands in the North American plains. This idea has been strongly opposed by ranchers with permits to graze cattle and sheep on federally managed lands.

The history of humanity's relationships to the environment provides many important lessons that can help us deal with today's environmental problems and avoid repeating past mistakes.

A continent ages quickly once we come.

ERNEST HEMINGWAY

This chapter addresses the following questions:

- What major beneficial and harmful effects have hunter–gatherer societies, agricultural societies, and industrialized societies had on the environment? What might be the environmental impact of the current information and globalization revolution?

- What are the major phases in the history of land and wildlife conservation, public health, and environmental protection in the United States?

- What is Aldo Leopold's land ethic?

2-1   CULTURAL CHANGES AND THE ENVIRONMENT

What Major Human Cultural Changes Have Taken Place? Agriculture, Industrialization, and Globalization

Since our hunter–gatherer days we have undergone three major cultural changes that have increased our impact on the environment.

Evidence from fossils, DNA analysis, and studies of ancient cultures suggests that the earliest form of the human (*Homo sapiens*) species was *Homo sapiens idaltu*, which existed about 160,000 years ago. The latest version of our species, *Homo sapiens sapiens*, has been around for only about 60,000 years. Thus the various versions of *Homo sapiens* have walked the earth for less than an eye blink of the estimated 3.7-billion-year existence of life on this marvelous planet. We are the planet's new infants.

Until about 12,000 years ago, we were mostly hunter–gatherers who typically moved as needed to find enough food for survival. Since then, three major cultural changes have occurred: the *agricultural revolution* (which began 10,000–12,000 years ago), the *industrial–medical revolution* (which began about 275 years ago), and the *information and globalization revolution* (which began about 50 years ago).

These changes have greatly increased our impact on the environment in three ways. They have given us much more energy and new technologies with which to alter and control more of the planet to meet our basic needs and increasing wants. They have also allowed expansion of the human population, mostly because of increased food supplies and longer life spans. In addition, they have greatly increased our resource use, pollution, and environmental degradation.

How Did Ancient Hunting-and-Gathering Societies Affect the Environment? Living Lightly on the Earth

Hunter–gatherers had a fairly small impact on their environment.

During most of their 60,000-year existence, *Homo sapiens sapiens* have been **hunter–gatherers.** They survived by collecting edible wild plant parts, hunting, fishing, and scavenging meat from animals killed by other predators. Our hunter–gatherer ancestors typically lived in small bands of fewer than 50 people who worked together to get enough food to survive. Many groups were nomadic, picking up their few possessions and moving seasonally from place to place to find enough food.

The earliest hunter–gatherers (and those still living this way today) survived through expert knowledge and understanding of their natural surroundings. Because of high infant mortality and an estimated average life span of 30–40 years, hunter–gatherer populations grew very slowly.

Advanced hunter–gatherers had greater environmental impacts than those of early hunter–gatherers. They used more advanced tools and fire to convert forests into grasslands. There is also some evidence that they probably contributed to the extinction of some large animals. They also altered the distribution of plants (and animals feeding on such plants) as they carried seeds and plants to new areas.

Early and advanced hunter–gatherers exploited their environment to survive. But their environmental impact usually was limited and local because of their small population, low resource use per person, migration that allowed natural processes to repair most of the damage they caused, and lack of technology that could have expanded their impact.

What Was the Agricultural Revolution? More Food, More People, Longer Lives, and an Increasing Ecological Footprint

Agriculture provided more food for more people who lived longer and in better health but also greatly increased environmental degradation.

Some 10,000–12,000 years ago, a cultural shift known as the **agricultural revolution** began in several regions of the world. It involved a gradual move from usually nomadic hunting-and-gathering groups to settled agricultural communities in which people domesticated wild animals and cultivated wild plants.

Plant cultivation probably developed in many areas, some including tropical forests of Southeast Asia, northeast Africa, and Mexico. People discovered how to grow various wild food plants from roots or tubers (fleshy underground stems). To prepare the land

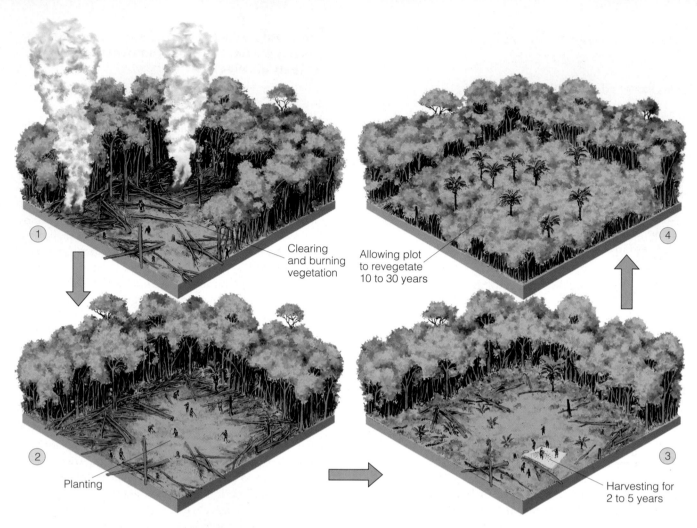

Figure 2-2 The first crop-growing technique may have been a combination of slash-and-burn and shifting cultivation in tropical forests. This method is sustainable only if small plots of the forest are cleared, cultivated for no more than 5 years, and then allowed to regenerate for 10–30 years to renew soil fertility. Indigenous cultures have developed many variations of this technique and have found ways to use some former plots nondestructively while they are being regenerated.

Image labels: ① Clearing and burning vegetation; ② Planting; ③ Harvesting for 2 to 5 years; ④ Allowing plot to revegetate 10 to 30 years

for planting, they cleared small patches of tropical forests by cutting down trees and other vegetation and then burning the underbrush (Figure 2-2). The ashes fertilized the often nutrient-poor tropical forest soils in this **slash-and-burn cultivation.**

Early growers also used various forms of **shifting cultivation** (Figure 2-2), primarily in tropical regions. After a plot had been used for several years, the soil became depleted of nutrients or reinvaded by the forest. Then the growers cleared a new plot. They learned that each abandoned patch normally had to be left fallow (unplanted) for 10–30 years before the soil became fertile enough to grow crops again. While patches were regenerating, growers used them for tree crops, medicines, fuelwood, and other purposes. In this manner, most early growers practiced *sustainable cultivation.*

These early farmers had fairly little impact on the environment. Their dependence mostly on human muscle power and crude stone or stick tools meant they could cultivate only small plots and their population size and density were low. In addition, normally enough land was available so they could move to other areas and leave abandoned plots unplanted for the several decades needed to restore soil fertility.

As more advanced forms of agriculture grew and spread they led to various beneficial and harmful effects (Figure 2-3).

What Is the Industrial–Medical Revolution? More People, Longer Lives, More Production, and an Even Larger Ecological Footprint

Because of the industrial–medical revolution more people live longer and healthier lives at a higher standard of living, but pollution, resource waste, and environmental degradation have increased.

The next cultural shift, the **industrial–medical revolution,** began in England in the mid-1700s and spread to the United States in the 1800s. It involved a shift from dependence on *renewable* wood (with supplies

Trade-Offs

Agricultural Revolution

Good News		Bad News

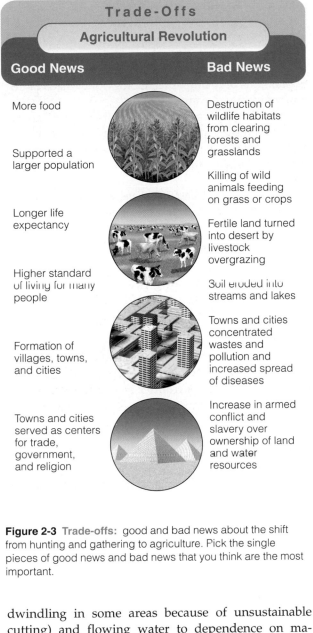

Good News	Bad News
More food	Destruction of wildlife habitats from clearing forests and grasslands
Supported a larger population	Killing of wild animals feeding on grass or crops
Longer life expectancy	Fertile land turned into desert by livestock overgrazing
Higher standard of living for many people	Soil eroded into streams and lakes
Formation of villages, towns, and cities	Towns and cities concentrated wastes and pollution and increased spread of diseases
Towns and cities served as centers for trade, government, and religion	Increase in armed conflict and slavery over ownership of land and water resources

Figure 2-3 Trade-offs: good and bad news about the shift from hunting and gathering to agriculture. Pick the single pieces of good news and bad news that you think are the most important.

dwindling in some areas because of unsustainable cutting) and flowing water to dependence on machines running on *nonrenewable* fossil fuels (first coal and later oil and natural gas). This led to a switch from small-scale, localized production of handmade goods to large-scale production of machine-made goods in centralized factories in rapidly growing industrial cities.

Factory towns grew into cities as rural people came to the factories for work. There they worked long hours under noisy, dirty, and hazardous conditions. Other workers toiled in dangerous coal mines.

In early industrial cities, coal smoke belching from chimneys was so heavy that many people died of lung ailments. Ash and soot covered everything, and some days the smoke was thick enough to blot out the sun.

Fossil fuel–powered farm machinery, commercial fertilizers, and new plant-breeding techniques in-

creased crop yields per acre. This helped protect biodiversity by reducing the need to expand the area of cropland. Because fewer farmers were needed, more people migrated to cities. With a larger and more reliable food supply and longer life spans, the human population began the sharp increase that continues today.

Figure 2-4 lists some of the beneficial (*good news*) and harmful (*bad news*) effects of the advanced industrial–medical revolution.

How Might the Information and Globalization Revolution Affect the Environment? Information Blessing or Information Overload?

Global access to information can help us understand and respond to environmental problems but can lead to confusion from information overload.

Trade-Offs

Industrial–Medical Revolution

Good News		Bad News

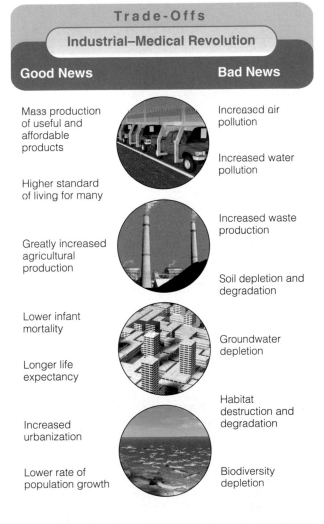

Good News	Bad News
Mass production of useful and affordable products	Increased air pollution
Higher standard of living for many	Increased water pollution
Greatly increased agricultural production	Increased waste production
Lower infant mortality	Soil depletion and degradation
Longer life expectancy	Groundwater depletion
Increased urbanization	Habitat destruction and degradation
Lower rate of population growth	Biodiversity depletion

Figure 2-4 Trade-offs: good and bad news about the effects of the advanced industrial revolution. Pick the single pieces of good news and bad news that you think are the most important.

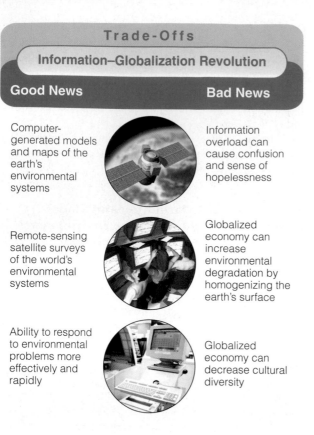

Trade-Offs

Information–Globalization Revolution

Good News	Bad News
Computer-generated models and maps of the earth's environmental systems	Information overload can cause confusion and sense of hopelessness
Remote-sensing satellite surveys of the world's environmental systems	Globalized economy can increase environmental degradation by homogenizing the earth's surface
Ability to respond to environmental problems more effectively and rapidly	Globalized economy can decrease cultural diversity

Figure 2-5 Trade-offs: good and bad news about the effects of this latest cultural revolution. Pick the single pieces of good news and bad news that you think are the most important.

Since 1950, and especially since 1970, we have begun making a new cultural shift called the **information and globalization revolution.** It is based on using new technologies for gaining rapid access to much more information on a global scale. These technologies include the telephone, radio, television, computers, the Internet, automated databases, and remote-sensing satellites. Figure 2-5 lists some of the possible beneficial and harmful effects of the information and globalization revolution.

2-2 ENVIRONMENTAL HISTORY OF THE UNITED STATES: THE TRIBAL AND FRONTIER ERAS

What Happened during the Tribal Era? Sustainable Living

Native Americans living in North America for at least 10,000 years had a fairly low environmental impact.

The environmental history of the United States can be divided into four eras: *tribal, frontier, conservation,* and *environmental.*

During the *tribal era,* North America was occupied by 5–10 million tribal people for at least 10,000 years be-

fore European settlers began arriving in the early 1600s. These indigenous people were called Indians by the Europeans and now are often called Native Americans. They practiced hunting and gathering, burned and cleared fields, and planted crops. Because of their small populations and simple technology, they had a fairly low environmental impact.

With some exceptions, most Native American cultures had a deep respect for the land and its animals and did not believe in land ownership, as indicated by the following quotation:

My people, the Blackfeet Indians, have always had a sense of reverence for nature that made us want to move through the world carefully, leaving as little mark behind as possible. (Jamake Highwater, Blackfoot)

What Happened During the Frontier Era (1607–1890)? Taking Over a Continent

European settlers saw the continent as a vast frontier to conquer and settle.

The *frontier era* began in the early 1600s when European colonists began settling North America. The early colonists developed a **frontier environmental worldview.** They viewed most of the continent as having vast and seemingly inexhaustible resources and as a hostile and dangerous wilderness to be conquered and managed for human use.

The government set up by European settlers conquered Native American tribes, took over the land, and urged people to spread across the continent. The transfer of vast areas of public land to private interests accelerated the settling of the continent.

This frontier environmental view prevailed for more than 280 years, until the government declared the frontier officially closed in 1890. However, this worldview that there is always another frontier to conquer remains as an important part of American culture today.

2-3 ENVIRONMENTAL HISTORY OF THE UNITED STATES: THE EARLY CONSERVATION ERA (1832–1960)

What Happened between (1832–70)? Ignored Warnings

A few people warned Americans that they were degrading their resource base, but few listened.

Between 1832 and 1870, some people became alarmed at the scope of resource depletion and degradation in the United States. They urged that part of the unspoiled wilderness on public lands—owned jointly by

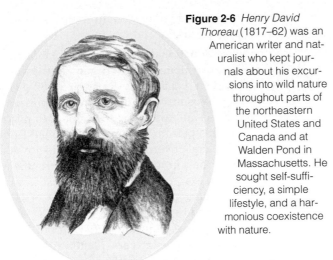

Figure 2-6 *Henry David Thoreau* (1817–62) was an American writer and naturalist who kept journals about his excursions into wild nature throughout parts of the northeastern United States and Canada and at Walden Pond in Massachusetts. He sought self-sufficiency, a simple lifestyle, and a harmonious coexistence with nature.

all people but managed by the government—be protected as a legacy to future generations.

Two early conservationists were *Henry David Thoreau* (Figure 2-6) and *George Perkins Marsh* (1801–1882). Thoreau was alarmed at the loss of numerous wild species from his native eastern Massachusetts. To gain a better understanding of nature, he built a cabin in the woods on Walden Pond near Concord, Massachusetts, lived there alone for 2 years, and wrote *Life in the Woods,* an environmental classic.*

In 1864, George Perkins Marsh, a scientist and member of Congress from Vermont, published *Man and Nature,* which helped legislators and citizens see the need for resource conservation. Marsh questioned the idea that the country's resources were inexhaustible. He also used scientific studies and case studies to show how the rise and fall of past civilizations were linked to the use and misuse of their resource base. Some of his resource conservation principles are still used today. Most of these warnings were not taken seriously.

What Happened Between 1870 and 1930? Government and Citizen Involvement

The government and newly formed private groups tried to protect more of the nation's natural resources and improve public health.

---

*I can identify with Thoreau. I spent 15 years living in the deep woods studying and thinking about how nature works and writing books such as the one you are reading. I lived in a school bus with an attached greenhouse. I used it as a scientific laboratory for evaluating things such as passive and active solar energy technologies, waste disposal (composting toilets), natural cooling (earth tubes), ways to save energy and water, and biological control of pests. It was great fun and I learned a lot. Since most of the world is urban I came out of the woods to find out more about this way of living.

Between 1870 and 1930, a number of actions increased the role of the federal government and private citizens in resource conservation and public health, as summarized in Figure 1 on p. A3 of Appendix 2. The *Forest Reserve Act of 1891* was a turning point in establishing the responsibility of the federal government for protecting public lands from resource exploitation.

In 1892, nature preservationist and activist *John Muir* (Figure 2-7) founded the Sierra Club. He became the leader of the *preservationist movement* that called for protecting large areas of wilderness on public lands from human exploitation, except for low-impact recreational activities such as hiking and camping. This idea was not enacted into law until 1964. Muir also proposed and lobbied for creation of a national park system on public lands.

Mostly because of political opposition, effective protection of forests and wildlife did not begin until *Theodore Roosevelt* (Figure 2-8, p. 26), an ardent conservationist, became president. His term of office, 1901–9, has been called the country's *Golden Age of Conservation.*

While in office he persuaded Congress to give the president power to designate public land as federal wildlife refuges. During his presidency he established wildlife reserves and more than tripled the size of the national forest reserves.

In 1905, Congress created the U.S. Forest Service to manage and protect the forest reserves. Roosevelt appointed *Gifford Pinchot* (1865–1946) as its first chief. Pinchot pioneered scientific management of forest resources on public lands. In 1906, Congress passed the *Antiquities Act,* which allows the president to protect areas of scientific or historical interest on federal lands as national monuments. Roosevelt used this act to protect the Grand Canyon and other areas that would later become national parks.

Figure 2-7 *John Muir* (1838–1914) was a geologist, explorer, and naturalist. He spent 6 years studying, writing journals, and making sketches in the wilderness of California's Yosemite Valley and then went on to explore wilderness areas in Utah, Nevada, the Northwest, and Alaska. He was largely responsible for establishing Yosemite National Park in 1890. He also founded the Sierra Club and spent 22 years lobbying actively for conservation laws.

Figure 2-8 *Theodore ("Teddy") Roosevelt* (1858– 1919) was a writer, explorer, naturalist, avid birdwatcher, and 26th president of the United States. He was the first national political figure to bring the issues of conservation to the attention of the American public. According to many historians, he has contributed more than any other president to natural resource conservation in the United States.

In 1907, Congress became upset because Roosevelt had added vast tracts to the forest reserves and banned further executive withdrawals of public forests. On the day before the bill became law, Roosevelt defiantly reserved another 6.5 million hectares (16 million acres). Most environmental historians view Roosevelt (a Republican) as the country's best environmental president.

In 1916, Congress passed the *National Park Service Act.* It declared that parks are to be maintained in a manner that leaves them unimpaired for future generations. The Act also established the National Park Service (within the Department of the Interior) to manage the system.

After World War I, the country entered a new era of economic growth and expansion. During the Harding, Coolidge, and Hoover administrations, the federal government promoted increased resource removal from public lands at low prices to stimulate economic growth.

President Hoover (a Republican) went even further and proposed that the federal government return all remaining federal lands to the states or sell them to private interests for economic development. But the Great Depression (1929–41) made owning such lands unattractive to state governments and private investors. The depression was bad news for the country. But some say that without it we might have little if any public lands left today.

What Happened between 1930 and 1960? Depression and War

During the economic depression of the 1930s the government bought land and hired many workers to restore the country's degraded environment and build dams to supply electricity and water.

A second wave of national resource conservation and improvements in public health began in the early 1930s as President *Franklin D. Roosevelt* (1882–1945) strove to bring the country out of the Great Depression. Figure 2 on p. A4 in Appendix 2 summarizes major events during this period. Roosevelt persuaded Congress to enact federal government programs to provide jobs and restore the country's degraded environment. The following are examples of these programs.

The government purchased large tracts of land from cash-poor landowners, and the *Civilian Conservation Corps* (CCC) was established in 1933. It put 2 million unemployed people to work planting trees and developing and maintaining parks and recreation areas. The CCC also restored silted waterways and built levees and dams for flood control.

The government built and operated many large dams in the Tennessee Valley and in the arid western states, including Hoover Dam on the Colorado River. The goals were to provide jobs, flood control, cheap irrigation water, and cheap electricity for industry.

Many environmental historians praise Roosevelt (a Democrat) for his efforts to get the country out of a major economic depression and restore past environmental degradation.

Federal resource conservation and public health policy during the 1940s and 1950s changed little, mostly because of preoccupation with World War II (1941–45) and economic recovery after the war.

2-4 ENVIRONMENTAL HISTORY OF THE UNITED STATES: THE ENVIRONMENTAL ERA (1960–2004)

What Happened during the 1960s? An Environmental Awakening

The modern environmental movement began and more citizens urged government to improve environmental quality.

A number of milestones in American environmental history occurred during the 1960s, as summarized in Figure 3 on p. A5 in Appendix 2. In 1962, biologist *Rachel Carson* (1907–64) published *Silent Spring,* which documented the pollution of air, water, and wildlife from pesticides such as DDT (Individuals Matter, right). This influential book helped broaden the concept of resource conservation to include preservation of wildlife and the *quality* of the air, water, and soil.

Many historians mark Carson's wake-up call as the beginning of the modern **environmental move-**

Rachel Carson (Figure 2-A) began her professional career as a biologist for the Bureau of U.S. Fisheries (later the U.S. Fish and Wildlife Service). In that capacity, she carried out research on oceanography and marine biology and wrote articles about the oceans and topics related to the environment.

In 1951, she wrote *The Sea Around Us*, which described in easily understandable terms the natural history of oceans and how humans were harming them. This book sold more than 2 million copies, was translated into 32 languages, and won a National Book Award.

During the late 1940s and throughout the 1950s, DDT and related compounds were increasingly used to kill insects that ate food crops, attacked trees, bothered people, and transmitted diseases such as malaria.

In 1958, DDT was sprayed to control mosquitoes near the home and private bird sanctuary of one of Rachel Carson's friends. After the spraying, her friend witnessed the agonizing deaths of several birds. She begged Carson to find someone to investigate the effects of pesticides on birds and other wildlife.

Carson decided to look into the issue herself and found that independent research on the environmental effects of pesticides was almost nonexistent. As a well-trained scientist, she surveyed the scientific literature, became convinced that pesticides could harm wildlife and humans, and methodically developed information about the harmful effects of widespread use of pesticides.

In 1962, she published her findings in *Silent Spring*, an allusion to the silencing of "robins, catbirds, doves, jays, wrens, and scores of other bird voices" because of their exposure to pesticides.

Many scientists, politicians, and policy makers read *Silent Spring*, and the public embraced it. But manufacturers of chemicals viewed the book as a serious threat to their booming pesticide sales and mounted a campaign to discredit her. A parade of critical reviewers and industry scientists claimed her book was full of inaccuracies, made selective and biased use of research findings, and failed to give a balanced account of the benefits of pesticides.

Some critics even claimed that, as a woman, she was incapable of understanding such a highly scientific and technical subject. Others charged that she was a hysterical woman and a radical nature lover trying to scare the public in order to sell books.

During these intense attacks, Carson was suffering from terminal cancer. Yet she strongly defended her research and countered her critics. She died in 1964—18 months after the publication of *Silent Spring*—without knowing that many historians consider her work an important contribution to the modern environmental movement then emerging in the United States.

Figure 2-A *Rachel Carson* (1907–1964)

ment in the United States. It flourished when a growing number of citizens organized to demand that political leaders enact laws and develop policies to curtail pollution, clean up polluted environments, and protect unspoiled areas from environmental degradation.

In 1964 Congress passed the *Wilderness Act*, inspired by the vision of John Muir more than 80 years earlier. It authorized the government to protect undeveloped tracts of public land as part of the National Wilderness System, unless Congress later decides they are needed for the national good. Land in this system is to be used only for nondestructive forms of recreation such as hiking and camping.

Between 1965 and 1970, the emerging science of *ecology* received widespread media attention. At the same time, the popular writings of biologists such as *Paul Ehrlich, Barry Commoner,* and *Garrett Hardin* awakened people to the interlocking relationships among population growth, resource use, and pollution.

During that period, a number of events increased public awareness of pollution. The public also became aware that pollution and loss of habitat were endangering well-known wildlife species such as the North American bald eagle, grizzly bear, whooping crane, and peregrine falcon.

During the 1969 U.S. *Apollo* mission to the moon, astronauts photographed the earth from space. This allowed people to see the earth as a tiny blue and white planet in the black void of space and led to the development of the *spaceship-earth environmental worldview.* It reminded us that we live on a marvelous planetary spaceship (*Terra I*) that we should not harm because it is the only home we have.

What Happened during the 1970s? The Environmental Decade

Increased awareness and public concern led Congress to pass a number of laws to improve environmental quality and conserve more of the nation's natural resources.

During the 1970s, media attention, public concern about environmental problems, scientific research, and action to address these concerns grew rapidly. Figure 4 on p. A6 of Appendix 2 summarizes major environmental events during this period, which is sometimes called the *first decade of the environment*

The first annual *Earth Day* was held on April 20, 1970. During this event, proposed by Senator *Gaylord Nelson* (born 1916), some 20 million people in more than 2,000 communities took to the streets to heighten awareness and to demand improvements in environmental quality.

Republican President *Richard Nixon* (1913–94) responded to the rapidly growing environmental movement. He established the *Environmental Protection Agency* (EPA) in 1970 and supported passage of the *Endangered Species Act of 1973*. This greatly strengthened the role of the federal government in protecting endangered species and their habitats.

In 1978, the *Federal Land Policy and Management Act* gave the *Bureau of Land Management* (BLM) its first real authority to manage the public land under its control, 85% of which is in 12 western states. This law angered a number of western interests whose use of these public lands was restricted for the first time.

In response, a coalition of ranchers, miners, loggers, developers, farmers, some elected officials, and others launched a political campaign known as the *sagebrush rebellion.* It had two major goals. One was to sharply reduce government regulation of the use of public lands. The other was to remove most public lands in the western United States from federal ownership and management and turn them over to the states. Then the plan was to persuade state legislatures to sell or lease the resource-rich lands at low prices to ranching, mining, timber, land development, and other private interests. This represented a return to President Hoover's plan to turn all public land over to private ownership that was thwarted by the Great Depression.

Jimmy Carter (a Democrat, born 1924), president between 1977 and 1981, was very responsive to environmental concerns. He persuaded Congress to create the *Department of Energy* to develop a long-range energy strategy to reduce the country's heavy dependence on imported oil. He appointed respected environmentalists to key positions in environmental and resource agencies and consulted with environmental leaders on environmental and resource policy matters.

In 1980, Carter helped create a *Superfund* as part of the *Comprehensive Environment Response, Compensation, and Liability Act* to clean up abandoned hazardous waste sites, including the Love Canal near Niagara Falls, New York. Carter also used the Antiquities Act of 1906 to triple the amount of land in the National Wilderness System and double the area in the National Park System (primarily by adding vast tracts in Alaska).

What Happened during the 1980s? Environmental Backlash

An anti-environmental movement formed to weaken or do away with many of the environmental laws passed in the 1960s and 1970s and to destroy the political effectiveness of the environmental movement.

Figure 5 on p. A6 in Appendix 2 summarizes some key environmental events during the 1980s that shaped U.S. environmental policy. During this decade, farmers and ranchers and leaders of the oil, automobile, mining, and timber industries strongly opposed many of the environmental laws and regulations developed in the 1960s and 1970s. They organized and funded a strong *anti-environmental movement* that persists today.

In 1981, *Ronald Reagan* (a Republican, born 1911), a self-declared *sagebrush rebel* and advocate of less federal control, became president. During his 8 years in office he angered environmentalists by appointing to key federal positions people who opposed most existing environmental and public land use laws and policies.

Reagan greatly increased private energy and mineral development and timber cutting on public lands. He also drastically cut federal funding for research on energy conservation and renewable energy resources and eliminated tax incentives for residential solar energy and energy conservation enacted during the Carter administration. In addition, he lowered automobile gas mileage standards and relaxed federal air and water quality pollution standards.

Although Reagan was immensely popular, many people strongly opposed his environmental and resource policies. This resulted in strong opposition in Congress, public outrage, and legal challenges by environmental and conservation organizations, whose memberships soared during this period.

In 1988, an industry-backed anti-environmental coalition called the *wise-use movement* was formed. Its major goals were to weaken or repeal most of the country's environmental laws and regulations and destroy the effectiveness of the environmental movement in the United States. Politically powerful coal, oil, mining, automobile, timber, and ranching interests helped back this movement.

Upon his election in 1989, *George H. W. Bush* (a Republican, born 1924) promised to be "the environmental president." But he received criticism from environmentalists for not providing leadership on such key environmental issues as population growth, global warming, and loss of biodiversity. He also continued support of exploitation of valuable resources on public lands at giveaway prices. In addition, he allowed some environmental laws to be undercut by the powerful influence of industry, mining, ranching, and real estate development industries.

What Happened from 1990 to 2004? Trying to Hold the Line

Since 1990 American environmentalists have spent most of their time and money trying to keep anti-environmentalists from weakening or eliminating most environmental laws passed in the 1960s and 1970s.

Figure 6 on p. A8 of Appendix 2 summarizes some key environmental events that took place between 1990 and 2004. In 1993, *Bill Clinton* (a Democrat, born 1946) became president and promised to provide national and global environmental leadership. During his 8 years in office he appointed respected environmentalists to key positions in environmental and resource agencies and consulted with environmentalists about environmental policy, as Carter did.

He also vetoed most of the anti-environmental bills (or other bills passed with anti-environmental riders attached) passed by a Republican-dominated Congress between 1995 and 2000. He announced regulations requiring sport utility vehicles (SUVs) to meet the same air pollution emission standards as cars. Clinton also used executive orders to make forest health the primary priority in managing national forests and to declare many roadless areas in national forests off limits to roads and logging. In addition, he used the Antiquities Act of 1906 to protect various parcels of public land in the West from development and resource exploitation by declaring them national monuments. He protected more public land as national monuments in the lower 48 states than any other president, including Teddy Roosevelt and Jimmy Carter..

Environmentalists criticized Clinton, however, for failing to push hard enough on key environmental issues such as global warming and global and national biodiversity protection.

Since 1990 environmentalists have had to spend much of their time and funds fighting efforts by the anti-environmental movement to discredit the environmental movement and weaken or eliminate most environmental laws passed during the 1960s and 1970s. They also had to counter claims by anti-environmental groups that problems such as global warming and ozone depletion are hoaxes or not very serious.

During the 1990s many small and mostly local grassroots environmental organizations sprang up to deal with environmental threats in their local communities. Interest in environmental issues increased on many college campuses and environmental studies programs at colleges and universities expanded. In addition, awareness of important but complex environmental issues such as sustainability, population growth, biodiversity protection, and threats from global warming increased.

In 2001, *George W. Bush* (a Republican, born 1946) became president. Like Reagan in the 1980s, he appointed to key federal positions people who opposed or wanted to weaken many existing environmental and public land use laws and policies. Also like Reagan, he did not consult with environmental groups and leaders in developing environmental policies, and he greatly increased private energy and mineral development and timber cutting on public lands. Bush weakened protections on almost as much public lands as Teddy Roosevelt protected.

Bush also opposed increasing automobile gas mileage standards as a way to save energy and reduce dependence on oil imports, and he supported relaxation of various federal air and water quality pollution standards. Like Reagan, he developed an energy policy that placed much greater emphasis on use of fossil fuels and nuclear power than on reducing energy waste and relying more on renewable energy resources.

In addition, he withdrew the U.S. from participation in the international Kyoto treaty designed to help reduce carbon dioxide emissions that can promote global warming. He also repealed or tried to weaken most of the pro-environmental measures established by Clinton.

In 2003, leaders of a dozen major environmental organizations charged that Bush, backed by a Republican-dominated Congress, was well on the way to compiling the worst environmental record of any president in the history of the country.

A few moderate Republican members of Congress have urged their party to return to its environmental roots, put down during Teddy Roosevelt's presidency, and shed its anti-environmental approach to legislation. Most Democrats agree and believe that the environmental problems we face are much too serious to be held hostage by political squabbling. They call for cooperation, not confrontation. They urge elected officials, regardless of party, to enter into a new pact to have the United States become the world leader in making this the *environmental century.*

2-5 CASE STUDY: ALDO LEOPOLD AND HIS LAND ETHIC

Who Was Aldo Leopold? Teacher, Conservationist, and Proponent of Land Ethics

Aldo Leopold played a major role in educating us about the need for conservation and providing ethical guidelines for our actions in nature.

Aldo Leopold (Figure 2-9) is best known as a strong proponent of *land ethics,* a philosophy in which humans as part of nature have an ethical responsibility to preserve wild nature.

After earning a master's degree in forestry from Yale University, he joined the U.S. Forest Service. He became alarmed by overgrazing and land deterioration on public lands where he worked, and was convinced the United States was losing too much of its mostly untouched wilderness.

In 1933, Leopold became a professor at the University of Wisconsin and founded the profession of game management. In 1935, he was one of the founders of the Wilderness Society.

He was a keen student of nature as he took long walks in the countryside. As years passed, he developed a deep understanding and appreciation for wildlife and urged us to include nature in our ethical concerns. Through his writings and teachings he became one of the founders of the *conservation* and *environmental movements* of the 20th century. In doing this, he laid important groundwork for the field of environmental ethics.

Leopold died in 1948 while fighting a brush fire at a neighbor's farm in central Wisconsin. His weekends of planting, hiking, and observing nature at his farm in

Figure 2-9 *Aldo Leopold* (1887–1948) was a forester, writer, and conservationist. His book *A Sand County Almanac* (published after his death) is considered an environmental classic that inspired the modern environmental movement. His *land ethic* expanded the role of humans as protectors of nature.

Wisconsin provided material for his most famous book, *A Sand County Almanac,* published after his death in 1949. Since then more than 2 million copies of this important book have been sold.

What Is Leopold's Concept of Land Ethics? Work with Nature

We need to become plain citizens of the earth instead of its conquerors.

The following quotations from his writings reflect Leopold's land ethic, and they form the basis for many of the beliefs of the modern *environmental wisdom worldview* (p. 17).

> All ethics so far evolved rest upon a single premise: that the individual is a member of a community of interdependent parts.
>
> That land is a community is the basic concept of ecology, but that land is to be loved and respected is an extension of ethics.
>
> The land ethic changes the role of Homo sapiens from conqueror of the land-community to plain member and citizen of it.
>
> We abuse land because we regard it as a commodity belonging to us. When we see land as a community to which we belong, we may begin to use it with love and respect.
>
> Anything is right when it tends to preserve the integrity, stability, and beauty of the biotic community. It is wrong when it tends otherwise.

Thank God, they cannot cut down the clouds!
HENRY DAVID THOREAU

CRITICAL THINKING

1. What three major things would you do to reduce the harmful environmental impacts of advanced industrial societies?

2. What one person do you believe has made the greatest and longest lasting contribution to the conservation and environmental movements in the United States (or in the country where you live)? Explain.

3. Public forests, grasslands, wildlife reserves, parks, and wilderness areas are owned by all citizens and managed for them by federal and state governments in the United States. In terms of the management policies for most of these lands, would you classify yourself as **(a)** a preservationist, **(b)** a conservationist, or **(c)** an advocate of transferring most public lands to private enterprise? Explain.

4. Do you favor or oppose efforts to greatly weaken or repeal most environmental laws in the United States (or in the country where you live)? Explain.

5. Some analysts believe the world's remaining hunter–gatherer societies should be given title to the land on which they and their ancestors have lived for centuries and should be left alone by modern civilization. They contend that we have created protected reserves for endangered wild species, so why not create reserves for these endangered human cultures? What do you think? Explain.

PROJECTS

1. What major changes (such as a change from agricultural to industrial, from rural to urban, or changes in population size, pollution, and environmental degradation) have taken place in your locale during the past 50 years? On balance, have these changes improved or decreased **(a)** the quality of your life and **(b)** the quality of life for members of your community as a whole?

2. Use the library or Internet to summarize the major accomplishments of the anti-environmental movement in the United States between 1980 and 2005.

3. Use the library or Internet to find bibliographic information about *Ernest Hemingway* and *Henry David Thoreau,* whose quotes appear at the beginning and end of this chapter.

4. Make a concept map of this chapter's major ideas using the section heads, subheads, and key terms (in boldface type). Look on the website for this book for information about making concept maps.

LEARNING ONLINE

The website for this book contains study aids and many ideas for further reading and research. They include a chapter summary, review questions for the entire chapter, flash cards for key terms and concepts, a multiple-choice practice quiz, interesting Internet sites, references, and a guide for accessing thousands of InfoTrac® College Edition articles. Log on to

http://biology.brookscole.com/miller14

Then click on the Chapter-by-Chapter area, choose Chapter 2, and select a learning resource.

CASE STUDY

An Environmental Lesson from Easter Island

Easter Island (Rapa Nui) is a small, isolated island in the great expanse of the South Pacific. Polynesians used double-hulled sea-going canoes to colonize this island about 2,500 years ago. They brought along their pigs, chickens, dogs, stowaway rats, taro roots, yams, bananas, and sugarcane.

Evidence from pollen grains found in the soil and lake sediments and in artifacts shows that the island was abundantly forested with a variety of trees—including basswoods (called hauhau) and giant palms. The Polynesians developed a civilization based on the island's trees. The towering palm trees were used for shelter, tools, and fishing boats. Hauhau trees were felled and burned to cook and keep warm in the island's cool winters, and rope was made from the tree's fibers. Land was also cleared of trees to plant taro, sugarcane, bananas, and yams.

Using these abundant tree resources, the Polynesians developed an impressive civilization. They also developed a technology capable of making and moving large stone structures, including their famous statues (Figure 3-1). The people flourished, with the population peaking at somewhere between 7,000 and 20,000 by 1400. However, they used up the island's precious trees faster than they were regenerated—an example of the tragedy of the commons.

By 1600, only a few small trees were left. Without large trees, the islanders could not build their traditional big canoes for hunting porpoises and catching fish in deeper offshore

waters, and no one could escape the island by boat. Without the once-great forests to absorb and slowly release water, springs and streams dried up, exposed soils eroded, crop yields plummeted, and famine struck. There was no firewood for cooking or keeping warm. The hungry islanders ate all of the island's birds. Then they began raising and eating rats, descendants of hitchhikers on the first canoes.

Both the population and the civilization collapsed as gangs fought one another for dwindling food supplies. Bone evidence indicates that the islanders began hunting and eating one another.

Dutch explorers reached the island on Easter Day, 1722, perhaps 1,000 years after the first Polynesians had landed. They found about 2,000 hungry Polynesians, living in caves on a shrubby grassland.

Like Easter Island at its peak, the earth is an isolated island in the vastness of space with no other suitable planet to migrate to. As on Easter Island, our population and resource consumption are growing and our resources are finite.

Will the humans on Earth Island re-create the tragedy of Easter Island on a grander scale, or will we learn how to live more sustainably on this planet that is our only home? Scientific knowledge is a key to learning how to live more sustainably. Thus we need to know what science is, understand the behavior of complex systems studied by scientists, and have a basic knowledge of the nature of the matter and energy that make up the earth's living and nonliving resources, as discussed in this chapter.

Figure 3-1 These massive stone figures on Easter Island are the remains of the technology created by an ancient civilization of Polynesians. Their civilization collapsed because the people used up the trees (especially large palm trees) that were the basis of their livelihood. More than 200 of these stone statues once stood on huge stone platforms lining the coast. At least 700 additional statues were found turned over, abandoned in rock quarries or on ancient roads between the quarries and the coast. No one knows how the early islanders (with no wheels, no draft animals, and no sources of energy other than their own muscles) transported these gigantic structures for miles before erecting them. We presume they accomplished it by felling large trees and using them to roll and erect the statues.

*Science is an adventure of the human spirit. It is essentially
an artistic enterprise, stimulated largely by curiosity, served
largely by disciplined imagination, and based largely on faith
in the reasonableness, order, and beauty of the universe.*

WARREN WEAVER

This chapter addresses the following questions:

- What is science, and what do scientists do?

- What are major components and behaviors of complex systems?

- What are the basic forms of matter? What makes matter useful to us as a resource?

- What are the major forms of energy? What makes energy useful to us as a resource?

- What scientific law governs changes of matter from one physical or chemical form to another?

- What three main types of nuclear changes can matter undergo?

- What are two scientific laws governing changes of energy from one form to another?

- How are the scientific laws governing changes of matter and energy from one form to another related to resource use and environmental degradation?

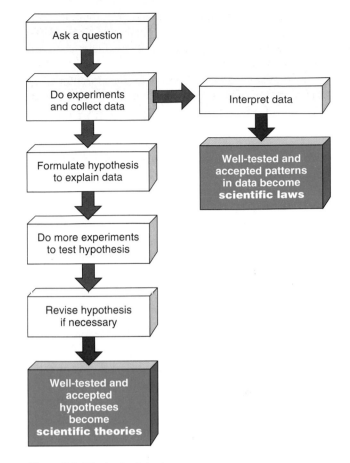

Figure 3-2 What scientists do.

3-1 THE NATURE OF SCIENCE

What Is Science and What Do Scientists Do?
Searching for Order in Nature

Scientists collect data, form hypotheses, and develop theories, models, and laws about how nature works.

Science is an attempt to discover order in the natural world and use that knowledge to describe what is likely to happen in nature. Its goal is to increase our understanding of the natural world. Science is based on the fundamental assumption that events in the natural world follow orderly patterns that can be understood through careful observation and experimentation.

Figure 3-2 summarizes the scientific process. Trace the pathways in this figure.

The first thing scientists do is ask a question or identify a problem to be investigated. Then they collect **scientific data,** or facts, related to the problem or question, by making observations and measurements. They often conduct **experiments** to study some phenomenon under known conditions. The resulting scientific data or facts must be confirmed by repeated observations and measurements, ideally by several different investigators.

The primary goal of science is not the data or facts themselves. Instead science seeks new ideas, princi-

ples, or models that connect and explain certain scientific data and descriptions of what is likely to happen in nature. Scientists working on a particular problem try to come up with a variety of possible explanations, or **scientific hypotheses,** of what they (or other scientists) observe in nature. A scientific hypothesis is an unconfirmed explanation of an observed phenomenon that can be tested by further research.

One method scientists use to test a hypothesis is to develop a **model,** an approximate or simplified representation or simulation of a system being studied. It may be an actual working model, a mental model, a pictorial model, a computer model, or a mathematical model.

Three important features of the scientific process are *skepticism, reproducibility, and peer review* of results by other scientists. Scientists tend to be highly skeptical of any new data or hypotheses until they can be confirmed or verified.

Peers, or scientists working in the same field, check for reproducibility by repeating and checking out one another's work to see if the data can be reproduced and whether proposed hypotheses are reasonable and useful.

Peer review happens when scientists openly publish details of the methods they used, the results of their experiments, and the reasoning behind their hypotheses. This process of publishing one's work for other scientists to examine and criticize helps keep scientists honest and reduces bias.

If repeated observations and measurements or tests using models support a particular hypothesis or a group of related hypotheses, the hypothesis becomes a **scientific theory.** In other words, a *scientific theory* is a verified, credible, and widely accepted scientific hypothesis or a related group of scientific hypotheses.

To scientists, scientific theories are not to be taken lightly. They are not guesses, speculations, or suggestions. Instead, they are useful explanations of processes or natural phenomena that have a high degree of certainty because they are supported by extensive evidence.

New evidence or a better explanation may modify, or in rare cases overturn, a particular scientific theory. But unless or until this happens, a scientific theory is the best and most reliable knowledge we have about how nature works.

Nonscientists often use the word *theory* incorrectly when they mean to refer to a *scientific hypothesis*, a tentative explanation or educated guess that needs further evaluation. The statement, "Oh, that's just a theory," made in everyday conversation, implies a lack of knowledge and careful testing—the opposite of the scientific meaning of the word.

Another important result of science is a **scientific,** or **natural, law:** a description of what we find happening in nature over and over in the same way. For example, after making thousands of observations and measurements over many decades, scientists formulated the *second law of thermodynamics.* Simply stated, this law says that heat always flows spontaneously from hot to cold—something you learned the first time you touched a hot object. A scientific law is no better than the accuracy of the observations or measurements upon which it is based. But if the data are accurate, a scientific law cannot be broken.

How Do Scientists Learn about Nature?
Follow Many Paths

There are many scientific methods.

We often hear about *the* scientific method. In reality, many **scientific methods** exist: they are ways in which scientists gather data and formulate and test scientific hypotheses, models, theories, and laws.

Here is an example of applying the scientific process to an everyday situation:

Observation: You switch on your trusty flashlight and nothing happens.

Question: Why did the light not come on?

Hypothesis: Maybe the batteries are bad.

Test the hypothesis: Put in new batteries and switch on the flashlight .

Result: Flashlight still does not work.

New hypothesis: Maybe the bulb is burned out.

Experiment: Replace bulb with a new bulb.

Result: Flashlight works when switched on.

Conclusion: Second hypothesis is verified.

Situations in nature are usually much more complicated than this. Many *variables* or *factors* influence most processes or parts of nature that scientists seek to understand. Ideally, scientists conduct a *controlled experiment* to isolate and study the effect of a single variable. To do such *single-variable analysis,* scientists set up two groups. One is an *experimental group* in which the chosen variable is changed in a known way. The other is a *control group* in which the chosen variable is not changed. If the experiment is designed properly, any difference between the two groups should result from the variable that was changed in the experimental group (see Connections, right).

A basic problem is that many of the problems environmental scientists investigate involve a huge number of interacting variables. This limitation is sometimes overcome by using *multivariable analysis—* running mathematical models on high-speed computers to analyze the interactions of many variables without having to carry out traditional controlled experiments.

What Types of Reasoning Do Scientists Use?
Bottom-Up and Top-Down Reasoning

Scientists use inductive reasoning to convert observations and measurements to a general conclusion and deductive reasoning to convert a generalization to a specific conclusion.

Scientists arrive at certain conclusions with varying degrees of certainty by using inductive and deductive reasoning. **Inductive reasoning** involves using specific observations and measurements to arrive at a general conclusion or hypothesis. It is a form of *"bottom-up" reasoning* that involves going from the specific to the general. For example, suppose we observe that a variety of different objects fall to the ground when we drop them from various heights. We might then use inductive reasoning to conclude that *all objects fall to the earth's surface when dropped.* Depending on the number of observations made, there may be a high degree of certainty in this conclusion. However, what we are really saying is that "All objects that we or other observers have dropped from various heights fall to the earth's surface." Although it is extremely unlikely, we

What Is Harming the Robins?

Suppose a scientist observes an abnormality in the growth of robin embryos in a certain area. She knows the area has been sprayed with a pesticide and suspects the chemical may be causing the abnormalities she has observed.

To test this hypothesis, the scientist carries out a *controlled experiment*. She maintains two groups of robin embryos of the same age in the laboratory. Each group is exposed to exactly the same conditions of light, temperature, food supply, and so on, except the embryos in the experimental group are exposed to a known amount of the pesticide in question.

The embryos in both groups are then examined over an identical period of time for the abnormality. If she finds a significantly larger number of the abnormalities in the experimental group than in the control group, the results support the idea that the pesticide is the culprit.

To be sure no errors occur during the procedure, the original researcher should repeat the experiment several times. Ideally one or more other scientists should repeat the experiment independently.

Critical Thinking

Can you find flaws in this experiment that might lead you to question the scientist's conclusions? (*Hint:* What other factors in nature—not the laboratory—and in the embryos themselves could possibly explain the results?)

---

cannot be *absolutely sure* someone will drop an object that does not fall to the earth's surface.

Deductive reasoning involves using logic to arrive at a specific conclusion based on a generalization or premise. It is a form of *"top-down"* reasoning that goes from the general to the specific. For example,

Generalization or premise: All birds have feathers.

Example: Eagles are birds.

Deductive conclusion: All eagles have feathers.

The conclusion of this *syllogism* (a series of logically connected statements) is valid as long as the premise is correct and we do not use faulty logic to arrive at the conclusion.

Deductive and inductive reasoning and critical thinking skills (p. 3) are important scientific tools. But scientists also try to come up with new or creative ideas

to explain some of the things they observe in nature. Often such ideas defy conventional logic and current scientific knowledge. According to physicist Albert Einstein, "There is no completely logical way to a new scientific idea." Intuition, imagination, and creativity are as important in science as they are in poetry, art, music, and other great adventures of the human spirit, as reflected in scientist Warren Weaver's quotation found at the opening of this chapter.

One of the exciting things about science is that it is never complete. Each discovery unearths new unanswered questions in an ongoing quest for knowledge about how the natural world works. This is one reason why people choose this profession.

How Valid Are the Results of Science? Very Reliable But Not Perfect

Scientists try to establish that a particular model, theory, or law has a very high probability of being true.

Scientists can do two major things. *First,* they can disprove things. *Second,* they can establish that a particular model, theory, or law has a very high probability or degree of certainty of being true. However, like scholars in any field, scientists cannot prove that their theories, models, and laws are *absolutely true.*

Although it may be extremely low, some degree of uncertainty is always involved in any scientific theory, model, or law. Most scientists rarely say something like, "Cigarettes cause lung cancer." Rather, the statement might be phrased, "There is overwhelming evidence from thousands of studies that indicate a significant relationship between cigarette smoking and lung cancer."

Most scientists also rarely use the word *proof.* When scientists hear someone say we should not take a scientific finding seriously because it has not been absolutely proven they know that this person either knows little about the nature of science or is using a debating or advertising trick to cast doubt on a widely accepted scientific finding. Scientists tend to use words like *projections* and *scenarios* to describe what is *likely* to happen in nature instead of making *predictions* or *forecasts* about what *will* happen.

What Is the Difference between Frontier Science and Sound Science? Preliminary and Well-Tested Results

Scientific results fall into those that have not been confirmed (frontier science) and those that have been well tested and widely accepted (sound science).

News reports about science often focus on two things: new so-called scientific breakthroughs, and disputes between scientists over the validity of preliminary and untested data, hypotheses, and models. These preliminary results, called **frontier science,** are often controversial because they have not been widely tested and accepted. At the frontier stage, it is normal and healthy for reputable scientists to disagree about the meaning and accuracy of data and the validity of various hypotheses.

By contrast, **sound science,** or **consensus science,** consists of data, theories, and laws that are widely accepted by scientists who are considered experts in the field involved. The results of sound science are based on a self-correcting process of open peer review. One way to find out what scientists generally agree on is to seek out reports by scientific bodies such as the U.S. National Academy of Sciences and the British Royal Society, which attempt to summarize consensus among experts in key areas of science.

What Is Junk Science and How Can We Detect It? Look Out for Baloney

Junk science is untested ideas presented as sound science.

Junk science consists of scientific results or hypotheses presented as sound science but not having undergone the rigors of the peer review process. Note that frontier science is not necessarily junk science. Instead, it represents tentative results or hypotheses that are in the process of being validated or rejected by peer review.

There are two problems in uncovering junk science. One is that some scientists, politicians, and other analysts label as junk science any science that does not support or further their particular agenda. The other is that reporters and journalists sometimes mislead us in presenting sound or consensus science along with a quote from a scientist in the field who disagrees with the consensus view or from one who is not an expert in the field being discussed. Such attempts to give a false sense of balance or fairness can mislead the public into distrusting well-established sound science.

Here are some critical thinking questions you can use to uncover junk science.

- How reliable are the sources making a particular claim? Do they have a hidden agenda? Are they experts in this field? What is their source of funding?

- Do the conclusions follow logically from the observations?

- Has the claim been verified by impartial peer review?

- How does the claim compare with the consensus view of experts in this field?

3-2 MODELS AND BEHAVIOR OF SYSTEMS

Why Are Models of Complex Systems Useful? Using Inputs, Throughputs, and Outputs to Make Projections

Scientists project the behavior of a complex system by developing a model of its inputs, throughputs (flows), and outputs of matter, energy, and information.

A **system** is a set of components that function and interact in some regular and theoretically understandable manner. Most *systems* have the following key components: **inputs** from the environment, **flows** or **throughputs** within the system at certain rates, and **outputs** to the environment (Figure 3-3).

Scientists use *models* or approximate representations or simulations to find out how systems work and to evaluate ideas or hypotheses. Some of the most powerful and useful technologies invented by humans are mathematical models, which are used to supplement our mental models. *Mathematical models* consist of one or more equations used to describe the behavior of a system and to describe how the system is likely to behave.

Making a mathematical model usually requires going many times through three steps. *First,* make a guess and write down some equations. *Second,* compute the likely behavior of the system implied by the equations. *Third,* compare the system's projected behavior with observations and behavior projected by mental models, existing experimental data, and scientific hypotheses, laws, and theories.

Mathematical models are important because they can give us improved perceptions and projections, especially in situations where our mental models are weak and unreliable. They are particularly useful when there are many interacting variables, when the time frame is long, and when controlled experiments are impossible, too slow, or too expensive to conduct.

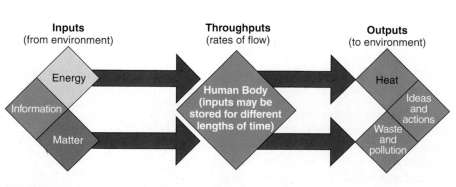

Figure 3-3 Major components of a *system* such as your body.

After building and testing a mathematical model, scientists use it to project what is *likely* to happen under a variety of conditions. In effect, they use mathematical models to answer *if–then* questions: "*If* we do such and such, *then* what is likely to happen now and in the future?" This process can give us a variety of projections or scenarios of possible futures or outcomes based on different assumptions.

Despite its usefulness, a mathematical model is nothing more than a set of hypotheses or assumptions about how we think a certain system works. Mathematical models (like all other models) are no better than the assumptions on which they are built and the data fed into them.

How Do Feedback Loops Affect Systems? Changing Direction

Outputs of matter, energy, or information fed back into a system can cause the system to do more of what it was doing (positive feedback) or less (negative feedback).

When someone asks you for feedback, they are asking for information that they can feed back into their mental processes to help them make a decision or carry out some action. All systems undergo change as a result of feedback loops. A **feedback loop** occurs when an output of matter, energy, or information is fed back into the system as an input and leads to changes in that system.

A **positive feedback loop** causes a system to change further in the same direction. One example involves depositing money in a bank at compound interest and leaving it there. The interest increases the balance, which through a positive feedback loop leads to more interest and an even higher balance.

A **negative,** or **corrective, feedback loop** causes a system to change in the opposite direction. An example is recycling aluminum cans. This involves melting aluminum and feeding it back into an economic system to make new aluminum products. This negative feedback loop of matter reduces the need to find, extract, and process virgin aluminum ore. It also reduces the flow of waste matter (discarded aluminum cans) into the environment.

The temperature-regulating system of your body is an example of a system governed by feedback. Normally a negative feedback loop prevents your body temperature from going too high. If you get hot, your brain receives this information and causes your body to sweat. The evaporation of sweat on your skin removes heat and cools your body. However, if your body temperature exceeds 42°C (108°F), your temperature control system breaks down as your body produces more heat than your sweat-dampened skin can get rid of.

Then a positive feedback loop caused by overloading the system overwhelms the negative feedback loop. These conditions produce a net gain in body heat, which produces even more body heat, and so on, until you die from heatstroke.

The tragedy on Easter Island discussed at the beginning of the chapter also involved the coupling of positive and negative feedback loops. As the abundance of trees turned to a shortage of trees, a positive feedback loop (more births than deaths) became weaker as death rates rose. Eventually a negative feedback loop (more deaths than births) dominated and caused a dieback of the island's human population.

How Do Time Delays Affect Complex Systems? Waiting for Something to Kick In

Sometimes corrective feedback takes so long to work that a system can cross a threshold and change its normal behavior.

Complex systems often show **time delays** between the input of a stimulus and the response to it. A long time delay can mean that corrective action comes too late. For example, a smoker exposed to cancer-causing chemicals in cigarette smoke may not get lung cancer for 20 years or more.

Time delays allow a problem to build up slowly until it reaches a *threshold level* and causes a fundamental shift in the behavior of a system. Prolonged delays dampen the negative feedback mechanisms that might slow, prevent, or halt environmental problems. Examples are population growth, leaks from toxic waste dumps, and degradation of forests from prolonged exposure to air pollutants.

What Is Synergy, and How Can It Affect Complex Systems? One Plus One Can Be Greater Than Two

Sometimes processes and feedbacks in a system can interact to amplify the results.

In arithmetic, 1 plus 1 always equals 2. However, in some of the complex systems found in nature, 1 plus 1 may add up to more than 2 because of synergistic interactions. A **synergistic interaction,** or **synergy,** occurs when two or more processes interact so that the combined effect is greater than the sum of their separate effects.

Synergy can result when two people work together to accomplish a task. For example, suppose you and I need to move a 140-kilogram (300-pound) tree that has fallen across the road. By ourselves, each of us can lift only, say, 45 kilograms (100 pounds). But if we work together and use our muscles properly, we can move the tree out of the way. That is using synergy to solve a problem. Research in the social sciences suggests that most political changes or changes in cultural beliefs are brought about by only about 5% (and rarely more than 10%) of a population working together (synergizing) and expanding their efforts to influence other people.

How Can We Anticipate Environmental Surprises? We Can Never Do Just One Thing

Because any action in a complex system has multiple and often unpredictable results, we should try to anticipate and plan for unintended results and surprises.

One basic principle of environmental science is *we can never do just one thing.* Any action in a complex system has multiple, unintended, and often unpredictable effects. Indeed, most of the harmful environmental problems we face today are unintended results of activities designed to increase the quality of human life (Figure 3-4).

One factor that can lead to an environmental surprise is a *discontinuity* or abrupt change in a previously stable system when some *environmental threshold* is crossed. For example, you may be able to lean back in a chair and balance yourself on two of its legs for a long time with only minor adjustments. But if you pass a certain threshold of movement, your balanced system suffers a discontinuity, or sudden shift, and you may find yourself on the floor. A similar change can happen when many trees in a forest start dying after being weakened and depleted of soil nutrients after decades of exposure to a cocktail of air pollutants.

3-3 MATTER

What Types of Matter Do We Find in Nature? Getting to the Bottom of Things

Matter exists in chemical forms as elements and compounds.

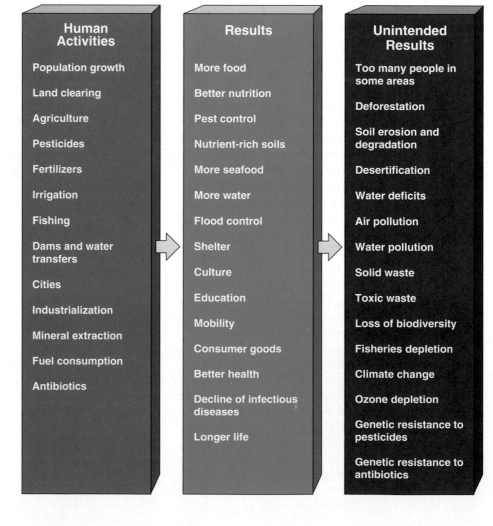

Figure 3-4 Natural capital degradation: human activities designed to improve the quality of life have had a number of unintended harmful environmental effects.

I am going to give you a brief introduction to some chemistry—a discussion of matter and energy. Some of you are saying "I hate chemistry. Why do I need to know this stuff?" The answer is that you and every other material thing on this planet are made up of chemicals and energy. To understand life and environmental problems, you need to know a wee bit of chemistry. I will try to make this journey as interesting and painless as possible. For those of you who have had some basic chemistry, this material will be a breeze.

Matter is anything that has mass (the amount of material in an object) and takes up space. Matter is found in two chemical forms. One is **elements**: the distinctive building blocks of matter that make up every material substance. The other consists of **compounds**: two or more different elements held together in fixed proportions by attractive forces called *chemical bonds*.

To simplify things, chemists represent each element by a one- or two-letter symbol. Examples used in this book are hydrogen (H), carbon (C), oxygen (O), nitrogen (N), phosphorus (P), sulfur (S), chlorine (Cl), fluorine (F), bromine (Br), sodium (Na), calcium (Ca), lead (Pb), mercury (Hg), arsenic (As), and uranium (U). Chemists have developed a way to classify elements in terms of their chemical behavior by arranging them in a *periodic table of elements,* as discussed in Appendix 3. *Good news.* The elements in the list above are the only ones you need to know to understand the material in this book.

From a chemical standpoint, how much are you worth? Not much. If we add up the market price per kilogram for each element in someone weighing 70 kilograms (154 pounds), the total value comes to about $120. Not very uplifting, is it?

But of course you are worth much more because your body is not just a bunch of chemicals enclosed in a bag of skin. Instead you are an incredibly complex system of air, water, soil nutrients, energy-storing chemicals, and food chemicals interacting in millions of ways to keep you alive and healthy. Feel better now?

What Are Nature's Building Blocks?
Matter's Bricks

Atoms, ions, and molecules are the building blocks of matter.

If you had a supermicroscope capable of looking at individual elements and compounds, you could see they are made up of three types of building blocks. The first is an **atom:** the smallest unit of matter that exhibits the characteristics of an element. The second is an **ion:** an electrically charged atom or combination of atoms. A third building block is a **molecule:** a combination of two or more atoms of the same or different elements held together by chemical bonds.

Some elements are found in nature as molecules. Examples are nitrogen and oxygen, which together make up about 99% of the volume of air you just inhaled. Two atoms of nitrogen (N) combine to form a gaseous molecule, with the shorthand formula N_2 (read as "N-two"). The subscript after the element's symbol indicates the number of atoms of that element in a molecule. Similarly, most of the oxygen gas in the atmosphere exists as O_2 (read as "O-two") molecules. A small amount of oxygen, found mostly in the second layer of the atmosphere (stratosphere), exists as O_3 (read as "O-three") molecules, a gaseous form of oxygen called *ozone.*

What Are Atoms Made Of? Looking Inside

Each atom has a tiny nucleus containing protons, and in most cases neutrons, and one or more electrons whizzing around somewhere outside the nucleus.

If you increased the magnification of your supermicroscope, you would find that each different type of atom contains a certain number of *subatomic particles.* There are three types of these atomic building blocks: positively charged **protons** (p), uncharged **neutrons** (n), and negatively charged **electrons** (e). Actually, there are other particles, but they need not concern us at this introductory level.

Each atom consists of an extremely small center, or **nucleus.** It contains one or more protons, and in most cases neutrons, and one or more electrons in rapid motion somewhere outside the nucleus. Atoms are incredibly small. More than 3 million hydrogen atoms could sit side by side on the period at the end of this sentence. Now that is tiny.

Each atom has a certain number of positively charged protons inside its nucleus and an equal number of negatively charged electrons outside its nucleus. Because these electrical charges cancel one another, *the atom as a whole has no net electrical charge.*

Each element has its own specific **atomic number,** equal to the number of protons in the nucleus of each of its atoms. The simplest element, hydrogen (H), has only 1 proton in its nucleus, so its atomic number is 1. Carbon (C), with 6 protons, has an atomic number of 6. Uranium (U), a much larger atom, has 92 protons and an atomic number of 92.

Because atoms are electrically neutral, the atomic number of an atom tells us the number of positively charged protons in its nucleus and the equal number of negatively charged electrons outside its nucleus. For example, an atom of uranium with an atomic number of 92 has 92 protons in its nucleus and 92 electrons outside, and thus no net electrical charge.

Because electrons have so little mass compared with the mass of a proton or a neutron, *most of an*

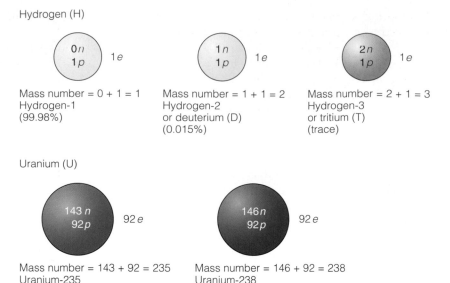

Figure 3-5 Isotopes of hydrogen and uranium. All isotopes of hydrogen have an atomic number of 1 because each has one proton in its nucleus; similarly, all uranium isotopes have an atomic number of 92. However, each isotope of these elements has a different mass number because its nucleus contains a different number of neutrons. Figures in parentheses indicate the percentage abundance by weight of each isotope in a natural sample of the element.

atom's mass is concentrated in its nucleus. The mass of an atom is described in terms of its **mass number:** the total number of neutrons and protons in its nucleus. For example, a hydrogen atom with 1 proton and no neutrons in its nucleus has a mass number of 1, and an atom of uranium with 92 protons and 143 neutrons in its nucleus has a mass number of 235 (92 + 143 = 235).

All atoms of an element have the same number of protons in their nuclei. But they may have different numbers of uncharged neutrons in their nuclei, and thus may have different mass numbers. Various forms of an element having the same atomic number but a different mass number are called **isotopes** of that element. Scientists identify isotopes by attaching their mass numbers to the name or symbol of the element. For example, hydrogen has three isotopes: hydrogen-1 (H-1), hydrogen-2 (H-2, common name *deuterium*), and hydrogen-3 (H-3, common name *tritium*). A natural sample of an element contains a mixture of its isotopes in a fixed proportion, or percentage abundance by weight (Figure 3-5).

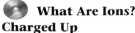 What Are Ions? Getting Charged Up

Atoms of some elements can lose or gain one or more electrons to form ions with positive or negative electrical charges.

Ions form when an atom of an element loses or gains one or more electrons. Thus an *ion* is an atom or groups of atoms with one or more net positive (+) or negative (−) electrical charges, one for each electron lost or gained. Atoms are neutral but ions are all charged up.

Some elements, known as *metals*, tend to lose one or more of their electrons and form positively charged ions. Like a quarterback passing a football, they are *electron givers.* For example, an atom of the metallic element sodium (Na, atomic number 11) with 11 positively charged protons and 11 negatively charged electrons can lose one of its electrons. It then becomes a sodium ion with a positive charge of 1 (Na^+) because it now has 11 positive charges (protons) but only 10 negative charges (electrons).

Other atoms, known as *nonmetals*, tend to gain one or more electrons and form negatively charged ions. Like a tight end waiting for a pass from a quarterback, they are *electron receivers.* For example, an atom of the nonmetallic element chlorine (Cl, with an atomic number of 17) can gain an electron and become a chlorine ion. The ion has a negative charge of 1 (Cl^-) because it has 17 positively charged protons and 18 negatively charged electrons.

The number of positive or negative charges on an ion is shown as a superscript after the symbol for an atom or a group of atoms. Examples of ions encountered in this book are positive ions such as hydrogen ions (H^+), calcium ions (Ca^{2+}), and ammonium ions (NH_4^+), and negative ions such as nitrate ions (NO_3^-), sulfate ions (SO_4^{2-}), and phosphate ions (PO_4^{3-}).

The amount of a substance in a unit volume of air, water, or other medium is called its **concentration.** It is like the number of people in a classroom or a swimming pool.

The concentration of hydrogen ions (H^+) in a water solution is a measure of its acidity or alkalinity, represented by a value called **pH.** On a *pH scale* of 0 to 14, *acids* have a pH less than 7, *bases* have a pH greater than 7, and a *neutral solution* has a pH of 7 (Figure 3-6).

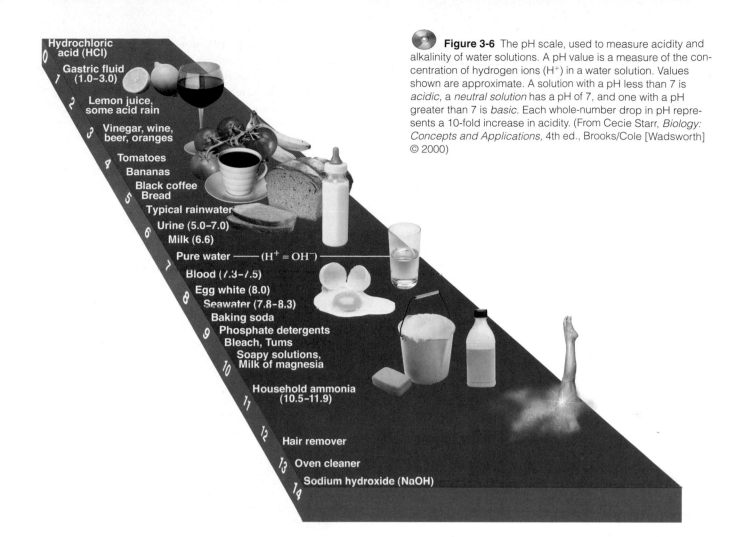

Figure 3-6 The pH scale, used to measure acidity and alkalinity of water solutions. A pH value is a measure of the concentration of hydrogen ions (H^+) in a water solution. Values shown are approximate. A solution with a pH less than 7 is *acidic*, a *neutral solution* has a pH of 7, and one with a pH greater than 7 is *basic*. Each whole-number drop in pH represents a 10-fold increase in acidity. (From Cecie Starr, *Biology: Concepts and Applications*, 4th ed., Brooks/Cole [Wadsworth] © 2000)

Labels on pH scale (from top/0 to bottom/14):

- 0 Hydrochloric acid (HCl)
- 1 Gastric fluid (1.0–3.0)
- 2 Lemon juice, some acid rain
- 3 Vinegar, wine, beer, oranges
- 4 Tomatoes, Bananas
- 5 Black coffee, Bread, Typical rainwater
- 6 Urine (5.0–7.0), Milk (6.6)
- 7 Pure water —— ($H^+ = OH^-$), Blood (7.3–7.5)
- 8 Egg white (8.0), Seawater (7.8–8.3), Baking soda
- 9 Phosphate detergents, Bleach, Tums, Soapy solutions, Milk of magnesia
- 10 Household ammonia (10.5–11.9)
- 12 Hair remover
- 13 Oven cleaner
- 14 Sodium hydroxide (NaOH)

What Holds the Atoms and Ions in Compounds Together? Giving, Receiving, and Sharing Electrons

Some compounds are made up of oppositely charged ions and others are made up of molecules.

Most matter exists as *compounds*, substances containing atoms or ions of more than one element that are held together by chemical bonds. Chemists use a shorthand **chemical formula** to show the number of atoms or ions of each type in a compound. The formula contains the symbols for each of the elements present and uses subscripts to represent the number of atoms or ions of each element in the compound's basic structural unit.

Some compounds are made up of oppositely charged ions and are called *ionic compounds*. Those made up of molecules of uncharged atoms are called *covalent* or *molecular compounds*.

Sodium chloride (table salt) is an *ionic compound* represented by the chemical formula NaCl. It consists of a three-dimensional array of oppositely charged *ions* (Na^+ and Cl^-). The forces of attraction between these oppositely charged ions are called *ionic bonds*, as

discussed in more detail in Appendix 3. They are formed when a metal atom (a giver) gives one or more electrons to a nonmetal atom (a receiver). Then the resulting positively and negatively charged ions attract one another. The result of this electron dating game, involving giving, receiving, and attraction between opposites, is an ionic compound.

Water, a *covalent* or *molecular compound*, consists of molecules made up of uncharged atoms of hydrogen (H) and oxygen (O). Each water molecule consists of two hydrogen atoms chemically bonded to an oxygen atom, yielding H_2O (read as "H-two-O") molecules. The bonds between the atoms in such molecules are called *covalent bonds*, as discussed in Appendix 3. Covalent compounds form when atoms of various elements share one or more electrons. It is electron dating by sharing.

What Are Organic Compounds? Think Carbon

Organic compounds contain carbon atoms combined with one another and with various other atoms such as hydrogen, nitrogen, or chlorine.

Table sugar, vitamins, plastics, aspirin, penicillin, and most of the chemicals in your body are **organic compounds.** If you could view these compounds with your supermicroscope, you would see that all (except one) have at least two carbon atoms (some have thousands) combined with each other and with atoms of one or more other elements such as hydrogen, oxygen, nitrogen, sulfur, phosphorus, chlorine, and fluorine. One exception, methane (CH_4), has only one carbon atom.

Almost all organic compounds are molecular compounds held together by covalent bonds. Organic compounds can be either *natural* (such as carbohydrates, proteins, and fats in natural foods) or *synthetic* (such as plastics and many drugs made by humans).

The millions of known organic (carbon-based) compounds include the following:

- *Hydrocarbons:* compounds of carbon and hydrogen atoms. An example is methane (CH_4), the main component of natural gas, and the simplest organic compound.

- *Chlorinated hydrocarbons:* compounds of carbon, hydrogen, and chlorine atoms. An example is the insecticide DDT ($C_{14}H_9Cl_5$).

- *Simple carbohydrates* (simple sugars): certain types of compounds of carbon, hydrogen, and oxygen atoms. An example is glucose ($C_6H_{12}O_6$), which most plants and animals break down in their cells to obtain energy.

What Are Genes, Chromosomes, and DNA Molecules? Linking Up Molecules

Simple organic molecules can link together to form more complex organic compounds.

Larger and more complex organic compounds, called *polymers,* consist of a number of basic structural or molecular units (*monomers*) linked by chemical bonds, somewhat like cars linked in a freight train. The three major types of organic polymers are *complex carbohydrates* consisting of two or more monomers of simple sugars (such as glucose) linked together, *proteins* formed by linking together monomers of amino acids, and *nucleic acids* (such as DNA and RNA) made by linked sequences of monomers called nucleotides, as discussed in Appendix 3.

Genes consist of specific sequences of nucleotides in a DNA molecule. Each gene carries codes (each consisting of three nucleotides) needed to make various proteins. These coded units of genetic information about specific traits are passed on from parents to offspring during reproduction.

Chromosomes are combinations of genes that make up a single DNA molecule, together with a number of proteins. Each chromosome typically contains thousands of genes. Genetic information coded in your chromosomal DNA is what makes you different

from an oak leaf, an alligator, or a flea and from your parents. The relationships of genetic material to cells are depicted in Figure 3-7, which you may find useful in getting genetic terms straight.

The total weight of the DNA needed to reproduce the world's 6.4 billion people is only about 50 milligrams—the weight of a small match. If the DNA coiled in your body were unwound, it would stretch about 960 million kilometers (600 million miles)—more than six times the distance between the sun and the earth.

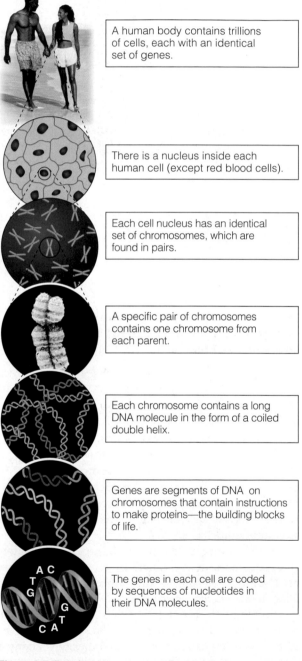

A human body contains trillions of cells, each with an identical set of genes.

There is a nucleus inside each human cell (except red blood cells).

Each cell nucleus has an identical set of chromosomes, which are found in pairs.

A specific pair of chromosomes contains one chromosome from each parent.

Each chromosome contains a long DNA molecule in the form of a coiled double helix.

Genes are segments of DNA on chromosomes that contain instructions to make proteins—the building blocks of life.

The genes in each cell are coded by sequences of nucleotides in their DNA molecules.

Figure 3-7 Relationships among cells, nuclei, chromosomes, DNA, genes, and nucleotides.

The different molecules of DNA that make up the millions of species found on the earth are like a vast and diverse genetic library. Each species is a unique book in that library.

The *genome* of a species is made up of the entire sequence of DNA "letters" or base pairs that combine to "spell out" the chromosomes in typical members of each species. In 2002, scientists were able to map out the genome for the human species.

What Are Inorganic Compounds? The Rest of the World's Compounds

Compounds without carbon–carbon and carbon–hydrogen bonds are called inorganic compounds.

Inorganic compounds do not have carbon–carbon or carbon–hydrogen bonds. Some of the inorganic compounds discussed in this book are sodium chloride (NaCl), water (H_2O), nitrous oxide (N_2O), nitric oxide (NO), carbon monoxide (CO), carbon dioxide (CO_2), nitrogen dioxide (NO_2), sulfur dioxide (SO_2), ammonia (NH_3), hydrogen sulfide (H_2S), sulfuric acid (H_2SO_4), and nitric acid (HNO_3). *Good news.* These are the only inorganic compounds you need to know to understand the material in this book.

Now you know about the fairly small cast of chemicals (atoms, ions, molecules, and compounds) you will encounter in this book.

What Are Four States of Matter? Let's Get Physical

Matter exists in solid, liquid, and gaseous physical states and a fourth state known as plasma.

The atoms, ions, and molecules that make up matter are found in three *physical states:* solid, liquid, and gas. For example, water exists as ice, liquid water, or water vapor depending on its temperature and the surrounding air pressure. The three physical states of any sample of matter differ in the spacing and orderliness of its atoms, ions, or molecules. A solid has the most compact and orderly arrangement and a gas the least compact and orderly arrangement. Liquids are somewhere in between.

A fourth state of matter is called **plasma.** It is a high-energy mixture of roughly equal numbers of positively charged ions and negatively charged electrons. A plasma forms when enough energy is applied to strip electrons away from the nuclei of atoms—somewhat like a blast of wind blowing the leaves off a tree.

Plasma is the most abundant form of matter in the universe. The sun and all stars consist mostly of plasma. There is little natural plasma on the earth, with most of it found in lightning bolts and flames.

But scientists have learned how to make artificial plasmas in fluorescent lights, arc lamps, neon signs, gas discharge lasers, and in TV and computer screens. They do this by running a high-voltage electric current through a gas. Scientists hope to be able to develop affordable plasma torches and use them to destroy toxic wastes, sterilize and clean water, remove soot from exhaust gases, and produce clean-burning hydrogen gas from diesel fuel, gasoline, or methane for use in fuel cells. Great stuff if we can do it.

What Is Matter Quality? Matter Usefulness

Matter can be classified as having high or low quality depending on how useful it is to us as a resource.

Matter quality is a measure of how useful a form of matter is to us as a resource, based on its availability and concentration, as shown in Figure 3-8. **High-quality**

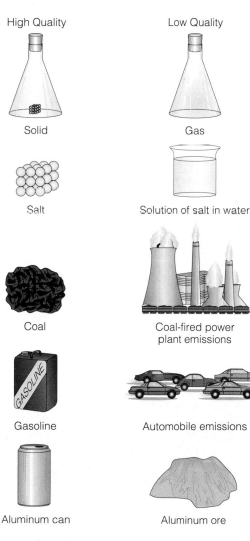

High Quality   Low Quality

Solid   Gas

Salt   Solution of salt in water

Coal   Coal-fired power plant emissions

Gasoline   Automobile emissions

Aluminum can   Aluminum ore

Figure 3-8 Examples of differences in matter quality. *High-quality matter* (left-hand column) is fairly easy to extract and concentrated; *low-quality matter* (right-hand column) is more difficult to extract and more dispersed than high-quality matter.

matter is concentrated, usually is found near the earth's surface, and has great potential for use as a matter resource. **Low-quality matter** is dilute, often is deep underground or dispersed in the ocean or the atmosphere, and usually has little potential for use as a material resource.

An aluminum can is a more concentrated, higher-quality form of aluminum than aluminum ore containing the same amount of aluminum. That is why it takes less energy, water, and money to recycle an aluminum can than to make a new can from aluminum ore.

Material efficiency, or **resource productivity,** is the total amount of material needed to produce each unit of goods or services. *Great news.* Business expert Paul Hawken and physicist Amory Lovins contend that resource productivity in developed countries could be improved by 75–90% within two decades using existing technologies.

3-4 ENERGY

What Is Energy? Doing Work and Transferring Heat

Energy is the work needed to move matter and the heat that flows from hot to cooler samples of matter.

Energy is the capacity to do work and transfer heat. Work is performed when an object such as a grain of sand, this book, or a giant boulder is moved over some distance. Work, or matter movement, also is needed to boil water or burn natural gas to heat a house or cook food. Energy is also the heat that flows automatically from a hot object to a cold object when they come in contact.

There are two major types of energy. One is **kinetic energy,** possessed by matter because of the matter's mass and its speed or velocity. Examples of this energy in motion are wind (a moving mass of air), flowing streams, heat flowing from a body at a high temperature to one at a lower temperature, and electricity (flowing electrons).

The second type is **potential energy,** which is stored and potentially available for use. Examples of this stored energy are a rock held in your hand, an unlit match, still water behind a dam, the chemical energy stored in gasoline molecules, and the nuclear energy stored in the nuclei of atoms.

Potential energy can be changed to kinetic energy. Drop a rock or this book on your foot and the book's potential energy when you held it changes into kinetic energy. When you burn gasoline in a car engine, the potential energy stored in the chemical bonds of its molecules changes into heat, light, and mechanical (kinetic) energy that propel the car.

What Is Electromagnetic Radiation? Think Waves

Some energy travels in waves at the speed of light.

Another type of energy is **electromagnetic radiation.** It is energy traveling in the form of a *wave* as a result of changing electric and magnetic fields.

There are many different forms of electromagnetic radiation, each with a different *wavelength* (distance between successive peaks or troughs in the wave) and *energy content*, as shown in Figure 3-9. Such radiation travels through space at the speed of light, which is about 300,000 kilometers (186,000 miles) per second. That is fast.

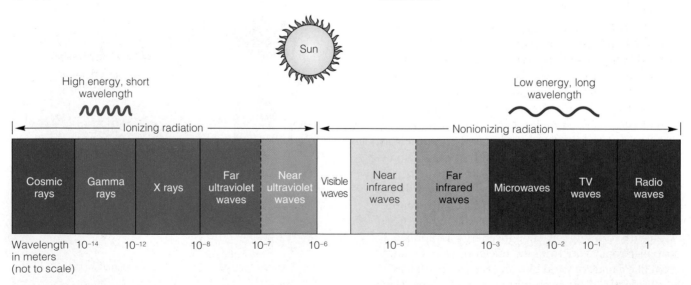

Figure 3-9 The *electromagnetic spectrum:* the range of electromagnetic waves, which differ in wavelength (distance between successive peaks or troughs) and energy content.

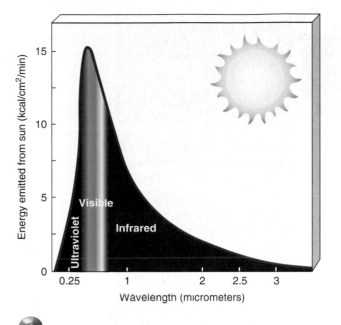

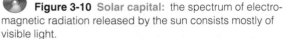

Figure 3-10 **Solar capital:** the spectrum of electro-magnetic radiation released by the sun consists mostly of visible light.

Cosmic rays, gamma rays, X rays, and ultraviolet radiation (Figure 3-9, left side) are called **ionizing radiation** because they have enough energy to knock electrons from atoms and change them to positively charged ions. The resulting highly reactive electrons and ions can disrupt living cells, interfere with body processes, and cause many types of sickness, including various cancers. The other forms of electromagnetic radiation (Figure 3-9, right side) do not contain enough energy to form ions and are called **nonionizing radiation.** No scientific consensus exists on whether nonionizing forms of electromagnetic radiation given off when an electric current passes through a wire or a motor is harmful to humans.

The visible light that you can detect with your eyes is a form of nonionizing radiation that occupies only a small portion of the full range, or spectrum, of different types of electromagnetic radiation. We are energy challenged because our senses can detect only a tiny amount of the different types of electromagnetic radiation that are all around us. Figure 3-10 shows that visible light makes up most of the spectrum of electro-magnetic radiation emitted by the sun.

What Is Heat and How Is It Transferred? Three Ways to Tango

Heat is the total kinetic energy of all the parts of a sample of matter and is transferred from one place to another by convection, conduction, and radiation.

Heat is the total kinetic energy of all the moving atoms, ions, or molecules within a given substance, excluding the overall motion of the whole object. For example, a glass of water consists of zillions of water molecules in constant motion. The heat stored in this sample of water is the total kinetic energy of all of these moving molecules. The term *heat* is also used to describe the energy that can be transferred between objects at different temperatures.

Temperature is the average speed of motion of the atoms, ions, or molecules in a sample of matter at a given moment. For example, the molecules in a sample of water in a glass have a certain average speed of motion. This is what we call its temperature.

Heat can be transferred from one place to another by three different methods. Study Figure 3-11 for an explanation of these methods.

Convection

Heating water in the bottom of a pan causes some of the water to vaporize into bubbles. Because they are lighter than the surrounding water, they rise. Water then sinks from the top to replace the rising bubbles.This up and down movement (convection) eventually heats all of the water.

Conduction

Heat from a stove burner causes atoms or molecules in the pan's bottom to vibrate faster. The vibrating atoms or molecules then collide with nearby atoms or molecules, causing them to vibrate faster. Eventually, molecules or atoms in the pan's handle are vibrating so fast it becomes too hot to touch.

Radiation

As the water boils, heat from the hot stove burner and pan radiate into the surrounding air, even though air conducts very little heat.

Figure 3-11 Three ways in which heat can be transferred from one place to another.

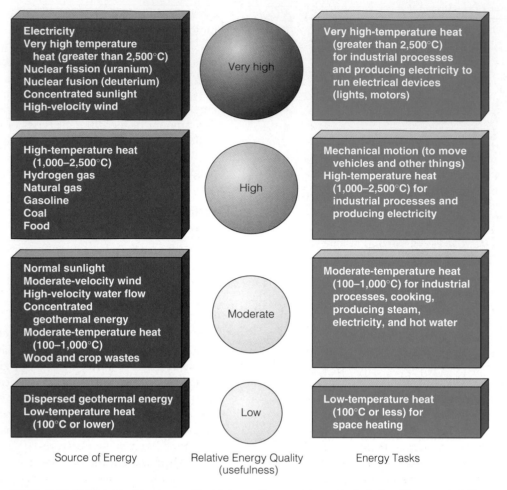

Figure 3-12 Categories of energy quality. *High-quality energy* is concentrated and has great ability to perform useful work. *Low-quality energy* is dispersed and has little ability to do useful work. To avoid unnecessary energy waste, it is best to match the quality of an energy source with the quality of energy needed to perform a task.

What Is Energy Quality? Energy Usefulness

Energy can be classified as having high or low quality depending on how useful it is to us as a resource.

Energy quality is a measure of an energy source's ability to do useful work, as seen in Figure 3-12. **High-quality energy** is concentrated and can perform much useful work. Examples are electricity, the chemical energy stored in coal and gasoline, concentrated sunlight, and nuclei of uranium-235 used as fuel in nuclear power plants.

By contrast, **low-quality energy** is dispersed and has little ability to do useful work. An example of low-quality energy is heat dispersed in the moving molecules of a large amount of matter (such as the atmosphere or a large body of water) so that its temperature is low.

For example, the total amount of heat stored in the Atlantic Ocean is greater than the amount of high-quality chemical energy stored in all the oil deposits of Saudi Arabia. Yet the ocean's heat is so widely dispersed, it cannot be used to move things or to heat things to high temperatures.

It makes sense to match the quality of an energy source with the quality of energy needed to perform a particular task (Figure 3-12) because doing so saves energy and usually money.

3-5 THE LAW OF CONSERVATION OF MATTER: A RULE WE CANNOT BREAK

What Is the Difference between a Physical and a Chemical Change? Changes in Form and in Chemical Makeup

Matter can change from one physical form to another or change its chemical composition.

When a sample of matter undergoes a **physical change,** its chemical composition is not changed. A piece of aluminum foil cut into small pieces is still aluminum foil. When solid water (ice) is melted or liquid water is boiled, none of the H_2O molecules involved are altered; instead, the molecules are organized in different spatial (physical) patterns.

In a **chemical change,** or **chemical reaction,** there is a change in the chemical composition of the elements or compounds. Chemists use shorthand chemical equations to represent what happens in a chemical reaction. For example, when coal burns completely, the solid carbon (C) it contains combines with oxygen gas (O_2) from the atmosphere to form the gaseous compound carbon dioxide (CO_2):

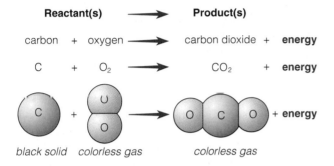

Energy is given off in this reaction, making coal a useful fuel. The reaction also shows how the complete burning of coal (or any of the carbon-containing compounds in wood, natural gas, oil, and gasoline) gives off carbon dioxide gas. This is a key gas that helps warm the lower atmosphere (troposphere).

The Law of Conservation of Matter: Why There Is No "Away"

When a physical or chemical change occurs, no atoms are created or destroyed.

We may change various elements and compounds from one physical or chemical form to another, but in no physical and chemical change can we create or destroy any of the atoms involved. All we can do is rearrange them into different spatial patterns (physical changes) or different combinations (chemical changes). This statement, based on many thousands of measurements, is known as the **law of conservation of matter.**

The law of conservation of matter means there is no "away" as in "to throw away." *Everything we think we have thrown away is still here somewhere on the planet with us in one form or another.* Dust and soot from the smokestacks of industrial plants, substances removed from polluted water at sewage treatment plants, and banned chemicals like DDT all have to go somewhere and can come back around to haunt us. For example, selling DDT abroad means it can return to the United States as residues in imported coffee, fruit, and other foods, or as fallout from air masses moved long distances by winds.

How Do Chemists Keep Track of Atoms? Atomic Bookkeeping

Chemical equations are used as an accounting system to verify that no atoms are created or destroyed in a chemical reaction.

In keeping with the law of conservation of matter, each side of a chemical equation must have the same number of atoms of each element involved. Minding this law leads to what chemists call a *balanced chemical equation.*

The equation for the burning of carbon (C + $O_2 \longrightarrow CO_2$) is balanced because one atom of carbon and two atoms of oxygen are on both sides of the equation. That is an easy one.

Now we try a slightly harder one. When electricity is passed through water (H_2O), the water molecule can be broken down into hydrogen (H_2) and oxygen (O_2). This chemical reaction can be represented by the following shorthand chemical equation:

$$H_2O \longrightarrow H_2 + O_2$$

| 2 H atoms | 2 H atoms | 2 O atoms |
| 1 O atom | | |

This equation is unbalanced because one atom of oxygen is on the left but two atoms are on the right.

We cannot change the subscripts of any of the formulas to balance this equation because that would make the substances different from those actually involved. That is a no-no according to the law of conservation of matter. Instead, we could use different numbers of the *molecules* involved to balance the equation. For example, we could use two water molecules:

$$2 H_2O \longrightarrow H_2 + O_2$$

| 4 H atoms | 2 H atoms | 2 O atoms |
| 2 O atoms | | |

This equation is still unbalanced because although the numbers of oxygen atoms on both sides are now equal, the numbers of hydrogen atoms are not.

We can correct this by having the reaction produce two hydrogen molecules:

$$2 H_2O \longrightarrow 2 H_2 + O_2$$

| 4 H atoms | 4 H atoms | 2 O atoms |
| 2 O atoms | | |

Now the equation is balanced, and the law of conservation of matter has been observed. We see that for every two molecules of water through which we pass electricity, two hydrogen molecules and one oxygen molecule are produced. By the way, this equation may change your life. It represents how we could use heat or electricity to decompose water and produce hydrogen gas that someday may replace oil as a fuel to power our societies—more on this hydrogen revolution in Chapter 18.

Ready for a little more practice? Try to balance the chemical equation for the reaction of nitrogen gas (N_2) with hydrogen gas (H_2) to form ammonia gas (NH_3).

How Harmful Are Pollutants? Too Much of Bad Things

The law of conservation of matter tells us that we will always produce some pollutants, but we can produce much less and clean up some of what we do produce.

We can make the environment cleaner and convert some potentially harmful chemicals into less harmful physical or chemical forms. But *the law of conservation of matter means we will always face the problem of what to do with some quantity of wastes and pollutants.*

Three factors determining the severity of a pollutant's harmful effects are its *chemical nature*, its *concentration*, and its *persistence*.

The second of these factors, *concentration*, is sometimes expressed in *parts per million (ppm)*; 1 ppm corresponds to 1 part pollutant per million parts of the gas, liquid, or solid mixture in which the pollutant is found. Smaller concentration units are parts per billion (ppb) and parts per trillion (ppt).

We can reduce the concentration of a pollutant by dumping it into the air or a large volume of water, but there are limits to the effectiveness of this dilution approach. For example, the water flowing in a river can dilute or disperse some of the wastes we dump into the river. But if we dump in too much waste this natural cleansing process does not work.

The third factor, **persistence**, is a measure of how long the pollutant stays in the air, water, soil, or body. Pollutants can be classified into four categories based on their persistence. **Degradable**, or **nonpersistent, pollutants** are broken down completely or reduced to acceptable levels by natural physical, chemical, and biological processes. Complex chemical pollutants that living organisms (usually specialized bacteria) break down into simpler chemicals are called **biodegradable pollutants.** Human sewage in a river, for example, is biodegraded fairly quickly by bacteria if the sewage is not added faster than it can be broken down.

Slowly degradable, or **persistent, pollutants** take decades or longer to degrade. Examples include the insecticide DDT and most plastics.

Nondegradable pollutants are chemicals that natural processes cannot break down. Examples include the toxic elements lead, mercury, and arsenic. Ideally, we should try not to use these chemicals, but if we do, we should figure out ways to keep them from getting into the environment.

3-6 NUCLEAR CHANGES

What Is Natural Radioactivity? Nuclei Losing Particles or Radiation

An atom can change from one isotope to another when its nucleus loses particles or gives off high-energy electromagnetic radiation.

In addition to physical and chemical changes, matter can undergo a third type of change known as a **nuclear change.** This occurs when nuclei of certain isotopes spontaneously change or are made to change into nuclei of different isotopes.

Three types of nuclear change are natural radioactive decay, nuclear fission, and nuclear fusion. **Natural radioactive decay** is a nuclear change in which unstable isotopes spontaneously emit fast-moving chunks of matter (called particles), high-energy radiation, or both at a fixed rate. The unstable isotopes are called **radioactive isotopes** or **radioisotopes.** Radioactive decay of these isotopes into various other isotopes continues until the original isotope is changed into a stable isotope that is not radioactive.

Radiation emitted by radioisotopes is damaging ionizing radiation. The most common form of ionizing energy released from radioisotopes is **gamma rays,** a form of high-energy electromagnetic radiation. You do not want to get exposed to these waves.

High-speed ionizing particles emitted from the nuclei of radioactive isotopes are most commonly of two types: **alpha particles** (fast-moving, positively charged chunks of matter that consist of two protons and two neutrons) and **beta particles** (high-speed electrons).

Each type of radioisotope spontaneously decays at a characteristic rate into a different isotope. This rate of decay can be expressed in terms of **half-life:** the time needed for *one-half* of the nuclei in a given quantity of a radioisotope to decay and emit their radiation to form a different isotope (Figure 3-13). The decay continues, often producing a series of different radioisotopes, until a nonradioactive isotope is formed.

Each radioisotope has a characteristic half-life, which may range from a few millionths of a second to several billion years (Table 3-1, right). An isotope's half-life cannot be changed by temperature, pressure, chemical reactions, or any other known factor.

Half-life can be used to estimate how long a sample of a radioisotope must be stored in a safe container before it decays to what is considered a safe level. A general rule is that such decay takes about 10 half-lives. Thus people must be protected from radioactive waste containing iodine-131 (which concentrates in the thyroid gland and has a half-life of 8 days) for 80 days (10 × 8 days). Plutonium-239, which is produced in nuclear reactors and used as the explosive in some

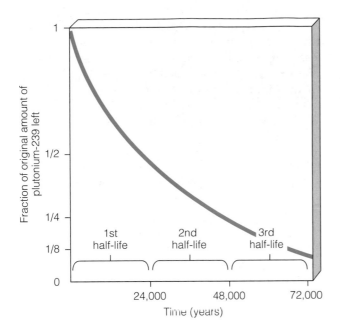

Figure 3-13 *Half-life.* The radioactive decay of plutonium-239, which is produced in nuclear reactors and used as the explosive in some nuclear weapons, has a half-life of 24,000 years. The amount of radioactivity emitted by a radioactive isotope decreases by one-half for each half-life that passes. Thus, after three half-lives, amounting to 72,000 years, one-eighth of a sample of plutonium-239 would still be radioactive.

nuclear weapons, can cause lung cancer when its particles are inhaled in minute amounts. Its half-life is 24,000 years. Thus it must be stored safely for 240,000 years (10 × 24,000 years)—about four times longer than the latest version of our species (*Homo sapiens*

Table 3-1 Half-Lives of Selected Radioisotopes

Isotope	Radiation Half-Life	Emitted
Potassium-42	12.4 hours	Alpha, beta
Iodine-131	8 days	Beta, gamma
Cobalt-60	5.27 years	Beta, gamma
Hydrogen-3 (tritium)	12.5 years	Beta
Strontium-90	28 years	Beta
Carbon-14	5,370 years	Beta
Plutonium-239	24,000 years	Alpha, gamma
Uranium-235	710 million years	Alpha, gamma
Uranium-238	4.5 billion years	Alpha, gamma

sapiens) has existed. Chapter 17 has more on the problem of what to do with the nuclear wastes we have created.

Exposure to ionizing radiation from alpha particles, beta particles, and gamma rays can damage cells in two ways. One is *genetic damage* from mutations or changes in DNA molecules that alter genes and chromosomes. If the mutation is harmful, it can lead to genetic defects in the next generation of offspring or several generations later.

The other is *somatic damage* to tissues, which causes harm during the victim's lifetime. Examples include burns, miscarriages, eye cataracts, and certain cancers.

According to the U.S. National Academy of Sciences, exposure over an average lifetime to average levels of ionizing radiation from natural and human sources causes about 1% of all fatal cancers and 5–6% of all normally encountered genetic defects in the U.S. population.

What Is Nuclear Fission? Splitting Heavy Nuclei

Neutrons can split apart the nuclei of certain isotopes with large mass numbers and release a large amount of energy.

Nuclear fission is a nuclear change in which nuclei of certain isotopes with large mass numbers (such as uranium-235) are split apart into lighter nuclei when struck by neutrons; each fission releases two or three more neutrons and energy (Figure 3-14, p. 50). Each of these neutrons, in turn, can cause an additional fission. For these multiple fissions to take place, enough fissionable nuclei must be present to provide the **critical mass** needed for efficient capture of these neutrons.

Multiple fissions within a critical mass form a **chain reaction,** which releases an enormous amount of energy (Figure 3-15, p. 50). This is somewhat like a room in which the floor is covered with spring-loaded mousetraps, each topped by a Ping-Pong ball. Open the door, throw in a single Ping-Pong ball, and watch the action in this simulated chain reaction of snapping mousetraps and balls flying around in every direction.

In an atomic bomb, an enormous amount of energy is released in a fraction of a second in an uncontrolled nuclear fission chain reaction. This reaction is initiated by an explosive charge, which pushes two masses of fissionable fuel together. This causes the fuel to reach the critical mass needed for a chain reaction and to give off a tremendous amount of energy in a gigantic explosion.

In the reactor of a nuclear power plant, the rate at which the nuclear fission chain reaction takes place is

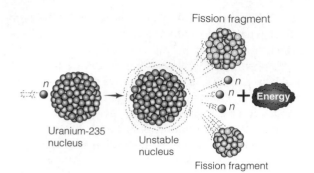

Figure 3-14 Fission of a uranium-235 nucleus by a neutron (*n*).

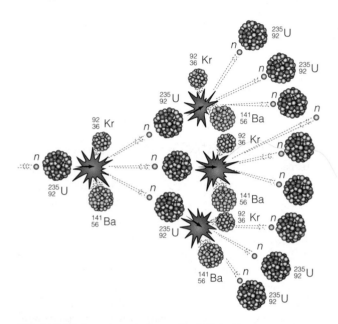

Figure 3-15 A *nuclear chain reaction* initiated by one neutron triggering fission in a single uranium-235 nucleus. This figure illustrates only a few of the trillions of fissions caused when a single uranium-235 nucleus is split within a critical mass of uranium-235 nuclei. The elements krypton (Kr) and barium (Ba), shown here as fission fragments, are only two of many possibilities.

controlled so that under normal operation only one of every two or three neutrons released is used to split another nucleus. In conventional nuclear fission reactors, the splitting of uranium-235 nuclei releases heat, which produces high-pressure steam to spin turbines and thus generate electricity.

What Is Nuclear Fusion? Forcing Light Nuclei to Combine

Extremely high temperatures can force the nuclei of isotopes of some lightweight atoms to fuse together and release large amounts of energy.

Nuclear fusion is a nuclear change in which two isotopes of light elements, such as hydrogen, are forced

together at extremely high temperatures until they fuse to form a heavier nucleus. Lots of energy is released when this happens. Temperatures of at least 100 million °C are needed to force the positively charged nuclei (which strongly repel one another) to fuse.

Nuclear fusion is much more difficult to initiate than nuclear fission, but once started it releases far more energy per unit of fuel than does fission. You would not be alive without nuclear fusion. Fusion of hydrogen nuclei to form helium nuclei is the source of energy in the sun and other stars.

After World War II, the principle of *uncontrolled nuclear fusion* was used to develop extremely powerful hydrogen, or thermonuclear, weapons. These weapons use the D–T fusion reaction, in which a hydrogen-2, or deuterium (D), nucleus and a hydrogen-3 (tritium, T) nucleus are fused to form a larger, helium-4 nucleus, a neutron, and energy, as shown in Figure 3-16.

Scientists have also tried to develop *controlled nuclear fusion*, in which the D–T reaction is used to pro-

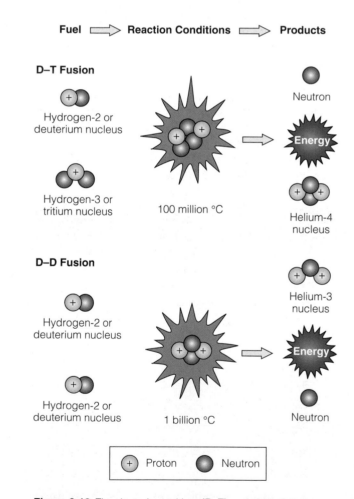

Figure 3-16 The deuterium–tritium (D–T) and deuterium–deuterium (D–D) *nuclear fusion* reactions, which take place at extremely high temperatures.

duce heat that can be converted into electricity. After more than 50 years of research, this process is still in the laboratory stage. Even if it becomes technologically and economically feasible, many energy experts do not expect it to be a practical source of energy until 2030, if then.

3-7 ENERGY LAWS: TWO RULES WE CANNOT BREAK

What Is the First Law of Thermodynamics? You Cannot Get Something for Nothing

In a physical or chemical change, we can change energy from one form to another but we can never create or destroy any of the energy involved.

Scientists have observed energy being changed from one form to another in millions of physical and chemical changes. But they have never been able to detect the creation or destruction of any energy (except in nuclear changes). The results of their experiments have been summarized in the **law of conservation of energy,** also known as the **first law of thermodynamics:** *In all physical and chemical changes, energy is neither created nor destroyed, but it may be converted from one form to another.*

This scientific law tells us that when one form of energy is converted to another form in any physical or chemical change, *energy input always equals energy output.* No matter how hard we try or how clever we are, we cannot get more energy out of a system than we put in; in other words, *we cannot get something for nothing in terms of energy quantity.* This is one of Mother Nature's basic rules that we have to live with.

What Is the Second Law of Thermodynamics? You Cannot Even Break Even

Whenever energy is changed from one form to another we always end up with less usable energy than we started with.

Because the first law of thermodynamics states that energy can be neither created nor destroyed, we may be tempted to think we will always have enough energy. Yet if we fill a car's tank with gasoline and drive around or use a flashlight battery until it is dead, something has been lost. If it is not energy, what is it? The answer is *energy quality* (Figure 3-12), the amount of energy available that can perform useful work.

Countless experiments have shown that when energy is changed from one form to another, a decrease in energy quality always occurs. The results of these experiments have been summarized in what is called the **second law of thermodynamics:** *When energy is changed from one form to another, some of the useful energy is always degraded to lower quality, more dispersed, less useful energy.* This degraded energy usually takes the form of heat given off at a low temperature to the surroundings (environment). There it is dispersed by the random motion of air or water molecules and becomes even less useful as a resource.

In other words, *we cannot even break even in terms of energy quality because energy always goes from a more useful to a less useful form when energy is changed from one form to another.* No one has ever found a violation of this fundamental scientific law. It is another one of Mother Nature's basic rules that we have to live with.

Consider three examples of the second law of thermodynamics in action. *First,* when a car is driven, only about 20–25% of the high-quality chemical energy available in its gasoline fuel is converted into mechanical energy (to propel the vehicle) and electrical energy (to run its electrical systems). The remaining 75–80% is degraded to low-quality heat that is released into the environment and eventually lost into space. Thus, most of the money you spend for gasoline is not used to get you anywhere.

Second, when electrical energy flows through filament wires in an incandescent lightbulb, it is changed into about 5% useful light and 95% low-quality heat that flows into the environment. In other words, this so-called *light bulb* is really a *heat bulb. Good news.* Scientists have developed compact fluorescent bulbs that are four times more efficient, and even more efficient bulbs are on the way. Do you use compact fluorescent bulbs?

Third, in living systems, solar energy is converted into chemical energy (food molecules) and then into mechanical energy (moving, thinking, and living). During each of these conversions, high-quality energy is degraded and flows into the environment as low-quality heat (Figure 3-17, p. 52). Trace the flows and energy conversions in this diagram.

The second law of thermodynamics also means that *we can never recycle or reuse high-quality energy to perform useful work.* Once the concentrated energy in a serving of food, a liter of gasoline, a lump of coal, or a chunk of uranium is released, it is degraded to low-quality heat that is dispersed into the environment.

Energy efficiency, or **energy productivity,** is a measure of how much useful work is accomplished by a particular input of energy into a system. *Good news.* There is plenty of room for improving energy efficiency. Scientists estimate that only about 16% of the energy used in the United States ends up performing useful work. The remaining 84% is either unavoidably wasted because of the second law of thermodynamics (41%) or unnecessarily wasted (43%).

Here is a lesson from thermodynamics. The cheapest and quickest way for us to get more energy is to

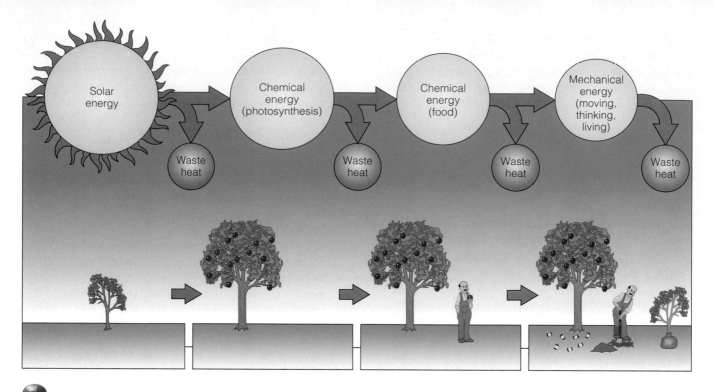

Figure 3-17 The second law of thermodynamics in action in living systems. Each time energy is changed from one form to another, some of the initial input of high-quality energy is degraded, usually to low-quality heat that is dispersed into the environment.

stop unnecessarily wasting almost half of the energy we use. We can do this by not driving gas-guzzling motor vehicles and by not living in poorly insulated and leaky houses. What are you doing to reduce your unnecessary waste of energy?

3-8 MATTER AND ENERGY LAWS AND ENVIRONMENTAL PROBLEMS

What Is a High-Throughput Economy? An Accelerating Treadmill

Most of today's economies increase economic growth by converting the world's resources to goods and services in ways that add large amounts of waste, pollution, and low-quality heat to the environment.

As a result of the law of conservation of matter and the second law of thermodynamics, individual resource use automatically adds some waste heat and waste matter to the environment. Most of today's advanced industrialized countries have **high-throughput (high-waste) economies** that attempt to sustain ever-increasing economic growth by increasing the one-way flow of matter and energy resources through their economic systems (Figure 3-18). These resources flow through their economies into planetary *sinks* (air, water, soil, organisms), where pollutants and wastes end up and can accumulate to harmful levels.

What happens if more and more people continue to use and waste more and more energy and matter resources at an increasing rate? In other words, what happens if most of the world's people get infected with the affluenza virus?

The law of conservation of matter and the two laws of thermodynamics discussed in this chapter tell us that eventually this consumption will exceed the capacity of the environment to dilute and degrade waste matter and absorb waste heat. However, they do not tell us how close we are to reaching such limits.

What Is a Matter-Recycling-and-Reuse Economy? Go in Circles Instead of Straight Lines

Recycling and reusing more of the earth's matter resources slows down depletion of nonrenewable matter resources and reduces our environmental impact.

There is a way to slow down the resource use and reduce our environmental impact in a high-throughput economy. We can convert such a linear high-throughput economy into a circular **matter-recycling-and-reuse economy** that recycles and reuses our matter outputs back instead of dumping them into the environment.

Changing to a matter-recycling-and-reuse economy is an important way to buy some time. But this does not allow more and more people to use more and more resources indefinitely, even if all of them were

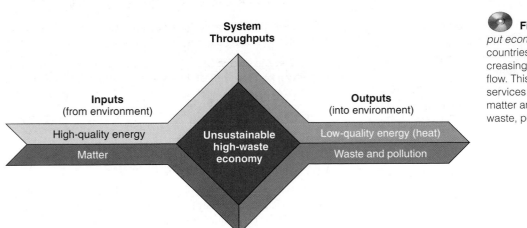

System Throughputs

Inputs
(from environment)

High-quality energy

Matter

Unsustainable high-waste economy

Outputs
(into environment)

Low-quality energy (heat)

Waste and pollution

Figure 3-18 The *high-through-put economies* of most developed countries are based on continually increasing the rates of energy and matter flow. This produces valuable goods and services but also converts high-quality matter and energy resources into waste, pollution, and low-quality heat.

somehow perfectly recycled and reused. The reason is that the two laws of thermodynamics tell us that recycling and reusing matter resources always requires using high-quality energy (which cannot be recycled) and adds waste heat to the environment.

What Is a Low-Throughput Economy? Learning from Nature

We can live more sustainably by reducing the throughput of matter and energy in our economies, not wasting matter and energy resources, recycling and reusing most of the matter resources we use, and stabilizing the size of our population.

Is there a better way out of the environmental situation that we have gotten ourselves into? You bet. The three scientific laws governing matter and energy changes suggest that the best long-term solution to our environmental and resource problems is to shift from an economy based on increasing matter and energy flow (throughput) to a more sustainable **low-throughput (low-waste) economy,** as summarized in Figure 3-19. This means building on the concept of recycling and reusing as much matter as possible by also reducing the throughput of matter and energy through an economy. This can be done by wasting less matter and energy, living more simply to decrease resource use per person, and slowing population growth to reduce the

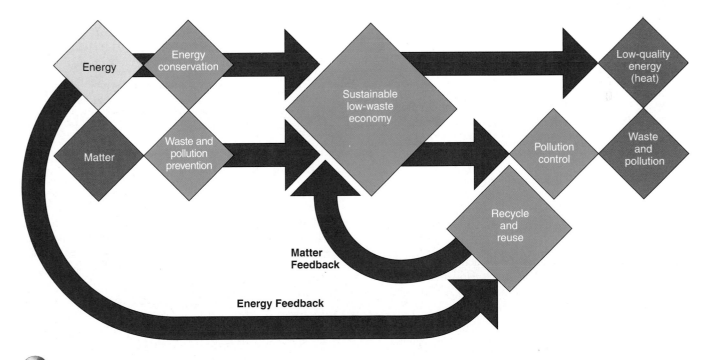

Figure 3-19 Solutions: lessons from nature. A *low-throughput economy,* based on energy flow and matter recycling, works with nature to reduce the throughput of matter and energy resources (items shown in green). This is done by **(1)** reusing and recycling most nonrenewable matter resources, **(2)** using renewable resources no faster than they are replenished, **(3)** using matter and energy resources efficiently, **(4)** reducing unnecessary consumption, **(5)** emphasizing pollution prevention and waste reduction, and **(6)** controlling population growth.

number of resource users. In other words, we learn to live more sustainably by heeding the *lessons from nature* revealed by the law of conservation of mass and the two laws of thermodynamics.

The next five chapters apply the three basic scientific laws of matter and thermodynamics to living systems and look at some *biological principles* that can also teach us how to live more sustainably by working with nature.

The second law of thermodynamics holds, I think, the supreme position among laws of nature. . . . If your theory is found to be against the second law of thermodynamics, I can give you no hope.

ARTHUR S. EDDINGTON

CRITICAL THINKING

1. Respond to the following statements:
   a. Scientists have not absolutely proven that anyone has ever died from smoking cigarettes.
   b. The greenhouse theory—that certain gases (such as water vapor and carbon dioxide) warm the atmosphere—is not a reliable idea because it is only a scientific theory.

2. See whether you can find an advertisement or an article describing some aspect of science in which **(a)** the concept of scientific proof is misused, **(b)** the term *theory* is used when it should have been *hypothesis,* and **(c)** a consensus scientific finding is dismissed or downplayed because it is "only a theory" or is not viewed as sound science.

3. How does a scientific law (such as the law of conservation of matter) differ from a societal law (such as maximum speed limits for vehicles)? Can each be broken? Explain.

4. A tree grows and increases its mass. Explain why this is not a violation of the law of conservation of matter

5. If there is no "away," why is the world not filled with waste matter?

6. Methane (CH_4) gas is the major component of natural gas. Write and balance the chemical equation for the burning of methane when it combines with oxygen gas in the atmosphere to form carbon dioxide and water.

7. Suppose you have 100 grams of radioactive plutonium-239 with a half-life of 24,000 years. How many grams of plutonium-239 will remain after **(a)** 12,000 years, **(b)** 24,000 years, and **(c)** 96,000 years?

8. Someone wants you to invest money in an automobile engine that will produce more energy than the energy in the fuel (such as gasoline or electricity) you use to run the motor. What is your response? Explain.

9. Use the second law of thermodynamics to explain why a barrel of oil can be used only once as a fuel.

PROJECTS

1. Use the library or Internet to find an example of junk science and explain why it is junk science. Compare your findings with those of your classmates.

2. **(a)** List two examples of negative feedback loops not discussed in this chapter, one that is beneficial and one that is detrimental. Compare your examples with those of your classmates. **(b)** Give two examples of positive feedback loops not discussed in this chapter. Include one that is beneficial and one that is detrimental. Compare your examples with those of your classmates.

3. If you have the use of a sensitive balance, try to demonstrate the law of conservation of mass in a physical change. Weigh a container with a lid (a glass jar will do), add an ice cube and weigh it again, and then allow the ice to melt and weigh it again. Explain how your results obey the law of conservation of matter.

4. Use the library or Internet to find examples of various perpetual motion machines and inventions that allegedly violate the two laws of thermodynamics by producing more high-quality energy than the high-quality energy needed to make them run. What has happened to these schemes and machines—many of them developed by scam artists to attract money from investors?

5. Use the library or the Internet to find bibliographic information about *Warren Weaver* and *Arthur S. Eddington,* whose quotes appear at the beginning and end of this chapter.

6. Make a concept map of this chapter's major ideas using the section heads, subheads, and key terms (in boldface). Look on the website for this book for information about making concept maps.

LEARNING ONLINE

The website for this book contains study aids and many ideas for further reading and research. They include a chapter summary, review questions for the entire chapter, flash cards for key terms and concepts, a multiple-choice practice quiz, interesting Internet sites, references, and a guide for accessing thousands of InfoTrac® College Edition articles. Log on to

http://biology.brookscole.com/miller14

Then click on the Chapter-by-Chapter area, choose Chapter 3, and select a learning resource.

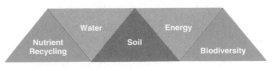
CASE STUDY

Have You Thanked the Insects Today?

Insects have a bad reputation. We classify many as *pests* because they compete with us for food, spread human diseases such as malaria, and invade our lawns, gardens, and houses. Some people have "bugitis," fear all insects, and think the only good bug is a dead bug. This view fails to recognize the vital roles insects play in helping sustain life on earth.

A large proportion of the earth's plant species (including many trees) depends on insects to pollinate their flowers (Figure 4-1, right). Without pollinating insects, we would have few fruits and vegetables to enjoy.

Insects that eat other insects help control the populations of at least half the species of insects we call pests. An example is the praying mantis (Figure 4-1, left). This free pest-control service is an important part of the natural capital that helps sustain us.

Insects have been around for at least 400 million years and are phenomenally successful forms of life. An estimated 10 quintillion (ten followed by 18 zeros) insects live with us on the earth—about 1.6 million insects for each one of us.

Some insects can reproduce at an astounding rate. For example, a single female housefly and her offspring can theoretically produce about 5.6 trillion houseflies in only one year.

Insects can rapidly evolve new genetic traits, such as resistance to pesticides. They also have an exceptional ability to evolve into new species when faced with new environmental conditions, and they are quite resistant to extinction.

For example, we can apply chemical pesticides to protect crops from pests. This can help grow more food. But there is a downside to pesticides. They can harm beneficial insects that help

protect crops from other insects. Widespread use of chemical pesticides can also accelerate the natural ability of rapidly reproducing insect pests to develop genetic resistance (immunity) to such chemicals. Thus in the long run, chemical pesticides can backfire and become less effective in reducing crop losses from insect pests—a glaring example of unintended consequences.

The environmental lesson is that although insects can thrive without newcomers such as us, we and most other land organisms would perish quickly without them. Learning about the roles insects play in nature requires us to understand how insects and other organisms living in a *biological community* such as a forest or pond interact with one another and with the nonliving environment. *Ecology* is the science that studies such relationships and interactions in nature, as discussed in this and the following six chapters.

Figure 4-1 Insects play important roles in helping sustain life on earth. The bright green caterpillar moth feeding on pollen in a crocus (right) and other insects pollinate flowering plants that serve as food for many plant eaters. The praying mantis eating a monarch butterfly (left) and many other insect species help control the populations of at least half of the insect species we classify as pests.

The earth's thin film of living matter is sustained by grand-scale cycles of energy and chemical elements.

G. EVELYN HUTCHINSON

This chapter addresses the following questions:

- What is ecology?

- What basic processes keep us and other organisms alive?

- What are the major components of an ecosystem?

- What happens to energy in an ecosystem?

- What are soils, and how are they formed?

- What happens to matter in an ecosystem?

- How do scientists study ecosystems?

- What are two principles of sustainability derived from learning how nature works?

4-1   THE NATURE OF ECOLOGY

What Is Ecology? Understanding Connections

Ecology is a study of connections in nature.

Ecology (from the Greek words *oikos,* "house" or "place to live," and *logos,* "study of") is the study of how organisms interact with one another and with their nonliving environment. In effect, it is a study of *connections in nature*—the house for the earth's life. Ecologists focus on trying to understand the interactions among organisms, populations, communities, ecosystems, and the biosphere (Figure 4-2).

An **organism** is any form of life. The **cell** is the basic unit of life in organisms. Organisms may consist of a single cell (bacteria, for instance) or many cells. Look in the mirror. What you see is about 10 trillion cells divided into about 200 different types.

On the basis of their cell structure, organisms can be classified as either *eukaryotic* or *prokaryotic.* Each cell of a **eukaryotic** organism is surrounded by a membrane and has a distinct *nucleus* (a membrane-bounded structure containing genetic material in the form of DNA), and several other internal parts called *organelles* (Figure 4-3a, p. 58). All organisms except bacteria and some algae are eukaryotic.

A membrane surrounds the cell of a **prokaryotic** organism, but the cell contains no distinct nucleus or other internal parts enclosed by membranes (Figure 4-3b). All bacteria are single-celled prokaryotic organisms. Most familiar organisms are eukaryotic, but they could not exist without hordes of prokaryotic organisms called *microbes* that can only be seen only with the aid of a microscope (see Case Study at right).

Organisms can be classified into **species,** groups of organisms that resemble one another in appearance, behavior, chemistry, and genetic makeup. Scientists use a special system to name each species. (See Appendix 4 to find out how biologists came up with the name *Homo sapiens sapiens* for our current species.)

Organisms that reproduce sexually by combining cells from both parents are classified as members of the same species if, under natural conditions, they can potentially breed with one another and produce live, fertile offspring.

How many species are on the earth? We do not know. Estimates range from 3.6 million to 100 million, many of them in tropical forests. Most are insects and microorganisms too small to be seen with the naked eye. A best guess is that we share the planet with 10–14 million other species.

So far biologists have identified, named, and briefly described about 1.4 million species, most of them insects (Figure 4-4, p. 58).

Case Study: What Species Rule the World? Small Matters!

Multitudes of tiny microbes such as bacteria, protozoa, fungi, and yeast help keep us alive.

They are everywhere and there are trillions of them. Billions are found inside your body, on your body, in a handful of soil, and in a cup of river water.

These mostly invisible rulers of the earth are *microbes,* a catchall term for many thousands of species of bacteria, protozoa, fungi, and yeasts—most too small to be seen with the naked eye.

Microbes do not get the respect they deserve. Most of us think of them as threats to our health in the form of infectious bacteria or "germs," fungi that cause athlete's foot and other skin diseases, and protozoa that cause diseases such as malaria. But these harmful microbes are in the minority.

You are alive because of multitudes of microbes toiling away, mostly out of sight. You can thank microbes for providing you with food by converting nitrogen gas in the atmosphere into forms that plants can take up from the soil as nutrients that keep you alive and well. Microbes also help produce foods such as bread, cheese, yogurt, vinegar, tofu, soy sauce, beer, and wine. You should also thank bacteria and fungi in the soil that decompose organic wastes into nutrients that can be taken up by plants that we and most other animals eat. Without these wee creatures we would be up to our eyeballs in waste matter.

Microbes, especially bacteria, help purify the water you drink by breaking down wastes. Bacteria in your intestinal tract break down the food you eat. Some microbes in your nose prevent harmful bacteria from reaching your lungs. Others are the source of disease-

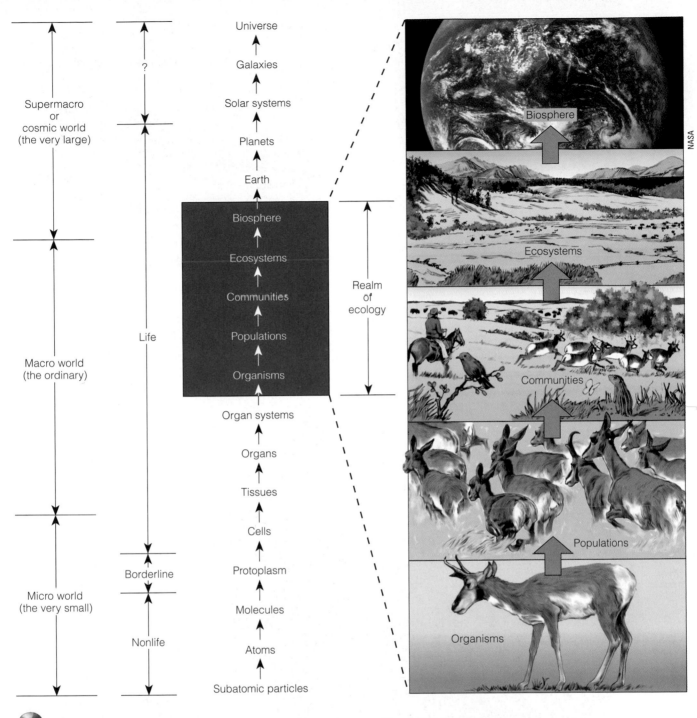

Figure 4-2 Natural capital: levels of organization of matter in nature. Note the five levels that ecology focuses on.

fighting antibiotics, including penicillin, erythromycin, and streptomycin. Genetic engineers are developing microbes that can extract metals from ores, break down various pollutants, and help clean up toxic waste sites.

Some microbes help control plant diseases and populations of insect species that attack our food crops. Relying more on these microbes for pest control can reduce the use of potentially harmful chemical pesticides.

What Is a Population? Life in a Group

Members of a species interact in groups called populations that live together in a particular place or habitat.

A **population** is a group of interacting individuals of the same species occupying a specific area (Figure 4-5, p. 59). Examples are all sunfish in a pond,

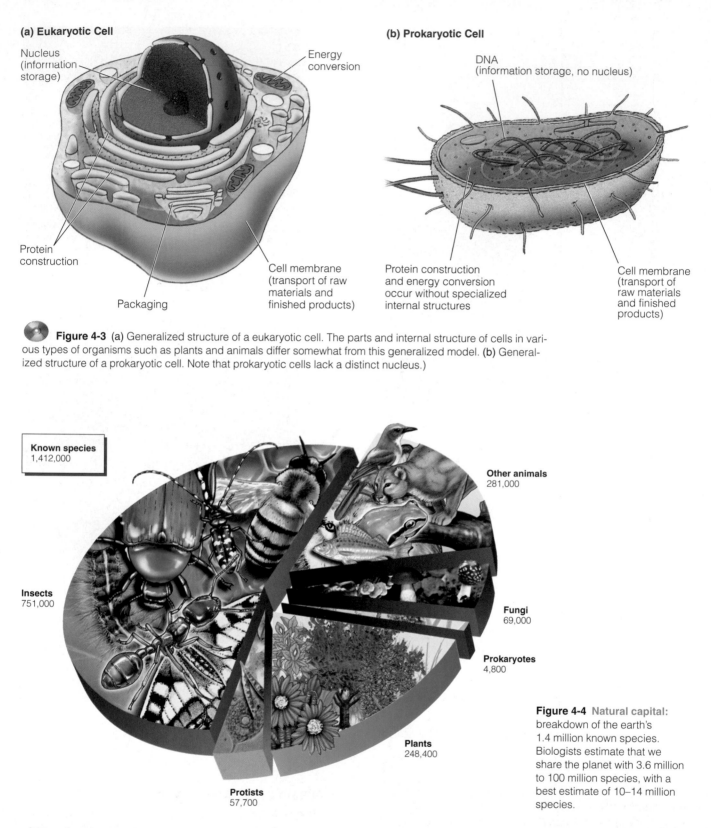

(a) Eukaryotic Cell

Nucleus (information storage)

Energy conversion

Protein construction

Packaging

Cell membrane (transport of raw materials and finished products)

(b) Prokaryotic Cell

DNA (information storage, no nucleus)

Protein construction and energy conversion occur without specialized internal structures

Cell membrane (transport of raw materials and finished products)

Figure 4-3 (a) Generalized structure of a eukaryotic cell. The parts and internal structure of cells in various types of organisms such as plants and animals differ somewhat from this generalized model. **(b)** Generalized structure of a prokaryotic cell. Note that prokaryotic cells lack a distinct nucleus.)

Known species 1,412,000

Insects 751,000

Protists 57,700

Plants 248,400

Other animals 281,000

Fungi 69,000

Prokaryotes 4,800

Figure 4-4 Natural capital: breakdown of the earth's 1.4 million known species. Biologists estimate that we share the planet with 3.6 million to 100 million species, with a best estimate of 10–14 million species.

white oak trees in a forest, and people in a country. In most natural populations, individuals vary slightly in their genetic makeup, which is why they do not all look or act alike. This is called a population's **genetic diversity** (Figure 4-6).

The place or environment where a population (or an individual organism) normally lives is its **habitat.** It may be as large as an ocean or as small as the intestine of a termite.

The area over which we can find a species is called its **distribution** or **range.** Many species, such as some tropical plants, have a small range and may be found on a single hillside. Other species such the grizzly bear have large ranges.

Figure 4-5 A population of monarch butterflies. The geographic distribution of this butterfly coincides with that of the milkweed plant, on which monarch larvae and caterpillars feed.

Figure 4-6 The *genetic diversity* among individuals of one species of Caribbean snail is reflected in the variations in shell color and banding patterns.

What Are Communities and Ecosystems? Interactions in Nature

A community consists of populations of different species living and interacting in an area, and an ecosystem is a community interacting with its physical environment of matter and energy.

A **community,** or **biological community,** consists of all the populations of the different species living and interacting in an area. It is a complex and interacting network of plants, animals, and microorganisms.

An **ecosystem** is a community of different species interacting with one another and with their physical environment of matter and energy. Ecosystems can range in size from a puddle of water to a stream, a patch of woods, an entire forest, or a desert. Ecosystems can be natural or artificial (human created). Examples of artificial ecosystems are crop fields, farm ponds, and reservoirs. All of the earth's ecosystems together make up what we call the **biosphere.**

Scientists studying ecosystems seek answers to several questions. *First,* how many members of each species are present? *Second,* how do these organisms capture energy and matter (nutrients) from their environment? *Third,* how do they transfer energy and matter among themselves and to other ecosystems? *Fourth,* how do they release energy into the environment and return nutrients to the environment for recycling?

4-2 THE EARTH'S LIFE-SUPPORT SYSTEMS

What Are the Major Parts of the Earth's Life-Support Systems? The Spheres of Life

The earth is made up of interconnected spherical layers that contain air, water, soil, minerals, and life.

We can think of the earth as being made up of several spherical layers, as diagrammed in Figure 4-7 (p. 60). Study this figure carefully. The **atmosphere** is a thin envelope or membrane of air around the planet. Its inner layer, the **troposphere,** extends only about 17 kilometers (11 miles) above sea level. It contains most of the planet's air, mostly nitrogen (78%) and oxygen (21%). The next layer, stretching 17–48 kilometers (11–30 miles) above the earth's surface, is the **stratosphere.** Its lower portion contains enough ozone (O_3) to filter out most of the sun's harmful ultraviolet radiation. This allows life to exist on land and in the surface layers of bodies of water.

The **hydrosphere** consists of the earth's water. It is found as *liquid water* (both surface and underground), *ice* (polar ice, icebergs, and ice in frozen soil layers called *permafrost*), and *water vapor* in the atmosphere. The **lithosphere** is the earth's crust and upper mantle; the crust contains nonrenewable fossil fuels (created from ancient fossils that were buried and subjected to intense pressure and heat) and minerals, and renewable soil chemicals (nutrients) needed for plant life.

The **biosphere** is the portion of the earth in which living (biotic) organisms exist and interact with one another and with their nonliving (abiotic) environment. The biosphere includes most of the hydrosphere and parts of the lower atmosphere and upper lithosphere. It reaches from the deepest ocean floor, 20 kilometers (12 miles) below sea level, to the tops of the highest mountains.

All parts of the biosphere are interconnected. If the earth were an apple, the biosphere would be no thicker than the apple's skin. *The goal of ecology is to understand the interactions in this thin, life-supporting global membrane of air, water, soil, and organisms.*

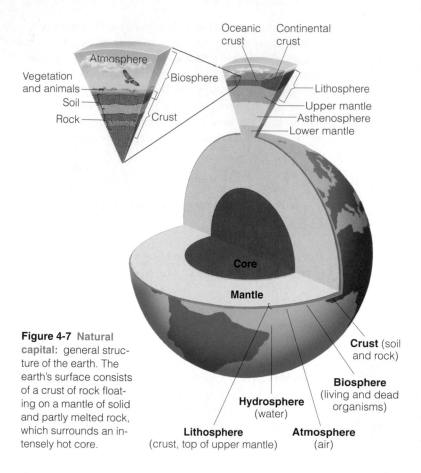

Figure 4-7 **Natural capital:** general structure of the earth. The earth's surface consists of a crust of rock floating on a mantle of solid and partly melted rock, which surrounds an intensely hot core.

Labels in figure:
Oceanic crust
Continental crust
Atmosphere
Vegetation and animals
Biosphere
Soil
Rock
Crust
Lithosphere
Upper mantle
Asthenosphere
Lower mantle
Core
Mantle
Crust (soil and rock)
Biosphere (living and dead organisms)
Hydrosphere (water)
Atmosphere (air)
Lithosphere (crust, top of upper mantle)

How Does the Sun Help Sustain Life on Earth? A Distant Nuclear Fusion Reactor

Energy released by the gigantic nuclear fusion reactor we call the sun provides the light and heat needed to sustain the earth's life.

Energy from the sun supports most life on the earth by lighting and warming the planet. It supports *photosynthesis,* the process in which green plants and some bacteria make compounds such as carbohydrates that keep them alive and feed most other organisms. And it powers the cycling of matter and drives the climate and weather systems that distribute heat and fresh water over the earth's surface.

About one-billionth of the sun's output of energy reaches the earth—a tiny sphere in the vastness of space—in the form of electromagnetic waves (Figure 3-10, p. 45). The amount of energy reaching the earth from the sun equals the amount of heat energy the earth reflects or radiates back into space. Otherwise, the earth would heat up to temperatures too hot for life as we know it.

Much of the sun's incoming energy is reflected away or absorbed by chemicals, dust, and clouds in the atmosphere as shown in Fig-

What Sustains Life on Earth? Sun, Cycles, and Gravity

Solar energy, the cycling of matter, and gravity sustain the earth's life.

Life on the earth depends on three interconnected factors, shown in Figure 4-8.

- The *one-way flow of high-quality energy* from the sun, through materials and living things in their feeding interactions, into the environment as low-quality energy (mostly heat dispersed into air or water molecules at a low temperature), and eventually back into space as heat. No round-trips are allowed because energy cannot be recycled.

- The *cycling of matter* (the atoms, ions, or molecules needed for survival by living organisms) through parts of the biosphere. Because the earth is closed to significant inputs of matter from space, the earth's essentially fixed supply of nutrients must be recycled again and again for life to continue. Nutrient trips in ecosystems are round-trips.

- *Gravity,* which allows the planet to hold onto its atmosphere and causes the downward movement of chemicals in the matter cycles.

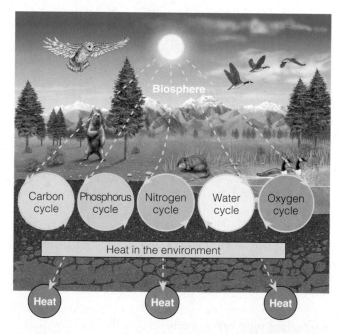

Figure 4-8 **Natural capital:** life on the earth depends on the *one-way flow of energy* (dashed lines) from the sun through the biosphere, the *cycling of crucial elements* (solid lines around circles), and *gravity,* which keeps atmospheric gases from escaping into space and draws chemicals downward in the matter cycles. This simplified model depicts only a few of the many cycling elements.

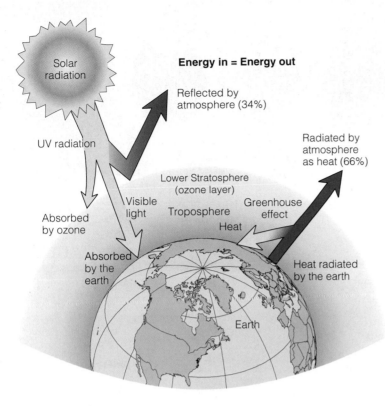

Figure 4-9 Solar capital: flow of energy to and from the earth.

Solar radiation

Energy in = Energy out

Reflected by atmosphere (34%)

Radiated by atmosphere as heat (66%)

UV radiation

Lower Stratosphere (ozone layer)

Visible light

Greenhouse effect

Absorbed by ozone

Troposphere

Heat

Absorbed by the earth

Heat radiated by the earth

Earth

ure 4-9. About 80% of the energy that gets through warms the troposphere and evaporates and cycles water through the biosphere. About 1% of this incoming energy generates winds, and green plants, algae, and bacteria use less than 0.1% to fuel photosynthesis.

Most of the solar radiation making it though the atmosphere hits the surface of the earth and is degraded into longer-wavelength infrared radiation. This infrared radiation interacts with so-called *greenhouse gases* (such as water vapor, carbon dioxide, methane, nitrous oxide, and ozone) in the troposphere. The radiation causes these gaseous molecules to vibrate and release infrared radiation with even longer wavelengths into the troposphere. As this radiation interacts with molecules in the air, it increases their kinetic energy, helping warm the troposphere and the earth's surface. Without this **natural greenhouse effect,** the earth would be too cold for life as we know it to exist and you would not be around to read this book.

4-3 ECOSYSTEM COMPONENTS

What Are Biomes and Aquatic Life Zones? Life on Land and at Sea

Life exists on land systems called biomes and in freshwater and ocean aquatic life zones.

Viewed from outer space, the earth resembles an enormous jigsaw puzzle consisting of large masses of land and vast expanses of ocean.

Biologists have classified the terrestrial (land) portion of the biosphere into **biomes** ("BY-ohms"). They are large regions such as forests, deserts, and grasslands characterized by a distinct climate and specific species (especially vegetation) adapted to it (Figure 4-10, p. 62).

Scientists divide the watery parts of the biosphere into **aquatic life zones,** each containing numerous ecosystems. Examples include *freshwater life zones* (such as lakes and streams) and *ocean* or *marine life zones* (such as coral reefs, coastal estuaries, and the deep ocean).

What Are the Major Components of Ecosystems? Matter, Energy, Life

Ecosystems consist of nonliving (abiotic) and living (biotic) components.

Two types of components make up the biosphere and its ecosystems. One type, called **abiotic,** consists of nonliving components such as water, air, nutrients, and solar energy. The other type, called **biotic,** consists of biological components—plants, animals, and microbes.

Figures 4-11 (p. 63) and 4-12 (p. 63) are greatly simplified diagrams of some of the biotic and abiotic components in a freshwater aquatic ecosystem and a terrestrial ecosystem.

How Tolerant Are Organisms to Environmental Conditions? Tolerance Limits

Populations of different species can thrive only under certain physical and chemical conditions.

Populations of different species thrive under different physical conditions. Some need bright sunlight, and

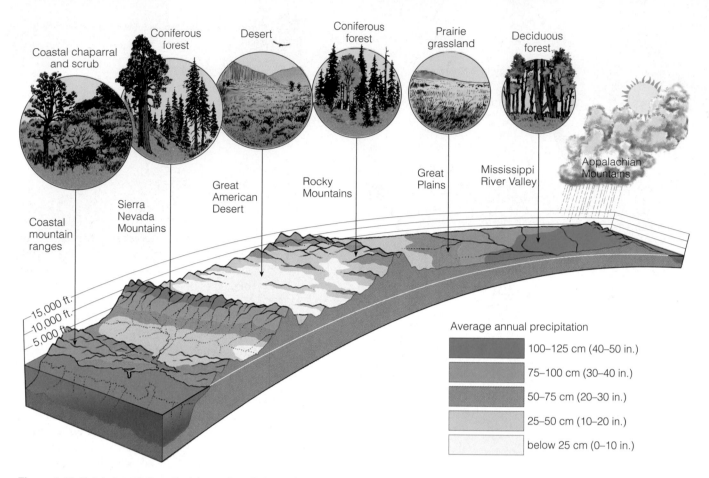

Figure 4-10 Natural capital: major biomes found along the 39th parallel across the United States. The differences reflect changes in climate, mainly differences in average annual precipitation and temperature (not shown).

others thrive better in shade. Some need a hot environment and others a cool or cold one. Some do best under wet conditions and others under dry conditions.

Each population in an ecosystem has a **range of tolerance** to variations in its physical and chemical environment (Figure 4-13, p. 64). Individuals within a population may also have slightly different tolerance ranges for temperature or other factors because of small differences in genetic makeup, health, and age. For example, a trout population may do best within a narrow band of temperatures (*optimum level* or *range*), but a few individuals can survive above and below that band (Figure 4-13). However, if the water becomes too hot or too cold, none of the trout can survive.

These observations are summarized in the **law of tolerance:** *The existence, abundance, and distribution of a species in an ecosystem are determined by whether the levels of one or more physical or chemical factors fall within the range tolerated by that species.* A species may have a wide range of tolerance to some factors and a narrow range of tolerance to others. Most organisms are least tolerant during juvenile or reproductive stages of their life cycles. Highly tolerant species can live in a variety of habitats with widely different conditions. Figure 4-14

(p. 64) shows how environmental physical conditions can limit the distribution of a particular species.

What Factors Limit Population Growth? There Are Always Limits in Nature

Availability of matter and energy resources can limit the number of organisms in a population.

A variety of factors can affect the number of organisms in a population. However, sometimes one factor, known as a **limiting factor,** is more important in regulating population growth than other factors. This ecological principle, related to the law of tolerance, is called the **limiting factor principle:** *Too much or too little of any abiotic factor can limit or prevent growth of a population, even if all other factors are at or near the optimum range of tolerance.*

On land, precipitation often is the limiting factor. Lack of water in a desert limits plant growth. Soil nutrients also can act as a limiting factor on land. Suppose a farmer plants corn in phosphorus-poor soil. Even if water, nitrogen, potassium, and other nutrients are at optimum levels, the corn will stop growing when it uses up the available phosphorus.

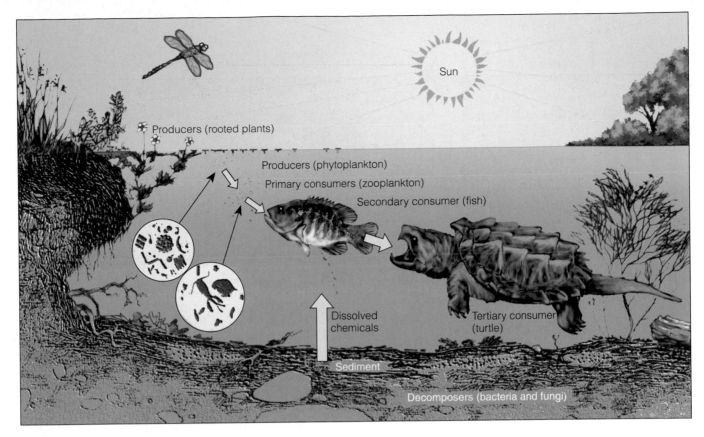

Figure 4-11 Major components of a freshwater ecosystem.

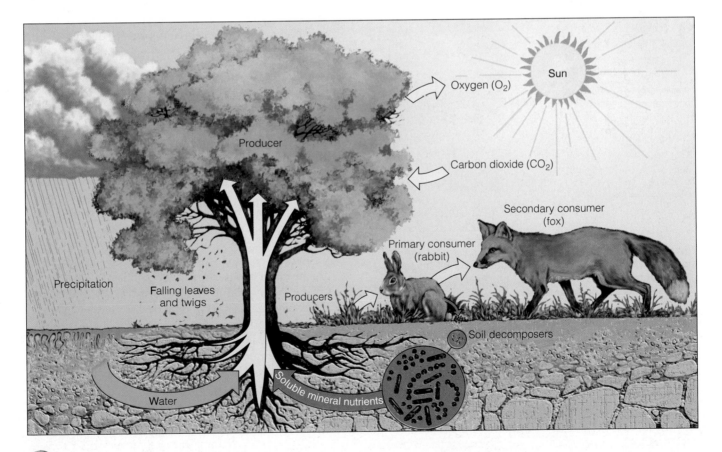

Figure 4-12 Major components of an ecosystem in a field.

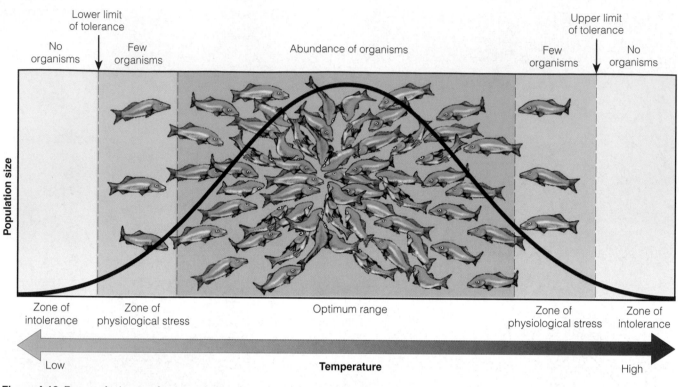

Figure 4-13 Range of tolerance for a population of organisms, such as fish, to an abiotic environmental factor—in this case, temperature.

Sugar Maple

Figure 4-14 The physical conditions of the environment can limit the distribution of a species. The green area shows the current range of sugar maple trees in eastern North America. (Data from U.S. Department of Agriculture).

Too much of an abiotic factor can also be limiting. For example, too much water or too much fertilizer can kill plants, a common mistake of many beginning gardeners.

Important limiting factors for aquatic ecosystems include temperature, sunlight, nutrient availability, and **dissolved oxygen (DO) content**—the amount of oxygen gas dissolved in a given volume of water at a particular temperature and pressure. Another limiting factor in aquatic ecosystems is **salinity**—the amounts of various inorganic minerals or salts dissolved in a given volume of water.

What Are the Major Biological Components of Ecosystems? Producers and Consumers

Some organisms in ecosystems produce food and others consume food.

The earth's organisms either produce or consume food. **Producers,** sometimes called **autotrophs** (self-feeders), make their own food from compounds obtained from their environment.

On land, most producers are green plants. In freshwater and marine ecosystems, algae and plants are the major producers near shorelines. In open water, the dominant producers are *phytoplankton*—mostly microscopic organisms that float or drift in the water.

Most producers capture sunlight to make carbohydrates (such as glucose, $C_6H_{12}O_6$) by **photosynthesis.** Although hundreds of chemical changes take place during photosynthesis, the overall chemical reaction can be summarized as follows:

carbon dioxide + water + **solar energy** $\longrightarrow$ glucose + oxygen

$6\,CO_2$ + $6\,H_2O$ + **solar energy** $\longrightarrow$ $C_6H_{12}O_6$ + $6\,O_2$

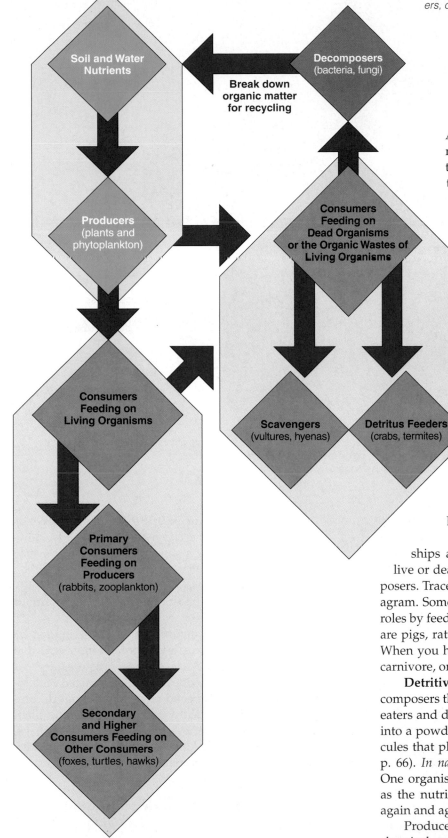

A few producers, mostly specialized bacteria, can convert simple compounds from their environment into more complex nutrient compounds *without sunlight.* This process is called **chemosynthesis.**

All other organisms in an ecosystem are **consumers,** or **heterotrophs** ("other feeders") that get the energy and nutrients they need by feeding on other organisms or their remains. **Decomposers** (mostly certain types of bacteria and fungi) are specialized consumers that recycle organic matter in ecosystems. They do this by breaking down (*biodegrading*) dead organic material or **detritus** ("di-TRI-tus", meaning "debris") to get nutrients. This releases the resulting simpler inorganic compounds into the soil and water, where producers can take them up as nutrients. Life works in circles.

Figure 4-15 shows the feeding relationships among producers, consumers feeding on live or dead organisms or their wastes, and decomposers. Trace the flows of matter and energy in this diagram. Some consumers, called **omnivores,** play dual roles by feeding on both plants and animals. Examples are pigs, rats, foxes, bears, cockroaches, and humans. When you had lunch today were you an herbivore, a carnivore, or an omnivore?

Detritivores consist of detritus feeders and decomposers that feed on detritus. Hordes of these waste eaters and degraders can transform a fallen tree trunk into a powder and finally into simple inorganic molecules that plants can absorb as nutrients (Figure 4-16, p. 66). *In natural ecosystems, there is little or no waste.* One organism's wastes serve as resources for others, as the nutrients that make life possible are recycled again and again. In nature *waste becomes food.*

Producers, consumers, and decomposers use the chemical energy stored in glucose and other organic compounds to fuel their life processes. In most cells

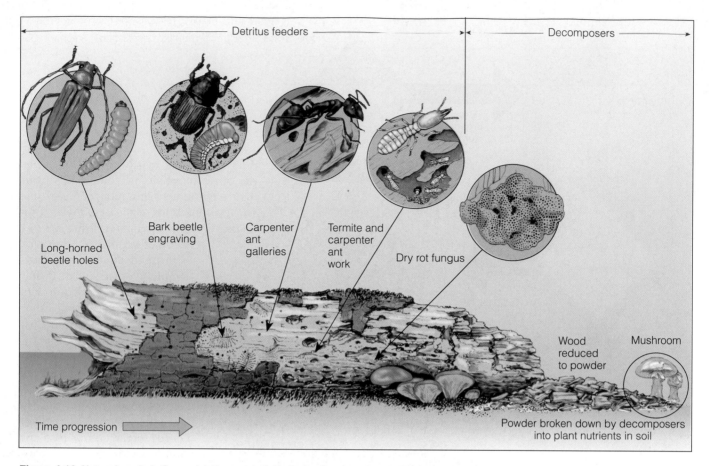

Detritus feeders —————————— ← — Decomposers —————————

Long-horned beetle holes

Bark beetle engraving

Carpenter ant galleries

Termite and carpenter ant work

Dry rot fungus

Wood reduced to powder

Mushroom

Time progression

Powder broken down by decomposers into plant nutrients in soil

Figure 4-16 **Natural capital:** Some detritivores, called *detritus feeders,* directly consume tiny fragments of this log. Other detritivores, called *decomposers* (mostly fungi and bacteria), digest complex organic chemicals in fragments of the log into simpler inorganic nutrients that can be used again by producers.

this energy is released by **aerobic respiration,** which uses oxygen to convert organic nutrients back into carbon dioxide and water. The net effect of the hundreds of steps in this complex process is represented by the following chemical reaction:

glucose + oxygen $\longrightarrow$ carbon dioxide + water + **energy**

$C_6H_{12}O_6 + 6 O_2 \longrightarrow 6 CO_2 + 6 H_2O +$ **energy**

Although the detailed steps differ, the net chemical change for aerobic respiration is the opposite of that for photosynthesis (p. 64).

Some decomposers get the energy they need by breaking down glucose (or other organic compounds) in the absence of oxygen. This form of cellular respiration is called **anaerobic respiration,** or **fermentation.** Instead of carbon dioxide and water, the end products of this process are compounds such as methane gas (CH_4, the main component of natural gas), ethyl alcohol (C_2H_6O), acetic acid ($C_2H_4O_2$, the key component of vinegar), and hydrogen sulfide (H_2S, when sulfur compounds are broken down). This is a plan B way to get a meal.

The survival of any individual organism depends on the *flow of matter and energy* through its body. However, an ecosystem as a whole survives primarily through a combination of *matter recycling* (rather than one-way flow) and *one-way energy flow* (Figure 4-17). Figure 4-17 sums up most of what is going on in our planetary home.

Decomposers complete the cycle of matter by breaking down detritus into inorganic nutrients that can be reused by producers. These waste eaters and nutrient recyclers provide us with this crucial ecological service and never send us a bill. Without decomposers, the entire world would be knee-deep in plant litter, dead animal bodies, animal wastes, and garbage, and most life as we know it would no longer exist. Have you thanked a decomposer today?

What Is Biodiversity? Variety Is the Spice of Life

A vital renewable resource is the biodiversity found in the earth's variety of genes, species, ecosystems, and ecosystem processes.

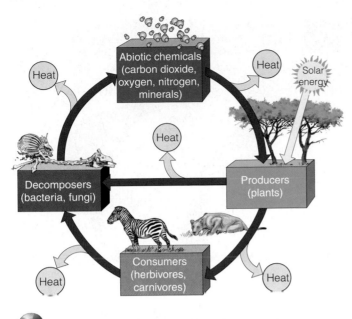

Figure 4-17 Natural capital: the main structural components of an ecosystem (energy, chemicals, and organisms). Matter recycling and the flow of energy from the sun, through organisms, and then into the environment as low-quality heat, link these components.

Biological diversity, or **biodiversity,** is one of the earth's most important renewable resources. Kinds of biodiversity include the following:

- **Genetic diversity**: the variety of genetic material within a species or a population

- **Species diversity**: the number of species present in different habitats

- **Ecological diversity**: the variety of terrestrial and aquatic ecosystems found in an area or on the earth

- **Functional diversity**: the biological and chemical processes such as energy flow and matter cycling needed for the survival of species, communities, and ecosystems (Figure 4-17)

Some people also include *human cultural diversity* as part of the earth's biodiversity. Each human culture has developed various ways to deal with changing environmental conditions.

The earth's biodiversity is the biological wealth or capital that helps keep us alive and supports our economies. Biologist Edward O. Wilson describes the earth's biodiversity as a "natural or biological Internet" that all living things are part of.

The earth's biodiversity supplies us with food, wood, fibers, energy, raw materials, industrial chemicals, and medicines—all of which pour hundreds of billions of dollars into the world economy each year. It also helps preserve the quality of the air and water, maintain

the fertility of soils, dispose of wastes, and control populations of pests that attack crops and forests.

Biodiversity is a renewable resource as long we live off the biological income it provides instead of the natural capital that supplies this income.

Some scientists say that the loss and degradation of biodiversity is the most important environmental problem we face. This is why understanding, protecting, and sustaining biodiversity is a major theme of ecology and of this book.

4-4 ENERGY FLOW IN ECOSYSTEMS

What Are Food Chains and Food Webs? Maps of Ecological Interdependence

Food chains and webs show how eaters, the eaten, and the decomposed are connected to one another in an ecosystem.

All organisms, whether dead or alive, are potential sources of food for other organisms. A caterpillar eats a leaf, a robin eats the caterpillar, and a hawk eats the robin. Decomposers consume the leaf, caterpillar, robin, and hawk after they die. As a result, *there is little matter waste in natural ecosystems.*

The sequence of organisms, each of which is a source of food for the next, is called a **food chain.** It determines how energy and nutrients move from one organism to another through an ecosystem (Figure 4-18, p. 68).

Ecologists assign each organism in an ecosystem to a feeding level, also called a **trophic level,** depending on whether it is a producer or a consumer and on what it eats or decomposes. Producers belong to the first trophic level, primary consumers to the second trophic level, secondary consumers to the third, and so on. Detritivores and decomposers process detritus from all trophic levels.

Real ecosystems are more complex than this. Most consumers feed on more than one type of organism, and most organisms are eaten or decomposed by more than one type of consumer. Because most species participate in several different food chains, the organisms in most ecosystems form a complex network of interconnected food chains called a **food web.** Trace the flows of matter and energy within the simplified food web in Figure 4-19 (p. 69). Trophic levels can be assigned in food webs just as in food chains. A food web shows how eaters, the eaten, and the decomposed are connected to one another. It is a map of life's interdependence.

How Can We Represent the Energy Flow in an Ecosystem? Think Pyramid

There is a decrease in the amount of energy available to each succeeding organism in a food chain or web.

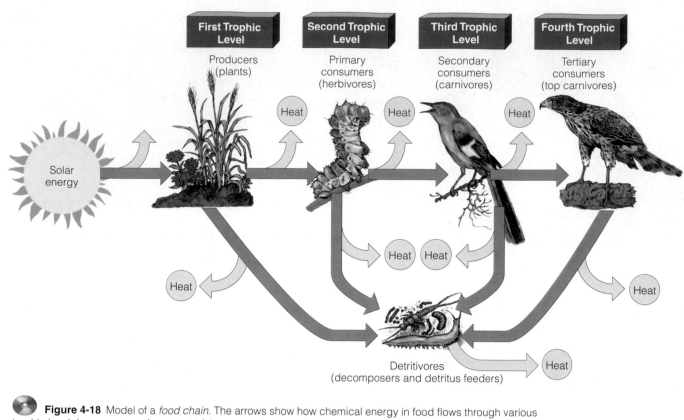

Figure 4-18 Model of a *food chain*. The arrows show how chemical energy in food flows through various *trophic levels* in energy transfers; most of the energy is degraded to heat, in accordance with the second law of thermodynamics. Food chains rarely have more than four trophic levels. Can you figure out why?

Each trophic level in a food chain or web contains a certain amount of **biomass,** the dry weight of all organic matter contained in its organisms. In a food chain or web, chemical energy stored in biomass is transferred from one trophic level to another.

Energy transfer through food chains and food webs is not very efficient. The reason is that with each transfer some usable energy is degraded and lost to the environment as low-quality heat, in accordance with the second law of thermodynamics. Thus only a small portion of what is eaten and digested is actually converted into an organism's bodily material or biomass, and the amount of usable energy available to each successive trophic level declines.

The percentage of usable energy transferred as biomass from one trophic level to the next is called **ecological efficiency.** It ranges from 2% to 40% (that is, a loss of 60–98%) depending on the types of species and the ecosystem involved, but 10% is typical.

Assuming 10% ecological efficiency (90% loss) at each trophic transfer, if green plants in an area manage to capture 10,000 units of energy from the sun, then only about 1,000 units of energy will be available to support herbivores and only about 100 units to support carnivores.

The more trophic levels in a food chain or web, the greater the cumulative loss of usable energy as energy flows through the various trophic levels. The **pyramid of energy flow** in Figure 4-20 (p. 70) illustrates this energy loss for a simple food chain, assuming a 90% energy loss with each transfer. How does this diagram help explain why there are not very many tigers in the world? Figure 4-21 (p. 70) shows the pyramid of energy flow during 1 year for an aquatic ecosystem in Silver Springs, Florida.

Energy flow pyramids explain why the earth can support more people if they eat at lower trophic levels by consuming grains, vegetables, and fruits directly rather than passing such crops through another trophic level and eating grain eaters such as cattle.

The large loss in energy between successive trophic levels also explains why food chains and webs rarely have more than four or five trophic levels. In most cases, too little energy is left after four or five transfers to support organisms feeding at these high trophic levels. This explains why there are so few top carnivores such as eagles, hawks, tigers, and white sharks. It also explains why such species usually are the first to suffer when the ecosystems supporting them are disrupted, and why these species are so vulnerable to extinction. Do you think we are on this list?

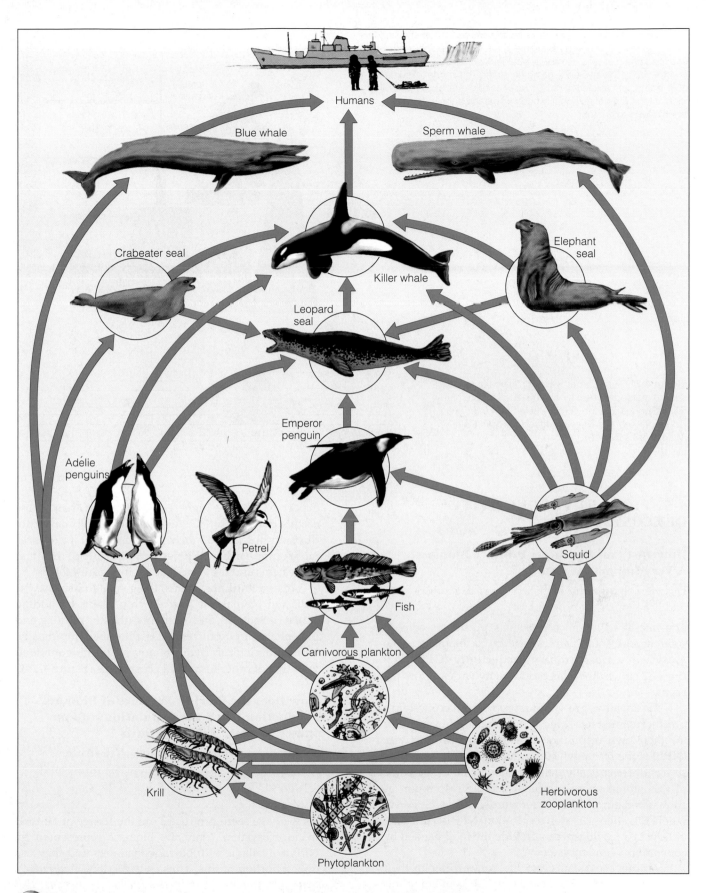

Figure 4-19 Greatly simplified *food web* in the Antarctic. Many more participants in the web, including an array of decomposer organisms, are not depicted here. This is part of life's internet.

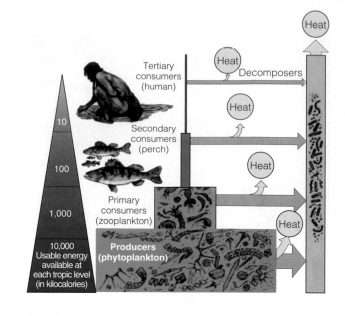

Figure 4-20 Generalized *pyramid of energy flow* showing the decrease in usable energy available at each succeeding trophic level in a food chain or web. In nature, ecological efficiency varies from 2% to 40%, with 10% efficiency being common. This model assumes a 10% ecological efficiency (90% loss of usable energy to the environment, in the form of low-quality heat) with each transfer from one trophic level to another.

Tertiary consumers (human)

Heat

Decomposers

Secondary consumers (perch)

Heat

Primary consumers (zooplankton)

Heat

Heat

Producers (phytoplankton)

10

100

1,000

10,000
Usable energy available at each tropic level (in kilocalories)

Figure 4-21 Annual energy flow (in kilocalories per square meter per year) for an aquatic ecosystem in Silver Springs, Florida. (From Cecie Starr, *Biology: Concepts and Applications*, 4th ed., Brooks/Cole [Wadsworth] © 2000)

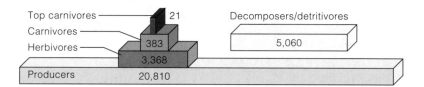

Top carnivores — 21

Carnivores —

Herbivores — 383

3,368

Producers 20,810

Decomposers/detritivores 5,060

4-5 PRIMARY PRODUCTIVITY OF ECOSYSTEMS

How Fast Can Producers Produce Biomass? A Very Important Rate

Different ecosystems use solar energy to produce and use biomass at different rates.

The *rate* at which an ecosystem's producers convert solar energy into chemical energy as biomass is the ecosystem's **gross primary productivity (GPP)**. Figure 4-22 shows how this productivity varies across the earth.

To stay alive, grow, and reproduce, an ecosystem's producers must use some of the biomass they produce for their own respiration. **Net primary productivity (NPP)** is the *rate* at which producers use photosynthesis to store energy *minus* the *rate* at which they use some of this stored energy through aerobic respiration as shown in Figure 4-23. In other words, NPP = GPP − R, where R is energy used in respiration. NPP is a measure of how fast producers can provide the food needed by consumers in an ecosystem.

Various ecosystems and life zones differ in their NPP, as graphed in Figure 4-24 (p. 72). Looking at this graph, what are the three most productive and the three least productive systems? Generally, would you

expect NPP to be higher at the equator than at the earth's poles? Why? Despite its low net primary productivity, there is so much open ocean that it produces more of the earth's NPP per year than any of the other ecosystems and life zones shown in Figure 4-24.

In agricultural systems, the goal is to increase the NPP and biomass of selected crop plants by adding water (irrigation) and nutrients (mostly nitrates and phosphates in fertilizers). Despite such inputs, the NPP of agricultural land is not very high compared with that of other terrestrial ecosystems (Figure 4-24).

How Does the World's Net Rate of Biomass Production Limit the Populations of Consumer Species? Nature's Limits

The number of consumer organisms the earth can support is determined by how fast producers can supply them with energy found in biomass.

As we have seen, producers are the source of all food in an ecosystem. Only the biomass represented by NPP is available as food for consumers and they use only a portion of this. Thus *the planet's NPP ultimately limits the number of consumers (including humans) that can survive on the earth.* This is an important lesson from nature.

Figure 4-22 Satellite data on the earth's gross primary productivity in terms of ocean and land concentrations of chlorophyll in producer organisms during the winter of 2004. On land rain forests and other highly productive areas are dark green and the least productive (mostly deserts) are brown or white in polar areas. At sea, the concentration of chlorophyll found in phytoplankton, a primary indicator of ocean productivity, ranges from red (highest) to orange, yellow, green, light blue, and dark blue (lowest). Where are the areas of highest and lowest productivity on land and at sea? (Image data from SeaWiFs Project/NASA and by kind permission of ORBIMAGE. All rights reserved.)

It is tempting to conclude from Figure 4-24 that a good way to feed the world's hungry millions would be to harvest plants in highly productive estuaries, swamps, and marshes. But people cannot eat most plants in these areas. In addition, these plants provide vital food sources (and spawning areas) for fish, shrimp, and other aquatic species that provide us and other consumers with protein.

We might also conclude from Figure 4-24 that we could grow more food for human consumption by clearing highly productive tropical rain forests and planting food crops. According to most ecologists, this is also a bad idea. Here is why. In tropical rain forests most nutrients are stored in the vegetation rather than in the soil, where they need to be to grow crops. When the trees are removed, frequent rains and growing crops rapidly deplete the nutrient-poor soils. Thus crops can be grown for only a short time without massive and expensive applications of commercial fertilizers.

Are the oceans a way out? Because the earth's vast open oceans provide the largest percentage of the earth's net primary productivity, why not harvest its primary producers (floating and drifting phytoplankton) to help feed the rapidly growing human population? The problem is that harvesting the widely

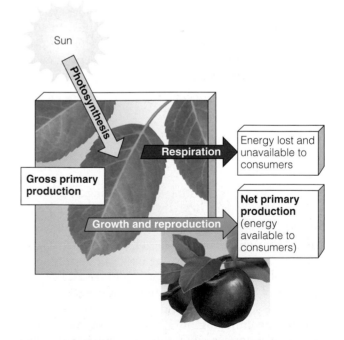

Figure 4-23 Distinction between gross primary productivity and net primary productivity. A plant uses some of its gross primary productivity to survive through respiration. The remaining energy is available to consumers.

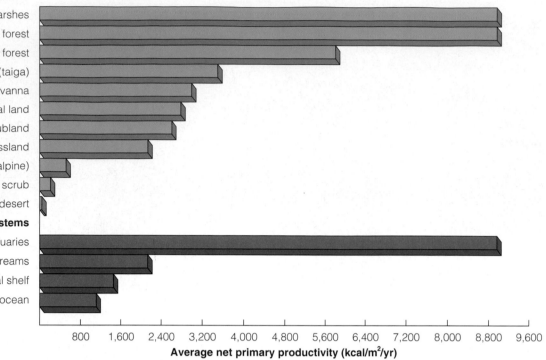

Terrestrial Ecosystems

Swamps and marshes
Tropical rain forest
Temperate forest
Northern coniferous forest (taiga)
Savanna
Agricultural land
Woodland and shrubland
Temperate grassland
Tundra (arctic and alpine)
Desert scrub
Extreme desert

Aquatic Ecosystems

Estuaries
Lakes and streams
Continental shelf
Open ocean

800 1,600 2,400 3,200 4,000 4,800 5,600 6,400 7,200 8,000 8,800 9,600

Average net primary productivity (kcal/m²/yr)

Figure 4-24 Natural capital: estimated annual average *net primary productivity* (NPP) per unit of area in major life zones and ecosystems, expressed as kilocalories of energy produced per square meter per year (kcal/m²/yr). (Data from *Communities and Ecosystems*, 2nd ed., by R. H. Whittaker, 1975. New York: Macmillan)

dispersed, tiny floating producers in the open ocean would take much more fossil fuel and other types of energy than the food energy we would get. In addition, this would disrupt the food webs of the open ocean that provide us and other consumer organisms with important sources of energy and protein from fish and shellfish.

How Much of the World's Net Rate of Biomass Production Do We Use? Let Them Eat Crumbs

Humans are using, wasting, or destroying a significant amount of the world's biomass faster than producers can make it.

Peter Vitousek, Stuart Roystaczer, and other ecologists estimate that humans now use, waste, or destroy about 27% of the earth's total potential NPP and 10–55% of the NPP of the planet's terrestrial ecosystems.

These scientists contend that this is the main reason we are crowding out or eliminating the habitats and food supplies of a growing number of other species. What might happen to us and to other consumer species if the human population doubles over the next 40–50 years and per capita consumption of re-

sources such as food, timber, and grassland rises sharply? This is an important question!

4-6 SOILS

What Is Soil and Why Is It Important? The Base of Life on Land

Soil is a slowly renewed resource that provides most of the nutrients needed for plant growth and also helps purify water.

Soil is a thin covering over most land that is a complex mixture of eroded rock, mineral nutrients, decaying organic matter, water, air, and billions of living organisms, most of them microscopic decomposers. Study the diagram in Figure 4-25 showing the profile of different aged soils. Soil is a renewable resource but it is renewed very slowly. Depending mostly on climate, the formation of just 1 centimeter (0.4 inch) of soil can take from 15 years to hundreds of years.

Soil is the base of life on land because it provides most of the nutrients needed for plant growth. Indeed, you are mostly soil nutrients imported into your body by the food you eat. Soil is also the earth's primary filter that cleanses water as it passes through. It is also a

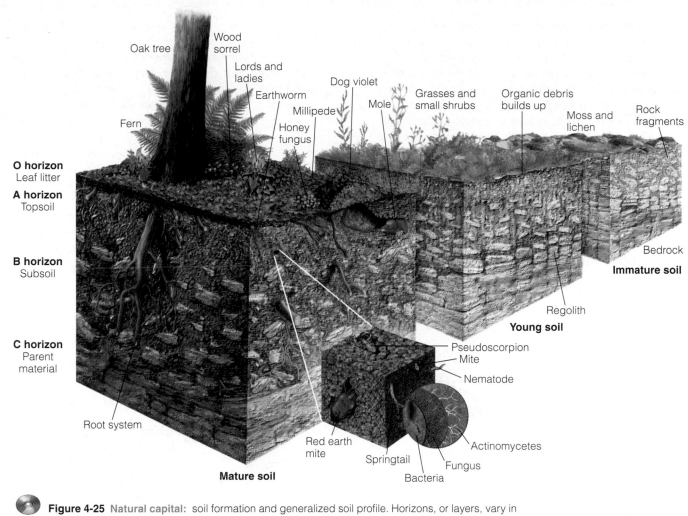

O horizon
Leaf litter

A horizon
Topsoil

B horizon
Subsoil

C horizon
Parent
material

Oak tree
Wood sorrel
Lords and ladies
Earthworm
Fern
Honey fungus
Millipede
Dog violet
Mole
Grasses and small shrubs
Organic debris builds up
Moss and lichen
Rock fragments

Bedrock

Immature soil

Regolith

Young soil

Root system

Mature soil

Red earth mite
Springtail
Bacteria
Pseudoscorpion
Mite
Nematode
Actinomycetes
Fungus

Figure 4-25 Natural capital: soil formation and generalized soil profile. Horizons, or layers, vary in number, composition, and thickness, depending on the type of soil. Soil is the base of life that provides the food you need to stay alive and healthy. (From Derek Elsom, *Earth*, 1992. Copyright © 1992 by Marshall Editions Developments Limited, New York: Macmillan. Used by permission.)

major component of the earth's water recycling and water storage processes. You can thank soil every time you drink a glass of water.

What Major Layers Are Found in Mature Soils? Layers Count

Most soils developed over a long time consist of several layers containing different materials.

Mature soils, or soils that have developed over a long time, are arranged in a series of horizontal layers called **soil horizons,** each with a distinct texture and composition that varies with different types of soils. A cross-sectional view of the horizons in a soil is called a **soil profile.** Most mature soils have at least three of the possible horizons (Figure 4-25). Think of them as floors in the building of life underneath your feet.

The top layer is the *surface litter layer,* or *O horizon.* It consists mostly of freshly fallen undecomposed or partially decomposed leaves, twigs, crop wastes, animal wastes, fungi, and other organic materials. Normally, it is brown or black.

The *topsoil layer,* or *A horizon*, is a porous mixture of partially decomposed organic matter, called **humus,** and some inorganic mineral particles. It is usually darker and looser than deeper layers. A fertile soil that produces high crop yields has a thick topsoil layer with lots of humus. This helps topsoil hold water and nutrients taken up by plant roots.

The roots of most plants and most of a soil's organic matter are concentrated in a soil's two upper layers. As long as vegetation anchors these layers, soil stores water and releases it in a nourishing trickle.

The two top layers of most well-developed soils teem with bacteria, fungi, earthworms, and small insects that interact in complex food webs such as the one shown in Figure 4-26 (p. 74). Bacteria and other decomposer microorganisms found by the billions in every handful of topsoil break down some of its complex organic compounds into simpler inorganic compounds soluble in water. Soil moisture carrying these dissolved nutrients is drawn up by the roots of plants and transported through stems and into leaves as part of the earth's chemical cycling processes.

The color of its topsoil tells us a lot about how useful a soil is for growing crops. Dark-brown or black topsoil is nitrogen-rich and high in organic matter. Gray, bright yellow, or red topsoils are low in organic matter and need nitrogen enrichment to support most crops. Pick up a handful of soil and look at its color. What did you learn?

The *B horizon (subsoil)* and the *C horizon (parent material)* contain most of a soil's inorganic matter, mostly broken-down rock consisting of varying mixtures of sand, silt, clay, and gravel. The C horizon lies on a base of unweathered parent rock called *bedrock*.

The spaces, or pores, between the solid organic and inorganic particles in the upper and lower soil layers contain varying amounts of air (mostly nitrogen and oxygen gas) and water. Plant roots need the oxygen for cellular respiration.

Some of the precipitation that reaches the soil percolates through the soil layers and occupies many of the soil's open spaces or pores. This downward movement of water through soil is called **infiltration.** As the water seeps down, it dissolves various minerals and organic matter in upper layers and carries them to lower layers in a process called **leaching.**

Most of the world's crops are grown on soils exposed when grasslands and deciduous (leaf-shedding) forests are cleared. Worldwide there are many thousands of different soil types—at least 15,000 in the United States. Five important soil types, each with a distinct profile, are shown in Figure 4-27.

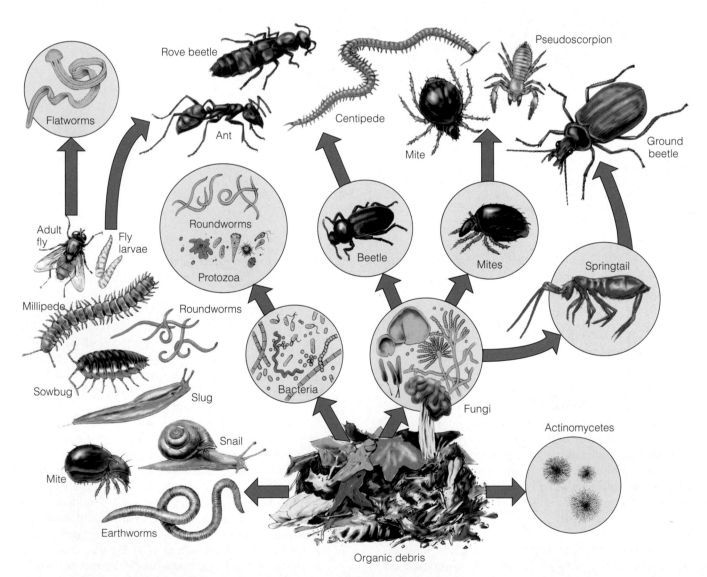

Figure 4-26 Natural capital: greatly simplified food web of living organisms found in soil.

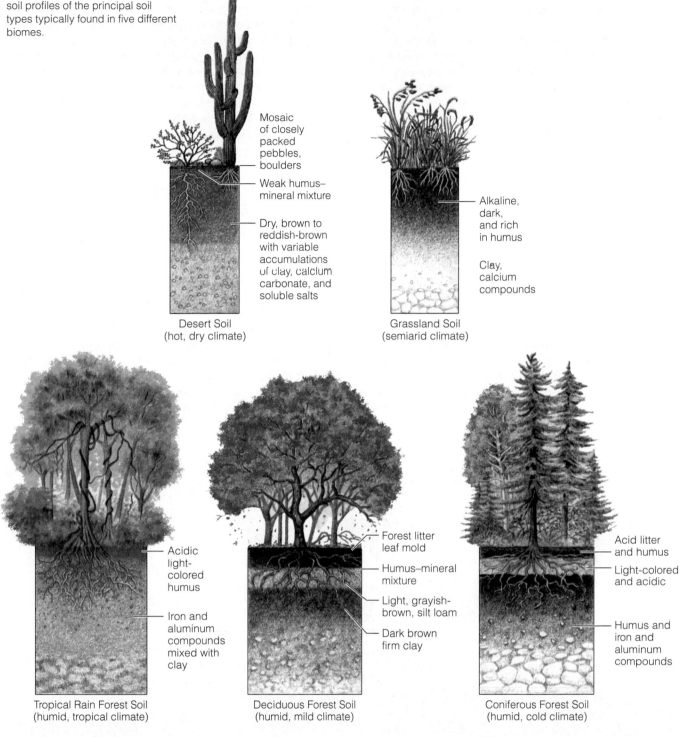

Figure 4-27 Natural capital: soil profiles of the principal soil types typically found in five different biomes.

Mosaic of closely packed pebbles, boulders

Weak humus– mineral mixture

Dry, brown to reddish-brown with variable accumulations of clay, calcium carbonate, and soluble salts

Desert Soil
(hot, dry climate)

Alkaline, dark, and rich in humus

Clay, calcium compounds

Grassland Soil
(semiarid climate)

Acidic light-colored humus

Iron and aluminum compounds mixed with clay

Tropical Rain Forest Soil
(humid, tropical climate)

Forest litter leaf mold

Humus–mineral mixture

Light, grayish-brown, silt loam

Dark brown firm clay

Deciduous Forest Soil
(humid, mild climate)

Acid litter and humus

Light-colored and acidic

Humus and iron and aluminum compounds

Coniferous Forest Soil
(humid, cold climate)

How Do Soils Differ in Texture and Porosity? Composition and Spaces Are Important

Soils vary in the size of the particles they contain and the amount of space between these particles.

Soils vary in their content of *clay* (very fine particles), *silt* (fine particles), *sand* (medium-size particles), and *gravel* (coarse to very coarse particles). The relative amounts of the different sizes and types of these mineral particles determine **soil texture.**

To get an idea of a soil's texture, take a small amount of topsoil, moisten it, and rub it between your fingers and thumb. A gritty feel means it contains a lot of sand. A sticky feel means a high clay content, and you should be able to roll it into a clump. Silt-laden soil feels smooth, like flour. A loam topsoil is best suited for plant growth. It has a texture between these extremes—a crumbly, spongy feeling—with many of its particles clumped loosely together.

Soil texture helps determine **soil porosity,** a measure of the volume of pores or spaces per volume of soil and of the average distances between those spaces. Fine particles are needed for water retention and coarse ones for air spaces. A porous soil has many pores and can hold more water and air than a less porous soil. The average size of the spaces or pores in a soil determines **soil permeability:** the rate at which water and air move from upper to lower soil layers.

4-7 MATTER CYCLING IN ECOSYSTEMS

What Are Biogeochemical Cycles? Going in Circles

Global cycles recycle nutrients through the earth's air, land, water, and living organisms and, in the process, connect past, present, and future forms of life.

All organisms are interconnected by vast global recycling systems made up of **nutrient cycles,** or **biogeochemical cycles** (literally, life–earth–chemical cycles). In these cycles, nutrient atoms, ions, and molecules that organisms need to live, grow, and reproduce are continuously cycled between air, water, soil, rock, and living organisms. These cycles, driven directly or indirectly by incoming solar energy and gravity, include the carbon, oxygen, nitrogen, phosphorus, and hydrologic (water) cycles (Figure 4-8).

The earth's chemical cycles connect past, present, and future forms of life. Some of the carbon atoms in your skin may once have been part of a leaf, a dinosaur's skin, or a layer of limestone rock. Your grandmother, Plato, or a hunter–gatherer who lived 25,000 years ago may have inhaled some of the oxygen molecules you just inhaled.

How Is Water Cycled in the Biosphere? The Water Cycle

A vast global cycle collects, purifies, distributes, and recycles the earth's fixed supply of water.

The **hydrologic cycle,** or **water cycle,** recycles the earth's fixed supply of water, as shown in Figure 4-28. Trace the flows and paths in this diagram. Solar energy evaporates water found on the earth's surface into the atmosphere. Some of this water returns to the earth as rain or snow, passes through living organisms, flows into bodies of water, and eventually is evaporated again to continue the cycle. The water cycle differs from most other nutrient cycles in that most of the water remains chemically unchanged and is transformed from one physical state to another.

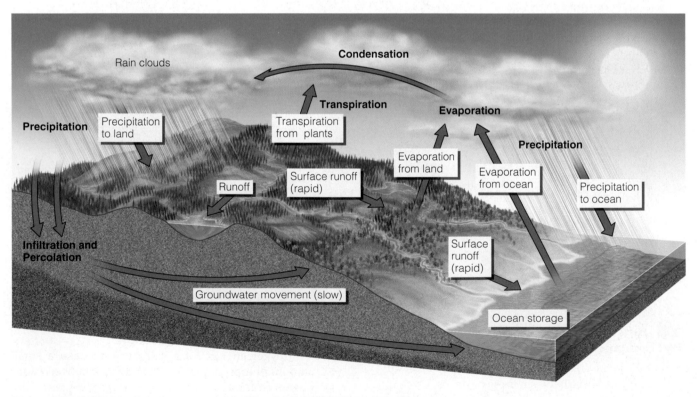

Figure 4-28 Natural capital: simplified model of the global *hydrologic cycle* that helps keep you alive.

The main processes in this water recycling and purifying cycle are *evaporation* (conversion of water into water vapor), *transpiration* (evaporation from plant leaves after water is extracted from soil by roots and transported throughout the plant), *condensation* (conversion of water vapor into droplets of liquid water), *precipitation* (rain, sleet, hail, and snow), *infiltration* (movement of water into soil), *percolation* (downward flow of water through soil and permeable rock formations to groundwater storage areas called aquifers), and *runoff* (surface movement down slopes to the sea to resume the cycle).

The water cycle is powered by energy from the sun, which evaporates water into the atmosphere, and by gravity, which draws the water back to the earth's surface as precipitation. About 84% of water vapor in the atmosphere comes from the oceans, and the rest comes from land. This should not surprise you since almost three-fourths of the earth is covered with water.

Winds and air masses transport water vapor over various parts of the earth's surface, often over long distances. Falling temperatures cause the water vapor to condense into tiny droplets that form clouds in the sky or fog near the surface. For precipitation to occur, air must contain **condensation nuclei:** tiny particles on which droplets of water vapor can collect. Sources of such particles include volcanic ash, soil dust, smoke, sea salts, and particulate matter emitted by factories, coal-burning power plants, and motor vehicles.

Currently about one-tenth of the fresh water returning to the earth's surface as precipitation becomes locked up in slowly flowing ice and snow called *glaciers*. But most precipitation falling on terrestrial ecosystems becomes *surface runoff*. This water flows into streams and lakes, which eventually carry water back to the oceans, where it can evaporate and cycle again.

Besides replenishing streams, lakes, and wetlands, surface runoff also causes soil erosion, which moves soil and weathered rock fragments from one place to another. Water is thus the primary sculptor of the earth's landscape. Because water dissolves many nutrient compounds, it is also a major medium for transporting nutrients within and between ecosystems and for removing and diluting wastes.

Throughout the hydrologic cycle, many natural processes purify water. Evaporation and subsequent precipitation act as a natural distillation process that removes impurities dissolved in water. Water flowing above ground through streams and lakes and below ground in aquifers is naturally filtered and purified by chemical and biological processes, mostly by the actions of decomposer bacteria. Thus *the hydrologic cycle can also be viewed as a cycle of natural renewal of water quality.*

How Are Human Activities Affecting the Water Cycle? Messing with Nature

We alter the water cycle by withdrawing large amounts of fresh water, clearing vegetation, eroding soils, polluting surface and underground water, and contributing to climate change.

During the past 100 years, we have been intervening in the earth's current water cycle in four major ways. *First,* we withdraw large quantities of fresh water from streams, lakes, and underground sources. In some heavily populated or heavily irrigated areas, withdrawals have led to groundwater depletion or intrusion of ocean salt water into underground water supplies.

Second, we clear vegetation from land for agriculture, mining, road and building construction, and other activities and sometimes cover the land with buildings, concrete, or asphalt. This increases runoff, reduces infiltration that recharges groundwater supplies, increases the risk of flooding, and accelerates soil erosion and landslides. We also increase flooding by destroying wetlands, which act like sponges to absorb and hold overflows of water.

Third, we modify water quality by adding nutrients (such as phosphates and nitrates found in fertilizers) and other pollutants. *Fourth,* according to a 2003 study by Ruth Curry and her colleagues, the earth's water cycle is speeding up as a result of a warmer climate caused partially by human inputs of carbon dioxide and other greenhouse gases into the atmosphere. This could change global precipitation patterns that affect the severity and frequency of droughts, floods, and storms. It can also intensify global warming by speeding up the input of water vapor—a powerful greenhouse gas—into the troposphere.

How Is Carbon Cycled in the Biosphere? Carbon Dioxide in Action

Carbon, the basic building block of organic compounds, recycles through the earth's air, water, soil, and living organisms.

Carbon is the basic building block of the carbohydrates, fats, proteins, DNA, and other organic compounds necessary for life. It is circulated through the biosphere by the **carbon cycle,** as shown in Figure 4-29 (p. 78). Trace the flows and paths in this diagram.

This cycle is based on carbon dioxide gas, which makes up about 0.038% of the volume of the troposphere and is also dissolved in water. The aerobic respiration of organisms, volcanic eruptions, the weathering of carbonate rocks, and the burning of carbon-containing compounds found in wood, grasses, and fossil fuels add carbon dioxide to the troposphere.

Figure 4-29 Natural capital: simplified model of the global *carbon cycle* that helps keep you alive. The left portion shows the movement of carbon through marine systems, and the right portion shows its movement through terrestrial ecosystems. Carbon reservoirs are shown as boxes; processes that change one form of carbon to another are shown in unboxed print. (From Cecie Starr, *Biology: Concepts and Applications,* 4th ed., Brooks/Cole [Wadsworth] © 2000)

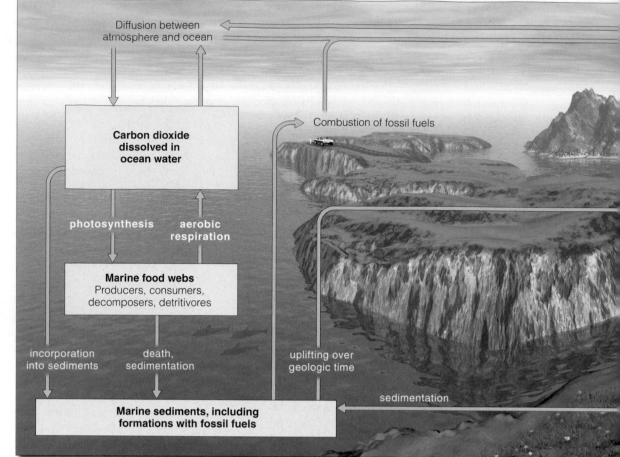

Diffusion between atmosphere and ocean

Combustion of fossil fuels

Carbon dioxide dissolved in ocean water

photosynthesis   aerobic respiration

Marine food webs
Producers, consumers, decomposers, detritivores

incorporation into sediments   death, sedimentation   uplifting over geologic time

sedimentation

Marine sediments, including formations with fossil fuels

Aerobic respiration in the cells of oxygen-using producers, consumers, and decomposers breaks down glucose and other complex organic compounds and converts the carbon back to CO_2, which is released into the troposphere and in water for reuse by producers.

Carbon dioxide is removed from the troposphere by terrestrial and aquatic producers, which use *photosynthesis* to convert it into complex carbohydrates such as glucose ($C_6H_{12}O_6$).

This linkage between *photosynthesis* in producers and *aerobic respiration* in producers, consumers, and decomposers circulates carbon in the biosphere and is a major part of the global carbon cycle. Oxygen and hydrogen, the other elements in carbohydrates, cycle almost in step with carbon.

Carbon dioxide is a key component of nature's thermostat. If the carbon cycle removes too much CO_2 from the atmosphere, the atmosphere will cool; if the cycle generates too much, the atmosphere will get warmer. Thus even slight changes in the carbon cycle can affect climate and ultimately the types of life that can exist on various parts of the planet.

Some carbon atoms take a long time to recycle. Over millions of years, buried deposits of dead plant matter and bacteria have been compressed between layers of sediment, where they form carbon-containing

fossil fuels such as coal and oil (Figure 4-29). This carbon is not released to the atmosphere as CO_2 for recycling until these fuels are extracted and burned, or until long-term geological processes expose these deposits to air. In only a few hundred years, we have extracted and burned huge quantities of fossil fuels that took millions of years to form. This is why fossil fuels are nonrenewable resources on a human time scale.

Oceans play important roles in the carbon cycle. Some of the atmosphere's carbon dioxide dissolves in ocean water, and the ocean's photosynthesizing producers remove some. On the other hand, as ocean water warms, some of its dissolved CO_2 returns to the atmosphere, just as carbon dioxide fizzes out of a carbonated beverage when it warms. The balance between these two processes plays a role in the earth's average temperature.

Some ocean organisms build their shells and skeletons by using dissolved CO_2 molecules in seawater to form carbonate compounds such as calcium carbonate ($CaCO_3$). When these organisms die, tiny particles of their shells and bone drift slowly to the ocean depths. There they are buried for eons (as long as 400 million years) in deep bottom sediments (Figure 4-29, left), where under immense pressure they are converted into limestone rock. Geological processes may eventu-

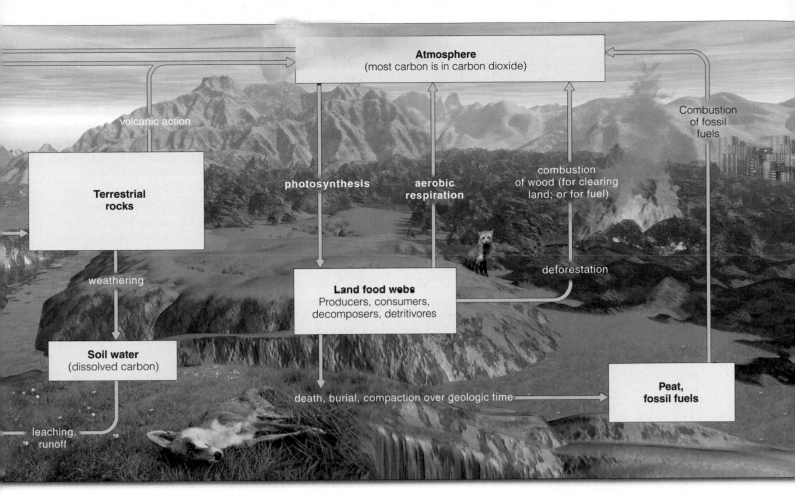

Atmosphere
(most carbon is in carbon dioxide)

volcanic action

Terrestrial rocks

photosynthesis

aerobic respiration

combustion of wood (for clearing land; or for fuel)

Combustion of fossil fuels

weathering

Land food webs
Producers, consumers, decomposers, detritivores

deforestation

Soil water
(dissolved carbon)

death, burial, compaction over geologic time

Peat, fossil fuels

leaching, runoff

ally expose the limestone to the atmosphere and acidic precipitation and make its carbon available to living organisms once again.

How Are Human Activities Affecting the Carbon Cycle? Messing with Nature's Thermostat

Carbon dioxide produced by burning fossil fuels and clearing photosynthesizing vegetation faster than it is replaced can increase the average temperature of the troposphere.

Since 1800 and especially since 1950, we have been intervening in the earth's carbon cycle in two ways that add carbon dioxide to the atmosphere. *First*, in some areas we clear trees and other plants that absorb CO_2 through photosynthesis faster than they can grow back. *Second*, we add large amounts of CO_2 by burning fossil fuels (Figure 4-30) and wood.

Computer models of the earth's climate systems suggest that increased concentrations of atmospheric CO_2 and other gases we are adding to the atmosphere could enhance the planet's *natural greenhouse effect* that helps warm the lower atmosphere (troposphere) and the earth's surface (Figure 4-9). The resulting *global warming* could disrupt global food production and

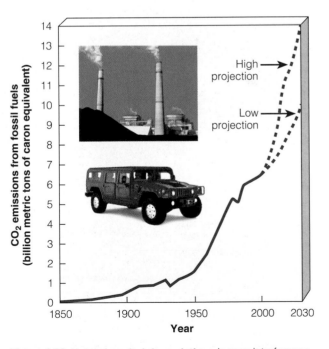

Figure 4-30 Natural capital degradation: human interference in the global carbon cycle from carbon dioxide emissions when fossil fuels are burned, 1850 to 2001 and projections to 2030 (dashed lines). (Data from UN Environment Programme, British Petroleum, International Energy Agency, and U.S. Department of Energy)

wildlife habitats, alter temperature and precipitation patterns, and raise the average sea level in various parts of the world—more about this in Chapter 21.

How Is Nitrogen Cycled in the Biosphere? Bacteria in Action

Different types of bacteria help recycle nitrogen through the earth's air, water, soil, and living organisms.

Nitrogen is the atmosphere's most abundant element, with chemically unreactive nitrogen gas (N_2) making up about 78% of the volume of the troposphere. Nitrogen is a crucial component of proteins, many vitamins, and the nucleic acids DNA and RNA. However, N_2 cannot be absorbed and used (metabolized) directly as a nutrient by multicellular plants or animals.

Fortunately, two natural processes convert N_2 gas in the atmosphere into compounds that can enter food webs as part of the **nitrogen cycle,** depicted in Fig-ure 4-31. Trace the flows and paths in this diagram. One of these processes is atmospheric electrical discharge in the form of lightning. This causes nitrogen (N_2) and oxygen (O_2) in the atmosphere to react and produce nitrogen oxide (NO). Try to write and balance the chemical equation for this reaction.

The other process is carried out by certain types of bacteria in aquatic systems, in the soil, and in the roots of some plants that can convert or "fix" N_2 into compounds useful as nutrients for plants and animals.

The nitrogen cycle consists of several major steps (Figure 4-31). In *nitrogen fixation,* specialized bacteria in the soil convert ("fix") gaseous nitrogen (N_2) to ammonia (NH_3) that can be used by plants. See if you can write a balanced chemical equation for the reaction of N_2 with H_2 to form NH_3.

Ammonia not taken up by plants may undergo *nitrification.* In this process, specialized aerobic bacteria convert most of the ammonia in soil to *nitrite ions* (NO_2^-), which are toxic to plants, and *nitrate ions*

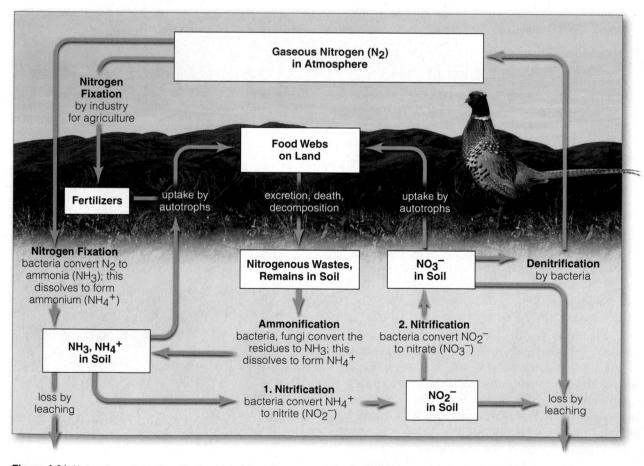

Figure 4-31 Natural capital: simplified model of the *nitrogen cycle* in a terrestrial ecosystem. Nitrogen reservoirs are shown as boxes; processes changing one form of nitrogen to another are shown in unboxed print. This cycle helps keep you alive. (From Cecie Starr and Ralph Taggart, *Biology: The Unity and Diversity of Life,* 9th ed., Wadsworth © 2001)

(NO_3^-), which are easily taken up by plants as a nutrient.

Nitrogen fixation and nitrification add inorganic ammonia, ammonium ions (NH_4^+), and nitrate ions to soil water. Then plant roots can absorb these dissolved substances in a step called *assimilation*. Plants use these ions to make nitrogen-containing organic molecules such as DNA, amino acids, and proteins. Animals in turn get their nitrogen by eating plants or plant-eating animals.

Plants and animals return nitrogen-rich organic compounds to the environment as wastes, cast-off particles, and dead bodies. In the *ammonification* step, vast armies of specialized decomposer bacteria convert this detritus into simpler nitrogen-containing inorganic compounds such as ammonia and water-soluble salts containing ammonium ions.

Nitrogen leaves the soil in the *denitrification* step in which other specialized anaerobic bacteria in water-logged soil and in the bottom sediments of lakes, oceans, swamps, and bogs convert NH_3 and NH_4^+ back into nitrite and nitrate ions and then into nitrogen gas (N_2) and nitrous oxide gas (N_2O). These gases are released to the atmosphere to begin the cycle again.

How Are Human Activities Affecting the Nitrogen Cycle? Altering Nature

Excessive inputs of various nitrogen-containing compounds into the environment from human activities is becoming a major regional and global environmental problem.

In the past 100 years, human activities have had several effects on the earth's current nitrogen cycle.

First, we add large amounts of nitric oxide (NO) to the atmosphere when we burn any fuel. In the atmosphere, this gas can be converted to nitrogen dioxide gas (NO_2) and nitric acid (HNO_3), which can return to the earth's surface as damaging *acid deposition,* commonly called *acid rain*—more on this in Chapter 20.

Second, we add nitrous oxide (N_2O) to the atmosphere through the action of anaerobic bacteria on livestock wastes and commercial inorganic fertilizers applied to the soil. This gas can warm the troposphere and deplete ozone in the stratosphere.

Third, we release large quantities of nitrogen stored in soils and plants as gaseous compounds into the troposphere through destruction of forests, grasslands, and wetlands. *Fourth,* we upset aquatic ecosystems by adding excess nitrates in agricultural runoff and discharges from municipal sewage systems—more on this in Chapter 22.

Fifth, we remove nitrogen from topsoil when we harvest nitrogen-rich crops, irrigate crops, and burn or clear grasslands and forests before planting crops.

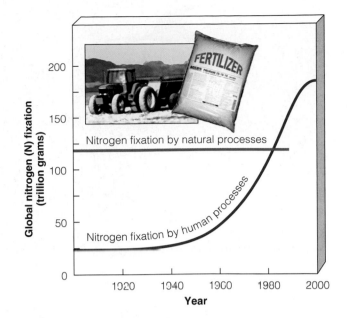

Figure 4-32 Natural capital degradation: human interference in the global nitrogen cycle. Human activities such as production of fertilizers now fix more nitrogen than all natural sources combined. (Data from UN Environment Programme, UN Food and Agriculture Organization, and U.S. Department of Agriculture)

Sixth, inputs of nitrogen into the air, soil, and water mostly from our activities is beginning to affect the biodiversity of terrestrial and aquatic systems by shifting their species composition towards species that can thrive on increased supplies of nitrogen nutrients.

Since 1950, human activities have more than doubled the annual release of nitrogen from the terrestrial portion of the earth into the rest of the environment (Figure 4-32). These excessive inputs of nitrogen into the air and water is a serious local, regional, and global environmental problem that so far has attracted fairly little attention compared to global environmental problems such as global warming and depletion of ozone in the stratosphere. Princeton University physicist Robert Socolow calls for the nations of the world to work out some type of international nitrogen management agreement to help prevent this problem from reaching crisis levels.

How Is Phosphorus Cycled in the Biosphere? Slow Cycling without Using the Atmosphere

Phosphorus cycles fairly slowly through the earth's water, soil, and living organisms.

Phosphorus circulates through water, the earth's crust, and living organisms in the phosphorus cycle, depicted

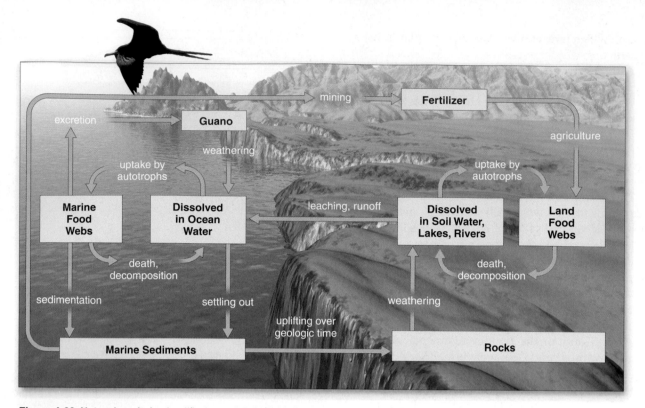

Figure 4-33 Natural capital: simplified model of the *phosphorus cycle*. Phosphorus reservoirs are shown as boxes; processes that change one form of phosphorus to another are shown in unboxed print. This cycle helps keep you alive. (From Cecie Starr and Ralph Taggart, *The Unity and Diversity of Life*, 9th ed., Wadsworth © 2001)

in Figure 4-33. Trace the flows and paths in this diagram. With the exception of small particles of phosphate in dust, very little phosphorus circulates in the atmosphere because soil conditions do not allow bacteria to convert chemical forms of phosphorus to gaseous forms of phosphorus. The phosphorus cycle is slow, and on a short human time scale much phosphorus flows one way from the land to the oceans.

Phosphorus is typically found as phosphate salts containing phosphate ions (PO_4^{3-}) in terrestrial rock formations and ocean bottom sediments. The weathering and erosion of phosphorus-containing rocks releases phosphorus into soil water, lakes, and rivers as phosphate ions, which are taken up by plant roots. Animals get most of the phosphorus they need from the food they eat. Decomposers break down organic phosphorus compounds in dead organisms into the soil where it can be reused by plants.

Phosphate can be lost from the cycle for long periods when it washes from the land into streams and rivers and is carried to the ocean. There it can be deposited as sediment on the sea floor and remain for millions of years. Someday geological uplift processes may expose these seafloor deposits from which

phosphate can be eroded to start the cyclical process again.

Because most soils contain little phosphate, it is often the *limiting factor* for plant growth on land unless phosphorus (as phosphate salts mined from the earth) is applied to the soil as a fertilizer. Phosphorus also limits the growth of producer populations in many freshwater streams and lakes because phosphate salts are only slightly soluble in water.

How Are Human Activities Affecting the Phosphorus Cycle? More Messing with Nature

We remove large amounts of phosphate from the earth to make fertilizer, reduce phosphorus in tropical soils by clearing forests, and add excess phosphates to aquatic systems.

We intervene in the earth's phosphorus cycle in three ways. *First,* we mine large quantities of phosphate rock to make commercial inorganic fertilizers. *Second,* we reduce the available phosphate in tropical soils when we cut down areas of tropical forests. *Third,* we disrupt aquatic systems with phosphates from runoff of animal wastes and fertilizers and discharges from sewage treatment systems—more on this in Chapter 22.

Scientists estimate that since 1900 human activities have increased the natural rate of phosphorus release into the environment about 3.7-fold.

How Is Sulfur Cycled in the Biosphere? The Sulfur Cycle

Sulfur cycles through the earth's air, water, soil, and living organisms.

Sulfur circulates through the biosphere in the **sulfur cycle**, shown in Figure 4-34. Trace the flows and paths in this diagram. Much of the earth's sulfur is stored underground in rocks and minerals, including sulfate (SO_4^{2-}) salts buried deep under ocean sediments.

Sulfur also enters the atmosphere from several natural sources. Hydrogen sulfide (H_2S)—a colorless, highly poisonous gas with a rotten-egg smell—is released from active volcanoes and from organic matter in swamps, bogs, and tidal flats broken down by anaerobic decomposers.

Sulfur dioxide (SO_2), a colorless, suffocating gas, also comes from volcanoes. Particles of sulfate (SO_4^{2-}) salts, such as ammonium sulfate, enter the atmosphere

from sea spray, dust storms, and forest fires. Plant roots absorb sulfate ions and incorporate the sulfur as an essential component of many proteins.

Certain marine algae produce large amounts of volatile dimethyl sulfide, or DMS (CH_3SCH_3). Tiny droplets of DMS serve as nuclei for the condensation of water into droplets found in clouds. Thus changes in DMS emissions can affect cloud cover and climate. In the atmosphere DMS is converted to sulfur dioxide.

In the atmosphere, sulfur dioxide (SO_2) from natural sources and human activities is converted to sulfur trioxide gas (SO_3) and to tiny droplets of sulfuric acid (H_2SO_4). Sulfur dioxide also reacts with other atmospheric chemicals such as ammonia to produce tiny particles of sulfate salts. These droplets and particles fall to the earth as components of *acid deposition*, which along with other air pollutants can harm trees and aquatic life—more on this in Chapter 20.

In the oxygen-deficient environments of flooded soils, freshwater wetlands, and tidal flats, specialized bacteria convert sulfate ions to sulfide ions (S^{2-}). The sulfide ions can then react with metal ions to form insoluble metallic sulfides, which are deposited as rock, and the cycle continues.

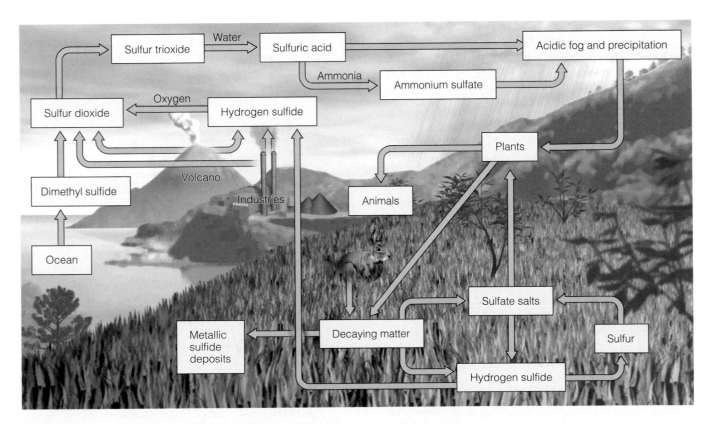

Figure 4-34 Natural capital: simplified model of the *sulfur cycle.*

How Are Human Activities Affecting the Sulfur Cycle? Overloading Nature

We add sulfur dioxide to the atmosphere by burning coal and oil, refining oil, and producing some metals from ores.

We add sulfur dioxide to the atmosphere in three ways. *First,* we burn sulfur-containing coal and oil to produce electric power. *Second,* we refine sulfur-containing petroleum to make gasoline, heating oil, and other useful products. *Third,* we convert sulfur-containing metallic mineral ores into free metals such as copper, lead, and zinc.

4-8 HOW DO ECOLOGISTS LEARN ABOUT ECOSYSTEMS?

What Is Field Research? Muddy Boots Ecology

Ecologists go into ecosystems and hang out in tree-tops to learn what organisms live there and how they interact.

Field research involves going into nature and observing and measuring the structure of ecosystems and what happens in them. Most of what we know about the structure and functioning of ecosystems described in this chapter has come from such research.

Ecologists trek through forests, deserts, and grasslands and wade or boat through wetlands, lakes, and streams collecting and observing species. Sometimes they carry out controlled experiments by isolating and changing a variable in part of an area and comparing the results with nearby unchanged areas.

Tropical ecologists erect tall construction cranes into the canopies of tropical forests to identify and observe the rich diversity of species living or feeding in these treetop habitats.

Increasingly, ecologists are using new technologies to collect field data. These include *remote sensing* from aircraft and satellites and *geographic information systems* (GISs), in which information gathered from broad geographic regions is stored in spatial databases (Figure 4-35). Then computers and GIS software can analyze and manipulate the data and combine them with ground and other data. These efforts can produce computerized maps of forest cover, water resources, air pollution emissions, coastal changes, relationships between cancers and sources of pollution, and changes in global sea temperatures.

Most satellite sensors use either reflected light or reflected infrared radiation to gather data. However, some new satellites have radar sensors that measure the reflection of microwave energy from the earth.

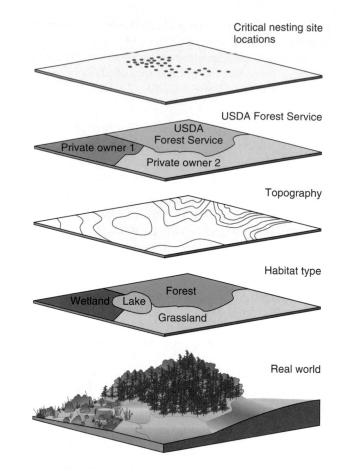

Figure 4-35 Geographic information systems (GISs) provide the computer technology for organizing, storing, and analyzing complex data collected over broad geographic areas. GISs enable scientists to overlay many layers of data (such as soils, topography, distribution of endangered populations, and land protection status).

These microwaves can "see" in the dark and penetrate smoke, clouds, haze, and water. This method is also being used to map the topography of the ocean floor and provide information about ocean currents and upward flows of nutrients from the ocean bottom (upwellings) that sustain fisheries.

How Are Ecosystems Studied in the Laboratory? Life under Glass

Ecologists use aquarium tanks, greenhouses, and controlled indoor and outdoor chambers to study ecosystems.

During the past 50 years, ecologists have increasingly supplemented field research by using *laboratory research* to set up, observe, and make measurements of model ecosystems and populations under laboratory conditions. Such simplified systems have been set up in containers such as culture tubes, bottles, aquarium tanks, and greenhouses and in indoor and outdoor chambers where temperature, light, CO_2, humidity, and other variables can be controlled carefully.

Such systems make it easier for scientists to carry out controlled experiments. In addition, such laboratory experiments often are quicker and cheaper than similar experiments in the field.

But there is a catch. We must consider whether what scientists observe and measure in a simplified, controlled system under laboratory conditions takes place in the same way in the more complex and dynamic conditions found in nature. Thus the results of laboratory research must be coupled with and supported by field research.

What Is Systems Analysis? Simulating Ecosystems

Ecologists develop mathematical and other models to simulate the behavior of ecosystems.

Since the late 1960s, ecologists have made increasing use of *systems analysis* to develop mathematical and other models that simulate ecosystems. Computer simulation of such models can help us understand large and very complex systems (such as rivers, oceans, forests, grasslands, cities, and climate) that cannot be adequately studied and modeled in field and laboratory research. Figure 4-36 outlines the major stages of systems analysis.

Researchers can change values of the variables in their computer models to project possible changes in environmental conditions, help anticipate environmental surprises, and analyze the effectiveness of various alternative solutions to environmental problems.

However, simulations and projections made using ecosystem models are no better than the data and assumptions used to develop the models. Thus careful field and laboratory ecological research must be used to provide the baseline data and determine the causal relationships between key variables needed to develop and test ecosystem models.

Why Do We Need Baseline Ecological Data? Understanding What We Have

We need baseline data on the world's ecosystems so we can see how they are changing and develop effective strategies for preventing or slowing their degradation.

According to a 2002 ecological study published by the Heinz Foundation, scientists have less than half of the basic ecological data they need to evaluate the status of ecosystems in the United States. Even fewer data are available for most other parts of the world.

Before we can understand what is happening to an ecosystem, community, or population and how best

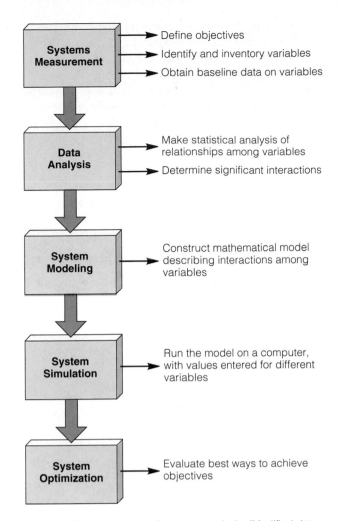

Figure 4-36 Major stages of systems analysis. (Modified data from Charles Southwick)

to prevent harmful environmental changes, we need to know its current condition. In other words, we need *baseline data* about its components, physical and chemical conditions, and how well it is functioning.

By analogy your doctor would like to have baseline data on your blood pressure, weight, and how well your organs and other systems are functioning as revealed by blood and other basic tests. Then when something happens to your health the doctor can run new tests and compare the results with the baseline data to determine what has changed and use this to come up with a treatment.

Ecologists call for a massive program to develop baseline data for the world's ecosystems. If we do not know how many elephants are in Africa, we cannot determine whether their populations are declining or increasing.

If we could see the bounty and beauty of nature the way it was 100 or 200 years ago, we would be outraged at what we have lost. But it is easy for us not to miss what we never saw or experienced.

For example, people visit degraded coastal environments such as a coral reef and call it beautiful because they are unaware how it used to look. Veteran divers say, "You should have seen it in the old days."

In this chapter we have seen that almost all natural ecosystems and the biosphere itself achieve *long-term* sustainability in two ways. *First,* they use *renewable solar energy* as their energy source. *Second,* they *recycle the chemical nutrients* their organisms need for survival, growth, and reproduction.

These two sustainability principles arise from the structure and function of natural ecosystems (Figures 4-8 and 4-17), the law of conservation of matter (p. 47), and the two laws of thermodynamics (p. 51). Thus the results of basic research in both the physical and biological sciences provide us with the same guidelines or lessons from nature on how we can live more sustainably on the earth, as summarized in Figure 3-19 (p. 53).

All things come from earth, and to earth they all return.
MENANDER (342–290 B.C.)

CRITICAL THINKING

1. **(a)** A bumper sticker asks, "Have you thanked a green plant today?" Give two reasons for appreciating a green plant. **(b)** Trace the sources of the materials that make up the bumper sticker, and decide whether the sticker itself is a sound application of the slogan.

2. Explain why microbes are the real rulers of the earth.

3. Explain how decomposers help keep you alive.

4. **(a)** How would you set up a self-sustaining aquarium for tropical fish? **(b)** Suppose you have a balanced aquarium sealed with a clear glass top. Can life continue in the aquarium indefinitely as long as the sun shines regularly on it? **(c)** A friend cleans out your aquarium and removes all the soil and plants, leaving only the fish and water. What will happen? Explain.

5. Make a list of the food you have eaten today and trace each type of food back to a particular producer.

6. Use the second law of thermodynamics (p. 51) to explain why there is such a sharp decrease in usable energy as energy flows through a food chain or web. Does an energy loss at each step violate the first law of thermodynamics (p. 51)? Explain.

7. Use the second law of thermodynamics (p. 51) to explain why many poor people in developing countries live on a mostly vegetarian diet.

8. Why do farmers not need to apply carbon to grow their crops but often need to add fertilizer containing nitrogen and phosphorus?

9. Carbon dioxide (CO_2) in the atmosphere fluctuates significantly on a daily and seasonal basis. Why are CO_2 levels higher during the day than at night?

10. What would happen to an ecosystem if **(a)** all its decomposers and detritus feeders were eliminated or **(b)** all its producers were eliminated? Are we necessary for the functioning of any natural ecosystem? Explain.

PROJECTS

1. Visit several types of nearby aquatic life zones and terrestrial ecosystems. For each site, try to determine the major producers, consumers, detritivores, and decomposers.

2. Write a brief scenario describing the sequence of consequences to us and to other forms of life if each of the following nutrient cycles stopped functioning: **(a)** carbon, **(b)** nitrogen, **(c)** phosphorus, and **(d)** water.

3. Use the library or the Internet to find out bibliographic information about *G. Evelyn Hutchinson* and *Menander,* whose quotes are found at the beginning and end of this chapter.

4. Make a concept map of this chapter's major ideas using the section heads, subheads, and key terms (in boldface). Look on the website for this book for information about making concept maps.

LEARNING ONLINE

The website for this book contains study aids and many ideas for further reading and research. They include a chapter summary, review questions for the entire chapter, flash cards for key terms and concepts, a multiple-choice practice quiz, interesting Internet sites, references, and a guide for accessing thousands of InfoTrac® College Edition articles. Log on to

http://biology.brookscole.com/miller14

Then click on the Chapter-by-Chapter area, choose Chapter 4, and select a learning resource.

5 Evolution and Biodiversity

Biodiversity

Earth: The Just-Right, Resilient Planet

Life on the earth (Figure 5-1) as we know it needs a certain temperature range: Venus is much too hot and Mars is much too cold, but the earth is *just right*. Otherwise, you would not be reading these words.

Life as we know it depends on the liquid water that dominates the earth's surface. Temperature is crucial because most life on the earth needs average temperatures between the freezing and boiling points of water.

The earth's orbit is the right distance from the sun to provide these conditions. If the earth were much closer, it would be too hot—like Venus—for water vapor to condense to form rain. If it were much farther away, its surface would be so cold—like Mars—that its water would exist only as ice. The earth also spins; if it did not, the side facing the sun would be too hot and the other side too cold for water-based life to exist.

The earth is also the right size: it has enough gravitational mass to keep its iron and nickel core molten and to keep the light gaseous molecules in its atmosphere (such as N_2, O_2, CO_2, and H_2O) from flying off into space.

On a time scale of millions of years, the earth is enormously resilient and adaptive. During the 3.7 billion years since life arose, the average surface temperature of the earth has remained within the narrow range of 10–20°C (50–68°F), even with a 30–40% increase in the sun's energy output. What a great temperature control system.

For several hundred million years oxygen has made up about 21% of the volume of earth's atmosphere. This is fortunate for us and most other forms of life. If the atmosphere's oxygen content dropped to about 15%, this would be lethal for most forms of life. If it increased to about 25%, oxygen in the atmosphere would

probably ignite into a giant fireball. And thanks to the development of photosynthesizing bacteria more than 2 billion years ago, an ozone sunscreen protects us and many other forms of life from an overdose of ultraviolet radiation. In short, this remarkable planet we live on is just right for life as we know it.

We can summarize the 3.7-billion-year biological history of the earth in one sentence: *Organisms convert solar energy to food, chemicals cycle, and a variety of species with different biological roles (niches) has evolved in response to changing environmental conditions.* Perhaps the two most astounding features of the planet are its incredibly rich diversity of life and its inherent ability to sustain life.

Here is the essence of this chapter. Each species here today represents a long chain of evolution and plays a unique ecological role (called its *niche*) in the earth's communities and ecosystems. These species, communities, and ecosystems also are essential for future evolution as populations of species continue to adapt to changes in environmental conditions.

Figure 5-1 Natural capital: the earth, a blue and white planet in the black void of space. Currently, it has the right physical and chemical conditions to allow the development of life as we know it.

NASA

There is grandeur to this view of life . . . that, whilst this planet has gone cycling on . . . endless forms most beautiful and most wonderful have been, and are being, evolved.

CHARLES DARWIN

This chapter addresses the following questions:

- How do scientists account for the emergence of life on the earth?

- What is evolution, and how has it led to the current diversity of organisms on the earth?

- What is an ecological niche, and how does it help a population adapt to changing environmental conditions?

- How do extinction of species and formation of new species affect biodiversity?

- What is the future of evolution and what should be our role in this future?

5-1 ORIGINS OF LIFE

How Did Life Emerge on the Primitive Earth? Chemistry First, Then Biology

Evidence indicates that the earth's life is the result of about 1 billion years of chemical evolution to form the first cells, followed by about 3.7 billion years of biological evolution to form the species we find on the earth today.

How did life on the earth evolve to its present incredible diversity of species? We do not know the full answers to these questions. But considerable evidence suggests that life on the earth developed in two phases over the past 4.6–4.7 billion years (Figure 5-2).

The first phase was *chemical evolution* of the organic molecules, biopolymers, and systems of chemical reactions needed to form the first cells. This took about 1 billion years.

This was followed by *biological evolution* from single-celled prokaryotic bacteria (Figure 4-3b, p. 58), to single-celled eukaryotic creatures (Figure 4-3a,

p. 58), and then to multicellular organisms. This second phase has been going on for about 3.7 billion years (Figure 5-3).

How Do We Know What Organisms Lived in the Past? Scientific Detective Work

Most of our knowledge about past life comes from fossils, chemical analysis, cores drilled out of buried ice, and DNA analysis.

Most of what we know of the earth's life history comes from **fossils:** mineralized or petrified replicas of skeletons, bones, teeth, shells, leaves, and seeds, or impressions of such items. Fossils give us physical evidence of organisms that lived long ago and reveal what their internal structures looked like.

Despite its importance, the fossil record is uneven and incomplete. Some forms of life left no fossils, some fossils have decomposed, and others are yet to be found. The fossils we have found so far are believed to represent only about 1% of all the species that have ever lived.

Evidence about the earth's early history also comes from chemical analysis and measurements of the half-lives of radioactive elements in primitive rocks and fossils. Information also comes from analysis of material in cores drilled out of buried ice and from comparisons of DNA of past and current organisms.

5-2 EVOLUTION AND ADAPTATION

What Is Evolution? Changing Genes

Evolution is the change in a population's genetic makeup over time.

According to scientific evidence, the major driving force of adaptation to changes in environmental conditions is **biological evolution,** or **evolution.** It is the change in a population's genetic makeup through successive generations. Note that *populations, not individuals, evolve by becoming genetically different.*

Chemical Evolution
(1 billion years)

Biological Evolution
(3.7 billion years)

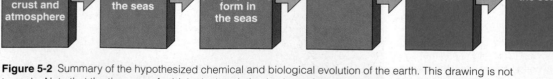

| Formation of the earth's early crust and atmosphere | Small organic molecules form in the seas | Large organic molecules (biopolymers) form in the seas | First protocells form in the seas | Single-cell prokaryotes form in the seas | Single-cell eukaryotes form in the seas | Variety of multicellular organisms form, first in the seas and later on land |

Figure 5-2 Summary of the hypothesized chemical and biological evolution of the earth. This drawing is not to scale. Note that the time span for biological evolution is almost four times longer than that for chemical evolution.

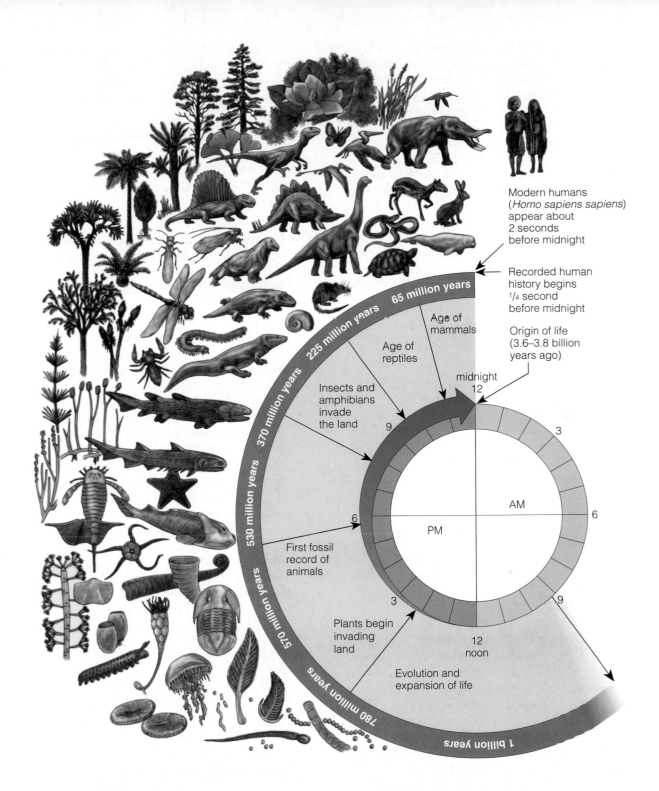

Modern humans
(*Homo sapiens sapiens*)
appear about
2 seconds
before midnight

**Recorded human
history begins**
¼ second
before midnight

Origin of life
(3.6–3.8 billion
years ago)

Age of
mammals

Age of
reptiles

65 million years

225 million years

370 million years

530 million years

570 million years

780 million years

1 billion years

Insects and
amphibians
invade
the land

First fossil
record of
animals

Plants begin
invading
land

Evolution and
expansion of life

midnight
12

9

6

3

AM

PM

6

3

9

12
noon

Figure 5-3 Natural capital: greatly simplified overview of the biological evolution of life on the earth, which was preceded by about 1 billion years of chemical evolution. Evidence indicates that microorganisms (mostly bacteria and, later, protists) that lived in water dominated the early span of biological evolution on the earth, between about 3.7 billion and 1 billion years ago. Plants and animals evolved first in the seas. Fossil and recent DNA evidence suggests that plants began moving onto land about 780 million years ago, and animals began invading the land about 370 million years ago. Humans arrived on the scene a very short time ago. We have been around for less than an eye blink of the earth's roughly 3.7-billion-year history of biological evolution. Although we are newcomers we have taken over much of the planet and now have the power to cause the premature extinction of many of the other species that travel with us as passengers on the earth as it hurtles through space.

According to the **theory of evolution,** all species descended from earlier, ancestral species. In other words, life comes from life. This widely accepted scientific theory explains how life has changed over the past 3.7 billion years and why life is so diverse today. Religious and other groups offer other explanations, but this is the accepted scientific explanation.

Biologists use the term **microevolution** to describe the small genetic changes that occur in a population. They use the term **macroevolution** to describe long-term, large-scale evolutionary changes through which new species form from ancestral species and other species are lost through extinction. Macroevolution also involves changes that occur at higher levels than species, such as the evolution of new genera, families, or classes of species (see Appendix 4).

How Does Microevolution Work? Changes in the Gene Pool

A population's gene pool changes over time when beneficial changes or mutations in its DNA molecules are passed on to offspring.

The first step in evolution is the development of *genetic variability* in a population. Every population of a species has a **gene pool:** its collection of genes or genetic resources potentially available to members' offspring in the next generation. *Microevolution* is a change in a population's gene pool over time.

Members of a population generally have the same number and kinds of genes. But a particular gene may have two or more different molecular forms, called **alleles.** Sexual reproduction leads to a random shuffling or recombination of alleles. As a result, each individual in a population (except identical twins) has a different combination of alleles.

Genetic variability in a population originates through **mutations:** random changes in the structure or number of DNA molecules in a cell that can be inherited by offspring. Mutations can occur in two ways. One is by exposure of DNA to external agents such as radioactivity, X rays, and natural and human-made chemicals (called *mutagens*). The other is the result of random mistakes that sometimes occur in coded genetic instructions when DNA molecules are copied each time a cell divides and whenever an organism reproduces.

Mutations can occur in any cells, but only those in reproductive cells are passed on to offspring. Some mutations are harmless but most are lethal. *Every so often, a mutation is beneficial.* The result is new genetic traits that give the bearer and its offspring better chances for survival and reproduction under existing environmental conditions or when conditions change.

Mutations are random and unpredictable, are the only source of totally new genetic raw material (alleles), and are rare. Life is a genetic shuffle. Once created by mutation, new alleles can be shuffled together or recombined *randomly* to create new combinations of genes in populations of sexually reproducing species. In addition to mutations, another source of evolutionary change is the trading of genes between bacteria.

What Role Does Natural Selection Play in Microevolution? Backing Reproductive Winners

Some members of a population may have genetic traits that enhance their ability to survive and produce offspring with these traits.

Natural selection occurs when some individuals of a population have genetically based traits that increase their chances of survival and their ability to produce offspring with the same traits. Three conditions are necessary for evolution of a population by natural selection.

There must be *genetic variability* for a trait in a population. The trait must be *heritable,* meaning it can be passed from one generation to another. And the trait must somehow lead to **differential reproduction.** This means it must enable individuals with the trait to leave more offspring than other members of the population.

An **adaptation,** or **adaptive trait,** is any heritable trait that enables organisms to better survive and reproduce under prevailing environmental conditions. Natural selection causes any allele or set of alleles that result in an adaptive trait to become more common in succeeding generations and alleles without the trait to become less common.

When faced with a critical change in environmental conditions, a population of a species has three possibilities: adapt to the new conditions through natural selection, migrate (if possible) to an area with more favorable conditions, or become extinct.

The process of microevolution can be summarized simply: *Genes mutate, individuals are selected, and populations evolve.*

What Is Coevolution? An Arms Race between Interacting Species

Interacting species can engage in a back-and-forth genetic contest in which each gains a temporary genetic advantage over the other.

Some biologists have proposed that *interactions between species* can also result in microevolution in each of their populations. According to this hypothesis, when populations of two different species interact over a long time, changes in the gene pool of one species can lead to changes in the gene pool of the other. This process is called **coevolution.** In this give-and-take evolutionary game, each species is in a genetic race to produce the largest number of surviving offspring.

One example is the interactions between bats and moths. Bats like to eat moths, and they hunt at night and use echolocation to navigate and locate their prey.

They do this by emitting extremely high-frequency and high-intensity pulses of sound. Then they analyze the returning echoes to create a sonic "image" of their prey. (We have copied this natural technology by using sonar to detect submarines, whale, and schools of fish.)

As a countermeasure, some moth species have evolved ears that are especially sensitive to the sound frequencies bats use to find them. When the moths hear the bat frequencies they try to escape by falling to the ground or flying evasively.

Some bat species then evolved ways to counter this defense by switching the frequency of their sound pulses. Some moths then evolved their own high-frequency clicks to jam the bats' echolocation system (we have also learned to jam radar). Some bat species then adapted by turning off their echolocation system and using the moth's clicks to locate their prey. Each species continues refining its adaptations in this ongoing coevolutionary contest. Coevolution is like an arms race between interacting populations of different species. Sometimes the predators are ahead and at other times the prey get the upper hand.

5-3 ECOLOGICAL NICHES AND ADAPTATION

What Is an Ecological Niche? How Species Coexist

Each species in an ecosystem has a specific role or way of life.

If asked what role a certain species such as an alligator plays in an ecosystem, an ecologist would describe its **ecological niche,** or simply **niche** (pronounced "nitch"). It is a species' way of life or functional role in a community or ecosystem and involves everything that affects its survival and reproduction.

A species' ecological niche includes the *adaptations* or *adaptive traits* its members have acquired through evolution. It also includes that species' range of tolerance for various physical and chemical conditions, such as temperature (Figure 4-13, p. 64) and water availability. In addition, it includes the types and amounts of resources the species uses (such as food or nutrients and space), how it interacts with other living and nonliving components of the ecosystems in which it is found, and the role it plays in the energy flow and matter cycling in an ecosystem. Finding out about these things for just one species takes a lot of research.

The ecological niche of a species is different from its **habitat,** the physical location where it lives. Ecologists often say that a niche is like a species' occupation, whereas habitat is like its address.

A species' **fundamental niche** is the full potential range of physical, chemical, and biological conditions and resources it could theoretically use if there were no direct competition from other species. But in a particular ecosystem, different species often compete with one another for one or more of the same resources. In other words, the niches of competing species overlap.

To survive and avoid competition for the same resources, a species usually occupies only part of its fundamental niche in a particular community or ecosystem—what ecologists call its **realized niche.** By analogy, you may be capable of being president of a particular company (your *fundamental professional niche*), but competition from others may mean you become only a vice president (your *realized professional niche*).

What are Generalist and Specialist Species? Broad and Narrow Niches

Some species have broad ecological roles and others have narrower or more specialized roles.

Scientists use the niches of species to broadly classify them as *generalists* or *specialists*. **Generalist species** have broad niches (Figure 5-4, right curve). They can live in many different places, eat a variety of foods, and tolerate a wide range of environmental conditions. Flies, cockroaches (Spotlight, p. 92), mice, rats, white-tailed deer, raccoons, coyotes, copperheads, channel catfish, and humans are generalist species.

Specialist species have narrow niches (Figure 5-4, left curve). They may be able to live in only one type of habitat, use only one or a few types of food, or tolerate only a narrow range of climatic and other environmental conditions. This makes them more prone to extinction when environmental conditions change.

For example, *tiger salamanders* are specialists because they can breed only in fishless ponds where their larvae will not be eaten. Threatened *red-cockaded woodpeckers* carve nest holes almost exclusively in

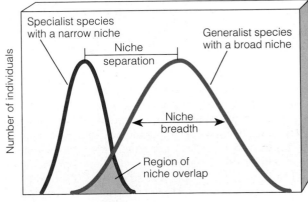

Figure 5-4 Overlap of the niches of two different species: a specialist and a generalist. In the overlap area the two species compete for one or more of the same resources. As a result, each species can occupy only a part of its fundamental niche; the part it occupies is its realized niche. Generalist species have a broad niche (right), and specialist species have a narrow niche (left).

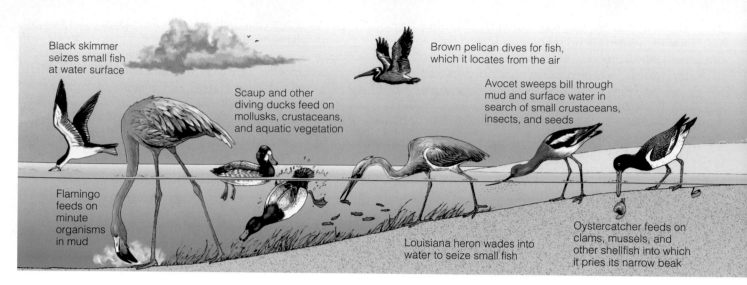

Figure 5-5 Specialized feeding niches of various bird species in a coastal wetland. Such resource partitioning reduces competition and allows sharing of limited resources.

longleaf pines that are at least 75 years old. China's highly endangered *giant pandas* feed almost exclusively on various types of bamboo. Figure 5-5 shows shorebirds that feed in specialized niches on crustaceans, insects, and other organisms on sandy beaches and their adjoining coastal wetlands.

Is it better to be a generalist than a specialist? It depends. When environmental conditions are fairly constant, as in a tropical rain forest, specialists have an advantage because they have fewer competitors. But under rapidly changing environmental conditions, the generalist usually is better off than the specialist.

Natural selection can lead to an increase in specialized species when the niches of species compete intensely for scarce resources. Over time one species may evolve into a variety of species with different adaptations that reduce competition and allow them to share limited resources.

This *evolutionary divergence* of a single species into a variety of similar species with specialized niches can

---

SPOTLIGHT

Cockroaches: Nature's Ultimate Survivors

Cockroaches, the bugs many people love to hate, have been around for about 350 million years and are one of the great success stories of evolution. They are so successful because they are *generalists*.

The earth's 4,000 cockroach species can eat almost anything including algae, dead insects, fingernail clippings, salts in tennis shoes, electrical cords, glue, paper, and soap. They can also live and breed almost anywhere except in polar regions.

Some cockroach species can go for months without food, survive for a month on a drop of water from a dishrag, and withstand massive doses of radiation. One species can survive being frozen for 48 hours.

They can usually evade their predators and a human foot in hot pursuit because most species have antennae that can detect minute movements of air, vibration sensors in their knee joints, and rapid response times (faster than you can blink). Some even have wings.

They also have high reproductive rates. In only a year, a single female Asian cockroach (especially prevalent in Florida) and its young can add about 10 million new cockroaches to the world. Their high reproductive rate also helps them quickly develop genetic resistance to almost any poison we throw at them.

Most cockroaches also sample food before it enters their mouth and learn to shun foul-tasting poi-

sons. They also clean up after themselves by eating their dead and, if food is scarce enough, their living.

The 25 different species of cockroach that live in homes can carry viruses and bacteria that cause diseases such as hepatitis, polio, typhoid fever, plague, and salmonella. They can also cause some people to have allergic reactions ranging from watery eyes to severe wheezing. About 60% of Americans suffering from asthma are allergic to dead or live cockroaches.

Critical Thinking

If you could, would you exterminate all cockroach species? What might be some ecological consequences of doing this?

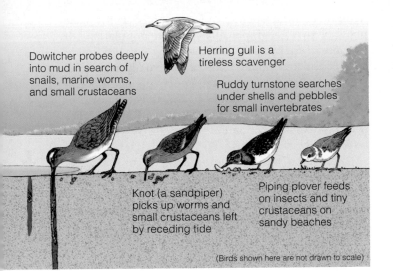

Dowitcher probes deeply into mud in search of snails, marine worms, and small crustaceans

Herring gull is a tireless scavenger

Ruddy turnstone searches under shells and pebbles for small invertebrates

Knot (a sandpiper) picks up worms and small crustaceans left by receding tide

Piping plover feeds on insects and tiny crustaceans on sandy beaches

(Birds shown here are not drawn to scale)

be illustrated by various species of honeycreepers on the island of Hawaii. Starting with a single ancestor, a variety of honeycreeper species evolved with different types of beaks specialized to feed on different types of food sources such as specific insects, nectar from different types of flowers, and certain types of seeds and fruit (Figure 5-6).

What Limits Adaptation? Life Is a Genetic Dice Game

A population's ability to adapt to new environmental conditions is limited by its gene pool and how fast it can reproduce.

Will adaptations to new environmental conditions in the not-to-distant future allow our skin to become more resistant to the harmful effects of ultraviolet radiation, our lungs to cope with air pollutants, and our liver to better detoxify pollutants?

The answer is *no* because of two limits to adaptations in nature. *First*, a change in environmental conditions can lead to adaptation only for genetic traits already present in the gene pool of a population. You must have dice to play the genetic dice game.

Second, even if a beneficial heritable trait is present in a population, the population's ability to adapt may be limited by its reproductive capacity. Populations of genetically diverse species that reproduce quickly— such as weeds, mosquitoes, rats, bacteria, or cockroaches—often adapt to a change in environmental conditions in a short time. In contrast, species that cannot produce large numbers of offspring rapidly, such as elephants, tigers, sharks, and humans, take a long time (typically thousands or even millions of years) to adapt through natural selection. You have to be able to throw the genetic dice fast.

Here is some *bad news* for most members of a population. Even when a favorable genetic trait is present

in a population, most of the population would have to die or become sterile so that individuals with the trait could predominate and pass the trait on. Thus most players get kicked out of the genetic dice game before they have a chance to win. This means that most members of the human population would have to die prematurely for hundreds of thousands of generations for a new genetic trait to predominate. This is hardly a desirable solution to the environmental problems we face.

What Are Two Commonly Misunderstood Aspects of Evolution? Strong Does Not Cut It and There Is No Grand Design

Evolution is about leaving the most descendants, and there is no master plan leading to genetic perfection.

There are two common misconceptions about evolution. One is that "survival of the fittest" means "survival of the strongest." To biologists, *fitness* is a measure of reproductive success, not strength. Thus the fittest individuals are those that leave the most descendants.

The other misconception is that evolution involves some grand plan of nature in which species become more perfectly adapted. From a scientific standpoint, no plan or goal of genetic perfection has been identified in the evolutionary process.

Fruit and seed eaters

Greater Koa-finch

Kona Grosbeak

Akiapolaau

Maui Parrotbill

Insect and nectar eaters

Kuai Akialaoa

Amakihi

Crested Honeycreeper

Apapane

Unkown finch ancestor

Figure 5-6 Evolutionary divergence of honeycreepers into a variety of specialized ecological niches. Each of these related species has a beak specialized to take advantage of certain types of food resources.

5-4 SPECIATION, EXTINCTION, AND BIODIVERSITY

How Do New Species Evolve? Moving Out and Moving On

A new species arises when members of a population are isolated from other members so long that changes in their genetic makeup prevent them from producing fertile offspring if they get together again.

Under certain circumstances, natural selection can lead to an entirely new species. In this process, called **speciation,** two species arise from one. For sexually reproducing species, a new species is formed when some members of a population can no longer breed with other members to produce fertile offspring.

The most common mechanism of speciation (especially among animals) is called *allopatric speciation.* It takes place in two phases: geographic isolation and reproductive isolation. **Geographic isolation** occurs when different groups of the same population become physically isolated from one another for long periods.

For example, part of a population may migrate in search of food and then begin living in another area with different environmental conditions (Figure 5-7). Populations also may become separated by a physical barrier (such as a mountain range, stream, lake, or road), by a change such as a volcanic eruption or earthquake, or when a few individuals are carried to a new area by wind or water. Other causes are the advance of glaciers during ice ages, changes in sea level that can create islands, shifts in ocean currents, a warmer or cooler climate that shifts vegetation northward or southward or up or down slopes, and a drier climate that divides a large lake into several small lakes.

The second phase of allopatric speciation is **reproductive isolation.** It occurs when mutation and natural selection operate independently in the gene pools of two geographically isolated populations. If this *divergence* process continues long enough, members of isolated populations of a sexually reproducing species may become so different in genetic makeup that when they get together again, they cannot produce live, fertile offspring. Then one species has become two, and speciation has occurred through *divergent evolution.*

For some rapidly reproducing organisms, this type of speciation may occur within hundreds of years. For most species it takes from tens of thousands to millions of years, which makes it difficult to observe and document the appearance of a new species.

A less common form of speciation is called *sympatric speciation.* It is the creation of a new species when groups in a population living close together are unable to interbreed because of a mutation or subtle behavioral changes. Some insects are candidates for this type of speciation perhaps when two populations experience different types of mutations by feeding on different types of plants.

What Is Extinction? Going, Going, Gone

A species becomes extinct when its populations cannot adapt to changing environmental conditions.

After speciation, the second process affecting the number and types of species on the earth is **extinction,** in which an entire species ceases to exist. Fossil records indicate that species tend to exist for 1-10 million years before becoming extinct, although catastrophic events accelerate the extinction process.

For most of the earth's geological history, species have faced incredible challenges to their existence. Continents have broken apart and moved over millions of years (Figure 5-8). The earth's land area has re-

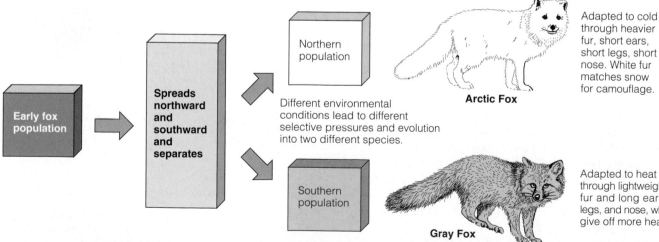

Figure 5-7 How geographic isolation can lead to reproductive isolation, divergence, and speciation.

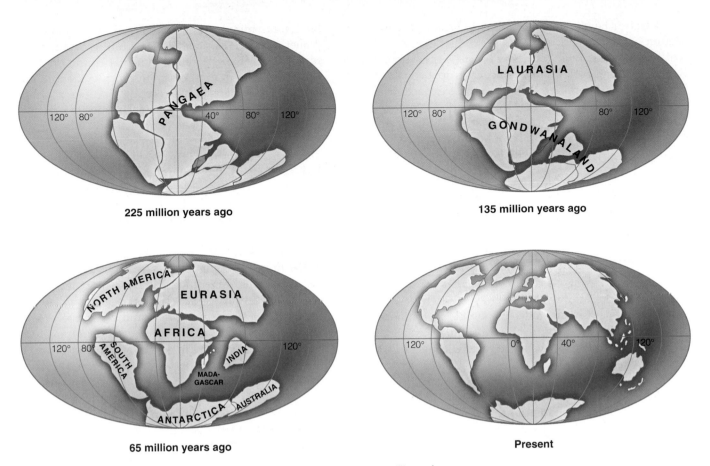

Figure 5-8 *Continental drift*, the extremely slow movement of continents over millions of years on several gigantic plates. This process plays a role in the extinction of species and the rise of new species. Populations are geographically and eventually reproductively isolated as land masses float apart and new coastal regions are created. Rock and fossil evidence indicates that about 200–250 million years ago all of the earth's present-day continents were locked together in a supercontinent called Pangaea (top left). About 180 million years ago, Pangaea began splitting apart as the earth's huge plates separated and eventually resulted in today's locations of the continents (bottom right).

peatedly shrunk as continents have been flooded, and expanded when the world's oceans have shrunk. At other times much of the planet's land has been covered with ice.

The earth's life has also had to cope with volcanic eruptions, meteorites and asteroids crashing onto the planet, and releases of large amounts of methane trapped beneath the ocean floor. Some of these events created dust clouds that shut down or sharply reduced photosynthesis long enough to eliminate huge numbers of producers and, soon thereafter, the consumers that fed on them.

Populations of existing species in some places have been also been reduced or eliminated by newly arrived migrant species or species that are accidentally or deliberately introduced into new areas. More recently our species arrived and began taking over or degrading more and more of the earth's resources and habitats. Today's biodiversity represents the species that have survived and thrived despite environmental upheavals.

What Is the Difference between Background Extinction, Mass Extinction, and Mass Depletion? Wiping Out Large Groups

All species eventually become extinct, but sometimes drastic changes in environmental conditions eliminate large groups of species.

Extinction is the ultimate fate of all species, just as death is for all individual organisms. Biologists estimate that 99.9% of all the species that ever existed are now extinct.

As local environmental conditions change, a certain number of species disappear at a low rate, called **background extinction.** Based on the fossil record and analysis of ice cores, biologists estimate that the average annual background extinction rate is one to five species for each million species on the earth.

In contrast, **mass extinction** is a significant rise in extinction rates above the background level. It is a catastrophic, widespread (often global) event in which large groups of existing species (perhaps 25–70%) are wiped out.

Scientists have also identified periods of **mass depletion** in which extinction rates are higher than normal but not high enough to classify as a mass extinction. Recent fossil and geological evidence casts doubt on the hypothesis that there have been five mass extinctions over the past 500 million years as is reported in many biology and environmental science textbooks. The new evidence suggests that there have been two mass extinctions and three mass depletions during this period.

What Is Adaptive Radiation? Take Advantage of Opportunities

Extinction of large groups of species opens up opportunities for new species to evolve and fill new or vacant niches.

A mass extinction or mass depletion crisis for some species is an opportunity for other species. The existence of millions of species today means that speciation, on average, has kept ahead of extinction, especially during the last 250 million years (Figure 5-9). Study this figure carefully.

Evidence shows that the earth's mass extinctions and depletions have been followed by periods of recovery called **adaptive radiations** in which numerous new species evolved to fill new or vacated ecological roles or niches in changed environments. Fossil records suggest that it takes 1–10 million years for adaptive radiations to rebuild biological diversity after a mass extinction or depletion.

How Are Human Activities Affecting the Earth's Biodiversity? The Earth Giveth and We Taketh

The scientific consensus is that human activities are decreasing the earth's biodiversity.

Speciation minus extinction equals *biodiversity*, the planet's genetic raw material for future evolution in response to changing environmental conditions. Extinction is a natural process. But much evidence indicates that humans have recently become a major force in the premature extinction of species. According to biologists Stuart Primm and Edward O. Wilson, three independent measures estimate that during the 20th century, extinction rates increased by 100–1,000 times the natural background extinction rate of about one to five species per million species per year. In other words, the annual estimated human-caused extinction rate is 100 to 1,000 species per million species.

Wilson and Primm warn that projected increases in the human population and resource consumption may cause the premature extinction of at least one-fifth of the earth's current species by 2030 and half by the end of the century. This could constitute a new *mass depletion* and possibly a new *mass extinction*. According to Wilson, if we make an "all-out effort to save the biologically richest parts of the world, the amount of loss can be cut at least by half."

On our short time scale, such major losses cannot be recouped by formation of new species; it took millions of years after each of the earth's past mass extinctions and depletions for life to recover to the previous level of biodiversity. If the recovery period lasts at least 5 million years, this would be 20 times longer than we have been around as a species.

We are also destroying or degrading ecosystems such as tropical forests, coral reefs, and wetlands that are centers for future speciation (See the Guest Essay on this topic by Norman Myers on the website for this chapter). Genetic engineering cannot stop this loss of biodiversity because genetic engineers rely on natural biodiversity for their genetic raw material.

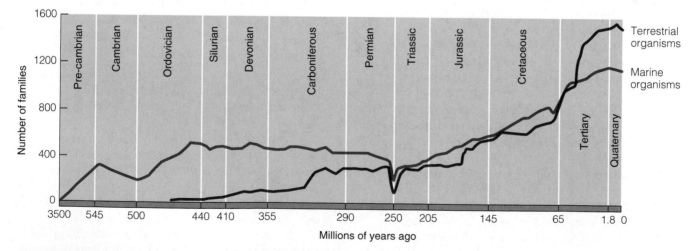

Figure 5-9 **Natural capital:** changes in the earth's biodiversity over geological time. The biological diversity of life on land and in the oceans has increased dramatically over the last 3.5 billion years, especially during the past 250 million years. Note that during the last 1.8 million years this increase has leveled off. During the next hundred years or so, will the human species be a major factor in decreasing the earth's biodiversity?

5-5 WHAT IS THE FUTURE OF EVOLUTION?

What Is Artificial Selection? Getting the Type of Fruit or Dog You Want

Humans pick members of a population with genetic traits they like and breed them to produce offspring with such traits.

We have used **artificial selection** to change genetic characteristics of populations. In this process, we select one or more desirable genetic traits in the population of a plant or animal, such as a type of wheat, fruit, or dog. Then we use *selective breeding* to end up with populations of the species containing large numbers of individuals with the desired traits.

For example, two crop plants such as a variety of a pear and of an apple can be crossbred with the goal of producing a pear with a more reddish color (Figure 5-10). This is repeated for a number of generations until the desired trait in the pear predominates.

Artificial selection results in many domesticated breeds or hybrids of the same species, all originally developed from one wild species. For example, despite their widely different genetic traits, the hundreds of different breeds of dogs are members of the same species because they can interbreed and produce fertile offspring.

Artificial selection has yielded food crops with higher yields, cows that give more milk, trees that grow faster, and a variety of types of dogs and cats.

But traditional crossbreeding is a slow process. And it can combine traits only from species that are close to one another genetically.

What Is Genetic Engineering? Transferring Genes between Species

Genetic engineers create genetically modified organisms by transplanting genes from one species to the DNA of another species.

Recently scientists have learned how to use genetic engineering to speed up our ability to manipulate genes. **Genetic engineering,** or **gene splicing,** is a set of techniques for isolating, modifying, multiplying, and recombining genes from different organisms. It enables scientists to transfer genes between different species that would never interbreed in nature. For example, genes from a fish species can be put into a tomato or strawberry.

The resulting organisms are called **genetically modified organisms (GMOs)** or **transgenic organisms.** Figure 5-11 (p. 98) outlines the steps involved in developing a genetically modified or transgenic plant. Study this figure carefully.

Gene splicing takes about half as much time to develop a new crop or animal variety and costs less

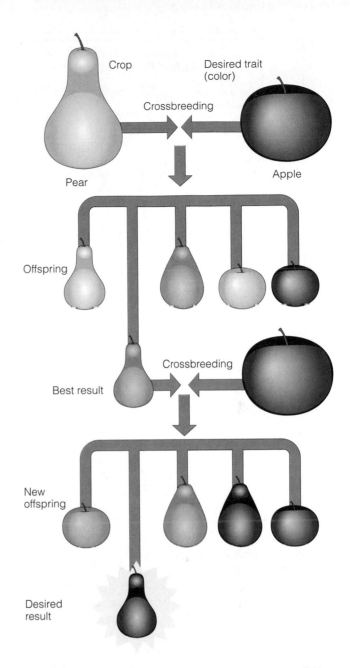

Figure 5-10 *Traditional crossbreeding* of species that are fairly close to one another genetically.

than traditional crossbreeding. While traditional crossbreeding involves mixing the genes of similar types of organisms through breeding, genetic engineering allows us to transfer traits between different types of organisms.

Scientists have used gene splicing to develop modified crop plants, genetically engineered drugs, and pest-resistant plants. They have also created genetically engineered bacteria to help clean up spills of oil and other toxic pollutants.

Genetic engineers have also learned how to produce a *clone* or genetically identical version of an individual in a population. Scientists have made clones of domestic animals such as sheep and cows and may

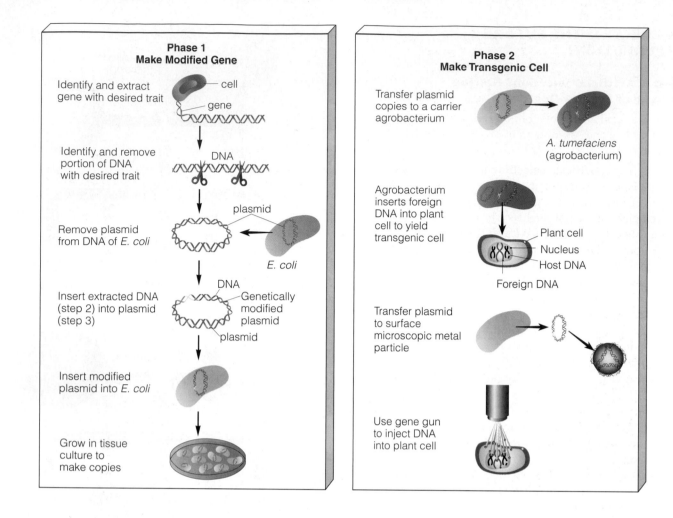

Phase 1
Make Modified Gene

Identify and extract gene with desired trait — cell, gene

Identify and remove portion of DNA with desired trait — DNA

Remove plasmid from DNA of *E. coli* — plasmid, *E. coli*

Insert extracted DNA (step 2) into plasmid (step 3) — DNA, Genetically modified plasmid, plasmid

Insert modified plasmid into *E. coli*

Grow in tissue culture to make copies

Phase 2
Make Transgenic Cell

Transfer plasmid copies to a carrier agrobacterium — *A. tumefaciens* (agrobacterium)

Agrobacterium inserts foreign DNA into plant cell to yield transgenic cell — Plant cell, Nucleus, Host DNA, Foreign DNA

Transfer plasmid to surface microscopic metal particle

Use gene gun to inject DNA into plant cell

someday be able to clone humans—a possibility that excites some people and horrifies others.

Bioengineers have developed chickens that lay low-cholesterol eggs, tomatoes with genes that can help prevent some types of cancer, and bananas and potatoes that contain oral vaccines to treat various viral diseases in developing countries where needles and refrigeration are not available.

Researchers envision using genetically engineered animals to act as biofactories for producing drugs, vaccines, antibodies, hormones, industrial chemicals such as plastics and detergents, and human body organs. This new field is called **biopharming.** For example, cows may be able to produce insulin for treating diabetes, perhaps more cheaply than making the insulin in laboratories. Have you considered this field as a career choice?

What Are Some Concerns about the Genetic Revolution? Genetic Wonderland or Genetic Wasteland?

Genetic engineering has great promise but it is an unpredictable process and raises a number of privacy, ethical, legal, and environmental issues.

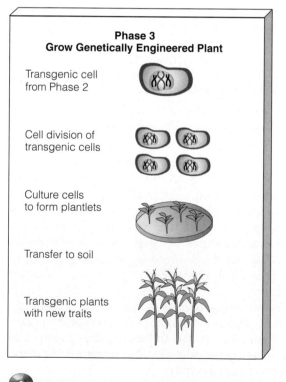

Phase 3
Grow Genetically Engineered Plant

Transgenic cell from Phase 2

Cell division of transgenic cells

Culture cells to form plantlets

Transfer to soil

Transgenic plants with new traits

Figure 5-11 *Genetic engineering.* Steps in genetically modifying a plant.

The hype about genetic engineering can lead us to believe that its results are controllable and predictable. In reality, most current forms of genetic engineering are messy and unpredictable. Genetic engineers can insert a gene into the nucleus of a cell but with current technology they do know whether the cell will incorporate the new gene into its DNA. They also do not know where the new gene will be located in the DNA molecule's structure and what effects this will have on the organism.

Thus, *genetic engineering is a trial and error process* with many failures and unexpected results. Indeed, the average success rate of current genetic engineering experiments is only about 1%.

Some people have genes that make them more likely to develop certain genetic diseases or disorders. We now have the power to detect these genetic deficiencies, even before birth. This raises some important issues. If gene therapy is developed for correcting these deficiencies, who will get it? Will it be mostly for the rich? Will this mean more abortions of genetically defective fetuses? Will health insurers refuse to insure people with certain genetic defects that could lead to health problems? Will employers refuse to hire them?

X *HOW WOULD YOU VOTE?* Should genetic screening be required as part of an application for a job, health insurance, or life insurance? Cast your vote online at http://biology.brookscole.com/miller14.

Some people dream of a day when our genetic prowess could eliminate death and aging altogether. As one's cells, organs, or other parts wear out or are damaged, they would be replaced with new ones. These replacement parts might be grown in genetic engineering laboratories or biopharms. Or people might choose to have a clone available for spare parts. Several countries have banned human cloning, but the research is taking place anyway and human clones may appear in the not too distant future.

This raises a number of questions: Is it moral to do this? Who decides? Who regulates it? Will genetically designed humans and clones have the same legal rights as other people?

X *HOW WOULD YOU VOTE?* Should we legalize the production of human clones if a reasonably safe technology for doing so becomes available? Cast your vote online at http://biology.brookscole.com/miller14.

What might be the environmental impacts of such genetic developments on resource use, pollution, and environmental degradation? If everyone could live with good health as long as they wanted for a price, sellers of body makeovers would encourage customers to line up. Each of these wealthy, long-lived people could have an enormous ecological footprint for perhaps centuries.

What Are Our Options? Time to Make Decisions

There are arguments over how much we should regulate genetic engineering research and development.

In the 1990s, a backlash developed against the increasing use of genetically modified food plants and animals. Some protesters are strongly against this new technology for a variety of reasons. Others advocate slowing down and taking a closer look at the short- and long-term advantages and disadvantages of this and other rapidly emerging genetic technologies.

Or at the very least, they say, we should require that all genetically modified crops and animal products and foods containing such components be clearly labeled as such. This would give consumers a more informed choice, as do the food labels that now require listing ingredients and nutritional information. Makers of genetically modified products strongly oppose such labeling (as food manufacturers opposed the other types of food labeling now in use) because they fear it would hurt sales.

Supporters of genetic engineering wonder why there is so much concern. After all, we have been genetically modifying plants and animals for centuries. Now we have a faster, better, and perhaps cheaper way to do it, so why not use it?

But proponents of more careful control of genetic engineering point out that most new technologies have had unintended harmful consequences (Figure 3-4, p. 38). The ecological lesson is that whenever we intervene in nature we must pause and ask, "What happens next?" This is why many analysts are cautious about rushing into genetic engineering and other forms of biotechnology without more careful evaluation of possible unintended consequences.

X *HOW WOULD YOU VOTE?* Should there be stricter government control over the development and use of genetic engineering technology? Cast your vote online at http://biology.brookscole.com/miller14.

How Did We Become Such a Powerful Species So Quickly? Brain and Thumb Power

We have thrived as a species mostly because of our complex brains and strong opposable thumbs.

Like other species, we have survived and thrived so far because we have certain adaptive traits. What are they?

First, look at the traits we do not have. We lack exceptional strength, speed, and agility. We do not have weapons such as claws or fangs, and we lack a protective shell or body armor.

Our senses are unremarkable. We see only visible light—a tiny fraction of the spectrum of electromagnetic radiation that bathes the earth. We cannot see

infrared radiation, as a rattlesnake can, or the ultraviolet light that guides some insects to their favorite flowers.

We cannot see as well or as far as an eagle or see well in the night like some owls and other nocturnal creatures. We cannot hear the high-pitched sounds that help bats maneuver in the dark. Our ears cannot pick up low-pitched sounds that are the songs of whales as they glide through the world's oceans. We cannot smell as keenly as a dog or a wolf. By such measures, our physical and sensory powers are pitiful.

Yet we have survived and flourished within less than a twitch of the earth's 3.7-billion year evolutionary history. Analysts attribute our success to two evolutionary adaptations: a *complex brain* and *strong opposable thumbs* that allow us to grip and use tools better than the few other animals that have thumbs. This has enabled us to develop technologies that extend our limited senses, weapons, and protective devices.

As a newly evolved infant species, we have quickly developed many powerful technologies to take over much of the earth's life-support systems to meet our basic needs and rapidly growing wants. We named ourselves *Homo sapiens sapiens*—the doubly wise species. If we keep degrading the life-support system for us and other species, some say we should be called *Homo ignoramus*—the unwise species. During this century we will probably learn which of these names is appropriate.

The *good news* is that we can change our ways. We can learn more about how to work with nature by understanding and copying the ways it has sustained itself for several billion years despite major changes in environmental conditions.

In the earth's ballet of life that has been playing on the global stage for about 3.7 billion years, life and death are interconnected. Some species appear and some disappear, but the show goes on. We only recently became members of this evolutionary ballet. Will we temporarily interrupt the show, take out some of the dancers, and get kicked off the stage? Or will we learn the rules of the ballet and have a long run? We live in interesting and challenging times.

All we have yet discovered is but a trifle in comparison with what lies hid in the great treasury of nature.

ANTONI VAN LEEUWENHOEK

CRITICAL THINKING

1. **(a)** How would you respond to someone who tells you that he or she does not believe in biological evolution because it is "just a theory"? **(b)** How would you respond to a statement that we should not worry about air pollution because through natural selection the human species will develop lungs that can detoxify pollutants?

2. How would you respond to someone who says that because extinction is a natural process we should not worry about the loss of biodiversity?

3. Describe the major differences between the ecological niches of humans and cockroaches. Are these two species in competition? If so, how do they manage to coexist?

4. Explain why you are for or against each of the following: **(a)** requiring labels indicating the use of genetically engineered components in any food item, **(b)** using genetic engineering to develop "superior" human beings, and **(c)** using genetic engineering to eliminate aging and death.

5. Suppose we could spray something in the air that would permanently give every person on earth unconditional love for all other people and unconditional love for nature. Explain why you would favor or oppose doing this. What professions, businesses, and educational subjects would this eliminate? How would this affect the entertainment, sports, and advertising businesses? How would it affect politics and environmental professions?

6. Congratulations! You are in charge of the future evolution on the earth. What are the three most important things you would do?

PROJECTS

1. An important adaptation of humans is a strong opposable thumb, which allows us to grip and manipulate things with our hands. As a demonstration of the importance of this trait, fold each of your thumbs into the palm of its hand and then tape them securely in that position for an entire day. After the demonstration, make a list of the things you could not do without the use of your thumbs.

2. Use the library or the Internet to find out what controls now exist on genetically engineered organisms in the United States, or the country where you live, and how well such controls are enforced.

3. Use the library or the Internet to find out bibliographic information about *Charles Darwin* and *Antoni van Leeuwenhoek*, whose quotes appear at the beginning and end of this chapter.

4. Make a concept map of this chapter's major ideas, using the section heads, subheads, and key terms (in boldface). Look on the website for this book for information about making concept maps.

LEARNING ONLINE

The website for this book contains study aids and many ideas for further reading and research. They include a chapter summary, review questions for the entire chapter, flash cards for key terms and concepts, a multiple-choice practice quiz, interesting Internet sites, references, and a guide for accessing thousands of InfoTrac® College Edition articles. Log on to

http://biology.brookscole.com/miller14

Then click on the Chapter-by-Chapter area, choose Chapter 5, and select a learning resource.

CASE STUDY

Blowing in the Wind: A Story of Connections

Wind, a vital part of the planet's circulatory system, connects most life on the earth. Without wind, the tropics would be unbearably hot and most of the rest of the planet would freeze.

Winds also transport nutrients from one place to another. Dust rich in phosphates and iron blows across the Atlantic from the Sahara Desert in Africa (Figure 6-1). This helps build up agricultural soils in the Bahamas and supplies nutrients for plants in the upper canopy of rain forests in Brazil. Iron-rich dust blowing from China's Gobi Desert falls into the Pacific Ocean between Hawaii and Alaska. This input of iron stimulates the growth of phytoplankton, the minute producers that support ocean food webs. This is the *good news.*

The *bad news* is that wind also transports harmful viruses, bacteria, fungi, and particles of long-lived pesticides and toxic metals. Particles of reddish-brown soil and pesticides banned in the United States are blown from Africa's deserts and eroding farmlands into the sky over Florida. This makes it difficult for the state to meet federal air pollution standards during summer months.

More *bad news.* Some types of fungi in this dust may play a role in degrading or killing coral reefs in the Florida Keys and the Caribbean. Scientists are currently studying possible links between contaminated African dust and a sharp rise in rates of asthma in the Caribbean since 1973.

Particles of iron-rich dust from Africa that enhance the productivity of algae have also been linked to outbreaks of toxic algal blooms—referred to as *red tides*—in Florida's coastal waters. People who eat shellfish contaminated by a toxin produced in red tides can become paralyzed or even die. Europe and the Middle East also receive contaminated African dust.

Pollution and dust from rapidly industrializing China and central Asia blow across the Pacific Ocean and degrade air quality over the western United States. In 2001, climate scientists reported that a huge dust storm of soil particles blown from northern China had blanketed areas from Canada to Arizona with a layer of dust. Studies show that Asian pollution contributes as much as 10% to West Coast smog, a threat expected to increase as China industrializes.

There is also *mixed news.* Particles from volcanic eruptions ride the winds, circle the globe, and change the earth's temperature for a while. Emissions from the 1991 eruption of Mount Pinatubo in the Philippines cooled the earth slightly for 3 years, temporarily masking signs of global warming. And volcanic ash, like the blowing desert dust, adds valuable trace minerals to the soil where it settles.

The lesson, once again, is that there is *no away* because *everything is connected.* Wind acts as part of the planet's circulatory system for heat, moisture, plant nutrients, and long-lived pollutants we put into the air. Movement of soil particles from one place to another by wind and water is a natural phenomenon. But when we disturb the soil and leave it unprotected, we hasten and intensify this process.

Wind is also an important factor in climate through its influence on global air circulation patterns. Climate, in turn, is crucial for determining what kinds of plant and animal life are found in the major biomes of the biosphere, as discussed in this chapter.

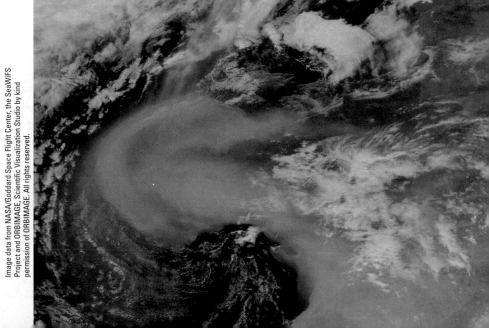

Figure 6-1 Some of the dust shown here blowing from Africa's Sahara Desert can end up as soil nutrients in Amazonian rain forests and particles of toxic air pollutants in Florida and the Caribbean.

To do science is to search for repeated patterns, not simply to accumulate facts, and to do the science of geographical ecology is to search for patterns of plant and animal life that can be put on a map.

ROBERT H. MACARTHUR

This chapter addresses the following broad questions about geographic patterns of ecology:

- What key factors determine the earth's weather?
- What key factors determine the earth's climate?
- How does climate determine where the earth's major biomes are found?
- What are the major types of desert biomes, and how do human activities affect them?
- What are the major types of grassland biomes, and how do human activities affect them?
- What are the major types of forest biomes, and how do human activities affect them?
- Why are mountain and arctic biomes important, and how do human activities affect them?

6-1 WEATHER: A BRIEF INTRODUCTION

What Is Weather? When Air Masses Meet

Weather is the result of the atmospheric conditions in a particular area over short time periods and is produced mostly by interacting air masses.

Weather is an area's short-term atmospheric conditions—typically over hours or days. Examples of atmospheric conditions are temperature, pressure, moisture content, precipitation, sunshine, cloud cover, and wind direction and speed.

Meteorologists use equipment on weather balloons, aircraft, ships, and satellites, as well as radar and stationary sensors, to obtain data on weather variables. They feed the data into computer models to draw weather maps. Other computer models project the weather for the next several days by calculating the probabilities that air masses, winds, and other factors will move and change in certain ways.

Much of the weather you experience is the result of interactions between the leading edges or fronts of moving masses of warm and cold air. Weather changes as one air mass replaces or meets another. The most dramatic changes in weather occur along a **front,** the boundary between two air masses with different temperatures and densities.

A **warm front** is the boundary between an advancing warm air mass and the cooler one it is replacing (Figure 6-2, top). Because warm air is less dense (weighs less per unit of volume) than cool air, an advancing warm front rises up over a mass of cool air. As the warm front rises, its moisture begins condensing into droplets

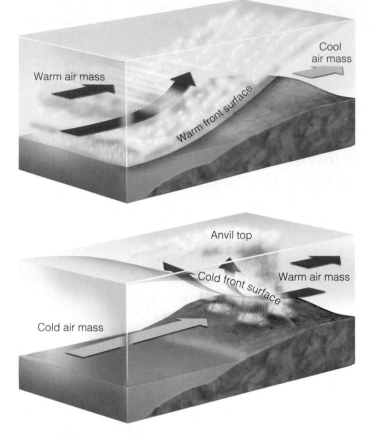

Figure 6-2 A *warm front* (top) occurs when an advancing mass of warm air meets and rises up over a retreating mass of denser cool air. A *cold front* (bottom) is the boundary formed when a mass of cold air wedges beneath a retreating mass of less dense warm air.

to form layers of clouds at different altitudes. Gradually the clouds thicken, descend to a lower altitude, and often release their moisture as rainfall. A moist warm front can bring days of cloudy skies and drizzle.

A **cold front** (Figure 6-2, bottom) is the leading edge of an advancing mass of cold air. Because cold air is denser than warm air, an advancing cold front stays close to the ground and wedges underneath less dense warmer air. An approaching cold front produces rapidly moving, towering clouds called *thunderheads.*

As a cold front passes through, we often experience high surface winds and thunderstorms. After the front passes through, we usually have cooler temperatures and a clear sky.

Further up near the top of the troposphere we find hurricane-force winds circling the earth, These powerful winds, called *jet streams,* follow rising and falling paths that have a strong influence on weather patterns.

What Are Highs and Lows? Pressure Changes

Weather is affected by up and down movements of masses of air with high and low atmospheric pressure.

Weather is also affected by changes in atmospheric pressure. *Air pressure* results from zillions of tiny molecules of gases (mostly nitrogen and oxygen) in the atmosphere zipping around at incredible speeds and hitting and bouncing off anything they encounter.

Atmospheric pressure is greater near the earth's surface because the molecules in the atmosphere are squeezed together under the weight of the air above. An air mass with high pressure, called a **high,** contains cool, dense air that descends toward the earth's surface and becomes warmer. Fair weather follows as long as the high-pressure air mass remains over an area.

In contrast, a low-pressure air mass, called a **low,** produces cloudy and sometimes stormy weather. Because of its low pressure and low density, the center of a low rises, and its warm air expands and cools. When the temperature drops below a certain level where condensation takes place, called the *dew point,* moisture in the air condenses and forms clouds. If the droplets in the clouds coalesce into large and heavy drops, precipitation occurs. Recall that the condensation of water vapor into water drops usually requires that the air contain suspended tiny particles of material such as dust, smoke, sea salts, or volcanic ash. These so-called condensation nuclei provide surfaces on which the droplets of water can form and coalesce. Now you know how rain forms.

What Are Tornadoes and Tropical Cyclones? Weather Godzillas

Tornadoes and tropical storms are weather extremes that can cause lots of damage but can sometimes have beneficial ecological effects.

Sometimes we experience *weather extremes.* Two examples are violent storms called *tornadoes* (which form over land) and *tropical cyclones* (which form over warm ocean waters and sometimes pass over coastal land).

Tornadoes or *twisters* are swirling funnel shaped clouds that form over land. They can destroy houses and cause other serious damage in areas when they touch down on the earth's surface. The United States is the world's most tornado-prone country, followed by Australia.

Tornadoes in the plains of the Midwest usually occur when a large, dry cold air front moving southward from Canada runs into a large mass of humid air moving northward from the Gulf of Mexico. Most tornadoes occur in the spring when fronts of cold air from the north penetrate deeply into the midwestern plains.

As the large warm-air mass moves rapidly over the more dense mass of cold air it rises rapidly and forms strong vertical convection currents that suck air upward, as shown in Figure 6-3. Trace the flows in this

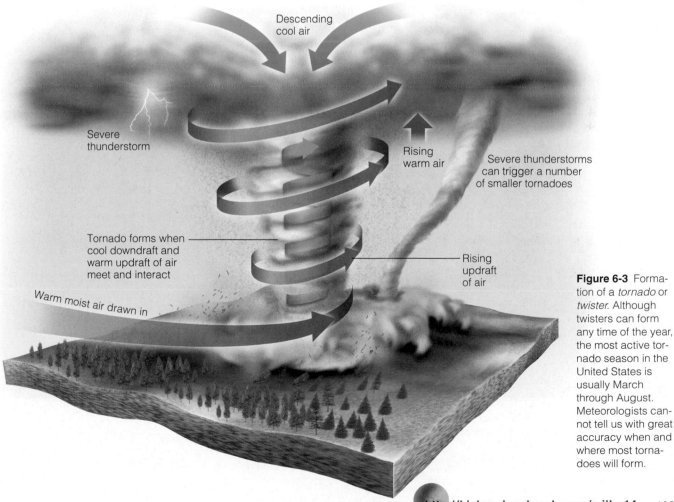

Descending cool air

Severe thunderstorm

Rising warm air

Severe thunderstorms can trigger a number of smaller tornadoes

Tornado forms when cool downdraft and warm updraft of air meet and interact

Rising updraft of air

Warm moist air drawn in

Figure 6-3 Formation of a *tornado* or *twister.* Although twisters can form any time of the year, the most active tornado season in the United States is usually March through August. Meteorologists cannot tell us with great accuracy when and where most tornadoes will form.

figure. Scientists hypothesize that the rising vortex of air starts spinning because the air near the ground in the funnel is moving slower than the air above. This rolls or spins the air ahead of the advancing front in a vertically rising air mass or vortex.

Large and dangerous storms called tropical cyclones are spawned by the formation of low-pressure cells of air over warm tropical seas. Figure 6-4 shows the formation and structure of a *tropical cyclone*. Trace the flows in this figure. *Hurricanes* are tropical cyclones that form in the Atlantic Ocean; those forming in the Pacific Ocean usually are called *typhoons*. Tropical cyclones take a long time to form and gain strength. As a result, meteorologists can track their paths and wind speeds and warn people in areas likely to be hit by these violent storms.

Hurricanes and typhoons can kill and injure people and damage property and agricultural production. But sometimes the long-term ecological and economic benefits of a tropical cyclone can exceed its short-term harmful effects.

For example, in parts of Texas along the Gulf of Mexico, coastal bays and marshes normally are closed off from freshwater and saltwater inflows. In August 1999, Hurricane Brett struck this coastal area. According to marine biologists, it flushed out excess nutrients from land runoff and dead sea grasses and rotting vegetation from the coastal bays and marshes. It also carved 12 channels through the barrier islands along the coast, allowing huge quantities of fresh seawater to flood the bays and marshes. This flushing out of the bays and marshes reduced brown tides consisting of explosive growths of algae feeding on excess nutrients. It also increased growth of sea grasses, which serve as nurseries for shrimp, crabs, and fish and food for millions of ducks wintering in Texas bays. Production of commercially important species of shellfish and fish also increased.

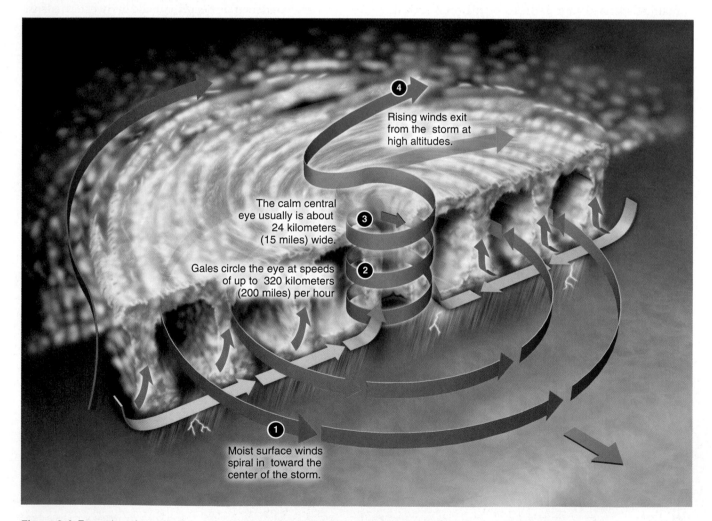

Figure 6-4 Formation of a *tropical cyclone*. Those forming in the Atlantic Ocean usually are called *hurricanes;* those forming in the Pacific Ocean usually are called *typhoons.*

6-2 CLIMATE: A BRIEF INTRODUCTION

What Is Climate? Long-Term Weather and Global Air Circulation

Climate is the average temperature and average precipitation of an area over long periods of time, which in turn are affected by global air circulation.

Climate is a region's long-term atmospheric conditions—typically over decades. *Average temperature* and *average precipitation* are the two main factors determining a region's climate and its effects on people, as shown in Figure 6-5.

Figure 6-6 (p. 106) is a generalized map of the earth's major climate zones. In what type of climate zone do you live?

The temperature and precipitation patterns that lead to different climates are caused primarily by the amount of incoming solar energy per unit area of land, air circulation over the earth's surface, and water circulation. Solar energy heats the atmosphere, evaporates water, helps create seasons, and causes air to circulate.

Four major factors determine global air circulation patterns. One is the *uneven heating of the earth's surface.* Air is heated much more at the equator, where the sun's rays strike directly throughout the year, than at the poles, where sunlight strikes at an angle and thus is spread out over a much greater area. You can observe this effect by shining a flashlight in a darkened room on the middle of a spherical object such as a basketball. These differences in the amount of incoming solar energy help explain why tropical regions near the equator are hot, polar regions are cold, and temperate regions in between generally have intermediate average temperatures.

A second factor is *seasonal changes in temperature and precipitation.* The earth's axis—an imaginary line connecting the north and south poles—is tilted. As a result, various regions are tipped toward or away from the sun as the earth makes its year-long revolution around the sun (Figure 6-7, p. 106). This creates opposite seasons in the northern and southern hemispheres.

A third factor is *rotation of the earth on its axis.* As the earth rotates, its surface turns faster beneath air masses at the equator and slower beneath those at the poles. This deflects air masses moving north and south to the west or east over different parts of the earth's surface (Figure 6-8, p. 107). The direction of air movement in these different areas sets up belts of *prevailing winds*—major surface winds that blow almost continually and distribute air and moisture over the earth's surface.

Fourth, *properties of air, water, and land* affect global air circulation. Heat from the sun evaporates ocean water and transfers heat from the oceans to the atmosphere, especially near the hot equator.

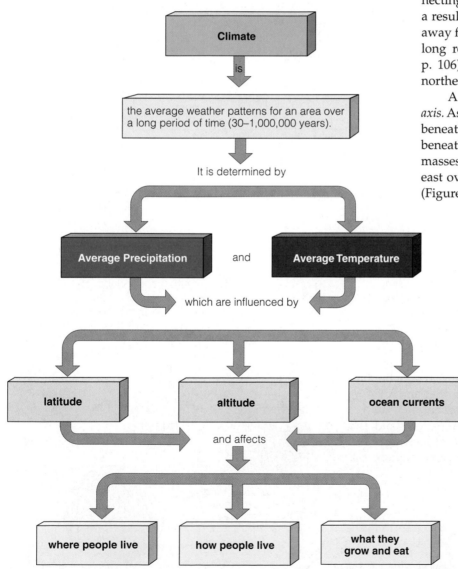

Figure 6-5 Climate and its effects. (Data from National Oceanic and Atmospheric Administration)

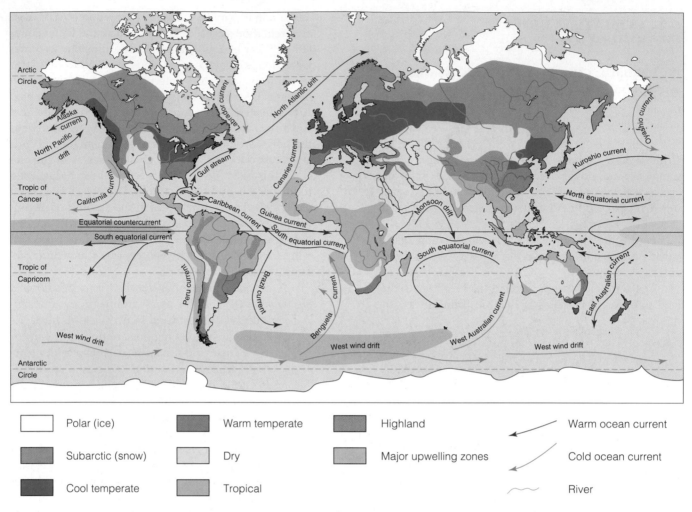

☐ Polar (ice)	▨ Warm temperate	▨ Highland
▨ Subarctic (snow)	▨ Dry	▨ Major upwelling zones
▨ Cool temperate	▨ Tropical	

Warm ocean current
Cold ocean current
River

Figure 6-6 Natural capital: generalized map of the earth's current climate zones, showing the major contributing ocean currents and drifts.

This evaporation of water creates cyclical convection cells that circulate air, heat, and moisture both vertically and from place to place in the troposphere, as shown in Figure 6-9. Trace the flows in this diagram. To understand this cycle, remember two things. *First,* hot air tends to rise, cool, and release moisture as precipitation. *Second,* cool air tends to sink, get warmer, and lose its moisture by evaporation (not precipitation).

The earth's air and water circulation patterns and its mixture of continents and oceans lead to an irregular distribution of climates and patterns of vegetation, as shown in Figure 6-10.

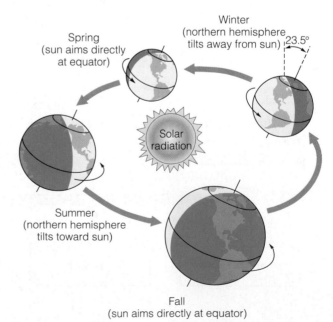

Figure 6-7 Seasons in the northern and southern hemispheres are caused by the tilt of the earth's axis. As the planet makes its annual revolution around the sun on an axis tilted about 23.5°, various regions are tipped toward or away from the sun. The resulting variations in the amount of solar energy reaching the earth create the seasons.

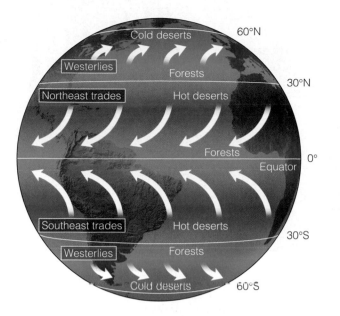

Figure 6-8 The earth's rotation deflects the movement of the air over different parts of the earth and creates global patterns of prevailing winds.

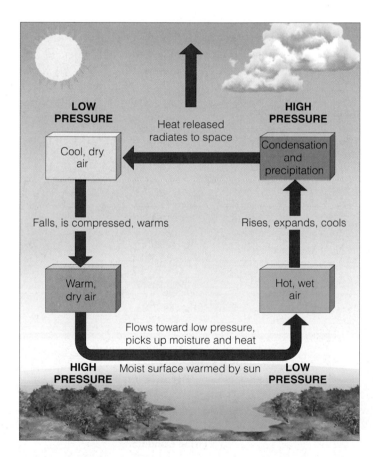

Figure 6-9 Transfer of energy by convection in the atmosphere. *Convection* occurs when matter warms, becomes less dense, and rises within its surroundings. This efficient means of heat transfer occurs in the planet's interior, oceans, and atmosphere (as shown here). Distribution of heat and water occurs in the atmosphere because vertical convection currents stir up air in the troposphere and transport heat and water from one area to another in circular convection cells called *Hadley cells.*

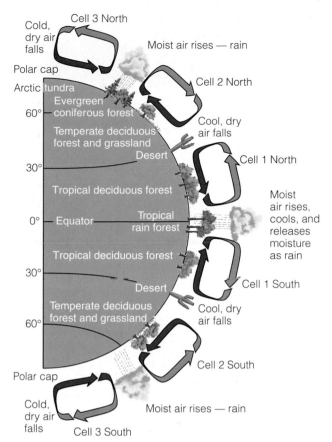

Figure 6-10 Natural capital: global air circulation and biomes. Heat and moisture are distributed over the earth's surface by vertical currents that form six large convection cells (Hadley cells) at different latitudes. The direction of airflow and the ascent and descent of air masses in these convection cells determine the earth's general climatic zones. The resulting uneven distribution of heat and moisture over the planet's surface leads to the forests, grasslands, and deserts that make up the earth's biomes.

How Do Ocean Currents and Winds Affect Regional Climates? Moving Heat and Air Around

Ocean currents and winds influence climate by redistributing heat received from the sun from one place to another.

The oceans absorb heat from the air circulation patterns just described, with the bulk of this heat absorbed near the warm tropical areas. This heat plus differences in water density create warm and cold ocean currents (Figure 6-6). These currents, driven by winds and the earth's rotation, redistribute heat received from the sun from one place to another and thus influence climate and vegetation, especially near coastal areas. They also help mix ocean waters and distribute nutrients and dissolved oxygen needed by aquatic organisms.

Winds can also affect regional climates and distribution of some forms of aquatic life. For example, wind blowing along steep western coasts of some continents pushes surface water away from the land. This outgoing surface water is replaced by an **upwelling** of cold, nutrient-rich bottom water, as shown in Figure 6-11.

Upwellings, whether far from shore or near shore, bring plant nutrients from the deeper parts of the ocean to the surface. In turn, these nutrients support large populations of phytoplankton, zooplankton, fish, and fish-eating seabirds.

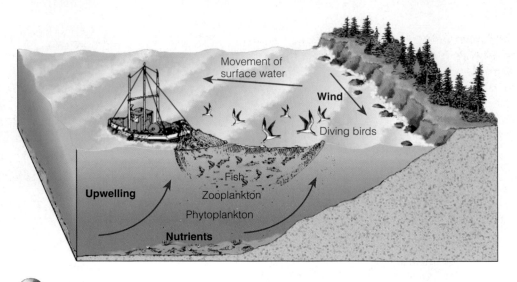

Figure 6-11 A *shore upwelling* (shown here) occurs when deep, cool, nutrient-rich waters are drawn up to replace surface water moved away from a steep coast by wind flowing along the coast toward the equator. Such areas support large populations of phytoplankton, zooplankton, fish, and fish-eating birds. *Equatorial upwellings* occur in the open sea near the equator (Figure 6-6) when northward and southward currents interact to push deep waters and their nutrients to the surface, thus greatly increasing primary productivity in such areas.

What Are El Niño and La Niña?
Changing Winds, Altered Upwellings, and Freaky Weather

El Niño occurs when a change in the direction of tropical winds warms coastal surface water, suppresses upwellings, and alters much of the earth's weather. La Niña is the reverse of this effect.

Every few years in the Pacific Ocean, normal shore upwellings (Figure 6-12, left) are affected by changes in climate patterns called the *El Niño–Southern Oscillation*, or *ENSO* (Figure 6-12, right). Trace the flows and components of the diagram in Figure 6-12.

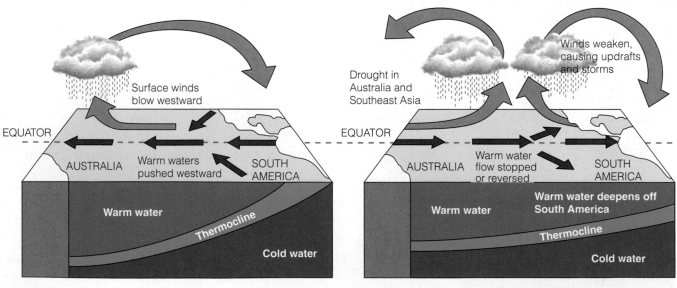

Figure 6-12 Normal trade winds blowing westward cause shore upwellings of cold, nutrient-rich bottom water in the tropical Pacific Ocean near the coast of Peru (left). A zone of gradual temperature change called the *thermocline* separates the warm and cold water. Every few years a climate shift known as the *El Niño–Southern Oscillation (ENSO)* disrupts this pattern. Westward-blowing trade winds weaken or reverse direction, which depresses the coastal upwellings and warms the surface waters off South America (right). When an ENSO lasts 12 months or longer, it severely disrupts populations of plankton, fish, and seabirds in upwelling areas and can trigger extreme weather changes over much of the globe (see Figure 6-13).

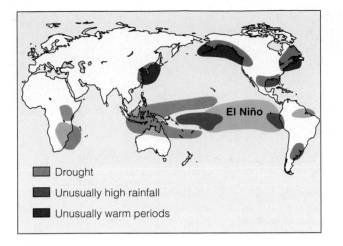

Figure 6-13 Typical global climatic effects of an El Niño–Southern Oscillation. During the 1996–98 ENSO, huge waves battered the California coast, and torrential rains caused widespread flooding and mudslides. In Peru, floods and mudslides killed hundreds of people, left about 250,000 people homeless, and ruined harvests. Drought in Brazil, Indonesia, and Australia led to massive wildfires in tinder-dry forests. India and parts of Africa also experienced severe drought. A catastrophic ice storm hit Canada and the northeastern United States, but the southeastern United States had fewer hurricanes. (Data from United Nations Food and Agriculture Organization)

In an ENSO, often called *El Niño*, prevailing tropical trade winds blowing westward weaken or reverse direction. This warms up surface water along the South and North American coasts, which suppresses the normal upwellings of cold, nutrient-rich water. The decrease in nutrients reduces primary productivity and causes a sharp decline in the populations of some fish species.

A strong ENSO can trigger extreme weather changes over at least two-thirds of the globe (Figure 6-13)—especially in lands along the Pacific and Indian Oceans—and distorts the fast-moving jet stream that flows high above North America.

La Niña, the reverse of El Niño, cools some coastal surface waters, and brings back upwellings. Typically La Niña means more Atlantic Ocean hurricanes, colder winters in Canada and the northeastern United States, and warmer and drier winters in the southeastern and southwestern United States. It also usually leads to wetter winters in the Pacific Northwest, torrential rains in Southeast Asia, lower wheat yields in Argentina, and more wildfires in Florida.

How Do Gases in the Atmosphere Affect Climate? The Natural Greenhouse Effect

Water vapor, carbon dioxide, and other gases influence climate by warming the lower troposphere and the earth's surface.

Small amounts of certain gases play a key role in determining the earth's average temperatures and thus its climates. These gases include water vapor (H_2O), carbon dioxide (CO_2), methane (CH_4), and nitrous oxide (N_2O).

Together these gases, known as **greenhouse gases**, allow mostly visible light and some infrared radiation and ultraviolet (UV) radiation from the sun to pass through the troposphere. The earth's surface absorbs much of this solar energy. This transforms it to longer-wavelength infrared radiation, which rises into the troposphere.

Some of this infrared radiation escapes into space and some is absorbed by molecules of greenhouse gases and emitted into the troposphere in all directions as even longer wavelength infrared radiation. Some of this released energy is radiated into space and some warms the troposphere and the earth's surface. This natural warming effect of the troposphere is called the **greenhouse effect**, diagrammed in Figure 6-14 (p. 110).

The basic principle behind the natural greenhouse effect is well established. Indeed, without its current concentrations of greenhouse gases (especially water vapor, which is found in the largest concentration), the earth would be a cold and mostly lifeless planet.

Human activities such as burning fossil fuels, clearing forests, and growing crops release carbon dioxide, methane, and nitrous oxide into the troposphere. There is concern that these large inputs of greenhouse gases can enhance the earth's natural greenhouse effect and lead to *global warming*—more on this in Chapter 21. This could alter precipitation patterns, shift areas where we can grow crops, raise average sea levels, and shift areas where some types of plants and animals can live.

How Does Topography of the Earth's Surface Affect Local Climate? Creating Deserts and Warming Cities

Mountains and cities affect local and regional climates.

Various topographic features of the earth's surface can create local and regional climatic conditions that differ from the general climate of a region. For example, mountains interrupt the flow of prevailing surface winds and the movement of storms. When moist air blowing inland from an ocean reaches a mountain range, it cools as it is forced to rise and expand. This causes the air to lose most of its moisture as rain and snow on the windward (wind-facing) slopes.

As the drier air mass flows down the leeward (away from the wind) slopes, it draws moisture out of the plants and soil over which it passes. The lower precipitation and the resulting semiarid or arid conditions

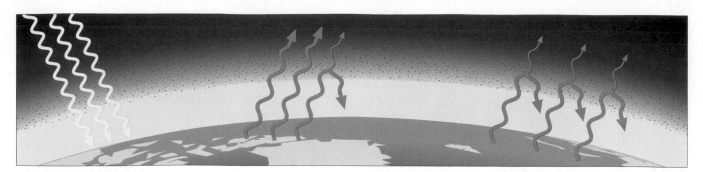

(a) Rays of sunlight penetrate the lower atmosphere and warm the earth's surface.

(b) The earth's surface absorbs much of the incoming solar radiation and degrades it to longer-wavelength infrared (IR) radiation, which rises into the lower atmosphere. Some of this IR radiation escapes into space as heat and some is absorbed by molecules of greenhouse gases and emitted as even longer wavelength IR radiation, which warms the lower atmosphere.

(c) As concentrations of greenhouse gases rise, their molecules absorb and emit more infrared radiation, which adds more heat to the lower atmosphere.

Figure 6-14 Natural capital: the *natural greenhouse effect.* Without the atmospheric warming provided by this natural effect, the earth would be a cold and mostly lifeless planet. According to the widely accepted greenhouse theory, when concentrations of greenhouse gases in the atmosphere rise, the average temperature of the troposphere rises. (Modified by permission from Cecie Starr, *Biology: Concepts and Applications,* 4th ed., Brooks/Cole [Wadsworth] 2000)

on the leeward side of high mountains are called the **rain shadow effect,** (Figure 6-15). This is one way some deserts form.

Cities also create distinct microclimates. Bricks, concrete, asphalt, and other building materials absorb and hold heat, and buildings block wind flow. Motor vehicles and the climate control systems of buildings release large quantities of heat and pollutants. As a result, cities tend to have more haze and smog, higher temperatures, and lower wind speeds than the surrounding countryside.

Now that you have the basics of climate we can examine how it applies to ecology.

6-3 BIOMES: CLIMATE AND LIFE ON LAND

Why Do Different Organisms Live in Different Places? Think Climate

Different climates lead to different communities of organisms, especially vegetation.

Why is one area of the earth's land surface a desert, another a grassland, and another a forest? Why do different types of deserts, grasslands, and forests exist? The general answer to these questions is differences in *climate* (Figure 6-6), caused mostly by differences in

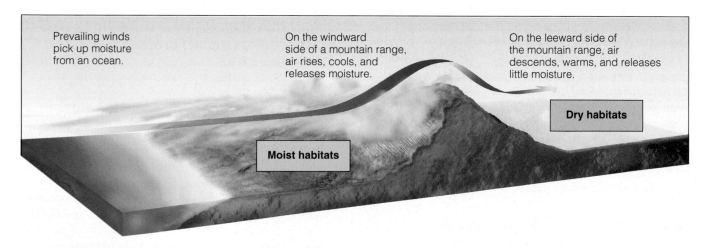

Prevailing winds pick up moisture from an ocean.

On the windward side of a mountain range, air rises, cools, and releases moisture.

On the leeward side of the mountain range, air descends, warms, and releases little moisture.

Dry habitats

Moist habitats

Figure 6-15 The *rain shadow effect* is a reduction of rainfall on the side of high mountains facing away from prevailing surface winds. It occurs when warm, moist air in prevailing onshore winds loses most of its moisture as rain and snow on the windward (wind-facing) slopes of a mountain range. This leads to semiarid and arid conditions on the leeward side of the mountain range and the land beyond. The Mojave Desert (east of the Sierra Nevada in California) and Asia's Gobi Desert are produced by this effect.

average temperature and precipitation due to global air and water circulation (Figure 6-10).

Figure 6-16 shows how scientists have divided the world into 12 major biomes. They are terrestrial regions with characteristic types of natural ecological communities adapted to the climate of each region. Study Figure 6-16 carefully and identify the type of biome you live in.

By comparing Figure 6-16 with Figure 6-6, you can see how the world's major biomes vary with climate. Make this comparison for the area where you live. Figure 4-10 (p. 62) shows major biomes in the United States as one moves through different climates along the 39th parallel.

Average annual precipitation and temperature (as well as soil type, Figure 4-27, p. 75) are the most important factors in producing tropical, temperate, or polar deserts, grasslands, and forests (Figure 6-17, p. 112).

On maps such as the one in Figure 6-16, biomes are presented as having boundaries and being covered with the same general type of vegetation. In reality, *biomes are not uniform.* They consist of a *mosaic of patches,* with somewhat different biological communities but with similarities unique to the biome. These patches occur mostly because the resources plants and animals need are not uniformly distributed. Go to a natural area in or near where you live and see if you can find patches with different vegetation.

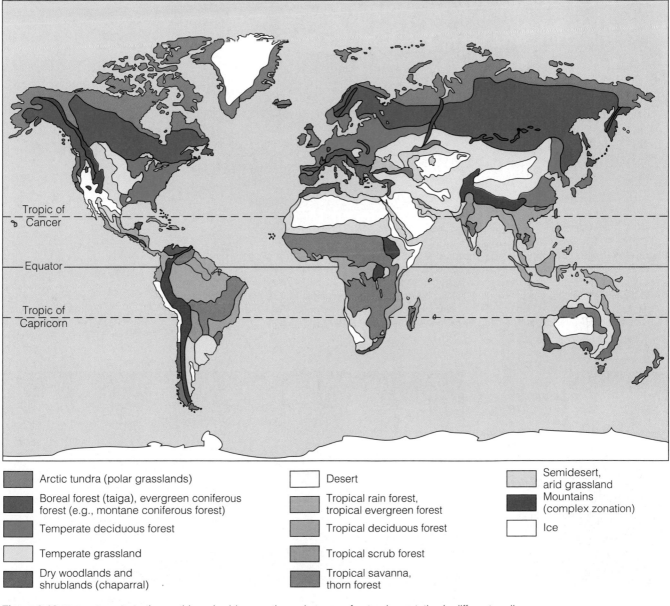

Arctic tundra (polar grasslands)

Boreal forest (taiga), evergreen coniferous forest (e.g., montane coniferous forest)

Temperate deciduous forest

Temperate grassland

Dry woodlands and shrublands (chaparral)

Desert

Tropical rain forest, tropical evergreen forest

Tropical deciduous forest

Tropical scrub forest

Tropical savanna, thorn forest

Semidesert, arid grassland

Mountains (complex zonation)

Ice

Figure 6-16 Natural capital: the earth's major *biomes*—the main types of natural vegetation in different undisturbed land areas—result primarily from differences in climate. Each biome contains many ecosystems whose communities have adapted to differences in climate, soil, and other environmental factors. In reality, people have removed or altered much of this natural vegetation in some areas for farming, livestock grazing, lumber and fuelwood, mining, and construction.

Figure 6-17 Natural capital: average precipitation and average temperature, acting together as limiting factors over a period of 30 or more years, determine the type of desert, grassland, or forest biome in a particular area. Although the actual situation is much more complex, this simplified diagram explains how climate determines the types and amounts of natural vegetation found in an area left undisturbed by human activities. (Used by permission of Macmillan Publishing Company. From Derek Elsom, *The Earth,* 1992. Copyright © 1992 by Marshall Editions Developments Limited. New York: Macmillan. Used by permission.)

Figure 6-18 Generalized effects of altitude (left) and latitude (right) on climate and biomes. Parallel changes in vegetation type occur when we travel from the equator to the poles or from lowlands to mountaintops. This generalized diagram shows only on of many possible sequences.

Figure 6-18 (facing page) shows how climate and vegetation vary with **latitude** (distance from the equator) and **altitude** (elevation above sea level). If you climb a tall mountain from its base to its summit, you can observe changes in plant life similar to those you would encounter in traveling from the equator to the earth's poles.

6-4 DESERT BIOMES

What Are the Major Types of Deserts? Hot, Medium, and Cold

Deserts have little precipitation and little vegetation and are found in tropical, temperate, and polar regions.

A **desert** is an area where evaporation exceeds precipitation. Annual precipitation is low and often scattered unevenly throughout the year. Deserts have sparse, widely spaced, mostly low vegetation.

Deserts cover about 30% of the earth's land surface and are found mostly in tropical and subtropical regions (Figure 6-16). The largest deserts are found in the interiors of continents, far from moist sea air and moisture-bearing winds. Other, more local deserts form on the downwind sides of mountain ranges because of the rain shadow effect (Figure 6-15).

During the day the baking sun warms the ground in the desert. At night, however, most of the heat stored in the ground radiates quickly into the atmosphere. This occurs because desert soils have little vegetation and moisture to help store the heat and the skies are usually clear. This explains why in a desert you may roast during the day but shiver at night.

A combination of low rainfall and different average temperatures creates tropical, temperate, and cold

deserts (Figures 6-17 and 6-19). Take a close look at the graphs in Figure 6-19.

Tropical deserts (Figure 6-19, left) are hot and dry most of the year. They have few plants and a hard, windblown surface strewn with rocks and some sand. They are the deserts we often see in movies.

In *temperate deserts,* daytime temperatures are high in summer and low in winter and there is more precipitation than in tropical deserts (Figure 6-19, center). The sparse vegetation consists mostly of widely dispersed, drought-resistant shrubs and cacti or other succulents adapted to the lack of water and temperature variations, as shown in Figure 6-20 (p. 114). Trace how nutrients and energy flow through the ecosystem in this diagram. In *cold deserts,* winters are cold, summers are warm or hot, and precipitation is low (Figure 6-19, right).

In the semiarid zones between deserts and grasslands, we find *semidesert.* This biome is dominated by thorn trees and shrubs adapted to long dry spells followed by brief, sometimes heavy rains.

How Do Desert Plants and Animals Survive? Stay Cool and Get Water Any Way You Can

Desert plants and animals have a number of strategies for staying cool and getting enough water to survive in hot and dry climates.

Adaptations for survival in the desert have two themes. One is *beat the heat* and the other is *every drop of water counts.* Desert plants exposed to the sunlight must conserve enough water for survival and lose enough heat so they do not overheat and die.

Desert plants have evolved a number of strategies for doing this. During long hot and dry spells plants

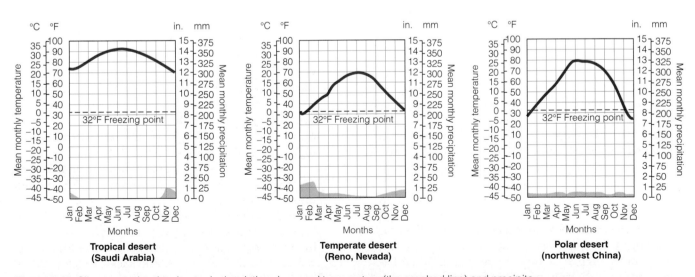

Figure 6-19 Climate graphs showing typical variations in annual temperature (the graphed line) and precipitation (the shaded area on the graph) in tropical, temperate, and polar (cold) deserts.

Figure 6-20 Natural capital: some components and interactions in a *temperate desert ecosystem*. When these organisms die, decomposers break down their organic matter into minerals that plants use. Colored arrows indicate transfers of matter and energy between producers, primary consumers (herbivores), secondary, or higher-level, consumers (carnivores), and decomposers. Organisms are not drawn to scale.

Red-tailed hawk

Gambel's quail

Yucca

Jack rabbit

Collared lizard

Agave

Prickly pear cactus

Roadrunner

Darkling beetle

Bacteria

Diamondback rattlesnake

Fungi

Kangaroo rat

| Producer to primary consumer | Primary to secondary consumer | Secondary to higher-level consumer | All producers and consumers to decomposers |

such as mesquite and creosote drop their leaves to survive in a dormant state. *Succulent (fleshy) plants,* such as the saguaro ("sah-WAH-ro") cactus, have three adaptations. They have no leaves, which can lose water by evapotranspiration. They store water and synthesize food in their expandable, fleshy tissue. And they reduce water loss by opening their pores (stomata) to take up carbon dioxide (CO_2) only at night. What a great evolutionary solution to a difficult problem.

Some desert plants use deep roots to tap into groundwater. Others such as prickly pear (Figure 6-20) and saguaro cacti use widely spread, shallow roots to collect water after brief showers and store it in their spongy tissue.

Evergreen plants conserve water by having wax-coated leaves that minimize evapotranspiration. Others, such as annual wildflowers and grasses, store much of their biomass in seeds that remain inactive, sometimes for years, until they receive enough water to germinate. Shortly after a rain these seeds germinate, grow, carpet some deserts with a dazzling array of colorful flowers, produce new seed, and die, all in only a few weeks.

Most desert animals are small. Some beat the heat by hiding in cool burrows or rocky crevices by day and coming out at night or in the early morning. Others become dormant during periods of extreme heat or drought.

Some desert animals have physical adaptations for conserving water. Insects and reptiles have thick outer coverings to minimize water loss through evaporation, and their wastes are dry feces and a dried concentrate of urine. Many spiders and insects get their water from dew or from the food they eat. Arabian oryxes survive by licking the dew that accumulates at night on rocks and on one another's hair.

Figure 6-21 shows major human impacts on deserts. What are the direct or indirect effects of your lifestyle on desert biomes?

Deserts take a long time to recover from disturbances because of their slow plant growth, low species diversity, slow nutrient cycling (because of little bacterial activity in their soils), and lack of water. Desert vegetation destroyed by livestock overgrazing and off-road vehicles may take decades to grow back.

Large desert cities

Soil destruction by off-road vehicles and urban development

Soil salinization from irrigation

Depletion of underground water supplies

Land disturbance and pollution from mineral extraction

Storage of toxic and radioactive wastes

Large arrays of solar cells and solar collectors used to produce electricity

Figure 6-21 Natural capital degradation: major human impacts on the world's deserts.

6-5 GRASSLAND, TUNDRA, AND CHAPARRAL BIOMES

What Are the Major Types of Grasslands? Hot, Mild, and Cold

Grasslands have enough precipitation to support grasses but not enough to support large stands of trees and are found in tropical, temperate, and polar regions.

Grasslands, or **prairies,** are regions with enough average annual precipitation to support grasses (and in some areas, a few trees). Most grasslands are found in the interiors of continents (Figure 6-16).

Grasslands persist because of a combination of seasonal drought, grazing by large herbivores, and occasional fires—all of which keep large numbers of shrubs and trees from growing. The three main types of grasslands—tropical, temperate, and polar (tundra)—result from combinations of low average precipitation and various average temperatures (Figures 6-17 and 6-22). Take a close look at the graphs in Figure 6-22.

What Are Tropical Grasslands and Savannas? Hot with On-and-Off Rain

Savannas are hot places that have rain, except during dry seasons, and enormous herds of hoofed animals occupying different ecological niches.

One type of tropical grassland, called a *savanna,* usually has warm temperatures year-round, two prolonged dry seasons, and abundant rain the rest of the year (Figure 6-22, left). African tropical savannas contain enormous herds of *grazing* (grass- and herb-eating) and *browsing* (twig- and leaf-nibbling) hoofed animals that feed on a wide variety of savanna plants, as shown in Figure 6-23 (p. 116). In this diagram, note the number of different niches that allow these animals to coexist.

As part of their niches, these and other large herbivores have evolved specialized eating habits that minimize competition between species for vegetation. For example, giraffes eat leaves and shoots from the tops of trees, elephants eat leaves and branches further down, Thompson's gazelles and wildebeests prefer short grass, and zebras graze on longer grass and stems.

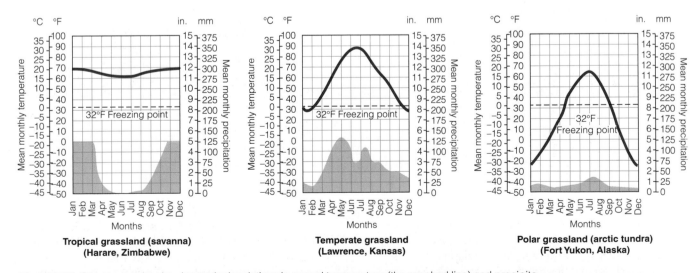

Figure 6-22 Climate graphs showing typical variations in annual temperature (the graphed line) and precipitation (the shaded area on the graph) in tropical, temperate, and polar (arctic tundra) grasslands.

Figure 6-23 Natural capital: some of the grazing animals found in different parts of the African savanna. These species share vegetation resources by occupying different feeding niches.

Many large savanna animal species are killed for their economically valuable coats and parts (tigers), tusks (rhinoceroses), and ivory tusks (elephants).

Humans have attempted to raise cattle in some savanna areas. Often these herds have helped convert savannas to deserts. One reason is that cattle require a lot more water than native herbivores (Figure 6-23) that are better adapted to the savanna climate. To get enough water the cattle must move back and forth between water holes. This frequent movement reduces their meat yield, tramples vegetation, and compacts large areas of soil.

Another problem is that cattle have moist droppings. As the droppings dry, they heat up in the sun, kill the grass underneath, and form patches of a nearly impenetrable soil covering (sometimes called *fecal pavement*). In contrast, native herbivores, such as antelope, produce dry fecal pellets that are readily decomposed with the nutrients returned to the soil. This illustrates a basic ecological rule: *Do not try to raise a plant or animal in an environment to which it is not adapted.*

What Are Temperate Grasslands? Good Weather and Soils for Crops and Livestock

Temperate grasslands with cold winters and hot and dry summers have deep and fertile soils that make them widely used for growing crops and grazing cattle.

Temperate grasslands cover vast expanses of plains and gently rolling hills in the interiors of North and South America, Europe, and Asia (Figure 6-16). In these grasslands, winters are bitterly cold, summers are hot and dry, and annual precipitation is fairly sparse and falls unevenly through the year (Figure 6-22, center).

Because the aboveground parts of most of the grasses die and decompose each year, organic matter accumulates to produce a deep, fertile soil (Figure 4-27, top right, p. 75). This soil is held in place by a thick network of intertwined roots of drought-tolerant grasses unless the topsoil is plowed up and allowed to blow away by prolonged exposure to high winds found in these biomes. The natural grasses are also adapted to fires ignited by lightning or set deliberately. The fires burn the plant above the ground but do not harm the roots, from which new life can spring.

Types of temperate grasslands in North America are the *tall-grass prairies* (Figure 6-24) and *short-grass prairies* of the midwestern and western United States and Canada. Trace the nutrient and energy flows and transfers in Figure 6-24. Here winds blow almost continuously, and evaporation is rapid, often leading to fires in the summer and fall.

Golden eagle
Pronghorn antelope
Coyote
Grasshopper sparrow
Grasshopper
Blue stem grass
Prairie dog
Bacteria
Fungi
Prairie coneflower

| ⇨ Producer to primary consumer | ⇨ Primary to secondary consumer | ⇨ Secondary to higher-level consumer | ⇨ All producers and consumers to decomposers |

Figure 6-24 Natural capital: some components and interactions in *a temperate tall-grass prairie ecosystem* in North America. When these organisms die, decomposers break down their organic matter into minerals that plants use. Colored arrows indicate transfers of matter and energy between producers, primary consumers (herbivores), secondary, or higher-level, consumers (carnivores), and decomposers. Organisms are not drawn to scale.

Figure 6-25 Natural capital degradation: replacement of a temperate grassland with a monoculture crop in California. When the tangled root network of natural grasses is removed, the fertile topsoil is subject to severe wind erosion unless it is covered with some type of vegetation.

National Archives/EPA Documerica

Many of the world's natural temperate grasslands have disappeared because they are great places to grow crops (Figure 6-25) and graze cattle. They are often flat, easy to plow, and have fertile, deep soils. However, plowing breaks up the soil and leaves it vulnerable to erosion by wind and water, and overgrazing can transform grasslands to semidesert or desert.

With the elimination of the bison (p. 20) and with plows that could break the dense turf, North America's long-grass prairies have become breadbaskets that produce huge quantities of grain. Only tiny protected remnants of the original prairies remain.

On the western short-grass prairie, cattle and sheep have replaced antelope and bison. Much of the sagebrush desert of the American West is the result of overgrazing of short-grass prairie.

What Are Polar Grasslands? Bitterly Cold Plains

Polar grasslands are covered with ice and snow except during a brief summer.

Polar grasslands, or *arctic tundra*, occur just south of the arctic polar ice cap (Figure 6-16). During most of the year these treeless plains are bitterly cold (Figure 6-22, right), swept by frigid winds, and covered with ice and snow. Winters are long and dark, and the scant precipitation falls mostly as snow, making this a "freezing desert."

This biome is carpeted with a thick, spongy mat of low-growing plants, primarily grasses, mosses, and dwarf woody shrubs, as shown in Figure 6-26. Trace the energy flows and nutrient transfers in this dia-

gram. Trees or tall plants cannot survive in the cold and windy tundra because they would lose too much of their heat. Most of the annual growth of the tundra's plants occurs during the 6- to 8-week summer, when sunlight shines almost around the clock.

One effect of the extreme cold is **permafrost,** a perennially frozen layer of the soil that forms not far below the surface when the water there freezes. During the brief summer the permafrost layer keeps melted snow and ice from soaking into the ground. Then the waterlogged tundra forms a large number of shallow lakes, marshes, bogs, ponds, and other seasonal wetlands. Hordes of mosquitoes, blackflies, and other insects thrive in these shallow surface pools. They feed large colonies of migratory birds (especially waterfowl) that return from the south to nest and breed in the bogs and ponds.

The arctic tundra's permanent animal residents are mostly small herbivores such as lemmings, hares, voles, and ground squirrels that burrow underground to escape the cold. Predators such as the lynx, weasel, snowy owl, and arctic fox eat them. Animals in this biome survive the intense winter cold through adaptations such as thick coats of fur (arctic wolf, arctic fox, and musk oxen), feathers (snowy owl), and living underground (arctic lemming).

Because of the cold, decomposition is slow. With low decomposer populations, the soil is poor in organic matter and in nitrates, phosphates, and other minerals. Human activities—mostly oil drilling sites, pipelines, mines, and military bases—leave scars that persist for centuries.

Another type of tundra, called *alpine tundra*, occurs above the limit of tree growth but below the per-

Figure 6-26 **Natural capital:** some components and interactions in an *arctic tundra (polar grassland) ecosystem.* When these organisms die, decomposers break down their organic matter into minerals that plants use. Colored arrows indicate transfers of matter and energy between producers, primary consumers (herbivores), secondary, or higher-level, consumers (carnivores), and decomposers. Organisms are not drawn to scale.

Long-tailed jaeger

Grizzly bear

Caribou

Mosquito

Snowy owl

Horned lark

Arctic fox

Willow ptarmigan

Dwarf willow

Lemming

Mountain cranberry

Moss campion

→ Producer to primary consumer

→ Primary to secondary consumer

→ Secondary to higher-level consumer

→ All producers and consumers to decomposers

manent snow line on high mountains (Figure 6-18, left). The vegetation is similar to that found in arctic tundra, but it gets more sunlight than arctic vegetation and has no permafrost layer.

Figure 6-27 (p. 120) lists major human impacts on grasslands. What are the direct and indirect effects of your lifestyle on grassland biomes?

What Is Chaparral? Great Place to Live but Beware of Fires and Mudslides

This temperate shrubland has a wonderful climate but is subject to fires in the fall followed by flooding and mudslides.

In many coastal regions that border on deserts, such as southern California and the Mediterranean, we find a

biome known as *temperate shrubland* or *chaparral.* Closeness to the sea provides a slightly longer winter rainy season than nearby temperate deserts have, and fogs during the spring and fall reduce evaporation.

Chaparral consists mostly of dense growths of low-growing evergreen shrubs and occasional small trees with leathery leaves that reduce evaporation. During the long, hot and dry summers, chaparral vegetation becomes very dry and highly flammable. In the fall, fires started by lightning or human activities spread with incredible swiftness.

Research reveals that chaparral is adapted to and maintained by periodic fires. Many of the shrubs store food reserves in their fire-resistant roots and have seeds that sprout only after a hot fire. With the first rain, annual grasses and wildflowers spring up and

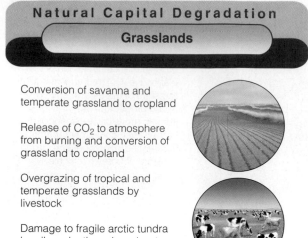

Natural Capital Degradation

Grasslands

Conversion of savanna and temperate grassland to cropland

Release of CO_2 to atmosphere from burning and conversion of grassland to cropland

Overgrazing of tropical and temperate grasslands by livestock

Damage to fragile arctic tundra by oil production, air and water pollution, and off-road vehicles

Figure 6-27 Natural capital degradation: major human impacts on the world's grasslands.

use nutrients released by the fire. New shrubs grow quickly and crowd out the grasses.

People like living in this biome because of its favorable climate. But those living in chaparral assume the high risk of losing their homes and possibly their lives to frequent fires and mudslides.

Some people in southern California learned this lesson in the fall of 2003 when fires raging through chaparral areas destroyed several thousand homes and killed 20 people. After fires often comes the hazard of flooding and mudslides; when heavy rains come, great

torrents of water pour off the unprotected burned hillsides to flood lowland areas. *Bottom line.* Chaparral has a great climate but is a risky place to live.

6-6 FOREST BIOMES

What Are the Major Types of Forests? Hot, Mild, and Cold

Forests have enough precipitation to support stands of trees and are found in tropical, temperate, and polar regions.

Undisturbed areas with moderate to high average annual precipitation tend to be covered with **forest,** which contains various species of trees and smaller forms of vegetation. The three main types of forest—tropical, temperate, and *boreal* (polar)—result from combinations of this precipitation level and various average temperatures (Figures 6-16 and 6-28). Take a close look at the graphs in Figure 6-28.

What Are Tropical Rain Forests? Centers of Biodiversity

These forests have heavy rainfall on most days and have a diversity of life forms occupying a variety of specialized niches in distinct layers.

Tropical rain forests are found near the equator, where hot, moisture-laden air rises and dumps its moisture (Figure 6-10). Trace the nutrient flows and energy transfers in Figure 6-29. These forests have a warm annual mean temperature (which varies little either daily or seasonally), high humidity, and heavy rainfall almost daily (Figure 6-28, left).

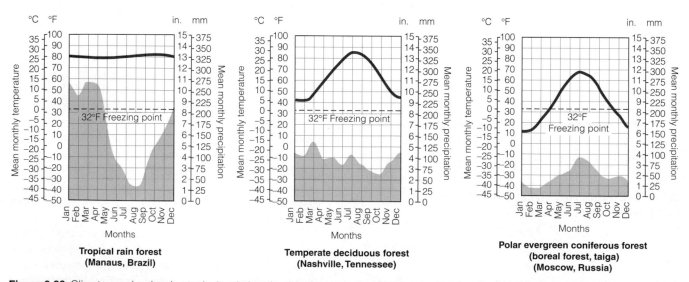

Figure 6-28 Climate graphs showing typical variations in annual temperature (the graphed line) and precipitation (the shaded area on the graph) in tropical, temperate, and polar forests.

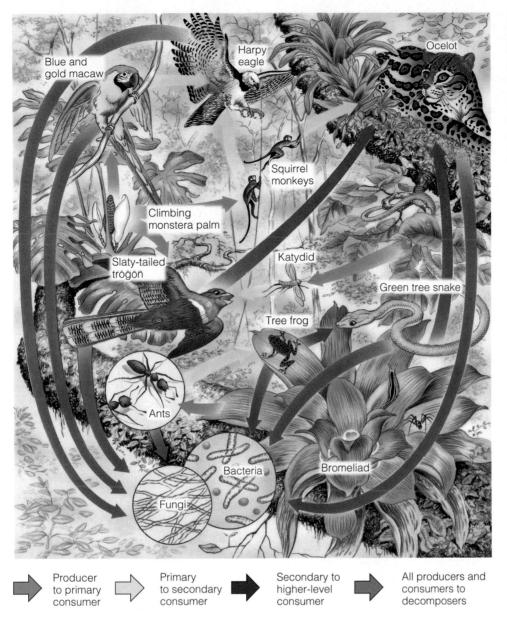

Figure 6-29 Natural capital: some components and interactions in a *tropical rain forest ecosystem*. When these organisms die, decomposers break down their organic matter into minerals that plants use. Colored arrows indicate transfers of matter and energy between producers, primary consumers (herbivores), secondary, or higher-level, consumers (carnivores), and decomposers. Organisms are not drawn to scale.

Blue and gold macaw	Harpy eagle	Ocelot
Climbing monstera palm	Squirrel monkeys	
Slaty-tailed trogon	Katydid	Green tree snake
	Tree frog	
Ants		
Fungi	Bacteria	Bromeliad

→ Producer to primary consumer →　Primary to secondary consumer →　Secondary to higher-level consumer →　All producers and consumers to decomposers

Tropical rain forests are dominated by **broadleaf evergreen plants,** which keep most of their broad leaves year-round. Typically, this biome's huge trees have shallow roots and wide bases that support their massive weight. The tops of the trees form a dense canopy which blocks most light from reaching the forest floor. Many of the plants that live at the ground level have enormous leaves to capture what little sunlight filters through to the dimly lit forest floor.

Individual trees may be draped with vines that reach to their tops to gain access to sunlight. Their trunks and branches may contain large numbers of *epiphytes* (or air plants), such as some types orchids and other plants. These plants grow without soil, get water from the humid air and rain, and receive nutrients falling from the trees' upper leaves and limbs.

Tropical rain forests are teeming with life and have incredible biological diversity. These diverse life forms occupy a variety of specialized niches in distinct layers, based in the plants' case mostly on their need for sunlight, as shown in Figure 6-30 (p. 122). Much of the animal life, particularly insects, bats, and birds, lives in the sunny *canopy* layer, with its abundant shelter and supplies of leaves, flowers, and fruits. To study life in the canopy, ecologists climb trees, use tall construction cranes, and build platforms and boardwalks in the upper canopy.

Stratification of specialized plant and animal niches in the layers of a tropical rain forest enables coexistence of a great variety of species. Although these forests cover only about 2% of the earth's land surface, they are habitats for at least half of the earth's terrestrial species.

Figure 6-30 Natural capital: stratification of specialized plant and animal niches in various layers of a *tropical rain forest*. The presence of these specialized niches enables species to avoid or minimize competition for resources and results in the coexistence of a great variety of species.

Dropped leaves, fallen trees, and dead animals decompose quickly because of the warm, moist conditions and hordes of bacteria, fungi, insects, and other detritivores. This rapid recycling of scarce soil nutrients is why little litter is found on the ground. Instead of being stored in the soil, most nutrients released by decomposition are taken up quickly and stored by trees, vines, and other plants. This explains why tropical rain forest soils have so few plant nutrients (Figure 4-27, bottom left, p. 75). Because of their poor soils, these are not good places to clear and grow crops or graze cattle.

What Are Temperate Deciduous Forests?
Seasonal Changes and Beautiful Colors

Most of the trees in these forests survive winter by dropping their leaves, which decay and produce a nutrient-rich soil.

Temperate deciduous forests grow in areas with moderate average temperatures that change significantly with the season. Trace the nutrient transfers and energy flows in Figure 6-31 (p. 123). These areas have long, warm summers, cold but not too severe winters, and abundant precipitation, often spread fairly evenly throughout the year (Figure 6-28, center).

This biome is dominated by a few species of broadleaf deciduous trees such as oak, hickory, maple, poplar, and beech. They survive cold winters by dropping their leaves in the fall and becoming dormant. Each spring they grow new leaves that change in the fall into an array of reds and golds before dropping.

Temperate deciduous forests have fewer tree species than tropical rain forests. But the penetration of more sunlight supports a richer diversity of plant life at ground level. Because of the fairly low rate of decomposition, these forests accumulate a thick layer

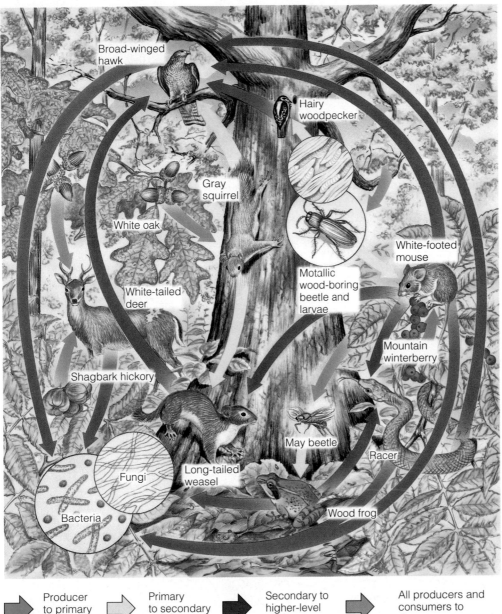

Figure 6-31 Natural capital: some components and interactions in a *temperate deciduous forest ecosystem.* When these organisms die, decomposers break down their organic matter into minerals that plants use. Colored arrows indicate transfers of matter and energy between producers, primary consumers (herbivores), secondary, or higher-level, consumers (carnivores), and decomposers. Organisms are not drawn to scale.

Broad-winged hawk

Hairy woodpecker

Gray squirrel

White oak

Metallic wood-boring beetle and larvae

White-footed mouse

White-tailed deer

Mountain winterberry

Shagbark hickory

May beetle

Racer

Fungi

Long-tailed weasel

Bacteria

Wood frog

→ Producer to primary consumer

→ Primary to secondary consumer

→ Secondary to higher-level consumer

→ All producers and consumers to decomposers

of slowly decaying leaf litter that is a storehouse of nutrients (Figure 4-27, bottom middle, p. 75).

The temperate deciduous forests of the eastern United States were once home for such large predators as bears, wolves, foxes, wildcats, and mountain lions (pumas). Today most of the predators have been killed or displaced, and the dominant mammal species often is the white-tailed deer, along with smaller mammals such as squirrels, rabbits, opossums, raccoons, and mice. Some of these diverse forests have been cleared and replaced with *tree plantations* consisting of only one tree species (see photo on p. vi).

Warblers, robins, and other bird species migrate to these forests during the summer to feed and breed. Many of these species are declining in numbers because of loss or fragmentation of their summer and winter habitats.

What Are Evergreen Coniferous Forests? Green and Cold

These forests in cold climates consist mostly of cone-bearing evergreen trees that keep their needles year-round to help the trees survive long and cold winters.

Evergreen coniferous forests, also called *boreal forests* and *taigas* ("TIE-guhs"), are found just south of the arctic tundra in northern regions across North America, Asia, and Europe (Figure 6-16). In this subarctic climate, winters are long, dry, and extremely cold, with sunlight available only 6–8 hours a day in the northernmost taiga. Summers are short, with mild to warm temperatures (Figure 6-28, right), and the sun typically shines 19 hours a day.

Most boreal forests are dominated by a few species of *coniferous* (cone-bearing) *evergreen trees* such as spruce, fir, cedar, hemlock, and pine that keep some of their narrow-pointed leaves (needles) all year long. This adaptation also allows the trees to take advantage of the brief summers without having to take time to grow new needles. Plant diversity is low because few species can survive the winters when soil moisture is frozen.

Beneath the stands of trees is a deep layer of partially decomposed conifer needles and leaf litter. Decomposition is slow because of the low temperatures, waxy coating of conifer needles, and high soil acidity. As the conifer needles decompose, they make the thin, nutrient-poor soil acidic and prevent most other plants (except certain shrubs) from growing on the forest floor (Figure 4-27, bottom right, p. 75).

These biomes contain a variety of wildlife, as depicted in Figure 6-32. Trace the flow of energy and nutrients in this diagram. During the brief summer the soil becomes waterlogged, forming acidic bogs, or *muskegs*, in low-lying areas of these forests. Warblers

Figure 6-32 Natural capital: some components and interactions in an *evergreen coniferous (boreal* or *taiga) forest ecosystem.* When these organisms die, decomposers break down their organic matter into minerals that plants use. Colored arrows indicate transfers of matter and energy between producers, primary consumers (herbivores), secondary, or higher-level, consumers (carnivores), and decomposers. Organisms are not drawn to scale.

Blue jay

Balsam fir

Great horned owl

Marten

Moose

White spruce

Wolf

Bebb willow

Pine sawyer beetle and larvae

Snowshoe hare

Fungi

Starflower

Bacteria

Bunchberry

Producer to primary consumer

Primary to secondary consumer

Secondary to higher-level consumer

All producers and consumers to decomposers

and other insect-eating birds feed on hordes of flies, mosquitoes, and caterpillars.

What Are Temperate Rain Forests? Dripping Green Giants and Mosses

Coastal areas support huge cone-bearing evergreen trees such as redwoods and Douglas fir in a cool and moist environment.

Coastal coniferous forests or *temperate rain forests* are found in scattered coastal temperate areas with ample rainfall or moisture from dense ocean fogs. Dense stands of large conifers such as Sitka spruce, Douglas fir, and redwoods dominate undisturbed areas of these biomes along the coast of North America, from Canada to northern California.

Most of the trees are evergreen because the abundance of water means that they have no need to shed their leaves. Tree trunks and the ground are frequently covered with mosses and ferns in this cool and moist environment. As in tropical rain forests, little light reaches the forest floor.

The ocean moderates the temperature so winters are mild and summers are cool. The trees in these moist forests depend on frequent rains and moisture from summer fog that rolls in off the Pacific.

Figure 6-33 lists major human impacts on the world's forests. What direct and indirect impacts does your lifestyle have on forest biomes?

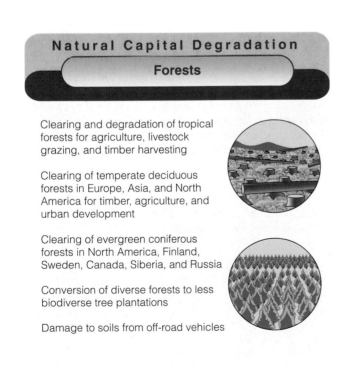

Natural Capital Degradation

Forests

Clearing and degradation of tropical forests for agriculture, livestock grazing, and timber harvesting

Clearing of temperate deciduous forests in Europe, Asia, and North America for timber, agriculture, and urban development

Clearing of evergreen coniferous forests in North America, Finland, Sweden, Canada, Siberia, and Russia

Conversion of diverse forests to less biodiverse tree plantations

Damage to soils from off-road vehicles

Figure 6-33 Natural capital degradation: major human impacts on the world's forests.

6-7 MOUNTAIN BIOMES

Why Are Mountains Ecologically Important? High-Altitude Biodiversity Sanctuaries

Mountains are high-elevation forested islands of biodiversity and often have snow-covered peaks that reflect solar radiation and gradually release water to lower-elevation streams and ecosystems.

Some of the world's most spectacular and important environments are mountains, which make up almost a fourth of the world's land surface. Mountains are places where dramatic changes in altitude, climate, soil, and vegetation take place over a very short distance (Figure 6-18, left).

Because of the steep slopes, mountain soils are especially prone to erosion when the vegetation holding them in place is removed by natural disturbances (such as landslides and avalanches) or human activities (such as timber cutting and agriculture). Many freestanding mountains are *islands of biodiversity* surrounded by a sea of lower-elevation landscapes transformed by human activities.

Mountains play important ecological roles. They contain the majority of the world's forests, which are habitats for much of the world's terrestrial biodiversity. They often are habitats for endemic species found nowhere else on earth, and they serve as sanctuaries for animal species driven from lowland areas.

They also help regulate the earth's climate when their snow- and ice-covered tops reflect solar radiation back into space. Mountains affect sea levels as a result of decreases or increases in glacial ice—most of which is locked up in Antarctica (the most mountainous of all continents). Finally, mountains play a critical role in the hydrologic cycle by gradually releasing melting ice, snow, and water stored in the soils and vegetation of mountainsides to small streams.

Despite their ecological, economic, and cultural importance, the fate of mountain ecosystems has not been a high priority of governments or many environmental organizations. Mountain ecosystems are coming under increasing pressure from several human activities (Figure 6-34, p. 126). What direct or indirect impacts does your lifestyle have on mountain biomes?

In this chapter we examined the connections among weather, climate, and the distribution of the earth's terrestrial biomes. Three general ecological lessons emerge from this study. *First*, different climates occur as a result of currents of air and water flowing over an unevenly heated planet spinning on a tilted axis. *Second*, different climates result in different communities of organisms, or biomes. *Third*, everything is connected.

Landless poor migrating uphill to survive

Timber extraction

Mineral resource extraction

Hydroelectric dams and reservoirs

Increasing tourism (such as hiking and skiing)

Air pollution from industrial and urban centers

Increased ultraviolet radiation from ozone depletion

Soil damage from off-road vehicles

Figure 6-34 Natural capital degradation: major human impacts on the world's mountains.

When we try to pick out anything by itself, we find it hitched to everything else in the universe.

JOHN MUIR

CRITICAL THINKING

1. List a limiting factor for each of the following ecosystems: **(a)** a desert, **(b)** arctic tundra, **(c)** alpine tundra, **(d)** the floor of a tropical rain forest, and **(e)** a temperate deciduous forest.

2. Why do deserts and arctic tundra support a much smaller biomass of animals than do tropical forests?

3. Some biologists have suggested restoring large herds of bison on public lands in the North American plains as a way of restoring remaining tracts of tall-grass prairie. Ranchers with permits to graze cattle and sheep on federally managed lands have strongly opposed this idea. Do you agree or disagree with the idea of restoring large numbers of bison to the plains of North America? Explain.

4. Why do most animals in a tropical rain forest live in its trees?

5. What biomes are best suited for **(a)** raising crops and **(b)** grazing livestock?

6. Congratulations! You are in charge of the world. What are the three most important features of your plan to help sustain the earth's terrestrial biodiversity?

PROJECTS

1. How has the climate changed in the area where you live during the past 50 years? Investigate the beneficial and harmful effects of these changes. How have these changes benefited or harmed you personally?

2. What type of biome do you live in? What have been the major effects of human activities over the past 50 years on the characteristic vegetation and animal life normally found in the biome you live in? How is your lifestyle affecting this biome?

3. Use the library or the Internet to find bibliographic information about *Robert H. MacArthur* and *John Muir*, whose quotes appear at the beginning and end of this chapter.

4. Make a concept map of this chapter's major ideas, using the section heads, subheads, and key terms (in boldface). Look on the website for this book for information about making concept maps.

LEARNING ONLINE

The website for this book contains study aids and many ideas for further reading and research. They include a chapter summary, review questions for the entire chapter, flash cards for key terms and concepts, a multiple-choice practice quiz, interesting Internet sites, references, and a guide for accessing thousands of InfoTrac® College Edition articles. Log on to

http://biology.brookscole.com/miller14

Then click on the Chapter-by-Chapter area, choose Chapter 6, and select a learning resource.

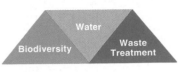

CASE STUDY

Why Should We Care about Coral Reefs?

In the shallow coastal zones of warm tropical and subtropical oceans we often find **coral reefs** (Figure 7-1, left). These stunningly beautiful natural wonders are among the world's oldest, most diverse, and most productive ecosystems.

Coral reefs are formed by massive colonies of tiny animals called *polyps*. They slowly build reefs by secreting a protective crust of limestone (calcium carbonate) around their soft bodies.

When the polyps die, their empty crusts remain as a platform for more reef growth. The result is an elaborate network of crevices, ledges, and holes that serve as calcium carbonate "condominiums" for a variety of marine animals.

Coral reefs involve a mutually beneficial relationship between the polyps and tiny single-celled algae called *zooxanthellae* ("zoh-ZAN-thel-ee") that live in the tissues of the polyps. The algae provide the polyps with color, food, and oxygen through photosynthesis and help produce calcium carbonate, which forms the coral skeleton. The polyps in turn provide the algae with a well-protected home and some of their nutrients. This is a win-win deal for both species.

Coral reefs provide a number of important ecological and economic services. When coral polyps form their limestone shells they remove carbon dioxide from the atmosphere as part of the carbon cycle. Coral reefs act as natural barriers that help protect about 15% of the world's coastlines from erosion by battering waves and storms. They also support at least one-fourth of all identified marine species and about two-thirds of marine fish species and produce about a tenth of the global fish catch.

The reefs provide jobs and building materials (which can help destroy the reefs) for some of the world's poorest countries and support fishing and tourism industries worth billions of dollars each year.

Finally, these biological treasures give us an underwater world to study and enjoy. Coral reefs provide these free ecological and economic services while occupying less than 0.1% of the world's ocean area—amounting to an area about half the size of France. They are true wonders of biodiversity!

Bad news. More than one-fourth of the world's coral reefs have been lost to coastal development, pollution, overfishing, warmer ocean temperatures, and other stresses. And these threats are increasing.

One problem is *coral bleaching* (Figure 7-1, right). It occurs when a coral becomes stressed and expels most of its colorful algae, leaving an underlying ghostly white skeleton of calcium carbonate. Two causes are increased water temperature and runoff of silt that covers the coral and prevents photosynthesis. Unable to grow or repair themselves, the corals eventually die unless the stress is removed and algae recolonize them.

Aquatic scientists view coral reefs as "aquatic biodiversity savings banks" and sensitive biological indicators of environmental conditions in their aquatic environment. The decline and degradation of these colorful oceanic sentinels should serve as a warning about the health of their habitats and the oceans that provide us with crucial ecological and economic services.

Figure 7-1 A healthy coral reef covered by colorful algae (left) and a bleached coral reef that has lost most of its algae (right) because of changes in the environment (such as cloudy water or too warm temperatures). With the algae gone, the white limestone of the coral skeleton becomes visible. If the environmental stress is not removed and no other alga species fill the abandoned niche, the corals die. These diverse and productive ecosystems are being damaged and destroyed at an alarming rate.

If there is magic on this planet, it is contained in water.
LOREN EISLEY

This chapter addresses the following questions:

- What are the basic types of aquatic life zones, and what factors influence the kinds of life they contain?
- What are the major types of saltwater life zones, and how do human activities affect them?
- What are the major types of freshwater life zones, and how do human activities affect them?
- How can we help sustain aquatic life zones?

7-1 AQUATIC ENVIRONMENTS

What Are the Two Major Types of Aquatic Life Zones? Salty and Fresh

Saltwater and freshwater aquatic life zones cover almost three-fourths of the earth's surface.

The aquatic equivalents of biomes are called *aquatic life zones*. The major types of organisms found in aquatic environments are determined by the water's *salinity*—the amounts of various salts such as sodium chloride (NaCl) dissolved in a given volume of water. As a result, aquatic life zones are divided into two major types: *saltwater* or *marine* (such as estuaries, coastlines, coral reefs, coastal marshes, mangrove swamps, and oceans) and *freshwater* (such as lakes, ponds, streams, rivers, and inland wetlands). Figure 7-2 shows the distribution of the world's major oceans, lakes, rivers, coral reefs, and mangroves.

The world's saltwater and freshwater aquatic systems cover about 71% of the earth's surface. We can think of the world's rivers, streams, lakes, and oceans as a giant circulatory system transporting water from one place to another as part of the earth's water cycle (Figure 4-28, p. 76). These aquatic systems also play vital roles in the earth's biological productivity, climate, biogeochemical cycles, and biodiversity.

What Kinds of Organisms Live in Aquatic Life Zones? Floaters, Swimmers, Crawlers, and Decomposers

Aquatic systems contain floating, drifting, swimming, bottom-dwelling, and decomposer organisms.

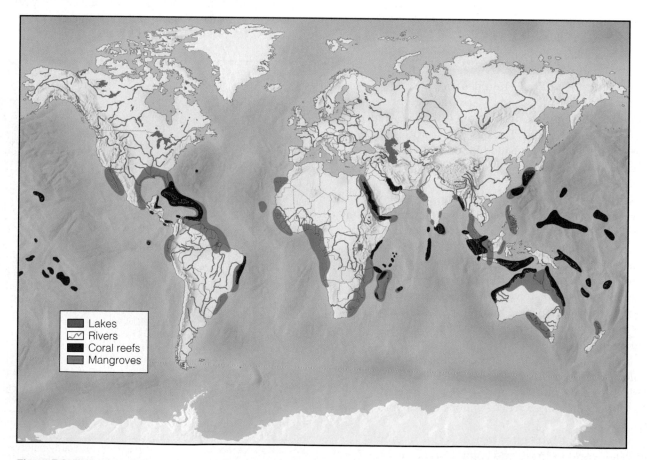

Figure 7-2 Natural capital: distribution of the world's major saltwater oceans, coral reefs, mangroves, and freshwater lakes and rivers.

Saltwater and freshwater life zones contain four major types of organisms. One group consists of weakly swimming, free-floating **plankton** carried by currents. There are three major types of plankton. One is **phytoplankton** ("FIE-toe-plank-ton," Greek for "drifting plants"), or *plant plankton*. They and many types of algae are the producers that support most aquatic food chains and food webs. Another type is **zooplankton** ("ZOE-oh-plank-ton", Greek for "drifting animals"), or *animal plankton*. They consist of primary consumers (herbivores) that feed on phytoplankton and secondary consumers that feed on other zooplankton. They range from single-celled protozoa to large invertebrates such as jellyfish.

Improved technologies for filtering and analyzing water have revealed huge populations of much smaller plankton called **ultraplankton**—photosynthetic bacteria no more than 2 micrometers (40 millionths of an inch) wide. Scientists estimate that these extremely small plankton may be responsible for 70% of the primary productivity near the ocean surface.

A *second group* of organisms consists of **nekton,** strongly swimming consumers such as fish, turtles, and whales. A *third group,* called **benthos,** dwells on the bottom. Examples are barnacles and oysters that anchor themselves to one spot, worms that burrow into the sand or mud, and lobsters and crabs that walk about on the bottom. A fourth group consists of **decomposers** (mostly bacteria) that break down the organic compounds in the dead bodies and wastes of aquatic organisms into simple nutrient compounds for use by producers.

How Do Aquatic Systems Differ from Terrestrial Systems? Living without Boundaries and Large Temperature Fluctuations

Living in water has its benefits and drawbacks.

Living in an aquatic environment has advantages and disadvantages (Figure 7-3). Several differences between aquatic and terrestrial systems also hinder the ability of researchers to understand aquatic systems. One is that aquatic systems have less pronounced and fixed physical boundaries than terrestrial ecosystems. This makes it difficult to count and manage populations of aquatic organisms.

Also, food chains and webs in aquatic systems are usually more complex and longer than those in terrestrial biomes. One reason is that the fluid medium of water systems and the variety of bottom habitats open up ways of getting food that are not available on land. Another difference is that aquatic systems (especially marine systems) are more difficult to monitor and study than terrestrial systems because of their size and because they are largely hidden from view.

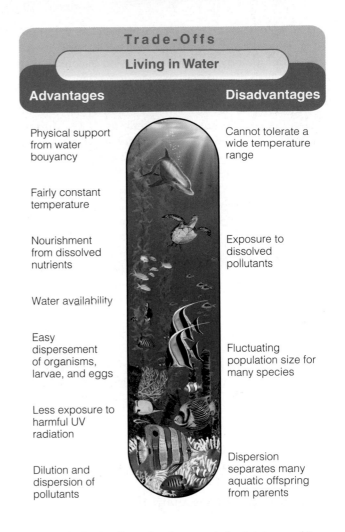

Trade-Offs

Living in Water

Advantages

Physical support from water bouyancy

Fairly constant temperature

Nourishment from dissolved nutrients

Water availability

Easy dispersement of organisms, larvae, and eggs

Less exposure to harmful UV radiation

Dilution and dispersion of pollutants

Disadvantages

Cannot tolerate a wide temperature range

Exposure to dissolved pollutants

Fluctuating population size for many species

Dispersion separates many aquatic offspring from parents

Figure 7-3 Trade-offs: advantages and disadvantages of living in water.

What Factors Limit Life at Different Depths in Aquatic Life Zones? Living in Layered Zones

Life in aquatic systems is found in surface, middle, and bottom layers.

Most aquatic life zones can be divided into three layers: *surface, middle,* and *bottom.* A number of environmental factors determine the types and numbers of organisms found in these layers. Examples are *temperature, access to sunlight for photosynthesis, dissolved oxygen content,* and *availability of nutrients* such as carbon (as dissolved CO_2 gas), nitrogen (as NO_3^-), and phosphorus (mostly as PO_4^{3-}) for producers.

In deep aquatic systems, photosynthesis is confined mostly to the upper layer, or **euphotic zone,** through which sunlight can penetrate. The depth of the euphotic zone in oceans and deep lakes can be reduced by excessive algal growth (algal blooms) clouding the water.

Dissolved O_2 levels are higher near the surface because oxygen-producing photosynthesis takes place there. Photosynthesis cannot take place below the sunlit layer. Thus at lower depths O_2 levels fall because of aerobic respiration by aquatic animals and decomposers. They also fall because less oxygen gas dissolves in the deeper and colder water than in warmer surface water.

In contrast, levels of dissolved CO_2 are *low* in surface layers because producers use CO_2 during photosynthesis and are *high* in deeper, dark layers where aquatic animals and decomposers produce CO_2 through aerobic respiration.

In shallow waters in streams, ponds, and oceans, ample supplies of nutrients for primary producers are usually available. By contrast, in the open ocean, nitrates, phosphates, iron, and other nutrients often are in short supply and limit net primary productivity (NPP) (Figure 4-24, p. 72). However, NPP is much higher in parts of the open ocean where upwellings bring such nutrients from the ocean bottom to the surface for use by producers.

Most creatures living on the bottom of the deep ocean and deep lakes depend on animal and plant plankton that die and fall into deep waters. Because this food is limited, deep-dwelling fish species tend to reproduce slowly. This makes them especially vulnerable to depletion from overfishing.

7-2 SALTWATER LIFE ZONES

Why Should We Care about the Oceans? Givers of Life

Although oceans occupy most of the earth's surface and provide many ecological and economic services, we know less about them than we do about the moon.

A more accurate name for Earth would be *Ocean* because saltwater oceans cover about 71% of the planet's surface (Figure 7-4). They contain about 250,000 known species of marine plants and animals and provide many important ecological and economic services, shown in Figure 7-5. Take a good look at this figure that describes part of your life-support system.

As landlubbers we have a distorted and limited view of our watery home. We know more about the surface of the moon than about the oceans that cover most of the earth. According to aquatic scientists, the scientific investigation of poorly understood marine and freshwater aquatic systems is a *research frontier* whose study could result in immense ecological and economic benefits.

What Is the Coastal Zone? Abundant Life near the Shore

The coastal zone makes up less than 10% of the world's ocean area but contains 90% of all marine species.

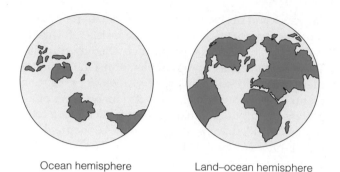

Ocean hemisphere   Land–ocean hemisphere

Figure 7-4 Natural capital: the ocean planet. The salty oceans cover about 71% of the earth's surface. About 97% of the earth's water is in the interconnected oceans, which cover 90% of the planet's mostly ocean hemisphere (left) and 50% of its land–ocean hemisphere (right). Freshwater systems cover less than 1% of the earth's surface.

Oceans have two major life zones: the *coastal zone* and the *open sea* (Figure 7-6). The **coastal zone** is the warm, nutrient-rich, shallow water that extends from the high-tide mark on land to the gently sloping, shal-

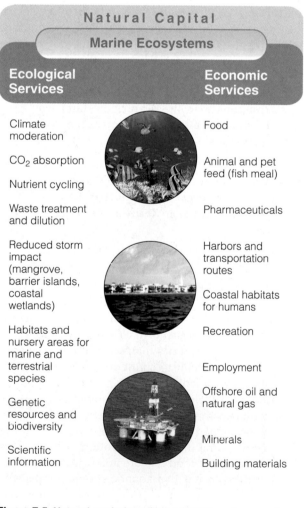

Natural Capital

Marine Ecosystems

Ecological Services	Economic Services
Climate moderation	Food
CO_2 absorption	Animal and pet feed (fish meal)
Nutrient cycling	
Waste treatment and dilution	Pharmaceuticals
Reduced storm impact (mangrove, barrier islands, coastal wetlands)	Harbors and transportation routes
	Coastal habitats for humans
Habitats and nursery areas for marine and terrestrial species	Recreation
	Employment
Genetic resources and biodiversity	Offshore oil and natural gas
	Minerals
Scientific information	Building materials

Figure 7-5 Natural capital: major ecological and economic services provided by marine systems.

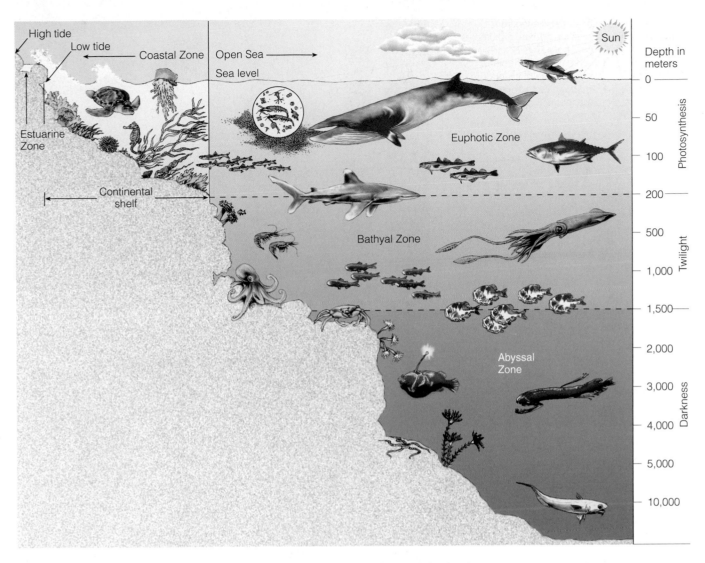

Figure 7-6 Natural capital: major life zones in an ocean (not drawn to scale). Actual depths of zones may vary.

low edge of the *continental shelf* (the submerged part of the continents). This zone has numerous interactions with the land and thus human activities easily affect it.

Although it makes up less than a tenth of the world's ocean area, the coastal zone contains 90% of all marine species and is the site of most large commercial marine fisheries. Most ecosystems found in the coastal zone have a high net primary productivity per unit of area. This occurs because of the zone's ample supplies of sunlight and plant nutrients flowing from land and distributed by wind and ocean currents.

What Are Estuaries, Coastal Wetlands, and Mangrove Swamps? Stressed Centers of Biological Productivity

Several highly productive coastal ecosystems are under increasing stress from human activities.

One highly productive area in the coastal zone is an **estuary,** a partially enclosed area of coastal water where seawater mixes with freshwater and nutrients from rivers, streams, and runoff from land (Figure 7-7, p. 132).

Estuaries and their associated **coastal wetlands** (land areas covered with water all or part of the year) include river mouths, inlets, bays, sounds, mangrove forest swamps in sheltered regions along tropical coasts, and salt marshes (Figure 7-8, p. 132) in temperate zones.

Temperature and salinity levels vary widely in estuaries and coastal wetlands because of the daily rhythms of the tides and seasonal variations in the flow of freshwater into the estuary. They are also affected by unpredictable flows of freshwater from coastal land and rivers after heavy rains and of salt water from the ocean as a result of storms, hurricanes, and typhoons. You have to be tough to survive in this harsh and variable environment.

Figure 7-7 View of an *estuary* taken from space. The photo shows the sediment plume at the mouth of Madagascar's Betsiboka River as it flows through the estuary and into the Mozambique Channel. Because of its topography, heavy rainfall, and the clearing of forests for agriculture, Madagascar is the world's most eroded country.

NASA

Herring gulls

Peregrine falcon

Snowy egret

Cordgrass

Short-billed dowitcher

Phytoplankton

Marsh periwinkle

Smelt

Zooplankton and small crustaceans

Soft-shelled clam

Bacteria

Clamworm

Figure 7-8 Natural capital: some components and interactions in a *salt marsh ecosystem* in a temperate area such as the United States. When these organisms die, decomposers break down their organic matter into minerals used by plants. Colored arrows indicate transfers of matter and energy between consumers (herbivores), secondary (or higher-level) consumers (carnivores), and decomposers. Organisms are not drawn to scale.

Producer to primary consumer

Primary to secondary consumer

Secondary to higher-level consumer

All consumers and producers to decomposers

The dominant organisms in *mangrove forest swamps* are trees that can grow in salt water. These trees have extensive roots that extend above the water, where they can obtain oxygen and provide support for the plant. These nutrient-rich forests are found in sheltered regions along tropical coasts (Figure 7-2). Because ocean waves do not reach these saltwater ecosystems, they collect mud and anaerobic sediment.

The constant water movement in estuaries and their associated coastal wetlands stirs up the nutrient-rich silt, making it available to producers. These systems filter toxic pollutants, excess plant nutrients, sediments, and other pollutants. They reduce storm damage by absorbing waves and storing excess water produced by storms. And they provide food, habitats, and nursery sites for a variety of aquatic species.

Bad news. We are degrading or destroying some of the ecological services that these important ecosystems provide at no cost. Researchers estimate that more than a third of the world's mangrove forests have been destroyed—mostly for developing aquaculture shrimp farms, growing crops, and coastal development projects.

What Niches Do Rocky and Sandy Shores Provide? Hold On, Dig In, or Hang Out in a Shell

Organisms in coastal areas experiencing daily low and high tides have evolved a number of ways to survive under harsh and changing conditions.

The area of shoreline between low and high tides is called the **intertidal zone.** This is not an easy place to live. Its organisms must be able to avoid being swept away or crushed by waves, and avoid or cope with being immersed during high tides and left high and dry (and much hotter) at low tides. They must also survive changing levels of salinity when heavy rains dilute salt water. To deal with such stresses, most intertidal organisms hold on to something, dig in, or hide in protective shells.

Some coasts have steep *rocky shores* pounded by waves. The numerous pools and other niches in the intertidal zone of rocky shores contain a great variety of species with different niches as shown in the top part of Figure 7-9 (p. 134).

Other coasts have gently sloping *barrier beaches,* or *sandy shores,* with niches for different marine organisms as shown in the bottom portion of Figure 7-9. Most of them are hidden from view and survive by burrowing, digging, and tunneling in the sand. These sandy beaches and their adjoining coastal wetlands are also home to a variety of shorebirds that feed in specialized niches on crustaceans, insects, and other organisms (Figure 5-5, p. 92).

What Are Barrier Islands? Natural Protectors of the Shore

Narrow islands off some shores help protect coastal zones from storm waves, but developing these islands reduces this natural protection and makes them risky places to live.

Barrier islands are low, narrow, sandy islands that form offshore from a coastline. Most run parallel to the shore. They are found along some coasts such as most of North America's Atlantic and Gulf coasts. These islands help protect the mainland, estuaries, and coastal wetlands from the onslaught of approaching storm waves.

These beautiful but very limited pieces of real estate are prime targets for real estate development. Almost one-fourth of the area of barrier islands in the United States has been developed.

Living on these islands can be risky. Sooner or later, many of the structures humans build on low-lying barrier islands (Figure 7-10, p. 135), such as Atlantic City, New Jersey, and Miami Beach, Florida, are damaged or destroyed by flooding, severe beach erosion, or major storms (including hurricanes).

Their low-lying beaches are constantly shifting, with gentle waves building them up and storms flattening and eroding them. Currents running parallel to the beaches constantly take sand from one area and deposit it in another. Some beach communities spend lots of money to replace the eroded sand. But sooner or later nature moves it somewhere else.

Figure 7-11 (p. 135) shows that undisturbed beaches on typical barrier islands have one or more rows of natural sand dunes with the sand held in place by the roots of grasses. These dunes serve as the first line of defense against the ravages of the sea at no cost. If we left it that way and built behind the second set of dunes, these beaches would be safer places for people and many of the populations of natural organisms that live there. Mainland coastal cities, beaches, and undisturbed salt marshes and estuaries behind these protective islands would also be safer from damage.

But this real estate is so scarce and valuable that developers want to cover it with buildings and roads. In doing this, they are rarely required to take into account the storm protection and other free ecological services the protective dunes provide. Thus the first thing most coastal developers do is to remove the protective dunes or build behind the first set of dunes. This means that large storms can flood and even sweep away seaside buildings and severely erode the sandy beaches. Then we call these human-influenced events natural disasters.

Governments often provide owners with fairly cheap property insurance, funds for replenishing eroded sand, and grants and low-cost loans to rebuild after damaging storms. This makes it less financially

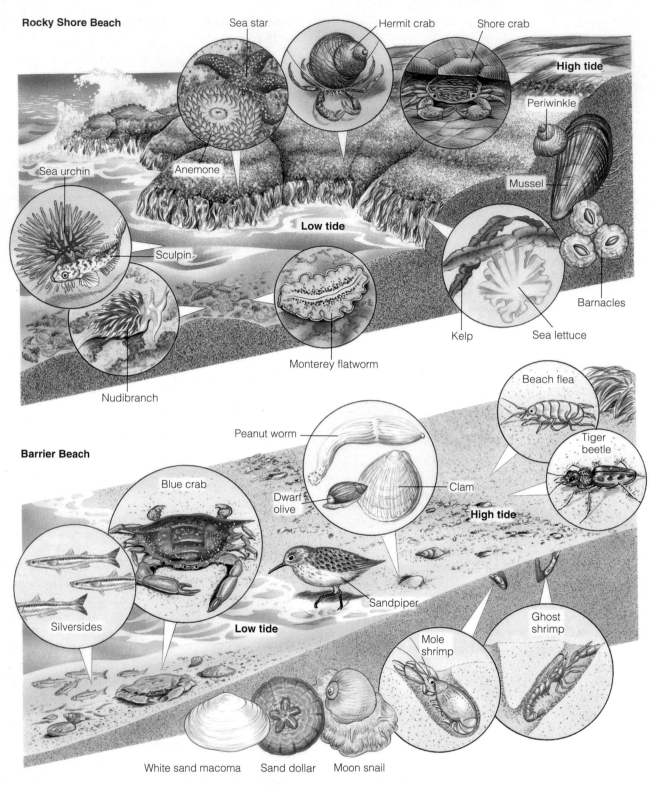

Rocky Shore Beach

Sea star

Hermit crab

Shore crab

High tide

Periwinkle

Sea urchin

Anemone

Sculpin

Mussel

Nudibranch

Monterey flatworm

Kelp

Sea lettuce

Barnacles

Low tide

Barrier Beach

Peanut worm

Beach flea

Blue crab

Dwarf olive

Clam

Tiger beetle

High tide

Silversides

Sandpiper

Low tide

Mole shrimp

Ghost shrimp

White sand macoma   Sand dollar   Moon snail

Figure 7-9 Living between the tides. Some organisms with specialized niches found in various zones on rocky shore beaches (top) and barrier or sandy beaches (bottom). Organisms are not drawn to scale.

Figure 7-10 A developed barrier island: Ocean City, Maryland, host to 8 million visitors a year. Rising sea levels from global warming may put this and many other barrier islands under water by the end of this century.

What Are Coral Reefs? Aquatic Oases of Biodiversity

Coral reefs are biologically diverse and productive ecosystems that are increasingly stressed by human activities.

Coral reefs (Figure 7-1 and photo on p. viii) form in clear, warm coastal waters of the tropics and subtropics (Figure 7-2). Coral reefs are ecologically complex in terms of the many interactions among the diverse organisms that live there as shown in Figure 7-12 (p. 136).

These reefs are vulnerable to damage because they grow slowly and are disrupted easily. They also thrive only in clear, warm, and fairly shallow water of constant high salinity. Corals can live only in water with a temperature of 18–30°C (64–86°F). Coral bleaching (Figure 7-1, right) can be triggered by an increase of just 1°C (1.8°F) above this maximum temperature.

The biodiversity of coral reefs can be reduced by natural disturbances such as severe storms, freshwater floods, and invasions of predatory fish. But throughout their long geologic history, coral reefs have been able to adapt to such natural environmental changes.

Today the biggest threats to the survival and biodiversity of many of the world's coral reefs come from sediment runoff and other human activities (Figure 7-13, p. 137). Scientists are concerned that these threats are occurring so rapidly (over decades) and over such a wide area that many of the world's coral reef systems may not have enough time to adapt.

Good news. There is growing evidence that coral reefs can recover when given a chance. When localities

risky to live in such places and hastens the destruction of their protective dune and vegetation systems.

Some argue that people who choose to live in these and other risky places should accept the full cost of such risks. They should not expect taxpayers to supplement their high property insurance costs or help them rebuild houses lost in hurricanes or other natural disasters that are expected risks in such areas.

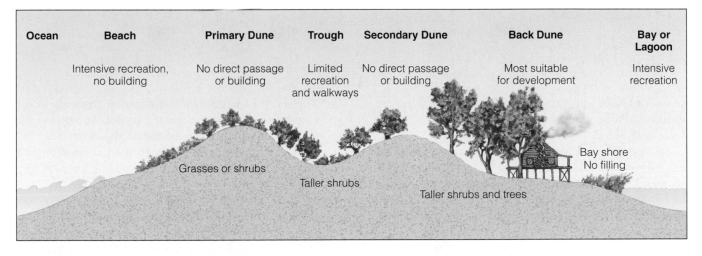

Ocean	Beach	Primary Dune	Trough	Secondary Dune	Back Dune	Bay or Lagoon
	Intensive recreation, no building	No direct passage or building	Limited recreation and walkways	No direct passage or building	Most suitable for development	Intensive recreation

Grasses or shrubs

Taller shrubs

Taller shrubs and trees

Bay shore
No filling

Figure 7-11 Natural capital: primary and secondary dunes on gently sloping sandy beaches help protect land from erosion by the sea. The roots of various grasses that colonize the dunes help hold the sand in place. Ideally, construction is allowed only behind the second strip of dunes, and walkways to the beach are built over the dunes to keep them intact. This helps preserve barrier beaches and protect buildings from damage by wind, high tides, beach erosion, and flooding from storm surges. Such protection is rare because the short-term economic value of oceanfront land is considered much higher than its long-term ecological value.

Figure 7-12 Natural capital: some components and interactions in a *coral reef ecosystem*. When these organisms die, decomposers break down their organic matter into minerals used by plants. Colored arrows indicate transfers of matter and energy between producers, primary consumers (herbivores), secondary (or higher-level) consumers (carnivores), and decomposers. Organisms are not drawn to scale.

Gray reef shark

Green sea turtle

Sea nettle

Fairy basslet

Blue tangs

Parrot fish

Sergeant major

Hard corals

Algae

Brittle star

Banded coral shrimp

Phytoplankton

Coney

Symbiotic algae

Zooplankton

Blackcap basslet

Sponges

Bacteria

Moray eel

| Producer to primary consumer | Primary to secondary consumer | Secondary to higher-level consumer | All consumers and producers to decomposers |

or nations have imposed restrictions on reef fishing or reduced inputs of nutrients and other pollutants, reefs have rebounded.

Some 300 coral reefs in 65 countries are protected as reserves or parks, and another 600 have been recommended for protection. But protecting reefs is difficult and expensive, and only half of the countries with coral reefs have set aside reserves that receive some protection from human activities.

What Biological Zones Are Found in the Open Sea? Where Is the Light?

The open ocean consists of a brightly lit surface layer, a dimly lit middle layer, and a dark bottom zone.

The sharp increase in water depth at the edge of the continental shelf separates the coastal zone from the vast volume of the ocean called the **open sea.** Primarily on the basis of the penetration of sunlight, it is divided into the three vertical zones shown in Figure 7-6.

The *euphotic zone* is the lighted upper zone where floating drifting phytoplankton carry out photosynthesis. Nutrient levels are low (except around upwellings), and levels of dissolved oxygen are high. Large, fast-swimming predatory fish such as swordfish, sharks, and bluefin tuna populate this zone.

The *bathyal zone* is the dimly lit middle zone that does not contain photosynthesizing producers because of a lack of sunlight. Various types of zooplankton and smaller fish, many of which migrate to the surface at night to feed, populate this zone.

The lowest zone, called the *abyssal zone,* is dark and very cold and has little dissolved oxygen. However, there are enough nutrients on the ocean floor to support about 98% of the species living in the ocean.

Most organisms of the deep waters and ocean floor get their food from showers of dead and decaying organisms (detritus) drifting down from upper lighted levels of the ocean. Some of these organisms, including many types of worms, are *deposit feeders,* which take mud into their guts and extract nutrients from it. Others such as oysters, clams, and sponges are *filter feeders,* which pass water through or over their bodies and extract nutrients from it. On parts of the dark, deep ocean floor near hydrothermal vents, scientists have found communities of organisms where specialized bacteria use chemosynthesis to produce their own food and food for other organisms feeding on them.

Average primary productivity and NPP per unit of area are quite low in the open sea except at an occasional equatorial upwelling, where currents bring up nutrients from the ocean bottom. However, because the open sea covers so much of the earth's surface, it makes the largest contribution to the earth's overall NPP.

Currently, about 40% of the world's population and more than half of the U.S. population live along

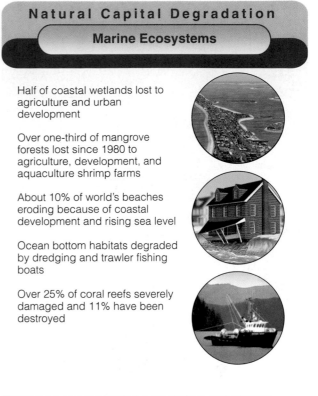

Natural Capital Degradation

Marine Ecosystems

Half of coastal wetlands lost to agriculture and urban development

Over one-third of mangrove forests lost since 1980 to agriculture, development, and aquaculture shrimp farms

About 10% of world's beaches eroding because of coastal development and rising sea level

Ocean bottom habitats degraded by dredging and trawler fishing boats

Over 25% of coral reefs severely damaged and 11% have been destroyed

Figure 7-14 Natural capital degradation: major human impacts on the world's marine systems.

coasts or within 100 kilometers (62 miles) of a coast. Some 13 of the world's 19 megacities with populations of 10 million or more people are in coastal zones. By 2030, at least 6.3 billion people—equal to the world's entire population in 2003—are expected to live in or near coastal zones. Figure 7-14 lists major human impacts on marine systems. Explain how your lifestyle can contribute directly or indirectly to these impacts.

7-3 FRESHWATER LIFE ZONES

What Are Freshwater Life Zones? Lakes, Wetlands, Rivers

Freshwater ecosystems provide important ecological and economic services even though they cover less than 1% of the earth's surface.

Freshwater life zones occur where water with a dissolved salt concentration of less than 1% by volume accumulates on or flows through the surfaces of terrestrial biomes. Examples are *standing* (lentic) bodies of freshwater such as lakes, ponds, and inland wetlands and *flowing* (lotic) systems such as streams and rivers. Although freshwater systems cover less than 1% of the earth's surface, they provide a number of important ecological and economic services (Figure 7-15, p. 138).

Natural Capital Degradation

Coral Reefs

Ocean warming

Soil erosion

Algae growth from fertilizer runoff

Mangrove destruction

Coral reef bleaching

Rising sea levels

Increased UV exposure from ozone depletion

Using cyanide and dynamite to harvest coral reef fish

Coral removal for building material, aquariums, and jewelery

Damage from anchors, ships, and tourist divers

Figure 7-13 Natural capital degradation: major threats to coral reefs.

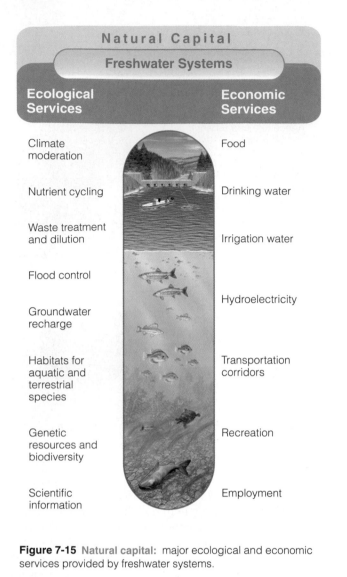

Natural Capital
Freshwater Systems

Ecological Services	Economic Services
Climate moderation	Food
Nutrient cycling	Drinking water
Waste treatment and dilution	Irrigation water
Flood control	
	Hydroelectricity
Groundwater recharge	
Habitats for aquatic and terrestrial species	Transportation corridors
Genetic resources and biodiversity	Recreation
Scientific information	Employment

Figure 7-15 Natural capital: major ecological and economic services provided by freshwater systems.

The volume of fresh water that we use provides us with free services worth about $3 trillion a year—equal to about 7% the value of all goods and services provided annually by the entire global economy.

What Life Zones Are Found in Freshwater Lakes? Life in Layers

Lakes consist of sunlit surface layers near and away from the shore, and at deeper levels a dark layer and a bottom zone.

Lakes are large natural bodies of standing fresh water formed when precipitation, runoff, or groundwater seepage fill depressions in the earth's surface. Causes of such depressions include *glaciation* (the Great Lakes of North America), *crustal displacement* (Lake Nyasa in East Africa) and *volcanic activity* (Crater Lake in Oregon; see photo on title page). Lakes are supplied with water from rainfall, melting snow, and streams that drain the surrounding watershed.

Lakes normally consist of four distinct zones that are defined by their depth and distance from shore as shown in Figure 7-16. The top layer is the *littoral zone* ("LIT-tore-el"). It consists of the shallow sunlit waters near the shore to the depth at which rooted plants stop growing, and it has a high biological diversity. It has adequate nutrients from bottom sediments. Bulrushes and cattails are plentiful near the shore and water lilies and entirely submerged plants flourish at the deepest depths of the littoral zone.

Next is the *limnetic zone* ("lim-NET-ic"): the open, sunlit water surface layer away from the shore that extends to the depth penetrated by sunlight. As the main photosynthetic body of the lake, its producers supply the food and oxygen that support most of the lake's consumers.

Next is the *profundal zone* ("pro-FUN-dahl"): the deep, open water where it is too dark for photosynthesis. Without sunlight and plants, oxygen levels are low. Fish adapted to its cooler and darker water are found in this zone.

Finally, at the bottom of the lake we find the *benthic zone* ("BEN-thic"). Mostly decomposers and detritus feeders and fish that swim from one zone to the other inhabit it. It is nourished mainly by detritus that falls from the littoral and limnetic zones and by sediment washing into the lake.

During the summer and winter, the water in deep temperate zone lakes becomes stratified into different temperature layers, which do not mix. Twice a year, in the fall and spring, the waters at all layers of these lakes mix in *overturns* that equalize the temperature at all depths. These overturns bring oxygen from the surface water to the lake bottom and nutrients from the lake bottom to the surface waters.

How Do Plant Nutrients Affect Lakes? Too Much of a Good Thing Is Not Good

A lake's supply of plant nutrients from its environment affect its physical and chemical conditions and the types and numbers of organisms it can support.

Ecologists classify lakes according to their nutrient content and primary productivity. A newly formed lake generally has a small supply of plant nutrients and is called an **oligotrophic** (poorly nourished) **lake** (Figure 7-17, left, p. 140). This type of lake is often deep, with steep banks. Glacier melt and mountain streams carrying little sediment supply water to such lakes. Because there is little sediment or microscopic life to cloud the water, such a lake usually has crystal-clear blue or green water and has small populations of phytoplankton and fish (such as smallmouth bass and trout). Because of their low levels of nutrients, these lakes have a low net primary productivity.

Over time, sediment, organic material, and inorganic nutrients wash into an oligotrophic lake, and

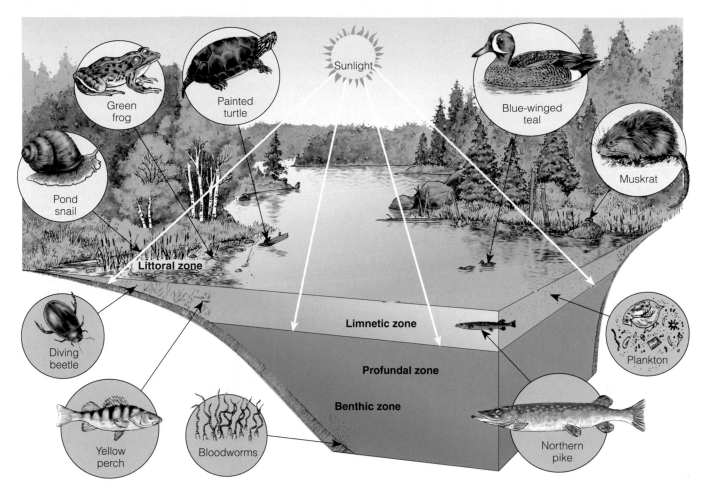

Figure 7-16 The distinct zones of life in a fairly deep temperate zone lake.

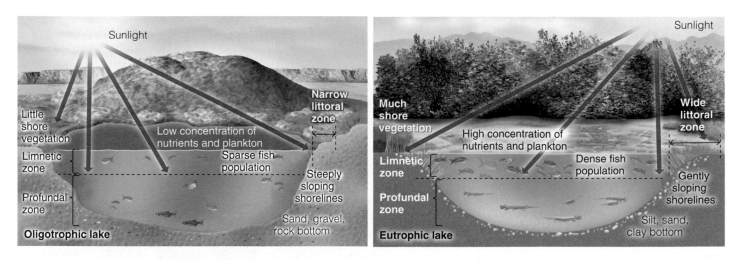

Figure 7-17 An *oligotrophic*, or nutrient-poor, lake (left) and a *eutrophic*, or nutrient-rich, lake (right). Mesotrophic lakes fall between these two extremes of nutrient enrichment. Nutrient inputs from human activities can accelerate eutrophication and lead to algae blooms and fish die-offs. This can reduce the fish populations shown for a europhic lake.

plants grow and decompose to form bottom sediments. A lake with a large or excessive supply of nutrients (mostly nitrates and phosphates) needed by producers is called a **eutrophic** (well-nourished) **lake**

(Figure 7-17, right). Such lakes typically are shallow and have murky brown or green water with poor visibility. Because of their high levels of nutrients, these lakes have a high net primary productivity.

In warm months the bottom layer of a eutrophic lake often is depleted of dissolved oxygen. Human inputs of nutrients from the atmosphere and from nearby urban and agricultural areas can accelerate the eutrophication of lakes, a process called *cultural eutrophication*. Many lakes fall somewhere between the two extremes of nutrient enrichment and are called **mesotrophic lakes.**

What Are the Major Characteristics of Freshwater Streams and Rivers? A Downhill Ride

Water flowing from mountains to the sea creates different aquatic conditions and habitats.

Precipitation that does not sink into the ground or evaporate is **surface water.** It becomes **runoff** when it flows into streams. The land area that delivers runoff, sediment, and dissolved substances to a stream is called a **watershed,** or **drainage basin.** Small streams join to form rivers, and rivers flow downhill to the ocean as shown in Figure 7-18.

In many areas, streams begin in mountainous or hilly areas that collect and release water falling to the earth's surface as rain or snow that melts during warm seasons. The downward flow of surface water and groundwater from mountain highlands to the sea takes place in three different aquatic life zones with different environmental conditions: the *source zone*, the *transition zone,* and the *floodplain zone* (Figure 7-18). Rivers in different areas can differ somewhat from this generalized model.

In the first, narrow *source zone* (Figure 7-18, top), headwaters, or mountain highland streams of cold, clear water, rush over waterfalls and rapids. As this turbulent water flows and tumbles downward, it dissolves large amounts of oxygen from the air. Since the water moves rapidly, floating plankton are less important. Instead, most plants such as algae and mosses are attached to rocks.

Since the water is shallow, light can usually penetrate to the bottom. But most of these streams are not very productive because of a lack of nutrients. Most nutrients come from organic matter (mostly leaves, branches, and the bodies of living and dead insects) that falls into the stream from nearby land.

This zone is populated by cold-water fish (such as trout in some areas), which need lots of dissolved oxygen. Many fish and other animals in fast-flowing headwater streams have compact and flattened bodies that allow them to live under stones.

In the *transition zone* (Figure 7-18, middle), the headwater streams merge to form wider, deeper streams that flow down gentler slopes with fewer

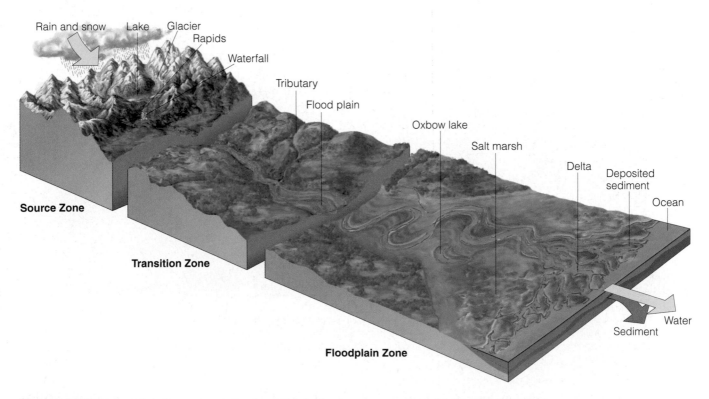

Figure 7-18 Natural capital: three zones in the downhill flow of water: *source zone* containing mountain (headwater) streams, *transition zone* containing wider, lower-elevation streams, and *floodplain zone* containing rivers, which empty into the ocean.

obstacles. The warmer water and other conditions in this zone support more producers (phytoplankton) and a variety of cool-water and warm-water fish species (such as black bass) with slightly lower oxygen requirements.

In the *floodplain zone* (see Figure 7-18, bottom), streams join into wider and deeper rivers that meander across broad, flat valleys. Water in this zone usually has higher temperatures and less dissolved oxygen than water in the first two zones. These slow-moving rivers sometimes support fairly large populations of producers such as algae and cyanobacteria and rooted aquatic plants along the shores. Because of increased erosion and runoff over a larger area, water in this zone often is muddy and contains high concentrations of suspended particulate matter (silt). The main channels of these slow-moving, wide, and murky rivers support distinctive varieties of fish (carp and catfish), whereas their backwaters support species similar to those present in lakes. They are nice places to fish. At its mouth, a river may divide into many channels as it flows through coastal wetlands and estuaries, where the river water mixes with ocean water (Figure 7-7).

As streams flow downhill, they become powerful shapers of land. Over millions of years the friction of moving water levels mountains and cuts deep canyons, and the rock and soil the water removes are deposited as sediment in low-lying areas.

Streams are fairly open ecosystems that receive many of their nutrients from bordering land ecosystems. We have established farmlands and constructed dams, power plants that need cooling water, sewage treatment plants, cities, recreation areas, and shipping terminals in the watersheds along the shores of rivers and streams, especially in their transition and floodplain zones. This greatly increases the flow of plant nutrients, sediment, and pollutants into these ecosystems. To protect a stream or river system from excessive inputs of nutrients and pollutants, we must protect the land around it.

What Are Freshwater Inland Wetlands? Natural Sponges

Inland wetlands absorb and store excess water from storms and provide a variety of wildlife habitats.

Inland wetlands are lands covered with fresh water all or part of the time (excluding lakes, reservoirs, and streams) and located away from coastal areas. Wetlands include *marshes* (dominated by grasses and reeds with a few trees), *swamps* (dominated by trees and shrubs), and *prairie potholes* (depressions carved out by glaciers). Other examples are *floodplains* (which receive excess water during heavy rains and floods)

and the wet *arctic tundra* in summer. Some wetlands are huge and some are small.

Some wetlands are covered with water year-round. Others, called *seasonal wetlands*, usually are underwater or soggy for only a short time each year. They include prairie potholes, floodplain wetlands, and bottomland hardwood swamps. Some stay dry for years before being covered with water again. In such cases, scientists must use the composition of the soil or the presence of certain plants (such as cattails, bulrushes, or red maples) to determine that a particular area is really a wetland. Inland wetlands provide a number of important and free ecological and economic services such as filtering toxic wastes and pollutants, absorbing and storing excess water from storms, and providing habitats for a variety of species.

What Are the Impacts of Human Activities on Freshwater Systems? Using and Abusing Rivers and Wetlands

We have built dams, levees, and dikes that reduce the flow of water and alter wildlife habitats in rivers; established nearby cities and farmlands that pollute streams and rivers; and filled in inland wetlands to grow food and build cities.

Human activities have four major impacts on freshwater systems. *First,* dams, diversions, or canals fragment almost 60% of the world's 237 large rivers. This alters and destroys wildlife habitats along rivers and in coastal deltas and estuaries by reducing water flow.

Second, flood control levees and dikes built along rivers alter and destroy aquatic habitats. *Third,* cities and farmlands add pollutants and excess plant nutrients to nearby streams and rivers.

Fourth, many inland wetlands have been drained or filled to grow crops or have been covered with concrete, asphalt, and buildings. In the United States, more than half of the inland wetlands estimated to have existed in the lower 48 states during the 1600s no longer exist. The total area of these wetland losses is greater than the size of California (which has lost 90% of its wetlands), Oregon, and Nevada combined. Such loss of important natural capital has been an important factor in increased flood and drought damage in the United States. Many other countries have suffered similar losses. For example, 80% of all wetlands in Germany and France have been destroyed.

In this chapter we have seen that human activities are severely stressing and overloading many of the world's aquatic systems. *Good news.* Research shows when such activities are reduced most of these systems can recover fairly quickly as long as aquatic life zones are not overfished or overloaded with pollutants and excessive nutrients.

According to scientists, we urgently need more research on how the world's aquatic systems work. With such information we will have a clearer picture of the impacts of our activities on the earth's aquatic biodiversity and how we can reduce these impacts.

All at last returns to the sea—to Oceanus, the ocean river, like the ever-flowing stream of time, the beginning and the end.

RACHEL CARSON

CRITICAL THINKING

1. List a limiting factor for each of the following: **(a)** the surface layer of a tropical lake, **(b)** the surface layer of the open sea, **(c)** an alpine stream, **(d)** a large, muddy river, **(e)** the bottom of a deep lake.

2. Why do terrestrial organisms evolve tolerances to broader temperature ranges than aquatic organisms do?

3. Why do aquatic plants such as phytoplankton tend to be very small, whereas most terrestrial plants such as trees tend to be larger and have more specialized structures such as stems and leaves for growth? Why are some aquatic animals, especially marine mammals such as whales, extremely large compared with terrestrial animals?

4. How would you respond to someone who proposes that we use the deep portions of the world's oceans to deposit our radioactive and other hazardous wastes because the deep oceans are vast and are located far away from human habitats? Give reasons for your response.

5. Someone tries to sell you several brightly colored pieces of dry coral. Explain in biological terms why this transaction is probably fraudulent.

6. Developers want to drain a large area of inland wetlands in your community and build a large housing development. List **(a)** the main arguments the developers would use to support this project and **(b)** the main arguments ecologists would use in opposing this project. If you were an elected city official, would you vote for or against this project? Can you come up with a compromise plan?

7. You are a defense attorney arguing in court for sparing an undeveloped old-growth tropical rain forest and a coral reef from severe degradation or destruction by de-velopment. Write your closing statement for the defense of each of these ecosystems. If the judge decides you can save only one of the ecosystems, which one would you choose, and why?

8. Congratulations! You are in charge of the world. What are the three most important features of your plan to help sustain the earth's aquatic biodiversity?

PROJECTS

1. Search for information about mangrove trees, using the Internet and library. Are the different species of mangrove trees closely related? What characteristics do they have in common? What characteristics do they have that make them a good place for fish to breed?

2. If possible, visit a nearby lake or reservoir. Would you classify it as oligotrophic, mesotrophic, or eutrophic? What are the primary factors contributing to its nutrient enrichment? Which of these factors are related to human activities?

3. Examine a topographic map for the area around a stream or lake near where you live to define the watershed for the stream. What human activities occur in the watershed? What influence, if any, do you expect these activities to have on the ecology of the stream or lake?

4. Use the library or the Internet to find bibliographic information about *Loren Eisley* and *Rachel Carson,* whose quotes appear at the beginning and end of this chapter.

5. Make a concept map of this chapter's major ideas, using the section heads, subheads. and key terms (in bold-face). Look on the website for this book for information about making concept maps.

LEARNING ONLINE

The website for this book contains study aids and many ideas for further reading and research. They include a chapter summary, review questions for the entire chapter, flash cards for key terms and concepts, a multiple-choice practice quiz, interesting Internet sites, references, and a guide for accessing thousands of InfoTrac® College Edition articles. Log on to

http://biology.brookscole.com/miller14

Then click on the Chapter-by-Chapter area, choose Chapter 7, and select a learning resource.

CASE STUDY
Flying Foxes: Keystone Species in Tropical Forests

The durian (Figure 8-1, top) is a prized fruit growing in Southeast Asian tropical forests. The odor of this football-sized fruit is so strong, it is illegal to have them on trains and in many hotel rooms in Southeast Asia. But its custardlike flesh has been described as "exquisite," "sensual," "intoxicating," and "the world's finest fruit."

Durian fruits come from a wild tree that grows in the tropical rain forest. Various species of nectar-, pollen-, and fruit-eating bats called *flying foxes* (Figure 8-1, bottom right) pollinate the flowers that hang high in durian trees (Figure 8-1, left). This pollination by flying foxes is an example of *mutualism:* an interaction between two species in which both species benefit.

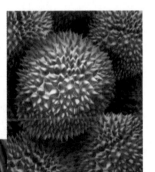

Many species of flying foxes are listed as endangered, and most populations are much smaller than historic numbers. One reason is deforestation. Another is hunting these bat species for their meat, which is sold in China and other parts of Asia. The bats also are killed to keep them from eating commercially grown fruits. Flying foxes are easy to hunt because they tend to congregate in large numbers when they feed or sleep.

Some ecologists classify flying foxes as *keystone species* because of the important roles they play in sustaining tropical forest communities. In addition to pollinating many plant species, the plant seeds they disperse in their droppings help maintain forest biodiversity and regenerate deforested areas.

Thus many other species depend on flying fox species. This explains why ecologists are concerned that the decline of flying fox populations could lead to a cascade of linked extinctions.

The story of flying foxes and durians illustrates the unique role (niche) of interacting species in a community. When populations of different species in a community interact with one another, they influence one another's ability to survive and reproduce. In this chapter we will look at these and other interactions and processes that occur in biological communities.

Figure 8-1 Flying foxes (bottom right) are bats that play key ecological roles in tropical rain forests in Southeast Asia by pollinating (left) and spreading the seeds of durian trees. Durians (top) are a highly prized tropical fruit.

Animal and vegetable life is too complicated a problem for human intelligence to solve, and we can never know how wide a circle of disturbance we produce in the harmonies of nature when we throw the smallest pebble into the ocean of organic life.

GEORGE PERKINS MARSH

This chapter addresses the following questions:

- What determines the number of species in a community?
- What different roles do species play in a community?
- How do species interact with one another in a community?
- How do communities change as environmental conditions change?
- Does high biodiversity increase the stability of a community?

8-1 COMMUNITY STRUCTURE AND SPECIES DIVERSITY

What Is Community Structure? Appearance, Diversity, and Niches

Biological communities differ in their physical appearance, the types and numbers of species they contain, and the ecological roles their species play.

Ecologists use three characteristics to describe a biological community. One is *physical appearance:* the relative sizes, stratification, and distribution of its populations and species, as shown in Figure 8-2 for various terrestrial communities. There are also differences in the physical structures and zones of communities in aquatic life zones such as oceans, rocky shores and sandy beaches, lakes, river systems, and inland wetlands.

The physical structure within a particular type of community or ecosystem can also vary. Most large terrestrial communities and ecosystems consist of a mosaic of vegetation patches of differing size. Life is patchy.

Community structure also varies around its *edges* where one type of community makes a transition to a different type of community. For example, the edge area between a forest and an open field may be sunnier, warmer, and drier than the forest interior and have a different combination of species than the forest and field interiors.

However, increased edge area from habitat fragmentation makes many species more vulnerable to stresses such as predators and fire. It also creates barriers that can prevent some species from colonizing new areas and finding food and mates.

A second characteristic of a community is its *species diversity:* a combination of its number of different species (*species richness*) and the abundance of individuals within each of its species (*species evenness*). For example, suppose we have two communities each with a total of 20 different species and 200 individuals and thus the same species diversity. But these communities could differ in their species richness and species evenness. For example, suppose community A has 10 individuals in each of its 20 species. And community B has 10 species with 2 individuals each and 10 other species, each with 18 individuals. Which community has the highest species evenness?

A third characteristic is a community's *niche structure:* the number of ecological niches, how they resemble or differ from each other, and how species interact

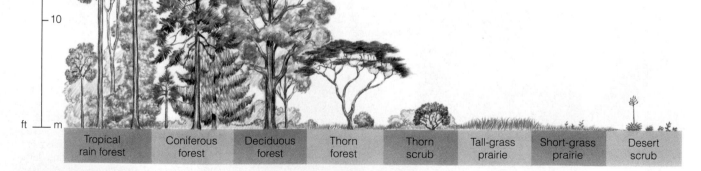

Figure 8-2 Natural capital: generalized types, relative sizes, and stratification of plant species in various terrestrial communities.

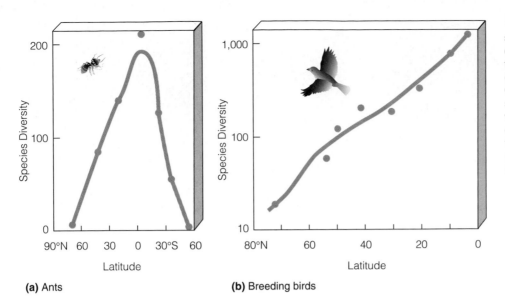

Figure 8-3 Changes in species diversity at different latitudes (distances from the equator) in terrestrial communities for **(a)** ants and **(b)** breeding birds of North and Central America. As a general rule, species diversity steadily declines as we go away from the equator toward either pole. (Modified by permission from Cecie Starr, *Biology: Concepts and Applications*, 4th ed., 2000, Brooks/Cole [Wadsworth])

(a) Ants   **(b)** Breeding birds

with one another. Studies indicate that the most species-rich environments are tropical rain forests, coral reefs, the deep sea, and large tropical lakes. Communities such as a tropical rain forest or a coral reef with a large number of different species (high species richness) generally have only a few members of each species (low species evenness).

Several factors affect the species diversity in communities. One is *latitude* (distance from the equator) in terrestrial communities (Figure 8-3). For most plants and animals, species diversity is highest in the tropics and declines from the equator to the poles. Another factor is *pollution* in aquatic systems (Figure 8-4). Other factors are habitat diversity, NPP, habitat disturbance, and time.

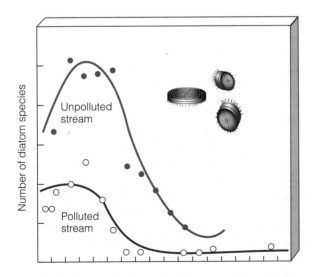

Figure 8-4 Changes in the species diversity and species abundance of diatom species in an unpolluted stream and a polluted stream. Both species diversity and species abundance decrease with pollution.

What Determines the Number of Species on Islands? Entrances and Exits

The number of species on an island is determined by how fast new species arrive and old species become extinct, the island's size, and how far it is from the mainland.

In the 1960s, Robert MacArthur and Edward O. Wilson began studying communities on islands to discover why large islands tend to have more species of a certain category such as insects, birds, or ferns than do small islands. To explain these differences in species richness with island size, MacArthur and Wilson proposed what is called the **species equilibrium model,** or the **theory of island biogeography.** According to this widely accepted model, a balance between two factors determines the number of different species found on an island: the rate at which new species immigrate to the island and the rate at which existing species become extinct on the island.

The model projects that at some point the rates of immigration and extinction should reach an equilibrium point (Figure 8-5a, p. 146) that determines the island's average number of different species. This is a fairly complex idea, so study Figure 8-5 carefully. The CD that comes with this book has a great animation of this model. Check it out.

According to the model, two features of an island affect its immigration and extinction rates and thus its species diversity. One is the island's *size*, with a small island tending to have fewer different species than a large one (Figure 8-5b). One reason is that a small island generally has a lower immigration rate because it is a smaller target for potential colonizers. In addition, a small island should have a higher extinction rate because it usually has fewer resources and less diverse habitats for its species.

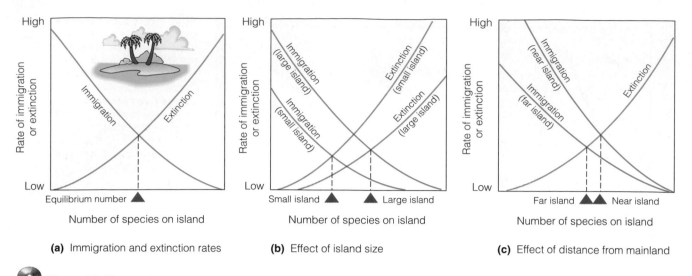

(a) Immigration and extinction rates	**(b)** Effect of island size	**(c)** Effect of distance from mainland

Figure 8-5 The *species equilibrium model* or *theory of island biogeography,* developed by Robert MacArthur and Edward O. Wilson. **(a)** The equilibrium number of species (blue triangle) on an island is determined by a balance between the immigration rate of new species and the extinction rate of species already on the island. **(b)** With time, large islands have a larger equilibrium number of species than smaller islands because of higher immigration rates and lower extinction rates on large islands. **(c)** Assuming equal extinction rates, an island near a mainland will have a larger equilibrium number of species than a more distant island because the immigration rate is greater for a near island than for a more distant one.

A second factor is an island's *distance from the nearest mainland* (Figure 8-5c). According to the model, if we have two islands about equal in size and other factors, the island closer to a mainland source of immigrant species should have the higher immigration rate and thus a higher species richness—assuming that extinction rates on both islands are about the same.

In recent years, scientists have used the model to help protect wildlife in *habitat islands* such as national parks surrounded by a sea of developed and fragmented land.

8-2 TYPES OF SPECIES

What Roles Do Various Species Play in Communities? A Biological Play with Many Parts

Communities can contain native, nonnative, indicator, keystone, and foundation species that play different ecological roles.

Ecologists often use labels—such as *native, nonnative, indicator, keystone,* or *foundation*—to describe the major ecological roles or niches various species play in communities. Any given species may play more than one of these five roles in a particular community.

Native species are those that normally live and thrive in a particular community. Others that evolved somewhere else and then migrate into or are deliber-

ately or accidentally introduced into a community are called **nonnative species, invasive species,** or **alien species.**

Throughout the earth's long history of life species in one part of the world have migrated from one community to another. But moving into a new community is not easy. Most organisms arriving in a new community do not survive because they are not able to find a suitable niche under their new environmental conditions.

Many people tend to think of nonnative or invasive species as villains. But most introduced and domesticated species of crops and animals such as chickens, cattle, and fish from around the world and many wild game species are beneficial to us. Indeed, most of the food crops and domesticated animals we depend on are not native to the communities where we raise them.

But sometimes a nonnative species can thrive and crowd out native species—an example of unintended and unexpected consequences. In 1957, for example, Brazil imported wild African bees to help increase honey production. Instead, the bees displaced domestic honeybees and reduced the honey supply.

Since then, these nonnative bee species, popularly known as "killer bees," have moved northward into Central America and parts of the southwestern United States, such as Texas, Arizona, New Mexico, and California. They are still heading north but should be stopped eventually by cold winters in the central United States unless they can adapt genetically to cold weather.

They are not the killer bees portrayed in some horror movies, but they are aggressive and unpredictable. They have killed thousands of domesticated animals and an estimated 1,000 people in the western hemisphere. Most of the people killed by these honeybees died because they were allergic to their stings or they fell down or became trapped and could not flee.

What Are Indicator Species? Smoke Alarms

Some species can alert us to harmful changes that are taking place in biological communities.

Species that serve as early warnings of damage or danger to a community are called **indicator species.** For example, the presence or absence of trout species in water at temperatures within their range of tolerance (Figure 4-13, p. 64) is an indicator of water quality because trout need clean water with high levels of dissolved oxygen.

Birds are excellent biological indicators because they are found almost everywhere and are affected quickly by environmental change such as loss or fragmentation of their habitats and introduction of chemical pesticides. Many bird species are declining. Butterflies are also good indicator species and in some areas are declining faster than bird species.

Using a living organism to monitor environmental quality is not new. Coal mining is a dangerous occupation, partly because of the underground presence of poisonous and explosive gases, many of which have no detectable odor. In the 1800s and early 1900s, coal miners took caged canaries into mines to act as early-warning sentinels. These birds sing loudly and often. If they quit singing for a long period and they appeared to be distressed, miners took this as an indicator for the presence of poisonous or explosive gases and got out of the mine.

The latest idea is to use indicator species to combat terrorism. Some scientists are trying to genetically engineer species of weedy plants to change their color rapidly when exposed to a harmful biological or chemical agent. If successful, genes from these plants could be inserted into evergreen trees, backyard shrubs, cheap houseplants, or even pond algae to turn them into early-warning systems for attacks using biological or chemical weapons.

Case Study: Why Are Amphibians Vanishing? Warnings from Frogs

The disappearance of many of the world's amphibian species may indicate a decline in environmental quality in many parts of the world.

Amphibians (frogs, toads, and salamanders) live part of their lives in water and part on land and are classified as indicator species. Frogs, for example, are good indicator species because they are especially vulnerable to environmental disruption at various points in their life cycle, shown in Figure 8-6. As tadpoles they live in water and eat plants, and as adults they live

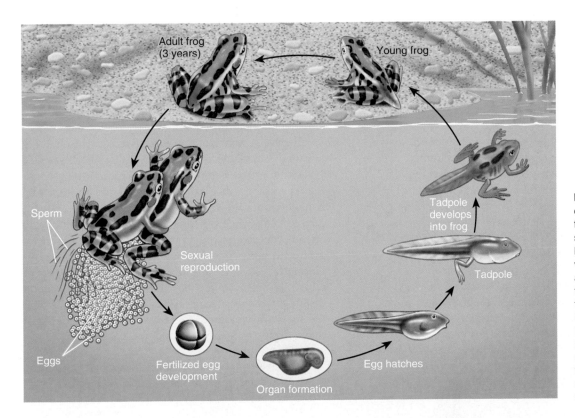

Figure 8-6 Typical life cycle of a frog. Populations of various frog species can decline because of the effects of various harmful factors at different points in their life cycle. Such factors include habitat loss, drought, pollution, increased ultraviolet radiation, parasitism, disease, overhunting for food (frog legs), and nonnative predators and competitors.

mostly on land and eat insects that can expose them to pesticides. Their eggs have no protective shells to block ultraviolet (UV) radiation or pollution. As adults, they take in water and air through their thin, permeable skins that can readily absorb pollutants from water, air, or soil.

Since 1980, populations of hundreds of the world's estimated 5,280 amphibian species have been vanishing or declining in almost every part of the world, even in protected wildlife reserves and parks.

No single cause has been identified to explain the amphibian declines. However, scientists have identified a number of factors that can affect frogs and other amphibians at various points in their life cycle.

One factor is *habitat loss and fragmentation,* especially because of the draining and filling of inland wetlands, deforestation, and development. Another is *prolonged drought,* which dries up breeding pools so that few tadpoles survive. Dehydration from lack of water can also weaken amphibians and make them more susceptible to fatal viruses, bacteria, fungi, and parasites.

Pollution can also play a role. Frog eggs, tadpoles, and adults are very sensitive to many pollutants, especially pesticides. Exposure to such pollutants may also make them more vulnerable to bacterial, viral, and fungal diseases, and cause an array of sexual abnormalities.

Increases in UV radiation caused by reductions in stratospheric ozone—caused when certain chemicals we make drift up into the stratosphere—can harm young embryos of amphibians found in shallow ponds. *Increased incidence of parasitism* by a flatworm (trematode) may account for deformities found in some frog species but not the worldwide decline of amphibians.

Overhunting can be a factor, especially in Asia and France, where frog legs are a delicacy. Populations of some amphibians can also be reduced by *immigration or introduction of nonnative predators and competitors* (such as fish) *and disease organisms.* A combination of such factors probably is responsible for the decline or disappearance of most amphibian species.

So why should we care if various amphibian species become extinct? Scientists give three reasons. *First,* it suggests that environmental quality is deteriorating in parts of the world because amphibians are sensitive biological indicators of changes in environmental conditions such as habitat loss and degradation, pollution, UV exposure, and climate change.

Second, adult amphibians play important ecological roles in biological communities. For example, amphibians eat more insects (including mosquitoes) than do birds. In some habitats, extinction of certain amphibian species could lead to extinction of other species, such as reptiles, birds, aquatic insects, fish, mammals, and other amphibians that feed on them or their larvae.

Third, from a human perspective, amphibians represent a genetic storehouse of pharmaceutical products waiting to be discovered. Compounds in hundreds of secretions from amphibian skin have been isolated and some are used as painkillers and antibiotics and in treating burns and heart disease.

The plight of some amphibian indicator species is a warning signal. They do not need us, but we and other species need them.

What Are Keystone Species? Major Players Who Help Keep Ecosystems Running Smoothly

Keystone species help determine the types and numbers of various other species in a community.

A keystone is the wedge-shaped stone placed at the top of a stone archway. Remove this stone and the arch collapses. In some communities, certain species called **keystone species** apparently play a similar role. They have a much larger effect on the types and abundances of many other species in a community than their numbers would suggest.

According to this hypothesis, keystone species play critical ecological roles. One is *pollination* of flowering plant species by bees, hummingbirds, bats, and other species. In addition, *top predator* keystone species feed on and help regulate the populations of other species. Examples are the wolf, leopard, lion, alligator (Connections, at right), and great white shark.

Have you thanked a *dung beetle* today? Maybe you should, because these keystone species rapidly remove, bury, and recycle animal wastes or dung. Without them we would be up to our eyeballs in such waste, and many plants would be starved for nutrients. These beetles also churn and aerate the soil, making it more suitable for plant life.

Ecologist Robert Paine conducted a controlled experiment along the rocky Pacific coast of the state of Washington that demonstrated the role of the sea star *Piaster orchaceus* as a keystone species in an intertidal zone community (Figure 7-9, top, p. 134). He removed sea stars from one community but not from an adjacent community, which served as a control group. In both communities he monitored the populations of 18 other species. In the community from which he removed the sea stars, all of the 18 species disappeared except mussels. In the community where the sea stars remained, they ate the mussels and kept them from multiplying and crowding out other species.

The loss of a keystone species can lead to population crashes and extinctions of other species in a com-

CONNECTIONS

The American alligator, North America's largest reptile, has no natural predators except humans. This species, which has been around for about 200 million years, has been able to adapt to numerous changes in the earth's environmental conditions.

This changed when hunters began killing large numbers of these animals for their exotic meat and their supple belly skin, used to make shoes, belts, and pocketbooks.

Other people considered alligators to be useless and dangerous and hunted them for sport or out of hatred. Between 1950 and 1960, hunters wiped out 90% of the alligators in Louisiana. By the 1960s, the alligator population in the Florida Everglades also was near extinction.

People who say "So what?" are overlooking the alligator's important ecological role or *niche* in subtropical wetland ecosystems. Alligators dig deep depressions, or gator holes. These holes hold fresh water during dry spells, serve as refuges for aquatic life, and supply fresh water and food for many animals.

In addition, large alligator nesting mounds provide nesting and feeding sites for herons and egrets. Alligators also eat large numbers of gar (a predatory fish) and thus help maintain populations of game fish such as bass and bream.

As alligators move from gator holes to nesting mounds, they help keep areas of open water free of invading vegetation. Without these free ecosystem services, freshwater ponds and shrubs and trees would fill in the coastal wetlands in the alligator's habitat, and dozens of species would disappear.

Some ecologists classify the North American alligator as a *keystone species* because of these important ecological roles in helping maintain the structure and function of its natural ecosystems. Some say it can also be classified as a *foundation species*.

In 1967, the U.S. government placed the American alligator on the endangered species list. Protected from hunters, the alligator population made a strong comeback in many areas by 1975—too strong, according to those who find alligators in their backyards and swimming pools, and to duck hunters, whose retriever dogs sometimes are eaten by alligators.

In 1977, the U.S. Fish and Wildlife Service reclassified the American alligator from an *endangered* species to a *threatened* species in Florida, Louisiana, and Texas, where 90% of the animals live. In 1987, this reclassification was extended to seven other states.

Alligators now number perhaps 3 million, most in Florida and Louisiana. It is generally illegal to kill members of a threatened species, but limited kills by licensed hunters are allowed in some areas of Texas, Florida, Louisiana, South Carolina, and Georgia to control the population.

To biologists, the comeback of the American alligator from near premature extinction by overhunting is an important success story in wildlife conservation.

The increased demand for alligator meat and hides has created a booming business in alligator farms, especially in Florida. Such farms reduce the rewards for illegal hunting of wild alligators.

Critical Thinking

Some homeowners in Florida believe they should have the right to kill any alligator found on their property. Others argue this should not be allowed because alligators are a threatened species, and housing developments have invaded the habitats of alligators, not the other way around. What is your opinion on this issue? Explain.

---

munity that depend on it for certain services. According to biologist Edward O. Wilson, "The loss of a keystone species is like a drill accidentally striking a power line. It causes lights to go out all over."

What Are Foundation Species? Players Who Create New Habitats and Niches

Foundation species can create and enhance habitats that can benefit other species in a community.

Some ecologists think the keystone species should be expanded to include the roles of *foundation species*, which play a major role in shaping communities by creating and enhancing habitat that benefits other species. For example, elephants push over, break, or uproot trees, creating forest openings in the savanna grasslands and woodlands of Africa. This promotes the growth of grasses and other forage plants that benefit smaller grazing species such as antelope. It also accelerates nutrient cycling rates. Some bat and bird foundation species can regenerate deforested areas and spread fruit plants by depositing plant seeds in their droppings.

Proponents of the foundation species hypothesis say that Paine's study of the role of the sea star *Piaster orchaceus* as a keystone species in an intertidal zone community did not take into account the role of mussels as *foundation species*.

According to this hypothesis, mussel beds are homes to hundreds of invertebrate species that do poorly in the presence of mussel competitors such as sea stars. When scientists measured the overall diversity of the species in a tide pool rather than just the 18 species observed by Paine they found that the overall diversity of species was greater when the keystone sea star species was absent. Its absence allowed the number of mussel species and the species they interact with to expand. From this point of view, the mussels should be viewed as a *foundation species* that expanded species richness.

8-3 SPECIES INTERACTIONS: COMPETITION AND PREDATION

How Do Species Interact? Ways to Get an Edge

Competition, predation, parasitism, mutualism, and commensalism are ways in which species can interact and increase their ability to survive.

When different species in a community have activities or resource needs in common, they may interact with one another. Members of these species may be harmed, helped, or unaffected by the interaction. Ecologists identify five basic types of interactions between species: *interspecific competition, predation, parasitism, mutualism,* and *commensalism.*

The most common interaction between species is *competition* for shared or scarce resources such as space and food. Ecologists call such competition between *species* **interspecific competition.**

When this occurs, parts of the fundamental niches of the competing species overlap (Figure 5-7, p. 94). With significant overlap, one of the competing species must migrate, if possible, to another area, shift its feeding habits or behavior through natural selection and evolution, suffer a sharp population decline, or become extinct in that area.

Humans are in competition with many other species for space, food, and other resources. As we convert more and more of the earth's land and aquatic resources and net primary productivity to our uses we deprive many other species of resources they need to survive.

How Have Some Species Reduced or Avoided Competition? Share the Wealth by Becoming More Specialized

Some species evolve adaptations that allow them to reduce or avoid competition for resources with other species.

Over a time scale long enough for evolution to occur, some species competing for the same resources evolve adaptations that reduce or avoid competition. One way this happens is through **resource partitioning.** It occurs when species competing for similar scarce resources evolve more specialized traits that allow them to use shared resources at different times, in different ways, or in different places.

Through evolution, the fairly broad niches of two competing species (Figure 8-7, top) can become more specialized (Figure 8-7, bottom) so that the species can share limited resources. When lions and leopards live in the same area, lions take mostly larger animals as prey, and leopards take smaller ones. Hawks and owls feed on similar prey, but hawks hunt during the day and owls hunt at night.

Ecologist Robert H. MacArthur studied the feeding habits of five species of North American warblers (small insect-eating birds) that hunt for insects and nest in the same type of spruce tree. He found that the bird species minimized the overlap of their niches and reduced competition among the species through resource partitioning. They did this by concentrating much of their hunting for insects in different parts of the spruce trees (Figure 8-8), employing different

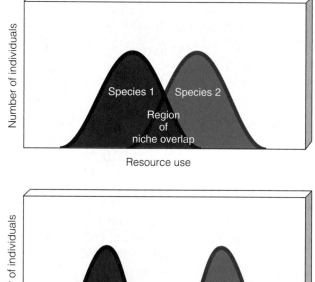

Figure 8-7 Natural capital: *resource partitioning* and *niche specialization* as a result of competition between two species. The top diagram shows the overlapping niches of two competing species. The bottom diagram shows that through evolution the niches of the two species become separated and more specialized (narrower) so that they avoid competing for the same resources.

hunting tactics, and nesting at slightly different times. Other examples are shown in Figure 5-5 (p. 92).

How Do Predator and Prey Species Interact? Eating and Being Eaten

Predator species feed on all or parts of other species called prey.

In **predation,** members of one species (the *predator*) feed directly on all or part of a living organism of another species (the *prey*). In this interaction, the predators benefit and the prey are harmed. The two kinds of organisms are said to have a *predator–prey relationship.* Such relationships are depicted in Figures 4-11 (p. 63), 4-12 (p. 63), 4-18 (p. 68), and 4-19 (p. 69).

Most of the world's predation is not the kind we see in many nature documentaries such as lions killing zebras or bears plucking salmon from streams. Instead, most predation occurs unseen at the microscopic level in soils and in the sediments of aquatic systems.

At the individual level, members of the prey species are clearly harmed. But at the population level, predation plays a role in evolution by natural selection. Predation can benefit the prey species because predators often kill the sick, weak, aged, and least fit members of a population (Case Study, below). This gives remaining prey better access to food supplies and prevents excessive population growth. Predation also helps successful genetic traits to become more dominant in the prey population through natural selection. This can enhance the reproductive success and long-term survival of the prey species.

Some people tend to view predators with contempt. When a hawk tries to capture and feed on a rabbit, some root for the rabbit. Yet the hawk like all predators is merely trying to get enough food to feed itself and its young; in the process, it is playing an important ecological role in controlling rabbit populations.

Case Study: Why Are Sharks Important Species? Culling the Oceans and Helping Improve Human Health

Some shark species eat and remove sick and injured ocean animals and some can help us learn how to fight cancer and immune system disorders.

The world's 370 shark species vary widely in size. The smallest is the dwarf dog shark, about the size of a large goldfish. The largest is the whale shark, the world's largest fish. It can grow to 15 meters (50 feet) long and weigh as much as two full-grown African elephants!

Various shark species, feeding at the top of food webs, cull injured and sick animals from the ocean and thus play an important ecological role. Without such shark species, the oceans would be overcrowded with dead and dying fish.

Figure 8-8 Sharing the wealth: *resource partitioning* of five species of common insect-eating warblers in the spruce forests of Maine. Each species minimizes competition with the others for food by spending at least half its feeding time in a distinct portion (shaded areas) of the spruce trees, and consuming somewhat different insect species. (After R. H. MacArthur, "Population Ecology of Some Warblers in Northeastern Coniferous Forests," *Ecology* 36 (1958): 533–36)

Many people—influenced by movies (such as *Jaws*), popular novels, and widespread media coverage of a fairly small number of shark attacks per year—think of sharks as people-eating monsters. However, the three largest species—the whale shark, basking shark, and megamouth shark—are gentle giants. They swim through the water with their mouths open, filtering out and swallowing huge quantities of *plankton.*

Every year, members of a few species of shark—mostly great white, bull, tiger, gray reef, lemon, hammerhead, shortfin mako, and blue—typically injure 60–100 people worldwide (60 in 2002). Between 1990 and 2003, sharks killed 8 people off U.S. coasts and 88 people worldwide—an average of 7 people per year. Most attacks are by great white sharks, which feed on sea lions and other marine mammals and sometimes mistake divers and surfers for their usual prey. Whose fault is this?

Media coverage of such attacks greatly distorts the danger from sharks. You are 30 times more likely to be killed by lightning than by a shark, and your chance of being killed by lightning is extremely small.

For every shark that kills or injures a person, we kill at least 1 million sharks, a total of about 100 million sharks each year. Sharks are caught mostly for their fins and then thrown back into the water to die.

Shark fins are widely used in Asia as a soup ingredient and as a pharmaceutical cure-all. They are worth as much as $563 per kilogram ($256 per pound). In high-end Hong Kong restaurants, a single bowl of shark fin soup can cost as much as $100!

According to a 2001 study by Wild Aid, shark fins sold in restaurants throughout Asia and in Chinese communities in cities such as New York, San Francisco, and London contain dangerously high levels of toxic mercury. Consumption of high levels of mercury is especially threatening for pregnant women, fetuses, and infants feeding on breast milk.

Sharks are also killed for their livers, meat (especially mako and thresher), hides (a source of exotic, high-quality leather), and jaws (especially great whites, whose jaws can sell for up to $10,000). They are also killed because we fear them. Some sharks (especially blue, mako, and oceanic whitetip) die when they are trapped in nets or lines deployed to catch swordfish, tuna, shrimp, and other commercially important species.

In addition to their important ecological roles, sharks can save human lives. They are helping us learn how to fight cancer, which sharks almost never get. Scientists are also studying their highly effective immune system because it allows wounds to heal without becoming infected.

Sharks have three natural traits that make them prone to population declines from overfishing. They take a long time to reach sexual maturity (10–24 years), have only a few offspring (between 2 and 10) once every year or two, and have long gestation (pregnancy) periods (up to 24 months for some species).

Sharks are among the most vulnerable and least protected animals on the earth. Eight of the world's shark species are in danger of extinction. In 2003, experts at the National Aquarium in Baltimore, Maryland, estimated that populations of a number of commercially valuable shark species have decreased by 90% since 1992.

In response to a public outcry over depletion of some shark species, the United States and several other countries have banned practices such as shark finning in their territorial waters. But such bans are difficult to enforce.

Some critics do not understand the concern because they see sharks as fish we should be able to catch and do what we want with. They also do not believe that sharks suffer from any harm we inflict on them so there is no need to treat them humanely. And they resent the United States and other countries forcing their ethical concerns about killing or harming sharks on nations and individuals that do not share these views. They also worry that such ethical concerns about species such as sharks and whales could spread to other aquatic and terrestrial species.

With more than 400 million years of evolution behind them, sharks have had a long time to get things right. Preserving their evolutionary genetic development begins with the knowledge that sharks do not need us, but we and other species need them.

> **X** *HOW WOULD YOU VOTE?* Do we have an ethical obligation to protect shark species from premature extinction and treat them humanely? Cast your vote online at http://biology .brookscole.com/miller14.

How Do Predators Increase Their Chances of Getting a Meal? Pursue, Ambush, and Immobilize

Some predators are fast enough to catch their prey, some hide and lie in wait, and some inject chemicals to paralyze their prey.

Predators have a variety of methods that help them capture prey. *Herbivores* can simply walk, swim, or fly up to the plants they feed on.

Carnivores feeding on mobile prey have two main options: *pursuit* and *ambush.* Some, such as the cheetah, catch prey by running fast; others, such as the American bald eagle, fly and have keen eyesight; still others, such as wolves and African lions, cooperate in capturing their prey by hunting in packs.

Other predators use camouflage—a change in shape or color—to hide in plain sight and ambush their prey. For example, praying mantises sit in flowers of a similar color and ambush visiting insects. White ermines (a type of weasel) and snowy owls hunt in snow-covered areas. The alligator snapping turtle, camouflaged in its stream-bottom habitat, dangles its worm-shaped tongue to entice fish into its powerful jaws. People camouflage themselves to hunt wild game and use camouflaged traps to ambush wild game.

Some predators use chemical warfare to attack their prey. For example, spiders and poisonous snakes use venom to paralyze their prey and to deter their predators.

How Do Prey Defend Themselves Against or Avoid Predators? Escape, Repel, Deceive, and Poison

Some prey escape their predators or have protective shells or thorns, some camouflage themselves, and some use chemicals to repel or poison predators

Prey species have evolved many ways to avoid predators, including the ability to run, swim, or fly fast, and a highly developed sense of sight or smell that alerts them to the presence of predators. Other avoidance adaptations are protective shells (as on armadillos, which roll themselves up into an armor-plated ball, and turtles), thick bark (giant sequoia), spines (porcupines), and thorns (cacti and rosebushes). Many lizards have brightly colored tails that break off when they are attacked, often giving them enough time to escape.

Other prey species use the camouflage of certain shapes or colors or the ability to change color (chameleons and cuttlefish). Some insect species have evolved shapes that look like twigs (Figure 8-9a), bark, thorns, or even bird droppings on leaves. A leaf insect may be almost invisible against its background (Figure 8-9b), and an arctic hare in its white winter fur blends into the snow. A spotted cheetah blends into the grass as it watches for grazing animals to chase down and kill.

Chemical warfare is another common strategy. Some prey species discourage predators with chemicals that are *poisonous* (oleander plants), *irritating* (stinging nettles and bombardier beetles, Figure 8-9c), *foul smelling* (skunks, skunk cabbages, and stinkbugs), or *bad tasting* (buttercups and monarch butterflies, Figure 8-9d). When attacked, some species

(a) Span worm

(b) Wandering leaf insect

(c) Bombardier beetle

(d) Foul-tasting monarch butterfly

(e) Poison dart frog

(f) Viceroy butterfly mimics monarch butterfly

(g) Hind wings of Io moth resemble eyes of a much larger animal.

(h) When touched, snake caterpillar changes shape to look like head of snake.

Figure 8-9 Some ways in which prey species avoid their predators by (a, b) camouflage, (c–e) chemical warfare, (d, e) warning coloration, (f) mimicry, (g) deceptive looks, and (h) deceptive behavior.

of squid and octopus emit clouds of black ink to confuse the predator and allow them to escape.

Scientists have identified more than 10,000 defensive chemicals made by plants. Some are herbivore poisons such as cocaine, caffeine, cyanide, opium, strychnine, peyote, nicotine, and rotenone, some of which we use as an insecticide. Others are herbivore repellents such as pepper, mustard, nutmeg, oregano, cinnamon, and mint, all of which we use to flavor or spice up our food. Major pharmaceutical companies view the plant world as a vast drugstore to study as a source for new medicines to treat a variety of human diseases and as a source of natural pesticides. Scientists going into nature to find promising natural chemicals are called *bioprospectors.* You might consider a career doing this fascinating work.

Many bad-tasting, bad-smelling, toxic, or stinging prey species have evolved *warning coloration,* brightly colored advertising that enables experienced predators to recognize and avoid them. They flash a warning, "Eating me is risky." Examples are brilliantly colored poisonous frogs (Figure 8-9e), red-, yellow-, and black-striped coral snakes, and foul-tasting monarch butterflies (Figure 8-9d) and grasshoppers.

Based on coloration, biologist Edward O. Wilson gives us two rules for evaluating possible danger from an unknown animal species we encounter in nature. *First,* if it is small and strikingly beautiful, it is probably poisonous. *Second,* if it is strikingly beautiful and easy to catch, it is probably deadly.

Other butterfly species, such as the nonpoisonous viceroy (Figure 8-9f), gain some protection by looking and acting like the monarch, a protective device known as *mimicry.* The harmless mountain king snake avoids predation by looking like the deadly and brilliantly colored coral snake.

Some prey species use *behavioral strategies* to avoid predation. Some attempt to scare off predators by puffing up (blowfish), spreading their wings (peacocks), or mimicking a predator (Figure 8-9h). Some moths have wings that look like the eyes of much larger animals (Figure 8-9g). Other prey species gain some protection by living in large groups (schools of fish, herds of antelope, flocks of birds).

Disease-carrying bacteria and fungi also attack species. Animals such as ants that live in complex social societies containing millions to trillions of individuals survive onslaughts by harmful infectious bacteria and fungi by evolving glands that secrete antibiotics and antifungals. Indeed, because of these secretions, the outer surface of an ant is almost free of bacteria and fungi and is much cleaner than most human skin. The ecological lessons from ants are simple and powerful. In a world ruled by bacteria, never bet against the bacteria and always rely on a variety of weapons!

Some biologists have begun exploring the use of antibiotics produced by ants to treat human infectious diseases. So far two antibiotic patents have been filed based on studying ants, and there are more to come. You might consider a career in this research frontier.

8-4 SPECIES INTERACTIONS: PARASITISM, MUTUALISM, AND COMMENSALISM

What Are Parasites, and Why Are They Important? Living On or In Another Species

Although parasites can harm their host organisms they can promote community biodiversity.

Parasitism occurs when one species (the *parasite*) feeds on part of another organism (the *host*), usually by living on or in the host. In this relationship, the parasite benefits and the host is harmed.

Parasitism can be viewed as a special form of predation. But unlike a conventional predator, a parasite usually is much smaller than its host (prey) and rarely kills its host. Also most parasites remain closely associated with, draw nourishment from, and may gradually weaken their hosts over time.

Tapeworms, disease-causing microorganisms, and other parasites live *inside* their hosts. Other parasites attach themselves to the *outside* of their hosts. Examples are ticks, fleas, mosquitoes, mistletoe plants, and fungi that cause diseases such as athlete's foot. Some parasites move from one host to another, as fleas and ticks do; others, such as tapeworms, spend their adult lives with a single host.

Some parasites have little contact with their host. For example, in North America cowbirds parasitize or take over the nests of other birds by laying their eggs in them and then letting the host birds raise their young.

From the host's point of view, parasites are harmful, but parasites play important ecological roles. Collectively, the matrix of parasitic relationships in a community acts somewhat like glue that helps hold the various species in a community together. Parasites also promote biodiversity by helping keep some species from becoming so plentiful that they eliminate other species.

How Do Species Interact So That Both Species Benefit? Win-Win Relationships

Pollination, fungi that help plant roots take up nutrients, and bacteria in your gut that help digest your food are examples of species interactions that benefit both species.

In **mutualism,** two species interact in a way that benefits both. The *pollination mutualism* between flowering plants and animals such as insects, birds, and bats is one of the most common forms of mutualism.

Some species benefit from *nutritional mutualism. Lichens,* hardy species that can grow on trees or barren rocks, consist of colorful photosynthetic algae and chlorophyll-lacking fungi living together. The fungi provide a home for the algae, and their bodies collect and hold moisture and mineral nutrients used by both species. The algae, through photosynthesis, provide sugars as food for themselves and the fungi.

A mutualistic relationship that combines nutrition and protection is *birds* that ride on the backs of large animals like African buffalo, elephants, and rhinoceroses (Figure 8-10a). The birds remove and eat parasites from the animal's body and often make noises warning the animal when predators approach.

Another example is *clownfish species*, which live within sea anemones, whose tentacles sting and paralyze most fish that touch them (Figure 8-10b). The clownfish, which are not harmed by the tentacles, gain protection from predators and feed on the detritus left from the meals of the anemones. The sea anemones benefit because the clownfish protect them from some of their predators.

Another example of nutritional mutualism is the highly specialized fungi that combine with plant roots to form mycorrhizae (from the Greek words for fungus and roots). The fungi get nutrition from the plant's roots. In turn the fungi benefit the plant by using their myriad networks of hairlike extensions to improve the plant's ability to extract nutrients and water from the soil (Figure 8-10c).

In *gut inhabitant mutualism*, vast armies of organisms such as bacteria live in an animal's digestive tract. The bacteria receive a sheltered habitat and food from their host. In turn, they help break down (digest) their host's food. Examples are the bacteria inside a termite's gut that digest wood or cellulose and provide the termite with food. Similarly, bacteria in your gut help digest the food you eat. Thank these little critters for helping keep you alive.

(a) Oxpeckers and black rhinoceros

(b) Clownfish and sea anemone

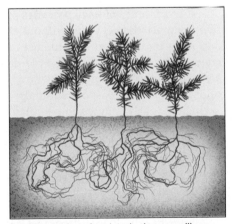

(c) Mycorrhizae fungi on juniper seedlings in normal soil

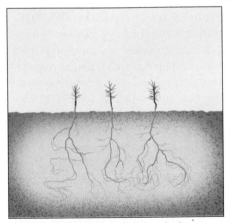

(d) Lack of mycorrhizae fungi on juniper seedlings in sterilized soil

Figure 8-10 Examples of *mutualism*. (a) Oxpeckers (or tickbirds) feed on and remove parasitic ticks that infest large thick-skinned animals such as a black rhinoceros. (b) A clownfish gains protection and food by living among deadly stinging sea anemones and helps protect the anemones from some of their predators. (c) Beneficial effects of mycorrhizal fungi attached to roots of juniper seedlings on plant growth compared to (d) growth of such seedlings in sterilized soil without mycorrhizal fungi.

It is tempting to think of mutualism as an example of cooperation between species, but actually it involves each species benefiting by exploiting the other.

How Do Species Interact So That One Benefits but the Other Is Not Harmed? Do No Harm

Some species interact in a way that helps one species but has little if any effect on the other.

Commensalism is a species interaction that benefits one species but has little, if any, effect on the other species. One example is a *redwood sorrel,* a small herb. It benefits from growing in the shade of tall redwood trees, with no known negative effects on the redwood trees.

Another example is plants called *epiphytes* (such as some types of orchids and bromeliads) that attach themselves to the trunks or branches of large trees in tropical and subtropical forests. These so-called *air plants* benefit by having a solid base on which to grow. They also live in an elevated spot that gives them better access to sunlight, water from the humid air and rain, and nutrients falling from the tree's upper leaves and limbs. This apparently does not harm the tree.

8-5 ECOLOGICAL SUCCESSION: COMMUNITIES IN TRANSITION

How Do Ecosystems Respond to Change? Shifting Community Composition

Over time new environmental conditions can cause changes in community structure that lead to one group of species being replaced by other groups.

All communities change their structure and composition over time in response to changing environmental conditions. The gradual change in species composition of a given area is called **ecological succession.** During succession some species colonize an area and their populations become more numerous, whereas populations of other species decline and may even disappear.

Ecologists recognize two types of ecological succession, depending on the conditions present at the beginning of the process. One is **primary succession,** which involves the gradual establishment of biotic communities on nearly lifeless ground. With the other, more common type, called **secondary succession,** biotic communities are established in an area where some type of biotic community is already present.

What Is Primary Succession? Establishing Life on Lifeless Ground

Over long periods, a series of communities with different species can develop in lifeless areas where there is no soil or bottom sediment.

Primary succession begins with an essentially lifeless area where there is no soil in a terrestrial ecosystem (Figure 8-11) or no bottom sediment in an aquatic ecosystem. Examples include bare rock exposed by a retreating glacier or severe soil erosion, newly cooled lava, an abandoned highway or parking lot, or a newly created shallow pond.

Primary succession usually takes an extremely long time. One reason is that before a community can become established on land, there must be soil. Depending mostly on the climate, it takes natural processes several hundred to several thousand years to produce fertile soil.

Soil formation begins when hardy **pioneer species** attach themselves to inhospitable patches of bare rock. Examples are wind-dispersed lichens and mosses, which can withstand the lack of moisture and soil nutrients and hot and cold temperature extremes found in such habitats.

These tough species start the soil formation process on patches of bare rock by trapping wind-blown soil particles and tiny pieces of detritus, producing tiny bits of organic matter, and secreting mild acids that slowly fragment and break down the rock. This chemical breakdown (weathering) is hastened by physical weathering such as the fragmentation of rock that occurs when water freezes in cracks and expands. This is a slow process. It may take a lichen 100 years to grow as large as a dinner plate.

As patches of soil build up and spread, a new plant community replaces the community of lichens and mosses. Most of these plants are tiny *annuals* that live for only a year. However, they produce flowers and seeds that fall to the ground and can germinate for the following growing season. These taller plants eliminate the lichens by depriving them of sunlight.

A community of *small perennial grasses* (plants that live for more than 2 years without having to reseed) and *herbs* or *ferns* normally replaces the annual plant community. The seeds of these plants germinate after arriving on the wind and in the rain, in the droppings of birds, or on the coats of mammals.

These **early successional plant species** grow close to the ground, can establish large populations quickly under harsh conditions, and have short lives. Some of their roots penetrate the rock and help break it up into more soil particles. The decay of their wastes and dead bodies also adds more nutrients to the soil.

After hundreds to a thousand or more years, the soil may be deep and fertile enough to store enough moisture and nutrients to support the growth of less hardy **midsuccessional plant species** of herbs, grasses, and low shrubs. Trees that need lots of sunlight and are adapted to the area's climate and soil usually replace these species.

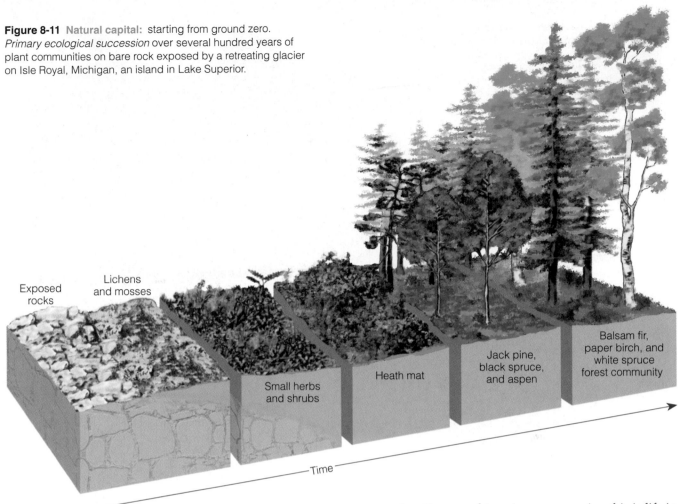

Figure 8-11 Natural capital: starting from ground zero. *Primary ecological succession* over several hundred years of plant communities on bare rock exposed by a retreating glacier on Isle Royal, Michigan, an island in Lake Superior.

Exposed rocks

Lichens and mosses

Small herbs and shrubs

Heath mat

Jack pine, black spruce, and aspen

Balsam fir, paper birch, and white spruce forest community

Time

As these tree species grow and create shade, they are replaced by **late successional plant species** (mostly trees) that can tolerate shade. Unless fire, flooding, severe erosion, tree cutting, climate change, or other natural or human processes disturb the area, what was once bare rock becomes a complex forest community.

Primary succession can also take place in newly created small ponds as a result of an influx of sediments and nutrients in runoff from the surrounding land. This sediment can support seeds or spores of plants reaching the pond by winds, birds, or other animals. Over time this process can transform the pond first into a marsh and eventually to dry land.

What Is Secondary Succession? Life Building on Life

A series of communities with different species can develop in places containing some soil or bottom sediment.

Secondary succession begins in an area where the natural community of organisms has been disturbed, removed, or destroyed but some soil or bottom sediment

remains. Compared to primary succession this is life in the fast lane. Candidates for secondary succession include abandoned farmlands, burned or cut forests, heavily polluted streams, and land that has been dammed or flooded. Because some soil or sediment is present, new vegetation can usually begin to germinate within a few weeks. Seeds can be present in soils, or they can be carried from nearby plants by wind or by birds and other animals.

European settlers cleared the mature native oak and hickory forests and planted the land with crops in the central or Piedmont region of North Carolina. Later they abandoned some of this farmland because of erosion and loss of soil nutrients. Figure 8-12 (p. 158) shows one way that such abandoned farmland has undergone secondary succession over 150–200 years.

Descriptions of ecological succession usually focus on changes in vegetation. But these changes in turn affect food and shelter for various types of animals. Thus as succession proceeds, the numbers and types of animals and decomposers also change.

As we have seen, primary and secondary succession involve changes in community structure. Thus the various stages of succession have different patterns of species diversity, trophic structure, niches, nutrient cycling, and energy flow and efficiency (Table 8-1, p. 158).

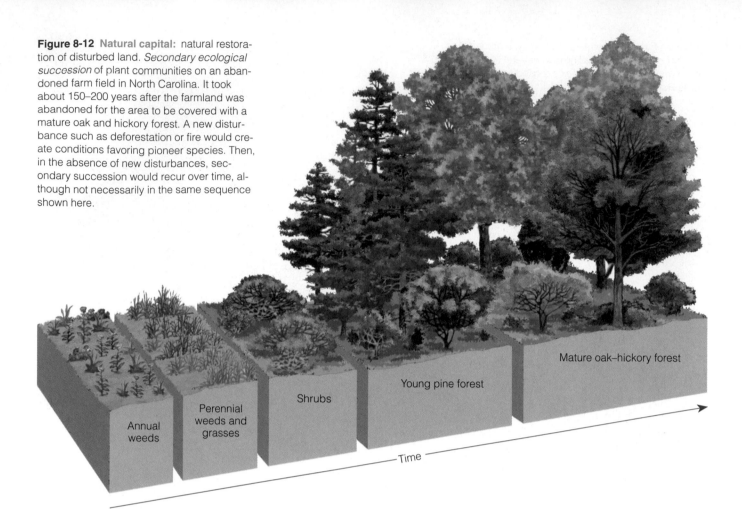

Figure 8-12 Natural capital: natural restoration of disturbed land. *Secondary ecological succession* of plant communities on an abandoned farm field in North Carolina. It took about 150–200 years after the farmland was abandoned for the area to be covered with a mature oak and hickory forest. A new disturbance such as deforestation or fire would create conditions favoring pioneer species. Then, in the absence of new disturbances, secondary succession would recur over time, although not necessarily in the same sequence shown here.

Annual weeds

Perennial weeds and grasses

Shrubs

Young pine forest

Mature oak–hickory forest

Time

Characteristic	Immature Ecosystem (Early Successional Stage)	Mature Ecosystem (Late Successional Stage)
Table 8-1 Ecosystem Characteristics at Immature and Mature Stages of Ecological Succession		
Ecosystem Structure		
Plant size	Small	Large
Species diversity	Low	High
Trophic structure	Mostly producers, few decomposers	Mixture of producers, consumers, and decomposers
Ecological niches	Few, mostly generalized	Many, mostly specialized
Community organization (number of interconnecting links)	Low	High
Ecosystem Function		
Biomass	Low	High
Net primary productivity	High	Low
Food chains and webs	Simple, mostly plant ⟶ herbivore with few decomposers	Complex, dominated by decomposers
Efficiency of nutrient recycling	Low	High
Efficiency of energy use	Low	High

How Do Species Replace One Another in Ecological Succession? Creating Beneficial and Hostile Conditions

Some species create conditions that favor the species that replace them; some create conditions that hinder their replacements; and some get along with the next group of species.

Ecologists have identified three factors that affect how and at what rate succession occurs. One is *facilitation*, in which one set of species makes an area suitable for species with different niche requirements. For example, as lichens and mosses gradually build up soil on a rock in primary succession, herbs and grasses can colonize the site. Similarly, plants such as legumes add nitrogen to the soil, making it more suitable for other plants found at later stages of succession.

A second factor is *inhibition*, in which early species hinder the establishment and growth of other species. Inhibition often occurs when plants release toxic chemicals that reduce competition from other plants. Succession then can proceed only when a fire, bulldozer, or other human or natural disturbance removes most of the inhibiting species.

A third factor is *tolerance*, in which late successional plants are largely unaffected by plants at earlier stages of succession. Tolerance may explain why late successional plants can thrive in mature communities without eliminating some early successional and mid-successional plants.

How Do Disturbances Affect Succession and Species Diversity? Setting Back the Community Clock

Changes in environmental conditions that disrupt a community can set back succession.

A **disturbance** is a change in environmental conditions that disrupts a community or ecosystem. Examples are fire, drought, flooding, mining, clear-cutting a forest, plowing a grassland, applying pesticides, climate change, and invasion by nonnative species. Environmental disturbances can range from catastrophic to mild and can be caused by natural changes or human activities. At any time during primary or secondary succession, such disturbances can convert a particular stage of succession to an earlier stage.

Many people think of all environmental disturbances as harmful. Large catastrophic disturbances can devastate communities and ecosystems. But many ecologists contend that in the long run some types of disturbances, even catastrophic ones such as fires and hurricanes, can be beneficial for the species diversity of some communities. Such disturbances create new conditions that can discourage or eliminate some species but encourage others by releasing nutrients and creating unfilled niches.

For example, when a large tree falls in a tropical forest, this local disturbance increases sunlight and nutrients for growth of plants in the understory. When a log hits a rock in an intertidal zone, it dislodges or kills many of the organisms that are growing on the rock and provides space for colonization by new intertidal organisms.

According to the *intermediate disturbance hypothesis,* communities that experience fairly frequent but moderate disturbances have the greatest species diversity. Researchers hypothesize that in such communities, moderate disturbances are large enough to create openings for colonizing species in disturbed areas but mild and infrequent enough to allow the survival of some mature species in undisturbed areas. Some field experiments support this hypothesis, but the scientific jury is still out on whether it applies to all types of communities.

Does Succession Proceed along an Expected Path, and Is Nature in Balance? Things Are Always Changing

Scientists cannot project the course of a given succession or view it as preordained progress toward a stable climax community that is in balance with its environment.

We may be tempted to conclude that ecological succession is an orderly sequence in which each stage leads automatically to the next, more stable stage. According to this classic view, succession proceeds along an expected path until a certain stable type of *climax community* occupies an area. Such a community is dominated by a few long-lived plant species and is in balance with its environment. This equilibrium model of succession is what ecologists meant years ago when they talked about the *balance of nature.*

Over the last several decades, many ecologists have changed their views about balance and equilibrium in nature. When these ecologists look at a community or ecosystem, such as a young forest, they see continuous change and instability instead of equilibrium and stability.

Under the old *balance-of-nature* view, a large terrestrial community undergoing succession eventually became covered with an expected type of climax vegetation. But a close look at almost any community reveals that it consists of an ever-changing mosaic of vegetation patches at various stages of succession. These patches result from a variety of mostly unexpected small and medium-sized disturbances.

Such research indicates that *we cannot project the course of a given succession or view it as preordained progress toward an ideally adapted climax community.* Rather, succession reflects the ongoing struggle by different species for enough light, nutrients, food, and space. This allows each to survive and gain reproductive

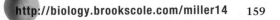

advantages over other species by occupying as much of its fundamental niche as possible.

This change in the way we view what is happening in nature explains why a growing number of ecologists prefer terms such as *biotic change* instead of *succession*—which implies an ordered and expected sequence of changes. Many ecologists have also replaced the term *climax community* with terms such as *mature community* or a mosaic of *vegetation patches* at various stages of succession.

8-6 ECOLOGICAL STABILITY, COMPLEXITY, AND SUSTAINABILITY

What Is Stability? Long-Term Sustainability through Continuous Change

Living systems maintain some degree of stability or sustainability through constant change in response to changing environmental conditions.

All systems from a cell to the biosphere are constantly changing in response to changing environmental conditions. Continents move, the climate changes, and disturbances and succession change the composition of communities. Even without human actions, ecological communities have always faced change.

So how do living organisms, communities, ecosystems, and the biosphere face changes and survive? All living systems, from single-celled organisms to the biosphere, contain complex networks of negative and positive feedback loops that interact to provide some degree of stability or sustainability over each system's expected life span.

This stability is maintained only by constant change in response to changing environmental conditions. For example, in a mature tropical rain forest, some trees die and others take their places. However, unless the forest is cut, burned, or otherwise destroyed, you would still recognize it as a tropical rain forest 50 or 100 years from now.

It is useful to distinguish among three aspects of stability or sustainability in living systems. One is **inertia,** or **persistence:** the ability of a living system to resist being disturbed or altered. A second is **constancy:** the ability of a living system such as a population to keep its numbers within the limits imposed by available resources. A third factor is **resilience:** the ability of a living system to repair damage after an external disturbance that is not too drastic.

Does Ecological Complexity Increase Ecological Stability? Mixed Results

Having many different species can provide some ecological stability or sustainability for communities, but we do not know whether this applies to all communities nor the minimum number of species needed for such stability.

In ecological terms, **complexity** refers to the number of species in a community (species richness) at each trophic level and the number of trophic levels in a community. It is one measure of a community's biodiversity.

In the 1960s, most ecologists believed the greater the species diversity and the accompanying web of feeding and biotic interactions in an ecosystem, the greater its stability. According to this hypothesis, a complex and biodiverse community with a diversity of species and feeding paths has more ways to respond to most environmental stresses because it does not have "all its eggs in one basket." This is a useful hypothesis, but recent research has found exceptions to this intuitively appealing idea.

Because no community can function without some producers and decomposers, there is a minimum threshold of species diversity below which communities and ecosystems cannot function. Beyond this, it is difficult to know whether simple communities are less stable than complex and biodiverse ones or to identify the threshold of species diversity needed to maintain community stability. Recent research by ecologist David Tilman and others suggests that communities with more species tend to have a higher net primary productivity and can be more resilient than simpler ones.

Many studies support the idea that some level of biodiversity provides insurance against catastrophe. But how much biodiversity is needed in various communities remains uncertain. For example, some recent research suggests that the average annual net primary productivity of an ecosystem reaches a peak with 10–40 producer species. Many ecosystems contain more producer species than this, but it is difficult to distinguish among those that are essential and those that are not. We need much more of this type of research.

Part of the problem is that ecologists disagree on how to define *stability*. Does an ecosystem need both high inertia and high resilience to be considered stable? Evidence suggests that some ecosystems have one of these properties but not the other. For example, tropical rain forests have high species diversity and high inertia; that is, they are resistant to significant alteration or destruction.

However, once a large tract of tropical forest is severely degraded, the community's resilience sometimes is so low that the forest may not be restored. Nutrients (which are stored primarily in the vegetation, not in the soil), and other factors needed for recovery may no longer be present. Such a large-scale loss of forest cover may also change the local or regional climate so that forests can no longer be supported.

By contrast, grasslands are much less diverse than most forests and have low inertia because they burn easily. However, because most of their plant matter is stored in underground roots, these ecosystems have high resilience and recover quickly. Grassland can be destroyed only if its roots are plowed up and something else is planted in its place, or if it is severely overgrazed by livestock or other herbivores. *Bad news.* We have been doing both of these things in some grassland areas for many decades.

Another difficulty is that populations, communities, and ecosystems are rarely, if ever, at equilibrium. Instead, nature is in a continuing state of disturbance, fluctuation, and change.

Why Should We Bother to Protect Natural Systems? The Precautionary Principle

Sometimes we should take precautionary measures to prevent serious harm even if some of the cause-and-effect relationships have not been established.

Some developers argue that if biodiversity does not necessarily lead to increased ecological stability, there is no point in trying to preserve and manage old-growth forests and other ecosystems. They conclude that we should cut down diverse old-growth forests, use the timber resources, and replace the forests with tree plantations of single tree species. Furthermore, they say, we should convert most of the world's grasslands to crop fields, drain and develop inland wetlands, dump our toxic and radioactive wastes into the deep ocean, and not worry about the premature extinction of species.

Ecologists point out that just because a system is not in equilibrium or balance does not mean that it cannot suffer from environmental degradation. They point to overwhelming evidence that human disturbances are disrupting some of the ecosystem services that support and sustain all life and all economies. They contend that our ignorance about the effects of our actions means we need to use great caution in making potentially harmful changes to communities and ecosystems on the fairly short-term time frame that concerns us.

As an analogy, we know that eating too much of certain types of foods and not getting enough exercise can greatly increase our chances of a heart attack, diabetes, and other disorders. But the exact connections between these health problems, chemicals in various foods, exercise, and genetics are still under study and often debated. We could use this uncertainty and unpredictability as an excuse to continue overeating and not exercising. But the wise course is to eat better and exercise more to help *prevent* potentially serious health problems.

This approach is based on the **precautionary principle:** When there is evidence that a human activity can harm our health or bring about changes in environmental conditions that can affect our economies or quality of life, we should take measures to prevent harm even if some of the cause-and-effect relationships have not been fully established scientifically. It is based on the commonsense idea behind many adages such as "Better safe than sorry," "Look before you leap," "First, do no harm," and "Slow down for speed bumps."

The precautionary principle makes sense, but it can be taken too far. If we do not take some risks, we would never learn much or discover what works and what does not.

The message is that we should take some risks but always think carefully about the possible short- and long-term expected and unintended effects (Figure 3-4, p. 38). Using the precautionary principle comes in when the potential risks seem too great or we have little information about the possible risks. Then it is time to step back, think about what we are doing, and do more research.

In this chapter we have seen how interactions among organisms in a community determine their abundances and distributions. Such interactions also serve as agents of natural selection on one another through *coevolution.* They also have significant effects on the structure and function of the ecosystems in which these organisms live. Everything is connected.

No part of the world is what it was before there were humans.
LAWRENCE B. SLOBODKIN

CRITICAL THINKING

1. How would you respond to someone who claims it is not important to protect areas of temperate and polar biomes because most of the world's biodiversity is in the tropics?

2. Why is the species diversity of a large island usually higher than that on a smaller island?

3. Why are predators generally less abundant than their prey?

4. What would you do if large numbers of cockroaches invaded your home? See whether you can come up with an ecological rather than a chemical (pesticide) approach to this problem.

5. How would you determine whether a particular species found in a given area is a keystone species?

6. Describe how evolution can affect predator–prey relationships.

7. How would you reply to someone who argues that **(a)** we should not worry about our effects on natural systems because succession will heal the wounds of human activities and restore the balance of nature, **(b)** efforts to preserve natural systems are not worthwhile because nature is largely unpredictable, and **(c)** because there is no balance in nature we should cut down diverse old-growth forests and replace them with tree plantations?

8. Suppose a hurricane blows down most of the trees in a forest. Timber company officials offer to salvage the fallen trees and plant a tree plantation to reduce the chances of fire and to improve the area's appearance, with an agreement that they can harvest the trees when they reach maturity and plant another tree plantation. Others argue that the damaged forest should be left alone because hurricanes and other natural events are part of nature, and the dead trees will serve as a source of nutrients for natural recovery through ecological succession. What do you think should be done, and why?

9. Congratulations! You are in charge of the world. What are the three most important features of your plan to help sustain the earth's biological communities?

PROJECTS

1. Make field studies, consult research papers, and interview people to identify and evaluate **(a)** the effects of the deliberate introduction of a beneficial nonnative species into the area where you live and **(b)** the effects of the deliberate or accidental introduction of a harmful nonnative species into the area where you live.

2. Use the library or Internet to find and describe two species not discussed in this textbook that are engaged in **(a)** a commensalistic interaction, **(b)** a mutualistic interaction, and **(c)** a parasite–host relationship.

3. Visit a nearby natural area and identify examples of **(a)** mutualism and **(b)** resource partitioning.

4. Use the library or Internet to identify the parasites likely to be found in your body.

5. Visit a nearby land area such as a partially cleared or burned forest or grassland or an abandoned crop field and record signs of secondary ecological succession. Study the area carefully to see whether you can find patches that are at different stages of succession because of various disturbances.

6. Use the library or the Internet to find bibliographic information about *George Perkins Marsh* and *Lawrence B. Slobodkin,* whose quotes appear at the beginning and end of this chapter.

7. Make a concept map of this chapter's major ideas, using the section heads, subheads, and key terms (in boldface). Look on the website for this book for information about making concept maps.

LEARNING ONLINE

The website for this book contains study aids and many ideas for further reading and research. They include a chapter summary, review questions for the entire chapter, flash cards for key terms and concepts, a multiple-choice practice quiz, interesting Internet sites, references, and a guide for accessing thousands of InfoTrac® College Edition articles. Log on to

http://biology.brookscole.com/miller14

Then click on the Chapter-by-Chapter area, choose Chapter 8, and select a learning resource.

CASE STUDY

Sea Otters: Are They Back from the Brink of Extinction?

Southern sea otters (Figure 9-1a) live in kelp forests (Figure 9-1c) in shallow waters along much of the Pacific coast of North America. Most remaining members of this species are found between California's coastal cities of Santa Cruz and Los Angeles.

These tool-using marine mammals use stones to pry shellfish off rocks underwater and to break open the shells while swimming on their backs and using their bellies as a table. Each day a sea otter consumes about a fourth of its weight in sea urchins (Figure 9-1b), clams, mussels, crabs, abalone, and about 40 other species of bottom-dwelling organisms.

Before European settlers arrived, about 1 million southern sea otters lived along the Pacific coastline of North America. By the early 1900s, the species was almost extinct because of overhunting for their thick and luxurious fur and because they competed with fishers for valuable abalone fish.

Good news. Between 1938 and 2003 the population of southern sea otters off California's coast increased from about 300 to 2,800. This partial recovery was helped when in 1977 the U.S. Fish and Wildlife Service declared the species endangered.

Why should we care about this species? One reason is that people love to look at these charismatic, cute, and cuddly animals as they play in the water. Anoher reason is biologists classify them as *keystone*

species that help keep sea urchins and other kelp-eating species from depleting kelp forests in offshore coastal waters. The third reason is *ethical.* Some people believe it is wrong to cause their premature extinction.

Why should we worry about preventing the loss of kelp forests? One reason is that they provide food, shelter, and protection for a variety of aquatic species. They also help reduce shore erosion and lessen the impact of storm waves on coastlines.

Wherever southern sea otters have returned or have been reintroduced, formerly deforested kelp areas recover within a few years and fish populations increase. This pleases biologists. But it upsets many commercial and recreational fishers, who argue that sea otters consume too many shellfish and dwindling stocks of abalone.

In 2003, the U.S. Fish and Wildlife Service published its recovery plan for the southern sea otter. It recommends reducing oil spills, ocean pollutants, and entanglements with fishing gear and boat motors. It also calls for more research to determine why otters have not recovered more fully after being named a threatened species in 1977. The plan says that the sea otter population would have to reach about 8,400 animals before it can be removed from the endangered species list.

Studying the population dynamics of southern sea otter populations and their interactions with other species has helped us to better understand the ecological importance of this keystone species. *Population dynamics* and *some basic ecological lessons* are the subjects of this chapter.

(a) Southern sea otter   **(b)** Sea urchin   **(c)** Kelp bed

Figure 9-1 Three of the species found in a kelp forest ecosystem. People visiting kelp forests marvel at their beauty and biodiversity. But old-timers say "You should have seen them before we started simplifying and degrading them."

In looking at nature ... never forget that every single organic being around us may be said to be striving to increase its numbers.

CHARLES DARWIN, 1859

This chapter addresses the following questions:

- How do populations change in size, density, makeup, and distribution in response to environmental stress?

- How do species differ in their reproductive patterns?

- What role does genetics play in the size and survival of a population?

- What are the major impacts of human activities on populations, communities, and ecosystems?

- What lessons can we learn from ecology about living more sustainably?

9-1 POPULATION DYNAMICS AND CARRYING CAPACITY

What Are the Major Characteristics of a Population? Changing and Clumping

Populations change in size, density, and age distribution, and most members of populations live together in clumps or groups.

Population dynamics is a study of how populations change in *size* (total number of individuals), *density* (number of individuals in a certain space), and *age distribution* (the proportion of individuals of each age in a population) in response to changes in environmental conditions.

Populations can also change in how they are distributed in their habitat. Three general patterns of *population distribution* or *dispersion* in a habitat are *clumping, uniform dispersion,* and *random dispersion* (Figure 9-2).

The populations of most species live in clumps or groups (Figure 9-2a). Examples are patches of vegetation, cottonwood trees clustered along streams, wolf

packs, flocks of geese, and schools of fish. Viewed from above, most of the world's landscapes are patchy with clumps of various plant and animal species found here and there. The same thing is found when we view the underwater world and the soil beneath our feet.

Why clumping? Four reasons. *First,* the resources a species needs vary greatly in availability from place to place. *Second,* living in herds, flocks, and schools can provide better protection from predators. *Third,* living in packs gives some predator species such as wolves a better chance of getting a meal. *Fourth,* some animal species form temporary groups for mating and caring for their young.

Some species maintain a fairly constant distance between individuals. By having this patter creosote bushes in a desert (Figure 9-2b) have better access to scarce water resources. Organisms with a random distribution (Figure 9-2c) are fairly rare. The world is mostly clumpy.

What Factors Govern Changes in Population Size? Entrances and Exits on the Global Stage

Populations increase through births and immigration and decrease through deaths and emigration.

Four variables—*births, deaths, immigration,* and *emigration*—govern changes in population size. A population increases by birth and immigration and decreases by death and emigration:

$$\text{Population change} = (\text{Births} + \text{Immigration}) - (\text{Deaths} + \text{Emigration})$$

These variables depend on changes in resource availability and other environmental changes (Figure 9-3).

A population's *age structure* can have a strong effect on how rapidly its size increases or decreases. Age structures are usually described in terms of organisms that are not mature enough to reproduce (the *prereproductive stage*), those that are capable of reproduction (the *reproductive stage*), and those that are too old to reproduce (the *postreproductive stage).*

Figure 9-2 Generalized *dispersion patterns* for individuals in a population throughout their habitat. The most common pattern is *clumps* of members of a population throughout their habitat, mostly because resources are usually found in patches.

(a) Clumped (elephants) **(b)** Uniform (creosote bush) **(c)** Random (dandelions)

The size of a population that includes a large proportion of young organisms in their reproductive stage or that will soon enter this stage is likely to increase. In contrast, the size of a population dominated by individuals past their reproductive stage is likely to decrease. The size of a population with a fairly even distribution between these three stages will likely remain stable because the reproduction by younger individuals will be roughly balanced by the deaths of older individuals.

What Limits Population Growth? Resources and Competitors

No population can grow indefinitely because resources such as light, water, and nutrients are limited and because of the presence of competitors or predators.

Populations vary in their capacity for growth, also known as the **biotic potential** of a population. The **intrinsic rate of increase (r)** is the rate at which a population would grow if it had unlimited resources. Most populations grow at a rate slower than this maximum.

Individuals in populations with a high rate of growth typically *reproduce early in life, have short generation times* (the time between successive generations), *can reproduce many times* (have a long reproductive life), and *have many offspring each time they reproduce.*

Some species have an astounding biotic potential. Without any controls on population growth, the descendants of a single female housefly could total about 5.6 trillion houseflies within about 13 months. If this exponential growth kept up, within a few years there would be enough houseflies to cover the earth's entire surface!

Fortunately, this is not realistic because *no population can grow indefinitely.* In the real world, a rapidly growing population reaches some size limit imposed by a shortage of one or more limiting factors, such as light, water, space, or nutrients, or by too many competitors or predators. *In nature there are always limits to population growth.* This important lesson from nature is the main message of this chapter.

Environmental resistance consists of all factors that act to limit the growth of a population. The size of the population of a particular species in a given place and time is determined by the interplay between its biotic potential and environmental resistance (Figure 9-3).

Together biotic potential and environmental resistance determine the **carrying capacity (K).** This is the maximum number of individuals of a given species that can be sustained indefinitely in a given space (area or volume). The growth rate of a population decreases as its size nears the carrying capacity of its environment because resources such as food and water begin to dwindle.

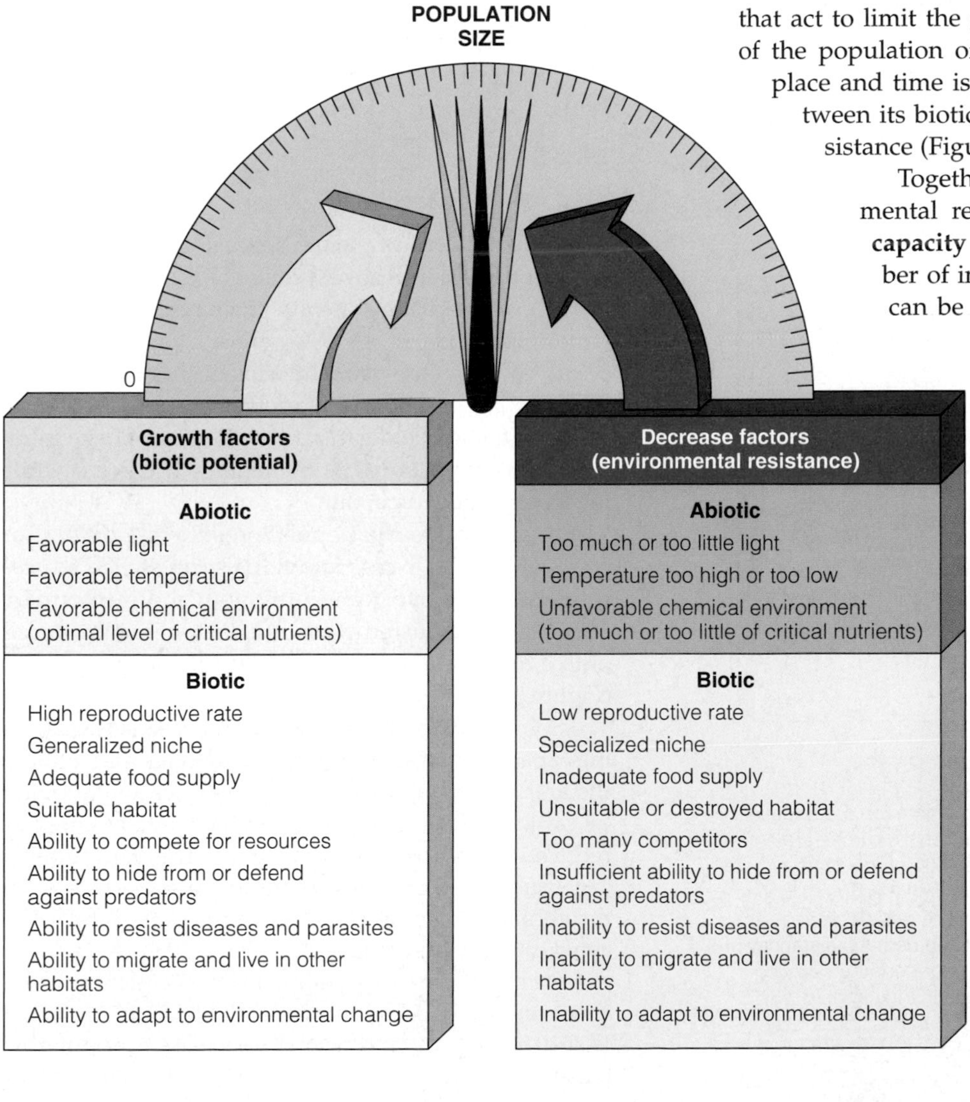

POPULATION SIZE

Growth factors (biotic potential)	Decrease factors (environmental resistance)
Abiotic	**Abiotic**
Favorable light	Too much or too little light
Favorable temperature	Temperature too high or too low
Favorable chemical environment (optimal level of critical nutrients)	Unfavorable chemical environment (too much or too little of critical nutrients)
Biotic	**Biotic**
High reproductive rate	Low reproductive rate
Generalized niche	Specialized niche
Adequate food supply	Inadequate food supply
Suitable habitat	Unsuitable or destroyed habitat
Ability to compete for resources	Too many competitors
Ability to hide from or defend against predators	Insufficient ability to hide from or defend against predators
Ability to resist diseases and parasites	Inability to resist diseases and parasites
Ability to migrate and live in other habitats	Inability to migrate and live in other habitats
Ability to adapt to environmental change	Inability to adapt to environmental change

Figure 9-3 Ecological trade-offs: factors that tend to increase or decrease the size of a population. Whether the size of a population grows, remains stable, or decreases depends on interactions between its growth factors (*biotic potential*) and its decrease factors (*environmental resistance*).

What Is the Difference between Exponential and Logistic Population Growth? J-Curves and S-Curves

With ample resources a population can grow rapidly, but as resources become limited its growth rate slows and levels off.

A population with few if any resource limitations grows exponentially. In *exponential growth*, a population grows at a fixed rate such as 1% or 2%. It starts slowly and grows faster as the population increases because the base size of the population is growing. Plotting the number of individuals against time yields a J-shaped growth curve (Figure 9-4, lower part of curve). Whether an exponential growth curve looks steep or "fast" depends on the time period of observation.

Logistic growth involves rapid exponential population growth followed by a steady decrease in population growth with time until the population size levels off. This occurs as the population encounters environmental resistance and its rate of growth decreases as it approaches the carrying capacity of its environment (Figure 9-4, top half of curve). After leveling off, such a population typically fluctuates slightly above and below the carrying capacity.

A plot of the number of individuals against time yields a sigmoid, or S-shaped, logistic growth curve (the whole curve in Figure 9-4). Figure 9-5 shows such a case involving sheep on the island of Tasmania, south of Australia, in the early 19th century.

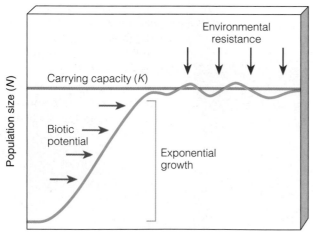

 Figure 9-4 No population can grow forever. *Exponential growth* (lower part of the curve) occurs when resources are not limiting and a population can grow at near its *intrinsic rate of increase* (*r*) or biotic potential. Such exponential growth is converted to *logistic growth*, in which the growth rate decreases as the population gets larger and faces environmental resistance. With time, the population size stabilizes at or near the *carrying capacity* (*K*) of its environment and results in the sigmoid (S-shaped) population growth curve shown in this figure. Depending on resource availability, the size of a population often fluctuates around its carrying capacity.

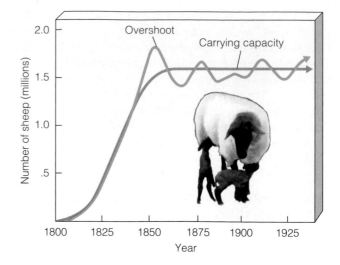

Figure 9-5 *Logistic growth* of a sheep population on the island of Tasmania between 1800 and 1925. After sheep were introduced in 1800, their population grew exponentially because of ample food. By 1855, they overshot the land's carrying capacity. Their numbers then stabilized and fluctuated around a carrying capacity of about 1.6 million sheep.

What Happens If the Population Size Exceeds the Carrying Capacity? Diebacks

When a population exceeds its resource supplies, many of its members die unless they can switch to new resources or move to an area with more resources.

The populations of some species do not make a smooth transition from exponential growth to logistic growth. Instead they use up their resource supplies and temporarily *overshoot*, or exceed, the carrying capacity of their environment. This occurs because of a *reproductive time lag:* the period needed for the birth rate to fall and the death rate to rise in response to resource overconsumption. Sometimes it takes a while for the message to get out.

In such cases the population suffers a *dieback*, or *crash*, unless the excess individuals can switch to new resources or move to an area with more resources. Such a crash occurred when reindeer were introduced onto a small island off the southwest coast of Alaska (Figure 9-6).

Sometimes when a population exceeds the carrying capacity of an area, it can cause damage that reduces the area's carrying capacity. For example, overgrazing by cattle on dry western lands in the United States has reduced grass cover in some areas. This has allowed sagebrush—which cattle cannot eat—to move in, thrive, and replace grasses. This reduces the land's carrying capacity for cattle.

Humans are not exempt from population overshoot and dieback, as shown by the tragedy on Easter Island (p. 32). Ireland also experienced a population crash after a fungus destroyed the potato crop in 1845.

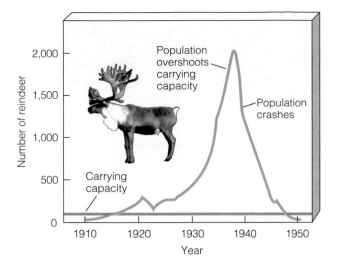

Figure 9-6 Exponential growth, overshoot, and population crash of reindeer introduced to a small island off the southwest coast of Alaska. When 26 reindeer (24 of them female) were introduced in 1910, lichens, mosses, and other food sources were plentiful. By 1935, the herd's population had soared to 2,000, overshooting the island's carrying capacity. This led to a population crash, with the herd plummeting to only 8 reindeer by 1950.

About 1 million people died, and 3 million others migrated to other countries.

Technological, social, and other cultural changes have extended the earth's carrying capacity for humans. We have increased food production and used large amounts of energy and matter resources to make normally uninhabitable areas habitable. A critical question is how long we will be able to keep doing this on a planet with a finite size and finite resources, and with a human population whose size and per capita resource use is growing exponentially.

✗ *How Would You Vote?* Can we continue expanding the earth's carrying capacity for humans? Cast your vote online at http://biology.brookscole.com/miller14.

How Does Population Density Affect Population Growth? Some Effects of Clumping

A population's density may or may not affect how rapidly it can grow.

Population density is the number of individuals in a population found in a particular space. *Density-independent population controls* affect a population's size regardless of its density. Such controls include floods, hurricanes, unseasonable weather, fire, habitat destruction (such as clearing a forest of its trees or filling in a wetland), pesticide spraying, and pollution. For example, a severe freeze in late spring can kill many individuals in a plant population, regardless of density.

Some factors that limit population growth have a greater effect as a population's density increases. Examples of such *density-dependent population controls* include competition for resources, predation, parasitism, and infectious disease.

Infectious disease is a classic type of density-dependent population control. An example is the *bubonic plague,* which swept through densely populated European cities during the 14th century. The bacterium causing this disease normally lives in rodents. It was transferred to humans by fleas that fed on infected rodents and then bit humans. The disease spread like wildfire through crowded cities, where sanitary conditions were poor and rats were abundant. At least 25 million people in European cities died from the disease.

What Kinds of Population Change Curves Do We Find in Nature? Variety Is the Spice of Life

Population sizes may stay about the same, suddenly increase and then decrease, vary in regular cycles, or change erratically.

In nature we find four general types of population fluctuations: *stable, irruptive, cyclic, and irregular* (Figure 9-7, p. 168). A species whose population size fluctuates slightly above and below its carrying capacity is said to have a fairly stable population size (Figures 9-5 and 9-7a). Such stability is characteristic of many species found in undisturbed tropical rain forests, where average temperature and rainfall vary little from year to year.

Some species, such as the raccoon and feral house mouse, normally have a fairly stable population. However, their population growth may occasionally explode, or *irrupt,* to a high peak and then crash to a more stable lower level or in some cases to a very low level (Figures 9-6 and 9-7b). Many short-lived, rapidly reproducing species such as algae and many insects have irruptive population cycles that are linked to seasonal changes in weather or nutrient availability. For example, in temperate climates, insect populations grow rapidly during the spring and summer and then crash during the hard frosts of winter.

The third type consists of *cyclic fluctuations* of population size over a regular time period (Figure 9-7c). Examples are lemmings, whose populations rise and fall every 3–4 years, and lynx and snowshoe hare, whose populations generally rise and fall in a 10-year cycle.

Finally, some populations appear to have *irregular behavior* in their changes in population size, with no recurring pattern (Figure 9-7d). Some scientists attribute this behavior to chaos in such systems. Other scientists contend that it may be orderly behavior whose details and interactions are still poorly understood.

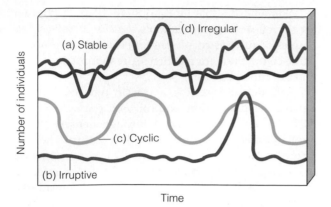

Figure 9-7 General types of simplified population change curves found in nature. **(a)** The population size of a species with a fairly *stable* population fluctuates slightly above and below its carrying capacity. **(b)** The populations of some species may occasionally explode, or *irrupt*, to a high peak and then crash to a more stable lower level. **(c)** Other species undergo sharp increases in their numbers, followed by crashes over fairly regular time intervals. Predators sometimes are blamed, but the actual causes of such boom-and-bust cycles are poorly understood. **(d)** The population sizes of some species change irregularly for mostly unknown reasons.

Do Predators Control Population Size? The Lynx–Hare Cycle

The population sizes of some predators and their prey change in cycles that appear to be caused by interaction between the two species, but other factors may be involved.

Some species that interact as predator and prey undergo cyclic changes in their numbers: sharp increases are followed by crashes (Figure 9-8).

For decades, predation has been the explanation for the 10-year population cycles of the snowshoe hare and its predator, the Canadian lynx. According to this *top-down control* hypothesis, lynx preying on hares periodically reduce the hare population. The shortage of hares then reduces the lynx population, which allows the hare population to build up again. At some point

the lynx population increases to take advantage of the increased supply of hares, starting the cycle again. Controlled laboratory experiments involving branconid wasp predators and bean weevil prey produce a similar pattern of out-of-phase fluctuations in the predator and prey populations.

Some research has cast doubt on this appealing hypothesis. Researchers have found that some snowshoe hare populations have similar 10-year boom-and-bust cycles on islands where lynx are absent. These scientists hypothesize that the periodic crashes in the hare population can also be influenced by their food supply. Large numbers of hares can die following a period when they consume food plants faster than the plants can be replenished, especially during winter.

Once the hare population crashes, the plants recover and the hare population begins rising again in a hare–plant cycle. If this *bottom-up control* hypothesis is correct, the lynx do not control hare populations. Instead, the changing hare population size may cause fluctuations in the lynx population.

The two hypotheses are not mutually exclusive. One problem is that the simple model of the predator–prey relationship assumes that the lynx are the only predators of hares and that the lynx prey only on hares. Neither is true in many of the real-life communities where these species are found. According to extensive research by Charles J. Krebs, the 10-year population cycle of snowshoe hares in boreal forests is caused by an interaction between predation (by lynx, coyotes, and other predators) and food supplies (especially in winter), and predation is the dominant process.

9-2 REPRODUCTIVE PATTERNS AND SURVIVAL

How Do Species Reproduce? Sexual Partners Are Not Always Needed

Some species reproduce without having sex and others reproduce by having sex.

Figure 9-8 Population cycles for the snowshoe hare and Canadian lynx. At one time scientists believed these curves provided circumstantial evidence that these predator and prey populations regulated one another. More recent research suggests that the periodic swings in the hare population are caused by a combination of predation by lynx and other predators (*top-down population control*) and changes in the availability of the food supply for hares. The rise and fall of the hare population apparently helps determine the lynx population (*bottom-up population control*). (Data from D. A. MacLulich)

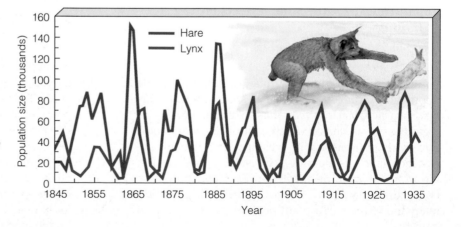

Reproductive individuals in populations of species have an inherent evolutionary drive to ensure that as many members of the next generation as possible will carry their genes. This increases the chance that their population will undergo evolution through natural selection.

Two types of reproduction can pass genes on to offspring. One is **asexual reproduction,** in which all offspring are exact genetic copies (clones) of a single parent. This is common in species such as bacteria that have only one cell. Each cell can divide to produce two identical cells that are genetic clones, or replicas of the original.

The second type is **sexual reproduction,** in which organisms produce offspring by combining sex cells or gametes (such as sperm and ovum) from both parents. This produces offspring with combinations of genetic traits from each parent.

Sexual reproduction has three disadvantages. *First,* males do not give birth. This means that females have to produce twice as many offspring to maintain the same number of young in the next generation as an asexually reproducing organism.

Second, there is an increased chance of genetic errors and defects during the splitting and recombination of chromosomes. *Third,* courtship and mating rituals consume time and energy, can transmit disease, and can inflict injury on males of some species as they compete for sexual partners.

So if sexual reproduction has some serious disadvantages, why do 97% of the earth's species use it? According to biologists, this happens because of two important advantages of sexual reproduction. One is that it provides a greater genetic diversity in offspring. A population with many different genetic possibilities has a greater chance of reproducing when environmental conditions change than does a brood of genetically identical clones. In addition, males of some species can gather food for the female and the young and protect and help train the young.

What Types of Reproductive Patterns Do Species Have? Opportunists and Competitors

Some species have a large number of small offspring and give them little parental care while other species have a few larger offspring and take care of them until they can reproduce.

In 1967, Robert H. MacArthur and Edward O. Wilson suggested that species could be classified into two fundamental reproductive patterns, *r-selected* and *K-selected species.* This classification depends on their position on the sigmoid (S-shaped) population growth curve (Figure 9-9) and the characteristics of their reproductive patterns (Figure 9-10, p. 170).

Species with a capacity for a high rate of population increase (*r*) are called **r-selected species** (Figure 9-9 and Figure 9-10, left). Such species reproduce early and put most of their energy into reproduction. Examples are algae, bacteria, rodents, annual plants (such as dandelions), and most insects.

These species have many, usually small offspring and give them little or no parental care or protection. They overcome the massive loss of offspring by producing so many that a few will survive to reproduce many more offspring to begin the cycle again.

Such species tend to be *opportunists.* They reproduce and disperse rapidly when conditions are favorable or when a disturbance opens up a new habitat or niche for invasion, as in the early stages of ecological succession.

Environmental changes caused by disturbances can allow opportunist species to gain a foothold. However, once established, their populations may crash because of unfavorable changes in environmental conditions or invasion by more competitive species. This helps explain why most r-selected or opportunist species go through irregular and unstable boom-and-bust cycles in their population size.

At the other extreme are *competitor* or **K-selected species** (Figure 9-9 and Figure 9-10, right). These species tend to reproduce late in life and have a small number of offspring with fairly long life spans.

Typically the offspring of such species develop inside their mothers (where they are safe), are born fairly large, mature slowly, and are cared for and protected by one or both parents until they reach reproductive age. This reproductive pattern results in a few big and strong individuals that can compete for resources and reproduce a few young to begin the cycle again.

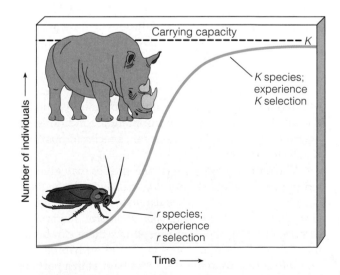

Figure 9-9 Positions of *r-selected* and *K-selected* species on the sigmoid (S-shaped) population growth curve.

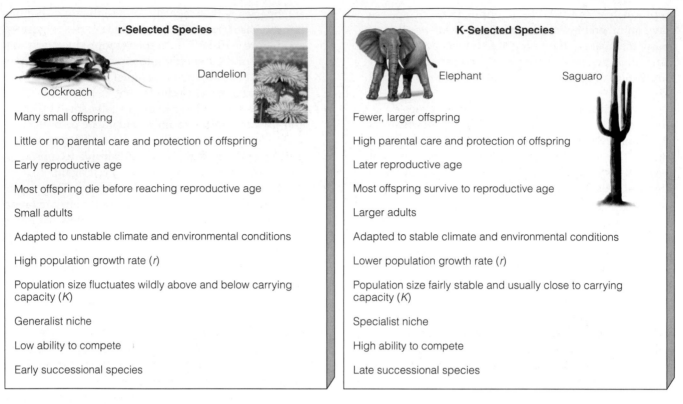

Figure 9-10 Generalized characteristics of *r-selected* or opportunist species and *K-selected* or competitor species. Many species have characteristics between these two extremes.

They are called K-selected species because they tend to do well in competitive conditions when their population size is near the carrying capacity (*K*) of their environment. Their populations typically follow a logistic growth curve.

Most large mammals (such as elephants, whales, and humans), birds of prey, and large and long-lived plants (such as the saguaro cactus, oak trees, and most tropical rain forest trees) are K-selected species. Many K-selected species—especially those with long generation times and low reproductive rates like elephants, rhinoceroses, and sharks—are prone to extinction.

Most organisms have reproductive patterns between the extremes of r-selected species and K-selected species, or they change from one extreme to the other under certain environmental conditions. In agriculture we raise both r-selected species (crops) and K-selected species (livestock).

The reproductive pattern of a species may give it a temporary advantage. But *the availability of suitable habitat for individuals of a population in a particular area is what determines its ultimate population size.* Regardless of how fast a species can reproduce, there can be no more dandelions than there is dandelion habitat and no more zebras than there is zebra habitat in a particular area.

What Are Survivorship Curves? At What Age Is Death Most Likely?

The populations of different species vary in how long individual members typically live.

Individuals of species with different reproductive strategies tend to have different *life expectancies.* One way to represent the age structure of a population is with a **survivorship curve,** which shows the percentages of the members of a population surviving at different ages. There are three generalized types of survivorship curves: *late loss, early loss,* and *constant loss* (Figure 9-11). Which type of curve applies to the human species?

A *life table* shows the numbers of individuals at each age on a survivorship curve. It shows the projected life expectancy and probability of death for individuals at each age.

Insurance companies use life tables of human populations to determine policy costs for customers. Life tables show that women in the United States survive an average of 6 years longer than men. This explains why a 65-year-old American man normally pays more for life insurance than a 65-year-old American woman.

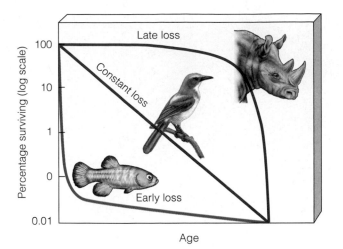

Figure 9-11 Three general survivorship curves for populations of different species, obtained by showing the percentages of the members of a population surviving at different ages. A *late loss* population (such as elephants, rhinoceroses, and humans) typically has high survivorship to a certain age, then high mortality. A *constant loss* population (such as many songbirds) shows a fairly constant death rate at all ages. For an *early loss* population (such as annual plants and many bony fish species), survivorship is low early in life. These generalized survivorship curves only approximate the behavior of species.

9-3 EFFECTS OF GENETIC VARIATIONS ON POPULATION SIZE

What Role Does Genetics Play in the Size of Populations? The Vulnerability of Small Isolated Populations

Variations in genetic diversity can affect the survival of small, isolated populations.

In most large populations genetic diversity is fairly constant. The loss or addition of individuals has little effect on the total gene pool.

However, genetic factors can affect the survival and genetic diversity of small, isolated populations. Several factors can play a role in the loss of genetic diversity and the survival of such populations. One is the *founder effect* when a few individuals in a population colonize a new habitat that is geographically isolated from other members of the population (Figure 5-7, p. 94). In such cases, limited genetic diversity or variability may threaten the survival of the colonizing population. Another problem is a *demograpic bottleneck.* It occurs when only a few individuals in a population survive a catastrophe such as a fire or hurricane. Lack of genetic diversity may limit the ability of these individuals to rebuild the population. A third factor is *genetic drift.* It involves random changes in the gene frequencies in a population that can lead to unequal reproductive success. For example, some individuals may breed more than others and their genes may eventually dominate the gene pool of the population. This change in gene frequency could help or hinder the survival of the population. The founder effect is one cause of genetic drift. A fourth factor is *inbreeding.* It occurs when individuals in a small population mate with one another. This can increase the frequency of defective genes within a population and affect its long-term survival.

What are Metapopulations? Exchanging Genes Now and Then

Variations in genetic diversity can affect the survival of small, isolated populations.

Some mobile populations that are geographically separated from one another can exchange genes when some of their members get together occasionally and mate. Such collections of interacting local populations of a species are called *metapopulations.*

Some local populations where birth rates are higher than death rates produce excess individuals that can migrate to other local populations. Other local populations where death rates are greater than birth rates can accept individuals from other populations. Conservation biologists can map out the locations of metapopulations and use this information to provide corridors and migration routes to enhance the overall population size, genetic diversity, and survial of related local populations.

9-4 HUMAN IMPACTS ON NATURAL SYSTEMS: LEARNING FROM NATURE

How Have Humans Modified Natural Ecosystems? Our Big Footprints

We have used technology to alter much of the rest of nature in ways that threaten the survival of many other species and could reduce the quality of life for our own species.

In this and the six preceding chapters we have looked at key concepts of science and ecology. It is time to review what lessons we can learn from this study of how nature operates and sustains itself. But first let us look at our environmental impact on the earth.

To survive and provide resources for growing numbers of people, we have modified, cultivated, built on, or degraded a large and increasing area of the earth's natural systems. Excluding Antarctica, our activities have directly affected to some degree about 83% of the earth's land surface (Figure 9-12, p. 172). Figure 9-13 (p. 172) compares some of the characteristics of natural and human-dominated systems.

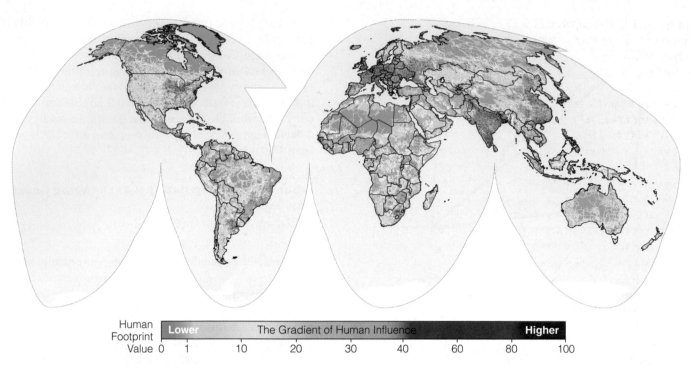

Figure 9-12 Natural capital degradation: the human footprint on the earth's land surface—in effect the sum of all ecological footprints (Figure 1-7, p. 10) of the human population. Colors represent the percentage of each area influenced by human activities. Excluding Antarctica and Greenland, human activities have directly affected to some degree about 83% of the earth's land surface and 98% of the area where it is possible to grow rice, wheat, or maize. (Data from Wildlife Conservation Society and the Center for International Earth Science Information Network at Columbia University [CIESIN]. Reprinted by permission.)

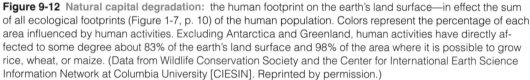

Property	Natural Systems	Human-Dominated Systems
Complexity	Biologically diverse	Biologically simplified
Energy source	Renewable solar energy	Mostly nonrenewable fossil fuel energy
Waste production	Little, if any	High
Nutrients	Recycled	Often lost or wasted
Net primary productivity	Shared among many species	Used, destroyed, or degraded to support human activities

Figure 9-13 Some typical characteristics of natural and human-dominated systems.

We have used technology to alter much of the rest of nature to meet our growing needs and wants in nine major ways. One is *reducing biodiversity by destroying, fragmenting, and degrading wildlife habitats.* This happens when we clear forests, dig up grasslands, and fill in wetlands to grow food or to construct buildings, highways, and parking lots.

A second is *reducing biodiversity by simplifying and homogenizing natural ecosystems.* Communities and ecosystems dominated by humans tend to have fewer species and fewer community interactions than do undisturbed ecosystems. When we plow grasslands and clear forests, we often replace thousands of interrelated plant and animal species with one crop or one kind of tree—called a *monoculture.* Then we spend a lot of time, energy, and money trying to protect such monocultures against threats such as invasions by *opportunist species* of plants (weeds) and *pests*—mostly insects, to which a monoculture crop is like an all-you-can-eat restaurant. Another threat is invasions by *pathogens*—fungi, viruses, or bacteria—that harm the plants and animals we want to raise.

A third type of alteration is *using, wasting, or destroying an increasing percentage of the earth's net primary productivity that supports all consumer species (including humans)*. This factor is the main reason we are crowding out or eliminating the habitats and food supplies of a growing number of other species.

A fourth type of intervention has unintentionally *strengthened some populations of pest species and disease-causing bacteria.* This has occurred through overuse of pesticides and antibiotics that has speeded up natural selection and caused genetic resistance to these chemicals.

A fifth effect has been to *eliminate some predators.* Some ranchers want to eliminate wolves, coyotes, eagles, and other predators that occasionally kill sheep. They also want to eradicate bison or prairie dogs that compete with their sheep or cattle for grass. A few big-game hunters push for elimination of predators that prey on game species.

Sixth, *we have deliberately or accidentally introduced new or nonnative species into ecosystems.* Most of these species, such as food crops and domesticated livestock, are beneficial to us but a few are harmful to us and other species.

Seventh, we have *overharvested some renewable resources.* Ranchers and nomadic herders sometimes allow livestock to overgraze grasslands until erosion converts these ecosystems to less productive semi-deserts or deserts. Farmers sometimes deplete soil nutrients by excessive crop growing. Some fish species are overharvested. Illegal hunting or poaching endangers wildlife species with economically valuable parts such as elephant tusks, rhinoceros horns, and tiger skins. In some areas, fresh water is being pumped out of underground aquifers faster than it is replenished.

Eighth, some human activities *interfere with the normal chemical cycling and energy flows in ecosystems.* Soil nutrients can erode from monoculture crop fields, tree plantations, construction sites, and other simplified ecosystems and overload and disrupt other ecosystems such as lakes and coastal ecosystems. Chemicals such as chlorofluorocarbons (CFCs) released into the atmosphere can increase the amount of harmful ultraviolet energy reaching the earth by reducing ozone levels in the stratosphere. Emissions of carbon dioxide and other greenhouse gases—from burning fossil fuels and from clearing and burning forests and grasslands—can trigger global climate change by altering energy flow through the troposphere.

Ninth, while most natural systems are powered by sunlight, *human-dominated ecosystems have become increasingly dependent on nonrenewable energy from fossil fuels.* Fossil fuel systems typically produce pollution, add more of the greenhouse gas carbon dioxide to the atmosphere, and waste much more energy than they need to.

To survive we must exploit and modify parts of nature. However, we are beginning to understand that any human intrusion into nature has multiple effects, most of them unintended and unpredictable (Figure 3-4, p. 38 and Connections, below).

We face two major challenges. *First*, we need to maintain a balance between simplified, human-altered ecosystems and the more complex natural ecosystems

Ecological Surprises

CONNECTIONS

Malaria once infected 9 out of 10 people in North Borneo, now known as Sabah. In 1955, the World Health Organization (WHO) began spraying the island with dieldrin (a DDT relative) to kill malaria-carrying mosquitoes. The program was so successful that the dreaded disease was nearly eliminated.

But unexpected things began to happen. The dieldrin also killed other insects, including flies and cockroaches living in houses. The islanders applauded. But then small insect-eating lizards that also lived in the houses died after gorging themselves on dieldrin-contaminated insects.

Next, cats began dying after feeding on the lizards. Then, in the absence of cats, rats flourished and overran the villages. When the people became threatened by sylvatic plague carried by rat fleas, the WHO parachuted healthy cats onto the island to help control the rats. Operation Cat Drop worked.

But then the villagers' roofs began to fall in. The dieldrin had killed wasps and other insects that fed on a type of caterpillar that either avoided or was not affected by the insecticide. With most of its predators eliminated, the caterpillar population exploded, munching its way through its favorite food: the leaves used in thatched roofs.

Ultimately, this episode ended happily: both malaria and the unexpected effects of the spraying program were brought under control. Nevertheless, this chain of unintended and unforeseen events emphasizes the unpredictability of interfering with an ecosystem. It reminds us that when we intervene in nature, we need to ask, "Now what will happen?"

Critical Thinking

Do you believe the beneficial effects of spraying pesticides on Sabah outweighed the resulting unexpected and harmful effects? Explain.

on which we and other species depend. *Second*, we need to slow down the rates at which we are altering nature for our purposes. If we simplify, homogenize, and degrade too much of the planet to meet our needs and wants, what is at risk is not the resilient earth but the quality of life for members of our species and other species we drive to premature extinction.

What Can We Learn from Ecology about Living More Sustainably? Copy Nature

We can develop more sustainable economies and societies by mimicking the four major ways that nature has adapted and sustained itself for several billion years.

So how can we live more sustainably? Ecologists say: Find out how nature has survived and adapted for several billion years and copy this strategy. Figure 9-14 (also found on the bottom half of the back cover) summarizes the four major ways in which life on earth has survived and adapted for several billion years. Figure 9-15 (left) gives an expanded description of these principles and Figure 9-15 (right) summarizes how we can live more sustainably by mimicking these fundamental but amazingly simple lessons from nature in designing our societies, products, and economies. *Figures 9-14 and 9-15 summarize the major message of this book. Study them carefully.*

Biologists have used these lessons from their ecological study of nature to formulate four guidelines for developing more sustainable societies and lifestyles:

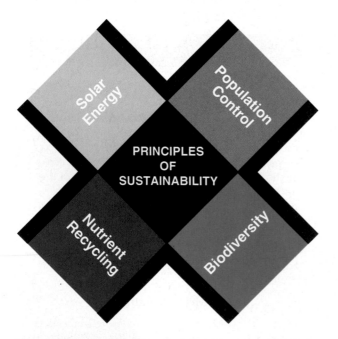

Figure 9-14 Sustaining natural capital: four interconnected principles of sustainability derived from learning how nature sustains itself. This diagram also appears on the bottom half of the back cover of this book.

Principles of Sustainability

How Nature Works   **Lessons for Us**

Runs on renewable solar energy.

Rely mostly on renewable solar energy.

Recycles nutrients and wastes. There is little waste in nature.

Prevent and reduce pollution and recycle and reuse resources.

Uses biodiversity to maintain itself and adapt to new environmental conditions.

Preserve biodiversity by protecting ecosystem services and preventing premature extinction of species.

Controls a species' population size and resource use by interactions with its environment and other species.

Reduce births and wasteful resource use to prevent environmental overload and depletion and degradation of resources.

Figure 9-15 Solutions: the four principles of sustainability (left) derived from observing nature have implications for the long-term sustainability of human societies (right). These four operating principles of nature are connected to one another. Failure of any single principle can lead to temporary or long-term unsustainability and disruption of ecosystems and human economies and societies.

- *Our lives, lifestyles, and economies are totally dependent on the sun and the earth.* We need the earth, but the earth does not need us. As a species, we are expendable.

- *Everything is connected to and interdependent with everything else.* The primary goal of ecology is to discover what connections in nature are the strongest, most important, and most vulnerable to disruption for us and other species.

- *We can never do merely one thing.* Any human intrusion into nature has unexpected and mostly unintended side effects (Figure 3-4, p. 38). When we alter nature we need to ask, "Now what will happen?"

- *We cannot indefinitely sustain a civilization that depletes and degrades the earth's natural capital, but we can sustain one that lives off the biological income provided by that capital.*

In the next chapter we will apply the principles of population dynamics and sustainability discussed in this chapter to the growth of the human population. In the three chapters after that we apply them to understanding the earth's terrestrial and aquatic biodiversity and how to help sustain them.

We cannot command nature except by obeying her.

Sir Francis Bacon

CRITICAL THINKING

1. **(a)** Why do biotic factors that regulate population growth tend to depend on population density and **(b)** Why do abiotic factors that regulate population tend to be independent of population density?

2. Why are pest species likely to be extreme r-selected species? Why are many endangered species likely to be extreme K-selected species?

3. Why is an animal that devotes most of its energy to reproduction likely to be small and weak?

4. Given current environmental conditions, if you had a choice, would you rather be an r-strategist or a K-strategist? Explain your answer.

5. List the type of survivorship curve you would expect given descriptions of the following organisms:
   a. This organism is an annual plant. It lives only 1 year. During that time, it sprouts, reaches maturity, produces many wind-dispersed seeds, and dies.
   b. This organism is a mammal. It reaches maturity after 10 years. It bears one young every 2 years. The parents and the rest of the herd protect the young.

6. Explain why a simplified ecosystem such as a cornfield usually is much more vulnerable to harm from insects and plant diseases than a more complex, natural ecosystem such as a grassland. Does this mean that we should never convert a grassland to a cornfield? Explain. What restrictions, if any, would you put on such conversions?

7. How has the human population generally been able to avoid environmental resistance factors that affect other populations? Is this likely to continue? Explain.

8. Explain why you agree or disagree with the four principles of sustainability listed in Figure 9-15 (left) and their lessons for human societies listed in Figure 9-15 (right). Identify aspects of your lifestyle that follow or violate each of these four sustainability principles. Would you be willing to change the aspects of your lifestyle that violate these sustainability principles? Explain.

PROJECTS

1. Use the principles of sustainability derived from the scientific study of how nature sustains itself (Figures 9-14 and 9-15) to evaluate the sustainability of the following parts of human systems: **(a)** transportation, **(b)** cities, **(c)** agriculture, **(d)** manufacturing, **(e)** waste disposal, and **(f)** your own lifestyle. Compare your analysis with those made by your classmates.

2. Use the library or the Internet to choose one wild plant species and one animal species and analyze the factors that are likely to limit the population of each species.

3. Use the library or the Internet to find bibliographic information about *Charles Darwin* and *Sir Francis Bacon*, whose quotes appear at the beginning and end of this chapter.

4. Make a concept map of this chapter's major ideas, using the section heads, subheads, and key terms (in boldface). Look on the website for this book for information about making concept maps.

LEARNING ONLINE

The website for this book contains study aids and many ideas for further reading and research. They include a chapter summary, review questions for the entire chapter, flash cards for key terms and concepts, a multiple-choice practice quiz, interesting Internet sites, references, and a guide for accessing thousands of InfoTrac® College Edition articles. Log on to

http://biology.brookscole.com/miller14

Then click on the Chapter-by-Chapter area, choose Chapter 9, and select a learning resource.

Slowing Population Growth in Thailand: A Success Story

Can a country sharply reduce its population growth in only 15 years? Thailand did.

In 1971, Thailand adopted a policy to reduce its population growth. When the program began, the country's population was growing at a very rapid rate of 3.2% per year, and the average Thai family had 6.4 children.

Fifteen years later in 1986, the country's population growth rate had been cut in half to 1.6%. By 2004, the rate had fallen to 0.8%, and the average number of children per family was 1.7.

There are a number of reasons for this impressive achievement. They include the creativity of the government-supported family planning program, a high literacy rate among women (90%), an increased economic role for women and advances in women's rights, and better health care for mothers and children. Other factors are the openness of the Thai people to new ideas and support of family planning by the country's religious leaders (95% of Thais are Buddhist). A key factor was the willingness of the government to encourage and financially support family planning and to work with the private, nonprofit Population and Community Development Association (PCDA).

Mechai Viravidaiya (Figure 10-1) led the way in reducing the country's population growth rate. This public relations genius and former government economist launched the PCDA in 1974 to help make family planning a national goal. PCDA workers handed out condoms at festivals, movie theaters, and even traffic jams, and they developed ads and witty songs about contraceptive use. Between 1971 and 2004, the percentage of married women using modern birth control rose from 15% to 70%— higher than the 58% usage in developed countries and the 51% usage in developing countries.

Viravidaiya helped establish a German-financed revolving loan plan to enable people participating in family planning programs to install toilets and drinking water systems. Low-rate loans were offered to farmers practicing family planning. The government also offers loans to individuals from a fund that increases as their village's level of contraceptive use rises. Education and economic rewards work.

All is not completely rosy. Although Thailand has done well in slowing population growth and raising per capita income, it has been less successful in reducing pollution and improving public health. Its capital, Bangkok, is plagued with notoriously high levels of traffic congestion and air pollution (Figure 10-2).

Figure 10-1 Individuals matter: *Mechai Viravidaiya,* a charismatic leader, played a major role in Thailand's successful efforts to reduce its population growth. In 1974, he established the private, nonprofit Population and Community Development Association (PCDA) to help implement family planning as a national goal.

Figure 10-2 This policeman and schoolchildren in Bangkok, Thailand, are wearing masks to reduce their intake of air polluted mainly by automobiles. Bangkok is one of the world's most car-clogged cities, with car commutes averaging 3 hours per day. Roughly one of every nine of its residents has a respiratory ailment.

The problems to be faced are vast and complex, but come down to this: 6.4 billion people are breeding exponentially. The process of fulfilling their wants and needs is stripping earth of its biotic capacity to produce life; a climactic burst of consumption by a single species is overwhelming the skies, earth, waters, and fauna.

PAUL HAWKEN

This chapter addresses the following questions:

- How is population size affected by birth, death, fertility, and migration rates?

- How is population size affected by age structure?

- How can we influence population size?

- What success have India and China had in slowing population growth?

- How can global population growth be reduced?

10-1 FACTORS AFFECTING HUMAN POPULATION SIZE

What Is Demography and Why Is it Important? How Population Change Affects Life, Death, and Economies

Changes in the size, composition, and distribution of human populations have important health, social, and economic effects.

How long are you likely to live and what will probably kill you? How many people are there in the country or area where you live and how many people are likely to live there in the future? How many children are you likely to have? How likely are you to get married or divorced? What kind of job will you probably have and how many times are you likely to change jobs? What are your chances of promotion? When are you likely to retire? About how many times will you move?

Demography is devoted to finding answers to these and other population-related questions. **Demography** is the study of the size, composition, and distribution of human populations and the causes and consequences of changes in these characteristics. Specialists in this field are called *demographers*.

How Is Population Size Affected by Birth Rates and Death Rates? Entrances and Exits

Population increases because of births and immigration, and decreases through deaths and emigration.

Human populations grow or decline through the interplay of three factors: *births, deaths,* and *migration*. **Population change** is calculated by subtracting the number of people leaving a population (through death and

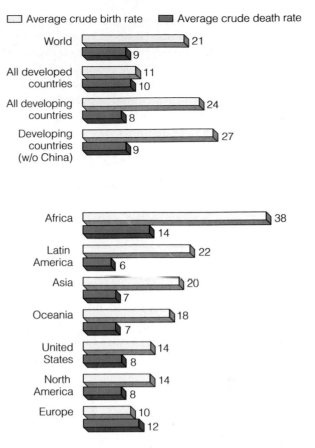

Figure 10-3 Average crude birth and death rates for various groupings of countries in 2004. (Data from Population Reference Bureau)

emigration) from the number entering it (through birth and immigration) during a specified period of time (usually a year):

Population change = (Births + Immigration) − (Deaths + Emigration)

When births plus immigration exceed deaths plus emigration, population increases; when the reverse is true, population declines.

Instead of using the total numbers of births and deaths per year, demographers use the **birth rate,** or **crude birth rate** (the number of live births per 1,000 people in a population in a given year), and the **death rate,** or **crude death rate** (the number of deaths per 1,000 people in a population in a given year). Figure 10-3 shows the crude birth and death rates for various groupings of countries in 2004.

How Fast Is the World's Population Growing? Good and Bad News

The rate at which the world's population increases has slowed, but the population is still growing fairly rapidly.

Birth rates and death rates are coming down worldwide, but death rates have fallen more sharply than birth rates. As a result, more births are occurring than deaths; every time your heart beats, 2.5 more babies are added to the world's population. At this rate, we share the earth and its resources with about 219,000 more people each day—97% of them in developing countries.

The rate of the world's annual population change is usually expressed as a percentage:

$$\text{Annual rate of natural population change (\%)} = \frac{\text{Birth rate} - \text{Death rate}}{1{,}000 \text{ persons}} \times 100$$

$$= \frac{\text{Birth rate} - \text{Death rate}}{10}$$

Exponential population growth has not disappeared but is occurring at a slower rate. The rate of the world's annual population growth (natural increase) dropped by almost half between 1963 and 2004, from 2.2% to 1.25%. This is *good news* but during the same period the population base doubled, from 3.2 billion to 6.4 billion. This drop in the rate of population increase is somewhat like learning that a truck heading straight at you has slowed from 100 kilometers per hour (kph) to 55 kph while its weight has doubled.

An exponential growth rate of 1.25% may seem small. But in 2004 it added about 80 million people to the world's population, compared to 69 million added in 1963, when the world's population growth reached its peak. An increase of 80 million people per year is roughly equal to adding another New York City every month, a Germany every year, and a United States every 3.7 years.

Also, there is a big difference between exponential population growth rates in developed and developing countries. In 2004, the population of developed countries was growing at a rate of 0.1%. That of the developing countries was 1.5%—almost 15 times faster.

As a result of these trends, the population of the developed countries, currently at 1.2 billion, is expected to change little in the next 50 years. In contrast, the population of the developing countries is projected to rise steadily from 5.2 billion in 2004 to 8 billion in 2050. The six nations expected to experience most of this growth are, in order: India, China, Pakistan, Nigeria, Bangladesh, and Indonesia.

What five countries have the largest numbers of people? Number 1 is China with 1.3 billion people—about one of every five people in the world. Number 2 is India with 1.1 billion people—about one of every six people. Together China and India—with roughly the same geographic areas—have 37% of the world's population. Number 3 is the United States, with 294 million people or 4.6% of the world's population.

Can you guess the next two most populous countries? What three countries are expected to have the most people in 2025? Look at Figure 10-4 to see if your answers are correct.

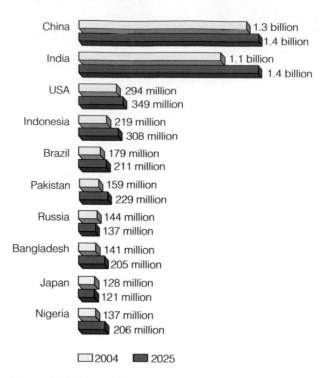

Figure 10-4 The world's 10 most populous countries in 2004, with projections of their population size in 2025. In 2004, more people lived in China than in all of Europe, Russia, North America, Japan, and Australia combined. (Data from World Bank and Population Reference Bureau)

How Long Does It Take to Double the Number of People on the Planet? The Rule of 70

Doubling time is how long it takes for a population growing at a specified rate to double its size.

One measure of population growth is **doubling time:** the time (usually in years) it takes for a population growing at a specified rate to double its size. A quick way to calculate doubling time is to use the **rule of 70:** 70/percentage growth rate = doubling time in years (a formula derived from the basic mathematics of exponential growth). For example, in 2004 the world's population grew by 1.2%. If that rate continues, the earth's population will double in about 56 years (70/1.25 = 56 years).

The population of the African country of Nigeria is increasing by 2.8% a year. How long will it take for its population to double?

How Have Global Fertility Rates Changed? Having Fewer Babies per Woman

The average number of children that a woman bears has dropped sharply since 1950, but the number is not low enough to stabilize the world's population in the near future.

Fertility is the number of births that occur to an individual woman or in a population. Two types of fertility rates affect a country's population size and growth rate. The first type, **replacement-level fertility,** is the number of children a couple must bear to replace themselves. It is slightly higher than two children per couple (2.1 in developed countries and as high as 2.5 in some developing countries), mostly because some female children die before reaching their reproductive years.

Does reaching replacement-level fertility mean an immediate halt in population growth? No, because so many future parents are alive. If each of today's couples had an average of 2.1 children and their children also had 2.1 children, the world's population would still grow for 50 years or more (assuming death rates do not rise).

The second type of fertility rate is the **total fertility rate (TFR):** the average number of children a woman typically has during her reproductive years. *Good news.* TFRs have dropped sharply since 1950 (Figure 10-5). In 2004, the average global TFR was 2.8 children per woman. It was 1.5 in developed countries (down from 2.5 in 1950) and 3.1 in developing countries (down from 6.5 in 1950). The highest TFRs are in Africa with a rate of 5.2 in 2004.

So how many of us are likely to be here in 2050? Answer: From 7.2 to 10.6 billion, depending on the world's projected average TFR (Figure 10-6). The medium projection is 8.9 billion people. About 97% of the growth in all three of these estimates is projected to take place in developing countries, where acute poverty (living on less than $1 per day) is a way of life for about 1.4 billion people.

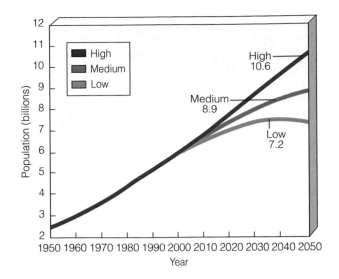

Figure 10-6 UN world population projections, assuming that by 2050 the world's total fertility rate is 2.5 (high), 2.0 (medium), or 1.5 (low) children per woman. The most likely projection is the medium one—8.9 billion by 2050. (Data from United Nations)

How Have Fertility and Birth Rates Changed in the United States? Ups and Downs

Population growth in the United States has slowed down but is not close to leveling off.

The population of the United States has grown from 76 million in 1900 to 294 million in 2004, despite oscillations in the country's TFR (Figure 10-7) and birth rate (Figure 10-8, p. 180). A sharp rise in the birth rate occurred after World War II. The period of high birth rates between 1946 and 1964 is known as the *baby-boom period.* This added 79 million people to the U.S. population. In 1957, the peak of the baby boom after World War II, the TFR reached 3.7 children per woman. Since then it has generally declined and remained at or below replacement level since 1972.

The drop in the TFR has led to a decline in the rate of population growth in the United States. But the

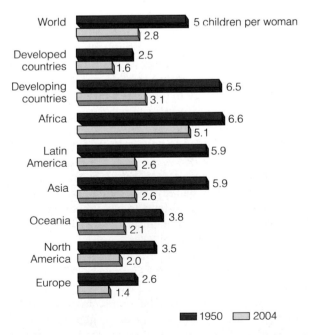

Figure 10-5 Good news: decline in total fertility rates for various groupings of countries, 1950–2004. (Data from United Nations)

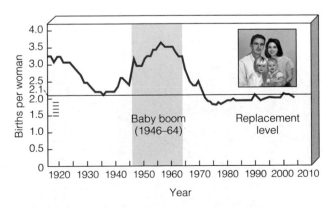

Figure 10-7 Total fertility rates for the United States between 1917 and 2004. (Data from Population Reference Bureau and U.S. Census Bureau)

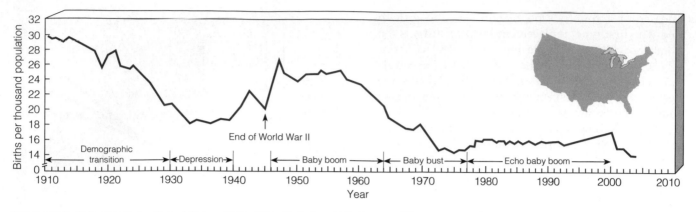

Figure 10-8 Birth rates in the United States, 1910–2004. (Data from U.S. Bureau of Census and U.S. Commerce Department)

country's population is still growing faster than that of any other developed country and is not close to leveling off.

About 2.9 million people were added to the U.S. population in 2004. Approximately 59% of this growth was the result of more births than deaths (1.7 million) and the rest came from legal and illegal immigration (1.2 million).

How many people are likely to be living in the United States in 2050 and 2100? No one knows, but the U.S. Bureau of Census makes projections using different assumptions about fertility rates and immigration rates. Its medium projection is that the U.S. population will increase from 294 million in 2004 to 420 million by 2050 and reach 571 million by 2100 (Figure 10-9). In contrast, population growth has slowed in other major developed countries since 1950 and most are expected to have declining populations after 2010. Because of a high per capita rate of resource use, each addition to the U.S. population has an enormous environmental impact (Figure 1-13, p. 15).

How do the population dynamics of the United States compare with those of its neighbors Canada and Mexico? Find out by looking at Figure 10-10. In addition to an almost fourfold increase in population growth,

some amazing changes in lifestyles took place in the United States during the 20th century (Figure 10-11).

What Factors Affect Birth Rates and Fertility Rates? Reducing Births

The number of children women have is affected by the cost of raising and educating children, educational and employment opportunities for women, infant deaths, marriage age, and availability of contraceptives and abortions.

Many factors affect a country's average birth rate and TFR. One is the *importance of children as a part of the labor force.* Proportions of children working tend to be higher in developing countries—especially in rural areas, where children begin working to help raise crops at an early age.

Another economic factor is the *cost of raising and educating children.* Birth and fertility rates tend to be lower in developed countries, where raising children is much more costly because they do not enter the labor force until they are in their late teens or 20s.

The *availability of private and public pension systems* affects how many children couples have. Pensions eliminate parents' need to have many children to help support them in old age.

Urbanization plays a role. Why? Because people living in urban areas usually have better access to family planning services and tend to have fewer children than those living in rural areas where children are needed to perform essential tasks.

Another important factor is *the educational and employment opportunities available for women.* TFRs tend to be low when women have access to education and paid employment outside the home. In developing countries, women with no education generally have two more children than women with a secondary school education.

Another factor is the *infant mortality rate.* In areas with low infant mortality rates, people tend to have a

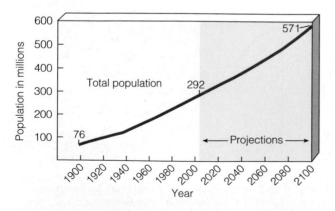

Figure 10-9 U.S. population growth, 1900–2004, and projections to 2100. (Data from U.S. Census Bureau)

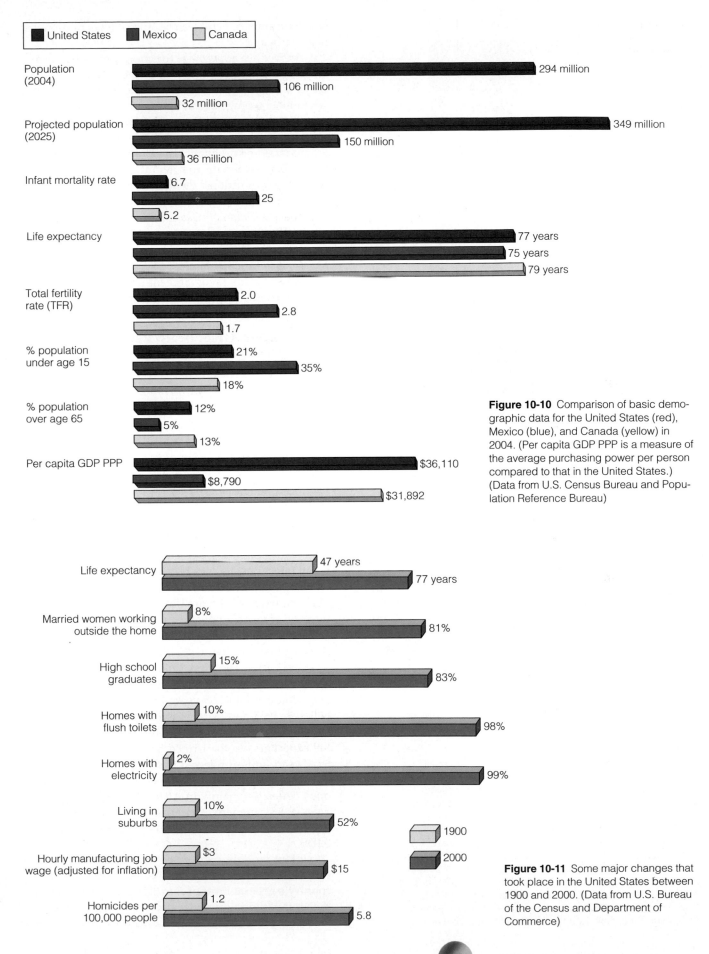

United States ■ **Mexico** ■ **Canada** □

Population (2004)
294 million
106 million
32 million

Projected population (2025)
349 million
150 million
36 million

Infant mortality rate
6.7
25
5.2

Life expectancy
77 years
75 years
79 years

Total fertility rate (TFR)
2.0
2.8
1.7

% population under age 15
21%
35%
18%

% population over age 65
12%
5%
13%

Per capita GDP PPP
$36,110
$8,790
$31,892

Figure 10-10 Comparison of basic demographic data for the United States (red), Mexico (blue), and Canada (yellow) in 2004. (Per capita GDP PPP is a measure of the average purchasing power per person compared to that in the United States.) (Data from U.S. Census Bureau and Population Reference Bureau)

Life expectancy
47 years
77 years

Married women working outside the home
8%
81%

High school graduates
15%
83%

Homes with flush toilets
10%
98%

Homes with electricity
2%
99%

Living in suburbs
10%
52%

Hourly manufacturing job wage (adjusted for inflation)
$3
$15

Homicides per 100,000 people
1.2
5.8

1900 □
2000 ■

Figure 10-11 Some major changes that took place in the United States between 1900 and 2000. (Data from U.S. Bureau of the Census and Department of Commerce)

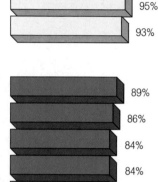

Extremely Effective

Total abstinence	100%
Sterilization	99.6%
Vaginal ring	98–99%

Highly Effective

IUD with slow-release hormones	98%
IUD plus spermicide	98%
Vaginal pouch ("female condom")	97%
IUD	95%
Condom (good brand) plus spermicide	95%
Oral contraceptive	93%

Effective

Cervical cap	89%
Condom (good brand)	86%
Diaphragm plus spermicide	84%
Rhythm method (Billings, Sympto-Thermal)	84%
Vaginal sponge impregnated with spermicide	83%
Spermicide (foam)	82%

Moderately Effective

Spermicide (creams, jellies, suppositories)	75%
Rhythm method (daily temperature readings)	74%
Withdrawal	74%
Condom (cheap brand)	70%

Unreliable

Douche	40%
Chance (no method)	10%

Figure 10-12 Typical effectiveness rates of birth control methods in the United States. Percentages are based on the number of undesired pregnancies per 100 couples using a specific method as their sole form of birth control for a year. For example, an effectiveness rating of 93% for oral contraceptives means that for every 100 women using the pill regularly for a year, 7 will get pregnant. Effectiveness rates tend to be lower in developing countries, primarily because of lack of education. Globally about 39% of the world's people using contraception rely on sterilization (32% of females and 7% of males), followed by IUDs (22%), the pill (14%), and male condoms (7%). Preferences in the United States are female sterilization (26%), the pill (25%), male condoms (19%), and male sterilization (10%). (Data from Alan Guttmacher Institute, Henry J. Kaiser Family Foundation, and the United Nations Population Division)

smaller number of children because fewer children die at an early age.

Average age at marriage (or, more precisely, the average age at which women have their first child) also plays a role. Women normally have fewer children when their average age at marriage is 25 or older.

Birth rates and TFRs are also affected by the *availability of legal abortions.* Each year about 190 million women become pregnant. The United Nations and the World Bank estimate that about 46 million of these women get abortions: 26 million of them legal and 20 million illegal (and often unsafe).

The *availability of reliable birth control methods* (Figure 10-12) allows women to control the number and spacing of the children they have. *Religious beliefs, traditions, and cultural norms* also play a role. In some countries, these factors favor large families and strongly oppose abortion and some forms of birth control.

What Factors Affect Death Rates? Reducing Deaths

Death rates have declined because of increased food supplies, better nutrition, advances in medicine, improved sanitation, and safer water supplies.

The rapid growth of the world's population over the past 100 years was not caused by a rise in the crude birth rate. Instead, it was caused largely by a decline in crude death rates, especially in developing countries.

More people started living longer and fewer infants died because of increased food supplies and distribution, better nutrition, medical advances such as vaccines and antibiotics, improved sanitation, and safer water supplies (which curtailed the spread of many infectious diseases).

Two useful indicators of overall health of people in a country or region are **life expectancy** (the average number of years a newborn infant can expect to live) and the **infant mortality rate** (the number of babies out of every 1,000 born who die before their first birthday).

Great news. The global life expectancy at birth increased from 48 years to 67 years (76 years in developed countries and 65 years in developing countries) between 1955 and 2004. It is projected to reach 74 in developing countries by 2050. Between 1900 and 2004, life expectancy in the United States increased from 47 to 77 years and is projected to reach 82 years by 2050.

Bad news. In the world's poorest and least developed countries, mainly in Africa, life expectancy is 49 years or less. In many African countries life expectancy is expected to fall further because of more deaths from AIDS.

Infant mortality is viewed as the best single measure of a society's quality of life because it reflects a country's general level of nutrition and health care. A high infant mortality rate usually indicates insufficient food (undernutrition), poor nutrition (malnutrition), and a high incidence of infectious disease (usually from contaminated drinking water and weakened disease resistance from undernutrition and malnutrition).

Good news. Between 1965 and 2004, the world's infant mortality rate dropped from 20 per 1,000 live births to 7 in developed countries and from 118 to 61 in developing countries. *Bad news.* At least 8 million infants (most in developing countries) die of preventable causes during their first year of life—an average of 22,000 mostly unnecessary infant deaths per day. This is equivalent to 55 jumbo jets, each loaded with 400 infants under age 1, crashing each day with no survivors!

The U.S. infant mortality rate declined from 165 in 1900 to 7 in 2004. This sharp decline was a major factor in the marked increase in U.S. average life expectancy during this period.

Still some 40 countries had lower infant mortality rates than the United States in 2004. Three factors keep the U.S. infant mortality rate higher than it could be: *inadequate health care for poor women during pregnancy and for their babies after birth, drug addiction among pregnant women,* and *a high teenage birth rate.*

Case Study: Should the United States Encourage or Discourage Immigration?

Immigration has played and continues to play a major role in the growth and cultural diversity of the U.S. population.

Only a few countries such as Canada, Australia, and the United States encourage immigration. And international migration to developed countries absorbs only about 1% of the annual population growth in developing countries

Since 1820 the United States has admitted almost twice as many immigrants as all other countries combined! However, the number of legal immigrants (including refugees) has varied during different periods because of changes in immigration laws and rates of economic growth (Figure 10-13). Currently, immigration accounts for about 41% of the country's annual population growth.

Between 1820 and 1960, most legal immigrants to the United States came from Europe. Since 1960, most have come from Latin America (51%) and Asia (30%), followed by Europe (13%).

What is the largest minority group in the United States? Answer: Latinos (67% of them from Mexico) made up 14% of the U.S. population in 2003. By 2050 Latinos are projected to make up one of every four people in the United States.

In 1995, the U.S. Commission on Immigration Reform recommended reducing the number of legal immigrants from about 900,000 to 700,000 per year for a transition period and then to 550,000 a year. Some analysts want to limit legal immigration to about 20% of the

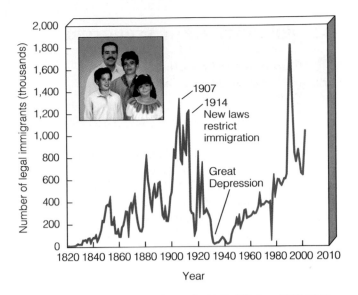

Figure 10-13 Legal immigration to the United States, 1820–2001. The large increase in immigration since 1989 resulted mostly from the Immigration Reform and Control Act of 1986, which granted legal status to illegal immigrants who could show they had been living in the country for several years. (Data from U.S. Immigration and Naturalization Service)

country's annual population growth. They would accept immigrants only if they can support themselves, arguing that providing immigrants with public services makes the United States a magnet for the world's poor.

There is also support for sharply reducing illegal immigration. But some are concerned that a crackdown on the country's 8–10 million illegal immigrants can also lead to discrimination against legal immigrants.

Proponents of reducing immigration argue that it would allow the United States to stabilize its population sooner and help reduce the country's enormous environmental impact. The public strongly supports reducing U.S. immigration levels. A January 2002 Gallup poll found that 58% of the people polled believed that immigration rates should be reduced (up from 45% in January 2001). A 1993 Hispanic Research Group survey found that 89% of Hispanic Americans supported an immediate moratorium on immigration.

Others oppose reducing current levels of legal immigration. They argue that this would diminish the historical role of the United States as a place of opportunity for the world's poor and oppressed. In addition, immigrants pay taxes, take many menial and low-paying jobs that other Americans shun, open businesses, and create jobs. Moreover, according to the U.S. Census Bureau, after 2020 higher immigration levels will be needed to supply enough workers as baby boomers retire.

X *HOW WOULD YOU VOTE?* Should immigration into the United States (or the country where you live) be reduced? Cast your vote online at http://biology.brookscole.com/miller14.

10-2 POPULATION AGE STRUCTURE

What Are Age Structure Diagrams? Sorting People by Age Groups

The number of people in young, middle, and older age groups determines how fast populations grow or decline.

As mentioned earlier, even if the replacement-level fertility rate of 2.1 were magically achieved globally tomorrow, the world's population would keep growing for at least another 50 years (assuming no large increase in death rates). The reason is a population's **age structure**: the distribution of males and females in each age group.

Demographers construct a population age structure diagram by plotting the percentages or numbers of males and females in the total population in each of three age categories: *prereproductive* (ages 0–14), *reproductive* (ages 15–44), and *postreproductive* (ages 45 and up). Figure 10-14 presents generalized age structure diagrams for countries with rapid, slow, zero, and negative population growth rates. Which of these figures best represents the country where you live?

Figure 10-15 shows how the age structure diagram for the United States changed between 1900 and 2000 and how it is projected to change by 2050.

How Does Age Structure Affect Population Growth? Teenagers Are the Population Wave of the Future

The number of people under age 15 is the major factor determining a country's future population growth.

Any country with many people below age 15 (represented by a wide base in Figure 10-14, left) has a powerful built-in momentum to increase its population size unless death rates rise sharply. The number of births will rise even if women have only one or two children, because a large number of girls will soon be moving into their reproductive years.

What is perhaps the world's most important population statistic? Answer: *30% of the people on the planet were under 15 years old in 2004.* These 1.9 billion young people are poised to move into their prime reproductive years. In developing countries the number is even higher: 33%, compared with 17% in developed countries.

We live in a *demographically divided world*. To see why, look at Figures 10-16 and 10-17 (p. 186).

How Can Age Structure Diagrams Be Used to Make Population and Economic Projections? Looking into a Crystal Ball

Changes in the distribution of a country's age groups have long-lasting economic and social impacts.

Between 1946 and 1964, the United States had a *baby boom* that added 79 million people to its population. Over time this group looks like a bulge moving up through the country's age structure (Figure 10-18, p. 186).

Baby boomers now make up nearly half of all adult Americans. As a result, they dominate the population's demand for goods and services. They also play an increasingly important role in deciding who gets elected and what laws are passed. Baby boomers who created the youth market in their teens and 20s are

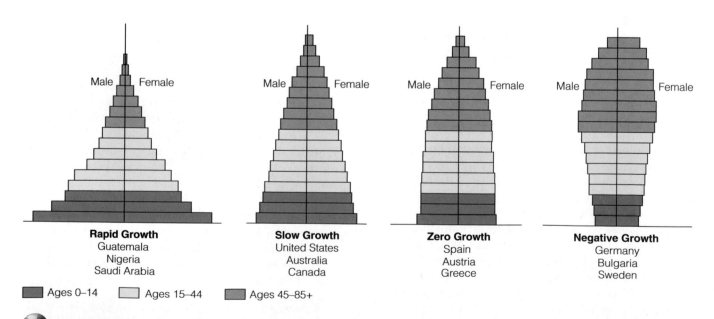

Rapid Growth	Slow Growth	Zero Growth	Negative Growth
Guatemala	United States	Spain	Germany
Nigeria	Australia	Austria	Bulgaria
Saudi Arabia	Canada	Greece	Sweden

■ Ages 0–14   □ Ages 15–44   ■ Ages 45–85+

Figure 10-14 Generalized population age structure diagrams for countries with rapid (1.5–3%), slow (0.3–1.4%), zero (0–0.2%), and negative population growth rates (a declining population). (Data from Population Reference Bureau)

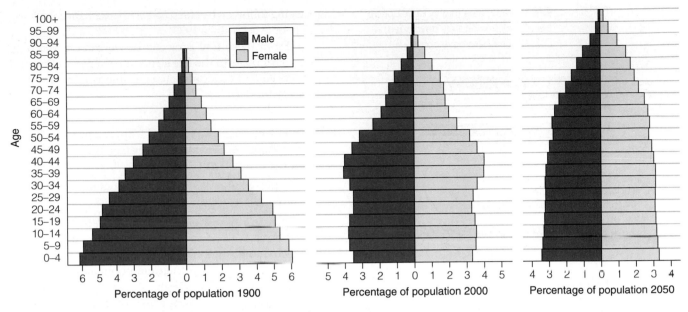

Figure 10-15 U.S. population by age and sex, 1900, 2000, and 2050 (projected). (Data from U.S. Census Bureau)

now creating the 50-something market and will soon move on to create a 60-something market. In 2011 the first baby boomers will turn 65, and the number of Americans over age 65 will grow sharply through 2029. According to some analysts, the retirement of baby boomers is likely to create a shortage of workers in the United States unless immigrant workers replace some of these retirees.

Much of the economic burden of helping support a large number of retired baby boomers will fall on the *baby-bust generation*. It consists of people born between 1965 and 1976 (when TFRs fell sharply to below 2.1; Figure 10-7).

Retired baby boomers are likely to use their political clout to force the smaller number of people in the baby-bust generation to pay higher income, healthcare, and Social Security taxes.

In other respects, the baby-bust generation should have an easier time than the baby-boom generation. Fewer people will be competing for educational opportunities, jobs, and services. Also, labor shortages may drive up their wages, at least for jobs requiring education or technical training beyond high school. However, this may not happen, if many American-owned companies operating at the global level (multinational companies) continue to export many low- and high-paying jobs to other countries.

Members of the baby-bust group may find it difficult to get job promotions as they reach middle age because members of the much larger baby-boom group will occupy most upper-level positions. Many baby boomers may delay retirement because of improved health, the need to accumulate adequate retirement funds, or extension of the retirement age needed to begin collecting Social Security. The baby-bust generation

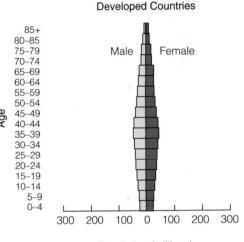

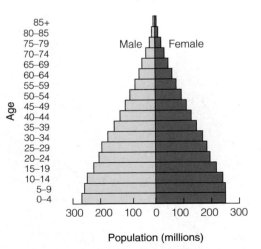

Figure 10-16 Population structure by age and sex in developing countries and developed countries, 2004. (Data from United Nations Population Division and Population Reference Bureau)

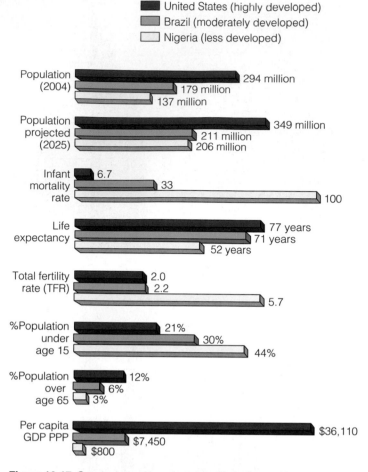

Population (2004)
- 294 million
- 179 million
- 137 million

Population projected (2025)
- 349 million
- 211 million
- 206 million

Infant mortality rate
- 6.7
- 33
- 100

Life expectancy
- 77 years
- 71 years
- 52 years

Total fertility rate (TFR)
- 2.0
- 2.2
- 5.7

%Population under age 15
- 21%
- 30%
- 44%

%Population over age 65
- 12%
- 6%
- 3%

Per capita GDP PPP
- $36,110
- $7,450
- $800

Figure 10-17 Comparison of key demographic indicators in 2004 for three countries, one highly developed (United States), one moderately developed (Brazil), and one less developed (Nigeria). (Data from Population Reference Bureau)

is being followed by the *echo-boom generation* consisting of people born since 1977.

From these few projections, we can see that any booms or busts in the age structure of a population create social and economic changes that ripple through a society for decades.

What Are Some Effects of Population Decline from Reduced Fertility? Sliding Down a Hill Too Fast Can Hurt

Rapid population decline as a result of more older people and fewer young people can lead to long-lasting economic and social problems.

The populations of most of the world's countries are projected to grow throughout most of this century. By 2004, however, 40 countries had populations that were either stable (annual growth rates at or below 0.3%) or declining. All are in Europe, except Japan. This means that about 14% of humanity (896 million people) lives in countries with stable or declining populations. By 2050, the United Nations projects, the population size of most developed countries (but not the United States) will have stabilized.

As the age structure of the world's population changes and the percentage of people age 60 or older increases (Figure 10-19), more countries will begin experiencing population declines. If population decline is gradual, its harmful effects usually can be managed.

But rapid population decline, like rapid population growth, can lead to serious economic and social problems. A country undergoing rapid population decline because of a "baby bust" or "birth dearth" has a

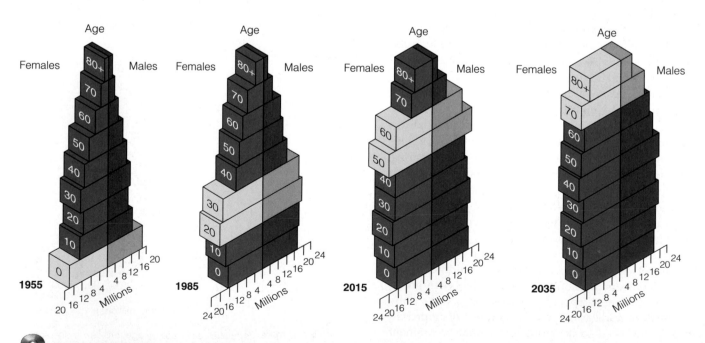

Figure 10-18 Tracking the baby-boom generation in the United States. (Data from Population Reference Bureau and U.S. Census Bureau)

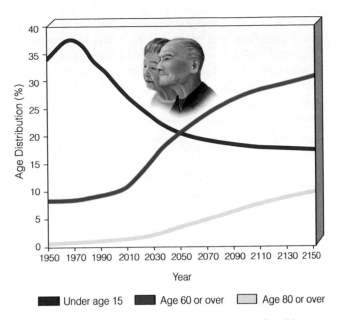

Figure 10-19 *Global aging.* Projected percentage of world population under age 15, age 60 or over, and age 80 or over, 1950–2150, assuming the medium fertility projection shown in Figure 10-6. Between 1998 and 2050 the number of people over age 80 is projected to increase from 66 million to 370 million. The cost of supporting a much larger elderly population will place enormous strains on the world's economy. (Data from the United Nations)

Legend: ■ Under age 15 ■ Age 60 or over □ Age 80 or over

Graph axes: Age Distribution (%) vertical (0 to 40); Year horizontal (1950 1970 1990 2010 2030 2050 2070 2090 2110 2130 2150)

sharp rise in the proportion of older people. They consume an increasingly larger share of medical care, social security funds, and other costly public services funded by a decreasing number of working taxpayers. Such countries can also face labor shortages unless they rely more on greatly increased automation or immigration of foreign workers.

Without babies to replenish the labor force and pay taxes, governments of countries such as Japan and some European countries facing rapid population decline will have a hard time funding the pensions of people who are living longer. To keep their finances in the black they will probably need to take unpopular steps, such as raising the retirement age, cutting retirement benefits, raising taxes, and increasing legal immigration.

What Are Some Effects of Population Decline from a Rise in Death Rates? The AIDS Tragedy

Large numbers of deaths from AIDS disrupt a country's social and economic structure by removing many young adults from its age structure.

Globally between 2000 and 2050, AIDS is projected to cause the premature deaths of 278 million people in 53 countries—38 of them in Africa. These premature deaths are almost equal to the entire current population of the United States. Read this paragraph again and think about the enormity of this tragedy.

Hunger and malnutrition kill mostly infants and children, but AIDS kills many young adults. This change in the age structure of a country has a number of harmful effects. One is a sharp drop in average life expectancy. In 16 African countries where up to a third of the adult population is infected with HIV, life expectancy could drop to 35–40 years of age.

Another effect is a loss of a country's most productive young adult workers and trained personnel such as scientists, farmers, engineers, teachers, and government, business, and health-care workers. This causes a significant increase in the number of orphans whose parents have died from AIDS—with 40 million orphans projected in Africa by 2010. It also causes a sharp drop in the number of productive adults available to support the young and elderly and to grow food.

Analysts call for the international community—especially developed countries—to develop and fund a massive program to help countries ravaged by AIDS in Africa and elsewhere. The program would have two major goals. One is to reduce the spread of HIV through a combination of improved education and health care. The other is to provide financial assistance for education and health care and volunteer teachers and health-care and social workers to help compensate for the missing young adult generation.

10-3 SOLUTIONS: INFLUENCING POPULATION SIZE

What Are the Advantages and Disadvantages of Reducing Births? An Important Controversy

There is disagreement over whether the world should encourage or discourage population growth.

The projected increase of the human population from 6.4 to 8.9 billion or more between 2004 and 2050 (Figure 10-6) raises an important question: *Can the world provide an adequate standard of living for 2.5 billion more people without causing widespread environmental damage?*

Controversy surrounds this and two related questions: whether the earth is overpopulated, and what measures, if any, should be taken to slow population growth. To some the planet is already overpopulated. To others we should encourage population growth to help stimulate economic growth by having more consumers.

Some analysts believe that asking how many people the world can support is the wrong question. They liken it to asking how many cigarettes one can smoke before getting lung cancer. Instead, they say, we should be asking what the *optimum sustainable population* of the earth might be, based on the planet's *cultural carrying capacity* See the Guest Essay by Garrett Hardin on this

topic on the website for this chapter. Such an optimum level would allow most people to live in reasonable comfort and freedom without impairing the ability of the planet to sustain future generations.

What is the optimum population size for the world (or for a particular country)? No one knows. Some consider it a meaningless concept; some put it at 20 billion, others at 8 billion, and others as low as 2 billion.

Those who do not believe the earth is overpopulated point out that the average life span of the world's 6.4 billion people is longer today than at any time in the past and is projected to get longer. They say that the world can support billions more people. They also see more people as the most valuable resource for solving the problems we face and stimulating economic growth by becoming consumers.

Some believe that all people should be free to have as many children as they want. And some view any form of population regulation as a violation of their religious beliefs. Others see it as an intrusion into their privacy and personal freedom. Some developing countries and some members of minorities in developed countries regard population control as a form of genocide to keep their numbers and power from rising.

Proponents of slowing and eventually stopping population growth have a different view. They point out that we fail to provide the basic necessities for one out of six people on the earth today. If we cannot or will not do this now, they ask, how will we be able to do this for the projected 2.5 billion more people by 2050?

Proponents of slowing population growth warn of two serious consequences if we do not sharply lower birth rates. One possibility is a higher death rate because of declining health and environmental conditions in some areas—something that is already happening in parts of Africa. Another is increased resource use and environmental harm as more consumers increase their already large ecological footprint in developed countries and in developing countries such as China and India that are undergoing rapid economic growth.

Population increase and the consumption that goes with it can increase environmental stresses such as *infectious disease, biodiversity losses, loss of tropical forests, fisheries depletion, increasing water scarcity, pollution of the seas,* and *climate change.*

Proponents of this view recognize that population growth is not the only cause of these problems. But they argue that adding several hundred million more people in developed countries and several billion more in developing countries can only intensify existing environmental and social problems.

These analysts believe people should have the freedom to produce as many children as they want, but only if it does not reduce the quality of other people's lives now and in the future, either by impairing the earth's ability to sustain life or by causing social disruption. They point out that limiting the freedom of individuals to do anything they want, in order to protect the freedom of other individuals, is the basis of most laws in modern societies.

How Can Economic Development Help Reduce Birth Rates? Economics Worked Once, But Will It Work Again?

History indicates that as countries become economically developed, their birth and death rates decline.

Demographers have examined the birth and death rates of western European countries that industrialized during the 19th century. From these data they developed a hypothesis of population change known as the **demographic transition**: as countries become industrialized, first their death rates and then their birth rates decline.

According to this hypothesis, the transition takes place in four stages (Figure 10-20):

First is the *preindustrial stage,* when there is little population growth because harsh living conditions lead to both a high birth rate (to compensate for high infant mortality) and a high death rate.

Next is the *transitional stage,* when industrialization begins, food production rises, and health care improves. Death rates drop and birth rates remain high, so the population grows rapidly (typically 2.5–3% a year).

During the third phase, called the *industrial stage,* the birth rate drops and eventually approaches the death rate as industrialization, medical advances, and modernization become widespread. Population growth continues, but at a slower and perhaps fluctuating rate, depending on economic conditions. Most developed countries and a few developing countries are in this third stage.

The last phase is the *postindustrial stage,* when the birth rate declines further, equaling the death rate and reaching zero population growth. Then the birth rate falls below the death rate and population size decreases slowly. Forty countries containing about 14% of the world's population have entered this stage and more of the world's developed countries are expected to enter this phase by 2050.

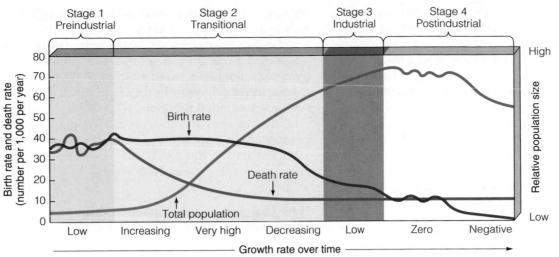

Figure 10-20
Generalized model of the demographic transition.

In most developing countries today, death rates have fallen much more than birth rates. In other words, these developing countries are still in the transitional stage, halfway up the economic development ladder, with high population growth rates.

Some economists believe that developing countries will make the demographic transition over the next few decades. But some population analysts fear that the still-rapid population growth in many developing countries will outstrip economic growth and overwhelm local life-support systems. This could cause many of these countries to be caught in a *demographic trap* at stage 2, the transition stage. This is now happening in a number of developing countries, especially in Africa. Indeed, some of the countries in Africa being ravaged by the HIV/AIDS epidemic are falling back to stage 1 as their death rates rise.

Analysts also point out that some of the conditions that allowed developed countries to develop are not available to many of today's developing countries. One problem is a shortage of skilled workers to produce the high-tech products necessary to compete in today's global economy. Another is a lack of the financial capital and other resources that allow rapid economic development.

Two other problems hinder economic development in many developing countries. One is a sharp rise in their debt to developed countries. Much of the income of such countries must be used to pay the interest on their debts. This leaves too little money for improving social, health, and environmental conditions.

Another problem is that since 1980 developing countries have been receiving less economic assistance from developed countries. Indeed, since the mid-1980s, developing countries have paid developed countries $40–50 billion a year (mostly in debt interest) more than they have received from these countries.

How Can Family Planning Help Reduce Birth and Abortion Rates and Save Lives? Planning for Babies Works

Family planning has been a major factor in reducing the number of births and abortions throughout most of the world.

Family planning provides educational and clinical services that help couples choose how many children to have and when to have them. Such programs vary from culture to culture, but most provide information on birth spacing, birth control, and health care for pregnant women and infants.

Family planning has helped raise the use of modern forms of contraception by married women in their reproductive years in developing countries from 10% of in the 1960s to 51% in 2004.

Studies also show that family planning has been responsible for at least 55% of the drop in TFRs in developing countries, from 6 in 1960 to 3.1 in 2004. Two examples of countries that have sharply reduced their population growth are Thailand (p. 176) and Iran (Case Study, p. 190). Family planning has also reduced the number of legal and illegal abortions per year and the risk of maternal and fetus death from pregnancy.

Despite such successes, there is also some *bad news*. *First,* according to John Bongaarts of the Population Council and the United Nations Population Fund, 42% of all pregnancies in the developing countries are unplanned and 26% end with abortion. *Second,* an estimated 150 million women in developing countries want to limit the number and determine the spacing of their children, but they lack access to contraceptive services. According to the United Nations, extending family planning services to these women and to those who will soon be entering their reproductive years could prevent an estimated 5.8 million births a year and more than 5 million abortions a year!

Some analysts call for expanding family planning programs to include teenagers and sexually active unmarried women, who are excluded in many existing programs. For teenagers, many advocate much greater emphasis on abstinence.

Another suggestion is to develop programs that educate men about the importance of having fewer children and taking more responsibility for raising them. Proponents also call for greatly increased research on developing new, more effective, and more acceptable birth control methods for men.

Finally, a number of analysts urge pro-choice and pro-life groups to join forces in greatly reducing unplanned births and abortions, especially among teenagers.

Case Study: Family Planning in Iran: A Success Story

Since 1989 Iran has used family planning and public education to cut its rate of population growth in half and reduce the average number of children per woman from 7 to 2.5.

When Ayatollah Khomeini assumed power in Iran in 1979, he did away with the family planning programs the Shah of Iran had put into place in 1967. He saw large families as a way to increase the size of his army.

Iranians responded and the country's population growth reached 4.4%—one of the world's highest rates. But this rapid growth in numbers began to overburden the country's economy and environment.

In 1989, the government reversed its policy and restored its family planning program. Government agencies were mobilized to raise public awareness of population issues, encourage smaller families, and provide free modern contraception. Religious leaders helped by mounting a crusade for smaller families. TV stations were used to provide family planning information throughout the country.

Iran became the first country to require couples to take a class on contraception before they could receive a marriage license. Between 1970 and 2000, the country also increased female literacy from 25% to 70% and female school enrollment from 60% to 90%.

These efforts paid off. Between 1989 and 2004, the country cut its population growth rate from 2.5% to 1.2% and its average family size from 7 children to 2.5—a remarkable change in 15 years.

How Can Empowering Women Help Reduce Birth Rates? Ensuring Education, Jobs, and Rights

Women tend to have fewer children if they are educated, have a paying job outside the home, and do not have their human rights suppressed.

What key factors lead women to have fewer and healthier children? Three things: education, paying jobs outside the home, and living in societies where their rights are not suppressed.

Roughly half of the world's people are female. Women do almost all of the world's domestic work and childcare, with little or no pay. Women also provide more unpaid health care than all the world's organized health services combined.

They also do 60–80% of the work associated with growing food, gathering fuelwood, and hauling water in rural areas of Africa, Latin America, and Asia (Figure 10-21). As one Brazilian woman put it, "For poor women the only holiday is when you are asleep."

Globally women account for two-thirds of all hours worked but receive only 10% of the world's income, and they own less than 2% of the world's land. In most developing countries, women do not have the legal right to own land or to borrow money. Women

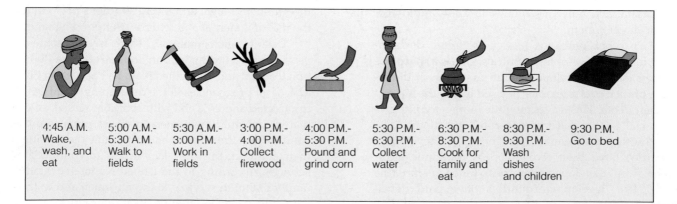

| 4:45 A.M. Wake, wash, and eat | 5:00 A.M.– 5:30 A.M. Walk to fields | 5:30 A.M.– 3:00 P.M. Work in fields | 3:00 P.M.– 4:00 P.M. Collect firewood | 4:00 P.M.– 5:30 P.M. Pound and grind corn | 5:30 P.M.– 6:30 P.M. Collect water | 6:30 P.M.– 8:30 P.M. Cook for family and eat | 8:30 P.M.– 9:30 P.M. Wash dishes and children | 9:30 P.M. Go to bed |

Figure 10-21 Typical workday for a woman in rural Africa. In addition to their domestic work, rural African women perform about 75% of all agricultural work. (Data from United Nations)

also make up 70% of the world's poor and 60% of the 875 million illiterate adults worldwide who can neither read nor write.

Women's representation in governments has been gradually increasing. Yet, only 11 of the world's 190 heads of state are women, and women hold just 14% of the seats in the world's parliaments.

According to United Nations Population Agency's executive director Thorya Obaid, "Many women in the developing world are trapped in poverty by illiteracy, poor health, and unwanted high fertility. All of these contribute to environmental degradation and tighten the grip of poverty. If we are serious about sustainable development, we must break this vicious cycle."

Breaking out of this trap means giving women everywhere full legal rights and the opportunity to become educated and earn income outside the home. Achieving this would slow population growth, promote human rights and freedom, reduce poverty, and slow environmental degradation—a win-win result.

Empowering women by seeking gender equality will take some major social changes. This will be difficult to achieve in male-dominated societies but it can be done.

Good news. An increasing number of women in developing countries are taking charge of their lives and reproductive behavior. They are not waiting around for the slow processes of education and cultural change. As it expands, such bottom-up change by individual women will play an important role in stabilizing population and providing women with equal rights.

10-4   CASE STUDIES: INDIA AND CHINA

What Success Has India Had in Controlling Its Population Growth? Some Progress but Not Enough

For over five decades India has tried to control its population growth with only modest success.

The world's first national family planning program began in India in 1952, when its population was nearly 400 million. In 2004, after 52 years of population control efforts, India was the world's second most populous country, with a population of 1.1 billion.

In 1952, India added 5 million people to its population. In 2004 it added 18 million. Figure 10-22 compares demographic data for India and China.

India faces a number of already serious poverty, malnutrition, and environmental problems that could worsen as its population continues to grow rapidly. By global standards, India's people are poor. Nearly half of India's labor force is unemployed or can find only occasional work.

India currently is self-sufficient in food grain production. Still, about 40% of its population and 53% of its children suffer from malnutrition, mostly because of poverty.

Furthermore, India faces serious resource and environmental problems. With 17% of the world's people, it has just 2.3% of the world's land resources and 2% of the world's forests. About half of the country's cropland is degraded as a result of soil erosion, waterlogging, salinization, overgrazing, and deforestation. In addition, over two-thirds of India's water is seriously polluted and sanitation services often are inadequate.

Without its long-standing family planning program, India's population and environmental problems would be growing even faster. Still, to its supporters the results of the program have been disappointing for several reasons: poor planning, bureaucratic inefficiency, the low status of women (despite constitutional guarantees of equality), extreme poverty, and lack of administrative and financial support.

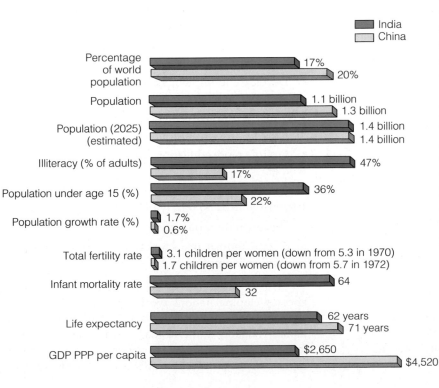

Figure 10-22 Comparison of basic demographic data for India (green) and China (yellow) in 2004. (Data from United Nations and Population Reference Bureau).

The government has provided information about the advantages of small families for years. Yet Indian women still have an average of 3.1 children. One reason is that most poor couples believe they need many children to do work and care for them in old age. Another is the strong cultural preference for male children, which means some couples keep having children until they produce one or more boys. These factors in part explain why even though 90% of Indian couples know of at least one modern birth control method, only 43% actually use one.

What Success Has China Had in Controlling Its Population Growth? Good Progress, Enforced with an Iron Hand

Since 1970 China has used a government-enforced program to cut its birth rate in half and sharply reduce its total fertility rate.

Since 1970, China has made impressive efforts to feed its people and bring its population growth under control. Between 1972 and 2004, China cut its crude birth rate in half and cut its total fertility rate from 5.7 to 1.7 children per woman (Figure 10-22).

To achieve its sharp drop in fertility, China has established the world's most extensive, intrusive, and strict population control program. Couples are strongly urged to postpone marriage and to have no more than one child. Married couples who pledge to have no more than one child receive extra food, larger pensions, better housing, free medical care, salary bonuses, free school tuition for their one child, and preferential treatment in employment when their child enters the job market. Couples who break their pledge lose such benefits.

The government also provides married couples with ready access to free sterilization, contraceptives, and abortion. This helps explain why about 83% of married women in China use modern contraception.

Government officials realized in the 1960s that the only alternative to strict population control was mass starvation. China is a dictatorship. Thus, unlike India, it has been able to impose a fairly consistent population policy throughout its society.

China has 21% of the world's population. But it has only 7% of the world's fresh water and cropland, 4% of its forests, and 2% of its oil. Soil erosion in China is serious and apparently getting worse. In the 1970s, the Chinese government had a system of health clinics that provided basic health care for its huge rural farm population. This system collapsed in the 1980s as China embraced capitalist economic reforms. According to government estimates, more than 800 million people—9 of every 10 rural Chinese—now have no health insurance or social safety net.

Population experts expect China's population to peak around 2040 and then begin a slow decline. This has led some members of China's parliament to call for amending the country's one-child policy so that some urban couples can have a second child. The goal would be to provide more workers to help support China's aging population.

What lesson can other countries learn from China? One possibility is to try to curb population growth before they must choose between mass starvation and coercive measures that severely restrict human freedom.

10-5 CUTTING GLOBAL POPULATION GROWTH

What Can We Do to Slow Population Growth? A New Vision

Experience indicates that the best way to slow population growth is a combination of investing in family planning, reducing poverty, and elevating the status of women.

In 1994, the United Nations held its third once-in-a-decade Conference on Population and Development in Cairo, Egypt. One of the conference's goals was to encourage action to stabilize the world's population at 7.8 billion by 2050 instead of the projected 8.9 billion. The major goals of the resulting population plan, endorsed by 180 governments, are to do the following by 2015:

- Provide universal access to family planning services and reproductive health care

- Improve health care for infants, children, and pregnant women

- Develop and implement national population polices

- Improve the status of women and expand education and job opportunities for young women

- Provide more education, especially for girls and women

- Increase the involvement of men in child-rearing responsibilities and family planning

- Sharply reduce poverty

- Greatly reduce unsustainable patterns of production and consumption

This is a tall order. But it can be done if developed and developing nations work together to implement such reforms. Some *good news* is that the experience of Japan, Thailand, South Korea, Taiwan, Iran, and China indicates that a country can achieve or come close to replacement-level fertility within a decade or two.

Such experience also suggests that the best way to slow population growth is a combination of investing in family planning, reducing poverty, and elevating the status of women.

In this chapter you have learned how the ecological principles of population dynamics can be applied to the human population. In the next three chapters, you will learn how various principles of ecology can be applied to help sustain the earth's biodiversity.

Our numbers expand but Earth's natural systems do not.
LESTER R. BROWN

CRITICAL THINKING

1. Why is it rational for a poor couple in a developing country such as India to have four or five children? What changes might lead such a couple to consider their behavior irrational?

2. Choose what you consider to be a major local, national, or global environmental problem, and describe the role of population growth in this problem. Compare your answer with those of your classmates.

3. Do you believe the population of **(a)** your own country and **(b)** the area where you live is too high? Explain.

4. Should everyone have the right to have as many children as they want? Explain.

5. Some people have proposed that the earth could solve its population problem by shipping people off to space colonies, each containing about 10,000 people. Assuming we could build such large-scale, self-sustaining space stations, how many people would have to be shipped off each day to provide living spaces for the 80 million people added to the earth's population this year? Current space shuttles can handle about 6 to 8 passengers. If this capacity could be increased to 100 passengers per shuttle, how many shuttles would have to be launched per day to offset the 80 million people added this year? According to your calculations, determine whether this proposal is a logical solution to the earth's population problem.

6. Some people believe the most important goal is to sharply reduce the rate of population growth in developing countries, where 97% of the world's population growth is expected to take place. Some people in developing countries agree that population growth in these countries can cause local environmental problems. But they contend that the most serious environmental problem the world faces is disruption of the global life-support system for the human species by high levels of resource consumption per person in developed countries, which use 80% of the world's resources. What is your view on this issue? Explain.

7. Suppose the cloning of human beings becomes possible without any serious health effects for cloned individuals. What effect might this have on the world's population size and growth rate? Explain.

8. Congratulations! You are in charge of the world. List the three most important features of your population policy.

PROJECTS

1. Assume your entire class (or manageable groups of your class) is charged with coming up with a plan for halving the world's population growth rate within the next 20 years. Develop a detailed plan that would achieve this goal, including any differences between policies in developing countries and those in developed countries. Justify each part of your plan. Try to anticipate what problems you might face in implementing the plan, and devise strategies for dealing with these problems.

2. Prepare an age structure diagram for your community. Use the diagram to project future population growth and economic and social problems.

3. Use the library or the Internet to find bibliographic information about *Paul Hawken* and *Lester R. Brown*, whose quotes appear at the beginning and end of this chapter.

4. Make a concept map of this chapter's major ideas, using the section heads, subheads, and key terms (in boldface). See material on the website for this book about how to prepare concept maps.

LEARNING ONLINE

The website for this book contains study aids and many ideas for further reading and research. They include a chapter summary, review questions for the entire chapter, flash cards for key terms and concepts, a multiple-choice practice quiz, interesting Internet sites, references, and a guide for accessing thousands of InfoTrac® College Edition articles. Log on to

http://biology.brookscole.com/miller14

Then click on the Chapter-by-Chapter area, choose Chapter 10, and select a learning resource.

Sustaining Terrestrial Biodiversity: Managing and Protecting Ecosystems

Biodiversity

CASE STUDY
Reintroducing Wolves to Yellowstone

At one time the gray wolf (Figure 11-1) ranged over most of North America. But between 1850 and 1900 an estimated 2 million wolves were shot, trapped, and poisoned by ranchers, hunters, and government employees. The idea was to make the West and the Great Plains safe for livestock and for big game animals prized by hunters.

It worked. When Congress passed the U.S. Endangered Species Act in 1973, only about 400–500 gray wolves remained in the lower 48 states, primarily in Minnesota and Michigan. In 1974 the U.S. Fish and Wildlife Service (USFWS) listed the gray wolf as endangered in all 48 lower states except Minnesota. Alaska was also not included because it had 6,000–8,000 gray wolves.

Ecologists recognize the important role this keystone predator species once played in parts of the West and the Great Plains. These wolves culled herds of bison, elk, caribou, and mule deer, and kept down coyote populations. They also provided uneaten meat for scavengers such as ravens, bald eagles, bears, ermines, and foxes.

In recent years, herds of elk, moose, and antelope have expanded. Their larger numbers have devastated some vegetation, increased erosion, and threatened the niches of other wildlife species. Reintroducing a keystone species such as the gray wolf into a terrestrial ecosystem is one way to help sustain its biodiversity and prevent environmental degradation.

In 1987, the USFWS proposed reintroducing gray wolves into the Yellowstone ecosystem. This brought angry protests. Some objections came from ranchers who feared the wolves would attack their cattle and sheep; one enraged rancher said that it was "like rein-

Figure 11-1 The *gray wolf* is a threatened species in the lower 48 states. Ranchers, hunters, miners, and loggers have vigorously opposed efforts to return this keystone species to its former habitat in the Yellowstone National Park area. However, wolves were reintroduced beginning in 1995 and now number several hundred.

troducing smallpox." Other protests came from hunters who feared the wolves would kill too many big game animals, and from mining and logging companies who worried the government would halt their operations on wolf-populated federal lands.

Since 1995, federal wildlife officials have caught gray wolves in Canada and relocated them in Yellowstone National Park and northern Idaho. By 2004 there were about 760 gray wolves in these two areas.

Their presence is causing a cascade of ecological changes in Yellowstone. With wolves around, elk are gathering less near streams and rivers. This has spurred the growth of aspen and willow trees that attract beavers, and elk killed by wolves are an important food source for grizzlies.

The wolves have cut coyote populations in half. This has increased populations of smaller animals such as ground squirrels and foxes hunted by coyotes, providing more food for eagles and hawks. Between 1995 and 2002 the wolves also killed 792 sheep, 278 cattle, and 62 dogs in the Northern Rockies.

In 2003, the U.S. Fish and Wildlife Service downgraded the gray wolf throughout most of the lower 48 states from endangered to threatened. In 2004, the agency proposed removing wolves from protection under the Endangered Species Act in Idaho and Montana. This would allow private citizens in these states to kill wolves that are attacking livestock or pets on private lands. Conservationists say this action is premature, warning that it could undermine one of the nation's most successful conservation efforts.

Population growth, economic development, and poverty are exerting increasing pressure on the world's forests, grasslands, parks, wilderness, and other terrestrial storehouses of biodiversity. This chapter and the two that follow are devoted to helping us understand and sustain the earth's biodiversity.

Forests precede civilizations, deserts follow them.
FRANÇOIS-AUGUSTE-RENÉ DE CHATEAUBRIAND

This chapter addresses the following questions:

- How have human activities affected the earth's biodiversity?

- What is conservation biology? What role does bioinfomatics play in helping sustain biodiversity?

- What are the major types of public lands in the United States, and how are they used?

- Why are forest resources important, and how are they used, managed, and sustained?

- How should forests in the United States be used, managed, and sustained?

- How serious is tropical deforestation, and how can we help sustain tropical forests?

- What problems do parks face, and how should we manage them?

- How should we establish, design, protect, and manage terrestrial nature reserves?

- What is wilderness, and why is it important?

- What is ecological restoration, and why is it important?

- What can we do to help sustain the earth's biodiversity?

11-1 HUMAN IMPACTS ON TERRESTRIAL BIODIVERSITY

How Have Human Activities Affected Global Biodiversity? Increasing Our Ecological Footprint

We have depleted and degraded some of the earth's biodiversity and these threats are expected to increase.

Figure 11-2 lists factors that tend to increase or decrease biodiversity. Many of our activities decrease biodiversity (Figure 11-3). According to biodiversity expert Edward O. Wilson, "The natural world is everywhere disappearing before our eyes—cut to pieces, mowed down, plowed under, gobbled up, replaced by human artifacts."

Consider a few examples of how human activities have decreased and degraded the earth's terrestrial biodiversity. According to the results of a 2002 study on the impact of the human ecological footprint on the earth's land (Figure 9-12, p. 172), we have disturbed to

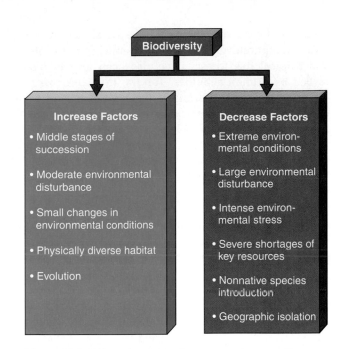

Figure 11-2 Factors that tend to increase or decrease the earth's biodiversity.

some extent at least half and probably about 83% of the earth's land surface (excluding Antarctica and Greenland).

About 82% of temperate deciduous forests have been cleared, fragmented, and dominated because their soils and climate are very favorable for growing

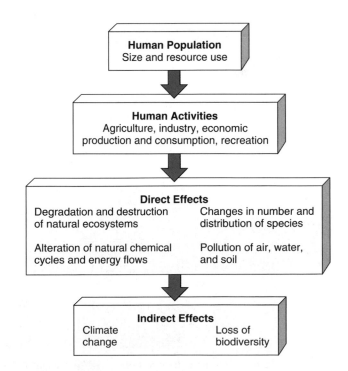

Figure 11-3 **Natural capital degradation:** major connections between human activities and the earth's biodiversity.

food and urban development. Chaparral and thorn scrub, temperate grasslands, temperate rain forests, and tropical dry forests have also been greatly disturbed by human activities. Tundra, tropical deserts, and land covered with ice are the least disturbed biomes because their harsh climates and poor soils make them unappealing to most human activities.

In the United States, at least 95% of the virgin forests in the lower 48 states have been logged for lumber and to make room for agriculture, housing, and industry. In addition, 98% of tallgrass prairie in the Midwest and Great Plains has disappeared, and 99% of California's native grassland and 85% of its original redwood forests are gone.

By some estimates, humans use, waste, or destroy about 10–55% of the net primary productivity of the planet's terrestrial ecosystems. And biologists estimate that the current global extinction rate of species is at least 100 times and probably 1,000 to 10,000 times what it was before humans existed.

These threats to the world's biodiversity are projected to increase sharply by 2018 (Figure 11-4). Study this figure carefully.

Figure 11-5 outlines the goals, strategies, and tactics for preserving and restoring the earth's terrestrial ecosystems (as discussed in this chapter) and preventing the premature extinction of species (as discussed in Chapter 12). Sustaining aquatic diversity is discussed in Chapter 13.

Why Should We Care About Biodiversity? Sustaining a Vital Part of the World's Life Support System

Biodiversity should be protected from degradation by human activities because it exists and because of its usefulness to us.

Biodiversity researchers contend that we should act to preserve the earth's overall diversity because its genes, species, ecosystems, and ecological processes have two types of value. One is **intrinsic** or **existence value** because these components of biodiversity exist, regardless of their use to us.

The other is **instrumental value** because of their usefulness to us. There are two major types of instrumental values. One consists of *use values* that benefit us in the form of economic goods and services, ecological services, recreation, scientific information, and preserving options for such uses in the future.

Another type consists of *nonuse values*. One is the *existence value* in knowing that a redwood forest, wilderness, or endangered species exists, even if we will never see it or get direct use from it. *Aesthetic value* is another nonuse value because many people appreciate a tree, forest, wild species, or a vista because of its beauty. *Bequest value* is a third type of nonuse value. It is based on a willingness of some people to pay to protect some forms of natural capital for use by future generations.

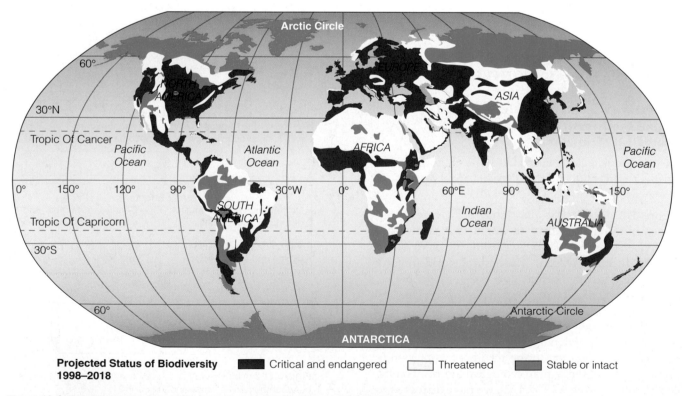

Projected Status of Biodiversity 1998–2018 ■ Critical and endangered ☐ Threatened ■ Stable or intact

Figure 11-4 Natural capital degradation: projected status of the earth's biodiversity between 1998 and 2018. (Data from World Resources Institute, World Conservation Monitoring Center, and Conservation International)

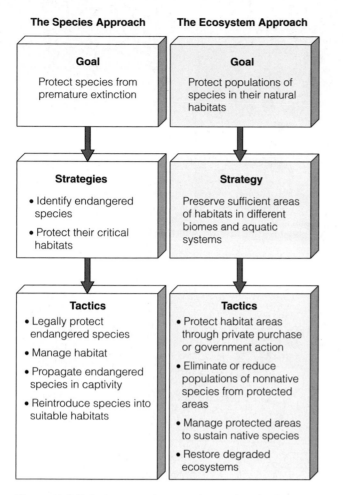

The Species Approach

Goal
Protect species from premature extinction

↓

Strategies
- Identify endangered species
- Protect their critical habitats

↓

Tactics
- Legally protect endangered species
- Manage habitat
- Propagate endangered species in captivity
- Reintroduce species into suitable habitats

The Ecosystem Approach

Goal
Protect populations of species in their natural habitats

↓

Strategy
Preserve sufficient areas of habitats in different biomes and aquatic systems

↓

Tactics
- Protect habitat areas through private purchase or government action
- Eliminate or reduce populations of nonnative species from protected areas
- Manage protected areas to sustain native species
- Restore degraded ecosystems

Figure 11-5 Solutions: goals, strategies, and tactics for protecting biodiversity.

11-2 CONSERVATION BIOLOGY

What Is Conservation Biology? Emergency Action to Sustain Biodiversity

Conservation biology uses rapid response strategies to stem the loss and degradation of the world's biodiversity.

Conservation biology is a multidisciplinary science that originated in the 1970s. Its goal is to use emergency responses to slow down the rate at which we are destroying and degrading the earth's biodiversity. Conservation biologists identify the most endangered and species-rich ecosystems, called *hot spots*. They then send in Rapid Assessment Teams of biologists to evaluate the situations, make recommendations, and take emergency action to stem the loss of biodiversity in such areas.

Conservation biology is based on Aldo Leopold's ethical principle that something is right when it tends to maintain the earth's life-support systems for us and other species and wrong when it does not. The Society of Conservation Biology is now one of the fastest

growing of all scientific societies and there are a dozen new scientific journals in this field. Have you considered a career in this field?

How Can Bioinfomatics Help Protect Biodiversity? Providing Good Information

Bioinfomatics analyzes and provides basic biological and ecological information to help us sustain biodiversity.

To understand and sustain biodiversity, we need basic biological and ecological information about the world's wild species. **Bioinfomatics** is the applied science of managing, analyzing, and communicating biological information.

Bioinfomatics uses tools such as high-resolution digitized images to photograph and analyze specimens of all known species and any new ones that are identified. It also builds computer databases to hold these images, DNA sequences for identifying bacteria and other microorganisms, and other biological information about the world's species and ecosystems. Such information is readily available through the Internet to anyone who wants it.

11-3 PUBLIC LANDS IN THE UNITED STATES

What Are the Major Types of U.S. Public Lands? Land for Current and Future Generations

More than a third of the land in the United States consists of publicly owned national forests, resource lands, parks, wildlife refuges, and protected wilderness areas.

No other nation has set aside as much of its land for public use, resource extraction, enjoyment, and wildlife as the United States. The federal government manages roughly 35% of the country's land, which belongs to every American. About 73% of this federal public land is in Alaska. Another 22% is in the western states (Figure 11-6, p. 198). The combined area of these public lands would cover up California and Alaska with land to spare.

Some federal public lands are used for many purposes. One example is the *National Forest System,* which consists of 155 forests and 22 grasslands. These forests, managed by the U.S. Forest Service, are used for logging, mining, livestock grazing, farming, oil and gas extraction, recreation, hunting, fishing, and conservation of watershed, soil, and wildlife resources.

A second example is the *National Resource Lands,* managed by the Bureau of Land Management (BLM). These lands are used primarily for mining, oil and gas extraction, and livestock grazing.

■ National parks and preserves	■ National forests	■ (and Xs) National wildlife refuges

Figure 11-6 Natural capital: national forests, national parks, and wildlife refuges managed by the U.S. federal government. U.S. citizens jointly own these and other public lands. (Data from U.S. Geological Survey)

A third system consists of 542 *National Wildlife Refuges* that are managed by the U.S. Fish and Wildlife Service (USFWS). Most refuges protect habitats and breeding areas for waterfowl and big game to provide a harvestable supply for hunters; a few protect endangered species from extinction. Permitted activities in most refuges include hunting, trapping, fishing, oil and gas development, mining, logging, grazing, some military activities, and farming.

Uses of other public lands are more restricted. One example is the *National Park System* managed by the National Park Service (NPS). It includes 56 major parks (mostly in the West) and 331 national recreation areas, monuments, memorials, battlefields, historic sites, parkways, trails, rivers, seashores, and lakeshores. Only camping, hiking, sport fishing, and boating can take place in the national parks, but sport hunting, mining, and oil and gas drilling is allowed in National Recreation Areas.

The most restricted public lands are 660 roadless areas that make up the *National Wilderness Preservation System.* These areas lie within the other types of public lands and are managed by agencies in charge of those lands. Most of these areas are open only for recreational activities such as hiking, sport fishing, camping, and nonmotorized boating.

How Should U.S. Public Lands Be Managed? An Ongoing Controversy

Since the 1800s there has been controversy over how U.S. public lands should be used because of the valuable resources they contain.

Federal public lands contain valuable oil, natural gas, coal, timber, and mineral resources. Since the 1800s there has been controversy over how the resources on these lands should be used and managed.

Most conservation biologists and environmental economists and many free-market economists believe the following four principles should govern use of public land:

- Protecting biodiversity, wildlife habitats, and the ecological functioning of public land ecosystems should be the primary goal.

- No one should receive government subsidies or tax breaks for using or extracting resources on public lands—a user-pays approach.

- The American people deserve fair compensation for extraction of any resources from their property.

- All users or extractors of resources on public lands should be responsible for any environmental damage they cause.

Aldo Leopold's land-use ethic (Section 2-5, p. 30) is the basis for most of these guiding principles.

There is strong and effective opposition to these ideas. Economists, developers, and resource extractors tend to view public lands in terms of their usefulness in providing mineral, timber, and other resources and their ability to increase short-term economic growth.

They have succeeded in blocking implementation of the four principles just listed. For example, in recent years the government has given more than $1 billion a year in subsidies to privately owned mining, logging, and grazing interests using U.S. public lands.

Some developers and resource extractors go further and have mounted a campaign to get the U.S. Congress to pass laws that would

- Sell public lands or their resources to corporations or individuals at less than fair market value.

- Slash federal funding for regulatory administration of public lands.

- Cut all old-growth forests in the national forests and replace them with tree plantations.

- Open all national parks, national wildlife refuges, and wilderness areas to oil drilling, mining, off-road vehicles, and commercial development.

- Do away with the National Park Service and launch a 20-year construction program of private concessions and theme parks run by private firms in the former national parks.

- Continue mining on public lands under the provisions of the 1872 Mining Law, which allows mining interests to pay no royalties to taxpayers for hard-rock minerals they remove.

- Repeal the Endangered Species Act or modify it to allow economic factors to override protection of endangered and threatened species.

- Redefine government-protected wetlands so that about half of them would no longer be protected.

- Prevent individuals or groups from legally challenging uses of public land for private financial gain.

> **X̄ HOW WOULD YOU VOTE?** Should much more of the U.S. public lands (or government-owned lands in the country where you live) be opened to extraction of timber, mineral, and energy resources? Cast your vote online at http://biology.brookscole.com/miller14.

11-4 MANAGING AND SUSTAINING FORESTS

What Are the Major Types of Forests? Old-Growth, Second-Growth, and Tree Plantations

Some forests have not been disturbed by human activities for several hundred years, others have grown back after being cut, and some consist of planted stands of a particular tree species.

Forests with at least 10% tree cover occupy about 30% of the earth's land surface (excluding Greenland and Antarctica). Figure 6-16 (p. 111) shows the distribution of the world's boreal, temperate, and tropical forests. These forests provide many important ecological and economic services (Figure 11-7, p. 200).

Forest managers and ecologists classify forests into three major types based on age and structure. One type is an **old-growth forest:** an uncut forest or regenerated forest that has not been seriously disturbed by human activities or natural disasters for at least several hundred years. Old-growth forests are storehouses of biodiversity because they provide ecological niches for a multitude of wildlife species (Figure 6-30, p. 122).

Natural Capital

Forests

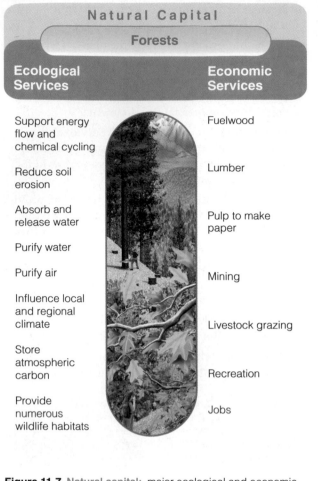

Ecological Services	Economic Services
Support energy flow and chemical cycling	Fuelwood
Reduce soil erosion	Lumber
Absorb and release water	Pulp to make paper
Purify water	
Purify air	Mining
Influence local and regional climate	Livestock grazing
Store atmospheric carbon	Recreation
Provide numerous wildlife habitats	Jobs

Figure 11-7 Natural capital: major ecological and economic services provided by forests.

A second type is a **second-growth forest:** a stand of trees resulting from secondary ecological succession (Figure 8-12, p. 158). They develop after the trees in an area have been removed by *human activities* (such as clear-cutting for timber or conversion to cropland) or by *natural forces* (such as fire, hurricanes, or volcanic eruption).

A **tree plantation,** also called a **tree farm,** is a third type (see photo on p. vi). It is a managed tract with uniformly aged trees of one species that are harvested by clear-cutting as soon as they become commercially valuable. They are then replanted and clear-cut again in a regular cycle (Figure 11-8).

Currently, about 63% of the world's forests are secondary-growth forests, 22% are old-growth forests, and 5% are tree plantations (that produce about one-fifth of the world's commercial wood). Five countries—Russia, Canada, Brazil, Indonesia, and Papua, New Guinea—have more than three-fourths of the world's remaining old-growth forests. Logging threatens about 39% of these forests. The rest are not threatened mostly because of their remoteness, not because laws protect them.

What Are the Major Types of Forest Management? Simple Tree Plantations and Diverse Forests

Some forests consist of one or two species of commercially important tree species that are cut down and replanted, and others contain diverse tree species harvested individually or in small groups.

There are two forest management systems. One is **even-aged management,** which involves maintaining trees in a given stand at about the same age and size. In this approach, sometimes called *industrial forestry,* a simplified *tree plantation* replaces a biologically diverse old-growth or second-growth forest. The plantation consists of one or two fast-growing and economically desirable species that can be harvested every 6–10 years, depending on the species (Figure 11-8).

A second type is **uneven-aged management,** which involves maintaining a variety of tree species in a stand at many ages and sizes to foster natural regeneration. Here the goals are biological diversity, long-term sustainable production of high-quality timber, selective cutting of individual mature or intermediate-aged trees, and multiple use of the forest for timber, wildlife, watershed protection, and recreation.

The fate of the world's remaining forests will be decided mostly by governments, which own about 80% of the remaining forests in developing countries. Governments in both developing and developed countries are under conflicting pressures from those wanting to log forests and convert them to agricultural land and urban development and conservationists who want to protect them—especially the world's remaining old-growth forests.

According to a 2001 study by the World Wildlife Fund, intensive but sustainable management of as little as one-fifth of the world's forests—an area twice the size of India—could meet the world's current and future demand for commercial wood and fiber. This intensive use of the world's tree plantations and some of its secondary forests would leave the world's remaining old-growth forest untouched.

How Are Trees Harvested? Be Selective or Chop Them All Down

Trees can be harvested individually from diverse forests, or an entire forest stand can be cut down in one or several phases.

The first step in forest management is to build roads for access and timber removal. Even carefully designed logging roads have a number of harmful effects (Figure 11-9). They include increased erosion and sediment runoff into waterways, habitat fragmentation, and biodiversity loss. Logging roads also expose

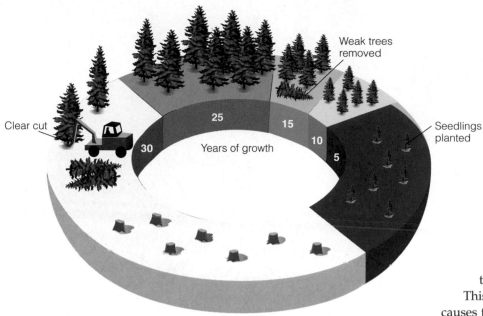

Figure 11-8 Short (25- to 30-year) rotation cycle of cutting and re-growth of a monoculture tree plantation in modern industrial forestry. In tropical countries, where trees can grow more rapidly year-round, the rotation cycle can be 6–10 years.

Clear cut

30

25

Years of growth

15

10

5

Weak trees removed

Seedlings planted

largest and best specimens of the most desirable species. Studies show that for every large tree cut down, 16 or 17 other trees are damaged or pulled down because a network of vines usually connects the trees in tropical forest canopies. This reduction of the forest canopy causes the forest floor to become warmer, drier, and more flammable and increases erosion of the forest's thin and usually nutrient-poor soil.

Some tree species grow best in full or moderate sunlight in medium to large clearings. Three major methods are used to harvest such species. One is **shelterwood cutting,** which removes all mature trees in an area in two or three cuttings over a period of time (Figure 11-10b).

Another is **seed-tree cutting** where loggers harvest nearly all of a stand's trees in one cutting but leave a few uniformly distributed seed-producing trees to regenerate the stand (Figure 11-10c).

The third approach is **clear-cutting,** which removes all trees from an area in a single cutting (Figure 11-10d). Figure 11-11 (p. 203) lists the advantages and disadvantages of clear-cutting. Shelterwood and seed-tree cutting are basically forms of clear-cutting carried out in two or more phases.

A clear-cutting variation that can provide a sustainable timber yield without widespread destruction is **strip cutting** (Figure 11-10e). It involves clear-cutting

forests to invasion by nonnative pests, diseases, and wildlife species. They also open once-inaccessible forests to farmers, miners, ranchers, hunters, and off-road vehicle users. In addition, logging roads on public lands in the United States disqualify the land for protection as wilderness.

Once loggers can reach a forest, they use various methods to harvest the trees (Figure 11-10, p. 202). With **selective cutting,** intermediate-aged or mature trees in an uneven-aged forest are cut singly or in small groups (Figure 11-10a). Selective cutting reduces crowding, encourages growth of younger trees, maintains an uneven-aged stand of trees of different species, and allows natural regeneration from surrounding trees. It can also help protect the site from soil erosion and wind damage, remove diseased trees, and allow a forest to be used for multiple purposes.

Sometimes loggers use a form of selective cutting called *high grading* to selectively cut trees in many tropical forests. It involves cutting and removing only the

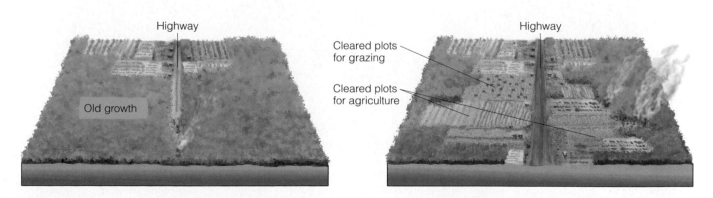

Highway

Old growth

Highway

Cleared plots for grazing

Cleared plots for agriculture

Figure 11-9 Natural capital degradation: building roads into previously inaccessible forests paves the way to their fragmentation, destruction, and degradation.

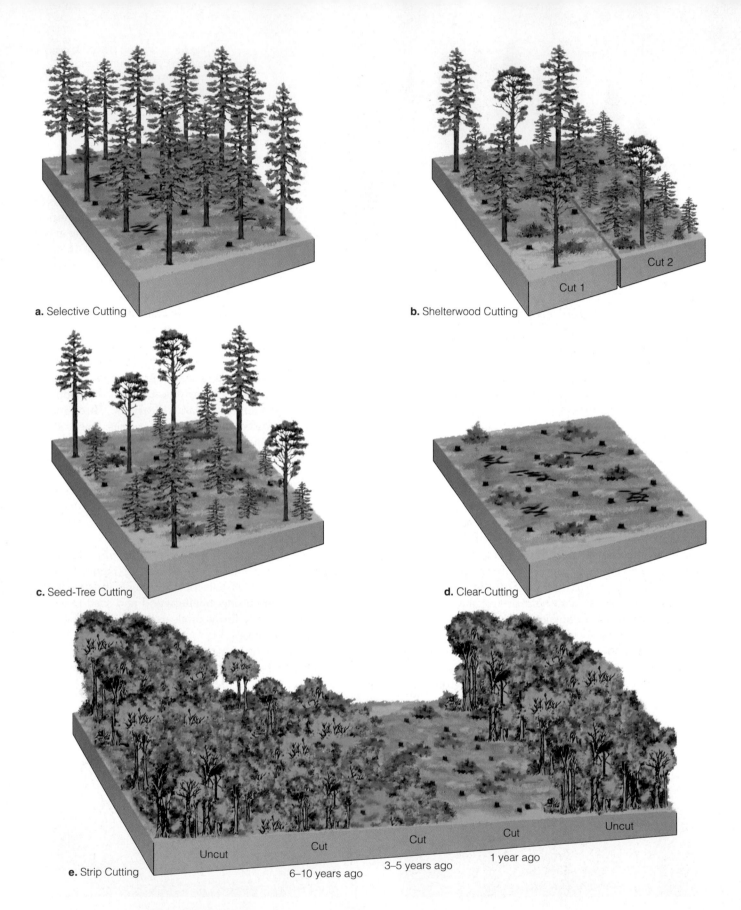

a. Selective Cutting

b. Shelterwood Cutting

Cut 1

Cut 2

c. Seed-Tree Cutting

d. Clear-Cutting

e. Strip Cutting

Uncut

Cut
6–10 years ago

Cut
3–5 years ago

Cut
1 year ago

Uncut

Figure 11-10 Tree harvesting methods.

Trade-Offs

Clear-Cutting Forests

Advantages	Disadvantages

Advantages	Disadvantages
Higher timber yields	Reduces biodiversity
Maximum economic return in shortest time	Disrupts ecosystem processes
Can reforest with genetically improved fast-growing trees	Destroys and fragments some wildlife habitats
Short time to establish new stand of trees	Leaves moderate to large openings
Needs less skill and planning	Increases soil erosion
Best way to harvest tree plantations	Increases sediment water pollution and flooding when done on steep slopes
Good for tree species needing full or moderate sunlight for growth	Eliminates most recreational value for several decades

Figure 11-11 Trade-offs: advantages and disadvantages of clear-cutting forests. Pick the single advantage and disadvantage that you think are the most important.

a strip of trees along the contour of the land, with the corridor narrow enough to allow natural regeneration within a few years. After regeneration, loggers cut another strip above the first, and so on. This allows clear-cutting of a forest in narrow strips over several decades with minimal damage.

What Are the Harmful Environmental Effects of Deforestation? Biodiversity Loss and Climate Change

Cutting down large forest areas reduces biodiversity and the ecological services forests provide and can contribute to regional and global climate change.

Deforestation is the temporary or permanent removal of large expanses of forest for agriculture or other uses. Harvesting timber and fuelwood from forests provides many economic benefits (Figure 11-7, right). However, deforestation can have many harmful environmental effects (Figure 11-12) that can reduce the ecological services provided by forests (Figure 11-7, left).

If left alone long enough, forests that have been logged or converted to cropland can revert to second-growth and even old-growth forests through secondary ecological succession. But this is not always the case. If deforestation occurs over a large enough area, it can cause a region's climate to become hotter and drier and prevent the return of a forest.

Deforestation can also contribute to projected global warming if trees are removed faster than they grow back. When forests are cleared for agriculture or other purposes and burned the carbon stored in the trees' biomass is released into the atmosphere as the greenhouse gas carbon dioxide (CO_2). Research indicates that when an old-growth forest is cut, it takes at least 200 years for a replacement forest to accumulate the same amount of carbon stored in the original forest.

What Is Happening to the World's Forests? Mixed News

Human activities have reduced the earth's forest cover by 20–50% and deforestation is continuing at a fairly rapid rate, except in most temperate forests in North America and Europe.

Global estimates of forest cover change are difficult to make because of lack of satellite and radar data, unmonitored land-use change, and different definitions of what constitutes a forest. But forest surveys are improving because of better satellite data.

Natural Capital Degradation

Deforestation

- Decreased soil fertility from erosion

- Runoff of eroded soil into aquatic systems

- Premature extinction of species with specialized niches

- Loss of habitat for migratory species such as birds and butterflies

- Regional climate change from extensive clearing

- Releases CO_2 into atmosphere from burning and tree decay

- Accelerates flooding

Figure 11-12 Natural capital degradation: harmful environmental effects of deforestation that can reduce the ecological services provided by forests. Which two of these effects do you think are the most serious?

Here are two pieces of *bad news. First,* surveys by the World Resources Institute (WRI) indicate that over the past 8,000 years human activities have reduced the earth's original forest cover by 20–50%.

Second, surveys by the UN Food and Agricultural Organization (FAO) and the World Resources Institute indicate that the global rate of forest cover loss during the 1990s was between 0.2% and 0.5% a year, and at least another 0.1–0.3% of the world's forests were degraded. If correct, the world's forests are being cleared and degraded at an exponential rate of 0.3–0.8% a year, with much higher rates in some areas. Over four-fifths of these losses are taking place in the tropics. The World Resources Institute estimates that if current deforestation rates continue, about 40% of the world's remaining intact forests will have been logged or converted to other uses within 10–20 years, if not sooner.

Here are two pieces of *good news. First,* the total area of many temperate forests in North America and Europe has increased slightly because of reforestation from secondary ecological succession on cleared forest areas and abandoned croplands.

Second, some of the cut areas of tropical forest have increased tree cover from regrowth and planting of tree plantations. But ecologists do not believe that tree plantations with their much lower biodiversity should be counted as forest any more than croplands should be counted as grassland. According to ecologist Michael L. Rosenzweig, "Forest plantations are just cornfields whose stalks have gotten very tall and turned to wood. They display nothing of the majesty of natural forests."

X *HOW WOULD YOU VOTE?* Should there be a global effort to sharply reduce the cutting of old-growth forests? Cast your vote online at http://biology.brookscole.com/miller14.

How Much Are the World's Ecological Services Worth? Putting a Price Tag on Mother Nature's Services

The huge economic value of the ecological services provided by the world's forests and other ecosystems is rarely counted in making decisions about how to use these ecosystems.

Currently forests are valued mostly for their economic services (Figure 11-7, right). But suppose we estimate and take into account the ecological services provided by forests (Figure 11-7, left). Then researchers say that the economic value of their long-term ecological services would be much greater than their short-term economic services.

In 1997, a team of ecologists, economists, and geographers attempted to estimate the monetary worth of the earth's natural ecological services (see top half of back cover). According to this crude appraisal led by ecological economist Robert Costanza of the University of Vermont, the economic value of income from the earth's ecological services is at least $36 trillion per year! This is fairly close to the $42 trillion value of all of the goods and services produced throughout the world in 2004. To provide an annual natural income of $36 trillion per year, the world's natural capital would have a value of at least $500 trillion—an average of about $82,000 for each person on earth!

Based on these estimates, *biodiversity is the world's biggest financial asset.* But unless this natural asset is given a financial value that is included in evaluating how we use forests and other ecological resources it will be used unsustainably and destroyed or degraded for short-term profit.

To estimate the monetary values of the ecological services provided by the world's natural capital, the researchers divided the earth's surface into 16 biomes (Figure 6-16, p. 111) and aquatic life zones. They omitted deserts and tundra because of a lack of data. Then they agreed on a list of 17 goods and services provided by nature and sifted through more than 100 studies that attempted to put a dollar value on such services in the 16 different types of ecosystems.

According to this appraisal, the world's forests provide us with ecological services worth about $4.7 trillion per year—equal in value to about one-tenth of all of the goods and services produced in the world in 2004.

The researchers say their estimates could easily be too low by a factor of 10 to 1 million or more. For example, their calculations included only estimates of the ecosystem services themselves, not the natural capital that generates them. They also omitted the value of nonrenewable minerals and fuels.

They hope such estimates will call people's attention to three important facts. The earth's ecosystem services are essential for all humans and their economies, their economic value is huge, and they are an ongoing source of ecological income as long as they are used sustainably.

Why have we not changed our accounting system to reflect these losses? One reason is that economic savings provided by conserving nature benefit everyone now and in the future, whereas profits made by exploiting nature are immediate and benefit a relatively small group of individuals. A second reason is that many current government subsidies and tax incentives support destruction and degradation of forests and other ecosystems for short-term economic gain. We get more of what we reward.

How Can We Manage Forests More Sustainably? Making Sustaining Forests Profitable

We can use forests more sustainably by including the economic value of their ecological services, harvesting trees no faster than they are replenished, and protecting old-growth and vulnerable areas.

Biodiversity researchers and a growing number of foresters call for more sustainable forest management. Figure 11-13 lists ways to do this. Which two of these solutions do you believe are most important?

Some conservation biologists suggest four ways to estimate how much of the world's remaining forests to protect. *First,* include estimates of the economic value of their ecological services in all decisions. *Second,* protect enough forest so that the rate of forest loss and degradation by human and natural factors in a particular area is balanced by the rate of forest renewal. *Third,* identify and protect forest areas that are centers of biodiversity and that are threatened by economic development. *Fourth,* establish and use methods to evaluate timber that has been grown sustainably, as discussed below.

Solutions: How Can We Certify Sustainably Grown Timber? Set Standards and Bring in Outside Evaluators

Organizations have developed standards for certifying that timber has been harvested sustainably and that wood products have been produced from sustainably harvested timber.

Collins Pine owns and manages a large area of productive timberland in northeastern California. Since 1940 the company has used selective cutting to help maintain ecological, economic, and social sustainability of its timberland.

Solutions

Sustainable Forestry

- Grow more timber on long rotations

- Rely more on selective cutting and strip cutting

- No clear-cutting, seed-tree, or shelterwood cutting on steeply sloped land

- No fragmentation of remaining large blocks of forest

- Sharply reduce road building into uncut forest areas

- Leave most standing dead trees and fallen timber for wildlife habitat and nutrient recycling

- Certify timber grown by sustainable methods

- Include ecological services of trees and forests in estimating economic value

Figure 11-13 Solutions: ways to manage forests more sustainably.

Since 1993 Scientific Certification Systems (SCS) has evaluated the company's timber production. SCS is part of the nonprofit Forest Stewardship Council (FSC). It was formed in 1993 to develop a list of environmentally sound practices for use in certifying timber and products made from such timber.

Each year SCS evaluates Collins's landholdings to ensure that cutting has not exceeded long-term forest regeneration, roads and harvesting systems have not caused unreasonable ecological damage, soils are not damaged, downed wood (boles) and standing dead trees (snags) are left to provide wildlife habitat, and the company is a good employer and a good steward of its land and water resources.

Another successful example of sustainable forestry certification involves the Menominee nation. Since 1890 it has selectively harvested trees of mixed species and ages from its tribal reservation land near Green Bay, Wisconsin—the state's single largest tract of virgin forest. Each year the Rain Forest Alliance's Smart Wood Program evaluates whether the Menominee harvest lumber from the tribal forest in an environmentally and socially responsible manner.

In 2001 the World Wildlife Fund (WWF) called on the world's five largest companies that harvest and process timber and buy wood products to adopt the FSC's sustainable management principles. According to the WWF, by doing this these five companies could essentially halt logging of old-growth forests and still meet the world's industrial wood and wood fiber needs using one-fifth of the world's forests.

Good news. In 2002, Mitsubishi, one of the world's largest forestry companies, announced that it would have third parties certify its forestry operations using standards developed by the Forest Stewardship Council. And Home Depot, Lowes, Andersen, and other major sellers of wood products in the United States have agreed to sell only wood certified as being sustainably grown by independent groups such as the Forest Stewardship Council (to the degree that certified wood is available).

11-5 FOREST RESOURCES AND MANAGEMENT IN THE UNITED STATES

What Is the Status of Forests in the United States? Encouraging News

U.S. forests cover more area than they did in 1920, more wood is grown than cut, and the country has set aside large areas of protected forests.

Forests cover about 30% of the U.S. land area, provide habitats for more than 80% of the country's wildlife species, and supply about two-thirds of the nation's total surface water.

Good news. Forests (including tree plantations) in the United States cover more area than they did in 1920. Many of the old-growth or frontier forests that were cleared or partially cleared between 1620 and 1960 have grown back naturally as fairly diverse second-growth (and in some cases third-growth) forest in every region of the United States, except much of the West. In 1995, environmental writer Bill McKibben cited forest regrowth in the United States—especially in the East—as "the great environmental story of the United States, and in some ways the whole world."

Also, every year more wood is grown in the United States than is cut, and each year the total area planted with trees increases. In addition, the United States was the world's first country to set aside large areas of forest in protected areas. By 2000, protected forests made up about 40% of the country's total forest area, mostly in the national forests (Figure 11-6).

Bad news. Since the mid-1960s, an increasing area of the nation's remaining old-growth and fairly diverse second-growth forests has been clear-cut and replaced with biologically simplified tree plantations. According to biodiversity researchers, this reduces overall forest biodiversity and disrupts ecosystem processes such as energy flow and chemical cycling. Some environmentally concerned citizens have protested the cutting

down of ancient trees and forests (Individuals Matter, below).

How Can We Reduce the Harmful Effects of Insects and Pathogens on U.S. Forests? Dealing with Bugs and Diseases

We can reduce tree damage from insects and diseases by inspecting imported timber, removing diseased and infected trees, and using chemicals and natural predators to help control insect pests.

Figure 11-14 shows some of the nonnative species of pests that are causing serious damage to certain tree species in parts of the United States. There are several ways to reduce the harmful impacts of tree diseases and of insects on forests. One is to ban imported timber that might introduce harmful new pathogens or insect pests. Another is to selectively remove or clear-cut infected and infested trees.

We can also develop tree species that are genetically resistant to common tree diseases. And we can control insect pests by applying conventional pesticides or using biological control (bugs that eat harmful bugs) combined with very small amounts of conventional pesticides.

INDIVIDUALS MATTER

Butterfly in a Redwood Tree

Butterfly is the nickname given to Julia Hill. This young woman spent 2 years of her life on a small platform near the top of a giant redwood tree in California to protest the clear-cutting of a forest of these ancient trees, some of them more than 1,000 years old.

She and other protesters were illegally occupying these trees as a form of *nonviolent civil disobedience* used decades ago by Mahatma Gandhi in his successful efforts to end the British occupation of India. Butterfly had never participated in any environmental protest or act of civil disobedience.

She went to the site to express her belief that it was wrong to cut down these ancient giants for short-term economic gain, even if you

own them. She planned to stay only for a few days.

But after seeing the destruction and climbing one of these magnificent trees she ended up staying in the tree for 2 years to bring publicity to what was happening and help save the surrounding trees. She became a media symbol of the protest and during her stay used a cell phone to communicate with members of the mass media throughout the world to help develop public support for saving the trees.

Can you imagine spending 2 years of your life in a tree on a platform not much bigger than a king-sized bed 55 meters (180 feet) above the ground and enduring high winds, intense rainstorms, snow, and ice? She was not living in a quiet pristine forest. All round her was the noise of trucks, chainsaws, and helicopters trying to scare her into returning to the ground.

She lost her courageous battle to save the surrounding forest but persuaded Pacific Lumber MAXXAM to save her tree (called Luna) and a 60-meter (200-foot) buffer zone around it. Not long after she descended from the tree someone used a chainsaw to seriously damage it, and cables and steel plates have been used to preserve it.

But maybe she and the earth did not lose. A book she wrote about her stand, and her subsequent travels to campuses all over the world, have inspired a number of young people to stand up for protecting biodiversity and other environmental causes.

She was leading by following in the tradition of Ghandi, who said, "My life is my message." Would you spend a day or a week of your life protesting something that you believed to be wrong?

Sudden oak death White pine blister rust Pine shoot beetle Beech bark disease Hemlock woolly adelgid

Figure 11-14 Some of the nonnative insect species that have invaded U.S. forests and are causing billions of dollars in damages and tree loss. The light green and orange colors show areas where green or red overlap with yellow. (Data from U.S. Forest Service)

How Do Fires Affect U.S. Forests? Surface, Crown, and Ground Fires

Forest fires can burn away flammable underbrush and small trees, burn large trees and leap from treetop to treetop, or burn flammable materials found under the ground.

Three types of fires can affect forest ecosystems. Some, called *surface fires* (Figure 11-15, left. p. 208), usually burn only undergrowth and leaf litter on the forest floor. These fires can kill seedlings and small trees but spare most mature trees and allow most wild animals to escape.

Occasional surface fires have a number of ecological benefits. They burn away flammable ground material and help prevent more destructive fires. They also release valuable mineral nutrients (tied up in slowly decomposing litter and undergrowth), stimulate the germination of certain tree seeds (such as those of the giant sequoia and jack pine), and help control pathogens and insects. In addition, some wildlife species such as deer, moose, elk, muskrat, woodcock, and quail depend on occasional surface fires to maintain their habitats and provide food in the form of vegetation that sprouts after fires.

Some extremely hot fires, called *crown fires* (Figure 11-15, right), may start on the ground but eventually burn whole trees and leap from treetop to treetop. They usually occur in forests that have had no surface fires for several decades. This allows dead wood, leaves, and other flammable ground litter to build up. These rapidly burning fires can destroy most vegetation, kill wildlife, increase soil erosion, and burn or damage human structures in their paths.

Figure 11-15 Surface fires (left) usually burn undergrowth and leaf litter on a forest floor and can help prevent more destructive crown fires (right) by removing flammable ground material. Sometimes carefully controlled surface fires are deliberately set to prevent buildup of flammable ground material in forests.

Surface fire **Crown fire**

Sometimes surface fires go underground and burn partially decayed leaves or peat. Such *ground fires* are most common in northern peat bogs. They may smolder for days or weeks and are difficult to detect and extinguish.

Solutions: How Can We Reduce Forest Damage from Fire? Set Little Fires, Allow Some Fires to Burn, and Clear Vegetation Near Buildings

We can reduce fire damage by setting controlled surface fires to prevent buildup of flammable material, allowing fires on public lands to burn unless they threaten human structures and lives, and clearing small areas around buildings.

Two ways to help protect forest resources from fire are *prevention* and *prescribed burning* (setting controlled ground fires to prevent buildup of flammable material). Ways to prevent forest fires include requiring burning permits, closing all or parts of a forest to travel and camping during periods of drought and high fire danger, and educating the public about the ecological effects of fire on forests.

In the United States, the Smokey Bear educational campaign of the Forest Service and the National Advertising Council has prevented countless forest fires. It has also saved many lives and prevented billions of dollars in losses of trees, wildlife, and human structures.

However, this educational program convinced much of the public that all forest fires are bad and should be prevented or put out. Ecologists warn that trying to prevent all forest fires increases the likelihood of destructive crown fires by allowing buildup of highly flammable underbrush and smaller trees in some forests.

According to the U.S. Forest Service, severe fires could threaten about 40% of all federal forest lands, mainly through fuel buildup from past rigorous fire protection programs (the Smokey Bear era), increased logging in the 1980s that left behind highly flammable

logging debris (called *slash*), and greater public use of federal forest lands. In addition, an estimated 40 million people now live in remote forested areas or areas with highly flammable chaparral vegetation with a high wildfire risk.

Ecologists and forest fire experts propose several strategies for reducing the harm from fires to forests and people. One is to set small prescribed surface fires or clear out (thin) flammable small trees and underbrush in the highest-risk forest areas. But prescribed fires require careful planning and monitoring to keep them from getting out of control.

During the spring of 2000, for example, a poorly planned prescribed fire got out of hand in an area managed by the Park Service near Los Alamos, New Mexico. The result was a 33-day fire that burned 19,000 hectares (47,000 acres), destroyed or damaged 280 homes, damaged 40 structures at the Los Alamos National Laboratory, and caused an estimated $1 billion in damages. In parts of fire-prone California, local officials are using goats as an alternative to prescribed burns (Solutions, right).

Another strategy is to allow many fires in national parks, national forests, and wilderness areas to burn and remove flammable underbrush and smaller trees as long as the fires do not threaten human structures and life. A third approach is to protect houses or other buildings by thinning a zone of 46–61 meters (150–200 feet) around such buildings and eliminating flammable materials such as wooden roofs.

In 2003, the U.S. Congress passed a law called the *Healthy Forests Initiative*. Under this law, timber companies are allowed to cut down economically valuable medium and large trees in most national forests for 10 years in return for clearing away smaller, more fire-prone trees and underbrush. The law also exempts most thinning projects from environmental reviews and appeals currently required by forest protection laws.

Will the law achieve the stated goal of reducing wildfires? According to biologists and many forest fire scientists, this law is likely to *increase* the chances of severe forest fires for two reasons. *First,* removing the most fire-resistant large trees—the ones that are valuable to timber companies—encourages dense growths of highly flammable young trees and rapidly growing underbrush. *Second,* removing the large and medium trees leaves behind highly flammable slash. Many of the worst fires in U.S. history—including some of those during the 1990s—burned through cleared forest areas containing slash.

Fire scientists agree that some forests on public lands need thinning to reduce the chances of catastrophic fires, but they believe a program to accomplish this should focus on two goals. One is to reduce ground-level fuel and vegetation in dry forest types and leave widely spaced medium and large trees that are the most fire resistant and thus can help forest recovery after a fire. These trees also provide critical wildlife habitat, especially as standing dead trees (snags) and logs where many animals live. The other goal would emphasize clearing of flammable vegetation around individual homes and buildings and near communities that are especially vulnerable to wildfire.

Critics of the Healthy Forests law say that these goals could be accomplished at a much lower cost to taxpayers by a law that would give grants to communities especially vulnerable to wildfires for thinning forests and protecting homes and buildings in their areas.

Goats to the Rescue

SOLUTIONS

California has thousands of wildfires every year. Prescribed burns are used to keep flammable underbrush down, but some officials are worried about such burns getting out of control.

Officials in California cities such as Monterey, Malibu, Berkeley, and Oakland are using goats to help reduce the flammable vegetation on surrounding hills during the state's six-month fire season.

This low-tech approach is working. It takes a herd of about 350 goats one day to eat their way through an acre of underbrush.

The goats are kept in movable pens, with electric fencing and water troughs. Before turning the herd loose, botanists put fences around small plants and trees that are rare or endangered. Dogs are typically used to herd the goats and help protect them from predators.

Critical Thinking

Can you think of any disadvantages of using goats to clear flammable underbrush? Are there any other animals that could do this job?

X *HOW WOULD YOU VOTE?* Do you support the Healthy Forests Act that allows timber companies to remove large and medium trees from most national forests without having to obey most environmental laws in exchange for thinning out flammable smaller trees and underbrush? Cast your vote online at http://biology.brookscole.com/miller14.

Case Study: How Should U.S. National Forests Be Managed? An Ongoing Controversy

There is controversy over whether U.S. national forests should be managed primarily for timber, their ecological services, recreation, or a mix of these uses.

For decades there has been controversy over the use of resources in the national forests. Timber companies push to cut as much of the timber in these forests as possible at low prices. Biodiversity experts and environmentalists call for sharply reducing or eliminating tree harvesting in national forests and using more sustainable forest management practices (Figure 11-13) for timber cutting in these forests. They believe that national forests should be managed primarily to provide recreation and to sustain biodiversity, water resources, and other ecological services.

Between 1930 and 1988, timber harvesting from national forests increased sharply. One reason is that timber companies pressured Congress to increase timber harvests. Also, Congress passed a law that allows the Forest Service to keep most of the money it makes on timber sales. This makes timber cutting a key way for the Forest Service to increase its budget.

In addition, a 1908 law gives counties within the boundaries of national forests one-fourth of the gross receipts from timber sales. This encourages county governments to push for increased timber harvesting.

By law, the Forest Service must sell timber for no less than the cost of reforesting the cleared land. But this price does not include the government-subsidized cost of building and maintaining access roads for timber removal by logging companies.

The Forest Service's timber-cutting program loses money because revenue from timber sales does not cover the costs of road building, timber sale preparation, administration, and other overhead costs. Because of such government subsidies, timber sales from U.S. federal lands have lost money for taxpayers in 97 of the last 100 years!

According to a 2000 study by the accounting firm Econorthwest, recreation, hunting, and fishing in national forests add 10 times more money to the national economy and provide 7 times more jobs than does extraction of timber and other resources. Figure 11-16 (p. 210) lists advantages and disadvantages of logging in national forests.

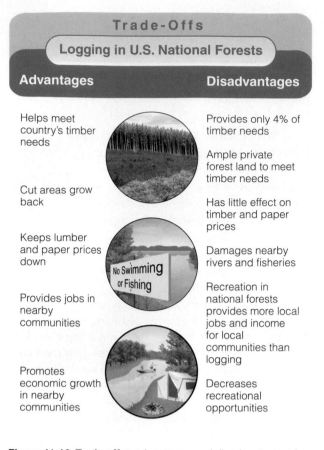

Trade-Offs

Logging in U.S. National Forests

Advantages	Disadvantages
Helps meet country's timber needs	Provides only 4% of timber needs
Cut areas grow back	Ample private forest land to meet timber needs
Keeps lumber and paper prices down	Has little effect on timber and paper prices
Provides jobs in nearby communities	Damages nearby rivers and fisheries
Promotes economic growth in nearby communities	Recreation in national forests provides more local jobs and income for local communities than logging
	Decreases recreational opportunities

Figure 11-16 Trade-offs: advantages and disadvantages of allowing logging in U.S. national forests. Pick the single advantage and disadvantage that you think are the most important. Explain.

How Can We Reduce the Need to Harvest Trees for Timber and Papermaking? Stop Waste and Make Paper from Tree-Free Fibers

Almost two-thirds of the wood consumed in the United States is wasted, and much of the paper we use could be made from agricultural residues and fast-growing crops such as kenaf.

One way to reduce the pressure to harvest trees on public and private land in the United States (and elsewhere) is to improve the efficiency of wood use. According to the Worldwatch Institute and forestry analysts, *up to 60% of the wood consumed in the United States is wasted unnecessarily.* This occurs because of inefficient use of construction materials, excess packaging, overuse of junk mail, inadequate paper recycling, and failure to reuse wooden shipping containers.

Only 4% of the total U.S. production of softwood timber comes from the national forests. Thus, reducing the waste of wood and paper products by only 4% could eliminate the need to remove any timber from national forests. This would allow these lands to be used primarily for recreation and biodiversity protection.

One way to reduce the pressure to harvest trees for paper production in national and private forests is to make paper by using fiber that does not come from trees. *Tree-free fibers* for making paper come from two sources: *agricultural residues* left over from crops (such as wheat, rice, and sugar) and *fast-growing crops* (such as kenaf and industrial hemp).

China uses tree-free pulp from rice straw and other agricultural wastes left after harvest to make almost two-thirds of its paper. Most of the small amount of tree-free paper produced in the United States is made from the fibers of a rapidly growing woody annual plant called *kenaf* (pronounced "kuh-NAHF"; see photo on p. vii).

Compared to pulpwood, kenaf needs less herbicide because it grows faster than most weeds and reduces insecticide use because its outer fibrous covering is nearly insect proof. Growing kenaf does not deplete soil nitrogen because it is a nitrogen fixer. And breaking down kenaf fibers takes less energy and fewer chemicals and thus produces less toxic wastewater than using conventional trees. According to the USDA kenaf is "the best option for tree-free papermaking in the United States" and could replace wood-based paper within 20–30 years.

11-6 TROPICAL DEFORESTATION

How Fast Are Tropical Forests Being Cleared and Degraded and Why Should We Care? Protecting the Priceless

Large areas of ecologically and economically important tropical forests are being cleared and degraded at a fast rate.

Tropical forests cover about 6% of the earth's land area—roughly the area of the lower 48 states. Climatic and biological data suggest that mature tropical forests once covered at least twice as much area as they do today, with most of the destruction occurring since 1950. Satellite scans and ground-level surveys used to estimate forest destruction indicate that large areas of tropical forests are being cut rapidly in parts of South America (especially Brazil), Africa, and Asia.

Studies indicate that more than half of the world's species of terrestrial plants and animals live in tropical rain forests. Brazil has about 40% the world's remaining tropical rain forest in the vast Amazon basin, which is about two-thirds the size of the continental United States. In 1970, deforestation affected only 1% of the area of the Amazon basin. By 2003, almost 20% had been deforested or degraded.

According to a 2001 study by Penn State researcher James Alcock, without immediate and aggressive action to reduce current forest destruction and degradation practices, Brazil's original Amazon rain forests may largely disappear within 40–50 years.

You probably have not heard about the loss of most of Brazil's Atlantic coastal rain forest. This less famous forest once covered about 12% of Brazil's land area. Now 93% of it has been cleared and most of what is left is recovering from previous cutting episodes. This represents a major loss of biodiversity because an area in this forest a little larger than two typical suburban house lots in the United States has 450 tree species! The entire United States has only about 865 native tree species.

There are disagreements about how rapidly tropical forests are being deforested and degraded because of three factors. *First*, it is difficult to interpret satellite images. *Second*, some countries hide or exaggerate deforestation rates for political and economic reasons. *Third*, governments and international agencies define forest, deforestation, and forest degradation in different ways.

For these reasons, estimates of global tropical forest loss vary from 50,000 square kilometers (19,300 square miles) to 170,000 square kilometers (65,600 square miles) per year. This is high enough to lose or degrade half of the world's remaining tropical forests in 35–117 years.

Most biologists believe that cutting and degrading most remaining old-growth tropical forests is a serious global environmental problem because of the important ecological and economic services they provide (Figure 11-7). For example, tropical forest plants provide chemicals used as blueprints for making most of the world's prescription drugs (Figure 11-17). And cutting these forests faster than they can grow back contributes to projected global warming because these forests are a storehouse for huge quantities of carbon, safely stored as organic compounds in plant biomass.

What Causes Tropical Deforestation and Degradation? The Big Five

The primary causes of tropical deforestation and degradation are population growth, poverty, environmentally harmful government subsidies, debts owed to developed countries, and failure to value ecological services.

Tropical deforestation results from a number of interconnected primary and secondary causes (Figure 11-18, p. 212). Population growth and poverty combine to drive subsistence farmers and the landless poor to tropical forests, where they try to grow enough food to survive. Government subsidies can accelerate deforestation by making timber or other tropical forest resources cheap, relative to the economic value of the ecological services they provide. Governments in Indonesia, Mexico, and Brazil also encourage the poor to colonize tropical forests by giving them title to land they clear. This can help reduce poverty but can lead to

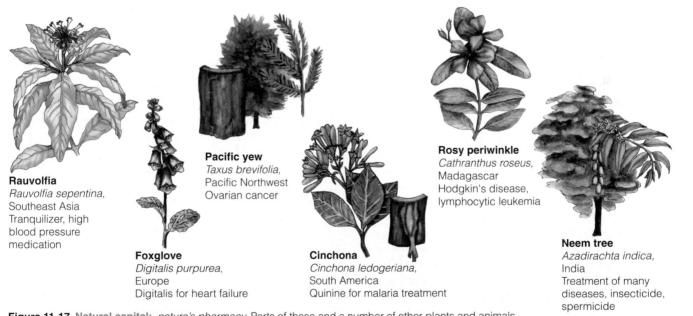

Rauvolfia
Rauvolfia sepentina,
Southeast Asia
Tranquilizer, high
blood pressure
medication

Foxglove
Digitalis purpurea,
Europe
Digitalis for heart failure

Pacific yew
Taxus brevifolia,
Pacific Northwest
Ovarian cancer

Cinchona
Cinchona ledogeriana,
South America
Quinine for malaria treatment

Rosy periwinkle
Cathranthus roseus,
Madagascar
Hodgkin's disease,
lymphocytic leukemia

Neem tree
Azadirachta indica,
India
Treatment of many
diseases, insecticide,
spermicide

Figure 11-17 Natural capital: *nature's pharmacy.* Parts of these and a number of other plants and animals (many of them found in tropical forests) are used to treat a variety of human ailments and diseases. Nine of the ten leading prescription drugs originally came from wild organisms. About 2,100 of the 3,000 plants identified by the National Cancer Institute as sources of cancer-fighting chemicals come from tropical forests. Despite their economic and health potential, fewer than 1% of the estimated 125,000 flowering plant species in tropical forests (and a mere 1,100 of the world's 260,000 known plant species) have been examined for their medicinal properties. Once the active ingredients in the plants have been identified, they can usually be produced synthetically. Many of these tropical plant species are likely to become extinct before we can study them.

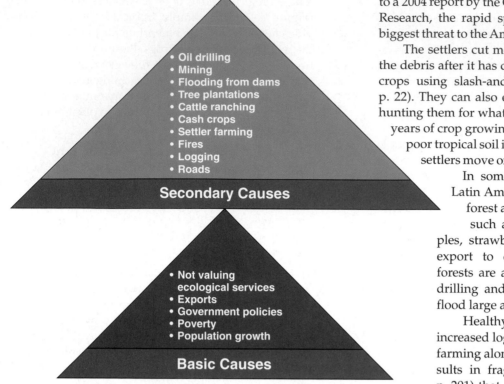

Figure 11-18 Natural capital degradation: major interconnected primary and secondary causes of the destruction and degradation of tropical forests. The importance of specific secondary causes varies in different parts of the world.

to a 2004 report by the Center for International Forestry Research, the rapid spread in cattle ranching is the biggest threat to the Amazon's tropical forests.

The settlers cut most of the remaining trees, burn the debris after it has dried for about a year, and plant crops using slash-and-burn agriculture (Figure 2-2, p. 22). They can also endanger some wild species by hunting them for what is called bushmeat. After a few years of crop growing and rain erosion, the nutrient-poor tropical soil is depleted of nutrients. Then the settlers move on to newly cleared land.

In some areas—especially Africa and Latin America—large sections of tropical forest are cleared for raising cash crops such as sugarcane, bananas, pineapples, strawberries, and coffee—mostly for export to developed countries. Tropical forests are also cleared for mining and oil drilling and to build dams on rivers that flood large areas of the forest.

Healthy rain forests do not burn. But increased logging, settlements, grazing, and farming along roads built in these forests results in fragments of forest (Figure 11-9, p. 201) that dry out. This makes such areas easier to ignite by lightning and for farmers and ranchers to burn. In addition to destroying and degrading biodiversity, this releases large amounts of carbon dioxide into the atmosphere.

Solutions: How Can We Reduce Deforestation and Degradation of Tropical Forests? Prevention Is Best

There are a number of ways to slow and reduce the deforestation and degradation of tropical forests.

Analysts have suggested various ways to protect tropical forests and use them more sustainably (Figure 11-19). One method is to help new settlers in tropical forests learn how to practice small-scale sustainable agriculture and forestry. The Lacandon Maya Indians of Chiapas, Mexico, for example, use a multi-layered system of agroforestry to cultivate as many as 75 crop species on 1-hectare (2.5-acre) plots for up to 7 years. After that, they plant a new plot to allow regeneration of the soil in the original plot.

In the lush rain forests of Peru's Palcazú Valley, Yaneshé Indians use strip cutting (Figure 11-10e) to harvest tropical trees for lumber. Tribe members also act as consultants to help other forest dwellers set up similar systems.

Another approach is to sustainably harvest some of the renewable resources such as fruits and nuts in rain forests. For example, about 6,000 families in the Petén region of Guatemala make a comfortable living by sustainably extracting various rain forest products.

environmental degradation unless the new settlers are taught how to use such forests more sustainably. In addition, international lending agencies encourage developing countries to borrow huge sums of money from developed countries to finance projects such as roads, mines, logging operations, oil drilling, and dams in tropical forests. Another cause is failure to value ecological services of forests (Figure 11-7, left).

The depletion and degradation of a tropical forest begins when a road is cut deep into the forest interior for logging and settlement and hunters are hired to kill wild animals to provide loggers and other work crews with meat.

Loggers typically use selective cutting to remove the best timber (high grading). This topples many other trees because of their shallow roots and the network of vines connecting trees in the forest's canopy. Timber exports to developed countries contribute significantly to tropical forest depletion and degradation. But domestic use accounts for more than 80% of the trees cut in developing countries.

After the best timber has been removed, timber companies often sell the land to ranchers. Within a few years they typically overgraze it and sell it to settlers who have migrated to the forest hoping to grow enough food to survive. Then they move their land-degrading ranching operations to another forest area. According

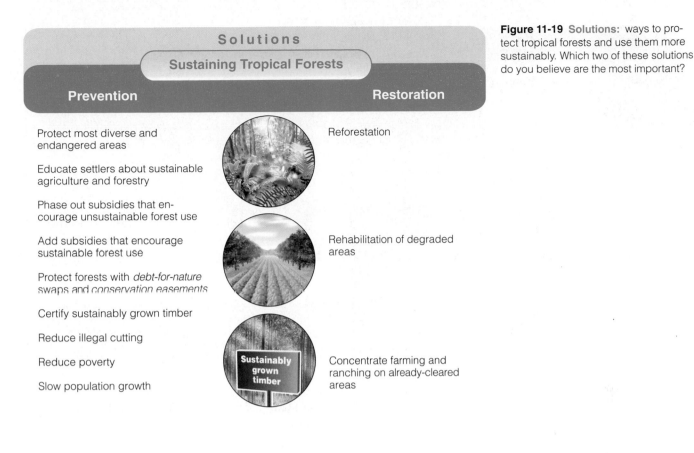

Figure 11-19 Solutions: ways to protect tropical forests and use them more sustainably. Which two of these solutions do you believe are the most important?

Solutions

Sustaining Tropical Forests

Prevention

Protect most diverse and endangered areas

Educate settlers about sustainable agriculture and forestry

Phase out subsidies that encourage unsustainable forest use

Add subsidies that encourage sustainable forest use

Protect forests with *debt-for-nature* swaps and *conservation easements*

Certify sustainably grown timber

Reduce illegal cutting

Reduce poverty

Slow population growth

Restoration

Reforestation

Rehabilitation of degraded areas

Sustainably grown timber

Concentrate farming and ranching on already-cleared areas

Another approach is to use *debt-for-nature swaps* to make it financially profitable for countries to protect tropical forests. In such a swap, participating countries act as custodians of protected forest reserves in return for foreign aid or debt relief. Since the first debt-for-nature swap in 1987, governments and private groups have carried out more than 20 such swaps in 10 countries.

Another important tool is using an international system for evaluating and certifying tropical timber produced by sustainable methods. Loggers can also use gentler methods for harvesting trees. For example, cutting canopy vines (lianas) before felling a tree can reduce damage to neighboring trees by 20–40%, and using the least obstructed paths to remove the logs can halve the damage to other trees. In addition, governments and individuals can mount efforts to reforest and rehabilitate degraded tropical forests and watersheds (Individuals Matter, p. 214). Another suggestion is to clamp down on illegal logging.

Solutions: The Incredible Neem Tree

The neem tree could eventually benefit almost everyone on the earth.

Suppose a single plant existed that could quickly reforest degraded land, supply fuelwood and lumber in dry areas, provide natural alternatives to toxic pesticides, be used to treat numerous diseases, and help

control human population growth? There is one: the *neem tree*, a broadleaf evergreen member of the mahogany family.

This remarkable tropical species, native to India and Burma, is ideal for reforestation because it can grow to maturity in only 5–7 years. It grows well in poor soil in semiarid lands such as those in Africa, providing abundant fuelwood, lumber, and lamp oil.

It also contains various natural pesticides. Chemicals from its leaves and seeds can repel or kill more than 200 insect species, including termites, gypsy moths, locusts, boll weevils, and cockroaches.

Extracts from neem seeds and leaves (Figure 11-17) can fight bacterial, viral, and fungal infections. Villagers call the tree a "village pharmacy" because its chemicals can relieve so many different health problems. People also use the tree's twigs as an antiseptic toothbrush and the oil from its seeds to make toothpaste and soap.

That is not all. Neem-seed oil evidently acts as a strong spermicide and may help in the development of a much-needed male birth control pill. According to a study by the U.S. National Academy of Sciences, the neem tree "may eventually benefit every person on the planet."

Despite its numerous advantages, ecologists caution against widespread planting of neem trees outside its native range. As a nonnative species, it could take over and displace native species because of its

Kenya's Green Belt Movement

In Kenya, Wangari Maathai founded the Green Belt Movement in 1977. The goals of this women's self-help group are to establish tree nurseries, raise seedlings, and plant and protect a tree for each of Kenya's 32 million people. By 2003, the 50,000 members of this grass-roots group had established 6,000 village nurseries and planted and protected more than 20 million trees.

The success of this project has sparked the creation of similar programs in more than 30 other African countries. She has said,

I don't really know why I care so much. I just have something inside me that tells me that there is a problem and I have to do something about it. And I'm sure it's the same

voice that is speaking to everyone on this planet, at least everybody who seems to be concerned about the fate of the world, the fate of this planet.

Figure 11-A Wangari Maathai, the first Kenyan woman to earn a Ph.D. (in anatomy) and to head an academic department (veterinary medicine) at the University of Nairobi, organized the internationally acclaimed Green Belt Movement in 1977. For her work in protecting the environment she has received many honors, including the Goldman Prize, the Right Livelihood Award, the UN Africa Prize for Leadership, and the Golden Ark Award. After years of being harrassed, beaten, and jailed for opposing government policies, she was elected to Kenya's parliament as a member of the Green Party in 2002. In 2003 she was also appointed Assistant Minister for Environment, Natural Resources, and Wildlife.

---

rapid growth and resistance to pests. But proponents say the benefits of neem trees far outweigh such risks. What do you think?

11-7 NATIONAL PARKS

What Are National Parks and How Are They Threatened? Under Assault

Countries have established over 1,100 national parks but most are threatened by human activities.

Today, more than 1,100 national parks larger than 10 square kilometers (4 square miles) are located in more than 120 countries.

According to a 1999 study by the World Bank and the World Wildlife Fund, only 1% of the parks in developing countries receive protection. Local people invade most of the unprotected parks in search of wood, cropland, game animals, and other natural products for their daily survival. Loggers, miners, and wildlife poachers (who kill animals to obtain and sell items such as rhino horns, elephant tusks, and furs) also invade many of these parks. Park services in developing countries typically have too little money and too few personnel to fight these invasions, either by force or by education.

Another problem is that most national parks are too small to sustain many large animal species. Also, many parks suffer from invasions by nonnative species that can reduce the populations of some native species and cause ecological disruption.

Case Study: National Parks in the United States

National parks in the United States face many threats.

The U.S. national park system, established in 1912, has 56 national parks (sometimes called the country's *crown jewels*), most of them in the West (Figure 11-6). State, county, and city parks supplement these national parks. Most state parks are located near urban areas and have about twice as many visitors per year as the national parks.

Popularity is one of the biggest problems of national and state parks in the United States. During the summer, users entering the most popular U.S. national and state parks often face hour-long backups and experience noise, congestion, eroded trails, and stress instead of peaceful solitude.

Many visitors expect parks to have grocery stores, laundries, bars, golf courses, video arcades, and other facilities found in urban areas. U.S. Park Service rangers spend an increasing amount of their time on law enforcement and crowd control instead of conservation, management, and education. Many overworked and underpaid rangers are leaving for better-paying jobs.

In some parks noisy dirt bikes, dune buggies, snowmobiles, and other off-road vehicles (ORVs) degrade the aesthetic experience for many visitors, destroy or damage fragile vegetation, and disturb wildlife.

Many parks suffer damage from the migration or deliberate introduction of nonnative species. European wild boars (imported to North Carolina in 1912 for hunting) threaten vegetation in part of the Great Smoky Mountains National Park. Nonnative mountain goats in Washington's Olympic National Park trample native vegetation and accelerate soil erosion. While some nonnative species have moved into parks, some economically valuable native species of animals and plants (including many threatened or endangered species) are killed or removed illegally in almost half of U.S. national parks.

Nearby human activities that threaten wildlife and recreational values in many national parks include mining, logging, livestock grazing, coal-burning power plants, water diversion, and urban development. Polluted air, drifting hundreds of kilometers, kills ancient trees in California's Sequoia National Park and often blots out the awesome views at Arizona's Grand Canyon. According to the National Park Service, air pollution affects scenic views in most national parks more than 90% of the time.

Analysts have made a number of suggestions for sustaining and expanding the national park system in the United States (Figure 11-20).

Private concessionaires provide campgrounds, restaurants, hotels, and other services for park visitors. Some analysts call for requiring concessionaires to com-

Solutions

National Parks

- Integrate plans for managing parks and nearby federal lands

- Add new parkland near threatened parks

- Buy private land inside parks

- Locate visitor parking outside parks and use shuttle buses for entering and touring heavily used parks

- Increase funds for park maintenance and repairs

- Survey wildlife in parks

- Raise entry fees for visitors and use funds for park management and maintenance

- Limit number of visitors to crowded park areas

- Increase number and pay of park rangers

- Encourage volunteers to give visitor lectures and tours

- Seek private donations for park maintenance and repairs

Figure 11-20 Solutions: suggestions for sustaining and expanding the national park system in the United States. Which two of these solutions do you believe are the most important? (Wilderness Society and National Parks and Conservation Association)

pete for contracts and pay franchise fees equal to 22% of their gross (not net) receipts. Currently concessionaires in national parks pay the government an average of only about 6–7% of their gross receipts in franchise fees. And many large concessionaires with long-term contracts pay as little as 0.75% of their gross receipts.

11-8 NATURE RESERVES

How Much of the Earth's Land Should We Protect from Human Exploitation? The Answer Is More

Ecologists believe that we should protect more land to help sustain the earth's biodiversity.

Most ecologists and conservation biologists believe the best way to preserve biodiversity is through a worldwide network of protected areas. Currently about 12% of the earth's land area has been protected strictly or partially in nature reserves, parks, wildlife refuges, wilderness, and other areas. In other words, we have reserved 88% of the earth's land for us, and most of the remaining 12% we have protected is ice, tundra, or desert where we do not want to live because it is too cold or too hot.

And this 12% figure is misleading because no more than 5% of these areas are actually protected. Thus, we have strictly protected only about 7% of the earth's terrestrial areas from potentially harmful human activities.

Conservation biologists call for protecting at least 20% of the earth's land area in a global system of biodiversity reserves that includes multiple examples of all the earth's biomes. Setting aside and helping sustain such a system will take action and funding by national governments (Case Study, p. 216), private groups (Solutions, p. 216), and cooperative ventures involving governments, businesses, and private conservation groups. Protection does not mean just drawing dotted lines around an area; it refers to ecologically sound management of areas.

Some progress is being made. In 2001, the Brazilian government launched a program to establish 80 parks in the Amazon River basin on government-owned lands and asked the World Wildlife Fund to help plan the system.

In 2002 Canada announced plans to create 10 huge new national parks and five new marine conservation areas by 2007. This will almost double the area occupied by the country's current 39 national parks. And in 2002 Gabon announced plans to set aside 10% of its land area for a system of national parks.

On the other hand, most developers and resource extractors oppose protecting even the current 12% of the earth's remaining undisturbed ecosystems. They contend that most of these areas contain valuable resources that would add to economic growth.

Ecologists and conservation biologists disagree. They view protected areas as islands of biodiversity that help sustain all life and economies and that serve as centers of future evolution See Norman Myer's Guest Essay on this topic on the website for this chapter.

Case Study: What Has Costa Rica Done to Protect Some of Its Land from Degradation? A Global Conservation Leader

Costa Rica has devoted a larger proportion of land than any other country to conserving its significant biodiversity.

Tropical forests once completely covered Central America's Costa Rica, which is smaller in area than West Virginia and about one-tenth the size of France. Between 1963 and 1983, politically powerful ranching families cleared much of the country's forests to graze cattle. They exported most of the beef produced to the United States and western Europe.

Despite such widespread forest loss, tiny Costa Rica is a superpower of biodiversity, with an estimated 500,000 plant and animal species. A single park in Costa Rica is home to more bird species than all of North America.

In the mid-1970s, Costa Rica established a system of reserves and national parks that by 2003 included about a quarter of its land—6% of it in reserves for indigenous peoples. Costa Rica now devotes a larger proportion of its land to biodiversity conservation than any other country!

The country's parks and reserves are consolidated into eight *megareserves* designed to sustain about 80% of Costa Rica's biodiversity (Figure 11-21). Each reserve contains a protected inner core surrounded by buffer zones that local and indigenous people use for sustainable logging, food growing, cattle grazing, hunting, fishing, and eco-tourism.

Costa Rica's biodiversity conservation strategy has paid off. Today, the $1 billion a year tourism business—almost two-thirds of it from eco-tourists—is the country's largest source of income.

To reduce deforestation the government has eliminated subsidies for converting forests to cattle grazing land. And it pays landowners to maintain or restore tree coverage. This helps stabilize the climate by absorbing carbon dioxide, controlling flooding, and purifying water. The goal is to make sustaining forests profitable. As a result Costa Rica has gone from having one of the world's highest deforestation rates to one of the lowest.

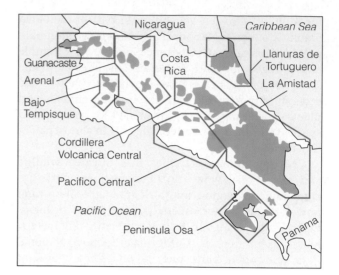

Figure 11-21 Solutions: Costa Rica has consolidated its parks and reserves into eight *megareserves* designed to sustain about 80% of the country's rich biodiversity.

A concern is that without careful government control, the 1 million tourists visiting Costa Rica each year could degrade some of the protected areas. Increased tourism could also stimulate the building of too many hotels, resorts, and other potentially harmful forms of development.

Solutions: The Nature Conservancy: Land Conservation through Private Action

The Nature Conservancy has used private and corporate donations to create the world's largest system of private natural areas and wildlife sanctuaries.

Since its founding by a group of professional ecologists in 1951, the *Nature Conservancy*—with more than 1 million members worldwide—has created the world's largest system of private natural areas and wildlife sanctuaries in 30 countries.

The organization uses private and corporate donations to maintain a fund for buying ecologically important pieces of land or wetlands threatened by development or other human activities. If it cannot buy land for habitat protection, the conservancy helps landowners obtain tax benefits in exchange for accepting legal restrictions or conservation easements preventing development. Landowners also receive sizable tax deductions by donating their land to the Nature Conservancy in exchange for lifetime occupancy rights.

According to John C. Sawhill, former president of the Nature Conservancy, "In the end, our society will be defined not only by what we create, but by what we refuse to destroy."

Should Reserves Be as Large as Possible? Generally, But Not Always

Large reserves usually are the best way to protect biodiversity, but in some places several well-placed, medium-sized, and isolated reserves can do the job.

Large reserves sustain more species and provide greater habitat diversity than do small reserves. They also minimize the area of outside edges exposed to natural disturbances (such as fires and hurricanes), invading species, and human disturbances from nearby developed areas.

However, research indicates that in some locales, several well-placed, medium-sized, and isolated reserves may better protect a wider variety of habitats and preserve more biodiversity than a single large reserve of the same area. A mixture of large and small reserves (Figure 11-21) may be the best way to protect a variety of species and communities against a number of different threats.

Establishing protected *habitat corridors* between reserves can help support more species and allow migration of vertebrates that need large ranges. They also permit migration of individuals and populations when environmental conditions in a reserve deteriorate and help preserve animals that must make seasonal migrations to obtain food. Corridors may also enable some species to shift their ranges if global climate change makes their current ranges uninhabitable.

On the other hand, corridors can threaten isolated populations by allowing movement of pest species, disease, fire, and exotic species between reserves. They also increase exposure of migrating species to natural predators, human hunters, and pollution. In addition, corridors can be costly to acquire, protect, and manage.

What Are Biosphere Reserves? A Great Idea

Biosphere reserves have an inner protected core surrounded by two buffer zones that can be used by local people for sustainable extraction of resources for food and fuel.

In 1971, the UN Educational, Scientific, and Cultural Organization (UNESCO) created the Man and the Biosphere (MAB) Programme. A major goal of the program is to establish at least one (and ideally five or more) *biosphere reserves* in each of the earth's 193 biogeographical zones. Today there are more than 425 biosphere reserves in 95 countries.

Each reserve must be large enough to contain three zones (Figure 11-22). The *core area* contains an important ecosystem that the government legally protects from all human activities except nondestructive research and monitoring.

A *buffer zone* surrounds and protects the core area. In this zone, emphasis is on nondestructive research, education, and recreation. Local people can also carry out sustainable logging, agriculture, livestock grazing, hunting, and fishing in this buffer zone, as long as such activities do not harm the core.

Finally, a second *buffer,* or *transition zone,* surrounds the inner buffer. In this zone local people can engage in more intensive but sustainable forestry, grazing, hunting, fishing, agriculture, and recreation than in the inner buffer zone. Doing this can enlist local people as partners in protecting a reserve from unsustainable uses.

So far, most biosphere reserves fall short of the ideal and receive too little funding for their protection and management. An international fund to help countries protect and manage biosphere reserves would cost about $100 million per year—about what the world's nations spend on weapons every 90 minutes.

What Is Adaptive Ecosystem Management? Cooperation and Flexibility

People with competing interests can work together to develop adaptable plans for managing and sustaining nature reserves.

Managing and sustaining a nature reserve is difficult. One problem is that reserves are constantly changing in response to environmental changes. Another is that they are affected by a variety of biological, cultural,

Biosphere Reserve

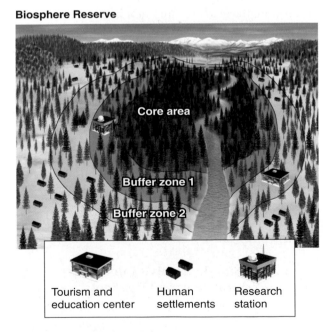

Figure 11-22 Solutions: a model *biosphere reserve.* In traditional parks and wildlife reserves, well-defined boundaries keep people out and wildlife in. By contrast, biosphere reserves recognize people's needs for access to sustainable use of various resources in parts of the reserve.

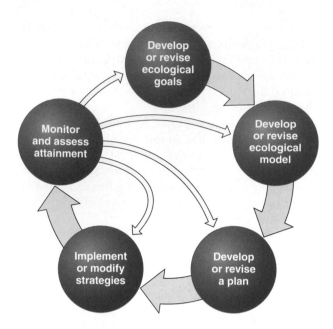

Figure 11-23 Solutions: the adaptive ecosystem management process.

economic, and political factors. In addition, their size, shape, and biological makeup often are determined by political, legal, and economic factors that depend on land ownership and conflicting public demands rather than by ecological principles and considerations.

One way to deal with these uncertainties and conflicts is through *adaptive ecosystem management*. It is based on using four principles. *First,* integrate ecological, economic, and social principles to help maintain and restore the sustainability and biological diversity of reserves while supporting sustainable economies and communities. *Second,* seek ways to get government agencies, private conservation organizations, scientists, business interests, and private landowners to reach a consensus on how to achieve common conservation objectives.

Third, view all decisions and strategies as scientific and social experiments and use failures as opportunities for learning and improvement. *Fourth,* emphasize continual information gathering, monitoring, reassessment, flexibility, adaptation, and innovation in the face of uncertainty and usually unpredictable change. Figure 11-23 summarizes the adaptive ecosystem management process.

What Areas Should Receive Top Priority for Establishing Reserves? Hot Spots

We can prevent or slow down losses of biodiversity by concentrating efforts on protecting hot spots where significant biodiversity is under immediate threat.

In reality, few countries are physically, politically, or financially able to set aside and protect large biodiversity reserves. To protect as much of the earth's remaining biodiversity as possible conservation biologists use

an *emergency action* strategy that identifies and quickly protects *biodiversity hot spots*. These "ecological arks" are areas especially rich in plant and animal species that are found nowhere else and are in great danger of extinction or serious ecological disruption.

Figure 11-24 shows 25 hot spots. They contain almost two-thirds of the earth's terrestrial biodiversity and are the only locations for more than one-third of the planet's known terrestrial plant and animal species. According to Norman Myers : "I can think of no other biodiversity initiative that could achieve so much at a comparatively small cost, as the hot spots strategy."

What Is Wilderness and Why Is It Important? Land Protected from Us

Wilderness is land legally set aside in a large enough area to prevent or minimize harm from human activities.

One way to protect undeveloped lands from human exploitation is by legally setting them aside as *wilderness*. According to the U.S. Wilderness Act of 1964, *wilderness* consists of areas "of undeveloped land affected primarily by the forces of nature, where man is a visitor who does not remain." U.S. President Theodore Roosevelt summarized what we should do with wilderness: "Leave it as it is. You cannot improve it."

The U.S. Wilderness Society estimates that a wilderness area should contain at least 4,000 square kilometers (1,500 square miles); otherwise it can be affected by air, water, and noise pollution from nearby human activities.

Wilderness supporters cite several reasons for preserving wild places. One is that they are areas where people can experience the beauty of nature and observe natural biological diversity. Such areas can also enhance the mental and physical health of visitors by allowing them to get away from noise, stress, development, and large numbers of people. Wilderness preservationist John Muir advised us,

> *Climb the mountains and get their good tidings.*
> *Nature's peace will flow into you as the sunshine into*
> *the trees. The winds will blow their freshness into you,*
> *and the storms their energy, while cares will drop off*
> *like autumn leaves.*

Even those who never use wilderness areas may want to know they are there, a feeling expressed by novelist Wallace Stegner:

> *Save a piece of country . . . and it does not matter in*
> *the slightest that only a few people every year will go*
> *into it. This is precisely its value . . . . We simply need*
> *that wild country available to us, even if we never do*
> *more than drive to its edge and look in. For it can be a*

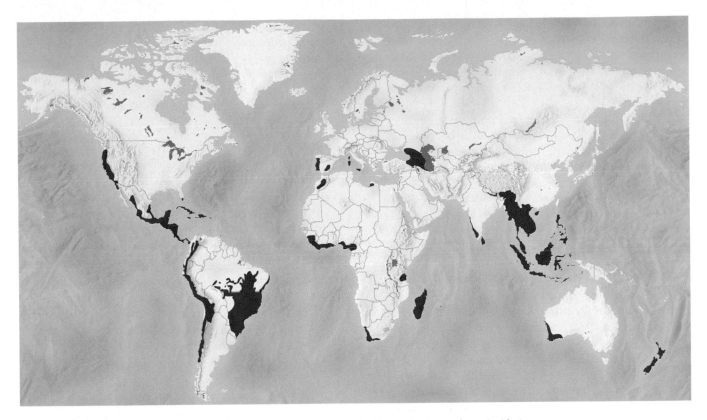

Figure 11-24 Endangered natural capital: twenty-five *hot spots* identified by ecologists as important but endangered centers of biodiversity that contain a large number of endemic plant and animal species found nowhere else. Research is adding new hot spots to this list. (Data from the Center for Applied Biodiversity Science at Conservation International)

means of reassuring ourselves of our sanity as creatures, a part of the geography of hope.

Some critics oppose protecting wilderness for its scenic and recreational value for a small number of people. They believe this is an outmoded concept that keeps some areas of the planet from being economically useful to humans.

Most biologists disagree. To them the most important reasons for protecting wilderness and other areas from exploitation and degradation are to *preserve their biodiversity* as a vital part of the earth's natural capital and to *protect them as centers for evolution* in response to mostly unpredictable changes in environmental conditions. In other words, wilderness is a biodiversity savings account and an eco-insurance policy.

Some analysts also believe wilderness should be preserved because the wild species it contains have a right to exist (or struggle to exist) and play their roles in the earth's ongoing saga of biological evolution and ecological processes, without human interference.

Case Study: How Much Wilderness Has Been Protected in the United States? Fighting for Crumbs and Losing

Only a small percentage of the land area of the United States has been set aside as wilderness.

In the United States, preservationists have been trying to save wild areas from development since 1900. Overall, they have fought a losing battle. Not until 1964 did Congress pass the Wilderness Act. It allowed the government to protect undeveloped tracts of public land from development as part of the National Wilderness Preservation System.

The area of protected wilderness in the United States increased tenfold between 1970 and 2000. Still, only about 4.6% of U.S. land is protected as wilderness—almost three-fourths of it in Alaska. Only 1.8% of the land area of the lower 48 states is protected, most of it in the West. In other words, Americans have reserved 98% of the continental United States to be used as they see fit and have protected only about 2% as wilderness. According to a 1999 study by the World Conservation Union (IUCN), the United States ranks 42nd among nations in terms of terrestrial area protected as wilderness, and Canada is in 36th place.

In addition, only 4 of the 413 wilderness areas in the lower 48 states are larger than 4,000 square kilometers (1,500 square miles). Also, the system includes only 81 of the country's 233 distinct ecosystems. Most wilderness areas in the lower 48 states are threatened habitat islands in a sea of development.

Almost 400,000 square kilometers (150,000 square miles) in scattered blocks of public lands could qualify for designation as wilderness—about 60% of it in the national forests. For two decades, these areas have been protected while they were evaluated for wilderness protection. Wilderness supporters would like to see all of these areas protected as part of the wilderness system.

This is unlikely because of the political strength of industries that see these areas as sources of resources for increased profits and short-term economic growth. The political efforts of these industries paid off when, in 2003, the Bush administration ceased protecting areas under consideration for classification as wilderness. This opens up many of these lands to road building, mining, oil drilling, logging, and off-road vehicle use. Such activities would also disqualify these areas for wilderness protection in the future.

Some wilderness advocates go further and call for creating *wilderness recovery areas.* They would do this by closing and obliterating nonessential roads in large areas of public lands, restoring wildlife habitats, allowing natural fires to burn, and reintroducing key species that have been driven from such areas.

Some ecologists and conservation biologists call for development of *The Wildlands Project (TWP)* to establish a network of protected wildlands and ecosystems throughout as much of the United States as possible. Accomplishing this would require cooperative efforts among government agencies, scientists, conservation groups, and private owners and users. It is unlikely that such projects will be implemented because of strong opposition to expansion of wilderness areas.

11-9 ECOLOGICAL RESTORATION

How Can We Rehabilitate and Restore Damaged Ecosystems? Making Amends for Our Actions

Scientists have developed a number of techniques for rehabilitating and restoring degraded ecosystems and creating artificial ecosystems.

Bad news. Almost every natural place on the earth has been affected or degraded to some degree by human activities. *Good news.* Much of the environmental damage we have inflicted on nature is at least partially reversible through **ecological restoration:** the process of repairing damage caused by humans to the biodiversity and dynamics of natural ecosystems. Examples include replanting forests, restoring grasslands, restoring wetlands, reclaiming urban industrial areas (brownfields), reintroducing native species, removing invasive species, and freeing river flows by removing dams.

Farmer and philosopher Wendell Berry says we should try to answer three questions in deciding whether and how to modify or rehabilitate natural ecosystems. *First,* what is here? *Second,* what will nature permit us to do here? *Third,* what will nature help us do here?

By studying how natural ecosystems recover, scientists are learning how to speed up repair operations using a variety of approaches. They include the following:

- *Restoration:* trying to return a particular degraded habitat or ecosystem to a condition as similar as possible to its natural state. However, we often lack knowledge about the previous composition of a degraded area and changes in climate, soil, and species composition can make it impossible to restore an area to its earlier state.

- *Rehabilitation:* attempts to turn a degraded ecosystem back into a functional or useful ecosystem without trying to restore it to its original condition. Examples include removing pollutants and replanting areas such as mining sites, landfills, and clear-cut forests to reduce soil erosion.

- *Remediation:* cleaning up chemical contaminants from a site by physical or chemical methods to protect human health and as a first step toward redevelopment of a site for human use. For example, an abandoned and polluted industrial plant—called a *brownfield*—may be cleaned up and then redeveloped into office buildings, apartments, a sports field, or a park.

- *Replacement:* replacing a degraded ecosystem with another type of ecosystem. For example, a productive pasture or tree farm may replace a degraded forest.

- *Creating artificial ecosystems:* Examples are the creation of artificial wetlands.

Researchers have suggested five basic science-based principles for carrying out ecological restoration.

- Mimic nature and natural processes and ideally let nature do most of the work, usually through secondary ecological succession.

- Recreate important ecological niches that have been lost.

- Rely on pioneer species, keystone species, foundation species, and natural ecological succession to facilitate the restoration process.

- Control or remove harmful nonnative species.

- If necessary, reconnect small patches to form larger ones and create corridors where existing patches are isolated.

Some analysts worry that environmental restoration could encourage continuing environmental destruction and degradation by suggesting any ecological harm we do can be undone. Some go further and say that we do not understand the incredible complexity of ecosystems well enough to restore or manage damaged natural ecosystems.

And ecologists point out that preventing ecosystem damage in the first place is cheaper and more effective than any form of ecological restoration. According to ecological restoration expert John Berger, "The purpose of ecological restoration is to repair previous damage, not legitimize further destruction."

Restorationists agree that restoration should not be used as an excuse for environmental destruction. But they point out that so far we have been able to protect or preserve no more than about 7% of nature from the effects of human activities. So ecological restoration is badly needed for much of the world's ecosystems that we have damaged.

They also point out that if a restored ecosystem differs from the original system this is better than nothing. And natural ecosystems are always changing anyway. They also contend that increased experience will improve the effectiveness of ecological restoration.

> **X** *HOW WOULD YOU VOTE?* Should we mount a massive effort to restore ecosystems we have degraded even though this will be quite costly? Cast your vote online at http://biology .brookscole.com/miller14.

Case Study: Ecological Restoration of a Tropical Dry Forest in Costa Rica

A degraded tropical dry forest in Costa Rica is being restored in a cooperative venture between tropical ecologists and local people.

Costa Rica is the site of one of the world's largest *ecological restoration* projects. In the lowlands of the country's Guanacaste National Park (Figure 11-21), a small tropical dry deciduous forest has been burned, degraded, and fragmented by large-scale conversion to cattle ranches and farms.

Now it is being restored and relinked to the rain forest on adjacent mountain slopes. The goal is to eliminate damaging nonnative grass and cattle and reestablish a tropical dry forest ecosystem over the next 100–300 years.

Daniel Janzen, professor of biology at the University of Pennsylvania and a leader in the field of restoration ecology, has helped galvanize international support and has raised more than $10 million for this restoration project. He recognizes that ecological restoration and protection of the park will fail unless the people in the surrounding area believe they will benefit from such efforts. Janzen's vision is to make the nearly 40,000 people who live near the park an essential part of the restoration of the degraded forest, a concept he calls *biocultural restoration.*

By actively participating in the project, local residents reap educational, economic, and environmental benefits. Local farmers make money by sowing large areas with tree seeds and planting seedlings started in Janzen's lab. Local grade school, high school, and university students and citizens' groups study the ecology of the park and go on field trips to the park. The park's location near the Pan American Highway makes it an ideal area for eco-tourism, which stimulates the local economy.

The project also serves as a training ground in tropical forest restoration for scientists from all over the world. Research scientists working on the project give guest classroom lectures and lead some of the field trips.

Janzen recognizes that in a few decades today's children will be running the park and the local political system. If they understand the ecological importance of their local environment, they are more likely to protect and sustain its biological resources. He believes that education, awareness, and involvement— not guards and fences—are the best ways to restore degraded ecosystems and protect largely intact ecosystems from unsustainable use.

11-10   WHAT CAN WE DO?

What Should Be Our Priorities? An Eight-Step Program

Biodiversity expert Edward O. Wilson has proposed eight priorities for protecting most of the world's remaining ecosystems and species.

In 2002, Edward O. Wilson, considered to be one of the world's foremost experts on biodiversity, published a book called *The Future of Life* (Knopf, New York). In this book, he proposed the following priorities for protecting most of the world's remaining ecosystems and species:

- *Take immediate action to preserve the world's biological hot spots* (Figure 11-24).

- *Keep intact the world's remaining old-growth forests and cease all logging of such forests.*

- *Complete the mapping of the world's terrestrial and aquatic biodiversity so we know what we have and ca make conservation efforts more precise and cost-effective.*

- *Determine the world's marine hot spots and assign them the same priority for immediate action as for those on land*—more on this in Chapter 13.

- *Concentrate on protecting and restoring everywhere the world's lakes and river systems, which are the most threatened ecosystems of all*—more on this Chapter 13.

- *Ensure that the full range of the earth's terrestrial and aquatic ecosystems are included in a global conservation strategy.*

- Plant trees and take care of them.

- Recycle paper and buy recycled paper products.

- Buy wood and wood products made from trees that have been grown sustainably.

- Help rehabilitate or restore a degraded area of forest or grassland near your home.

- When building a home, save all the trees and as much natural vegetation and soil as possible.

- Landscape your yard with a diversity of plants natural to the area instead of having a monoculture lawn.

Figure 11-25 What can you do? Ways to help sustain terrestrial biodiversity.

■ *Make conservation profitable.* This involves finding ways to raise the income of people who live in or near nature reserves so they can become partners in their protection and sustainable use. It also requires providing financial help from private and government sources to governments that protect their forests and other nature reserves.

■ *Initiate ecological restoration products worldwide* to heal some of the damage we have done and increase the share of the earth's land and water allotted to the rest of nature.

According to Wilson, such a conservation strategy would cost about $30 billion per year—an amount that could be provided by a tax of one cent per cup of coffee. According to biologist David Suzuhi, "We must try our best in everything we do not to disrupt the natural systems around us because, ultimately, we are completely dependent on them. That is what sustainability all about."

This strategy for protecting the earth's precious biodiversity will not be implemented without bottom-up political pressure on elected officials from individual citizens and groups. It will also require cooperation among key people in government, the private sector, science, and engineering using adaptive management (Figure 11-23). Figure 11-25 lists some ways you can help sustain the earth's terrestrial biodiversity.

We abuse land because we regard it as a commodity belonging to us. When we see land as a community to which we belong, we may begin to use it with love and respect.

ALDO LEOPOLD

CRITICAL THINKING

1. Do you agree or disagree with the program that reintroduced populations of the gray wolf in the Yellowstone ecosystem? Explain. Do you favor reintroducing grizzly bears to Yellowstone or other public lands in the western United States? Explain.

2. Explain why you agree or disagree with **(a)** the four principles that biologists and some economists have suggested for using public land in the United States (p. 199) and **(b)** the nine suggestions made by developers and resource extractors for managing and using U.S. public land (p. 199).

3. Explain why you agree or disagree with each of the proposals for providing more sustainable use of forests throughout the world, listed in Figure 11-13, p. 205.

4. Should there be a ban on the use of off-road motorized vehicles and snowmobiles on all public lands? Explain.

5. Should the U.S. government (or the government of the country where you live) continue providing private companies that harvest timber from public lands with subsidies for reforestation and for building and maintaining access roads? Explain.

6. In the early 1990s, Miguel Sanchez, a subsistence farmer in Costa Rica, was offered $600,000 by a hotel developer for a piece of land that he and his family had been using sustainably for many years. The land contained an old-growth rain forest and a black sand beach in an area under rapid development. Sanchez refused the offer. What would you have done if you were a poor subsistence farmer in Miguel Sanchez's position? Explain your decision.

7. Should developed countries provide most of the money to preserve remaining tropical forests in developing countries? Explain.

8. If ecosystems are undergoing constant change, why should we **(a)** establish and protect nature reserves and **(b)** carry out ecological restoration?

9. Congratulations! You are in charge of protecting and sustaining the world's terrestrial biodiversity. List the three most important features of your policies for using and managing **(a)** forests and **(b)** parks.

PROJECTS

1. Obtain a topographic map of the region where you live and use it to identify local, state, and federally owned lands in the form of parks, rangeland, forests, and wilderness areas. Identify the government agency or agencies responsible for managing each of these areas, and try to evaluate how well these agencies are preserving the natural resources on this public land on your behalf.

2. What has happened to the biome in which you live during the past 50 years? How much, if any, of it remains in its original state? How much of it should be protected from further degradation? How much of it could be restored?

3. If possible, try to visit **(a)** a diverse old-growth forest, **(b)** an area that has been recently clear-cut, and **(c)** an area that was clear-cut 5–10 years ago. Compare the biodiversity, soil erosion, and signs of rapid water runoff in each of the three areas.

4. For many decades, New Zealand has had a policy of meeting all its demand for wood and wood products by growing timber on intensively managed tree plantations. Use the library or Internet to evaluate the effectiveness of this approach and its major advantages and disadvantages.

5. Use the library and Internet to find one example of a successful ecological restoration project not discussed in this chapter and one that failed. For your example, describe the strategy used, the ecological principles involved, and why the project succeeded or failed.

6. Use the library or the Internet to find bibliographic information about *François-Auguste-René de Chateaubriand*

and *Aldo Leopold,* whose quotes appear at the beginning and end of this chapter.

7. Make a concept map of this chapter's major ideas, using the section heads, subheads, and key terms (in bold-face). Look on the website for this book for information about making concept maps.

LEARNING ONLINE

The website for this book contains study aids and many ideas for further reading and research. They include a chapter summary, review questions for the entire chapter, flash cards for key terms and concepts, a multiple-choice practice quiz, interesting Internet sites, references, and a guide for accessing thousands of InfoTrac® College Edition articles. Log on to

http://biology.brookscole.com/miller14

Then click on the Chapter-by-Chapter area, choose Chapter 11, and select a learning resource.

12 Sustaining Biodiversity: The Species Approach

CASE STUDY
The Passenger Pigeon: Gone Forever

In 1813, bird expert John James Audubon saw a single flock of passenger pigeons that he estimated was 16 kilometers (10 miles) wide and hundreds of kilometers long, and contained perhaps a billion birds. The flock took three days to fly past him and was so dense that it darkened the skies.

By 1914, the passenger pigeon (Figure 12-1) had disappeared forever. How could a species that was once the most common bird in North America and probably the world become extinct in only a few decades? The answer is, humans wiped them out. The main reasons for the extinction of this species were uncontrolled commercial hunting and loss of the bird's habitat and food supply as forests were cleared to make room for farms and cities.

Passenger pigeons were good to eat, their feathers made good pillows, and their bones were widely used for fertilizer. They were easy to kill because they flew in gigantic flocks and nested in long, narrow colonies.

Commercial hunters would capture one pigeon alive, sew its eyes shut, and tie it to a perch called a stool. Soon a curious flock would land beside this "stool pigeon"—a term we now use to describe someone who turns in another person for breaking the law. Then the birds would be shot or ensnared by nets that might trap more than 1,000 birds at once.

Beginning in 1858, passenger pigeon hunting became a big business. Shotguns, traps, artillery, and even dynamite were used. People burned grass or sulfur below their roosts to suffocate the birds. Shooting galleries used live birds as targets. In 1878, one professional pigeon trapper made $60,000 by killing 3 million birds at their nesting grounds near Petoskey, Michigan!

By the early 1880s, only a few thousand birds remained. At that point, recovery of the species was doomed because the females laid only one egg per nest each year. On March 24, 1900, a young boy in Ohio shot the last known wild passenger pigeon. The last passenger pigeon on earth, a hen named Martha after Martha Washington, died in the Cincinnati Zoo in

John James Audubon/The New York Historical Society

Figure 12-1 **Lost natural capital:** passenger pigeons have been extinct in the wild since 1900. The last known passenger pigeon died in the Cincinnati Zoo in 1914.

1914. Her stuffed body is now on view at the National Museum of Natural History in Washington, D.C.

Eventually all species become extinct or evolve into new species. But biologists estimate that human activities have increased the natural rate of extinction by a factor of 1,000 to 10,000—perhaps more. Studies indicate that this rate of loss of biodiversity is expected to increase as the human population grows, consumes more resources, disturbs more of the earth's land and aquatic systems, and uses more of the earth's net plant productivity that supports all species.

The last word in ignorance is the person who says of an animal or plant: "What good is it?" . . . If the land mechanism as a whole is good, then every part of it is good, whether we understand it or not . . . . Harmony with land is like harmony with a friend; you cannot cherish his right hand and chop off his left.

ALDO LEOPOLD

This chapter addresses the following questions:

- How do biologists estimate extinction rates, and how are human activities affecting these rates?

- Why should we care about biodiversity and species extinction?

- What human activities endanger wildlife?

- How can we help prevent premature extinction of species?

- What is reconciliation ecology, and how can it be used to help prevent premature extinction of species?

12-1 SPECIES EXTINCTION

What Are Three Types of Species Extinction? Local, Ecological, and Biological

Species can become extinct locally, ecologically, or globally.

Biologists distinguish among three levels of species extinction. One is *local extinction*. It occurs when a species is no longer found in an area it once inhabited but is still found elsewhere in the world. Most local extinctions involve losses of one or more populations of a species.

The second type is *ecological extinction*. It occurs when so few members of a species are left that it can no longer play its ecological roles in the biological communities where it is found.

The third type is *biological extinction*, when a species is no longer found anywhere on the earth (Figures 12-1 and 12-2). Biological extinction is forever.

What Are Endangered and Threatened Species? Ecological Smoke Alarms

An endangered species could soon become extinct and a threatened species is likely to become extinct.

Biologists classify species heading toward biological extinction as either *endangered* or *threatened* (Figure 12-3, p. 226). An **endangered species** has so few individual survivors that the species could soon become extinct over all or most of its natural range. A **threatened,** or **vulnerable, species** is still abundant in its natural range but because of declining numbers is likely to become endangered in the near future.

Some species have characteristics that make them more vulnerable than others to ecological and biological extinction (Figure 12-4, p. 228). As biodiversity expert Edward O. Wilson puts it, "the first animal species to go are the big, the slow, the tasty, and those with valuable parts such as tusks and skins."

A 2000 joint study by the World Conservation Union and Conservation International and a 1999 study by the World Wildlife Fund found that human activities threaten several types of species with premature extinction (Figure 12-5, p. 228). And a 2000 survey by the Nature Conservancy and the Association for Biodiversity Information found that about one-third of 21,000 animal and plant species in the United States are vulnerable to premature extinction.

Passenger pigeon   Great auk   Dodo   Dusky seaside sparrow   Aepyornis (Madagascar)

Figure 12-2 Lost natural capital: some animal species that have become prematurely extinct largely because of human activities, mostly habitat destruction and overhunting.

Grizzly bear
(threatened)

Kirkland's warbler

White top pitcher plant

Arabian oryx
(Middle East)

African elephant
(Africa)

Mojave desert tortoise
(threatened)

Swallowtail butterfly

Humpback chub

Golden lion tamarin
(Brazil)

Siberian tiger
(Siberia)

West Virginia spring
salamander

Giant panda
(China)

Whooping crane

Knowlton cactus

Blue whale

Mountain gorilla
(Africa)

Pine barrens
tree frog (male)

Swamp pink

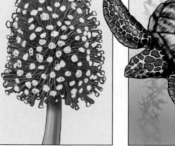
Hawksbill sea turtle

El Segunda blue butterfly

Figure 12-3 Endangered natural capital: species that are endangered or threatened with premature extinction largely because of human activities. Almost 30,000 of the world's species and 1,200 of those in the United States are officially listed as in danger of becoming extinct. Most biologists believe the actual number of species at risk is much larger.

Florida manatee

Northern spotted owl
(threatened)

Gray wolf

Florida panther

Bannerman's turaco
(Africa)

Devil's hole pupfish

Snow leopard
(Central Asia)

Symphonia
(Madagascar)

Black-footed ferret

Utah prairie dog

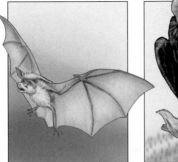

Ghost bat
(Australia)

California condor

Black lace cactus

Black rhinoceros
(Africa)

Oahu tree snail

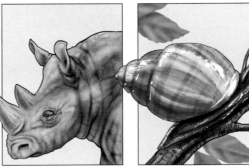

How Do Biologists Estimate Extinction Rates? Peering into a Cloudy Looking Glass

Scientists use measurements and models to estimate extinction rates.

Evolutionary biologists estimate that 99.9% of all species that ever existed are now extinct because of a combination of background extinction, mass extinctions, and mass depletions taking place over thousands to millions of years. Biologists also talk of an *extinction spasm,* in which large numbers of species are lost over a period of a few centuries or at most 1,000 years.

Biologists trying to catalog extinctions have three problems. *First,* the extinction of a species typically takes such a long time that it is not easy to document. *Second,* we have identified only about 1.4–1.8 million of the world's estimated 5–100 million species. *Third,* we know little about most of the species we have identified.

The truth is we do not know how many species are becoming extinct each year mostly because of our activities. But scientists do the best they can with the tools they have to estimate past and projected future extinction rates.

One approach is to study past records documenting the rate at which mammals and birds have become extinct since we came on the scene and comparing this with the fossil records of such extinctions prior to our arrival. For example, there is a detailed study on extinction of Pacific island birds by early human colonists. Since the 1960s the International Union for the Conservation of Nature and Natural Resources (IUCN)—also known as the World Conservation Union—has kept

Characteristic	Examples
Low reproductive rate (K-strategist)	Blue whale, giant panda, rhinoceros
Specialized niche	Blue whale, giant panda, Everglades kite
Narrow distribution	Many island species, elephant seal, desert pupfish
Feeds at high trophic level	Bengal tiger, bald eagle, grizzly bear
Fixed migratory patterns	Blue whale, whooping crane, sea turtles
Rare	Many island species, African violet, some orchids
Commercially valuable	Snow leopard, tiger, elephant, rhinoceros, rare plants and birds
Large territories	California condor, grizzly bear, Florida panther

Figure 12-4 Characteristics of species that are prone to ecological and biological extinction.

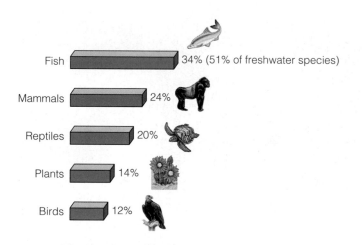

Fish 34% (51% of freshwater species)
Mammals 24%
Reptiles 20%
Plants 14%
Birds 12%

Figure 12-5 **Endangered natural capital:** percentage of various types of species threatened with premature extinction because of human activities. (Data from World Conservation Union, Conservation International, and World Wildlife Fund)

Red Lists that have become the world standard for listing all threatened species throughout the world. These lists provide baseline information on how some of the earth's biodiversity changes over time. Biologists use these lists to identify species that have become extinct and to determine shifts in the types and numbers of species endangered by human activities.

Another way that biologists project future extinction rates is to observe how the number of species present increases with the size of an area. This *species–area*

relationship suggests that on average, a 90% loss of habitat causes the extinction of about 50% of the species living in that habitat. For example, scientists estimate that about 50% of the world's existing terrestrial species live in tropical forests and that about one-third of the remaining tropical forests will be cut or burned during the next few decades. If these assumptions are valid, the species–area relationship suggests about 1 million species in these tropical forests will become extinct during this period.

The methods just described give similar estimates of past and future extinction rates. They also provide strong evidence that human actions have caused recent extinctions and that the situation will get worse. Scientists also use models to estimate the risk that a particular population of a species will become endangered or extinct within a certain time.

Estimates of future extinction rates vary because of differing assumptions about the earth's total number of species, the proportion of these species found in tropical forests, the rate at which tropical forests are being cleared, and the reliability of the methods used to make these estimates.

How Are Human Activities Affecting Extinction Rates? Taking Out More Species

Biologists estimate that the current rate of extinction is at least 1,000 to 10,000 times the rate before we arrived.

In due time all species become extinct, but there is considerable evidence that we are hastening the final exit for a growing number of species. Before we came on the scene the estimated extinction rate was roughly one species per million annually. This amounted to an annual extinction rate of about 0.0001% per year.

Using the methods just described, biologists conservatively estimate that the current rate of extinction is at least 1,000 to 10,000 times the rate before we arrived. This amounts to an annual extinction rate of 0.1% to 1% per year.

So how many species are we probably losing prematurely each year? This depends on how many species are on the earth. Assuming that the extinction rate is 0.1%, each year we are losing 5,000 species per year if there are 5 million species, 14,000 if there are 14 million species (biologists' current best guess), and 100,000 if there are 100 million species.

Most biologists would consider the premature loss of 1 million species over 100–200 years an extinction crisis or spasm that if kept up would lead to a mass depletion or even a mass extinction. At an extinction rate of 0.1% a year, the time it would take to lose 1 million species would be 200 years if there were a total of 5 million species, 71 years with a total of 14 million species, and 10 years with 100 million species. How many years would it take to lose 1 million species for each of these three species estimates if the extinction rate is 1% a year?

According to researchers Edward O. Wilson and Stuart Primm, at a 1% extinction rate at least 20% of the world's current animal and plant species could be gone by 2030 and 50% could vanish by the end of this century. In the words of biodiversity expert Norman Myers, "Within just a few human generations, we shall—in the absence of greatly expanded conservation efforts—impoverish the biosphere to an extent that will persist for at least 200,000 human generations or twenty times longer than the period since humans emerged as a species."

Most biologists consider extinction rates of 0.1–1% to be conservative estimates for several reasons. *First*, both the rate of species loss and the extent of biodiversity loss are likely to increase during the next 50–100 years because of the projected exponential growth of the world's human population and per capita resource use. In other words, the size of our already large ecological footprint (Figure 1-7, p. 10 and Figure 9-12, p. 172) is likely to increase.

Second, current and projected extinction rates are much higher than the global average in parts of the world that are endangered centers of biodiversity. Conservation biologists estimate that such biologically rich areas could lose one-fourth to one-half of their estimated species within a few decades. They urge us to focus our efforts on slowing the much higher rates of extinction in such *hot spots* (Figure 11-24, p. 219) as the best and quickest way to protect much of the earth's biodiversity from being lost prematurely.

Third, we are eliminating, degrading, and simplifying many biologically diverse environments—such as tropical forests, tropical coral reefs, wetlands, and estuaries—that serve as potential colonization sites for the emergence of new species. Thus, in addition to increasing the rate of extinction, we may also be limiting long-term recovery of biodiversity by reducing the rate of speciation for some types of species. In other words, we are also creating a *speciation crisis*. See Norman Myers Guest Essay on this topic on the website for this chapter.

Philip Levin, Donald Levin, and other biologists also argue that the increasing fragmentation and disturbance of habitats throughout the world may increase the speciation rate for rapidly reproducing opportunist species such as weeds, rodents, and cockroaches and other insects. Thus the real threat to biodiversity from current human activities may not be a permanent decline in the number of species but a long-term erosion in the earth's variety of species and habitats.

Some people, most of them not biologists, say the current estimated extinction rates are too high and are based on inadequate data and models. Researchers agree that their estimates of extinction rates are based on inadequate data and sampling. They continually strive to get better data and improve the models they use to estimate extinction rates.

However, they point to clear evidence that human activities have increased the rate of species extinction and that this rate is likely to rise. According to these biologists, arguing over the numbers and waiting to get better data and models should not be used as excuses for inaction. They call for us to implement a *precautionary strategy* now to help prevent a significant decrease in the earth's genetic, species, ecological, and functional diversity.

To these biologists, we are not heeding the warning of Aldo Leopold about preserving biodiversity as we tinker with the earth: "To keep every cog and wheel is the first precaution of intelligent tinkering."

12-2 IMPORTANCE OF WILD SPECIES

Why Should We Preserve Wild Species? They Have Value

We should not cause the premature extinction of species because of the economic and ecological services they provide.

So what is all the fuss about? If all species eventually become extinct, why should we worry about losing a few more because of our activities? Does it matter that the passenger pigeon, the 80–100 remaining Florida panthers, or some unknown plant or insect in a tropical forest becomes prematurely extinct because of our activities?

We know that new species eventually evolve to take the place of ones lost through extinction spasms, mass depletions, or mass extinctions. So why should we care if we speed up the extinction rate over the next 50–100 years? The answer is that *it will take at least 5 million years for speciation to rebuild the biodiversity we are likely to destroy during this century!*

Conservation biologists and ecologists say we should act now to prevent the premature extinction of species because of their **instrumental value** based on their usefulness to us in the form of economic and ecological services. For example, species provide economic value in the form of food crops, fuelwood and lumber, paper, and medicine (Figure 11-17, p. 211).

Another instrumental value is the *genetic information* in species. Genetic engineers use this information to produce new types of crops (Figure 5-11, p. 98) and foods and edible vaccines for viral diseases such as hepatitis B. Carelessly eliminating many of the species making up the world's vast genetic library is like burning books before we read them. Wild species also provide a way for us to learn how nature works and sustains itself.

The earth's wild plants and animals also provide us with *recreational pleasure*. Each year Americans spend over three times as many hours watching wildlife—doing nature photography and bird watching, for example—as they spend on watching movies or professional sporting events.

Wildlife tourism, or *eco-tourism,* generates at least $500 billion per year worldwide, and perhaps twice that much. Conservation biologist Michael Soulé estimates that one male lion living to age 7 generates $515,000 in tourist dollars in Kenya but only $1,000 if killed for its skin. Similarly, over a lifetime of 60 years a Kenyan elephant is worth about $1 million in eco-tourist revenue—many times more than its tusks are worth when sold illegally for their ivory.

Ideally, eco-tourism should not cause ecological damage. In addition, it should provide income for local people to motivate them to preserve wildlife and funds for the purchase and maintenance of wildlife preserves and conservation programs. Much eco-tourism does not meet these standards, and excessive and unregulated eco-tourism can destroy or degrade fragile areas and promote premature species extinction. The website for this chapter lists some guidelines for evaluating eco-tours.

Case Study: Why Should We Care about Bats? Ecological Allies

Because of the important ecological and economic roles bats play, we should view them as valuable allies, not as enemies to kill.

Worldwide there are 950 known species of bats—the only mammals that can fly. However, bats have two traits that make them vulnerable to extinction. *First,* they reproduce slowly. *Second,* many bat species live in huge colonies in caves and abandoned mines, which people sometimes block. This prevents them from leaving to get food and can disturb their hibernation.

Bats play important ecological roles. About 70% of all bat species feed on crop-damaging nocturnal insects and other insect pest species such as mosquitoes. This makes them the major nighttime SWAT team for such insects.

In some tropical forests and on many tropical islands, *pollen-eating bats* pollinate flowers, and *fruit-eating bats* distribute plants throughout tropical forests by excreting undigested seeds.

As keystone species, such bats are vital for maintaining plant biodiversity and for regenerating large areas of tropical forest cleared by human activities. If you enjoy bananas, cashews, dates, figs, avocados, or mangos, you can thank bats.

Many people mistakenly view bats as fearsome, filthy, aggressive, rabies-carrying bloodsuckers. But most bat species are harmless to people, livestock, and crops. In the United States, only 10 people have died of bat-transmitted disease in four decades of record keeping; more Americans die each year from falling coconuts.

Because of unwarranted fears of bats and lack of knowledge about their vital ecological roles, several bat species have been driven to extinction. Currently, about one-fourth of the world's bat species, including the ghost bat (Figure 12-3), are listed as endangered or threatened. Conservation biologists urge us to view bats as valuable allies, not as enemies.

What Is the Intrinsic Value of Species? Existence Rights

Some people believe that each wild species has an inherent right to exist.

Some people believe that each wild species also has *intrinsic* or *existence* value based on its inherent right to exist and play its ecological roles regardless of its usefulness to us. Biologist Edward O. Wilson believes most people feel obligated to protect other species and the earth's biodiversity because most humans seem to have a natural affinity for nature that he calls *biophilia* (Connections, right). As novelist Fyodor Dostoevsky said in his 1889 novel *The Brothers Karamazov,* "Love the animals, love the plants, love everything. If you love everything, you will perceive the divine mystery in things. Once you perceive it, you will begin to comprehend it better every day. And you will come at last to love the whole world with an all-embracing love."

Some people distinguish between the survival rights of plants and those of animals, mostly for practical reasons. Poet Alan Watts once said he was a vegetarian "because cows scream louder than carrots."

Other people distinguish among various types of species. For example, they might think little about getting rid of the world's mosquitoes, cockroaches, rats, or disease-causing bacteria.

Some proponents of existence rights such as Nobel Prize winner Albert Schweitzer go further and assert that each individual organism has a right to survive without human interference. Others apply this to individuals of some species but not to those of other species. Unless they are strict vegetarians, for example, some people see no harm in having others kill domesticated animals in slaughterhouses to provide them with meat, leather, and other products. But these same people might deplore the killing of wild animals such as deer, squirrels, or rabbits. Where do you stand on this issue?

Some conservation biologists also caution us not to focus primarily on protecting relatively big organisms—the plants and animals we can see and are familiar with. They remind us that the true foundation of the earth's ecosystems and ecological processes are the invisible bacteria, and the algae, fungi, and other *microorganisms* that decompose the bodies of larger organisms and recycle the nutrients needed by all life (Case Study, p. 56).

Biophilia

Biologist Edward O. Wilson contends that because of the billions of years of biological connections leading to the evolution of the human species, we have an inherent affinity for the natural world. He calls this phenomenon *biophilia* (love of life).

Evidence of this natural and emotional affinity for life is seen in the preference most people have for almost any natural scene over one from an urban environment. Given a choice, most people prefer to live in an area where they can see water, grassland, or a forest. More people visit zoos and aquariums than attend all professional sporting events combined.

In the 1970s I was touring the space center at Cape Canaveral in Florida. During our bus ride the tour guide pointed out each of the abandoned multimillion-dollar launch sites and gave a brief history of each launch. Most of us were utterly bored. Suddenly people started rushing to the front of the bus and staring out the window with great excitement. What they were looking at was a baby alligator—a dramatic example of how *biophilia* can triumph over *technophilia*.

Not everyone has biophilia. Some have the opposite feeling about many or most forms of life. This fear of life is called *biophobia*. Biophobia varies in intensity and degree with individuals based on heredity and experience with various forms of life. For example, some movies, books, and TV programs condition us to fear or be repelled by certain species such as snakes, spiders, insects (especially ones that bite, sting, or crawl around our houses such as cockroaches, bats, sharks, rats, and bacteria). Throughout this book I have tried to show you the important ecological roles such species play.

But I understand that many of you will fear many of these species regardless of how useful they are to us and the functioning of ecosystems. Fear is a difficult emotion to overcome.

Critical Thinking

Do you have an affinity for wildlife and wild ecosystems (biophilia)? If so, how do you display this love of wildlife in your daily actions? What patterns of your consumption help destroy and degrade wildlife?

12-3 EXTINCTION THREATS FROM HABITAT LOSS AND DEGRADATION

What Is the Role of Habitat Loss and Degradation? Creating Homeless Species

The greatest threat to a species is the loss and degradation of the place where it lives.

Figure 12-6 shows the basic and secondary causes of the endangerment and premature extinction of wild species. Conservation biologists sometimes summarize the main secondary factors leading to premature extinction using the acronym **HIPPO** for **h**abitat destruction and fragmentation, **i**nvasive (alien) species, **p**opulation growth (too many people consuming too many resources), **p**ollution, and **o**verharvesting.

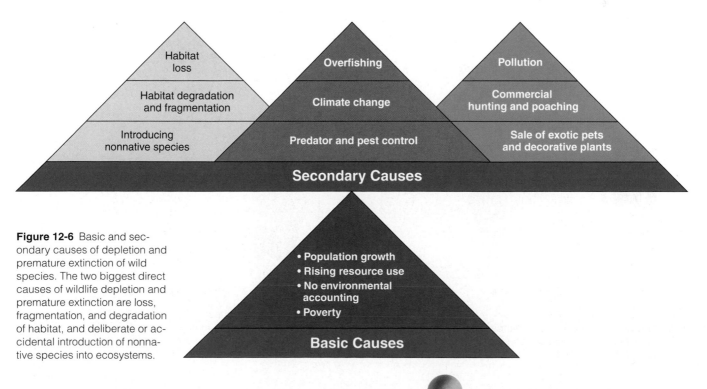

Figure 12-6 Basic and secondary causes of depletion and premature extinction of wild species. The two biggest direct causes of wildlife depletion and premature extinction are loss, fragmentation, and degradation of habitat, and deliberate or accidental introduction of nonnative species into ecosystems.

According to biodiversity researchers, the greatest threat to wild species is habitat loss (Figure 12-7), degradation, and fragmentation. In other words, many species have a hard time surviving after we take over their ecological "house" and food supplies and make them homeless.

Deforestation of tropical forests is the greatest eliminator of terrestrial species followed by the destruction of wetlands and plowing of grasslands. Globally, temperate biomes have been affected more by habitat loss and degradation than have tropical biomes because of widespread development in temperate countries over the past 200 years. Emphasis is now shifting to many tropical biomes.

According to the Nature Conservancy, the major types of habitat disturbance threatening endangered species in the United States are, in order of importance: agriculture, commercial development, water development, outdoor recreation (including off-road vehicles), livestock grazing, and pollution.

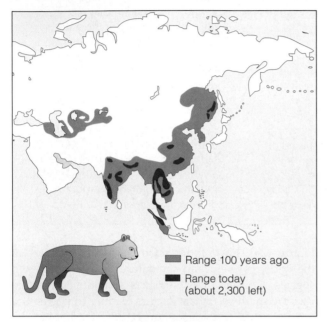

Range 100 years ago
Range today (about 2,300 left)

Indian Tiger

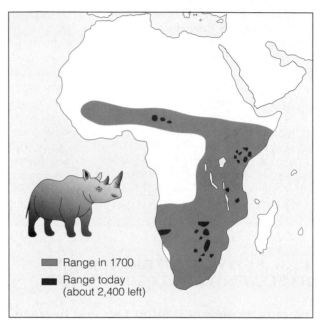

Range in 1700
Range today (about 2,400 left)

Black Rhino

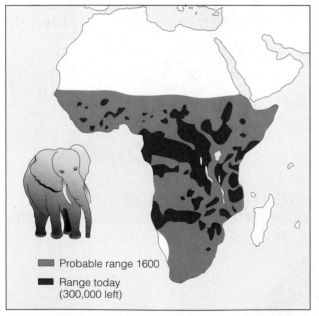

Probable range 1600
Range today (300,000 left)

African Elephant

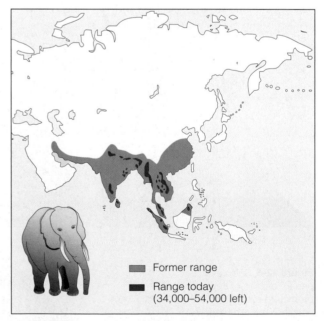

Former range
Range today (34,000–54,000 left)

Asian or Indian Elephant

Figure 12-7 Degraded natural capital: reductions in the ranges of four wildlife species, mostly the result of habitat loss and hunting. What will happen to these and millions of other species when the world's human population doubles and per capita resource consumption rises sharply in the next few decades? (Data from International Union for the Conservation of Nature and World Wildlife Fund)

Island species, many of them endemic species found nowhere else on earth, are especially vulnerable to extinction when their habitats are destroyed, degraded, or fragmented.

What Is the Role of Habitat Fragmentation? Isolating and Weakening Populations of Species

Species are more vulnerable to extinction when their habitats are divided into smaller, more isolated patches.

Habitat fragmentation occurs when a large, continuous area of habitat is reduced in area and divided into smaller, more scattered, and isolated patches or "habitat islands." This divides populations of a species into smaller and more isolated groups that are more vulnerable to predators, invasion by more competitive species, disease, and catastrophic events such as a storm or fire. Also, it creates barriers that can hinder some species from dispersing and colonizing new areas, getting enough to eat, and finding mates.

Certain types of species are especially vulnerable to local and regional extinction because of habitat fragmentation. They include species that are rare, that need to roam unhindered over large areas, and that cannot rebuild their population because of a low reproductive capacity. Also included are species with specialized niches and species that are sought by people for furs, food, medicines, or other uses.

The theory of island biogeography (p. 145) has been used to understand the effects of fragmentation on species extinction and to develop ways to help prevent such extinction.

Case Study: How Do Human Activities Affect Bird Species? A Disturbing Message from the Birds

Our activities are causing serious declines in the populations of many bird species.

Approximately 70% of the world's 9,800 known bird species are declining in numbers and about one of every six bird species is threatened with extinction, mostly because of habitat loss and fragmentation. A 2002 National Audubon Society study found that a quarter of all U.S. bird species are declining in numbers or are at risk of disappearing. Figure 12-8 shows the 10 most threatened U.S. songbird species according to a 2002 study by the National Audubon Society.

Cerulean warbler   Sprague's pipit   Bichnell's thrush   Black-capped vireo   Golden-cheeked warbler

Florida scrub jay   California gnatcatcher   Kirtland's warbler   Henslow's sparrow   Bachman's warbler

Figure 12-8 Threatened natural capital: ten most threatened species of U.S. songbirds according to a 2002 study by the National Audubon Society. Most of these species are threatened because of habitat loss and fragmentation from human activities. Almost 1,200 species—about 12% of the world's 9,800 known bird species—may face premature extinction during this century.

Nonnative species are the second greatest threat to birds. They include bird-eating cats, rats, brown-tree snakes, and mongooses.

Birds can also be loved to death. A third of the world's 330 parrot species are threatened from a combination of habitat loss and capture for the pet trade (often illegal), especially in Europe and the United States.

At least 23 species of seabirds face extinction because they are being drowned after becoming hooked on miles of baited lines put out by fishing boats. Millions of migrating birds are also killed each year when they collide with power lines, communications towers, and skyscrapers that we have erected in the middle of their migration routes. For example, each year U.S. hunters kill about 121 million birds. But about 1 billion birds are killed in the U.S. each year by flying into glass windows.

Other threats to birds are oil spills, exposure to pesticides, herbicides that destroy their habitats, and swallowing toxic lead shotgun pellets left in wetlands and lead sinkers left by anglers. Poorly regulated illegal hunting and capture also take a heavy toll.

Conservation biologists view this decline of bird species as an early warning of the greater loss of biodiversity to come. The reason is that birds are excellent *environmental indicators* because they live in every climate and biome, respond quickly to environmental changes in their habitats, and are easy to track and count.

Besides serving as indicator species, birds play important ecological roles. These include helping control populations of rodents and insects (which decimate many tree species), pollinating a variety of flowering plants, spreading plants throughout their habitats by consuming and excreting plant seeds, and scavenging dead animals. Conservation biologists urge us to listen more carefully to what birds are telling us about the state of the environment.

12-4 EXTINCTION THREATS FROM NONNATIVE SPECIES

What Is the Role of Deliberately Introduced Species? Good and Bad News

Many nonnative species provide us with food, medicine, and other benefits but a few can wipe out some native species, disrupt ecosystems, and cause large economic losses.

We depend heavily on nonnative organisms for ecosystem services, food, shelter, medicine, and aesthetic enjoyment.

According to a 2000 study by ecologist David Pimentel, introduced species such as corn, wheat, rice, other food crops, cattle, poultry, and other livestock provide more than 98% of the U.S. food supply. Similarly, nonnative tree species are grown in about 85% of the world's tree plantations. Some deliberately introduced species have also helped control pests.

The problem is that some introduced species have no natural predators, competitors, parasites, or pathogens to help control their numbers in their new habitats. Such species can reduce or wipe out populations of many native species and trigger ecological disruptions. Figure 12-9 shows some of the estimated 50,000 nonnative species deliberately or accidentally introduced into the United States that have caused ecological and economic harm.

After habitat loss and degradation, the deliberate or accidental introduction of nonnative species into ecosystems is the biggest cause of animal and plant extinctions. Nonnative species threaten almost half of the more than 1,260 endangered and threatened species in the United States and 95% of those in the state of Hawaii, according to the U.S. Fish and Wildlife Service. They are also blamed for about two-thirds of fish extinctions in the United States between 1900 and 2000. One example of a deliberately introduced plant species is the *kudzu* ("CUD-zoo") *vine*, which grows rampant in the southeastern United States (see Case Study below).

Deliberately introduced animal species have also caused ecological and economic damage. An example is the estimated 1 million European *wild (feral) boars*, or *hogs* (Figure 12-9), found in parts of Florida, Texas, and other states. They breed like rabbits, have razor-sharp tusks, compete for food with endangered animals, root up farm fields, and cause traffic accidents. Game and wildlife officials have had little success in controlling their numbers with hunting and trapping and say there is no way to stop them. Another example is the estimated 30 million *feral cats* and 41 million *outdoor pet cats* introduced into the United States; they kill about 568 million birds per year!

Case Study: Deliberate Introduction of the Kudzu Vine: Unintended Consequences

The rapidly growing kudzu vine has spread throughout much of the southern United States and is almost impossible to control.

In the 1930s the *kudzu vine* was imported from Japan and planted in the southeastern United States to help control soil erosion. It does control erosion. But it is so prolific and difficult to kill that it engulfs hillsides, gardens, trees, abandoned houses and cars, stream banks,

Figure 12-9 (facing page) **Threats to natural capital:** some nonnative species that have been deliberately or accidentally introduced into the United States.

Deliberately Introduced Species

Purple loosestrife

European starling

African honeybee
("Killer bee")

Nutria

Salt cedar
(Tamarisk)

Marine toad
(Giant toad)

Water hyacinth

Japanese beetle

Hydrilla

European wild boar
(Feral pig)

Accidentally Introduced Species

Sea lamprey
(attached to lake trout)

Argentina fire ant

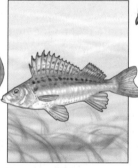
Brown tree snake

Eurasian ruffe

Common pigeon
(Rock dove)

Formosan termite

Zebra mussel

Asian long-horned beetle

Asian tiger mosquito

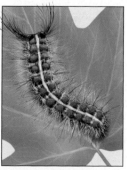
Gypsy moth larvae

Figure 12-10 Kudzu taking over a house and a truck. This vine can grow 5 centimeters (2 inches) per hour and is now found from east Texas to Florida and as far north as southeastern Pennsylvania and Illinois. Kudzu was deliberately introduced into the United States for erosion control, but it cannot be stopped by being dug up or burned. Grazing by goats and repeated doses of herbicides can destroy it, but goats and herbicides also destroy other plants, and herbicides can contaminate water supplies. Recently, scientists have found a common fungus (*Myrothecium verrucaria*) that can kill kudzu within a few hours, apparently without harming other plants.

patches of forest, and anything else in its path (Figure 12-10).

This vine, sometimes called "the vine that ate the South," has spread throughout much of the southern United States. It could spread as far north as the Great Lakes by 2040 if projected global warming occurs.

Kudzu is considered a menace in the United States. But Asians use a powdered kudzu starch in beverages, gourmet confections, and herbal remedies for a range of diseases. A Japanese firm has built a large kudzu farm and processing plant in Alabama and ships the extracted starch to Japan.

Although kudzu can engulf and kill trees, it could eventually help save trees from loggers. Research at the Georgia Institute of Technology indicates that kudzu may be used as a source of tree-free paper.

What Is the Role of Accidentally Introduced Species? Aliens Taking Over

A growing number of accidentally introduced species are causing serious economic and ecological damage.

Many unwanted nonnative invaders arrive from other continents as stowaways on aircraft, in the ballast wa-

ter of tankers and cargo ships, and as hitchhikers on imported products such as wooden packing crates. Cars and trucks can spread seeds of nonnative species imbedded in tire treads.

In the late 1930s, the extremely aggressive *Argentina fire ant* (Figure 12-9) was introduced accidentally into the United States in Mobile, Alabama. The ants may have arrived on shiploads of lumber or coffee imported from South America or by hitching a ride in the soil-containing ballast water of cargo ships.

These ants spawn and spread rapidly. Bother them, and up to 100,000 ants can swarm out of their nest to attack you with their painful and burning stings.

Without natural predators, fire ants have spread rapidly by land and water (they can float) throughout the South, from Texas to Florida and as far north as Tennessee and North Carolina (Figure 12-11). They are also found in Puerto Rico and recently have invaded California and New Mexico.

Wherever fire ants have gone, they have sharply reduced or wiped out up to 90% of native ant populations. Their extremely painful stings have killed deer fawns, birds, livestock, pets, and at least 80 people allergic to their venom. These ants have invaded cars and caused accidents by attacking drivers, damaged crops (such as soybeans, corn, strawberries, and potatoes), disrupted phone service and electrical power, caused fires by chewing through underground cables, and cost the United States an estimated $600 million per year. Their large mounds, which raise large boils on the land, can ruin crop fields, and their painful stings can make backyards uninhabitable.

Widespread pesticide spraying in the 1950s and 1960s temporarily reduced fire ant populations. But this chemical warfare hastened the advance of the rapidly multiplying fire ant by reducing populations of many native ant species. Worse, it promoted devel-

1918

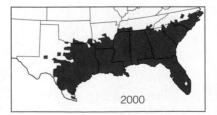

2000

Figure 12-11 Natural capital degradation: expansion of the *Argentina fire ant* in southern states, 1918–2000. This invader is also found in Puerto Rico, New Mexico, and California. (Data from U.S. Department of Agriculture)

The Termite from Hell

CASE STUDY

Forget killer bees and fire ants. The homeowner's nightmare is the Formosan termite (Figure 12-9). It is the most voracious, aggressive, and prolific of more than 2,000 known termite species.

These termites probably arrived on the U.S. mainland from Hawaii during or soon after World War II. They were stowaways in wooden packing materials on military cargo ships that docked in southern ports such as New Orleans, Louisiana, and Houston, Texas.

Formosan termites consume wood nine times faster than domestic termites. Their huge colonies can contain up to 73 million insects compared to about 1 million in the colonies of most native termites.

Domestic termite colonies have to be in contact with soil, which can be chemically treated around the outside of a building to reduce infestation. But Formosan termites

can establish a colony in an attic and in trees. This makes applying pesticides around the perimeter of a building virtually worthless in fighting these pests.

Over the past decade, the Formosan termite has caused more damage in New Orleans than hurricanes, floods, and tornadoes combined. Infestations affect as many as 90% of the houses and one-third of the oak trees in the city. The famous French Quarter has one of the world's most concentrated infestations.

Once confined to Louisiana, these termites have invaded at least a dozen other states, including Alabama, Florida, Mississippi, North and South Carolina, Texas, and California. They cause at least $1.1 billion in damage each year and the damage is increasing.

In New Orleans, the U.S. Department of Agriculture is using a variety of techniques in an attempt to control the species in a heavily infested 15-block area of the French

Quarter. They hope to develop techniques for dealing with these invaders elsewhere.

One method is to bait the termites by putting out blocks of wood to detect their presence. Once the termites are found in a block of wood it is replaced by another block that is baited with a pesticide toxic to termites. Termites feeding on this wood carry the pesticide back to their nest, where it is spread to other members of the nest.

Scientists have also found a cottony mold that can kill 100% of the termites in contact with it within a week. They are working on a method for producing the mold and using it as part of the bait approach.

Critical Thinking

What important ecological roles do termites play in nature? If the Formosan termite and other termite species could be eradicated (a highly unlikely possibility), would you favor doing this? Explain.

---

opment of genetic resistance to pesticides in the rapidly multiplying fire ants through natural selection. In other words, we helped wipe out their competitors and make them genetically stronger.

Researchers at the U.S. Department of Agriculture are experimenting with use of biological controls such as a tiny parasitic Brazilian fly and a pathogen imported from South America to reduce fire ant plantations. Tests are underway to see if sending in these stealth agents will work. But before widespread use of biological control agents, researchers must be sure they will not cause problems for native ant species or become pests themselves. Fire ants are not all bad. They prey on some other insect pests, including ticks and horse fly larvae. Another unplanned for harmful invader is the *Formosan termite* (Case Study, above).

Solutions: How Can We Reduce Threats from Nonnative Species? Prevention Pays

Prevention is the best way to reduce the threats from nonnative species because once they have arrived it is difficult and expensive to slow their spread.

Once a nonnative species gets established in an ecosystem, its wholesale removal is almost impossible—somewhat like trying to get smoke back into a chimney or trying to unscramble an egg. Thus the best way to limit the harmful impacts of nonnative species is to prevent them from being introduced and becoming established.

There are several ways to do this. One is to identify major characteristics that allow species to become successful invaders and the types of ecosystems that are vulnerable to invaders (Figure 12-12, p. 238). Such information can be used to screen out potentially harmful invaders. In 2003, marine ecologist Kevin Lafferty and his colleagues reported that many invading animal species gained a competitive advantage in their new homes because they leave behind about half of their native parasites and diseases.

We can also inspect imported goods that are likely to contain invader species. A third strategy is to identify major harmful invader species and pass international laws banning their transfer from one country to another, as is now done for endangered species.

Prevention and control can help. But many of these invaders are tiny, hard to detect, and able to

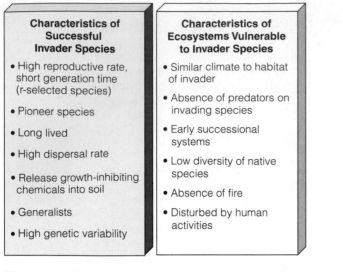

Characteristics of Successful Invader Species	Characteristics of Ecosystems Vulnerable to Invader Species
• High reproductive rate, short generation time (r-selected species)	• Similar climate to habitat of invader
• Pioneer species	• Absence of predators on invading species
• Long lived	• Early successional systems
• High dispersal rate	• Low diversity of native species
• Release growth-inhibiting chemicals into soil	• Absence of fire
• Generalists	• Disturbed by human activities
• High genetic variability	

Figure 12-12 Threats to natural capital: some general characteristics of successful invader species and ecosystems vulnerable to invading species.

breed rapidly. We also need to remind ourselves that the globalization of our economies and lifestyles is what helps bring these new and unwanted biological immigrants into countries throughout the world.

Case Study: Exploding Deer Populations in the United States: Should We Put Bambi on Birth Control?

In suburban areas we can trap and move deer somewhere else, put them on birth control, sterilize them, or not plant their favorite foods around houses.

A related problem is the explosion of deer populations in suburban areas. In this case we are the invader species. Americans have increasingly moved into the woods habitat of deer and provided them with flowers, garden crops, and other plants they like to eat.

Deer are edge species that like to live in the woods for security and venture into nearby fields, lawns, or gardens for food. Suburbanization has created an all-you-can-eat edge paradise for deer. The deer also raid nearby farmers' fields and orchards, threaten rare plants and animals in some areas, and spread Lyme disease (carried by deer ticks) to humans.

You may be surprised to learn that deer kill and injure more people each year in the United States than any other wild species. Collisions between deer and vehicles occur more than 1.5 million times each year, injure thousands of people, typically kill at least 200 people annually, and cause more than $1 billion in additional damages.

There are no easy answers to the deer population problem in the suburbs. Increased hunting—by changing hunting rules to allow killing of more female deer (does)—can cut down the overall deer population. But

this will have little effect on deer living near suburban areas because it is too dangerous to allow hunting there. Deer could also be trapped and moved somewhere else, but this is expensive and must be repeated every few years. And where are we going to take them?

Darts loaded with deer contraceptive could be fired into does each year to hold down the birth rate. But this is also expensive and must be repeated each year. One possibility is an experimental single-shot contraceptive vaccine that causes does to stop producing eggs for several years. Another approach, being tested by state biologists in Connecticut, is to trap dominant males and use chemical injections to sterilize them. However, both these approaches will require years of testing.

Meanwhile, if you live in the suburbs, expect deer to chow down on your shrubs, flowers, and garden plants. They have to eat every day like you do. You might consider not planting their favorite foods around your house.

12-5 EXTINCTION THREATS FROM POACHING AND HUNTING

How Serious Is the Illegal Taking or Killing of Wild Species? Making Big Money

Some protected species are killed for their valuable parts or are sold live to collectors.

Organized crime has moved into illegal wildlife smuggling because of the huge profits involved. Smuggling wildlife—including many endangered species—is the third largest and most lucrative illegal cross-border smuggling activity after arms and drugs. At least two-thirds of all live animals illegally smuggled around the world die in transit.

Poverty is one reason behind the illegal smuggling of wild species. Some poor people struggling to survive in areas with rich stores of wildlife kill or trap such species to make enough money to survive and feed their families. What would you do in the same situation? Others are professional poachers.

To poachers, a *live mountain gorilla* is worth $150,000, a *panda pelt* $100,000 (only about 1,500 pandas are left in the wild), a *chimpanzee* $50,000, and an *Imperial Amazon macaw* $30,000. A *rhinoceros horn* is worth as much as $28,600 per kilogram ($13,000 per pound) because of its use in dagger handles in the Middle East and as a fever reducer and alleged aphrodisiac in China—the world's largest consumer of wildlife—and other parts of Asia.

In 1950, an estimated 100,000 tigers existed in the world. Despite international protection, today fewer than 7,500 tigers remain in the wild (about 4,000 in India), mostly because of habitat loss and poaching for fur and bones. Bengal tigers are at risk because a tiger fur sells for $100,000 in Tokyo. With the body parts of a single tiger worth $5,000–20,000, it is not surprising that illegal hunting has skyrocketed, especially in India. Without emergency action, few or no tigers may be left in the wild within 20 years.

As commercially valuable species become endangered, their black market demand soars. This increases their chances of premature extinction from poaching. Most poachers are not caught. And the money they can make far outweighs the small risk of being caught, fined, or imprisoned.

Case Study: The Rising Demand for Bushmeat in Africa: Hungry People Trying to Survive

Rapid population growth in parts of Africa has increased the number of people hunting wild animals for food or for sale of their meat to restaurants.

Indigenous people in much of West and Central Africa have sustainably hunted wildlife for *bushmeat* as a source of food for centuries. But in the last two decades the level of hunting for bushmeat in some areas has skyrocketed.

In forests throughout West and Central Africa virtually every type of wild animal is being hunted by local people, frequently illegally, for food or to supply restaurants (Figure 12-13). The bushmeat trade is also increasing in Southeast Asia, the Caribbean, and Central and South America.

Figure 12-13 *Bushmeat*, such as this gorilla head, is consumed as a source of protein by local people in parts of West Africa and sold in the national and international marketplace. You can find bushmeat on the menu in Cameroon and the Congo in West Africa as well as in Paris, France, and Brussels, Belgium—often supplied by illegal poaching.

Killing wild animals for bushmeat has become more widespread for four reasons. *First*, an eightfold increase in Africa's population during the last century has led more people to survive by hunting wild animals. *Second*, logging roads have allowed miners, ranchers, and settlers to move into once inaccessible forests. *Third*, restaurants in many parts of the world have begun serving bushmeat dishes. *Fourth*, many people living in poverty find that selling wild animals or their valuable parts to collectors, meat suppliers, and poachers is a way to make enough money to survive.

So what is the big deal? After all, people have to eat. And for most of the time our species has been around we survived by hunting and gathering wild species.

The problem is that the current depletion of bushmeat species in some areas has ecological impacts. It has caused the local extinction of many animals in West Africa and has driven one species—Miss Waldron's red colobus monkey—to complete extinction. It is also a factor in greatly reducing gorilla, orangutan, and chimpanzee populations. For example, wealthy patrons of some restaurants regard gorilla meat as a source of status and power.

It also threatens forest carnivores such as crowned eagles and leopards by depleting their main prey species. The forest itself is also changed because of the decrease in seed-dispersing animals.

12-6 OTHER EXTINCTION THREATS

What Is the Role of Predator Control? If They Bother You, Kill Them

Killing predators that bother us or cause economic losses threatens some species with premature extinction.

People try to exterminate species that compete with them for food and game animals. For example, U.S. fruit farmers exterminated the Carolina parakeet around 1914 because it fed on fruit crops. The species was easy prey because when one member of a flock was shot, the rest of the birds hovered over its body, making themselves easy targets.

African farmers kill large numbers of elephants to keep them from trampling and eating food crops. Each year, U.S. government animal control agents shoot, poison, or trap thousands of coyotes, prairie dogs, wolves, bobcats, and other species that prey on livestock, on species prized by game hunters, and on crops or fish raised in aquaculture ponds. Since 1929 U.S. ranchers and government agencies have poisoned 99% of North America's prairie dogs because horses and cattle sometimes step into the burrows and break their legs. This has also nearly wiped out the endangered black-footed ferret (Figure 12-3; about 600 are left in the wild), which preyed on the prairie dog. This is another example of unintended consequences because of not understanding the connections between species.

What Is the Role of the Market for Exotic Pets and Decorative Plants? Are We Really Pet and Plant Lovers?

Legal and illegal trade in wildlife species used as pets or for decorative purposes threatens some species with extinction.

The global legal and illegal trade in wild species for use as pets is a huge and very profitable business. However, for every live animal captured and sold in the pet market, an estimated 50 others are killed.

About 25 million U.S. households have exotic birds as pets, 85% of them imported. More than 60 bird species, mostly parrots, are endangered or threatened because of this wild bird trade. According to the U.S. Fish and Wildlife Service, collectors of exotic birds may pay $10,000 for a threatened hyacinth macaw smuggled out of Brazil; however, during its lifetime a single macaw left in the wild might yield as much as $165,000 in tourist income. A 1992 study suggested that keeping a pet bird indoors for more than 10 years doubles a person's chances of getting lung cancer from inhaling tiny particles of bird dander.

Other wild species whose populations are depleted because of the pet trade include amphibians, reptiles, mammals, and tropical fish (taken mostly from the coral reefs of Indonesia and the Philippines). Divers commonly catch tropical fish by using plastic squeeze bottles of cyanide to stun them. For each fish caught alive, many more die. In addition, the cyanide solution kills the coral animals that create the reef, which is a center for marine biodiversity.

Things do not have to be this way. Pilai Poonswad decided to do something about poachers taking hornbills—large, beautiful, and rare birds—from a rain forest in Thailand. She visited the poachers in their villages and showed them why the birds are worth more alive than dead. Now she has a number of ex-poachers earning much more money than they did before by taking eco-tourists into the forest to see these magnificent birds. Because of their vested financial interest in preserving the hornbills, they also help protect them from poachers.

Some exotic plants, especially orchids and cacti, are endangered because they are gathered (often illegally) and sold to collectors to decorate houses, offices, and landscapes. The United States imports about 75% of all orchids and 99% of all live cacti sold each year. A collector may pay $5,000 for a single rare orchid, and a single rare mature crested saguaro cactus can earn cactus rustlers as much as $15,000.

In other words, collecting exotic pets and plants kills large numbers of them and endangers many of these species and others that depend on them. Are such collectors lovers or haters of the species they collect? Should we leave most exotic species in the wild?

What Are the Roles of Climate Change and Pollution? Speeding Up the Treadmill and Poisoning Species

Projected climate change and exposure to pollutants such as pesticides can threaten some species with premature extinction.

Most natural climate changes in the past have taken place over long periods of time. This gave species more time to adapt or evolve into new species to cope. But considerable evidence indicates that human activities such as greenhouse gas emissions and deforestation may bring about rapid climate change during this century. This could change the habitats many species and accelerate extinction of some species.

According to a 2000 study by the World Wildlife Fund, global warming could increase extinction by altering one-third of the world's wildlife habitats by 2100. This includes 70% of the habitat in high-altitude arctic and boreal biomes. Ten of the world's 17 penguin species are endangered or threatened mostly because of higher temperatures in their polar habitats. Another problem is that some species may not have enough time to adapt or migrate to areas with more favorable climates.

Pollution threatens populations and species in a number of ways. A major extinction threat is from the unintended effects of pesticides. According to the U.S. Fish and Wildlife Service, each year in the United States, pesticides kill about one-fifth of the country's beneficial honeybee colonies, more than 67 million birds, and 6–14 million fish. They also threaten about one-fifth of the country's endangered and threatened species.

12-7 PROTECTING WILD SPECIES: THE RESEARCH AND LEGAL APPROACH

How Can International Treaties Help Protect Endangered Species? Some Success

International treaties have helped reduce the international trade of endangered and threatened species, but enforcement is difficult.

Several international treaties and conventions help protect endangered or threatened wild species. One of the most far-reaching is the 1975 *Convention on Interna-*

tional Trade in Endangered Species (CITES). This treaty, now signed by 160 countries, lists some 900 species that cannot be commercially traded as live specimens or wildlife products because they are in danger of extinction. The treaty also restricts international trade of 29,000 other species because they are at risk of becoming threatened.

CITES has helped reduce international trade in many threatened animals, including elephants, crocodiles, and chimpanzees. However, the effects of this treaty are limited because enforcement is difficult and varies from country to country, and convicted violators often pay only small fines. Also, member countries can exempt themselves from protecting any listed species. And much of the highly profitable illegal trade in wildlife and wildlife products goes on in countries that have not signed the treaty.

The *Convention on Biological Diversity (CBD)*, ratified by 186 countries, legally binds signatory governments to reversing the global decline of biological diversity. The treaty requires each signatory nation to inventory its biodiversity and develop a *national conservation strategy*—a detailed plan for managing and preserving its biodiversity.

Implementing this treaty has been slow because some key countries such as the United States have not ratified it. Also, it has no severe penalties or other enforcement mechanisms.

How Can National Laws Help Protect Endangered Species? A Tough and Controversial Act in the United States

One of the world's most far-reaching and controversial environmental laws is the U.S. Endangered Species Act passed in 1973.

The United States controls imports and exports of endangered wildlife and wildlife products through two laws. One is the *Lacey Act of 1900.* It prohibits transporting live or dead wild animals or their parts across state borders without a federal permit.

The other is the *Endangered Species Act of 1973 (ESA),* which was amended in 1982, 1985, and 1988. It was designed to identify and legally protect endangered species in the United States and abroad. This act is probably the most far-reaching environmental law ever adopted by any nation, which has made it controversial. Canada and a number of other countries have similar laws.

The National Marine Fisheries Service (NMFS) is responsible for identifying and listing endangered and threatened ocean species, and the U.S. Fish and Wildlife Services (USFWS) identifies and lists all other endangered and threatened species. Any decision by either agency to add or remove a species from the list must be based on biological factors alone, without consideration of economic or political factors.

The act also forbids federal agencies to carry out, fund, or authorize projects that would jeopardize an endangered or threatened species or destroy or modify the critical habitat it needs to survive. However, in 2003 Congress exempted the Defense Department from this requirement and from the Marine Mammal Protection Act.

On private lands, fines up to $100,000 and one-year imprisonment can be imposed to ensure protection of the habitats of endangered species. This part of the act has been controversial because many of the listed species live totally or partially on private land.

The act also makes it illegal for Americans to sell or buy any product made from an endangered or threatened species. These species cannot be hunted, killed, collected, or injured in the United States, and this protection has been extended to threatened and endangered foreign species.

In 2003, however, the Bush administration proposed eliminating protection of foreign species, causing an uproar by conservationists. With this rule change, American hunters, circuses, and the pet industry could pay individuals or governments to kill, capture, and import animals that are on the brink of extinction in other countries.

X HOW WOULD YOU VOTE? Should the U.S. Endangered Species Act no longer protect threatened and endangered species in other countries? Cast your vote online at http://biology.brookscole.com/miller14.

Between 1973 and 2004, the number of U.S. species on the official endangered and threatened list increased from 92 to about 1,260 species—60% of them plants and 40% animals. According to a 2000 study by the Nature Conservancy, about one-third of the country's species are at risk of extinction, and 15% of all species are at high risk. This amounts to about 30,000 species, compared to the 1,260 species currently protected under the ESA. The study also found that many of the country's rarest and most imperiled species are concentrated in a few hot spots (Figure 12-14, p. 242).

What Are Critical Habitat Designations and Recovery Plans? How to Rebuild Populations

The Endangered Species Act requires protecting the critical habitat and developing a recovery plan for each listed species, but lack of funding and political opposition hinder these efforts.

The ESA generally requires the secretary of the interior to designate and protect the *critical habitat* needed for

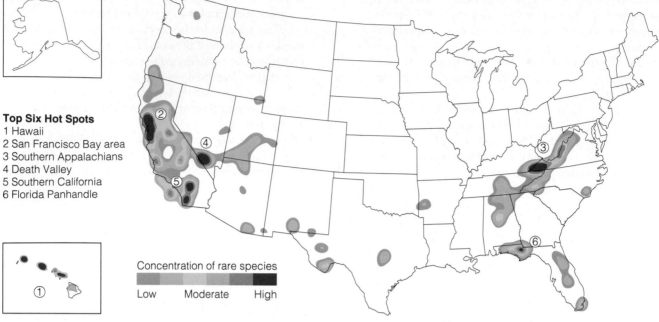

Top Six Hot Spots
1 Hawaii
2 San Francisco Bay area
3 Southern Appalachians
4 Death Valley
5 Southern California
6 Florida Panhandle

Concentration of rare species

Low   Moderate   High

Figure 12-14 Threatened natural capital: biodiversity hot spots in the United States. This map shows areas that contain the largest concentrations of rare and potentially endangered species. (Data from State Natural Heritage Programs, the Nature Conservancy, and Association for Biodiversity Information)

the survival and recovery of each listed species. So far critical habitats have been established for only about one-third of the species on the ESA list, mostly because of political pressure and a lack of funds. Beginning in 2001 the Bush administration stopped listing new species and designating critical habitat for listed species unless required by court order.

Getting listed is only half the battle. Next, the USFWS or the NMFS is supposed to prepare a plan to help the species recover. By 2004, final recovery plans had been developed and approved for 79% of the listed endangered or threatened species in the United States. Examples of successful recovery plans include those for the American alligator, the gray wolf, the bald eagle, and the peregrine falcon. *Bad news.* About half of current recovery plans exist only on paper, mostly because of political opposition and limited funds.

Should the Government Compensate Landowners When Endangered Species Decrease the Economic Value of Their Land? Private Versus Public Property Rights

There is controversy over whether the government should compensate private property owners who suffer financial losses when it restricts how they can use their land because of the presence of threatened or endangered species.

Critical habitats for more than half of the listed endangered and threatened species in the United States are found on private land. One controversy over the ESA is the political and legal issue of whether federal and state governments must compensate private property owners when government laws or regulations limit how the owners can use their property and decrease its financial value. Many people who own land on which threatened or endangered species live think the law goes too far and infringes on their property rights.

The Fifth Amendment of the U.S. Constitution gives the government the power, known as *eminent domain*, to force a citizen to sell property needed for a public good. For example, suppose the government needs some or all of land you own for a road. It can legally take your land but must reimburse you based on the land's fair market value.

The current controversy is over whether the Constitution requires the government to compensate you if instead of taking your property (a *physical taking*) it reduces its value by not allowing you to do certain things with it (a *regulatory taking*). For example, you might not be allowed to build on some or all of your property or to harvest trees from your land because these areas are habitats for an endangered species.

This can decrease the value of one's property. For example, a property owner in Travis, Texas, saw her land decrease in value from $830,000 to $38,000 because it contained two endangered bird species: the black-capped vireo and the golden-cheeked warbler. Should she be compensated for this loss by the federal government?

Most people would say yes. The problem is that requiring government compensation for regulatory takings would cost so much that it could cripple the financial ability of state and federal governments to protect the public good by enforcing existing or future environmental, land-use, health, and safety laws. The high costs involved could also hinder passage of any new environmental land-use, environmental, health, and safety laws because of lack of funds to compensate citizens and businesses. Some anti-environmentalists and elected officials opposed to the ESA push for requiring compensation for regulatory takings as a way to weaken or gut the ESA.

The controversy over regulatory takings is a continuation of the long-standing conflict over two types of freedoms: the right to be protected by law from the damaging actions of others and the right to do as one pleases without undue government interference. Achieving a balance between these conflicting types of individual rights is a difficult problem that governments have been wrestling with for centuries.

How Can We Encourage Private Landowners to Protect Endangered Species? Trying to Find Win-Win Compromises

Congress has amended the Endangered Species Act to help landowners protect endangered species on their land.

The ESA has encouraged some developers, timber companies, and other private landowners to avoid government regulation and possible loss of economic value by managing their land to reduce its use by endangered species. The National Association for Homebuilders, for example, has published practical tips for developers and other landowners to avoid ESA issues. Suggestions include planting crops, plowing fields between crops to prevent native vegetation and endangered species from occupying the fields, clearing forests, and burning or managing vegetation to make it unsuitable for local endangered species. Some landowners who discover small populations of endangered animals may also be tempted to use the "shoot, shovel, and shut up" solution.

Congress changed the ESA in several ways to help deal with these and other problems associated with the regulatory takings issue. In 1982, Congress amended the ESA to allow the secretary of the interior to use *habitat conservation plans (HCPs)*. They are designed to strike a compromise between the interests of private landowners and those of endangered and threatened species.

With an HCP, landowners, developers, or loggers are allowed to destroy some critical habitat or kill all or part of an endangered or threatened species population on private land in exchange for taking steps to protect that species. Such measures might include setting aside a part of the species' habitat as a protected area, protecting critical nesting sites, maintaining travel corridors for the species involved, paying to relocate the species to another suitable habitat, removing competitors and predators, or paying money to have the government buy suitable habitat elsewhere.

Once the plan is approved it cannot be changed, even if new data show that the plan cannot protect a species and help it recover. By 2004, some 400 HCPs had been developed.

Some wildlife conservationists support this approach because it can help head off use of evasive techniques and reduce political pressure to weaken or eliminate the ESA. However, there are two major criticisms of HCPs. One is that many of them have been approved without enough scientific evaluation of their effects on a species' recovery. Another problem is that many plans are political compromises that do not protect the species or make inadequate provisions for its recovery.

In 1999, the USFWS approved two new approaches for encouraging private landowners to protect threatened or endangered species. One is *safe harbor agreements* in which landowners voluntarily agree to take specified steps to restore, improve, or maintain habitat for threatened or endangered species located on their land. In return, landowners get technical help. They also receive government assurances that the natural resources involved will not face future restrictions once the agreement is over, and that after the agreement has expired landowners can return the property to its original condition without penalty.

Another method is the use of *voluntary candidate conservation agreements* in which landowners agree to take specific steps to help conserve a species whose population is declining but is not yet listed as endangered or threatened. Participating landowners receive technical help and assurances that no additional resource-use restrictions will be imposed on the land covered by the agreement if the species is listed as endangered or threatened in the future.

Should the Endangered Species Act Be Weakened? One Side of the Story

Some believe that the Endangered Species Act should be weakened or repealed because it has been a failure, tramples on private property rights, and hinders economic development of private land.

Since 1992 Congress has been debating the reauthorization of the ESA with proposals ranging from eliminating the act, to weakening it, to strengthening it.

The strongest opposition to the act is in the western United States where most public lands are located. Many westerners view the federal regulatory agencies managing these public lands as an enemy that wants to take away private property rights and restrict industries from having access to rich biological and mineral resources found on public lands. Many of these people believe that the ESA puts the rights and welfare of endangered plants and animals above those of people.

Many opponents of the ESA also contend that it has not worked and has caused severe economic losses by hindering development on private land that contains endangered or threatened species. Since 1995, efforts to weaken the ESA have included the following suggested changes:

▪ Make protection of endangered species on private land voluntary.

▪ Have the government compensate landowners if it forces them to stop using part of their land to protect endangered species (the regulatory takings issue).

▪ Make it harder and more expensive to list newly endangered species by requiring government wildlife officials to navigate through a series of hearings and peer review panels.

▪ Eliminate the need to designate critical habitats because developing and implementing a recovery plan is more important. And designating critical habitats is a lengthy, complex, and costly process that delays development of recovery plans. Also, dealing with lawsuits for failure to develop critical habitats takes up most of the limited funds for carrying out the ESA.

▪ Allow the secretary of the interior to permit a listed species to become extinct and to determine whether a species should be listed.

▪ Allow the secretary of the interior to give any state, county, or landowner permanent exemption from the law.

Other critics want do away with the ESA entirely. But since this is politically unpopular with the American public, most efforts are designed to weaken the act and reduce its already meager funding.

Should the Endangered Species Act Be Strengthened? The Other Side of the Story

According to most conservation biologists, the Endangered Species Act should be strengthened and modified to develop a new system to protect and sustain the country's biodiversity.

Most conservation biologists and wildlife scientists agree that the ESA has some deficiencies and needs to be simplified and streamlined. But they contend that the ESA has not been a failure (see Case Study, below).

They also contest the charge that the ESA has caused severe economic losses. Government records show that since 1979 only about 0.05% of the almost 200,000 projects evaluated by the USFWS have been blocked or canceled as a result of the ESA. In addition,

What Has the Endangered Species Act Accomplished?

CASE STUDY

Critics of the ESA call it an expensive failure because only 37 species have been removed from the endangered list. *Fourteen of these species recovered, 8 became extinct, and the rest were removed because of technical errors or discovery of new populations.*

Most biologists agree that the act needs strengthening and modification. But they disagree that the act has been a failure, for four reasons.

First, species are listed only when they are in serious danger of extinction. This is like setting up a poorly funded hospital emergency room that takes only the most desperate cases, often with little hope for recovery, and saying it should be shut down because it has not saved enough patients.

Second, it takes decades for most species to become endangered or threatened. Thus it usually takes decades to bring a species in critical condition back to the point where it can be removed from the list. Expecting the ESA—which has been in existence only since 1973—to quickly repair the biological depletion of many decades is unrealistic.

Third, the most important measure of the law's success is that the condition of almost 40% of the listed species is stable or improving. A hospital emergency room taking only the most desperate cases and then stabilizing or improving the condition of 40% of its patients would be considered an astounding success!

Fourth, the federal endangered species budget was only $58 million in 2005—about what the Department of Defense spends in a little more than an hour or 20¢ a year per

U.S. citizen. To supporters of the ESA, it is amazing that so much has been accomplished in stabilizing or improving the condition of almost 40% of the listed species on a shoestring budget.

Yes, the act can be improved and federal regulators have sometimes been too heavy-handed in enforcing it. But instead of gutting or doing away with this important act, biologists call for it to be strengthened and modified to help protect ecosystems and the nation's overall biodiversity.

Some critics say that only 20% of the endangered species are stable or improving. If correct, this is still an incredible bargain.

Critical Thinking

Should the budget for the Endangered Species Act be drastically increased? Explain.

the act allows for economic concerns. By law, a decision to list a species must be based solely on science. But once a species is listed, economic considerations can be weighed against species protection in protecting critical habitat and designing and implementing recovery plans. Also, private lands designated as critical habitats are not affected by the ESA unless the landowner plans an action that requires a federal permit.

Furthermore, the act authorizes a special cabinet-level panel, nicknamed the "God Squad," to exempt any federal project from having to comply with the act if the economic costs are too high.

Finally, the act allows the government to issue permits and exemptions to landowners with listed species living on their property and use habitat conservation plans, safe harbor agreements, and candidate conservation agreements to bargain with private landowners.

A study by the U.S. National Academy of Sciences recommended three major changes to make the ESA more scientifically sound and effective.

- Greatly increase the meager funding for implementing the act.

- Develop recovery plans more quickly.

- When a species is first listed, establish a core of its survival habitat as a temporary emergency measure that could support the species for 25–50 years.

Some suggest concentrating limited ESA funds on protecting species that have the best chances of surviving and that play important ecological and economic roles. Some say this is unethical because all species should have a right to exist. But proponents argue that because of limited funding we are already deciding which species to save and that this is a better use of limited funds. What do you think?

Most biologists and wildlife conservationists believe the United States should modify the act to emphasize protecting and sustaining biological diversity and ecological functioning rather than attempting to save individual species. This new ecosystems approach would follow three principles:

- Find out what species and ecosystems the country has.

- Locate and protect the most endangered ecosystems and species within such systems.

- Put more emphasis on preventing species from becoming threatened and ecosystems from becoming degraded.

- Provide private landowners who agree to help protect endangered ecosystems with significant financial incentives (tax breaks, write-offs) and technical help.

X HOW WOULD YOU VOTE? Should the Endangered Species Act be modified to protect the nation's overall biodiversity? Cast your vote online at http://biology.brookscole.com/miller14.

12-8 PROTECTING WILD SPECIES: THE SANCTUARY APPROACH

What Is the Role of Wildlife Refuges and Other Protected Areas? Protect the Homes of Species in Trouble

The United States has set aside 542 federal refuges for wildlife, but many refuges are suffering from environmental degradation.

In 1903, President Theodore Roosevelt established the first U.S. federal wildlife refuge at Pelican Island, Florida. Since then the National Wildlife Refuge System has grown to 542 refuges. Since 1995 visits to national parks have leveled off while those to wildlife refuges have almost doubled. More than 35 million Americans visit these refuges each year to hunt, fish, hike, or watch birds and other wildlife.

More than three-fourths of the refuges are concentrated along major bird migration corridors or *flyways* (Figure 12-15, p. 246). They serve as vital wetland sanctuaries for protecting millions of migratory waterfowl as they journey north and south each year to find food, suitable climate, and other conditions necessary for reproduction.

About one-fifth of U.S. endangered and threatened species have habitats in the refuge system, and some refuges have been set aside for specific endangered species. These have helped Florida's key deer, the brown pelican, and the trumpeter swan to recover.

Conservation biologists call for setting aside more refuges to help protect endangered plants. They also urge Congress and state legislatures to allow abandoned military lands that contain significant wildlife habitat to become national or state wildlife refuges.

Bad news. According to a General Accounting Office study, activities considered harmful to wildlife occur in nearly 60% of the nation's wildlife refuges. A 2002 study by the National Wildlife Refuge Association found that invasions by nonnative species are wreaking havoc on many of the nation's wildlife refuges. Also, too much hunting and fishing and use of powerboats and off-road vehicles can take their toll on wildlife populations in heavily used refuges.

Can Gene Banks, Botanical Gardens, and Farms Help Save Most Endangered Plant Species? Important but Limited Solutions

Establishing gene banks and botanical gardens and using farms to raise threatened species can help protect species from extinction, but these options lack funding and storage space.

Gene or *seed banks* preserve genetic information and endangered plant species by storing their seeds in refrigerated, low-humidity environments. The world's more

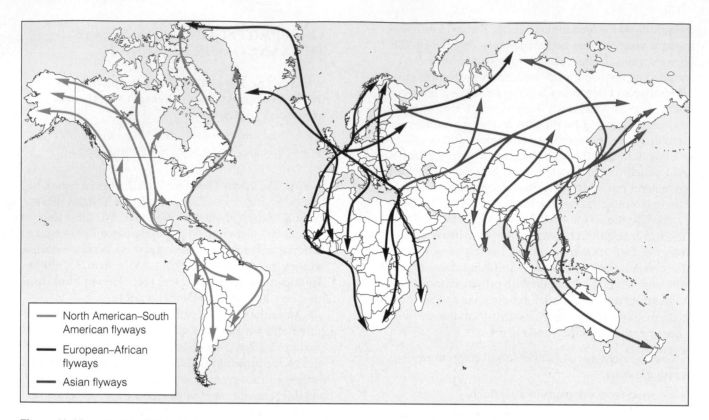

Figure 12-15 Natural capital: major *flyways* used by migratory birds, mostly waterfowl. Each route has a number of subroutes. Some countries along such flyways have entered into agreements and treaties to protect crucial habitats, especially wetlands, needed by such species, both along their migration routes and at each end of their journeys.

Legend:
— North American–South American flyways
— European–African flyways
— Asian flyways

than 100 seed banks have focused on storing seeds of the approximately 100 plant species that provide us with food—with just 14 plants providing about 90% of the calories in our food. Some banks are beginning to store seeds for a wider range of species that may be threatened with extinction or a loss of genetic diversity. Scientists urge the establishment of many more such banks, especially in developing countries.

But seed banks are expensive to operate and can be destroyed by accidents. Also, because stored seeds do not evolve, they may not survive when used in the future.

The world's 1,600 *botanical gardens* and *arboretums* contain living plants, representing almost one-third of the world's known plant species. However, they contain only about 3% of the world's rare and threatened plant species.

Botanical gardens also help educate an estimated 150 million visitors a year about the need for plant conservation. But these sanctuaries have too little storage capacity and too little funding to preserve most of the world's rare and threatened plants.

We can take pressure off some endangered or threatened species by raising them on *farms* for com-

mercial sale. One example is the use of farms in Florida to raise alligators for meat and hides. Another example is *butterfly farms* in Papua New Guinea, where many butterfly species are threatened by habitat destruction and fragmentation, commercial overexploitation, and environmental degradation.

Can Zoos and Aquariums Help Protect Most Endangered Animal Species? Important but Expensive and Limited

Zoos and aquariums can help protect endangered animal species, but lack funding and storage space.

Zoos, aquariums, game parks, and animal research centers are being used to preserve some individuals of critically endangered animal species, with the long-term goal of reintroducing the species into protected wild habitats. Two techniques for preserving endangered terrestrial species are egg pulling and captive breeding. *Egg pulling* involves collecting wild eggs laid by critically endangered bird species and then hatching them in zoos or research centers. In *captive breeding,* some or all of the wild individuals of a critically endangered species are captured for breeding in captivity, with the aim of reintroducing the offspring into the wild.

Other techniques for increasing the populations of captive species include artificial insemination, surgical implantation of eggs of one species into a surrogate mother of another species (embryo transfer), use of incubators, and cross-fostering (in which the young of a rare species are raised by parents of a similar species). Scientists also use computer databases of the family lineages of species in zoos and DNA analysis to match individuals for mating—a computer dating service for zoo animals—and to prevent genetic erosion through inbreeding.

Proponents urge zoos and wildlife managers to collect and freeze cells of endangered species for possible cloning. They believe that such miniature *frozen zoos* could play a role in bringing back depleted species in the future.

The ultimate goal of captive breeding programs is to build up populations to a level where they can be reintroduced into the wild. However, before conservation biologists attempt a reintroduction they study the factors that originally caused the species to become endangered, whether these factors still exist, and whether there is enough suitable habitat available.

After more than two decades of captive breeding efforts, only a handful of endangered species have been returned to the wild. Examples shown in Figure 12-3 include the black-footed ferret, California condor, Arabian oryx, and golden lion tamarin. Most reintroductions fail because of lack of suitable habitat, inability of individuals bred in captivity to survive in the wild, or renewed overhunting or capture of some returned species.

Lack of space and money limits efforts to maintain populations of endangered animal species in zoos and research centers. The captive population of each species must number 100–500 individuals to avoid extinction through accident, disease, or loss of genetic diversity through inbreeding. Recent genetic research indicates that 10,000 or more individuals are needed for an endangered species to maintain its capacity for biological evolution.

According to one estimate, using all the space in the 201 accredited U.S. zoos for captive breeding could sustain only about 100 large animal species on a long-term basis. Thus the major conservation role of zoos will be to help educate the public about the ecological importance of the species they display and the need to protect habitat.

Public aquariums that exhibit unusual and attractive fish and some marine animals such as seals and dolphins also help educate the public about the need to protect such species. In the United States, more than 35 million people visit aquariums each year. However, public aquariums have not served as effective gene banks for endangered marine species, especially marine mammals that need large volumes of water.

Instead of seeing zoos and aquariums as sanctuaries, some critics see most of them as prisons for once wild animals. They also contend that zoos and aquariums foster the false notion that we do not need to preserve large numbers of wild species in their natural habitats.

Some people criticize zoos and aquariums for putting on shows with animals wearing clothes, riding bicycles, or performing tricks. They see this as fostering the idea that the animals are there primarily to entertain us by doing things people do and in the process raising money for their keepers.

Conservation biologists point out that zoos, aquariums, and botanical gardens, regardless of their benefits and drawbacks, are not biologically or economically feasible solutions for most of the world's current endangered species and the much larger number expected over the next few decades.

12-9 RECONCILIATION ECOLOGY

What Is Reconciliation Ecology? Rethinking Conservation Strategy

Reconciliation ecology involves finding ways to share the places we dominate with other species.

In 2003, ecologist Michael L. Rosenzweig wrote the book *Win-Win Ecology: How Earth's Species Can Survive in the Midst of Human Enterprise* (Oxford University Press). He strongly supports the eight-point program of Edward O. Wilson to help save the earth's natural habitats by establishing and protecting nature reserves (p. 221). He also supports the species protection strategies discussed in this chapter.

But he contends that in the long run these approaches will fail for two reasons. One is that current reserves are devoted to saving only about 7% of nature. To Rosenzweig the real challenge is to help sustain wild species in the human-dominated portion of nature that makes up 93% of the planetary ecological "cake"

The other problem is that setting aside funds and refuges and passing laws to protect endangered and threatened species are essentially desperate attempts to save species that are in deep trouble. This can help a few species, but the real challenge is learning how to keep species from getting to such a point in the first place.

Rosenzweig suggests that we develop a new form of conservation biology called **reconciliation ecology.** It is the science of inventing, establishing, and maintaining new habitats to conserve species diversity in places where people live, work, or play. In other words, we need to learn how to share the spaces we dominate with other species.

How Can We Implement Reconciliation Ecology? Observe, Be Creative, and Cooperate with Your Neighbors

Some people are finding creative ways to practice reconciliation ecology in their neighborhoods and cities.

Practicing reconciliation ecology begins by looking at the habitats we prefer. Given a choice, most people prefer a grassy and fairly open habitat with a few scattered trees. We also like water and prefer to live near a stream, lake, river, or ocean. We also love flowers.

The problem is that most species do not like what we like or cannot survive in the habitats we prefer. No wonder so few of them live with us.

So what do we do? Reconciliation ecology goes beyond efforts to attract birds to backyards. For example, providing a self-sustaining habitat for a butterfly species may require 20 or so neighbors to band together. Doing this for an insect-eating bat species could help keep down mosquitoes and other pesky insects in a neighborhood.

The safe harbor agreements and voluntary candidate conservation agreements that are part of the Endangered Species Act are examples of reconciliation ecology in action. They reward responsible stewardship by private landowners who take voluntary actions to help protect endangered or threatened species

or species that may soon become threatened. For example, people have worked together to help preserve bluebirds within human dominated habitats (Case Study, below).

Another form of restoration ecology involves replacing some monoculture yards in neighborhoods with diverse yards using plant species adapted to local climates that are selected to attract certain species. This would make neighborhoods more biologically diverse and interesting, keep down insect pests, and require less use of noisy and polluting lawnmowers.

Communities could have contests and awards for people designing the most biodiverse and species-friendly yards and gardens. Signs could describe the type of ecosystem being mimicked and the species being protected as a way to educate and encourage experiments by other people.

In Berlin, Germany, people have planted gardens on many large rooftops. These can be designed to support a variety of species by varying the depth and type of soil and their exposure to sun. Such roofs also save energy by providing insulation, help cool cities, and conserve water by reducing evapotranspiration. Reconciliation ecology proponents call for a global campaign to use the roofs of the world to help sustain biodiversity.

San Francisco's Golden Gate Park is a 410-hectare (1,012-acre) oasis of gardens and trees in the midst of a large city. It is a good example of reconciliation ecology because it was designed and planted by humans who transformed it from a system of sand dunes.

CASE STUDY

Using Reconciliation Ecology to Protect Bluebirds?

Let me tell you a story about bluebirds. *Bad news.* Populations of bluebirds in much of the eastern United States are declining

There are two reasons. One is that these birds nest in tree holes of a certain size. Dead and dying trees once provided plenty of these holes. But today timber companies often cut down all of the trees, and homeowners manicure their property by removing dead and dying trees.

A second reason is that two aggressive, abundant, and nonnative bird species—starlings and house sparrows—also like to nest in tree holes and take them away from

bluebirds. To make matters worse, starlings eat the blueberries the bluebirds need to survive during the winter.

Good news. People have come up with a creative way to help save the bluebird. They have designed nest boxes with holes large enough to accommodate bluebirds but too small for starlings. They also found that house sparrows like shallow boxes, so they made the bluebird boxes deep enough to make them unattractive nesting sites for the sparrows.

In 1979, the North American Bluebird Society was founded to spread the word and encourage people to use bluebird boxes on their property and to keep house

cats away from nesting bluebirds. Now bluebird numbers are building back up.

Properly designed nest boxes are also being used to boost the population of red-cockaded woodpeckers on Florida's Elgin Air Force Base. Nest boxes in swamplands have done the same thing for America's wood ducks.

Restoration ecology works! Perhaps you might want to consider a career in this exciting new field.

Critical Thinking

See if you can come up with a reconciliation project to help protect threatened bird or other species in your neighborhood or on the grounds of your school.

- Do not but furs, ivory products, and other materials made from endangered or threatened animal species.

- Do not buy wood and paper products produced by cutting remaining old-growth forests in the tropics.

- Do not buy birds, snakes, turtles, tropical fish, and other animals that are taken from the wild.

- Do not buy orchids, cacti, and other plants that are taken from the wild.

Figure 12-16 What can you do? Ways to help premature extinction of species.

The Department of Defense controls large areas of land in the United States. Rosenzweig and other reconciliation ecologists believe that some of this land could serve as laboratories for developing and testing reconciliation ecology ideas.

Some college campuses and schools might also serve as reconciliation ecology laboratories. How about your campus or school?

In this chapter, we have seen that protecting the terrestrial species that make up part of the earth's biodiversity from premature extinction is a difficult, controversial, and challenging responsibility. Figure 12-16 lists some things you can do to help prevent the premature extinction of species.

We know what to do. Perhaps we will act in time.

EDWARD O. WILSON

CRITICAL THINKING

1. How do **(a)** population growth, **(b)** poverty, and **(c)** climate change affect biodiversity?

2. Discuss your gut-level reaction to the following statement: "Eventually all species become extinct. Thus it does not really matter that the passenger pigeon is extinct and that the whooping crane, the California condor, and the world's remaining rhinoceros and tiger species are endangered mostly because of human activities." Be honest about your reaction, and give arguments for your position.

3. **(a)** Do you accept the ethical position that each species has the inherent right to survive without human interference, regardless of whether it serves any useful purpose for humans? Explain. Would you extend this right to *Anopheles* mosquito species, which transmit malaria, and

to infectious bacteria? **(b)** Do you believe each individual of an animal species has an inherent right to survive? Explain. Would you extend such rights to individual plants and microorganisms and to tigers that have killed people? Explain.

4. Explain why you agree or disagree with **(a)** using animals for research, **(b)** keeping animals captive in a zoo or aquarium, and **(c)** killing surplus animals produced by a captive-breeding program in a zoo when no suitable habitat is available for their release.

5. What would you do if **(a)** your yard and house are invaded by fire ants, **(b)** you find bats flying around your yard at night, and **(c)** deer invade your yard and eat your shrubs and vegetables?

6. Which of the following statements best describes your feelings toward wildlife? **(a)** As long as it stays in its space, wildlife is OK. **(b)** As long as I do not need its space, wildlife is OK. **(c)** I have the right to use wildlife habitat to meet my own needs. **(d)** When you have seen one redwood tree, fox, elephant, or some other form of wildlife, you have seen them all, so lock up a few of each species in a zoo or wildlife park and do not worry about protecting the rest. **(e)** Wildlife should be protected.

7. List your three favorite wild species. Examine why they are your favorites. Are they cute and cuddly looking, like the giant panda and the koala? Do they have humanlike qualities, like apes or penguins that walk upright? Are they large, like elephants or blue whales? Are they beautiful, like tigers and monarch butterflies? Are any of them plants? Are any of them species such as bats, sharks, snakes, or spiders that most people are afraid of? Are any of them microorganisms that help keep you alive? Reflect on what your choice of favorite species tells you about your attitudes toward most wildlife.

8. Environmental groups in a heavily forested state want to restrict logging in some areas to save the habitat of an endangered squirrel. Timber company officials argue that the well-being of one type of squirrel is not as important as the well-being of the many families affected if the restriction causes them to lay off hundreds of workers. If you had the power to decide this issue, what would you do and why? Can you come up with a compromise?

9. Explain why some developers and extractors of resources on public land oppose the development of national databases of the species found in a particular country.

10. Congratulations! You are in charge of preventing the premature extinction of the world's existing species from human activities. What would be the three major components of your program to accomplish this goal?

PROJECTS

1. Make a log of your own consumption of all products for a single day. Relate your level and types of consumption to the **(a)** decline of wildlife species and **(b)** increased

destruction, degradation, and fragmentation of wildlife habitats in the United States (or the country where you live) and in tropical forests.

2. Identify examples of habitat destruction or degradation in your community that have had harmful effects on the populations of various wild plant and animal species. Develop a management plan for rehabilitating these habitats and species.

3. Choose a particular animal or plant species that interests you and use the library or the Internet to find out **(a)** its numbers and distribution, **(b)** whether it is threatened with extinction, **(c)** the major future threats to its survival, **(d)** actions that are being taken to help sustain this species, and **(e)** a type of reconciliation ecology that might be useful in sustaining this species.

4. Work with your classmates to develop an experiment in reconciliation ecology for your campus.

5. Use the library or the Internet to find bibliographic information about *Aldo Leopold* and *Edward O. Wilson,* whose quotes appear at the beginning and end of this chapter.

6. Make a concept map of this chapter's major ideas, using the section heads, subheads, and key terms (in boldface). Look on the website for this book for information about making concept maps.

LEARNING ONLINE

The website for this book contains study aids and many ideas for further reading and research. They include a chapter summary, review questions for the entire chapter, flash cards for key terms and concepts, a multiple-choice practice quiz, interesting Internet sites, references, and a guide for accessing thousands of InfoTrac® College Edition articles. Log on to

http://biology.brookscole.com/miller14

Then click on the Chapter-by-Chapter area, choose Chapter 12, and select a learning resource.

CASE STUDY

A Biological Roller Coaster Ride in Lake Victoria

Lake Victoria, shared by Kenya, Tanzania, and Uganda in East Africa, is the world's second largest freshwater lake (Figure 13-1, left). It has been in ecological trouble for more than two decades.

Until the early 1980s, Lake Victoria had more than 500 species of fish found nowhere else. About 80% of them were small algae-eating fishes known as cichlids (pronounced "SIK-lids"), each with a slightly different ecological niche.

Since 1980 some 200 of the cichlid species have become extinct, and some of those that remain are in trouble.

Four factors caused this dramatic loss of aquatic biodiversity. *First,* there was a large increase in the population of the Nile perch (Figure 13-1, right). This fish was deliberately introduced into the lake during the 1950s and 1960s to stimulate local economies and the fishing industry, which exports large amounts of the fish to several European countries. The population of this large, prolific, and ravenous fish exploded by displacing the cichlids and by 1985 had wiped out many of them.

Also, the native people who depended on the cichlids for protein cannot afford the perch, and the mechanized fishing industry has put most small-scale fishers and fish vendors out of business. This has increased poverty and protein malnutrition.

Second, in the 1980s the lake began experiencing frequent algal blooms because of nutrient runoff from surrounding farms, deforested land, untreated sewage, and declines in the populations of the algae-eating cichlids. This greatly decreased oxygen levels in the lower depths of the lake and drove remaining native cichlids and other fish species to shallower waters, where they are more vulnerable to fishing nets. The turbid water caused by eutrophication also made it hard for female cichlids to select mates by color and can lead to the extinction of some species.

Third, since 1987 the nutrient-rich lake has been invaded by the water hyacinth (Figure 12-9, p. 235). This rapidly growing plant carpeted large areas of the lake, blocked sunlight, deprived fish and plankton of oxygen, and reduced the diversity of important aquatic plant species. *Good news.* The population of water hyacinths has been reduced sharply by introducing two weevils for biological control and mechanical removal at strategic locations.

Fourth, the Nile perch now shows signs of being overfished. This may allow a gradual return of some of the remaining cichlids.

This ecological story shows the dynamics of large aquatic systems and illustrates that we can never do just one thing when we intrude into an ecosystem of connected species. There are always unintended consequences.

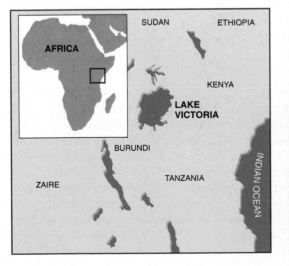

Figure 13-1 Although the Nile perch (right) is a fine food fish, it has played a key role in a major loss of biodiversity in East Africa's Lake Victoria (left).

The coastal zone may be the single most important portion of our planet. The loss of its biodiversity may have repercussions far beyond our worst fears.

G. CARLETON RAY

This chapter addresses the following questions:

- What is aquatic biodiversity, and what is its economic and ecological importance?

- How are human activities affecting aquatic biodiversity?

- How can we protect and sustain marine biodiversity?

- How can we manage and sustain the world's marine fisheries?

- How can we protect, sustain, and restore wetlands?

- How can we protect, sustain, and restore lakes, rivers, and freshwater fisheries?

13-1   AN OVERVIEW OF AQUATIC BIODIVERSITY

What Do We Know about the Earth's Aquatic Biodiversity? Some but Not Nearly Enough

We know fairly little about the biodiversity of the world's marine and freshwater systems.

The world's interconnected oceans cover 71% of the planet's surface. The oceans support a variety of species at different depths (Figure 7-6, p. 131, and Figure 13-2).

About 63% of the roughly 25,000 known fish species exist in marine systems—50% in coastal waters, 12% in the deep sea, and 1% in open ocean. The remaining 37% live in freshwater systems.

Freshwater aquatic systems also contain a variety of visible plant and animal species (Figure 13-3, p. 254) and microscopic species. Many lakes contain an assortment of species found in different layers (Figure 7-16, p. 139). As they flow from their mountain (headwater) streams to the ocean, river systems have a variety of ecological habitats that support different aquatic species (Figure 7-18, p. 140).

We have explored only about 5% of the earth's global ocean and know fairly little about its biodiversity and how it works. We also know fairly little about freshwater biodiversity. Scientific investigation of poorly understood marine and freshwater aquatic systems is a *research frontier* whose study could result in immense ecological and economic benefits.

In 2003, a three-year study by the Pew Oceans Commission found that the coastal waters of the United States are in deep trouble and that laws protecting them need fundamental reforms. Here are four of the commission's major recommendations. *First,* pass a National Ocean Policy Act that commits the country to protect, sustain, and restore the living oceans. *Second,* double the federal budget for ocean research. *Third,* base fisheries management on preserving aquatic ecosystems and habitats rather than relying mostly on catch limits for individual species. *Fourth,* set up a network of marine reserves, linked by protected corridors, to help protect fish breeding and nursery grounds.

What Are Some General Patterns of Marine Biodiversity? Abundant Life near the Water's Edge and in the Deep

Coral reefs, coastal areas, and the ocean bottom are centers of marine biodiversity.

Despite the lack of knowledge about overall marine biodiversity, scientists have established several general patterns. *First,* the greatest marine biodiversity occurs in *coral reefs* (Figure 7-12, p. 136), *estuaries,* and the *deep-ocean floor. Second,* biodiversity is higher near coasts than in the open sea because of the greater variety of producers, habitats, and nursery areas in coastal areas.

Third, biodiversity is higher in the bottom (benthic) region of the ocean than in the surface (pelagic) region because of the greater variety of habitats and food sources on the ocean bottom. *Fourth,* the lowest marine biodiversity probably is found in the middle depths of the open ocean. Can you explain why?

Why Should We Care about Aquatic Biodiversity? Keeping Us Alive and Supporting Our Economies

The world's marine and freshwater systems provide important ecological and economic services.

Marine systems provide a variety of important ecological and economic services (Figure 7-5, p. 130). Globally, we get about 6% of our total protein and almost a fifth of our animal protein from marine fish and shellfish.

Seaweed and other organisms provide chemicals used in cosmetics and pharmaceuticals worth $400 million per year. Chemicals from several types of al-

Kelp

Hogfish

Cobia

Carrageen

Pacific sailfish

Batfish

Yellow jack

Moray

Red snapper

Red algae

Striped drum

Angelfish

Bladder kelp

Sea lettuce

Orange roughy

Chinook salmon

Devilfish

Great barracuda

Laminaria

Porcupine fish

Sockeye salmon

Grouper

Dulse

Chilean sea bass

Figure 13-2 Natural capital: marine biodiversity. Some ocean inhabitants.

gae, sea anemones (Figure 8-10b, p. 155), sponges, and mollusks have antibiotic and anticancer properties. Anticancer chemicals have also been extracted from porcupine fish, puffer fish, and shark liver. We use chemicals from seaweeds and octopuses to treat hypertension and coral material to reconstruct our bones. These are only a few of many examples.

Freshwater systems also provide important ecological and economic services (Figure 7-15, p. 138). Although lakes, rivers, and wetlands occupy only 1% of the earth's surface, they provide ecological and economic services worth trillions of dollars per year.

13-2 HUMAN IMPACTS ON AQUATIC BIODIVERSITY

How Has Habitat Loss and Degradation Affected Marine Biodiversity? Our Large Aquatic Footprints

Human activities have destroyed or degraded a large proportion of the world's coastal wetlands, coral reefs, mangroves, and ocean bottom.

The greatest threat to the biodiversity of the world's oceans is loss and degradation of habitats. Here are four

Figure 13-3 Natural capital: freshwater biodiversity. Some inhabitants of freshwater rivers and lakes.

examples. *First,* during the last century we lost about half of the world's coastal wetlands. This can decrease fish catches in coastal waters because coastal wetlands and marshes provide essential spawning, feeding, and nursery areas for major commercial fish species.

Second, more than one-fourth of the world's diverse coral reefs have been severely damaged, mostly by human activities (Figure 7-13, p. 137). By 2050 another 70% of the world's coral reefs may be severely damaged or eliminated.

Third, more than a third of the world's original mangrove forest swamps have disappeared, mostly because of clearing for coastal development, growing

crops, and aquaculture shrimp farms. Such activities threaten many of the world's remaining mangrove systems.

Fourth, many bottom habitats are being degraded and destroyed by dredging operations and trawler boats, which like giant submerged bulldozers drag huge nets weighted down with heavy chains and steel plates over ocean bottoms to harvest bottom fish and shellfish. Each year thousands of trawlers scrape and disturb an area of ocean bottom equal to the combined size of Brazil and India and about 150 times larger than the area of forests clear-cut each year. Recovery in heavily trawled areas rarely is possible because of repeated scraping. Slow-growing, long-lived corals, sponges, and fish are particularly vulnerable to trawling. In 2004, some 1,134 scientists signed a statement urging the United Nations to declare a moratorium on bottom trawling on the high seas.

How Have Human Activities Affected Marine Fish Populations and Species? Gone Fishing, Fish Gone

About three-fourths of the world's commercially valuable marine fish species are overfished or fished near their limits.

Studies indicate that about three-fourths of the world's 200 commercially valuable marine fish species (40% in U.S. waters) are either overfished or fished to their estimated sustainable yield. Overfishing is the greatest threat to populations of fish that live in surface waters. Populations of bottom-dwelling fish are affected by a combination of overfishing and disruption of habitat by trawler fishing.

In most cases, overfishing leads to *commercial extinction.* This is usually only a temporary depletion of fish stocks, as long as depleted areas and fisheries are allowed to recover. But this is changing. Today fish are hunted throughout the world's oceans by a global fleet of millions of fishing boats—some of them longer than a football field. These fleets, most supported by government subsidies, use sonar, satellite global positioning systems, and aircraft to find fish. Then they catch them by deploying gigantic nets or lines containing many thousands of hooks that can stretch as far as 80 kilometers (50 miles). Modern industrial fishing can cause 80% depletion of a target fish species in only 10–15 years.

One result of the increasingly efficient global hunt for fish is that *big fish in many populations of commercially valuable species are becoming scarce.* In 2003, fishery scientists Ransom Myers and Boris Worm looked at fishing data for 13 commercial fisheries since 1952. Their data indicate that during the last 45 years the abundance of large open-ocean fish such as swordfish, marlins, tunas, and sharks and bottom-dwelling groundfish such as cod plummeted by 90%! A 2004 study by Jeffrey Hutchings and John Reynolds found that 230 populations of marine fish have suffered an 83% drop in breeding population size from known historic levels.

Many depleted species, like the bottom-dwelling North Atlantic cod, may never recover because too much of their habitat has been destroyed or degraded or there are too few survivors to find mates. For example, after stocks had dropped by 97–99% since the early 1960s, Canada closed its cod fishery, putting thousands out of work. After a decade there is no sign of recovery.

The smaller fish are next. As the fishing industry has depleted its most valuable and larger species, it has begun working its way down marine food webs to exploit smaller and faster-growing varieties at lower trophic levels (Figure 13-4). If this process continues, it will begin to unravel food webs, disrupt marine ecosystems, and hinder the recovery of fish feeding at higher trophic levels because the species they eat have also been overfished.

If this happens, the most abundant remaining species will be jellyfish, barnacles, and plankton. If we keep vacuuming the seas, McDonald's may begin serving barnacle burgers instead of fish sandwiches.

Most fishing boats are after one or a small number of commercially valuable species. However, their gigantic nets and incredibly long lines of hooks also catch nontarget species, called *bycatch.* Almost one-third of the world's annual fish catch consists of such species that are thrown overboard dead or dying. In addition to wasting potential sources of food, this can deplete the populations of bycatch species that play important ecological roles in oceanic food webs.

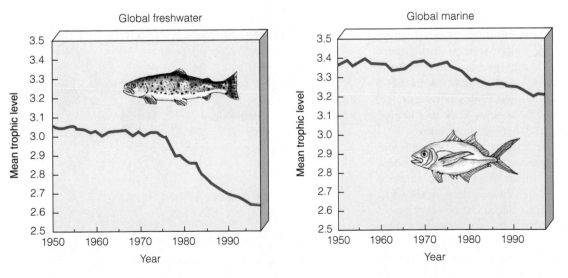

Figure 13-4 Mean trophic levels of the global marine (right) and freshwater (left) fish catch have declined since 1950. (Daniel Pauly)

To sum it up: *Many species are overfished, big fish are becoming scarce, smaller fish are next, and we throw away 30% of the fish we catch.*

How Have Human Activities Affected the Survival of Aquatic Species? Many Extinctions on the Horizon

Marine and aquatic fish are threatened with premature extinction by human activities more than any other group of species.

Human activities such as overfishing, habitat destruction and degradation, invasions by nonnative species, and pollution are endangering a number of aquatic species. According to marine biologists, at least 1,200 marine species have become extinct in the past few hundred years, and many thousands of additional marine species could disappear during this century. Indeed, *fish are threatened with extinction by human activities more than any other group of species* (Figure 12-5, p. 228).

Also, according to the UN Food and Agriculture Organization and the World Wildlife Fund, at least a fifth of the world's 10,000 known freshwater fish species (37% in the United States) are threatened with extinction or have already become extinct. Indeed, freshwater animals are disappearing five times faster than land animals.

The tiny seahorse is vulnerable to global extinction chiefly because in dried form it is used in traditional Chinese medicine to treat heart disease, asthma, impotence, and a host of other ills.

How Have Nonnative Species Affected Fish Populations and Species? Alien Invaders Have Hit the Water

Nonnative species are an increasing threat to marine and freshwater biodiversity.

Another problem is the deliberate or accidental introduction of hundreds of nonnative species (Figure 12-9, p. 235) into coastal waters, wetlands, and lakes throughout the world. These bioinvaders can displace or cause the extinction of native species and disrupt ecosystem functions, as happened to Lake Victoria (p. 251). Bioinvaders are blamed for about two-thirds of fish extinctions in the United States between 1900 and 2000. Invasive aquatic species cost the United States an average of about $16 million per hour!

Many aquatic invaders arrive in the ballast water of ships when it is discharged in the waters of ports. One way to reduce this threat is to require ships to discharge their ballast water and replace it with salt water at sea before entering ports. Other ways are to require ships to sterilize their ballast water or pump nitrogen into it (Individuals Matter, p. 257).

Let us take a look at two aquatic invader species. The *Asian swamp eel* has invaded the waterways of south Florida, probably after escaping from a home aquarium. This rapidly reproducing eel eats almost anything—including many prized fish species—by sucking them in like a vacuum cleaner. It can elude cold weather, drought, fires, and predators (including humans with nets) by burrowing into mud banks. It is also resistant to waterborne poisons because it can breathe air and can wriggle across dry land to invade new waterways, ditches, canals, and marshes. Eventually it could take over much of the waterways of the southeastern United States as far north as Chesapeake Bay. You have to admire a species with such an array of survival skills.

The *purple loosestrife* (Figure 12-9, p. 235) is a perennial plant that grows in wetlands in parts of Europe. In the early 1880s, it was imported into the United States as an ornamental plant. It was also released accidentally into U.S. waterways in ballast water contaminated with its seeds.

A single plant can produce more than 2.5 million seeds a year. The seeds are spread by water, in mud, and by becoming attached to wildlife, livestock, people, and tire treads.

Few native plants can compete with this prolific and highly productive plant. This explains why it has spread to temperate and boreal wetlands in 35 states (Figure 13-5) and into southeastern Canada.

As it spreads, it reduces wetland biodiversity by displacing native vegetation and reducing habitat for some forms of wetland wildlife. Some conservationists call this plant the "purple plague."

Hopeful news. Some states have recently introduced two natural predators of loosestrife from Europe: a weevil species and a leaf-eating beetle. It will take some time to determine the effectiveness of this biological

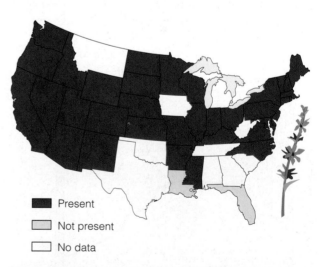

Present

Not present

No data

Figure 13-5 Natural capital degradation: distribution of purple loosestrife in wetlands in the lower 48 states. This nonnative species from Eurasia was introduced deliberately into the United States in the early 1980s and has spread to wetlands in 35 states. (U.S. Department of Agriculture)

control approach and to be sure the introduced species themselves do not become pests.

13-3 PROTECTING AND SUSTAINING MARINE BIODIVERSITY

Why Is It Difficult to Protect Marine Biodiversity? Out of Sight, Out of Mind

Coastal development, the invisibility and vastness of the world's oceans, and lack of legal jurisdiction hinder protection of marine biodiversity.

There are several reasons why protecting marine biodiversity is difficult. One is rapidly growing coastal development and the accompanying massive inputs of sediment and other wastes from land into coastal waters. This harms shore-hugging species and threatens biologically diverse and highly productive coastal ecosystems such as coral reefs, marshes, and mangrove forest swamps.

Another factor is that much of the damage to the oceans and other bodies of water is not visible to most people. And many people incorrectly view the seas as an inexhaustible resource that can absorb an almost infinite amount of waste and pollution.

In addition, most of the world's ocean area lies outside the legal jurisdiction of any country. Thus it is an open-access resource, subject to overexploitation because of the tragedy of the commons.

How Can We Protect Endangered and Threatened Marine Species? Legal Agreements and Awareness

We can use laws, international treaties, and education to help reduce the premature extinction of marine species.

One widely used method for protecting biodiversity is identifying and protecting endangered, threatened, and rare species, as has been done to help save a number of endangered terrestrial species (Chapter 12).

This strategy has also been used to protect a number of endangered and threatened marine reptiles (turtles) and mammals (especially whales, seals, and sea lions; Figure 13-6, p. 258). Each year plastic items dumped from ships and left as litter on beaches threaten the lives of millions of marine mammals, turtles, and seabirds that ingest, become entangled in, choke on, or are poisoned by such debris (see photo on p. viii).

Three of eight major sea turtle species (Figure 13-7, p. 258) are endangered (Kemp's ridley, leatherbacks, and hawksbills), and the rest are threatened, mostly because of four factors. *First* is loss or degradation of beach habitat where they come ashore to lay eggs. *Second* is the legal and illegal taking of eggs. *Third* is the increased use of turtles as sources of food, medicinal

Killing Invader Species and Saving Shipping Companies Money

INDIVIDUALS MATTER

A large cargo ship typically has a dozen or more ballast tanks below deck. Each tank is the size of a high school gymnasium and holds millions of gallons of water.

When a ship takes on cargo and leaves port it pumps water into the ballast tanks to keep it low in the water, submerge its rudder, and help maintain stability. This water also contains large numbers of fish, crabs, clams, and other species (many of them microscopic) found in the port's local waters.

When the ship's cargo is removed at its destination its ballast water is released until the ship is loaded again. This dumps millions of foreign organisms into rivers and bays. Thus cargo ships moving about 80% of the goods traded internationally play the primary role in the release of nonnative aquatic organisms into various parts of the world.

In 2002, researchers Mario Tamburri and Kerstin Wasson found that pumping nitrogen gas into ballast tanks while a ship is at sea virtually eliminates dissolved oxygen in the ballast water. This saves the shipping industry money by reducing corrosion of a ship's steel compartments. In addition, within three days it kills most fish, crabs, clams, and other potential invader species lurking in the ballast tanks.

ingredients, tortoiseshell (for jewelry), and leather from their flippers. *Fourth*, many turtles are unintentionally captured and drowned by commercial fishing boats—especially shrimp trawlers and those using long lines of hooks. Conservationists estimate that each year the global longline fishing industry unintentionally hooks and kills as many as 40,000 sea turtles as bycatch. In 2004 the United States banned long-line swordfish fishing off the Pacific coast to save dwindling sea turtle populations.

Until recently, U.S. shrimp trawling boats killed as many as 55,000 sea turtles (mostly endangered loggerheads and Kemp's ridleys) each year. To reduce this slaughter, since 1989 the U.S. government has required offshore shrimp trawlers to use turtle exclusion devices (TEDs). In 2004 researchers at the National Marine Fisheries Service reported that longline fishing boats using a rounder hook with a smaller opening and baited with mackerel instead of squid could reduce the sea turtle bycatch by 65–90%.

National and international laws and treaties to help protect marine species include the 1975 Convention on International Trade in Endangered Species

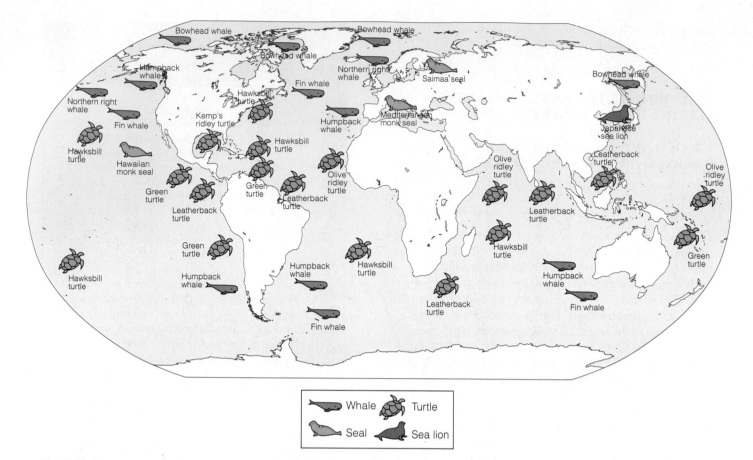

Figure 13-6 Natural capital degradation: endangered and threatened marine mammals (whales, seals, and sea lions) and reptiles (turtles). Many marine fish, seabirds, and invertebrate species are also threatened.

Figure 13-7 Natural capital degradation: major species of sea turtles that have roamed the seas for 150 million years, showing their relative adult sizes. Three of these species (Kemp's ridley, leatherbacks, and hawksbills) are endangered, and the rest are threatened as a result of human activities.

(CITES), the 1979 Global Treaty on Migratory Species, the U.S. Marine Mammal Protection Act of 1972, the U.S. Endangered Species Act of 1973 (p. 241), the U.S. Whale Conservation and Protection Act of 1976, and the 1995 International Convention on Biological Diversity.

Public aquariums that exhibit unusual and attractive fish and some marine animals such as seals and dolphins have played an important role in educating the public about the need to protect such species.

Case Study: Should Commercial Whaling Be Resumed? An Ongoing Controversy

After many of the world's whale species were overharvested, commercial whaling was banned in 1970, but there are efforts to overturn this ban.

Cetaceans are an order of mostly marine mammals ranging in size from the 0.9-meter (3-foot) porpoise to the giant 15- to 30-meter (50- to 100-foot) blue whale. They are divided into two major groups: *toothed whales* and *baleen whales* (Figure 13-8, p. 260).

Toothed whales, such as the porpoise, sperm whale, and killer whale (orca), bite and chew their food and feed mostly on squid, octopus, and other marine animals. *Baleen whales*, such as the blue, gray, humpback, and finback, are filter feeders. Instead of teeth they have several hundred horny plates made of baleen, or whalebone, that hang down from the upper jaw. These plates filter plankton from the seawater, especially tiny shrimplike krill (Figure 4-19, p. 69). Baleen whales are the more abundant of the two cetacean groups.

Whales are fairly easy to kill because of their large size and their need to come to the surface to breathe. Mass slaughter has become efficient with the use of radar and airplanes to locate them, fast ships, harpoon guns, and inflation lances that pump dead whales full of air and make them float.

Whale harvesting, mostly in international waters, has followed the classic pattern of a tragedy of the commons, with whalers killing an estimated 1.5 million whales between 1925 and 1975. This overharvesting reduced the populations of 8 of the 11 major species to the point at which it no longer paid to hunt and kill them (*commercial extinction*). It also drove some commercially prized species such as the giant blue whale to the brink of biological extinction (see Case Study, p. 260).

In 1946, the International Convention for the Regulation of Whaling established the International Whaling Commission (IWC), which now has 49 nation members. Its mission was to regulate the whaling industry by setting annual quotas to prevent overharvesting and commercial extinction.

This did not work well for two reasons. *First*, IWC quotas often were based on inadequate data or ig-

nored by whaling countries. *Second*, without powers of enforcement the IWC was not able to stop the decline of most commercially hunted whale species.

In 1970, the United States stopped all commercial whaling and banned all imports of whale products. Under pressure from environmentalists, the U.S. government, and governments of many nonwhaling countries in the IWC, the IWC has imposed a moratorium on commercial whaling since 1986. It worked. The estimated number of whales killed commercially worldwide dropped from 42,480 in 1970 to about 1,200 in 2004.

Despite the ban, IWC members Japan and Norway have continued to hunt certain whale species, and Iceland resumed hunting whales in 2002—stating that a certain number of whales needed to be harvested for scientific purposes. Japan, Norway, Iceland, Russia, and a growing number of small tropical island countries—which Japan brought into the IWC to support its position—continue working to overthrow the IWC ban on commercial whaling and reverse the international ban on buying and selling whale products.

They argue that commercial whaling should be allowed because it has long been a traditional part of the economies and cultures of countries such as Japan, Iceland, and Norway. They also contend that the ban is based on emotion, not updated scientific estimates of whale populations.

The moratorium on commercial whaling has led to a sharp rebound in the estimated populations of sperm, pilot, and minke whales. Proponents of resuming whaling see no scientific reason for not resuming controlled and sustainable hunting of these species and other whale species with populations of at least 1 million.

Conservationists disagree. Some argue that whales are peaceful, intelligent, sensitive, and highly social mammals that pose no threat to humans and should be protected for ethical reasons. Others question IWC estimates of the allegedly recovered whale species, noting the inaccuracy of past IWC estimates of whale populations. Also, many conservationists fear that opening the door to any commercial whaling may eventually lead to widespread harvests of most whale species by weakening current international disapproval and legal sanctions against commercial whaling.

Proponents of resuming whaling say that people in other countries have no right to tell Japanese, Norwegians, and other whaling culture countries that because we like whales they must not eat them. This would be like people in India who consider cows sacred telling Americans and Europeans that they should not be allowed to eat beef.

X *HOW WOULD YOU VOTE?* Should commercial whaling be resumed? Cast your vote online at http://biology .brookscole.com/miller14.

Case Study: Near Extinction of the Blue Whale: Big Species are Easy to Kill

Commercial whaling almost drove the blue whale to extinction and it may never recover.

The biologically endangered blue whale (Figure 13-8) is the world's largest animal. Fully grown, it is longer than three train boxcars and weighs more than 25 elephants. The adult has a heart the size of a Volkswagen

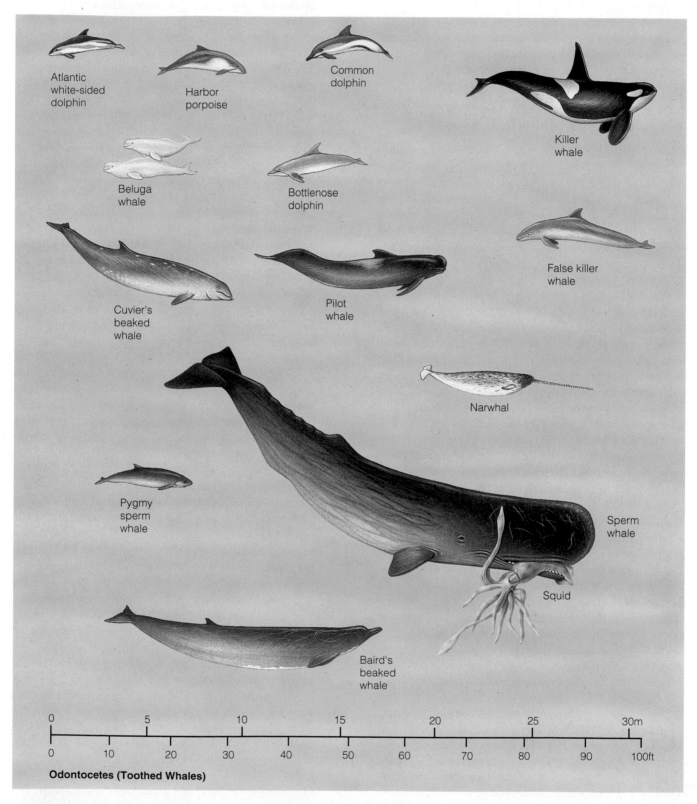

Figure 13-8 Natural capital: examples of cetaceans, which can be classified as either toothed whales or baleen whales.

Beetle car, and some of its arteries are so big that a child could swim through them.

Blue whales spend about 8 months a year in Antarctic waters. There they find an abundant supply of krill (Figure 4-19, p. 69), which they filter by the trillions daily from seawater. During the winter they migrate to warmer waters where their young are born.

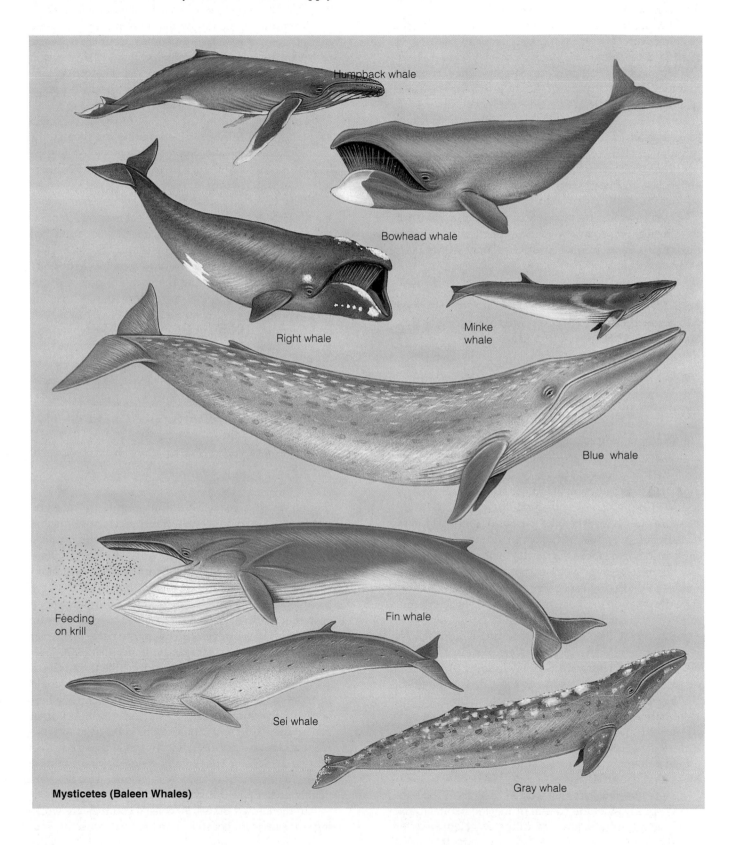

Humpback whale

Bowhead whale

Right whale

Minke whale

Blue whale

Feeding on krill

Fin whale

Sei whale

Gray whale

Mysticetes (Baleen Whales)

Before commercial whaling began an estimated 200,000 blue whales roamed the Antarctic Ocean. Today the species has been hunted to near biological extinction for its oil, meat, and bone. There are probably fewer than 10,000 of these whales left.

A combination of prolonged overharvesting and certain natural characteristics of blue whales caused its decline. Their huge size made them easy to spot. They were caught in large numbers because they grouped together in their Antarctic feeding grounds. They also take 25 years to mature sexually and have only one offspring every 2–5 years. This low reproductive rate makes it difficult for the species to recover once its population falls beneath a certain threshold.

Blue whales have not been hunted commercially since 1964 and have been classified as an endangered species since 1975. Despite this protection, some marine biologists fear that too few blue whales remain for the species to recover and avoid extinction. Others believe that with continued protection they will make a slow comeback.

What Is the Role of International Agreements and Protected Marine Sanctuaries? Hopeful but Limited Progress

Nations have established various types of marine sanctuaries, but most receive only partial protection and fully protected areas make up less than 0.01% of the world's ocean area.

Under the United Nations Law of the Sea, all coastal nations have sovereignty over the waters and seabed up to 19 kilometers (12 miles) offshore. They also have almost total jurisdiction over their Exclusive Economic Zone (EEZ), which extends 320 kilometers (200 miles) offshore. Taken together, the nations of the world have jurisdiction over 36% of the ocean surface and 90% of the world's fish stocks. However, instead of using this law to protect their fishing grounds, many governments promoted overfishing, subsidized new fishing fleets, and failed to establish and enforce stricter regulation of fish catches

Since 1986 the World Conservation Union (IUCN) has helped establish a global system of *marine protected areas* (MPAs), mostly at the national level. An MPA is an area of ocean protected from some or all human activities. The 1,300 existing MPAs provide partial protection for about 0.2% of the earth's total ocean area.

In addition, the United Nations Environment Programme has spearheaded efforts to develop 12 regional agreements to protect large marine areas shared by several countries.

And about 90 of the world's 350 biosphere reserves (p. 217) include coastal or marine habitats. *Marine reserves*—also known as fully protected areas or no-take MPAs—are areas where no extraction and alteration of any living or nonliving resources is allowed. More than 20 coastal nations, including the United States, have established marine reserves that vary widely in size. In 2002, the Australian government established the world's largest marine reserve.

Scientific studies show that within fully protected marine reserves, fish populations double, fish size grows by almost a third, fish reproduction triples, and species diversity increases by almost one-fourth. Furthermore, this improvement happens within 2–4 years after strict protection begins and lasts for decades.

However, less than 0.01% of the world's ocean area consists of fully protected marine reserves. In the United States the total area of fully protected marine habitat is only about 130 square kilometers (50 square miles). In other words, *we have failed to strictly protect 99.99% of the world's ocean area from human exploitation.*

In 1997, a group of international marine scientists called for governments to increase fully protected marine reserves to at least 20% of the ocean's surface by 2020. The 2003 Pew Fisheries Commission study recommended establishing many more protected marine reserves in U.S. coastal waters and connecting them with protected corridors so fish can move back and forth. In 2004 marine biologist Elliott Norse proposed establishment of *moveable marine reserves* that move with the animals as they migrate through the oceans.

What Is the Role of Integrated Coastal Management? Cooperation Can Work

Some communities have worked together to develop integrated plans for managing their coastal areas.

Integrated coastal management is a community-based effort to develop and use coastal resources more sustainably. The overall aim is for groups competing for the use of coastal resources to identify shared problems and goals. Then they attempt to develop workable, cost-effective, and adaptable solutions that preserve biodiversity and environmental quality while meeting economic and social needs. In other words, develop and implement integrated plans using the principles of *adaptive ecosystem management* (Figure 11-23, p. 218).

Ideally, the overall goal is to zone the land and sea portions of an entire coastal area. Such zoning would include some fully protected marine reserves where no exploitive human activities are allowed and other zones where different kinds and levels of human activities are permitted. Australia's huge Great Barrier Reef Marine Park is managed this way. Currently, more than 100 integrated coastal management programs are being developed throughout the world.

In the United States, 90 coastal counties are working to establish coastal management systems, but fewer than 20 of these plans have been implemented.

What Role Can Reconciliation Ecology Play? Share the Spaces We Dominate

We can greatly increase the use of reconciliation ecology in protecting, sustaining, and restoring aquatic systems.

Reconciliation ecology (p. 247) has a role to play in protecting, sustaining, and restoring aquatic systems. Here are two examples given by Michael L. Rosenzweig.

One involves large numbers of *American crocodiles* taking up residence in 38 long canals dug to provide cooling water for Florida's Turkey Point electric power plant. The sides of each canal are topped with berms of the dirt dredged to dig them out. The berms support a variety of plants and small trees, including red mangroves. In addition, a group of American crocodiles showed up and began living and breeding in the canals. This was not a planned experiment in reconciliation ecology. It just happened.

Now we zoom to the city of Eliat, Israel, at one tip of the Red Sea. There we find a magnificent coral reef at the water's edge, which is a major tourist attraction. To help protect the reef from excessive development and destructive tourism, Israel set aside part of the reef as a nature reserve.

The bad news is that most of the rest of the reef is gone as a result of tourism, industry, and inadequate sewage treatment. Enter Reuven Yosef, a pioneer in reconciliation ecology, who has developed an underwater restaurant called the Red Sea Star Restaurant. Take an elevator down two floors and walk into a restaurant surrounded with windows looking out into a beautiful coral reef.

This reef was created from broken pieces of coral. When divers find broken pieces of coral in the nearby reserve they bring them to Yosef's coral hospital.

Most pieces of broken coral soon become infected and die. But researchers have learned how to treat the coral fragments with antibiotics and store them while they are healing in large tanks of fresh seawater.

After several months of healing, divers bring the fragments to the watery area outside the Red Sea Star Restaurant's windows. There they are wired to panels of iron mesh cloth. The coral grow and cover the iron matrix. Then fish and other creatures show up.

Using his creativity and working with nature, Yosef has helped create a small coral reef and provides a beautiful place for restaurant customers to see the reef without having to be divers or snorklers.

We need to greatly increase experiments in reconciliation ecology in the world's aquatic systems. Perhaps you can begin such an experiment in your area. You also might want to consider this field as a career choice.

13-4 MANAGING AND SUSTAINING THE WORLD'S MARINE FISHERIES

How Can We Manage Fisheries to Sustain Stocks and Protect Biodiversity? Many Ideas

There are a number of ways to manage marine fisheries more sustainably and protect marine biodiversity.

Overfishing is a serious threat to biodiversity in coastal waters and to some marine species in open-ocean waters. Figure 13-9 lists measures that analysts

Solutions

Managing Fisheries

Fishery Regulations

Set catch limits well below the maximum sustainable yield

Improve monitoring and enforcement of regulations

Economic Approaches

Sharply reduce or eliminate fishing subsidies

Charge fees for harvesting fish and shellfish from publicly owned offshore waters

Certify sustainable fisheries

Protected Areas

Establish no-fishing areas

Establish more marine protected areas

Rely more on integrated coastal management

Consumer Information

Label sustainably harvested fish

Publicize overfished and threatened species

Bycatch

Use wide-meshed nets to allow escape of smaller fish

Use net escape devices for seabirds and sea turtles

Ban throwing edible and marketable fish back into the sea

Aquaculture

Restrict coastal locations for fish farms

Control pollution more strictly

Depend more on herbivorous fish species

Nonative Invasions

Kill organisms in ship ballast water

Filter organisms from ship ballast water

Dump ballast water far at sea and replace with deep-sea water

Figure 13-9 Solutions: ways to manage fisheries more sustainably and protect marine biodiversity.

have suggested for managing global fisheries more sustainably and protecting marine biodiversity. Most of these approaches rely on some sort of government regulation.

But in nature there are almost always surprises because we still have little understanding of how ecosystems work. For example, researchers have found that reducing fishing to protect fish stocks can harm populations of some seabirds.

In the North Sea, the bycatch tossed back into the sea is eaten by a seabird species called the great skua. Because of such an abundance of food the size of the great skua population rose sharply. But reducing fish quotas has meant fewer discards for these seabirds. To make up for the loss these birds have been preying in other seabirds such as kittiwakes and puffins—another example of unintended consequences from our actions.

One way to reduce overfishing is to develop better measurements and models for projecting fish populations. Until recently, management of commercial fisheries has been based primarily on the *maximum sustained yield (MSY)*. It involves using a mathematical model to project the maximum number of fish that can be harvested annually from a fish stock without causing a population drop.

But experience has shown that the MSY concept has helped hasten the collapse of most commercially valuable stocks for several reasons. *First*, populations and growth rates of fish stocks are difficult to measure. *Second*, population sizes of fish stocks usually are based on unreliable and sometimes underreported catch figures by fishers. *Third*, harvesting a particular species at its estimated maximum sustainable level can affect the populations of other target and nontarget fish species and other marine organisms—those pesky connections again. *Fourth*, fishing quotas are difficult to enforce and many groups managing fisheries have ignored projected MSYs for short-term political or economic reasons.

In recent years, fishery biologists and managers have begun placing more emphasis on the *optimum sustained yield (OSY)* concept. This approach attempts to take into account interactions with other species and to provide more room for error. But it still depends on the poorly understood biology of fish and changing ocean conditions. Also, many bodies governing fisheries ignore OSY estimates for short-term political and economic reasons.

Another approach is *multispecies management* of a number of interacting species, which takes into account their competitive and predator–prey interactions. Such models are still in the development and testing stage.

A more ambitious approach is to develop complex computer models for managing multispecies fisheries in *large marine systems*. However, it is a political challenge to get groups of nations to cooperate in planning and managing them.

Despite the scientific and political difficulties, some limited management of several large marine systems is under way. Examples include the Mediterranean Sea by 17 of the 18 nations involved and the Great Barrier Reef under the exclusive control of Australia.

A basic problem is the uncertainties built into using any of these approaches. As a result, many fishery scientists and environmentalists are increasingly interested in using the *precautionary principle* for managing fisheries and large marine systems. This means sharply reducing fish harvests and closing some overfished areas until they recover and we have more information about what levels of fishing can be sustained.

Should Governments Control Access to Fisheries? Regulation and Cooperation Can Work

Some fishing communities regulate fish harvests on their own and others work with the government to regulate them.

By international law, a country's offshore fishing zone extends to 370 kilometers (200 nautical miles, or 230 statute miles) from its shores. Foreign fishing vessels can take certain quotas of fish within such zones, called *exclusive economic zones*, but only with a government's permission.

Ocean areas beyond the legal jurisdiction of any country are known as the *high seas*. International maritime law and international treaties set some limits on the use of the living and mineral common-property resources in the high seas. But such laws and treaties are difficult to monitor and enforce.

Traditionally, many coastal fishing communities have developed allotment and enforcement systems that have sustained their fisheries, jobs, and communities for hundreds and sometimes thousands of years. An example is Norway's Lofoten fishery, one of the world's largest cod fisheries. For 100 years it has been self-regulated, with no government quota regulations and no participation by the Norwegian government. Cooperation can work.

However, the influx of large modern fishing boats and fleets has weakened the ability of many coastal communities to regulate and sustain local fisheries. Many community management systems have been replaced by *comanagement,* in which coastal communities and the government work together to manage fisheries. In this approach, a central government typically sets quotas for various species, divides the quotas among communities, and may limit fishing seasons and regulate the type of fishing gear that can be used to harvest a particular species.

Each community then allocates and enforces its quota among its members based on its own rules. Often communities focus on managing inshore fisheries and the central government manages the offshore fisheries. When it works, community-based comanagement illustrates that the tragedy of the commons is not inevitable.

Should We Use the Marketplace to Control Access to Fisheries? Good and Bad News about an Interesting Idea

Some countries try to prevent overfishing by giving each fishing vessel quotas that can be bought or sold in the marketplace.

Some countries are using a market-based system called *individual transfer quotas (ITQs)* to help control access to fisheries. The government gives each fishing vessel owner a specified percentage of the total allowable catch (TAC) for a fishery in a given year.

Owners are permitted to buy, sell, or lease their quotas like private property. Currently about 50 of the world's fisheries are managed by the ITQ system. It was introduced in New Zealand in 1986 and in Iceland in 1990. In these countries there has been some reduction in overfishing and the overall fishing fleet and an end to government fishing subsidies that encourage overfishing.

But enforcement has been difficult, some fishers illegally exceed their quotas, and the wasteful bycatch has not been reduced.

Environmentalists have identified four problems with the ITQ approach and have made suggestions for its improvement. *First,* in effect it transfers ownership of publicly owned fisheries to private commercial fishers but still makes the public responsible for the costs of enforcing and managing the system. *Remedy:* Collect fees (not to exceed 5% of the value of the catch) from quota holders to pay for the costs of government enforcement and management of the ITQ system.

Second, it can squeeze out small fishing vessels and companies because they do not have the capital to buy ITQs from others. For example, 5 years after the ITQ system was implemented in New Zealand, three companies controlled half the ITQs. *Remedy:* Do not allow any fisher or fishing company to accumulate more than a fifth of the total quota of a fishery.

Third, the ITQ system can increase poaching and sales of illegally caught fish on the black market, as has happened to some extent in New Zealand. Some of this comes from small-scale fishers who receive no quota or too small a quota to make a living. Some also comes from larger-scale fishers who deliberately exceed their quotas. *Remedy:* Require strict record keeping and have well-trained observers on all fishing vessels with quotas.

Fourth, the fishing quotas (TACS) are often set too high to prevent overfishing. *Remedy:* Leave 10–50% of the estimated MSY of an ITQ fishery as a buffer to protect the fishery from unexpected decline.

13-5 PROTECTING, SUSTAINING, AND RESTORING WETLANDS

How Are Wetlands Protected in the United States? Some Progress

Requiring government permits for filling or destroying U.S. wetlands has slowed their loss, but there are continuing attempts to weaken this protection.

Coastal and inland wetlands are important reservoirs of aquatic biodiversity that provide many vital ecological and economic services. In the United States, a federal permit is required to fill or to deposit dredged material into wetlands occupying more than 1.2 hectares (3 acres). According to the U.S. Fish and Wildlife Service, this law helped cut the average annual wetland loss by 80% between 1969 and 2002.

There are continuing attempts to weaken wetlands protection by using unscientific criteria to classify areas as wetlands. Also, only about 8% of remaining inland wetlands is under federal protection, and federal, state, and local wetland protection is weak.

The stated goal of current U.S. federal policy is zero net loss in the function and value of coastal and inland wetlands. A policy known as *mitigation banking* allows destruction of existing wetlands as long as an equal area of the same type of wetland is created or restored.

Some wetland restoration projects have been successful (Individuals Matter, p. 266).

However, a 2001 study by the National Academy of Sciences found that at least half of the attempts to create new wetlands fail to replace lost ones and most of the created wetlands do not provide the ecological functions of natural wetlands. In addition, wetland creation projects often fail to meet the standards set for them and are not adequately monitored.

Figure 13-10 lists ways to help sustain wetlands in the United States and elsewhere. Many developers,

---

Solutions

Protecting Wetlands

Legally protect existing wetlands

Steer development away from existing wetlands

Use mitigation banking only as a last resort

Require creation and evaluation of a new wetland before destroying an existing wetland

Restore degraded wetlands

Try to prevent and control invasions by nonnative species

Figure 13-10 Solutions: ways to help sustain the world's wetlands.

Restoring a Wetland

Humans have drained, filled in, or covered over swamps, marshes, and other wetlands for centuries. They have done this to create rice fields and other land to grow crops, create land for urban development and highways, reduce disease such as malaria caused by mosquitoes, and extract minerals, oil, and natural gas.

Some people have begun to question such practices as we learn more about the ecological and economic importance of coastal and inland wetlands. Can we turn back the clock to restore or rehabilitate lost marshes?

California rancher Jim Callender decided to try. In 1982, he bought 20 hectares (50 acres) of a Sacramento Valley rice field that had been a marsh until the early 1970s. To grow rice, the previous owner had destroyed the marsh by bulldozing, draining, leveling, uprooting the native plants, and spraying with chemicals to kill the snails and other food of the waterfowl.

Callender and his friends set out to restore the marshland. They hollowed out low areas, built up islands, replanted bulrushes and other plants that once were there, reintroduced smartweed and other plants needed by birds, and planted fast-growing Peking willows. After years of care, hand planting, and annual seeding with a mixture of watergrass, smartweed, and rice, the marsh is once again a part of the Pacific flyway used by migratory waterfowl (Figure 12-15, p. 246).

Jim Callender and others have shown that at least part of the continent's degraded or destroyed wetlands can be reclaimed with scientific knowledge and hard work. Such restoration is useful, but to most ecologists the real challenge is to protect remaining wetlands from harm in the first place.

farmers, and resource extractors vigorously oppose these suggestions.

Case Study: Restoring the Florida Everglades: Will It Work?

The world's largest ecological restoration project involves trying to undo some of the damage inflicted on Florida's Everglades by human activities.

South Florida's Everglades was once a 100-kilometer-wide (60-mile-wide), knee-deep sheet of water flowing slowly south from Lake Okeechobee to Florida Bay (Figure 13-11). As this shallow body of water trickled south it created a vast network of wetlands with a variety of wildlife habitats.

But since 1948 much of the southward natural flow of the Everglades has been diverted and disrupted by a system of canals, levees, spillways, and pumping stations. The most devastating blow came in the 1960s, when the U.S. Army Corps of Engineers transformed the meandering 103-mile-long Kissimmee River (Figure 13-11) into a straight 84-kilometer (56-mile) canal. The canal provided flood control by speeding the flow of water but drained large wetlands north of Lake Okeechobee, which farmers then turned into cow pastures.

Below Lake Okeechobee, farmers planted and fertilized vast agricultural fields of sugarcane and vegetables. Historically the Everglades has been a nutrient–poor aquatic system, with low phosphorus levels. But runoff of phosphorus from fertilizers has greatly increased phosphorus levels. This large nutrient input has stimulated the growth of nonnative plants such as cattails, which have taken over and displaced saw grass, choked waterways, and disrupted food webs in a vast area of the Everglades.

Mostly as a result of these human alterations, the natural Everglades has shrunk to half its original size and dried out, leaving large areas vulnerable to summer wildfires. Urbanization has also contributed to the loss of biodiversity in the Everglades by fragmenting much of its habitat.

To help preserve the lower end of the system, in 1947 the U.S. government established Everglades National Park, which contains about a fifth of the remaining Everglades. But this did not work—as conservationists had predicted—because the massive plumbing and land development project to the north cut off much of the water flow needed to sustain the park's wildlife.

As a result, 90% of the park's wading birds have vanished, and populations of other vertebrates, from deer to turtles, are down 75–95%. Florida Bay, south of the Everglades is a shallow estuary with many tiny islands or keys. Large volumes of fresh water that once flowed through the park into Florida Bay have been diverted for crops and cities, causing the bay to become saltier and warmer. This and increased nutrient input from crop fields and cities have stimulated the growth of large algae blooms that sometimes cover 40% of the bay. This has threatened the coral reefs and the diving, fishing, and tourism industries of the bay and Florida Keys—another example of unintended consequences.

By the 1970s, state and federal officials recognized that this huge plumbing project threatens wildlife—a major source of tourism income for Florida—and the water supply for the 6 million residents of south

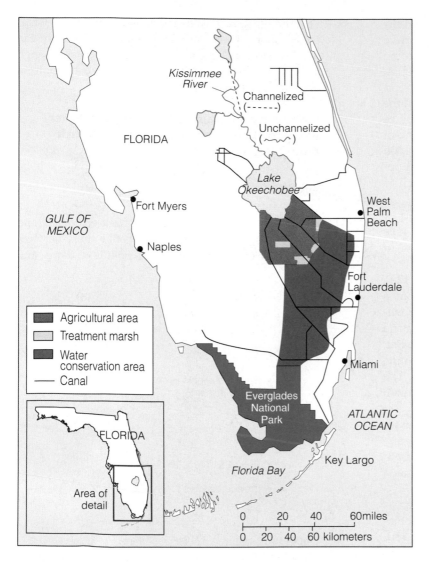

Figure 13-11 The world's largest ecological restoration project is an attempt to undo and redo an engineering project that has been destroying Florida's Everglades and threatening water supplies for south Florida's growing population.

Florida and the 6 million more people projected to be living there by 2050.

After more than 20 years of political haggling, in 1990 Florida's state government and the federal government agreed on the world's largest ecological restoration project. It is to be carried out by the U.S. Army Corps of Engineers between 2000 and 2038, with Florida and the federal government sharing its projected $7.8 billion cost.

The project has several ambitious goals. *First,* restore the curving flow of more than half of the Kissimmee River. *Second,* remove 400 kilometers (250 miles) of canals and levees blocking water flow south of Lake Okeechobee. *Third,* buy 240 square kilometers (93 square miles) of farmland and allow it to flood to create artificial marshes to filter agricultural runoff before it reaches Everglades National Park. *Fourth,* create a network of artificial marshes. *Fifth,* create 18 large reservoirs to ensure an adequate water supply for south

Florida's current and projected population and the lower Everglades. *Sixth,* build new canals, reservoirs, and huge pumping systems to capture 80% of the water currently flowing out to sea and return it to the Everglades.

Will this huge ecological restoration project work? It depends not only on the abilities of scientists and engineers but also on prolonged political and economic support from citizens, Florida's politically powerful sugarcane and agricultural industries, and elected state and federal officials.

Bad news. The carefully negotiated plan is unraveling. In 2003, sugarcane growers persuaded the Florida legislature to increase the amount of phosphorus they could discharge and extend the deadline for doing this from 2006 to 2016.

According to critics, the main goal of the Everglades restoration plan is to provide water for urban and agricultural development with ecological restoration as a secondary goal. Also, the plan does not specify how much of the water rerouted toward south and central Florida will go to the parched park instead of to increased industrial, agricultural, and urban development. In 2002, a National Academy of Sciences panel said that the plan would probably not clear up Florida Bay's nutrient enrichment problems.

The need to make expensive and politically controversial efforts to undo some of the damage to the Everglades caused by 120 years of agricultural and urban development is another example of failure to heed two fundamental lessons from nature: Prevention is the cheapest and best way to go and there are almost always unintended consequences because we can never do one thing when we intervene in nature.

13-6 PROTECTING, SUSTAINING, AND RESTORING LAKES AND RIVERS

Case Study: Can the Great Lakes Survive Repeated Invasions by Alien Species? They Keep Coming

Invasions by nonnative species have upset the ecological functioning of the Great Lakes for decades, and more invaders keep arriving.

Invasions by nonnative species is a major threat to the biodiversity and ecological functioning of lakes, as illustrated by what has happened to the Great Lakes.

Collectively, the Great Lakes are the world's largest body of fresh water. Since the 1920s, they have

been invaded by at least 162 nonnative species and the number keeps rising. Many of the alien invaders arrive on the hulls or in bilge water discharges of oceangoing ships that have been entering the Great Lakes through the St. Lawrence seaway for over 40 years.

One of the biggest threats, *sea lampreys,* reached the western lakes through the Welland Canal as early as 1920. This parasite attaches itself to almost any kind of fish and kills the victim by sucking out its blood (Figure 12-9, p. 235). Over the years it has depleted populations of many important sport fish species such as lake trout.

The United States and Canada keep the lamprey population down by applying a chemical that kills their larvae in their spawning streams—at a cost of about $15 million a year.

In 1986, larvae of the *zebra mussel* (Figure 12-9, p. 235) arrived in ballast water discharged from a European ship near Detroit, Michigan. This thumbnail-sized nonnative species reproduces rapidly and has no known natural enemies in the Great Lakes. As a result, it has displaced other mussel species and depleted the food supply for some other Great Lakes species. The mussels have also clogged irrigation pipes, shut down water intake systems for power plants and city water supplies, fouled beaches, and grown in huge masses on boat hulls, piers, pipes, rocks, and almost any exposed aquatic surface. This mussel has also spread to freshwater communities in parts of southern Canada and 18 states in the United States.

Possible *good news.* In 2001, scientists at Indiana's Purdue University Calumet reported on the use of an oscillating dipole apparatus to produce what is popularly known as a "zebra mussel death ray." It emits extremely low frequency electromagnetic waves that can kill zebra mussels, apparently without harming the environment or other species. This approach is being evaluated. New methods for treating ballast water have also been developed (Individuals Matter, p. 257).

Zebra mussels may not be good for us and some fish species but they can benefit a number of aquatic plants. By consuming algae and other microorganisms, the mussels increase water clarity, which permits deeper penetration of sunlight and more photosynthesis. This allows some native plants to thrive and return the plant composition of Lake Erie (and presumably other lakes) closer to what it was 100 years ago. Because the plants provide food and increase dissolved oxygen, their comeback may benefit certain aquatic animals.

More *bad news.* In 1991, a larger and potentially more destructive species, the *quagga mussel,* invaded the Great Lakes, probably discharged in the ballast water of a Russian freighter. It can survive at greater depths and tolerate more extreme temperatures than the zebra mussel. There is concern that it may eventually colonize areas such as Chesapeake Bay and waterways in parts of Florida.

The *Asian carp* is also expected to reach the Great Lakes soon. These highly prolific fish, which can quickly grow as long as 1.2 meters (4 feet) and weigh up to 50 kilograms (110 pounds), have no natural predators in the Great Lakes.

Case Study: Managing the Columbia River Basin for People and Salmon: A Difficult Balancing Act

Constructing a large number of dams along the Columbia River has provided many human benefits but has threatened wild salmon populations.

Rivers and streams provide important ecological and economic services (Figure 13-12). But these services can be disrupted by overfishing, pollution, dams, and

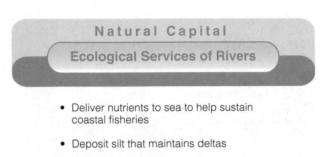

Natural Capital

Ecological Services of Rivers

- Deliver nutrients to sea to help sustain coastal fisheries

- Deposit silt that maintains deltas

- Purify water

- Renew and renourish wetlands

- Provide habitats for wildlife

Figure 13-12 Natural capital: important ecological services provided by rivers.

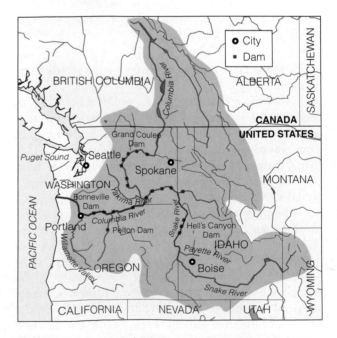

Figure 13-13 Degraded natural capital: the Columbia River basin. (Data from Northwest Power Planning Council)

water withdrawal for irrigation. An example of such disruption is the Columbia River—North America's fourth largest river (Figure 13-13).

This basin has the world's largest hydroelectric power system. It has 119 dams, 19 of which are major generators of inexpensive hydroelectric power. It also supplies municipal and industrial water for several major urban areas and is a source of water for irrigating large areas of agricultural land.

Salmon are migratory fish that spawn in the upper reaches of streams and rivers. Their offspring migrate downstream to the ocean, where they spend most of their adult lives. The adults complete their life cycle by returning to their place of birth to spawn and die (Figure 13-14, left). A series of dams and extensive forest clearing of land adjacent to stream banks can severely disrupt the salmon life cycle because salmon need free flowing rivers to return, spawn, and lay eggs at the sites where they were hatched. This requires not cutting nearby forests that can cloud salmon spawning sites with silt and cover spawned eggs. In other words, salmon need nearby intact forests.

In 2001 research by ecologist Thomas Reimichen indicated that salmon can improve the health of the nearby forests. This occurs because bears feeding on large amounts of the salmon strew half-eaten salmon

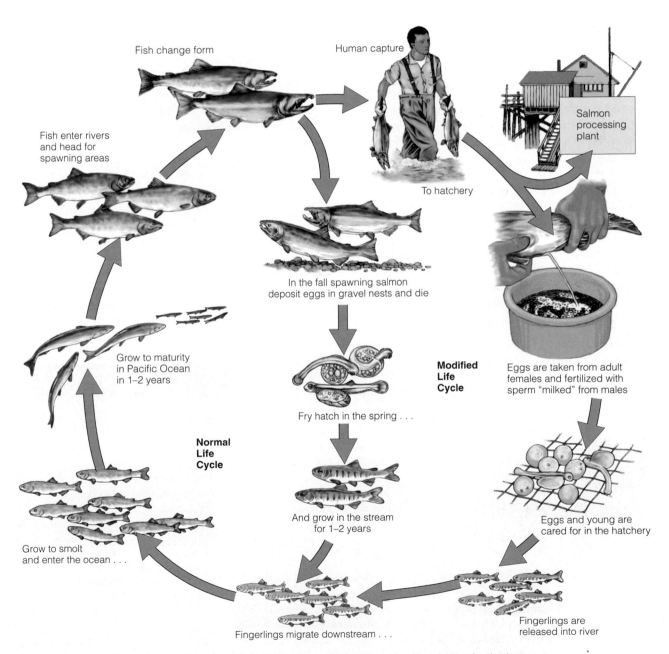

Figure 13-14 Normal life cycle of wild salmon (left) and human-modified life cycle of hatchery-raised salmon (right). Salmon spend part of their lives in fresh water and part in salt water.

carcasses on the forest floor. As they decompose the carcasses help fertilize the forest and provide food for a variety of insects and other scavengers. Another study indicated that trees in forests along streams in the Pacific Northwest with healthy salmon populations grow up to three times faster than those along streams without salmon.

Thus such forests and salmon need each other for good health. Since the dams were built, the Columbia River's wild Pacific salmon population has dropped by 94% and nine Pacific Northwest salmon species are listed as endangered or threatened. The dams are not the only cause of this decline. Other factors include overfishing of salmon in the Pacific Ocean, destruction of salmon spawning grounds in streams by sediment from logging and mining, and withdrawals of water for irrigation and other human uses. Also, the lack of shade in salmon spawning streams where all of the trees have been cut makes the water too hot for survival of salmon eggs.

Commercial fishing operations have modified the wild salmon's natural cycle by using salmon ranching, a form of aquaculture in which salmon eggs and young are raised in a hatchery and then released (Figure 13-14, right). But ranch salmon that escape and interbreed with wild ones reduce the genetic diversity of the wild fish and their ability to survive.

In 1980, the U.S. Congress passed the Northwest Power Act. It has two main goals, which often conflict with one another. One was to develop and implement long-range plans to meet the region's electricity needs. The other was to rebuild wild and hatchery-raised salmon and other fish populations.

Figure 13-15 lists some of the strategies that have been used to help restore wild salmon populations.

The federal government has spent over $3 billion in efforts to save the salmon but none have been effective. Environmentalists, Native American tribes, and commercial salmon fishers want the government to remove four small hydroelectric dams on the lower Snake River in Washington to restore salmon spawning habitat. Farmers, barge operators, and aluminum workers argue that removing the dams would hurt local economies by reducing irrigation water, eliminating cheap transportation of commodities by ship in the affected areas, and reducing the supply of cheap electricity for industries and consumers.

Can this wild salmon restoration project work? No one knows because it will take decades to see whether the salmon populations can be rebuilt. Despite problems, this program demonstrates that people with diverse and often conflicting economic, political, and environmental interests can work together to try new ideas and develop potentially sustainable solutions to complex resource management issues. It is an example of a large-scale reconciliation ecology project.

Critics of this expensive salmon restoration program argue that populations of wild salmon are stable in Alaska, so we should not care that wild salmon are declining in the Pacific Northwest. They also contend that the economic costs to the hydroelectric power, shipping, and timber industries and to farmers and consumers exceed the value of saving wild salmon. Some have worked to restore salmon populations in specific streams (Individuals Matter, below).

X ***How Would You Vote?*** Should federal efforts to rebuild wild salmon populations in the Columbia River Basin be abandoned? Cast your vote online at http://biology.brookscole .com/miller14.

INDIVIDUALS MATTER

The Man Who Planted Trees to Restore a Stream

In 1980 heart problems forced John Beal, an engineer with the Boeing Company, to take some time off. To improve his health he began taking daily walks. His strolls took him by a small stream called Hamm Creek that flows from the southwest hills of Seattle, Washington, into the Duwamish River that empties into Puget Sound.

He remembered when the stream was a spawning ground for salmon and evergreen trees lined its banks. Now the polluted stream had no fish and the trees were gone.

He decided to restore Hamm Creek. He persuaded companies to stop polluting the creek and hauled out many truckloads of garbage. Then he began a 15-year project of planting thousands of trees along the stream's banks. He also restored natural waterfalls and ponds and salmon spawning beds.

At first he worked alone, but word spread and other people joined him. TV news reports and newspaper articles about the restoration project brought more volunteers.

The creek's water now runs clear, its vegetation has been restored, and salmon have returned to spawn. His reward is the personal satisfaction he feels about having made a difference for Hamm Creek and his community. His dedication to making the world a better place is an outstanding example of the idea that *all sustainability is local.*

Building upstream hatcheries

Releasing juvenile salmon from hatcheries to underpopulated streams

Releasing extra water from dams to wash juvenile salmon downstream

Building fish ladders so adult salmon can bypass dams during upstream migration

Using trucks and barges to transport salmon around dams

Reducing silt runoff from logging roads above salmon spawning streams

Banning dams from some stream areas

Figure 13-15 Solutions: Some strategies used to rebuild salmon populations in the Columbia River basin.

How Can Freshwater Fisheries Be Managed and Sustained? Encourage Some Species and Discourage Others

Freshwater fisheries can be sustained by building and protecting populations of desirable species, preventing overfishing, and decreasing populations of less desirable species.

Sustainable management of freshwater fish involves encouraging populations of commercial and sport fish species, preventing such species from being overfished, and reducing or eliminating populations of less desirable species. Ways to do this include regulating the time and length of fishing seasons and the number and size of fish that can be taken.

Other techniques include building reservoirs and farm ponds and stocking them with fish, fertilizing nutrient-poor lakes and ponds, and protecting and creating fish spawning sites. In addition, fishery managers can protect fish habitats from sediment buildup and other forms of pollution, prevent excessive growth of aquatic plants from large inputs of plant nutrients, and build small dams to control water flow.

Improving habitats, breeding genetically resistant fish varieties, and using antibiotics and disinfectants can control predators, parasites, and diseases. Hatcheries can be used to restock ponds, lakes, and streams with prized species such as trout and salmon, and entire river basins can be managed to protect valued species such as salmon. Some individuals have worked to help restore degraded streams.

How Can Wild and Scenic Rivers Be Protected and Restored? Let More of Them Run Free

A federal law helps protect a tiny fraction of U.S. wild and scenic rivers from dams and other forms of development.

In 1968, the U.S. Congress passed the National Wild and Scenic Rivers Act. It established the National Wild and Scenic Rivers System to protect rivers and river segments with outstanding scenic, recreational, geological, wildlife, historical, or cultural values.

Congress established a three-tiered classification scheme. *Wild rivers* are rivers or segments of rivers that are relatively inaccessible and untamed and that are not permitted to be widened, straightened, dredged, filled, or dammed. The only activities allowed are camping, swimming, nonmotorized boating, sport hunting, and sport and commercial fishing.

Scenic rivers are free of dams, mostly undeveloped, accessible in some places by roads, and of great scenic value. *Recreational rivers* are rivers or sections of rivers that are readily accessible by roads and that may have some dams or development along their shores.

Currently only 0.2% of the country's 6 million kilometers (3.5 million miles) are protected by the Wild and Scenic Rivers System. In contrast, dams and reservoirs are found on 17% of the country's total river length.

Environmentalists urge Congress to add 1,500 additional river segments to the system, a goal vigorously opposed by some local communities and anti-environmental groups. Achieving this goal would protect about 2% of the country's river systems. That is still 98% for us and only 2% for the wild rivers and their species.

In this chapter we have seen that threats to aquatic biodiversity are real and growing and are even greater than threats to terrestrial biodiversity. Keys to sustaining life in the earth's aquatic systems include greatly increasing research to learn more about aquatic life, greatly expanding efforts to protect and restore aquatic biodiversity, and promoting integrated ecological management of connected terrestrial and aquatic systems. These things can be done if enough people insist on it.

To promote conservation, fishers and officials need to view fish as a part of a larger ecological system, rather than simply as a commodity to extract.

ANNE PLATT MCGINN

CRITICAL THINKING

1. What three actions would you take to deal with the ecological and economic problems of Africa's Lake Victoria?

2. Why is marine biodiversity higher **(a)** near coasts than in the open sea and **(b)** on the ocean's bottom than at its surface?

3. Why is it more difficult to identify and protect endangered marine species than to protect such species on land?

4. List three methods for using the precautionary principle as a way to manage marine fisheries and help protect marine biodiversity.

5. Should fishers harvesting fish from a country's publicly owned waters be required to pay the government (taxpayers) fees for the fish they catch? Explain. If your livelihood depended on commercial fishing, would you be for or against such fees?

6. Are you for or against using Individual Transfer Quotas (p. 265) as the major method for managing fisheries? Explain. What are the alternatives?

7. Are you for or against using mitigation banking to help sustain wetlands? Explain. What restrictions, if any, would you put on such activities?

8. Congratulations! You are in charge of protecting the world's aquatic biodiversity. List the three most important points of your policy to accomplish this goal.

PROJECTS

1. Survey the condition of a nearby wetland, coastal area, river, or stream. Has its condition improved or deteriorated during the last 10 years? What local, state, or national efforts are being used to protect this aquatic system? Develop a plan for protecting it.

2. Work with your classmates to develop an experiment in aquatic reconciliation ecology for your campus or local community.

3. Use the library or the Internet to find bibliographic information about *G. Carleton Ray* and *Anne Platt McGinn*, whose quotes appear at the beginning and end of this chapter.

4. Make a concept map of this chapter's major ideas, using the section heads, subheads, and key terms (in boldface). Look on the website for this book about how to prepare concept maps.

LEARNING ONLINE

The website for this book contains study aids and many ideas for further reading and research. They include a chapter summary, review questions for the entire chapter, flash cards for key terms and concepts, a multiple-choice practice quiz, interesting Internet sites, references, and a guide for accessing thousands of InfoTrac® College Edition articles. Log on to

http://biology.brookscole.com/miller14

Then click on the Chapter-by-Chapter area, choose Chapter 13, and select a learning resource.

CASE STUDY
Growing Perennial Crops on the Kansas Prairie by Copying Nature

Think about farms in Kansas and you probably picture seemingly endless fields of wheat or corn plowed up and planted each year. By 2040, the picture might change, thanks to pioneering research at the nonprofit *Land Institute* near Salina, Kansas.

The institute, headed by plant geneticist Wes Jackson, is experimenting with an ecological approach to agriculture on the midwestern prairie. It relies on planting a mixture of different crops in the same area, a technique called *polyculture*. This involves planting a mix of perennial grasses (Figure 14-1, right), legumes (a source of nitrogen fertilizer, Figure 14-1, left), sunflowers, grain crops, and plants that provide natural insecticides in the same field.

The goal is to raise food by mimicking many of the natural conditions of the prairie without losing fertile grassland soil. Institute researchers believe that

perennial polyculture can be blended with *modern monoculture* to help reduce the latter's harmful environmental effects.

Because these plants are perennials, there is no need to plow up and prepare the soil each year to replant them. This takes much less labor than conventional monoculture or diversified organic farms that grow annual crops. It also reduces soil erosion because the unplowed soil is not exposed to wind and rain. And it reduces the need for irrigation because the deep roots of such perennials retain more water than annuals. There is also less pollution from chemical fertilizers and pesticides. This sounds like a win-win solution.

Thirty-six years of research by the institute have shown that various mixtures of perennials grown in parts of the midwestern prairie could be used as important sources of food. One such mix of perennial crops includes *eastern grama grass* (a warm-season grass that is a relative of corn with three times as much protein as corn and twice as much as wheat; Figure 14-1, right), *mammoth wildrye* (a cool-season grass distantly related to rye, wheat, and barley), *Illinois bundleflower* (a wild nitrogen-producing legume that can enrich the soil and whose seeds can serve as livestock feed; Figure 14-1, left), and *Maximilian sunflower* (which produces seeds with as much protein as soybeans).

This important research may eventually help us come closer to producing and distributing enough food to meet everyone's basic nutritional needs and doing this without degrading the soil, water, air, and biodiversity that support all food production. However, this will require more evaluation of the costs involved and the feasibility of integrating such practices into conventional agricultural production systems.

Figure 14-1 Solutions: The Land Institute in Salina, Kansas, is a farm, a prairie laboratory, and a school dedicated to changing the way we grow food. It advocates growing a diverse mixture (polyculture) of edible perennial plants to supplement traditional annual monoculture crops. Two of these perennial crops are eastern grama grass (bottom) and the Illinois bundleflower (top left).

This chapter analyzes the world's crop, meat, and fish production systems and how they can be made more sustainable. It addresses the following questions:

- How is the world's food produced?

- How are green revolution and traditional methods used to raise crops?

- How are soils being degraded and eroded, and what can be done to reduce these losses?

- How much has food production increased, how serious is malnutrition, and what are the environmental effects of producing food?

- How can we increase production of crops, meat, and fish and shellfish?

- How do government policies affect food production?

- How can we design and shift to more sustainable agricultural systems?

14-1 HOW IS FOOD PRODUCED?

What Systems Provide Us with Food? The Challenges Ahead

Croplands, rangelands, and ocean fisheries supply most of our food and since 1950 production from all three systems has increased dramatically.

Historically, humans have depended on three systems for their food supply. *Croplands* mostly produce grains, and provide about 77% of the world's food. *Rangelands* produce meat, mostly from grazing livestock, and supply about 16% of the world's food. *Ocean fisheries* supply about 7% of the world's food.

Since 1950 there has been a staggering increase in global food production from all three systems. This occurred because of technological advances such as increased use of tractors and farm machinery and high-tech fishing boats and gear; inorganic chemical fertilizers; irrigation; pesticides; high-yield varieties of wheat, rice, and corn; densely populated feedlots and enclosed pens for raising cattle, pigs, and chickens; and aquaculture ponds and ocean cages for raising some types of fish and shellfish.

We face important challenges in increasing food production without causing serious environmental harm. To feed the world's 8.9 billion people projected to exist in 2050, we must produce and equitably dis-tribute more food than has been produced since agriculture began about 10,000 years ago, and do it in an environmentally sustainable manner.

Can we do this? Some analysts say we can, mostly by using genetic engineering (Figure 5-11, p. 98). Others have doubts. They are concerned that environmental degradation, pollution, lack of water for irrigation, overgrazing by livestock, overfishing, and loss of vital ecological services may limit future food production. A key problem is that human activities continue to take over or degrade more of the planet's *net primary productivity*, which supports all life.

We also face the challenge of sharply reducing poverty because about one out of five people do not have enough land to grow their own food or enough money to buy sufficient food—regardless of how much is available.

What Plants and Animals Feed the World? Our Three Most Important Crops

Wheat, rice, and corn provide more than half of the calories in the food consumed by the world's people.

The earth has perhaps 30,000 plant species with parts that people can eat. Since the beginning of agriculture about 10,000 of the species have been used as a source of food for people and livestock. Today only 14 plant and 8 terrestrial animal species supply an estimated 90% of our global intake of calories. Just three grain crops—*wheat, rice,* and *corn*—provide more than half the calories people consume. These three grains, and most other food crops, are *annuals,* whose seeds must be replanted each year. This dependence on just a few plant species for food represents a dramatic reduction in agricultural biodiversity.

Two-thirds of the world's people survive primarily on traditional grains (mainly rice, wheat, and corn), mostly because they cannot afford meat. As incomes rise most people consume more meat and other products of domesticated livestock, which in turn means more grain consumption by those animals.

Fish and shellfish are an important source of food for about 1 billion people, mostly in Asia and in coastal areas of developing countries. But on a global scale fish and shellfish supply only 7% of the world's food and about 6% of the protein in the human diet.

What Are the Major Types of Food Production? High-Input and Low-Input Agriculture

About 80% of the world's food supply is produced by industrialized agriculture and 20% by subsistence agriculture.

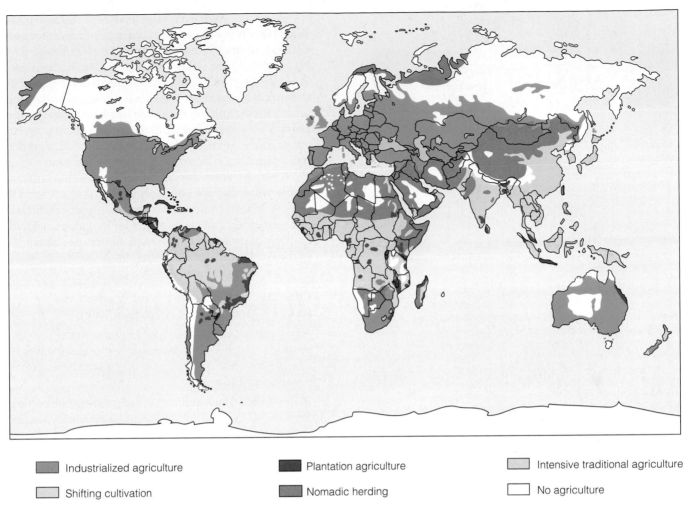

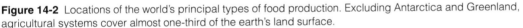

| ■ Industrialized agriculture | ■ Plantation agriculture | ■ Intensive traditional agriculture |
| ■ Shifting cultivation | ■ Nomadic herding | □ No agriculture |

Figure 14-2 Locations of the world's principal types of food production. Excluding Antarctica and Greenland, agricultural systems cover almost one-third of the earth's land surface.

There are two major types of agricultural systems: *industrialized* and *traditional*. **Industrialized agriculture,** or **high-input agriculture,** uses large amounts of fossil fuel energy, water, commercial fertilizers, and pesticides to produce single crops (monocultures) or livestock animals for sale. Practiced on about a fourth of all cropland, mostly in developed countries (Figure 14-2), high-input industrialized agriculture has spread since the mid-1960s to some developing countries.

Plantation agriculture is a form of industrialized agriculture used primarily in tropical developing countries. It involves growing *cash crops* (such as bananas, coffee, soybeans, sugarcane, cocoa, and vegetables) on large monoculture plantations, mostly for sale in developed countries.

An increasing amount of livestock production in developed countries is industrialized. Large numbers of cattle are brought to densely populated *feedlots,* where they are fattened up for about 4 months before slaughter. Most pigs and chickens in developed coun-

tries spend their lives in densely populated pens and cages and eat mostly grain grown on cropland.

Traditional agriculture consists of two main types, which together are practiced by about 2.7 billion people (42% of the world's people) in developing countries and provide about a fifth of the world's food supply. **Traditional subsistence agriculture** typically uses mostly human labor and draft animals to produce only enough crops or livestock for a farm family's survival. Examples of this very low-input type of agriculture include numerous forms of shifting cultivation in tropical forests and nomadic livestock herding.

In **traditional intensive agriculture,** farmers increase their inputs of human and draft labor, fertilizer, and water to get a higher yield per area of cultivated land. They produce enough food to feed their families and to sell for income.

Croplands, like natural ecosystems, provide important ecological and economic services listed in Figure 14-3 (p. 276). Indeed, agriculture is the world's

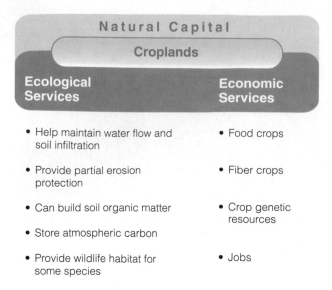

Figure 14-3 Natural capital: ecological and economic services provided by croplands.

largest industry, providing a living for one of every five (1.3 billion) people.

14-2 PRODUCING FOOD BY GREEN REVOLUTION AND TRADITIONAL TECHNIQUES

How Have Green Revolutions Increased Food Production? High-Input Monocultures in Action

Since 1950, most of the increase in global food production has come from using high-input agriculture to produce more crops on each unit of land.

Farmers can produce more food by farming more land or getting higher yields per unit of area from existing cropland. Since 1950, most of the increase in global food production has come from increased yields per unit of area of cropland in a process called the **green revolution.**

The green revolution involves three steps. *First,* develop and plant monocultures (Figure 6-25, p. 118) of selectively bred or genetically engineered high-yield varieties of key crops such as rice, wheat, and corn. *Second,* produce high yields by using large inputs of fertilizer, pesticides, and water. *Third,* increase the number of crops grown per year on a plot of land through *multiple cropping.*

This high-input approach dramatically increased crop yields in most developed countries between 1950 and 1970 in what is called the *first green revolution* (Figure 14-4, blue areas).

A *second green revolution* has been taking place since 1967. It involves introducing fast-growing dwarf varieties of rice (Figure 14-5) and wheat (developed by Norman Bourlag, who later received a Nobel Peace Prize for his work) into several developing countries in tropical and subtropical climates (Figure 14-4, green areas). Producing more food on less land is also an important way to protect biodiversity by saving large areas of forests, grasslands, wetlands, and easily eroded mountain terrain from being used to grow food.

Yield increases depend not only on fertile soil and ample water but also on high inputs of fossil fuels to run machinery, produce and apply inorganic fertilizers and pesticides, and pump water for irrigation. All told, high-input green revolution agriculture uses about 8% of the world's oil output.

Case Study: Industrial Food Production in the United States: A Success Story

America's industrialized agricultural system produces about 17% of the world's grain but has a larger environmental impact than any other American industry.

In the United States industrialized farming has become *agribusiness* as big companies and larger family-owned farms have taken control of almost three-fourths of U.S. food production. According to environmental educator David Orr, "the U.S. food system is increasingly dominated by 'superfarms', which are roughly to farming what WalMart is to retailing."

In total annual sales, agriculture is bigger than the automotive, steel, and housing industries combined. It generates about 18% of the country's gross national income and almost a fifth of all jobs in the private sector, employing more people than any other industry. With only 0.3% of the world's farm labor force, U.S. farms produce about 17% of the world's grain and nearly half of the world's corn and soybean exports.

Since 1950, U.S. farmers have used green revolution techniques to more than double the yield of key crops such as wheat, corn, and soybeans without cultivating more land. Such increases in the yield per hectare of key crops have kept large areas of forests, grasslands, wetlands, and easily erodible land from being converted to farmland.

In addition, the country's agricultural system has become increasingly efficient. While the U.S. output of crops, meat, and dairy products has been increasing steadily since 1975, the major inputs of labor and resources—with the exception of pesticides—to produce each unit of that output have fallen steadily since 1950.

U.S. consumers now spend only about 2% of their income on domestically produced food, compared to about 11% in 1948. Adjusted for inflation, U.S. farm products now cost about one-third of what they did in

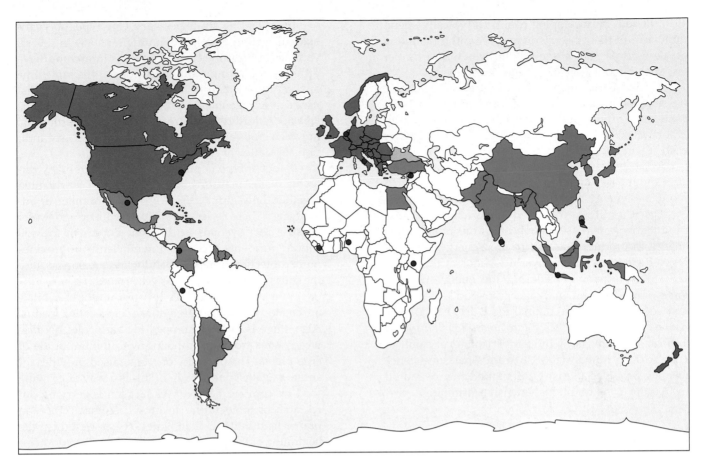

First green revolution (developed countries)

Second green revolution (developing countries)

● Major international agricultural research centers and seed banks

Figure 14-4 Countries whose crop yields per unit of land area increased during the two green revolutions. The first (blue) took place in developed countries between 1950 and 1970; the second (green) has occurred since 1967 in developing countries with enough rainfall or irrigation capacity. Several agricultural research centers and gene or seed banks (red dots) play a key role in developing high-yield crop varieties.

International Rice Research Institute, Manila

Figure 14-5 Solutions: high-yield, semidwarf variety of rice called IR-8 (left), developed as part of the second green revolution. Crossbreeding two parent strains of rice produced it: PETA from Indonesia (center) and DGWG from China (right). The shorter and stiffer stalks of the new variety allow the plants to support larger heads of grain without toppling over and increase the benefit of applying more fertilizer.

1910. People in developing countries typically spend up to 40% of their income on food. And the 1.1 billion of the world's poor struggling to live on $1 a day or less spend about 70% of their income on food.

The industrialization of agriculture has been made possible by the availability of cheap energy, most of it from oil. Putting food on the table consumes about 17% of all commercial energy used in the United States each year (Figure 14-6). *Good news.* The input of energy needed to produce a unit of food has fallen considerably and most plant crops in the United States provide more food energy than the energy used to grow them.

Bad news. If we include livestock, the U.S. food production system uses about three units of fossil fuel energy to produce one unit of food energy. That energy efficiency is much lower if we look at the whole U.S. food system. Considering the energy used to grow, store, process, package, transport, refrigerate, and cook all plant and animal food, *about 10 units of nonrenewable fossil fuel energy are needed to put 1 unit of food energy on the table.* By comparison, every unit of energy from human labor in traditional subsistence farming provides at least 1 unit of food energy and up to 10 units using traditional intensive farming.

What Growing Techniques Are Used in Traditional Agriculture? Low-Input Agrodiversity in Action

Many traditional farmers in developing countries use low-input agriculture to produce a variety of different crops on each plot of land.

Traditional farmers in developing countries grow about one-fifth of the world's food on about three-fourths of its cultivated land. Many traditional farmers simultaneously grow several crops on the same plot, a practice known as **interplanting.** Such crop diversity reduces the chance of losing most or all of the year's food supply to pests, bad weather, and other misfortunes.

Interplanting strategies vary. One type, **polyvarietal cultivation,** involves planting a plot with several varieties of the same crop. Another is **intercropping**—growing two or more different crops at the same time on a plot (for example, a carbohydrate-rich grain that uses soil nitrogen and a protein-rich legume that puts it back). A third type is **agroforestry,** or **alley cropping,** in which crops and trees are grown together (see Individuals Matter, at right).

A fourth type is **polyculture,** in which many different plants maturing at various times are planted together. Low-input *polyculture* has a number of advantages. There is less need for fertilizer and water because root systems at different depths in the soil capture nutrients and moisture efficiently. It provides more protection from wind and water erosion because the soil is covered with crops year-round. There is little or no need for insecticides because multiple habitats are created for natural predators of crop-eating insects. Also, there is little or no need for herbicides because weeds have trouble competing with the multitude of crop plants. The diversity of crops raised provides insurance against bad weather. This is a way of growing food by copying nature. Wes Jackson is carrying out research on polyculture to grow perennial crops on prairie land in the United States (see case study at the beginning of this chapter).

Recent ecological research found that on average, low-input polyculture produces higher yields per hectare of land than high-input monoculture. For example, a 2001 study by ecologists Peter Reich and David Tilman found that carefully controlled polyculture plots with 16 different species of plants consistently outproduced plots with 9, 4, or only 1 type of plant species.

Traditional farmers in arid and semiarid areas with low natural soil fertility have developed innovative methods to boost crop production. For example, in the African countries of Niger and Burkina Faso, a

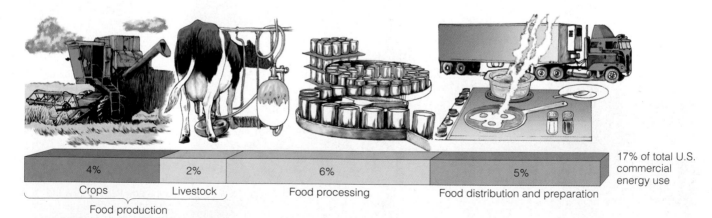

| 4% | 2% | 6% | 5% |
| Crops | Livestock | Food processing | Food distribution and preparation |

Food production

17% of total U.S. commercial energy use

Figure 14-6 In the United States, industrialized agriculture uses about 17% of all commercial energy. In the United States, food travels an average of 2,400 kilometers (1,500 miles) from farm to table. (Data from David Pimentel and Worldwatch Institute)

Low-Tech Sustainable Agriculture in Africa

In 2003, Cuban-born soil scientist Pedro Sanchez was awarded the World Food Prize, the "Nobel Prize" of agriculture. He received it for developing a low-tech and sustainable form of agriculture that has increased crop yields fourfold, restored depleted soils, and helped more than 150,000 Africans (most in sub-Saharan Africa) escape from hunger and poverty.

Since 1992 Sanchez has been working with scientists in Kenya, Africa, to develop an effective cultivation system that grows crops and trees together.

A basic problem is that many African soils are thin and often depleted of nitrogen and phosphorus after several years of intense crop growing. Adding commercial inorganic fertilizers would help, but most African farmers cannot afford them.

Sanchez and his colleagues developed the following soil replenishment and crop-growth system to deal with these problems. *First*, at the beginning of the rainy season farmers plant corn in rows between local varieties of fast-growing trees (Figure 14-14c, p. 285). The corn is harvested and the trees are allowed to grow for a year.

Second, just before the second corn-planting season the trees are cut down and their leaves are dug into the soil to add nitrogen. In addition, the trees can be used for firewood, which also helps prevent deforestation.

Third, phosphorus is added to the soil by crushing small deposits of phosphate rock found throughout much of Africa. Africa's mildly acidic soils help dissolve the phosphate fertilizer into the soluble form of phosphate needed by corn plants.

Fourth, farmers chop up the leaves and stems of a weedy shrub called the Mexican sunflower that is found along many roadside and farm boundaries. They place the plant pieces in planting holes with corn seed to provide micronutrients needed for healthy crop growth.

This four-part technologically simple system can then be used to provide food and restore depleted soil on a more sustainable basis. The system also helps empower the women who raise most of the crops in Africa by bringing in extra income.

---

farmer innovation called *tassas* has tripled yields on at least 100,000 hectares (250,000) acres of unproductive land. *Tassas* are small pits dug in the soil, filled with manure, and then planted with crops once they fill with infrequent rain.

14-3 SOIL EROSION AND DEGRADATION

What Causes Soil Erosion? The Big Three

Water, wind, and people cause soil erosion.

Most people in developed countries get their food from grocery stores, fast-food chains, and restaurants. But we need to remind ourselves that *all food comes from the earth or soil*—the base of life. This explains why preserving the world's topsoil (Figure 4-25, p. 73) is the key to producing enough food to feed the world's growing population.

Land degradation occurs when natural or human-induced processes decrease the future ability of land to support crops, livestock, or wild species. One type of land degradation is **soil erosion:** the movement of soil components, especially surface litter and topsoil from one place to another. The two main agents of erosion are *flowing water* and *wind,* with water causing most soil erosion (see photo on p. ix, right).

Some soil erosion is natural and some is caused by human activities. In undisturbed vegetated ecosystems, the roots of plants help anchor the soil, and usually soil is not lost faster than it forms. Soil becomes more vulnerable to erosion through human activities that destroy plant cover, including farming, logging, construction, overgrazing by livestock, off-road vehicle use, and deliberate burning of vegetation.

Soil erosion has two major harmful effects. One is *loss of soil fertility* through depletion of plant nutrients in topsoil. The other harmful effect occurs when eroded soil ends up as sediment in nearby surface waters, where it can pollute water, kill fish and shellfish, and clog irrigation ditches, boat channels, reservoirs, and lakes.

Soil, especially topsoil, is classified as a renewable resource because natural processes regenerate it. However, if topsoil erodes faster than it forms on a piece of land, it eventually becomes nonrenewable.

How Serious Is Global Soil Erosion? Mostly Bad News

Soil is eroding faster than it is forming on more than a third of the world's cropland, and much of this land also suffers from salt buildup and waterlogging.

A 1992 joint survey by the United Nations (UN) Environment Programme and the World Resources Institute estimated that topsoil is eroding faster than it forms on about 38% of the world's cropland (Figure 14-7, p. 280). According to a 2000 study by the Consultative Group on International Agricultural Research, soil erosion and degradation has reduced food production on about 16% of the world's cropland.

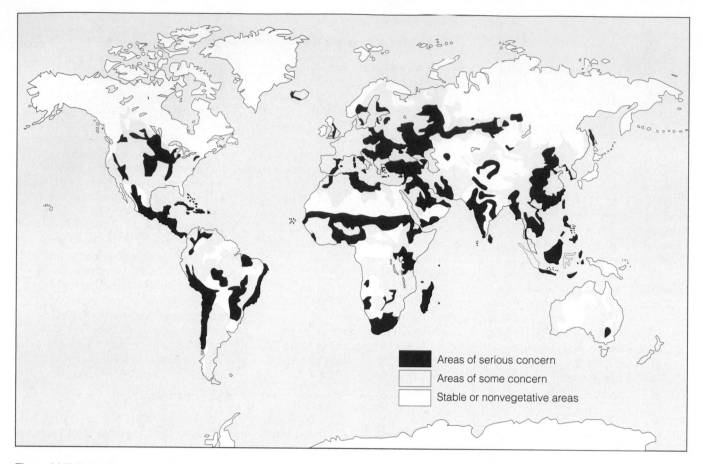

Figure 14-7 **Natural capital degradation:** global soil erosion. (Data from UN Environment Programme and the World Resources Institute)

Soil expert David Pimentel estimates that worldwide soil erosion causes damages of at least $375 billion per year (an average of $42 million per hour), including direct damage to agricultural lands and indirect damage to waterways, infrastructure, and human health. See his Guest Essay on this subject on the website for this chapter.

Some analysts contend that erosion estimates are overstated because they underestimate the abilities of some local farmers to restore degraded land. The UN Food and Agriculture Organization (FAO) also points out that much of the eroded topsoil does not go far and is deposited further down a slope, valley, or plain. In some places, the loss in crop yields in one area could be offset by increased yields elsewhere.

Case Study: Soil Erosion in the United States Today; *Some* Hopeful *News*

Soil in the United States erodes faster than it forms on most cropland, but since 1987 erosion has been cut by about two-thirds.

In the 1930s, several midwestern states lost large amounts of topsoil as a result of poor cultivation prac-

tices and prolonged drought (Case Study, next page). This taught the country some important lessons about the need for soil conservation.

The situation has improved dramatically since then. But according to the Natural Resources Conservation Service, soil on cultivated land in the United States is still eroding about 16 times faster than it can form. Erosion rates are even higher in heavily farmed regions. An example is the Great Plains, which has lost one-third or more of its topsoil in the 150 years since it was first plowed.

Good news. Of the world's major food-producing nations, only the United States is sharply reducing some of its soil losses through a combination of planting crops without disturbing the soil and government-sponsored soil conservation programs.

The 1985 Food Security Act (Farm Act) established a strategy for reducing soil erosion in the United States. In the first phase of this program, farmers receive a subsidy for taking highly erodible land out of production and replanting it with soil-saving grass or trees for 10–15 years. In 2003, roughly one-tenth of U.S. cropland was in this Conservation Reserve Program (CRP).

According to the U.S. Department of Agriculture, since 1985 this program has cut soil losses on cropland in the United States by about two-thirds. And between 1982 and 1997, the area of U.S. farmland with the greatest potential for wind erosion and water erosion decreased by nearly one-third. If lawmakers continue to support this program, it could eventually cut such soil losses as much as 80%.

A second provision of the Farm Act authorizes the government to forgive all or part of farmers' debts to the Farmers Home Administration if they agree not to farm highly erodible cropland or wetlands for 50 years. The farmers must plant trees or grass on this land or restore it to wetland.

These efforts to slow soil erosion are important. But effective soil conservation is practiced today on only about half of all U.S. agricultural land and on about half of the country's most erodible cropland.

Case Study: The Dust Bowl: An Environmental Lesson from Nature

In the 1930s, a large area of cropland in the midwestern United States had to be abandoned because of severe soil erosion caused by a combination of poor cultivation practices and prolonged drought.

In the 1930s, Americans learned a harsh environmental lesson when much of the topsoil in several dry and windy midwestern states was lost through a combination of poor cultivation practices and prolonged drought.

Before settlers began grazing livestock and planting crops there in the 1870s, the deep and tangled root systems of native prairie grasses anchored the fertile topsoil firmly in place. But plowing the prairie tore up these roots, and the agricultural crops the settlers planted annually in their place had less extensive root systems.

After each harvest, the land was plowed and left bare for several months, exposing it to high winds. Overgrazing by livestock in some areas also destroyed large expanses of grass, denuding the ground.

The stage was set for severe wind erosion and crop failures; all that was needed was a long drought. One came between 1926 and 1937 when the annual precipitation dropped by about almost two-thirds. In the 1930s, dust clouds created by hot, dry windstorms blowing across the barren exposed soil darkened the sky at midday in some areas; rabbits and birds choked to death on the dust.

During May 1934, a cloud of topsoil blown off the Great Plains traveled some 2,400 kilometers (1,500 miles) and blanketed most of the eastern United States with dust. Laundry hung out to dry by women in Georgia quickly became covered with dust blown in from the Midwest. Journalists gave the worst-hit

Figure 14-8 The *Dust Bowl* of the Great Plains, where a combination of extreme drought and poor soil conservation practices led to severe wind erosion of topsoil in the 1930s.

part of the Great Plains a new name: the *Dust Bowl* (Figure 14-8).

During the "dirty thirties," large areas of cropland were stripped of topsoil and severely eroded. This triggered one of the largest internal migrations in U.S. history. Thousands of farm families from Oklahoma, Texas, Kansas, and Colorado abandoned their dust-choked farms and dead livestock and migrated to California or to the industrial cities of the Midwest and East. Most found no jobs because the country was in the midst of the Great Depression.

In May 1934, Hugh Bennett of the U.S. Department of Agriculture (USDA) went before a congressional hearing in Washington to plead for new programs to protect the country's topsoil. Lawmakers took action when Great Plains dust began seeping into the hearing room.

In 1935, the United States passed the *Soil Erosion Act*, which established the Soil Conservation Service (SCS) as part of the USDA. With Bennett as its first head, the SCS (now called the Natural Resources Conservation Service) began promoting sound soil conservation practices, first in the Great Plains states and later elsewhere. Soil conservation districts were formed throughout the country, and farmers and ranchers were given technical assistance in setting up soil conservation programs.

What Is Desertification, and How Serious Is It? Decreasing Land Productivity

About one-third of the world's land has lost some of its productivity from a combination of drought and human activities that reduce or degrade topsoil.

In **desertification,** the productive potential of arid or semiarid land falls by 10% or more because of a combination of natural climate change that causes prolonged drought and human activities that reduce or degrade

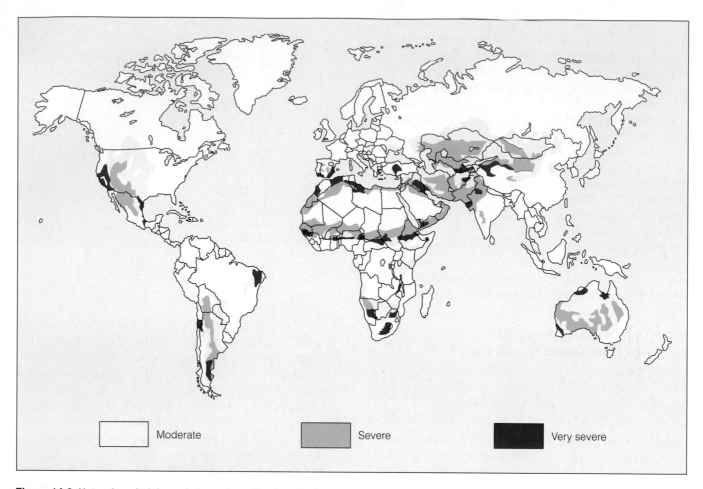

| | Moderate | | Severe | | Very severe |

Figure 14-9 Natural capital degradation: desertification of arid and semiarid lands. It is caused by a combination of prolonged drought and human activities that expose soil to erosion. (Data from UN Environment Programme and Harold E. Drengue)

topsoil. The process can be *moderate* (a 10–25% drop in productivity), *severe* (a 25–50% drop), or *very severe* (a drop of 50% or more, usually creating huge gullies and sand dunes). Note that only in extreme cases does desertification lead to what we call desert.

Over thousands of years the earth's deserts have expanded and contracted, mostly because of natural climate changes. However, human activities can accelerate desertification in some parts of the world (Figure 14-9). Study Figure 14-9 to find out the areas of the world most affected by desertification. Is it a problem where you live?

According to a 2003 UN conference on desertification, about a third of the world's land and 70% of all drylands is suffering from the effects of desertification. UN officials estimate that this loss of soil productivity threatens the livelihoods of at least 250 million people in 110 countries (70 in Africa). China is facing serious desertification, as its portion of the Gobi Desert expanded by an area half the size of Pennsylvania between 1994 and 1999.

Figure 14-10 summarizes the major causes and consequences of desertification. We cannot control when or where prolonged droughts may occur, but we can reduce overgrazing, deforestation, and destructive forms of planting, irrigation, and mining that leave soil barren. We can also restore land suffering from desertification by planting trees and grasses that anchor soil and hold water.

How Do Excess Salts and Water Degrade Soils? Crop Losses from Too Much Salt and Water

Repeated irrigation can cause loss of crop productivity by salt buildup in the soil and waterlogging of crop plants.

The one-fifth of the world's cropland that is irrigated produces almost 40% of the world's food. But irrigation has a downside. Most irrigation water is a dilute solution of various salts, picked up as the water flows over or through soil and rocks. Irrigation water not ab-

Causes		Consequences
Overgrazing		Worsening drought
Deforestation		Famine
Erosion		Economic losses
Salinization		Lower living standards
Soil compaction		Environmental refugees
Natural climate change		

Figure 14-10 Causes and consequences of desertification.

sorbed into the soil evaporates, leaving behind a thin crust of dissolved salts (such as sodium chloride) in the topsoil.

Repeated annual applications of irrigation water lead to the gradual accumulation of salts in the upper soil layers. This accumulation of salts is called **salinization** (Figure 14-11). It stunts crop growth, lowers crop yields, and eventually kills plants and ruins the land (see figure on p. ix, left).

According to a 1995 study, severe salinization has reduced yields on about a fifth of the world's irrigated cropland, and almost another third has been moderately salinized. The most severe salinization occurs in Asia, especially in China, India, and Pakistan.

Salinization affects almost one-fourth of irrigated cropland in the United States. But the proportion is much higher in some heavily irrigated western states.

We know how to prevent and deal with soil salinization, as summarized in Figure 14-12. But some of these remedies are expensive.

Another problem with irrigation is **waterlogging** (Figure 14-11). Farmers often apply large amounts of irrigation water to leach salts deeper into the soil. But without adequate drainage, water accumulates underground and gradually raises the water table. Saline

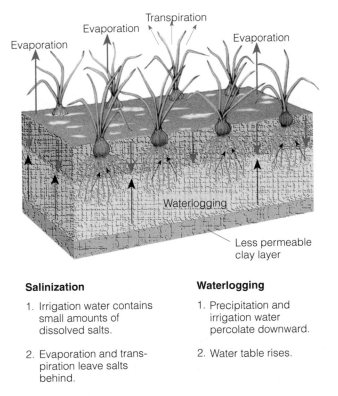

Salinization

1. Irrigation water contains small amounts of dissolved salts.

2. Evaporation and transpiration leave salts behind.

3. Salt builds up in soil.

Waterlogging

1. Precipitation and irrigation water percolate downward.

2. Water table rises.

Figure 14-11 Natural capital degradation: *salinization* and *waterlogging* of soil on irrigated land without adequate drainage can decrease crop yields.

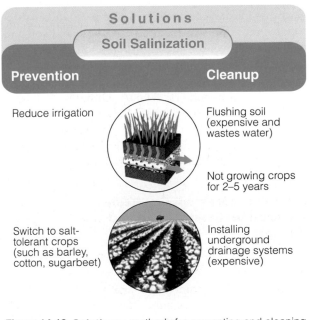

Solutions

Soil Salinization

Prevention	Cleanup
Reduce irrigation	Flushing soil (expensive and wastes water)
	Not growing crops for 2–5 years
Switch to salt-tolerant crops (such as barley, cotton, sugarbeet)	Installing underground drainage systems (expensive)

Figure 14-12 Solutions: methods for preventing and cleaning up soil salinization.

water then envelops the deep roots of plants, lowering their productivity and killing them after prolonged exposure. At least one-tenth of the world's irrigated cropland suffers from waterlogging, and the problem is getting worse.

14-4 SOIL CONSERVATION

How Can Conservation Tillage Reduce Soil Erosion? Do Not Disturb the Soil

Modern farm machinery can plant crops without disturbing the soil.

Soil conservation involves using ways to reduce soil erosion and restore soil fertility. For hundreds of years, farmers have used various methods to reduce soil erosion, mostly by keeping the soil covered with vegetation.

In **conventional-tillage farming**, farmers plow the land and then break up and smooth the soil to make a planting surface. In areas such as the midwestern United States, harsh winters prevent plowing just before the spring growing season. Thus crop fields often are plowed in the fall. This leaves the soil bare during the winter and early spring and makes it vulnerable to erosion.

Many U.S. farmers use **conservation-tillage farming** to disturb the soil as little as possible while planting crops. With *minimum-tillage farming,* the soil is not disturbed over the winter. Then at planting time special tillers break up and loosen the subsurface soil without turning over the topsoil, previous crop residues, or any cover vegetation. In *no-till farming,* special planting machines inject seeds, fertilizers, and weed killers (herbicides) into thin slits made in the unplowed soil and then smooth over the cut. Figure 14-13 lists the advantages and disadvantages of conservation tillage.

In 2003, farmers used conservation tillage on about 45% of U.S. cropland. The USDA estimates that using conservation tillage on 80% of U.S. cropland would reduce soil erosion by at least half. Conservation tillage also has great potential to reduce soil erosion and raise crop yields in the Middle East and in Africa.

What Other Methods Can Reduce Soil Erosion? Several Tried and True Methods

Farmers have developed a number of ways to grow crops that reduce soil erosion.

Figure 14-14 show some of the methods farmers have used to reduce soil erosion. One is **terracing,** which can reduce soil erosion on steep slopes by converting the land into a series of broad, nearly level terraces that run across the land contour (Figure 14-14a). This

retains water for crops at each level and reduces soil erosion by controlling runoff.

Another method is **contour farming**, which involves plowing and planting crops in rows across the slope of the land rather than up and down (Figure 14-14b). Each row acts as a small dam to help hold soil and to slow water runoff.

Farmers also use **strip cropping** to reduce soil erosion (Figure 14-14b). It involves planting alternating strips of a row crop (such as corn or cotton) and another crop that completely covers the soil (such as grass or a grass and legume mixture). The cover crop traps soil that erodes from the row crop, catches and reduces water runoff, and helps prevent the spread of pests and plant diseases.

One way to reduce erosion is to leave crop residues on the land after the crops are harvested. Another is to plant **cover crops** such as alfalfa, clover, or rye immediately after harvest to help protect and hold the soil.

Another method for slowing erosion is **alley cropping** or **agroforestry,** in which several crops are planted together in strips or alleys between trees and shrubs that can provide fruit or fuelwood (Figure 14-14c). The trees or shrubs provide shade (which reduces water loss by evaporation) and help retain and slowly release soil moisture. They also can provide fruit, fuelwood, and trimmings that can be used as mulch (green manure) for the crops and as fodder for livestock.

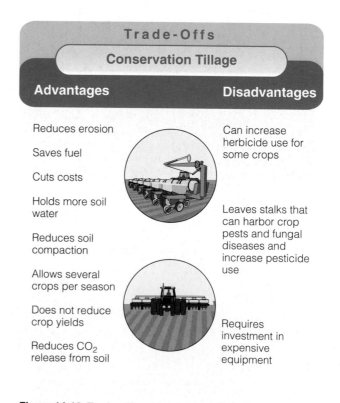

Trade-Offs

Conservation Tillage

Advantages	Disadvantages
Reduces erosion	Can increase herbicide use for some crops
Saves fuel	
Cuts costs	
Holds more soil water	Leaves stalks that can harbor crop pests and fungal diseases and increase pesticide use
Reduces soil compaction	
Allows several crops per season	
Does not reduce crop yields	Requires investment in expensive equipment
Reduces CO_2 release from soil	

Figure 14-13 Trade-offs: advantages and disadvantages of using *conservation tillage*. Pick the single advantage and disadvantage that you think are the most important.

(a) Terracing

(b) Contour planting and strip cropping

(c) Alley cropping

(d) Windbreaks

Figure 14-14 Solutions: in addition to conservation tillage, soil conservation methods include (a) terracing, (b) contour planting and strip cropping, (c) alley cropping or agroforestry, and (d) windbreaks.

Some farmers establish **windbreaks,** or **shelterbelts,** of trees (Figure 14-14d) to reduce wind erosion, help retain soil moisture, supply wood for fuel, and provide habitats for birds, pest-eating and pollinating insects, and other animals.

Some governments use *land classification* to identify easily erodible (marginal) land that should be neither planted in crops nor cleared of vegetation. In the United States, the Natural Resources Conservation Service has set up a classification system to identify types of land that are suitable or unsuitable for cultivation.

How Can We Maintain and Restore Soil Fertility? Conservation and Fertilizers

Soil conservation can reduce loss of soil nutrients, and applying inorganic and organic fertilizers can help restore lost nutrients.

The best way to maintain soil fertility is through soil conservation. The next best thing to do is to restore some of the plant nutrients that have been washed, blown, or leached out of soil or removed by repeated crop harvesting.

Fertilizers can partially restore lost plant nutrients. Farmers can use **organic fertilizer** from plant and animal materials or **commercial inorganic fertilizer** produced from various minerals.

There are several types of *organic fertilizer.* One is **animal manure:** the dung and urine of cattle, horses, poultry, and other farm animals. It improves soil structure, adds organic nitrogen, and stimulates beneficial soil bacteria and fungi.

Manure use in the United States has decreased because most farmers no longer raise crops and livestock on the same farm and it costs too much to transport animal manure from feedlots near urban areas to distant rural crop-growing areas. Also, tractors and other motorized farm machinery have largely replaced horses and other draft animals that added manure to the soil.

Burning poultry wastes to produce electricity leaves a phosphorus-rich ash. Researchers at the U.S. Department of Agriculture are evaluating its value as an organic fertilizer. Also, Canada-based International

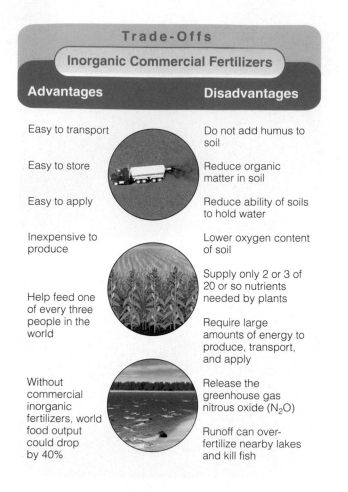

Trade-Offs

Inorganic Commercial Fertilizers

Advantages	Disadvantages
Easy to transport	Do not add humus to soil
Easy to store	Reduce organic matter in soil
Easy to apply	Reduce ability of soils to hold water
Inexpensive to produce	Lower oxygen content of soil
Help feed one of every three people in the world	Supply only 2 or 3 of 20 or so nutrients needed by plants
	Require large amounts of energy to produce, transport, and apply
Without commercial inorganic fertilizers, world food output could drop by 40%	Release the greenhouse gas nitrous oxide (N_2O)
	Runoff can over-fertilize nearby lakes and kill fish

Figure 14-15 Trade-offs: advantages and disadvantages of using *inorganic commercial fertilizers* to enhance or restore soil fertility. Pick the single advantage and disadvantage that you think are the most important.

Bio-Recovery Corporation has developed a bacterial process that converts biodegradable human and animal wastes into pathogen free, nutrient rich organic fertilizer in only 72 hours.

A second type of organic fertilizer called **green manure** consists of freshly cut or growing green vegetation plowed into the soil to increase the organic matter and humus available to the next crop.

A third type is **compost,** produced when microorganisms in soil break down organic matter such as leaves, food wastes, paper, and wood in the presence of oxygen.

Some farmers also use *spores of mushrooms,* including puffballs and truffles, as organic fertilizer. The spores take in more moisture and nutrients from the soil. Unlike typical fertilizers that farmers must apply every few weeks, one application of mushroom fungi lasts all year and costs just pennies per plant.

Crops such as corn, tobacco, and cotton can deplete nutrients (especially nitrogen) in the topsoil if planted on the same land several years in a row. One way to reduce such losses is **crop rotation.** Farmers plant areas or strips with nutrient-depleting crops one year. The next year they plant the same areas with legumes (whose root nodules add nitrogen to the soil). A typical rotation is corn → soybeans (a legume) → oats → alfalfa (a legume). In addition to helping restore soil nutrients, this method reduces erosion by keeping the soil covered with vegetation. It also helps reduce crop losses to insects by presenting them with a changing target.

Can Inorganic Fertilizers Save the Soil? A Partial Solution

Inorganic fertilizers can help restore soil fertility if they are used with organic fertilizers and their harmful environmental effects are controlled.

Many farmers (especially in developed countries) rely on *commercial inorganic fertilizers. The active ingredients typically are inorganic compounds* containing *nitrogen, phosphorus,* and *potassium.* Other plant nutrients may also be present in low or trace amounts. These fertilizers account for about one-fourth of the world's crop yield. According to Canadian geographer Vaclav Smil, without synthetic inorganic fertilizer we could only feed 2–3 million people.

Figure 14-15 lists the advantages and disadvantages of using inorganic fertilizers to enhance or restore soil fertility. Inorganic chemical fertilizers can replace depleted inorganic nutrients, but they do not replace organic matter. Thus for healthy soil, both inorganic and organic fertilizers should be used.

14-5 FOOD PRODUCTION, NUTRITION, AND ENVIRONMENTAL EFFECTS

How Much Has Food Production Increased? Impressive Gains that Are Slowing Down

After increasing significantly since 1950, global grain production has mostly leveled off since 1985, and per capita grain production has declined since 1978.

After almost tripling between 1950 and 1985, world grain production has essentially leveled off (Figure 14-16, left). And after rising by about 36% between 1950 and 1978, per capita food production has declined (Figure 14-16, right). The sharpest drops in per capita food production have occurred in Africa since 1970, in the former Soviet Union since 1990, and in China since 1998.

Good news. We produce more than enough food to meet the basic nutritional needs of every person on the earth. *Bad news:* one out of six people in developing countries are not getting enough to eat because food is not distributed equally among the world's people. This

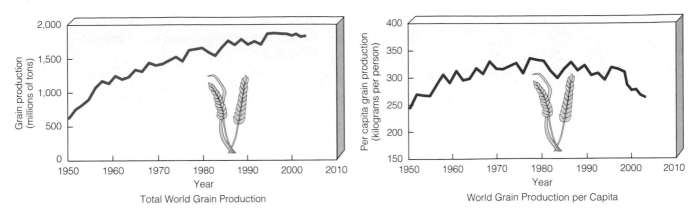

Figure 14-16 Total worldwide grain production of wheat, corn, and rice (left), and per capita grain production (right), 1950–2003. In order, the world's three largest grain-producing countries are China, the United States, and India. (U.S. Department of Agriculture, Worldwatch Institute, UN Food and Agriculture Organization, and Earth Policy Institute)

occurs because of differences in soil, climate, political and economic power, and average per capita income.

Most agricultural experts agree that *the root causes of hunger and malnutrition are and will continue to be poverty and inequality,* which prevent poor people from growing or buying enough food regardless of how much is available. Other factors are war, corruption, and tariffs and subsidies that make it hard for poor people to see excess food they produce.

How Serious Are Undernutrition and Malnutrition? Some Progress

Some people cannot grow or buy enough food to meet their basic energy needs, and others do not get enough protein and other key nutrients.

To maintain good health and resist disease, we need fairly large amounts of *macronutrients* (such as protein, carbohydrates, and fats), and smaller amounts of *micronutrients* consisting of various vitamins (such as A, C, and E) and minerals (such as iron, iodine, and calcium).

People who cannot grow or buy enough food to meet their basic energy needs suffer from **chronic undernutrition.** Chronically undernourished children are likely to suffer from mental retardation and stunted growth. They are also susceptible to infectious diseases such as diarrhea and measles that rarely kill children in developed countries.

Many of the world's poor can only afford to live on a low-protein, high-carbohydrate diet consisting of grains such as wheat, rice, or corn. Many suffer from **malnutrition** resulting from deficiencies of protein and other key nutrients.

The two most common nutritional deficiency diseases are marasmus and kwashiorkor. *Marasmus* (from the Greek word *marasmos*, "to waste away") occurs when a diet is low in both calories and protein. Most victims are either nursing infants of malnourished mothers or children who do not get enough food after being weaned from breast-feeding. A child suffering from severe marasmus is usually very thin and shriveled and looks like a very old miniature starving person (Figure 1-12, p. 13). *Good news.* If the child is treated in time with a balanced diet, most of these effects can be reversed.

Kwashiorkor (meaning "displaced child" in a West African dialect) is a severe protein deficiency occurring in infants and children ages 1–3, usually after the arrival of a new baby deprives them of breast milk. The displaced child's diet changes to grain or sweet potatoes, which provide enough calories but not enough protein. Such children typically have a bloated belly, reddish-orange hair, and discolored and puffy skin. *Good news.* If caught soon enough, most of the harmful effects can be cured with a balanced diet. Otherwise, children who survive their first year or two suffer from stunted growth and mental retardation.

Good news. According to the UN Food and Agriculture Organization (FAO), the average daily food intake in calories per person in the world and in developing countries rose sharply between 1961 and 2003, and is projected to continue rising through 2030 (Figure 14-17). Also, the estimated number of chronically undernourished or malnourished people fell from 918 million in 1970 to 825 million in 2001, about 95% of them in developing countries.

Bad news. About one of every six people in developing countries (including about one of every three children below age 5) is chronically undernourished or malnourished. The FAO estimates that at least 5.5 million people die prematurely from a combination of poverty, undernutrition, malnutrition, and increased susceptibility to normally nonfatal infectious diseases

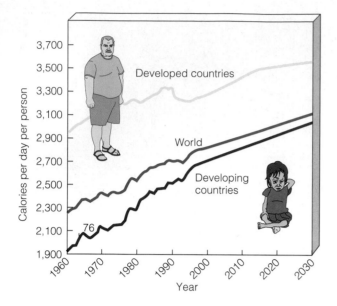

Figure 14-17 The average daily food intake in calories per person in the world, developing countries, and developed countries: 1961–2003 and projected increases to 2030. The average adult male needs about 2,500 calories per day for good health. (Data from UN Food and Agriculture Organization)

(such as measles and diarrhea) because of their weakened condition. This means that each day an average of 15,100 people—80% of them children under age 5—die prematurely from these causes related to poverty.

Studies by the United Nations Children's Fund (UNICEF) indicate that one-half to two-thirds of childhood deaths from nutrition-related causes could be prevented at an average annual cost of $5–10 per child by taking the following measures:

- Immunizing children against childhood diseases such as measles

- Encouraging breast-feeding (except for mothers with AIDS)

- Preventing dehydration from diarrhea by giving infants a mixture of sugar and salt in a glass of water

- Preventing blindness by giving children a vitamin A capsule twice a year at a cost of about 75¢ per child or fortifying common foods with vitamin A and other micronutrients at a cost of about 10¢ per child annually

- Providing family planning services to help mothers space births at least 2 years apart

- Increasing education for women, with emphasis on nutrition, drinking water sterilization, and child care

Some people in developed countries also suffer from lack of access to enough food for good health. In the United States, about 11 million people (half of them children under age 5) do not have access to enough food on a regular basis for good health.

How Serious Are Micronutrient Deficiencies? Important but Limited Progress

One of every three persons has a deficiency of one or more vitamins and minerals, especially vitamin A, iron, and iodine.

According to the World Health Organization (WHO), about one out of three people suffer from a deficiency of one or more vitamins and minerals. The most widespread micronutrient deficiencies in developing countries involve *vitamin A, iron,* and *iodine.*

According to the WHO, 120–140 million children in developing countries are deficient in vitamin A. Globally about 250,000 children under age 6 go blind each year from a lack of vitamin A and up to 80% of them die within a year.

Scientists recently spliced genes into rice to make it rich in beta-carotene, the source of vitamin A. Eating normal amounts of this vitamin-fortified rice—called Golden Rice—should provide 20–40% of the daily requirements of vitamin A. But the beta-carotene in this rice is not converted to vitamin A in the body of a poorly nourished person.

Other nutritional deficiency diseases are caused by lack of minerals. Too little *iron*—a component of hemoglobin that transports oxygen in the blood—causes *anemia.* According to a 1999 survey by the WHO, one of every three people in the world, mostly women and children in tropical developing countries, suffers from too little iron. Iron deficiency causes fatigue, makes infection more likely, and increases a woman's chances of dying in childbirth and an infant's chances of dying of infection during its first year of life.

Elemental *iodine* is essential for proper functioning of the thyroid gland, which produces a hormone that controls the body's rate of metabolism. Chronic lack of iodine, found in seafood and crops grown in iodine-rich soils, can cause stunted growth, mental retardation, and goiter—an abnormal enlargement of the thyroid gland that can lead to deafness. According to the United Nations, about 26 million children suffer brain damage each year from lack of iodine and 600 million—mostly in South and Southeast Asia—suffer from goiter.

How Serious Is Overnutrition? Bad and Getting Worse

In developed countries overnutrition is a major cause of preventable deaths.

Overnutrition occurs when food energy intake exceeds energy use and causes excess body fat. Overnourished people are classified as *overweight* if they are roughly 4.5–14 kilograms (10–30 pounds) over a healthy body weight and *obese* if they are more than 14 kilograms (30 pounds) over a healthy weight. Too many calories, too little exercise, or both can cause overnutrition.

People who are underfed and underweight and those who are overfed and overweight face similar health problems: *lower life expectancy, greater susceptibility to disease and illness,* and *lower productivity and life quality.* We live in a world where 1 billion people have health problems because they do not get enough to eat and 1.7 billion worry about health problems from eating too much. According to a 2004 study by the International Obesity Task Force, about 1 of every 4 people in the world are overweight and 5% are obese.

In developed countries, overnutrition is the second leading cause of preventable deaths after smoking, mostly from heart disease, cancer, stroke, and diabetes. About one out of seven adults in developed countries suffer from overnutrition. According to the Centers for Disease Prevention and Control, about two-thirds of Americans adults are overweight and almost one-third is obese—the highest overnutrition rate of any developed country. The $40 billion Americans spend each year trying to lose weight is 1.7 times more than the $19 billion per year needed to eliminate undernutrition and malnutrition in the world. More than half of all adults are overweight in Russia, the United Kingdom, and Germany compared to 15% in China.

In 2004, the World Health Organization urged governments to discourage food and beverage ads that exploit children; tax less-healthy foods; and limit high-fat and high-sugar foods in schools.

What Are the Environmental Effects of Producing Food? Agriculture Is Number One

Modern agriculture has a greater harmful environmental impact than any other human activity, and these effects may limit future food production.

Modern agriculture has significant harmful effects on air, soil, water, and biodiversity, as Figure 14-18 (p. 290) shows. According to many analysts, agriculture has a greater harmful environmental impact than any other human activity!

Some analysts believe these harmful environmental effects can be overcome and will not limit future food production. Other analysts disagree. For example, according to Norman Myers, a combination of environmental factors may limit future food production. They include *soil erosion, salt buildup and waterlogging of soil on irrigated lands, water deficits and droughts,* and *loss of wild species* that provide the genetic resources for improved forms of foods.

According to a 2002 study by the UN Department for Economic and Social Affairs, close to 30% of the world's cropland has been degraded to some degree by soil erosion, salt buildup, and chemical pollution, and 17% has been seriously degraded. Such environmental factors may limit food production in India and China (Case Study, below), the world's two most populous countries.

Case Study: Can China's Population Be Fed? A Precarious Situation

Population growth, economic growth, lack of resources, and the harmful environmental effects of food production may limit crop production in China.

Since 1970, China has made significant progress in feeding its people and slowing its rate of population growth. But there is concern that crop yields may not be able to keep up with demand because of its growing population and economic development. A basic problem is that with 20% of the world's people, China has only 7% of the world's cropland and fresh water, 4% of its forests, and 2% of its oil.

Between 1998 and 2003, China's grain production fell by 18%. This decline in grain production occurred mostly because of a drop in cropland because of a loss of irrigation water, desert expansion, and conversion of cropland to nonfarm uses. Another factor was a decline in planting two crops a year because of a loss of farm labor as more Chinese migrated from rural areas to cities in search of jobs. According to projections by the Worldwatch Institute and the U.S. Central Intelligence Agency, China's grain production could fall much more between 2003 and 2030, mostly because of water shortages, degraded cropland, diversion of water from cropland to cities, and continued conversion of cropland to nonfarm uses.

As incomes in China have risen, so has meat consumption. Even if China's currently booming economy resulted in no increases in meat consumption, the projected drop in grain production would mean that by 2030 China would need to import more than the world's total grain exports (roughly half of which come from the United States).

But suppose the increased demand for meat led to a rise in per capita grain consumption equal to one-half the current U.S. level. Then China would need to import more than the entire current grain output of the United States.

The Earth Policy Institute and the U.S. Central Intelligence Agency warn that if either of these scenarios turns out to be correct, no country or combination of countries has the potential to supply even a small fraction of China's potential food supply deficit. This does not take into account huge grain deficits that are projected in other parts of the world by 2025, especially in Africa and India.

To food expert Lester Brown, China is facing an ecological meltdown by exceeding the carrying capacity of its land. It is "overplowing its land, overgrazing its rangelands, depleting its soils, expanding its deserts,

Biodiversity Loss

Loss and degradation of habitat from clearing grasslands and forests and draining wetlands

Fish kills from pesticide runoff

Killing of wild predators to protect livestock

Loss of genetic diversity from replacing thousands of wild crop strains with a few monoculture strains

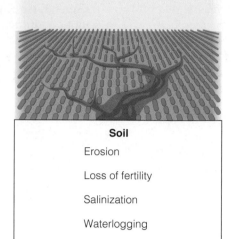

Soil

Erosion

Loss of fertility

Salinization

Waterlogging

Desertification

Air Pollution

Greenhouse gas emissions from fossil fuel use

Other air pollutants from fossil fuel use

Pollution from pesticide sprays

Water

Water waste

Aquifer depletion

Increased runoff and flooding from land cleared to grow crops

Sediment pollution from erosion

Fish kills from pesticide runoff

Surface and groundwater pollution from pesticides and fertilizers

Overfertilization of lakes and slow-moving rivers from runoff of nitrates and phosphates from fertilizers, livestock wastes, and food processing wastes

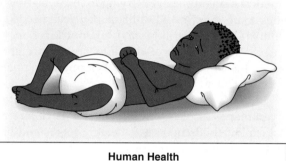

Human Health

Nitrates in drinking water

Pesticide residues in drinking water, food, and air

Contamination of drinking and swimming water with disease organisms from livestock wastes

Bacterial contamination of meat

Figure 14-18 Natural capital degradation: major environmental effects of food production. According to UN studies, land degradation reduced cumulative food production worldwide by about 13% on cropland and 4% on pastureland between 1950 and 2000.

overcutting its forests, and overpumping its aquifers." See his guest essay on the website for this chapter.

Other analysts disagree. According to a 1997 study by the International Food Policy Institute, China should be able to feed its population and begin exporting grain again by 2020 if the government invests in expanding irrigation, using more water-efficient forms of irrigation, and increasing agricultural research. Also, recent satellite surveys show that China

has about 40% more potential cropland than previously thought. In addition, the World Bank concluded in 1997 that China's domestic food production should keep up with its population growth for the next two or three decades without having to import large amounts of grain. China's food production dilemma illustrates how the problems of population growth, economic development, and environmental degradation can interact.

14-6 INCREASING CROP PRODUCTION

What Is the Gene Revolution? From Crossbreeding to Mixing Genes in a New Way

We can increase crop yields by using crossbreeding to mix the genes of similar types of organisms and genetic engineering to mix those of different organisms.

For centuries, farmers and scientists have used *crossbreeding* through *artificial selection* to develop genetically improved varieties of crop strains (Figure 5-10, p. 97). Such selective breeding has had amazing results. Ancient ears of corn were about the size of your little finger and wild tomatoes were once the size of a grape.

But traditional crossbreeding is a slow process, typically taking 15 years or more to produce a commercially valuable new variety and can combine traits only from species that are close to one another genetically. It also provides varieties that are useful for only about 5–10 years before pests and diseases reduce their effectiveness.

Scientists are creating a *third green revolution*—actually a *gene revolution*—by using genetic engineering to develop genetically improved strains of crops and livestock animals. It involves splicing a gene from one species and transplanting it into the DNA of another species (Figure 5-11, p. 98). Compared to traditional crossbreeding, gene splicing takes about half as long to develop a new crop, cuts costs, and allows the insertion of genes from almost any other organism into crop cells.

Ready or not, the world is entering the *age of genetic engineering*. More than two-thirds of the food products on U.S. supermarket shelves contain ingredients made from genetically engineered crops, and the proportion is increasing rapidly. Currently, genetically engineered crops account for about 5% of the world's crop area. But by 2020 more cropland may be devoted to genetically engineered crops than to conventional crossbred crops.

Bioengineers are developing or plan to develop new varieties of crops resistant to heat, cold, herbicides, insect pests, parasites, viral diseases, drought, and salty or acidic soil. They also hope to develop crop plants that that can grow faster and survive with little or no irrigation and with less fertilizer and pesticides.

For example, bioengineers have altered citrus trees (that normally take 6 years to produce fruit) to yield fruit in only one year. They hope to go further and use *advanced tissue culture* techniques to mass-produce only orange juice sacs. This would eliminate the need for citrus orchards and would free large amounts of land for other purposes such as biodiversity protection.

A team of research scientists at Washington State University is experimenting with cell cultures to produce a variety of food and medical products in fermentation tanks or bioreactors. These cell factories would contain mixtures of various plant and animal cells suspended in nutrient solutions of salts and carbohydrates. If successful and affordable, such food factory systems could produce food independent of local weather in environmentally controlled buildings in local areas. This would reduce the environmental impacts of food production and greatly reduce long-distance shipping costs.

However, critics note that so far most genetically modified crops have been for use in temperate areas rather than on subsistence crops in the tropics where food needs are the greatest. Also two-thirds of available transgenic crops have been engineered to tolerate more herbicides (and thus increase herbicide sales) rather than for pest resistance and improved food quality. This has occurred because seed companies understandably concentrate on developing crops that will give them high profits in countries where farmers can afford the new varieties.

How Safe Are Genetically Modified Foods? Savior or Frankenfood?

There is controversy over whether the benefits of genetically engineered food outweigh its unintended and potentially harmful effects.

Despite the promise, there is considerable controversy over the use of *genetically modified food (GMF)* and other forms of genetic engineering. Such food is seen by its producers and investors as a potentially sustainable way to solve world food problems but critics consider it potentially dangerous "Frankenfood". Figure 14-19 (p. 292) summarizes the projected advantages and disadvantages of this new technology. Study this figure carefully.

Critics recognize the potential benefits of genetically modified crops. But they warn that we know too little about the potential harm to human health and ecosystems from widespread use of such crops. Also, genetically modified organisms cannot be recalled if they cause unintended harmful genetic and ecological effects—as some scientists expect to happen.

In 2002 biologist Barry Commoner warned, "The genetically engineered crops now being grown represent a massive uncontrolled experiment whose outcome is inherently unpredictable. The results could be catastrophic." Until we have more information, such critics call for more controlled field experiments, more research and long-term safety testing to better understand the risks, and stricter regulation of this technology.

Trade-Offs

Genetically Modified Crops and Foods

Projected Advantages	Projected Disadvantages
Need less fertilizer	Irreversible and unpredictable genetic and ecological effects
Need less water	
More resistant to insects, plant disease, frost, and drought	Harmful toxins in food from possible plant cell mutations
Faster growth	New allergens in food
Can grow in slightly salty soils	Lower nutrition
Less spoilage	Increased evolution of pesticide-resistant insects and plant diseases
Better flavor	
Less use of conventional pesticides	Creation of herbicide-resistant weeds
Tolerate higher levels of herbicide use	Harm beneficial insects
Higher yields	Lower genetic diversity

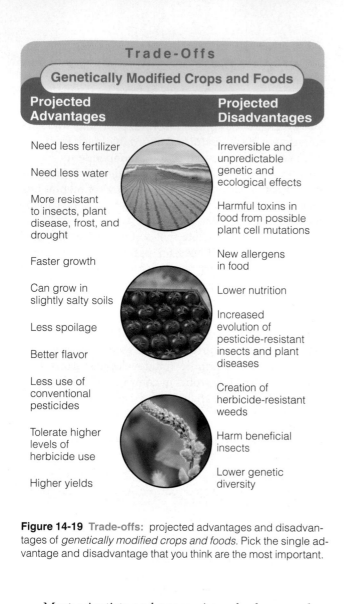

Figure 14-19 Trade-offs: projected advantages and disadvantages of *genetically modified crops and foods.* Pick the single advantage and disadvantage that you think are the most important.

Most scientists and economists who have evaluated genetic engineering of crops believe that its potential benefits outweigh the potential risks. A statement signed in 2000 by over 2,100 scientists—including Nobel laureates James Watson (codiscoverer of DNA) and Norman Bourlag (founder of the second green revolution)—supported the use of food modified by genetic engineering. According to a 2004 report by the U.N. Food and Agriculture Organization, geneticaly modified (GM) crops hold great promise for farmers in developing countries. But the study pointed out that so far the technology has not been focused on developing GM crops for the poor. The report also called for more research and government regulation to assess any harmful environmental effects from this technology.

A 2004 study by the Ecological Society of America recommended more caution in releasing genetically engineered organisms into the environment. Also, agroecologists Miguel Altieri and Peter Rosset point out that the idea of using genetic engineering to pro-

vide enough food to feed everyone is based on two faulty assumptions. One is that world hunger is caused by a global shortage of food. The other is that genetic engineering is the only and best way to increase food production. The reality is that poverty and inequality not food production are primary causes of hunger and malnutrition. Research also shows that polyculture using perennial crops can produce higher crop yields than current green revolution and genetic revolution techniques.

✗ *How Would You Vote?* Do the potential advantages of genetically engineered foods outweigh their potential disadvantages? Cast your vote online at http://biology.brookscole.com/miller14.

Many analysts and consumer advocates believe governments should require mandatory labeling of genetically modified foods. This would provide consumers with information to help them make informed choices about the foods they buy. Such labeling is required in Japan, Europe, South Korea, Canada, Australia, and New Zealand and is favored by 81% of Americans polled in 1999.

Industry representatives and the U.S. Department of Agriculture oppose this because they claim that genetically modified foods are not substantially different from foods developed by conventional crossbreeding methods. Also, they fear—probably correctly—that labeling such foods would hurt sales by arousing suspicion.

✗ *How Would You Vote?* Should all genetically engineered foods be so labeled? Cast your vote online at http://biology.brookscole.com/miller14.

Can We Continue Expanding the Green Revolution? Maybe, Maybe Not

Lack of resources such as water and fertile soil, and environmental factors may limit our ability to continue increasing crop yields.

Many analysts believe we can produce all the food we need in the future by spreading the use of existing high-yield green revolution crops and genetically engineered crops to more of the world.

Other analysts disagree. They point to several factors that have limited the success of the green and gene revolutions to date and may continue to do so. One problem is that without huge amounts of fertilizer and water, most green revolution crop varieties produce yields that are no higher (and are sometimes lower) than those from traditional strains. Another problem is that green revolution and genetically engineered crop strains and their high inputs of water, fertilizer, and pesticides cost too much for most subsistence farmers in developing countries. Scientists also point out that continuing to increase fertilizer, water, and pesticide inputs eventually produces no additional increase in crop yields. For example, grain yields rose about 2.1% a year between 1950 and 1990, but the increase dropped to 1.1% per year between 1990 and 2000 and to 0.5% between 1997 and 2002. No one knows whether this downward trend will continue.

There is also concern that crop yields in some areas may start dropping as soil erodes and loses fertility, irrigated soil becomes salty and waterlogged, underground and surface water supplies become depleted and polluted with pesticides and nitrates from fertilizers, and populations of rapidly breeding pests develop genetic immunity to widely used pesticides. We do not know how close we are to such environmental limits.

Also, according to Indian economist Vandana Shiva, overall gains in crop yields from new green and gene revolution varieties may be much lower than claimed. The yields are based on comparisons between the output per hectare of old and new *monoculture* varieties rather than between the even higher yields per hectare for *polyculture* cropping systems and the new monoculture varieties that often replace polyculture crops.

There is also concern that the projected increased loss of biodiversity can limit the genetic raw material needed for future green and gene revolutions. The UN Food and Agriculture Organization estimates that two-thirds of all seeds planted in developing countries are of uniform strains. Such genetic uniformity increases the vulnerability of food crops to pests, diseases, and harsh weather.

Will People Try New Foods? Changing Eating Habits Is Difficult

A variety of plants and insects could be used as sources of food, but most consumers are reluctant to try new foods.

Some analysts recommend greatly increased cultivation of less widely known plants to supplement or replace such staples as wheat, rice, and corn. One of many possibilities is the *winged bean* common in New Guinea and Southeast Asia. This fast-growing bean is a good source of protein and has so many edible parts it has been called a supermarket on a stalk. It also needs little fertilizer because of nitrogen-fixing nodules in its roots.

Some edible insects—called *microlivestock*—are also important potential sources of protein, vitamins, and minerals in many parts of the world. There are about 1,500 edible insect species. Examples include black ant larvae (served in tacos in Mexico), giant wa-

terbugs (crushed into vegetable dip in Thailand), *Mopani,* or emperor moth caterpillars (eaten in South Africa), cockroaches (eaten by Kalahari desert dwellers), lightly toasted butterflies (a favorite food in Bali), and fried ants (sold on the streets of Bogota, Colombia). Most of these insects are 58–78% protein by weight—three to four times as protein-rich as beef, fish, or eggs. One problem is getting farmers to take the financial risk of cultivating new types of food crops. Another is convincing consumers to try new foods. Would you try a bug soup?

Some plant scientists believe we should rely more on *polycultures of perennial crops* (p. 278), which are better adapted to regional soil and climate conditions than most annual crops. Using perennials would also eliminate the need to till soil and replant seeds each year, greatly reducing energy use, saving water, and reducing soil erosion and water pollution from eroded sediment. Not surprisingly, large seed companies that make their money selling seeds each year for annual crops generally oppose this idea.

Is Irrigating More Land the Answer? A Limited Solution

The amount of irrigated land per person has been falling since 1978 and is projected to fall much more during the next few decades.

About 40% of the world's food production and two-thirds of the world's rice and wheat comes from the 20% of the world's cropland that is irrigated. *Good news.* Between 1950 and 2003, the world's irrigated area tripled, with most of the growth occurring from 1950 to 1978.

Bad news. The amount of irrigated land per person has been falling since 1978 and is projected to fall much more between 2004 and 2050. One reason is that since 1978 the world population has grown faster than irrigated agriculture. Other factors are depletion of underground water supplies (aquifers), inefficient use of irrigation water, and salt buildup in soil on irrigated cropland. In addition, the majority of the world's farmers do not have enough money to irrigate their crops.

Is Cultivating More Land the Answer? Another Limited Solution

Significant expansion of cropland is unlikely over the next few decades because of poor soils, limited water, high costs, and harmful environmental effects.

Theoretically, the world's cropland could be more than doubled by clearing tropical forests and irrigating arid land. But much of this is *marginal land* with poor soil fertility, steep slopes, or both. Cultivation of such land is unlikely to be sustainable.

Much of the world's potentially cultivable land lies in dry areas, especially in Australia and Africa. Large-scale irrigation in these areas would require expensive dam projects, use large inputs of fossil fuel to pump water long distances, and deplete groundwater supplies by removing water faster than it is replenished. It would also require expensive efforts to prevent erosion, groundwater contamination, salinization, and waterlogging, all of which reduce crop productivity.

Furthermore, these potential increases in cropland would not offset the projected loss of almost one-third of today's cultivated cropland caused by erosion, overgrazing, waterlogging, salinization, and urbanization.

Such expansion of cropland would also reduce wildlife habitats and thus the world's biodiversity. According to the FAO, cultivating all potential cropland in developing countries would reduce the areas of forests, woodlands, and permanent pasture by almost half. Clearing forests would also release a huge amount of carbon dioxide into the atmosphere and accelerate global warming, which is expected to cause shifts in the areas where some crops could be grown.

Bottom line: *Many analysts believe that significant expansion of cropland is unlikely over the next few decades.*

Can We Grow More Food in Urban Areas? Some Untapped Potential

People in urban areas could save money by growing more of their food.

According to the United Nations Development Program, urban gardens provide about 15% of the world's food supply. Food experts believe that people in urban areas could live more sustainably and save money by growing more of their food. Such food could be grown in empty lots, in backyards, on rooftops and balconies, and by raising fish in tanks and sewage lagoons.

Growing food in urban areas reduces stresses on soil and biodiversity in nonurban areas. It can also provide food and jobs for low-income urban residents. However, it can lead to conflicts over how urban land should be used. And urban soil needs to be checked for traces of toxic pollutants such as lead and mercury.

A study by the UN Center for Human Settlements estimated that up to half of the total area in many cities in developing countries is vacant public land that could be used to produce food. Is there a vacant lot or a rooftop in your neighborhood that could be used to grow food?

How Much Food Is Wasted? Way Too Much

Up to 70% of the food we produce is wasted through spoilage, inefficient processing and preparation, and plate waste.

We have greatly increased the efficiency of food production. But the efficiency of food consumption is still low. According to the FAO, as much as 70% of the food we produce is lost through spoilage, inefficient processing and preparation, and plate waste. Even affluent countries such as the United States, Canada, Switzerland, Italy, and Belgium waste nearly 60% of their food. Cutting such losses in half would go a long way in meeting global food needs and reducing the environmental impact of agriculture. How much of the food on your plate is wasted?

14-7 PRODUCING MORE MEAT

How Are Rangelands Used to Produce Meat? Grass and Shrubs for Livestock

Much of the rangeland that makes up 40% of the world's ice-free land is used to raise livestock.

Most analysts call for an increase in meat production because we will need more meat to feed the projected increase in the world's population. Also, when incomes rise, meat consumption per person usually increases.

Rangelands are grasslands in temperate and tropical climates that supply forage or vegetation for grazing (grass-eating) and browsing (shrub-eating) animals. Almost 4 billion cattle, sheep, and goats graze on about 42% of the world's rangeland. Livestock also graze in **pastures:** managed grasslands or enclosed meadows usually planted with domesticated grasses or other forage.

Most rangeland grasses have a deep and complex network of roots (Figure 4-27, top right, p. 75) that help anchor the plants. Blades of rangeland grass grow from the base, not the tip. Thus as long as only the upper half is eaten, rangeland grass is a renewable resource that can be grazed again and again.

Moderate levels of grazing are healthy for grasslands because removal of mature vegetation stimulates rapid regrowth and encourages greater plant diversity. If not overdone, disturbance of the soil surface by the hooves of grazing animals allows more rainfall to reach the roots of rangeland grasses. The key to the health of rangeland is to prevent both overgrazing and undergrazing by domesticated livestock and wild herbivores.

Is Producing More Meat the Answer? More Protein at the Expense of the Environment

Meat and meat products are important sources of protein, but meat production has many harmful environmental effects.

Meat and meat products are good sources of high-quality protein. Between 1950 and 2003, world meat production increased more than fivefold, and per capita meat production more than doubled. It is likely to more than double again by 2050 as affluence rises in middle-income developing countries and people begin consuming more meat.

Some analysts expect most future increases in meat production to come from densely populated *feedlots*, where animals are fattened for slaughter by feeding on grain grown on cropland or meal produced from fish. Feedlots account for about 43% of the world's beef production, half of pork production, and almost three-fourths of poultry production.

In the United States, most production of cattle, pigs, and poultry is concentrated in increasingly large, factory-like production facilities in only a few areas. As many as 100,000 cattle may be confined to a single feedlot complex and 10,000 hogs may be crowded almost shoulder to shoulder in a giant barn.

This industrialized approach increases meat productivity. But it has a number of harmful environmental effects. Animal wastes from such facilities are typically stored in enormous open lagoons, which can rupture or leak and contaminate groundwater and nearby streams and rivers. In 1999, for example, torrential rains from Hurricane Floyd caused a number of hog and poultry waste lagoons in southeastern North Carolina to over-

flow and spill their wastes into local rivers. Living near a feedlot or animal waste lagoon is also a nasal assault.

Expanding feedlot production of meat will increase pressure on the world's grain supply because feedlot livestock consume grain produced on cropland instead of feeding on natural grasses. It will also increase pressure on the world's fish supply because about one-third of the world's fish catch is used to feed livestock. Livestock production also has an enormous environmental impact (Connections, below).

What Are the Effects of Overgrazing? Eroding Soil and Fewer Livestock

Overgrazing can lead to soil erosion and limit livestock production.

Overgrazing occurs when too many animals graze too long and exceed the carrying capacity of a grassland area. It lowers the net primary productivity of grassland vegetation, reduces grass cover, and when combined with prolonged drought can cause desertification. It also exposes the soil to erosion by water and wind (Figure 14-20, left) and compacts the soil (which diminishes its capacity to hold water). Overgrazing also enhances invasion of exposed land by woody

---

Some Environmental Consequences of Meat Production

CONNECTIONS

The meat-based diet of affluent people in developed and developing countries has a number of harmful environmental effects. More than half of the world's cropland (19% in the United States) is used to produce livestock feed grain (mostly field corn, sorghum, and soybeans). Livestock and fish raised for food also consume about 37% of the world's grain production and 70% of grain production in the United States.

Meat production uses more than half the water withdrawn from the world's rivers and aquifers each year. Most of this water is used to irrigate crops fed to livestock and to wash away animal wastes.

According to Canadian scientist Vaclav Smil, producing one calorie of energy in the flesh of a cow, pig, or chicken requires 11–15 calories of feed. The energy needed to produce a single hamburger is enough to

drive a small car about 32 kilometers (20 miles).

About 14% of U.S. topsoil loss is directly associated with livestock grazing. Cattle belch out about 16% of the methane (a greenhouse gas about 25 times more potent than carbon dioxide) released into the atmosphere. Also, some of the nitrogen in commercial inorganic fertilizer used to grow livestock feed is converted to nitrous oxide, a greenhouse gas released from the soil into the atmosphere.

Livestock in the United States produce about 20 times more waste (manure) than is produced by the country's human population. A single cow produces as much waste as 16 humans. Only about half of this nutrient-rich livestock waste is recycled into the soil. Manure washing off the land or leaking from lagoons used to store animal wastes can kill fish by depleting dissolved oxygen.

Chickens, pigs, and cows use about 70% of the antibiotics con-

sumed in the United States. According to the World Health Organization and the FAO, widespread use of antibiotics in the livestock industry is increasing the development of microbes that are genetically resistant to widely used antibiotics. This makes it harder to fight infectious diseases in both humans and livestock animals. In 2004, McDonalds began requiring its chicken suppliers to stop giving their birds antibiotics to promote growth.

Producing meat can also endanger wildlife species. According to a 2002 report by the National Public Lands Grazing Campaign, livestock grazing in the United States has contributed to population declines of almost a fourth of the country's threatened and endangered species.

Critical Thinking

Are you willing to eat less meat or not eat any meat? Explain.

shrubs such as mesquite and prickly pear cactus. Finally, overgrazing can limit livestock production.

We do not know the condition of much of the world's rangeland because of a lack of detailed surveys. However, limited data from surveys in various countries by the FAO indicate that overgrazing by livestock has caused as much as a fifth of the world's rangeland to lose productivity, mostly by desertification (Figure 14-9).

How Can Rangelands Be Managed More Sustainably to Produce More Meat?
Control and Restore

We can sustain rangeland productivity by controlling the number and distribution of livestock and by restoring degraded rangeland.

The most widely used method for more sustainable management of rangeland is to control the number of grazing animals and the duration of their grazing in a given area so the carrying capacity of the area is not exceeded. However, determining the carrying capacity of a range site is difficult and costly.

Livestock tend to aggregate around natural water sources especially thin strips of lush vegetation along streams or rivers known as *riparian zones* (Figure 14-21, left) and ponds established to provide water for livestock. As a result, areas around such water sources tend to be overgrazed and other areas can be undergrazed.

Studies indicate that 65–75% of the wildlife in the western United States depends totally on riparian habitats. According to a 1999 study in the *Journal of Soil*

Figure 14-20 Rangeland: overgrazed (left) and lightly grazed (right).

and Water Conservation, livestock grazing has damaged approximately 80% of stream and riparian ecosystems in the United States.

To help prevent such damage, livestock can be moved from one grazing area to another and riparian areas can be fenced off. Sometimes protected areas can recover in a few years (Figure 14-21, right). Ranchers can also provide supplemental feed at selected sites and locate water holes and tanks and salt blocks in strategic places.

Figure 14-21 Solutions: cattle on a riparian zone of a public rangeland along Arizona's San Pedro River (left) in the mid-1980s just before this section of waterway was protected by banning domestic livestock grazing for 15 years, eliminating sand and gravel operations and water pumping rights in nearby areas, and limiting access by off-highway vehicles. The right photo shows the recovery of this riparian area at the same time of year after 10 years of protection.

A more expensive and less widely used method of rangeland management is to suppress the growth of unwanted invader plants by herbicide spraying, mechanical removal, or controlled burning. A cheaper way to discourage unwanted vegetation is controlled, short-term trampling by large numbers of livestock.

Replanting barren areas with native grass seeds and applying fertilizer can increase growth of desirable vegetation and reduce soil erosion. But this is an expensive way to restore severely degraded rangeland.

How Can We Produce Meat More Sustainably? Shifting Our Meat Priorities

We can reduce the environmental impacts of meat production by relying more on fish and chicken and less on beef and pork.

Livestock and fish vary widely in the efficiency with which they convert grain into animal protein (Figure 14-22). A more sustainable form of meat production and consumption would involve shifting from less grain-efficient forms of animal protein, such as beef and pork, to more grain-efficient ones, such as poultry and farmed fish (Figure 14-22).

Some environmentalists have called for reducing livestock production (especially cattle) to decrease its environmental effects and to feed more people. This would decrease the environmental impact of livestock production, but it would not free up much land or grain to feed more of the world's hungry people.

Cattle and sheep that graze on rangeland use a resource (grass) that humans cannot eat, and most of this land is not suitable for growing crops. Moreover, because of poverty, insufficient economic aid, and the nature of global economic and food distribution systems, very little if any additional grain grown on land once used to raise livestock or livestock feed would reach the world's hungry people.

14-8 CATCHING AND RAISING MORE FISH AND SHELLFISH

Where Do We Get the Fish and Shellfish We Eat? Oceans and Fish Farms

About 88% of the fish and shellfish we eat comes from the ocean or is produced by aquaculture in aquatic feedlots.

The world's third major food-producing system consists of **fisheries**: concentrations of particular aquatic species suitable for commercial harvesting in a given ocean area or inland body of water. About 55% of the annual commercial catch of fish and shellfish comes from the ocean, mostly from plankton-rich coastal waters.

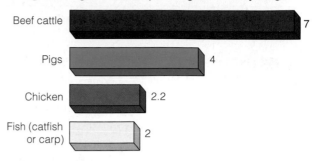

Kilograms of grain needed per kilogram of body weight

Figure 14-22 Efficiency of converting grain into animal protein. Data in kilograms of grain per kilogram of body weight added. (U.S. Department of Agriculture)

The rest of the catch comes from using *aquaculture* to raise marine and freshwater fish like livestock animals in feedlots in ponds and underwater cages and from inland freshwater fishing from lakes, rivers, reservoirs, and ponds. About one-third of the world's marine fish harvest is used as animal feed, fishmeal, and oils.

Some commercially important marine species of fish and shellfish are shown in Figure 14-23 (p. 298). Fish and shellfish supply about 7% of the global food supply and are the primary source of animal protein for about 1 billion people, mostly in developing countries.

How Are Fish and Shellfish Harvested? Hunt and Gather As Much As You Can

High-tech global fishing fleets roam the world's oceans to find and harvest most of the fish and shellfish we eat.

The world's commercial marine fishing industry is dominated by industrial fishing fleets using global satellite positioning equipment, sonar, huge nets and long fishing lines, spotter planes, and large factory ships that can process and freeze their catches. Figure 14-24 (p. 299) shows the major methods used for the commercial harvesting of various marine fish and shellfish.

Let us look at a few of these methods. *Trawler fishing* is used to catch fish and shellfish—especially shrimp, cod, flounder, and scallops—that live on or near the ocean floor. It involves dragging a funnel-shaped net held open at the neck along the ocean bottom and weighed down with chains or metal plates. This scrapes up almost everything that lies on the ocean floor and often destroys bottom habitats—somewhat like clear-cutting the ocean floor. Newer trawling nets are large enough to swallow 12 jumbo jets and even larger ones are on the way! The large mesh of the net allows most small fish to escape but can capture and kill other species such as seals and

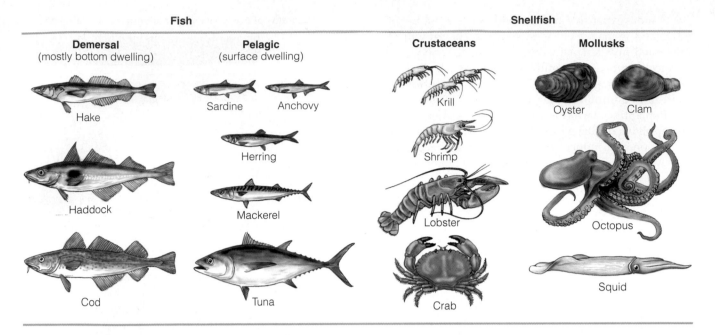

Figure 14-23 Some major types of commercially harvested marine fish and shellfish.

endangered and threatened sea turtles. Only the large fish are kept. Most of the fish and other aquatic species—called *bycatch*—are thrown back into the ocean dead or dying.

Another method, *purse-seine fishing,* involves catching surface-dwelling species such as tuna, mackerel, anchovies, and herring, which tend to feed in schools near the surface or in shallow areas. After locating a school the fishing vessel surrounds it with a large net called a purse seine. Then they close the net like a drawstring purse to trap the fish. Nets used to capture yellowfin tuna in the eastern tropical Pacific Ocean have killed large numbers of dolphins that swim on the surface above schools of tuna.

Fishing vessels also use *longlining*. It involves putting out lines up to 130 kilometers (80 miles) long, hung with thousands of baited hooks. The depth of the lines can be adjusted to catch open-ocean fish species such as swordfish, tuna, and sharks or bottom fishes such as halibut and cod. Longlines also hook endangered sea turtles, sea-feeding albatross birds, and pilot whales and dolphins.

With *drift-net fishing*, fish are caught by huge drifting nets that can hang as much as 15 meters (50 feet) below the surface and be up to 64 kilometers (40 miles) long. This method can lead to overfishing of the desired species and may trap and kill large quantities of unwanted fish and marine mammals (such as dolphins, porpoises, and seals), marine turtles, and seabirds. Since 1992, a UN ban on the use of drift nets longer than 2.5 kilometers (1.6 miles) in international waters has sharply reduced use of this technique. But

longer nets continue to be used because compliance is voluntary and it is difficult to monitor fishing fleets over vast ocean areas. Also, the decrease in drift nets has led to increased use of longlines, which often have similar effects on marine wildlife.

Figure 14-25 shows the effects of these now common efforts to increase the seafood harvest. After increasing fourfold between 1960 and 1982, the annual commercial fish catch (marine plus freshwater harvest but excluding aquaculture) has declined and leveled off (Figure 14-25, left). After doubling between 1950 and 1956, the per capita catch leveled off until 1980 and since then has been declining (Figure 14-25, right) and may continue to decline because of overfishing, pollution, habitat loss, and population growth.

Connections: How Are Overfishing and Habitat Degradation Affecting Fish Harvests? Dropping Yields

About three-fourths of the world's commercially valuable marine fish species are overfished or fished at their biological limit.

Fish are renewable resources as long as the annual harvest leaves enough breeding stock to renew the species for the next year. **Overfishing** is the taking of so many fish that too little breeding stock is left to maintain numbers.

Prolonged overfishing leads to *commercial extinction,* when the population of a species declines to the point at which it is no longer profitable to hunt for them. Fishing fleets then move to a new species or a

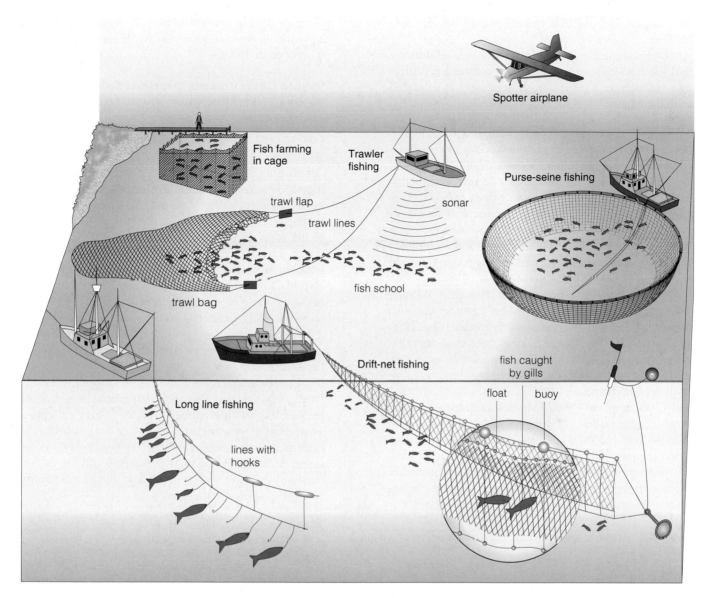

Figure 14-24 Major commercial fishing methods used to harvest various marine species. These methods have become so effective that many fish have become commercially extinct.

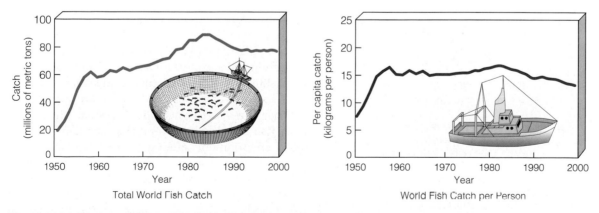

Total World Fish Catch

World Fish Catch per Person

Figure 14-25 Natural capital degradation: world fish catch (left) and world fish catch per person (right), 1950–2000. The total catch and per capita catches since 1990 may be about 10% lower than shown here because of the discovery in 2000 that since 1990 China had apparently been inflating its fish catches. (Data from UN Food and Agriculture Organization and Worldwatch Institute)

new region, hoping that the overfished species will recover.

Overfishing is not new. Historical studies indicate that some species were overfished beginning centuries ago. However, overfishing has greatly accelerated with the expansion of today's large and efficient global fishing fleets.

According to the UN Food and Agriculture Organization, about three-fourths of the world's 200 commercially valuable marine fish species are either overfished or fished to their estimated maximum sustainable yield. According to the Ocean Conservancy, "we are spending the principal of our marine fish resources rather than living off the interest they provide." Analysts warn that some of these fisheries are so depleted that even if all fishing stopped immediately it would take up to 20 years for stocks to recover.

Studies by the U.S. National Fish and Wildlife Foundation show that 14 major commercial fish species in U.S. waters such as some groundfishes (Figure 14-26) have been severely depleted. Also, degradation, destruction, and pollution of wetlands, estuaries, coral reefs, salt marshes, and mangroves threaten populations of fish and shellfish.

Good news. In 1995, fisheries biologists studied population data for 128 depleted fish stocks and concluded that 125 of them could recover with careful management. This involves establishing fishing quotas, restricting use of certain types of fishing gear and methods, limiting the number of fishing boats, closing fisheries during spawning periods, and setting aside networks of no-take reserves. So far we are not doing most of these things.

Should Governments Continue Subsidizing Fishing Fleets? Too Many Boats Chasing Too Few Fish

Government subsidies given to the fishing industry are a major cause of overfishing.

Overfishing is a big and growing problem because we have too many commercial fishing boats and fleets trying to hunt and gather a dwindling supply of the most desirable fish.

It costs the global fishing industry about $120 billion a year to catch $70 billion worth of fish. Government subsidies such as fuel tax exemptions, price controls, low-interest loans, and grants for fishing gear make up most of the $50 billion annual deficit of the industry. Without such subsidies, some of the world's fishing boats and fleets would have to go out of business and the number of fish caught would approach their sustainable yield.

Continuing to subsidize excess fishing allows some fishers to keep their jobs and boats a little longer while making less and less money until the fishery collapses. Then all jobs are gone, and fishing communities suffer even more—another example of the tragedy of the commons in action. Some of the money could be shifted from subsidies to programs to buy out some fishing boats and retrain their crews.

✗ *How Would You Vote?* Should governments eliminate all fishing subsidies? Cast your vote online at http://biology .brookscole.com/miller14.

What Is Aquaculture? Feedlots of the Sea

Raising large numbers of fish and shellfish in ponds and cages is the world's fastest growing type of food production.

Aquaculture involves raising fish and shellfish for food like crops instead of going out and hunting and gathering them. It is the world's fastest-growing type of food production and accounts for about one-third of the fish and shellfish we eat. China, the world leader, produces over two-thirds of the world's aquaculture output.

There are two basic types of aquaculture. One called **fish farming** involves cultivating fish in a controlled environment (often a coastal or inland pond, lake, reservoir, or rice paddy) and harvesting them when they reach the desired size.

The other is **fish ranching.** It involves holding anadromous species such as salmon that live part of their lives in fresh water and part in salt water in captivity for the first few years of their lives, usually in fenced-in areas or floating cages in coastal lagoons and

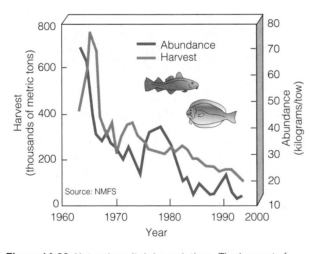

Figure 14-26 Natural capital degradation: The harvest of groundfishes (yellowtail flounder, haddock, and cod) in the Georges Bank off the coast of New England in the North Atlantic, once one of the world's most productive fishing grounds, has declined sharply since 1965. Stocks dropped to such low levels that since December 1994 the National Marine Fisheries Services has banned fishing of these species in the Georges Bank. The closure helped. By 1999 populations of the major groundfishes began recovering. However, the fishery still remains closed because it has been so severely depleted. (Data from U.S. National Marine Fisheries Service)

estuaries. Then the fish are released, and adults are harvested when they return to spawn (Figure 13-14, right, p. 269).

Figure 14-27 lists the major advantages and disadvantages of aquaculture. Some analysts project that freshwater and saltwater aquaculture production could provide at least half of the world's seafood by 2020. Some also propose increased use of aquaculture to grow single-cell algae such as *Spirulina*, which is 70% protein.

But other analysts warn that the harmful environmental effects of aquaculture (Figure 14-27, right) could limit future production. Also, some trends in aquaculture could harm ocean fisheries. For example, intensive farming of large carnivorous fish like salmon and trout is replacing traditional aquaculture in which farmed fish such as carp and tilapia eat plants and detritus. This increases overfishing of smaller marine species used to feed farmed carnivorous species. If kept up, this depletion of the seas to feed aquaculture farms could cause the collapse of both marine fisheries and carnivorous aquaculture. Figure 14-28 lists some ways to make aquaculture more sustainable and to reduce its environmental effects.

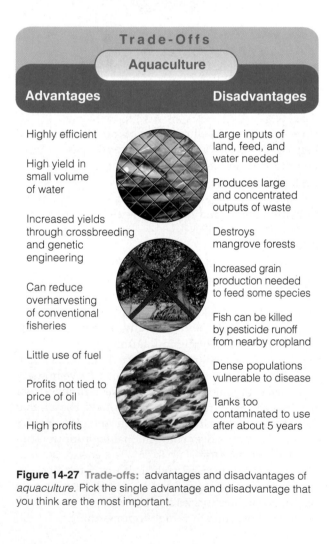

Figure 14-27 Trade-offs: advantages and disadvantages of *aquaculture.* Pick the single advantage and disadvantage that you think are the most important.

Solutions

More Sustainable Aquaculture

- Reduce use of fishmeal as a feed to reduce depletion of other fish

- Improve pollution management of aquaculture wastes

- Reduce escape of aquaculture species into the wild

- Restrict location of fish farms to reduce loss of mangrove forests and other threatened areas

- Farm some aquaculture species (such as salmon and cobia) in deeply submerged cages to protect them from wave action and predators and allow dilution of wastes into the ocean

- Set up a system for certifying sustainable forms of aquaculture

Figure 14-28 Solutions: Ways to make aquaculture more sustainable and reduce its harmful environmental effects.

However, even under the most optimistic projections, increasing both the wild catch and aquaculture will not increase world food supplies significantly. The reason is that currently fish and shellfish supply only about 1% of the energy and 6% of the total protein in the human diet.

X HOW WOULD YOU VOTE? Do the advantages of aquaculture outweigh its disadvantages? Cast your vote online at http://biology.brookscole.com/miller14.

14-9 GOVERNMENT AGRICULTURAL POLICY

How Do Government Agricultural Policies Affect Food Production? To Interfere or Not to Interfere

Governments can use price controls to keep food prices artificially low, give farmers subsidies to encourage food production, or eliminate food price controls and subsidies and let farmers respond to market demand.

Agriculture is a financially risky business. Whether farmers have a good year or a bad year depends on factors over which they have little control: weather, crop prices, crop pests and diseases, interest rates, and the global market. Because of the need for reliable food supplies despite fluctuations in these factors, most governments provide various forms of assistance to farmers and consumers.

Governments use three main approaches to do this. One is to use price controls to *keep food prices artificially*

low. This makes consumers happy but means farmers may not be able to make a living.

Another is to *give farmers subsidies and tax breaks to keep them in business and encourage them to increase food production.* Globally, government price supports and other subsidies for agriculture total more than $300 billion per year (about $100 billion per year in the United States)—an average of more than half a million dollars per minute! If government subsidies are too generous and the weather is good, farmers may produce more food than can be sold. The resulting surplus depresses food prices, which reduces the financial incentive for farmers in developing countries to increase domestic food production—those connections again.

A third approach is to *eliminate most or all price controls and subsidies and let farmers respond to market demand without government interference.* However, some analysts urge that any phaseout of farm subsidies should be coupled with increased aid for the poor and the lower middle class, who would suffer the most from any increase in food prices. Many environmentalists say that instead of eliminating all subsidies we should use them to reward farmers and ranchers who protect the soil, conserve water, reforest degraded land, protect and restore wetlands, conserve wildlife, and practice more sustainable agriculture and fishing.

X HOW WOULD YOU VOTE? Should governments phase out subsidies for conventional industrialized agriculture and phase in subsidies for more sustainable agriculture? Cast your vote online at http://biology.brookscole.com/miller14.

14-10 SUSTAINABLE AGRICULTURE

What Is More Sustainable Agriculture? Learn From Nature

We can produce food more sustainably by reducing resource throughput and working with nature.

There are three main ways to reduce hunger and malnutrition and the harmful environmental effects of agriculture. One is to *slow population growth.* Another is to *reduce poverty* so people can grow or buy enough food for their survival and good health.

The third is to develop and phase in systems of more **sustainable** or **low-input agriculture**—also called **organic farming** or **agroecology**—over the next few decades. Figure 14-29 lists the major components of more sustainable agriculture. This method of food production uses technologies based on ecological knowledge to increase yields, control pests, and build soil fertility. It relies more on a variety of perennial crops (polyculture) rather than monoculture of annual crops. It recognizes that it is unwise to overuse pesti-

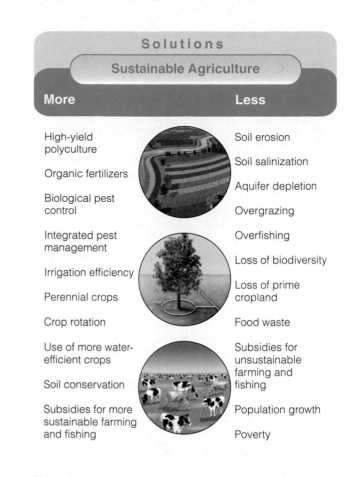

Solutions		
Sustainable Agriculture		
More		**Less**

More	Less
High-yield polyculture	Soil erosion
Organic fertilizers	Soil salinization
Biological pest control	Aquifer depletion
Integrated pest management	Overgrazing
Irrigation efficiency	Overfishing
Perennial crops	Loss of biodiversity
Crop rotation	Loss of prime cropland
Use of more water-efficient crops	Food waste
Soil conservation	Subsidies for unsustainable farming and fishing
Subsidies for more sustainable farming and fishing	Population growth
	Poverty

Figure 14-29 Solutions: components of more sustainable, low-throughput agriculture.

cides because this eliminates natural predators that help control pest populations and causes pest populations to become genetically resistant to widely used pesticides. Organic farmers or agroecologists also rely more or totally on manure and tilled-in crop residues to help maintain and build soil fertility by increasing its carbon content. This can help reduce runoff and improve water quality.

Studies have shown that low-input organic farming produces roughly equivalent yields with lower carbon dioxide emissions, uses about half as much energy per unit of yield than conventional farming, improves soil fertility, and generally is more profitable for the farmer than high-input farming.

In 2002, agricultural scientists Paul Mader and David Dubois reported the results of a 21-year study comparing organic and conventional farming. Their results and those from other studies have shown that for most crops low-input organic farming has a number of advantages over conventional high-input farming. They include use of up to 56% less energy per unit of yield, and improved soil health and fertility. Organic farming also provides more habitats for wild plant and animal species and generally is more profitable for the farmer than high-input farming.

Currently, organic farming is used on less than 1% of the world's cropland (0.2% in the United States) but on 6–10% of the cropland in many European countries. In 2002, global sales of organic foods amounted to about $23 billion and such sales are growing rapidly in the United States, Canada, and much of Europe.

Most proponents of more sustainable agriculture are not opposed to high-yield agriculture. Instead, they see it as vital for protecting the earth's biodiversity by reducing the need to cultivate new and often marginal land. They call for using environmentally sustainable forms of both high-yield polyculture (pp. 273 and 278) and high-yield monoculture for growing crops.

How Can We Make the Transition to More Sustainable Agriculture? Get Serious

More research, demonstration projects, government subsidies, and training can promote a shift to more sustainable agriculture.

Analysts suggest four major strategies to help farmers make the transition to more sustainable agriculture. *First,* greatly increase research on sustainable agriculture and improving human nutrition. *Second,* set up demonstration projects throughout each country so farmers can see how more sustainable agricultural systems work. *Third,* provide subsidies and increased foreign aid to encourage its use. *Fourth,* establish training programs in sustainable agriculture for farmers and government agricultural officials and encourage the creation of college curricula in sustainable agriculture and human nutrition.

Phasing in more sustainable agriculture involves applying the four principles of sustainability (Figure 9-15, p. 174) to producing food. The goal is to feed the world's people while sustaining the earth's natural capital and living off the natural income it provides. This will not be easy, but it can be done. Figure 14-30 lists some ways you can promote more sustainable agriculture.

The sector of the economy that seems likely to unravel first is food. Eroding soils, deteriorating rangelands, collapsing fisheries, falling water tables, and rising temperatures are converging to make it difficult to expand food production fast enough to keep up with the demand.

LESTER R. BROWN

CRITICAL THINKING

1. Summarize the major economic and ecological advantages and limitations of each of the following proposals for increasing world food supplies and reducing hunger over the next 30 years: **(a)** cultivating more land by clearing tropical forests and irrigating arid lands, **(b)** catching more fish in the open sea, **(c)** producing more fish and

What Can You Do?
Sustainable Agriculture

- Waste less food

- Reduce or eliminate meat consumption

- Feed pets balanced grain foods instead of meat

- Use organic farming to grow some of your food

- Buy organic food

- Compost your food wastes

Figure 14-30 What can you do? Ways to promote more sustainable agriculture.

shellfish with aquaculture, and **(d)** increasing the yield per area of cropland.

2. List five ways in which your lifestyle directly or indirectly contributes to soil erosion.

3. What are the three most important actions you would take to reduce hunger **(a)** in the country where you live and **(b)** in the world?

4. Some have suggested that rangelands could be used to raise wild grazing animals for meat instead of conventional livestock. Others consider it unethical to raise and kill wild herbivores for food. What do you think? Explain.

5. Should governments phase in agricultural tax breaks and subsidies to encourage farmers to switch to more sustainable farming? Explain your answer.

6. Explain why you support or oppose greatly increased use of **(a)** genetically modified food, **(b)** perennial food crops, and **(c)** polyculture.

7. Suppose you live near a coastal area and a company wants to use a fairly large area of coastal marshland for an aquaculture operation. If you were an elected local official, would you support or oppose such a project? Explain. What safeguards or regulations would you impose on the operation?

8. Congratulations! You are in charge of the world. List the three most important features of **(a)** your agricultural policy, **(b)** your policy to reduce soil erosion, and **(c)** your policy for more sustainable harvesting and farming of fish and shellfish.

PROJECTS

1. Conduct a survey of soil erosion and soil conservation in and around your community on cropland, construction sites, mining sites, grazing land, and deforested land. Use these data to develop a plan for reducing soil erosion in your community.

2. If possible, visit both a conventional industrialized farm and an organic or low-input farm. Compare **(a)** soil erosion and other forms of land degradation, **(b)** use and costs of energy, **(c)** use and costs of pesticides and inorganic fertilizer, **(d)** use and costs of natural pest control and organic fertilizer, **(e)** yields per hectare for the same crops, and **(f)** overall profit per hectare for the same crops.

3. Try to gather data evaluating the harmful environmental effects of nearby agriculture on your local community. What is being done to reduce these effects?

4. Use health and other local government records to estimate how many people in your community suffer from undernutrition or malnutrition. Has this problem increased or decreased since 1980? What are the basic causes of this hunger problem, and what is being done to alleviate it? Share the results of your study with local officials, and present your own plan for improving efforts to reduce hunger in your community.

5. Make a list of all the food you eat in one day and read all labels or look up the amount of calories, fat, protein, and carbohydrates in each food. Then determine how many calories you took in that day and the percentage of your diet from fat, protein, and carbohydrates. Rate your diet as healthy, borderline healthy, or unhealthy. Compare your results with those of your classmates.

6. Use the library or the Internet to learn about the four types of vegetarians and the advantages and disadvantages of a vegetarian diet in terms of your health and the environment.

7. Use the library or the Internet to find bibliographic information about *Aldo Leopold* and *Lester R. Brown,* whose quotes appear at the beginning and end of this chapter.

8. Make a concept map of this chapter's major ideas, using the section heads, subheads, and key terms (in boldface type). Look on the website for this book for information about making concept maps.

LEARNING ONLINE

The website for this book contains study aids and many ideas for further reading and research. They include a chapter summary, review questions for the entire chapter, flash cards for key terms and concepts, a multiple-choice practice quiz, interesting Internet sites, references, and a guide for accessing thousands of InfoTrac® College Edition articles. Log on to

http://biology.brookscole.com/miller14

Then click on the Chapter-by-Chapter area, choose Chapter 14, and select a learning resource.

CASE STUDY

Water Conflicts in the Middle East

In the near future, water-short countries in the Middle East are likely to engage in conflicts over access to water resources. Most water in this dry region comes from three shared river basins: the Nile, Jordan, and Tigris–Euphrates (Figure 15-1).

Three countries—Ethiopia, Sudan, and Egypt—use most of the water that flows in Africa's Nile River, with Egypt being last in line along the river.

To meet the water needs of its rapidly growing population, Ethiopia plans to divert more water from the Nile. So does Sudan. Such upstream diversions would reduce the amount of water available to Egypt, which cannot exist without irrigation water from the Nile.

Egypt can go to war with Sudan and Ethiopia for more water, cut population growth, or improve irrigation efficiency. Other options are to import more grain to reduce the need for irrigation water, work out water-sharing agreements with other countries, or suffer the harsh human and economic consequences of hydrological poverty.

The Jordan basin is by far the most water-short region, with fierce competition for its water among Jordan, Syria, Palestine (Gaza and the West Bank), and Israel.

Syria plans to build dams and withdraw more water from the Jordan River, decreasing the downstream water supply for Jordan and Israel. Israel warns that it may destroy the largest dam that Syria plans to build.

Turkey, located at the headwaters of the Tigris and Euphrates rivers, controls how much water flows downstream to Syria and Iraq before emptying into the Persian Gulf. Turkey is building 24 dams along the upper Tigris and Euphrates to generate electricity and irrigate a large area of land.

If completed, these dams will reduce the flow of water downstream to Syria and Iraq by up to 35% in normal years and much more in dry years. Syria also plans to build a large dam along the Euphrates to divert water arriving from Turkey. This will leave little water for Iraq and could lead to a water war between it and Syria.

Resolving these water distribution problems will require a combination of regional cooperation in allocating water supplies, slowed population growth, improved efficiency in water use, higher water prices to help improve irrigation efficiency, and increased grain imports to reduce water needs. This will not be easy.

To many analysts, emerging water shortages in many parts of the world—along with the related problems of biodiversity loss and climate change—are the three most serious environmental problems the world faces during this century.

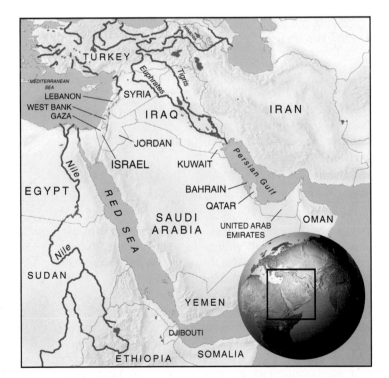

Figure 15-1 The Middle East, whose countries have some of the highest population growth rates in the world. Because of the dry climate, food production depends heavily on irrigation. Existing conflicts between countries in this region over access to water may soon overshadow both long-standing religious and ethnic clashes and take over valuable oil supplies.

Our liquid planet glows like a soft blue sapphire in the hard-edged darkness of space. There is nothing else like it in the solar system. It is because of water.

JOHN TODD

This chapter discusses the water problems we face and ways to use this wonderful but irreplaceable resource more sustainably. It addresses the following questions:

- What are water's unusual physical properties?
- How much fresh water is available to us, and how much of it are we using?
- What causes freshwater shortages, and what can be done about this problem?
- What are the advantages and disadvantages of using dams and reservoirs to supply more water?
- What are the advantages and disadvantages of transferring large amounts of water from one place to another?
- What are the advantages and disadvantages of withdrawing groundwater and converting salt water to fresh water?
- How can we waste less water?
- What are the causes of flooding, and what can be done to reduce the risk of flooding and flood damage?
- How can we use the earth's water more sustainably?

15-1 WATER'S IMPORTANCE AND UNIQUE PROPERTIES

Why Is Water So Important? Liquid Natural Capital

Water keeps us alive, moderates climate, sculpts the land, and removes and dilutes wastes and pollutants.

We live on the water planet, with a precious film of water—most of it salt water—covering about 71% of the earth's surface. Look in the mirror. What you see is about 60% water, most of it inside your cells.

No species can do without water. You could survive several weeks without food but only a few days without water. It takes huge amounts of water to supply you with food, shelter, and other needs and wants. Water also plays a key role in sculpting the earth's surface, moderating climate, and removing and diluting wastes and pollutants.

What Are Some Important Properties of Water? An Amazing Molecule

Water's unique and important properties arise mostly from attractive forces between its molecules.

Water is a remarkable substance with a unique combination of properties:

- *There are strong forces of attraction (called hydrogen bonds, Appendix 3, Figure 4) between molecules of water.* These attractive forces are the major factor determining water's distinctive properties.

- *Water exists as a liquid over a wide temperature range because of the strong forces of attraction between water molecules.* Without its high boiling point the oceans would have evaporated a long time ago.

- *Liquid water changes temperature slowly because it can store a large amount of heat without a large change in temperature.* This high heat capacity helps protect living organisms from temperature fluctuations. It also moderates the earth's climate and makes water an excellent coolant for car engines and power plants.

- *Evaporating liquid water takes large amounts of energy because of the strong forces of attraction between its molecules.* Water absorbs large amounts of heat as it changes into water vapor and releases this heat as the vapor condenses back to liquid water. This helps distribute heat throughout the world and determine the climates of various areas (Figure 6-9, p. 107 and Figure 6-10, p. 107. This property also makes evaporation cooling process—explaining why you feel cooler when perspiration evaporates from your skin.

- *Liquid water can dissolve a variety of compounds.* This enables it to carry dissolved nutrients into the tissues of living organisms, flush waste products out of those tissues, serve as an all-purpose cleanser, and help remove and dilute the water-soluble wastes of civilization. This property also means that water-soluble wastes can easily pollute water.

- *Water filters out wavelengths of the sun's ultraviolet (UV) radiation that would harm some aquatic organisms.*

- *Attractive forces between the molecules of liquid water cause its surface to contract and to adhere to and coat a solid.* These strong cohesive forces allow narrow columns of water to rise through a plant from its roots to its leaves (capillary action).

- *Unlike most liquids, water expands when it freezes.* This means that ice floats on water because it has a lower density (mass per unit of volume) than liquid water. Otherwise lakes and streams in cold climates would freeze solid and lose most of their current forms of aquatic life. Because water expands upon freezing, it can break pipes, crack a car's engine block (which is why we use antifreeze), break up streets, and fracture rocks that end up as soil particles.

Without these unique properties of water, you and most other forms of life on this planet would not exist. As water endlessly recycles through the biosphere, it physically connects us to one another, to other forms of life, and to the entire planet.

Despite its importance, water is one of our most poorly managed resources. We waste it and pollute it. We also charge too little for making it available. This encourages still greater waste and pollution of this resource, for which we have no substitute. As Benjamin Franklin said many decades ago: "It is not until the well runs dry that we know the worth of water."

15-2 SUPPLY, RENEWAL, AND USE OF WATER RESOURCES

How Much Fresh Water Is Available? Natural Recycling to the Rescue

Only about 0.01% of the earth's water supply is available to us as fresh water, but this supply is recycled.

Only a tiny fraction of the planet's abundant water is available to us as fresh water (Figure 15-2). Study this figure carefully to see where the world's water is found. About 97.4% of the world's total volume of water is found in oceans and saline lakes and is too salty for drinking, irrigation, or industry (except as a coolant). Most of the remaining 2.6% that is fresh water is locked up in ice caps or glaciers or in groundwater too deep or salty to be used.

Thus only about 0.014% of the earth's total volume of water is easily available to us as soil moisture, usable groundwater, water vapor, and lakes and streams. If the world's water supply were only 100 liters (26 gallons), our usable supply of fresh water would be only 0.014 liter, or 2.5 teaspoons!

Fortunately, the world's fresh water supply is continuously collected, purified, recycled, and distributed in the solar-powered *hydrologic cycle* (Figure 4-28, p. 76). You might want to review this figure to trace the circulation of the earth's water as it evaporates from water, land, and organisms into the atmosphere; con-

denses and falls as precipitation back to the earth's land and water; and flows across the earth's surfaces as runoff into streams, rivers, lakes, wetlands, and the seas and infiltrates underground.

This magnificent water recycling and purification system works only as long as we do not overload water systems with slowly degradable and nondegradable wastes or withdraw water from underground supplies faster than it is replenished. *Bad news.* In parts of the world we are doing both of these things.

Differences in average annual precipitation divide the world's continents, countries, and people into water *haves* and *have-nots.* Some places get lots of rain (the dark green and light green areas in Figure 6-6, p. 106) and others very little (the yellow areas in Figure 6-6, p. 106). For example, Canada, with only 0.5% of the world's population, has 20% of the world's fresh water, whereas China, with 21% of the world's people, has only 7% of the supply.

What Is Surface Water? Water on Top

Water that does not sink into the ground or evaporate into the air runs off into bodies of water.

One of our most precious resources is fresh water that flows across the earth's land surface and into the world's rivers, streams, lakes, wetlands, and estuaries. The precipitation that does not infiltrate the ground or return to the atmosphere by evaporation (including transpiration from plants) is called **surface runoff.** The region from which surface water drains into a river, lake, wetland, or other body of water is called its **watershed** or **drainage basin.**

The renewable supplies of freshwater we depend on consist of accessible surface runoff and freshwater that infiltrates into aquifers found fairly near the earth's surface. About two-thirds of the world's annual runoff is lost by seasonal floods and is not available for human use. The remaining one-third is **reliable runoff:** the amount of runoff that we can generally count on as a stable source of water from year to year.

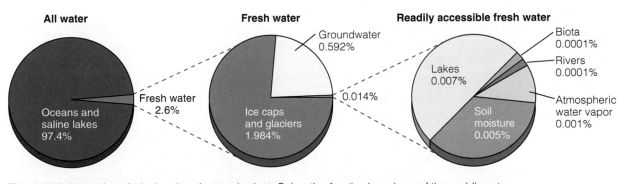

Figure 15-2 Natural capital: the planet's water budget. Only a tiny fraction by volume of the world's water supply is fresh water available for human use.

What Is Groundwater? Water Down Below

Some precipitation infiltrates the ground and is stored in spaces in soil and rock.

Some precipitation infiltrates the ground and percolates downward through voids (pores, fractures, crevices, and other spaces) in soil and rock (Figure 15-3). The water in these spaces is called **groundwater**—one of our most important sources of fresh water. Groundwater found within 1 kilometer (0.6 miles) of the earth's surface contains more than 100 times all the water found in the world's rivers, streams, freshwater lakes, and reservoirs combined.

Close to the surface in the **zone of aeration,** the pores of the soil contain a mixture of air and some water. Lower layers of soil where the spaces are completely filled with water make up the **zone of satura-** tion. We drill shallow wells to tap into groundwater in this zone. The **water table** is located at the top of the zone of saturation. It falls in dry weather or when we remove groundwater faster than it is replenished, and rises in wet weather.

Deeper down are geological layers called **aquifers**: porous, water-saturated layers of sand, gravel, or bedrock through which groundwater flows. They are like large elongated sponges through which groundwater seeps. Fairly watertight layers of rock or clay below an aquifer keep the water from seeping out. About one of every three people on the earth depends on water pumped out of aquifers for drinking and other uses.

Most aquifers are replenished naturally by precipitation that percolates downward through soil and rock in what is called **natural recharge.** But some are recharged from the side by *lateral recharge* from nearby streams. Most aquifers recharge extremely slowly.

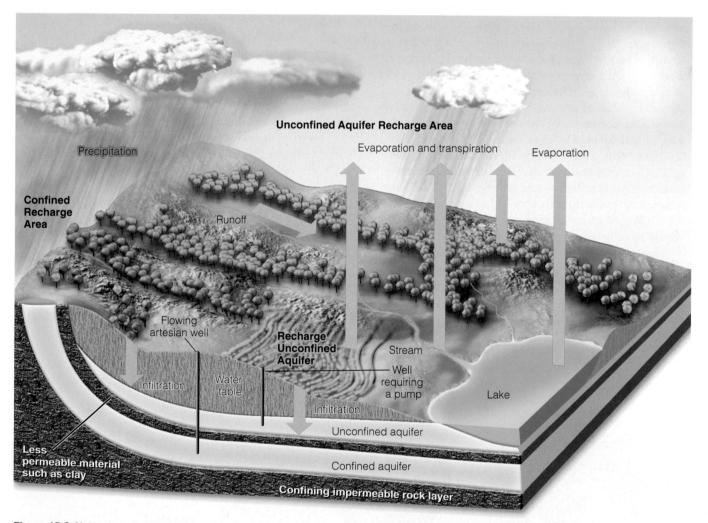

Figure 15-3 Natural capital: groundwater system. An *unconfined aquifer* is an aquifer with a water table. A *confined aquifer* is bounded above and below by less permeable beds of rock. Groundwater in this type of aquifer is confined under pressure. Some aquifers are replenished by precipitation and others are not.

Groundwater normally moves from points of high elevation and pressure to points of lower elevation and pressure. This movement is quite slow, typically only a meter or so (about 3 feet) per year and rarely more than 0.3 meter (1 foot) per day.

There is a *hydrological connection* between groundwater and surface water because eventually most groundwater flows into rivers, lakes, estuaries, and wetlands. Thus if we disrupt the hydrological cycle by removing groundwater faster than it is replenished, some nearby streams, lakes, and wetlands can dry up.

Some aquifers get very little, if any, recharge and on a human time scale are nonrenewable resources. Typically these *nonreplenishable aquifers*—also called *fossil aquifers*—are found fairly deep underground and were formed tens of thousands of years ago. Withdrawals from them amount to *water mining* that, if kept up, will deplete these ancient deposits.

How Much of the World's Reliable Water Supply Are We Withdrawing? Taking Half Now and More Later

We are using more than half of the world's reliable runoff of surface water and could be using 70–90% by 2025.

Withdrawal is the total amount of water we remove from a river, lake, or aquifer for any purpose. Some of this water may be returned to its source. For example, most water withdrawn from a river or lake to help cool power plants may be returned to its source. However, this input of heated water can disrupt aquatic life, a phenomenon known as *thermal pollution*.

Some of the water withdrawn from a source may be returned to that source for reuse. *Consumptive water use* occurs when water withdrawn is not available for reuse in the basin from which it was removed—mostly because of losses such as evaporation, seepage into the ground, transport to another area, or contamination.

During the last century, the human population tripled, global water withdrawal increased sevenfold, and per capita withdrawal quadrupled. As a result, we now withdraw about 34% of the world's reliable runoff. We leave another 20% in streams to transport goods by boats, dilute pollution, and sustain fisheries and wildlife. Thus *we directly or indirectly use about 54% of the world's reliable runoff of surface water*.

Because of increased population growth alone, global withdrawal rates of surface water could reach more than 70% of the reliable surface runoff by 2025—90% if per capita withdrawal of water continues rising at the current rate. This is a global average, with withdrawal rates already exceeding the reliable runoff in a growing number of areas.

How Do We Use the World's Fresh Water? Watering Crops Is Number One

Irrigation is the biggest user of water (70%) followed by industries (20%), and cities and residences (10%).

Worldwide, we use about 70% of the water we withdraw each year from surface waters and aquifers to irrigate one-fifth of the world's cropland. This produces about 40% of the world's food, including two-thirds of the rice and wheat. About 85% of the water withdrawn for irrigation is consumed and not returned to its water basin, mostly because of evaporation and seepage into the ground. Some of this water is also contaminated with salts and pesticides.

Industry uses about 20% of the water withdrawn each year, and cities and residences use the remaining 10%. Uses of withdrawn water vary from one region or country to another (Figure 15-4)..

Just about anything you do uses water. Reading a newspaper, driving a car, eating a hamburger or a bowl of rice, wearing cotton clothing, or drinking a beverage out of an aluminum can involve processes or products that require large amounts of water (Figure 15-5, p. 310).

There is also a difference in water priorities between developed and developing nations. According to the United Nations, for example, the daily minimum amount of water needed to support three fourths of the world's people is equal to the amount of water used each day to irrigate the world's golf courses.

Case Study: Freshwater Resources in the United States—Unequal Distribution

The United States has plenty of fresh water, but supplies vary in different areas depending on climate.

The United States has more than enough renewable fresh water. But much of it is in the wrong place at the

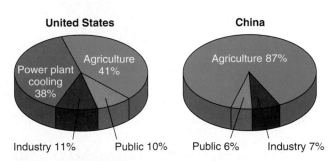

Figure 15-4 Use of water withdrawn in the United States and China. In the United States, about two-thirds of irrigation water comes from surface sources and the rest comes from underground. (Worldwatch Institute and World Resources Institute)

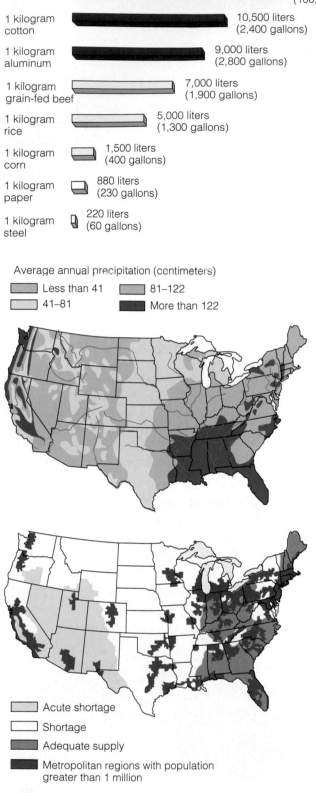

1 automobile — 400,000 liters (106,000 gallons)

1 kilogram cotton — 10,500 liters (2,400 gallons)

1 kilogram aluminum — 9,000 liters (2,800 gallons)

1 kilogram grain-fed beef — 7,000 liters (1,900 gallons)

1 kilogram rice — 5,000 liters (1,300 gallons)

1 kilogram corn — 1,500 liters (400 gallons)

1 kilogram paper — 880 liters (230 gallons)

1 kilogram steel — 220 liters (60 gallons)

Figure 15-5 Amount of water needed to produce some common agricultural and manufactured products. (U.S. Geological Survey)

Average annual precipitation (centimeters)

- Less than 41
- 41–81
- 81–122
- More than 122

- Acute shortage
- Shortage
- Adequate supply
- Metropolitan regions with population greater than 1 million

Figure 15-6 Average annual precipitation and major rivers (top) and water-deficit regions in the continental United States and their proximity to metropolitan areas having or projected to have populations greater than 1 million (bottom). (U.S. Water Resources Council and U.S. Geological Survey)

wrong time or is contaminated by agricultural and industrial practices. The eastern states usually have ample precipitation, whereas many western states have too little (Figure 15-6, top).

In the East, most water is used for energy production, cooling, and manufacturing. The largest use by far in the West (85%) is for irrigation.

In many parts of the eastern United States, the most serious water problems are flooding, occasional urban shortages, and pollution. For example, the 3 million residents of Long Island, New York, get most of their water from an increasingly contaminated aquifer.

The major water problem in the arid and semiarid areas of the western half of the country (Figure 15-6, bottom) is a shortage of runoff, caused by low precipitation (Figure 15-6, top), high evaporation, and recurring prolonged drought. Water tables in many areas are dropping rapidly as farmers and cities deplete aquifers faster than they are recharged.

In 2003, the U.S. Department of the Interior mapped out water *hot spots* in 17 western states (Figure 15-7). In these areas, competition for scarce water to support growing urban areas, irrigation, recreation,

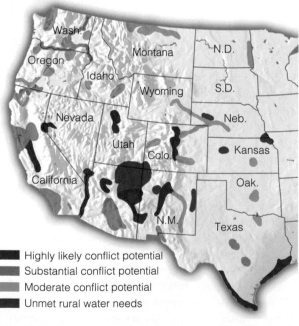

- Highly likely conflict potential
- Substantial conflict potential
- Moderate conflict potential
- Unmet rural water needs

Figure 15-7 Water *hot spot* areas in 17 western states that by 2025 could be faced with intense conflicts and "water wars" over competition for scarce water for urban growth, irrigation, recreation, and wildlife. Some analysts suggest that this is a map of places not to live over the next 25 years. (U.S. Department of the Interior)

and wildlife could trigger "water wars"—intense political and legal conflicts—in the next 20 years.

15-3 TOO LITTLE WATER

What Causes Freshwater Shortages? Climate and Demand

Dry climate, drought, dry soil, and too many people using the reliable supply cause water scarcity.

According to Swedish hydrologist Malin Falkenmark, there are four causes of water scarcity: *dry climate, drought* (a prolonged period in which precipitation is at least 70% lower and evaporation is higher than normal), *desiccation* (drying of exposed soil because of activities such as deforestation and overgrazing by livestock), and *water stress* (low per capita availability of water caused by increasing numbers of people relying on limited runoff).

Figure 15-8 shows the degree of stress on the world's major river systems, based on the amount of water available compared to the amount used by humans. A country is said to be *water stressed* when the volume of reliable runoff per person drops below about 1,700 cubic meters (60,000 cubic feet) per year. This usually occurs when water withdrawal is more than 20% higher than the reliable supply. A country

suffers from *water scarcity* when per capita water availability falls below 1,000 cubic meters (35,000 cubic feet) per year.

According to the United Nations, about 41% of the world's population lives in river basins located in 20 countries that suffer from water stress or water scarcity. Look at the red and orange areas in Figure 15-8 to see where these areas are located. The number of countries suffering from water stress or water scarcity could grow to 40 countries by 2020 and 60 countries by 2050.

Some areas have lots of water, but the largest rivers carrying most of the runoff are far from agricultural and population centers. For example, South America has the largest annual water runoff of any continent, but 60% of the runoff flows through the Amazon River in remote areas where few people live.

In some areas, overall precipitation is plentiful but arrives mostly during short periods or cannot be collected and stored because of a lack of storage capacity. For example, only a few hours of rain provide over half of India's rainfall during a four-month monsoon season.

The volumes of some of the world's lakes and rivers have shrunk drastically, mostly because of human withdrawals of water for irrigation and industry. Siberia's Aral Sea, once the world's fourth-largest freshwater lake, has shrunk in area to less than half its former size and lost 83% of its volume of water since

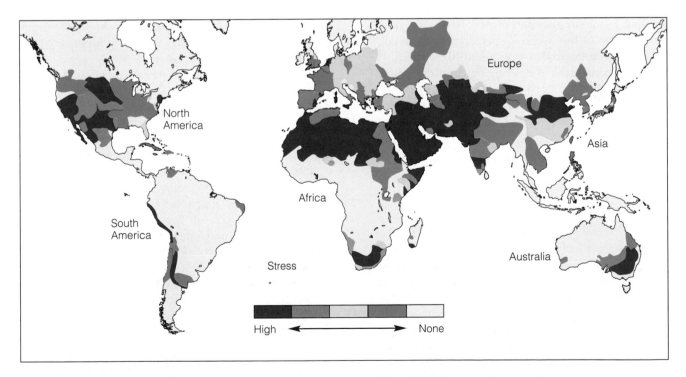

Figure 15-8 Natural capital degradation: stress on the world's major river basins, based on a comparison of the amount of water available with the amount used by humans. (Data from World Commission on Water Use in the 21st Century)

1960. The area of West Africa's Lake Chad, once the world's sixth largest lake, has shrunk by 92% since 1960 because of a combination of water diversion for irrigation and periods of prolonged drought. In the dry season, water flowing in the Colorado River and a number of other rivers throughout the world rarely reaches the ocean.

How Many of the World's People Do Not Have Access to Enough Fresh Water? Aquatic Inequality

About one out of six people do not have regular access to an adequate and affordable supply of clean water, and this could increase to at least one out of four people by 2050.

Good news. During the last century about 816 million people gained access to fresh water that is safe enough to drink. And in rural areas the percentage of families with access to safe drinking water rose from 10% to nearly 75%.

Bad news. A 2003 study by the United Nations found that about one out of six people do not have regular access to an adequate and affordable supply of safe drinking water and this could increase to at least one out of four people by 2050.

Even when a plentiful supply of water exists, most of the 1.1 billion poor people living on less than $1 a day cannot afford a safe supply of drinking water and must live in *hydrological poverty.* Most are cut off from municipal water supplies and must collect water from unsafe sources or buy water—often coming from polluted rivers—from private vendors at high prices. In water-short rural areas in developing countries, many women and children must walk long distances each day, carrying heavy jars or cans, to get a meager and sometimes contaminated supply of water.

How Can We Increase Freshwater Supplies? Withdraw More and Waste Less

We can increase water supplies by building dams, bringing in water from elsewhere, withdrawing groundwater, converting salt water to fresh water, wasting less water, and importing food.

There are several ways to increase the supply of fresh water in a particular area. One is to build dams and reservoirs to store runoff for release as needed. Another is to bring in surface water from another area. We can also withdraw groundwater and convert salt water to fresh water (desalination). Other strategies are to reduce water waste and import food to reduce water use in growing producing crops and raising livestock.

In *developed countries,* people tend to live where the climate is favorable and bring in water from an-

other watershed. In *developing countries,* most people (especially the rural poor) must settle where the water is and try to capture the precipitation they need.

Who Should Own and Manage Freshwater Resources? Government versus Private Ownership

There is controversy over whether water supplies should be owned and managed by governments or by private corporations.

Is access to enough clean water to meet one's basic needs a basic human right, or is water a commodity to be sold in the marketplace? Most people would say that everyone should have a right to clean water. The problem is, who will pay for making water available to everyone?

Most water resources are owned by governments and managed as publicly owned resources for their citizens. However, an increasing number of governments are retaining ownership of these public resources but hiring private companies to manage them. In addition, three large transnational companies—Vivendi, Suez, and RWE—based in Europe have a long-range strategy to buy up as much of the world's water supplies as possible, especially in Europe and North America.

Currently, about 85% of Americans get their water from publicly owned utilities. This may change. Within 10 years the three European-based water companies aim to control 70% of the water supply in the United States by buying up American water companies and entering into agreements with most cities to manage their water supplies.

The argument is that private companies have the money and expertise to manage these resources better and more efficiently than government bureaucracies. Experience with this *public–private partnership* approach is mixed. Some companies hired to manage water resources have improved efficiency, done a good job, and in a few cases lowered rates.

But in the late 1980s, Prime Minister Margaret Thatcher placed England's water management in private hands. The result was financial mismanagement, skyrocketing water rates, deteriorating water quality, and company executives giving themselves generous financial compensation packages. In the late 1990s, Prime Minister Tony Blair brought the system under control by imposing much stricter government oversight. *The message:* Governments hiring private companies to manage water resources must set standards and maintain strict oversight of such contracts.

Some government officials want to go further and sell public water resources to private companies. Many people oppose full privatization of water resources because they believe that water is a public resource too important to be left solely in private hands. Also, once a city's water systems have been taken over by a foreign-based corporation, efforts to return the systems to public control can lead to severe economic penalties under the rules of the World Trade Organization (WTO).

In the Bolivian town of Cochabamba, 60% of the water was being lost through leaky pipes. With no money to fix the pipes, the Bolivian government sold the town's water system to a subsidiary of the Bechtel Corporation. Within 6 months, the company doubled water rates. This led to a general strike and violent street clashes between protesters and government troops that led to 10,000 injured people and 7 deaths. The Bolivian government ended up tearing up the contract. But the town's current government–private cooperative water management system is in shambles and most of the leaks have not been stopped.

Some analysts point to two possible problems in a fully privatized water system. *First,* since private companies make money by delivering water, they have more incentive to sell as much water as they can rather than to conserve it. *Second,* because of lack of money to pay water bills, the poor will continue to be left out. There are no easy answers for managing the water that everyone needs.

X *How Would You Vote?* Should private companies own and manage most of the world's water resources? Cast your vote online at http://biology.brookscole.com/miller14.

15-4 USING DAMS AND RESERVOIRS TO SUPPLY MORE WATER

What Are the Advantages and Disadvantages of Large Dams and Reservoirs? Mixed Blessings

Large dams and reservoirs can produce cheap electricity, reduce downstream flooding, and provide year-round water for irrigating cropland, but they also displace people and disrupt aquatic systems.

An estimated 800,000 dams of all sizes now restrict the flow of the world's rivers. Large dams and reservoirs have benefits and drawbacks (Figure 15-9). Their main purpose is to capture and store runoff and release it as needed to control floods; to generate electricity; and to supply water for irrigation and for towns and cities. Reservoirs also provide recreational activities such as swimming, fishing, and boating.

The more than 45,000 large dams (22,000 of them in China) built on the world's 227 largest rivers have increased the annual reliable runoff available for human use by nearly one-third. But a series of dams on a river, especially in arid areas, can reduce downstream flow to a trickle and prevent it from reaching the sea as a part of the hydrologic cycle. According to the World Commission on Water in the 21st Century, half of the world's

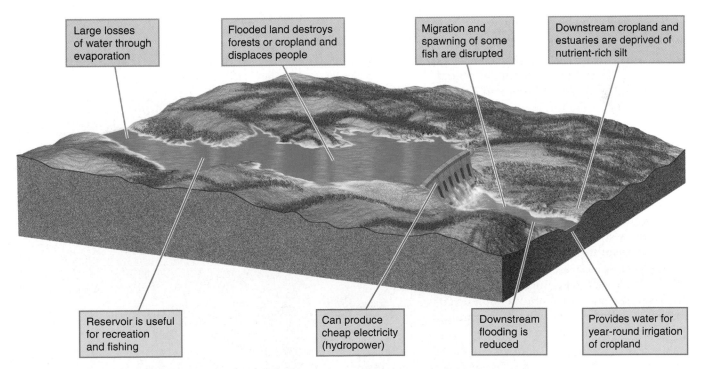

Large losses of water through evaporation

Flooded land destroys forests or cropland and displaces people

Migration and spawning of some fish are disrupted

Downstream cropland and estuaries are deprived of nutrient-rich silt

Reservoir is useful for recreation and fishing

Can produce cheap electricity (hydropower)

Downstream flooding is reduced

Provides water for year-round irrigation of cropland

Figure 15-9 Trade-offs: advantages (green) and disadvantages (orange) of large dams and reservoirs. The world's 45,000 large dams (higher than 15 meters or 50 feet) capture and store about 14% of the world's runoff, provide water for about 45% of irrigated cropland, and supply more than half the electricity used by 65 countries. The United States has more than 70,000 large and small dams, capable of capturing and storing half of the country's entire river flow. Pick the single advantage and disadvantage that you think are the most important.

major rivers are going dry part of the year because of flow reduction by dams (see the case study below).

This engineering approach to river management has displaced between 40 and 80 million people from their homes and has flooded an area of mostly productive land roughly equal to the area of California. In addition, this approach often impairs some of the important ecological and economic services rivers provide (Figure 13-12, p. 269). In 2003 the World Resources Institute estimated that dams and reservoirs have strongly or moderately fragmented and disturbed 60% of the world's major river basins. According to water-resource expert Peter H. Gleck, at least a fourth of the world's freshwater fish species are threatened or endangered, primarily because dams and water withdrawals have destroyed many free-flowing rivers.

Because of evaporation and seepage from their reservoirs, some dams lose more water than they provide. The reservoirs behind dams also eventually fill up with silt, which makes the dams useless for storing water or producing electricity.

How Would You Vote? Do the advantages of large dams outweigh their disadvantages? Cast your vote online at http://biology.brookscole.com/miller14.

Case Study: The Colorado River Basin—An Overtapped Resource

The Colorado River has so many dams and withdrawals that it often does not reach the ocean.

The Colorado River flows 2,300 kilometers (1,400 miles) from the mountains of central Colorado to the Mexican border and eventually to the Gulf of California (Figure 15-10).

During the past 50 years, this once free-flowing river has been tamed by a gigantic plumbing system consisting of 14 major dams and reservoirs, hundreds of smaller dams, and a network of aqueducts and canals that supply water to farmers, ranchers, and cities.

This domesticated river provides electricity from hydroelectric plants at major dams, water for more than 25 million people in seven states, and water used to grow about 15% of the nation's produce and livestock. The river also supports a multibillion-dollar recreation industry of whitewater rafting, boating, fishing, camping, and hiking.

The river supplies water to some of the nation's driest and hottest cities. Take away this tamed river and Las Vegas, Nevada, would be a mostly uninhabited desert area; San Diego, California, could not support its present population; and California's Imperial Valley, which grows a major portion of the nation's vegetables, would consist mostly of cactus and mesquite plants.

There are three major problems associated with use of this river's water. One is that the Colorado River basin includes some of the driest lands in the United States and Mexico (Figure 15-6). In addition, legal pacts in 1922 and 1944 allocated more water for human use in the U.S. and Mexico than the river can supply—even in years without a drought. The pacts also allocated no water for environmental purposes.

Finally, because of so many withdrawals, since 1960 the river has rarely made it to the Gulf of California except during a few years with higher than normal precipitation. This threatens the survival of species that spawn in the river, destroys estuaries that serve as breeding grounds for numerous aquatic species, and increases saltwater contamination of aquifers near the coast.

Traditionally, about 80% of the water withdrawn from the Colorado has been used to irrigate crops and raise cattle. This large-scale use of water for agriculture was made possible because the government paid

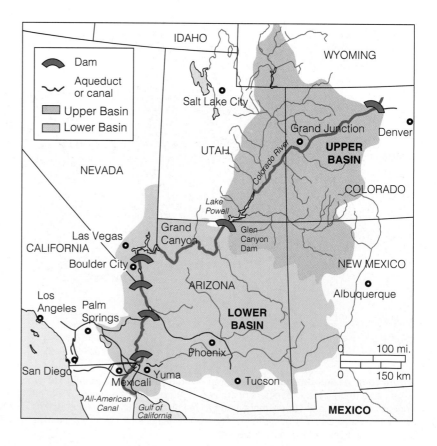

Figure 15-10 The *Colorado River basin*. The area drained by this basin is equal to more than one-twelfth of the land area of the lower 48 states.

for the dams and reservoirs and has supplied many of the farmers and ranchers with water at a low price. This has led to inefficient use of irrigation water, including growing crops such as rice, cotton, and alfalfa that need a lot of water in a water-short area.

It is estimated that improving overall irrigation efficiency by 10–15% would provide enough water to support projected urban growth in the areas served by the river until 2020.

Other problems are evaporation, leakage, and siltation from large reservoirs created by some of the dams along the Colorado River. For example, there are huge losses of river water in the Lake Mead and Lake Powell dam reservoirs from a combination of evaporation and seepage of water into porous rock beds under the reservoirs.

In addition, as the flow of the Colorado River slows in large reservoirs behind dams it drops much of its load of suspended silt. The amount of silt being deposited on the bottoms of the Lake Powell and Lake Mead reservoirs is roughly equivalent to that from having 20,000 dump trucks dump dirt into the reservoirs every day. Probably sometime during this century these reservoirs will be too full of silt to store water for generating hydroelectric power or controlling floods.

These problems illustrate the dilemmas that governments and people living in semiarid regions with shared river systems face as population and economic growth place increasing demands on limited supplies of surface water. Currently there are no cooperative agreements for use of 158 of the world's 263 water basins shared by two or more countries.

Case Study: China's Three Gorges Dam— A Controversial Project

There is debate over whether the advantages of the world's largest dam and reservoir will outweigh its disadvantages.

When completed, China's Three Gorges project on the mountainous upper reaches of the Yangtze River will include the world's largest hydroelectric dam and reservoir. Two kilometers (1.2 miles) long, the dam is supposed to be completed by 2013 at a cost of at least $25 billion. Figure 15-11 lists major advantages and disadvantages of this controversial project.

Bad news. About 1.9 million people are being relocated from the area to be flooded to form a gigantic 600-kilometer-long (385-mile-long) reservoir behind the dam—long enough to stretch from San Francisco to Los Angeles and as large as Lake Superior. Many people are being uprooted from their ancestral homes and relocated on land too barren to grow much food. The reservoir will also flood one of China's most beau-

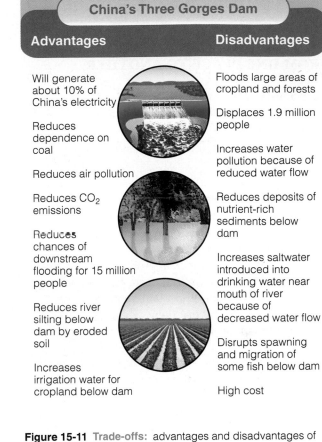

Trade-Offs

China's Three Gorges Dam

Advantages	Disadvantages
Will generate about 10% of China's electricity	Floods large areas of cropland and forests
Reduces dependence on coal	Displaces 1.9 million people
Reduces air pollution	Increases water pollution because of reduced water flow
Reduces CO_2 emissions	Reduces deposits of nutrient-rich sediments below dam
Reduces chances of downstream flooding for 15 million people	Increases saltwater introduced into drinking water near mouth of river because of decreased water flow
Reduces river silting below dam by eroded soil	Disrupts spawning and migration of some fish below dam
Increases irrigation water for cropland below dam	High cost

Figure 15-11 Trade-offs: advantages and disadvantages of the Three Gorges dam being built across the Yangtze River in China. Pick the single advantage and disadvantage that you think are the most important.

tiful areas, 1,350 cities and villages, and thousands of archeological and cultural sites.

Good news. When completed, the dam will have the electric output of 18 large coal-burning or nuclear power plants and will help reduce China's dependence on coal and its emissions of the greenhouse gas CO_2. It will also help hold back the Yangtze River's floodwaters, which have killed more than 500,000 people during the past 100 years—including 4,000 people in 1998. In addition, it will enable large cargo-carrying ships to travel deep into China's interior, greatly reducing transportation costs.

Mixed news. Because the dam is built over a seismic fault, geologists worry that the dam might collapse and cause a major flood that would kill millions of people. Engineers claim that the dam can withstand the maximum projected earthquake. Others are not so confident, noting that since 1949 more than 3,200 dams in China have collapsed and killed several hundred thousand people. About 80 small cracks have already been discovered in the dam. Critics claim that it would have been cheaper, less disruptive, and safer to build a series of smaller dams.

15-5 TRANSFERRING WATER FROM ONE PLACE TO ANOTHER

Case Study: The Aral Sea Disaster— A Glaring Example of Unintended Consequences

Diverting water from the Aral Sea and its two feeder rivers mostly for irrigation has created a major ecological, economic, and health disaster.

Tunnels, aqueducts, and underground pipes can transfer stream runoff collected by dams and reservoirs from water-rich areas to water-poor areas. However, they also create environmental problems. Indeed, most of the world's dam projects and large-scale water transfers illustrate the important ecological principle that *you cannot do just one thing*. There are almost always a number of unintended environmental consequences (Figure 3-4, p. 38).

An example is the shrinking of the Aral Sea (Figure 15-12). It is a result of a large-scale water transfer project in an area of the former Soviet Union with the driest climate in central Asia. Since 1960, enormous amounts of irrigation water have been diverted from the inland Aral Sea and its two feeder rivers to create one of the world's largest irrigated areas, mostly for raising cotton and rice. The irrigation canal, the world's longest, stretches over 1,300 kilometers (800 miles)—equivalent to one-third the width of the continental United States.

This large-scale water diversion project, coupled with droughts and high evaporation rates in this area's hot and dry climate, has caused a regional ecological, economic, and health disaster. Since 1960 the sea's salinity has tripled, its surface area has decreased by 58%, and it has lost 83% of its water. In effect it has been transformed from a single large lake (Figure 15-12, left) into three smaller lakes (Figure 15-12, right). Water withdrawal for agriculture has reduced the sea's two supply rivers to mere trickles.

About 85% of the area's wetlands have been eliminated and roughly half the area's bird and mammal species have disappeared. In addition, a huge area of former lake bottom has been converted to a human-made desert covered with glistening white salt. The increased salt concentration caused the presumed extinction of 20 of the area's 24 native fish species. This

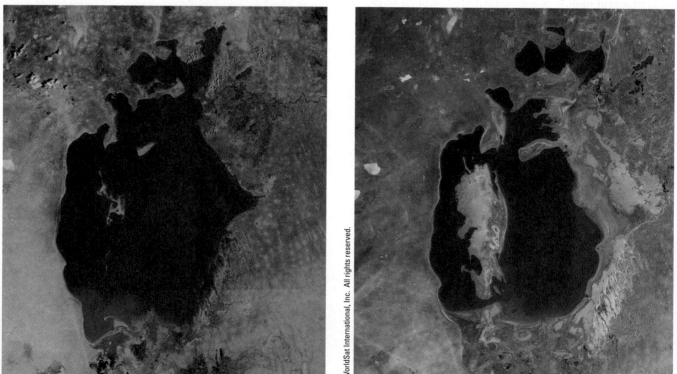

Figure 15-12 Natural capital degradation: the *Aral Sea* was once the world's fourth largest freshwater lake. Since 1960 it has been shrinking and getting saltier because most of the water from the rivers that replenish it has been diverted to grow cotton and food crops. These satellite photos show the sea in 1976 and in 1997. As the lake shrinks, it leaves behind a salty desert, economic ruin, increasing health problems, and severe ecological disruption.

has devastated the area's fishing industry, which once provided work for more than 60,000 people. Fishing villages and boats once on the sea's coastline now are abandoned in the middle of a salt desert.

Winds pick up the salty dust that encrusts the lake's now-exposed bed and blow it onto fields as far as 300 kilometers (190 miles) away. As the salt spreads, it pollutes water and kills wildlife, crops, and other vegetation. Aral Sea dust settling on glaciers in the Himalayas is causing them to melt at a faster than normal rate—another example of connections and unintended consequences.

To raise yields, farmers have increased use of herbicides, insecticides, fertilizers, and irrigation water on some crops. Many of the chemicals have percolated downward and accumulated to dangerous levels in groundwater, from which most of the region's drinking water comes.

Shrinkage of the Aral Sea has altered the area's climate. The once-huge sea acted as a thermal buffer that moderated the heat of summer and the extreme cold of winter. Now there is less rain, summers are hotter and drier, winters are colder, and the growing season is shorter. The combination of such climate change and severe salinization has reduced crop yields by 20–50% on almost a third of the area's cropland.

Finally, there have been increasing health problems from a combination of toxic dust, salt, and contaminated water for a growing number of the 58 million people living in the Aral Sea's watershed.

Can the Aral Sea be saved, and can the area's serious ecological and human health problems be reduced? There is agreement that the sea will never return to its former volume. Efforts have focused primarily on stopping further shrinkage and undoing some of the ecological and health damage caused by its shrinkage.

Encouraging news. Since 1999, the United Nations and the World Bank have spent about $600 million to purify drinking water and upgrade irrigation and drainage systems, which improves irrigation efficiency and flushes salts from croplands. In addition, some artificial wetlands and lakes have been constructed to help restore aquatic vegetation, wildlife, and fisheries.

The five countries surrounding the lake and its two feeder rivers have worked to improve irrigation efficiency and to partially replace water-thirsty crops such as rice and cotton, with other crops that require less water. As a result, the total annual volume of water in the Aral Sea basin has stabilized.

In 2004 the United Nations Environment Programme warned that Lake Balkhash in Kazakhstan could meet a fate like that of the Aral Sea if pollution and water withdrawals from the river Ili flowing into it from northwestern China continue increasing.

Case Study: The California Water Transfer Project—Bringing Water to a Desert

There is controversy over the massive transfer of water from water-rich northern California to water-poor southern California.

One of the world's largest water transfer projects is the *California Water Project* (Figure 15-13). It uses a maze of giant dams, pumps, and aqueducts (cement-lined artificial rivers) to transport water from water-rich northern California to southern California's heavily populated arid and semiarid agricultural regions and cities.

For decades, northern and southern Californians have feuded over how the state's water should be allocated under this project. Southern Californians say they need more water from the north to grow more crops and to support Los Angeles, San Diego, and other growing urban areas. Agriculture uses three-fourths of the water withdrawn in California, much of it used inefficiently for growing water-thirsty crops such as alfalfa under desertlike conditions.

Opponents in the north say that sending more water south would degrade the Sacramento River, threaten fisheries, and reduce the flushing action that helps clean San Francisco Bay of pollutants. They also argue that much of the water sent south is wasted. They point to studies showing that making irrigation just 10% more efficient would provide enough water for domestic and industrial uses in southern California.

Mono Lake, a salt lake in southern California's desert just east of Yosemite National Park, is an ecological causality of water transfers from one water basin to another. Diversion of water from rivers feeding the lake has shrunk the lakes volume by one-third. This

Figure 15-13 Solutions: California Water Project and the Central Arizona Project. These projects involve large-scale water transfers from one watershed to another. Arrows show the general direction of water flow.

has reduced populations of resident and migrating gulls, ducks, and wading birds that use the lake for food and shelter and led to lawsuits over greatly reducing diversion of water flowing into the lake. In 1994, the California Water Resources Control Board required Los Angeles to restrict its water diversions and allow water to be restored to the lake.

According to a 2002 joint study by a team of scientists and engineers, projected global warming is likely to sharply reduce water availability in California (especially southern California) and other water-short states in the western United States even under the study's best-case scenario. The study projects that overall precipitation levels are likely to remain constant, but warmer temperatures will cause what would have fallen as snow during the winter to come down as rain.

Currently the annual snow pack in California's northern Sierra Nevada mountains acts as a natural reservoir by storing water as snow through the winter and then slowly releasing it as water during spring and summer when water demand is high. If winter precipitation falls as rain instead of as snow, it will fill rivers and streams at a time of year when the demand for water is low, and will lead to shortages during spring and summer when demand is high. Some analysts project that sometime during this century many of the people living in arid southern California cities (such as Los Angeles and San Diego), and farmers in the area, will have to move elsewhere because of a lack of water.

Pumping out more groundwater is not the answer because it is already being withdrawn faster than it is replenished throughout much of California and desalinating ocean water is too expensive for irrigation. To most analysts, quicker and cheaper solutions are to improve irrigation efficiency, stop growing water-thirsty crops in a desert climate, and allow farmers to sell cities the legal rights to withdraw certain amounts of water from rivers.

Case Study: Canada's James Bay Watershed Transfer Project—Rearranging Nature

The first phase of a gigantic 50-year project to produce hydroelectric power for Canada and the United States has been completed, but the second phase has been postponed.

Another major watershed transfer project is Canada's James Bay project. It is a $60 billion, 50-year scheme to harness the wild rivers that flow into Quebec's James and Hudson Bays to produce electric power for Canadian and U.S. consumers (Figure 15-14).

If completed, this megaproject by Hydro-Quebec involves building 600 dams and dikes that will reverse or alter the flow of 19 giant rivers covering a watershed three times the size of New York state. The project will flood an area of boreal forest and tundra equal in area to Washington state or Germany. It will also

Figure 15-14 If completed, the *James Bay project* in northern Quebec will alter or reverse the flow of 19 major rivers and flood an area the size of the state of Washington to produce hydropower for consumers in Quebec and the United States, especially in New York State. Phase I of this 50-year project is completed.

displace thousands of indigenous Cree and Inuit, who for 5,000 years have lived off James Bay by subsistence hunting, fishing, and trapping.

After 20 years, the $16 billion first phase of the project has been completed. It diverted three major rivers, flooded a huge area of tundra and forest, and built dams that generate electricity equal to that from 26 large coal-burning or nuclear power plants.

The second phase was postponed indefinitely in 1994 because the first phase produced more power than could be sold. Opposition by the Cree, whose ancestral hunting grounds would have been flooded, and by Canadian and U.S. environmentalists, along with New York state's cancellation of two contracts to buy electricity, also helped to postpone phase II.

15-6 TAPPING GROUNDWATER, CONVERTING SALTWATER TO FRESHWATER, SEEDING CLOUDS, AND TOWING ICEBERGS AND BIG BAGGIES

What Are the Advantages and Disadvantages of Withdrawing Groundwater? Avoid Too Many Straws in the Glass

Most aquifers are renewable sources unless the water is removed faster than it is replenished or becomes contaminated.

Aquifers provide drinking water for about one-fourth of the world's people. In the United States, water pumped from aquifers supplies almost all of the drinking water in rural areas, one-fifth of that in urban areas, and 43% of irrigation water.

Relying more on groundwater has advantages and disadvantages (Figure 15-15). The *good news* is that aquifers are widely available and are renewable sources of water as long as the water is not withdrawn faster than it is replaced and as long as the aquifers do not become contaminated.

The *bad news* is that water tables are falling in many areas of the world as the rate of pumping out water (mostly to irrigate crops) exceeds the rate of natural recharge from precipitation. The problem of falling water tables sneaks up on us because we cannot see it happening. The first sign is shallow wells going dry, followed by loss of water from deeper wells if the water mining process continues. Covering aquifer recharge areas with urban development also contributes to aquifer depletion.

The world's three largest grain-producing countries—China, India, and the United States—are overpumping many of their aquifers. In 2002, China announced a massive project to pump surface water

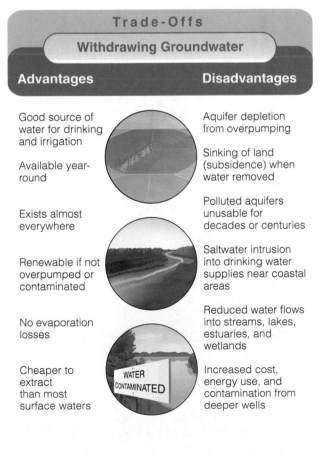

Figure 15-15 Trade-offs: advantages and disadvantages of withdrawing groundwater. Pick the single advantage and disadvantage that you think are the most important.

through three huge aqueducts from the Yangtze River in China's water-rich south to the country's arid north. The water will help grow more food and slow aquifer depletion in the North China Plain. This project is to begin in 2005 but will not be completed until 2050. Environmentalists are concerned about its potentially harmful ecological impacts.

In the United States, groundwater is being withdrawn at four times its replacement rate. The most serious overdrafts are in parts of the huge Ogallala Aquifer, underlying eight states in the arid high plains from southern South Dakota to central Texas (Case Study, p. 320) and in parts of the arid Southwest (Figure 15-16, p. 320). Serious groundwater depletion is also taking place in California's water-short Central Valley, which supplies about half the country's vegetables and fruits. Do you live in any of the blue areas in Figure 15-16 or in a similar area in another country where groundwater is being withdrawn faster than it is replenished?

Saudi Arabia is as water-poor as it is oil-rich. It gets about 70% of its drinking water at a high cost from the world's largest desalination complex on its eastern coast. The rest of the country's water is pumped from deep aquifers, most as nonrenewable as the country's oil. Yet this water-short nation wastes much of its scarcest resource with large numbers of fountains, swimming pools, and countless irrigation sprinklers that suck nonrenewable water from deep underground and let precious water evaporate into the hot, dry desert air. Hydrologists estimate that because of the rapid depletion of its fossil aquifers, most irrigated agriculture in Saudi Arabia may disappear within 10 to 20 years.

According to water resource expert Sandra Postel, about 480 million people are being fed with grain produced with eventually unsustainable water mining from aquifers. This example of the tragedy of the commons is expected to increase as irrigated areas are expanded to help feed 2.5 billion more people projected to join the ranks of humanity between 2004 and 2050.

In addition to limiting future food production, overpumping aquifers is increasing the gap between the rich and poor in some areas. As water tables drop, farmers must drill deeper wells, buy larger pumps, and use more electricity to run the pumps. Poor farmers cannot afford to do this and end up losing their land and either working for richer farmers or migrating to cities already crowded with poor people struggling to survive.

Withdrawing lots of water sometimes allows the sand and rock in aquifers to collapse, causing the land above the aquifer to *subside* or sink. Once an aquifer becomes compressed, recharge is impossible. Since 1950, some spots above aquifers in California's heavily farmed San Joaquin Valley has sunk or subsided more than 50 meters (160 feet).

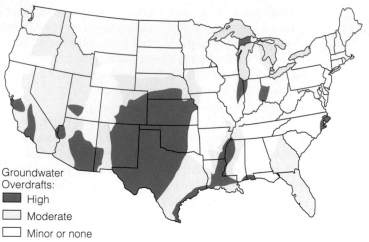

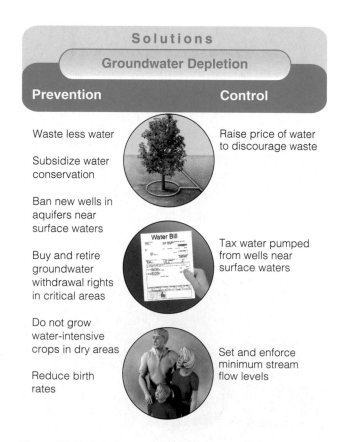

Figure 15-16 Natural capital degradation: areas of greatest aquifer depletion from groundwater overdraft in the continental United States. Aquifer depletion is also high in Hawaii and Puerto Rico (not shown on map). It causes the land above the aquifer to subside or sink in spots in most of these areas. (U.S. Water Resources Council and U.S. Geological Survey)

Groundwater Overdrafts:
- ■ High
- ▢ Moderate
- ☐ Minor or none

coastal areas of Florida, California, South Carolina, and Texas.

Figure 15-18 lists ways to prevent or slow the problem of groundwater depletion. Study this figure carefully.

Case Study: The Shrinking Ogallala Aquifer— Using Up Natural Capital

Pumping water from the world's largest aquifer has greatly increased food production, but overpumping is a serious problem in some areas.

Mexico City, built on a lakebed, has one of the world's worse subsidence problems because of an increase in groundwater overdrafts from rapid population growth and urbanization. In recent years, some parts of the city have sunk as much as 8 meters (26 feet).

Excessive withdrawal of groundwater can cause the roof of a cavern or underground conduit to collapse suddenly and create a large crater. Such *sinkholes* can form suddenly without warning and swallow houses, cars, and trees. Subsidence and the formation of sinkholes usually compress the rock particles in aquifers and thus lead to a permanent loss of the aquifer.

Finally, groundwater overdrafts near coastal areas can contaminate groundwater supplies by causing intrusion of salt water into freshwater aquifers used to supply water for irrigation and domestic purposes (Figure 15-17). This is an especially serious problem in

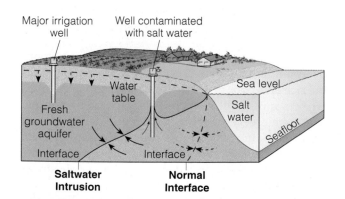

Figure 15-17 Natural capital degradation: *saltwater intrusion* along a coastal region. When the water table is lowered, the normal interface (dashed line) between fresh and saline groundwater moves inland (solid line), making groundwater supplies unusable for irrigation and domestic purposes.

Solutions

Groundwater Depletion

Prevention	Control
Waste less water	Raise price of water to discourage waste
Subsidize water conservation	
Ban new wells in aquifers near surface waters	Tax water pumped from wells near surface waters
Buy and retire groundwater withdrawal rights in critical areas	
Do not grow water-intensive crops in dry areas	Set and enforce minimum stream flow levels
Reduce birth rates	

Figure 15-18 Solutions: ways to prevent or slow groundwater depletion. Which two of these solutions do you believe are the most important?

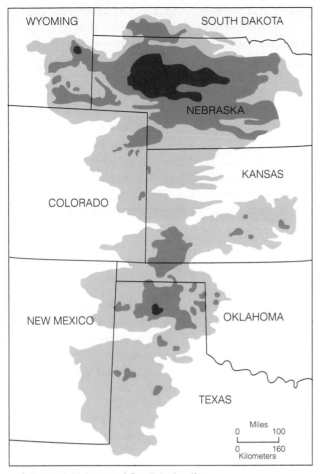

Saturated thickness of Ogallala Aquifer

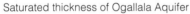 Less than 61 meters (200 ft.)

61–183 meters (200–600 ft.)

More than 183 meters (600 ft.)
(as much as 370 meters or 1,200 ft. in places)

Figure 15-19 Natural capital degradation: the Ogallala is the world's largest known aquifer. If the water in this aquifer were above ground, it could cover all 50 states with 0.5 meter (1.5 feet) of water. Water withdrawn from this aquifer is used to grow crops, raise cattle, and provide cities and industries with water. As a result, this aquifer, which is renewed very slowly, is being depleted, especially at its thin southern end in parts of Texas, New Mexico, Oklahoma, and Kansas. (Data from U.S. Geological Survey)

During the retreat of the last ice age about 15,000–30,000 years ago, vast amounts of water were deposited underground. This created the world's largest known aquifer, the Ogallala (Figure 15-19).

Pumping large amounts of water from this fossil aquifer has helped transform vast areas of arid high-plains prairie into one of the largest and most productive agricultural regions in the United States. Mostly because of irrigated farming, this region produces one-fifth of U.S. agricultural output (including 40% of its feedlot beef).

However, this aquifer is being overpumped and gradually depleted in some areas. Although it is gigantic, the Ogallala is essentially a one-time deposit of ancient water with an extremely slow recharge rate. In some areas, water is being pumped out 8–10 times faster than the slow natural recharge rate.

The northernmost states (Wyoming, North Dakota, South Dakota, and parts of Colorado) still have ample water supplies from the aquifer. But in parts of the southern states, where the aquifer is thinner, the water is being depleted rapidly—especially in the Texas High Plains.

Government subsidies designed to increase crop production also increase depletion of the Ogallala. These subsidies encourage farmers to grow water-thirsty cotton in the lower basin, give crop-disaster payments, and provide tax breaks in the form of groundwater depletion allowances (with larger breaks for heavier groundwater use).

Can Deep Aquifers Supply More Water? Solution or Pipe Dream?

Scientists are evaluating huge, deep aquifers as a source of water.

With global water shortages looming, scientists are evaluating deep aquifers—found at depths of 0.8 kilometer (0.5 mile) or more—as future water sources. Some of these are gigantic nonrenewable aquifers containing deposits of water as much as a million years old.

Seismic and core-drilling technologies used by the oil industry are being used to locate and evaluate the largest of these deep aquifers, some of which run underneath several countries. Preliminary results suggest that some of these aquifers hold enough water to support billions of people for centuries. They also indicate that the water quality is much higher than that in most of the world's rivers and lakes.

There are two major concerns about tapping these mostly one-time deposits. One is that we know little about the geological and ecological impacts of pumping from deep aquifers. The other is that no international water treaties govern the rights to and ownership of water that underlies several countries. Without such treaties, there could be legal and physical conflicts over who has the right to tap into and use these resources.

How Useful Is Desalination? A Costly Option

Removing salt from seawater will probably not be done widely because of high costs and what to do with the resulting salt.

Desalination involves removing dissolved salts from ocean water or from brackish (slightly salty) water in aquifers or lakes. It is another way to increase supplies of fresh water.

One method for desalinating water is *distillation*—heating salt water until it evaporates, leaves behind salts in solid form, and condenses as fresh water. Another method is *reverse osmosis*—pumping salt water at high pressure through a thin membrane with pores that allow water molecules, but not most dissolved salts, to pass through. In effect, high pressure pushes fresh water out of salt water.

There are about 13,500 desalination plants in 120 countries, mostly the desert nations of the Middle East, North Africa, the Caribbean, and the Mediterranean. These plants meet less than 0.3% of the world's water needs.

Oil-rich and water-short Middle Eastern countries produce about 60% of the world's desalinated water. Saudi Arabia is the largest producer and accounts for more than a third of the world's output, followed by the United States, which produces about a fifth of the world's desalinated water.

Water-short Israel plans to get as much half of its water from desalination by 2008. Some water-short coastal cities in the United States, such as Tampa, Florida, have built desalination plants to supplement water supplies. In California, coastal cities such as Los Angeles, San Diego, and Monterey may build such plants.

There are two major problems with the widespread use of desalination. One is the high cost because it takes a lot of energy to desalinate water. Currently, desalinating water costs two to three times as much as the conventional purification of fresh water, although recent advances in reverse osmosis have brought the energy costs down somewhat.

The second problem is that desalination produces large quantities of briny wastewater that contains lots of salt and other minerals. Dumping concentrated brine into a nearby ocean increases the salinity of the ocean water, which threatens food resources and aquatic life in the vicinity. Dumping it on land could contaminate groundwater and surface water.

Bottom line: Currently, significant desalination is practical only for water-short wealthy countries and cities that can afford its high cost.

Scientists are working to develop new membranes for reverse osmosis that can separate water from salt more efficiently and under less pressure. If successful, this strategy could bring down the cost of desalination. Even so, it probably will not be cheap enough to irrigate conventional crops or meet much of the world's demand for fresh water unless scientists can figure out how to use solar energy or other means to desalinate seawater cheaply and how to safely dispose of the salt left behind.

Can Cloud Seeding and Towing Icebergs or Gigantic Water Bags Improve Water Supplies? Solutions or Pipe Dreams?

Seeding clouds with tiny particles of chemicals to increase rainfall, or towing icebergs or huge bags filled with fresh water to dry coastal areas, probably will not provide significant amounts of fresh water in the future.

For decades, 10 states, mostly in the water-short western United States, and 24 other countries have experimented with seeding clouds with dry ice or tiny particles of chemicals such as silver iodide. The hypothesis is that the particles become nuclei around which raindrops form and thus produce more rain or snow over dry regions and more snow over mountains.

Bad news. First, cloud seeding does not work well in very dry areas where rain is needed most, because there are few clouds to seed. *Second,* although some proponents in the multimillion-dollar cloud-seeding industry say the technology works, a 2003 report by the U.S. National Academy of Sciences says there is no compelling scientific evidence that it does. *Third,* it introduces large amounts of the cloud-seeding chemicals into soil and water systems, possibly harming people, wildlife, and agricultural productivity.

Fourth, seeding has led to legal disputes over the ownership of cloud water. For example, during a 1977 drought the attorney general of Idaho accused officials in neighboring Washington state of "cloud rustling" and threatened to file suit in federal court.

Some analysts have proposed towing huge icebergs from Antarctic or Arctic waters to arid coastal areas such as Saudi Arabia and southern California and pumping fresh water from the melting bergs ashore. But nobody is sure how to do it, and even if they did it might cost too much, especially for water-short developing countries.

In 2002, a company proposed collecting spring runoff water from several rivers in northern California and piping it offshore to gigantic plastic (fiberpoly) bags as long as three football fields. Then tugboats would carry the floating bags to southern California where the water would be piped to shore.

The California Coastal Commission opposes this scheme until the environmental impact can be assessed. The costs are unknown but would probably be high enough that only wealthy areas could afford it. Stay tuned.

15-7 REDUCING WATER WASTE

What Are the Benefits of Reducing Water Waste? A Win-Win Solution

We waste about two-thirds of the water we use but using water more efficiently could reduce wastage to about 15%.

Mohamed El-Ashry of the World Resources Institute estimates that *65–70% of the water people use throughout the world is lost through evaporation, leaks, and other losses.* The United States, the world's largest user of water, does slightly better but still loses about half of the water it withdraws. El-Ashry believes it is economically and technically feasible to reduce such water losses to 15%, thereby meeting most of the world's water needs for the foreseeable future.

This win-win solution will also decrease the burden on wastewater plants and reduce the need for expensive dams and water transfer projects that destroy wildlife habitats and displace people. It will also slow depletion of groundwater aquifers and save energy and money.

According to water resource experts, the main cause of water waste is that *we charge too little for water.* Such *underpricing* is mostly the result of government subsidies that provide irrigation water, electricity, and diesel fuel for farmers to pump water from rivers and aquifers at below-market prices.

Subsidies keep the price of water so low that users have little or no financial incentive to invest in well-known water-saving technologies. According to water resource expert Sandra Postel, "By heavily subsidizing water, governments give out the false message that it is abundant and can afford to be wasted—even as rivers are drying up, aquifers are being depleted, fisheries are collapsing, and species are going extinct."

For example, U.S. farmers typically pay only one-fifth of the true cost of water provided by federally financed dams and other water supply projects. The low price for water in arid areas of the western United States encourages farmers there to grow crops that need a lot of water.

But farmers, industries, and others benefiting from government water subsidies argue that they promote settlement and agricultural production in arid and semiarid areas, stimulate local economies, and help lower prices of food, manufactured goods, and electricity for consumers.

Most water resource experts believe that in this century's era of water scarcity in many areas, governments will have to make the unpopular decision to raise water prices. China did this in 2002 because it faced water shortages in most of its major cities, rivers running dry, and falling water tables in key agricultural areas.

Higher water prices encourage water conservation but make it harder for low-income farmers and city dwellers to buy enough water to meet their needs. On the other hand, when South Africa raised water prices it established *life-line* rates that give each household a set amount of water at a low price to meet basic needs. When users exceed this amount, the price rises.

The second major cause of water waste is *lack of government subsidies for improving the efficiency of water use.* A basic rule of economics is that you get more of

what you reward. Subsidies for efficient water use would sharply reduce water waste.

✗ HOW WOULD YOU VOTE? Should water prices be raised sharply to help reduce water waste? Cast your vote online at http://biology.brookscole.com/miller14.

Solutions: How Can We Waste Less Irrigation Water? Deliver Water Precisely When and Where Needed

About 60% of the world's irrigation water is wasted, but several irrigation techniques could reduce the waste to 5–20%.

About 60% of the irrigation water applied throughout the world does not reach targeted crops and does not contribute to food production. Most irrigation water comes from groundwater wells or surface water sources and flows by gravity through unlined ditches in crop fields so the water can be absorbed by crops (Figure 15-20, left, p. 324). This *flood irrigation* method delivers far more water than the crops need and typically loses 40% of the water through evaporation, seepage, and runoff.

More efficient and environmentally sound irrigation technologies exist that can greatly reduce water demands and waste by delivering water more precisely to crops. One of these technologies is a *center-pivot low-pressure sprinkler* (Figure 15-20, right), which uses pumps to spray water on crops. Typically, it allows 80% of the water to reach crops and thus cuts water use by one-fourth compared to conventional gravity flow systems.

Another method is *low-energy precision application (LEPA) sprinklers,* a form of center-pivot irrigation that puts 90–95% of the water where crops need it by spraying the water closer to the ground and in larger droplets than the center-pivot low-pressure system. LEPA sprinklers use 20–30% less energy than low-pressure sprinklers and typically use 37% less water than conventional gravity flow systems.

Farmers also use *surge valves* or *time-controlled valves* on conventional gravity flow irrigation systems (Figure 15-20, left). These valves send water down irrigation ditches in pulses instead of a continuous stream. This can raise irrigation efficiency to 80% and cut water use by a fourth.

Another method is to use *soil moisture detectors* to water crops only when they need it. Some farmers in Texas bury a $1 cube of gypsum, the size of a lump of sugar, at the root zone of crops. Wires embedded in the gypsum run back to a small portable meter that indicates soil moisture, enabling the farmers to cut their irrigation water use by 33–66% in many cases.

Drip irrigation or *microirrigation systems* (Figure 15-20, center) are the most efficient ways to deliver small amounts of water precisely to crops. These systems consist of a network of perforated plastic tubing installed at or below the ground level. Small holes or

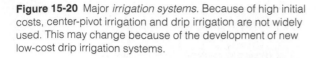

Figure 15-20 Major *irrigation systems*. Because of high initial costs, center-pivot irrigation and drip irrigation are not widely used. This may change because of the development of new low-cost drip irrigation systems.

Center pivot
(efficiency 80% with low-pressure sprinkler and 90–95% with LEPA sprinkler)

Water usually pumped from underground and sprayed from mobile boom with sprinklers.

Drip irrigation
(efficiency 90–95%)

Above- or below-ground pipes or tubes deliver water to individual plant roots.

Gravity flow
(efficiency 60% and 80% with surge valves)

Water usually comes from an aqueduct system or a nearby river.

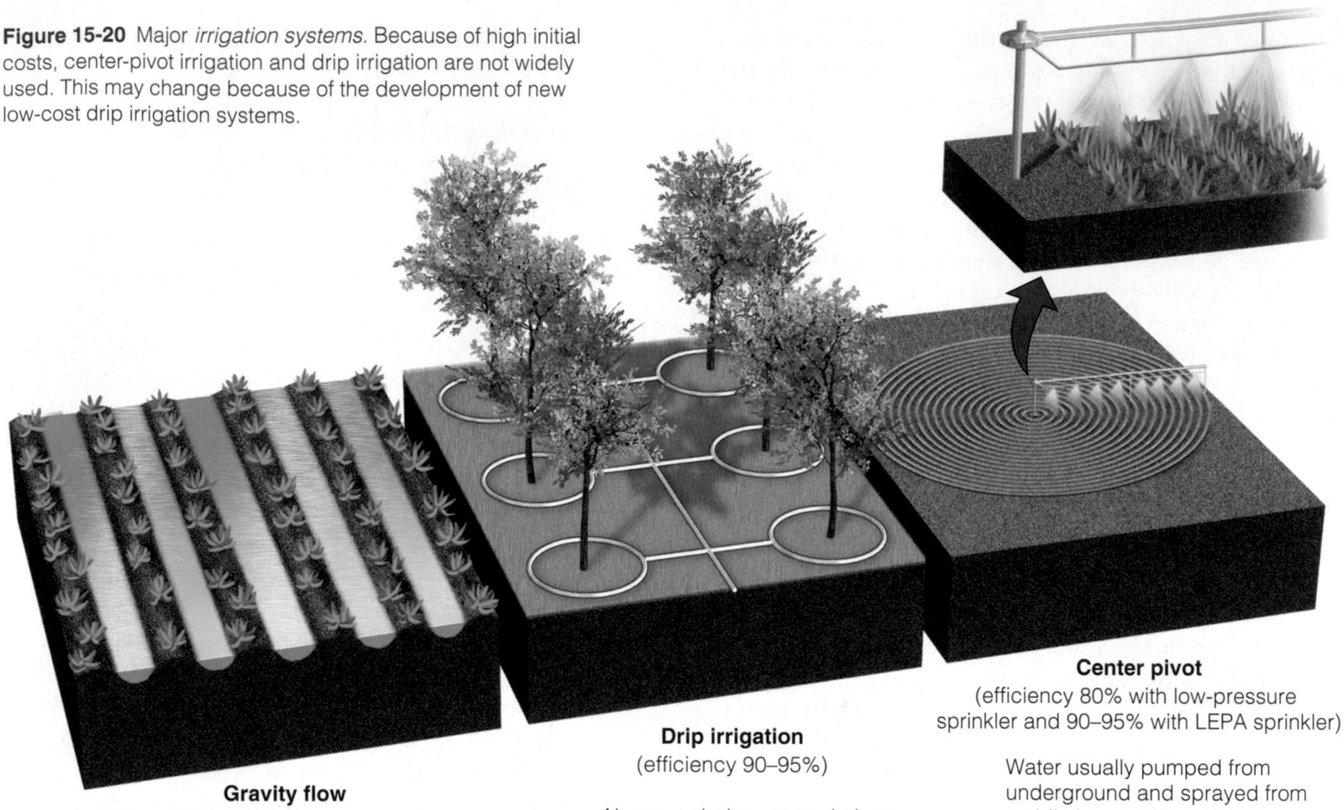

Solutions

Reducing Irrigation Water Waste

- Lining canals bring water to irrigation ditches

- Leveling fields with lasers

- Irrigating at night to reduce evaporation

- Using soil and satellite sensors and computer systems to monitor soil moisture and add water only when necessary

- Polyculture

- Organic farming

- Growing water-efficient crops using drought-resistant and salt-tolerant crop varieties

- Irrigating with treated urban waste water

- Importing water-intensive crops and meat

Figure 15-21 Solutions: methods for reducing water waste in irrigation. Which two of these solutions do you believe are the most important?

emitters in the tubing deliver drops of water at a slow and steady rate close to the plant roots.

Another innovation is DRiWATER®, called "drip irrigation in a box." It consists of 1-liter (1.1-quart) packages of gel-encased water that is released slowly into the soil after being buried near plant roots. It wastes almost no water and lasts about 3 months.

Drip irrigation is very efficient, with 90—95% of the water reaching the crops. It is also adaptable because the flexible and lightweight plastic tubing can be fitted to match the patterns of crops in a field and left in place or moved around.

Bad news. Drip irrigation is used on just over 1% of the world's irrigated crop fields and 4% of those in the United States. However, this percentage rises to 90% in Cyprus, 66% in Israel, and 13% in California. The main problem is that the capital cost of conventional drip irrigation systems is too high for most poor farmers and for use on low-value row crops.

Also, irrigation water is underpriced because of government subsidies. Raise water prices enough and drip irrigation would quickly be used to irrigate most of the world's crops. *Good news.* The capital cost of a new type of drip irrigation system is one-tenth as much per hectare as conventional drip systems.

Figure 15-21 lists other ways to reduce water waste in irrigating crops. Since 1950, water-short Israel has used many of these techniques to slash irrigation water waste by about 84% while irrigating 44% more land. Israel now treats and reuses 30% of its municipal sewage water for crop production and plans to increase this to 80% by 2025. The government also gradually removed most government water subsidies to raise the price of irrigation water to one of the highest in the world. Israelis also import most of their wheat and meat and concentrate on growing fruits, vegetables, and flowers that need less water. Researchers have also found that farmers can often sustain grain yields by using one fourth less irrigation water as long as crops receive enough water during their critical stages of growth.

Many of the world's poor farmers cannot afford to use most of the modern technological methods for increasing irrigation and its efficiency. Instead they use small-scale and low-cost traditional technologies.

Some (in Bangladesh, for example) use pedal-powered treadle pumps to move water through irrigation ditches. Others use buckets or small tanks with holes for drip irrigation.

How Can We Waste Less Water in Industry, Homes, and Businesses? Copy Nature, Raise Prices, Improve Efficiency

We can save water by using yard plants that need little water, using drip irrigation, raising water prices, fixing leaks, and using water-saving toilets and other appliances.

Figure 15-22 lists ways to use water more efficiently in industries, homes, and businesses. Many homeowners and businesses in water-short areas are copying nature by replacing green lawns with vegetation adapted to a dry climate (Figure 15-23). This win-win approach is called *Xeriscaping* (pronounced "ZEER-i-scaping"), and reduces water use by 30–85% and sharply reduces the need for labor, fertilizer, and fuel. It also reduces polluted runoff, air pollution, and yard wastes. Some people in dry areas collect rainwater from gutters in large plastic barrels on wheels and use it to water their flowers and gardens.

About one-fifth of all U.S. public water systems do not have water meters and charge a single low rate for almost unlimited use of high-quality water. Many apartment dwellers have little incentive to conserve water because water use is included in their rent. In Boulder, Colorado, introducing water meters reduced water use by more than one-third.

Because of laws requiring water conservation, the desert city of Tucson, Arizona, consumes half as much water per person as Las Vegas, a desert city with even less rainfall and less emphasis on water conservation (Spotlight, p. 326).

In the United States, flushing toilets with water clean enough to drink is the single largest use of

---

- Redesign manufacturing processes

- Landscape yards with plants that require little water

- Use drip irrigation

- Fix water leaks

- Use water meters and charge for all municipal water use

- Raise water prices

- Use waterless composting toilets

- Require water conservation in water-short cities

- Use water-saving toilets, showerheads, and front-loading clothes washers

- Collect and reuse household water to irrigate lawns and nonedible plants

- Purify and reuse water for houses, apartments, and office buildings

Figure 15-22 Solutions: methods of reducing water waste in industries, homes, and businesses. Which two of these solutions do you believe are the most important?

Figure 15-23 Solutions: *Xeriscaping.* This technique can reduce water use by as much as 85% by landscaping with rocks and plants that need little water and are adapted to the growing conditions in arid and semiarid areas. The term Xeriscape® was first used in 1978 in Denver, Colorado, and means "water conservation through creative landscaping."

Running Short of Water in Las Vegas, Nevada

Las Vegas, located in the Mojave Desert, is an artificial aquatic wonderland of large trees, green lawns and golf courses, waterfalls, and swimming pools. According to water experts, *Las Vegas uses more water per person than any other city in the world*. It is also one of the fastest-growing cities in the United States.

Tucson, Arizona, in the Sonoran Desert, is a model of water conservation. It began a strict water conservation program in 1976 that included raising water rates 500% for some residents. As a result, low-flush toilets, low-flow showerheads, Xeriscaping, and use of drip irrigation have become the norm.

In contrast, Las Vegas, which gets one-third less rainfall than Tucson, only recently started to encourage water conservation. It has raised water rates, but they are still less than half those in Tucson and have not gone up. Las Vegas is almost wholly dependent on water from the Colorado River stored in Lake Mead. Since 1999, a drought has reduced its water level to 59% of capacity, and it could drop to 42% by 2008.

Water experts project that even if these recent water conservation efforts are successful, Las Vegas may begin running short of water by 2007.

Critical Thinking

If you were an elected official in charge of Las Vegas, what three actions would you take to improve water conservation? What might be the political implications of doing these things?

domestic water. It accounts for about 38% of domestic water use, followed by bathing (32%), and washing clothes and dishes (20%). New models of low-flush toilets can remove the equivalent of two dozen golf balls in one flush without clogging and use fewer than 6.1 liters (1.6 gallons) of water—less than half the amount required by EPA standards.

About 50–75% of the water from bathtubs, showers, bathroom sinks, and clothes washers in a typical house could be stored and reused as *gray water* for irrigating lawns and nonedible plants. About two-thirds of the wastewater in Israel is reused this way. All of Singapore's sewage is treated at reclamation plants for reuse by industry.

Another problem is leakage and other losses from systems that supply water to homes, businesses, and industries. According to UN studies, 15–40% of the water supplied in urban areas is lost mostly through leakage of water mains, pipes, pumps, and valves, and illegal water hook-ups. In some parts of Africa such

losses can reach 50–70% of the water extracted from surface waters and aquifers. Even in an advanced industrialized country such as the United States such losses average 10–30%. However, such losses have been reduced to about 3% in Copenhagen and Denmark and 5% in Fukuoka, Japan.

How Can We Reduce the Use of Water in Removing Industrial and Household Wastes? Changing the Way We Deal with Wastes

We can mimic the way nature deals with wastes instead of using large amounts of high-quality water to wash away and dilute our industrial and animal wastes.

Industrialized countries use large amounts of water good enough to drink to dilute and wash or flush away industrial, animal, and household wastes. Sewage treatment plants remove valuable plant nutrients and dump most of them into rivers, lakes, and oceans. This overloads aquatic systems with plant nutrients that should be recycled to the soil, as nature does.

We also use high-quality water to flush toxic industrial wastes into sewers and send them to treatment plants that do not remove most of the harmful chemicals. The FAO estimates that if current trends continue, within 40 years we will need the world's entire reliable flow of river water just to dilute and transport the wastes we produce.

We can use five principles to redesign the way we manage sewage and industrial wastes while saving enormous amounts of water.

- Use pollution prevention and waste reduction to sharply decrease the amount of industrial wastes we produce.

- Ban discharge of industrial toxic wastes into municipal sewer systems.

- Rely more on waterless composting toilets that convert human fecal matter to a small amount of dry and odorless soil-like humus material that can be removed from a composting chamber every year or so and returned to the soil as fertilizer. They work. I used one for almost two decades without any problems.

- Return the nutrient-rich sludge produced by conventional waste treatment plants to the soil as a fertilizer. Banning the input of toxic industrial chemicals into sewage treatment plants will make this feasible.

- Shift to new ways to treat sewage that mimic the way nature breaks down and recycles the nutrients in organic waste material. Examples include solar-powered waste treatment systems and using wetlands to treat sewage, as discussed in Chapter 22.

15-8 TOO MUCH WATER

🔘 What Causes Flooding? Rain and People

Heavy rainfall, rapid melting of snow, removing vegetation, and destroying wetlands cause flooding.

Heavy rain and rapid snowmelt are the major causes of natural flooding by streams. A flood happens when water in a stream overflows its normal channel and spills into the adjacent area, called a **floodplain** (Figure 15-24).

People settle on floodplains because of their many advantages: fertile soil, ample water for irrigation, availability of nearby rivers for transportation and recreation, and flat land suitable for crops, buildings, highways, and railroads.

Floods have several benefits. They provide the world's most productive farmland because the land is regularly covered with nutrient-rich silt left after floodwaters recede. They also recharge groundwater and help refill wetlands.

But each year, floods kill up to 25,000 people and cause tens of billions of dollars in property damage. Floods, like droughts, are usually considered natural disasters.

Since the 1960s several types of human activities have contributed to the sharp rise in flood deaths and damages. One is *removal of water-absorbing vegetation,* especially on hillsides (Figure 15-25, p. 328). Another is *draining wetlands* that absorb floodwaters and reduce the severity of flooding. *Living on floodplains* also increases the threat of damage from flooding. Flooding also increases when we *pave or build* and replace water-absorbing vegetation, soil, and wetlands with highways, parking lots, and buildings that cannot absorb rainwater.

In developed countries, people deliberately settle on floodplains and then expect dams, levees, and other devices to protect them from floodwaters. However, when heavier-than-normal rains occur, these devices can be overwhelmed. In many developing countries, the poor have little choice but to try to survive in flood-prone areas (see the Case Study below).

Case Study: Living on Floodplains in Bangladesh—Danger for the Poor

Bangladesh has increased flooding because of upstream deforestation of Himalayan mountain slopes and the clearing of mangrove forests on its coastal floodplains.

Bangladesh is one of the world's most densely populated countries, with 141 million people (projected to reach 280 million by 2050) packed into an area roughly the size of Wisconsin. It is also one of the world's poorest countries.

The people of Bangladesh depend on moderate annual flooding during the summer monsoon season to grow rice and help maintain soil fertility in the delta basin. The annual floods deposit eroded Himalayan soil on the country's crop fields.

In the past, great floods occurred every 50 years or so. But since the 1970s they have come about every 4 years. Bangladesh's increased flood problems begin in the Himalayan watershed, where several factors—rapid population growth, deforestation, overgrazing,

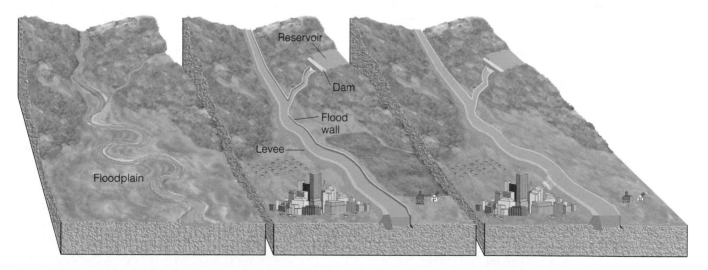

Figure 15-24 Land in a natural floodplain (left) often is flooded after prolonged rains. When the floodwaters recede, deposits of silt are left behind, creating a nutrient-rich soil. To reduce the threat of flooding and thus to allow people to live in floodplains, rivers have been narrowed and straightened (channelized), equipped with protective levees and walls, and dammed to create reservoirs that store and release water as needed. These alterations can give a false sense of security to floodplain dwellers. In the long run, such measures can greatly increase flood damage because they can be overwhelmed by prolonged rains (right), as happened along the Mississippi River in the midwestern United States during the summer of 1993.

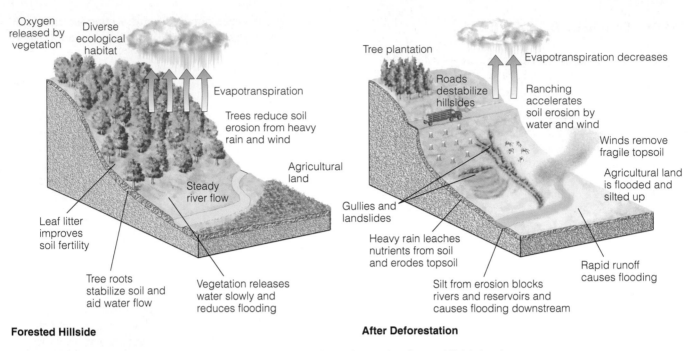

Forested Hillside

Oxygen released by vegetation

Diverse ecological habitat

Evapotranspiration

Trees reduce soil erosion from heavy rain and wind

Agricultural land

Steady river flow

Leaf litter improves soil fertility

Tree roots stabilize soil and aid water flow

Vegetation releases water slowly and reduces flooding

After Deforestation

Tree plantation

Evapotranspiration decreases

Roads destabilize hillsides

Ranching accelerates soil erosion by water and wind

Winds remove fragile topsoil

Agricultural land is flooded and silted up

Gullies and landslides

Heavy rain leaches nutrients from soil and erodes topsoil

Silt from erosion blocks rivers and reservoirs and causes flooding downstream

Rapid runoff causes flooding

Figure 15-25 Natural capital degradation: hillside before and after deforestation. Once a hillside has been deforested for timber and fuelwood, livestock grazing, or unsustainable farming, water from precipitation rushes down the denuded slopes, erodes precious topsoil, and floods downstream areas. A 3,000-year-old Chinese proverb says, "To protect your rivers, protect your mountains."

and unsustainable farming on steep and easily erodible mountain slopes—have greatly diminished the ability of its mountain soils to absorb water.

Think of a forest as a complex living sponge for catching, storing, using, and recycling water and releasing it in small amounts. Cut down the forests on the Himalayan mountains and what happens? Instead of being absorbed and released slowly, water from the monsoon rains runs off the denuded Himalayan foothills, carrying vital topsoil with it (Figure 15-25, right).

This increased runoff of soil, combined with heavier-than-normal monsoon rains, has increased the severity of flooding along Himalayan rivers and downstream in Bangladesh. A disastrous flood in 1998 covered two-thirds of Bangladesh's land area for 9 months, leveled 2 million homes, drowned at least 2,000 people, and left 30 million people homeless. It also destroyed more than one-fourth of the country's crops, which caused thousands of people to die of starvation. In 2002, another flood left 5 million people homeless and inundated large areas of rice fields.

Living on Bangladesh's coastal floodplain also carries dangers from storm surges and cyclones. In 1970, as many as 1 million people drowned in one storm, and another surge killed an estimated 139,000 people in 1991.

In their struggle to survive, the poor in Bangladesh have cleared many of the country's coastal mangrove forests for fuelwood, farming, and aquaculture ponds for raising shrimp. This has led to more severe flooding because these coastal wetlands help shelter Bangla-desh's low-lying coastal areas from storm surges and cyclones. Damages and deaths from cyclones in areas of Bangladesh still protected by mangrove forests have been much lower than in areas where the forests have been cleared.

How Can We Reduce Flood Risks? Think about Where You Want to Live

We can reduce flooding risks by controlling river water flows, preserving and restoring wetlands, identifying and managing flood-prone areas, and if possible choosing not to live in such areas.

We can use several methods to reduce the risk from flooding. One is to *straighten and deepen streams*, a process called *channelization* (Figure 15-24, middle). Channelization can reduce upstream flooding but it removes bank vegetation and increases stream velocity. This increased flow of water can promote upstream bank erosion, increase downstream flooding and sediment deposition, and reduce habitats for aquatic wildlife.

Another approach is to *build levees or floodwalls* along the sides of streams (Figure 15-24, middle). Levees contain and accelerate stream flow, but this increases the water's capacity for doing damage downstream. They also do not protect against unusually high and powerful floodwaters, as occurred in 1993 when two-thirds of the levees built along the Mississippi River in the United States were damaged or destroyed.

Building dams can also reduce the threat of flooding by storing water in a reservoir and releasing it gradually. Another way to reduce flooding is to *preserve existing wetlands and restore degraded wetlands* to take advantage of the natural flood control provided by floodplains.

We can also *identify and manage flood-prone areas.* Actions include prohibiting certain types of buildings or activities in high-risk flood zones, and elevating or floodproofing buildings that are allowed on floodplains. In addition, we can construct a floodway that allows floodwater to flow through a community with minimal damage. These *prevention,* or *precautionary,* approaches are based on thousands of years of experience that can be summed up in one idea: *Sooner or later the river (or ocean) always wins.*

On a personal level we can use the precautionary approach to *think carefully about where we live.* Many of the poor live in flood-prone areas because they have nowhere else to go. But most people can do some research and choose not to live in areas subject to flooding.

15-9 A MORE SUSTAINABLE WATER FUTURE

How Can We Use Water More Sustainably? A Blue Revolution

We can use water more sustainably by cutting waste, raising water prices, preserving forests on water basins, and slowing population growth.

Sustainable water use is based on the commonsense principle stated in an old Inca proverb: "The frog does not drink up the pond in which it lives." Figure 15-26 lists ways to implement this principle.

Historically, our response to water scarcity has been to expand the supply by building more dams, transporting water from one area to another, and drilling more wells. These measures will continue to some degree, but water resource experts project that the emphasis will begin shifting from increasing the water supply to reducing the water demand by using water more efficiently and stabilizing population.

The challenge in developing such a *blue revolution* is to implement a mix of strategies. One involves using technology to irrigate crops more efficiently and to save water in industries and homes. A second approach uses economic and political decisions to remove subsidies that cause water to be underpriced and thus wasted while guaranteeing low prices for low-income consumers, and adding subsidies that reward reduced water waste.

A third component is to switch to new waste production and treatment systems that accept only nontoxic wastes, use less or no water to treat wastes, return nutrients in plant and animal wastes to the soil, and mimic the ways that nature decomposes and recycles organic wastes.

Solutions

Sustainable Water Use

- Not depleting aquifers
- Preserving ecological health of aquatic systems
- Preserving water quality
- Integrated watershed management
- Agreements among regions and countries sharing surface water resources
- Outside party mediation of water disputes between nations
- Marketing of water rights
- Raising water prices
- Wasting less water
- Decreasing government subsidies for supplying water
- Increasing government subsidies for reducing water waste
- Slowing population growth

Figure 15-26 Solutions: methods for achieving more sustainable use of the earth's water resources.

A fourth strategy is to leave enough water in rivers to protect wildlife, ecological processes, and the natural ecological services provided by rivers (Figure 13-12, p. 269). We now control the flow rates of most rivers. As a result, some analysts say that we have an ethical and ecological responsibility to manage rivers for wildlife and restore those we have damaged. Ecological restoration efforts show that when we restore a river's flow and reconnect it with its floodplain its wildlife and ecological health can return. One way to do this is to follow South Africa's lead in legally establishing a freshwater reserve for all rivers designed to sustain their biodiversity and the valuable ecosystem services they provide for other species and society.

Each of us can help bring about this blue revolution by using and wasting less water. We can also support government policies that result in more sustainable use of the world's water and better ways to treat our industrial and household wastes. Figure 15-27 (p. 330) lists ways you can reduce your water use and waste. Since virtually everything we use requires water to produce, cutting down on unnecessary consumption is a key to reducing water use and water pollution. Reducing meat consumption is another way to lower water use and waste. The typical meat-intensive U.S. diet

- Use water-saving toilets, showerheads, and faucet aerators.

- Shower instead of taking baths, and take short showers.

- Repair water leaks.

- Turn off sink faucets while brushing teeth, shaving, or washing.

- Wash only full loads of clothes or use the lowest possible water-level setting for smaller loads.

- Wash a car from a bucket of soapy water, and use the hose for rinsing only.

- If you use a commercial car wash, try to find one that recycles its water.

- Replace your lawn with native plants that need little if any watering.

- Water lawns and gardens in the early morning or evening.

- Use drip irrigation and mulch for gardens and flowerbeds.

- Use recycled (gray) water for watering lawns and houseplants and for washing cars.

Figure 15-27 What can you do? Ways you can reduce your use and waste of water.

requires about twice as much water per person as a nutritious vegetarian diet.

As air is a sacred gas, so is water a sacred liquid that links us to all the oceans of the world and ties us back in time to the very birthplace of life.

DAVID SUZUKI

CRITICAL THINKING

1. How do human activities increase the harmful effects of prolonged drought? How can we reduce these effects?

2. Put the following users in order of how much water you would allocate to them from the Colorado River (if the legal system allowed it): farmers, ranchers, cities, Native Americans, and Mexico. Explain your choices.

3. What role does population growth play in water supply problems?

4. Explain why you are for or against **(a)** gradually phasing out government subsidies of irrigation projects in the western United States (or in the country where you live)

and **(b)** providing government subsidies to farmers for improving irrigation efficiency.

5. Should we use up slowly renewable underground water supplies such as the Ogallala aquifer (Figure 15-19, p. 321) or save much of such supplies for future generations? Explain.

6. Calculate how many liters and gallons of water are wasted in 1 month by a toilet that leaks 2 drops of water per second (1 liter of water equals about 3,500 drops and 1 liter equals 0.265 gallon).

7. How do human activities contribute to flooding and flood damage? How can these effects be reduced?

8. Congratulations! You are in charge of managing the world's water resources. What are the three most important things you would do?

PROJECTS

1. In your community,
   a. What are the major sources of the water supply?
   b. How is water use divided among agricultural, industrial, power plant cooling, and public uses?
   c. Who are the biggest consumers of water?
   d. What has happened to water prices (adjusted for inflation) during the past 20 years? Are they too low to encourage water conservation and reuse?
   e. What water supply problems are projected?

2. Develop a water conservation plan for your school and submit it to school officials.

3. Consult with local officials to identify any floodplain areas in your community. Develop a map showing these areas and the types of activities (such as housing, manufacturing, roads, and recreational use) found on these lands. Evaluate management of such floodplains in your community and come up with suggestions for improvement.

4. Use the library or the Internet to find bibliographic information about *John Todd* and *David Suzuki*, whose quotes appear at the beginning and end of this chapter.

5. Make a concept map of this chapter's major ideas, using the section heads, subheads, and key terms (in bold-face type). See material on the website for this book about how to prepare concept maps.

LEARNING ONLINE

The website for this book contains study aids and many ideas for further reading and research. They include a chapter summary, review questions for the entire chapter, flash cards for key terms and concepts, a multiple-choice practice quiz, interesting Internet sites, references, and a guide for accessing thousands of InfoTrac® College Edition articles. Log on to

http://biology.brookscole.com/miller14

Then click on the Chapter-by-Chapter area, choose Chapter 15, and select a learning resource.

16 Geology and Nonrenewable Mineral Resources

CASE STUDY
The General Mining Law of 1872

Some people have gotten rich by using the little-known U.S. General Mining Law of 1872. It was designed to encourage mineral exploration and the mining of *hardrock minerals* (such as gold, silver, copper, zinc, nickel, and uranium) on U.S. public lands and to help develop the then sparsely populated West.

Under this law, a person or corporation can get mining rights for hardrock minerals or assume legal ownership for parcels of land on essentially all U.S. public land except parks and wilderness by *patenting* it. They state that they believe the land contains valuable hardrock minerals and promise to spend $500 to improve the land for mineral development. They can also pay the federal government $6–12 per hectare ($2.50–5.00 an acre) to buy the land. Then they can use it for essentially any purpose, mine it, lease it, build on it, or sell it. People have built golf courses, hunting lodges, hotels, and housing subdivisions on public land that they bought from taxpayers at 1872 prices.

So far, public lands containing an estimated $240–385 billion (adjusted for inflation) of publicly owned mineral resources have been transferred to private interests at 1872 prices. Domestic and foreign mining companies operating under this law remove mineral resources worth at least $2–3 billion per year on once-public land they bought this way.

In addition, the Congressional Budget Office estimates that mining companies remove hardrock minerals worth at least $650 million per year from public land that has not been transferred to private ownership. About 20% of the mining rights for U.S. public lands are owned by foreign countries. Hardrock mining companies pay no royalties on the minerals they extract from such lands.

In 1992, the 1872 law was modified to require mining companies to post bonds to cover 100% of the estimated cleanup costs in case they go bankrupt. In the past this was not required. As a result, cleaning up land and streams (Figure 16-1) damaged by more than 550,000 abandoned hardrock mines (mostly in the West), would cost U.S. taxpayers $33–72 billion! Mining companies are lobbying Congress to overturn or greatly weaken the pollution bond requirement.

Mining companies defend the 1872 law, pointing out that they must invest large sums (often $100 million or more) to locate and develop an ore site before they make any profits from mining hardrock minerals. In addition, their mining operations provide high-paying jobs to miners, supply vital resources for industry, stimulate the national and local economies, reduce trade deficits, and save American consumers money on products produced from minerals.

Environmentalists call for revising the law to ban the patenting (sale) of public lands but allow 20-year leases of designated public land for hardrock mining. They would also require mining companies to pay a *gross* royalty of 8–12% on the wholesale value of all minerals removed from public land—similar to what oil, natural gas, and coal companies pay. Environmentalists also want much stricter requirements for cleanup of any environmental damage caused by mining companies.

Canada, Australia, South Africa, and other countries that are major extractors of hardrock minerals have laws that require royalty payments and full responsibility for environmental damage. What is your opinion on this issue?

Figure 16-1 This polluted creek in Montana is one example of how gold mining on public or nonpublic land can contaminate water with highly toxic cyanide or mercury used to extract gold from its ore. In addition, air and water convert the sulfur in gold ore to sulfuric acid, which releases toxic metals such as cadmium and copper into streams and groundwater. Until recently, companies mining for hardrock minerals on public lands were not required to clean up such environmental harm.

© Bryan Peterson

Civilization exists by geological consent, subject to change without notice.

WILL DURANT

This chapter discusses the earth's basic geological processes and the nonrenewable mineral resources we use. It addresses the following questions:

- What major geologic processes occur within the earth and on its surface?

- What are the hazards from earthquakes and volcanic eruptions?

- What are rocks, and how are they recycled by the rock cycle?

- How do we find and extract mineral resources from the earth's crust?

- Will there be enough nonrenewable mineral resources for future generations?

16-1 GEOLOGIC PROCESSES

Is the Earth a Stable or a Dynamic Planet? Change Is the Norm

The planet we live on is constantly changing as a result of processes taking place on and below its surface.

Geology, the subject of this chapter, is the science devoted to the study of dynamic processes occurring on the earth's surface and in its interior. Geologists study and analyze rocks and the features and processes of the earth's interior and surface. Some of these processes lead to geologic hazards such as earthquakes and volcanic eruptions, and others produce the renewable soil and nonrenewable mineral and energy resources that support life and economies.

You probably think of the ground you stand on as stable and unmoving. Wrong. It is part of a dynamic planet whose surface and interior are constantly changing. Fortunately, most of these geologic changes take place very slowly on our short human time scale and most occur out of sight within the earth's interior.

Over eons continents have moved to new positions, breaking apart and crunching into one another (Figure 5-8, p. 95). Inside the earth, huge cyclical flows of molten rock break the earth's surface into a series of gigantic plates that move very slowly across the planet's surface. Go outside and you are standing on one of these moving plates. Hard to believe, isn't it?

Energy from the sun and from the earth's interior, coupled with the erosive power of flowing water, have created continents, mountains, valleys, plains, and ocean basins in an ongoing process that continues to change the landscape. And every now and then the solid earth under our feet shakes, rattles, and rolls during an earthquake, or erupts like a punctured boil when a volcano forms or awakens after a long geological sleep.

What Is the Earth's Structure? Living on a Layered Sphere

The earth's three major zones are its core, mantle, and crust.

As the primitive earth cooled over eons, its interior separated into three major concentric zones, layers which geologists identify as the *core,* the *mantle,* and the *crust* (Figure 4-7, p. 60). In other words, beneath your feet is a crust of soil and rock floating on a mantle of partly melted and solid rock, which surrounds an intensely hot core. What we know about the earth's interior comes mostly from indirect evidence such as density measurements, seismic (earthquake) wave studies, measurements of interior heat flow, lava analyses, and research on meteorite composition.

The **core** is the earth's innermost zone. It is intensely hot and has a solid inner part, surrounded by a liquid core of molten or semisolid material.

A thick solid zone called the **mantle** surrounds the core. Most of the mantle is solid rock, but under its rigid outermost part is a zone—the *asthenosphere*—of very hot, partly melted rock that flows slowly and can be deformed like soft plastic.

The outermost and thinnest zone of the earth is called the **crust.** It consists of the *continental crust,* which underlies the continents (including the continental shelves extending into the oceans), and the *oceanic crust,* which underlies the ocean basins and covers 71% of the earth's surface (Figure 16-2).

16-2 INTERNAL AND EXTERNAL GEOLOGIC PROCESSES

What Geologic Processes Occur within the Earth's Interior? Welcome to the Hot Zone

Huge volumes of heated and molten rock move around within the earth's interior.

We tend to think of the earth's crust, mantle, and core as fairly static. But geologic processes taking place within the earth and on its surface, mostly over thousands to millions of years, bring about changes in these components (Figure 16-3, p. 334).

Geologic changes originating from the earth's interior, called *internal processes,* generally build up the planet's surface. Heat from the earth's interior provides the energy for these processes, but gravity also plays a role.

Residual heat from the earth's formation is still being given off as the inner core cools and the outer core

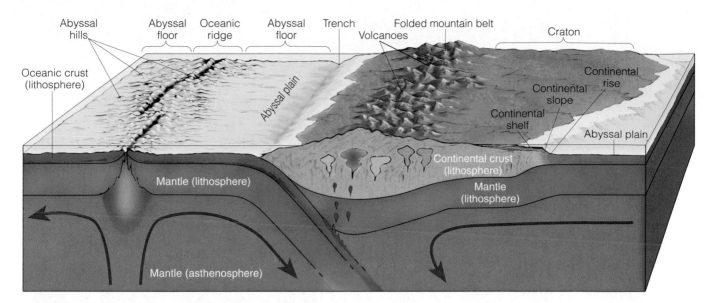

Abyssal hills
Abyssal floor
Oceanic ridge
Abyssal floor
Trench
Volcanoes
Folded mountain belt
Craton
Oceanic crust (lithosphere)
Continental rise
Continental slope
Continental shelf
Abyssal plain
Abyssal plain
Mantle (lithosphere)
Continental crust (lithosphere)
Mantle (lithosphere)
Mantle (asthenosphere)

Figure 16-2 **Natural capital:** major features of the earth's crust and upper mantle. The *lithosphere*, composed of the crust and outermost mantle, is rigid and brittle. The *asthenosphere*, a zone in the mantle, can be deformed by heat and pressure.

cools and solidifies. Continued decay of radioactive elements in the crust, especially the continental crust, provides much of this heat flow from within. Heat from the core causes much of the mantle to deform and flow slowly like heated plastic.

Measurements of heat flow within the earth suggest that two kinds of movement occur in the mantle's asthenosphere. One is *convection cells* or *currents* that move large volumes of rock and heat in loops within the mantle like a giant conveyer belt (Figure 16-3). Internal heat pushes soft rock in the mantle's asthenosphere upward, then downward as the rock cools. This pattern resembles convection in the atmosphere (Figure 6-9, p. 107) or in a pot of boiling water (Figure 3-11, left, p. 45).

The second type of movement involves *mantle plumes,* in which mantle rock flows slowly upward in a column (Figure 16-3), like smoke from a chimney on a cold, calm morning. When the moving rock reaches the top of the plume, it moves out in a radial pattern, as if it were flowing up an umbrella through the handle and then spreading out in all directions from the tip of the umbrella to the rim. There is a lot going on beneath your feet.

What Is Plate Tectonics? The Earth Is Moving

Huge solid plates—called tectonic plates—move extremely slowly across the earth's surface.

The flows of energy and heated material in the mantle convection cells cause about 15 huge rigid plates, called **tectonic plates,** to move extremely slowly across the earth's surface (Figure 16-3 and Figure 16-4,

p. 335). These plates are about 100 kilometers (60 miles) thick. They are composed of the continental and oceanic crust and the rigid, outermost part of the mantle (above the asthenosphere), a combination called the **lithosphere.**

These plates move constantly, supported by the slowly flowing asthenosphere. They are somewhat like large icebergs floating extremely slowly on the surface of an ocean or like the world's largest and slowest-moving surfboards. Some plates move about 1 centimeter (slightly more than a third of an inch) a year. Others at a seafloor spreading zone can move as much as 18 centimeters (7 inches) a year. You are riding or surfing on one of these plates throughout your entire life, but the motion is too slow for you to notice.

The theory explaining the movements of the plates and the processes that occur at their boundaries is called **plate tectonics.** The concept, which became widely accepted by geologists in the 1960s, was developed from an earlier idea called *continental drift.* Throughout the earth's history, continents have split and joined as plates have very slowly drifted thousands of kilometers back and forth across the planet's surface (Figure 5-8, p. 95).

As these plates collide, break apart, and slide by one another over millions of years, they produce mountains on land (such as the Himalayas and the Appalachian Mountains of the eastern United States), huge ridges and trenches on the ocean floor, and other features of the earth's surface (Figures 16-2 and 16-3). These movements and geological processes continue today. Natural hazards such as volcanoes and earthquakes are likely to be found at plate boundaries. In addition, plate movements and interactions affect the

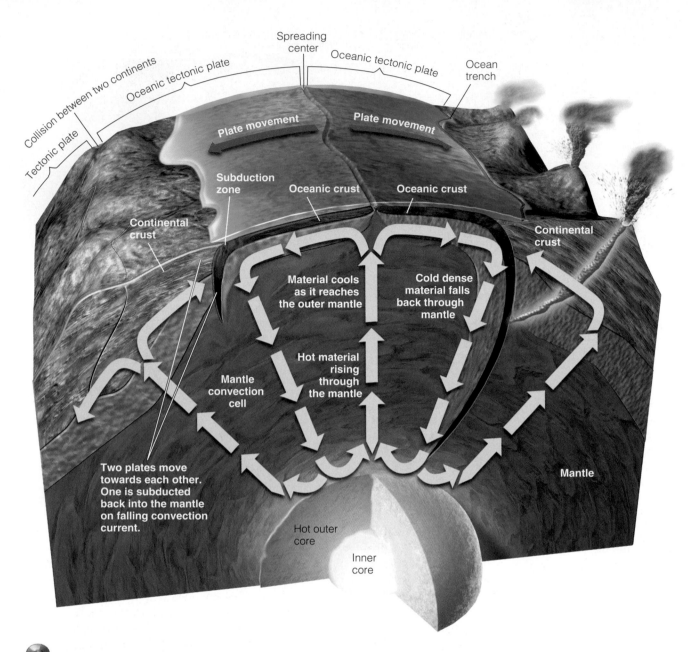

Spreading center

Oceanic tectonic plate

Ocean trench

Collision between two continents

Oceanic tectonic plate

Tectonic plate

Plate movement

Plate movement

Subduction zone

Oceanic crust

Oceanic crust

Continental crust

Material cools as it reaches the outer mantle

Cold dense material falls back through mantle

Continental crust

Mantle convection cell

Hot material rising through the mantle

Two plates move towards each other. One is subducted back into the mantle on falling convection current.

Mantle

Hot outer core

Inner core

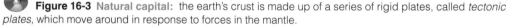

Figure 16-3 Natural capital: the earth's crust is made up of a series of rigid plates, called *tectonic plates*, which move around in response to forces in the mantle.

earth's climate and concentrate many of the minerals we extract and use.

The theory of plate tectonics also helps explain how certain patterns of biological evolution occurred. By reconstructing the course of continental drift over millions of years, scientists can trace how species migrated from one area to another when continents that are now far apart were still joined together. As the continents separated, populations became geographically and reproductively isolated, and speciation occurred. In other words, tectonic plates have played a major but mostly unnoticed role in the drama of life that continues to unfold on our planetary home.

What Types of Boundaries Occur between the Earth's Plates? Moving Apart, Colliding, and Sliding

The earth's tectonic plates move apart, push together, and slide past one another.

Lithospheric plates have three types of boundaries (Figure 16-5, p. 336). One is a **divergent plate** boundary, where the plates move apart in opposite directions. A second type is a **convergent plate boundary**, where the plates are pushed together by internal forces (Figure 16-5, middle). When an oceanic plate collides with a continental plate, the continental plate usually rides up over the denser oceanic plate and pushes it down into the mantle in a process called *subduction.* The area where this collision and subduction takes place is called a *subduction zone.* Over time the subducted plate melts and rises back to the earth's surface as molten rock or magma. A *trench* ordinarily forms at the boundary between the two converging plates. Stresses in the plate undergoing subduction cause earthquakes at convergent plate boundaries.

The third type of boundary is a **transform fault,** which occurs where plates slide and grind past one another along a fracture (fault) in the lithosphere (Figure 16 5, bottom). Most transform faults are on the ocean floor but a few are found on land. An example is the North American Plate and the Pacific Plate that slide and rub past each other along California's San Andreas Fault. As a result, southern California is slowly moving along this transform fault toward northern California. Sometime perhaps 30 million years from now the geographical area we now call Los Angeles will probably slowly grind and slide by the geographical area now known as San Francisco. But by

then neither of these areas will exist as we know them today. This is fascinating stuff from a geological perspective of the earth's history but is not a problem that we need worry about.

What Geologic Processes Occur on the Earth's Surface? Erosion and Weathering

Water and wind move large amounts of soil and broken-down pieces of rock from one place to another.

Geologic changes based directly or indirectly on energy from the sun and on gravity (rather than on heat in the earth's interior) are called *external processes.* Internal processes generally build up the earth's surface. In contrast, external processes tend to wear it down and produce a variety of landforms and environments formed by the buildup of eroded sediment.

A major external process is **erosion:** the process by which material is dissolved, loosened, or worn away from one part of the earth's surface and deposited elsewhere. Flowing streams cause most erosion. Wind

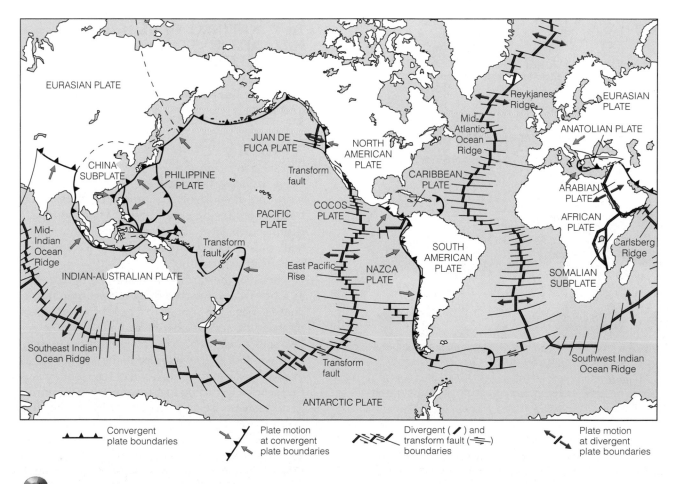

Figure 16-4 The earth's major tectonic plates. These bands correspond to the patterns for the types of lithospheric plate boundaries shown in Figure 16-5. What plate do you live on?

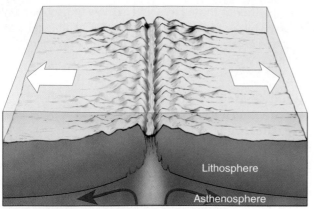

Oceanic ridge at a divergent plate boundary

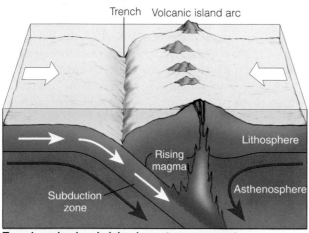

Trench and volcanic island arc at a convergent plate boundary

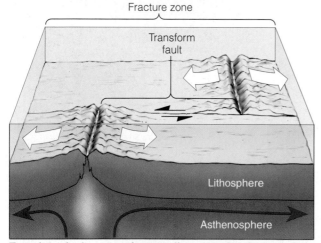

Transform fault connecting two divergent plate boundaries

Figure 16-5 Types of boundaries between the earth's lithospheric plates. All three types occur both in oceans and on continents.

blowing particles of soil from one area to another is another cause. Human activities, particularly those that destroy vegetation that holds soil in place, accelerate erosion, as discussed in Section 14-3 (p. 279).

Weathering consists of the physical, chemical, and biological processes that break down rocks and minerals into smaller particles that can be eroded. There are three types of weathering processes. One is *physical* or *mechanical weathering,* in which a large rock mass is broken into smaller fragments. This is similar to what happens when you hammer a rock into pieces. The most important agent of mechanical weathering is *frost wedging,* in which water collects in pores and cracks of rock, expands upon freezing, and splits off pieces of the rock. This is also what causes most of the potholes we complain about in our roads and streets.

A second process is *chemical weathering,* in which one or more chemical reactions decompose a mass of rock. Most chemical weathering involves a reaction of rock material with oxygen, carbon dioxide, and moisture in the atmosphere and on the ground.

A third process is *biological weathering,* the conversion of rock or minerals into smaller particles through the action of living things. For example, lichens produce acids that can chemically weather rocks. And roots growing into and rubbing against rock can physically break it into small pieces.

16-3 NATURAL GEOLOGIC HAZARDS: EARTHQUAKES AND VOLCANIC ERUPTIONS

What Are Earthquakes? Shake, Rattle, and Roll

Earthquakes occur when a part of the earth's crust suddenly fractures, shifts to relieve stress, and releases energy as shock waves.

Stress in the earth's crust can cause solid rock to deform until it suddenly fractures and shifts along the fracture, producing a fault (Figure 16-5, bottom). The faulting or a later abrupt movement on an existing fault causes an **earthquake** that has certain features and effects (Figure 16-6).

Relief of the earth's internal stress releases energy as shock waves, which move outward from the earthquake's focus like ripples in a pool of water. Scientists measure the severity of an earthquake by the *magnitude* of its shock waves. The magnitude is a measure of the amount of energy released in the earthquake, as indicated by the amplitude (size) of the vibrations when they reach a recording instrument (seismograph).

Scientists use the **Richter scale,** on which each unit represents an amplitude 10 times greater than the next smaller unit. Thus a magnitude 5.0 earthquake is 10 times greater than a magnitude 4.0, and a magnitude 6.0 quake is 100 times greater than a magnitude 4.0 quake. Seismologists rate earthquakes as *insignifi-*

cant (less than 4.0 on the Richter scale), *minor* (4.0–4.9), *damaging* (5.0–5.9), *destructive* (6.0–6.9), *major* (7.0–7.9), and *great* (over 8.0).

Earthquakes often have *aftershocks* that gradually decrease in frequency over a period of up to several months, and some have *foreshocks* from seconds to weeks before the main shock.

The *primary effects of earthquakes* include shaking and sometimes a permanent vertical or horizontal displacement of the ground. These effects may have serious consequences for people and for buildings, bridges, freeway overpasses, dams, and pipelines. An earthquake is a very large rock-and-roll event.

Secondary effects of earthquakes include rockslides, urban fires, and flooding caused by *subsidence* (sinking) of land. Coastal areas can be severely damaged by earthquakes at sea that can generate huge water waves, called *tsunamis* (also called tidal waves, although they have nothing to do with tides) that travel as fast as 950 kilometers (590 miles) per hour. You cannot outrun one of these waves.

One way to reduce loss of life and property from earthquakes is to examine historical records and make geologic measurements to locate active fault zones. We can also map high-risk areas (Figure 16-7), establish building codes that regulate the placement and design of buildings in areas of high risk, and increase research on projecting when and where earthquakes will occur. Then people can decide how high the risk might be and whether they want to accept the risk and live in areas subject to earthquakes.

Engineers know how to make homes, large buildings, bridges, and freeways more earthquake resistant. But this can be expensive, especially the reinforcement of existing structures.

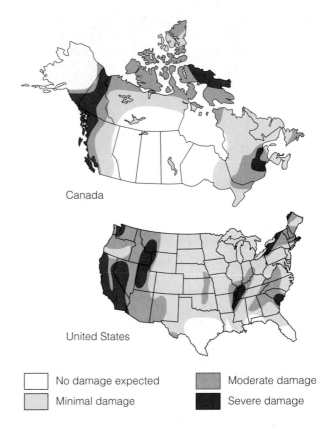

Canada

United States

No damage expected | Moderate damage
Minimal damage | Severe damage

Figure 16-7 Expected damage from earthquakes in Canada and the contiguous United States. This map is based on earthquake records. (U.S. Geological Survey)

What Are Volcanoes? Popping an Earth Cork

Some volcanoes erupt quietly and release flows of molten rock but others erupt explosively and spew large chunks of lava rock, ash, and harmful gases into the atmosphere.

An active **volcano** occurs where magma (molten rock) reaches the earth's surface through a central vent or a long crack (*fissure*; Figure 16-8, p. 338). Volcanic activity can release *ejecta* (debris ranging from large chunks of lava rock to ash that may be glowing hot), liquid lava, and gases (such as water vapor, carbon dioxide, and sulfur dioxide) into the environment.

Volcanic activity is concentrated for the most part in the same areas as seismic activity. Some volcanoes erupt explosively and eject large quantities of gases and particulate matter (soot and mineral ash) high into the troposphere. Most of the particles of soot and ash soon fall back to the earth's surface. But gases such as sulfur dioxide remain in the atmosphere and are converted to tiny droplets of sulfuric acid, many of which stay above the clouds and may not be washed out by rain for up to 3 years. These tiny droplets reflect some of the sun's energy and can cool the atmosphere for 1–4 years.

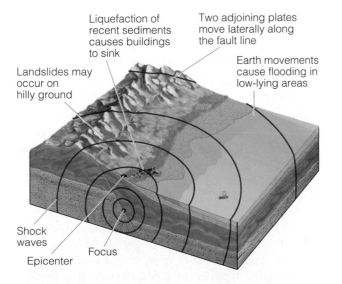

Liquefaction of recent sediments causes buildings to sink

Two adjoining plates move laterally along the fault line

Landslides may occur on hilly ground

Earth movements cause flooding in low-lying areas

Shock waves

Focus

Epicenter

Figure 16-6 Major features and effects of an *earthquake*.

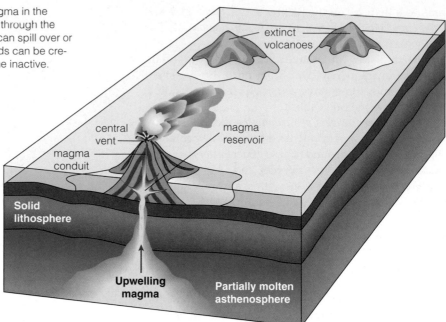

Figure 16-8 A *volcano* erupts when molten magma in the partially molten asthenosphere rises in a plume through the lithosphere to erupt on the surface as lava that can spill over or be ejected into the atmosphere. Chains of islands can be created by eruptions of volcanoes that then become inactive.

Other volcanoes erupt more quietly. They involve primarily lava flows, which can cover roads and villages and ignite brush, trees, and homes.

We tend to think negatively of volcanic activity, but it also provides some benefits. One is outstanding scenery in the form of majestic mountains, some lakes (such as Crater Lake in Oregon; see title page photo), and other landforms. Perhaps the most important benefit of volcanism is the highly fertile soils produced by the weathering of lava.

We can reduce the loss of human life and sometimes property from volcanic eruptions. One way is to use historical records and geologic measurements to identify high-risk areas so that people can try to avoid living in them. Other methods involve developing effective evacuation plans and trying to develop measurements that warn us when volcanoes are likely to erupt.

Scientists are studying phenomena that precede an eruption. Examples include tilting or swelling of the cone, changes in magnetic and thermal properties of the volcano, changes in gas composition, and increased seismic activity.

16-4 MINERALS, ROCKS, AND THE ROCK CYCLE

What Are Minerals and Rocks? Hard Stuff

The earth's crust consists of solid inorganic elements and compounds called minerals and masses of one or more minerals called rock.

The earth's crust, still forming in various places, is composed of minerals and rocks. It is the source of almost all the nonrenewable resources we use: fossil fuels, metallic minerals, and nonmetallic minerals (Figure 1-6, p. 9). It is also the source of soil (Figure 4-25, p. 73) and of the elements that make up your body and the bodies of other living organisms. You are mostly water mixed with chemically transformed particles of minerals and rock and a small amount of air.

A **mineral** is an element or inorganic compound that occurs naturally and is solid with a regular internal crystalline structure. Some minerals consist of a single element, such as gold, silver, diamond (carbon), and sulfur. But most of the more than 2,000 identified minerals occur as inorganic compounds formed by various combinations of elements. Examples are salt, mica, and quartz.

Rock is a solid combination of one or more minerals that is part of the earth's crust. Some kinds of rock, such as limestone (calcium carbonate, or $CaCO_3$) and quartzite (silicon dioxide, or SiO_2), contain only one mineral, but most rocks consist of two or more minerals.

What Are the Major Rock Types and How Are They Recycled? Really Slow Recycling

The earth's crust contains igneous, sedimentary, and metamorphic rocks that are recycled by the rock cycle.

Based on the way it forms, rock is placed in three broad classes. One is **igneous rock,** formed below or on the earth's surface when molten rock (magma) wells up from the earth's upper mantle or deep crust, cools, and hardens. Examples are granite (formed underground) and lava rock (formed aboveground when molten lava cools and hardens). Although often covered by sedimentary rocks or soil, igneous rocks form the bulk of the earth's crust. They also are the main source of many nonfuel mineral resources.

A second type is **sedimentary rock.** It is formed from sediment produced when existing rocks are weathered and eroded into small pieces, and transported from their sources by water, wind, or gravity to downstream, downwind, or downhill sites. These sediments are deposited in layers that accumulate over time and increase the weight and pressure on underlying layers. A combination of pressure and dissolved minerals seeping through the layers of sediment crystallize and bind sediment particles together to form *sedimentary rock.*

Examples are sandstone and shale (formed from pressure created by deposited layers of sediment), dolomite and limestone (formed from the compacted shells, skeletons, and other remains of dead organisms), and lignite and bituminous coal (derived from plant remains). Some types of sedimentary rock are the result of crystals precipitating out of or growing in solutions and then being compacted in layers or drying out to leave crystals behind. An example is rock salt, which we know as table salt or sodium chloride (NaCl).

The third type is **metamorphic rock,** produced when a preexisting rock is subjected to high temperatures (which may cause it to melt partially), high pressures, chemically active fluids, or a combination of these agents. These forces can change or transform a rock by reshaping its internal crystalline structure and its physical properties and appearance. Examples are anthracite (a form of coal), slate (formed when shale and mudstone are heated), and marble (produced when limestone is exposed to heat and pressure).

The interaction of physical and chemical processes that changes rocks from one type to another is called the **rock cycle** (Figure 16-9). It recycles the earth's three types of rocks over millions of years and is the slowest of the earth's cyclic processes. It also concentrates the planet's nonrenewable mineral resources on which we depend. Without the incredibly slow rock cycle you would not exist.

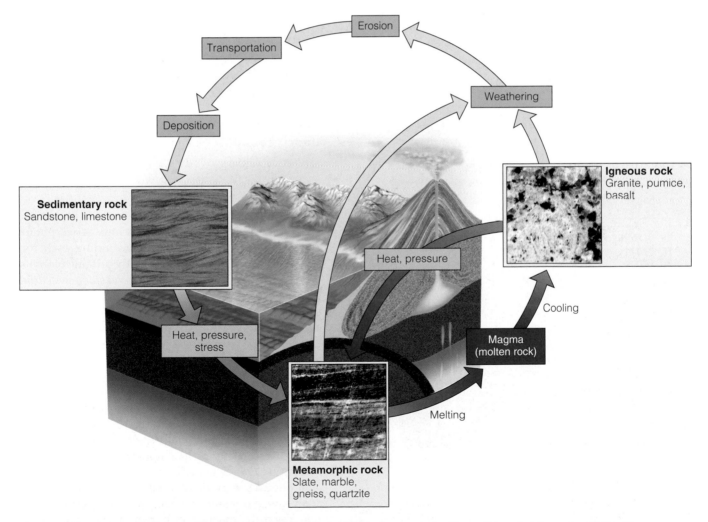

Figure 16-9 Natural capital: the *rock cycle* is the slowest of the earth's cyclic processes. The earth's materials are recycled over millions of years by three processes: *melting, erosion,* and *metamorphism,* which produce *igneous, sedimentary,* and *metamorphic* rocks. Rock of any of the three classes can be converted to rock of either of the other two classes, or can be recycled within its own class.

What Are Nonrenewable Mineral Resources? Useful Rock Resources

Mineral resources are nonrenewable materials that we can extract from the earth's crust.

A **nonrenewable mineral resource** is a concentration of naturally occurring material in or on the earth's crust that can be extracted and processed into useful materials at an affordable cost. Over millions to billions of years the earth's internal and external geologic processes have produced numerous nonfuel mineral resources and fossil fuel energy resources. Because they take so long to produce, they are classified as *nonrenewable resources*.

We know how to find and extract more than 100 nonrenewable minerals from the earth's crust. They include *metallic mineral resources* (iron, copper, aluminum), *nonmetallic mineral resources* (salt, clay, sand, phosphates, and soil), and *energy resources* (coal, oil, natural gas, and uranium).

Ore is rock containing enough of one or more metallic minerals to be mined profitably. We convert about 40 metals extracted from ores into many everyday items that we either use and discard (Figure 3-18, p. 53) or learn to reuse, recycle, or use less wastefully (Figure 3-19, p. 53).

The U.S. Geological Survey divides nonrenewable mineral resources into four major categories (Figure 16-10):

- **Identified resources**: deposits of a nonrenewable mineral resource with a *known* location, quantity, and quality, or whose existence is based on direct geologic evidence and measurements

- **Reserves**: identified resources from which a usable nonrenewable mineral can be extracted profitably at current prices

- **Undiscovered resources**: potential supplies of a nonrenewable mineral resource assumed to exist on the basis of geologic knowledge and theory but with unknown specific locations, quality, and amounts

- **Other resources**: undiscovered resources and identified resources not classified as reserves

Most published estimates of the supply of a given nonrenewable resource refer to *reserves*. Reserves can increase when new deposits are found or when higher prices or improved mining technology make it profitable to extract deposits that previously were too expensive to extract. Theoretically, all *other resources* could eventually be converted to reserves, but this is highly unlikely.

16-5 FINDING, REMOVING, AND PROCESSING NONRENEWABLE MINERAL RESOURCES

How Are Buried Mineral Deposits Found? Underground Detective Work

Promising underground deposits of minerals are located by a variety of physical and chemical methods.

Mining companies use several methods to find promising mineral deposits. One is using aerial photos and satellite images to reveal protruding rock formations (outcrops) associated with certain minerals. Also, planes can be equipped with *radiation-measuring equipment* to detect deposits of radioactive minerals such as uranium ore, and a *magnetometer* to measure changes in the earth's magnetic field caused by magnetic minerals such as iron ore. Another method uses a *gravimeter* to measure differences in gravity caused by differences in density between an ore deposit and the surrounding rock.

Underground methods include drilling a deep well and extracting core samples. Scientists can also put sensors in existing wells to detect electrical resistance or radioactivity to pinpoint the location of oil and natural gas. Scientists also make *seismic surveys* on land and at sea by detonating explosive charges and analyzing the resulting shock waves to get information about the makeup of buried rock layers. Yet another method is to perform *chemical analysis* of water and plants to detect deposits of underground minerals that have leached into nearby bodies of water or have been absorbed by plant tissues.

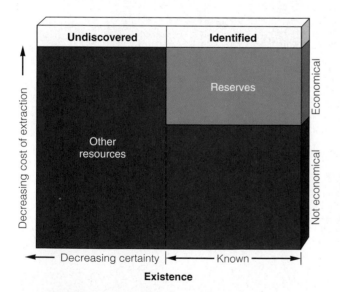

Figure 16-10 Natural capital: general classification of mineral resources. (The area shown for each class does not represent its relative abundance.) In theory, all mineral resources classified as *other resources* could become reserves because of rising mineral prices or improved mineral location and extraction technology. In practice, geologists expect only a fraction of other resources to become reserves.

How Are Buried Mineral Deposits Removed? Heavy Digging and Lifting

Shallow deposits of mineral deposits are removed by surface mining, and deep deposits are removed by subsurface mining.

After suitable mineral deposits are located, several different types of mining techniques are used to remove them, depending on their location and type. Shallow deposits are removed by **surface mining,** and deep deposits are removed by **subsurface mining.**

In surface mining, mechanized equipment strips away the **overburden** of soil and rock and usually discards it as waste material called **spoils.** Surface mining extracts about 90% of the nonfuel mineral and rock resources and 60% of the coal (by weight) that are used in the United States.

The type of surface mining used depends on the resource being sought and on local topography. There are several types of surface mining. One is **open-pit mining** (Figure 16-11a), in which machines dig holes and remove ores (such as iron and copper), sand, gravel, and stone (such as limestone and marble). A second method is **dredging** (Figure 16-11b), in which chain buckets and draglines scrape up underwater mineral deposits.

A third method used where the terrain is fairly flat is **area strip mining** (Figure 16-11c). A gigantic earth-mover strips away the overburden, and a huge power shovel digs a cut to remove the mineral deposit. Then the trench is filled with overburden and a new cut is made parallel to the previous one. The process is repeated over the entire site. If filling it does not restore the land, area strip mining leaves a wavy series of highly erodible hills of rubble called *spoil banks.*

Contour strip mining (Figure 16-11d) is used on hilly or mountainous terrain. A power shovel cuts a series of terraces into the side of a hill. An earthmover removes the overburden, a power shovel extracts the coal, and the overburden from each new terrace is

(a) Open Pit Mine

(b) Dredging

(c) Area Strip Mining

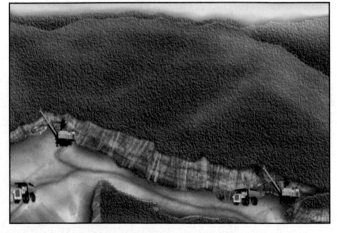

(d) Contour Strip Mining

Figure 16-11 Major mining methods used to extract *surface deposits* of solid mineral and energy resources.

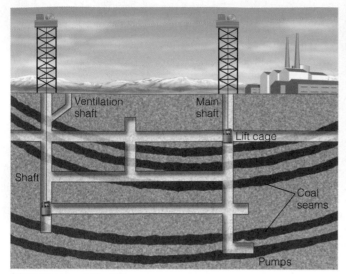

(a) Underground Coal Mine

(b) Room-and-Pillar

Figure 16-12 Major mining methods used to extract *underground deposits* of solid mineral and energy resources (primarily coal). **(a)** Mine shafts and tunnels are dug and blasted out. **(b)** In *room-and-pillar* mining, machinery is used to gouge out coal and load it onto a shuttle car in one operation, and pillars of coal are left to support the mine roof. **(c)** In *longwall coal mining*, movable steel props support the roof and cutting machines shear off the coal onto a conveyor belt. As the mining proceeds, roof supports are moved forward and the roof behind is allowed to fall (often causing the land above to sink or subside).

(c) Longwall Mining of Coal

dumped onto the one below. Unless the land is restored, a wall of dirt is left in front of a highly erodible bank of soil and rock called a *highwall*.

Another method is **mountaintop removal.** It uses explosives, massive shovels, and huge machinery called draglines to remove the top of a mountain and expose seams of coal underneath. The resulting waste rock and dirt is pushed down into the nearest streams and valleys below. This form of surface mining—increasingly used in West Virginia—causes considerable environmental damage.

Although surface-mined land can be restored (except in arid and semiarid areas), it is expensive and not done in many countries. In the United States, the *Surface Mining Control and Reclamation Act of 1977* requires mining companies to restore most surface-mined land so it can be used for the same purpose as before it was mined. The act also levied a tax on mining companies

to restore land that was disturbed by surface mining before the law was passed. But reclamation efforts are only partially successful and coal companies continue lobbying elected officials to have the law weakened or to choke off funds for its enforcement.

Subsurface mining (Figure 16-12) removes coal and various metal ores that are too deep to be extracted by surface mining. Miners dig a deep vertical shaft, blast subsurface tunnels and chambers to get to the deposit, and use machinery to remove the ore or coal and transport it to the surface.

Subsurface mining disturbs less than one-tenth as much land as surface mining and usually produces less waste material. But it leaves much of the resource in the ground and is more dangerous and expensive than surface mining. Hazards include cave-ins, explosions, and lung diseases (such as black lung) caused by prolonged inhalation of mining dust.

16-6 ENVIRONMENTAL EFFECTS OF USING MINERAL RESOURCES

What Are the Environmental Impacts of Nonrenewable Mineral Resources? Degradation, Waste, and Pollution

Extracting, processing, and using mineral resources can disturb the land, kill miners, erode soils, produce large amounts of solid waste, and pollute the air, water, and soil.

The mining, processing, and use of mineral resources takes enormous amounts of energy and often causes land disturbance, soil erosion, and air and water pollution (Figure 16-13).

Mining can harm the environment in a number of ways. One is *scarring and disruption of the land surface.* The Department of the Interior estimates that at least 500,000 surface-mined sites dot the U.S. landscape, mostly in the West. Cleanup costs are estimated in the tens of billions of dollars.

Another problem is collapse of land above underground mines. This *subsidence* can cause houses to tilt, sewer lines to crack, gas mains to break, and groundwater systems to be disrupted.

Toxin-laced mining wastes can be blown or deposited elsewhere by wind or water erosion. Another problem is *acid mine drainage.* It occurs when rainwater seeping through a mine or mine wastes carries sulfuric acid (H_2SO_4, produced when aerobic bacteria act on iron sulfide minerals in spoils) to nearby streams and groundwater (Figures 16-1 and 16-14, p. 344). This contaminates water supplies and can destroy some forms of aquatic life. According to the U.S. Environmental Protection Agency, mining has polluted about 40% of western watersheds.

Mining can also result in emission of toxic chemicals into the atmosphere. In the United States, the mining industry produces more toxic emissions than any other industry—typically accounting for almost half of such emissions. Finally, some forms of wildlife can be exposed to toxic mining wastes stored in holding ponds and leaking from such ponds.

What Is the Typical Life Cycle of a Nonrenewable Metal Resource? Going Around in Circles

Metal ores are extracted from the earth's crust, purified, smelted to extract the desired metal, and converted to the desired products.

Figure 16-15 (p. 345) depicts the typical life cycle of a metal resource. It begins with extracting ore from the earth's crust.

Ore typically has two components. One is the *ore mineral* containing the desired metal and the other is

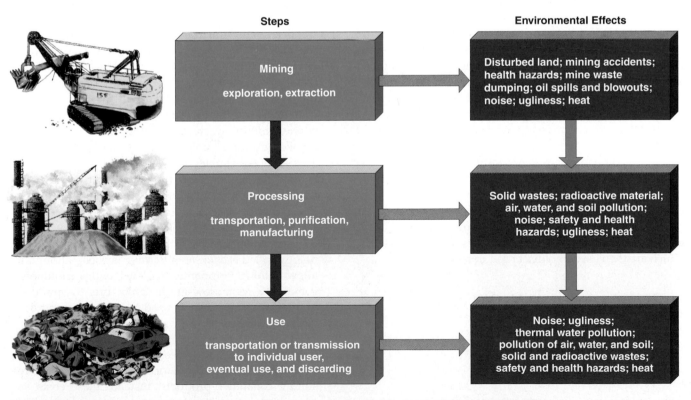

Figure 16-13 Natural capital degradation: some harmful environmental effects of extracting, processing, and using nonrenewable mineral and energy resources. The energy used to carry out each step causes additional pollution and environmental degradation.

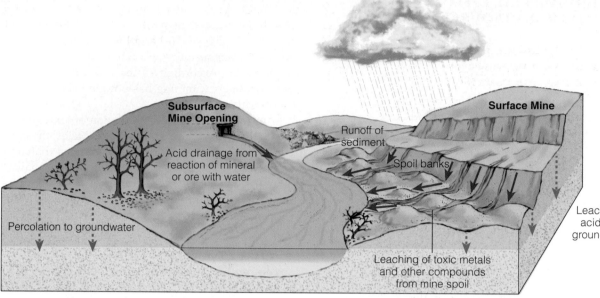

Figure 16-14 Natural capital degradation: pollution and degradation of a stream and groundwater by runoff of acids—called *acid mine drainage*—and by toxic chemicals from surface and subsurface mining. These substances can kill fish and other aquatic life. Acid mine drainage has damaged more than 19,000 kilometers (12,000 miles) of streams in the United States, mostly in Appalachia and the West.

waste material called *gangue* (pronounced "gang"). Removing the gangue from ores produces piles of waste called *tailings*. Particles of toxic metals blown by the wind or leached from tailings by rainfall can contaminate surface water and groundwater.

After gangue has been removed, **smelting** is used to separate the metal from the other elements in the ore mineral. Without effective pollution control equipment, smelters emit enormous quantities of air pollutants, which damage vegetation and soils in the surrounding area. They also cause water pollution and produce liquid and solid hazardous wastes that must be disposed of safely.

Some companies are using improved technology to reduce pollution from smelting, thereby lowering production costs, saving costly cleanup bills, and decreasing liability for damages.

Once smelting has produced the pure metal, it is usually melted and converted to desired products, which are then used and discarded or recycled.

Are There Environmental Limits to Resource Extraction and Use? A Serious Problem

Environmental damage caused by extraction, processing, and use of mineral resources can limit their availability.

Some environmentalists and resource experts do not believe that exhaustion of supplies is the greatest danger from continually increasing consumption of nonrenewable mineral resources. Instead, it is more likely to be the environmental damage caused by their extraction, processing, and conversion to products (Figure 16-13).

The environmental impacts from mining an ore are affected by its percentage of metal content, or *grade.* The more accessible and higher-grade ores are usually exploited first. As they are depleted, it takes more money, energy, water, and other materials to exploit lower-grade ores. This in turn increases land disruption, mining waste, and pollution.

For example, gold miners typically remove an amount of ore equal to the weight of 50 automobiles to extract a piece of gold that would fit inside your clenched fist. Most newlyweds would be surprised to know that about 5.5 metric tons (6 tons) of mining waste was created to make their two gold wedding rings. In Australia and North America, a mining technology called *cyanide heap leaching* is cheap enough to allow mining companies to level entire mountains containing very low-grade gold ore. Cyanide—a highly toxic chemical—is used to separate about 85% of the world's gold from waste ore.

Currently, most of the harmful environmental costs of mining and processing minerals are not included in the prices for processed metals and the resulting consumer products. Instead, these costs are passed on to society and future generations, which gives mining companies and manufacturers little in-

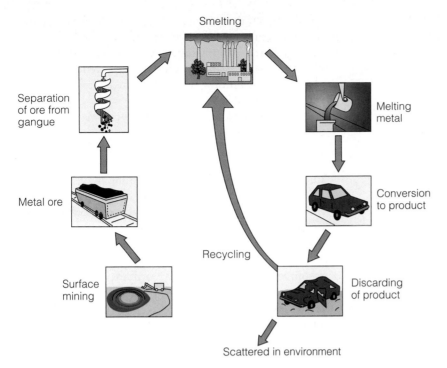

Figure 16-15 Typical *life cycle of a metal resource*. Each step in this process uses energy and produces some pollution and waste heat.

centive to reduce resource waste and pollution. Environmentalists and some economists call for phasing in *full-cost pricing*—including the cost of environmental harm done in the price of goods made from minerals.

> **✗ How Would You Vote?** Should market prices of goods made from minerals include their harmful environmental costs? Cast your vote online at http://biology.brookscole.com/miller14

16-7 SUPPLIES OF MINERAL RESOURCES

Will We Have Enough Nonrenewable Mineral Resources? Making Educated Guesses

The future supply of a resource depends on its available and affordable supply and how rapidly that supply is used.

The future supply of nonrenewable minerals depends on two factors. One is the actual or potential supply of the mineral and the other is the rate at which we use it.

We never completely run out of any mineral. However, a mineral becomes *economically depleted* when it costs more to find, extract, transport, and process the remaining deposit than it is worth. At that point, there are five choices: *recycle or reuse existing supplies, waste less, use less, find a substitute,* or *do without.*

Depletion time is how long it takes to use up a certain proportion—usually 80%—of the reserves of a mineral at a given rate of use (Figure 1-8, p. 11). When ex-

perts disagree about depletion times, they are often using different assumptions about supply and rate of use (Figure 16-16).

The shortest depletion time assumes no recycling or reuse and no increase in reserves (curve A, Figure 16-16). A longer depletion time assumes that recycling will stretch existing reserves and that better mining technology, higher prices, and new discoveries will increase reserves (curve B, Figure 16-16). An even longer depletion time assumes that new discoveries will further expand reserves and that recycling, reuse, and reduced consumption will extend supplies (curve C, Figure 16-16). Finding a substitute for a resource leads to a new set of depletion curves for the new resource.

We can use geological methods to make fairly good estimates of the reserves of most resources (Figure 16-10, blue) and less accurate measurements of potential other supplies of mineral resources (Figure 16-10, red). Rising prices and improved mining technology can convert some of the other resources to reserves, but it is difficult to make accurate projections of how much this will add to the usable supply.

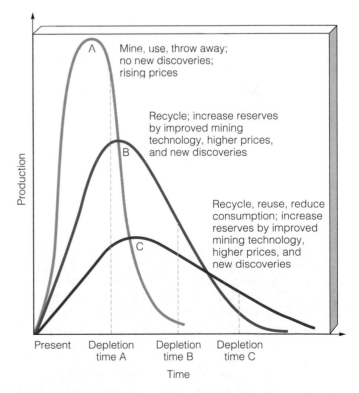

Figure 16-16 *Depletion curves* for a nonrenewable resource (such as aluminum or copper) using three sets of assumptions. Dashed vertical lines represent times when 80% depletion occurs.

We do know that the demand for mineral resources is increasing at a rapid rate as more people consume more and more stuff. For example, since 1940 Americans alone have used up as large a share of the earth's mineral resources as all previous generations put together, and this resource-use treadmill is speeding up.

Will we run out of affordable supplies of a particular mineral resource? No one knows. If we do, can we find an acceptable substitute? Some think we can. Are there environmental limits to the use of mineral resources? Many environmentalists think so unless we can use microorganisms, or other less environmentally harmful ways, to extract and process minerals, or nanotechnology to construct materials we need from atoms and molecules.

How Does Economics Affect Supplies of Nonrenewable Minerals? Prices Can Make a Difference If the Market Is Truly Free

A rising price for a scarce mineral resource can increase supplies and encourage more efficient use.

Geologic processes determine the quantity and location of a mineral resource in the earth's crust. But economics determines what part of the known supply is extracted and used.

According to standard economic theory, in a competitive free market a plentiful mineral resource is cheap when its supply exceeds demand. And when a resource becomes scarce its price rises. This can encourage exploration for new deposits, stimulate development of better mining technology, and make it profitable to mine lower-grade ores. It can also encourage a search for substitutes and promote resource conservation.

But according to some economists, this price effect may no longer apply very well in most developed countries. One reason is that industry and government in such countries often control the supply, demand, and prices of minerals to such an extent that a truly competitive free market does not exist.

Most mineral prices are artificially low because governments subsidize development of their domestic mineral resources to help promote economic growth and national security. In the United States, for instance, mining companies get depletion allowances amounting to 5–22% of their gross income (depending on the mineral). They can also reduce the taxes they pay by deducting much of their costs for finding and developing mineral deposits. In addition, hardrock mining companies operating in the United States can buy public land at 1872 prices or use public land and pay no royalties to the government on the minerals they extract (see the Case Study that opens this chapter).

Between 1982 and 2004, U.S. mining companies received more than $6 billion in government subsidies.

Critics argue that taxing rather than subsidizing the extraction of nonfuel mineral resources would provide governments with revenue, create incentives for more efficient resource use, promote waste reduction and pollution prevention, and encourage recycling and reuse of mineral resources.

Mining company representatives say they need subsidies and low taxes to keep the prices of minerals low for consumers and to encourage the companies not to move their mining operations to other countries with no such taxes and less stringent mining regulations.

Economic problems can also hinder the development of new supplies of mineral resources because exploring for them takes lots of increasingly scarce investment capital and is risky financially. Typically, if geologists identify 10,000 possible deposits of a given resource, only 1,000 sites are worth exploring; only 100 justify drilling, trenching, or tunneling; and only 1 becomes a producing mine or well. If you had lots of financial capital, would you invest it in developing a nonrenewable mineral resource?

Should More Mining Be Allowed on Public Lands in the United States?

There is controversy over whether to extract more mineral resources from public lands.

About one-third of the land in the United States is public land owned jointly by all U.S. citizens. This land, consisting of national forests, parks, resource lands, and wilderness (Figure 11-6, p. 198), is managed by various government agencies under laws passed by Congress.

For decades, resource developers, environmentalists, and conservationists have argued over how this land should be used. Extractors of mineral resources complain that three-fourths of the country's vast public lands, with many areas containing rich deposits of mineral resources, are off limits to mining.

In recent decades, they have stepped up efforts to have Congress open up most of these lands to mineral development, sell off mineral-rich public lands to private interests, or turn their management over to state and local governments (which often can be more readily influenced by mining and development interests). Since 2002, the Bush administration and Congress have expanded the extraction of mineral, timber, and fossil fuel resources on U.S. public lands.

Conservation biologists and environmentalists strongly oppose such efforts. They argue that this increases environmental degradation and decreases biodiversity.

Can We Get Enough Minerals by Mining Lower-Grade Ores? Taking It to the Limit

New technologies can increase the mining of low-grade ores at affordable prices, but harmful environmental effects can limit this.

Some analysts contend that all we need to do to increase supplies of a mineral is to extract lower grades of ore. They point to the development of new earth-moving equipment, improved techniques for removing impurities from ores, and other technological advances in mineral extraction and processing.

In 1900, the average copper ore mined in the United States was about 5% copper by weight. Today it is 0.5%, and copper costs less (adjusted for inflation). New methods of mineral extraction may allow even lower-grade ores of some metals to be used.

But several factors can limit the mining of lower-grade ores. One is the increased cost of mining and processing larger volumes of ore. Another is the availability of fresh water needed to mine and process some minerals—especially in arid and semiarid areas. A third limiting factor is the environmental impact of the increased land disruption, waste material, and pollution produced during mining and processing (Figure 16-13).

One way to improve mining technology is to use microorganisms for in-place (*in situ*, pronounced "in-SY-too") mining. This biological approach removes desired metals from ores while leaving the surrounding environment undisturbed. It also reduces air pollution associated with the smelting of metal ores and reduces water pollution associated with using hazardous chemicals such as cyanides and mercury to extract gold.

Once a commercially viable ore deposit has been identified, wells are drilled into it and the ore is fractured. Then the ore is inoculated with natural or genetically engineered bacteria to extract the desired metal. Next the well is flooded with water, which is pumped to the surface, where the desired metal is removed. Then the water is recycled.

This technique permits economical extraction from low-grade ores that are used more as high-grade ores are depleted. Currently, more than 30% of all copper produced worldwide, worth more than $1 billion a year, comes from such *biomining*. If naturally occurring bacteria cannot be found to extract a particular metal, genetic engineering techniques could be used to produce such bacteria.

However, microbiological ore processing is slow. It can take decades to remove the same amount of material that conventional methods can remove within months or years. So far, biological mining methods are economically feasible only with low-grade ore for which conventional techniques are too expensive.

Can We Use Nanotechnology to Produce New Materials? Evaluating a New Technology That May Change the World

Building new materials from the bottom up by assembling atoms and molecules has enormous potential but could have potentially harmful unintended effects.

Nanotechnology is using science and engineering at the atomic and molecular level to build materials with specified properties from the bottom up. It involves finding ways to manipulate atoms and molecules as small as 1–100 nanometers—billionths of a meter—wide. For comparison, your unaided eye cannot see things smaller than 10,000 nanometers across and the width of a typical human hair is 50,000 nanometers.

This atomic and molecular approach to manufacturing uses raw materials of abundant atoms such as carbon, oxygen, and hydrogen and arranges them to create everything from medicines and solar cells to automobile bodies, hopefully with little environmental harm and without depleting nonrenewable resources. One example is soccer-ball-shaped forms of carbon called *buckyballs*.

Nanotechnology scientists entice us with visions of a *molecular economy*. They include a supercomputer the size of a sugar cube that could store all the information in the U.S. Library of Congress, biocomposite materials smaller than a human cell that would make your bones and tendons super strong, designer molecules that could seek out and kill only cancer cells, foam with nanoparticles that could provide superthermal insulation, and windows, kitchens, and bathrooms that never need cleaning. The list could go on.

This research is in the early stages and tangible results are a decade away. But already nanotechnology has been used to develop stain-resistant and wrinkle-free materials for pants and sunscreens that block ultraviolet light.

Nobel laureate Horst Stormer says, "Nanotechnology has given us the tools . . . to play with the ultimate toy box of nature—atoms and molecules. . . . The possibilities to create new things appear limitless." You might want to consider this rapidly emerging field as a career choice.

So what is the catch? What are some possible unintended harmful consequences of nanotechnology? One concern is that as particles get smaller they become more reactive and potentially more toxic because they have large surface areas relative to their mass. Another is that nanosize particles can get through the natural defenses of our bodies. They could easily reach the lungs and from there migrate to other organs, including the central nervous system, and to the bloodstream.

In 2004, Eva Olberdorster, an environmental toxicologist at Southern Methodist University, found that fish swimming in water loaded with buckyballs experience brain damage within 48 hours. Little is known about how buckyballs and other nanoparticles behave in the human body. But factories are churning out buckyballs and these and other nanoparticles are starting to show up in products from cosmetics to sunscreens and in the environment.

Many analysts say we need to do two things before unleashing widespread use of nanotechnology.

First, carefully investigate its potential ecological, health, and societal risks. *Second*, develop guidelines and regulations for controlling and guiding its spread until we have better answers to many of the "What happens next?" questions about this technology.

Can We Get More Minerals from Seawater and by Mining the Ocean Floor? Some Problems

Most minerals in seawater cost too much to extract, and mineral resources found on the deep ocean floor are not being removed because of high costs and squabbles over who owns them.

Ocean mineral resources are found in seawater, sediments and deposits on the shallow continental shelf, hydrothermal ore deposits (Figure 16-17), and manganese-rich nodules on the deep-ocean floor.

Most of the chemical elements found in seawater occur in such low concentrations that recovering them takes more energy and money than they are worth. Only magnesium, bromine, and sodium chloride are abundant enough to be extracted profitably at current prices with existing technology.

Deposits of minerals (mostly sediments) along the continental shelf and near shorelines are significant sources of sand, gravel, phosphates, sulfur, tin, copper, iron, tungsten, silver, titanium, platinum, and diamonds.

Rich hydrothermal deposits of gold, silver, zinc, and copper are found as sulfide deposits in the deep-ocean floor and around black smokers (Figure 16-17). Currently, it costs too much to extract these minerals even though

some of these deposits contain large concentrations of important metals.

Manganese-rich nodules found on the deep-ocean floor at various sites may be a future source of manganese and other key metals. They might be sucked up from the ocean floor by giant vacuum pipes or scooped up by buckets on a continuous cable operated by a mining ship.

So far these nodules and resource-rich mineral beds in international waters have not been developed because of high costs and squabbles over who owns them and how any profits from extracting them should be distributed among the world's nations.

Some environmentalists believe seabed mining probably would cause less environmental harm than mining on land. But they are concerned that removing seabed mineral deposits and dumping back unwanted material will stir up ocean sediments, destroy seafloor organisms, and have potentially harmful effects on poorly understood ocean food webs and marine biodiversity. They call for more research to help evaluate such possible effects.

Figure 16-17 Natural capital:
hydrothermal ore deposits form when mineral-rich superheated water shoots out of vents in solidified magma on the ocean floor. After mixing with cold seawater, black particles of metal ore precipitate out and build up as chimneylike ore deposits around the vents. A variety of organisms, supported by bacteria that produce food by chemosynthesis, exist in the dark ocean around these black smokers.

Can We Find Substitutes for Scarce Nonrenewable Mineral Resources? The Materials Revolution

Scientists and engineers are developing new types of materials that can serve as substitutes for many metals.

Some analysts believe that even if supplies of key minerals become too expensive or scarce, human ingenuity will find substitutes. They point to the current *materials revolution* in which silicon and new materials, particularly ceramics and plastics, are being developed and used as replacements for metals.

Ceramics have many advantages over conventional metals. They are harder, stronger, lighter, and longer lasting than many metals, and they withstand intense heat and do not corrode. Within a few decades we may have high-temperature ceramic superconductors in which electricity flows without resistance. Such a development may lead to faster computers, more efficient power transmission, and affordable electromagnets for propelling high-speed magnetic levitation trains.

High-strength plastics and composite materials strengthened by lightweight carbon and glass fibers are beginning to transform the automobile and aerospace industries. They cost less to produce than metals because they take less energy, do not need painting, and can be molded into any shape. New plastics and gels are also being developed to provide superinsulation without taking up much space. And nanotechnology may result in many new materials.

Substitutes can be found for many scarce mineral resources. But finding substitutes for some key materials may be difficult or impossible. Examples are helium, phosphorus for phosphate fertilizers, manganese for making steel, and copper for wiring motors and generators.

In addition, some substitutes are inferior to the minerals they replace. For example, aluminum could replace copper in electrical wiring. But producing aluminum takes much more energy than producing copper, and aluminum wiring is a greater fire hazard than copper wiring.

Mineral resources are the building blocks on which modern society depends. Knowledge of their physical nature and origins, the web they weave between all aspects of human society and the physical earth, can lay the foundations for a sustainable society.

ANN DORR

CRITICAL THINKING

1. Explain why you support or oppose each of the following proposals concerning extraction of hardrock minerals on public land in the United States: **(a)** not granting title to public lands in the United States for actual or claimed hardrock mining, **(b)** requiring mining companies to pay a royalty of 8–12% on the *gross* revenues they earn from hardrock minerals they extract from public lands, and **(c)** making hardrock mining companies legally responsible for restoring the land and cleaning up environmental damage caused by their activities.

2. Describe what would probably happen if **(a)** plate tectonics stopped and **(b)** erosion and weathering stopped. If you could, would you eliminate either group of processes? Explain.

3. Imagine you are an igneous rock. Act as a microscopic reporter and send in a written report on what you experience as you move through various parts of the rock cycle (Figure 16-9, p. 339). Repeat this experience, assuming in turn you are a sedimentary rock and then a metamorphic rock.

4. In the area where you live, are you more likely to experience an earthquake or a volcanic eruption? What can you do to escape or reduce the harm if such a disaster strikes? What actions can you take when it occurs?

5. Congratulations! You are in charge of the world. What are the three most important features of your policy for developing and sustaining the world's nonrenewable mineral resources?

PROJECTS

1. Write a brief scenario describing the series of consequences to us and to other forms of life if the rock cycle stopped functioning.

2. What mineral resources are extracted in your area? What mining methods are used, and what have been their harmful environmental impacts? How has mining these resources benefited the local economy?

3. Use the library or the Internet to find out where earthquakes and volcanic eruptions have occurred during the past 30 years, and then stick small flags on a map of the world or place dots on Figure 16-4 (p. 335) and compare their locations with the plate boundaries shown in this figure.

4. Use the library or the Internet to find bibliographic information about *Will Durant* and *Ann Dorr,* whose quotes appear at the beginning and end of this chapter.

5. Make a concept map of this chapter's major ideas using the section heads, subheads, and key terms (in boldface). Look on the website for this book for information about making concept maps.

LEARNING ONLINE

The website for this book contains study aids and many ideas for further reading and research. They include a chapter summary, review questions for the entire chapter, flash cards for key terms and concepts, a multiple-choice practice quiz, interesting Internet sites, references, and a guide for accessing thousands of InfoTrac® College Edition articles. Log on to

http//biology.brookscole.com/miller14

Then click on the Chapter-by-Chapter area, choose Chapter 16, and select a learning resource.

CASE STUDY
Bitter Lessons from Chernobyl

Chernobyl is known around the globe as the site of the world's most serious nuclear power plant accident (Figure 17-1). On April 26, 1986, a series of explosions in one of the reactors in a nuclear power plant in Ukraine—then part of the Soviet Union— blew the massive roof off a reactor building and flung radioactive debris and dust high into the atmosphere. A huge radioactive cloud spread over much of Belarus, Russia, Ukraine, and other parts of Europe and eventually encircled the planet. Clouds of radioactive material escaped into the atmosphere for 10 days. The surrounding environment and people were exposed to radiation levels about 100 times higher than those caused by the atomic bomb the United States dropped on Hiroshima, Japan, near the end of World War II.

According to various UN studies, the disaster was caused by poor reactor design and human error. Around 30 people near the accident site died from radiation exposure and nearly 2,000 children developed thyroid cancer. Thousands of others may have also developed various cancers and died. But because of poor recordkeeping no one knows the exact death toll, with

the estimated number of premature deaths ranging from 8,000 to 15,000. Regardless of numbers, this was a major human-caused tragedy.

More than 100,000 people had to leave their homes. Most were not evacuated until at least 10 days after the accident. These environmental refugees had to leave their possessions behind. They also had to say goodbye to lush green wheat fields and blossoming apple trees, land their families had farmed for generations, cows and goats that would be shot because the grass they ate was radioactive, and their cats and dogs poisoned with radioactivity.

In 2003, Ukraine officials downgraded the area in a 27-square kilometer (17-square mile) radius from the reactor to a "zone with high risk" to allow those willing to accept the health risk to return home.

The accident exposed more than half a million people to dangerous levels of radioactivity and has caused several thousand cases of thyroid cancer. The total cost of the accident is at least $140 billion according to the U.S. Department of Energy and could eventually reach at least $358 billion according to Ukrainian officials— many times more than the value of all nuclear electricity ever generated in the former Soviet Union.

Chernobyl taught us that *a major nuclear accident anywhere has effects that reverberate throughout much of the world.*

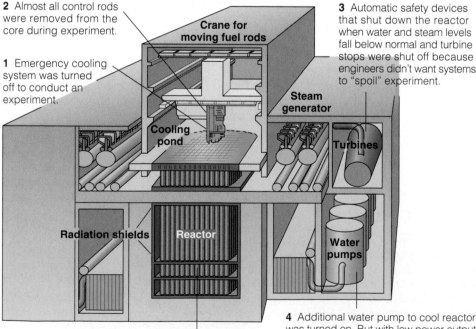

Figure 17-1 Major events leading to the Chernobyl nuclear power plant accident on April 26, 1986, in the former Soviet Union. The accident happened because engineers turned off most of the reactor's automatic safety and warning systems to keep them from interfering with an unauthorized safety experiment. Also, the safety design of the reactor was inadequate (there was no secondary containment shell, as in Western-style reactors), and a design flaw led to unstable operation at low power. After the reactor exploded, crews exposed themselves to lethal levels of radiation to put out fires and encase the shattered reactor in a hastily constructed concrete tomb. This 19-story tomb is crumbling and leaking radioactive materials into the surrounding area. In 2003, the European Bank of Reconstruction and Development provided $85 million to make emergency repairs to the tomb and will provide an additional $750 million to build a new protective shield around the damaged reactor.

2 Almost all control rods were removed from the core during experiment.

1 Emergency cooling system was turned off to conduct an experiment.

Crane for moving fuel rods

3 Automatic safety devices that shut down the reactor when water and steam levels fall below normal and turbine stops were shut off because engineers didn't want systems to "spoil" experiment.

Steam generator

Cooling pond

Turbines

Radiation shields

Reactor

Water pumps

5 Reactor power output was lowered too much, making it too difficult to control.

4 Additional water pump to cool reactor was turned on. But with low power output and extra drain on system, water didn't actually reach reactor.

Typical citizens of advanced industrialized nations each consume as much energy in six months as typical citizens in developing countries consume during their entire life.

MAURICE STRONG

This chapter evaluates fossil fuel and nuclear power energy resources. It addresses the following questions:

- How should we evaluate energy alternatives?
- What are the advantages and disadvantages of conventional and nonconventional oil?
- What are the advantages and disadvantages of natural gas?
- What are the advantages and disadvantages of coal and converting coal to gaseous and liquid fuels?
- What are the advantages and disadvantages of conventional nuclear fission, breeder nuclear fission, and nuclear fusion?

17-1 EVALUATING ENERGY RESOURCES

What Types of Energy Do We Use? Supplementing Free Solar Capital

About 99% of the energy that heats the earth and our homes comes from the sun, and the remaining 1% comes mostly from burning fossil fuels.

Some 99% of the energy that heats the earth and all of our buildings comes directly from the sun at no cost to us. Without this essentially inexhaustible solar energy (*solar capital*), the earth's average temperature would

be −240°C (−400°F), and life as we know it would not exist.

Solar energy comes from the nuclear fusion of hydrogen atoms that make up the sun's mass. Thus *life on earth is made possible by a gigantic nuclear fusion reactor safely located in space about 150 million kilometers (93 million miles) away.*

This direct input of solar energy also produces several other *indirect forms of renewable solar energy*. Examples are wind, falling and flowing water (hydropower), and biomass (solar energy converted to chemical energy stored in chemical bonds of organic compounds in trees and other plants).

Commercial energy sold in the marketplace makes up the remaining 1% of the energy we use. Most commercial energy comes from extracting and burning *nonrenewable mineral resources* obtained from the earth's crust, primarily carbon-containing fossil fuels—oil, natural gas, and coal—as shown in Figure 17-2.

What Types of Commercial Energy Does the World Depend On? The Fossil Fuel Era

About 78% of the commercial energy used worldwide comes from nonrenewable fossil fuels.

About 84% of the commercial energy consumed in the world comes from *nonrenewable* energy resources (78% from fossil fuels and 6% from nuclear power; Figure 17-3, left, p. 352). The remaining 16% comes from *renewable* energy resources—biomass (10%), hydropower (5%), and a combination of geothermal, wind, and solar energy (1%).

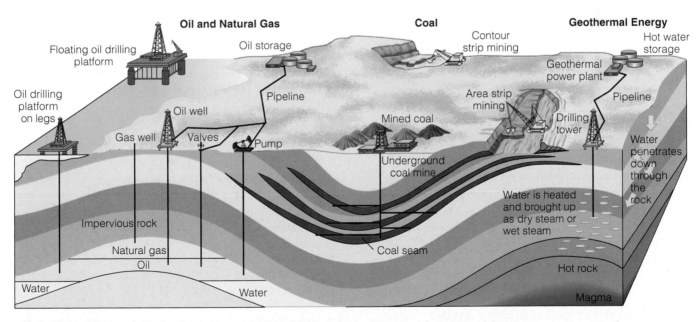

Figure 17-2 Natural capital: important nonrenewable energy resources that can be removed from the earth's crust are coal, oil, natural gas, and some forms of geothermal energy. Nonrenewable uranium ore is also extracted from the earth's crust and then processed to increase its concentration of uranium-235, which can be used as a fuel in nuclear reactors to produce electricity.

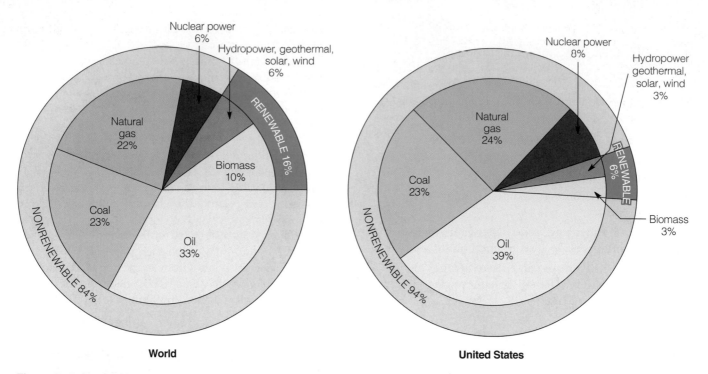

Figure 17-3 Commercial energy use by source for the world (left) and the United States (right) in 2002. Commercial energy amounts to only 1% of the energy used in the world; the other 99% is direct solar energy received from the sun and is not sold in the marketplace. (U.S. Department of Energy, British Petroleum, Worldwatch Institute, and International Energy Agency)

Figure 17-4 shows the global consumption of energy by fuel type between 1970 and 2003, with projections to 2020. Note that oil predominates, followed by natural gas.

Roughly half the world's people in developing countries burn wood and charcoal to heat their dwellings and cook their food. This *biomass energy* is renewable as long as wood supplies are not harvested faster than they are replenished. Most of this biomass is collected by users and not sold in the marketplace.

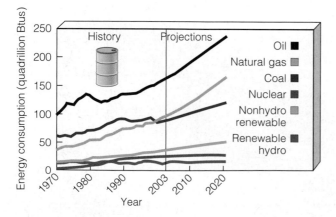

Figure 17-4 Global energy consumption by fuel type, 1970–2003, with projections to 2020. (A Btu is a British thermal unit, a standard measure of heat for value comparison of various fuels.) (Data from U.S. Department of Energy, *Annual Energy Review, 2003 and 2004*)

Thus the actual percentage of renewable biomass energy used in the world is higher than the 10% figure shown in Figure 17-3 (left).

Bad news. Many people in developing countries face a *fuelwood shortage* that is expected to get worse because of unsustainable harvesting of fuelwood. Also, people die prematurely from breathing particles emitted by burning wood indoors in open fires and poorly designed primitive stoves.

What Is the Energy Future of the United States? Searching for Fossil Fuel Substitutes

There is debate over whether U.S. energy policy for this century should continue its dependence on oil and coal or depend more on natural gas, hydrogen, and solar cells.

The United States is the world's largest energy user, with the average American consuming as much energy in one day as a person in the poorest countries consumes in a year. In 2004, with only 4.6% of the population, the United States used 24% of the world's commercial energy. In contrast, India, with 16% of the world's people, used about 3% of the world's commercial energy.

About 94% of the commercial energy used in the United States comes from *nonrenewable* energy resources (86% from fossil fuels and 8% from nuclear power; Figure 17-3, right). The remaining 6% comes mostly from renewable biomass and hydropower.

Figure 17-5 shows energy consumption by fuel in the United States from 1970 to 2003, with projections to 2020. Note that the main projected trends between 2003 and 2020 are increased use of oil and natural gas and a leveling off of coal use.

An important environmental, economic, and political issue is what energy resources the United States might be using by 2050 and 2100. Figure 17-6 shows shifts in use of various commercial sources of energy in the United States since 1800 and one scenario projecting changes to a solar–hydrogen energy age by 2100. According to the U.S. Department of Energy and the Environmental Protection Agency, burning fossil fuels causes more than 80% of U.S. air pollution and 80% of U.S. carbon dioxide emissions. For many energy experts the need to use cleaner and less climate-disrupting (noncarbon) energy resources—not the depletion of fossil fuels—is the driving force behind the projected transition to a solar–hydrogen energy age in the United States and in other parts of the world before the end of this century.

Whether the shift shown in Figure 17-6, or some other scenario, occurs depends primarily on energy resources the U.S. government decides to *promote* by use of subsidies and tax breaks. If we want energy alternatives such as solar energy and hydrogen to become main dishes instead of side orders on our energy menu, they must be nurtured by subsidies and tax breaks.

A political problem is that the fossil fuel and nuclear power industries that have been receiving government subsidies for over 50 years understandably do not want to give them up. And they use their considerable political power to keep them, even though they are mature industries that do not need such nurturing.

Thus the energy path of the United States (or any country) is primarily a political decision made by government officials with pressure from officials of energy companies and from citizens. As a citizen, you can

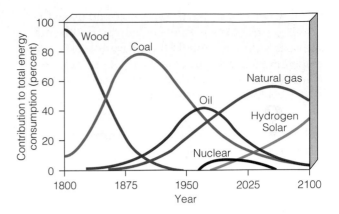

Figure 17-6 Shifts in the use of commercial energy resources in the United States since 1800, with projected changes to 2100. Shifts from wood to coal and then from coal to oil and natural gas have each taken about 50 years. The projected shift to 2100 is only one of many possible scenarios that depend on a variety of assumptions. (U.S. Department of Energy)

play an important role in helping decide the energy future for yourself and future generations. Indeed, it is one of the most important political acts you can undertake. This explains why you should have an understanding of the advantages and disadvantages of our major energy options, as discussed in this chapter and the one that follows.

How Can We Decide Which Energy Resources to Use? Evaluating Alternative Resources

We need to answer several questions in deciding which energy resources to promote.

Energy policies must be developed with the future in mind because experience shows that it usually takes at least 50 years and huge investments to phase in new energy alternatives to the point where they provide 10–20% of total energy use. Making projections such as those in Figure 17-6 and converting them into energy policy involves answering the following questions for *each* alternative:

- How much of the energy resource is likely to be available in the near future (the next 15–25 years) and the long term (the next 25–50 years)?

- What is the net energy yield for the resource?

- How much will it cost to develop, phase in, and use the resource?

- What government research and development subsidies and tax breaks will be used to help develop the resource?

- How will dependence on the resource affect national and global economic and military security?

- How vulnerable is the resource to terrorism?

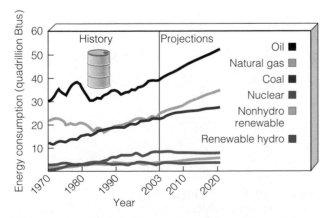

Figure 17-5 Energy consumption by fuel in the United States, 1970–2003, with projections to 2020. (U.S. Department of Energy, *Annual Energy Review, 2003 and 2004*)

■ How will extracting, transporting, and using the resource affect the environment, human health, and the earth's climate? Should these harmful costs be included in the market price of the resource through a combination of taxes and phasing out environmentally harmful subsidies (full-cost pricing)?

What Is Net Energy? The Only Energy That Really Counts

Net energy is the amount of high-quality usable energy available from a resource after subtracting the energy needed to make it available for use.

It takes energy to get energy. For example, before oil is useful to us it must be found, pumped from beneath the ground or ocean floor, transferred to a refinery and converted to useful fuels (such as gasoline, diesel fuel, and heating oil), transported to users, and burned in furnaces and cars. Each step uses high-quality energy. The second law of thermodynamics tells us that some of it will always be wasted and degraded to lower-quality energy.

The usable amount of *high-quality energy* available from a given quantity of a resource is its **net energy.** It is the total amount of energy available from the resource minus the energy needed to find, extract, process, and get it to consumers. It is calculated by estimating the total energy available from the resource over its lifetime minus the amount of energy *used* (the first law of thermodynamics), *automatically wasted* (the second law of thermodynamics), and *unnecessarily wasted* in finding, processing, concentrating, and transporting the useful energy to users.

Net energy is like your net spendable income—your wages minus taxes and job-related expenses. For example, suppose that for every 10 units of energy in oil in the ground we have to use and waste 8 units of energy to find, extract, process, and transport the oil to users. Then we have only 2 units of *useful energy* available from every 10 units of energy in the oil.

We can express net energy as the ratio of useful energy produced to the useful energy used to produce it. In the example just given, the *net energy ratio* would be 10/8, or 1.25. The higher the ratio, the greater the net energy. When the ratio is less than 1, there is a net energy loss.

Figure 17-7 shows estimated net energy ratios for various types of space heating, high-temperature heat for industrial processes, and transportation. In terms of net energy, how do the energy resources used to heat your home and propel your car (if you have one) stack up compared to other alternatives?

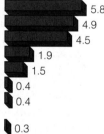

Space Heating

Passive solar	5.8
Natural gas	4.9
Oil	4.5
Active solar	1.9
Coal gasification	1.5
Electric resistance heating (coal-fired plant)	0.4
Electric resistance heating (natural-gas-fired plant)	0.4
Electric resistance heating (nuclear plant)	0.3

High-Temperature Industrial Heat

Surface-mined coal	28.2
Underground-mined coal	25.8
Natural gas	4.9
Oil	4.7
Coal gasification	1.5
Direct solar (highly concentrated by mirrors, heliostats, or other devices)	0.9

Transportation

Natural gas	4.9
Gasoline (refined crude oil)	4.1
Biofuel (ethyl alcohol)	1.9
Coal liquefaction	1.4
Oil shale	1.2

Figure 17-7 *Net energy ratios* for various energy systems over their estimated lifetimes. The higher the net energy ratio, the greater the net energy available. (U.S. Department of Energy and Colorado Energy Research Institute, *Net Energy Analysis*, 1976; and Howard T. Odum and Elisabeth C. Odum, *Energy Basis for Man and Nature*, 3rd ed., New York: McGraw-Hill, 1981)

Currently, oil has a high net energy ratio because much of it comes from large, accessible, and cheap-to-extract deposits such as those in the Middle East. When those are depleted, the net energy ratio of oil will decline and prices will rise.

Conventional nuclear energy has a low net energy ratio because of the large amounts of energy needed to make it available. We have to extract and process uranium ore, convert it into nuclear fuel, build and operate nuclear power plants, dismantle the highly radioactive plants after their 15–60 years of useful life, and store the resulting highly radioactive wastes safely for 10,000–240,000 years depending on the types of radioisotopes they contain. Each of these steps in what is called the *nuclear fuel cycle* uses energy and costs money. Some analysts estimate that ultimately the conventional nuclear fuel cycle will lead to a net energy loss; we will have to put more energy into it than we will ever get out of it.

17-2 OIL

What Is Crude Oil, and How Is It Extracted and Processed? Gooey Stuff to Which We Are Addicted

Crude oil is a thick liquid containing hydrocarbons that we extract from underground deposits and separate into products such as gasoline, heating oil, and asphalt.

Petroleum, or **crude oil** (oil as it comes out of the ground), is a thick and gooey liquid consisting of hundreds of combustible hydrocarbons along with small amounts of sulfur, oxygen, and nitrogen impurities. We have oil today because of a series of three lucky geological events taking place over millions of years. The first event occurred when sediments buried dead organic material raining down onto seafloors faster than it could decay. The next event took place eons later when the seafloor sediments ended up with the right depth for pressure and heat to slowly "cook" or convert the buried organic material into *oil*. The third geological break came about because the oil was able to collect in porous limestone or sandstone rock covered by an impermeable cap of shale or silt to keep it from escaping (Figure 17-2) and thus making it and other fossil fuels part of the carbon cycle (Figure 4-29, p. 78).

Any change in this fortunate chain of events would have meant no oil, which provides about a third of the energy we use today to heat our homes and other buildings and to run our motor vehicles. Oil and its chemical cousin natural gas also provide us with food grown with the help of hydrocarbon-based fertilizers and pesticides. This type of oil is also known as *conventional oil* or *light oil*.

Today's global oil industry is a marvel of technology and management skills. Satellites help find promising oil deposits. Sophisticated computers and software programs analyze seismic data to create 3-D images of the earth's interior. High-tech equipment can drill oil and natural gas wells to a depth of almost 6 kilometers (4 miles). Drilling platforms on the high seas are engineering marvels that can withstand major hurricanes. The incredibly complex process of managing and coordinating the discovery, production, marketing, and distribution of oil throughout the world to billions of users is an amazing process.

Deposits of crude oil and natural gas often are trapped together under a dome deep within the earth's crust on land or under the seafloor (Figure 17-2). The crude oil is dispersed in pores and cracks in underground rock formations, somewhat like water saturating a sponge. To extract the oil, a well is drilled into the deposit. Then oil drawn by gravity out of the rock pores and into the bottom of the well is pumped to the surface.

On average, producers get only about 35–50% of the oil out of an oil deposit—although some believe that improved drilling technology may increase the recovery rate to 75%. The remaining *heavy crude oil* is too difficult or expensive to recover. As oil prices rise, it can become economical to remove about 10–25% of this remaining heavy oil by flushing the well with steam and water. But this lowers the net energy yield for the recovered oil.

Drilling for oil causes only moderate damage to the earth's land because the wells occupy fairly little land area. But drilling for oil and transporting it around the world results in oil spills on land and in aquatic systems. In addition, harmful environmental effects are associated with the extraction, processing, and use of any nonrenewable resource from the earth's crust (Figure 16-13, p. 343).

According to oil producers, improved extraction technologies can increase oil production without serious damage to environmentally sensitive areas. One method allows oil and natural gas producers to drill deeper in most locations. In addition, oil producers can now use one drilling rig (derrick) on a pad to drill several gas or oil pockets at the same time. Another new technology allows oil or gas extraction from distances as far away as 8 kilometers (5 miles) by drilling at angles (slant drilling).

After it is extracted, crude oil is transported to a *refinery* by pipeline, truck, or ship (oil tanker). There it is heated and distilled in gigantic columns to separate it into components with different boiling points (Figure 17-8, p. 356)—another technological marvel based on complex chemistry and chemical engineering. However, refining oil decreases its net energy yield. In the United States, for example, petroleum refining accounts for about 8% of all U.S. energy consumption.

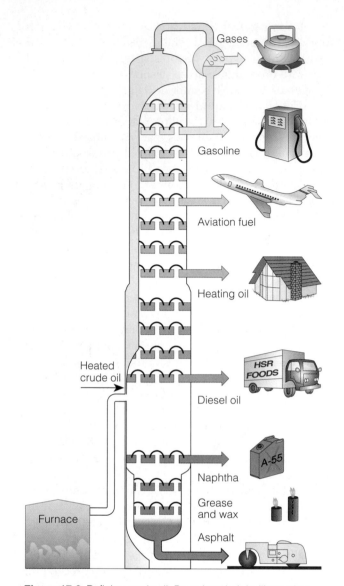

Figure 17-8 Refining crude oil. Based on their boiling points, components are removed at various levels in a giant distillation column. The most volatile components with the lowest boiling points are removed at the top of the column.

Some products of oil distillation, called **petrochemicals,** are used as raw materials in manufacturing pesticides, plastics, synthetic fibers, paints, medicines, and many other products. Look at your clothes and other items around you and try to figure out how many of these things were made from chemicals produced by distilling oil.

Who Has the World's Oil Supplies? OPEC Rules

Eleven OPEC countries—most of them in the Middle East—have 78% of the world's proven oil reserves and most of the world's unproven reserves.

The oil industry is the world's largest business. Thus control of current and future oil reserves is the single greatest source of global economic and political power.

The 11 countries that make up the Organization of Petroleum Exporting Countries (OPEC) have 78% of the world's estimated crude oil reserves. This explains why OPEC is expected to have long-term control over the supplies and prices of the world's conventional oil. Today OPEC's members are Algeria, Indonesia, Iran, Iraq, Kuwait, Libya, Nigeria, Qatar, Saudi Arabia, United Arab Emirates, and Venezuela.

Saudi Arabia, with 25%, has by far the largest proportion of the world's proven crude oil reserves, followed by Canada (15%)—because its huge supply of oil sand was recently classified as a conventional source of oil. Other countries with large proven reserves are Iraq (11%), the United Arab Emirates (9.3%), Kuwait (9.2%), and Iran (8.6%).

Most analysts say it is only a matter of time before the Middle Eastern share of global oil production increases from its current 30% to at least 50%. This is why the world's other nations have such vital economic and military security interests in helping preserve political stability in the often-volatile Middle East.

Here is the problem in a nutshell. Oil is the most widely used energy resource in the world and in the United States. Some call the people in developed countries *oilaholics,* and the world's largest suppliers to them are Canada, Saudi Arabia, and several other Persian Gulf Middle Eastern countries. To some the world economy is built largely on how long Saudia Arabia's House of Saud rulers can continue. There is also concern that terrorist assaults on a few key parts of the country's oil system could put the Saudis out of the oil business for up to 2 years and create global economic chaos.

Case Study: How Much Oil Does the United States Have? Rapidly Dwindling Supplies

The United States—the world's largest oil user—has only 2.9% of the world's proven oil reserves and only a small percentage of its unproven reserves.

Figure 17-9 shows the locations of the major known deposits of fossil fuels in the United States and Canada and ocean areas where more crude oil and natural gas might be found. About one-fourth of U.S. domestic oil production comes from offshore drilling (mostly off the coasts of Texas and Louisiana, Figure 17-10) and 17% from Alaska's North Slope.

The United States has only 2.9% of the world's oil reserves. But it uses about 26% of the crude oil extracted worldwide each year (over two-thirds of that for transportation), mostly because oil is an abundant, convenient, and cheap fuel (Figure 17-11, p. 358). Despite an upsurge in exploration and test drilling, U.S. oil extraction has declined since 1985, and most geologists do not expect a significant increase in domestic supplies (Figure 17-12, p. 358). And the United States produces most of its dwindling supply of oil at a high

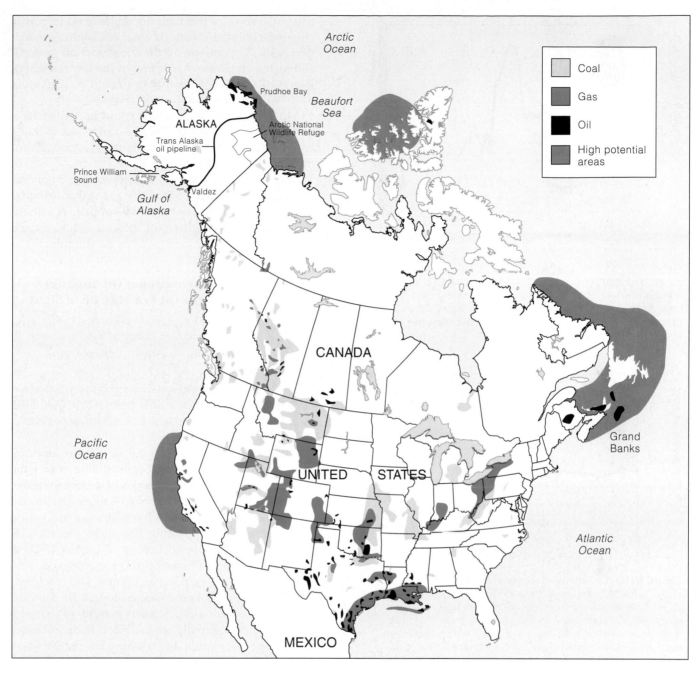

Figure 17-9 Natural capital: locations of the major known deposits of oil, natural gas, and coal in North America and offshore areas where more crude oil and natural gas might be found. Geologists do not expect to find very much new oil and natural gas in North America. (Council on Environmental Quality and U.S. Geological Survey)

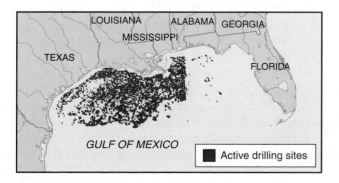

Figure 17-10 Offshore drilling for oil accounts for about one-fourth of U.S. oil production. About nine of every ten barrels of this oil comes from the Gulf of Mexico, where there are 4,000 oil drilling platforms and 53,000 kilometers (33,000 miles) of underwater pipeline. (U.S. Geological Survey)

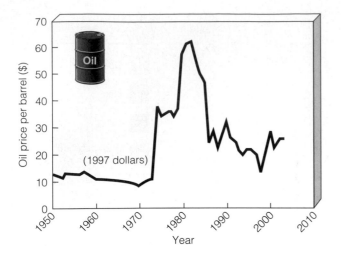

Figure 17-11 Inflation-adjusted price of oil, 1950–2003. When adjusted for inflation, oil costs about the same as it did in 1975. Although low oil prices have stimulated economic growth, they have discouraged improvements in energy efficiency and use of renewable energy resources. (U.S. Department of Energy and Department of Commerce)

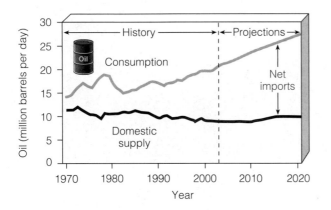

Figure 17-12 U.S. petroleum supply, consumption, and imports, 1970–2003, with projections to 2020. During 2003, the United States imported most of its oil from four nations in the following order of importance: the non-OPEC nations of Canada and Mexico and the OPEC nations of Venezuela and Saudi Arabia. In the not-too-distant future, the Department of Energy projects that the U.S. will have to depend more on the Middle East for oil, because it contains by far most of the world's discovered and undiscovered oil. (U.S. Department of Energy, *Annual Energy Review, 2003* and *2004*)

cost of $7.50–$10 per barrel compared to about $2.50 in Saudi Arabia.

Bottom line: If you think of U.S. oil reserves as a six-pack of oil, four of the cans are empty. Geologists estimate that if the country opens up virtually all of its public lands and coastal regions to oil exploration, it may find at best about half a can of new oil at a high economic and environmental cost.

In 2003, the United States imported about 55% of the oil it used (up from 36% in 1973 when OPEC imposed an oil embargo against the U.S. and other na-

tions). Reasons for this high dependence on imported oil are declining domestic oil reserves, higher production costs for domestic oil than for most oil imports, and increased oil use. According to the Department of Energy (DOE), the United States could be importing 64–70% of the oil it uses by 2020 (Figure 17-12).

In 1970, a bushel of wheat could be traded for a barrel of oil. Now it takes 9 bushels of wheat to buy a barrel of oil. In 2003, grain exports paid for only 11% of the U.S. oil import bill of $99 billion.

Some analysts contend that depending on oil imports is not necessarily bad. They argue that using up limited and declining domestic oil supplies is a drain-America-first policy that will increase future dependence on foreign oil supplies. What do you think?

How Long Will Conventional Oil Supplies Last? The End of the Oil Era May Be in Sight

Known and projected global oil reserves should last for 42–93 years and U.S. reserves for 10–48 years depending on how rapidly we use oil.

Production of the world's estimated oil reserves is expected to peak between a little before 2010 and 2030, and production of estimated U.S. reserves peaked in 1975.

We are not yet running out of oil. But once oil production peaks, we will begin sliding down the bell-shaped oil production curve of a nonrenewable resource (Figure 1-8, p. 11) from 50% depletion toward 80% depletion, when it costs too much to extract what is left. At some point during this slide, we will shift from an abundant supply of cheap oil (Figure 17-11) to a dwindling supply of increasingly expensive oil.

According to geologists, known and projected global reserves of oil are expected to be 80% depleted within 42–93 years and U.S. reserves in 10–48 years depending on how rapidly we use oil. If these estimates are correct, oil should be reaching its sunset years sometime this century. Appendix 5 (p. A12) summarizes milestones in the Age of Oil.

Can We Meet the World's Growing Demand for Oil? Rapid Exponential Growth Is a Hungry Beast

Just to keep using conventional oil at the current rate, we must discover global oil reserves equivalent to a new Saudi Arabian supply every 10 years.

Even if much more oil is somehow found, we are ignoring the consequences of the high (1–5% per year) exponential growth in oil consumption in the world, especially in developing countries (Figure 17-13). It is hard to get a grip on the incredible amount of oil we

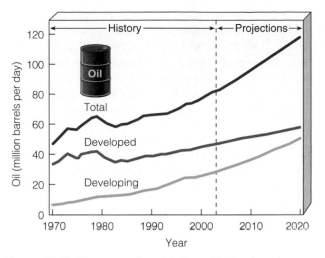

Figuro 17-13 Oil consumption globally and in developed and developing regions, 1970–2003, with projections to 2020. In order, the world's three largest consumers of oil are the United States, China, and Japan—all with limited domestic oil supplies. China imports about a third of its oil and could be importing 50% by 2010. (U.S. Department of Energy, *Annual Energy Review, 2003 and 2004*)

consume. Maybe this will help. *Stretched end to end, the number of barrels of oil the world used in 2004 would wrap around the earth's equator 636 times, and projected oil use in 2020 would circle the equator 913 times!*

Suppose we continue to use oil at the current rate with no increase in oil consumption—a highly unlikely assumption. Even under this conservative no-growth estimate:

- Saudi Arabia, with the world's largest crude oil reserves, could supply world oil needs for about 10 years.

- The estimated reserves under Alaska's North Slope—the largest ever found in North America—would meet current world demand for only 6 months or U.S. demand for 3 years.

- The estimated reserves in Alaska's Arctic National Wildlife Refuge would meet the world's current oil demand for only 1–5 months or U.S. oil demand for 7–24 months (see the Case Study, at right).

Thus for the world just to keep using conventional oil at the current rate, we must discover reserves equivalent to a new Saudi Arabian supply every 10 years. According to most geologists, this is highly unlikely.

And many developing countries such as China and India are rapidly expanding their use of oil. By 2005 China could be using as much oil as the United States and the two countries would be competing to import dwindling supplies of increasingly expensive oil. Indeed, if everyone in the world consumed as much oil as the average American, the world's proven oil reserves would be gone in a decade. Exponential growth is an incredibly powerful force.

Case Study: Should Oil and Gas Development Be Allowed in the Arctic National Wildlife Refuge? To Drill or Not to Drill

There is controversy between oil companies and environmentalists over whether to drill for oil and natural gas in Alaska's Arctic National Wildlife Refuge.

The Arctic National Wildlife Refuge (ANWR) on Alaska's North Slope (Figure 17-9) contains more than one-fifth of all land in the U.S. National Wildlife Refuge System. The refuge's coastal plain is the only stretch of Alaska's arctic coastline not open to oil and gas development.

This tundra biome is home to a diverse community of species, including polar bears, arctic foxes, musk oxen, and peregrine falcons. During the brief arctic summer it serves as a nesting ground for millions of tundra swans, snow geese and other migratory birds, and as a calving ground for a herd of about 130,000 caribou. Partly because of its harsh climate, this is an extremely fragile ecosystem.

Since 1980, U.S. oil companies have been lobbying Congress for permission to carry out exploratory drilling in the coastal plain because they believe it might contain oil and natural gas deposits. Alaska's elected representatives in Congress strongly support such drilling because the state uses revenue from oil production to finance most of its budget and to provide annual dividends to citizens. Environmentalists and conservationist strongly oppose drilling in this area. These polarized positions are summarized in Figure 17-14 (p. 360). Study this figure carefully.

According to drilling opponents, the potential ecological risks are not worth the estimated one-in-five chance of finding enough oil to meet all of the country's needs for only 7–24 months. They point out that improving motor vehicle fuel efficiency is a much faster, cheaper, cleaner, and more secure way to increase future oil supplies. For example, improving fuel efficiency by just 0.4 kilometer per liter (1 mile per gallon) for new cars, SUVs, and light trucks in the United States would save more oil than is ever likely to be produced from the ANWR. In addition, it would be cheaper for the United States to join with Canada in building a pipeline to import some of its potentially abundant oil produced from its oil sands.

In their efforts to either use or protect ANWR, both sides have probably exaggerated their positions. But this issue is symbolic of the fundamental clash between people with different environmental world views.

✗ HOW WOULD YOU VOTE? Do you support opening up Alaska's Arctic National Wildlife Refuge to oil development? Cast your vote online at http://biology.brookscole.com/miller14.

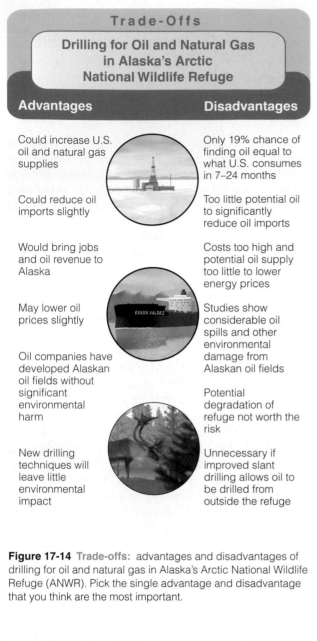

Trade-Offs

Drilling for Oil and Natural Gas in Alaska's Arctic National Wildlife Refuge

Advantages	Disadvantages
Could increase U.S. oil and natural gas supplies	Only 19% chance of finding oil equal to what U.S. consumes in 7–24 months
Could reduce oil imports slightly	Too little potential oil to significantly reduce oil imports
Would bring jobs and oil revenue to Alaska	Costs too high and potential oil supply too little to lower energy prices
May lower oil prices slightly	Studies show considerable oil spills and other environmental damage from Alaskan oil fields
Oil companies have developed Alaskan oil fields without significant environmental harm	Potential degradation of refuge not worth the risk
New drilling techniques will leave little environmental impact	Unnecessary if improved slant drilling allows oil to be drilled from outside the refuge

Figure 17-14 Trade-offs: advantages and disadvantages of drilling for oil and natural gas in Alaska's Arctic National Wildlife Refuge (ANWR). Pick the single advantage and disadvantage that you think are the most important.

What Are the Major Advantages and Disadvantages of Conventional Oil? A Difficult Choice

Conventional oil is versatile fuel and reserves can last for at least 50 years, but burning it produces air pollution and releases the greenhouse gas carbon dioxide.

Figure 17-15 lists the advantages and disadvantages of using conventional crude oil as an energy resource. A serious problem is that burning oil or any carbon-containing fossil fuel releases CO_2 into the atmosphere and thus can help promote climate change through global warming. Currently, burning oil mostly as gasoline and diesel fuel for transportation accounts for about 43% of global CO_2 emissions. Figure 17-16 com-

pares the relative amounts of CO_2 emitted per unit of energy by the major fossil fuels and nuclear power.

In 1999 Mike Bowling, CEO of ARCO Oil, said, "We are embarked on the beginning of the last days of the Age of Oil." He went on to discuss the need for the world to shift from a carbon-based to a hydrogen-based energy economy during this century (Figure 17-6).

How Useful Are Heavy Oils from Oil Sand and Oil Shale? Can Heavier Substitutes Save the Day?

Heavy oils from oil sand and oil shale could supplement conventional oil, but there are environmental problems.

Oil sand, or **tar sand,** is a mixture of clay, sand, water, and a combustible organic material called *bitumen*—a

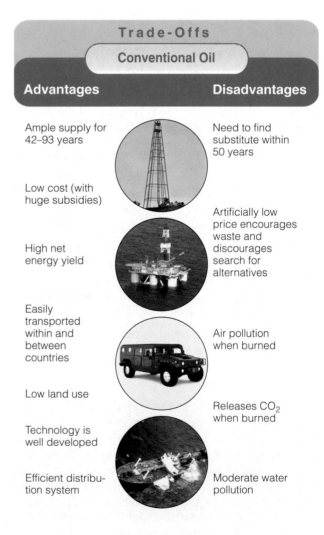

Trade-Offs

Conventional Oil

Advantages	Disadvantages
Ample supply for 42–93 years	Need to find substitute within 50 years
Low cost (with huge subsidies)	Artificially low price encourages waste and discourages search for alternatives
High net energy yield	
Easily transported within and between countries	Air pollution when burned
Low land use	Releases CO_2 when burned
Technology is well developed	
Efficient distribution system	Moderate water pollution

Figure 17-15 Trade-offs: advantages and disadvantages of using conventional crude oil as an energy resource. Pick the single advantage and disadvantage that you think are the most important.

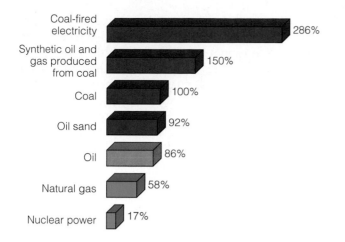

Figure 17-16 CO_2 emissions per unit of energy produced by various fuels, expressed as percentages of emissions produced by burning coal directly. (Data from U.S. Department of Energy)

thick and sticky heavy oil with a high sulfur content and that smells like asphalt. It was created when conventional oil escaping from its birthplace was degraded into tar by bacteria and groundwater.

Oil sands nearest the surface are dug up in what looks like a war zone by gigantic electric shovels and loaded into house-sized trucks that carry them to upgrading plants. There they are mixed with hot water and steam to extract the bitumen, which is heated in huge cookers to convert it into a low-sulfur synthetic crude oil suitable of refining. Heating the cookers requires vast amounts of natural gas that reduces the net energy yield for the oil. Two tons of oil sand are strip mined for each barrel of oil and 3 barrels of water are needed to extract each barrel of bitumen.

Bitumen in deeper deposits of oil sand can be removed by underground processing. This involves using one well to inject steam into the underground oil sands to loosen the bitumen and another well to suck the bitumen out. This leaves the land largely undisturbed and eliminates the need for giant tailings ponds to store water, sand, and clay left over from surface mining.

Northeastern Alberta in Canada has about three-fourths of the world's oil sand reserves, about a tenth of them close enough to the surface to be recovered by surface or underground mining. Improved technology may allow extraction of twice that amount.

Currently these deposits supply about a fifth of Canada's oil needs and this proportion is expected to increase. Because of the dramatic reductions in development and production costs, in 2003 the oil industry began counting Canada's oil sands as reserves of conventional oil. This means that Canada has 15% of the world's oil reserves, second only to Saudi Arabia.

If a pipeline is built to transfer some of this crude synthetic crude oil from western Canada to the northwestern United States, Canada could greatly reduce future U.S. dependence on oil imports from the Middle East and add to its income. Other fairly large deposits of oil sands are in Utah, Venezuela, Colombia, and Russia.

Bad news. Extracting and producing oil sands has a severe impact on the land and produces more water pollution, much more air pollution (especially sulfur dioxide), and more CO_2 per unit of energy than conventional crude oil. Also, costs are skyrocketing because it takes so many highly skilled workers to build and operate an oil sands site.

Another potential source of oil are deposits of *oil shale,* which are neither oil nor shale rock. Instead, *oil shales* are fine-grained sedimentary rocks (Figure 17-17, left) containing a solid combustible mixture of hydrocarbons called *kerogen.* It can be distilled from crushed oil shale rock by heating it in a large container to yield **shale oil** (Figure 17-17, right). Before the thick shale oil can be sent by pipeline to a refinery, it must be heated to increase its flow rate and processed to remove sulfur, nitrogen, and other impurities.

Estimated potential global supplies of shale oil are about 240 times larger than estimated global supplies of conventional oil. But most deposits are of such a low grade that with current oil prices and technology it takes more energy and money to mine and convert kerogen to crude oil than the resulting fuel is worth. Producing and using shale oil also has a much higher environmental impact than conventional oil.

Figure 17-18 (p. 362) lists the advantages and disadvantages of using heavy oil from oil sand and oil shale as energy resources. Overall, do you believe the advantages outweigh the disadvantages?

Figure 17-17 Natural capital: oil shale rock (left) and the shale oil (right) extracted from it. Big U.S. oil shale projects have been canceled because of excessive cost.

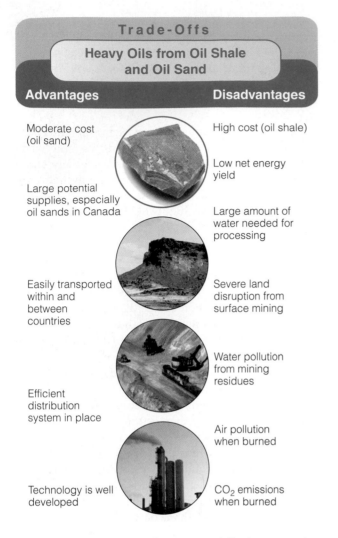

Trade-Offs

Heavy Oils from Oil Shale and Oil Sand

Advantages	Disadvantages
Moderate cost (oil sand)	High cost (oil shale)
Large potential supplies, especially oil sands in Canada	Low net energy yield
Easily transported within and between countries	Large amount of water needed for processing
Efficient distribution system in place	Severe land disruption from surface mining
	Water pollution from mining residues
Technology is well developed	Air pollution when burned
	CO_2 emissions when burned

Figure 17-18 Trade-offs: advantages and disadvantages of using heavy oils from oil shale and oil sand as energy resources. Pick the single advantage and disadvantage that you think are the most important.

17-3 NATURAL GAS

What Is Natural Gas? Mostly Methane

Natural gas, consisting mostly of methane, is often found above reservoirs of crude oil.

In its underground gaseous state, **natural gas** is a mixture of 50–90% by volume of methane (CH_4), the simplest hydrocarbon. It also contains smaller amounts of heavier gaseous hydrocarbons such as ethane (C_2H_6), propane (C_3H_8), and butane (C_4H_{10}), and small amounts of highly toxic hydrogen sulfide (H_2S).

Conventional natural gas lies above most reservoirs of crude oil (Figure 17-2). Like oil, natural gas was formed from fossil deposits of phytoplankton and animals buried on the seafloor for millions of years and subjected to high temperatures and pressures.

However, unless a natural gas pipeline has been built deposits of natural gas found above oil deposits (Figure 17-2) cannot be used. Indeed, the natural gas found above oil reservoirs in deep sea and remote land areas is often viewed as an unwanted byproduct and is burned off. This wastes a valuable energy resource and releases carbon dioxide into the atmosphere.

Unconventional natural gas is found in other underground sources. One is *methane hydrate,* in which small bubbles of natural gas are trapped in ice crystals deep under the arctic permafrost and beneath deep-ocean sediments. Globally the amount of energy in methane hydrates is about twice that in the earth's oil, natural gas, and coal resources combined.

So far it costs too much to get natural gas from methane hydrates and unconventional sources of natural gas, but the extraction technology is being developed rapidly, especially by Japan that has large deposits off its coast and few deposits of conventional oil and natural gas. One problem is that when methane hydrate is brought to the surface it warms up and releases methane (a greenhouse gas) into the atmosphere.

When a natural gas field is tapped, propane and butane gases are liquefied and removed as **liquefied petroleum gas (LPG).** LPG is stored in pressurized tanks for use mostly in rural areas not served by natural gas pipelines. The rest of the gas (mostly methane) is dried to remove water vapor. Then it is cleansed of poisonous hydrogen sulfide and other impurities and pumped into pressurized pipelines for distribution.

At a very low temperature natural gas can be converted to **liquefied natural gas (LNG).** This highly flammable liquid can then be shipped to other countries in refrigerated tanker ships.

How Is Natural Gas Used? A Versatile Fuel

Natural gas can be burned to heat space and water, generate electricity, and propel vehicles.

Natural gas is a versatile fuel that can be burned to heat water and buildings and to generate electricity. It can also be used as a fuel for cars and trucks with fairly inexpensive engine modifications. Natural gas is especially useful for running fleets of taxis and delivery and work vehicles operating from garages and maintenance facilities that can be used to supply them with this fuel.

Increasingly, natural gas is used to run medium-sized turbines that produce electricity. These clean-burning turbines have a much higher energy efficiency (50–60%) than coal-burning power plants (24–35%). They are cheaper to build per kilowatt-hour, require less time to install, and are easier and cheaper to maintain than large-scale coal and nuclear power plants.

Burning natural gas emits CO_2 but at a lower rate per unit of energy than other fossil fuels (Figure 17-16). For these reasons, natural gas use is expected to grow worldwide (Figure 17-4) and in the United States (Figures 17-5 and 17-6).

Who Has the World's Natural Gas Supplies and How Long Will the Supplies Last? More Abundant than Oil

Russia and Iran have almost half of the world's reserves of conventional natural gas, and global reserves should last 62–125 years.

Russia has about 31% of the world's proven natural gas reserves, followed by Iran (15%) and Qatar (9%). About 36% of the world's natural gas reserves are in Middle Eastern countries. The United States has only 3% of the world's proven reserves. Geologists expect to find more natural gas, especially in unexplored developing countries.

The long-term global outlook for natural gas supplies is better than for conventional oil. At the current consumption rate, geologists estimate that known reserves and undiscovered potential reserves of conventional natural gas should last the world for 62–125 years and the United States for 55–80 years, depending on how rapidly it is used.

They project that *conventional* and *unconventional* supplies of natural gas (the latter available at higher prices) should last at least 200 years at the current consumption rate and 80 years if consumption rates rise 2% per year.

Figure 17-19 lists the advantages and disadvantages of natural gas as an energy resource. Energy experts project greatly increased global use of natural gas during this century because of its fairly abundant supply, and lower pollution and CO_2 rates per unit of energy compared to other fossil fuels (Figure 17-16).

Because of its advantages over oil, coal, and nuclear energy, some analysts see natural gas as the best fuel to help make the transition to improved energy efficiency and greater use of solar energy and hydrogen over the next 50 years.

What Is the Future of Natural Gas in the United States? Declining Supplies and Rising Imports

Natural gas production in the United States is expected to continue declining, resulting in increased dependence on imports from Canada, Russia, and the Middle East.

In 2002, natural gas was burned to provide 53% of the heat in U.S. homes and 16% of the country's electricity. By 2020, the U.S. Department of Energy projects that natural gas will be burned to produce about one-third

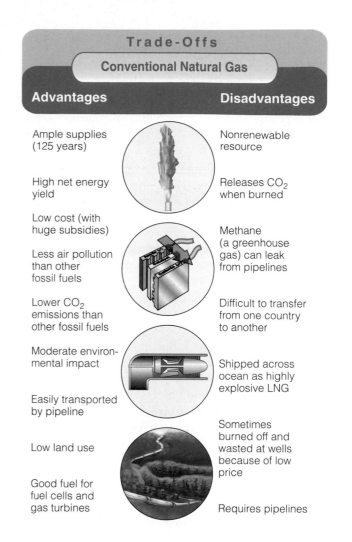

Trade-Offs

Conventional Natural Gas

Advantages	Disadvantages
Ample supplies (125 years)	Nonrenewable resource
High net energy yield	Releases CO_2 when burned
Low cost (with huge subsidies)	Methane (a greenhouse gas) can leak from pipelines
Less air pollution than other fossil fuels	
Lower CO_2 emissions than other fossil fuels	Difficult to transfer from one country to another
Moderate environmental impact	Shipped across ocean as highly explosive LNG
Easily transported by pipeline	
Low land use	Sometimes burned off and wasted at wells because of low price
Good fuel for fuel cells and gas turbines	Requires pipelines

Figure 17-19 Trade-offs: advantages and disadvantages of using conventional natural gas as an energy resource. Pick the single advantage and disadvantage that you think are the most important.

of the country's electricity, if the natural gas pipeline distribution system is greatly expanded.

Bad news. U.S. production of natural gas has been declining for a long time, and most geologists do not believe this situation will be reversed. More natural gas could be imported from Canada, but this will require building a major pipeline between the two countries. Also, production in Canada is expected to peak between 2020 and 2030. Then the United States and the rest of the world would have to rely increasingly on Russia and the Middle East for supplies of natural gas.

More liquefied natural gas could be imported by ship. But this requires cooling the gas to a very low temperature to liquefy it, shipping it in special tankers, and building special LNG receiving terminals. This is quite expensive and reduces the net energy yield for natural gas. Also, LNG is highly flammable and could lead to large-scale fires at receiving terminals.

17-4 COAL

What Is Coal, and How Is It Extracted? A Mostly Carbon Fuel

Coal, which can be extracted by surface and underground mining, consists mostly of carbon plus small amounts of sulfur and trace amounts of mercury and radioactive material.

Coal is a solid fossil fuel formed in several stages as buried remains of land plants that lived 300–400 million years ago were subjected to intense heat and pressure over many millions of years (Figure 17-20). Coal is mostly carbon and contains small amounts of sulfur, released into the atmosphere as SO_2 when coal is burned. Burning coal also releases trace amounts of toxic mercury and radioactive materials.

Anthracite (which is about 98% carbon) is the most desirable type of coal because of its high heat content and low sulfur content. However, because it takes much longer to form, it is less common and therefore more expensive than other types of coal.

Some coal is extracted underground by miners working in tunnels and shafts (Figure 16-12, p. 342). This is one of the world's most dangerous occupations because of accidents and black lung disease caused by prolonged inhalation of coal dust particles. *Area strip mining* (Figure 16-11c, p. 341) is used to extract coal found close to the earth's surface on flat terrain, and *contour strip mining* (Figure 16-11d) is used on hilly or mountainous terrain. In some cases, entire mountaintops are removed and dumped into the valleys below to expose seams of coal. The scarred land from the surface mining of coal is not restored in most countries and only partially restored in parts of the United States.

Fly over parts of West Virginia and you will see some mountains looking as if their tops had been sliced off with a machete and others so deeply mined that they look like ugly miniature Grand Canyons. Enormous slurry ponds containing mining waste are sandwiched between the remains of these mountains. Mountaintop mining has polluted some 760 kilometers (470

miles) of West Virginia's streams and displaced thousands of families.

After coal is removed, trains usually transport it to a processing plant, where it is broken up, crushed, and washed to remove impurities. After the coal is dried it is shipped (again usually by train) to users, mostly power plants and industrial plants.

How Is Coal Used, and How Long Will Supplies Last?

Coal is burned mostly to produce electricity and steel, and reserves in the United States, Russia, and China could last hundreds to thousands of years.

Coal is burned to generate 62% of the world's electricity (52% in the United States) and make three-fourths of its steel. Coal is by far the world's most abundant fossil fuel, with deposits containing ten times more energy than oil and natural gas resources combined. According to the U.S. Geological Survey, identified and unidentified supplies of coal could last the world for 214–1,125 years, depending on the rate of usage.

The United States has one-fourth of the world's proven coal reserves. Russia has 16% and China 12%. In 2002, just over half of global coal consumption was split almost evenly between China and the United States.

China has enough proven coal reserves to last 300 years at its current rate of consumption. According to the U.S. Geological Survey, identified U.S. coal reserves should also last about 300 years at the current consumption rate, and unidentified U.S. coal resources could extend those supplies for perhaps another 100 years, at a higher cost. However, if U.S. coal use should increase by 4% a year—as the coal industry projects—the country's proven coal reserves would last only 64 years.

Figure 17-21 lists the advantages and disadvantages of using coal as an energy resource. *Bottom line.*

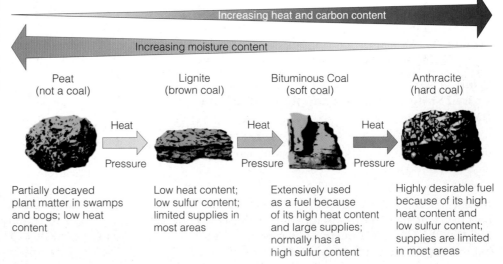

Figure 17-20 Natural capital: stages in coal formation over millions of years. Peat is a soil material made of moist, partially decomposed organic matter. Lignite and bituminous coal are sedimentary rocks, whereas anthracite is a metamorphic rock (Figure 16-9, p. 339).

Increasing heat and carbon content →

← Increasing moisture content

| Peat (not a coal) | | Lignite (brown coal) | | Bituminous Coal (soft coal) | | Anthracite (hard coal) |

Heat / Pressure

Partially decayed plant matter in swamps and bogs; low heat content

Low heat content; low sulfur content; limited supplies in most areas

Extensively used as a fuel because of its high heat content and large supplies; normally has a high sulfur content

Highly desirable fuel because of its high heat content and low sulfur content; supplies are limited in most areas

Trade-Offs

Coal

Advantages		Disadvantages
Ample supplies (225–900 years)		Very high environmental impact
High net energy yield		Severe land disturbance, air pollution, and water pollution
Low cost (with huge subsidies)		High land use (including mining)
Mining and combustion technology well-developed		Severe threat to human health
		High CO_2 emissions when burned
Air pollution can be reduced with improved technology (but adds to cost)		Releases radioactive particles and toxic mercury into air

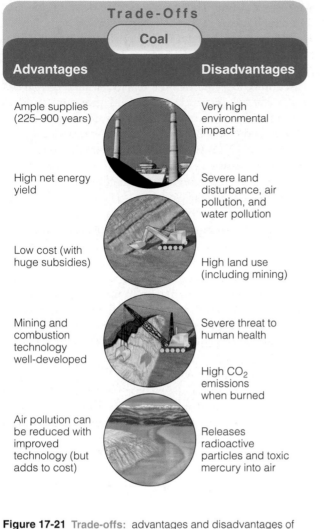

Figure 17-21 Trade-offs: advantages and disadvantages of using coal as an energy resource. Pick the single advantage and disadvantage that you think are the most important.

Coal is the world's most abundant fossil fuel, but mining and burning it has a severe environmental impact on air, water, and land and accounts for over a third of the world's annual CO_2 emissions. Each year in the United States alone, air pollutants—such as sulfur dioxide, particulates, and toxic metals such as mercury, arsenic, and lead—released when coal is burned kill thousands of people prematurely (estimates range from 65,000 to 200,000), cause at least 50,000 cases of respiratory disease, and result in several billion dollars of property damage. Many people are unaware that burning coal is also responsible for about one-fourth of atmospheric mercury pollution in the United States and releases far more radioactive particles into the air than normally operating nuclear power plants.

In China, millions of people burning coal in unvented stoves for heat and cooking are exposed to dangerous levels of particulate matter and toxic metals such as mercury and arsenic.

What Are the Advantages and Disadvantages of Converting Solid Coal into Gaseous and Liquid Fuels? Better for the Air, Worse for the Climate

Coal can be converted to gaseous and liquid fuels that burn cleaner than coal, but costs are high, and producing and burning them add more carbon dioxide to the atmosphere than burning coal.

Solid coal can be converted into **synthetic natural gas (SNG)** by **coal gasification** or into a liquid fuel such as methanol or synthetic gasoline by **coal liquefaction.** Figure 17-22 lists the advantages and disadvantages of using these *synfuels*.

Trade-Offs

Synthetic Fuels

Advantages		Disadvantages
Large potential supply		Low to moderate net energy yield
		Higher cost than coal
Vehicle fuel		Requires mining 50% more coal
		High environmental impact
Moderate cost (with large government subsidies)		Increased surface mining of coal
		High water use
Lower air pollution when burned than coal		Higher CO_2 emissions than coal

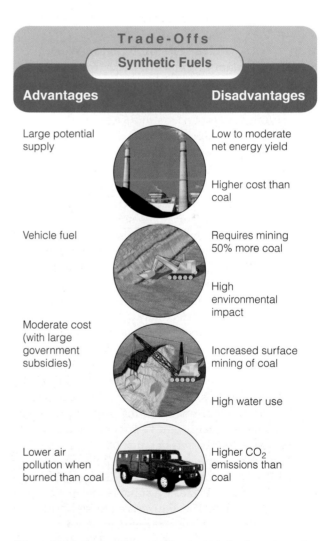

Figure 17-22 Trade-offs: advantages and disadvantages of using synthetic natural gas (SNG) and liquid synfuels produced from coal. Pick the single advantage and disadvantage that you think are the most important.

Without huge government subsidies, most analysts expect these synthetic fuels to play only a minor role as energy resources in the next 20–50 years. Compared with burning conventional coals, they require mining 50% more coal and their production and burning add 50% more carbon dioxide to the atmosphere. Also, they cost more to produce.

However, the U.S. Department of Energy and a consortium of major oil companies are working on ways to reduce CO_2 emissions during the coal gasification process. They hope to develop metal-ceramic membranes that trap carbon dioxide gas. The CO_2 could then be compressed and piped off to underground repositories or other permanent storage sites. If this works, burning gasified coal could be a cheaper and cleaner way to produce electricity than burning coal, oil, or natural gas. Stay tuned.

17-5 NUCLEAR ENERGY

How Does a Nuclear Fission Reactor Work? Splitting Nuclei to Produce Electricity

In a conventional nuclear reactor, isotopes of uranium and plutonium undergo controlled nuclear fission and the resulting heat is used to produce steam that spins turbines to generate electricity.

To evaluate the advantages and disadvantages of nuclear power, we must know how a conventional nuclear power plant and its accompanying nuclear fuel cycle work. In a nuclear fission chain reaction, neutrons split the nuclei of atoms such as uranium-235 and plutonium-239 and release energy mostly as high-temperature heat as a result of the chain reaction (Figure 3-15, p. 50). In the reactor of a nuclear power plant, the rate of fission is controlled and the heat generated is used to produce high-pressure steam, which spins turbines that generate electricity.

Light-water reactors (LWRs) like the one in the diagram in Figure 17-23 produce about 85% of the world's nuclear-generated electricity (100% in the United States). The *core* of an LWR contains 35,000– 70,000 long, thin fuel rods, each packed with fuel pellets. Each pellet is about one-third the size of a cigarette and contains the energy equivalent of 0.9 metric ton (1 ton) of coal or four barrels of crude oil.

The *uranium oxide fuel* in each pellet consists of about 97% nonfissionable uranium-238 and 3% fissionable uranium-235. To create a suitable fuel, the concentration of uranium-235 in the ore is increased (enriched) from 0.7% (its natural concentration in uranium ore) to 3% by removing some of the uranium-238.

Control rods made of neutron-absorbing materials, such as boron or cadmium, are moved in and out of the spaces between the fuel assemblies in the core to absorb neutrons. This regulates the rate of fission and amount of power the reactor produces.

A material called a *moderator* slows down the neutrons emitted by the fission process to keep the chain reaction going. The moderator can be liquid water (used in 75% of the world's reactors, called *pressurized water reactors,* Figure 17-23), solid graphite (used in 20% of all reactors, mostly in France, the former Soviet Union, and Great Britain), or heavy water (deuterium oxide or D_2O, used in 5% of all reactors). Graphite-moderated reactors (used in the ill-fated Chernobyl plant; Figure 17-1) can also produce fissionable plutonium-239 for nuclear weapons.

A *coolant,* usually water, circulates through the reactor's core to remove heat to keep fuel rods and other materials from melting and to produce steam for generating electricity. In Great Britain, gaseous carbon dioxide is blown into the core to keep the fuel assemblies cool. The greatest danger in water-cooled reactors is a loss of coolant that would allow the fuel to quickly overheat, melt down, and possibly release radioactive materials to the environment. An LWR reactor has an emergency core-cooling system as a backup to help prevent such meltdowns.

As a further safety backup, a *containment vessel* with very thick and strong walls surrounds the reactor core. It is designed to keep radioactive materials from escaping into the environment in case of an internal explosion or core meltdown within the reactor and to protect the core from external threats such as a plane crash. Containment vessels typically consist of a 1.2-meter (4-foot) steel-reinforced concrete wall with a steel liner.

Water-filled pools or *dry casks* with thick steel walls are used for on-site storage of highly radioactive spent fuel rods removed when reactors are refueled. Most spent fuel rods are stored in 6-meter- (20-foot-) deep pools of boron-treated water to shield against radiation and to keep the fuel from heating up, catching fire, and releasing radioactive materials into the environment. The long-term goal is to transport spent fuel rods and other long-lived radioactive wastes to an underground facility where they must be stored safely for 10,000–240,000 years until their radioactivity falls to safe levels.

The overlapping and multiple safety features of a modern nuclear reactor greatly reduce the chance of a serious nuclear accident. But these safety features make nuclear power plants very expensive to build and maintain.

What Is the Nuclear Fuel Cycle? Looking at the Whole Picture

The nuclear fuel cycle includes the mining of uranium, processing it to make a satisfactory

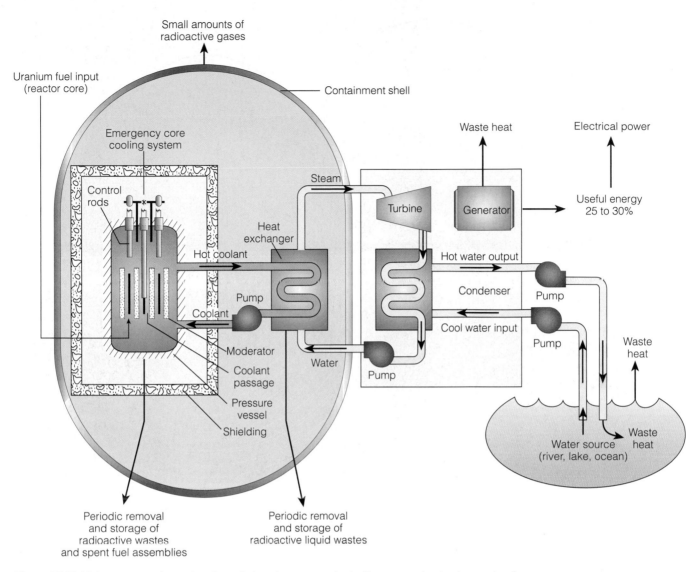

Figure 17-23 Light-water–moderated and –cooled nuclear power plant with a pressurized water reactor. Some plants use huge cooling towers to transfer some of the waste heat to the atmosphere.

fuel, using it in a reactor, safely storing the resulting highly radioactive wastes for thousands of years, and dealing with the highly radioactive reactor after its useful life.

Nuclear power plants, each with one or more reactors, are only one part of the nuclear fuel cycle (Figure 17-24, p. 368). Unlike other energy resources, nuclear energy produces *high-level radioactive wastes* that give off large amounts of harmful ionizing radiation for a short time and small amounts for a long time. Such wastes, consisting mainly of spent fuel rods from commercial nuclear power plants and assorted wastes from the production of nuclear weapons, must be stored safely for thousands of years.

After approximately 15–60 years of operation, a nuclear reactor becomes dangerously contaminated with radioactive materials, and many of its parts be-

come brittle or corroded and worn out. Unless the plant's life can be extended by expensive renovation, it must be *decommissioned* or retired.

Once a nuclear reactor comes to the end of its useful life it cannot be shut down and abandoned like a coal-burning plant. It contains large quantities of intensely radioactive materials that must be kept out of the environment for many thousands of years.

In the *closed nuclear fuel cycle* (Figure 17-24, dotted lines), the fissionable isotopes uranium-235 and plutonium-239 are removed from spent fuel assemblies for reuse as nuclear fuel. Their removal means that the remaining radioactive wastes must be stored safely for about 10,000 years. Currently, these isotopes are rarely removed from spent fuel rods and other nuclear wastes because of high costs and the potential use of the removed isotopes in nuclear weapons.

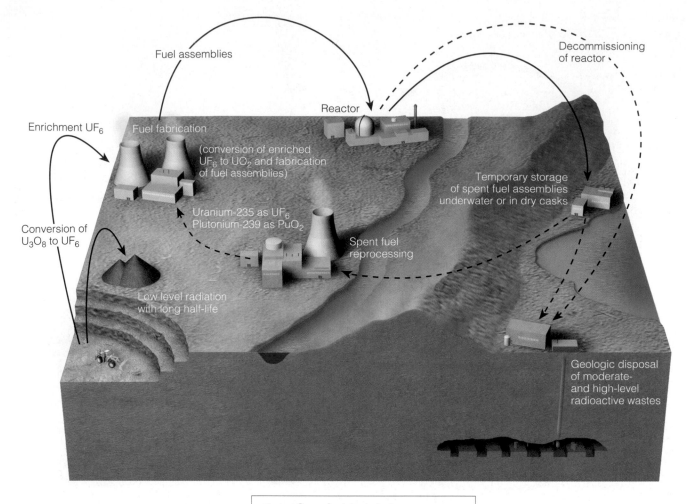

Fuel assemblies

Decommissioning
of reactor

Enrichment UF$_6$

Reactor

Fuel fabrication

(conversion of enriched
UF$_6$ to UO$_2$ and fabrication
of fuel assemblies)

Temporary storage
of spent fuel assemblies
underwater or in dry casks

Conversion of
U$_3$O$_8$ to UF$_6$

Uranium-235 as UF$_6$
Plutonium-239 as PuO$_2$

Spent fuel
reprocessing

Low level radiation
with long half-life

Geologic disposal
of moderate-
and high-level
radioactive wastes

⟵ Open fuel cycle today
⟵ - - Prospective "closed" end fuel cycle

Figure 17-24 The nuclear fuel cycle.

In the *open nuclear fuel cycle* (solid lines, Figure 17-24) the isotopes are not removed by reprocessing the nuclear wastes and are eventually buried in an underground disposal facility. These wastes must be stored safely for about 240,000 years—several times longer than the latest version of our species has been around.

In evaluating the safety, economic feasibility, and overall environmental impact of nuclear power, energy experts and economists caution us to look at this entire cycle, not just the nuclear plant itself.

How Did We Get into Nuclear Power and How Successful Has It Been? A Faded Dream

After more than 50 years of development and enormous government subsidies, nuclear power has not lived up to its promise.

U.S. utility companies began developing nuclear power plants in the late 1950s for three reasons. *First*, the Atomic Energy Commission (which had the conflicting

roles of promoting and regulating nuclear power) promised utility executives that nuclear power would produce electricity at a much lower cost than coal and other alternatives. Indeed, President Dwight D. Eisenhower declared in a 1953 speech that nuclear power would be "too cheap to meter."

Second, the government (taxpayers) paid about one-fourth of the cost of building the first group of commercial reactors and guaranteed there would be no cost overruns. *Third*, after insurance companies refused to insure nuclear power, Congress passed the Price–Anderson Act to protect the U.S. nuclear industry and utilities from significant liability in case of accidents.[*]

In the 1950s, researchers projected that by the year 2000 at least 1,800 nuclear power plants would supply 21% of the world's commercial energy (25% in the United States) and most of the world's electricity.

---

[*]This act limits the nuclear industry's liability for any accident to $9.5 billion. According to the U.S. Nuclear Regulatory Commission, a worst-case accident would cause more than $300 billion in damages.

After more than 50 years of development, enormous government subsidies, and an investment of $2 trillion worldwide, these goals have not been met. Instead, by 2002, 441 commercial nuclear reactors in 30 countries were producing only 6% of the world's commercial energy and 19% of its electricity.

Since 1989, electricity production from nuclear power has increased only slightly and is now the world's slowest-growing energy source. According to the U.S. Department of Energy, the percentage of the world's electricity produced by nuclear power will fall to 12% by 2025 because the retirement of aging existing reactors is expected to exceed construction of new ones.

No new nuclear power plants have been ordered in the United States since 1978, and all 120 plants ordered since 1973 have been canceled. In 2004, there were 103 licensed and operating commercial nuclear power reactors in 31 states—most in the eastern half of the country (Figure 17-25). Is there a nuclear reactor in your vicinity? These reactors generate about 21% of the country's electricity and 8% of its total energy. This percentage is expected to decline over the next two to three decades as existing plants wear out and are retired.

According to energy analysts and economists, there are several major reasons for the failure of nuclear power to grow as projected. They include multibillion-dollar construction cost overruns, higher operating costs and more malfunctions than expected, and poor management. Two other major setbacks have been public concerns about safety and stricter government safety regulations, especially after the accidents in 1979 at the Three Mile Island nuclear plant in Pennsylvania and in 1986 at Chernobyl (p. 350).

Another problem is investor concerns about the economic feasibility of nuclear power that take into account the entire nuclear fuel cycle. At Three Mile Island, investors lost over a billion dollars in one hour from damaged equipment and repair, even though the

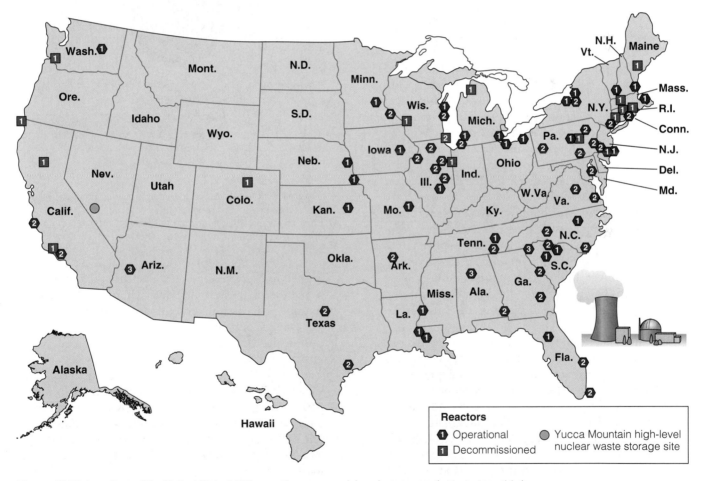

Figure 17-25 Locations of the United States' 103 operating commercial nuclear power plant reactors, 14 decommissioned reactors (with highly radioactive used fuel stored on site), and the recently approved site in Nevada for storage of highly radioactive used fuel from operating and decommissioned nuclear reactors. Numbers refer to the number of reactors at each nuclear power plant site. There are at least 30 other sites (mostly in the West) containing high-level nuclear wastes produced mostly by making nuclear weapons that will also ship wastes to Nevada's Yucca Mountain underground storage site. (Data from U.S. Nuclear Regulatory Commission and U.S. Department of Energy)

reactor core did not melt down and no human lives were lost. Also, concern has risen about the vulnerability of nuclear power plants to terrorist attacks after the events of September 11, 2001, in the United States. Experts are especially concerned about the vulnerability of poorly protected and intensely radioactive spent fuel rods stored in pools or casks outside of reactor buildings.

What Are the Advantages and Disadvantages of the Conventional Nuclear Fuel Cycle? Better than Coal but Much More Costly and Vulnerable to Terrorist Attack

The nuclear fuel cycle has a fairly low environmental impact and a very low risk of an accident, but costs are high, radioactive wastes must be stored safely for thousands of years, and facilities are vulnerable to terrorist attack.

Figure 17-26 lists the major advantages and disadvantages of the conventional nuclear fuel cycle. Using nuclear power to produce electricity has some important advantages over coal-burning power plants (Figure 17-27).

Some proponents of nuclear power in the United States claim it will help reduce dependence on imported oil. But other analysts point out that nuclear power has little effect on U.S. oil use because burning oil typically produces only 2–3% of the electricity in the United States. The major use for oil is to produce gasoline and diesel fuel for transportation, which would not be affected by increasing the use of nuclear power to produce electricity.

Proponents say we should increase the use of nuclear power because its use does not release the greenhouse gas carbon dioxide into the atmosphere. It is true that nuclear power plants do not release carbon dioxide. However, the nuclear fuel cycle does release this gas into the atmosphere, although emissions are less per unit of energy than burning fossil fuels (Figure 17-16).

Because of multiple built-in safety features, the risk of exposure to radioactivity from nuclear power plants in the United States and most other developed countries is extremely low. However, a partial or complete meltdown or explosion is possible, as the Chernobyl and Three Mile Island accidents have taught us.

The U.S. Nuclear Regulatory Commission (NRC) estimates there is a 15–45% chance of a complete core meltdown at a U.S. reactor during the next 20 years. The NRC also found that 39 U.S. reactors have an 80% chance of failure in the containment shell from a meltdown or an explosion of gases inside containment structures.

Throughout the world, nuclear scientists and government officials urge the shutdown of 35 poorly de-

Trade-Offs

Conventional Nuclear Fuel Cycle

Advantages	Disadvantages
Large fuel supply	High cost even with large subsidies
Low environmental impact (without accidents)	Low net energy yield
Emits 1/6 as much CO_2 as coal	High environmental impact (with major accidents)
Moderate land disruption and water pollution (without accidents)	Catastrophic accidents can happen (Chernobyl)
Moderate land use	No widely acceptable solution for long-term storage of radioactive wastes and decommissioning worn-out plants
Low risk of accidents because of multiple safety systems (except in 35 poorly designed and run reactors in former Soviet Union and eastern Europe)	Subject to terrorist attacks
	Spreads knowledge and technology for building nuclear weapons

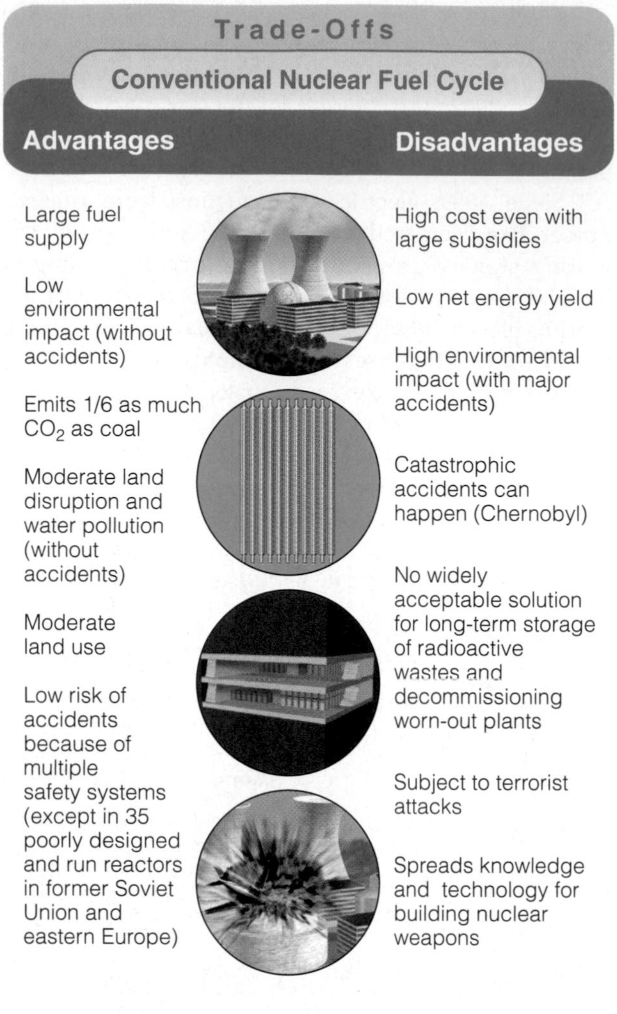

Figure 17-26 Trade-offs: advantages and disadvantages of using the conventional nuclear fuel cycle (Figure 17-24) to produce electricity. Pick the single advantage and disadvantage that you think are the most important.

signed and poorly operated nuclear reactors in some republics of the former Soviet Union and in eastern Europe. This is unlikely without economic aid from developed countries.

In the United States, there is widespread public distrust of the ability of the NRC and the Department of Energy (DOE) to enforce nuclear safety in commercial (NRC) and military (DOE) nuclear facilities. In 1996, George Galatis, a respected senior nuclear engineer, said, "I believe in nuclear power but after seeing the NRC in action I'm convinced a serious accident is not just likely, but inevitable. . . . They're asleep at the wheel."

Concerns about the safety of some U.S. nuclear power plants grew in 2002 when inspectors found that leaking boric acid had eaten a softball-size hole through nearly the entire reactor lid at a nuclear plant near Toledo, Ohio. The only thing preventing a rupture of the high-pressure reactor vessel and a possible

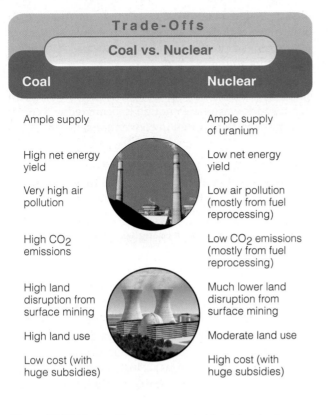

Trade-Offs
Coal vs. Nuclear

Coal	Nuclear
Ample supply	Ample supply of uranium
High net energy yield	Low net energy yield
Very high air pollution	Low air pollution (mostly from fuel reprocessing)
High CO_2 emissions	Low CO_2 emissions (mostly from fuel reprocessing)
High land disruption from surface mining	Much lower land disruption from surface mining
High land use	Moderate land use
Low cost (with huge subsidies)	High cost (with huge subsidies)

Figure 17-27 Trade-offs: comparison of the risks of using nuclear power (based on the nuclear fuel cycle) and coal-burning plants to produce electricity. If you had to choose, would you rather live next door to a coal-fired power plant or a nuclear power plant?

meltdown was a 1-centimeter (0.44-inch) thick stainless steel liner.

How Vulnerable Are U.S. Nuclear Power Plants to Terrorist Attack? A Serious Concern

There is great concern about the vulnerability of U.S. nuclear power plants and the nuclear wastes they store to terrorist attack.

The 2001 destruction of New York City's World Trade Center towers raised fears that a similar attack could break open a reactor's containment shell and set off a reactor meltdown that could create a major radioactive disaster.

Nuclear officials say such concerns are overblown and that U.S. nuclear plants could survive such an attack because of the thickness and strength of the containment walls. But a 2002 study by the Nuclear Control Institute found that the plants were not designed to withstand the crash of a large jet traveling at the impact speed of the two hijacked airliners that hit the World Trade Center.

This is not surprising because in 1982 the U.S. Nuclear Regulatory Commission ruled that owners of nuclear power plants did not have to design the plants to survive threats such as suicidal airliner crashes. According to the NRC, requiring such construction would make nuclear electricity too expensive to be competitive.

An even greater concern is insufficient security at U.S. nuclear power plants against ground-level attacks by terrorists. During a series of ground-based security exercises by the NRC between 1991 and 2001, mock attackers were able to simulate the destruction of enough equipment to cause a meltdown of nearly half of U.S. nuclear plants. And according to a 2002 study by the nonprofit Project on Government Oversight (POGO), these tests did not realistically represent a terrorist attack scenario. This study also found that that many security guards at nuclear power plants have low morale and are overworked, underpaid, undertrained, and not equipped with sufficient firepower to repel a serious ground attack by terrorists.

The NRC contends that the security weaknesses revealed by earlier mock tests have been corrected. But many analysts are unconvinced and note that since September 2001 the NRC has stopped staging such tests.

According to critics, the problem is that the NRC is reluctant to require utilities to significantly upgrade plant security because this would increase the costs of nuclear power and make it less competitive in the marketplace.

How Safe Is High-Level Radioactive Waste Stored at U.S. Nuclear Power Plants? Vulnerable to Terrorists

Spent fuel rods stored underwater in pools or in dry casks outside of the containment shells at nuclear plants are vulnerable to attack by terrorists.

Most high-level radioactive wastes are spent fuel rods. A spent-fuel storage pool typically holds five to ten times more long-lived radioactivity than the radioactive core inside a plant's reactor.

Suppose that water drains out of a spent-fuel pool or a dry storage cask ruptures because of unlikely but possible events such as earthquake, airplane impact, or terrorist act. Then, according to NRC studies, the highly radioactive and thermally hot fuel would be exposed to air and steam. This would cause the zirconium outer cover of the fuel assemblies to catch fire and burn fiercely.

The NRC acknowledges that such a fire could not be extinguished and would burn for days. This would release significant amounts of radioactive materials into the atmosphere, contaminate large areas for many decades, and create economic and psychological havoc.

Unlike the reactor core with its thick concrete protective dome, spent-fuel pools have little protective cover. The pools have backup cooling systems to help prevent a fire, but these could malfunction or be destroyed by a terrorist attack or a deliberate crash by a small airplane. For example, studies in 2002 by the Institute for Resource and Security Studies and the Federation of American Scientists estimated that release of all radioactive material in the spent-fuel rods in the storage pool at the Millstone Unit 3 reactor in Connecticut because of an accident or terrorist attack would put five times more radioactive material into the atmosphere than the 1986 Chernobyl accident. And an area larger than New York state would be uninhabitable for at least 30 years because of radioactive contamination.

According to these studies, about 161 million people—57% of the U.S. population—live within 121 kilometers (75 miles) of an aboveground spent-fuel site. There are 127 such sites in 44 states, mostly in the eastern half of the country (Figure 17-25).

U.S. nuclear power officials consider such events to be highly unlikely worst-case scenarios and question some of the estimates. They also contend that nuclear power facilities are safe from attack. Critics are not convinced and call for constructing much more secure structures to protect spent-fuel storage sites. They accuse the NRC of failure to require this because it would impose additional costs on utility companies, raise the cost of nuclear power, and make it a less attractive energy alternative.

What Do We Do with Low-Level Radioactive Waste? Dump It in the Ocean, Bury It in Special Landfills, or Mix It with Ordinary Trash

The nuclear fuel cycle and other nuclear facility processes produce low-level radioactive wastes that must be stored safely for 100–500 years before they decay to safe levels.

Each part of the nuclear fuel cycle (Figure 17-24) produces low-level and high-level solid, liquid, and gaseous radioactive wastes with various half-lives (Table 3-1, p. 49). Wastes classified as *low-level radioactive wastes* give off small amounts of ionizing radiation and must be stored safely for 100–500 years before decaying to safe levels. Such wastes include tools, building materials, clothing, glassware, and other items that have been contaminated by radioactivity.

From the 1940s to 1970, most low-level radioactive waste produced in the United States and most other countries was put into steel drums and dumped into the ocean; the United Kingdom and Pakistan still dispose of them this way.

Today, low-level waste materials from commercial nuclear power plants, hospitals, universities, industries, and other producers in the United States are put in steel drums and shipped to the two regional land-

fills run by federal and state governments. Attempts to build new regional dumps for low-level radioactive waste using improved technology have met with fierce public opposition.

To lower costs, nuclear industry and utility officials have been lobbying Congress and the NRC to declare such waste safe enough to be mixed with ordinary trash and deposited in conventional landfills.

What Should We Do with High-Level Radioactive Waste? A Dangerous and Long-Lasting Unintended Consequence

There is disagreement among scientists over methods for the long-term storage of high-level radioactive waste.

After more than 50 years of research, scientists still do not agree on whether there is a safe method for storing high-level radioactive waste. Some believe the long-term safe storage or disposal of high-level radioactive wastes is technically possible. Others disagree, pointing out that it is impossible to demonstrate that any method will work for 10,000–240,000 years.

Here are some of the proposed methods and their possible drawbacks.

Bury it deep underground. This favored strategy is under study by all countries producing nuclear waste. In 2001, the U.S. National Academy of Sciences concluded that the geologic repository option is the only scientifically credible long-term solution for safely isolating such wastes. However, according to an earlier 1990 report by the U.S. National Academy of Sciences, "Use of geological information to pretend to be able to make very accurate predictions of long-term site behavior is scientifically unsound."

Shoot it into space or into the sun. Costs would be very high, and a launch accident—like the explosion of the space shuttle *Challenger*—could disperse high-level radioactive wastes over large areas of the earth's surface. This strategy has been abandoned for now.

Bury it under the Antarctic ice sheet or the Greenland ice cap. The long-term stability of the ice sheets is not known. They could be destabilized by heat from the wastes, and retrieving the wastes would be difficult or impossible if the method failed. This strategy is prohibited by international law.

Dump it into descending subduction zones in the deep ocean (Figure 16-5, middle, p. 336). But wastes eventually might be spewed out somewhere else by volcanic activity, and containers might leak and contaminate the ocean before being carried downward. Also, retrieval would be impossible if the method did not work. This strategy is prohibited by international law.

Bury it in thick deposits of mud on the deep-ocean floor in areas that tests show have been geologically stable for 65 million years. The waste containers eventually would corrode and release their radioactive contents. This approach is prohibited by international law.

Change it into harmless, or less harmful, isotopes. Currently no way exists to do this. Scientists are investigating the use of a linear accelerator to speed up the normal rates of radioactive decay. But even if this or other methods are developed, costs would probably be very high, and the resulting toxic materials and low-level (but very long-lived) radioactive wastes would still need to be disposed of safely.

Case Study: The Yucca Mountain Storage Site for High-Level Radioactive Wastes— Controversy over Desert Burial

Scientists disagree over the decision to store high-level nuclear wastes at an underground storage site in Nevada.

In 1985, the U.S. Department of Energy (DOE) announced plans to build a repository for underground storage of high-level radioactive wastes from commercial nuclear reactors and some nuclear weapons facilities. The site is to be built on federal land in the Yucca Mountain desert region, 160 kilometers (100 miles) northwest of Las Vegas, Nevada (Figure 17-25).

The proposed facility (Figure 17-28) is expected to cost at least $58 billion to build (financed partly by a tax on nuclear power). It is scheduled to open by 2010 and begin taking in high-level radioactive waste now stored at 127 sites in 44 states. But officials concede that it is not likely to open until 2015. After the site is filled with waste it is supposed to be monitored for 300 years and then sealed

The wastes are to be buried in tunnels deep below the surface of the almost 1,500-meter- (5,000-foot-) high mountain and well above the current water table. They will be inside containers made of a special metal alloy designed to withstand the high temperatures of the radioactive waste and covered with a shield to protect the metal from corrosion by dripping water.

Currently, the area gets only 15 centimeters (6 inches) of rainfall per year and most of this evaporates in the desert heat before it can seep underground. But no one knows whether the climate of this area will get wetter over the next 10,000 to 240,000 years.

A number of scientists and energy analysts have serious concerns about the safety of this site. For one, they are concerned that rock fractures and tiny cracks may allow water to leak into the site and eventually corrode casks holding radioactive waste. DOE computer models said that water would not flow into the site, but a scientist found evidence that at one time water had flowed deep into the mountain through tiny cracks in a matter of decades. In 1998, Jerry Szymanski, formerly the DOE's top geologist at Yucca Mountain and now an outspoken opponent of the site, said that if water flooded the site it could cause an explosion so large that "Chernobyl would be small potatoes."

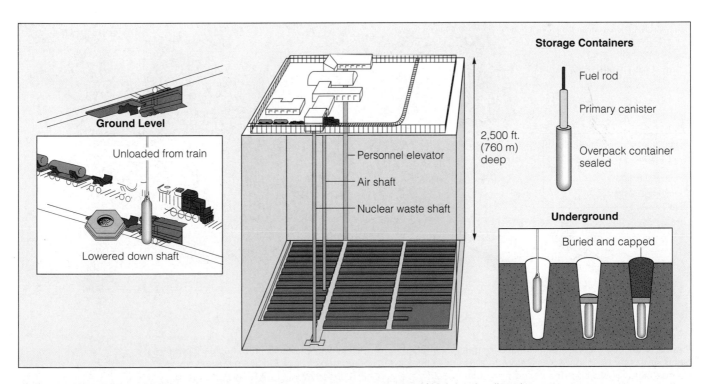

Figure 17-28 Solutions: general design for deep underground permanent storage of high-level radioactive wastes from commercial nuclear power plants in the United States. (U.S. Department of Energy)

In 2004, physicist Paul Craig resigned from a federal panel of experts evaluating the Yucca Mountain project so he could speak more freely about the project. He said that the metal canisters used to store the waste and their protective drip shields are badly designed and that they "would corrode and that would eventually lead to leakage of nuclear waste."

In addition, geologists point out a nearby active volcano and 32 active earthquake fault lines running through the site—an unusually high number. Other scientists claim that data show that the site should be safe from water, earthquakes, and volcanic eruptions.

Despite such concerns, in January 2002 the U.S. energy secretary found that the site is scientifically sound and recommended to President Bush and Congress that highly radioactive waste from the nation's nuclear power plants and some nuclear weapons sites be deposited under Yucca Mountain. The secretary cited this as an important way to help protect wastes now stored at nuclear plants from possible terrorist attack.

This decision raised a storm of protest from Nevada's elected officials and citizens (80% of them opposed to the site) and others concerned about safety. The governor of Nevada charged that the DOE lowered its scientific standards for evaluating the site's geologic integrity. Opponents charge that politics, not sound geology, played a major role in the decision.

Opponents also contend that the Yucca Mountain waste site should not be opened because *it can decrease national security.* One reason is that it would require at least 19,000 shipments of wastes over much of the country (Figure 17-29)—an average of a shipment each day for the estimated 38 years before the site is filled. These wastes would be put into specially designed casks and shipped in trucks and rail cars. Critics contend that it is much more difficult to protect such a large number of shipments from terrorist attack than to provide more secure ways to store such wastes at nuclear power plant sites.

Also shipping nuclear wastes to the Yucca Mountain site would not decrease the possibilities of sabotage of wastes stored at the country's nuclear plant sites in pools and casks (unless their security is significantly upgraded) because the plants will be producing new wastes about as fast as the old wastes are shipped out. By 2036 to 2041, when the Yucca Mountain site may be filled, there will be about as much nuclear waste stored at nuclear plant sites as there is today.

The DOE and proponents of nuclear power say the risks of an accident or sabotage of waste shipments are negligible. They point out that the shipments are packed in thick metal casks and protected by armed guards in urban areas. Opponents believe the risks are underestimated, especially after the events of September 11, 2001. They also call for armed guards through-

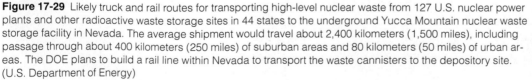

★ Nuclear power plants
△ Yucca Mountain
— Railroads
— Highways

Figure 17-29 Likely truck and rail routes for transporting high-level nuclear waste from 127 U.S. nuclear power plants and other radioactive waste storage sites in 44 states to the underground Yucca Mountain nuclear waste storage facility in Nevada. The average shipment would travel about 2,400 kilometers (1,500 miles), including passage through about 400 kilometers (250 miles) of suburban areas and 80 kilometers (50 miles) of urban areas. The DOE plans to build a rail line within Nevada to transport the waste cannisters to the depository site. (U.S. Department of Energy)

out the entire rail or truck trip instead of only in urban areas.

For example, a terrorist hidden along a busy highway or railway or atop an urban building could use a shoulder-mounted missile launcher to fire one or more antitank missiles that could penetrate the thick walls of a shipping cask on a truck or train car (Figure 17-29) and release radioactive materials.

There is disagreement over the possible effects of such an event. Some analysts say it would spread radioactive particles over no more than a 1.6-kilometer (1-mile) radius and that the area and the people affected could be decontaminated by being hosed down. Other analysts say the affected area could be five to thirty times that estimate. And if such an event occurred in an urban area, the spreading radioactivity could cause 300–18,000 fatal cancers, and result in at least $10 billion in damages. According to a 2002 DOE study, in a worst-case scenario such an urban attack could release enough radioactivity to expose 96,000 people and cause 48 fatal cancers.

In 2002, the U.S. National Academy of Sciences, in collaboration with Harvard and University of Tokyo scientists, urged the U.S. government to slow down and rethink the nuclear waste storage process. They contend that storing spent-fuel rods in dry-storage casks in well-protected buildings at nuclear plant sites is an adequate solution for at least 100 years in terms of safety and national security. This would buy time to carry out more research on this complex problem and to evaluate other sites and storage methods that might be more acceptable scientifically and politically.

Despite these suggestions and many objections from scientists and citizens, during the summer of 2002 Congress approved Yucca Mountain as the official site for storing the country's commercial nuclear wastes. Opponents want the law repealed. Meanwhile, Nevada is still fighting the project in the courts. This story illustrates how science, politics, and economics can interact as people attempt to solve a difficult and controversial problem.

X *How Would You Vote?* Should highly radioactive spent fuel be stored in well-protected buildings at nuclear power plant sites instead of shipping them to a single site for underground burial? Cast your vote online at http://biology .brookscole.com/miller14.

What Can We Do with Worn-out Nuclear Plants? A Costly Dilemma

When a nuclear reactor reaches the end of its useful life we have to keep its highly radioactive materials from reaching the environment for thousands of years.

When a nuclear plant comes to the end of its useful life, it must be decommissioned. Scientists have proposed three ways to do this.

One is to dismantle the plant and store its large volume of highly radioactive materials in a high-level nuclear waste storage facility (Figure 17-28), whose safety is questioned by a number of scientists.

A second approach is to put up a physical barrier around the plant and set up full-time security for 30–100 years before the plant is dismantled. This allows time for some of the radioactive material to decay to levels that make dismantlement safer. A third option is to enclose the entire plant in a tomb that must last and be monitored for several thousand years.

Regardless of the method chosen, decommissioning adds to the total costs of nuclear power as an energy option. So far, only a few plants have been torn down. But doing this cost two to ten times as much as it did to build them. The total estimated costs for decommissioning the 103 reactors now in operation in the United States range from $200 billion to $1 trillion. This further decreases the net energy yield of nuclear power and adds to its already high cost.

At least 228 large commercial reactors worldwide (20 in the United States) are scheduled for retirement by 2012. However, the Nuclear Regulatory Commission has approved extending the life of at least 40 reactors to 60 years. Opponents contend that this could increase the risk of nuclear accidents in aging reactors. In 2003, congressional auditors reported that the owners of almost half the nuclear power reactors in the United States are not setting aside enough money to decommission them when they are retired, which will saddle taxpayers with the bill.

What Are "Dirty" Radioactive Bombs? A Serious Threat

Terrorists could wrap conventional explosives around small amounts of various radioactive materials that are fairly easy to get, detonate such bombs, and contaminate an area with radioactivity for decades.

Since the terrorist attacks in the United States on September 11, 2001, there has been growing concern about threats from explosions of so-called *dirty bombs*. Such a bomb consists of an explosive such as dynamite mixed with or wrapped around some form of radioactive material—an amount that could fit in a coffee cup.

Radioactive materials can be stolen from thousands of poorly guarded and difficult-to-protect sources or bought on the black market. Sources might be hospitals that use radioisotopes (such as cobalt-60) to treat cancer, diagnose diseases, and sterilize some types of medical equipment. Another source could be university research labs. Some industries also use radioisotopes to detect leaks in underground pipes, irradiate food, examine mail and other materials, and detect flaws in pipe welds and boilers. Radioactive materials such as americium-241 are also found in smoke detectors.

Detonating a dirty bomb at street level or on a rooftop does not cause a nuclear blast. But such an explosion and subsequent cancers in a densely populated city could kill a dozen to 1,000 people, spread radioactive material over several to hundreds of blocks, and contaminate buildings and soil in the affected area for up to 10 times the half-life of the isotope used. Cleaning up such an area would cost billions of dollars. In addition, detonating a dirty bomb would cause intense psychological terror and panic throughout much of a country. As a result, terrorists would succeed in their primary objective.

Since 1986, the NRC has recorded 1,700 incidents in the United States in which radioactive materials used by industrial, medical, or research facilities have been stolen or lost. And since 1991, the International Atomic Energy Agency (IAEA) has detected 671 incidents of illicit trafficking in dirty-bomb materials.

Can We Afford Nuclear Power? Burning Money

Even with massive government subsidies, the nuclear power fuel cycle is an expensive way to produce electricity compared to a number of other energy alternatives.

Experience has shown that the nuclear power fuel cycle is an expensive way to produce electricity, even when huge government subsidies partially shield it from free-market competition with other energy sources.

In the United States, costs rose dramatically in the 1970s and 1980s because of unanticipated safety problems and stricter regulations after the Three Mile Island and Chernobyl accidents. In 1995, the World Bank said that nuclear power is too costly and risky. *Forbes* business magazine has called the failure of the U.S. nuclear power program "the largest managerial disaster in U.S. business history, involving $1 trillion in wasted investment and $10 billion in direct losses to stockholders." And the *Economist* says, "Not one [nuclear power plant], anywhere in the world, makes commercial sense."

In recent years, the operating costs of many U.S. nuclear power plants have dropped, mostly because of less downtime. But environmentalists and economists point out that the true cost of nuclear power must be based on the entire nuclear power fuel cycle, not merely the operating cost of individual plants. According to them, when these costs are included the overall cost of nuclear power is very high (even with huge government subsidies) compared to many other energy alternatives.

Partly to address cost concerns, the U.S. nuclear industry hopes to persuade Congress and utility companies to build hundreds of smaller second-generation plants using standardized designs, which they claim are safer and can be built more quickly (in 3–6 years).

These *advanced light-water reactors (ALWRs)* have built-in *passive safety features* designed to make explosions or the release of radioactive emissions almost impossible. However, according to *Nucleonics Week*, an important nuclear industry publication, "Experts are flatly unconvinced that safety has been achieved—or even substantially increased—by the new designs." In addition, these new designs do not eliminate the threats and the expense and hazards of long-term radioactive waste storage and power plant decommissioning.

Each new plant will cost up to $2 billion. Nuclear power proponents want Congress to provide the industry with up to $350 million in taxpayer subsidies between 2004 and 2009 for new advanced reactor start-up costs.

Is Breeder Nuclear Fission a Feasible Alternative? A Failed Technology

Because of very high costs and bad safety experiences with several nuclear breeder reactors, this technology has essentially been abandoned.

Some nuclear power proponents urge the development and widespread use of **breeder nuclear fission reactors,** which generate more nuclear fuel than they consume by converting nonfissionable uranium-238 into fissionable plutonium-239. Because breeders would use more than 99% of the uranium in ore deposits, the world's known uranium reserves would last at least 1,000 years, and perhaps several thousand years.

However, if the safety system of a breeder reactor fails, the reactor could lose some of its liquid sodium coolant, which ignites when exposed to air and reacts explosively if it comes into contact with water. This could cause a runaway fission chain reaction and perhaps a nuclear explosion powerful enough to blast open the containment building and release a cloud of highly radioactive gases and particulate matter. Leaks of flammable liquid sodium can also cause fires, which have happened with all experimental breeder reactors built so far.

In addition, existing experimental breeder reactors produce plutonium so slowly that it would take 100–200 years for them to produce enough to fuel a significant number of other breeder reactors. In 1994, the United States ended government-supported research for breeder technology after providing about $9 billion in research and development funding.

In December 1986, France opened a commercial-size breeder reactor. It was so expensive to build and operate that after spending $13 billion, the government spent another $2.75 billion to shut it down permanently in 1998. Because of this experience, other countries have abandoned their plans to build full-size commercial breeder reactors.

Is Nuclear Fusion a Feasible Alternative? A Costly 50-Year Dream Still at the Laboratory Stage

Nuclear fusion has a number of advantages, but after more than five decades of research and billions of dollars in government research and development subsidies, this technology is still at the laboratory stage.

For decades, scientists have hoped that controlled nuclear fusion will provide an almost limitless source of high-temperature heat and electricity to supply most of the world's commercial energy. Research has focused on the D–T nuclear fusion reaction, in which two isotopes of hydrogen—deuterium (D) and tritium (T)—fuse at about 100 million °C (180 million °F; Figure 3-16, p. 50).

According to a 2001 Department of Energy task force, fusion energy has a number of important advantages. They include no emissions of conventional air pollutants or carbon dioxide, an almost infinite fuel supply (water), and wastes that are much less radioactive so they would need to be stored for only about 100 years.

There would be no risk of meltdown or release of large amounts of radioactive materials from a terrorist attack and little risk from additional proliferation of nuclear weapons because bomb-grade materials (such as enriched uranium-235 and plutonium-239) are not required for fusion energy.

Fusion power might also be used to destroy toxic wastes, supply electricity for ordinary use, and decompose water and produce the hydrogen gas needed to run a hydrogen economy by the end of this century.

This sounds great. So what is holding up fusion energy? After more than 50 years of research and expenditures of more than $25 billion of mostly government funds in the United States, controlled nuclear fusion is still in the laboratory stage. None of the approaches tested so far have produced more energy than they use.

If researchers can eventually get more energy out of nuclear fusion than they put in, the next step would be to build a small fusion reactor and then scale it up to commercial size. This is an extremely difficult engineering problem. Also, the estimated cost of building and operating a commercial fusion reactor (even with huge government subsidies) is several times that of a comparable conventional fission reactor.

Proponents contend that with greatly increased federal funding, a commercial nuclear fusion power plant might be built by 2030 or perhaps by 2020 with emphasis on developing a new technique called muon-catalyzed fusion. However, many experts do not expect nuclear fusion to be a significant energy source until 2100, if then.

What Should Be the Future of Nuclear Power in the United States? Phase Out or Keep Options Open

There is disagreement over whether the United States should phase out nuclear power or keep this option open in case other alternatives do not pan out.

Since 1948, nuclear energy (fission and fusion) has received about 58% of all federal energy research and development funds in the United States—compared to 22% for fossil fuels, 11% for renewable energy, and 8% for energy efficiency and conservation. Because the results of such a huge investment of taxpayer dollars in nuclear power have been disappointing, some analysts call for phasing out all or most government subsidies and tax breaks for nuclear power and using the money to subsidize and accelerate the development of other, more promising energy technologies.

To these analysts, nuclear power is a complex, expensive, inflexible, and centralized way to produce electricity that is too vulnerable to terrorist attack. They believe it is a technology whose time has passed in a world where electricity will increasingly be provided by small, decentralized, easily expandable power plants such as natural gas turbines, farms of wind turbines on land and at sea, arrays of solar cells, and hydrogen-powered fuel cells. According to investors and World Bank economic analysts, conventional nuclear power simply cannot compete in today's increasingly open, decentralized, and unregulated energy market unless it is artificially shielded from free-market competition by huge government subsidies.

Proponents of nuclear power argue that governments should continue funding research and development and pilot plant testing of smaller and potentially safer and cheaper reactor designs along with breeder fission and nuclear fusion. They say we need to keep nuclear options available for use in the future if various renewable energy options fail to keep up with electricity demands and reduce CO_2 emissions to acceptable levels. Germany does not buy these arguments and has plans to phase out nuclear power over the next two decades.

X *HOW WOULD YOU VOTE?* Should nuclear power be phased out in the country where you live over the next 20 to 30 years? Cast your vote online at http://biology.brookscole.com/miller14.

Civilization as we know it will not survive unless we can find a way to live without fossil fuels.

DAVID GOLDSTEIN

CRITICAL THINKING

1. Just to continue using oil at the current rate (not the projected higher exponential rate), we must discover and add to global oil reserves the equivalent of a new Saudi Arabian supply (the world's largest) *every 10 years.* Do you believe this is possible? If not, what effects might this have on your life and on the life of a child or grandchild you might have?

2. List five actions you can take to reduce your dependence on oil and gasoline derived from it. Which do you actually plan to do?

3. Explain why you are for or against continuing to increase oil imports in the United States or in the country where you live. If you favor reducing dependence on oil imports, list the three best ways to do this.

4. Explain why you agree or disagree with the following proposals by various energy analysts to help solve U.S. energy problems: **(a)** find and develop more domestic supplies of oil, **(b)** place a heavy federal tax on gasoline and imported oil to help reduce the waste of oil resources, **(c)** increase dependence on nuclear power, and **(d)** phase out all nuclear power plants by 2025.

5. What do you believe should be done with high-level radioactive wastes? Explain.

6. Would you favor having high-level nuclear waste transported by truck or train through the area where you live to a centralized underground storage site? Explain. What are the options?

7. Explain why you agree or disagree with each of the following proposals made by the U.S. nuclear power industry: **(a)** provide at least $100 billion in government subsidies to build a large number of better-designed nuclear fission power plants to reduce dependence on imported oil and slow global warming, **(b)** prevent the public from participating in hearings on licensing new nuclear power plants and on safety issues at the nation's nuclear reactors, **(c)** restore government subsidies to develop a breeder nuclear fission reactor program, and **(d)** greatly increase federal subsidies for developing nuclear fusion.

8. Should the United States and other developed countries provide economic and technical aid for closing 35 poorly designed and poorly operated nuclear reactors in some republics of the former Soviet Union and in eastern Europe? Explain.

9. Congratulations! You are in charge of the world. List the three most important features of your policy to develop nonrenewable energy resources during the next 50 years.

PROJECTS

1. How is the electricity in your community produced? How has the inflation-adjusted cost of that electricity changed since 1970?

2. Write a two-page scenario of what your life might be like without oil. Compare and discuss the scenarios developed by members of your class.

3. Use the library or the Internet to find information about the accident that took place at the Three Mile Island (TMI) nuclear power plant near Harrisburg, Pennsylvania, in 1979. According to the nuclear power industry, the TMI accident showed that its safety systems work because the accident caused no known deaths. Other analysts argue that the accident was a wake-up call about the potential dangers of nuclear power plants that led to tighter and better safety regulations. Use the information you find to determine which of these positions you support, and defend your choice.

4. Use the library or the Internet to find bibliographic information about *Maurice Strong* and *David Goldstein,* whose quotes appear at the beginning and end of this chapter.

5. Make a concept map of this chapter's major ideas, using the section heads, subheads, and key terms (in boldface). Look on the website for this book for information about making concept maps.

LEARNING ONLINE

The website for this book contains study aids and many ideas for further reading and research. They include a chapter summary, review questions for the entire chapter, flash cards for key terms and concepts, a multiple-choice practice quiz, interesting Internet sites, references, and a guide for accessing thousands of InfoTrac® College Edition articles. Log on to

http://biology.brookscole.com/miller14

Then click on the Chapter-by-Chapter area, choose Chapter 17, and select a learning resource.

18 Energy Efficiency and Renewable Energy

Energy

The Coming Energy-Efficiency and Renewable-Energy Revolution

Energy analyst Amory Lovins built a large, solar heated, superinsulated, partially earth-sheltered home and office in Snowmass, Colorado (Figure 18-1), with severely cold winter temperatures.

This structure also houses the research center for the Rocky Mountain Institute (cofounded in 1982 by Amory and Hunter Lovins), an office used by 45 people. This office–home gets 99% of its space and water heating and 95% of its daytime lighting from the sun, and uses one-tenth the usual amount of electricity for a structure of its size.

With today's superinsulating windows a house can have many windows without much heat loss in cold weather or heat gain in hot weather. Thinner insulation now being developed will allow roofs and walls to be insulated far better than in today's best superinsulated houses.

A small but growing number of people in developed and developing countries get their electricity from *solar cells* that convert sunlight directly into electricity. They can be attached like shingles to a roof, used as roofing, or applied to window glass as a coating. Solar-cell prices are high but falling.

According to many scientists and executives of oil and automobile companies, we are in the beginning stages of a *hydrogen revolution* to be phased in during this century as the Age of Oil (Appendix 5) begins winding down (Figure 17-6, p. 353). Because there is little hydrogen gas (H_2) around, we have to use another energy resource to produce it from water or various organic compounds such as methane. We could do this by passing electricity produced by renewable energy from wind turbines, hydroelectric power plants, solar cells, biomass, and geothermal energy from the earth's interior through water to make H_2 gas. Energy-efficient *fuel cells* could use the hydrogen to produce electricity to run cars and appliances, heat water, and heat and cool buildings.

Burning hydrogen in a fuel cell by combining it with oxygen produces water vapor and no carbon dioxide. Thus shifting to hydrogen as our primary source of energy would eliminate most air pollution and also greatly slow global warming—as long as the hydrogen is produced from water and not carbon-containing fossil fuels and the nuclear fuel cycle that emit the greenhouse gas CO_2 into the atmosphere.

Figure 18-1 The Rocky Mountain Institute in Colorado. This facility is a home and a center for the study of energy efficiency and sustainable use of energy and other resources. It is also an example of energy-efficient passive solar design.

Robert Millman/Rocky Mountain Institute

If the United States wants to save a lot of oil and money and increase national security, there are two simple ways to do it: Stop driving Petropigs and stop living in energy sieves.

AMORY B. LOVINS

This chapter evaluates the use of energy efficiency and renewable energy as energy alternatives. It addresses the following questions:

- How can we improve energy efficiency and what are the advantages of doing so?

- What are the advantages and disadvantages of using solar energy to heat buildings and water and produce electricity?

- What are the advantages and disadvantages of using flowing water to produce electricity?

- What are the advantages and disadvantages of using wind to produce electricity?

- What are the advantages and disadvantages of burning plant material (biomass) to heat buildings and water, produce electricity, and propel vehicles (biofuels)?

- What are the advantages and disadvantages of extracting heat from the earth's interior (geothermal energy) and using it to heat buildings and water and to produce electricity?

- What are the advantages and disadvantages of producing hydrogen gas and burning it to produce electricity, heat buildings and water, and propel vehicles?

- What are the advantages and disadvantages of using smaller, decentralized micropower sources to heat buildings and water, produce electricity, and propel vehicles?

- How can we make a transition to a more sustainable energy future?

18-1 THE IMPORTANCE OF IMPROVING ENERGY EFFICIENCY

What Is Energy Efficiency and How Much Energy Do We Waste? Saving Money by Not Wasting Energy

Using less energy to do useful work reduces the environmental impact of energy use and saves money.

Energy efficiency is a measure of the useful energy produced by an energy conversion device compared to the energy that ends up being converted to low-quality, essentially useless heat. For example, the light produced by a light bulb is useful energy, while the heat it produces is wasted energy.

If you replace an incandescent bulb that is only 5% efficient with a compact fluorescent bulb that is 20% efficient, you get the same amount of light using one-fourth as much energy. This reduces pollution and carbon dioxide emissions and saves money on your electric bill—a win-win solution for you and the earth. Figure 18-2 lists major economic and environmental advantages of reducing energy waste.

Some critics like to paint proponents of conserving energy as calling for personal sacrifice, giving up cars, freezing in winter, wearing sweaters, and burning up in the summer. This is an incorrect and misleading view of energy conservation, which is implemented mainly by using existing technologies and developing new ones that waste less energy.

You may be surprised to learn that about 84% of all commercial energy used in the United States is wasted (Figure 18-3). About 41% of the energy used is wasted automatically because of the degradation of energy quality imposed by the second law of thermodynamics. But about 43% of the energy used in the United States is wasted unnecessarily, mostly by using fuel-wasting motor vehicles, furnaces, and other devices and living and working in leaky, poorly insulated, poorly designed buildings. See the Guest Essay on this topic by Amory Lovins on the website for this chapter. The U.S. Department of Energy (DOE) estimates that the United States unnecessarily wastes as much energy as two-thirds of the world's population consumes.

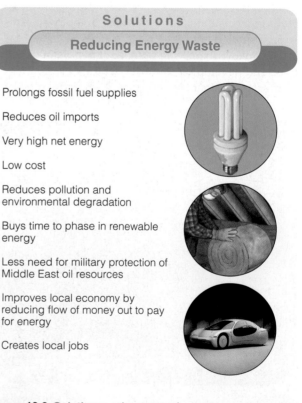

Solutions

Reducing Energy Waste

Prolongs fossil fuel supplies

Reduces oil imports

Very high net energy

Low cost

Reduces pollution and environmental degradation

Buys time to phase in renewable energy

Less need for military protection of Middle East oil resources

Improves local economy by reducing flow of money out to pay for energy

Creates local jobs

Figure 18-2 Solutions: advantages of reducing energy waste. Global improvements in energy efficiency could save the world about $1 trillion per year—an average of $114 million per hour!

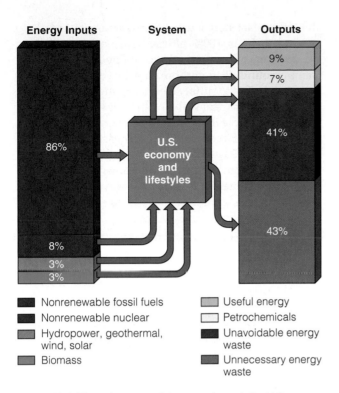

Energy Inputs | **System** | **Outputs**

- ■ Nonrenewable fossil fuels
- ■ Nonrenewable nuclear
- ▨ Hydropower, geothermal, wind, solar
- ▨ Biomass
- ▨ Useful energy
- □ Petrochemicals
- ■ Unavoidable energy waste
- ▨ Unnecessary energy waste

Figure 18-3 Flow of commercial energy through the U.S. economy. Note that only 16% of all commercial energy used in the United States ends up performing useful tasks or being converted to petrochemicals; the rest is unavoidably wasted because of the second law of thermodynamics (41%) or is wasted unnecessarily (43%).

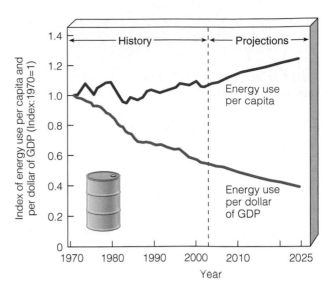

Figure 18-4 *Good news.* Energy use per person needed to produce a dollar of the U.S. gross domestic product, 1970–2003 with projections to 2025 (Index: 1970 = 1). However, energy use per capita is rising. (U. S. Department of Energy)

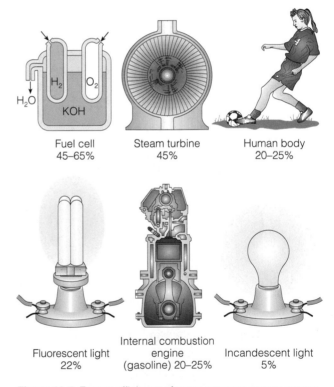

Fuel cell 45–65%
Steam turbine 45%
Human body 20–25%

Fluorescent light 22%
Internal combustion engine (gasoline) 20–25%
Incandescent light 5%

Figure 18-5 Energy efficiency of some common energy conversion devices.

We can save energy and money by buying more energy-efficient cars, lighting, heating systems, water heaters, air conditioners, and appliances. Some energy-efficient models may cost more initially, but in the long run they usually save money by having a lower **life cycle cost:** initial cost plus lifetime operating costs.

Good news. In the United States, the amount of energy used per person to produce a dollar of gross domestic product declined sharply between 1970 and 2003 and is projected to continue dropping through 2025.(Figure 18-4). Such improvements in energy efficiency have cut U.S. energy bills by $275 billion a year.

Bad news. Unnecessary energy waste still costs the United States about $300 billion per year—an average of $570,000 per minute.

The energy conversion devices we use vary in their energy efficiencies (Figure 18-5). Four widely used devices waste large amounts of energy. One is the *incandescent light bulb*, which wastes 95% of its energy input of electricity. In other words, it is a *heat bulb.* The second is a *nuclear power plant* producing electricity for space heating or water heating. Such a plant wastes about 86% of the energy in its nuclear fuel and probably 92% when we include the energy needed to deal with its radioactive wastes for thousands of years and to retire the plant. Third is a motor vehicle with an

internal combustion engine, which wastes 75–80% of the energy in its fuel.

The fourth is a *coal-burning power plant,* in which two-thirds of the energy released by burning coal ends up as waste heat in the environment. Energy experts call for us to replace these four energy-wasting technologies or greatly improve their energy efficiency over the next few decades.

What Is Net Energy Efficiency? Honest Energy Accounting

Net energy efficiency is a measure of how much useful energy we get from an energy resource after subtracting the energy used and wasted in making the energy available.

Recall that the only energy that really counts is *net energy* (p. 354). The *net energy efficiency* of a system used to heat your house, for example, is determined by the efficiency of each step in the energy conversion for the entire system.

Figure 18-6 shows the net energy efficiency for heating two well-insulated homes. One is heated with electricity produced at a nuclear power plant, transported by wire to the home, and converted to heat (electric resistance heating). The other is heated passively: direct solar energy enters through high-efficiency windows facing the sun and strikes heat-absorbing materials that store the heat for slow release.

This analysis shows that converting the high-quality energy in nuclear fuel to high-quality heat at several thousand degrees in the power plant, converting this heat to high-quality electricity, transmitting the electricity to users, and using the electricity to provide low-quality heat for warming a house to only about 20°C (68°F) is very wasteful of high-quality energy. Although the last step of using the incoming electricity to produce heat is 100% efficient, the numerous steps needed to get the electricity to the house waste enormous amounts of energy. Burning coal or any fossil fuel at a power plant to supply electricity and transmitting it long distances to heat water or space is also inefficient.

This example illustrates two general principles for saving energy. First, *keep the number of steps in an energy conversion process as low as possible.* Each time we convert energy from one form to another or transmit it, some useful energy is almost always lost. Second, *strive to have the highest possible energy efficiency for each step in an energy conversion process.*

18-2 WAYS TO IMPROVE ENERGY EFFICIENCY

How Can We Save Energy in Industry? Cogenerate, Buy New Motors, and Use Efficient Lighting

Industries can save energy and money by producing both heat and electricity from an energy source and by using energy-efficient electric motors and lighting.

Some industries save energy and money by using **cogeneration,** or *combined heat and power (CHP)* systems. In such a system two useful forms of energy (such as steam and electricity) are produced from the same fuel

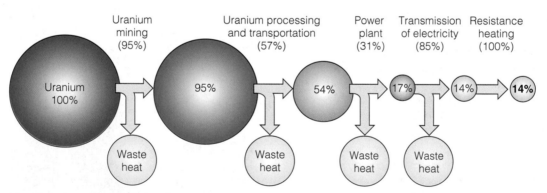

Electricity from Nuclear Power Plant

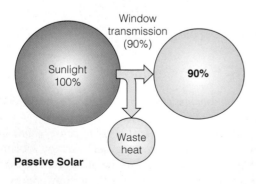

Passive Solar

Figure 18-6 Comparison of net energy efficiency for two types of space heating. The cumulative net efficiency is obtained by multiplying the percentage shown inside the circle before each step by the energy efficiency for that step (shown in parentheses). So 100 × 0.95 = 95%; 95 × 0.57 = 54%; and so on. Because of the second law of thermodynamics, in most cases the greater the number of steps in an energy conversion process, the lower its net energy efficiency. About 86% of the energy used to provide space heating by electricity produced at a nuclear power plant is wasted. If the additional energy needed to deal with nuclear wastes and to retire highly radioactive nuclear plants after their useful life is included, then the net energy yield for a nuclear plant is only about 8% (or 92% waste). By contrast, with passive solar heating, only about 10% of incoming solar energy is wasted.

source. These systems have an energy efficiency of 80–90% (compared to about 30–40% for coal-fired boilers and nuclear power plants) and emit two-thirds less CO_2 per unit of energy produced than conventional coal-fired boilers do.

Cogeneration has been widely used in western Europe for years. Its use in the United States (where it now produces 9% of the country's electricity) and China is growing.

Another way to save energy and money in industry is to *replace energy-wasting electric motors,* which consume about one-fourth of the electricity produced in the United States. Most of these motors are inefficient because they run only at full speed with their output throttled to match the task—somewhat like driving a car fast with your foot on the brake pedal. Each year a heavily used electric motor consumes 10 times its purchase cost in electricity—equivalent to using $200,000 worth of gasoline each year to fuel a $20,000 car! The costs of replacing such motors with new adjustable-speed drive motors would be paid back in about 1 year and save an amount of energy equal to that generated by 150 large (1,000-megawatt) power plants.

A third way to save energy is to *switch from low-efficiency incandescent lighting to higher-efficiency fluorescent lighting.*

How Can We Save Energy in Transportation? Replace Gas Guzzlers with Gas Sippers

The best way to save energy in transportation is to increase the fuel efficiency of motor vehicles.

Good news. Between 1973 and 1985, the average fuel efficiency rose sharply for new cars sold in the United States and to a lesser degree for pickup trucks, minivans, and sport utility vehicles (SUVs) (Figure 18-7). This occurred primarily because of government-mandated *Corporate Average Fuel Economy (CAFE)* standards. *Bad news.* Between 1985 and 2004, the average fuel efficiency for new passenger cars sold in the United States leveled off or declined slightly.

Fuel-efficient cars are available, but account for less than 1% of all car sales. One reason is that the inflation-adjusted price of gasoline today in the United States is low (Figure 18-8 and Connections, p. 384). A second reason is that two-thirds of U.S. consumers prefer SUVs, pickup trucks, minivans, and other large, inefficient vehicles. A third reason is the failure of elected officials to raise CAFE standards since 1985 because of opposition from automakers and oil companies.

Suppose that Congress required the average motor vehicle in the United States to get 17 kilometers per liter (kpl) [40 miles per gallon (mpg)] within 10 years. According to energy analysts, this would cut gasoline consumption in half, save more than three times the amount of oil in the nation's current proven oil re-

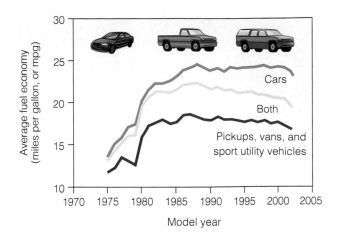

Figure 18-7 Average fuel economy of new vehicles sold in the United States, 1975–2004. The largest and most inefficient vehicles, like the Hummer, are not covered by fuel economy regulations. (U.S. Environmental Protection Agency and National Highway Traffic Safety Administration)

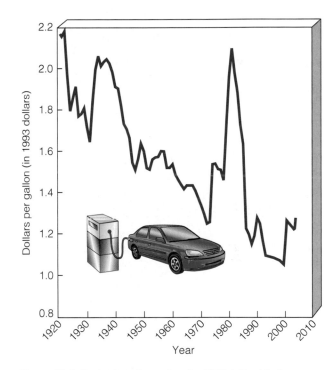

Figure 18-8 Real price of gasoline (in 1993 dollars) in the United States, 1920–2004. The 225 million motor vehicles in the United States use about 40% of the world's gasoline. Gasoline is one of the cheapest items American consumers buy and costs less per liter than bottled water. (U.S. Department of Energy)

serves, and also save more than enough oil to eliminate all current oil imports to the United States from the Middle East. In 2003, China announced plans to impose much stricter fuel-efficiency standards than the United States. The goals are (1) to enourage car companies to develop hybrid, fuel cell, and other fuel-saving vehicles, (2) reduce the country's dependence on oil imports, and (3) reduce carbon dioxide emissions.

The Real Cost of Gasoline in the United States

Economists and environmentalists point out that gasoline costs U.S. consumers much more than it appears. This is because the real cost of gasoline is not paid directly at the pump.

According to a 1998 study by the International Center for Technology Assessment, the hidden costs of gasoline to U.S. consumers is about $1.30–3.70 per liter ($5–14 per gallon), depending on how the costs are estimated. These hidden costs include the following:

- Government subsidies and tax breaks for oil companies and road builders.

- Pollution cleanup.

- Military protection of oil supplies in the Middle East (at least $30 billion a year not including the Iraq War).

- Environmental, health, and social costs such as increased medical bills and insurance premiums, time wasted in traffic jams, noise pollution, increased mortality from air and water pollution, urban sprawl, and harmful effects on wildlife species and habitats.

Economists point out that if these harmful costs were included as taxes in the market price of gasoline, we would have much more energy-efficient and less polluting cars. However, gasoline and car companies benefit financially by being able to pass these hidden costs on to consumers and future generations.

This is basically an education and political problem. Most consumers are unaware that they are paying these harmful costs and do not connect them with gasoline use. Also, politicians running on a platform of raising gasoline prices in the United States 3–11-fold would be committing political suicide.

Critical Thinking

Some economists have suggested that U.S. consumers might be willing to pay much more for gasoline if (1) they understood they are already paying these hidden costs indirectly and (2) the tax revenues from gasoline sales were used to reduce taxes on wages, income, and wealth and provide a safety net for low- and middle-class consumers. Would you support or oppose such a proposal? Explain.

X HOW WOULD YOU VOTE? Should the government greatly increase fuel efficiency standards for all vehicles in the United States or the country where you live? Cast your vote online at http://biology.brookscole.com/miller14.

Are Hybrid-Electric Vehicles the Answer? A New Option

Fuel-efficient hybrid-electric vehicles are powered by a battery and a small internal combustion engine that recharges the battery.

There is rapidly growing interest in developing *super-efficient cars* that could eventually get 34–128 kpl (80–300 mpg). This concept was pioneered and developed in detail in the 1980s by physicist Amory Lovins. See his Guest Essay on the website for this chapter.

One type of energy-efficient car uses a *hybrid-electric internal combustion engine*. It runs on gasoline, diesel fuel, or natural gas and uses a small battery (recharged by the internal combustion engine) to provide the energy needed for acceleration and hill climbing (Figure 18-9).

Toyota introduced its first hybrid vehicle in 1997 and Honda and Nissan have been selling several models of hybrid vehicles in the United States since 2000. Carmakers plan to introduce at least 20 hybrid models, including cars, trucks, SUVs, and vans, in the next 4–5

years. In 2004 Toyota (Lexus) and Ford (its Escape model) began selling hybrid SUVs with the fuel efficiency of a compact car.

Toyota has a strong lead in developing such vehicles but in 2003 General Motors announced it would have the manufacturing capability to build as many as 1 million hybrid cars, trucks, and SUVs by 2006. Sales of hybrid motor vehicles are projected to grow rapidly and probably dominate motor vehicle sales between 2010 and 2030.

Some people buy trucks or SUVs because they believe they are safer than midsize automobiles. But safety studies reveal that SUVs and pickups are more dangerous to people in them and to those in vehicles they may run into than most midsize and large automobiles. And they are no safer than some models of compact cars. The main reason is that SUVs and trucks are taller and heavier than most other vehicles. This makes them more likely to roll over and harder to control in emergency stops.

Are Fuel-Cell Cars the Answer? Possible Star of the Future

Automakers are developing fuel-efficient cars powered by fuel cells running on hydrogen and producing little pollution.

A **Combustion engine:**
Small, efficient internal combustion engine powers vehicle with low emissions.

B **Fuel tank:**
Liquid fuel such as gasoline, diesel, or ethanol runs small combustion engine.

C **Electric motor:**
Traction drive provides additional power, recovers braking energy to recharge battery.

D **Battery bank:**
High-density batteries power electric motor for increased power.

E **Regulator:**
Controls flow of power between electric motor and battery bank.

F **Transmission:**
Efficient 5-speed automatic transmission.

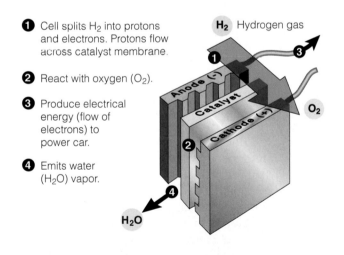

Fuel
Electricity

Figure 18-9 Solutions: general features of a car powered by a *hybrid gas–electric engine*. A small internal combustion engine recharges the batteries, thus reducing the need for heavy banks of batteries and solving the problem of the limited range of conventional electric cars. The bodies of future models will probably be made of lightweight composite plastics that offer more protection in crashes, do not need to be painted, do not rust, can be recycled, and have fewer parts than conventional car bodies. (Concept information from DaimlerChrysler, Ford, Honda, and Toyota)

Another type of superefficient car is an electric vehicle that uses a *fuel cell*—a device that combines hydrogen gas (H_2) and oxygen gas (O_2) fuel to produce electricity and water vapor ($2\ H_2 + O_2 \longrightarrow 2\ H_2O$) (Figure 18-10).

Fuel cells are at least twice as efficient as internal combustion engines, have no moving parts, require little maintenance, and produce little or no pollution depending on how their hydrogen fuel is produced. Most major automobile companies have developed prototype fuel-cell cars. They hope to have a variety of affordable fuel-cell vehicles on the market by 2020 (with a few models available by 2010) and greatly increase their use by 2050. Until then hybrids will probably have an advantage because they are available now and get their fuel from regular filling stations instead of having to depend on building a new network of hydrogen filling stations.

In 2001, Bill Ford, grandson of Henry Ford and chairman of the Ford Motor Company, said, "I believe

1 Cell splits H_2 into protons and electrons. Protons flow across catalyst membrane.

2 React with oxygen (O_2).

3 Produce electrical energy (flow of electrons) to power car.

4 Emits water (H_2O) vapor.

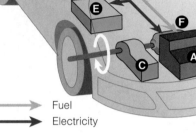

H_2 Hydrogen gas

O_2

H_2O

A **Fuel cell stack:**
Hydrogen and oxygen combine chemically to produce electricity.

B **Fuel tank:**
Hydrogen gas or liquid or solid metal hydride stored on board or made from gasoline or methanol.

C **Turbo compressor:**
Sends pressurized air to fuel cell.

D **Traction inverter:**
Module converts DC electricity from fuel cell to AC for use in electric motor.

E **Electric motor / transaxle:**
Converts electrical energy to mechanical energy to turn wheels.

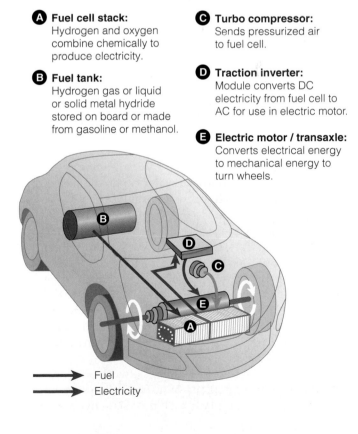

Fuel
Electricity

Figure 18-10 Solutions: general features of an electric car powered by a *fuel cell* running on hydrogen gas. The hydrogen can be produced from natural gas, gasified coal, or methanol, or by using renewable energy sources such as wind turbines or solar cells to produce electricity needed to decompose water into hydrogen and oxygen gas. Prototype models are on the road now and manufacturers hope to have some models of such cars on the market within a decade. (Concept information from DaimlerChrysler, Ford, Ballard, Toyota, and Honda)

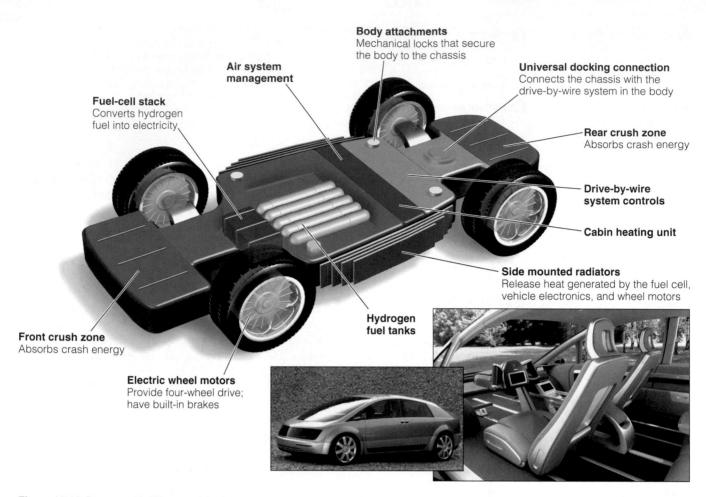

Figure 18-11 Prototype Hy-Wire car of the future developed by General Motors. It combines a hydrogen fuel cell with drive-by-wire technology. It consists of a skateboard-like chassis and a variety of snap-on bodies. The company claims the car could be on the road within a decade but some analysts believe that it will be 2020–2030 before a variety of such cars from various manufacturers will be mass produced. (Basic information from General Motors)

fuel cells will finally end the 100-year reign of the internal combustion engine."

In 2002, General Motors developed a prototype of the hydrogen-fuel car of the future (Figure 18-11). This Hy-Wire (for hydrogen-by-wire) car handles like a high-speed sports car, zips along with no engine noise, and emits only wisps of warm water vapor and heat—no smelly exhaust, no smog, no greenhouse gases.

The heart of this car is a thin flat aluminum chassis that looks like a skateboard. It houses a stack of fuel cells, hydrogen fuel tanks, electronic controls, and wheels with built-in electric motors and brakes. Such a car should have a fuel efficiency equivalent of more than 43 kpl (100 mpg).

The car's fiberglass body plugs into the chassis much like a laptop computer connects to a docking station. The chassis can come in compact, medium, and large sizes for different models.

The car would be refueled by a network of hydrogen gas stations or perhaps by a fuel-cell system in your garage or workplace that produces hydrogen from natural gas and also provides electricity, heating, and air conditioning for your home or workplace.

At first these cars will be expensive, but prices should come down from increased mass production. General Motors says that because these hydrogen-powered fuel-cell vehicles have so few components they will eventually be cheaper (and safer) than vehicles with internal combustion engines.

General Motors hopes to be selling such cars within a decade. Many analysts believe it will be around 2020–2030 before a large number of affordable fuel cell cars will be available but this depends on how rapidly new innovations in this emerging technology can be developed. Stay tuned.

How Can We Design Buildings to Save Energy? Work with Nature

We can save energy in buildings by getting heat from the sun, superinsulating them, and using plant-covered ecoroofs.

Atlanta's 13-story Georgia Power Company building uses 60% less energy than conventional office buildings of the same size. The largest surface of the building faces south to capture solar energy. Each floor extends out over the one below it, which blocks out the higher summer sun to reduce air conditioning costs but allows warming by the lower winter sun. Energy-efficient lights focus on desks rather than illuminating entire rooms. In contrast the conventional Sears Tower building in Chicago consumes more energy in a day than does a city of 150,000 people.

Green architecture is beginning to catch on in Europe, the United States, and Japan. In the Netherlands, the ING Bank built an energy-efficient headquarters that cost no more than a conventional building but uses 92% less energy. This saves the bank $2.9 million a year.

Since 2001 the U.S. Green Building Council has certified 89 office or apartment buildings, condos, manufacturing plants, convention centers, schools, libraries, and college buildings (such as the environmental studies building at the University of California at Santa Barbara) as meeting strict environmental design standards. More than 1,000 other buildings have applied for the council's sought-after seal of approval.

In 2000 the 4,000-member Green Building Council's Leadership in Energy and Environmental Design program (LEED) established building standards and a silver, gold, and platinum scoring system that is used by an increasing number of architects, developers, and elected officials across the United States. In 2004, the two buildings with the most platinum points were the National Resources Defense Council's three-story office building in Santa Monica, California, and the Audubon Society's office building in Los Angeles, California. The mayors of New York City and Chicago have vowed to

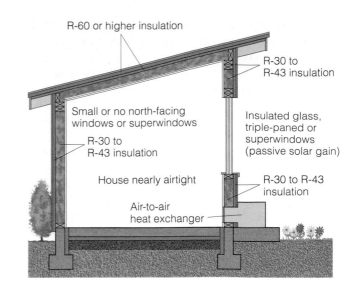

Figure 18-12 Solutions: major features of a *superinsulated house.* Such a house is so heavily insulated and so airtight that heat from direct sunlight, appliances, and human bodies can warm it with little or no need for a backup heating system. An air-to-air heat exchanger prevents buildup of indoor air pollution.

make their cities the greenest in the United States. Have you considered a career in green architecture?

Another energy-efficient design is a *superinsulated house* (Figure 18-12). Such houses typically cost 5% more to build than conventional houses of the same size. But this extra cost is paid back by energy savings within about 5 years and can save a homeowner $50,000–100,000 over a 40-year period. Superinsulated houses in Sweden use 90% less energy for heating and cooling that the typical American home.

Since the mid-1980s there has been growing interest in building superinsulated houses called *strawbale houses* (Figure 18-13). The walls are made by stacking

Figure 18-13 Solutions: energy-efficient, environmentally healthy, and affordable Victorian-style *strawbale house* designed and built by Alison Gannett in Crested Butte, Colorado. The left photo was taken during construction, and the right photo shows the completed house. Depending on the thickness of the bales, plastered strawbale walls have an insulating value of R-35 to R-60, compared to R-12 to R-19 in a conventional house.(The R-value is a measure of resistance to heat flow.) Such houses are also great sound insulators.

compacted bales of low-cost straw and then covering the bales on the outside and inside with plaster or adobe. The main problem is getting banks and other moneylenders to recognize the potential of this and other unconventional types of housing and to provide homeowners with construction loans. (See the Guest Essay about strawbale and solar energy houses by Nancy Wicks on the website for this chapter.)

Ecoroofs or *green roofs* covered with plants have been used in Germany, in other parts of Europe, and in Iceland for decades. With proper design, these plant-covered roof gardens provide good insulation, absorb storm water and release it slowly, outlast conventional roofs, and make a building or home more energy efficient. Designing and installing such systems could be an interesting career.

How Can We Save Energy in Existing Buildings? Stop Leaks and Use Energy-Efficient Devices

We can save energy in existing buildings by insulating them, plugging leaks, and using energy-efficient heating and cooling systems, appliances, and lighting.

Here are some ways to save energy in existing buildings.

- *Insulate and plug leaks.* About one-third of heated air in U.S. homes and buildings escapes through closed windows and holes and cracks (Figure 18-14)— roughly equal to the energy in all the oil flowing through the Alaska pipeline every year. During hot weather these windows and cracks also let heat in, increasing the use of air conditioning. Although not very sexy, adding insulation and plugging leaks in a house are two of the quickest, cheapest, and best ways to save energy and money.

- *Use energy-efficient windows.* Replacing all windows in the United States with low-E (low-emissivity) windows would cut expensive heat losses from houses by two-thirds and reduce CO_2 emissions. Widely available superinsulating windows insulate as well as 8–12 sheets of glass. Although they cost 10–15% more than double-glazed windows, this cost is paid back rapidly by the energy they save. Even better windows will reach the market soon.

- *Stop other heating and cooling losses.* Leaky heating and cooling ducts in attics and unheated basements allow 20–30% of a home's heating and cooling energy to escape and draw unwanted moisture and heat into the home. Careful sealing can reduce this loss. Some designs for new homes keep the ducts inside the home's thermal envelope so that escaping hot or cool air feeds back into the living space.

- *Heat houses more efficiently* (Figure 18-15). In order, the most energy-efficient ways to heat a space are: superinsulation, a geothermal heat pump, passive solar heating, a conventional heat pump (in warm climates only), small cogenerating microturbines, and a high-efficiency (85–98%) natural gas furnace. The most wasteful and expensive way is to use electric resistance heating with the electricity produced by a coal-fired or nuclear power plant (Figure 18-6). In Germany and the United States there is increasing use

Figure 18-14 An infrared photo (thermogram) showing heat loss (red, white, and orange) around the windows, doors, roofs, and foundations of houses and stores in Plymouth, Michigan. Many homes and buildings in the United States and in most other countries are so full of leaks that their heat loss in cold weather and heat gain in hot weather are equivalent to having a large window-sized hole in the wall of the house.

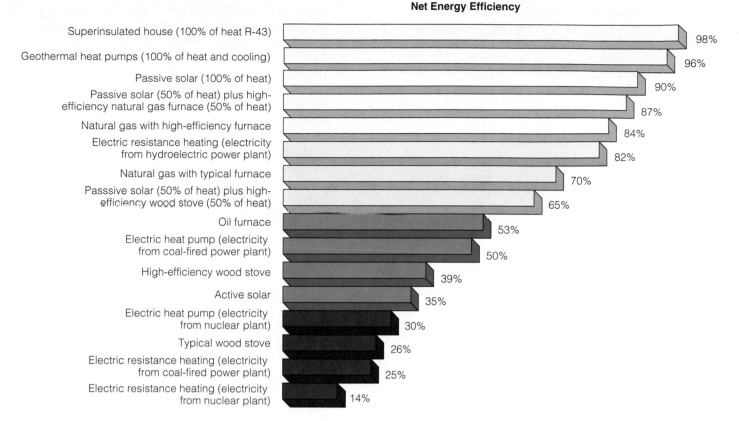

Net Energy Efficiency

Superinsulated house (100% of heat R-43) — 98%

Geothermal heat pumps (100% of heat and cooling) — 96%

Passive solar (100% of heat) — 90%

Passive solar (50% of heat) plus high-efficiency natural gas furnace (50% of heat) — 87%

Natural gas with high-efficiency furnace — 84%

Electric resistance heating (electricity from hydroelectric power plant) — 82%

Natural gas with typical furnace — 70%

Passsive solar (50% of heat) plus high-efficiency wood stove (50% of heat) — 65%

Oil furnace — 53%

Electric heat pump (electricity from coal-fired power plant) — 50%

High-efficiency wood stove — 39%

Active solar — 35%

Electric heat pump (electricity from nuclear plant) — 30%

Typical wood stove — 26%

Electric resistance heating (electricity from coal-fired power plant) — 25%

Electric resistance heating (electricity from nuclear plant) — 14%

Figure 18-15 Solutions: ways to heat an enclosed space such as a house, ranked by net energy efficiency. (Data from Howard T. Odum)

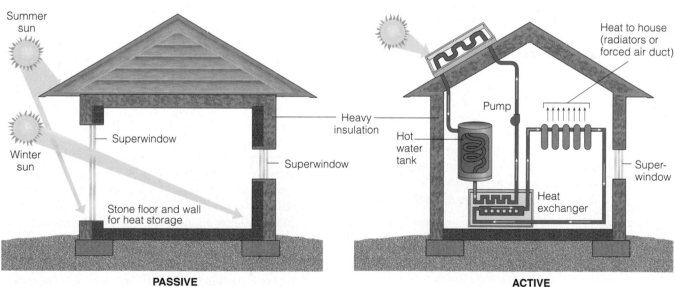

PASSIVE

ACTIVE

Figure 18-16 Solutions: passive and active solar heating for a home.

of cogeneration units or *microturbines* about the size of a refrigerator. They run on natural gas or liquefied petroleum gas (LPG) to produce heat and electricity for businesses, small apartment buildings, neighborhood groups of four or five energy-efficient houses, and small government facilities such as police stations. In 6–8 years, they pay for themselves in saved fuel and electricity.

■ *Heat water more efficiently.* One way to do this is to use a *tankless instant water heater* (about the size of a small suitcase) fired by natural gas or LPG but not by electricity. These devices, widely used in many parts of Europe, heat water instantly as it flows through a small burner chamber, provide hot water only when it is needed, and cost 30–50% less to heat water than traditional heaters.* They cost 2–4 times more than conventional water heaters, but save money because they last 3–4 times longer and cost less to operate than conventional tank heaters.

A well-insulated, conventional natural gas or LPG water heater is also fairly efficient. But all conventional natural gas and electric resistance heaters waste energy by keeping a large tank of water hot all day and night and can run out after a long shower or two—like running your car all night until you drive it.

■ *Use energy-efficient appliances.*** Since 1978 the Department of Energy (DOE) has set federal energy-efficiency standards for more than 20 appliances used in the United States, and similar programs exist in 43 other countries. A 2001 study by the National Academy of Sciences found that between 1978 and 2000, the $7 billion spent by the DOE on this program saved consumers more than $30 billion in energy costs and provided environmental benefits valued conservatively at $60–80 billion.

If all households in the United States used the most efficient frost-free refrigerator now available, 18 large (1,000-megawatt) power plants could close. Microwave ovens can cut electricity use for cooking by 25–50% (but not if used for defrosting food). Clothes dryers with moisture sensors cut energy use by 15%, and front-loading washers use 50% less energy than top-loading models but cost about the same.

■ *Use energy-efficient lighting.* Americans spend about a quarter of their electricity budget on lighting. But many do not realize that they could cut these costs 30–60% by replacing energy-wasting incandescent bulbs and halogen torchiere bulbs (which because of their high heat output have caused fires and increased air conditioning costs) with much more efficient fluorescent bulbs. They cost $5–10 but last 6–10 times longer than an incandescent and pay for themselves in a year or two. Three-way and dimmable versions are now available (see http://www.tcpi .com). Replacing 25 incandescent bulbs in a house or building with energy-efficient fluorescent bulbs typically saves about $1,125. What a great investment payoff.

Students in Brown University's environmental studies program showed that the school could save more than $40,000 per year just by replacing the incandescent light bulbs in exit signs with compact fluorescent bulbs. What is your school doing to save electricity and money in lighting?

However, these and other fluorescent bulbs contain toxic mercury than when discarded can contaminate landfills and groundwater or get into the atmosphere if incinerated. A Florida company collects used bulbs and extracts the toxic mercury for reuse.

Within the next two decades, both incandescent and fluorescent bulbs may be replaced by even more efficient white-light LEDs (light-emitting diodes) and organic LEDs (OLEDs). Westinghouse is selling a 20-watt LED bulb with a light output equal to a 100-watt incandescent bulb. It costs $40, but saves money because it lasts 80 times longer than incandescents (see http://westinghouselighting.com).

■ *Cut off electrical devices when not using them.* Cutting off lights, computers, TVs, and other appliances when they are not needed and cutting off their instant-on feature can make a big difference in energy use and bills. At 9 P.M. one weekday evening, major TV stations in Bangkok, Thailand, cooperated with the government in showing a dial that gave the city's current use of electricity. Viewers were asked to turn off unnecessary lights and appliances. They then watched the dial register a 735-megawatt drop in electricity use—a decrease equal to the output of two medium-sized coal-burning power plants. This visual experience showed individuals that reducing their unnecessary electricity use could cut their bills and close down power plants.

■ *Set strict energy-efficiency standards for new buildings.* Building codes could require new houses use 60–80% less energy than conventional houses of the same size, as has been done in Davis, California. Because of tough national energy-efficiency standards, the average home in Sweden consumes about one-third as much energy as the average American home of the same size.

---

*They work very well. I used them in a passive solar office and living space for 15 years. Models are available for $500–1,000 from companies such as Rinnai, Bosch, Takagi, and Envirotech. For information visit http://foreverhotwater.com.

**Each year the American Council for an Energy-Efficient Economy (ACEEE) publishes a list of the most energy-efficient major appliances mass-produced for the U.S. market. A copy can be obtained from the council at 1001 Connecticut Avenue NW, Suite 801, Washington, DC 20036, or on its website at http://www.aceee.org/consumerguide/index.htm.

Why Are We Still Wasting So Much Energy? We Get What We Reward

Low-priced oil and gasoline and lack of government tax breaks for saving energy promote energy waste.

With such an impressive array of benefits (Figure 18-2), why is there so little emphasis on improving energy efficiency? One reason is a glut of low-cost oil and gasoline. As long as energy is artificially cheap because its market price does not include its harmful costs (Connections, p. 384), people are more likely to waste it and not make investments in improving energy efficiency.

Another reason is a lack of sufficient government tax breaks and other economic incentives for consumers and businesses to invest in improving energy efficiency.

Would you like to earn about 20% a year on your money, tax-free and risk-free? Invest it in improving the energy efficiency of your home and in energy-efficient lights and appliances. You get your investment back in a few years and then make about 20% a year by having lower heating, cooling, and electricity bills. This is a win-win deal for you and the earth.

X *HOW WOULD YOU VOTE?* Should the United States or the country where you live greatly increase its emphasis on improving energy efficiency? Cast your vote online at http://biology.brookscole.com/miller14.

18-3 USING RENEWABLE ENERGY TO PROVIDE HEAT AND ELECTRICITY

What Are the Main Types of Renewable Energy? Solar Capital

Six types of renewable energy are solar, flowing water, wind, biomass, geothermal, and hydrogen.

One of the four keys to sustainability (bottom half of back cover) based on learning from nature is to *rely mostly on renewable solar energy*. We can get renewable solar energy directly from the sun or indirectly from moving water, wind, and biomass. Two other forms of renewable energy are geothermal energy from the earth's interior and using renewable energy to produce hydrogen fuel from water. Like fossil fuels and nuclear power, each of these renewable energy alternatives has advantages and disadvantages, as discussed in the remainder of this chapter.

If renewable energy is so great, why does it provide only 16% of the world's energy and 6% of the energy in the United States? One reason is that renewable energy resources have received and continue to receive much lower government tax breaks, subsidies, and research and development (R & D) funding than fossil fuels and nuclear power have received for

decades. The other reason is that the prices we pay for fossil fuels and nuclear power do not include their harm to the environment and to human health.

In other words, the economic dice have been loaded against solar, wind, and other forms of renewable energy. If the economic playing field was made more even, energy analysts say that many of these forms of renewable energy would take over—another example of the *you-get-what-you-reward* economic principle in action.

Here are four encouraging developments favoring increased use of renewable energy. *First*, in 2001 the European Union (EU) adopted nonbinding agreements for its member countries to get 12% of their total energy and 22% of their electricity from renewable energy by 2010.

Second, California gets about 12% of its electricity from renewable resources (7% of it from wind turbines) and wants to get 20% from such resources by 2010. *Third*, a 2001 joint study by the American Council for an Energy-Efficient Economy, the Tellus Institute, and the Union of Concerned Scientists showed how renewable energy could provide 20% of U.S. energy by 2020 if given sufficient government R & D subsidies and tax breaks. *Fourth*, according to the Worldwatch Institute, all U.S. electricity could be provided by farms of wind turbines operating in just three states—Kansas, North Dakota, and South Dakota—or with solar energy on a 260-square-kilometer (100-square-mile) plot in the Nevada or southern California desert.

How Can We Use Direct Solar Energy to Heat Houses and Water? Face the Sun and Store Its Heat

We can heat buildings by orienting them toward the sun (passive solar heating) or by pumping a liquid such as water through rooftop collectors (active solar heating).

Buildings and water can be heated by direct solar energy using two methods: passive and active (Figure 18-16). A **passive solar heating system** absorbs and stores heat from the sun directly within a structure (Figure 18-1, Figure 18-16, left, and Figure 18-17, p. 392). See the Guest Essay by Nancy Wicks on this topic on the website for this chapter.

Using passive solar energy is not new. For thousands of years, many people have intuitively followed the first principle of sustainability (bottom half of back cover). They have oriented their dwellings to take advantage of heat from the sun and used adobe and thick stone walls to collect and store heat during the day and gradually release it at night.

In today's passively heated buildings, energy-efficient windows and attached greenhouses face the

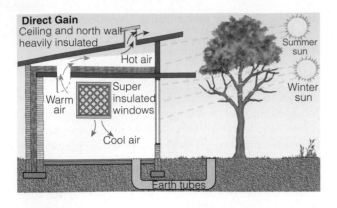

Direct Gain
Ceiling and north wall heavily insulated
Hot air
Warm air
Super insulated windows
Cool air
Summer sun
Winter sun
Earth tubes

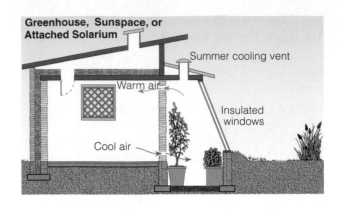

Greenhouse, Sunspace, or Attached Solarium
Summer cooling vent
Warm air
Insulated windows
Cool air

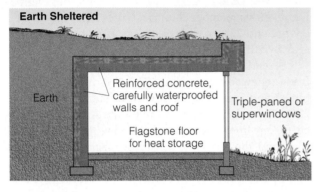

Earth Sheltered
Earth
Reinforced concrete, carefully waterproofed walls and roof
Triple-paned or superwindows
Flagstone floor for heat storage

Figure 18-17 Solutions: three examples of *passive solar design* for houses.

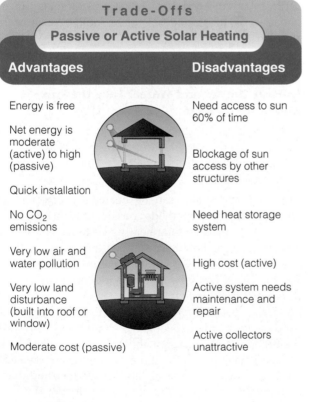

Trade-Offs

Passive or Active Solar Heating

Advantages	Disadvantages
Energy is free	Need access to sun 60% of time
Net energy is moderate (active) to high (passive)	Blockage of sun access by other structures
Quick installation	
No CO$_2$ emissions	Need heat storage system
Very low air and water pollution	High cost (active)
Very low land disturbance (built into roof or window)	Active system needs maintenance and repair
Moderate cost (passive)	Active collectors unattractive

Figure 18-18 Trade-offs: advantages and disadvantages of heating a house with passive or active solar energy. Pick the single advantage and the single disadvantage that you think are the most important.

sun to collect solar energy by direct gain. Walls and floors of concrete, adobe, brick, stone, salt-treated timber, and water in metal or plastic containers store much of the collected solar energy as heat and release it slowly throughout the day and night. A small backup heating system such as a vented natural gas or propane heater may be used but is not necessary in many climates.

On a life cycle cost basis, good passive solar and superinsulated design is the cheapest way to heat a home or small building in regions with access to ample sunlight. Such a system usually adds 5–10% to the construction cost, but the life cycle cost of operating such a house is 30–40% lower. The typical payback time for passive solar features is 3–7 years.

An **active solar heating system** absorbs energy from the sun by pumping a heat-absorbing fluid (such as water or antifreeze solution) through special collectors usually mounted on a roof or on special racks to face the sun (Figure 18-16, right). A typical active collector has a flat black surface, a coil through which the heat-absorbing medium such as water is pumped, and a cover consisting of two or three layers of glass.

Some of the collected heat can be used directly. The rest can be stored in a large insulated container filled with gravel, water, clay, or a heat-absorbing chemical for release as needed. Often these insulated heat storage containers are located under a house.

Active solar collectors can also supply hot water and are widely used in areas of the world with sunny climates. More than 1 million homes in Florida and California heat all or some of their water with one or more active solar collectors.

Figure 18-18 lists the major advantages and disadvantages of using passive or active solar energy for heating buildings. Passive solar energy is great for new homes in sunny areas but cannot be used to heat existing homes and buildings not oriented to receive sunlight or where trees or other buildings block access to sunlight. Active solar collectors are good for heating water in sunny areas. But most analysts do not expect widespread use of active solar collectors for heating houses because of their high costs, maintenance requirements, and unappealing appearance.

How Can We Cool Houses Naturally? Insulate and Work with Nature

We can cool houses by superinsulating them, taking advantage of breezes, shading them, having light-colored roofs, and using geothermal cooling.

Here are some ways to have a cooler house. Use superinsulation and superinsulating windows, open windows to take advantage of breezes, and use fans to keep air moving. Block the high summer sun with deciduous trees and window overhangs, (Figure 18-17, top left), or awnings.

Use a light-colored roof to reflect up to 80% of the sun's heat, compared to only 8% for a roof colored dark gray. Suspend reflective insulating foil in an attic to block heat from radiating down into the house.

Another option is to place plastic *earth tubes* underground where the earth is cool year-round. In this geothermal cooling system, a tiny fan can pipe cool and partially dehumidified air into an energy-efficient house (Figure 18-17, top left).* In warm climates you can also use high-efficiency heat pumps for air conditioning.

Toronto, Canada's largest city, cools downtown buildings by pumping cold water from the depths of Lake Ontario and passing it through building air conditioning systems. This reduces the use of coal for producing electricity, cuts greenhouse gas emissions, and slashes summer use of electricity for air conditioning by 90%.

How Can We Use Solar Energy to Generate High-Temperature Heat and Electricity? Desert Power

Large arrays of solar collectors in sunny deserts can produce high-temperature heat to spin turbines and produce electricity, but costs are high.

Several *solar thermal systems* can collect and transform radiant energy from the sun into high-temperature thermal energy (heat), which can be used directly or converted to electricity. These systems are used mostly in desert areas with ample sunlight.

One method uses a *central receiver system*, called a *power tower*. Huge arrays of computer-controlled mirrors called *heliostats* track the sun and focus sunlight on a central heat collection tower (top drawing in Figure 18-19).

Australia is building a different type of power tower in its sunny outback. It will consist of a concrete thermal chimney twice the height of the world's tallest building surrounded by a gigantic sloped solar greenhouse with a diameter of 5 kilometers (3 miles). As the hot air collected by the huge greenhouse flows up into the tower it will spin 32 giant turbines and produce enough electricity to serve 200,000 homes. Some of the heat collected during the day will be stored in tubes filled with water. The heat released from this water after dark should keep the power plant working throughout the night. This project is a miniature version of how the earth makes wind from solar energy.

Another approach is a *solar thermal plant* in which sunlight is collected and focused on arrays of oil-filled pipes running through the middle of a large area of curved solar collectors (bottom drawing in Figure 18-19). This concentrated sunlight can generate temperatures high enough for producing steam to run turbines and generate electricity. At night or on cloudy days, high-efficiency combined-cycle natural gas turbines can supply backup electricity as needed.

On an individual scale, inexpensive *solar cookers* can focus and concentrate sunlight and cook food, especially in rural villages in sunny developing countries. They can be made by fitting an insulated box big enough to hold three or four pots with a transparent, removable top. Solar cookers reduce deforestation for fuelwood and the time and labor needed to collect firewood. They also reduce indoor air pollution from smoky fires.

Figure 18-19 lists the advantages and disadvantages of concentrating solar energy to produce high-temperature heat or electricity. Most analysts do not expect widespread use of such technologies over the next few decades because of high costs, limited suitable sites,

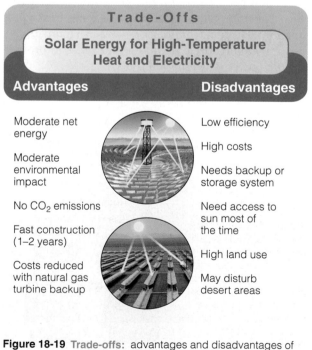

Trade-Offs

Solar Energy for High-Temperature Heat and Electricity

Advantages	Disadvantages
Moderate net energy	Low efficiency
Moderate environmental impact	High costs
No CO_2 emissions	Needs backup or storage system
Fast construction (1–2 years)	Need access to sun most of the time
Costs reduced with natural gas turbine backup	High land use
	May disturb desert areas

Figure 18-19 Trade-offs: advantages and disadvantages of using solar energy to generate high-temperature heat and electricity. Pick the single advantage and the single disadvantage that you think are the most important.

---

*They work. I used them in a passively heated and cooled office and home for 15 years. People allergic to pollen and molds should add an air purification system, but this is also necessary with a conventional cooling system.

and availability of cheaper ways to produce electricity such as combined-cycle natural gas and wind turbines.

How Can We Produce Electricity with Solar Cells? Use Your Roof or Windows as a Power Plant

Solar cells that convert sunlight to electricity can be incorporated into roofing materials or windows, and the high costs of doing this are expected to fall.

Solar energy can be converted directly into electrical energy by **photovoltaic (PV) cells,** commonly called **solar cells** (Figure 18-20). A typical solar cell is a transparent wafer containing a semiconductor material with a thickness ranging from less than that of a human hair to a sheet of paper. Sunlight energizes and causes electrons in the semiconductor to flow, creating an electrical current. These devices have no moving parts, require little maintenance, produce no pollution during operation, and last as long as a conventional fossil fuel or nuclear power plant.

The semiconductor material used in solar cells can be made into lightweight paper-thin rigid or flexible sheets and incorporated into traditional-looking roofing materials (blue in Figure 18-20). Glass walls and windows of buildings can also have built-in solar cells. In 2004, energy giant British Petroleum (BP) began building the world's largest factory to produce windows and cladding and roofing materials that will incorporate BP's power-producing solar cells.

Easily expandable banks of solar cells can be used to provide electricity in developing countries for 1.7 billion people in rural villages without electricity. Such banks of cells can also produce electricity at a small power plant (bottom drawing in Figure 18-21), using combined-cycle natural gas turbines to provide backup power when the sun is not shining. Another possibility is to use arrays of solar cells to convert water to hydrogen gas that can be distributed

to energy users by pipeline, as natural gas is. With financing from the World Bank, India (the world's number-one market for solar cells) is installing solar-cell systems in 38,000 villages, and Zimbabwe is bringing solar electricity to 2,500 villages. By 2004, more than 1 million homes in the world, most of them in villages in developing countries (and about 200,000 in the United States), were getting some or all of their electricity from solar cells mostly because they were long distances from a power grid.

Figure 18-21 lists the advantages and disadvantages of solar cells. Current costs of producing electricity from solar cells are high but are expected to drop because of savings from mass production and new designs. Solar cells can also be incorporated into carbon-based polymers similar to Teflon that can be applied to surfaces in thin layers. The first generation of such *organic solar cells* that can convert 20–35% of the sun's energy into electricity could enter the marketplace within a few years. These solar cells could be printed on a sheet of paper, stuck onto your house or car windows, painted on your house, or even incorporated into your clothing—making you a walking tiny power plant.

Some envision incorporating tiny rods of semiconductors with a thickness of several nanometers (a tiny fraction of the thickness of a hair on your head) in plastic materials. Such *nano solar cells* can be manufactured in extremely high volumes at a very low cost. Stay tuned.

Currently solar cells supply only about 0.05% of the world's electricity. But with increased government and private R & D and greater government tax breaks and other subsidies they could provide over a quarter of the world's electricity by 2040. If such projections are correct, the production, sale, and installation of solar cells could become one of the world's largest and fastest

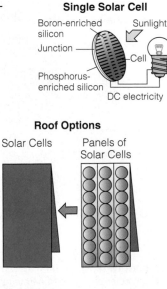

Figure 18-20 Solutions: photovoltaic (PV) (solar) cells can provide electricity for a house or building using new solar-cell roof shingles or PV panel roof systems that look like a blue metal roof. Arrays of such cells can also produce electricity for a village or at a small power plant.

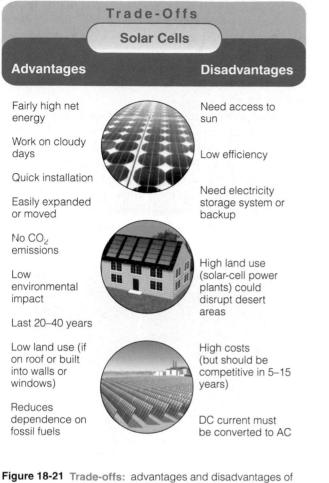

Advantages	Disadvantages
Fairly high net energy	Need access to sun
Work on cloudy days	Low efficiency
Quick installation	
Easily expanded or moved	Need electricity storage system or backup
No CO$_2$ emissions	High land use (solar-cell power plants) could disrupt desert areas
Low environmental impact	
Last 20–40 years	
Low land use (if on roof or built into walls or windows)	High costs (but should be competitive in 5–15 years)
Reduces dependence on fossil fuels	DC current must be converted to AC

Figure 18-21 Trade-offs: advantages and disadvantages of using solar cells to produce electricity. Pick the single advantage and the single disadvantage that you think are the most important.

growing businesses. This is another exciting field to consider as a career choice.

18-4 PRODUCING ELECTRICITY FROM THE WATER CYCLE

How Can We Produce Electricity from Flowing Water? Renewable Hydropower

Water flowing in rivers and streams can be trapped in reservoirs behind dams and released as needed to spin turbines and produce electricity.

Solar energy evaporates water and deposits it as water and snow in other areas as part of the water cycle (Figure 4-28, p. 76). Water flowing from high elevations to lower elevations in rivers and streams can be controlled by dams and reservoirs and used to produce electricity. This indirect form of renewable solar energy is called *hydropower* (Figure 15-9, p. 313).

Three methods are used to produce such electricity. One is *large-scale hydropower,* in which a high dam is built across a large river to create a reservoir. Some of the water stored in the reservoir is allowed to flow through huge pipes at controlled rates, spinning turbines and producing electricity.

Another method is *small-scale hydropower.* A low dam with no reservoir or only a small one is built across a small stream, and the stream's flow of water is used to spin turbines and produce electricity. Submerging small high-efficiency turbines in a stream without impeding stream navigation or fish movements can also produce electricity. A smaller turbine called a micro-hydrogenerator can be used to provide affordable electricity for a single home.

A third method is *pumped-storage hydropower.* Pumps use surplus electricity from a conventional power plant to pump water from a lake or a reservoir to another reservoir at a higher elevation. When more electricity is needed, water in the upper reservoir is released, flows through turbines, and generates electricity on its return to the lower reservoir.

In 2002, hydropower supplied 20% of the world's electricity, 99% in Norway, 75% in New Zealand, 25% in China, and 7% in the United States (but about 50% on the West Coast).

Figure 18-22 (p. 396) lists the advantages and disadvantages of using large-scale hydropower plants to produce electricity.

✗ HOW WOULD YOU VOTE? Do the advantages of using large-scale hydropower plants to produce electricity outweigh the disadvantages? Cast your vote online at http://biology .brookscole.com/miller14.

According to the United Nations, only about 13% of the world's technically exploitable potential for hydropower has been developed. Much of this untapped potential is in China (p. 315), India, South America, Central Africa, and parts of the former Soviet Union.

Because of increasing concern about the harmful environmental and social consequences of large dams, there has been growing pressure on the World Bank and other development agencies to stop funding new large-scale hydropower projects. Also, according to a 2000 study by the World Commission on Dams, hydropower in tropical countries is a major emitter of greenhouse gases. This occurs because reservoirs that power the dams can trap rotting vegetation, which can emit greenhouse gases such as carbon dioxide and methane.

Small-scale hydropower projects eliminate most of the harmful environmental effects of large-scale projects. But their electrical output can vary with seasonal changes in stream flow.

We can also produce electricity from water flows by tapping into the energy from tides and waves. Most analysts expect these sources to make little contribution to world electricity production because of high costs and lack of enough areas with the right conditions.

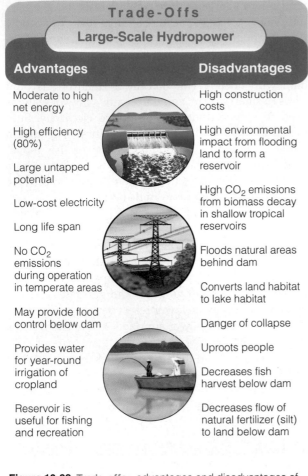

Trade-Offs

Large-Scale Hydropower

Advantages	Disadvantages
Moderate to high net energy	High construction costs
High efficiency (80%)	High environmental impact from flooding land to form a reservoir
Large untapped potential	
Low-cost electricity	High CO_2 emissions from biomass decay in shallow tropical reservoirs
Long life span	
No CO_2 emissions during operation in temperate areas	Floods natural areas behind dam
	Converts land habitat to lake habitat
May provide flood control below dam	Danger of collapse
Provides water for year-round irrigation of cropland	Uproots people
	Decreases fish harvest below dam
Reservoir is useful for fishing and recreation	Decreases flow of natural fertilizer (silt) to land below dam

Figure 18-22 Trade-offs: advantages and disadvantages of using large dams and reservoirs to produce electricity. Pick the single advantage and the single disadvantage that you think are the most important.

18-5 PRODUCING ELECTRICITY FROM WIND

What Is the Global Status of Wind Power? A Star Is Born

Since 1995 the use of wind turbines to produce electricity has increased almost sevenfold.

The greater heating of the earth at the equator than at the poles and the earth's rotation (Figure 6-8, p. 107) set up flows of air called *wind*. This indirect form of solar energy can be captured by wind turbines (Figure 18-23) and converted into electricity.

Since 1990, wind power has been by far the world's fastest growing source of energy, with its use increasing almost sevenfold between 1995 and 2004.

Europe is leading the world into the age of wind energy and out of the age of coal and other fossil fuels. About three-fourths of the world's wind power is produced in Europe in inland and offshore wind farms. And European companies manufacture about 80% of

the wind turbines sold in the global marketplace. Denmark has banned coal and gets 90% of its electricity from wind. Nine of the world's 10 leading wind turbine manufacuring companies are in three countries—Denmark, Germany, and Spain—mostly because of strong and consistent government subsidies and tax breaks.

Wind power is also being developed rapidly in India (the world's number-two market for wind energy) and to a lesser degree in China. By 2030, India could use wind to generate a fourth of its electricity.

Much of the world's potential wind power remains untapped. According to the 2003 Wind Force 12 report, wind parks on only one-tenth of the earth's land could produce twice the world's projected demand for electricity by 2020.

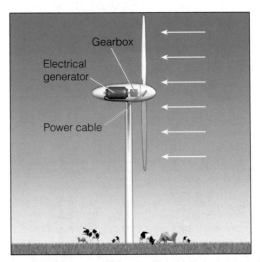

Wind turbine

Wind farm

Figure 18-23 Solutions: wind turbines can be used to produce electricity individually or in clusters, called wind farms or wind parks. Since 1990, wind power has been the world's fastest growing source of energy. Our energy future may be blowing in the wind.

The DOE calls the Great Plains states of Oklahoma, South Dakota, North Dakota, Kansas, Nebraska, and Texas the "Saudi Arabia of wind" and points out that they have enough wind resources to more than meet all the nation's electricity needs. According to the American Wind Energy Association, with increased and consistent government subsidies and tax breaks, wind power could produce almost a fourth of U.S. electricity by 2025.

A growing number of U.S. farmers and ranchers make more money by leasing their land for wind power production than by growing crops or raising cattle. This explains why many of them are joining environmentalists and wind industry executives in urging political leaders to increase government research and development and tax breaks for wind power.

Figure 18-24 lists the advantages and disadvantages of using wind to produce electricity. According to energy analysts, wind power has more advantages and fewer serious disadvantages than any other energy resource. Between the 1980s and 2004, the cost of wind-generated electricity dropped ninefold from 36¢ to about 4¢ per kilowatt-hour at favorable wind sites. This is about the same price as using coal, natural gas, and hydropower (at highly favorable sites) to produce electricity and three times cheaper than nuclear power. Wind power is like an underdog racehorse that is beginning to break out of a pack of other, more pampered (subsidized) energy racehorses.

If wind turbines are mass-produced like automobiles, the cost for a kilowatt of wind-generated energy could drop to 1–2 cents, making it by far the cheapest way to produce electricity. Many governments and corporations are recognizing that *there is money in wind*. Do the math. The average price of electricity in the United States in 2003 was 7.2¢ per kilowatt-hour. If wind companies can produce a kilowatt of electricity at about 4¢ now, and in the not-too-distant future, for 1.5–2¢, the potential profits are huge. This explains why General Electric, one of the world's largest multinational companies, recently decided to get into wind power.

Some critics allege that wind turbines suck large numbers of birds into their wind stream. However, as long as wind farms are not located along bird migration routes most birds learn to fly around them. Wind power developers now make sophisticated studies of bird migration paths to help them locate onshore and offshore wind parks and are designing new turbines to reduce this problem.

Also, studies have shown that much larger numbers of birds die when they are sucked into jet engines, killed by domesticated and feral cats, and crash into skyscrapers, plate glass windows, communications towers, and car windows.

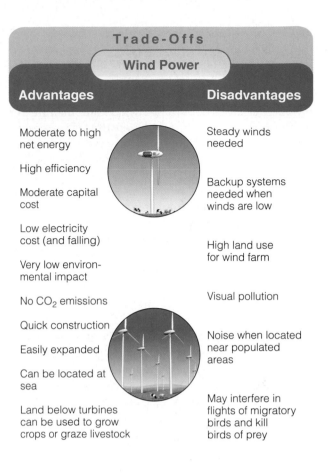

Figure 18-24 **Trade-offs:** advantages and disadvantages of using wind to produce electricity. Wind power experts project that by 2025 wind power could supply more than 10% of the world's electricity and 10–25% of the electricity used in the United States. Pick the single advantage and the single disadvantage that you think are the most important.

Even larger numbers of birds, fish, and other forms of wildlife are killed by oil spills, air pollution, water pollution, and release of toxic wastes from use of fossil fuels such as coal and oil. The key questions are, Which types of energy resources lead to the lowest loss of wildlife? and How can we minimize loss of wildlife from use of any energy resource?

X *HOW WOULD YOU VOTE?* Should we greatly increase our dependence on wind power? Cast your vote online at http://biology.brookscole.com/miller14.

18-6 PRODUCING ENERGY FROM BIOMASS

How Is Biomass Used to Provide Energy? Burning Carbon Compounds

Plant materials and animal wastes can be burned to provide heat or electricity or converted into gaseous or liquid biofuels.

Biomass consists of plant materials and animal wastes that can be burned directly as a solid fuel or converted into gaseous or liquid **biofuels** (Figure 18-25). Most biomass is burned *directly* for heating, cooking, and industrial processes or *indirectly* to drive turbines and produce electricity. Burning wood and manure for heating and cooking supplies about 10% of the world's energy and about 30% of the energy used in developing countries (90% in the poorest countries such as Bangladesh, Ethiopia, Burundi, and Bhutan).

In 2002, about 350 biomass power plants supplied about 3% of the commercial energy and 2% of the electricity used in the United States. The U.S. government has a goal of increasing the use of biomass energy to 9% of the country's total commercial energy by 2010.

One way to produce biomass fuel is to plant, harvest, and burn large numbers of fast-growing trees (especially cottonwoods, poplars, sycamores, willows, and leucaenas), shrubs, perennial grasses (such as switchgrass), and water hyacinths in *biomass plantations.*

In agricultural areas, *crop residues* (from sugarcane, rice, cotton, and coconuts) and *animal manure* can be collected and burned or converted into biofuels. In some developing countries the poor gather animal manure or dung by hand, dry it, and burn it for heat and cooking. On the surface, this appears to be a free and logical use of wasted biomass.

But some ecologists argue that it makes more sense to use animal manure as a fertilizer and crop residues to feed livestock, retard soil erosion, and fertilize the soil. Not allowing these animal and crop wastes to return to the soil as natural fertilizer can reduce food production and food supplies in poor countries. Also burning dried dung in open fires wastes about 90% of its heat content.

Figure 18-26 lists the general advantages and disadvantages of burning solid biomass as a fuel. One problem is that burning biomass produces CO_2. However, if the rate of use of biomass does not exceed the rate at which it is replenished by new plant growth (which takes up CO_2), there is no net increase in CO_2 emissions. But repeated cycles of growing and harvesting biomass plantations can deplete the soil of key nutrients.

How Can Gaseous Fuels Be Produced from Biomass? Bacteria and Chemistry to the Rescue

Some forms of biomass can be converted into gaseous and liquid biofuels.

Bacteria and various chemical processes can convert some forms of biomass into gaseous biofuels (Figure 18-25). One of them is *biogas*—a mixture of 60% methane and 40% CO_2.

In rural China, anaerobic bacteria in more than 500,000 *biogas digesters* on farms and in homes convert

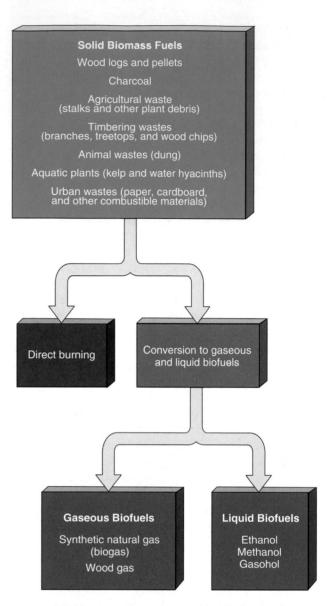

Figure 18-25 Principal types of biomass fuel.

plant and animal wastes into methane gas that is used for heating and cooking. After the biogas has been removed, the almost odorless solid residue is used as fertilizer on food crops or, if it is contaminated, on trees. When they work, biogas digesters are very efficient and burning natural gas produced from dung produces much more heat than burning the dung itself. But they are slow and unpredictable, a problem that could be corrected by developing more reliable models. They also add CO_2 to the atmosphere.

In some places in the United States, bacteria convert livestock wastes from cattle, hogs, and chickens to biogas. One way to do this is to put the wastes in a long, lined, insulated pit. A flexible liner stretching across the digester pit inflates like a balloon as it collects the biogas, which can then be burned to heat the digester or nearby farm buildings or to produce electricity.

Trade-Offs

Solid Biomass

Advantages	Disadvantages
Large potential supply in some areas	Nonrenewable if harvested unsustainably
Moderate costs	Moderate to high environmental impact
No net CO_2 increase if harvested and burned sustainably	CO_2 emissions if harvested and burned unsustainably
Plantation can be located on semiarid land not needed for crops	Low photosynthetic efficiency
	Soil erosion, water pollution, and loss of wildlife habitat
Plantation can help restore degraded lands	Plantations could compete with cropland
	Often burned in inefficient and polluting open fires and stoves
Can make use of agricultural, timber, and urban wastes	

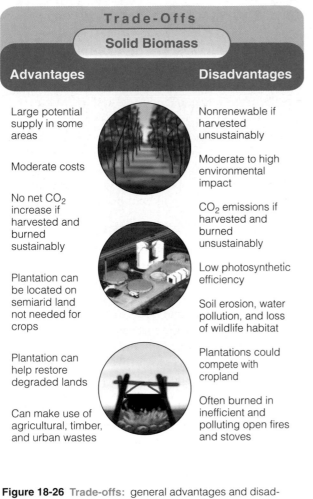

Figure 18-26 Trade-offs: general advantages and disadvantages of burning solid biomass as a fuel. Pick the single advantage and single disadvantage that you think are the most important.

Bacteria in large digesters can also convert municipal garbage and sewage to methane gas. Wells drilled into about 300 large landfills in the United States recover methane produced by the decomposition of organic wastes and burn it to produce enough electricity to meet the needs of a million homes. BMW's automobile factory near Spartanburg, S.C. gets more than a fourth of its electricity and a tenth of its heat by burning methane gas piped in from a nearby landfill owned by Waste Management.

What Are the Advantages and Disadvantages of Using Liquid Ethanol and Methanol Produced from Biomass as a Fuel? Mixed Signals

Some believe we can rely much more on ethanol and methanol as a fuel, but others disagree.

Some analysts believe that liquid ethanol produced from biomass could replace gasoline and diesel fuel when oil becomes too scarce or expensive. *Ethanol* can be made from sugar and grain crops (sugarcane, sugar beets, sorghum, sunflowers, and corn) by fermentation and distillation. Gasoline mixed with 10–23% pure ethanol makes *gasohol*, which can be burned in conventional gasoline engines.

Figure 18-27 lists the advantages and disadvantages of using ethanol as a vehicle fuel compared to gasoline. Ethanol could be produced from surplus grain crops. But industrialized agriculture uses more energy in the form of petroleum-based vehicle fuel, fertilizers, and pesticides than the energy obtained by burning ethanol produced by such crops. Thus, there is a *net energy loss* from growing grain crops, converting the grain to ethanol, distilling the ethanol, and distributing and burning it as a motor vehicle fuel. The U.S. government gives large subsidies to corn growers to produce ethanol as part of the national energy policy. Critics see this as a politically motivated giveaway and waste of money that could be used to support more promising renewable energy alternatives.

X HOW WOULD YOU VOTE? Do the advantages of using liquid ethanol as a fuel outweigh the disadvantages? Cast your vote online at http://biology.brookscole.com/miller14.

Trade-Offs

Ethanol Fuel

Advantages	Disadvantages
High octane	Large fuel tank needed
	Lower driving range
Some reduction in CO_2 emissions	Net energy loss
	Much higher cost
Reduced CO emissions	Corn supply limited
	May compete with growing food on cropland
Can be sold as gasohol	Higher NO emissions
	Corrosive
Potentially renewable	Hard to start in cold weather

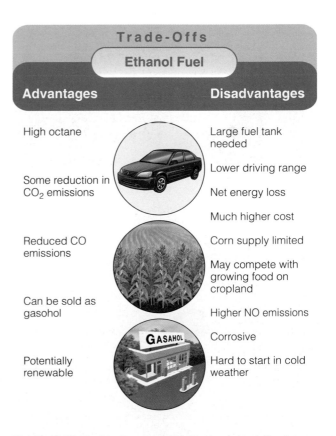

Figure 18-27 Trade-offs: general advantages and disadvantages of using ethanol as a vehicle fuel compared to gasoline. Pick the single advantage and single disadvantage that you think are the most important.

Some analysts believe that liquid methanol produced from biomass could replace gasoline and diesel fuel when oil becomes too scarce or expensive. *Methanol* is made mostly from natural gas but can also be produced at a higher cost from coal and biomass such as wood, wood wastes, agricultural wastes, sewage sludge, and garbage.

Figure 18-28 lists the advantages and disadvantages of using methanol as a vehicle fuel compared to gasoline. According to a 1997 analysis by David Pimentel and two other researchers, "Large-scale biofuel production is not an alternative to the current use of oil and is not even an advisable option to cover a significant fraction of it."

However, chemist George A. Olah believes that establishing a *methanol economy* is preferable to the highly publicized hydrogen economy. He points out that methanol can be produced chemically from carbon dioxide in the atmosphere, which could also help slow projected global warming. In addition, methanol can be converted to other hydrocarbon compounds that can be used to produce a variety of useful chemicals like those made from petroleum and natural gas.

✗ *How Would You Vote?* Do the advantages of using liquid methanol as a fuel outweigh the disadvantages? Cast your vote online at http://biology.brookscole.com/miller14.

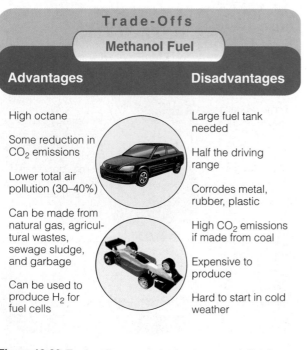

Trade-Offs

Methanol Fuel

Advantages	Disadvantages
High octane	Large fuel tank needed
Some reduction in CO_2 emissions	Half the driving range
Lower total air pollution (30–40%)	Corrodes metal, rubber, plastic
Can be made from natural gas, agricultural wastes, sewage sludge, and garbage	High CO_2 emissions if made from coal
	Expensive to produce
Can be used to produce H_2 for fuel cells	Hard to start in cold weather

Figure 18-28 Trade-offs: general advantages and disadvantages of using methanol as a vehicle fuel compared to gasoline. Pick the single advantage and the single disadvantage that you think are the most important.

18-7 GEOTHERMAL ENERGY

What Is Geothermal Energy? Tapping the Earth's Internal Heat

We can use geothermal energy stored in the earth's mantle to heat and cool buildings and to produce electricity.

Geothermal energy consists of heat stored in soil, underground rocks, and fluids in the earth's mantle. Examples are volcanic rock, geysers, and hot springs. Scientists have developed several ways to tap into this stored energy to heat and cool buildings and to produce electricity.

Throughout most of the world (except tundra areas with permafrost) the temperature of the earth at a depth of about 3 meters (10 feet) is 10–16°C (50–60°F). *Geothermal heat pumps* can tap into this difference between underground and surface temperatures in most places and use a system of pipes and ducts to heat or cool a building. These devices use the earth as a heat source in winter and as a heat sink during summer. They are a very efficient and cost-effective way to heat or cool a space.

A related way to heat or cool a building is *geothermal exchange* or *geoexchange*. It involves using buried pipes filled with a fluid to move heat in or out of the ground depending on the season and the heating or cooling requirements. In the winter, for example, heat is removed from fluid in pipes buried in the ground and blown through house ducts. In the summer this process is reversed. According to the U.S. Environmental Protection Agency, geothermal exchange is the most energy-efficient, cost-effective, and environmentally clean way to heat or cool a building.

We have also learned to tap into deeper and more concentrated underground reservoirs of geothermal energy. One type of reservoir contains *dry steam* with water vapor but no water droplets. Another consists of *wet steam*, a mixture of water vapor and water droplets. The third is *hot water* trapped in fractured or porous rock at various places in the earth's crust.

If such geothermal reservoirs are close to the surface, wells can be drilled to extract the dry steam, wet steam, or hot water (Figure 17-2, p. 351), which can be used to heat homes and buildings or to spin turbines and produce electricity.

There are three other nearly nondepletable sources of geothermal energy. One is *molten rock* (magma). Another is *hot dry-rock zones*, where molten rock that has penetrated the earth's crust heats subsurface rock to high temperatures. A third source is low- to moderate-temperature *warm-rock reservoir deposits*. Heat from such deposits could be used to preheat water and run heat pumps for space heating and air conditioning. Hot dry-rock zones can be found almost anywhere about 8–10 kilometers (5–6 miles) below the earth's surface. Researchers in several countries are exploring whether these zones can provide affordable geothermal energy.

Currently, about 22 countries (most of them in the developing world) are extracting energy from geothermal sites to produce about 1% of the world's electricity. Geothermal energy is used to heat about 85% of Iceland's buildings, produce electricity, and provide heat to grow most of its fruits and vegetables in greenhouses heated by geothermal energy. The world's largest operating geothermal system, called *The Geysers*, extracts energy from a dry steam reservoir north of San Francisco, California. It provides electricity for about 1.7 million homes. In 1999, Santa Monica, California, became the first city in the world to get all its electricity from geothermal energy.

Figure 18-29 lists the advantages and disadvantages of using geothermal energy. It generally has a much lower environmental impact than fossil fuel energy resources.

But geothermal energy has two main problems. One is that the cost of tapping large-scale reservoirs of geothermal energy is too high for all but the most concentrated and accessible sources. New technologies may bring these costs down.

The other is that some dry- or wet-steam geothermal reservoirs can be depleted if heat is removed faster than natural processes renew it. Thus geothermal resources can be nonrenewable on a human time scale, but the potential supply is so vast that it is usually classified as a renewable energy resource. Recirculating all of the hot water back into the underground reservoir can also slow heat depletion from such reservoirs.

18-8 HYDROGEN

Can Hydrogen Replace Oil? Good-bye Oil, Smog, and CO_2 Emissions, Hello Hydrogen

Some energy analysts view hydrogen gas as the best fuel to replace oil during the last half of this century.

When oil is gone or what is left costs too much to use, how will we fuel vehicles, industry, and buildings? Many scientists and executives of major oil companies and automobile companies say the fuel of the future is hydrogen gas (H_2)—envisioned in 1874 by science fiction writer Jules Verne in his book *The Mysterious Island.*

Figure 18-30 (p. 402) lists the advantages and disadvantages of using hydrogen as an energy resource. Electricity (electrolysis) or high temperatures (thermolysis) can be used to split water molecules into gaseous hydrogen and oxygen ($2\,H_2O \longrightarrow 2\,H_2 + O_2$). And

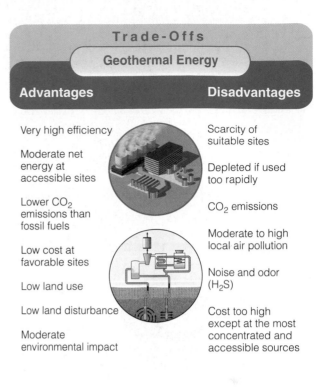

Figure 18-29 Trade-offs: advantages and disadvantages of using geothermal energy for space heating and to produce electricity or high-temperature heat for industrial processes. Pick the single advantage and the single disadvantage that you think are the most important.

when the hydrogen gas is used as a fuel it combines with oxygen gas in the air and produces nonpolluting water vapor ($2\,H_2 + O_{29} \longrightarrow 2\,H_2O$).*

Proponents envision using hydrogen in energy-efficient and nonpolluting fuel cells to provide electricity for running buses, cars (Figures 18-10 and 18-11), houses, and other buildings. Widespread use of hydrogen could provide most of the energy needed to run an economy (Figure 18-31, p. 403). Proponents believe that such systems can be available by 2020–2030 and then be phased in during this century.

So what is the catch? There are three problems in turning the vision of widespread use of hydrogen as a fuel into reality. *First,* hydrogen is chemically locked up in water and organic compounds such as methane and gasoline. *Second,* it takes energy and money to produce hydrogen from water and organic compounds. In other words, *hydrogen is not a source of energy. It is a fuel produced by using energy—lots of it. Third,* fuel cells are the best way to use hydrogen to produce electricity, but current versions are expensive.

---

*Water vapor is a potent greenhouse gas. However, because there is already so much of it in the atmosphere, human additions of this gas are insignificant.

Trade-Offs

Hydrogen

Advantages	Disadvantages
Can be produced from plentiful water	Not found in nature
Low environmental impact	Energy is needed to produce fuel
	Negative net energy
Renewable if produced from renewable energy resources	CO_2 emissions if produced from carbon-containing compounds
No CO_2 emissions if produced from water	Nonrenewable if generated by fossil fuels or nuclear power
Good substitute for oil	High costs (but may eventually come down)
Competitive price if environmental and social costs are included in cost comparisons	Will take 25 to 50 years to phase in
Easier to store than electricity	Short driving range for current fuel cell cars
Safer than gasoline and natural gas	No fuel distribution system in place
Nontoxic	Excessive H_2 leaks may deplete ozone
High efficiency (45–65%) in fuel cells	

Figure 18-30 Trade-offs: advantages and disadvantages of using hydrogen as a fuel for vehicles and for providing heat and electricity. Pick the single advantage and the single disadvantage that you think are the most important.

We could use electricity from coal-burning and conventional nuclear power plants to electrolyze water. But doing this is expensive and subjects us to the harmful environmental effects associated with using these fuels (Figure 17-21, p. 365, and Figure 17-26, p. 370). We can also use a reforming process that involves using high temperatures and chemical processes to separate hydrogen from carbon atoms in organic compounds found in conventional fuels such as natural gas, methanol, ethanol, or gasoline. A problem is that getting hydrogen from organic compounds such as methane (CH_4) produces carbon dioxide ($CH_4 + 2\ H_2O \longrightarrow 4\ H_2 + CO_2$). And according to a 2002 study by physicist Marin Hoffer and a team of other scientists, these reforming processes add more

CO_2 to the atmosphere per unit of heat generated than does burning these carbon-containing fuels directly. Thus using this approach could accelerate projected global warming unless we can develop affordable ways to store (sequester) the CO_2 underground or in the deep ocean. We can also gasify coal or biomass to produce hydrogen, but this is more expensive than using natural gas and also releases CO_2.

Most proponents of hydrogen believe that if we are to get its very low pollution and low CO_2 emission benefits, the energy used to produce H_2 by decomposing water must come from low-polluting, renewable sources that emit little or no CO_2. The most likely sources are electricity generated by wind farms, hydropower, geothermal energy, solar cells (when their prices come down), or biological processes in bacteria and algae (Spotlight, p. 404).

In 1999, DaimlerChrysler, Royal Dutch Shell, Norsk Hydro, and Icelandic New Energy announced government-approved plans to turn the tiny country of Iceland into the world's first "hydrogen economy" by 2040—the brainchild of chemist Bragi Árnason, known as "Professor Hydrogen." The country's abundant renewable geothermal energy, hydropower, and offshore winds will be used to produce hydrogen from seawater and the H_2 will be used to run its buses, passenger cars, fishing vessels, and factories. Iceland's first hydrogen service station opened in 2003.

Once hydrogen is produced we must have a way to store it for use as needed. Here are some of the ways that scientists and engineers are investigating for hydrogen storage.

Store it in compressed gas tanks either above or below the ground or aboard motor vehicles. In 2002, General Motors developed a lightweight high-pressure hydrogen storage tank that can be used on cars and can store enough hydrogen to provide a range of nearly 480 kilometers (300 miles) before refueling.

Store it as more dense liquid hydrogen. But the liquid hydrogen must be stored in tanks kept at very low temperatures. This is costly, takes almost a third of the hydrogen's original fuel energy, and requires a large amount of insulation.

Store it in solid metal hydride compounds. Certain metals can absorb and chemically bond hydrogen in their latticework of atoms. Heating such metal hydride compounds releases the hydrogen gas as needed. DaimlerChrysler has found a way to store hydrogen as *sodium borohydride* in a nontoxic and nonflammable solution that can be pumped in and out of the vehicle safely and cleanly and without leaks of hydrogen gas.

Absorb hydrogen gas on activated charcoal or graphite nanofibers. Like hydrides, this is a safe and efficient way to store hydrogen, but an input of energy is needed to release the hydrogen. *Trap and store hydrogen*

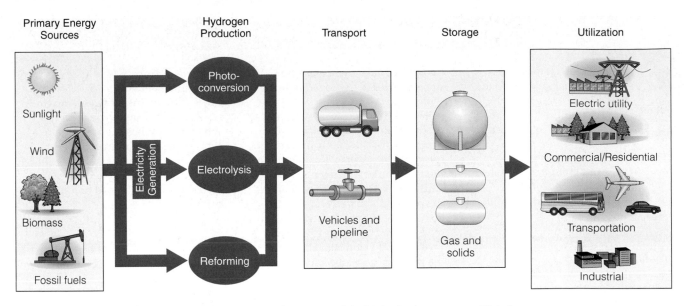

Figure 18-31 Solutions: hydrogen energy system used to run a sustainable hydrogen economy. (Data from U.S. Department of Energy and the Worldwatch Institute)

gas in a framework of water molecules called *clathrate hydrates* or in tiny glass *microspheres.* Stay tuned for further developments from this research.

Hydrogen is highly flammable and burns with an invisible flame. But it may be safer than gasoline for two reasons. *First*, when this light gas is released it quickly disperses into the atmosphere instead of posing a fire hazard by puddling on the ground like gasoline. *Second*, metal hydrides, charcoal powders, graphite, nanofibers, and glass microspheres containing hydrogen will not explode or burn if a vehicle's tank is ruptured in an accident.

Will Widespread Use of Hydrogen Decrease Protective Ozone in the Stratosphere? Probably Not With with Careful Use of Hydrogen

A preliminary study suggests that widespread use of hydrogen could decrease the concentration of protective ozone in the stratosphere over Antarctica for a few months each year.

In 2003, researchers Tracey Tromp and John Eiler at the California Institute of Technology published a paper that sent shivers down the back of hydrogen proponents. On the basis of computer models, they projected that if hydrogen eventually replaces all fossil fuels, hydrogen gas leaking from such a global system could rise into the stratosphere, be oxidized to form water vapor, increase depletion of the ozone layer over Antarctica during part of the year, and allow more harmful ultraviolet radiation to reach the earth's surface.

Most press reports failed to note that the authors and other scientists gave several reasons why this problem may not be as serious as this preliminary

study suggests. *First*, the authors' model is based on still poorly understood atmospheric chemical interactions involved in the hydrogen fuel cycle. This includes the possibility that excess hydrogen in the troposphere would be absorbed by soils or removed by reactions with other chemicals in the atmosphere before most of it can reach the stratosphere. *Second*, the assumptions about leakage of hydrogen may be much too high because of improved technology and vigilance to reduce such leaks. *Third*, global efforts are in place to drastically reduce ozone depletion in the stratosphere by 2050, mostly from chlorine and bromine compounds we have been putting into the atmosphere (more on this in Chapter 21). Since widespread use of hydrogen is not expected until after 2050, this potential threat would be greatly diminished.

What Are Some Possible Potholes in the Hydrogen Highway? Getting Diverted

Because large-scale use of hydrogen is probably 25–50 years away, we should not let its potential divert us from the immediate priorities of sharply reducing greenhouse gas emissions by increasing fuel efficiency and encouraging the use of renewable energy to help us produce hydrogen and phase out fossil fuels.

Some analysts urge the United States to spend about $100 billion over the next two decades to spur the development of a renewable-energy hydrogen revolution that would be phased in during this century. The media hype about hydrogen can divert us from the fact that it will probably not be in widespread use for 25–50 years.

Producing Hydrogen from Green Algae Found

In a few decades we may be able to use large-scale cultures of green algae to produce hydrogen gas. This simple plant grows almost everywhere and is commonly found in pond scum.

When living in air and sunlight, green algae carry out photosynthesis like other plants and produce carbohydrates and oxygen gas. However, in 2000, Tasios Melis, a researcher at the University of California at Berkeley, found a way to modify the photosynthesis process to make these algae produce bubbles of hydrogen rather than oxygen.

First, he grew cultures of hundreds of billions of the algae in the normal way with plenty of sunlight, nutrients, and water. Then he cut off their supply of two key nutrients: sulfur and oxygen. Within 20 hours, the plant cells underwent a metabolic change and switched from an oxygen-producing to a hydrogen-producing metabolism, allowing the researcher to collect hydrogen gas bubbling from the culture.

Melis believes he can increase the efficiency of this hydrogen---

producing process 10-fold. If so, sometime in the future a *biological hydrogen plant* might cycle a mixture of algae and water through a system of clear tubes exposed to sunlight to produce hydrogen. The gene responsible for producing the hydrogen might even be transferred to other plants to produce hydrogen.

Critical Thinking

What might be some ecological problems related to the widespread use of this method for producing hydrogen?

While we are working to develop a renewable-energy hydrogen revolution, energy analysts call for us to focus on two more immediate and important priorities. One is to begin sharply reducing our dependence on carbon-containing fossil fuels—especially oil and coal, which emit large quantities of carbon dioxide. The other is the related challenge of sharply reducing our emissions of carbon dioxide and other greenhouse gases to help slow global warming and climate change. Analysts suggest that we do this by

- Greatly improving fuel-efficiency standards for motor vehicles through a combination of mandatory government standards and much higher taxes on gasoline and diesel fuels, coupled with a corresponding reduction in income and payroll taxes. This could be done within 10 years. Some energy analysts accuse car companies of misleading the public by saying that we do not need to increase government (CAFE) fuel-efficiency standards because we can depend on hydrogen.

- Providing large tax breaks for people and businesses using fuel-efficient cars, buildings, heating systems, and household appliances and keeping such breaks in place for at least 25 years

- Investing much more in public transit running on less polluting natural gas as an alternative to the car and using at least half of the money collected by gasoline taxes to promote this change

- Greatly increasing research and development subsidies for development and phasing in of renewable-energy technologies, such as wind power, solar cells, biomass, and geothermal energy, and providing such subsidies for at least 25 years. Such non-carbon energy technologies will be needed to produce hydrogen.

- Providing very large tax breaks for people and businesses using renewable-energy technologies and keeping such breaks in place for at least 25 years.

It will take large amounts of fossil fuel energy and money to phase in the use of hydrogen during the last two-thirds of this century. If we do not conserve fossil fuels their prices might rise to the point where we could not afford to use them to help us make the transition to a renewable-energy hydrogen economy by the end of this century. Also, failing to reduce the threat of climate change is likely to divert huge amounts of money from hydrogen to help us deal with the harmful effects of climate change.

18-9 ENTERING THE AGE OF DECENTRALIZED MICROPOWER

What Is Micropower? Think Small and Dispersed

Energy analysts expect dispersed, small-scale energy-generating units to replace centralized, large-scale power plants over the next few decades.

According to Chuck Linderman, director of energy supply policy for the Edison Electric Institute, the era of big central power plant systems is coming to a close. He and other energy analysts believe the chief feature of electricity production over the next few decades will be

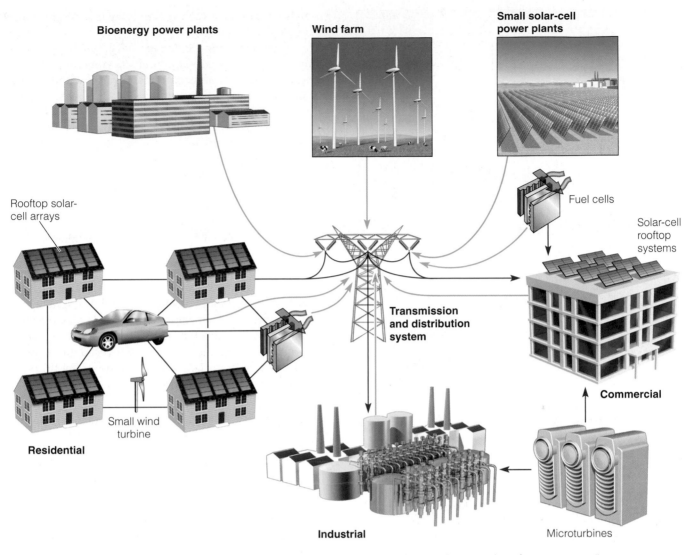

Bioenergy power plants

Wind farm

Small solar-cell power plants

Rooftop solar-cell arrays

Fuel cells

Solar-cell rooftop systems

Transmission and distribution system

Commercial

Small wind turbine

Residential

Industrial

Microturbines

Figure 18-32 Solutions: *decentralized power system* in which electricity is produced by a large number of dispersed, small-scale *micropower systems*. Some would produce power on site and others would feed the power they produce into a conventional electrical distribution system. Over the next few decades, many energy and financial analysts expect a shift to this type of power system.

decentralization to dispersed, small-scale **micropower systems** that generate 1–10,000 kilowatts (Figure 18-32). This shift from centralized *macropower* to dispersed *micropower* is analogous to the computer industry's shift from large centralized mainframes to increasingly smaller, widely dispersed PCs, laptops, and handheld computers.

Figure 18-33 (p. 406) lists some of the advantages of decentralized micropower systems over traditional macropower systems. The potential for financial gain by companies and investors in micropower systems is huge, with a $10 trillion market projected by 2018. Decentralized micropower systems could also work well for 1.7 billion people in isolated villages in developing countries.

18-10 A SUSTAINABLE ENERGY STRATEGY

What Roles Will Economics and Politics Play in Our Energy Future? Rewards Pay Off

Governments can use a combination of subsidies, tax breaks, and taxes to promote or discourage use of various energy alternatives.

To most analysts the key to making a shift to more sustainable energy resources and societies is economics and politics. Governments can use two basic economic and political strategies to help stimulate or discourage use of a particular energy resource.

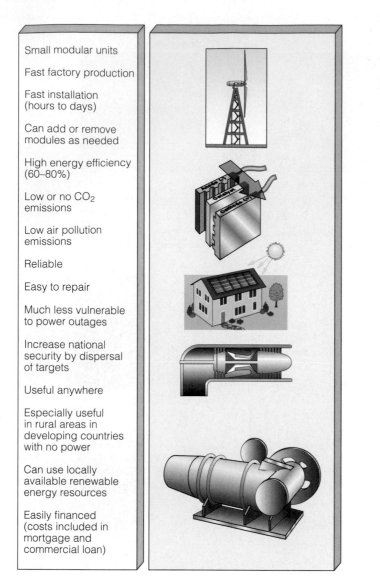

Figure 18-33 Solutions: advantages of micropower systems.

- Small modular units

- Fast factory production

- Fast installation (hours to days)

- Can add or remove modules as needed

- High energy efficiency (60–80%)

- Low or no CO_2 emissions

- Low air pollution emissions

- Reliable

- Easy to repair

- Much less vulnerable to power outages

- Increase national security by dispersal of targets

- Useful anywhere

- Especially useful in rural areas in developing countries with no power

- Can use locally available renewable energy resources

- Easily financed (costs included in mortgage and commercial loan)

One approach is to *keep energy prices artificially low to encourage use of selected energy resources.* This is done mostly by providing research and development subsidies and tax breaks, and by enacting regulations that help stimulate the development and use of energy resources receiving such support. For decades, this approach has been used to help the fossil fuel and nuclear power industries in the United States (Figure 18-34) and in most other developed countries. This approach has created an uneven economic playing field that encourages energy waste and rapid depletion of nonrenewable energy resources and discourages the development of other energy alternatives and improvements in energy efficiency. For example, in the United States people who buy the biggest SUVs for business cars get a tax deduction of up to $100,000. People buying an energy-efficient hybrid car got a $1,500 tax deduction in 2004, but this is being reduced to $500 by 2006. Energy

analysts say this reward system is the reverse of what it should be.

A second option is to *keep energy prices artificially high to discourage use of a resource.* Governments can raise the price of an energy resource by withdrawing existing tax breaks and other subsidies, enacting restrictive regulations, or adding taxes on its use. This increases government revenues, encourages improvements in energy efficiency, reduces dependence on imported energy, and decreases use of an energy resource that has a limited future supply.

Many economists favor *increasing taxes on fossil fuels* as a way to reduce air and water pollution, slow greenhouse gas emissions, and encourage improvements in energy efficiency and greater use of renewable energy. For example in Germany and Great Britain, where gasoline costs more than $1.30 per liter ($5 per gallon), overall oil and gasoline consumption has fallen since the 1970s. Some economists believe the public might accept these higher taxes if income and payroll taxes were lowered as gasoline or other fossil fuel taxes were raised. And energy assistance would be provided for the poor and lower middle class who would bear the brunt of taxes on gasoline and other energy-intensive goods.

X *HOW WOULD YOU VOTE?* Should the government increase taxes on fossil fuels and offset this by reducing income and payroll taxes and providing an energy safety net for the poor and lower middle class? Cast your vote online at http://biology .brookscole.com/miller14.

How Can We Develop a More Sustainable Energy Future? Stop the Waste, Use the Sun, and Cut Pollution

A more sustainable energy policy would improve energy efficiency, rely more on renewable energy, and reduce the harmful environmental effects of using fossil fuels and nuclear energy.

Figure 18-35 lists strategies for making the transition to a more sustainable energy future over the next few decades. Energy analysts also call for the United States to modernize its aging electrical grid system. Energy analysts describe the United States as a first-world nation with a third-world electrical grid system. This system is highly vulnerable to disruption from unforeseen power outages, sabotage by terrorists, and attacks by cyber-terrorists on the computer programs that run it. They call for the country to waste no time in transforming it into a smart, flexible, responsive, and digitally controlled network.

Energy analysts estimate that implementing policies such as those shown in Figure 18-35 over the next several decades could save money, create a net gain in jobs, reduce greenhouse gas emissions, and sharply reduce air and water pollution. According to proponents, these policies would also increase national secu-

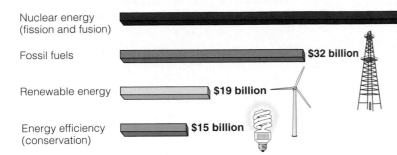

Nuclear energy
(fission and fusion) .... $73 billion

Fossil fuels .... $32 billion

Renewable energy .... $19 billion

Energy efficiency
(conservation) .... $15 billion

Figure 18-34 U.S. energy policy priorities. U.S. Department of Energy research and development funding for various sources of energy, 1948–2003. If other government subsidies and tax breaks are included, the figures for nuclear power and fossil fuels are much higher, and fossil fuels receive approximately 60% of these benefits, nuclear energy 30%, and renewable energy and energy conservation only 10%. (U.S. Department of Energy)

rity in two ways: first by reducing dependence on imported oil, and second by decreasing dependence on large and centralized nuclear power and coal plants that are vulnerable to terrorist attacks.

We have the technology, creativity, and wealth to make the transition to a more sustainable energy future. But making this transition depends primarily on *politics*, and thus on pressure individuals and groups can put on elected officials and officials of energy resource companies by voting with their ballots and pocketbooks (by refusing to buy some products and

letting company executives know why). Figure 18-36 (p. 408) lists some ways you can contribute to making this transition by reducing the amount of energy you use and waste.

A transition to renewable energy is inevitable, not because fossil fuel supplies will run out—large reserves of oil, coal, and gas remain in the world—but because the costs and risks of using these supplies will continue to increase relative to renewable energy.

MOHAMED EL-ASHRY

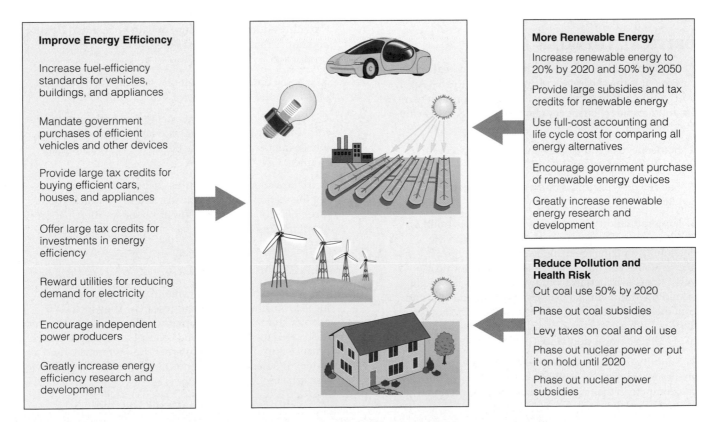

Improve Energy Efficiency

Increase fuel-efficiency standards for vehicles, buildings, and appliances

Mandate government purchases of efficient vehicles and other devices

Provide large tax credits for buying efficient cars, houses, and appliances

Offer large tax credits for investments in energy efficiency

Reward utilities for reducing demand for electricity

Encourage independent power producers

Greatly increase energy efficiency research and development

More Renewable Energy

Increase renewable energy to 20% by 2020 and 50% by 2050

Provide large subsidies and tax credits for renewable energy

Use full-cost accounting and life cycle cost for comparing all energy alternatives

Encourage government purchase of renewable energy devices

Greatly increase renewable energy research and development

Reduce Pollution and Health Risk

Cut coal use 50% by 2020

Phase out coal subsidies

Levy taxes on coal and oil use

Phase out nuclear power or put it on hold until 2020

Phase out nuclear power subsidies

Figure 18-35 Solutions: suggestions of various energy analysts to help make the transition to a more sustainable and less risky energy future.

- Drive a car that gets at least 15 kilometers per liter (35 miles per gallon) and join a carpool.

- Use mass transit, walking, and bicycling.

- Superinsulate your house and plug all air leaks.

- Turn off lights, TV sets, computers, and other electronic equipment when they are not in use.

- Wash laundry in warm or cold water.

- Use passive solar heating.

- For cooling, open windows and use ceiling fans or whole-house attic or window fans.

- Turn thermostats down in winter and up in summer.

- Buy the most energy-efficient homes, lights, cars, and appliances available.

- Turn down the thermostat on water heaters to 43–49°C (110–120°F) and insulate hot water heaters and pipes.

Figure 18-36 What can you do? Ways to reduce your use and waste of energy.

CRITICAL THINKING

1. A home builder installs electric baseboard heat and claims, "It is the cheapest and cleanest way to go." Apply your understanding of the second law of thermodynamics and net energy efficiency chain (Figure 18-6) to evaluate this claim.

2. Someone tells you we can save energy by recycling it. How would you respond?

3. Should gas-guzzling motor vehicles be taxed heavily? Explain.

4. Congratulations! You have won $250,000 to build a house of your choice anywhere you want. With the goal of maximizing energy efficiency, what type of house would you build? Where would you locate it? What types of materials would you use? What types of materials would you *not* use? How would you heat and cool the house? How would you heat water? What type of lighting, stove, refrigerator, washer, and dryer would you use? Which of these appliances could you do without?

5. Should government subsidies and tax breaks for all energy alternatives be eliminated so all energy choices can compete in the marketplace on an even economic footing? Explain.

6. Should government tax breaks and other subsidies for fossil fuels and nuclear power be phased out and re-placed with subsidies and tax breaks for improving energy efficiency and renewable energy alternatives? Explain.

7. Explain why you agree or disagree with each of the proposals suggested in Figure 18-35 (p, 407) as ways to promote a more sustainable energy future.

8. List the parts of your daily life that depend on the electrical grid system.

9. Congratulations! You are in charge of the U.S. Department of Energy (or the energy agency in the country where you live). What proportions of your research and development budget would you devote to fossil fuel, nuclear power, renewable energy, and improving energy efficiency? How would you distribute your funds among the various types of renewable energy?

10. Congratulations! You are in charge of the world. List the five most important features of your energy policy.

PROJECTS

1. Make a study of energy use in your school and use the findings to develop an energy-efficiency improvement program. Present your plan to school officials.

2. Learn how easy it is to produce hydrogen gas from water using a battery, some wire for two electrodes, and a dish of water. Hook a wire to each of the poles of the battery, immerse the electrodes in the water, and observe bubbles of hydrogen gas being produced at the negative electrode and bubbles of oxygen at the positive electrode. Carefully add a small amount of battery acid to the water and notice that this increases the rate of hydrogen production.

3. Use the library or the Internet to compare the energy policies of the United States, Germany, and China.

4. Use the library or the Internet to find bibliographic information about *Amory B. Lovins* and *Mohamed El-Ashry*, whose quotes appear at the beginning and end of this chapter.

5. Make a concept map of this chapter's major ideas, using the section heads, subheads, and key terms (in boldface). Look on the website for this book for information about making concept maps.

LEARNING ONLINE

The website for this book contains study aids and many ideas for further reading and research. They include a chapter summary, review questions for the entire chapter, flash cards for key terms and concepts, a multiple-choice practice quiz, interesting Internet sites, references, and a guide for accessing thousands of InfoTrac® College Edition articles. Log on to

http://biology.brookscole.com/miller14

Then click on the Chapter-by-Chapter area, choose Chapter 18, and select a learning resource.

CASE STUDY
The Big Killer

What is roughly the diameter of a 30-caliber bullet, can be bought almost anywhere, is highly addictive, and kills about 13,700 people every day, or one every 6 seconds? It is a cigarette. *Cigarette smoking is the world's most preventable major cause of suffering and premature death among adults.*

According to the World Health Organization (WHO), tobacco helped kill about 80 million people between 1950 and 2004. This is 2.6 times more than the 30 million people killed in battle in all wars during the 20th century!

The WHO estimates that each year tobacco contributes to the premature deaths of at least 5 million people (about half from developed countries and half from developing countries) from 34 illnesses including *heart disease, lung cancer, other cancers, bronchitis, emphysema,* and *stroke.* By 2030 the annual death toll from smoking-related diseases is projected to reach 10 million (Figure 1-15, p. 17)—an average of about 27,400 preventable deaths per day or 1 death every 3 seconds. About 70% of these deaths are expected to occur in developing countries.

According to a 2002 study by the Centers for Disease Control and Prevention, smoking kills about 442,000 Americans per year prematurely, an average of 1,211 deaths per day (Figure 19-1). This death toll is roughly equivalent to three fully loaded 400-passenger jumbo jets crashing *every day* with no survivors!

Yet, this ongoing major human tragedy rarely makes the news.

The overwhelming consensus in the scientific community is that the nicotine inhaled in tobacco smoke is highly addictive. Only 1 in 10 people who try to quit smoking succeed, about the same relapse rate as for recovering alcoholics and those addicted to heroin or crack cocaine. A British government study showed that adolescents who smoke more than one cigarette have an 85% chance of becoming smokers. People can also be exposed to secondhand smoke from others, called *passive smoking.*

According to a 2002 study by the Centers for Disease Control and Prevention, smoking in the United States costs about $158 billion a year for medical bills, increased insurance costs, disability, lost earnings and productivity because of illness, and property damage from smoking-caused fires. This is an average of about $7 per pack of cigarettes sold in the United States.

Many health experts urge that a $3–5 federal tax be added to the price of a pack of cigarettes in the United States. Such a tax would mean that the users of cigarettes (and other tobacco products), not the rest of society, would pay a much greater share of the health, economic, and social costs associated with their smoking.

Other suggestions for reducing the death toll and health effects of smoking in the United States (and in other countries) include banning all cigarette advertising, prohibiting the sale of cigarettes and other tobacco products to anyone under 21 (with strict penalties for violators), and banning cigarette vending machines. Analysts also call for classifying and regulating the use of nicotine as an addictive and dangerous drug, eliminating all federal subsidies and tax breaks to tobacco farmers and tobacco companies, and using cigarette tax income to finance an aggressive antitobacco advertising and education program.

So far, the U.S. Congress has not enacted such reforms. Critics say this is mostly because tobacco companies donated tens of millions of dollars to candidates running for Congress and the presidency.

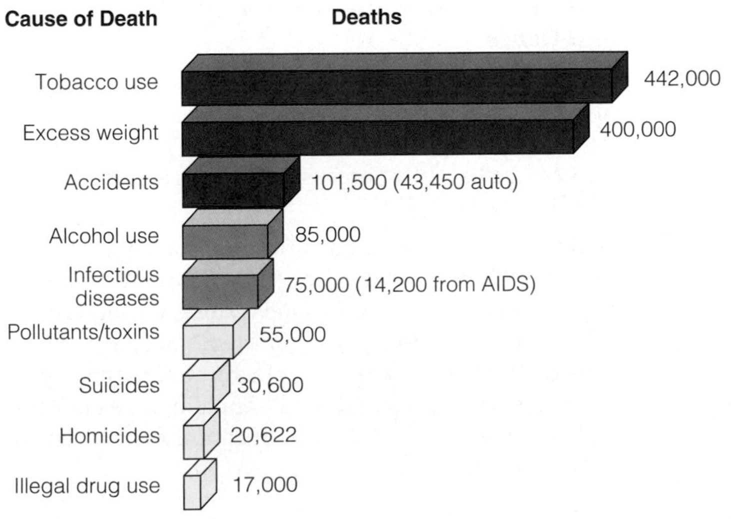

Cause of Death   **Deaths**

- Tobacco use — 442,000
- Excess weight — 400,000
- Accidents — 101,500 (43,450 auto)
- Alcohol use — 85,000
- Infectious diseases — 75,000 (14,200 from AIDS)
- Pollutants/toxins — 55,000
- Suicides — 30,600
- Homicides — 20,622
- Illegal drug use — 17,000

Figure 19-1 Annual deaths in the United States from tobacco use and eight other causes in 2003. Smoking is by far the nation's leading cause of preventable death. (U.S. National Center for Health Statistics and Centers for Disease Control and Prevention and U.S. Surgeon General)

The dose makes the poison.

PARACELSUS, 1540

This chapter addresses the following questions:

- What types of hazards do people face?

- What is toxicology, and how do scientists measure toxicity?

- What chemical hazards do people face, and how can they be measured?

- What types of disease (biological hazards) threaten people in developing countries and developed countries?

- How can risks be estimated, managed, and reduced?

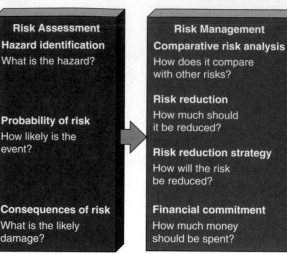

Figure 19-2 *Risk assessment* and *risk management.*

19-1 RISK, PROBABILITY, AND HAZARDS

What Is Risk? The Chances We Take

Risk is a measure of the likelihood that you will suffer harm from a hazard.

Risk is the *possibility* of suffering harm from a hazard that can cause injury, disease, death, economic loss, or environmental damage. **Risk assessment** is the scientific process of estimating how much harm a particular hazard can cause to human health. **Risk management** involves deciding whether or how to reduce a particular risk to a certain level and at what cost.

Risk is usually expressed in terms of *probability:* a mathematical statement about how likely one is to suffer harm from a hazard. Scientists often state probability in terms such as "The lifetime probability of developing lung cancer from smoking a pack of cigarettes a day is 1 in 250." This means that 1 of every 250 people who smoke a pack of cigarettes a day will develop lung cancer over a typical lifetime (usually considered 70 years).

It is important to distinguish between *possibility* and *probability.* When we say that it is *possible* that a smoker can get lung cancer we are saying that this event could happen. *Probability* gives us an estimate of the likelihood of such an event. Figure 19-2 summarizes how risks are assessed and managed.

What Are the Major Types of Hazards? They Are All Around Us, But How Risky Are They?

We can suffer harm from cultural hazards, chemical hazards, physical hazards, and biological hazards, but determining the risks involved is difficult.

We can suffer harm from four major types of hazards:

- *Cultural hazards* such as unsafe working conditions, smoking, poor diet, drugs, drinking, driving, criminal assault, unsafe sex, and poverty

- *Physical hazards* such as ionizing radiation, fire, tornado (Figure 6-3, p. 103), hurricane (Figure 6-4, p. 104), flood (Figure 15-24, p. 327), volcanic eruption (Figure 16-8, p. 338), and earthquake (Figure 16-6, p. 337)

- *Chemical hazards* from harmful chemicals in the air, water, soil, and food

- *Biological hazards* from pathogens (bacteria, viruses, and parasites), pollen and other allergens, and animals such as bees and poisonous snakes

19-2 TOXICOLOGY: ASSESSING CHEMICAL HAZARDS

What Determines Whether a Chemical Is Harmful? How Much, How Often, and Genes

The harm caused by exposure to a chemical depends on the amount of exposure (dose), frequency of exposure, who is exposed, how well the body's detoxification systems work, and one's genetic makeup.

Toxicity measures how harmful a substance is in causing injury, illness, or death to a living organism. This depends on several factors. One is **dose,** the amount of a substance a person has ingested, inhaled, or absorbed through the skin. Other factors are frequency of exposure, who is exposed (adult or child, for example), and how well the body's detoxification systems (such as the liver, lungs, and kidneys) work.

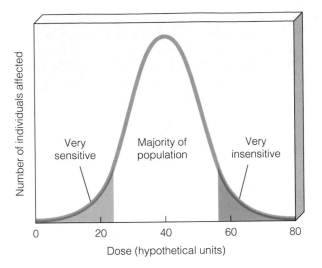

Figure 19-3 Typical variations in sensitivity to a toxic chemical within a population, mostly because of differences in genetic makeup. Some individuals in a population are very sensitive to small doses of a toxin (left), and others are very insensitive (right). Most people fall between these two extremes (middle).

The y-axis is labeled "Number of individuals affected". The x-axis is labeled "Dose (hypothetical units)" with values 0, 20, 40, 60, 80. The curve labels are "Very sensitive", "Majority of population", and "Very insensitive".

Toxicity also depends on *genetic makeup* that determines an individual's sensitivity to a particular toxin (Figure 19-3). This genetic variation in individual responses to exposure to various toxins raises a difficult ethical, political, and economic question. When regulating levels of a toxic substance in the environment, should the allowed level be set to protect the most sensitive individuals (at great cost) or the average person? What is your view on this issue? Why?

Five major factors can affect the harm caused by a substance. One is its *solubility*. *Water-soluble toxins* (which are often inorganic compounds) can move throughout the environment and get into water supplies and the aqueous solutions that surround the cells in our bodies.

Oil- or fat-soluble toxins (which are usually organic compounds) can penetrate the membranes surrounding an organism's cells because the membranes allow similar oil-soluble chemicals to pass through them. Thus, oil- or fat-soluble toxins can accumulate in body tissues and cells.

A second factor is a substance's *persistence*. Many chemicals, such as the pesticide DDT (banned in many countries but still used in some), are often used because of their persistence or resistance to breakdown. They do their job for a long time. But this persistence also means they can have long-lasting harmful effects on the health of wildlife and people.

A third factor for some substances is **bioaccumulation,** in which some molecules are absorbed and stored in specific organs or tissues at higher than normal levels. This means that a chemical found at a fairly low concentration in the environment can build up to a harmful level in certain organs and tissues.

A related factor is **biomagnification,** in which levels of some potential toxins in the environment are magnified as they pass through food chains and webs. Organisms at low trophic levels might ingest only small amounts of a toxin, but each animal on the next level up that eats many of those organisms will take in larger amounts of that toxin. As the toxin moves through higher trophic levels, organisms at each level consume increasingly greater amounts of the toxin. Figure 19-4 provides an illustration of this effect. Examples of chemicals that can be biomagnified include long-lived, fat-soluble organic compounds such as DDT,

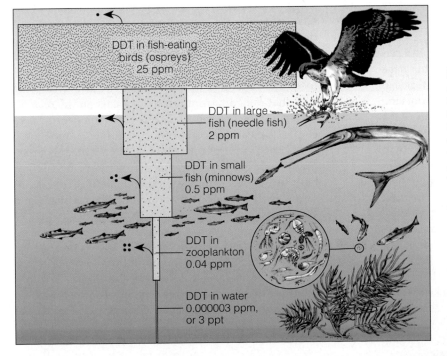

DDT in fish-eating birds (ospreys) 25 ppm

DDT in large fish (needle fish) 2 ppm

DDT in small fish (minnows) 0.5 ppm

DDT in zooplankton 0.04 ppm

DDT in water 0.000003 ppm, or 3 ppt

Figure 19-4 *Bioaccumulation* and *biomagnification.* DDT is a fat-soluble chemical that can accumulate in the fatty tissues of animals. In a food chain or web, the accumulated DDT can be biologically magnified in the bodies of animals at each higher trophic level. This diagram shows that the concentration of DDT in the fatty tissues of organisms was biomagnified about 10 million times in this food chain in an estuary near Long Island Sound in New York. If each phytoplankton organism takes up from the water and retains one unit of DDT, a small fish eating thousands of zooplankton (which feed on the phytoplankton) will store thousands of units of DDT in its fatty tissue. Then each large fish that eats 10 of the smaller fish will ingest and store tens of thousands of units, and each bird (or human) that eats several large fish will ingest hundreds of thousands of units. Black dots represent DDT, and arrows show small losses of DDT through respiration and excretion.

PCBs (oily chemicals used in electrical transformers), and some radioactive isotopes (such as strontium-90).

A fifth factor is *chemical interactions* that can decrease or multiply the harmful effects of a toxin. An *antagonistic interaction* can reduce harmful effects. For example, vitamins E and A apparently interact to reduce the body's response to some cancer-causing chemicals.

A *synergistic interaction* multiplies harmful effects. For instance, workers exposed to tiny fibers of asbestos increase their chances of getting lung cancer 20-fold. But asbestos workers who also smoke have a 400-fold increase in lung cancer rates. In such cases, one plus one can be a lot greater than two.

The effects of exposure to a chemical can be acute or chronic. The type and amount of health damage resulting from exposure to a chemical or other agent are called the **response**. An *acute effect* is an immediate or rapid harmful reaction to an exposure—ranging from dizziness to death. A *chronic effect* is a permanent or long-lasting consequence (kidney or liver damage, for example) from exposure to a single dose or to repeated sublethal doses of a harmful substance.

What Are Some Basic Principles of Toxicology? The Dose Makes the Poison—Or Does It?

Any substance can be harmful if ingested in a large enough quantity, but the critical question is, what is the lowest level of exposure that causes harm?

A basic concept of toxicology is that *any synthetic or natural chemical can be harmful if ingested in a large enough quantity.* In other words, every chemical is harmful at some level of exposure. For example, drinking 100 cups of strong coffee one after another would expose most people to a lethal dosage of caffeine. Similarly, downing 100 tablets of aspirin or 1 liter (1.1 quarts) of pure alcohol (ethanol) would kill most people.

The critical question is, *how much exposure to a particular toxic chemical causes a harmful response?* This is the meaning of the chapter-opening quote by the German scientist Paracelsus about the dose making the poison.

A basic problem is that people vary in terms of the dose of a toxin they can tolerate without significant harm, because of differences in their genetic makeup (Figure 19-3). Because of this variation in how individuals respond to exposure to a toxic chemical, a better way to state Paracelsus' principle of toxicology is: The dose makes the poison, *but differently for different individuals.*

Your body has three major mechanisms for reducing the harmful effects of some chemicals. *First,* it can break down (usually by enzymes found in the liver), dilute, or excrete—for example, in your breath, sweat, and urine—small amounts of most toxins to keep them from reaching harmful levels. However, accumulations of high levels of toxins can overload the ability of your liver and kidneys to degrade and excrete such substances.

Second, your cells have enzymes that can sometimes repair damage to DNA and protein molecules. *Third,* cells in some parts of your body (such as your skin and the linings of your gastrointestinal tract, lungs, and blood vessels) can reproduce fast enough to replace damaged cells. However, such high rates of cell reproduction can be altered by exposure to ionizing radiation and certain chemicals so that cell growth accelerates and creates a nonmalignant or malignant (cancerous) tumor.

Should We Be Concerned about Trace Levels of Toxic Chemicals? It Depends on the Chemical.

Trace amounts of chemicals in the environment or your body may or may not be harmful.

Should we be concerned about trace amounts of various chemicals in air, water, food, and our bodies? The honest answer is that we do not know in most cases because of a lack of data and the difficulty of determining the effects of exposures to low levels of chemicals.

Some scientists think that trace levels of most chemicals are not harmful. They point to the dramatic increase in average life expectancy in the United States since 1950. They say we should concentrate limited research funds on much greater health risks such as smoking, obesity, and infectious diseases (especially those that affect people in developing countries).

Other scientists are not so sure and believe that much more research is needed to help us evaluate the possible long-term harm caused by exposure to low levels of thousands of new synthetic chemicals that we have put into the environment during the past few decades.

Chemists are able to detect increasingly small amounts of potentially toxic chemicals in air, water, and food. This is good news, but it can give the false impression that dangers from toxic chemicals are increasing when in some cases all we are doing is uncovering levels of chemicals that have been around for a long time.

Some people also have the mistaken idea that natural chemicals are safe and synthetic chemicals are harmful. In fact, many synthetic chemicals are quite safe if used as intended, and many natural chemicals are deadly.

The average person, for instance, is far more likely to be killed by aflatoxin, a carcinogen produced by molds in peanut butter and corn, than to be killed by lightning or by a shark. However, the chance of dying of cancer from eating several spoonfuls of peanut butter a day is quite small.

How Can We Estimate the Toxicity of a Chemical? Kill Half of the Animals in a Test Population

Chemicals vary widely in their toxicity to humans and other animals.

A **poison** or **toxin** is a chemical that adversely affects the health of a living human or animal by causing injury, illness, or death. One method for determining the relative toxicity of various chemicals is to measure its effects on test animals. A widely used method for estimating the toxicity of a chemical is to determine its *lethal dose* (*LD*). This is often done by measuring a chemical's **median lethal dose** or **LD50**: the amount received in one dose that kills 50% of the animals (usually rats and mice) in a test population within a 14-day period (Figure 19-5).

Chemicals vary widely in their toxicity (Table 19-1, p. 414). Some poisons can cause serious harm or death after a single acute exposure at very low dosages. Others cause such harm only at dosages so huge that it is nearly impossible to get enough into the body. Most chemicals fall between these two extremes. In 2004, the U.S. Environmental Protection Agency listed arsenic, lead, mercury, vinyl chloride (used to make polyvinylchloride or PVC plastics), and polychlorinated biphyenyls (PCBs) in order as the five top toxic substances in terms of human and environmental health in the list of 276 substances it regulates under the Comprehensive Environmental Response, Compensation, and Liability Act, commonly known as the Superfund Act.

How Do Scientists Use Case Reports and Epidemiological Studies to Estimate Toxicity? Reports from the Front Lines and Controlled Experiments

We can estimate toxicity by using case reports about the harmful effects of chemicals on human health and by comparing the health of a group of people exposed to a chemical with that of a similar group not exposed to the chemical.

Scientists use various methods to get information about the harmful effects of chemicals on human health. One is *case reports,* usually made by physicians. They provide information about people suffering some adverse health effect or death after exposure to a chemical. Such information often involves accidental poisonings, drug overdoses, homicides, or suicide attempts.

Most case reports are not reliable sources for estimating toxicity because the actual dosage and the exposed person's health status are often not known. But such reports can provide clues about environmental hazards and suggest the need for laboratory investigations.

Another source of information is *epidemiological studies.* They involve comparing the health of people exposed to a particular chemical (the *experimental group*) with the health of another group of statistically similar people not exposed to the agent (the *control group*). The goal is to determine whether the statistical association between exposure to a toxic chemical and a health problem is strong, moderate, weak, or undetectable.

Three factors can limit the usefulness of epidemiological studies. One problem is that in many cases too few people have been exposed to high enough levels of a toxic agent to detect statistically significant differences. Another limitation is that conclusively linking an observed effect with exposure to a particular chemical is difficult because people are exposed to many different toxic agents throughout their lives. Another limitation is that we cannot use epidemiological studies to evaluate hazards from new technologies or chemicals to which people have not been exposed.

How Do Scientists Use Laboratory Experiments to Estimate Toxicity? Controversial Animal Testing

Exposing a population of live laboratory animals (especially mice and rats) to known amounts of a chemical is the most widely used method for determining its toxicity.

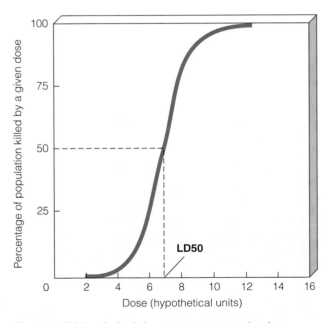

Figure 19-5 Hypothetical *dose-response curve* showing determination of the LD50, the dosage of a specific chemical that kills 50% of the animals in a test group. This is one method that toxicologists use to determine and compare the toxicities of different chemicals.

Table 9-1 Toxicity Ratings and Average Lethal Doses for Humans

Toxicity Rating	LD50 (milligrams per kilogram of body weight)*	Average Lethal Dose†	Examples
Supertoxic	Less than 0.01	Less than 1 drop	Nerve gases, botulism toxin, mushroom toxins, dioxin (TCDD)
Extremely toxic	Less than 5	Less than 7 drops	Potassium cyanide, heroin, atropine, parathion, nicotine
Very toxic	5–50	7 drops to 1 teaspoon	Mercury salts, morphine, codeine
Toxic	50–500	1 teaspoon to 1 ounce	Lead salts, DDT, sodium hydroxide, sodium fluoride, sulfuric acid, caffeine, carbon tetrachloride
Moderately toxic	500–5,000	1 ounce to 1 pint	Methyl (wood) alcohol, ether, phenobarbital, amphetamines (speed), kerosene, aspirin
Slightly toxic	5,000–15,000	1 pint to 1 quart	Ethyl alcohol, Lysol, soaps
Essentially nontoxic	15,000 or greater	More than 1 quart	Water, glycerin, table sugar

*Dosage that kills 50% of individuals exposed
†Amounts of substances in liquid form at room temperature that are lethal when given to a 70.4-kilogram (155-pound) human

The most widely used method for determining toxicity is to expose a population of live laboratory animals (especially mice and rats) to measured doses of a specific substance under controlled conditions. Animal tests take 2–5 years and cost $200,000 to $2 million per substance tested. Such tests also kill or harm and can be painful to the test animals. The goal is to develop data on the response of the test animals to various doses of a chemical (called a dose-response curve). But estimating the effects of low doses is difficult.

Animal welfare groups want to limit or ban use of test animals or ensure that experimental animals are treated in the most humane manner possible. More humane methods for carrying out toxicity tests are available. They include computer simulations and using tissue cultures of cells and bacteria, chicken egg membranes, and measurements of changes in the electrical properties of individual animal cells.

These alternatives can greatly decrease the use of animals for testing toxicity. But many scientists contend that some animal testing is needed because the alternative methods cannot adequately mimic the complex biochemical interactions of a live animal.

Acute toxicity tests are run to develop a **dose-response curve,** which shows the effects of various dosages of a toxic agent on a group of test organisms (Figure 19-6). Such tests are *controlled experiments* in which the effects of the chemical on a *test group* are compared with the responses of a *control group* of organisms not exposed to the chemical. Care is taken that organisms in both groups are as identical as possible in age, health status, and genetic makeup, and that all are exposed to the same environmental conditions.

Fairly high dosages are used to reduce the number of test animals needed, obtain results quickly, and lower costs. Otherwise, tests would have to be run on millions of laboratory animals for many years, and manufacturers could not afford to test most chemicals.

For the same reasons, scientists usually use mathematical models to extrapolate the results of high-dose exposures to low-dose levels. Then they extrapolate the low-dose results on the test organisms to humans to estimate LD50 values for acute toxicity (Table 19-1).

According to the *nonthreshold dose-response model* (Figure 19-6, left), any dosage of a toxic chemical or ionizing radiation causes harm that increases with the

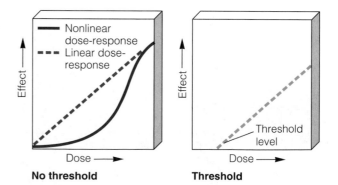

Figure 19-6 Two types of *dose-response curves*. The linear and nonlinear curves in the left graph apply if even the smallest dosage of a chemical or ionizing radiation has a harmful effect that increases with the dosage. The curve on the right applies if a harmful effect occurs only when the dosage exceeds a certain *threshold level*. Which model is better for a specific harmful agent is uncertain because of the difficulty in estimating the response to very low dosages.

dosage. With the *threshold dose-response model* (Figure 19-6, right), a threshold dosage must be reached before any detectable harmful effects occur, presumably because the body can repair the damage caused by low dosages of some substances. Establishing which of these models applies at low dosages is extremely difficult and controversial. To be on the safe side, scientists usually use the nonthreshold dose-response model.

Some scientists challenge the validity of extrapolating data from test animals to humans because human physiology and metabolism often differ from those of the test animals. Other scientists say that such tests and models work fairly well (especially for revealing cancer risks) when the correct experimental animal is chosen or when a chemical is toxic or harmful to several different test animal species.

The problem of estimating toxicities is difficult. One problem is that in real life each of us is exposed to a variety of chemicals, some of which can interact in ways to decrease or enhance their individual effects over short and long times. Thus we could further modify Paracelsus' original idea as follows: The dose *of a usually unknown mixture of chemicals* makes the poison, but differently for different individuals.

There are more problems. Toxicologists have great difficulty in estimating the toxicity of a single substance. Adding the problem of evaluating mixtures of potentially toxic substances, separating out which ones are the culprits, and determining how they can interact with one another is overwhelming from a scientific and economic standpoint. For example, just studying the interactions of all possible combinations of three of the 500 most widely used industrial chemicals would take 20.7 million experiments—a physical and financial impossibility.

The effects of a particular chemical can also depend upon when exposure occurs. For example, children can be much more susceptible to toxic substances than an adult for several reasons. On a per weight basis children breathe more air, drink more water, and eat more food than do adults. They are also exposed to toxins in dust or soil when they frequently put their fingers, toys, or other objects in their mouths. In addition, immune systems and processes for degrading or excreting toxins and repairing damage are usually less well developed in children than in adults.

In 2003, the U.S. Environmental Protection Agency proposed that in determining risk, regulators should assume that the risk of children getting cancer from exposure to chemicals that can cause cancer is 10 times the exposure risk of adults. Some health scientists contend that these guidelines are too weak and to be on the safe side regulators should assume that the risk of harm from toxins for children should be 100 times that of adults.

In 2003, the U.S. government initiated a National Children's Study that will follow the health and exposure levels to key toxins for 100,00 children from birth to age 18. As you can see, toxicologists have important but difficult jobs.

Can a Little Bit of Arsenic or Radiation Be Good for You? Controversy over Hormesis

There is controversy over the hypothesis that very small doses of radiation and some toxins may have beneficial health effects.

There is a hypothesis that radiation and some toxic substances that can harm or kill us at high doses may have beneficial health effects at very low doses. This phenomenon is called *hormesis*. A possible explanation for this effect is that very small doses of some substances may stimulate cellular repair or other beneficial responses.

Edward Calabrese, a highly respected toxicologist at the University of Massachusetts, Amherst, has made a thorough examination of the literature on this subject. He has concluded that the idea has merit and needs more research to test its validity and discover the possible mechanisms involved.

Poor Paracelsus. If this idea turns out to have validity for some substances, we must further modify his original hypothesis as follows: The dose of the mixture of chemicals usually makes the poison—differently for different individuals—but in some cases a tiny bit of a poison may be good for you. The various possible revisions of the original hypothesis proposed by Paracelsus are a good example of how scientific hypotheses are modified to account for new data.

Scientists are waiting for more evidence to come in before accepting the hormesis hypothesis. Stay tuned for more developments about this fascinating idea.

How Good Are Estimates of Toxicity? Taking Uncertainty into Account

Because all methods of estimating toxicity have serious limitations, allowed exposure levels are usually set well below the estimated harmful levels.

As we have seen, all methods for estimating toxicity levels and risks have serious limitations. But they are all we have. To take this uncertainty into account and minimize harm, scientists and regulators typically set allowed exposure levels to toxic substances and ionizing radiation at 1/100 or even 1/1,000 of the estimated harmful levels.

Despite their many limitations, carefully conducted and evaluated toxicity studies are important sources of information for understanding dose-response effects and estimating and setting exposure standards. But citizens, lawmakers, and regulatory officials must recognize the huge uncertainties involved in all such studies.

19-3 CHEMICAL HAZARDS

What Are Toxic and Hazardous Chemicals? Causing Death and Harm

Toxic chemicals can kill, and hazardous chemicals can cause various types of harm.

A **toxic chemical** is a chemical, which through its chemical action on life processes, can cause temporary or permanent harm or death to humans or animals. Its toxicity is often measured in terms of its medium lethal dose (Figure 19-5). A **hazardous chemical** can harm humans or other animals because it is flammable or explosive or because it can irritate or damage the skin or lungs, interfere with oxygen uptake, or induce allergic reactions.

There are three major types of potentially toxic agents. One consists of **mutagens,** chemicals or ionizing radiation that cause or increase the frequency of random *mutations,* or changes, in the DNA molecules found in cells. An example is nitrous acid (HNO_2) formed by digestion of nitrite preservatives in foods. *Most mutations are harmless.* One reason is that organisms have biochemical repair mechanisms that can correct mistakes or changes in the DNA code.

But harmful mutations occurring in reproductive cells can be passed on to offspring and to future generations. It is generally accepted that there is no safe threshold for exposure to harmful mutagens.

A second type consists of **teratogens,** chemicals that cause harm or birth defects to a fetus or embryo. Ethyl alcohol is an example of a teratogen. Drinking during pregnancy can lead to offspring with a low birth weight and a number of physical, developmental, and mental problems. Thalidomide is also a potent teratogen.

The third group is **carcinogens,** chemicals or ionizing radiation that cause or promote **cancer**—the growth of a malignant (cancerous) tumor, in which certain cells multiply uncontrollably. An example is benzene, a widely used chemical solvent. Many cancerous tumors spread by **metastasis** when malignant cells break off from tumors and travel in body fluids to other parts of the body. There they start new tumors, making treatment much more difficult. Typically, 10–40 years may elapse between the initial exposure to a carcinogen and the appearance of detectable symptoms. Partly because of this time lag, many healthy teenagers and young adults have trouble believing their smoking, drinking, eating, and other lifestyle habits today could lead to some form of cancer before they reach age 50.

What Effects Can Some Chemicals Have on Immune, Nervous, and Endocrine Systems? Possible Harm from Small Doses

Long-term exposure to some chemicals at low doses may disrupt the body's immune, nervous, and endocrine systems.

Since the 1970s a growing body of research on wildlife and laboratory animals, along with some epidemiological studies of humans, indicates that long-term exposure to low doses of some chemicals in the environment can disrupt the body's immune, nervous, and endocrine systems.

The *immune system* consists of specialized cells and tissues that protect the body against disease and harmful substances by forming antibodies that make invading agents harmless. Ionizing radiation and some chemicals can weaken the human immune system and leave the body vulnerable to attacks by allergens, infectious bacteria, viruses, and protozoans. Examples are arsenic and dioxins.

Some natural and synthetic chemicals in the environment, called *neurotoxins,* can harm the human *nervous system* (brain, spinal cord, and peripheral nerves). For example, many poisons and the venom of poisonous snakes are neurotoxins, which inhibit, damage, or destroy nerve cells (neurons) that transmit electrochemical messages throughout the body. Effects can include behavioral changes, paralysis, and death. Other examples of neurotoxins are PCBs, mercury, and certain pesticides.

The *endocrine system* is a complex network of glands that release very small amounts of *hormones* in the bloodstream of humans and other vertebrate animals. Low levels of these chemical messengers turn on and off bodily systems that control sexual reproduction, growth, development, learning ability, and behavior.

Each type of hormone has a specific molecular shape that allows it to attach only to certain cell receptors (Figure 19-7, left). Once bonded together, the hormone and its receptor molecule can signal cell mechanisms to execute the chemical message carried by the hormone.

Case Study: Are Hormonally Active Agents a Human Health Threat? Serious Concern but Inconclusive Evidence

Exposure to low levels of certain synthetic chemicals may disrupt the effects of natural hormones in animals, but more research is needed to determine the effects of these chemicals on humans.

There is concern that human exposure to low levels of certain synthetic chemicals can mimic and disrupt the effects of natural hormones. Over the last 25 years, experts from a number of disciplines have been piecing together field studies on wildlife, studies on labora-

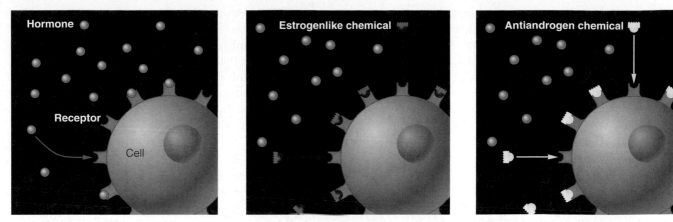

Figure 19-7 Hormones are molecules that act as messengers in the endocrine system to regulate various bodily processes, including reproduction, growth, and development. Each type of hormone has a unique molecular shape that allows it to attach to specially shaped receptors on the surface of, or inside, cells and to transmit its chemical message (left). Molecules of certain pesticides and other synthetic chemicals have shapes similar to those of natural hormones and can affect the endocrine system in people and various other animals. These molecules are called *hormonally active agents* (HAAs). Some HAAs, sometimes called *hormone mimics,* disrupt the endocrine system by attaching to estrogen receptor molecules (center) and giving too-strong, too-weak, or mistimed signals. Other HAAs, sometimes called *hormone blockers,* prevent natural hormones such as androgens from attaching to their receptors (right) so that no signal is given. Some pollutants, called *thyroid disrupters,* may disrupt hormones released by thyroid glands and cause growth and weight disorders and brain and behavioral disorders. Because of the difficulty in determining the harmful effects of long-term exposure to low levels of HAAs, there is uncertainty over their effects on human health.

tory animals, and epidemiological studies of human populations. This analysis suggests that a variety of human-made chemicals can act as *hormone* or *endocrine disrupters,* known as *hormonally active agents (HAAs).* Examples of hormone disrupters are DDT, PCBs, and certain herbicides.

Some, called *hormone mimics,* are chemicals similar to estrogens (female sex hormones). They can disrupt the endocrine system by attaching to estrogen receptor molecules (Figure 19-7, center). Others, called *hormone blockers,* disrupt the endocrine system by preventing natural hormones such as androgens (male sex hormones) from attaching to their receptors (Figure 19-7, right). Estrogen mimics and hormone blockers are sometimes called *gender benders* because of their possible effects on sexual development and reproduction. There is also growing concern about still another group of HAAs—pollutants that can act as *thyroid disrupters* and cause growth, weight, brain, and behavioral disorders.

Is long-term exposure to low levels of HAAs a threat to human health? A 1999 study of the possible effects of hormonally active agents on humans by a U.S. National Academy of Sciences panel of scientists came to three major conclusions. *First,* "adverse reproductive and developmental effects have been observed in human populations, wildlife, and laboratory animals as a consequence of exposure to HAAs." *Second,* "there have been only a few studies of the effects of HAAs in humans, but the results of laboratory and wildlife studies suggest that HAAs have the potential to affect human immune functions." *Third,* greatly increased research is needed to come to a more definitive conclusion about whether low levels of most HAAs in the environment pose a threat to human health.

Bottom line: We do not know whether exposure to trace amounts of various hormonally active chemicals introduced into the environment have harmful effects on humans and other animals. Some scientists say there is no definitive evidence for harm from HAAs to humans and dismiss it as a minor threat. But others say there is enough preliminary evidence to warrant greatly increased research on their possible effects. This will take decades.

Some scientists say we need to wait for the results of more research before banning or severely restricting HAAs. Other scientists believe that as a precaution, we should sharply reduce the use of potential hormone disrupters.

Why Do We Know So Little about the Harmful Effects of Chemicals? Establishing Guilt Is Difficult

Under existing laws most chemicals are considered innocent until shown to be guilty, and estimating their toxicity to establish guilt is difficult, uncertain, and expensive.

According to risk assessment expert Joseph V. Rodricks, "Toxicologists know a great deal about a few chemicals, a little about many, and next to nothing about most." The U.S. National Academy of Sciences estimates that only about 10% of at least 80,000 chemicals in commercial use have been thoroughly screened for toxicity, and only 2% have been adequately tested to determine whether they are carcinogens, teratogens, or mutagens. Hardly any of the chemicals in commercial use have been screened for possible damage to the human nervous, endocrine, and immune systems.

Currently, federal and state governments do not regulate about 99.5% of the commercially used chemicals in the United States. There are several reasons for this lack of regulation. One is that under existing U.S. laws, most chemicals are considered *innocent until shown to be guilty.* Some analysts think this is the opposite of the way it should be. They ask why chemicals should have the same legal rights as people.

A second reason is that there are not enough funds, personnel, facilities, and test animals available to provide such information for more than a small fraction of the many individual chemicals we encounter in our daily lives. A third limitation is that it is difficult and expensive to analyze the combined effects of multiple exposures to various chemicals and the possible interactions of such chemicals.

X *How Would You Vote?* Should chemicals be regulated based on their effects on the nervous, immune, and endocrine systems? Cast your vote online at http://biology.brookscole .com/miller14.

Is Pollution Prevention the Answer?
Taking Precautions

Preliminary but not conclusive evidence that a chemical causes significant harm should spur preventive action, some say.

So where does this leave us? We do not know a lot about the potentially toxic chemicals around us and inside of us, and estimating their effects is very difficult, time consuming, and expensive. Is there a way out of this dilemma?

Some scientists and health officials, especially those in European Union countries, are pushing for much greater emphasis on *pollution prevention.* They say we should not release into the environment chemicals that we know or suspect can cause significant harm. This means looking for harmless or less harmful substitutes for toxic and hazardous chemicals or recycling them within production processes so they do not reach the environment.

This prevention strategy greatly reduces the expenditures of huge amounts of money on statistically uncertain and controversial toxicity studies and exposure standards. It also lowers the risk from exposure to potentially hazardous chemicals and products and their possible but poorly understood multiple interactions.

This approach is based on the **precautionary principle:** When there is plausible but incomplete scientific evidence (frontier science evidence) of significant harm to humans or the environment from a proposed or existing chemical or technology, we should take action to prevent or reduce the risk instead of waiting for more conclusive (sound or consensus science) evidence. This principle is based on familiar axioms: "Look before you leap." "Better safe than sorry." "An ounce of prevention is worth a pound of cure."

Under this approach, those proposing to introduce a new chemical or technology would bear the burden of establishing its safety. This means two major changes in the way we evaluate risks. *First,* new chemicals and technologies would be assumed harmful until scientific studies can show otherwise. *Second,* existing chemicals and technologies that appear to have a strong chance of causing significant harm would be removed from the market until their safety can be established.

Some movement is being made in this direction, especially in the European Union. In 2000, negotiators agreed to a global treaty that would ban or phase out use of 12 of the most notorious *persistent organic pollutants (POPs),* also called the *dirty dozen.* The list included DDT and eight other persistent pesticides, PCBs, and dioxins and furans. New chemicals would be added to the list when the harm they cause is seen as outweighing their usefulness. This treaty went into effect in 2004.

Manufacturers and businesses agree that some chemicals are too dangerous for widespread use and that some technologies such as coal-burning plants carry high health risks. But they contend that widespread application of the precautionary principle would make it too expensive and almost impossible to introduce any new chemical or technology. Strict application of the precautionary principle would stifle chemical and technological innovation and risk taking. We can never have a risk-free society. For example, if we had strictly applied the precautionary principle would we have automobiles, antibiotics, or plastics?

On the other hand, proponents of increased reliance on the precautionary principle say that it will encourage innovation in developing less harmful alternative chemicals and technologies and in finding ways to prevent as much pollution as possible instead of relying mostly on pollution control. It is true that we cannot have a risk-free society. But proponents believe we should make greater use of the precautionary principle effort to reduce many of the risks we face. As you can see, there are no easy answers for knowing when to apply the precautionary principle.

X *How Would You Vote?* Should we assume that new chemicals that can end up in the environment are guilty of causing harm until proven innocent? Cast your vote online at http://biology.brookscole.com/miller14.

19-4 BIOLOGICAL HAZARDS: DISEASE IN DEVELOPED AND DEVELOPING COUNTRIES

What Are Nontransmissible and Transmissible Diseases? To Spread or Not to Spread

Diseases not caused by living organisms do not spread from one person to another, and those caused by living organisms such as bacteria and viruses can spread from person to person.

A **nontransmissible disease** is caused by something other than a living organism and does not spread from one person to another. Such diseases tend to develop slowly and have multiple causes. Examples are cardiovascular (heart and blood vessel) disorders, most cancers, diabetes, asthma, emphysema, and malnutrition.

A **transmissible disease** is caused by a living organism and can spread from one person to another. *Infectious agents* or *pathogens* (such as a bacterium, virus, protozoa, or parasite; Figure 19-8) cause such diseases. These agents are spread by air, water, food, and body fluids, and by some insects (such as mosquitoes; Figure 19-9, p. 420) and other nonhuman carriers. All such pathways are called *vectors*.

Typically, a *bacterium* is a one-celled microorganism that can replicate (clone) itself by simple cell division. A *virus* is a microscopic, noncellular infectious agent. Its DNA or RNA contains instructions for making more viruses, but it has no apparatus to do this. To replicate, a virus must invade a host cell and take over the cell's DNA to create a factory for producing more viruses (Figure 19-10, p. 421). A *parasite* is an organism that feeds off another organism (p. 154). *Protozoans* are a diverse assortment of microscopic or near-microscopic organisms that live as single cells or in simple colonies. Examples are *Giardia lamblia* that causes giardiasis, a gastrointestinal disease transmitted by water, and several species of *Plasmodium* that transmit malaria.

According to the World Health Organization, about 30% of all deaths per year are caused by nontransmissible cardiovascular disease, 26% by transmissible infectious disease (Figure 19-11, p. 421), and 12% by nontransmissible cancers.

As a country industrializes, it usually makes an *epidemiological transition* in which deaths from the infectious diseases of childhood *decrease* and those from the chronic diseases of adulthood (heart disease and stroke, cancer, and respiratory conditions) *increase.*

Good news. Since 1900, and especially since 1950, the incidence of infectious diseases and the death rates from such diseases have been greatly reduced. This has been done mostly by a combination of better health care, using antibiotics to treat infectious disease caused by bacteria, and developing vaccines to prevent the spread of some infectious viral diseases.

Bad news. Many disease-carrying bacteria have developed genetic immunity to widely used antibiotics (Case Study, below). Also, many disease-transmitting species of insects such as mosquitoes have become immune to widely used pesticides that once helped control their populations.

Case Study: Are We Losing Ground in Our Struggle against Infectious Bacteria? Growing Germ Resistance to Antibiotics

Rapidly producing infectious bacteria can undergo natural selection and become genetically resistant to widely used antibiotics.

We may be falling behind in our efforts to prevent infectious bacterial diseases because of the astounding reproductive rate of bacteria, which can produce 16,777,216 offspring in 24 hours. Their high reproductive rate allows them to become genetically resistant to an increasing number of antibiotics through natural selection. They can also transfer such resistance to nonresistant bacteria.

Other factors play a role in the potentially serious rise in the incidence of some infectious bacterial diseases—such as tuberculosis (Case Study, p. 421)—once controlled by antibiotics. One is that harmful bacteria are spread around the globe by human travel and the trade of goods. Another is that overuse of pesticides

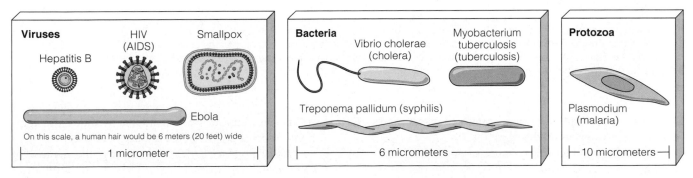

Figure 19-8 Examples of *pathogens* or agents that can cause transmissible diseases. A micrometer is one-millionth of a meter.

Dengue Fever

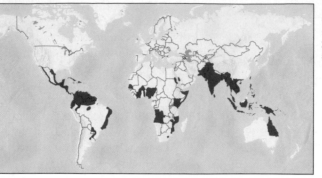

Painful and sometimes fatal.
Carried by four related viruses and
strikes during rainy season.
**2.5 million people at risk;
50 million new cases a year.**

Malaria

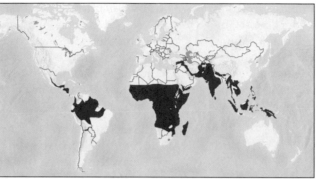

Endemic in more than 100 countries.
Caused by four protozoa species.
**270–500 million new cases and
1 million deaths per year.**

Yellow Fever

Dreaded far more than 400 years.
Viral disease that causes symptoms from
mild to severe illness and death.
**200,000 new cases and
30,000 deaths a year.**

Figure 19-9 A few species of mosquito act as *vectors* to transmit pathogens for a number of infectious diseases, including the three whose normal ranges are mapped here. Female mosquitoes feed on blood and male mosquitoes on plant juices. Thus, only the female mosquito bites people and animals to feed on their blood, and in the process transmits pathogens from one victim to later victims. *Throughout human history, disease transmission by female mosquitoes has probably killed more people than any other single factor.* However, most mosquito species do not transmit infectious diseases and mosquitoes play important ecological roles. Their eggs are a major food source for fish, various insects, and frogs and other amphibians. Adult mosquitoes are an important source of food for bats, spiders, and many insect and bird species. Mosquitoes locate us by the CO_2 we give off, the odor of lactic acid secreted by our skin, and our body heat, which is why heat-absorbing dark clothes attract mosquitoes more than light-colored clothes do. Once a female mosquito finds us she pierces our skin, injects an anticoagulant mixed with saliva that keeps our blood flowing and also causes an itchy bump to rise, and drinks her fill of our blood. (World Health Organization and the U.S. Centers for Disease Control and Prevention)

increases populations of pesticide-resistant insects and other carriers of bacterial diseases.

An additional factor is overuse of antibiotics. According to a 2000 study by Richard Wenzel and Michael Edward, at least half of all antibiotics used to treat humans are prescribed unnecessarily. In many countries antibiotics are available without prescriptions, which also promotes unnecessary use.

According to a 2001 study by the Union of Concerned Scientists, nearly 75% of all antibiotics manufactured in the United States are used mostly in feed additives to boost livestock production. Recent

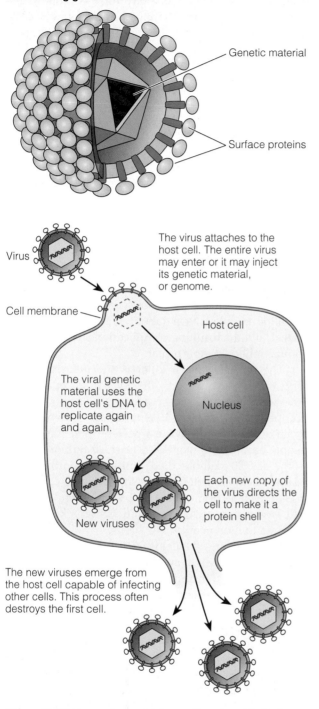

A typical virus consists of a shell of proteins surrounding genetic material

Genetic material

Surface proteins

Virus

The virus attaches to the host cell. The entire virus may enter or it may inject its genetic material, or genome.

Cell membrane

Host cell

The viral genetic material uses the host cell's DNA to replicate again and again.

Nucleus

Each new copy of the virus directs the cell to make it a protein shell

New viruses

The new viruses emerge from the host cell capable of infecting other cells. This process often destroys the first cell.

Figure 19-10 How a virus reproduces. (American Medical Association)

studies show that resistant strains of infectious diseases that develop in livestock animals can spread to humans through contact with infected animals or water and through food webs. *Good news.* Because of public pressure, efforts are being made to phase out the use of antibiotics to boost livestock. Some fast food

chains now refuse to buy meat from livestock treated with antibiotics.

The result of these factors acting together is that every major disease-causing bacterium now has strains that resist at least one of the roughly 160 antibiotics we use to treat bacterial infections. Consequently, the United States and other countries are seeing an increase in the number of patients who contract infectious bacterial disease while they are in a hospital or other medical facility. According to a 2002 study by the Joint Commission for the Accreditation of Healthcare Organizations, each year nearly 2 million Americans leave hospitals with mostly preventable infections they acquired there, and at least 90,000 of them died prematurely because of such infections.

Biologist Paul Ewald suggests that we should stop trying to obliterate lethal microbes and instead focus on how to weaken their effects by forcing them to mutate in certain ways. Maybe we can get the forces of evolution on our side by causing viruses to become less virulent as they spread among the population.

Case Study: The Global Tuberculosis Epidemic—A Growing Threat

Tuberculosis (TB) kills about 1.7 million people a year and could kill 28 million people by 2020.

Since 1990, one of the world's most underreported stories has been the rapid spread of tuberculosis (TB). According to the World Health Organization, this highly infectious bacterial disease infects about

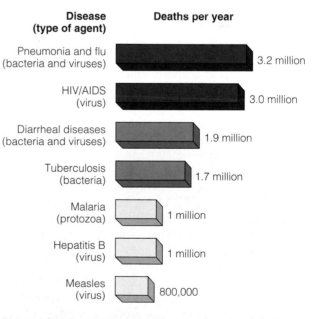

Disease (type of agent)	Deaths per year
Pneumonia and flu (bacteria and viruses)	3.2 million
HIV/AIDS (virus)	3.0 million
Diarrheal diseases (bacteria and viruses)	1.9 million
Tuberculosis (bacteria)	1.7 million
Malaria (protozoa)	1 million
Hepatitis B (virus)	1 million
Measles (virus)	800,000

Figure 19-11 Each year the world's seven deadliest infectious diseases kill about 12.6 million people—most of them poor people in developing countries. This amounts to about 34,500 mostly preventable deaths every day. (World Health Organization)

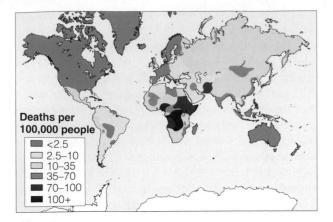

Figure 19-12 The current global tuberculosis epidemic. This easily transmitted disease is spreading rapidly and now kills about 1.7 million people a year—about 84% of them in developing countries. (World Health Organization)

Deaths per 100,000 people

- <2.5
- 2.5–10
- 10–35
- 35–70
- 70–100
- 100+

9 million people per year and kills about 1.7 million of them—mostly in developing countries (Figure 19-12). The WHO projects that between 2004 and 2020, about 28 million people will die of the disease, unless current efforts and funding to control TB are greatly strengthened and expanded.

The bacterium causing TB infection moves from person to person mainly in airborne droplets produced by coughing, sneezing, singing, or even talking. At the rate of 9 million people per year, the TB bacillus has now infected about one of every three people in the world. To be infected means to have the organisms in your body, whether or not you are sick.

During their lifetime about 5–10% of all infected people will become sick or infectious (able to spread the disease) with active TB, especially when their immune system is weakened. Left untreated, each person with active TB typically infects 10–15 other people.

Most infected people do not appear to be sick, and about half of them do not even know they are infected. As a result, this serious health problem has been called a *silent global epidemic.*

Several factors account for the recent increase in TB. One is the lack of TB screening and control programs, especially in developing countries, where about 95% of the new cases occur. A second problem is that most strains of the TB bacterium have developed genetic resistance to almost all effective antibiotics.

Another factor is increased population growth and urbanization that have increased contacts between people and have spread TB, especially in areas where large numbers of the poor are crowded together. In addition, the spread of AIDS greatly weakens the immune system and allows TB bacteria to multiply in AIDS victims.

Slowing the spread of the disease involves early identification and treatment of people with active TB, especially those with a chronic cough. Treatment with a combination of four inexpensive drugs can cure 90%

of those with active TB. However, to be effective, the drugs must be taken every day for 6–8 months. Because the symptoms disappear after a few weeks, many patients think they are cured and stop taking the drugs. This allows the disease to recur in a hard-to-treat form. It then spreads to other people, and drug-resistant strains of TB bacteria develop.

How Serious Is the Threat from Viral Diseases? Watch Out for HIV, Flu, and Hepatitis B

HIV, flu, and hepatitis B viruses infect and kill many more people each year than the highly publicized Ebola, West Nile, and SARS viruses.

What are the world's three most widespread and dangerous viruses? The biggest killer is the *human immunodeficiency virus (HIV)* that is transmitted by unsafe sex, sharing of needles by drug users, infected mothers to offspring before or during birth, and exposure to infected blood. On a global scale, HIV infects at least 5 million people a year (about 41,000 in the United States), and the resulting complications from AIDS kill about 3 million people a year.

The second biggest killer is the *influenza* or *flu* virus that is transmitted by the body fluids or airborne emissions of an infected person and kills about 1 million people per year. In 1919 a highly virulent global strain of the influenza virus infected up to 500 million people around the world. At least 20 million people (some say 50 million) died within 6 months. Many health scientists believe that sooner or later such a mass infection from a new and very potent flu virus will sweep the world again and perhaps kill several hundred million people.

The third largest killer is the *hepatitis B virus (HBV)* that damages the liver and kills about 1 million a year. Like HIV, it is transmitted by unsafe sex, sharing of needles by drug users, infected mothers to offspring before or during birth, and exposure to infected blood.

In recent years, three other viruses have received widespread coverage in the media. One is the *Ebola virus* transmitted by the blood or other body fluids of an infected person. Another is the *West Nile virus,* transmitted by the bite of a common mosquito that has become infected by feeding on birds carrying the virus. A third is the *severe acute respiratory syndrome (SARS) virus.* This easily transmitted virus first emerged in southern China in 2002. It infected more than 8,000 in 30 countries and killed nearly 800 before the outbreak was brought under control.

Health officials are concerned about the emergence and spread of these three and other emerging viral diseases and are working hard to control the spread of these diseases. But in terms of annual infection rates and deaths, the three most dangerous viruses by far are HIV, flu, and hepatitis B.

For example, the West Nile virus has spread throughout most of the lower 48 states but the chances of being infected and killed by it is low (about 1 in 2,500). In 2003, the flu killed more Americans in two days than the West Nile virus killed during the entire year.

You can greatly reduce your chances of getting infectious diseases such as flu, the common cold, and SARS that spread from person to person by practicing good old-fashioned hygiene. Wash your hands thoroughly and often, and avoid touching your mouth, nose, and eyes.

It is much harder to fight viral infections than infections caused by bacteria and protozoa. One problem is that most drugs that can kill a virus also harm the cells of its host. Treating viral infections such as colds, flu, and most mild coughs and sore throats with antibiotics is useless and increases genetic resistance in disease-causing bacteria.

The best weapons against viruses are *vaccines* that stimulate the body's immune system to produce antibodies to ward off viral infections. Immunization with vaccines has helped reduce the spread of viral diseases such as smallpox, polio, rabies, influenza, measles, and hepatitis B. But vaccines are not available for many viral diseases.

Case Study: How Serious Is the Global Threat from HIV and AIDS? A Rapidly Growing Health Threat

The spread of acquired immune deficiency syndrome (AIDS), caused by HIV, is one of the world's most serious and rapidly growing health threats.

Sex can be hazardous to your health. Worldwide, almost 400 million people are infected with a *sexually transmitted disease (STD)* each year. According to the U.S. Centers for Disease Control, almost one of every four Americans is walking around with an STD and at least one in every three sexually active persons in the United States will contract an STD by age 24. STDs are rampant in high schools and colleges, where many students think, "It cannot happen to me." Polls indicate that 50–66% of sexually active students do not use condoms and more than 40% have two or more sex partners. Some STDs can cause infertility in men and women. Others can cause genital warts and genital cancers or, in the case of HIV, eventually death.

The global spread of *acquired immune deficiency syndrome (AIDS)*, caused by HIV, a serious and rapidly growing health threat. The virus itself is not deadly, but it kills immune cells and leaves the body defenseless against infectious bacteria and other viruses. According to the WHO, by the beginning of 2004 some 38 million people worldwide (96% of them in developing countries, especially African countries south of the

Sahara Desert) were infected with HIV. Every day about 14,000 more people—most of them between the ages of 15 and 24—get infected with HIV. According to U.S. Secretary of State Colin Powell, "AIDS is . . . now more destructive than any army, any conflict, and any weapon of mass destruction."

The news is going to get worse. Infection rates are increasing rapidly in five countries—Nigeria, Ethiopia, Russia, Indonesia, Vietnam, India, and China—that together have 40% of the world's population.

Within 7–10 years, at least half of those with HIV develop AIDS. This long incubation period means that infected people often spread the virus for several years without knowing they are infected. So far, *there is no vaccine to prevent HIV and no cure for AIDS. Once you get AIDS, you will eventually die,* although drugs may help some infected people live longer. However, only a tiny fraction of those suffering from AIDS can afford to use these costly drugs.

Between 1980 and 2004, more than 20 million people (460,000 in the United States) died of AIDS-related diseases. An estimated 280,000 of the roughly 900,000 Americans infected with HIV do not know it and there are about 42,000 new infections a year. In the United States, free or low-cost confidential testing for HIV exposure is available at many public health offices and at many doctors' offices. However, it takes a few weeks to 6 months or more before enough antibodies form in response to an HIV infection for any test to show that the virus is present. In addition, the conventional tests require several hours of lab time, often at another location, so that results may not be available for 1 to 2 weeks. In 2003, the CDC began purchasing and pilot testing nationwide a new device called OraQuick. It can use a small amount of blood from a finger prick to test for HIV within 20 minutes. The accuracy rate is 99.6%, roughly the same as more conventional blood tests.

AIDS has caused the life expectancy of 700 million people living in sub-Saharan Africa to drop from 62 to 47 years. The premature deaths of teachers, health care workers, and other young, productive adults in such countries leads to diminished education and health care, decreased food production and economic development, and disintegrating families. Such deaths drastically alter a country's age structure diagram (Figure 19-13, p. 424). AIDS has left 15 million orphans— roughly equal to every child under age 5 in America.

Between 2004 and 2020, the WHO estimates 60 million more deaths from AIDS and a death toll reaching as high as 5 million a year by 2020.

According to the WHO, a global strategy to slow the spread of AIDS should have five major priorities. *First,* shrink the number of people capable of infecting others by quickly reducing the number of new infections below the number of deaths. *Second,* concentrate on the groups in a society that are most

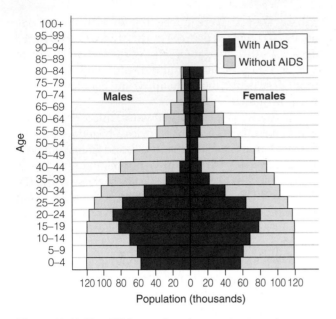

Figure 19-13 How AIDS can affect the age structure of a population. This figure shows the projected age structure of Botswana's population in 2020 with and without AIDS. (U.S. Census Bureau)

likely to spread the disease, such as truck drivers, sex workers, and soldiers. *Third,* provide free HIV testing and pressure people to get tested.

Fourth, use a mass advertising and education program for adults and schoolchildren to help prevent the disease with emphasis on abstinence and condom use. *Fifth,* provide free or low-cost drugs to slow the progress of the disease.

Senegal acted early to check the spread of the virus and has kept the proportion of its young adults infected with HIV below 1%. Within a decade, Botswana cut its HIV infection rates in half with good leadership from its government, health-care facilities that are better than average, and vast wealth from diamonds. It provided free testing for HIV and pressured people to get tested. Those with HIV or AIDS were provided with free or low-cost drugs to slow the disease's progress.

> **X̶ HOW WOULD YOU VOTE?** Should developed and developing nations mount a global campaign to reduce the spread of AIDS and to help countries affected by this disease? Cast your vote online at http://biology.brookscole.com/miller14.

💿 Case Study: Malaria: A Deadly Parasitic Disease That Is Making a Comeback

Malaria kills about 1 million people a year and has probably killed more people than all of the wars ever fought.

About one of every five people in the world—most of them living in poor African countries—is at risk of malaria (Figure 19-9, middle). Worldwide, an estimated 300–500 million people are infected with the protozoan parasites that cause malaria, and there are 270–500 million new cases each year. Malaria is not just a concern for the people living in the areas where it occurs (Figure 19-9, middle), but also for anyone traveling to these areas—including many unsuspecting tourists—because there is no vaccine for this disease.

Malaria is caused by a parasite that is spread by the bites of certain mosquito species. It infects and destroys red blood cells, causing fever, chills, drenching sweats, anemia, severe abdominal pain and headaches, vomiting, extreme weakness, and greater susceptibility to other diseases. The disease kills about 1 million people each year (about 900,000 of them children under age 5)—an average of 2,700 deaths per day. Many children who survive severe malaria have brain damage or impaired learning ability.

Malaria is caused by four species of protozoan parasites in the genus *Plasmodium*. Most cases of the disease are transmitted when an uninfected female mosquito from any one of about 60 *Anopheles* mosquito species bites an infected person, ingests blood that contains the parasite, and later bites an uninfected person (Figure 19-14). When this happens, *Plasmodium* parasites move out of the mosquito and into the human's bloodstream, multiply in the liver, and enter blood cells to continue multiplying. Malaria can also be transmitted by blood transfusions or by sharing needles.

The malaria cycle repeats itself until immunity develops, treatment is given, or the victim dies. *Over the course of human history, malarial protozoa probably have killed more people than all the wars ever fought.*

The mosquitoes that transmit malaria breed in shallow pools and puddles—often in tire ruts and hoof prints—near human dwellings and apparently are attracted to smelly feet. During the 1950s and 1960s, the spread of malaria was sharply curtailed by draining swamplands and marshes, spraying breeding areas with insecticides, and using drugs to kill the parasites in the bloodstream. Since 1970, malaria has come roaring back. Most species of the malaria-carrying *Anopheles* mosquito have become genetically resistant to widely used insecticides. Worse, the *Plasmodium* parasites have become genetically resistant to common antimalarial drugs.

Researchers are working to develop new antimalarial drugs (such as artemisinins derived from the Chinese herbal remedy qinghaosu), vaccines, and biological controls for *Anopheles* mosquitoes. But such approaches receive too little funding and have proved more difficult than originally thought.

In 2002, scientists announced they had broken the genetic codes for both the mosquito and the parasite responsible for most malaria cases. Eventually this important information could uncover genetic vulnerabil-

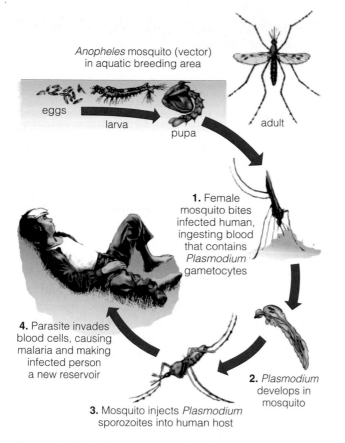

Anopheles mosquito (vector) in aquatic breeding area

eggs

larva

pupa

adult

1. Female mosquito bites infected human, ingesting blood that contains *Plasmodium* gametocytes

4. Parasite invades blood cells, causing malaria and making infected person a new reservoir

3. Mosquito injects *Plasmodium* sporozoites into human host

2. *Plasmodium* develops in mosquito

Figure 19-14 The life cycle of malaria. *Plasmodium* circulates from mosquito to human and back to mosquito.

ities in these organisms. This could allow scientists to alter the genetic makeup of mosquitoes so they cannot carry and transmit the parasite to humans. It could also lead to more effective drugs, vaccines, insecticides, and insect repellents to counter the disease.

Meanwhile, health experts say prevention is the best approach to slowing the spread of malaria. Methods include increasing water flow in irrigation systems to prevent mosquito larvae from developing (an expensive solution that uses much more water than required for irrigation) and fixing leaking water pipes.

A problem is that poor villagers in malarial regions cannot afford screens on their homes and mosquito nets for their beds. The WHO calls for countries to do away with all taxes and tariffs on insecticide-treated bed nets, and to give such bed nets to the poor.

Other approaches include cultivating fish that feed on mosquito larvae (biological control), clearing vegetation around houses, planting trees that soak up water in low-lying marsh areas where mosquitoes thrive (a method that can degrade or destroy ecologically important wetlands), and using zinc and vitamin A supplements to boost resistance to malaria in children.

Spraying the inside of homes with low concentrations of DDT about twice a year greatly reduces the number of malaria cases. But under an international treaty enacted in 2002, DDT and five of its chlorinated-hydrocarbon cousins are being phased out in developing countries. However, the treaty allows 25 countries to continue using DDT for malaria control until other alternatives are available.

Health officials in developing countries call for much greater funding for research on finding ways to prevent and treat malaria. Each year more than $70 billion is spent on research on disease. If you look at the number of people dying each year from malaria, a fair share of the global research funding for malaria would be about $1.75 billion a year. The actual figure spent annually for malaria research is about $85 million a year.

Solutions: How Can We Reduce the Incidence of Infectious Diseases? More Money and Assistance

We can sharply reduce the incidence of infectious diseases if the world is willing to provide the necessary funds and assistance.

Bad news. First, death rates from infectious diseases in developing countries are unacceptably high.

Second, only about 10% of global medical research and development money is spent on infectious diseases in developing countries, even though more people worldwide suffer and die from these diseases than from all other diseases combined. *Third,* major drug companies have greatly decreased research on developing antibiotics and vaccines because they are difficult and costly to develop. They also produce lower profits because patients take them for only a short time compared to medicines for treating chronic diseases such as diabetes and hypertension that must be taken every day for years.

Fourth, about one-third of the world's people, mostly in developing countries, lack adequate access to clean drinking water and sanitation facilities.

Fifth, the WHO estimates that children under 5 make up only 10% of the world's population but account for 40% of global illness. Eleven million children a year die before their fifth birthday from causes that are mostly preventable and treatable.

Good news. According to the WHO, the global death rate from infectious diseases dropped by about two-thirds between 1970 and 2000 and is projected to continue dropping. Also, between 1971 and 2000, the percentage of children in developing countries immunized with vaccines to prevent tetanus, measles, diphtheria, typhoid fever, and polio increased from 10% to 84%—saving about 10 million lives a year.

Figure 19-15 (p. 426) lists measures that health scientists and public health officials suggest to help prevent or reduce the incidence of infectious diseases that affect humanity (especially in developing countries).

Solutions

Infectious Diseases

Increase research on tropical diseases and vaccines

Reduce poverty

Decrease malnutrition

Improve drinking water quality

Reduce unnecessary use of antibiotics

Educate people to take all of an antibiotic prescription

Reduce antibiotic use to promote livestock growth

Careful hand washing by all medical personnel

Immunize children against major viral diseases

Oral rehydration for diarrhea victims

Global campaign to reduce HIV/AIDS

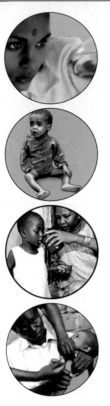

Figure 19-15 Solutions: ways to prevent or reduce the incidence of infectious diseases, especially in developing countries. Which two of these solutions do you believe are the most important?

An important breakthrough has been the development of simple *oral rehydration therapy* to help prevent death from dehydration for victims of diarrheal diseases, which cause about one-fourth of all deaths of children under age 5. It involves administering a simple solution of boiled water, salt, and sugar or rice, at a cost of only a few cents per person. It has been the major factor in reducing the annual number of deaths from diarrhea from 4.6 million in 1980 to 1.9 million in 2002. Few investments have saved so many lives at such a low cost.

In 2001, the WHO began promoting a do-it-yourself technique that uses sunlight to disinfect water. The process is simple: fill a transparent plastic bottle with contaminated water and lay it horizontally on a flat black surface (which absorbs more heat and kills more pathogens than a lighter surface can) in the sunlight. After several hours, the heat and ultraviolet rays of the sun kill most illness-causing microorganisms in polluted water. This simple method is especially useful in tropical countries where there is intense sunlight.

How Serious Is the Threat of Bioterrorism? A Growing Concern

Bioterrorism that involves releasing infectious organisms into the air, water supply, or food supply is a serious and growing threat.

One of the threats in our increasingly interconnected global society is *bioterrorism*. It involves the deliberate release of disease-causing bacteria or viruses into the air, water supply, or food supply of concentrated urban populations.

According to antiterrorism experts, bioterrorism is a much easier, cheaper, and more effective way to cause illness, death, and mass terror and chaos than crashing planes into buildings or setting off dirty nuclear weapons. The materials and tools to make biological weapons are inexpensive and easy to get. A state-of-the-art laboratory for making biological warfare agents requires about $10,000 of off-the-shelf equipment such as a beer fermenter, a protein-based culture of the disease to be produced, protective plastic clothing, and a gas mask. The lab could be housed in a space about the size of a small bathroom. Now that the sequencing of the genome of the flu virus is nearly complete, bioterrorists can develop more lethal flu viruses and easily transmit them through the air in tiny droplets.

Since the end of World War II, the United States and the former Soviet Union both have spent billions of dollars developing, producing, and stockpiling large quantities of biological weapons of mass destruction. Figure 19-16 provides information about some of the common bacterial and viral agents these countries have studied and developed.

Both countries have used recombinant DNA techniques to produce more dangerous versions of these organisms that act faster, are more virulent, and are resistant to antibiotics used to treat them. They have also created new and even more dangerous infectious organisms with properties that are classified as top secret.

Both countries have promised to destroy their biological weapons. But because of the secrecy of these programs there is no way to know how many weapons remain.

Thousands of former Soviet scientists with knowledge about how to develop these weapons are living in poverty. There is fear that countries interested in developing biological weapons will hire some of them. In 1995, the U.S. Central Intelligence Agency (CIA) identified 16 nations suspected of having programs to develop and stockpile biological warfare agents. In addition, thousands of molecular biologists and graduate-school students around the world have enough knowledge about recombinant DNA and cloning technology to design and mass produce biological warfare agents.

Once made, the bacteria or viruses can be carried in a small vial or aerosol container not detectable by conventional security equipment. They could be released

Agent	Contagious	Symptoms	Mortality (if untreated)	Existence of vaccine	Treatment
Smallpox (virus)	Yes	Fever, aches, headache, red spots on face and torso	30%	Yes	Vaccination within 4 days after exposure, IV hydration
Hemorrhagic fever (viruses)	Yes	Vary but include fever, bleeding, shock, and coma	Varies	No	Ebola has no cure, antiviral riboflavin and some antibiotics may help
Inhalation anthrax (bacterium)	No	Fever, chest pain, difficulty breathing, respiratory failure	90–100%	Yes	Early treatment with Cipro and other antibiotics
Botulism (bacterium)	No	Blurred vision, progressive paralysis, death within 24 hours if not treated	60–100%	Yes	Equine antitoxin given early. Intensive care, respirator
Pneumonic plague (bacterium)	Yes	High fever, chills, headache, coughing blood, difficulty breathing, respiratory failure	90–100%	No	Antibiotics
Tularemia (bacterium)	No	Fever, sore throat, weakness, respiratory stress, pneumonia	30–60%	Yes (in testing)	Antibiotics

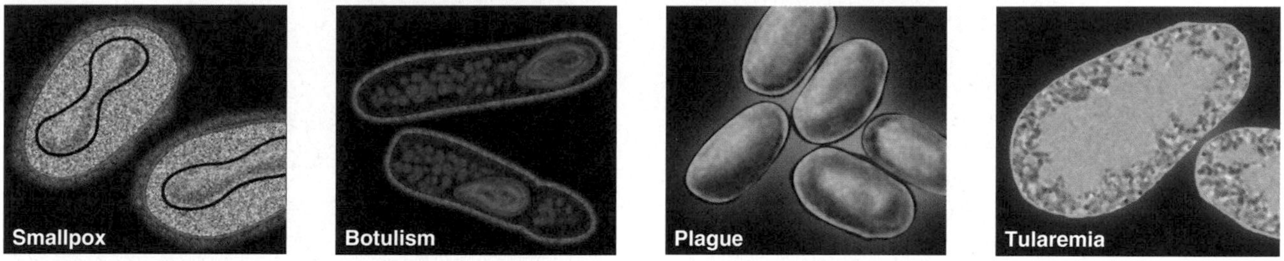

Smallpox · Botulism · Plague · Tularemia

Figure 19-16 Characteristics of common agents that might be used by terrorists as biological weapons.

in a crowded subway car, into a public water supply, or into the unprotected, ground-level air intakes found in most office buildings. A terrorist organization with volunteers willing to die for their cause could infect volunteers with a normally fatal disease organism that is easily transmitted from one human to another. After waiting until they are contagious, the volunteers could be sent on airplane trips throughout the world. Millions could die and the social and economic fabric of affected societies would unravel.

According to a 2003 *worst-case scenario* published in the *Proceedings of the National Academy of Sciences*, if terrorists release 1 kilogram (2.2 pounds) of anthrax spores in a city of 10 million people, at least 123,000 people would die, even if everyone took the appropriate antibiotics within 48 hours after exposure.

A more likely version of this scenario is that the attack might go unnoticed until a few victims turned up sick at hospitals. Then waves of very sick people would overwhelm hospitals, most of which lack enough stocks of antibiotics, vaccines, equipment, and staffing to handle such a big surge in emergency patients. Casualties among medical workers would compound the crisis and chaos would reign.

I know you are thinking: Whoa, enough already. This stuff is depressing and scary. But it is a reality in today's world. Let us look at some more hopeful news about bioterrorism.

Early detection of biological agents is a key to treating exposed victims and preventing the spread of diseases to others. Some scientists are trapping common insects such as bees, beetles, moths, and crickets to see whether they can be used as environmental monitors of chemical and biological agents. Others are trying to develop inexpensive and easy-to-use DNA detectors to quickly and accurately diagnose any infectious disease such as smallpox. For example, MIT biologist Todd Rider has developed a biological sensor to detect within minutes dangerous biological agents such as anthrax. He made the sensor out of mouse immune cells by inserting a gene for antibodies for a particular biological agent (such as anthrax) along with a gene that causes a jellyfish to glow. When a biological agent activates the antibody, the immune cells of the mouse light up.

Also, treatments are available for the most common biological agents (Figure 19-16)—unless they have been genetically modified to make such treatments fail. And outbreaks can be kept under control if hospitals stock large supplies of antibiotics and vaccines for treatment of common diseases, provide emergency and hospital workers with detection systems and protective gear, and alert doctors to the symptoms of the most common biological warfare agents.

19-5 RISK ANALYSIS

How Can We Estimate Risks? Evaluate, Compare, Decide

Scientists have developed ways to evaluate and compare risks, decide how much risk is acceptable, and find affordable ways to reduce them.

Risk analysis involves identifying hazards and evaluating their associated risks (*risk assessment*, Figure 19-2, left), ranking risks (*comparative risk analysis*), determining options and making decisions about reducing or eliminating risks (*risk management*, Figure 19-2, right), and informing decision makers and the public about risks (*risk communication*).

Statistical probabilities based on past experience, animal testing and other tests, and epidemiological studies are used to estimate risks from older technologies and chemicals. To evaluate new technologies and products, risk evaluators use more uncertain statistical probabilities, based on models rather than actual experience and testing.

Figure 19-17 lists the results of a *comparative risk analysis,* summarizing the greatest ecological and health risks identified by a panel of scientists acting as advisers to the U.S. Environmental Protection Agency (EPA).

The greatest risks many people face today are rarely dramatic enough to make the daily news. In terms of the number of premature deaths per year (Figure 19-18) and reduced life span, *the greatest risk by far is poverty* (Figure 19-19 (p. 430) and Figure 1-15, p. 17). Its high death toll is a result of malnutrition, increased susceptibility to normally nonfatal infectious diseases, and often fatal infectious diseases from lack of access to a safe water supply.

Thus *the sharp reduction or elimination of poverty would do far more to improve longevity and human health than any other measure.* It would also greatly improve human rights, provide more people with income to stimulate economic development, and reduce environmental degradation and the threat of terrorism. Sharply reducing poverty is a win-win situation for people, economies, and the environment.

After the health risks associated with poverty and gender, the greatest risks of premature death are mostly the result of unhealthful choices that people make about their lifestyles—what I referred to early in this chapter as cultural hazards (Figures 19-18 and 19-19).

By far the best ways to reduce one's risk of premature death and serious health problem are to avoid

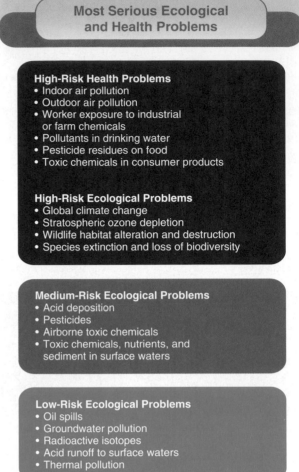

Figure 19-17 *Comparative risk analysi*s of the most serious ecological and health problems according to scientists acting as advisers to the U.S. Environmental Protection Agency. Risks under each category are not listed in rank order. (Science Advisory Board, *Reducing Risks*, Washington, D.C.: Environmental Protection Agency, 1990)

smoking and exposure to smoke, lose excess weight, reduce consumption of foods containing cholesterol and saturated fats, eat a variety of fruits and vegetables, exercise regularly, avoid alcohol or drink no more than two drinks a day, avoid excess sunlight (which ages skin and may cause skin cancer), and have only safe sex.

How Can We Estimate Risks of Using Increasingly Complex Technology in Our Lives? A Difficult Task

Estimating risks from using certain technologies is difficult because of the unpredictability of human behavior, human error, and sabotage.

The more complex a technological system and the more people needed to design and run it, the more difficult it is to estimate the risks. The overall reliability of

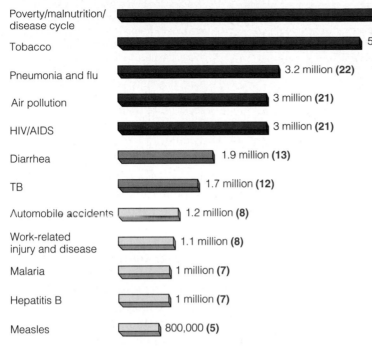

Cause of Death	Annual Deaths
Poverty/malnutrition/disease cycle	11 million (75)
Tobacco	5 million (34)
Pneumonia and flu	3.2 million (22)
Air pollution	3 million (21)
HIV/AIDS	3 million (21)
Diarrhea	1.9 million (13)
TB	1.7 million (12)
Automobile accidents	1.2 million (8)
Work-related injury and disease	1.1 million (8)
Malaria	1 million (7)
Hepatitis B	1 million (7)
Measles	800,000 (5)

Figure 19-18 Number of deaths per year in the world from various causes. Numbers in parentheses give these deaths in terms of the number of fully loaded 400-passenger jumbo jets crashing *every day of the year* with no survivors. Because of sensational media coverage, most people have a distorted view of the largest annual causes of death. (World Health Organization and U.S. Center for Disease Control and Prevention)

a technological system (expressed as a percentage) or the probability expressed as a percentage that a device will complete a task without failing is the product of two factors:

$$\text{System reliability (\%)} = \frac{\text{Technology}}{\text{reliability}} \times \frac{\text{Human}}{\text{reliability}}$$

With careful design, quality control, maintenance, and monitoring, a highly complex system such as a nuclear power plant or space shuttle can achieve a high degree of technology reliability. But human reliability usually is much lower than technology reliability and is almost impossible to predict: To err is human.

Suppose the technology reliability of a nuclear power plant is 95% (0.95) and human reliability is 75% (0.75). Then the overall system reliability is 71% (0.95 × 0.75 × 100 = 0.71 = 71%). Even if we could make the technology 100% reliable (1.0), the overall system reliability would still be only 75% (1.0 × 0.75 × 100 = 75%). The crucial dependence of even the most carefully designed systems on unpredictable human reliability helps explain essentially "impossible" tragedies such as the Chernobyl nuclear power plant accident and the *Challenger* and *Columbia* space shuttle accidents.

One way to make a system more foolproof or fail-safe is to move more of the potentially fallible elements from the human side to the technical side. However, chance events such as a lightning bolt can knock out an automatic control system, and no machine or computer program can completely replace human judgment. Also, the parts in any automated control system are manufactured, assembled, tested, certified, and maintained by fallible human beings. In addition, computer software programs used to monitor and control complex systems can also contain human error or can be deliberately modified by computer viruses to malfunction.

How Useful Is Risk Analysis? A Very Difficult Task

The results of risk analysis are usually very uncertain.

Here are some of the key questions involved in evaluating the reliability of risk analysis:

■ How reliable are risk assessment data and models? Sections 19-2 and 19-3)

■ Who profits from a risk analysis that allows certain levels of harmful chemicals into the environment, and who suffers?

■ Should estimates emphasize short-term risks, or should more weight be put on long-term risks? Who should make this decision?

■ Who should do a particular risk analysis, and who should review the results? A government agency? Independent scientists? The public?

■ Should cumulative effects of various risks be considered, or should risks be considered separately, as is usually done? Suppose a pesticide is found to have an annual risk of killing 1 out of 1 million through cancer, the current EPA limit. Cumulatively, however, effects from 40 such pesticides might kill 40, or 400 of every million people because of synergistic effects.

■ How widespread is each risk? About how many people are likely to be affected?

■ Should risk levels be higher for workers (as is almost always the case) than for the general public? What say should workers and their families have in this decision?

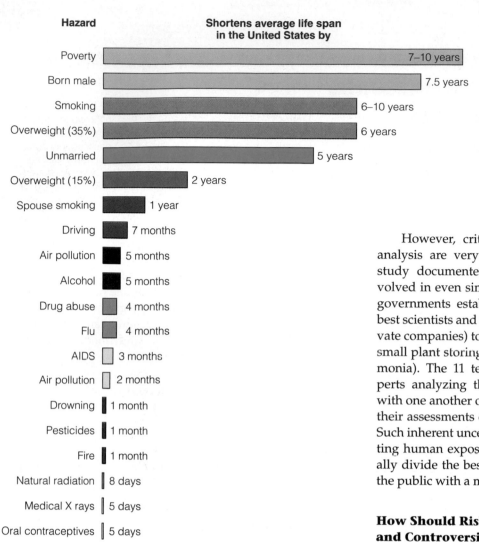

Hazard	Shortens average life span in the United States by
Poverty	7–10 years
Born male	7.5 years
Smoking	6–10 years
Overweight (35%)	6 years
Unmarried	5 years
Overweight (15%)	2 years
Spouse smoking	1 year
Driving	7 months
Air pollution	5 months
Alcohol	5 months
Drug abuse	4 months
Flu	4 months
AIDS	3 months
Air pollution	2 months
Drowning	1 month
Pesticides	1 month
Fire	1 month
Natural radiation	8 days
Medical X rays	5 days
Oral contraceptives	5 days
Toxic waste	4 days
Flying	1 day
Hurricanes, tornadoes	1 day
Living lifetime near nuclear plant	10 hours

Figure 19-19 Comparison of risks people face, expressed in terms of shorter average life span. After poverty and gender, the greatest risks people face are mostly from the lifestyle choices they make. These are only generalized relative estimates. Individual response to some of these risks can vary with factors such as genetic variation, family medical history, emotional makeup, stress, and social ties and support. (Data from Bernard L. Cohen)

However, critics point out that results of risk analysis are very uncertain. For example, a recent study documented the significant uncertainties involved in even simple risk analysis. Eleven European governments established 11 different teams of their best scientists and engineers (including those from private companies) to assess the hazards and risks from a small plant storing only one hazardous chemical (ammonia). The 11 teams, consisting of world-class experts analyzing this very simple system, disagreed with one another on fundamental points and varied in their assessments of the hazards by a factor of 25,000. Such inherent uncertainty explains why regulators setting human exposure levels for toxic substances usually divide the best results by 100 to 1,000 to provide the public with a margin of safety.

How Should Risks Be Managed? A Complex and Controversial Process

Risk management involves trying to answer a number of difficult and controversial questions about whether and how to reduce a particular societal risk to a certain level and at what cost.

Risk management includes the administrative, political, and economic actions taken to decide whether and how to reduce a particular societal risk to a certain level and at what cost.

Risk management involves answering the following questions:

- How reliable is the risk analysis for each risk?
- Which risks to human health should be given the highest priority?
- How much risk is acceptable and to whom?
- How much is a life worth, and how much money should we spend per life saved? Most government risk analyses set the value of a life at about $3.7 million for all people and about $1.4 million for people over 70. How much do you believe your life is worth?
- How much will it cost to reduce each risk to an acceptable level?

- How much risk is acceptable, and to whom is it acceptable? According to the National Academy of Sciences, exposure to toxic chemicals is responsible for 2–4% of the 521,000 cancer deaths in the United States; this amounts to 10,400–20,800 premature cancer deaths per year. Is this acceptable and to whom?

Proponents point to the numerous advantages of risk analysis. It is a useful way to organize and analyze available scientific information, identify significant hazards, and focus on areas that need more research. It can also help regulators to decide how money for reducing risks should be allocated and to stimulate people to make more informed decisions about health and environmental goals and priorities.

- How should limited funds be spent to provide the greatest benefit?

- How will the risk management plan be monitored, enforced, and communicated to the public?

Each step in this process involves making value judgments and weighing trade-offs to find some reasonable compromise among often conflicting political, economic, health, and environmental interests.

How Well Do We Perceive Risks? Most of Us Flunk

Most individuals are poor at evaluating the relative risks they face, mostly because of misleading information and irrational fears.

Most of us are not good at assessing the relative risks from the hazards that can affect us. Also, many people deny or shrug off the high-risk chances of death (or injury) from voluntary activities they enjoy, such as *motorcycling* (1 death in 50 participants), *smoking* (1 in 300 participants by age 65 for a pack-a-day smoker), *hang gliding* (1 in 1,250), and *driving* (1 in 3,300 without a seatbelt and 1 in 6,070 with a seatbelt). Indeed, the most dangerous thing most people in many countries do each day is drive or ride in a car.

Yet some of these same people may be terrified about the possibility of being killed by a *gun* (1 in 28,000 in the United States), *flu* (1 in 130,000), *nuclear power plant accident* (1 in 200,000), *West Nile virus* (1 in 1 million), *lightning* (1 in 3 million), *commercial airplane crash* (1 in 9 million), *snakebite* (1 in 36 million), or *shark attack* (1 in 281 million).

What Factors Distort Our Perceptions of Risk? Irrational Fears and Perceptions Can Take Over

Several factors can give people a distorted sense of risk.

Here are four factors that can cause people to see a technology or a product as being riskier than experts judge it to be. *First* is the *degree of control* we have. Most of us have a greater fear of things over which we do not have personal control. For example, some individuals feel safer driving their own car for long distances through heavy traffic than traveling the same distance on a plane. But look at the math. The risk of dying in a car accident while using your seatbelt is 1 in 6,070 whereas the risk of dying in a commercial airliner crash is 1 in 9 million. Can you think of another example?

Second is *fear of the unknown.* Most people have greater fear of a new, unknown product or technology than they do of an older and more familiar one. Examples include a greater fear of genetically modified food than of food produced by traditional plant breeding techniques, and a greater fear of nuclear power plants than of more familiar coal-fired power plants. Can you think of another example?

Third is whether or not we voluntarily take the risk. For example, we might perceive that the risk from driving, which is largely *voluntary,* is less than that from a nuclear power plant, which is mostly imposed on us whether we like it or not. Can you come up with another example?

Fourth is whether a risk is *catastrophic, not chronic.* We usually have a much greater fear of a well-publicized death toll from a single catastrophic accident rather than the same or an even larger death toll spread out over a longer time. Examples include a severe nuclear power plant accident, an industrial explosion, or an accidental plane crash, as opposed to coal-burning power plants, automobiles, and smoking. Can you think of another example?

There is also concern over the *unfair distribution of risks* from the use of a technology or certain chemicals. Citizens are outraged when government officials decide to put a hazardous waste landfill or incinerator in or near their neighborhood. Even when the decision is based on careful risk analysis, it is usually seen as politics, not science. Residents will not be satisfied by estimates that the lifetime risks of cancer death from the facility are not greater than, say, 1 in 100,000. Instead, they point out that living near the facility means that they will have a much higher risk of dying from cancer than would people living farther away.

How Can You Become Better at Risk Analysis? Analyze, Compare, and Evaluate Your Lifestyle

To become better at risk analysis you can carefully evaluate the barrage of bad news, compare risks, and concentrate on reducing risks over which we have some control.

You can do three things to become better at estimating risks. *First,* carefully evaluate what the media presents. Recognize that the media often give an exaggerated view of risks to capture our interest and thus sell newspapers or gain TV viewers.

Second, compare risks. Do you risk getting cancer by eating a charcoal-broiled steak once or twice a week? Yes, because in theory anything can harm you. The question is whether this danger is great enough for you to worry about. In evaluating a risk the question is not, "Is it safe?" but rather, "How risky is it compared to other risks?"

Third, concentrate on the most serious risks to your life and health over which you have some control over and stop worrying about smaller risks and those over which you have little or no control. When you worry about something, the most important question to ask is, "Do I have any control over this?"

For example, the top four killers of Americans (and people in many countries) are heart attacks, strokes, cancer, and accidents (many of them involving motor vehicles). You have control over major ways to reduce these risks because you decide whether to smoke, what to eat, how much exercise you get, how much alcohol you consume, your exposure to the sun's ultraviolet rays, how safely you drive, and whether or not you practice safe sex. Concentrate on evaluating these important choices, and you will have a much greater chance of living a healthy, longer, happier, and less fearful life.

The burden of proof imposed on individuals, companies, and institutions should be to show that pollution prevention options have been thoroughly examined, evaluated, and used before lesser options are chosen.

JOEL HIRSCHORN

CRITICAL THINKING

1. Explain why you agree or disagree with the proposals for reducing the death toll and other harmful effects of smoking listed on p. 409. Do you believe that there should be a ban on smoking indoors in all public places? Explain.

2. Do you believe the precautionary approach should be used to deal with the *potential* harm from hormonally active agents (HAAs) while more definitive research is carried out over the next two decades? Explain. What harmful effects could using this approach have on the economy and on your lifestyle?

3. Should we have zero pollution levels for all toxic and hazardous chemicals? Explain. What are the alternatives?

4. Evaluate the following statements:
   a. We should not get worked up about exposure to toxic chemicals because almost any chemical in a large enough dosage can cause some harm.
   b. We should not worry so much about exposure to toxic chemicals because through genetic adaptation we can develop immunity to such chemicals.
   c. We should not worry so much about exposure to toxic chemicals because we can use genetic engineering to reduce or eliminate such problems.

5. Should pollution levels be set to protect the most sensitive people in a population (Figure 19-3, left, p. 411) or the average person (Figure 19-3, middle)? Explain.

6. Should laboratory-bred animals be used in laboratory experiments in toxicology? Explain. What are the alternatives?

7. What are the five major risks you face from **(a)** your lifestyle, **(b)** where you live, and **(c)** what you do for a living? Which of these risks are voluntary and which are involuntary? List the five most important things you can do to reduce these risks. Which of these things do you actually plan to do?

8. Congratulations! You are in charge of a global risk–benefit analysis board to evaluate whether certain chemicals or technologies should be approved for widespread use. Explain why you would approve or disapprove each of the following: **(a)** drugs to slow the aging process, **(b)** drugs that would cause people to have unconditional love for everyone and thus have the potential to do away with hate, violence, and war, **(c)** genetic engineering advances that would allow parents to have genes inserted into lab-produced fetuses to produce designer babies with their desired checklist of enhanced genetic traits, **(d)** allowing people to have a genetic clone that they can use for spare parts to help them live longer, and **(e)** putting everyone in the world under constant electronic surveillance to help prevent bioterrorism.

9. Congratulations! You are in charge of the world. List the three most important features of your program to reduce the risk from exposure to **(a)** toxic and hazardous chemicals, **(b)** infectious disease organisms, and **(c)** viruses.

PROJECTS

1. Use the library or the Internet to find recent articles that support or refute the hormesis hypothesis.

2. Use the library or the Internet to find recent articles describing the increasing genetic resistance in disease-causing bacteria to commonly used antibiotics. Evaluate the evidence and claims in these articles.

3. Pick a specific viral disease and use the library or Internet to find out **(a)** how it spreads, **(b)** its effects, **(c)** strategies for controlling its spread, and **(d)** possible treatments.

4. Use the library or the Internet to find bibliographic information about *Paracelsus* and *Joel Hirschorn*, whose quotes appear at the beginning and end of this chapter.

5. Make a concept map of this chapter's major ideas, using the section heads, subheads, and key terms (in boldface). Look on the website for this book for information about making concept maps.

LEARNING ONLINE

The website for this book contains study aids and many ideas for further reading and research. They include a chapter summary, review questions for the entire chapter, flash cards for key terms and concepts, a multiple-choice practice quiz, interesting Internet sites, references, and a guide for accessing thousands of InfoTrac® College Edition articles. Log on to

http://biology.brookscole.com/miller14

Then click on the Chapter-by-Chapter area, choose Chapter 19, and select a learning resource.

20 Air Pollution

CASE STUDY
When Is a Lichen Like a Canary?

Nineteenth-century coal miners took canaries with them into the mines—not for their songs but for the moment when they stopped singing. Then the miners knew it was time to get out of the mine because the air contained methane, which could ignite and explode.

Today we use sophisticated equipment to monitor air quality, but living things such as lichens (Figure 20-1) can also warn us of bad air. Lichens, which are not plants, consist of a fungus and an alga living together, usually in a mutually beneficial (mutualistic) partnership.

You have probably seen lichens growing as crusts or leafy growths on rocks (Figure 20-1, right), walls, tombstones, and tree trunks or as beards hanging down from twigs and branches (Figure 20-1, left).

These hardy pioneer species are good biological indicators of air pollution because they are always absorbing air as a source of nourishment. A highly polluted area around an industrial plant may have no lichens or only gray-green crusty lichen. An area with moderate air pollution may have orange crusty lichens on walls. Walls and trees in areas with fairly clean air may have leafy lichens.

Some lichen species are sensitive to specific air-polluting chemicals. Old man's beard (*Usnea trichodea*) and yellow *Evernia* lichens, for example, sicken or die in the presence of too much sulfur dioxide.

Because lichens are widespread, long lived, and anchored in place, they can also help track pollution to its source. The scientist who discovered sulfur dioxide pollution on Isle Royale in Lake Superior, where no car or smokestack had ever intruded, used *Evernia* lichens to point the finger northward to coal-burning facilities at Thunder Bay, Canada.

In 1986, the Chernobyl nuclear power plant in Ukraine (Figure 17-1, p. 350) exploded and spewed radioactive particles into the atmosphere. Some of these particles fell to the ground over northern Scandinavia and were absorbed by lichens that carpet much of Lapland. The area's Saami people depend on reindeer meat for food, and the reindeer feed on lichens. After Chernobyl, more than 70,000 reindeer had to be killed and the meat discarded because it was too radioactive to eat. Scientists helped the Saami identify where to find the remaining uncontaminated reindeer by analyzing lichens to pinpoint the most contaminated areas.

We all must breathe air from a global atmospheric commons in which air currents and winds can transport some pollutants long distances. Lichens can alert us to the danger, but as with all forms of pollution, the best solution is prevention.

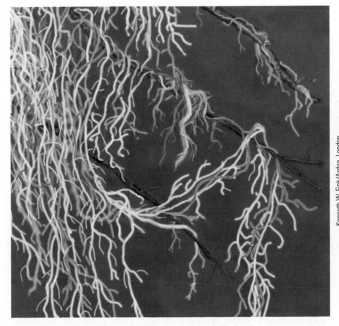

Kenneth W. Fink/Ardea, London

Figure 20-1 Red and yellow crustose lichens growing on slate rock in the foothills of the Sierra Nevada near Merced, California (right), and *Usnea trichodea* lichen growing on a branch of a larch tree in Gifford Pinchot National Park, Washington (left). The vulnerability of various lichen species to specific air pollutants can help researchers detect levels of these pollutants and track down their sources.

I thought I saw a blue jay this morning. But the smog was so bad that it turned out to be a cardinal holding its breath.

MICHAEL J. COHEN

This chapter discusses the pollutants found in outdoor and indoor air and how they can be reduced. It addresses the following questions:

- What layers are found in the atmosphere?
- What are the major outdoor air pollutants, and where do they come from?
- What are two types of smog?
- What is acid deposition, and how can it be reduced?
- What are the harmful effects of air pollutants?
- How can we prevent and control air pollution?

20-1 STRUCTURE AND SCIENCE OF THE ATMOSPHERE

What Are Key Characteristics of the Atmosphere? Several Layers with Different Properties

The atmosphere consists of several layers with different temperatures, pressures, and composition.

We live at the bottom of a thin layer of gases surrounding the earth, called the *atmosphere*. It is divided into several spherical sublayers (Figure 20-2), each characterized by abrupt changes in temperature as a result of differences in the absorption of incoming solar energy.

Density and atmospheric pressure also vary throughout the atmosphere. Gravitational forces pull the gas molecules in the atmosphere toward the earth's surface. This means that the air we breathe at sea level has a higher *density* (more molecules per liter) than the air we inhale on top of the world's highest mountain.

Atmospheric pressure is a measure of the mass per unit area of air. It is caused by the bombardment of a surface by the molecules in air. The pressure of the atmosphere increases as the density of air increases because a volume of air with a high density has more gas molecules than a volume with a lower density. Thus, *atmospheric pressure decreases with altitude*. This explains why the pressure exerted on each square centimeter of your body by the bombardment of hordes of gas molecules is greater at sea level than at the top of a tall mountain.

What Is the Troposphere? Weather Breeder

The atmosphere's innermost layer is made up mostly of nitrogen and oxygen, with smaller amounts of water vapor and carbon dioxide.

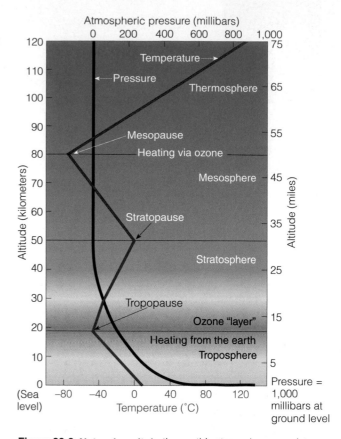

Figure 20-2 Natural capital: the earth's atmosphere consists of several layers. The average temperature of the atmosphere varies with altitude (red line). The average temperature of the atmosphere at the earth's surface is determined by a combination of two factors. One is *natural heating* by incoming sunlight and certain greenhouse gases that release absorbed energy as heat into the lower troposphere (the natural *greenhouse effect*; Figure 6-14, p. 110). The other is *natural cooling* by surface evaporation of water and convection processes that transfer heat to higher altitudes and latitudes (Figure 6-10, p. 107). Most UV radiation from the sun is absorbed by ozone (O_3), found primarily in the stratosphere in the *ozone layer* 17–26 kilometers (10–16 miles) above sea level.

About 75–80% of the earth's air mass is found in the *troposphere*, the atmospheric layer closest to the earth's surface. This layer extends only about 17 kilometers (11 miles) above sea level at the equator and about 8 kilometers (5 miles) over the poles. If the earth were the size of an apple, this lower layer containing the air we breathe would be no thicker than the apple's skin.

Take a deep breath. About 99% of the volume of the air you just inhaled from the troposphere consists of two gases: nitrogen (78%) and oxygen (21%). The remainder consists of water vapor (varying from 0.01% at the frigid poles to 4% in the humid tropics), slightly less than 1% argon (Ar), 0.038% carbon dioxide (CO_2), and trace amounts of several other gases.

The troposphere is also the layer of the atmosphere involved in the chemical cycling of the earth's vital nutrients. In addition, this thin and turbulent layer of rising and falling air currents and winds is largely responsible for the planet's short-term *weather* and long-term *climate*.

To biologist and environmental scientist David Suzuki, "Air is a matrix or universal glue that joins all life together. . . . Every breath is an affirmation of our connection with other living things, a renewal of our link with our ancestors, and a contribution to generations yet to come."

What Is the Stratosphere? Earth's Global Sunscreen

Ozone in the atmosphere's second layer filters out most of the sun's UV radiation that is harmful to us and most other species.

The atmosphere's second layer is the *stratosphere*, which extends from about 17 to 48 kilometers (11–30 miles) above the earth's surface (Figure 20-2). Although the stratosphere contains less matter than the troposphere, its composition is similar, with two notable exceptions: its volume of water vapor is about 1/1,000 as much and its concentration of ozone (O_3) is much higher (Figure 20-3).

Stratospheric ozone is produced when some of the oxygen molecules there interact with ultraviolet (UV) radiation emitted by the sun ($3 O_2 + UV \longrightarrow 2 O_3$). This "global sunscreen" of ozone in the stratosphere keeps about 95% of the sun's harmful UV radiation from reaching the earth's surface.

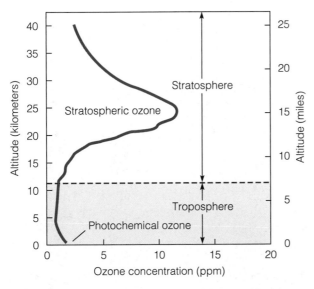

Figure 20-3 Natural capital: average distribution and concentrations of ozone in the troposphere and stratosphere. *Beneficial ozone* that forms in the stratosphere protects life on earth by filtering out most of the incoming harmful UV radiation emitted by the sun. *Harmful* or *photochemical ozone* forms in the troposphere when various air pollutants undergo chemical reactions under the influence of sunlight. Ozone in this atmosphere near the earth's surface damages plants, lung tissues, and some materials such as rubber.

This UV filter of "good" ozone in the lower stratosphere allows us and other forms of life to exist on land and helps protect us from sunburn, skin and eye cancer, cataracts, and damage to our immune systems. It also prevents much of the oxygen in the troposphere from being converted to photochemical ozone, a harmful air pollutant.

Much evidence indicates that some human activities are *decreasing* the amount of beneficial or "good" ozone in the stratosphere and *increasing* the amount of harmful or "bad" ozone in the troposphere—especially in some urban areas.

20-2 OUTDOOR AIR POLLUTION

What Are the Major Types and Sources of Air Pollution? Burning Fossil Fuels Is the Major Culprit

Outdoor air pollutants come mostly from natural sources and burning fossil fuels in motor vehicles and power and industrial plants.

Air pollution is the presence of chemicals in the atmosphere in concentrations high enough to affect climate and harm organisms and materials. The effects of airborne pollutants range from annoying to lethal.

Table 20-1 (p. 436) lists the major classes of pollutants commonly found in outdoor (ambient) air. The majority comes from natural sources. They include dust particles blowing off the earth's surface (Figure 6-1, p. 102), volatile organic chemicals released by some plants, the decay of plants, forest fires, volcanic eruptions, and sea spray. Most natural sources of air pollution are spread out and, except for those from volcanic eruptions and some forest fires, rarely reach harmful levels.

Air pollution is not new (Spotlight, p. 437). Throughout human history, beginning with the discovery of fire, we have added various types of pollutants to the troposphere. Our inputs increased when we began extracting and burning coal, first for heat and later for generating electricity and producing materials such as steel.

Burning oil, gasoline, and natural gas also adds pollutants to the atmosphere. Pollutants from our activities can reach harmful levels in the troposphere, especially in urban areas where people, cars, and industrial activities are concentrated.

Most outdoor pollutants in today's urban areas enter the atmosphere from the burning of fossil fuels in power plants and factories (*stationary sources*) and in motor vehicles (*mobile sources*). Scientists classify outdoor air pollutants into two categories. **Primary pollutants** are those emitted directly into the troposphere in a potentially harmful form. Examples are soot

Table 20-1 Major Classes of Air Pollutants

Class	Examples
Carbon oxides	Carbon monoxide (CO) and carbon dioxide (CO_2)
Sulfur oxides	Sulfur dioxide (SO_2) and sulfur trioxide (SO_3)
Nitrogen oxides	Nitric oxide (NO), nitrogen dioxide (NO_2), nitrous oxide (N_2O) (NO and NO_2 often are lumped together and labeled NO_x)
Volatile organic compounds (VOCs)	Methane (CH_4), propane (C_3H_8), chlorofluorocarbons (CFCs)
Suspended particulate matter (SPM)	Solid particles (dust, soot, asbestos, lead, nitrate, and sulfate salts), liquid droplets (sulfuric acid, PCBs, dioxins, and pesticides)
Photochemical oxidants	Ozone (O_3), peroxyacyl nitrates (PANs), hydrogen peroxide (H_2O_2), aldehydes
Radioactive substances	Radon-222, iodine-131, strontium-90, plutonium-239 (Table 3-1, p. 49)
Hazardous air pollutants (HAPs), which cause health effects such as cancer, birth defects, and nervous system problems	Carbon tetrachloride (CCl_4), methyl chloride (CH_3Cl), chloroform ($CHCl_3$), benzene (C_6H_6), ethylene dibromide ($C_2H_2Br_2$), formaldehyde (CH_2O_2)

and carbon monoxide. While in the troposphere, some of these primary pollutants may react with one another or with the basic components of air to form new pollutants, called **secondary pollutants** (Figure 20-4).

With their concentration of cars and factories, cities normally have higher outdoor air pollution lev-

els than rural areas. However, prevailing winds can spread long-lived primary and secondary air pollutants from urban and industrial areas to the countryside and to other urban areas.

Indoor air pollutants come from infiltration of polluted outside air and various chemicals used or pro-

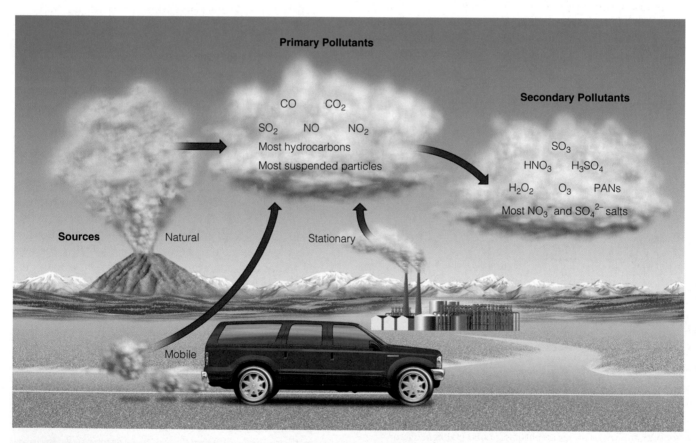

Figure 20-4 Natural capital degradation: sources and types of air pollutants. Human inputs of air pollutants may come from *mobile sources* (such as cars) and *stationary sources* (such as industrial and power plants). Some *primary air pollutants* may react with one another or with other chemicals in the air to form *secondary air pollutants.*

Air Pollution in the Past: The Bad Old Days

Modern civilization did not invent air pollution. It probably began when humans discovered fire and used it to burn wood in poorly ventilated caves for warmth and cooking and inhaled unhealthy smoke and soot.

During the Middle Ages, a haze of wood smoke hung over densely packed urban areas. The industrial revolution brought even worse air pollution as coal was burned to power factories and heat homes.

By the 1850s, London had become well known for its "pea soup" fog, a mixture of coal smoke and fog that blanketed the city. In 1880, a prolonged coal fog killed an estimated 2,200 people. Another in 1911 killed more than 1,100 Londoners. The authors of a report on this disaster coined the word *smog* for the deadly mixture of smoke and fog that enveloped the city.

In 1952, an even worse yellow fog lasted for 5 days and killed an estimated 4,000–12,000 Londoners, prompting Parliament to pass the Clean Air Act of 1956. Additional air pollution disasters in 1956, 1957, and 1962 killed 2,500 more people. Because of strong air pollution laws, London's air today is much cleaner, and "pea soup" fogs are a thing of the past. Now the major threat is from air pollutants emitted by motor vehicles.

The industrial revolution, powered by coal-burning factories and homes, brought air pollution to the United States. Large industrial cities such as Pittsburgh, Pennsylvania, and St. Louis, Missouri, were known for their smoky air. By the 1940s, the air over some cities was so polluted that people had to use their automobile headlights during the day.

The first documented air pollution disaster in the United States occurred on October 29, 1948, at the small industrial town of Donora in Pennsylvania's Monongahela River Valley south of Pittsburgh. Pollutants from the area's coal-burning industries became trapped in a dense fog that stagnated over the valley for 5 days. About 6,000 of the town's 14,000 inhabitants became sick, and 22 of them died. This killer fog resulted from a combination of mountainous terrain surrounding the valley and weather conditions that trapped and concentrated deadly pollutants emitted by the community's steel mill, zinc smelter, and sulfuric acid plant.

In 1963, high concentrations of air pollutants accumulated in the air over New York City, killing about 300 people and injuring thousands. Other episodes in New York, Los Angeles, and other large cities in the 1960s led to much stronger air pollution control programs in the 1970s.

In 1952, Oregon became the first state to pass a law controlling air pollution. Congress passed the original version of the Clean Air Act in 1963. But it did not have much effect until a stronger version was enacted in 1970 and the Environmental Protection Agency was created and empowered to set and enforce national air pollution standards. Even stricter emission standards were imposed by amendments to the Clean Air Act in 1977 and 1990. Mostly as a result of these laws and actions by states and local areas, there has been a dramatic improvement in air quality throughout the United States.

Critical Thinking

Explain why you agree or disagree with the statement that air pollution in the United States should not be a major concern because of the significant progress in reducing outdoor air pollution since 1970.

---

duced inside buildings, as discussed in Section 20-5. Experts in risk analysis rate indoor and outdoor air pollution as high-risk human health problems.

According to the World Health Organization (WHO), one of every six people on the earth or more than 1.1 billion people live in urban areas where outdoor air is unhealthy to breathe. Most of them live in densely populated cities in developing countries where air pollution control laws do not exist or are poorly enforced. In other words, poverty can mean poor air for the poor.

In the United States and most other developed countries, government-mandated standards set maximum allowable atmospheric concentrations, or criteria, for six *criteria* or *conventional air pollutants* commonly found in outdoor air (Table 20-2, p. 438). Most scientists would also add volatile organic compounds (VOCs, Table 20-1) to this list because of their role in the formation of photochemical smog that plagues many cities. Most air pollutants are gases. But some are *aerosols*, which consist of tiny particles of solids or droplets of liquids suspended in the air. *Good news.* Regulating these six criteria pollutants has helped sharply reduce their levels in most developed countries.

Should Carbon Dioxide Be Classified as an Air Pollutant? Most Scientists Say Yes

Carbon dioxide can be classified as an air pollutant because it can warm the atmosphere and contribute to global climate change.

Most scientists would add CO_2 to the gang of six criteria air pollutants (Table 20-2) despite the fact that the

Table 20-2 Major Outdoor Air Pollutants*

CARBON MONOXIDE (CO)

Description: Colorless, odorless gas that is poisonous to air-breathing animals; forms during the incomplete combustion of carbon-containing fuels ($2 C + O_2 \longrightarrow 2 CO$).

Major human sources: Cigarette smoking (p. 409), incomplete burning of fossil fuels. About 77% (95% in cities) comes from motor vehicle exhaust.

Health effects: Reacts with hemoglobin in red blood cells and reduces the ability of blood to bring oxygen to body cells and tissues. This impairs perception and thinking; slows reflexes; causes headaches, drowsiness, dizziness, and nausea; can trigger heart attacks and angina; damages the development of fetuses and young children; and aggravates chronic bronchitis, emphysema, and anemia. At high levels it causes collapse, coma, irreversible brain cell damage, and death.

NITROGEN DIOXIDE (NO₂)

Description: Reddish-brown irritating gas that gives photochemical smog its brownish color; in the atmosphere can be converted to nitric acid (HNO_3), a major component of acid deposition.

Major human sources: Fossil fuel burning in motor vehicles (49%) and power and industrial plants (49%).

Health effects: Lung irritation and damage; aggravates asthma and chronic bronchitis; increases susceptibility to respiratory infections such as the flu and common colds (especially in young children and older adults).

Environmental effects: Reduces visibility; acid deposition of HNO_3 can damage trees, soils, and aquatic life in lakes.

Property damage: HNO_3 can corrode metals and eat away stone on buildings, statues, and monuments; NO_2 can damage fabrics.

SULFUR DIOXIDE (SO₂)

Description: Colorless, irritating; forms mostly from the combustion of sulfur-containing fossil fuels such as coal and oil ($S + O_2 \longrightarrow SO_2$); in the atmosphere can be converted to sulfuric acid (H_2SO_4), a major component of acid deposition.

Major human sources: Coal burning in power plants (88%) and industrial processes (10%).

Health effects: Breathing problems for healthy people; restriction of airways in people with asthma; chronic exposure can cause a permanent condition similar to bronchitis. According to the WHO, at least 625 million people are exposed to unsafe levels of sulfur dioxide from fossil fuel burning.

Environmental effects: Reduces visibility; acid deposition of H_2SO_4 can damage trees, soils, and aquatic life in lakes.

Property damage: SO_2 and H_2SO_4 can corrode metals and eat away stone on buildings, statues, and monuments; SO_2 can damage paint, paper, and leather.

SUSPENDED PARTICULATE MATTER (SPM)

Description: Variety of particles and droplets (aerosols) small and light enough to remain suspended in atmosphere for short periods (large particles) to long periods (small particles; Figure 20-6, p. 441); cause smoke, dust, and haze.

Major human sources: Burning coal in power and industrial plants (40%), burning diesel and other fuels in vehicles (17%), agriculture (plowing, burning off fields), unpaved roads, construction.

Health effects: Nose and throat irritation, lung damage, and bronchitis; aggravates bronchitis and asthma; shortens life; toxic particulates (such as lead, cadmium, PCBs, and dioxins) can cause mutations, reproductive problems, cancer.

Environmental effects: Reduces visibility; acid deposition of H_2SO_4 droplets can damage trees, soils, and aquatic life in lakes.

Property damage: Corrodes metal; soils and discolors buildings, clothes, fabrics, and paints.

OZONE (O₃)

Description: Highly reactive, irritating gas with an unpleasant odor that forms in the troposphere as a major component of photochemical smog (Figures 20-3 and 20-5).

Major human sources: Chemical reaction with volatile organic compounds (VOCs, emitted mostly by cars and industries) and nitrogen oxides to form photochemical smog (Figure 20-5).

Health effects: Breathing problems; coughing; eye, nose, and throat irritation; aggravates chronic diseases such as asthma, bronchitis, emphysema, and heart disease; reduces resistance to colds and pneumonia; may speed up lung tissue aging.

Environmental effects: Ozone can damage plants and trees; smog can reduce visibility.

Property damage: Damages rubber, fabrics, and paints.

LEAD

Description: Solid toxic metal and its compounds, emitted into the atmosphere as particulate matter.

Major human sources: Paint (old houses), smelters (metal refineries), lead manufacture, storage batteries, leaded gasoline (being phased out in developed countries).

Health effects: Accumulates in the body; brain and other nervous system damage and mental retardation (especially in children); digestive and other health problems; some lead-containing chemicals cause cancer in test animals.

Environmental effects: Can harm wildlife.

*Data from U.S. Environmental Protection Agency.

EPA, under pressure from most U.S. oil and coal companies, says it is not.

These scientists give three reasons for classifying CO_2 as an air pollutant. *First,* in high enough concentrations any chemical in the air can become a pollutant. *Second,* we have been increasing the concentration of CO_2 in the troposphere by burning fossil fuels and clearing CO_2-absorbing trees faster than they are growing back in many areas. *Third,* the troposphere is warming and there is considerable evidence that the additional CO_2 (a greenhouse gas) added to the troposphere by human activities plays a role in this change.

So who cares if it gets a little warmer? Warmer winters would be nice. The problem is that enhancement of the earth's natural greenhouse effect (Figure 6-14, p. 110)—called *global warming*—can change where and how much precipitation falls, affect where we can grow food, and flood some areas of the world because of rising sea levels. Thus because higher atmospheric levels of CO_2 can harm some people, economies, and ecosystems, most scientists say that it meets the criteria for being classified as an air pollutant.

In 2003, 12 states sued the U.S. Environmental Protection Agency (EPA) for failure to regulate emissions of carbon dioxide as allowed under the Clean Air Act. The EPA contends that the Clean Air Act does not give the agency the power to control gases that can cause global warming. Meanwhile, some states are passing laws that regulate carbon dioxide emissions.

20-3 PHOTOCHEMICAL AND INDUSTRIAL SMOG

What Is Photochemical Smog? Brown-Air Smog

Photochemical smog is a mixture of air pollutants formed by the reaction of nitrogen oxides and volatile organic hydrocarbons under the influence of sunlight.

A *photochemical reaction* is any chemical reaction activated by light. Air pollution known as **photochemical smog** is formed when a mix of nitrogen oxides (NO and NO_2, both called NO_x) and volatile organic hydrocarbon compounds from natural and human sources chemically react under the influence of UV radiation from the sun to produce a mixture of more than 100 primary and secondary pollutants (Figure 20-5, p. 440). In greatly simplified terms,

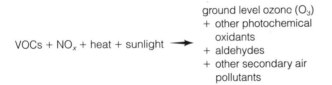

The formation of photochemical smog involves a complex series of chemical reactions. It begins inside automobile engines and in the boilers of coal-burning power and industrial plants. At the high temperatures found there, nitrogen and oxygen in air react to produce colorless nitric oxide ($N_2 + O_2 \longrightarrow 2\ NO$). In the atmosphere, some of the NO is converted to nitrogen dioxide (NO_2), a yellowish-brown gas with a choking odor (Table 20-2). NO_2 is the cause of the brownish haze that hangs over many cities during the afternoons of sunny days, explaining why photochemical smog sometimes is called *brown-air smog*.

When exposed to ultraviolet radiation from the sun, some of the NO_2 engages in a complex series of reactions with hydrocarbons (mostly released by vegetation, motor vehicles, and other human activities) that produce photochemical smog—a mixture of ozone, nitric acid, aldehydes, peroxyacyl nitrates (PANs), and other secondary pollutants.

Collectively, NO_2, O_3, and PANs are called *photochemical oxidants* because they can react with and oxidize certain compounds in the atmosphere or inside your lungs that normally are not oxidized. Mere traces of these oxidants (especially ozone, Table 20-2) and aldehydes in photochemical smog can irritate the respiratory tract and damage crops and trees.

Hotter days lead to higher levels of ozone and other components of smog. As traffic increases on a sunny day, smog builds up to peak levels by early afternoon, irritating people's eyes and respiratory tracts.

All modern cities have some photochemical smog, but it is much more common in cities with sunny, warm, dry climates and lots of motor vehicles. Examples are Los Angeles, Denver, and Salt Lake City in the United States; Sydney, Australia; Mexico City; São Paulo, Brazil; and Buenos Aires, Argentina.

What might happen if 400 million Chinese drive gasoline-fueled cars by 2050 as projected? According to a 1999 article in *Geophysical Research Letters*, the resulting photochemical smog could cover the entire western Pacific in ozone, extending to the United States.

How Can Trees Contribute to Photochemical Smog? Hydrocarbon Emitters

Some hydrocarbon-emitting tree species can contribute to the formation of photochemical smog.

Trees have many environmental benefits. They emit oxygen, absorb CO_2, provide shade (which reduces energy needed for air conditioning), and help absorb and remove various pollutants from the air.

When Ronald Reagan was running for president in 1980 he was ridiculed when he implied that we did not need tougher air pollution laws because trees and plants caused 80% of all air pollution. This was certainly an exaggeration.

But recently scientists have found that some species of trees and plants (such as some oak species, sweet gums, poplars, and kudzu) in and around urban areas play a larger role in smog formation than was once thought. They do so by emitting volatile organic compounds (VOCs, especially hydrocarbons such as isoprene) that are ingredients in the development of photochemical smog.

Unless forests of such trees are close to urban areas, most of their VOC emissions occur in nonurban areas and disperse into the atmosphere. Thus they do not make a significant contribution to the formation of photochemical smog except in forested areas near urban areas with large sources of NO_x and plenty of sunlight. This is in contrast to cars, which along with refineries and other sources emit most of their VOCs, NO_x, and other pollutants into urban air.

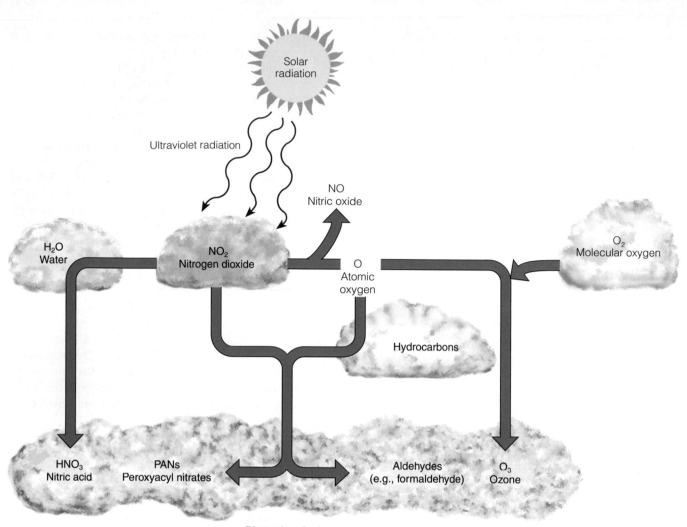

Photochemical smog

Figure 20-5 **Natural capital degradation:** simplified scheme of the formation of photochemical smog. The severity of smog generally is associated with atmospheric concentrations of ozone at ground level.

Because of trees' ecological and aesthetic benefits, environmentalists support their widespread planting in urban areas. But they say the emphasis should be on tree species that emit low levels of VOCs.

How Does Industrial Smog Form, and How Big a Problem Is It? Gray-Air Smog Is a Danger in Some Developing Countries

Industrial smog is a mixture of sulfur dioxide, droplets of sulfuric acid, and a variety of suspended solid particles emitted by burning coal and oil.

Fifty years ago, cities such as London, England, and Chicago and Pittsburgh in the United States burned large amounts of coal and heavy oil (which contain sulfur impurities) in power plants and factories and for heating homes and cooking food. During winter, people in such cities were exposed to **industrial smog** consisting mostly of sulfur dioxide, aerosols contain-ing suspended droplets of sulfuric acid formed from sulfur dioxide, and a variety of suspended solid particles (Figure 20-6).

The chemistry of industrial smog is fairly simple. When burned, the carbon in coal and oil is converted to carbon dioxide ($C + O_2 \longrightarrow CO_2$) and carbon monoxide ($2\ C + O_2 \longrightarrow 2\ CO$). Some of the un-burned carbon also ends up in the atmosphere as suspended particulate matter (soot), another ingredient of industrial smog.

The sulfur compounds in coal and oil also react with oxygen to produce sulfur dioxide (SO_2), a color-less, suffocating gas. Sulfur dioxide also is emitted into the troposphere when metal sulfide ores are roasted or smelted to convert a metal ore (such as lead sulfide, PbS) to a free metal (such as Pb).

In the troposphere, some of the sulfur dioxide reacts with oxygen to form sulfur trioxide (SO_3), which then reacts with water vapor in the air to produce tiny

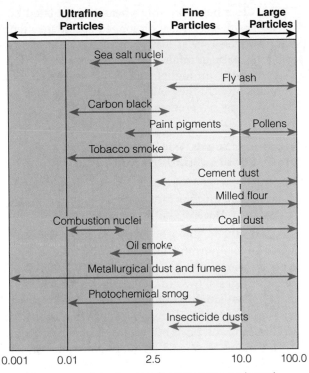

Ultrafine Particles | Fine Particles | Large Particles

Sea salt nuclei
Fly ash
Carbon black
Paint pigments | Pollens
Tobacco smoke
Cement dust
Milled flour
Combustion nuclei | Coal dust
Oil smoke
Metallurgical dust and fumes
Photochemical smog
Insecticide dusts

0.001   0.01   2.5   10.0   100.0

Average particle diameter (micrometers or microns)

Figure 20-6 Suspended particulate matter consists of particles of solid matter and droplets of liquid that are small and light enough to remain suspended in the atmosphere for some period (the larger the particle, the sooner it usually falls to earth). Suspended particles are found in a wide variety of types and sizes, ranging in diameter from 0.001 micrometer to 100 micrometers (a micrometer, or micron, is one-millionth of a meter, or about 0.00004 inch). Since 1987, the EPA has focused on *fine particles* smaller than 10 microns (known as *PM-10*). In 1997, the agency began focusing on reducing emissions of *ultrafine particles* with diameters less than 2.5 microns (known as *PM-2.5*) because these particles are small enough to reach the lower part of human lungs and contribute to respiratory diseases.

suspended droplets of sulfuric acid (H_2SO_4). Some of these droplets react with ammonia in the atmosphere to form solid particles of ammonium sulfate [$(NH_4)_2SO_4$]. The tiny suspended particles of such salts and carbon (soot) give the resulting industrial smog a gray color, explaining why it is sometimes called *gray-air smog*.

Today urban industrial smog is rarely a problem in most developed countries. This has happened because coal and heavy oil are burned only in large boilers with reasonably good pollution control or with tall smokestacks that transfer the pollutants to downwind rural areas.

However, industrial smog is a problem in industrialized urban areas of China, India, Ukraine, and some eastern European countries (especially the "black triangle" region of Slovakia, Poland, Hungary, and the Czech Republic), where large quantities of coal are burned with inadequate pollution controls.

In addition to providing electricity and running industries, coal is burned for heating homes and cooking by millions of poor families. As a result, China has some of the world's most polluted indoor and outdoor air. This explains why many residents of Beijing develop serious respiratory problems and some die prematurely from the coal-generated air pollution. After only a few days in this region visitors often suffer from coughs and bronchial irritation.

Some Chinese cities have so many smokestacks and home chimneys belching coal smoke that residents see the sun for only a few weeks a year. This is similar to conditions in many U.S. cities a hundred years ago and earlier in many European cities (Spotlight, p. 437).

What Factors Influence the Formation of Photochemical and Industrial Smog? Rain, Wind, Buildings, Mountains, and Temperature

Outdoor air pollution can be reduced by precipitation, sea spray, and winds and increased by urban buildings, mountains, and high temperatures.

The frequency and severity of smog in an area depend on local climate and topography, population density, the amount of industry, and the fuels used in industry, heating, and transportation.

Three natural factors help reduce outdoor air pollution. One is *rain and snow*, which help cleanse the air of pollutants. This helps explain why cities with dry climates are more prone to photochemical smog than cities with wet climates. A second factor is *salty sea spray from the oceans*, which can wash out particulates and other water-soluble pollutants from air that flows from land onto the oceans.

A third factor is *winds*, which can help sweep pollutants away, dilute them by mixing them with cleaner air, and bring in fresh air. However, these pollutants are blown somewhere else or are deposited from the sky onto surface waters, soil, and buildings. There is no away.

Four other factors can *increase* outdoor air pollution. One is *urban buildings*, which can slow wind speed and reduce dilution and removal of pollutants. Another is *hills and mountains*. They can reduce the flow of air in valleys below them, allowing pollutant levels to build up at ground level. In addition, *high temperatures* found in most urban areas promote the chemical reactions leading to photochemical smog formation.

A fourth factor is called the *grasshopper effect* based on atmospheric distillation, which transfers volatile air pollutants from tropical and temperate areas to the earth's poles. It occurs when volatile compounds evaporate from warm terrestrial areas at low latitudes and rise high into the atmosphere. Then they are deposited in the oceans or carried to higher latitudes at or near the earth's poles by atmospheric currents and oceanic currents of water.

This explains why for decades pilots have reported dense layers of reddish-brown haze over the Arctic. It also explains why polar bears, whales, sharks, and other top carnivores and native peoples in the Arctic have high levels of DDT and other long-lived pesticides, toxic metals (such as lead and mercury), and polychlorinated biphenyls (PCBs) in their bodies even though there are no concentrations of industrial facilities and cars in these remote areas.

How Can Temperature Inversions Increase Outdoor Air Pollution? Trapping Pollutants near the Ground

A layer of warm air sitting on top of a layer of cool air near the ground can prevent outdoor pollutants from rising and dispersing.

During daylight, the sun warms the air near the earth's surface. Normally, this warm air and most of the pollutants it contains rise to mix with the cooler air above it. This mixing of warm and cold air creates turbulence, which disperses the pollutants.

Under certain atmospheric conditions, however, a layer of warm air can lie atop a layer of cooler air nearer the ground, a situation known as a **temperature inversion.** Because the cooler air is denser than the warmer air above it, the air near the surface does not rise and mix with the air above it. Pollutants can concentrate in this stagnant layer of cool air near the ground.

Areas with two types of topography and weather conditions are especially susceptible to prolonged temperature inversions (Figure 20-7). One such area is a town or city located in a valley surrounded by mountains that experiences cloudy and cold weather during part of the year (Figure 20-7, top). In such cases, the surrounding mountains along with the clouds block much of the winter sun that causes air to heat and rise, and the mountains block air from being blown away. As long as these stagnant conditions persist, concentrations of pollutants in the valley below will build up to harmful and even lethal concentrations. This is what happened during the 1948 air pollution disaster in the valley town of Donora, Pennsylvania (Spotlight, p. 437).

The second type of area typically is a city with several million people and motor vehicles in an area with a sunny climate, light winds, mountains on three sides, and the ocean on the other. Here, the conditions are ideal for photochemical smog worsened by frequent thermal inversions (Figure 20-7, bottom).

This describes California's heavily populated Los Angeles basin, which has prolonged subsidence temperature inversions at least half of the year, mostly during the warm summer and fall. When a thermal inversion persists throughout the day, the surrounding mountains prevent the polluted surface air from being blown away by sea breezes.

Case Study: South Asia's Massive Brown Cloud—Choking in China and India

A huge dark brown cloud of industrial smog, caused by coal burning in countries such as China and India, stretches over much of southeastern Asia.

A 2002 study by the UN Environment Programme warned of the harmful effects of a huge blanket of mostly industrial smog—called the *Asian Brown Cloud.* Satellite images show a dark brown cloud, 3 kilometers (2 miles) thick, stretching nearly continuously across much of India, Bangladesh, and the industrial heart of China and parts of the open sea in this area.

This cloud is caused by huge emissions of ash, smoke, dust, and acidic compounds produced by people burning coal in industries and homes and clearing and burning forests for planting crops, along with dust blowing off deserts in western Asia. As the cloud travels, it picks up many toxic pollutants.

In a way, the rapid industrialization of parts of southeastern Asia, especially China and India, is repeating on a much larger scale the smoky, unhealthy coal-burning past of the industrial revolution in Europe and the United States during the 19th and early 20th centuries (Spotlight, p. 437).

Here are some of the possible harmful effects of the Asian pollution. One is that the cloud reduces the amount of solar energy hitting the earth's surface underneath it by 2% to as much as 15% in some areas. This may be reducing India's winter rice harvests by 3–10%. And acids in the haze falling to the surface can damage crops, trees, and life in lakes.

The cloud may also be an important contributor to illnesses and premature deaths from respiratory diseases for people who live under it. In Beijing, China, atmospheric concentrations of airborne particulates are routinely five times higher than those in the Los Angeles basin. Not one of India's 22 cities with more than 1 million people meets WHO air pollution standards. Instead of blue skies, many of the inhabitants living under the Asian brown cloud see gray skies much of the year.

Another problem is that the aerosols of fine particles and droplets in this huge cloud appear to be causing changes in regional climate, warming some areas, cooling other areas, and shifting patterns of rainfall including India's vital winter monsoon sea-

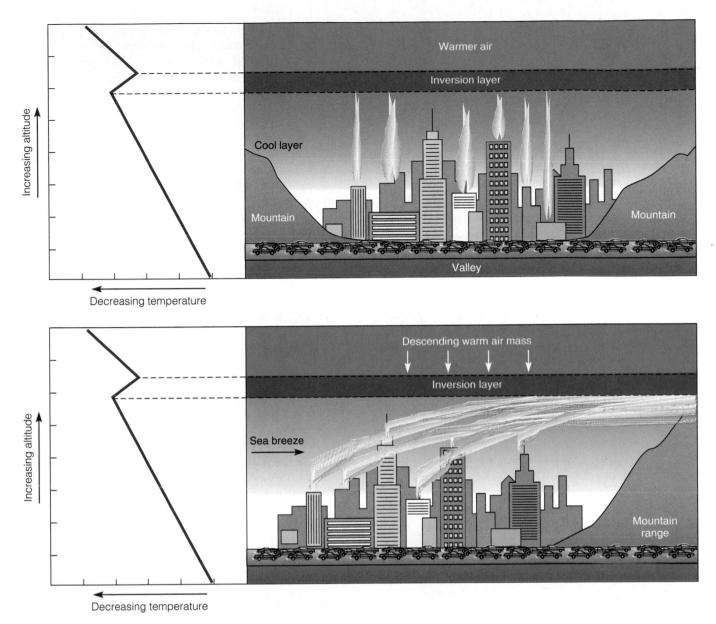

Figure 20-7 Two sets of topography and weather conditions that lead to prolonged *temperature inversions,* in which a warm air layer sits atop a cooler air layer. Air pollutants can build to harmful levels during an inversion. Top, a temperature inversion can occur during cold, cloudy weather in a valley surrounded by mountains (top). Frequent and prolonged temperature inversions can also occur in an area with a sunny climate, light winds, mountains on three sides, and the ocean on the other. A layer of descending warm air from a high-pressure system prevents ocean-cooled air near the ground from ascending enough to disperse and dilute pollutants. Because of their topography, Los Angeles and Mexico City have frequent temperature inversions, many of them prolonged during the summer.

son. It may even play a role in the intensity and frequency of the El Niño–Southern Oscillation (ENSO) climate cycle (Figure 6-12, p. 108) and may thus affect North and South America and other parts of the world (Figure 6-13, p. 109).

The history of air pollution control in Europe and the United States shows that atmospheric hazes can be cleared up fairly quickly. This is done by setting standards for coal-burning industries and utilities, shifting from coal to cleaner-burning natural gas in industries and dwellings in urban areas, relying more on renewable energy resources such as wind and hydropower to produce electricity, and requiring catalytic converters on cars. China is beginning to do this and hopes to restore clear blue skies to Beijing by the 2008 Olympics. India's capital city of Delhi has also made progress in reducing air pollution under orders from India's Supreme Court.

20-4 REGIONAL OUTDOOR AIR POLLUTION FROM ACID DEPOSITION

What Is Acid Deposition, and Where Does It Occur? Acids Falling on Your Head

Sulfur dioxide, nitrogen oxides, and particulates can react in the atmosphere to produce acidic chemicals that can travel long distances before returning to the earth's surface.

Most coal-burning power plants, ore smelters, and other industrial plants in developed countries use tall smokestacks to emit sulfur dioxide, suspended particles, and nitrogen oxides high into the troposphere where wind can mix, dilute, and disperse them.

These tall smokestacks reduce *local* air pollution, but they can increase *regional* air pollution downwind. This occurs because the primary pollutants, sulfur dioxide and nitrogen oxides, emitted into the atmosphere above the inversion layer are transported as much as 1,000 kilometers (600 miles) by prevailing winds. During their trip, they form secondary pollutants such as nitric acid vapor, droplets of sulfuric acid, and particles of acid-forming sulfate and nitrate salts.

These acidic substances remain in the atmosphere for 2–14 days, depending mostly on prevailing winds, precipitation, and other weather patterns. During this period they descend to the earth's surface in two forms. One is *wet deposition* as acidic rain, snow, fog, and cloud vapor with a pH less than 5.6 (Figure 3-6, p. 41). The other is *dry deposition* as acidic particles. The resulting mixture is called **acid deposition** (Figure 20-8), sometimes termed *acid rain*. Most dry deposition occurs within about 2–3 days fairly near the emission sources, whereas most wet deposition takes place within 4–14 days in more distant downwind areas.

Acid deposition is a regional air pollution problem in most parts of the world that are downwind from coal-burning facilities and from urban areas with large numbers of cars. Such areas include the eastern United States (Figure 20-9) and other parts of the world (Figure 20-10, p. 446). Look at the maps to see if you live in a major acid deposition area.

In the United States, coal-burning power and industrial plants in the Ohio Valley emit the largest quantities of sulfur dioxide and other pollutants that can cause acid deposition. Mostly as a result of these emissions along with those by other industries and motor vehicles in urban areas, typical precipitation in the eastern United States has a pH of 4.4–4.8 (Figure 20-9). This is about 10 or more times the acidity of natural precipitation, which has a pH of 5.6. Some mountaintop forests in the eastern United States and east of Los Angeles, California, are bathed in fog and dews as acidic as lemon juice, with a pH of 2.3—about 1,000 times the acidity of normal precipitation.

In some areas, soils contain basic compounds such as calcium carbonate ($CaCO_3$) or limestone that can react with and neutralize, or *buffer*, some inputs of acids.

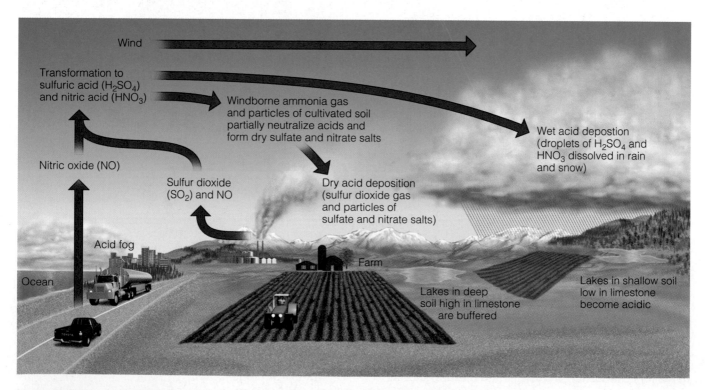

Wind

Transformation to sulfuric acid (H_2SO_4) and nitric acid (HNO_3)

Nitric oxide (NO)

Windborne ammonia gas and particles of cultivated soil partially neutralize acids and form dry sulfate and nitrate salts

Wet acid deposition (droplets of H_2SO_4 and HNO_3 dissolved in rain and snow)

Sulfur dioxide (SO_2) and NO

Dry acid deposition (sulfur dioxide gas and particles of sulfate and nitrate salts)

Acid fog

Ocean

Farm

Lakes in deep soil high in limestone are buffered

Lakes in shallow soil low in limestone become acidic

Figure 20-8 **Natural capital degradation:** *acid deposition*, which consists of rain, snow, dust, or gas with a pH lower than 5.6, is commonly called acid rain. Soils and lakes vary in their ability to buffer or remove excess acidity.

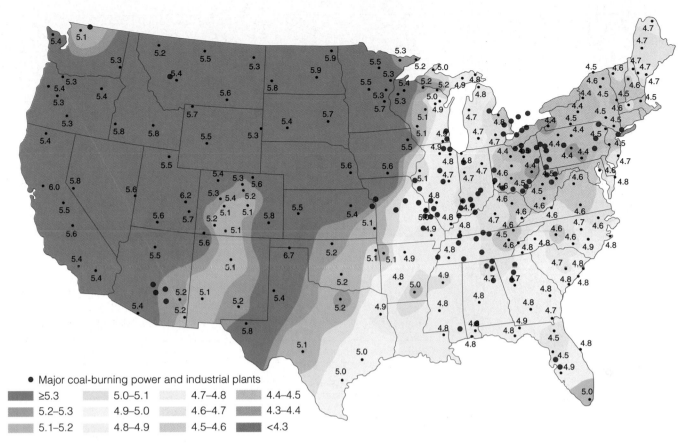

Figure 20-9 *Natural capital degradation:* pH values from field measurements at 250 sites (black dots) in 48 states in 2002. Yellow, tan, and rose indicate areas with lowest pH (highest acidity). Red dots show major sources of sulfur dioxide (SO_2) emissions, mostly large coal-burning power plants. The damaging acidic rain and snow that falls in the Northeast results from coal-fired power and industrial plants and cars in the region and pollution that blows in from such plants in the Midwest. In the East the primary component of acid deposition is H_2SO_4 (formed from SO_2 emitted by coal-burning plants). In the West HNO_3 predominates (formed mostly from NO_x emissions from motor vehicles). According to the EPA, about two-thirds of the SO_2 and a fourth of the NO_x that are the primary causes of acid deposition come from coal-burning power plants. (National Atmospheric Deposition Program/National Trends Network, 2003)

Two types of *sensitive areas* are vulnerable to acid deposition because they lack the buffering capacity needed to neutralize the acidic compounds falling from the sky. One has areas with thin acidic soils derived mostly from granite rock without such natural buffering (Figure 20-10, green and most red areas). The other is areas where the buffering capacity of soils has been depleted by decades of acid deposition (some red areas in Figure 20-10).

Many acid-producing chemicals generated in one country are exported to other countries by prevailing winds. For example, acidic emissions from industrialized areas of western Europe (especially the United Kingdom and Germany) and eastern Europe blow into Norway, Switzerland, Austria, Sweden, the Netherlands, and Finland. And some SO_2 and other emissions from coal-burning power and industrial plants in the Ohio Valley of the United States (Figure 20-9) end up in southeastern Canada. In addition, some acidic emissions in China end up in Japan and North and South Korea.

The worst acid deposition is in Asia, especially China, which gets about 59% of its energy from burning coal. Scientists estimate that by 2025, China will emit more sulfur dioxide than the United States, Canada, and Japan combined.

What Are Some Harmful Effects of Acid Deposition? Lung Disease, Corrosion, Haze, and Dead Fish

Acid deposition can cause or worsen respiratory disease, attack metallic and stone objects, decrease atmospheric visibility, and kill fish.

Acid deposition has a number of harmful effects. It contributes to human respiratory diseases such as bronchitis and asthma, and can leach toxic metals (such as lead and copper) from water pipes into drinking water. It also damages statues, national monuments, buildings, metals, and car finishes. Limestone, marble (which is a form of limestone), and sandstone, which are used in statues and as building materials,

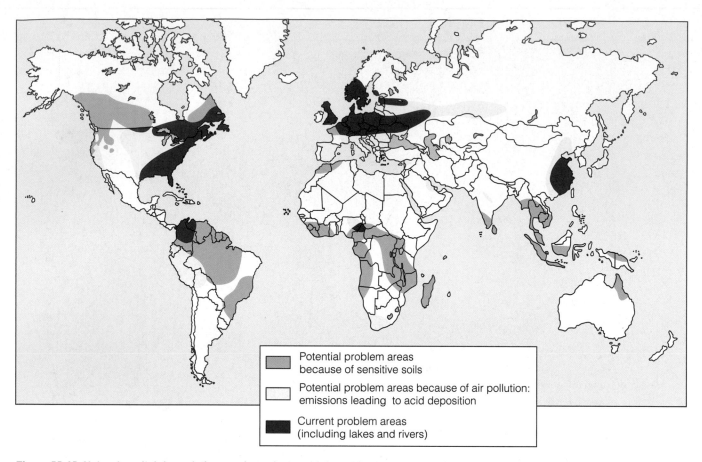

Figure 20-10 Natural capital degradation: regions where acid deposition is now a problem (red) and regions with the potential to develop this problem (yellow and green). Such regions have large inputs of air pollution (mostly from power plants, industrial plants, and ore smelters) or are sensitive areas with soils and bedrock that cannot neutralize (buffer) inputs of acidic compounds (green areas and most red areas). (World Resources Institute and U.S. Environmental Protection Agency)

are especially susceptible because they dissolve even in weak acid solutions. Large amounts of money are spent each year to clean and repair monuments and buildings that have been attacked by acid deposition. Acid deposition also decreases atmospheric visibility, mostly because of the sulfate particles it contains.

For most aquatic systems, acid deposition has harmful effects when the pH falls below 6 and especially below 4.5, which kills most fish. Another effect is the release into lakes of aluminum ions (Al^{3+}) attached to minerals in nearby soil. These ions asphyxiate many kinds of fish by stimulating excessive mucus formation, which clogs their gills.

Lakes vary in their sensitivity to inputs of acids. Those on sand or igneous rocks, such as granite, generally have little buffering capacity to neutralize acids and thus are more susceptible to acidification. Much of the damage to aquatic life in such sensitive areas is a result of *acid shock*. This is caused by the sudden runoff of large amounts of highly acidic water and aluminum ions into lakes and streams when snow melts in the spring or after unusually heavy rains.

Because of excess acidity, several thousand lakes in Norway and Sweden contain no fish, and many more lakes there have lost most of their acid-neutralizing capacity. In Canada, at least 1,200 acidified lakes contain few if any fish, and some fish populations in many thousands of other lakes are declining because of increased acidity. In the United States, several hundred lakes (most in the Northeast) are threatened with excess acidity.

Acid deposition is not always the main culprit. Some lakes are acidic because they are surrounded by naturally acidic soils.

What Are the Effects of Acid Deposition on Plants and Soils? Depleting Nutrients and Damaging and Weakening Plants

Acid deposition can deplete some soil nutrients, release toxic ions into the soil, and weaken plants so they become more susceptible to other stresses.

Acid deposition (often along with other air pollutants such as ozone) can harm forests and crops, especially when the soil pH falls below 5.1. Effects of acid deposition on trees and other plants are caused partly by chemical interactions in forest and cropland soils (Figure 20-11).

At first, sustained acid precipitation adds nitrogen and sulfur to the soils, which stimulates plant growth. But continued acid inputs can cause several problems. One is that the acids leach essential plant nutrients and calcium and magnesium salts (which can reduce acidity) from soils. This reduces plant productivity and the ability of the soils to buffer or neutralize acidic inputs.

Another problem is that calcium deficiencies in plants produced in such nutrient-depleted soils can be passed on to herbivores. For example, birds eating calcium-deficient plant material could have less calcium for egg production, mammals could have

weaker bones, and insects could have weaker exoskeletons—another example of connections and unintended consequences.

In addition, some air pollutants can harm some types of trees through *synergistic effects*. For example, studies show that no visible injury occurs to white pine seedlings when they are exposed individually to low concentrations of sulfur dioxide and ozone. However, if the seedlings are exposed to the same concentrations of both pollutants simultaneously, visible damage occurs, apparently because the two pollutants interacted synergistically.

Acid inputs can also dissolve insoluble soil compounds and release ions of metals such as aluminum, lead, cadmium, and mercury. When absorbed from soil, these ions are highly toxic to plants and animals. In addition, acid deposition can promote growth of

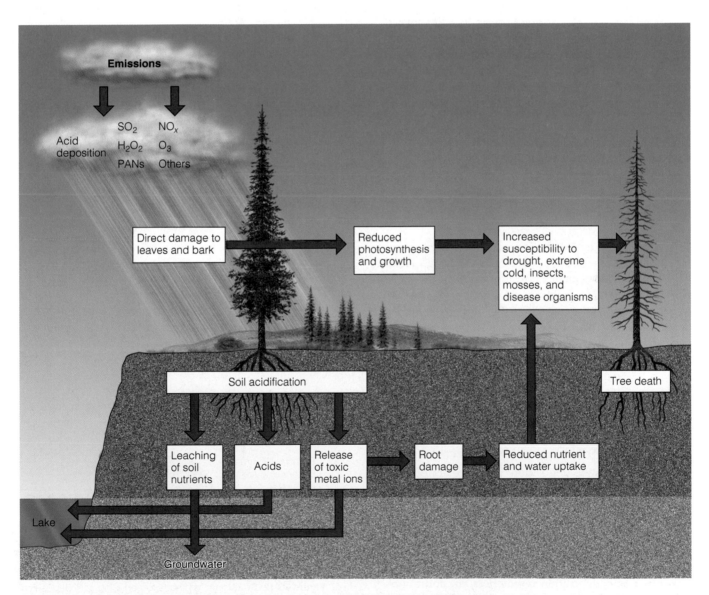

Figure 20-11 **Natural capital degradation:** air pollutants are one of several interacting stresses that can damage, weaken, or kill trees and pollute surface and groundwater.

acid-loving mosses that can kill trees. These mosses hold enough water to drown tree roots and can kill mycorrhizae fungi that help the roots absorb nutrients (Figure 8-10c, p. 155).

A major problem is that acid deposition weakens trees and other plants so they become more susceptible to other types of damage such as severe cold, diseases, insect attacks, drought, and harmful mosses. Thus acid deposition rarely kills trees directly but can weaken them and make them more susceptible to other stresses.

Mountaintop forests are the terrestrial areas hardest hit by acid deposition. These areas tend to have thin soils without much buffering capacity, and trees on mountaintops (especially conifers such as red spruce and balsam fir that keep their leaves year-round) are bathed almost continuously in very acidic fog and clouds.

Note that *most of the world's forests and lakes are not being destroyed or seriously harmed by acid deposition*. It is a regional problem that can harm forests and lakes downwind from coal-burning facilities and from large car-dominated cities without adequate pollution controls. Do you live in an area affected by acid deposition?

How Serious Is Acid Deposition in the United States? Some Hopeful Signs

Much progress has been made in understanding and reducing acid deposition in the United States, but there is a long way to go.

Between 1980 and 1990, the federal government sponsored a massive study called the National Acid Precipitation Assessment Program (NAPAP). Its goals were to coordinate government acid deposition research and assess the costs, benefits, and effectiveness of the country's acid deposition legislation and control programs.

Good news. Acid deposition has not led to a decline in overall tree growth in the vast majority of forests in the United States and Canada. Also, the 1990 amendments to the Clean Air Act have lead to significant reductions in SO_2 and NO_x emissions from coal-fired power and industrial plants, and further reductions are projected.

Bad news. Acid deposition has accelerated the leaching of plant nutrients—such as ions of calcium and magnesium—from soils in some areas and this could eventually decrease tree growth. Acid deposition has also increased concentrations of toxic forms of aluminum in some soil and in lakes and streams.

According to a 2001 study by Gene Likens and nine other acid deposition researchers, an additional 80% reduction in SO_2 emissions from coal-burning power and industrial plants in the midwestern United States (red dots in Figure 20-9) will be needed for northeastern lakes, forests, and streams to recover from past and projected effects of acid deposition.

Bottom line. The 1990 amendments to the Clean Air Act have helped reduce some of the harmful impacts of acid deposition in the United States but there is still a long way to go.

What Can Be Done to Reduce Acid Deposition? Plenty, But There Is Political Opposition

A number of prevention and control methods can reduce acid deposition, but implementing these solutions is politically difficult.

Figure 20-12 summarizes ways to reduce acid deposition. According to most scientists studying the problem, the best solutions are *prevention approaches* that

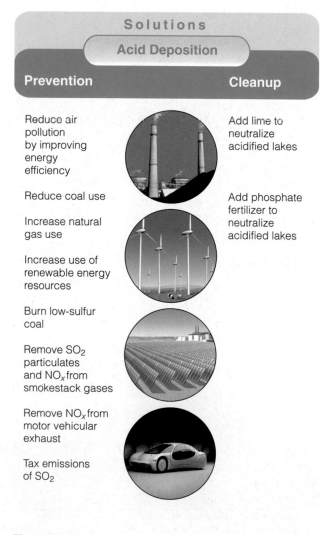

Figure 20-12 Solutions: methods for reducing acid deposition and its damage.

reduce or eliminate emissions of SO_2, NO_x, and particulates, as discussed in Section 20-7.

To help reduce SO_2 emissions and thus acid deposition, some coal-burning power plants in the United States have increased their use of low-sulfur lignite coal (Figure 17-20, p. 364). However, because low-sulfur lignite coal has a low heating value, more coal must be burned to generate the same amount of electricity. This increases air pollution by emitting more CO_2, toxic mercury, and radioactive particles into the troposphere—connections again.

Controlling acid deposition is a political hot potato. One problem is that the people and ecosystems it affects often are quite distant from those who cause the problem. Also, countries with large supplies of coal (such as China, India, Russia, and the United States) have a strong incentive to use it as a major energy resource. And owners of coal-burning power plants say the costs of adding pollution control equipment, using low-sulfur coal, or removing sulfur from coal are too high and would increase the cost of electricity for consumers.

Environmentalists respond that affordable and much cleaner ways are available to produce electricity—including wind turbines and burning natural gas in turbines. They also point out that the largely hidden health and environmental costs of burning coal are roughly twice its market cost. They urge inclusion of these costs in the price of producing electricity from coal. Then consumers would have more realistic knowledge about the harmful effects of burning coal.

Large amounts of limestone or lime can be used to neutralize acidified lakes or surrounding soil—the primary cleanup approach now being used. But there are several problems with liming. One is that it is an expensive and temporary remedy that usually must be repeated annually. Also, it can kill some types of plankton and aquatic plants and can harm wetland plants that need acidic water. Finally, it is difficult to know how much lime to put where (in the water or at selected places on the ground).

In 2002, researchers in England found that adding a small amount of phosphate fertilizer can neutralize excess acidity in a lake. The effectiveness of this approach is being evaluated.

20-5 INDOOR AIR POLLUTION

How Serious Is Indoor Air Pollution? Being Indoors Can Be Hazardous to Your Health

Indoor air pollution usually is a much greater threat to human health than outdoor air pollution.

If you are reading this book indoors, you may be inhaling more air pollutants with each breath than if you were outside. Figure 20-13 (p. 450) shows some typical sources of indoor air pollution. Which are you exposed to?

EPA studies have revealed some alarming facts about indoor air pollution in the United States. *First,* levels of 11 common pollutants generally are two to five times higher inside homes and commercial buildings than outdoors and as much as 100 times higher in some cases. *Second,* pollution levels inside cars in traffic-clogged urban areas can be up to 18 times higher than outside. *Third,* the health risks from exposure to such chemicals are magnified because people typically spend 70–98% of their time indoors or inside vehicles.

As a result of these studies, in 1990 the EPA placed indoor air pollution at the top of the list of 18 sources of cancer risk—causing as many as 6,000 premature cancer deaths per year. At greatest risk are smokers, infants and children under age 5, the old, the sick, pregnant women, people with respiratory or heart problems, and factory workers.

Danish and U.S. EPA studies have linked various air pollutants found in buildings to dizziness, headaches, coughing, sneezing, shortness of breath, nausea, burning eyes, chronic fatigue, irritability, skin dryness and irritation, and flu-like symptoms, known as the *sick-building syndrome.* New buildings are more commonly "sick" than old ones because of reduced air exchange (to save energy) and chemicals released from new carpeting and furniture. EPA studies indicate that almost one in five of the 4 million commercial buildings in the United States are considered "sick" (including EPA headquarters).

According to the EPA and public health officials, the four most dangerous indoor air pollutants in developed countries are *cigarette smoke, formaldehyde, radioactive radon-222 gas,* and *very small fine and ultrafine particles. Good news.* Between 1985 and 2004, the number of U.S. cities banning indoor smoking in facilities used by the public increased from 202 to almost 1,700.

In developing countries, the indoor burning of wood, charcoal, dung, crop residues, and coal in open fires or in unvented or poorly vented stoves for cooking and heating exposes inhabitants to high levels of particulate air pollution. According to the World Bank, as many as 2.8 million people (most of them women and children) in developing countries die prematurely each year from breathing elevated levels of such indoor smoke. *Thus indoor air pollution for the poor is by far the world's most serious air pollution problem*—a glaring example of the relationship between poverty and environmental quality.

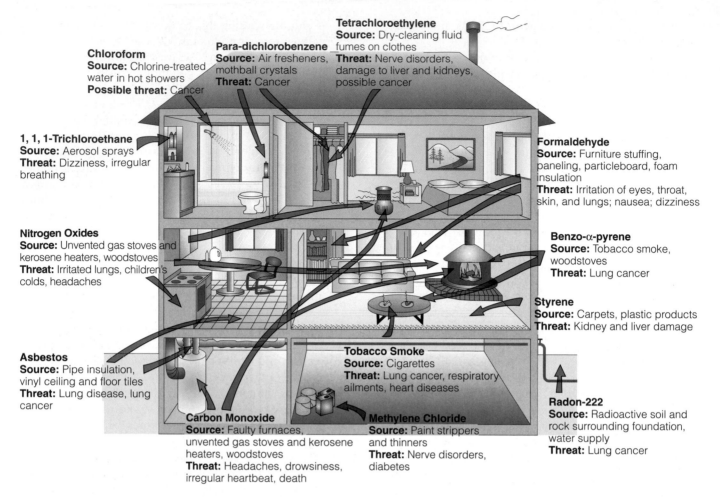

Chloroform
Source: Chlorine-treated water in hot showers
Possible threat: Cancer

Para-dichlorobenzene
Source: Air fresheners, mothball crystals
Threat: Cancer

Tetrachloroethylene
Source: Dry-cleaning fluid fumes on clothes
Threat: Nerve disorders, damage to liver and kidneys, possible cancer

1, 1, 1-Trichloroethane
Source: Aerosol sprays
Threat: Dizziness, irregular breathing

Formaldehyde
Source: Furniture stuffing, paneling, particleboard, foam insulation
Threat: Irritation of eyes, throat, skin, and lungs; nausea; dizziness

Nitrogen Oxides
Source: Unvented gas stoves and kerosene heaters, woodstoves
Threat: Irritated lungs, children's colds, headaches

Benzo-α-pyrene
Source: Tobacco smoke, woodstoves
Threat: Lung cancer

Styrene
Source: Carpets, plastic products
Threat: Kidney and liver damage

Asbestos
Source: Pipe insulation, vinyl ceiling and floor tiles
Threat: Lung disease, lung cancer

Tobacco Smoke
Source: Cigarettes
Threat: Lung cancer, respiratory ailments, heart diseases

Carbon Monoxide
Source: Faulty furnaces, unvented gas stoves and kerosene heaters, woodstoves
Threat: Headaches, drowsiness, irregular heartbeat, death

Methylene Chloride
Source: Paint strippers and thinners
Threat: Nerve disorders, diabetes

Radon-222
Source: Radioactive soil and rock surrounding foundation, water supply
Threat: Lung cancer

Figure 20-13 Some important indoor air pollutants. (Data from U.S. Environmental Protection Agency)

Are You Exposed to Formaldehyde? A Serious Problem

Formaldehyde, found in a variety of common materials and household products, can cause a number of health problems.

The chemical that causes most people in developed countries difficulty is *formaldehyde,* a colorless, extremely irritating gas widely used to manufacture common household materials. According to the EPA and the American Lung Association, 20–40 million Americans suffer from chronic breathing problems, dizziness, rash, headaches, sore throat, sinus and eye irritation, wheezing, and nausea caused by daily exposure to low levels of formaldehyde emitted from common household materials. Are you one of these people?

There are many sources of formaldehyde. They include building materials (such as plywood, particleboard, paneling, and high-gloss wood used in floors and cabinets), furniture, drapes, upholstery, adhesives in carpeting and wallpaper, urethane-formaldehyde insulation, fingernail hardener, and wrinkle-free coating on permanent-press clothing (Figure 20-13). The EPA estimates that as many as 1 of every 5,000 people who live in manufactured homes for more than 10 years will develop cancer from formaldehyde exposure.

Case Study: Are You Being Exposed to Radioactive Radon Gas? Test the Air in Your House

Radon-222, a radioactive gas found in some soil and rocks, can seep into some houses and increase the risk of lung cancer.

Radon-222—a naturally occurring radioactive gas that you cannot see, taste, or smell—is produced by the radioactive decay of uranium-238. Most soil and rock contain small amounts of uranium-238. But this isotope is much more concentrated in underground deposits of minerals such as uranium, phosphate, granite, and shale.

When radon gas from such deposits seeps upward through the soil and is released outdoors, it disperses quickly in the atmosphere and decays to harmless levels. However, in buildings above such deposits radon gas can enter through cracks in foundations and walls, openings around sump pumps and drains, and hollow concrete blocks (Figure 20-14). It tends to be pulled into a house because of the slightly lower atmospheric

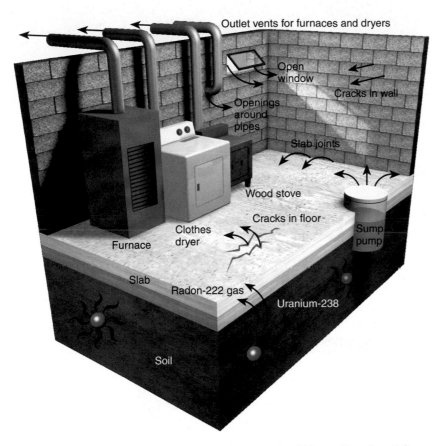

Figure 20-14 Sources and paths of entry for indoor radon-222 gas. (Data from U.S. Environmental Protection Agency)

Because radon hot spots can occur almost anywhere, we do not know which buildings have unsafe levels of radon without conducting tests. In 1988, the EPA and the U.S. Surgeon General's Office recommended that everyone living in a detached house, a townhouse, a mobile home, or on the first three floors of an apartment building test for radon.

Ideally, radon levels should be monitored continuously in the main living areas (not basements or crawl spaces) for 2 months to a year. By 2003, only about 6% of U.S. households had conducted radon tests (most lasting only 2–7 days and costing $20–100 per home). In 2003, the EPA again urged Americans to test their home for indoor radon gas. Have you tested your home for radon?

For information about radon testing visit the EPA website at http://www.epa.gov/iaq/radon or call the radon hotline at 800-SOS-RADON. According to the EPA, radon control could add $350–500 to the cost of a new home, and correcting a radon problem in an existing house could run $800–2,500. Remedies include sealing cracks in the foundation and walls, increasing ventilation by cracking a window or installing vents, and using a fan to create cross ventilation.

pressure inside most homes. Once inside it can build up to high levels, especially in unventilated lower levels of homes and buildings.

Radon-222 gas quickly decays into solid particles of other radioactive elements such as polonium-210 that if inhaled expose lung tissue to a large amount of ionizing radiation from alpha particles. According to the National Academy of Sciences and the EPA, prolonged exposure for a lifetime of 70 years to low levels of radon or radon acting together with smoking is responsible for 7,000–30,000 (with a best estimate of 14,000) of the approximately 140,000 lung cancer deaths each year in the United States. This makes radon the second leading cause of lung cancer after smoking. Most of the deaths are among smokers or former smokers. Each year nonsmokers account for about 2,100–2,900 deaths from lung cancer.

Scientists made two assumptions in estimating risks from radon. One is that there is no safe threshold dose for radon exposure. The other is that the incidence of lung cancer in uranium miners exposed to high levels of radon in mines can be extrapolated to estimate lung cancer deaths for people in homes exposed to much lower levels of radon. Some scientists question the validity of these assumptions.

20-6 EFFECTS OF AIR POLLUTION ON LIVING ORGANISMS AND MATERIALS

How Does Your Respiratory System Help Protect You from Air Pollution? Your Air Pollution Security System

Your respiratory system has several ways to help protect you from air pollution, but some air pollutants can overcome these defenses.

Your respiratory system (Figure 20-15, p. 452) has a number of mechanisms that help protect you from much air pollution. Hairs in your nose filter out large particles. Sticky mucus in the lining of your upper respiratory tract captures smaller (but not the smallest) particles and dissolves some gaseous pollutants. Sneezing and coughing expel contaminated air and mucus when pollutants irritate your respiratory system.

In addition, hundreds of thousands of tiny mucus-coated hairlike structures called *cilia* line your upper respiratory tract. They continually wave back

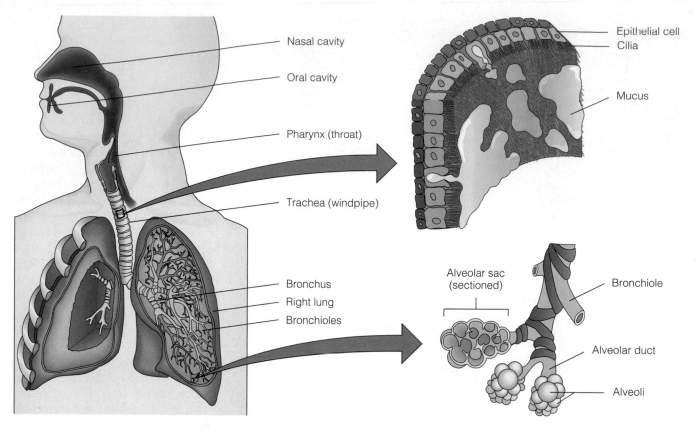

Figure 20-15 Major components of the human respiratory system.

and forth and transport mucus and the pollutants they trap to your throat (where they are swallowed or expelled).

But prolonged or acute exposure to air pollutants including tobacco smoke can overload or break down these natural defenses. This can cause or contribute to various respiratory diseases. One that can start early in life is *asthma*, typically an allergic reaction causing muscle spasms in the bronchial walls and acute shortness of breath. According to a 2004 report by Harvard Medical School's Center for Health and the Global Environment, asthma among U.S. preschool children (ages 3 to 5) grew 160% between 1980 and 1994.

Years of smoking and breathing air pollutants can lead to other respiratory disorders including *lung cancer* and *chronic bronchitis,* which involves persistent inflammation and damage to the cells lining the bronchi and bronchioles. The results are mucus buildup, loud, painful coughing, and shortness of breath. Damage deeper in the lung can cause *emphysema,* which is irreversible damage to air sacs or alveoli leading to loss of lung elasticity and acute shortness of breath (Figure 20-16).

People with respiratory diseases are especially vulnerable to air pollution, as are older adults, infants, pregnant women, and people with heart disease.

Figure 20-16 Normal human lungs (left) and the lungs of a person who died of emphysema (right). Prolonged smoking and exposure to air pollutants can cause emphysema in anyone, but about 2% of emphysema cases result from a defective gene that reduces the elasticity of the air sacs in the lungs. Anyone with this hereditary condition, for which testing is available, should not smoke and should not live or work in a highly polluted area.

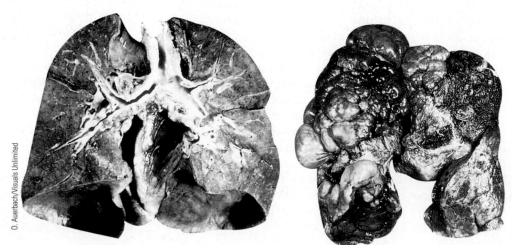

O. Auerbach/Visuals Unlimited

How Many People Die Prematurely from Air Pollution? A Major Killer

Each year air pollution prematurely kills about 3 million people, mostly from indoor air pollution in developing countries.

Table 20-2 lists some of the harmful health effects of prolonged or chronic exposure to six major air pollutants. According to the World Health Organization, worldwide at least 3 million people (most of them in Asia) die prematurely each year from the effects of air pollution—an average of 8,200 deaths per day. About 2.8 million of these deaths (93%) are from *indoor* air pollution, mostly from burning wood or coal inside dwellings in developing countries. *This explains why the World Health Organization and the World Bank consider indoor air pollution one of the world's most serious environmental problems.*

In the United States, the EPA estimates that annual deaths related to indoor and outdoor air pollution range from 150,000 to 350,000 people—equivalent to one to two fully loaded 400-passenger jumbo jets crashing *each day* with no survivors. Millions more become ill and lose work time. Most of these deaths are related to inhalation of fine and ultrafine particulates in indoor air.

According to recent studies by the EPA, each year more than 125,000 Americans get cancer from breathing soot-laden diesel fumes from buses, trucks, tractors, bulldozers and other construction equipment, and portable generators. The EPA says that in one year a large diesel-powered bulldozer produces as much air pollution as 26 cars.

The EPA has proposed emission standards for diesel-powered vehicles that become effective in 2007 with full compliance by 2012. The EPA estimates that these standards should reduce diesel-fuel emissions by more than 90% and prevent as many as 12,000 premature deaths. Manufacturers of diesel-powered engines and vehicles dispute the EPA findings and hope to relax or delay the standards. Environmentalists contend that the standards will encourage the use of hybrid diesel-electric buses or better buses and other large vehicles powered by natural gas.

20-7 PREVENTING AND REDUCING AIR POLLUTION

How Have Laws Helped Reduce Air Pollution in the United States? Clean Air Acts to the Rescue

Clean Air Acts in the United States have greatly reduced outdoor air pollution from six major pollutants.

The U.S. Congress passed Clean Air Acts in 1970, 1977, and 1990. With these laws the federal government established air pollution regulations for key pollutants that are enforced by each state and by major cities.

Congress directed the EPA to establish *national ambient air quality standards (NAAQS)* for six outdoor criteria pollutants (Table 20-2). The EPA regulates these chemicals by using *criteria* developed from risk assessment methods to set maximum permissible levels in outdoor air.

One limit, called a *primary standard,* is set to protect human health, and another, called a *secondary standard,* is intended to prevent environmental and property damage. Each standard specifies the maximum allowable level, averaged over a specific period, for a certain pollutant in outdoor (ambient) air. The EPA has also established national emission standards for more than 188 *hazardous air pollutants* (HAPs) that may cause serious *health* and *ecological effects.* These chemicals include neurotoxins, carcinogens, mutagens, teratogens, endocrine system disrupters, and other toxic compounds (Section 19-3, p. 416). Most of these chemicals are chlorinated hydrocarbons, volatile organic compounds, or compounds of toxic metals.

Measurements of HAPs in outdoor air are made in about 50 locations in the United States. One of the best sources of information about these chemicals in your local area is the annual *Toxic Release Inventory* (TRI) collected and released to the public as part of community right to know laws enacted by Congress in 1986. This TRI law requires 23,000 refineries, power plants, hard rock mines, chemical manufacturers, and factories to report their releases above certain minimum amounts and their waste management methods for 667 toxic chemicals.

Great news. According to a 2003 EPA report, combined emissions of the six criteria air pollutants decreased by 48% between 1970 and 2002 even with significant increases in gross domestic product, vehicle miles traveled, energy consumption, and population.

Also, between 1983 and 2002, emissions of each of the six major outdoor air pollutants decreased—by 93% for lead, 41% for carbon monoxide, 40% for volatile organic compounds (VOCs), 34% for suspended particulate matter with a diameter less than 10 microns (PM-10), 33% for sulfur dioxide, and 15% for NO_x. During this same period the ground-level atmospheric concentration for ozone averaged over 8 hours decreased by 14%. Between 1983 and 2002, the average atmospheric concentration of very fine suspended particulate matter with a diameter less than 2.5 microns (PM-2.5) decreased by 8%. And nationwide emissions of toxic chemicals into the air dropped by about 21% between 1990 and 1999.

However, releases of two HAPs—mercury and dioxins which are toxic at very low levels—have increased in recent years. According to the EPA, about 100 million Americans live in areas where the estimated cancer from HAPs—mostly formaldehyde, acetaldehyde, benzene, and 1,3 butadiene—is 10 in 1 million or ten times the normally accepted standard of 1 death per 1 million persons.

After dropping in the 1980s, smog levels did not drop between 1993 and 2003 mostly because reducing smog requires much bigger cuts in emissions of nitrogen oxides from power and industrial plants and motor vehicles. Also, according to the EPA, in 2003 more than 170 million people lived in 474 of the nation's 2,700 counties in 31 states where air is unhealthy to breathe during part of the year because of high levels of air pollutants—primarily ozone and fine particles. However, for most urban areas such conditions exist for only a few days a year.

How Can U.S. Air Pollution Laws Be Improved? We Can Do Better

Environmentalists applaud the success of U.S. air pollution control laws but have suggested several ways to make them more effective.

The reduction of outdoor air pollution in the United States since 1970 has been a remarkable success story. This occurred because of two factors. *First,* U.S. citizens insisted that laws be passed and enforced to improve air quality. *Second,* the country was affluent enough to afford such controls and improvements.

But more can be done. Environmentalists point to several deficiencies in the Clean Air Act. *One is continuing to rely mostly on pollution cleanup rather than prevention.* An example of the power of prevention is that in the United States, the air pollutant with the largest drop (98% between 1970 and 2002) in its atmospheric level was lead, which was largely banned in gasoline. This is viewed as one of the greatest pollution success stories in the country's history.

Second is the failure of Congress to increase fuel-efficiency standards for cars, sport utility vehicles (SUVs), and light trucks. According to environmental scientists, increased fuel efficiency would reduce air pollution from motor vehicles more quickly and effectively than any other method, reduce CO_2 emissions, save energy, and save consumers enormous amounts of money.

Third, there has also been *inadequate regulation of emissions from inefficient two-cycle gasoline engines.* These engines are used in lawn mowers, leaf blowers, chain saws, jet skis, outboard motors, and snowmobiles. According to the California Air Resources Board, a 1-hour ride on a typical jet ski creates more air pollution than the average U.S. car does in a year, and operating a 100-horsepower boat engine for 7 hours emits more air pollutants than a new car driven 160,000 kilometers (100,000 miles).

In 2001, the EPA announced plans to reduce emissions from most of these sources by 2007. But manufacturers push for extending these deadlines.

Fourth, there is little or no regulation of air pollution from oceangoing ships in American ports. According to the Earth Justice Legal Defense Fund, a single ship emits more air pollution than 2,000 diesel trucks.

Fifth, the Clean Air Acts *have failed to do much about reducing emissions of carbon dioxide and other greenhouse gases.* Also, the laws *have failed to deal seriously with indoor air pollution* even though it is by far the most serious air pollution problem in terms of poorer health, premature death, and economic losses from lost work time and increased health costs.

Sixth, finally, there is a need for *better enforcement of the Clean Air Acts.* According to a 2002 government study, doing this would save about 6,000 lives and prevent 140,000 asthma attacks each year in the United States.

Executives of companies affected by implementing such policies claim that correcting these deficiencies in the country's Clean Air Act would cost too much, harm economic growth, and cost jobs. Proponents contend that history has shown that most industry cost estimates of implementing various air pollution control standards in the United States were many times the actual cost. In addition, implementing such standards has helped increase economic growth and create jobs by stimulating companies to develop new technologies for reducing air pollution emissions. Many of these technologies are sold in the international marketplace.

✗ HOW WOULD YOU VOTE? Should the U.S. Clean Air Act be strengthened? Cast your vote online at http://biology .brookscole.com/miller14.

Case Study: Should We Use the Marketplace to Reduce Pollution? Emissions Trading

Allowing producers of air pollutants to buy and sell government air pollution allotments in the marketplace can help reduce emissions.

To help reduce SO_2 emissions, the Clean Air Act of 1990 allows an *emissions trading policy,* which enables the 110 most polluting power plants in 21 states (primarily in the Midwest and East, red dots in Figure 20-9) to buy and sell SO_2 pollution rights.

This process begins with each of the coal-burning plants measuring the sulfur dioxide emitted from their smokestacks. Each year, a coal-burning power plant is given a certain number of pollution credits, or rights to emit a certain amount of SO_2. A utility that emits less SO_2 than its limit has a surplus of pollution credits. It can use these credits to avoid reductions in SO_2 emissions at another of its plants, keep them for future plant expansions, or sell them to other utilities, private citizens, or environmental groups.

Proponents argue that this system allows the marketplace to determine the cheapest, most efficient way to get the job done instead of having the government dictate how to control air pollution. Some environmentalists see this *cap-and-trade* market approach as an improvement over the regulatory *command-and-control* approach, as long as it achieves a net reduction in SO_2 pollution. This would be accomplished by limiting the total number of credits and gradually lowering the *emissions cap* or annual number of credits, as has been done since 2000.

One of the neat things about the SO_2 emissions market is that anyone can participate. Environmental groups can buy up such rights to pollute and not use them. You could personally reduce air pollution by buying a certificate allowing you to add 0.9 metric ton (1 ton) of SO_2 to the atmosphere and hanging it on the wall. You can purchase these certificates and give them away as birthday or holiday gifts. See http://www.epa.gov/airmarkets/ for a list of brokers and other sellers of SO_2 permits.

Some environmentalists criticize the cap-and-trade program. They contend that it allows utilities with older, dirtier power plants to buy their way out and keep on emitting unacceptable levels of SO_2. This could lead (as it has) to continuing high levels of air pollution in certain areas or "hot spots."

This approach also creates incentives to cheat because air quality regulation is based largely on self-reporting of emissions. To help keep the system honest, the environmentalists call for unannounced spot monitoring by the government and high fines for cheaters.

In addition, the success of any emissions trading approach depends on how low the initial cap is set and then on how much it is reduced annually to promote continuing innovation in air pollution prevention and control. Without these elements, these critics say that emissions trading programs mostly move air pollutants from one area to another without achieving an overall reduction in air quality.

Good news. Between 1980 and 2002, the emissions trading system helped reduce SO_2 emissions from electric power plants in the United States by 40%. And the cost of doing this was less than one-tenth the cost projected by industry because this market-based system motivated companies to reduce emissions in more efficient ways.

The EPA has also created an emissions trading program for smog-forming nitrogen oxides (NO_x) in a number of states in the East and Midwest. Emissions trading may also be implemented for particulate emissions and volatile organic compounds (VOCs) and for the combined emissions of SO_2, NO_x, and mercury from coal-burning power plants.

This combined emissions trading scheme, proposed by the Bush administration in 2001 under its *Clear Skies* initiative, is strongly supported by the electric power industry. The initiative aims for a 70% reduction in emissions of mercury, sulfur dioxide, and nitrogen oxides by 2018. But critics call this the *Dirty Skies* plan, largely crafted by the country's biggest air polluters. According to Frank O'Donnell, executive director of the Clean Air Trust, the *Clear Skies* plan was "drawn up by and for the big polluters." Critics claim that the plan will set caps too low, allow emissions of toxic mercury to triple by 2013, cause a 50% increase in SO_2 emissions, and require no controls for reducing emissions of the greenhouse gas CO_2.

Environmentalists and health scientists are particularly opposed to using a cap-and-trade program to control emissions of mercury by coal-burning power plants and industries because it is highly toxic, falls out of the atmosphere fairly near such facilities, and does not breakdown in the environment. Coal-burning plants choosing to buy permits instead of sharply reducing their mercury emissions would create hot spots with unacceptably high levels of mercury.

Bad news. In 2002, the EPA reported results from evaluation of the country's oldest and largest emissions trading program, in effect since 1993 in southern California. The EPA study found that this cap-and-trade model "produced far less emissions reductions than were either projected for the program or could have been expected from" the command-and-control system it replaced. The study also found accounting abuses, including emissions caps set 60% higher than current emissions levels. Cap-and-trade programs need to be carefully monitored.

X HOW WOULD YOU VOTE? Should emissions trading be used to help control emissions of all major air pollutants? Cast your vote online at http://biology.brookscole.com/miller14.

How Can We Reduce Outdoor Air Pollution from Coal-Burning Facilities? Prevention Is Best

There are a number of ways to prevent and control air pollution from coal-burning facilities.

Figure 20-17 (p. 456) summarizes ways to reduce emissions of sulfur oxides, nitrogen oxides, and particulate matter from stationary sources such as electric power plants and industrial plants that burn coal.

Good news. Between 1980 and 2002, emissions of SO_2 from U.S. electric power plants decreased by 40%, emissions of NO_x emissions by 30%, and soot emissions by 75%. Emphasis has been on output approaches that add equipment to remove some of the particulate, NO_x, and SO_2 pollutants after they are produced (Figure 20-18, p. 457). Pollution can also be reduced at the

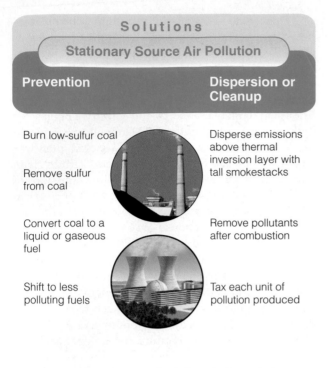

Solutions

Stationary Source Air Pollution

Prevention	Dispersion or Cleanup
Burn low-sulfur coal	Disperse emissions above thermal inversion layer with tall smokestacks
Remove sulfur from coal	
Convert coal to a liquid or gaseous fuel	Remove pollutants after combustion
Shift to less polluting fuels	Tax each unit of pollution produced

Figure 20-17 Solutions: methods for reducing emissions of sulfur oxides, nitrogen oxides, and particulate matter from stationary sources such as coal-burning electric power plants and industrial plants.

input stage by using cleaner coal technologies. One is *fluidized-bed combustion,* which reduces pollutant emissions and burns coal more efficiently by blowing a stream of hot air into a boiler to burn a mixture of powdered coal and crushed limestone. Another is *coal gasification,* which has a mixture of advantages and disadvantages (Figure 17-22, p. 365). Environmentalists argue for much greater emphasis on prevention approaches to decrease the levels of these pollutants reaching the troposphere.

Case Study: What Should We Do about Air Pollution from Older Coal-Burning Facilities? A Burning Controversy

There is controversy over whether older coal-burning plants in the United States should have to meet the same air pollution standards as new plants.

For several decades, environmental scientists, environmentalists, and the attorneys general of a number of states have been fighting to require about 20,000 older coal-burning power plants and industrial plants (red dots in Figure 20-9) and oil refineries in the United States to meet the air pollution standards required for new facilities under the Clean Air Act. Such plants were not required to do this because they were already in existence when the law was passed in 1970.

A 1977 rule in the Clean Air Act—called the *New Source Review*—requires these older facilities to upgrade pollution control equipment when they expand or modernize their facilities. But for almost three decades their owners have been getting around the rules by expanding the plants and calling it maintenance.

In addition they have lobbied elected officials to have this rule overturned. In 2002, their efforts paid off when the Bush administration announced that it was easing the New Source Review restrictions and thus making it easier for older facilities to expand and modernize without having to add expensive pollution control equipment.

Opponents say the revised rule would gut the only provision in the Clean Air Act that could be used to force such facilities to reduce air pollution emissions. They believe it is naive to think that refineries and coal-burning power and industrial plants are going to spend billions of dollars installing modern pollution control equipment when they have not done so in almost three decades and now do not have to (unless they can somehow persuade Congress to have taxpayers foot the bill). A 2004 study by Abt Associates (a consulting firm that the EPA has used to evaluate air pollution policies) found that strengthening the pollution standards for these aging coal-fired plants would prevent 22,000 premature deaths per year.

In 2003, the attorneys general of nine northeastern states challenged this change in the New Source Review in federal court. Their lawsuit alleges that the EPA is exceeding its power in overturning the rule. In 2003, the National Academy of Public Administration, an independent advisory body set up by Congress in 1984, opposed easing the New Source Review rules. Instead, the group advised Congress to give the dirtiest U.S. coal-fired power and industrial plants a 10-year deadline to either install the latest air pollution control equipment or shut down to protect human health. The group said that when Congress established the new source rules in 1977 it did not intend for dirty plants to run indefinitely.

✗ HOW WOULD YOU VOTE? Should older coal-burning power and industrial plants have to meet the same air pollution standards as new plants? Cast your vote online at http://biology.brookscole.com/miller14.

How Can We Reduce Outdoor Air Pollution from Motor Vehicles? Emphasize Prevention

There are a number of ways to prevent and control air pollution from motor vehicles.

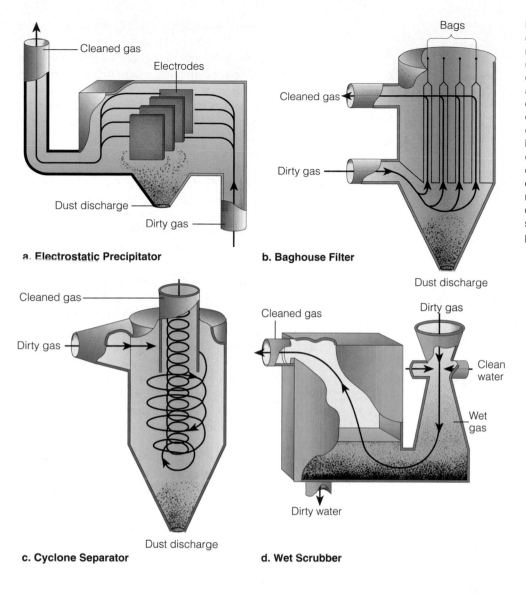

a. Electrostatic Precipitator

Cleaned gas

Electrodes

Dust discharge

Dirty gas

b. Baghouse Filter

Bags

Cleaned gas

Dirty gas

Dust discharge

c. Cyclone Separator

Cleaned gas

Dirty gas

Dust discharge

d. Wet Scrubber

Cleaned gas

Dirty gas

Clean water

Wet gas

Dirty water

Figure 20-18 Solutions: four commonly used output or control methods for removing particulates (a, b, c, d), and SO_2 (d) from the exhaust gases of electric power and industrial plants. Of these, only baghouse filters remove many of the more hazardous fine particles. All these methods produce hazardous materials that must be disposed of safely, and except for cyclone separators, all of them are expensive. Modern wet scrubbers remove 98% of the SO_2 and 98% of the particulate matter in smokestack emissions, but they are expensive to install and maintain.

Figure 20-19 (p. 458) lists ways to reduce emissions from motor vehicles, the primary culprits in producing photochemical smog. One way to make significant reductions is to get older, high-polluting vehicles off the road. According to EPA estimates, 10% of the vehicles on the road in the United States emit 50–70% of vehicular air pollutants. But people who cannot afford to buy a newer car often own old cars. One suggestion is to pay people to take their old cars off the road, which would result in huge savings in health and air pollution control costs.

In 2003, a team of research engineers working with grants from the National Science Foundation and an oil company found a way for oil refineries, power plants, and vehicles to use a class of chemicals called zeolites to do a better job than other alternatives in removing sulfur impurities from diesel fuel, jet fuel, and gasoline. This *adsorption process*, which is carried out at room temperature and pressure, should be cheaper, easier, and more effective than using expensive catalytic converters that operate at high temperatures and pressures to remove sulfur. Stay tuned to see if this technology works as projected and is implemented.

Good news. Over the next 10–20 years air pollution from motor vehicles should decrease from increased use of *partial zero-emission vehicles (PZEVs)* that emit almost no air pollutants because of improved engine and emission systems, *hybrid-electric vehicles* (Figure 18-9, p. 385), and vehicles powered by fuel cells running on hydrogen (Figure 18-10, p. 385).

Bad news. The growing number of motor vehicles in urban areas of many developing countries is contributing to the already poor air quality there. Many of these vehicles are 10 or more years old, have no pollution control devices, and continue to burn leaded gasoline.

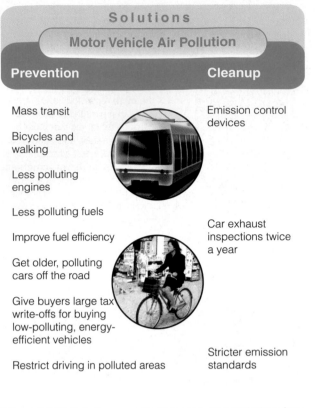

Solutions

Motor Vehicle Air Pollution

Prevention	Cleanup

Prevention

Mass transit

Bicycles and walking

Less polluting engines

Less polluting fuels

Improve fuel efficiency

Get older, polluting cars off the road

Give buyers large tax write-offs for buying low-polluting, energy-efficient vehicles

Restrict driving in polluted areas

Cleanup

Emission control devices

Car exhaust inspections twice a year

Stricter emission standards

Figure 20-19 Solutions: methods for reducing emissions from motor vehicles.

What Should We Do about Ultrafine Particles? Another Controversy

There is controversy over reducing emissions of ultrafine particles that pose a serious threat to human health.

Research indicates that invisible particles—especially *fine particles* with diameters less than 10 microns (PM-10) and *ultrafine particles* with diameters less than 2.5 microns (PM-2.5)—pose a significant health hazard. Such particles come from a variety of sources (Figure 20-6).

They are not effectively captured by most air pollution control equipment and are small enough to penetrate the respiratory system's natural defenses against air pollution. They can also bring with them droplets or other particles of toxic or cancer-causing pollutants that become attached to their surfaces.

Once they are lodged deep within the lungs, these fine particles can cause chronic irritation that can trigger asthma attacks, aggravate other lung diseases, and cause lung cancer. These lung problems interfere with the blood's uptake of oxygen and release of CO_2, which strains the heart and increases the risk of death from heart disease. According to several recent studies of air pollution in U.S. cities, fine and ultrafine parti-

cles prematurely kill 65,000–200,000 Americans each year.

Exposure to particulate air pollution is much worse in most developing countries, where urban air quality has generally deteriorated. The World Bank estimates that reducing particulate levels globally to WHO guidelines would prevent 300,000–700,000 premature deaths per year!

In 1997, the EPA announced stricter emission standards for ultrafine particles in the United States. The EPA estimates the cost of implementing the standards at $7 billion per year, with the resulting health and other benefits estimated at $120 billion per year.

According to industry officials, the new standard is based on flimsy scientific evidence and its implementation will cost $200 billion per year. EPA officials say that their review of the scientific evidence—one of the most exhaustive reviews ever undertaken by the agency—supports the need for the new standard for ultrafine particles. Furthermore, a 2000 study by the Health Effects Institute of 90 large American cities supported the link between fine and ultrafine particles and higher rates of death and disease. Industries affected by these new standards are lobbying Congress to have them weakened or overturned or to extend deadlines for their implementation.

✗ HOW WOULD YOU VOTE? Do you support establishing a stricter standard for emissions of ultrafine particles? Cast your vote online at http://biology.brookscole.com/miller14.

How Can We Reduce Indoor Air Pollution? Emphasize Prevention

Little effort has been spent on reducing indoor air pollution even though it is a much greater threat to human health than outdoor air pollution.

Reducing indoor air pollution does not require setting indoor air quality standards and monitoring the more than 100 million homes and buildings in the United States (or the buildings in any country). Instead, air pollution experts suggest several ways to prevent or reduce indoor air pollution (Figure 20-20). Another possibility for cleaner indoor air in some high-rise buildings is rooftop greenhouses through which building air can be circulated.

In developing countries, indoor air pollution from open fires and leaky and inefficient stoves that burn wood, charcoal, or coal could be reduced if governments gave people inexpensive clay or metal stoves, which burn biofuels more efficiently while venting their exhaust to the outside, or stoves that use solar energy to cook food (solar cookers) in sunny areas. Doing this would also reduce deforestation by using less fuelwood and charcoal.

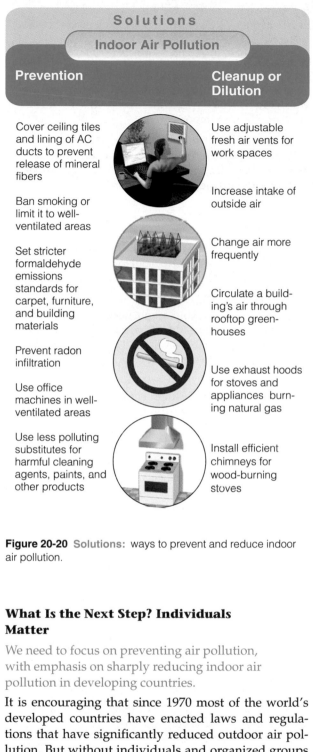

Solutions

Indoor Air Pollution

Prevention	Cleanup or Dilution
Cover ceiling tiles and lining of AC ducts to prevent release of mineral fibers	Use adjustable fresh air vents for work spaces
Ban smoking or limit it to well-ventilated areas	Increase intake of outside air
Set stricter formaldehyde emissions standards for carpet, furniture, and building materials	Change air more frequently
	Circulate a building's air through rooftop greenhouses
Prevent radon infiltration	
Use office machines in well-ventilated areas	Use exhaust hoods for stoves and appliances burning natural gas
Use less polluting substitutes for harmful cleaning agents, paints, and other products	Install efficient chimneys for wood-burning stoves

Figure 20-20 Solutions: ways to prevent and reduce indoor air pollution.

What Is the Next Step? Individuals Matter

We need to focus on preventing air pollution, with emphasis on sharply reducing indoor air pollution in developing countries.

It is encouraging that since 1970 most of the world's developed countries have enacted laws and regulations that have significantly reduced outdoor air pollution. But without individuals and organized groups putting strong political pressure on elected officials in the 1970s and 1980s these laws and regulations would not have been enacted, funded, and implemented. In turn, these legal requirements spurred companies, scientists, and engineers to come up with better ways to control outdoor pollution.

The current laws are a useful *output approach* to controlling pollution. To environmentalists, however,

the next step is to shift to *preventing air pollution*. With this approach, the question is not *What can we do about the air pollutants we produce?* but *How can we avoid producing such pollutants in the first place?*

Figure 20-21 shows ways to prevent outdoor and indoor air pollution over the next 30–40 years. Like the shift to *controlling outdoor air pollution* between 1970 and 2000, this new shift to *preventing outdoor and indoor air pollution* will not take place without political pressure on elected officials by individual citizens and groups. Figure 20-22 (p. 460) lists some ways that you can reduce your exposure to indoor air pollution.

Turning the corner on air pollution requires moving beyond patchwork, end-of-pipe approaches to confront pollution at its sources. This will mean reorienting energy, transportation, and industrial structures toward prevention.

HILARY F. FRENCH

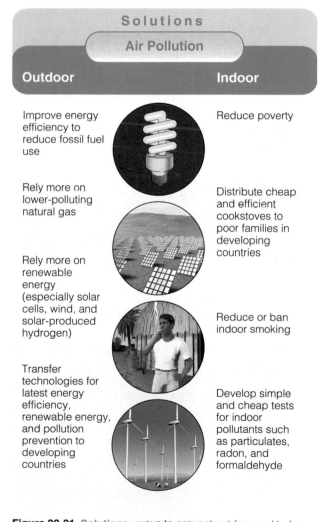

Solutions

Air Pollution

Outdoor	Indoor
Improve energy efficiency to reduce fossil fuel use	Reduce poverty
Rely more on lower-polluting natural gas	Distribute cheap and efficient cookstoves to poor families in developing countries
Rely more on renewable energy (especially solar cells, wind, and solar-produced hydrogen)	Reduce or ban indoor smoking
Transfer technologies for latest energy efficiency, renewable energy, and pollution prevention to developing countries	Develop simple and cheap tests for indoor pollutants such as particulates, radon, and formaldehyde

Figure 20-21 Solutions: ways to prevent outdoor and indoor air pollution over the next 30–40 years.

- Test for radon and formaldehyde inside your home and take corrective measures as needed.

- Do not buy furniture and other products containing formaldehyde.

- Remove your shoes before entering your house to reduce inputs of dust, lead, and pesticides.

- Test your house or workplace for asbestos fiber levels and for any crumbling asbestos materials if it was built before 1980.

- Don't live in a pre-1980 house without having its indoor air tested for asbestos and lead.

- Do not store gasoline, solvents, or other volatile hazardous chemicals inside a home or attached garage.

- If you smoke, do it outside or in a closed room vented to the outside.

- Make sure that wood-burning stoves, fireplaces, and kerosene- and gas-burning heaters are properly installed, vented, and maintained.

- Install carbon monoxide detectors in all sleeping areas.

Figure 20-22 What can you do? Ways to reduce your exposure to air pollution.

CRITICAL THINKING

1. Explain why you agree or disagree with the following statement: "Because we have not proved absolutely that anyone has died or suffered serious disease from nitrogen oxides, current federal emission standards for this pollutant should be relaxed."

2. Identify climate and topographic factors in your local community that (a) intensify air pollution and (b) help reduce air pollution.

3. Should all tall smokestacks be banned? Explain.

4. Explain how sulfur in coal can increase the acidity of rainwater.

5. Do you agree or disagree with the possible weaknesses of the U.S. Clean Air Act listed on p. 454? Defend each of your choices.

6. Explain why you agree or disagree with each of the proposals listed in Figure 20-21 (p. 459) for shifting the emphasis to preventing air pollution over the next several decades. Which two of these proposals do you believe are the most important?

7. Congratulations! You are in charge of reducing air pollution in the country where you live. List the three most important features of your policy for (a) outdoor air pollution and (b) indoor air pollution.

PROJECTS

1. Evaluate your exposure to some or all of the indoor air pollutants in Figure 20-13 (p. 450) in your school, workplace, and home. Come up with a plan for reducing your exposure to these pollutants.

2. Have buildings at your school been tested for radon? If so, what were the results? What has been done about areas with unacceptable levels? If this testing has not been done, talk with school officials about having it done.

3. Use the library or the Internet to find bibliographic information about *Michael J. Cohen* and *Hilary F. French*, whose quotes appear at the beginning and end of this chapter.

4. Make a concept map of this chapter's major ideas, using the section heads, subheads, and key terms (in boldface). Look on the website for this book for information about making concept maps.

LEARNING ONLINE

The website for this book contains study aids and many ideas for further reading and research. They include a chapter summary, review questions for the entire chapter, flash cards for key terms and concepts, a multiple-choice practice quiz, interesting Internet sites, references, and a guide for accessing thousands of InfoTrac® College Edition articles. Log on to

http://biology.brookscole.com/miller14

Then click on the Chapter-by-Chapter area, choose Chapter 20, and select a learning resource.

21 Climate Change and Ozone Loss

Climate Control

CASE STUDY

A.D. 2060: Green Times on Planet Earth

Mary Wilkins sat in the living room of the solar-powered and earth-sheltered house (Figure 21-1) she shared with her daughter Jane and her family. It was July 4, 2060: Independence Day. She heard the hum of solar-powered pumps trickling water to rows of organically grown vegetables and glanced at the fish in the aquaculture and waste treatment tank in the greenhouse that provided much of her home's heat. Mary began putting the finishing touches on her grandchildren's costumes for this afternoon's pageant in Rachel Carson Park. It would honor earth heroes who began the Age of Ecology in the 20th century and those who continued this tradition in the 21st century.

She was delighted that her 12-year-old grandson Jeffrey had been chosen to play Aldo Leopold (Figure 2-9, p. 30), who in the late 1940s began urging people to work with the earth. Her pride swelled when her 10-year-old granddaughter Lynn was chosen to play Rachel Carson (Figure 2-A, p. 27), who in the 1960s warned people about threats from their increasing exposure to pesticides and other potentially harmful chemicals. Her neighbor's son Manuel had been chosen to play biologist Edward O. Wilson, who in the last third of the 20th century explained the need to preserve the earth's biodiversity.

The transition to more sustainable societies and economies began around 2010 when people and governments began to mimic the way the earth has sustained itself for billions of years (Figure 9-15, p. 174). By 2060 the loss of global biodiversity had been cut in half. Most air pollution began gradually disappearing when energy from the sun, wind, and hydrogen began replacing that from oil and coal. Most food was now produced by more sustainable agriculture.

Preventing pollution and reducing resource waste had become important money-saving priorities for businesses and households based on the four Rs of resource consumption: *reduce, reuse, recycle,* and *refuse.* Walking and bicycling had increased in a growing number of cities and towns designed for people instead of cars.

World population had stabilized at 8 billion in 2028 and then had begun a slow decline. Significant atmospheric warming had occurred by 2050. But the rate of additional warming began decreasing by 2050 as hydrogen produced by using electricity produced by wind farms, solar cells, and geothermal energy was being phased in to replace carbon-containing fossil fuels. International treaties enacted in the 1990s banned the chemicals that had begun depleting ozone in the stratosphere during the last quarter of the 20th century. By 2050, ozone levels in the stratosphere had returned to 1980 levels.

Two hours later Mary, her daughter Jane, and her son-in-law Gene watched with pride as 40 children honored the leaders of the Age of Ecology. At the end, Lynn stepped forward and said, "Today we have honored many earth heroes, but the real heroes are the people in this audience and around the world who have worked to help sustain the earth's life-support systems for us and other species. We thank you for giving us such a wonderful gift and promise to leave the earth even better for our children and grandchildren and all living creatures."

This hopeful scenario describes the more sustainable type of world we could have by 2060 if enough of us work to help implement such a vision. This is an exciting challenge. Jump in.

Figure 21-1 An *earth-sheltered house* in the United States. Solar cells on the roof provide most of the house's electricity. About 13,000 families across the United States have built such houses. Mary Wilkins's fictional house in 2060 could be similar to this one.

We are embarked on the most colossal ecological experiment of all time—doubling the concentration in the atmosphere of an entire planet of one of the most important gases in the earth's atmosphere—and we really have little idea of what might happen.

PAUL A. COLINVAUX

This chapter discusses how our activities are changing the world's climate and depleting ozone in the stratosphere, and what we can do about these threats. It addresses the following questions:

- How have the earth's temperature and climate changed in the past?

- How might the earth's temperature change in the future?

- What factors can affect changes in the earth's average temperature?

- What are some possible beneficial and harmful effects of a warmer earth?

- What can we do to slow or adapt to projected increases in the earth's temperature?

- How have human activities depleted ozone in the stratosphere, and why should we care?

- What can we do to slow and eventually reverse ozone depletion in the stratosphere caused by human activities?

21-1 PAST CLIMATE CHANGE

How Have the Earth's Temperature and Climate Changed in the Past? Climate Change Is Not New

Temperature and climate have been changing throughout the earth's history.

The earth's climate—determined mostly by its average temperature and average precipitation—is not fixed. Therefore, climate change is neither new nor unusual. Over the past 4.7 billion years it has shifted due to volcanic emissions, changes in solar input, continents moving as a result of shifting tectonic plates, strikes by large meteorites, and other factors.

At some times (over hundreds to millions of years), the troposphere's average temperature has changed gradually and at other times fairly quickly (over a few decades to 100 years) as shown in the Figure 21-2 graphs. Over the past 900,000 years, the average temperature of the troposphere has undergone prolonged periods of *global cooling* and *global warming* (Figure 21-2, top left). These alternating cycles of freezing and thawing are known as *glacial and interglacial* (between ice ages) *periods*.

During each cold period, thick glacial ice covered much of the earth's surface for about 100,000 years. Most of it melted during a *warmer interglacial period*

lasting 10,000–12,500 years that followed each glacial period.

For roughly 12,000 years, we have had the good fortune to live in an interglacial period with a fairly stable climate and a moderate average global surface temperature (Figure 21-2, top right and bottom left). However, even during this generally stable period, regional climates have changed significantly. For example, about 7,000 years ago, most of the current Sahara desert received almost 20 times more annual rainfall than it does today.

How Do Scientists Study Climate Change? Drill Holes and Make Measurements

Geologic records and atmospheric measurements provide a wealth of information about past atmospheric temperatures and climate.

Scientific clues about the earth's past temperatures and climate are found deep within its glaciers and ice caps, such as those in Greenland and Antarctica. Scientists drill into these museums of atmospheric history and extract long cores of ice (Figure 21-3). In 2004, data from cores drilled in antarctic ice indicated that the current interglacial period could last for another 15,000 years before a new ice age occurs—unless our activities seriously alter the earth's climate.

Scientists analyze air bubbles trapped in different segments of these *ice cores* to uncover information about past tropospheric composition, temperature trends such as those in Figure 21-2, greenhouse gas concentrations, solar activity, snowfall, and forest fire frequency (from trapped layers of soot particles).

Scientists also study past climates by drilling cores into the bottoms of lakes, ponds, and swamps. Then they analyze different zones of the sediment for pollen, fossils, and other clues about what types of plants lived in the past and trends in plant life over time. For those who like detective work, finding out about the earth's climate history is a fascinating activity.

Scientists also make direct measurements to get current information about tropospheric temperature, composition, and trends. They measure temperatures using thermometers on land and at sea and on weather balloons at various altitudes. Direct temperature records go back to 1861. Scientists have also been using infrared sensors on satellites to get temperature information about the troposphere.

Finally, scientists collect air samples at different locations and altitudes and analyze them to detect changes in the chemical composition of the troposphere. For example, since 1958 environmental chemist Charles Keeling has analyzed CO_2 levels in the troposphere at the Mauna Loa observatory in Hawaii.

Once they have accumulated a certain amount of data, scientists get together to try to reach a consensus. In 1988, the United Nations and the World Meteorolog-

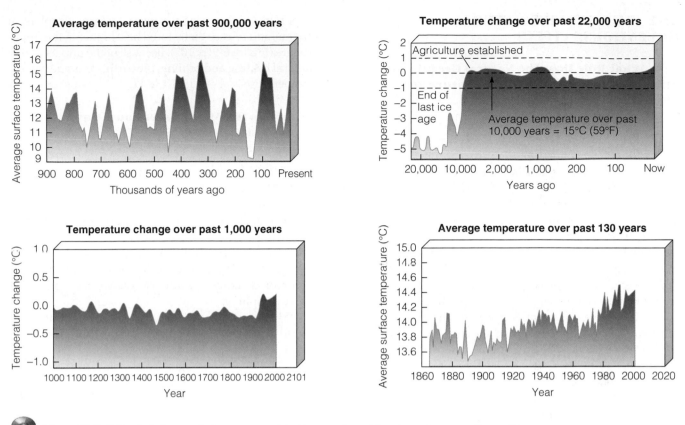

Average temperature over past 900,000 years

Average surface temperature (°C) / Thousands of years ago

Temperature change over past 22,000 years

Temperature change (°C)

Agriculture established
End of last ice age
Average temperature over past 10,000 years = 15°C (59°F)

Years ago

Temperature change over past 1,000 years

Temperature change (°C) / Year

Average temperature over past 130 years

Average surface temperature (°C) / Year

Figure 21-2 Estimated changes in the average global temperature of the atmosphere near the earth's surface over different periods of time. Past temperature changes are estimated by analysis of radioisotopes in rocks and fossils, plankton and radioisotopes in ocean sediments, ice cores from ancient glaciers, temperature measurements at different depths in boreholes drilled deep into the earth's surface, pollen from lake bottoms and bogs, tree rings, historical records, and temperature measurements (since 1861). (Data from Goddard Institute for Space Studies, Intergovernmental Panel on Climate Change, National Academy of Sciences, National Aeronautics and Space Agency, National Center for Atmospheric Research, and National Oceanic and Atmospheric Administration)

ical Organization established the Intergovernmental Panel on Climate Change (IPCC) to document past climate change and project future climate change. The IPCC is a network of over 2,000 leading climate experts from 70 nations.

Panels of scientists from the U.S. National Academy of Sciences and the American Geophysical Union (AGU) have also evaluated possible future climate changes. In addition, the U.S. Congress created the *U.S. Global Change Research Program* (*USGCRP*) in 1990 to project future climate changes and the potential impacts.

Recall that science can never give us absolute certainty or proof. Instead, it establishes levels of certainty or probability that a scientific model or theory is true. The IPCC expresses its conclusions and projections in probabilities using several levels of certainty: *virtually certain* (more than 99% probability), *very likely* (90–99% probability), and *likely* (66–90% probability). Throughout this chapter I use these categories to describe IPCC conclusions and projections about atmospheric temperature changes and their possible affects on climate.

Figure 21-3 Ice cores such as this one extracted by drilling deep holes in ancient glaciers at various sites in antarctica and Greenland can be analyzed to obtain information about past climates.

21-2 THE EARTH'S NATURAL GREENHOUSE EFFECT

What Role Does the Natural Greenhouse Effect Play in the Earth's Temperature and Climate? A Giver of Life

Certain gases in the atmosphere absorb heat and warm the lower atmosphere.

In addition to incoming sunlight, a natural process called the *greenhouse effect* (Figure 6-14, p. 110) warms the earth's lower troposphere and surface. Some of the energy from the sun warms the earth's surface, causing it to radiate infrared energy back toward space. Clouds, water vapor, carbon dioxide, and other gases in the lower troposphere are heated when they absorb some of this outgoing infrared energy. These clouds and gases (called *greenhouse gases*) then radiate heat as longer-wavelength infrared radiation in all directions. Some of the released energy is radiated into space and some warms the troposphere and the earth's surface.

Swedish chemist Svante Arrhenius first recognized this natural tropospheric heating effect in 1896. Since then numerous laboratory experiments and measurements of atmospheric temperatures at different altitudes have confirmed this relationship. As a result, it is one of the most widely accepted theories in the atmospheric sciences.

A *natural cooling process* also takes place at the earth's surface. Large quantities of heat are absorbed by the evaporation of liquid surface water, and the water vapor molecules rise, condense to form droplets in clouds, and release their stored heat higher in the troposphere (Figure 6-9, p. 107). Because of the impact of this natural heating and cooling, the earth's average surface temperature is about 15°C (59°F).

What Are the Major Greenhouse Gases? Two Important Molecules

The two major greenhouse gases are water vapor and carbon dioxide.

Table 21-1 shows the major sources, average time in the troposphere, and relative warming potential of various greenhouse gases in the troposphere. The two greenhouse gases with the largest concentrations are *water vapor*, controlled by the hydrologic cycle, and *carbon dioxide* (CO_2), controlled by the carbon cycle. Carbon dioxide is the greenhouse gas we have added to the troposphere.

The coal, oil, and natural gas that support the world's economy all contain carbon that plants and sunshine converted to organic compounds hundreds of millions of years ago. Under high pressures and temperatures these buried organic compounds were converted to fossil fuels. Extracting and burning these storehouses of carbon releases carbon dioxide into the atmosphere.

According to the measurements of CO_2 concentrations in glacial ice, estimated changes in tropospheric CO_2 levels correlate fairly closely with estimated variations in the average global temperature near the

Table 21-1 Major Greenhouse Gases from Human Activities

Greenhouse Gas	Human Sources	Average Time in the Troposphere	Relative Warming Potential (compared to CO_2)
Carbon dioxide (CO_2)	Fossil fuel burning, especially coal (70–75%), deforestation, and plant burning	100–120 years	1
Methane (CH_4)	Rice paddies, guts of cattle and termites, landfills, coal production, coal seams, and natural gas leaks from oil and gas production and pipelines	12–18 years	23
Nitrous oxide (N_2O)	Fossil fuel burning, fertilizers, livestock wastes, and nylon production	114–120 years	296
Chlorofluorocarbons (CFCs)*	Air conditioners, refrigerators, plastic foams	11–20 years (65–110 years in the stratosphere)	900–8,300
Hydrochloro-fluorocarbons (HCFCs)	Air conditioners, refrigerators, plastic foams	9–390	470–2,000
Hydrofluorocarbons (HFCs)	Air conditioners, refrigerators, plastic foams	15–390	130–12,700
Halons	Fire extinguishers	65	5,500
Carbon tetrachloride	Cleaning solvent	42	1,400

*CFC use is being phased out, but they remain in the troposphere for 1–2 decades.

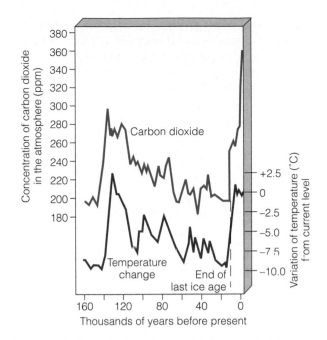

Figure 21-4 Atmospheric carbon dioxide levels and global temperature. Estimated long-term variations in average global temperature of the atmosphere near the earth's surface are graphed along with average tropospheric CO_2 levels over the past 160,000 years. The rough correlation between CO_2 levels in the troposphere and temperature shown in these estimates based on ice core data suggests a connection between these two variables, although no definitive causal link has been established. In 1999, the world's deepest ice core sample revealed a similar correlation between air temperatures and the greenhouse gases CO_2 and CH_4 going back 460,000 years. (Data from Intergovernmental Panel on Climate Change and National Center for Atmospheric Research)

earth's surface during the past 160,000 years (Figure 21-4). Trace the curves in this figure.

21-3 CLIMATE CHANGE AND HUMAN ACTIVITIES

How Have Human Activities Affected Tropospheric Concentrations of Greenhouse Gases? Messing with the Carbon Cycle

Humans have increased concentrations of greenhouse gases in the troposphere by burning fossil fuels, clearing and burning forests and grasslands, raising large numbers of livestock such as cattle, planting rice, and using inorganic fertilizers.

Figure 21-5 shows that since 1861, the concentrations of the greenhouse gases CO_2, CH_4, and N_2O in the troposphere have risen sharply, especially since 1950. Current CO_2 levels in the troposphere (Figure 21-5, top) appear to be higher than they have been in at least

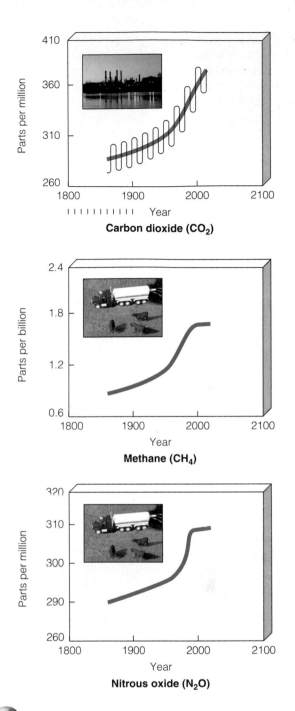

Figure 21-5 Increases in average concentrations of the greenhouse gases carbon dioxide, methane, and nitrous oxide in the troposphere between 1861 and 2003. The fluctuations in the CO_2 curve represent seasonal changes in photosynthetic activity that cause small differences between summer and winter concentrations of CO_2. (Data from Intergovernmental Panel on Climate Change, National Center for Atmospheric Research, and World Resources Institute)

160,000 years (Figure 21-4, blue curve). According to the IPCC, three human activities have emitted large amounts of greenhouse gases into the troposphere at a faster rate than natural processes can remove them.

One has been the sharp rise in the use of fossil fuels, which release large amounts of CO_2 and CH_4 into the troposphere. Electricity generated by coal is responsible for about 42% of this input, transportation 24%, industrial processes 20%, and residential and commercial uses 14%. Exhale, start a car, turn up the thermostat, turn on a light, burn leaves or a fireplace log, or do just about anything, and you add carbon dioxide to the troposphere. Burning a gallon of gasoline (which weighs about 2.7 kilograms, or 6 pounds) produces about 9 kilograms (20 pounds) of CO_2.

A second process is deforestation and clearing and burning of grasslands to raise crops and build cities, which release CO_2 and N_2O. Third is the raising of an increasing number of cattle and other livestock that release methane as a result of their digestive processes. A fourth process is cultivation of rice in paddies and use of inorganic fertilizers that release N_2O into the troposphere.

What Role Does the United States Play in Greenhouse Gas Emissions? Number One

The United States emits more greenhouse gases as a nation and on a per person basis than any other country.

The United States is by far the world's largest emitter of CO_2. Although the United Staates has only 4.6% of the world's population it produces an estimated 24% of the annual global emissions. The U.S. is followed by the European Union (12%), China (11%), Russia (7%), Japan (5%), and India (5%). However, the combined CO_2 emissions of the Asian countries of China, India, Japan, and South Korea are over twice the emissions of European Union countries and are approaching those of the United States.

According to the IPCC, emissions of CO_2 from U.S. coal-burning power and industrial plants are very likely to exceed the combined CO_2 emissions of 146 nations where three-fourths of the world's people live. CO_2 emissions from U.S. motor vehicles are roughly equivalent to those produced by everything that powers the Japanese economy.

The U.S. also emits large quantities of CH_4. Most comes from landfills (35% of all U.S. CH_4 emissions), domesticated livestock and their manure (26%), natural gas and oil systems (20%), and coal mining (10%).

In addition, the United States has the world's highest per capita CO_2 emissions, followed by Australia, Canada, and the Netherlands. During a 70-year lifetime, each U.S. citizen typically emits about 454 metric tons (500 tons) of CO_2 into the troposphere.

Is the Troposphere Warming? Very Likely

There is considerable evidence that the earth's troposphere is warming.

Here are five of many IPCC findings that support the scientific consensus that it is *very likely* (90–99% probability) that the troposphere is getting warmer. *First,* the 20th century was the hottest century in the past 1,000 years (Figure 21-2, bottom left). *Second,* since 1861 the average global temperature of the troposphere near the earth's surface has risen 0.6°C (1.1°F) over the entire globe and about 0.8°C (1.4°F) over the continents. Most of this increase has taken place since 1980.

Third, the 16 warmest years on record have occurred since 1980 and the 10 warmest years since 1990. The hottest year was 1998, followed in order by 2002, 2001, and 2003. Based on climate records going back to 1500, the summer of 2003 was the hottest Europe experienced in 500 years. More than 19,000 deaths were attributed to the heat. *Fourth,* glaciers and floating sea ice in some parts of the world are melting and shrinking (Case Study, below). *Fifth,* during the last century the world's average sea level rose by 0.1–0.2 meter (4–8 inches), partly from runoff from melting ice and partly because of the volume of ocean water expands when its temperature increases.

A few scientists have been skeptical of atmospheric warming. They pointed to a 1990 study showing that since 1979 temperature measurements near the earth's surface have been rising while satellite and other measurements showed no appreciable warming of the mid and upper troposphere. However, in 2002 and 2004 researchers analyzed these data and found much the same warming in these areas as thermometers show at te earth's surface.

The terms global warming and global climate change are often used interchangeably but they are not the same. **Global warming** refers to temperature increases in the troposphere, which in turn can cause climate change. **Global climate change** is a broader term that refers to changes in any aspects of the earth's climate, including temperature, precipitation, and storm intensity. It can involve global warming or cooling, but our focus will be on global warming. Global warming should not be confused with the problem of *ozone depletion* (Table 21-2), as discussed in Section 21-9.

Case Study: Warning Signals from the Earth's Ice and Snow: Meltdowns Are Under Way

Some of the world's floating ice and land-based glaciers are slowly melting, reflecting less incoming sunlight back into space, and helping warm the troposphere further.

The average temperature of the troposphere is strongly affected by the vast amounts of frozen water found as ice and snow near the earth's poles and in most of the world's mountain glaciers. In the Arctic region, this water is locked up in ice caps that cover Greenland and floating sea ice in the Arctic Ocean. The

Table 21-2 Major Characteristics of Global Warming and Ozone Depletion

Characteristic	Global Warming	Ozone Depletion
Region of atmosphere involved	Troposphere.	Stratosphere.
Major substances involved	CO_2, CH_4, N_2O (greenhouse gases).	O_3, O_2, chlorofluorocarbons (CFCs).
Interaction with radiation	Molecules of greenhouse gases absorb infared (IR) radiation from the earth's surface, vibrate, and release longer-wavelength IR radiation (heat) into the lower troposphere. This natural greenhouse effect helps warm the lower troposphere.	About 95% of incoming ultraviolet (UV) radiation from the sun is absorbed by O_3 molecules in the stratosphere and does not reach the earth's surface.
Nature of problem	There is a high (90–99%) probability that increasing concentrations of greenhouse gases in the troposphere from burning fossil fuels,deforestation, and agriculture are enhancingthe natural greenhouse effect and raisingthe earth's average surface temperature (Figure 21-2, bottom right, and Figure 21-11, p. 471).	CFCs and other ozone-depleting chemicals released into the troposphere by human activities have made their way to the stratosphere, where they decrease O_3 concentration. This can allow more harmful UV radiation to reach the earth's surface.
Possible consequences	Changes in climate, agricultural productivity, water supplies, and sea level.	Increased incidence of skin cancer, eye cataracts, and immune system suppression and damage to crops and phytoplankton.
Possible responses	Decrease fossil fuel use and deforestation; prepare for climate change.	Eliminate or find acceptable substitutes for CFCs and other ozone-depleting chemicals.

South Pole is covered by Antarctica, which contains about 70% of the earth's ice.

As the atmosphere warms, it causes more convection that transfers surplus heat from equatorial to polar areas (Figure 6-10, p. 107). Thus, temperature increases tend to be greater in polar regions. This explains why scientists regard the ice- and snow-covered areas at or near the earth's poles and as *early warning sentinels* of changes in the average temperature of earth's troposphere. Measurements from the Arctic Sea, Greenland, and the northwestern shores of Alaska

(Figure 17-9, p. 357) show that floating sea ice around the North Pole (Arctic) and Greenland is melting and thinning faster than it is being formed. For example, less ice covered the Arctic Ocean at the end of 2003 than in any year since 1979 when satellites began keeping track of such ice (Figure 21-6).

Why should we care if there is less ice in the Arctic? The answers lies in the **albedo** or reflectivity of different parts of the earth's surface (Figure 21-7, p. 467). Light-colored surfaces of ice and snow help cool the earth by reflecting 80–90% of incoming sunlight back

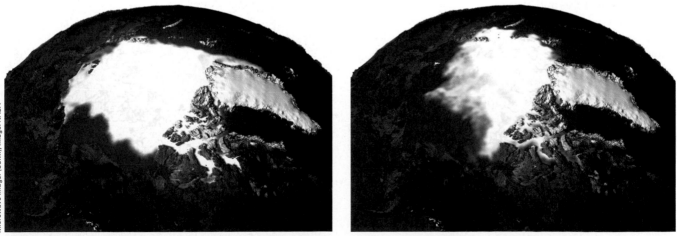

Data collected by Defense Satellite Program (DMSP) Special Sensor Microwave Imager (SSMI), Image: NASA

Figure 21-6 Satellite data showing Arctic sea ice in 1979 (left) and in 2003 (right). according to NASA, the ice cover shrunk by 9% during this period. [Defense Meteorological Satellite Program (DMSP) Special Sensor Microwave Imager (SSMI)]

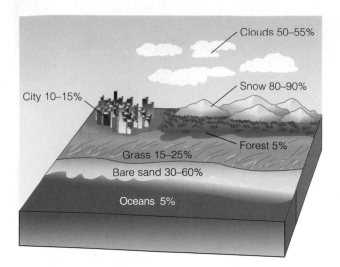

Figure 21-7 The *albedo*, or reflectivity of incoming solar energy, of different parts of the earth's surface varies greatly. (Data from NOAA)

into space. Much less sunlight is reflected by darker surfaces such as forests, grass, cities, and oceans. Thus the world's coldest regions are part of the earth's air-conditioning system.

A rise in the earth's temperature can cause gradual melting of some of the earth's ice caps, floating ice, and mountain glaciers to melt. This would expose darker and less reflective surfaces of water and land and result in a warmer troposphere. As more ice melts, the troposphere can become warmer, which melts more ice and increases the tropospheric temperature even more

It is not known whether this shrinkage and thinning of floating sea ice is the result of natural polar climate fluctuations, global warming caused by human-caused increases in greenhouse gases, or a combination of both factors—the last being the most likely explanation, according to many climate scientists. Regardless of the cause, such changes can affect the earth's temperatures and climate.

Because it is floating, large-scale melting of Arctic Ocean ice will not raise global sea levels—just as an ice cube in a glass of water does not raise the water level when it melts. However, according to researchers at the University of California at Santa Cruz, as much as half of the Arctic sea ice could disappear by 2050. If this happens, it would shift the course of the storm-guiding jet stream northward. The researchers estimate this would reduce wintertime rain and snowfall by nearly a third over an area stretching from southern British Columbia to Mexico. The resulting drop in the Sierra Nevada snow pack would sharply reduce the spring and summer water supply for states such as California.

Many scientists believe that the biggest long-term climate danger comes from Greenland. They are espe-

cially concerned about partial or eventually complete melting of the land-based glaciers or ice sheets that cover Greenland.

If this occurred, as it did in a previous interglacial warm period 110,000–130,000 years ago (Figure 21-8), average sea levels would rise by 7 meters (23 feet). In 2002, glaciologist Konrad Steffen reported that ice covering about a third of Greenland's total area is melting at a much faster rate than at any time since records have been kept. Researchers have calculated that a 3°C (5°F) rise in the earth's average atmospheric temperature—within the range projected during this century—would be enough eventually to melt the entire Greenland ice sheet. They estimate this would take about 1,000 years but partial melting could accelerate an increase in average sea level during this century. This is an area that scientists will be watching closely.

Would you like a preview of some of the effects of rapid atmospheric warming over the next 25–30 years? Visit Alaska, where average winter temperatures have increased by 4°C (8°F)—since 1960 and year-round temperatures have risen by 3°C (5°F). Most of this increase occurred since 1976. The hottest year in Alaskan history was 2002, and the winter of 2003 was the second warmest on record.

These warmer temperatures are melting glaciers and snow in parts of Alaska. Some of the permafrost under arctic tundra soils is warming and melting. This releases large amounts of CO_2 and CH_4 into the troposphere, which can accelerate tropospheric warming. The melting permafrost has caused buildings, roads, telephone and utility lines, and parts of the Trans-Alaska pipeline (Figure 17-9, p. 357) to sink, shift, and in some cases break up. In some parts of Alaska trees

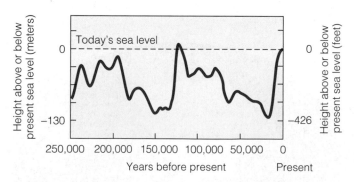

Figure 21-8 Changes in *average sea level* over the past 250,000 years based on data from cores removed from the ocean. The coming and going of glacial periods (ice ages) largely determine the rise and fall of sea level. As glaciers melted and retreated since the peak of the last glacial period about 18,000 years ago, the earth's average sea level has risen about 125 meters (410 feet). (Adapted from Tom Garrison, *Oceanography: An Invitation to Marine Science*, 3/E, © 1998. Brooks/Cole.)

are dying because the permafrost underneath them is melting. According to University of Alaska scientists, Alaskan forests are also threatened by greatly increased populations of the spruce-bark beetle (which can kill spruce trees) because of a lack of cold spells that help keep them under control.

However, there are economic benefits from a warmer Alaska. They include a longer growing season, ice-free ports, more people moving to the state, and more tourists visiting and spending money year-round.

During the last 25 years glaciers have also been melting and shrinking at accelerating rates on many of the world's mountaintops. Only 27 of the 150 glaciers found during the middle of the last century in Montana's Glacier National Park remain. Tanzania's Mount Kilimanjaro—Africa's tallest peak—may be ice-free within 15 years. Other evidence indicates that 80% of South American glaciers could disappear within 15 years.

The disappearance of mountain glaciers means a loss of frozen water reservoirs that partially thaw out during warm months and release water for use by farms and city-dwellers in the valleys below. This is *bad news* for countries like Peru, Ecuador, and Bolivia, which rely on annual water release by mountain glaciers for irrigation and household use.

21-4 PROJECTING FUTURE CHANGES IN THE EARTH'S TEMPERATURE

How Do Scientists Model Changes in the Earth's Temperature and Climate? Computer Models as Crystal Balls

Scientists have developed complex mathematical models of the earth's climate systems, and they use them to project future changes in the earth's average temperature.

To project the effects of increases in greenhouse gases on average global temperature, scientists develop models of how interactions among solar energy and the earth's land, oceans, ice, and greenhouse gases determine the average temperature of the troposphere. Figure 21-9 (p. 470) gives a greatly simplified summary of some of these interactions. Trace the flows and connections in this figure.

Scientists use this information to develop global climate models (also know as *coupled global circulation models*) that are applied to the atmosphere to project the effects of increases in greenhouse gases on average global temperature. Currently 14 research laboratories are operating, evaluating, and improving coupled general circulation models.

These modelers develop a three-dimensional representation of how energy, air masses, and moisture flow through the atmosphere, based on the laws of physics and the major factors affecting the earth's temperature and climate shown in Figure 21-9.

Computer simulations begin by covering the earth's surface with a grid of several hundred huge squares (Figure 21-10, p. 471). Each square provides the base for a stacked of gigantic imaginary cells, each several hundred kilometers on a side and about 3 kilometers (2 miles) high. These layers of cells extend down into the ocean and up into the atmosphere. Data on variables such as solar energy, sunlight, air pressure, temperature, water vapor, and winds or currents that affect climate in each cell are fed into the model. Then a complex set of mathematical equations simulates flows of matter and energy among the cells and the entire climate model is fed into a supercomputer. New climate data can be added to the model to improve its accuracy.

Such models provide scenarios of what is *very likely* or *likely* to happen based on various assumptions and data fed into the model. How well the results correspond to the real world depends on the assumptions of the model (based on current knowledge about the systems making up the earth, oceans, and atmosphere) and the accuracy of the data used.

What Is the Scientific Consensus about Future Changes in the Earth's Temperature? Hotter Times Ahead

Most climate scientists agree that human activities have influenced recent temperature increases and will lead to further significant temperature increases during this century.

In 1990, 1995, and 2001, the IPCC published reports that evaluate how global temperatures changed in the past (Figure 21-2) and are likely to change during this century. The IPCC reached its conclusions on the basis of scientific principles governing climate, data from past events, human emissions of CO_2 and other greenhouse gases, current temperature measurements, and global climate models.

Here are three major findings of the 2001 report.

- Despite many uncertainties, the latest climate models match the records of global temperature changes since 1850 very closely.

- "There is new and stronger evidence that most of the warming observed over the last 50 years is attributable to human activities."

- It is very likely (90–99% probability) that the earth's mean surface temperature will increase by 1.4–5.8°C (2.5–10.4°F) between 2000 and 2100 (Figure 21-11, p. 471).

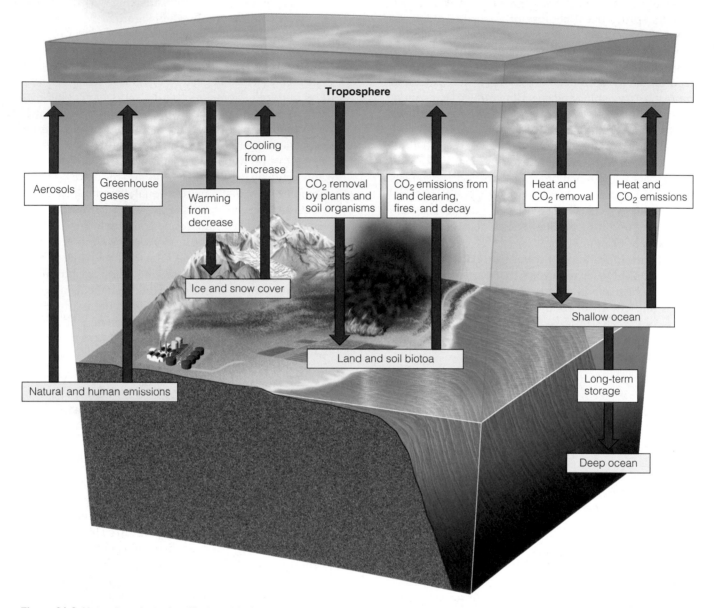

Sun

Troposphere

Aerosols

Greenhouse gases

Cooling from increase

Warming from decrease

CO_2 removal by plants and soil organisms

CO_2 emissions from land clearing, fires, and decay

Heat and CO_2 removal

Heat and CO_2 emissions

Ice and snow cover

Natural and human emissions

Land and soil biotoa

Shallow ocean

Long-term storage

Deep ocean

Figure 21-9 Natural capital: simplified model of some of the major processes that interact to determine the average temperature and greenhouse gas content of the troposphere and thus the earth's climate.

A 2001 report by the National Academy of Sciences and a 2002 Bush administration report, prepared by various U.S. government agencies for the United Nations, reached similar conclusions. In 2004, the American Geophysical Union released a position statement that said, "Scientific evidence strongly indicates that humans have played a role in the rapid warming of the past half century." And it is "virtually certain" that increasing greenhouse gases will warm the planet.

IPCC and other climate scientists holding the consensus view agree that current climate models need to be improved. This is why the *very likely* projected range of average atmospheric temperature during this century is quite broad (Figure 21-11). Scientists are hard at work trying to develop better models and narrow down such uncertainties.

A few climate scientists disagree with the consensus view about future temperature changes in the earth's atmosphere. They say we know too little about how the earth's climate works to make reliable projections about such changes. They also point out that some of the projected climate changes can be beneficial to some regions. And they believe we can use our inge-

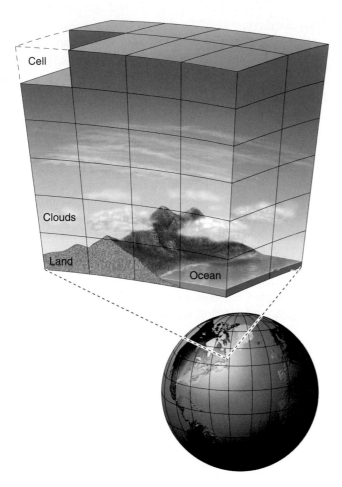

Figure 21-10 Global circulation model (GCM) of climate divides the earth's atmosphere into large numbers of gigantic boxes or cells stacked many layers high. The laws of physics and our understanding of global air circulation patterns and other factors that can affect climate are used to describe numerically what happens to major variables affecting climate in each cell and how they change from one cell to another.

nuity to offset most of the undesirable effects of climate change.

X | HOW WOULD YOU VOTE? Do you believe that we will experience significant global warming during this century? Cast your vote online at http://biology.brookscole.com/miller14.

Why Should We Be Concerned about a Warmer Earth? The Speed of Change Is What Counts

A rapid increase in the temperature of the troposphere would give humans and other species little time to deal with its effects.

Climate scientists warn that the concern is not just a temperature change but how rapidly it occurs, regardless of cause. Past temperature changes often took place over thousands to a hundred thousand years (Figure 21-2, top left). The problem we face is a fairly sharp projected increase in the temperature of the troposphere during this century (Figure 21-11).

According to the IPCC, it is *very likely* that this will be the fastest temperature change of the past 1,000 years. Such rapid temperature change can affect the availability of water resources by altering rates of evaporation and precipitation. It can also change wind patterns and weather, dry some areas, add moisture to others, alter some ocean currents, shift areas where crops can be grown, increase average sea levels and flood some coastal wetlands and cities and low-lying islands, and alter the structure and location of some of the world's biomes. These are major changes in the earth's atmospheric conditions. An increase in the earth's average temperature within a few decades or a century gives us little time to deal with its effects.

In 2002, the U.S. National Academy of Sciences issued a study, which raised the possibility that the temperature of the troposphere could rise drastically in only a decade or two. The report cited abrupt and long-lasting changes in tropospheric temperatures that have occurred during the last 100,000 years.

The report lays out a nightmarish worst-case scenario in which ecosystems suddenly collapse, low-lying cities are flooded, forests are consumed in vast fires, grasslands die out and turn into dust bowls, wildlife disappears, and tropical waterborne and insect-transmitted infectious diseases spread rapidly beyond their current ranges.

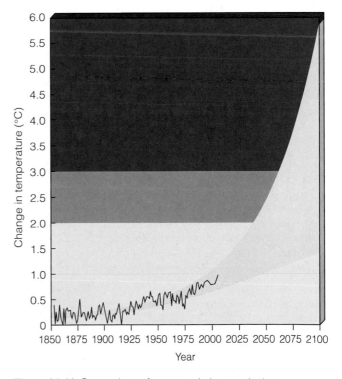

Figure 21-11 Comparison of measured changes in the average temperature of the atmosphere at the earth's surface between 1860 and 2003 and the projected range of temperature increase during the rest of this century. (Data from U.S. National Academy of Sciences, National Center for Atmospheric Research, and Intergovernmental Panel on Climate Change)

These possibilities were affirmed by a 2003 analysis carried out by Peter Schwartz and Doug Randall for the Department of Defense. They also projected widespread rioting and regional conflict in some countries faced with dwindling food, water, and energy supplies. The authors concluded that global warming must "be viewed as a serious threat to global stability and should be elevated beyond a scientific debate to a U.S. national security concern."

21-5 FACTORS AFFECTING THE EARTH'S TEMPERATURE

Scientists have identified a number of natural and human-influenced factors that might *amplify* (positive feedback) or *dampen* (negative feedback) projected changes in the average temperature of the troposphere. The fairly wide range of projected future temperature changes shown in Figure 21-11 results from including what is known about these factors in climate models. Let us examine some possible wild cards that could help or make matters worse or better during this century.

Can the Oceans Store More CO₂ and Heat? We Do Not Know

There is uncertainty about how much CO_2 and heat the oceans can remove from the troposphere and how long they might remain in the oceans.

The oceans help moderate the earth's average surface temperature by removing about 29% of the excess CO_2 we pump into the atmosphere as part of the global carbon cycle. They also absorb heat from the atmosphere and slowly transfer some of it to the deep ocean, where it is removed from the climate system for long but unknown periods of time (Figure 21-9).

Oocean currents on the surface and deep down are connected and act like a gigantic conveyor belt to store CO_2 and heat in the deep sea and to transfer hot and cold water from the tropics to the poles (Figure 21-12 and Figure 6-6, p. 106).

Scientists do not know how rapidly heat absorbed by the ocean from the troposphere can be transferred to the deep ocean by such currents and other mixing processes. They also do not know whether, over the next few decades, the oceans will release some of their stored heat and dissolved CO_2 into the troposphere, thereby amplifying its global warming.

Evidence suggests that large changes in the speed of the ocean currents in this conveyor belt, and its stopping and starting, contributed to wild swings in northern hemisphere temperatures during past ice ages. Scientists are trying to learn more about how these currents operate to evaluate the likelihood of the loop slowing down or stalling during this century and the effects this might have on regional and global atmospheric temperatures.

In 2003 a group of physical oceanographers, including Sydney Levitus of the National Oceanic and Atmospheric Administration (NOAA) and Ruth Curry of the Woods Hole Oceanographic Institute, announced the results of a compilation of millions of observations and measurements of the Atlantic Ocean from pole to pole. Their analysis indicated that tropical oceans are now much saltier and oceans closer to the poles are less salty than they were 40 years ago. In other words, during this period fresh water has been lost from the low latitudes and added at high latitudes.

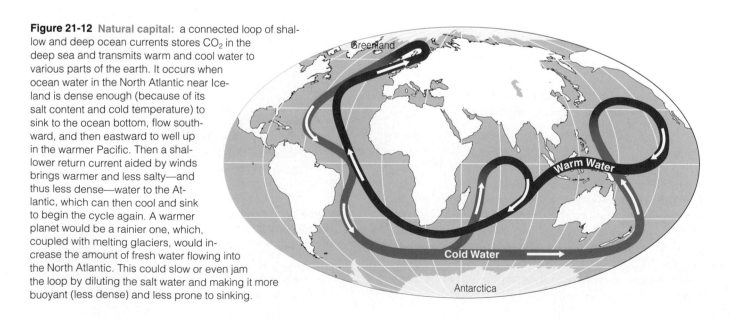

Figure 21-12 Natural capital: a connected loop of shallow and deep ocean currents stores CO_2 in the deep sea and transmits warm and cool water to various parts of the earth. It occurs when ocean water in the North Atlantic near Iceland is dense enough (because of its salt content and cold temperature) to sink to the ocean bottom, flow southward, and then eastward to well up in the warmer Pacific. Then a shallower return current aided by winds brings warmer and less salty—and thus less dense—water to the Atlantic, which can then cool and sink to begin the cycle again. A warmer planet would be a rainier one, which, coupled with melting glaciers, would increase the amount of fresh water flowing into the North Atlantic. This could slow or even jam the loop by diluting the salt water and making it more buoyant (less dense) and less prone to sinking.

They suggest that this indicates that global warming may be playing a key role in accelerating the global water cycle. If this hypothesis is correct, global warming may affect global precipitation patterns and alter the distribution, severity, and frequency of droughts, floods, and storms. Such an accelerated hydrologic cycle could also intensify global warming by increasing the rate of evaporation of water—a potent greenhouse gas—into the troposphere.

It could also slow down the conveyor belt (Figure 21-12) that helps draw warm Gulf Stream waters northward in the Atlantic, pumping heat into northern regions and moderating wintertime air temperatures, especially in western Europe.

The large loop of shallow and deep ocean currents shown in Figure 21-12 helps keep much of the northern hemisphere (especially Europe) fairly warm by pulling warm tropical water north, pushing cold water south, and releasing much of the heat stored in the water into the troposphere.

If this loop of currents should slow sharply or shut down, northern Europe and the northeast coast of North America would experience severe regional cooling. In other words, *global warming can lead to significant global cooling in some parts of the world,* with the climate of western Europe possibly resembling that of Siberia. Disruption or significant slowing of the loop would also disrupt other parts of the world with floods, droughts, severe storms, and searing heat.

How Might Changes in Cloud Cover Affect the Troposphere's Temperature? Another Uncertainty

Warmer temperatures create more clouds that could warm or cool the troposphere, but we do not know which effect might dominate.

One of the largest unknowns in global climate models is the effect of changes in the global distribution of clouds or the temperature of the troposphere. Warmer temperatures increase evaporation of surface water and create more clouds. These additional clouds can have a *warming effect* (positive feedback), by absorbing and releasing heat into the troposphere, or a *cooling effect* (negative feedback) by reflecting more sunlight back into space.

The net result of these two opposing effects depends on several factors. One is how much water vapor will enter the troposphere as the earth's surface warms. In 2004, measurements by researchers Andrew Dessler and Ken Minschwaner verified that water vapor is increasing in the troposphere as the earth warms. However, they found that increases in water vapor in the upper troposphere were not as high as many global circulation climate models have assumed.

The effects of clouds on atmospheric temperatures also depend on whether it is day or night. Other factors include the type (thin or thick), coverage (continuous or discontinuous), and altitude of the cloud, and the size and number of water droplets or ice crystals formed in clouds.

For example, an increase in thick and continuous clouds at low altitudes can decrease surface warming by reflecting and blocking more sunlight. However, an increase in thin and discontinuous cirrus clouds at high altitudes can warm the lower troposphere and increase surface warming. What climate scientists know about the effects of clouds has been included in the latest climate models., but much uncertainty remains.

In 1999, researchers at the University of Colorado reported that the wispy condensation trails (contrails) left behind by jet planes might have a greater impact on the temperature of the troposphere than scientists had thought. Using infrared satellite images, they found that jet contrails expand and turn into large cirrus clouds that tend to release heat into the upper troposphere. If these preliminary results are confirmed, emissions from jet planes could be responsible for as much as half of the tropospheric warming in the northern hemisphere.

How Might Outdoor Air Pollution Affect the Troposphere's Temperature? A Temporary Effect

Aerosol pollutants and soot produced by human activities can warm or cool the troposphere, but such effects will decrease with any decline in such outdoor air pollution.

Aerosols (microscopic droplets and solid particles) of various air pollutants are released or formed in the troposphere by volcanic eruptions and human activities, and they can increase cloud cover.

Some of the resulting clouds have a high albedo and reflect more incoming sunlight back into space during the day. This could help counteract the heating effects of increased greenhouse gases.

Nights are warmer becaue the presence of clouds prevents some of the heat stored in the earth's land and water during the day from being radiated into space. These pollutants may explain why most of the recent warming in the northern hemisphere occurs at night.

But these interactions are complex. Aerosol pollutants in the lower troposphere can either warm or cool the air, depending on factors such as their size and the reflectivity of the underlying surface.

Most tropospheric aerosols, such as sulfate particles produced by fossil fuel combustion, tend to cool the troposphere and thus can temporarily slow global warming. However, a recent study by Mark Jacobson of Stanford University indicated that tiny particles of *soot* or *black carbon aerosols*—produced mainly from incomplete combustion in coal burning, diesel engines, and open fires—may be responsible for 15–30% of global warming during the past 50 years. If so, soot would be the second biggest human contribution to global warming, after the greenhouse gas CO_2.

One possible effect of increased aerosols in the troposphere is *global* or *solar dimming*. In 2004, scientists reported that measurements showed a drop in the amount of sunshine reaching the earth's surface by as much as 10% between 1960 and 1990, with a 37% drop in Hong Kong. Satellite measurements showed that a decrease in the amount of energy from the sun—solar radiation—could not account for this effect. Scientists hypothesize that pollution can dim sunlight in two ways. One is that soot particles in the atmosphere reflect some of the sunlight back into space. Another possibility is that the airborne particles cause more water droplets to condense out of the air, leading to thicker and darker clouds, which can reduce incoming sunlight. However, preliminary measurements showed that the amount of sunlight reaching the earth's surface increased slightly between 2001 and 2003. Scientists are trying to sort out the complexities and causes of these phenomena.

Climate scientists do not expect aerosol pollutants to counteract or enhance projected global warming very much in the next 50 years for two reasons. One is that aerosols and soot fall back to the earth or are washed out of the lower atmosphere within weeks or months, whereas CO_2 and other greenhouse gases remain in the troposphere for decades to several hundred years. The other is that aerosol inputs into the troposphere are being reduced—especially in developed countries.

Can Increased CO_2 Levels Stimulate Photosynthesis and Remove More CO_2 from the Air? A Temporary and Limited Effect

Increased CO_2 in the troposphere could increase plant photosynthesis, but several factors can limit or offset this effect.

Some studies suggest that more CO_2 in the troposphere could increase the rate of plant photosynthesis in areas with adequate water and soil nutrients. This would remove more CO_2 from the troposphere and help slow atmospheric warming.

However, recent studies indicate that this CO_2 removal would be temporary for two reasons. One is that it would slow as the plants reach maturity and take up less CO_2 from the troposphere. The other is that carbon stored by the plants as organic compounds would be returned to the troposphere as CO_2 when the plants die and decompose or burn.

A 2004 study by a team of U.S. and Brazilian scientists showed that undisturbed old-growth Amazon rainforests are experiencing rapid changes in species composition apparently because of rising atmospheric levels of carbon dioxide. Higher CO_2 levels are fertilizing many species of trees and fast growing larger trees are out competing smaller younger trees. This is changing the mix of tree and wildlife species or biodiversity makeup of theses rainforests. Initially, this can increase the uptake of CO_2 from the atmosphere. But as these larger trees mature and die out sooner the reduction in denser wood and foliage could eventually lead to a drop in the amount of carbon dioxide these rainforests remove from the atmosphere. In addition, plant-eating insects that breed more rapidly and year-round in warmer temperatures could offset much of the increased plant growth.

How Might a Warmer Troposphere Affect Methane Emissions? Accelerated Warming

Warmer air can release methane gas stored in bogs, wetlands, and tundra soils, causing a feedback loop that makes the air warmer.

Global warming could be accelerated by an increased release of methane (a potent greenhouse gas) from two major sources. One is bogs and other wetlands and the other is ice-like compounds called *methane hydrates* trapped beneath the arctic permafrost. Significant amounts of methane would be released into the troposphere if the permafrost in tundra and boreal forest soils partially or completely melts, as is occurring in parts of Canada, Alaska, China, and Mongolia. The resulting tropospheric warming could lead to more methane release and still more warming.

21-6 POSSIBLE EFFECTS OF A WARMER WORLD

What Are Some Possible Effects of a Warmer Troposphere? Winners and Losers

A warmer troposphere would have beneficial and harmful effects, but poor nations in the tropics will suffer the most.

A warmer troposphere could have a number of beneficial and harmful effects, listed in Figures 21-13 and 21-14, for humans, other species, and ecosystems, depending mostly on their locations and on how rapidly the temperature changes. Study these figures carefully. However, betting on living in an area with favorable climate change in the future is like playing a game of Russian roulette. Global climate models are improving, but so far we cannot make reliable projections about how the climates of particular regions are likely to change.

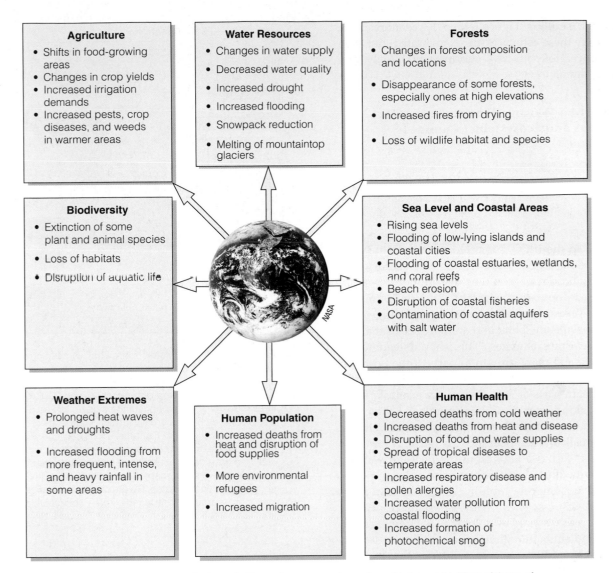

Agriculture
- Shifts in food-growing areas
- Changes in crop yields
- Increased irrigation demands
- Increased pests, crop diseases, and weeds in warmer areas

Water Resources
- Changes in water supply
- Decreased water quality
- Increased drought
- Increased flooding
- Snowpack reduction
- Melting of mountaintop glaciers

Forests
- Changes in forest composition and locations
- Disappearance of some forests, especially ones at high elevations
- Increased fires from drying
- Loss of wildlife habitat and species

Biodiversity
- Extinction of some plant and animal species
- Loss of habitats
- Disruption of aquatic life

Sea Level and Coastal Areas
- Rising sea levels
- Flooding of low-lying islands and coastal cities
- Flooding of coastal estuaries, wetlands, and coral reefs
- Beach erosion
- Disruption of coastal fisheries
- Contamination of coastal aquifers with salt water

Weather Extremes
- Prolonged heat waves and droughts
- Increased flooding from more frequent, intense, and heavy rainfall in some areas

Human Population
- Increased deaths from heat and disruption of food supplies
- More environmental refugees
- Increased migration

Human Health
- Decreased deaths from cold weather
- Increased deaths from heat and disease
- Disruption of food and water supplies
- Spread of tropical diseases to temperate areas
- Increased respiratory disease and pollen allergies
- Increased water pollution from coastal flooding
- Increased formation of photochemical smog

Figure 21-13 *Winners and losers.* Projected effects of a warmer atmosphere for the world. Most of these effects could be harmful or beneficial depending on where one lives. Current models of the earth's climate cannot make reliable projections about where such effects might take place at a regional level and how long they might last. (Data from Intergovernmental Panel on Climate Change, U.S. Global Climate Change Research Program, U.S. National Academy of Sciences)

According to the IPCC, the largest burden of the harmful effects of moderate global warming will fall on people and economies in poorer tropical and subtropical nations without the economic and technological resources needed to adapt to its harmful impacts.

In 2003, scientists at the World Health Organization estimated that each year about 150,000 people—mostly children in developing countries in Asia and Africa—die prematurely from side effects of global warming ranging from increases in malaria to malnutrition. They estimated that this death toll could double by 2020. Some researchers estimate that by the end of this century the annual death toll from global warming

- Less severe winters
- More precipitation in some dry areas
- Less precipitation in some wet areas
- Increased food production in some areas
- Expanded population and range for some plant and animal species adapted to higher temperatures

Figure 21-14 *Winners:* Possible *beneficial effects* of a warmer atmosphere for some countries and people.

could reach 6 million (Figure 1-15, p. 17) or more. Some analysts say these estimates are exaggerated. But even with a lower toll this is a serious and largely preventable human tragedy.

How Might a Warmer Troposphere Affect Organisms and Ecosystems? Change is upon Them.

A warmer troposphere will change the distribution and population sizes of wild species, shift locations of some of the world's ecosystems, and threaten some protected reserves and coral reefs.

According to the IPCC, projected change in the temperature of the troposphere during this century will have a significant effect on the "distributions, population sizes, population density, and behavior of wildlife." A warmer climate could expand ranges and populations of some plant and animal species that can adapt to warmer climates. This should lead to increased tree and plant growth in parts of the northern United States, Canada, Russia, central Asia, and northern Europe. In parts of Scandinavia, for example, birch trees are taking over traditional reindeer lichen pastures. And the reindeer are having to compete for lichen with elk and red deer moving north.

There is also *bad news* from the IPCC. A warmer troposphere would threaten plant and animal species that could not migrate rapidly enough to new areas (Figure 21-15), species with specialized niches, and those with a narrow tolerance for temperature change. And shifts in regional climate would threaten many parks, wildlife reserves, wilderness areas, wetlands, and coral reefs—thwarting some current efforts to stem the loss of biodiversity. Also, species likely to do better in a warmer world include certain rapidly multiplying weeds, insect pests, and disease-carrying organisms such as mosquitoes and water-borne bacteria.

The IPCC says it is *very likely* that tree deaths will increase from more disease and higher pest populations that would thrive in areas with a warmer climate. It is also *very likely* that wildfires in forest and grassland areas with drier climates will increase, destroying wildlife habitats and, as a result, releasing large amounts of CO_2 into the troposphere.

A 2004 report by the UN Environment Programme estimated that at least 1 million species (especially plant, mammal, butterfly, and bird species) could face premature extinction by 2050 unless greenhouse gas emissions are drastically reduced. According to the IPCC, ecosystems *most likely* to be disrupted and lose species are coral reefs, polar seas, coastal wetlands, arctic and alpine tundra, and high-elevation mountaintops.

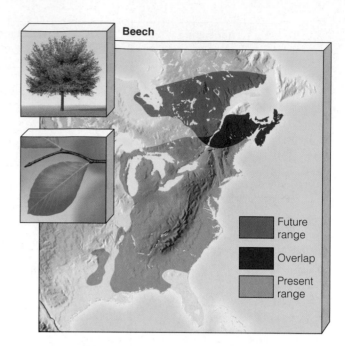

Figure 21-15 Possible effects of global warming on the geographic range of beech trees based on ecological evidence and computer models. According to one projection, if CO_2 emissions doubled between 1990 and 2050, beech trees (now common throughout the eastern United States) would survive only in a greatly reduced range in northern Maine and southeastern Canada. This is only one of a number tree species whose geographic ranges could be changed drastically by increased atmospheric warming. For example, native sugar maples are likely to disappear in the northeastern United States. On the other hand, ranges of some tree species adapted to a warm climate would spread. (Data from Margaret B. Davis and Catherine Zabinski, University of Minnesota)

How Might a Warmer Troposphere Affect Agriculture? Winners and Losers

Food production may increase in some areas and decrease in others.

In a warmer world, agricultural productivity may increase in some areas and decrease in others. For example, some analysts project that warmer temperatures and increased precipitation at northern latitudes may lead to a northward shift of some agricultural production from the midwestern United States to Canada. But overall food production could decrease because soils in these areas of Canada are generally less fertile than those in the midwestern United States.

A decrease in high-elevation snow packs could lead to a sharp decline in agricultural productivity in some heavily irrigated areas. For example, water experts project increasing water shortages in areas such as central and southern California that receive most of their water in summer months from snow melting on the Sierra Nevada as well as snowmelt in the Rockies that feeds the Colorado River (Figure 15-10, p. 314). Warmer winter temperatures in the Sierra Nevada and the Rockies would cause most precipitation to fall as rain rather than snow. This would increase flooding during the winter months and sharply reduce the summer supply of water for central and southern California.

Even larger effects would occur if snow mass in the Himalayas decreased. Such a change could reduce water available in summer for irrigation from the Yellow, Indus, and Ganges Rivers. Irrigation water from these rivers is vital. It is currently used to produce the world's two largest wheat harvests in China and in India. Also, reduced water flow in the summer from the Yangtze River in China would harm the world's largest rice harvest.

Crop and fish production in some areas could be reduced by rising sea levels that would flood river deltas, which are home to some of the world's most productive agricultural lands and coastal aquaculture ponds.

What Are Some Possible Effects of Rising Sea Levels? Seek Higher Ground

Rising sea levels could flood low-lying coastal wetlands and islands, coral reefs, and parts of some of the world's coastal cities.

Another problem with a warmer world is a rise in global sea level caused by runoff from melting snow and ice and by the fact that water expands slightly when heated. In their 2001 IPCC report, climate scientists projected that global sea levels are *very likely* to rise during this century (Figure 21-16).

The high projected rise in sea level of about 88 centimeters (35 inches) would have a number of harmful effects. They include the following:

- Threatening half of the world's coastal estuaries, wetlands (one-third of those in the United States, especially in southern Louisiana and southern Florida), and coral reefs

- Disrupting many of the world's coastal fisheries

- Flooding low-lying barrier islands and causing gently sloping coastlines (especially along the U.S. East Coast) to erode and retreat inland by about 1.3 kilometers (0.8 mile)

- Flooding agricultural lowlands and deltas in parts of Bangladesh, India, and China, where much of the world's rice is grown

- Contaminating freshwater coastal aquifers with salt water

- Submerging some low-lying islands in the Pacific Ocean (the Marshall Islands) and the Indian Ocean (the Maldives, a chain of 1,200 small islands)

One comedian jokes that he plans to buy land in Kansas because it will probably become valuable beachfront

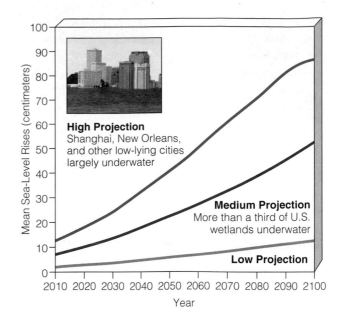

Figure 21-16 It is very likely that global sea levels will rise from 9–88 centimeters (4–35 inches) during this century. If this occurs, flooding and coastal erosion would be especially severe in heavily populated coastal areas of the tropics and warm temperate regions. (Data from Intergovernmental Panel on Climate Change, 2000)

property. Another boasts that she is not worried because she lives in a houseboat—the "Noah strategy." On a more serious note, a Netherlands architectural firm has designed a prototype for an energy-efficient, floating home for use in areas subject to flooding.

21-7 DEALING WITH THE THREAT OF GLOBAL WARMING

What Are Our Options? The Great Climate Debate

There is disagreement over what we should do about the threat of global warming.

As we have seen, nearly all climate scientists agree that the earth's temperature is very likely to increase during this century and that human activities play a part in this change. Despite this scientific consensus, there is debate among scientists over the causes of these changes (natural or human), how rapidly they might occur, the effects on humans and ecosystems, and how we should respond to this potentially serious long-term global threat.

Economists and policymakers also disagree over how we should respond to the threat of climate change. They disagree on whether

- the economic costs of reducing greenhouse gas emissions are higher than the economic benefits.

- developed countries, developing countries, or both should take responsibility for reducing greenhouse gas emissions,

- actions to reduce greenhouse gas emissions should be voluntary or required as a result of national laws and an international treaty.

As a result of these scientific, economic, and political disagreements, there are three schools of thought concerning what we should do about projected global warming. One is to *do more research before acting.* With this *wait-and-see-strategy,* many scientists and economists call for more research and a better understanding of the earth's climate system before making far-reaching and controversial economic and political decisions such as phasing out fossil fuels. This is the current position of the U.S. government.

A second and rapidly growing group of scientists, economists, business leaders, and political leaders especially in the European Union believe that we should *act now to reduce the risks from climate change brought about by global warming.* They argue that the potential for harmful economic, ecological, and social consequences is so great that action should not be delayed. They believe that current evidence indicates that global warming is occurring and that if we delay by waiting for even more conclusive evidence, it will be too late to slow down the degree and rate of such warming. In other words, global warming is a good candidate for applying the *precautionary principle.*

In 1997, more than 2,500 scientists from a variety of disciplines signed a Scientists' Statement on Global Climate Disruption and concluded, "We endorse those [IPCC] reports and observe that the further accumulation of greenhouse gases commits the earth irreversibly to further global climatic change and consequent ecological, economic, and social disruption. The risks associated with such changes justify preventive action through reductions in emissions of greenhouse gases." Also in 1997, 2,700 economists led by eight Nobel laureates declared, "As economists, we believe that global climate change carries with it significant environmental, economic, social, and geopolitical risks and that preventive steps are justified."

A third strategy is to *act now as part of a no-regrets strategy.* Scientists and economists supporting this approach say we should take the key actions needed to slow global warming—even if the threat does not materialize—because such actions lead to other important environmental, health, and economic benefits. For example, a reduction in the combustion of fossil fuels, especially coal, will lead to sharp reductions in air pollution that lowers food and timber productivity, decreases biodiversity, and prematurely kills large numbers of people. Reducing oil use would also de-

crease dependence on imported oil, which threatens economic and military security. And improving energy efficiency has numerous economic and environmental advantages (Figure 18-2, p. 380).

> **✗ HOW WOULD YOU VOTE?** Should we act now to help slow global warming? Cast your vote online at http://biology.brookscole.com/miller14.

What Can We Do to Reduce the Threat? Conserve Energy, Use Renewable Energy, and Intercept Greenhouse Gas Emissions

We can improve energy efficiency, rely more on carbon-free renewable energy resources, and find ways to keep much of the CO_2 we produce out of the troposphere.

Figure 21-17 presents a variety of prevention and cleanup solutions that climate analysts have suggested for slowing the rate and degree of global warming. The solutions come down to three major strategies: *improve energy efficiency to reduce fossil fuel use, shift from carbon-based fossil fuels to a mix of carbon-free renewable energy resources,* and *sequester or store as much CO_2 as possible in soil, in vegetation, underground, and in the deep ocean.* The effectiveness of these three strategies would be enhanced by *reducing population* to decrease the number of fossil fuel consumers and CO_2 emitters and by *reducing poverty* to decrease the need of the poor to clear more land for crops and wood.

Scientists are also developing new power plant designs that would eliminate smokestack emissions of CO_2 and other pollutants. One approach is to develop modified forms of coal gasification to increase the energy efficiency of coal-fired power plants from 35% to 70%. There is also research on using metal-ceramic membranes in coal gasification plants to trap CO_2 for sequestering. However, if such plants can be developed, they are likely to be quite costly and it would take many decades for them to replace existing power plants.

> **✗ HOW WOULD YOU VOTE?** Should we phase out the use of fossil fuels over the next fifty years? Cast your vote online at http://biology.brookscole.com/miller14.

Can We Remove and Store (Sequester) Enough CO_2 to Slow Global Warming? Are Output Approaches the Answer?

We can prevent some of the CO_2 we produce from circulating in the troposphere, but the costs may be high and the effectiveness of various approaches is unknown.

Figure 21-18 (p. 480) shows several potential techniques to remove CO_2 from the troposphere or from

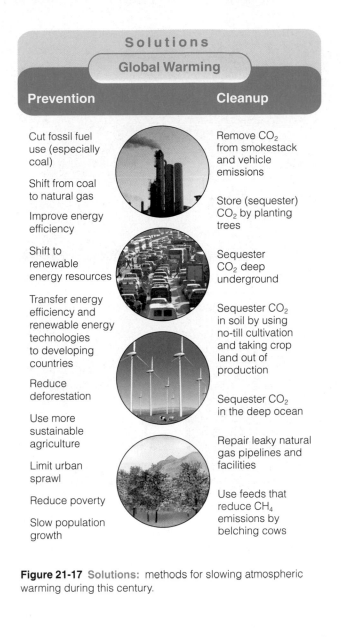

Solutions

Global Warming

Prevention	Cleanup
Cut fossil fuel use (especially coal)	Remove CO_2 from smokestack and vehicle emissions
Shift from coal to natural gas	Store (sequester) CO_2 by planting trees
Improve energy efficiency	
Shift to renewable energy resources	Sequester CO_2 deep underground
Transfer energy efficiency and renewable energy technologies to developing countries	Sequester CO_2 in soil by using no-till cultivation and taking crop land out of production
Reduce deforestation	Sequester CO_2 in the deep ocean
Use more sustainable agriculture	Repair leaky natural gas pipelines and facilities
Limit urban sprawl	
Reduce poverty	Use feeds that reduce CH_4 emissions by belching cows
Slow population growth	

Figure 21-17 Solutions: methods for slowing atmospheric warming during this century.

smokestacks and store (sequester) it in other parts of the environment.

One possible way to remove CO_2 from the atmosphere is to plant trees that store (sequester) it in biomass. But studies indicate that this is a temporary approach because trees release their stored CO_2 back into the atmosphere when they die and decompose or if they are burned (for example, by forest fires or to clear land for crops).

A second approach is *soil sequestration* in which plants such as switchgrass are used to remove CO_2 from the air and store it in the soil. But warmer temperatures can increase decomposition in soils and return some of the stored CO_2 to the atmosphere.

A third strategy is to *reduce the release of carbon dioxide and nitrous oxide from soil.* Ways to do this include *conservation cultivation* (Figure 14-13, p. 284) and

retiring depleted crop fields, leaving them untouched as conservation reserves.

A fourth approach is to remove CO_2 from smokestacks and *pump it deep underground* into unminable coal seams and abandoned oil fields or *inject it into the deep ocean,* as shown in Figure 21-18.

There are several problems with this strategy. One is that current methods can remove only about 30% of the CO_2 from smokestack emissions and would double or triple the cost of producing electricity by burning coal. The U.S. Department of Energy estimates that the cost of sequestering carbon dioxide in various underground and deep ocean repositories will have to be reduced at least 10-fold to make this approach economically feasible. In addition, injecting large quantities of CO_2 into the ocean could upset the global carbon cycle, seawater acidity, and some forms of deep-sea life in unpredictable ways.

Some scientists have suggested that we add iron to the oceans (especially in Antarctic waters) to stimulate the growth of marine algae, which could remove more CO_2 through photosynthesis. But the algae would return it to the atmosphere a short time later when they died unless the carbon is somehow deposited in the deep ocean. Furthermore, we do not know the potential effects of applying large amounts of iron to the ocean's ecosystems.

How Can Governments Reduce the Threat of Global Warming? Use Sticks and Carrots

Governments can tax greenhouse gas emissions and energy use, increase subsidies and tax breaks for saving energy and using renewable energy, and decrease subsidies and tax breaks for fossil fuels.

Governments could use three major methods to promote the solutions to slowing global warming listed in Figure 21-17. One is to phase in output-based *carbon taxes* on each unit of CO_2 emitted by fossil fuels (especially coal and gasoline) or input-based *energy taxes* on each unit of fossil fuel (especially coal and gasoline) that is burned. Decreasing taxes on income, labor, and profits to offset increases in consumption taxes on carbon emissions or fossil fuel use could help make such a strategy more politically acceptable.

A second strategy is to *level the economic playing field* by greatly increasing government subsidies for energy-efficiency and carbon-free renewable-energy technologies, carbon sequestration, and more sustainable agriculture, and by phasing out subsidies and tax breaks for using fossil fuels.

The third strategy is *technology transfer.* Governments of developed countries could fund the transfer of energy-efficiency, carbon-free renewable-energy,

Figure 21-18 Solutions: methods for removing carbon dioxide from the atmosphere or from smokestacks and storing (sequestering) it in plants, soil, deep underground reservoirs, and the deep ocean.

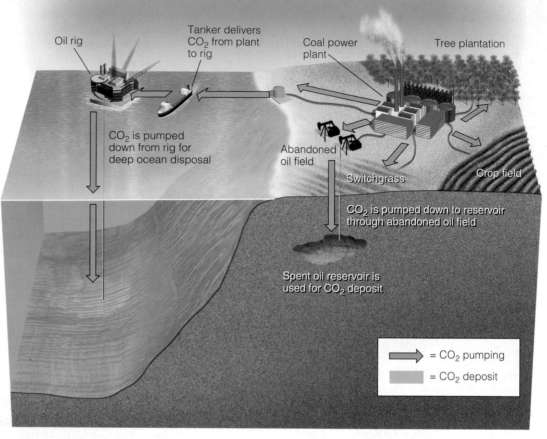

Oil rig

Tanker delivers CO$_2$ from plant to rig

Coal power plant

Tree plantation

CO$_2$ is pumped down from rig for deep ocean disposal

Abandoned oil field

Switchgrass

Crop field

CO$_2$ is pumped down to reservoir through abandoned oil field

Spent oil reservoir is used for CO$_2$ deposit

= CO$_2$ pumping

= CO$_2$ deposit

carbon-sequestration, and more sustainable agriculture technologies to developing countries. Increasing the current tax on each international currency transaction by a quarter of a penny could finance this technology transfer, which would then generate wealth for developing countries.

How Can We Use the Marketplace to Reduce or Prevent Greenhouse Gas Emissions? Emissions Trading

Establishing a global emissions trading program could help reduce greenhouse gas emissions.

An economic approach to slow global warming is to agree to global and national limits on greenhouse gas emissions and encourage industries and countries to meet these limits by selling and trading greenhouse gas emission permits in the marketplace. This approach stimulates companies to develop new technologies to reduce greenhouse gas emissions and increase profits. In the United States, this market approach has been used to reduce SO$_2$ emissions ahead of target goals at a fraction of the projected cost (p. 453).

In a greenhouse gas emissions trading program, industries and countries could earn greenhouse gas emission credits by improving energy efficiency, switching from coal to natural gas, and adopting certain farming, ranching, and soil-building and conser-

vation practices. Credits could also be earned by switching from coal and other fossil fuels to forms of carbon-free renewable energy such as solar, wind, hydrogen, and geothermal. For example, a coal-burning power plant in Illinois could earn emission credits by building a wind farm in Oregon.

In addition, credits could be earned by sequestering CO$_2$ from the atmosphere by reforestation or by injecting it into the deep ocean or secure underground reservoirs (Figure 21-18). For example, a coal-burning power plant in Ohio might earn credits by financing a CO$_2$-removing reforestation project in Costa Rica.

Companies or countries that manage to produce fewer emissions than their permits allowed could sell some of their credits to other participants. As a result, participants that devise innovative ways to reduce greenhouse gas production are rewarded by increased profit. And participants that produce excess amounts of greenhouse gas face increased costs because they are fined or have to buy extra permits from participants who have earned credits by reducing their emissions.

Some analysts believe this market-based approach is more politically and economically feasible than relying primarily on government regulation to impose taxes on carbon emissions or fuel used. Other analysts point to some problems with emissions trading. One is that carbon fuels are burned in so many homes, vehicles, factories, and crop fields that it would be difficult to monitor compliance. For that reason, many analysts think that emissions trading programs should be used in conjunction with other approaches, such as taxes on fossil fuel use, significant government subsidies for energy efficiency and renewable energy, and removal of subsidies for fossil fuels.

Another problem is that it is politically difficult for the world's countries to agree on what should count as credits or how any such credits should be divided among nations.

Can We Afford to Reduce the Threat of Global Warming? Not Acting Will Probably Cost More

It will very likely cost us less to help slow and adapt to global warming now than to deal with its harmful effects later.

According to a 2001 study by the UN Environment Programme, projected global warming will cost the world economy more than $300 billion annually by 2050 ($30 billion per year in the United States) unless nations make strong efforts to curb greenhouse gas emissions.

According to a number of economic studies, implementing the strategies listed in Figure 21-16 would boost the global and U.S. economy, provide much-needed jobs (especially in developing countries with large numbers of unemployed and underemployed people), and cost much less than trying to deal with the harmful effects of these problems.

However, according to some widely publicized economic models developed by economist William Nordhaus and others, the projected costs of reducing CO_2 emissions will greatly exceed the projected benefits. Other economists criticize these models as being unrealistic and too gloomy for two reasons. *First*, they do not include the huge cost savings from implementing many of the strategies listed in Figure 21-17 such as improving energy efficiency. *Second*, they underestimate the ability of the marketplace to act rapidly when money is to be made from reducing greenhouse gas emissions.

21-8 WHAT IS BEING DONE TO REDUCE GREENHOUSE GAS EMISSIONS?

What Is the Kyoto Protocol? A Controversial International Agreement

Getting countries to agree on reducing their greenhouse gas emissions is difficult.

In December 1997, more than 2,200 delegates from 161 nations met in Kyoto, Japan, to negotiate a treaty to help slow global warming. The resulting *Kyoto Protocol* would require 39 developed countries to cut emissions of CO_2, CH_4, and N_2O to an average of about 5.2% below 1990 levels by 2012. The initial steps of the protocol were directed at these 39 countries because they are responsible for a majority of the world's CO_2 emissions (58% in 1999) and thus should take the lead in reducing their emissions.

The protocol would not require poorer developing countries to make cuts in their greenhouse gas emissions until a later version of the treaty. It would also allow greenhouse gas emissions trading among participating countries. By mid-2004, the Kyoto Protocol had been ratified by more than 120 countries.

Some climate analysts praise the Kyoto agreement as a small but important step in attempting to slow projected global warming. But according to computer models, the 5.2% reduction goal of the Kyoto Protocol would shave only about 0.06°C (0.1°F) off the 0.7–1.7°C (1–3°F) temperature rise projected by 2060.

In 2001, President George W. Bush withdrew U.S. participation from the Kyoto Protocol because he argued that it was too expensive and did not require emissions reductions by developing countries such as China and India that have large and increasing emissions of greenhouse gases. This decision set off strong protests by many scientists, citizens, and leaders throughout most of the world who pointed out that strong leadership is needed by the United States because it has the highest total and per capita CO_2 emissions of any country. According to most climate analysts, the Kyoto Protocol will accomplish little without the full participation of the United States, Russia, China, and India. However, Scott Barnett, an expert on environmental treaties, believes that the Kyoto Protocol is a badly thought out agreement that will not work.

How Can We Move Beyond the Kyoto Protocol Stalemate? Forging a New Strategy

Countries could work together to develop a new international approach to slowing global warming.

In 2004, Richard B. Stewart and Jonathan B. Wiener proposed that countries work together to develop a new strategy for slowing global warming.

They urge the development of a new climate treaty by the United States; China, India, Russia, and other major emitters among developing countries, and Australia and any other developed countries not participating in the Kyoto Protocol. The treaty would include participation by developing countries, develop an effective emissions trading program that includes developing countries omitted from such trading by the Kyoto Protocol, set achievable targets for reducing emissions for each 10 of the next 40 years, and evaluate global and national strategies for adapting to the harmful ecological and economic effects of global warming.

Scott Bennett suggests starting again with a new approach that sets technological goals and standards, not targets and timetables. This or other alternative new approaches would allow the United States to provide much-needed leadership on this important global issue instead of being seen as a spoiler. Such a parallel treaty could be used as a basis for overhauling the Kyoto Protocol. Or countries participating in the protocol could agree to join the new parallel treaty.

What Are Some Countries, Businesses, States, and Cities Doing to Help Delay Global Warming? Good News

Many countries, companies, cities, states, and provinces are reducing their greenhouse gas emissions, improving energy efficiency, and increasing their use of carbon-free renewable energy.

Many countries are reducing their greenhouse gas emissions. For example, by 2000 Great Britain had reduced its CO_2 emissions to its 1990 level, well ahead of its Kyoto target goal. It did this mostly by relying more on natural gas than on coal, improving energy efficiency in industry and homes, and reducing gasoline use by raising its tax on gasoline. Between 2000 and 2050, Great Britain aims to cut its CO_2 emissions by 60%, mostly by improving energy efficiency and by relying on renewable resources for 20% of its energy by 2030. To help accomplish this goal, the government has greatly increased research and development spending for renewable energy and tax breaks for renewable energy.

According to a 2001 study by the Natural Resources Defense Council, China reduced its CO_2 emissions by 17% between 1997 and 2000, a period during which CO_2 emissions in the United States rose by 14%. The government did this by phasing out coal subsidies, shutting down inefficient coal-fired electric plants, speeding up its 20-year commitment to increase energy efficiency, and restructuring its economy to increase use of renewable energy resources.

A growing number of major global companies, such as Alcoa, DuPont, IBM, Toyota, BP Amoco, and Shell, have established targets to reduce their greenhouse gas emissions by 10–65% from 1990 levels by 2010. For example, BP Amoco has already met goals that exceed those in the Kyoto Protocol at no net cost to the company.

Since 1990, governments in more than 500 cities around the world (including 110 in the United States) have established programs to reduce their greenhouse gas emissions. What is your community doing?

What Are Some Individuals and Schools Doing to Help Delay Global Warming? Change from the Bottom Up

Some individuals and schools are reducing their greenhouse gas emissions, wasting less energy, and relying more on carbon-free renewable energy.

Each of us leaves a *climate change legacy* because at least half the greenhouse gases emitted by our daily activities will still be in the troposphere a century from now. *Good news.* About 400,000 U.S. households are buying carbon-free electricity from their utility companies. Figure 21-19 lists some things you can do to cut your CO_2 emissions. How many of these things are you doing?

Some universities and colleges around the United States (and in some other countries) are taking steps to reduce CO_2 emissions. For example, students and faculty at Oberlin College in Ohio have asked their board of trustees to reduce its CO_2 emissions to zero by 2020 by buying renewable energy or producing its own. Twenty-five Pennsylvania colleges have joined to purchase wind power and other forms of carbon-free renewable energy. What is your school doing to help slow global warming?

How Can We Prepare for Global Warming? Get Ready for Change

A growing number of countries and cities are looking for ways to cope with the harmful effects of climate change.

According to the latest global climate models, the world needs to cut current emissions of greenhouse gases (not just CO_2) by at least 50% by 2018 to stabilize concentrations of such gases in the air at their present levels. Such a large reduction in emissions is extremely unlikely for political and economic reasons because it would require rapid, widespread changes in industrial processes, energy sources, transportation options, and individual lifestyles.

As a result, a growing number of climate analysts and economists suggest that we should also begin preparing for the possible effects of long-term atmospheric warming and climate change. Figure 21-20 shows some ways to implement this *adaptation* strategy.

Reducing CO₂ Emissions

- Drive a fuel-efficient car, walk, bike, carpool, and use mass transit

- Use energy-efficient windows

- Use energy-efficient appliances and lights

- Heavily insulate your house and seal all drafts

- Reduce garbage by recycling and reuse

- Insulate hot water heater

- Use compact fluorescent bulbs

- Plant trees to shade your house during summer

- Set water heater no higher than 49°C (120°F)

- Wash laundry in warm or cold water

- Use low-flow shower head

Figure 21-19 **What can you do?** Ways to reduce your annual emissions of CO_2.

Why Are Global Warming and Climate Change Such Difficult Problems to Deal with? A Complex, Long-Term, and Controversial Challenge

Global warming and climate change are hard to deal with because they have many causes (some poorly understood); their effects are long-term and uneven; and there is controversy over how they should be addressed

Several characteristics of global warming and climate change pose difficult and often controversial scientific, economic, political, and ethical questions about how to address these threats.

First, these problems *have many complex and still poorly understood causes and effects.* Second, they *are long-term problems.* Elected officials who have to make tough decisions about dealing with these issues will be long gone when the beneficial or harmful effects of their actions occur. The long-term effects of climate change also raise an important ethical question. How much are we willing to change or sacrifice *now* for benefits that may not be realized in our lifetimes but could greatly benefit our children, grandchildren, and the plants and animals that we share the planet with?

Third *the harmful and beneficial effects of climate change are uneven.* There will be winners and losers.

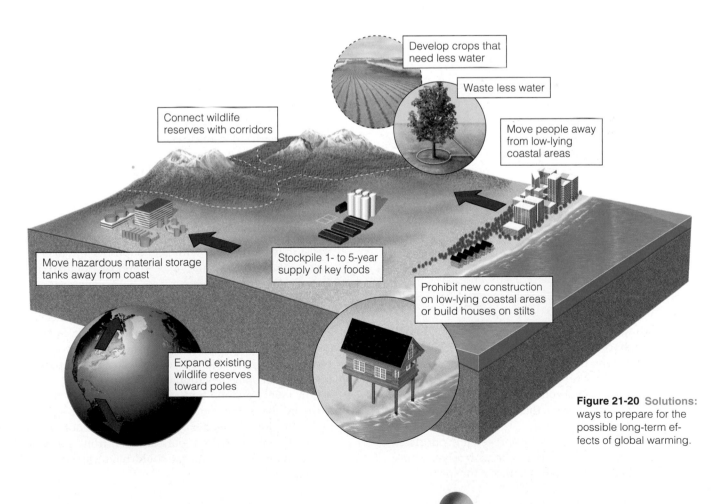

Develop crops that need less water

Waste less water

Connect wildlife reserves with corridors

Move people away from low-lying coastal areas

Move hazardous material storage tanks away from coast

Stockpile 1- to 5-year supply of key foods

Prohibit new construction on low-lying coastal areas or build houses on stilts

Expand existing wildlife reserves toward poles

Figure 21-20 **Solutions:** ways to prepare for the possible long-term effects of global warming.

Winning nations are less likely to bring about controversial changes or spend large sums of money to slow down something that will benefit them. The catch: We do not know which countries and parts of countries will be winners and losers until it is too late to avoid harmful effects.

Fourth, *reducing greenhouse gas emissions will take unprecedented international respoinse to a global problem that is of uncertain magnitude.* And this must be done using political and economic systems not designed to deal with long-term threats.

Because of these characteristics, you can see why so many analysts believe that responding to this threat is one of the most important and challenging dilemmas we face.

21-9 OZONE DEPLETION IN THE STRATOSPHERE

What Is the Threat from Ozone Depletion? A Clear Danger

Less ozone in the stratosphere will allow more harmful UV radiation to reach the earth's surface.

A layer of ozone in the lower stratosphere (Figure 20-2, p. 434, and Figure 20-3, p. 435) keeps about 95% of the sun's harmful ultraviolet (UV) radiation from reaching the earth's surface. Measuring instruments on balloons, aircraft, and satellites show considerable seasonal depletion (thinning) of ozone concentrations in the stratosphere above Antarctica and the Arctic. Similar measurements reveal a lower overall loss of stratospheric ozone everywhere except over the tropics.

Based on these measurements and on mathematical and chemical models, the overwhelming consensus of researchers in this field is that ozone depletion (thinning) in the stratosphere is a serious threat to humans, other animals, and some of the sunlight-driven primary producers (mostly plants) that support the earth's food.

What Causes Ozone Depletion? From Dream Chemicals to Nightmare Chemicals

Widespread use of a number of useful and long-lived chemicals has reduced ozone levels in the stratosphere.

Thomas Midgley, Jr., a General Motors chemist, discovered the first *chlorofluorocarbon (CFC)* in 1930, and chemists developed similar compounds to create a family of highly useful CFCs. The two most widely used are CFC-11 (trichlorofluoromethane, CCl_3F) and CFC-12 (dichlorodifluoromethane, CCl_2F_2), known by their trade name, Freons.

These chemically stable (nonreactive), odorless, nonflammable, nontoxic, and noncorrosive compounds seemed to be dream chemicals. Inexpensive to manufacture, they became popular as coolants in air conditioners and refrigerators (replacing toxic sulfur dioxide and ammonia), propellants in aerosol spray cans, cleaners for electronic parts such as computer chips, fumigants for granaries and ship cargo holds, and bubbles in plastic foam used for insulation and packaging. Between 1960 and the early 1990s, CFC production rose sharply.

But it turned out that CFCs were too good to be true. In 1974, calculations by chemists Sherwood Rowland and Mario Molina at the University of California-Irvine indicated that CFCs were lowering the average concentration of ozone in the stratosphere. They shocked both the scientific community and the $28-billion-per-year CFC industry by calling for an immediate ban of CFCs in spray cans (for which substitutes were available).

Rowland and Molina's research led them to four major conclusions. *First,* CFCs remain in the troposphere because they are insoluble in water and chemically unreactive. *Second,* over 11–20 years these heavier-than-air chemicals are lifted into the stratosphere mostly through convection, random drift, and the turbulent mixing of air in the troposphere.

Third, once they reach the stratosphere, the CFC molecules break down under the influence of high-energy UV radiation. This releases highly reactive chlorine atoms (Cl), as well as atoms of fluorine (F), bromine (Br) and Iodine (I), which accelerate the breakdown of ozone (O_3) into O_2 and O in a cyclic chain of chemical reactions, one of which is shown in Figure 21-21. This causes ozone in various parts of the stratosphere to be destroyed faster than it is formed.

Finally, each CFC molecule can last in the stratosphere for 65–385 years, depending on its type. During that time, each chlorine atom released from these molecules can convert hundreds of molecules of O_3 to O_2.

Overall, according to Rowland and Molina's calculations and later models and atmospheric measurements of CFCs in the stratosphere, these dream molecules had turned into global ozone destroyers.

The CFC industry (led by DuPont), a powerful, well-funded adversary with a lot of profits and jobs at stake, attacked Rowland and Molina's calculations and conclusions. The researchers held their ground, expanded their research, and explained the meaning of their calculations to other scientists, elected officials, and the media. After 14 years of delaying tactics, DuPont officials acknowledged in 1988 that CFCs were depleting the ozone layer and agreed to stop producing them once they found substitutes.

In 1995, Rowland and Molina received the Nobel Prize in chemistry for their work. In awarding the prize, the Royal Swedish Academy of Sciences said that they contributed to "our salvation from a global environmental problem that could have catastrophic consequences."

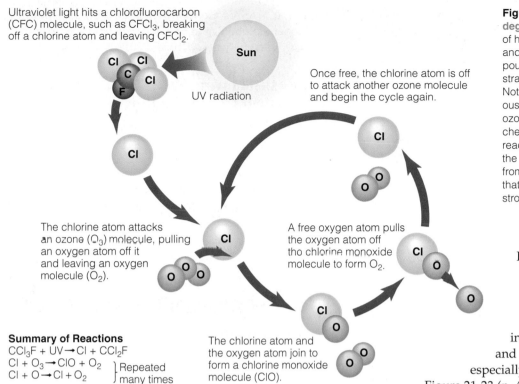

Ultraviolet light hits a chlorofluorocarbon (CFC) molecule, such as CFCl₃, breaking off a chlorine atom and leaving CFCl₂.

Sun

UV radiation

Once free, the chlorine atom is off to attack another ozone molecule and begin the cycle again.

Cl

Cl

The chlorine atom attacks an ozone (O₃) molecule, pulling an oxygen atom off it and leaving an oxygen molecule (O₂).

Cl

A free oxygen atom pulls the oxygen atom off the chlorine monoxide molecule to form O₂.

Cl O

Cl O

The chlorine atom and the oxygen atom join to form a chlorine monoxide molecule (ClO).

Summary of Reactions
CCl₃F + UV → Cl + CCl₂F
Cl + O₃ → ClO + O₂   ⎫ Repeated
Cl + O → Cl + O₂   ⎬ many times

Figure 21-21 Natural capital degradation: simplified summary of how chlorofluorocarbons (CFCs) and other chlorine-containing compounds can destroy ozone in the stratosphere faster than it is formed. Note that chlorine atoms are continuously regenerated as they react with ozone. Thus they act as catalysts, chemicals that speed up chemical reactions without being used up by the reaction. Bromine atoms released from bromine-containing compounds that reach the stratosphere also destroy ozone by a similar mechanism.

What Other Chemicals Deplete Stratospheric Ozone? More Culprits

A number of chemicals can end up in the stratosphere and deplete ozone there for up to several hundred years.

CFCs are not the only ozone-depleting compounds (ODCs). Others are *halons* and *hydrobromofluorocarbons* (*HBFCs*) (used in fire extinguishers), *methyl bromide* (a widely used fumigant), *hydrogen chloride* (emitted into the stratosphere by space shuttles), and cleaning solvents such as *carbon tetrachloride, methyl chloroform,* n-*propyl bromide,* and *hexachlorobutadiene.*

The oceans and occasional volcanic eruptions also release chlorine compounds into the troposphere. But most of these do not make it to the stratosphere because they dissolve easily in water and wash out of the troposphere in rain. Bromine compounds may be less likely to wash out of the troposphere, but further study is needed to confirm this possibility. Measurements and models indicate that 75–85% of the observed ozone losses in the stratosphere since 1976 are the result of ozone-depleting chemicals released into the atmosphere by human activities beginning in the 1950s.

What Happens to Ozone Levels over the Earth's Poles Each Year? It Drops Each Winter and Spring.

During four months of each year up to half of the ozone in the stratosphere over Antarctica is depleted.

In 1984, researchers analyzing satellite data discovered that 40–50% of the ozone in the upper stratosphere over Antarctica disappeared during the Antarctic late winter and spring (August–November), especially since 1976 (Figure 21-22).

Figure 21-23 (p. 486) shows the seasonal variation of ozone with altitude over Antarctica during 2003. The observed loss of ozone above Antarctica often is called an *ozone hole.* A more accurate term is *ozone thinning* because the ozone depletion varies with altitude and location.

The total area of the stratosphere above Antarctica that suffers from ozone thinning during the peak season varies from year to year and in some recent years has covered an area greater than that of North America In 2003, the area of thinning was the second largest size ever.

Measurements indicate that CFCs and other ODCs are the primary culprits. Each winter, steady winds blow in a circular pattern over the earth's poles. This creates a *polar vortex:* a huge swirling mass of very cold air that is isolated from the rest of the atmosphere until the sun returns a few months later.

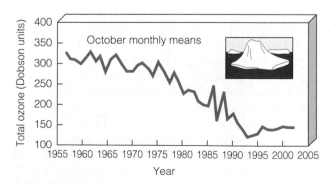

Figure 21-22 Mean total level of ozone for October over the Halley Bay measuring station in Antarctica, 1956–2003. (Data from British Antarctic Survey and World Meteorological Organization)

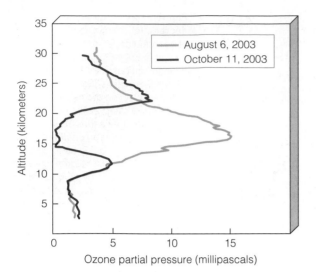

Figure 21-23 Seasonal variation of ozone level with altitude over Antarctica during 2003. Note the severe depletion of ozone during October (during the Antarctic spring, red line) and its return to more normal levels in August (during the Antarctic winter, green line). (Data from National Oceanic and Atmospheric Administration)

When water droplets in clouds enter this circling stream of extremely frigid air, they form tiny ice crystals. The surfaces of these ice crystals collect CFCs and other ozone depleting chemicals in the stratosphere, setting up conditions for the formation of ClO, the molecule most responsible for the seasonal loss of ozone over the Antarctic.

When partial sunlight returns in October, the light stimulates ClO molecules which reduce ozone (Figure 21-21). Within weeks, this cyclic reaction typically destroys 40–50% of the ozone above Antarctica (100% in some places).

As summer approaches and temperatures warm, the polar vortex begins to break up and mix again with the rest of the atmosphere. Then new ozone forms over Antarctica until the next dark winter.

When the vortex breaks up, huge masses of ozone-depleted air above Antarctica flow northward and linger for a few weeks over parts of Australia, New Zealand, South America, and South Africa. This raises biologically damaging UV-B levels in these areas by 3–10%, and in some years as much as 20%.

In 1988, scientists discovered that similar but usually less severe ozone thinning occurs in the stratosphere over the Arctic during the arctic spring and early summer (February–May), with a seasonal ozone loss of 11–38% (compared to a typical 50% loss above Antarctica). When this mass of air above the Arctic breaks up each spring, large masses of ozone-depleted air flow south to linger over parts of Europe, North America, and Asia. In 2002, models indicated that the Arctic is unlikely to develop the large-scale ozone thinning found over the Antarctic. Unfortunately, however, according to a 1998 model developed by scien-

tists at NASA's Goddard Institute for Space Studies, ozone depletion over the Antarctic and Arctic will be at its worst between 2010 and 2019.

Why Should We Be Worried about Ozone Depletion? Life in the Ultraviolet Zone

Increased UV radiation reaching the earth's surface from ozone depletion in the stratosphere is harmful to human health, crops, forests, animals, and materials.

Why should we care about ozone loss? Figure 21-24 lists some of the expected effects of decreased levels of

Natural Capital Degradation

Effects of Ozone Depletion

Human Health

- Worse sunburn
- More eye cataracts
- More skin cancers
- Immune system suppression

Food and Forests

- Reduced yields for some crops
- Reduced seafood supplies from reduced phytoplankton
- Decreased forest productivity for UV-sensitive tree species

Wildlife

- Increased eye cataracts in some species
- Decreased population of aquatic species sensitive to UV radiation
- Reduced population of surface phytoplankton
- Disrupted aquatic food webs from reduced phytoplankton

Air Pollution and Materials

- Increased acid deposition
- Increased photochemical smog
- Degradation of outdoor paints and plastics

Global Warming

- Accelerated warming because of decreased ocean uptake of CO_2 from atmosphere by phytoplankton and CFCs acting as greenhouse gases

Figure 21-24 Natural capital degradation: Expected effects of decreased levels of ozone in the stratosphere.

ozone in the stratosphere. From a human standpoint the answer is that with less ozone in the stratosphere, more biologically damaging UV-A and UV-B radiation will reach the earth's surface. This will give humans worse sunburns, more eye cataracts (a clouding of the eye's lens that reduces vision and can cause blindness if not corrected), and more skin cancers (Figure 21-25 and Connections, at right).

Humans can make cultural adaptations to increased UV radiation by staying out of the sun, protecting their skin with clothing, and applying sunscreens. However, plants and animals that help support us and other forms of life cannot make such changes except through biological evolution, a process that can take a long time.

Connections: What Cancer Are You Most Likely to Get? Look in the Mirror

Exposure to UV radiation is a major cause of skin cancers.

Research indicates that years of exposure to UV-B ionizing radiation in sunlight is the primary cause of *squamous cell* (Figure 21-25, left) and *basal cell* (Figure 21-25, center) *skin cancers*. Together these two types make up 95% of all skin cancers. Typically there is a 15- to 40-year lag between excessive exposure to UV-B and development of these cancers.

Caucasian children and adolescents who experience only one severe sunburn double their chances of getting these two types of cancers. Some 90–95% of

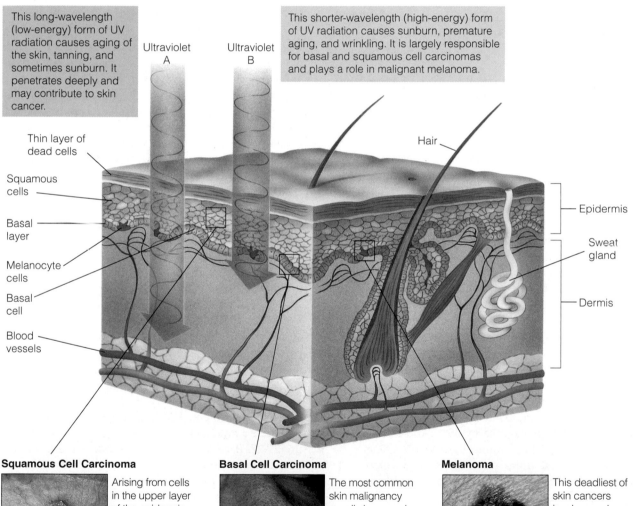

This long-wavelength (low-energy) form of UV radiation causes aging of the skin, tanning, and sometimes sunburn. It penetrates deeply and may contribute to skin cancer.

This shorter-wavelength (high-energy) form of UV radiation causes sunburn, premature aging, and wrinkling. It is largely responsible for basal and squamous cell carcinomas and plays a role in malignant melanoma.

Ultraviolet A

Ultraviolet B

Thin layer of dead cells

Squamous cells

Basal layer

Melanocyte cells

Basal cell

Blood vessels

Hair

Epidermis

Sweat gland

Dermis

Squamous Cell Carcinoma

Arising from cells in the upper layer of the epidermis, this cancer is also caused by exposure to sunlight or tanning lamps. It is usually curable if treated early. It grows faster than basal cell carcinoma and can metastasize.

Basal Cell Carcinoma

The most common skin malignancy usually is caused by excessive exposure to sunlight or tanning lamps. It develops slowly, rarely metastasizes and is nearly 100% curable if diagnosed early and treated properly.

Melanoma

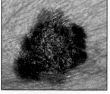

This deadliest of skin cancers involves melanocyte cells, which produce pigment. It can develop from a mole or on blemished skin, grows quickly, and can spread to other parts of the body (metastasize).

Figure 21-25 Structure of the human skin and the relationships between ultraviolet (UV-A and UV-B) radiation and the three types of skin cancer. (The Skin Cancer Foundation)

- Stay out of the sun, especially between 10 A.M. and 3 P.M.

- Do not use tanning parlors or sunlamps.

- When in the sun, wear protective clothing and sunglasses that protect against UV-A and UV-B radiation.

- Be aware that overcast skies do not protect you.

- Do not expose yourself to the sun if you are taking antibiotics or birth control pills.

- Use a sunscreen with a protection factor of 15 or 25 if you have light skin.

- Examine your skin and scalp at least once a month for moles or warts that change in size, shape, or color or sores that keep oozing, bleeding, and crusting over. If you observe any of these signs, consult a doctor immediately.

Figure 21-26 **What can you do?** Ways to reduce your exposure to harmful UV radiation.

these types of skin cancer can be cured if detected early enough, although their removal may leave disfiguring scars. These cancers kill 1–2% of their victims, which amounts to about 2,300 deaths in the United States each year.

A third type of skin cancer, *malignant melanoma* (Figure 21-25, right), occurs in pigmented areas such as moles anywhere on the body. Within a few months, this type of cancer can spread to other organs.

It kills about one-fourth of its victims (most under age 40) within 5 years, despite surgery, chemotherapy, and radiation treatments. Each year it kills about 100,000 people (including more than 7,400 Americans), mostly Caucasians. It can be cured if detected early enough, but recent studies show that some melanoma survivors have a recurrence more than 15 years later.

A 2003 study found that women who used tanning parlors once a month or more increased their chance of developing malignant melanoma by 55%. And a 2004 study at Dartmouth College found that people using tanning beds were 2.5 times more likely to develop basal cell carcinoma and 1.5 times more susceptible to squamous cell carcinoma.

Recent evidence suggests that about 90% of sunlight's melanoma-causing effect may come from exposure to UV-A (which is not blocked by window glass) and 10% from UV-B. Tanning booth lights and sunlamps emit mostly UV-A. Some sunscreens provide little or no protection from UV-A unless they contain chemicals such as zinc oxide or avobenzene (also

called Parasol 1789). Read the fine print on the tube to see if such chemicals are present.

Evidence indicates that people (especially Caucasians) who experience three or more blistering sunburns before age 20 are five times more likely to develop malignant melanoma than those who have never had severe sunburns. About 10% of those who get malignant melanoma have an inherited gene that makes them especially susceptible to the disease. Figure 21-26 lists ways for you to protect yourself from harmful UV radiation.

21-10 PROTECTING THE OZONE LAYER

How Can We Protect the Ozone Layer? Say No

To reduce ozone depletion we must stop producing ozone-depleting chemicals.

The consensus of researchers in this field is that we should immediately stop producing all ozone-depleting chemicals. However, even with immediate and consistent action, models indicate it will take about 50 years for the ozone layer to return to 1980 levels and about 100 years for recovery to pre-1950 levels. *Good news.* Substitutes are available for most uses of CFCs, and others are being developed (Individuals Matter, at right).

In 1987, representatives of 36 nations meeting in Montreal, Canada, developed a treaty, commonly known as the *Montreal Protocol.* Its goal was to cut emissions of CFCs (but not other ozone depleters) into the atmosphere by about 35% between 1989 and 2000. After hearing more bad news about seasonal ozone thinning above Antarctica in 1989, representatives of 93 countries met in London in 1990 and in Copenhagen, Denmark (1992), and adopted the *Copenhagen Protocol,* an amendment which accelerated the phasing out of key ozone-depleting chemicals.

These landmark international agreements, now signed by 177 countries, are important examples of global cooperation in response to a serious global environmental problem. Without them, ozone depletion would be a much more serious threat, as shown in Figure 21-27. If nations continue to follow these treaties, ozone levels should return to 1980 levels by 2050 and 1950 levels by 2100.

However, according to a 1998 study by the World Meteorological Organization, ozone depletion in the stratosphere has been cooling the troposphere and has helped offset or disguise as much as 30% of the global warming from our emissions of greenhouse gases. Thus restoring the ozone layer could lead to an increase in global warming. But the alternative is worse. Environmental choices such as these are not easy.

Ray Turner and His Refrigerator

Ray Turner, an aerospace manager at Hughes Aircraft in California, made an important low-tech ozone-saving discovery by using his head—and his refrigerator. His concern for the environment led him to look for a cheap and simple substitute for the CFCs used as cleaning agents to remove films of oxidation from the electronic circuit boards manufactured at his plant.

He started by looking in his refrigerator. He decided to put drops of various substances on a corroded penny to see whether any of them removed the film of oxidation. Then he used his soldering gun to see whether solder would stick to the surface of the penny, indicating the film had been cleaned off.

First, he tried vinegar. No luck. Then he tried some ground-up lemon peel, also a failure. Next he tried a drop of lemon juice and watched as the solder took hold. The rest, as they say, is history.

Today, Hughes Aircraft uses inexpensive citrus-based solvents that are CFC-free to clean circuit boards. This new cleaning technique has reduced circuit board defects by about 75% at Hughes. And Turner got a hefty bonus. Now other companies, such as AT&T, clean computer boards and chips using acidic chemicals extracted from cantaloupes, peaches, and plums. Maybe you can find a solution to an environmental problem in your refrigerator, grocery store, drugstore, or backyard.

The ozone protocols set an important precedent for global cooperation and action to avert potential global disaster by using *prevention* to solve a serious environmental problem. Nations and companies agreed

to work together to solve this problem for three reasons. *First*, there was convincing and dramatic scientific evidence of a serious problem. *Second*, CFCs were produced by a small number of international companies. *Third*, the certainty that CFC sales would decline over a period of years unleashed the economic and creative resources of the private sector to find even more profitable substitute chemicals.

The atmosphere is the key symbol of global interdependence. If we can't solve some of our problems in the face of threats to this global commons, then I can't be very optimistic about the future of the world.

MARGARET MEAD

CRITICAL THINKING

1. In preparation for the 1992 UN Conference on the Human Environment in Rio de Janeiro, President George H. W. Bush's top economic adviser gave an address in Williamsburg, Virginia, to representatives of governments from a number of countries. He told his audience not to worry about global warming because the average temperature increases scientists are projecting were much less than the temperature increase he experienced in coming from Washington, D.C., to Williamsburg, Virginia. What is the fundamental flaw in this reasoning?

2. What changes might occur in **(a)** the global hydrologic cycle (Figure 4-28, p. 76) and **(b)** the global carbon cycle (Figure 4-29, p. 78) if the troposphere experiences significant warming? Explain.

3. What will be the likely effect of clearing forests and converting them to grasslands and crops on **(a)** the earth's reflectivity (albedo) and **(b)** the earth's average surface temperature? Explain.

4. One way to help slow the rate of CO_2 emissions is to reduce the clearing of forests—especially in developing countries where intense deforestation is taking place. Should the United States and other developed countries pay poorer countries to stop cutting their forests? Explain.

5. Of the three schools of thought on what should be done about possible global warming (p. 478), which do you favor? Explain.

6. Of the proposals in Figure 21-17 (p. 479) for reducing emissions of greenhouse gas emissions into the troposphere, with which do you disagree? Why?

7. Of the proposals in Figure 21-20 (p. 483) for preparing for the effects of global warming, with which do you disagree? Why?

8. What consumption patterns and other features of your lifestyle directly add greenhouse gases to the atmosphere? Which, if any, of these things would you be willing to give up to slow global warming and reduce other forms of air pollution?

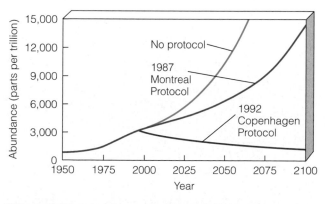

Figure 21-27 Solutions: projected concentrations of ozone-depleting chemicals (ODCs) in the stratosphere under three scenarios: no action, the 1987 Montreal Protocol, and the 1992 Copenhagen Protocol. (Data from World Meteorological Organization)

9. Congratulations! You are in charge of the world. List your three most important actions for dealing with the problems of **(a)** global warming and **(b)** depletion of ozone in the stratosphere.

PROJECTS

1. As a class, conduct a poll of students at your school to determine **(a)** whether they understand the difference between global warming of the troposphere and ozone depletion in the stratosphere (Table 21-2, p. 467) and **(b)** whether they believe global warming from an enhanced greenhouse effect is a very serious problem, a moderately serious problem, or of little concern. Tally the results to see whether there are differences related to each poll participant's year in school, political leaning (liberal, conservative, independent), and sex.

2. As a class, conduct a poll of students at your school to determine whether they believe stratospheric ozone depletion is a very serious problem, a moderately serious problem, or of little concern. Tally the results to see whether there are differences related to each poll participant's year in school, political leaning (liberal, conservative, independent), and sex.

3. Use the library or the Internet to determine how the current government policy on global warming in the country where you live compares with the policy suggestions made by various analysts and listed in Figures 21-17 (p. 479) and 21-20 (p. 483).

4. Write a 1- to 2-page scenario of what your life could be like by 2060 if nations, companies, and individuals do not take steps to reduce projected global warming caused at least partly by human activities. Contrast your scenario with the positive scenario at the opening of this chapter. Compare and critique scenarios written by other members of your class.

5. If you drive a car, calculate how much CO_2 it emits per day by taking the number of miles you drive, multiplying it by 20, and dividing the result by the number of miles per gallon your car gets. If you use the metric system, multiply the kilometers driven by the number of liters of gasoline it takes to drive your car 100 kilometers and divide the result by 42 to get your daily CO_2 emissions in kilograms. In either case, add another 20% to include the CO_2 emitted in manufacturing the gasoline you used. Compare your results with other members of your class.

6. Use the library or the Internet to find bibliographic information about *Paul A. Colinvaux* and *Margaret Mead*, whose quotes appear at the beginning and end of this chapter.

7. Make a concept map of this chapter's major ideas, using the section heads, subheads, and key terms (in boldface). Look on the website for this book for information about making concept maps.

LEARNING ONLINE

The website for this book contains study aids and many ideas for further reading and research. They include a chapter summary, review questions for the entire chapter, flash cards for key terms and concepts, a multiple-choice practice quiz, interesting Internet sites, references, and a guide for accessing thousands of InfoTrac® College Edition articles. Log on to

http://biology.brookscole.com/miller14

Then click on the Chapter-by-Chapter area, choose Chapter 21, and select a learning resource.

CASE STUDY

Learning Nature's Ways to Purify Sewage

Some communities and individuals are seeking better ways to purify sewage by working with nature. Ecologist John Todd designs, builds, and operates innovative ecological wastewater treatment systems called *living machines* (Figure 22-1).

This ecological purification process begins when sewage flows into a passive solar greenhouse or outdoor sites containing rows of large open tanks populated by an increasingly complex series of organisms. In the first set of tanks, algae and microorganisms decompose organic wastes, with sunlight speeding up the process. Water hyacinths, cattails, bulrushes, and other aquatic plants growing in the tanks take up the resulting nutrients.

After flowing though several of these natural purification tanks, the water passes through an artificial marsh of sand, gravel, and bulrush plants to filter out algae and remaining organic wastes. Some of the plants also absorb (sequester) toxic metals such as lead and mercury and secrete natural antibiotic compounds that kill pathogens.

Next the water flows into aquarium tanks. Snails and zooplankton in these tanks consume microorganisms and are in turn consumed by crayfish, tilapia, and other fish that can be eaten or sold as bait. After 10 days, the clear water flows into a second artificial marsh for final filtering and cleansing.

The water can be made pure enough to drink by exposing it to ultraviolet light or by passing it through an ozone generator, usually immersed out of sight in an attractive pond or wetland habitat. Selling the ornamental plants, trees, and baitfish produced as by-products of such living machines helps reduce costs. Operating costs are about the same as for a conventional sewage treatment plant.

Some communities and industries are working with nature by using natural and artificial wetlands to purify wastewater, as discussed later in this chapter.

Water pollution is related to air pollution, land-use practices, climate change, energy use, solid and hazardous waste, and the number of people, farms, and industries producing sewage and other wastes. These connections explain the need to solve water pollution problems by integrating them with policies for the problems just listed. Otherwise, environmentalists warn we will continue to shift environmental problems from one part of the environment to another.

Ocean Arks International

Figure 22-1 Solution: Ecological wastewater purification by a *living machine.* At the Providence, Rhode Island, Solar Sewage Treatment Plant, biologist John Todd demonstrates how ecological waste engineering in a greenhouse can be used to purify wastewater in an ecological process he invented. Todd and others are conducting research to perfect solar-aquatic sewage treatment systems based on working with nature.

Today everybody is downwind or downstream from somebody else.

WILLIAM RUCKELSHAUS

This chapter addresses the following questions:

- What pollutes water, where do the pollutants come from, and what effects do they have?
- What are the major water pollution problems of streams and lakes?
- What causes groundwater pollution, and how can it be prevented?
- What are the major water pollution problems of oceans?
- How can we prevent and reduce surface water pollution?
- How safe is drinking water, and how can it be made safer?

22-1 TYPES, EFFECTS, AND SOURCES OF WATER POLLUTION

What Are the Major Types and Effects of Water Pollutants? Unseen Threats

Infectious bacteria, inorganic and organic chemicals, and excess heat pollute water.

Water pollution is any chemical, biological, or physical change in water quality that has a harmful effect on living organisms or that makes water unsuitable for desired uses. Table 22-1 lists the major classes of water pollutants along with their major human sources and harmful effects. Study this table carefully. Note that excessive heat is considered a water pollutant.

Table 22-2 lists some common diseases that can be transmitted to humans through drinking water contaminated with infectious agents. The World Health Organization (WHO) estimates that 3.4 million people

---

Table 22-1 Major Categories of Water Pollutants

INFECTIOUS AGENTS

Examples: Bacteria, viruses, protozoa, and parasitic worms
Major Human Sources: Human and animal wastes
Harmful Effects: Disease

OXYGEN-DEMANDING WASTES

Examples: Organic waste such as animal manure and plant debris that can be decomposed by aerobic (oxygen-requiring) bacteria
Major Human Sources: Sewage, animal feedlots, paper mills, and food processing facilities
Harmful Effects: Large populations of bacteria decomposing these wastes can degrade water quality by depleting water of dissolved oxygen. This causes fish and other forms of oxygen-consuming aquatic life to die.

INORGANIC CHEMICALS

Examples: Water-soluble (1) acids, (2) compounds of toxic metals such as lead (Pb), arsenic (As), and selenium (Se), and (3) salts such as sodium chloride (NaCl) in ocean water and fluorides (F$^-$) found in some soils

Major Human Sources: Surface runoff, industrial effluents, and household cleansers
Harmful Effects: Can (1) make fresh water unusable for drinking or irrigation, (2) cause skin cancers and crippling spinal and neck damage (F$^-$), (3) damage the nervous system, liver, and kidneys (Pb and As), (4) harm fish and other aquatic life, (5) lower crop yields, and (6) accelerate corrosion of metals exposed to such water.

ORGANIC CHEMICALS

Examples: Oil, gasoline, plastics, pesticides, cleaning solvents, detergents
Major Human Sources: Industrial effluents, household cleansers, surface runoff from farms and yards
Harmful Effects: Can (1) threaten human health by causing nervous system damage (some pesticides), reproductive disorders (some solvents), and some cancers (gasoline, oil, and some solvents) and (2) harm fish and wildlife.

PLANT NUTRIENTS

Examples: Water-soluble compounds containing nitrate

(NO$_3^-$), phosphate (PO$_4^{3-}$), and ammonium (NH$_4^+$) ions
Major Human Sources: Sewage, manure, and runoff of agricultural and urban fertilizers
Harmful Effects: Can cause excessive growth of algae and other aquatic plants, which die, decay, deplete water of dissolved oxygen, and kill fish. Drinking water with excessive levels of nitrates lowers the oxygen-carrying capacity of the blood and can kill unborn children and infants ("blue-baby syndrome").

SEDIMENT

Examples: Soil, silt
Major Human Sources: Land erosion
Harmful Effects: Can (1) cloud water and reduce photosynthesis, (2) disrupt aquatic food webs, (3) carry pesticides, bacteria, and other harmful substances, (4) settle out and destroy feeding and spawning grounds of fish, and (5) clog and fill lakes, artificial reservoirs, stream channels, and harbors.

RADIOACTIVE MATERIALS

Examples: Radioactive isotopes of iodine, radon,

uranium, cesium, and thorium
Major Human Sources: Nuclear and coal-burning power plants, mining and processing of uranium and other ores, nuclear weapons production, natural sources
Harmful Effects: Genetic mutations, miscarriages, birth defects, and certain cancers

HEAT (THERMAL POLLUTION)

Examples: Excessive heat
Major Human Sources: Water cooling of electric power plants and some types of industrial plants. Almost half of all water withdrawn in the United States each year is for cooling electric power plants.
Harmful Effects: Lowers dissolved oxygen levels and makes aquatic organisms more vulnerable to disease, parasites, and toxic chemicals. When a power plant first opens or shuts down for repair, fish and other organisms adapted to a particular temperature range can be killed by the abrupt change in water temperature—known as *thermal shock.*

Table 22-2 Common Diseases Transmitted to Humans Through Contaminated Drinking Water

Type of Organism	Disease	Effects
Bacteria	Typhoid fever	Diarrhea, severe vomiting, enlarged spleen, inflamed intestine; often fatal if untreated
	Cholera	Diarrhea, severe vomiting, dehydration; often fatal if untreated
	Bacterial dysentery	Diarrhea; rarely fatal except in infants without proper treatment
	Enteritis	Severe stomach pain, nausea, vomiting; rarely fatal
Viruses	Infectious hepatitis	Fever, severe headache, loss of appetite, abdominal pain, jaundice, enlarged liver; rarely fatal but may cause permanent liver damage
Parasitic protozoa	Amoebic dysentery	Severe diarrhea, headache, abdominal pain, chills, fever; if not treated can cause liver abscess, bowel perforation, and death
	Giardiasis	Diarrhea, abdominal cramps, flatulence, belching, fatigue
Parasitic worms	Schistosomiasis	Abdominal pain, skin rash, anemia, chronic fatigue, and chronic general ill health

die prematurely each year from waterborne diseases. This means that during your lunch hour about 400 people died from such diseases. Each year, diarrhea alone kills about 1.9 million people—about 90% of them children under 5 in developing countries. The number of children killed by largely preventable diarrhea in the past 10 years is greater than the number of people killed in all armed conflicts since World War II.

In the United States, an estimated 1.5 million people a year become ill from infectious agents found in water and food. For example, in 1993 a protozoan parasite called *Cryptosporidium* contaminated the public drinking water supply in Milwaukee, Wisconsin. About 370,000 people developed severe diarrhea and at least 100 people with weakened immune systems died.

How Do We Measure Water Quality? Biology and Chemistry in Action

Scientists monitor water quality by using bacterial counts, chemical analysis, and indicator organisms.

Scientists use a number of biological and chemical methods to measure water quality. One involves measuring the number of colonies of *fecal coliform bacteria* (such as various strains of *Escherichia coli*) present in a water sample (Figure 22-2). Various strains of these bacteria live in the colon or intestines of humans and other animals and thus are present in their fecal wastes. Although most strains of coliform bacteria do not cause disease, their presence indicates that water has been exposed to human or animal wastes that are likely to contain disease-causing agents.

To be considered safe for drinking, water should contain no colonies of coliform bacteria in a sample of 100 milliliters (about 1/2 cup). To be considered safe for swimming, it should have no more than 200 colonies per 100 milliliters. By contrast, raw sewage may contain several million coliform bacterial colonies in 100 milliliters of water.

When dangerous levels of fecal coliform bacteria are detected scientists try to determine whether the source is from humans, various types of livestock, or wild animals such as birds or raccoons. A new field of science called *bacterial source tracking (BST)* uses molecular biology techniques to determine subtle differences in strains of *E. coli* based on their animal host.

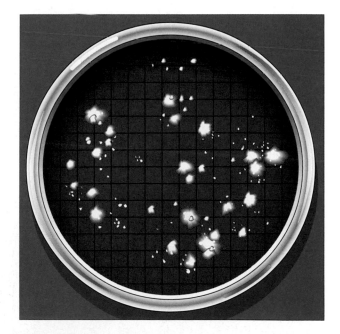

Figure 22-2 *Fecal coliform bacteria* test is used to indicate the likely presence of disease-causing bacteria in water. It is carried out by passing a water sample through a filter, placing the filter disk on a growth medium that supports coliform bacteria (such as *E. coli*) for 24 hours, and then counting the number of colonies of coliform bacteria (shown as clumps in the figure).

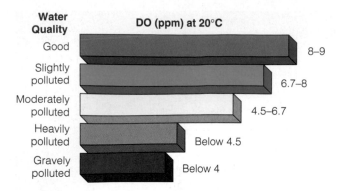

Figure 22-3 Water quality and dissolved oxygen (DO) content in parts per million (ppm) at 20°C (68°F). Only a few fish species can survive in water with less than 4 ppm of dissolved oxygen at this temperature.

The level of *dissolved oxygen* is related to the amount of oxygen-demanding wastes, so called because they are broken down by oxygen-requiring bacteria, and plant nutrients in a sample of water (Figure 22-3). Scientists also measure the *biological oxygen demand (BOD)*, the amount of dissolved oxygen consumed by aquatic decomposers.

They also use *chemical analysis* to determine the presence and concentrations of inorganic and organic chemicals that pollute water. They measure sediment content by evaporating the water in a sample and weighing the resulting sediment. Suspended sediment clouds water. Scientists use an instrument called a col-

orimeter to measure the turbidity or clarity (transparency) of a water sample.

Scientists can also monitor water pollution by using living organisms as *indicator species*. For example, they remove aquatic plants such as cattails and analyze them to determine pollution in areas contaminated with fuels, solvents, and other organic chemicals. Bottom-dwelling species such as mussels that feed by filtering water through their bodies can also be analyzed to determine water quality.

Genetic engineers are working to develop bacteria and yeasts (single-celled fungi) that fluoresce or glow in the presence of specific pollutants such as toxic heavy metals in the ocean, toxins in the air from chemical weapons, and carcinogens in food. This development of *biomonitors* or *biosensors* is a rapidly growing field that might interest you as a career choice.

What Are Point and Nonpoint Sources of Water Pollution? Concentrated and Diffuse Sources

Water pollution can come from single sources or a variety of dispersed sources.

Point sources discharge pollutants at specific locations through drain pipes, ditches, or sewer lines into bodies of surface water (Figure 22-4). Examples include factories, sewage treatment plants (which remove some but not all pollutants), underground mines, and oil tankers.

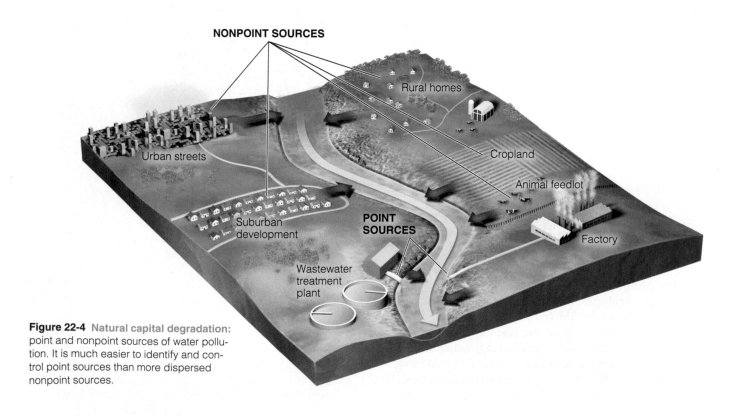

Figure 22-4 Natural capital degradation: point and nonpoint sources of water pollution. It is much easier to identify and control point sources than more dispersed nonpoint sources.

Because point sources are at specific places, they are easy to identify, monitor, and regulate. Most developed countries control point-source discharges of many harmful chemicals into aquatic systems. But there is little control of such discharges in most developing countries.

Nonpoint sources are scattered and diffuse and cannot be traced to any single site of discharge (Figure 22-4). Examples include acid deposition and runoff of chemicals into surface water from croplands, livestock feedlots, logged forests, urban streets, lawns, golf courses, and parking lots. There has been little progress in controlling water pollution from nonpoint sources because of the difficulty and expense of identifying and controlling discharges from so many diffuse sources.

What Are the Major Sources of Water Pollution? Supplying Food and Goods

The leading sources of water pollution are agriculture, industries, and mining.

Agricultural activities are by far the leading cause of water pollution. Sediment eroded from agricultural lands and overgrazed rangeland is the largest source. Other major agricultural pollutants include fertilizers and pesticides, bacteria from livestock and food processing wastes, and excess salt from soils of irrigated cropland.

Industrial facilities are another large source of water pollution. *Mining* is a third source. Surface mining disturbs the earth's surface, creating a major source of eroded sediments and runoff of toxic chemicals. Acidic compounds draining from active and abandoned subsurface and surface mines into streams can kill fish and other aquatic life (Figure 16-14, p. 344).

Is the Water Safe to Drink? For Most but Not All

One of every five people in the world lack access to safe drinking water.

Good news. About 95% of the people in developed countries and 74% of those in developing countries have access to clean drinking water.

Bad news. According to the WHO, about 1.4 billion people in developing countries do not have access to clean drinking water. As a result, each day about 9,300 people die prematurely from infectious diseases spread by contaminated water or the lack of water for adequate hygiene.

The United Nations estimates it would cost about $23 billion a year over 8–10 years to bring low-cost safe water and sanitation to the people in the world who do not have it. If developed countries paid half of that cost, it would amount to an average of $19 a year for each person in such countries.

Connections: How Might Projected Climate Change Affect Water Quality? More Pollution

In a warmer world, too much rain and too little rain can increase water pollution.

Global warming projections include changes in precipitation: some areas will get much more precipitation and other areas will get less. A moisture-laden atmosphere generates more intense downpours, which can flush more harmful chemicals, plant nutrients, and microorganisms into waterways. Massive flooding can spread disease-carrying pathogens by contaminating water treatment facilities and wells. It can also cause lagoons that store animal wastes, as well sewer lines that carry both sewage and storm runoff, to overflow and release raw sewage into rivers and streams.

Prolonged drought can reduce river flows that dilute wastes. It can also spread infectious diseases more rapidly among people who lack enough water to stay clean. Warmer water temperatures can threaten aquatic life by reducing dissolved oxygen levels, and can increase the growth rates of populations of harmful bacteria.

22-2 POLLUTION OF FRESHWATER STREAMS

What Are the Water Pollution Problems of Streams? Pollution Overload and Low Flow

Flowing streams can recover from a moderate level of degradable water pollutants if their flows are not reduced.

Rivers and other flowing streams can recover rapidly from moderate levels of degradable, oxygen-demanding wastes and excess heat through a combination of dilution and biodegradation of such wastes by bacteria. But this natural recovery process does not work when streams are overloaded with pollutants or when drought, damming, or water diversion for agriculture and industry reduce their flows. Also, these natural dilution and biodegradation processes do not eliminate slowly degradable and nondegradable pollutants.

In a flowing stream, the breakdown of degradable wastes by bacteria depletes dissolved oxygen and creates an *oxygen sag curve* (Figure 22-5, p. 496). This reduces or eliminates populations of organisms with high oxygen requirements until the stream is cleansed of wastes.

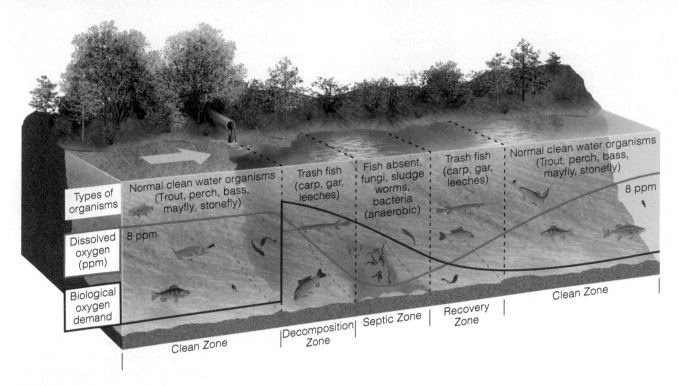

Figure 22-5 Natural capital: dilution and decay of degradable, oxygen-demanding wastes and heat in a stream, showing the oxygen sag curve (blue) and the curve of oxygen demand (red). Depending on flow rates and the amount of pollutants, streams recover from oxygen-demanding wastes and heat if they are given enough time and are not overloaded.

The depth and width of the oxygen sag curve and thus the time and distance needed for a stream to recover depend on several factors. They include the volume of incoming degradable wastes and the stream's volume, flow rate, temperature, and pH level. Similar oxygen sag curves can be plotted when heated water from industrial and power plants is discharged into streams.

What Have Developed Countries Done to Reduce Stream Pollution? Good and Bad News

Most developed countries have sharply reduced point-source pollution, but toxic chemicals and pollution from nonpoint sources are still problems.

Water pollution control laws enacted in the 1970s have greatly increased the number and quality of waste-water treatment plants in the United States and most other developed countries. Such laws also require industries to reduce or eliminate point-source discharges into surface waters.

These efforts have enabled the United States to hold the line against increased pollution by disease-causing agents and oxygen-demanding wastes in most of its streams. This is an impressive accomplishment given the rise in the country's economic activity, re-source consumption, and population since passage of these laws.

One success story is the cleanup of Ohio's Cuyahoga River. It was so polluted that in 1959 and again in 1969 it caught fire and burned for several days as it flowed through Cleveland. The highly publicized image of this burning river prompted elected officials to enact laws limiting the discharge of industrial wastes into the river and sewage systems and provide funds to upgrade sewage treatment facilities. Today the river is cleaner and is widely used by boaters and anglers. This accomplishment illustrates the power of bottom-up pressure by citizens to spur elected officials to change a severely polluted river into an economically and ecologically valuable public resource. Individuals matter!

Another spectacular cleanup occurred in Great Britain. In the 1950s, the Thames River was little more than a flowing anaerobic sewer. Now, after more than 45 years of effort and hundreds of millions of dollars spent by British taxpayers and private industry, the Thames has made a remarkable recovery. Commercial fishing is thriving and the number of fish species has increased 20-fold since 1960. In addition, many species of waterfowl and wading birds have returned to their former feeding grounds.

There is also some *bad news*. Large fish kills and drinking water contamination still occur in parts of developed countries. Two causes of these problems are accidental or deliberate releases of toxic inorganic and organic chemicals by industries or mines and malfunctioning sewage treatment plants. A third cause is nonpoint runoff of pesticides and excess plant nutrients from cropland and animal feedlots.

What Have Developing Countries Done to Reduce Stream Pollution? Little Progress

Stream pollution in most developing countries is a serious and growing problem.

Available data indicate that stream pollution from discharges of untreated sewage and industrial wastes is a serious and growing problem in most developing countries. According to a 2003 report by the World Commission on Water in the 21st Century, half of the world's 500 major rivers are heavily polluted, most of them running through developing countries. Most of these countries cannot afford to build waste treatment plants and do not have or do not enforce laws for controlling water pollution.

Industrial wastes and sewage pollute more than two-thirds of India's water resources (Case Study, below) and 54 of the 78 streams monitored in China. Only about 10% of the sewage produced in Chinese cities is treated. In Latin America and Africa, most streams passing through urban or industrial areas suffer from severe pollution.

Case Study: India's Ganges River: Religion, Poverty, and Health

Religious beliefs, cultural traditions, poverty, little economic development, and a large population interact to cause severe pollution of the Ganges River in India.

To India's Hindu people, the Ganges is a holy river. Each day more than 1 million Hindus bathe or take a "holy dip" in the river. Many people also drink its water and use it to wash their clothes.

Bad news. The Ganges is highly polluted. About 350 million people—one-third of the country's population—live in the Ganges River basin. Very little of the sewage produced by these people and by the industries and 29 large cities in the basin is treated.

This situation is complicated by the Hindu belief in cremating the dead to free the soul and throwing the ashes in the holy Ganges to increase the chances of the soul getting into heaven. Traditionally, wood fires burn most bodies in the open air. This creates air pollution and helps deplete India's forests.

It also causes water pollution because many people cannot afford enough wood for cremation. As a result, many bodies are dumped into the river without cremation or are only partially burned. Decomposition of these bodies depletes dissolved oxygen and adds disease-carrying bacteria and viruses to the water. This problem is expected to get worse because about 19 million people are added to India's population each year—about a third of them to the Ganges River basin.

Good news. The Indian government has launched a plan to help clean up the river. It involves building waste treatment plants in the basin's 29 large cities and constructing 32 electric crematoriums along the banks of the river that can burn bodies more efficiently and at a lower cost than wood cremation. The government also introduced 25,000 snapping turtles to devour corpses.

Bad news. Most of the sewage treatment plants are not completed or do not work very well and only a few of the crematoriums have been completed. There is also concern that many Hindus will not abandon the traditional ritual of wood cremation or will not be able to afford any type of cremation.

This situation shows how religious and cultural conditions and poverty can affect environmental problems and solutions to such problems.

22-3   POLLUTION OF FRESHWATER LAKES

Why Are Lakes and Reservoirs More Vulnerable to Pollution than Most Streams? Too Little Flow and Mixing

Dilution of pollutants in lakes is less effective than in most streams because most lake water is not mixed well and has little flow.

In lakes and reservoirs, dilution of pollutants often is less effective than in streams for two reasons. One is that lakes and reservoirs often contain stratified layers that undergo little vertical mixing. The other is that they have little flow. The flushing and changing of water in lakes and large artificial reservoirs can take from 1 to 100 years, compared with several days to several weeks for streams.

This means that lakes and reservoirs are more vulnerable than streams to contamination by runoff or discharge of plant nutrients, oil, pesticides, and toxic substances such as lead, mercury, and selenium. These contaminants can kill bottom life and fish and birds that feed on contaminated aquatic organisms. Many toxic chemicals and acids also enter lakes and reservoirs from the atmosphere.

As they pass through food webs in lakes, the concentrations of some chemicals can be biologically magnified. Examples include DDT (Figure 19-4, p. 411), PCBs (Figure 22-6, p. 498), some radioactive isotopes, and some mercury compounds.

Figure 22-6 *Biological magnification* of PCBs (polychlorinated biphenyls) in an aquatic food chain in the Great Lakes. Most of the 209 different PCBs are insoluble in water, soluble in fats, and resistant to biological and chemical degradation—properties that result in their accumulation in the tissues of organisms and their biological amplification in food chains and webs. Although the long-term health effects on people exposed to low levels of PCBs are unknown, high doses of PCBs in laboratory animals produce liver and kidney damage, gastric disorders, birth defects, skin lesions, hormonal changes, smaller penis size, and tumors. Boys in Taiwan exposed to PCBs while in their mothers' wombs developed abnormally small penises. In the United States, manufacture and use of PCBs have been banned since 1976. Before then, millions of metric tons of these long-lived chemicals were released into the environment. Many of them still exist in bottom sediments of lakes, streams, and oceans.

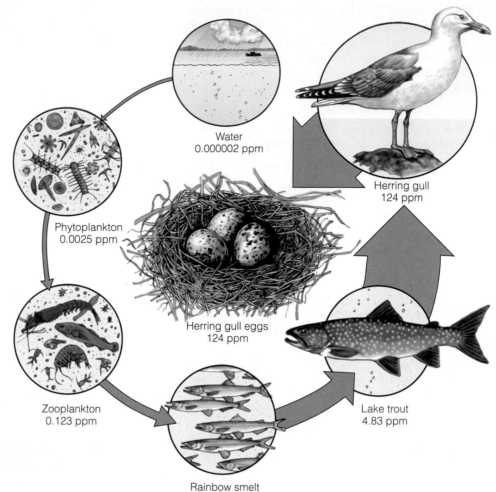

Water
0.000002 ppm

Phytoplankton
0.0025 ppm

Herring gull
124 ppm

Herring gull eggs
124 ppm

Zooplankton
0.123 ppm

Lake trout
4.83 ppm

Rainbow smelt
1.04 ppm

What Is Cultural Eutrophication and How Can It Be Reduced? Too Much of a Good Thing

Various human activities can overload lakes with plant nutrients, which decrease dissolved oxygen and kill some aquatic species.

Eutrophication is the name given to the natural nutrient enrichment of lakes, mostly from runoff of plant nutrients such as nitrates and phosphates from surrounding land. Over time, some lakes become more eutrophic (Figure 7-17, right, p. 139), but others do not because of differences in the surrounding drainage basins.

An increase in plant nutrients can be beneficial to populations of floating phytoplankton that feed aquatic organisms. In turn, this can increase the growth rate and abundance of some fish and other desirable species.

But excessive inputs of nutrients can upset aquatic ecosystems. Near urban or agricultural areas, human activities can greatly accelerate the input of plant nutrients to a lake. This process is called **cultural eutrophication.** It is mostly nitrate- and phosphate-containing effluents from various sources that cause such a change (Figure 22-7).

During hot weather or drought, this nutrient overload produces dense growths or "blooms" of organisms such as algae and cyanobacteria and thick growths of water hyacinths, duckweed, and other aquatic plants. These dense colonies of plant life can reduce lake productivity and fish growth by decreasing the input of solar energy needed for photosynthesis.

In addition, when the algae die, their decomposition by swelling populations of aerobic bacteria depletes dissolved oxygen in the surface layer of water near the shore and in the bottom layer. This oxygen depletion can kill fish and other aerobic aquatic animals. If excess nutrients continue to flow into a lake, anaerobic bacteria take over and produce gaseous decomposition products such as smelly, highly toxic hydrogen sulfide and flammable methane.

According to the U.S. Environmental Protection Agency (EPA), about one-third of the 100,000 medium to large lakes and about 85% of the large lakes near major population centers in the United States have some

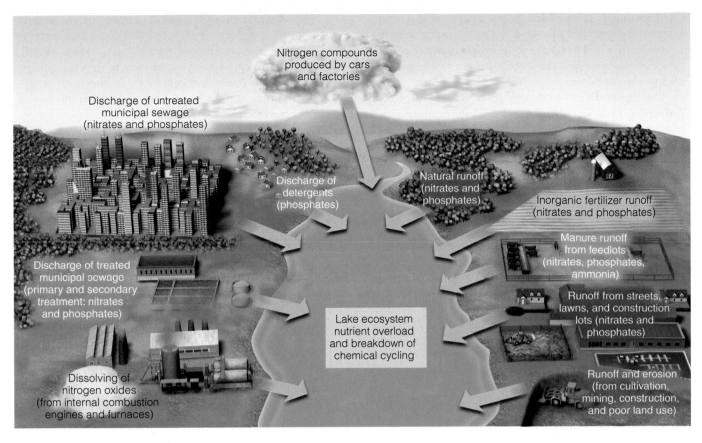

Figure 22-7 Natural capital degradation: principal sources of nutrient overload causing *cultural eutrophication* in lakes and coastal areas. The amount of nutrients from each source varies according to the types and amounts of human activities occurring in each airshed and watershed. The enlarged populations of algae and plants (stimulated by increased nutrient input) die. Then their decomposition by aerobic bacteria lowers levels of dissolved oxygen. This can kill fish and other aquatic life and reduce biodiversity and the aesthetic and recreational value of the lake.

degree of cultural eutrophication. One-fourth of the lakes in China also suffer from cultural eutrophication.

Cultural eutrophication also occurs in marine ecosystems, especially in coastal waters and partially enclosed estuaries and bays. It also affects enclosed seas, such as the Mediterranean, Baltic, and Black Seas.

There are several ways to *prevent* or *reduce* cultural eutrophication. They include using advanced (but expensive) waste treatment systems to remove nitrates and phosphates before wastewater enters lakes, banning or limiting the use of phosphates in household detergents and other cleaning agents, and using soil conservation and land-use control to reduce nutrient runoff.

There are also several ways to *clean up* lakes suffering from cultural eutrophication. Examples are mechanically removing excess weeds, controlling undesirable plant growth with herbicides and algicides, and pumping air through lakes and reservoirs to avoid oxygen depletion (an expensive and energy-intensive method).

As usual, pollution prevention is more effective and usually is cheaper in the long run than cleanup.

Good news. If excessive inputs of plant nutrients are stopped, a lake can usually return to its previous state (see the two case studies that follow).

Case Study: Lake Washington—A Success Story

Lake Washington near Seattle has recovered from severe cultural eutrophication.

Lake Washington in the metropolitan area of Seattle is a success story of recovery from severe cultural eutrophication caused by decades of sewage and other inputs.

Recovery took place within about 4 years after the sewage was diverted into the nearby Puget Sound. This worked for three reasons. *First,* a large body of water (Puget Sound) with a rapid rate of exchange involving Pacific Ocean waters was available to receive and dilute the sewage wastes. *Second,* the lake had not yet filled with weeds and sediment, because of its large size and depth. *Third,* preventive action was taken before the lake had become a shallow, highly eutrophic lake. Today, the lake's water quality is good.

Bad news. Now the Puget Sound is in trouble. There is growing concern about increased urban runoff caused by the area's rapidly growing population, overflows of raw sewage, and large inputs of toxic materials into the sound.

Taking pollution from one place (Lake Washington) and putting it somewhere else (Puget Sound) is an output approach that can be overwhelmed by a combination of more people and more wastes. The way out is to prevent most wastes from reaching either of these two bodies of water.

Case Study: Pollution in the Great Lakes— Hopeful Progress

Pollution of the Great Lakes has dropped significantly but there is a long way to go.

The five interconnected Great Lakes of North America (Figure 22-8) formed about 10,500 years ago when retreating glaciers melted and poured water into the land basins carved out by the slowly moving glaciers. These lakes contain at least 95% of the fresh surface water in the United States and one-fifth of the world's fresh surface water.

The Great Lakes basin is also home for about 30% of the Canadian population and 14% of the U.S. population. At least 38 million people obtain their drinking water from these lakes.

Despite their enormous size, these lakes are vulnerable to pollution from point and nonpoint sources. One reason is that less than 1% of the water entering these lakes flows out to the St. Lawrence River each year. Another reason is that in addition to land runoff these lakes get atmospheric deposition of large quanti-

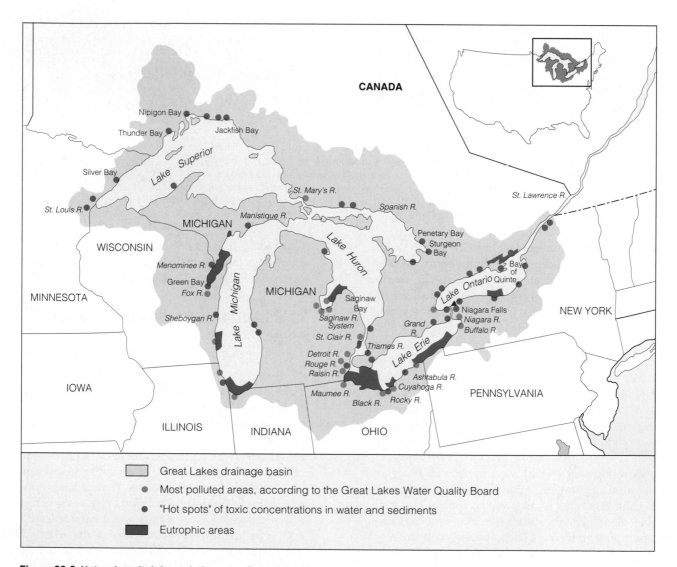

Figure 22-8 Natural capital degradation: the *Great Lakes basin* and the locations of some of its water quality problems. The Great Lakes region is dotted with several hundred abandoned toxic waste sites listed by the EPA as Superfund sites to receive cleanup priority. (Data from U.S. Environmental Protection Agency)

ties of acids, pesticides, and other toxic chemicals, often blown in from hundreds or thousands of kilometers away.

By the 1960s, many areas of the Great Lakes were suffering from severe cultural eutrophication, huge fish kills, and contamination from bacteria and a variety of toxic industrial wastes. The impact on Lake Erie was particularly intense because it is the shallowest of the Great Lakes and has the highest concentrations of people and industrial activity along its shores. Many bathing beaches had to be closed, and by 1970 the lake had lost most of its native fish.

Since 1972, Canada and the United States have joined forces and spent more than $20 billion on a Great Lakes pollution control program. This program has decreased algal blooms, increased dissolved oxygen levels and sport and commercial fishing catches in Lake Erie, and allowed most swimming beaches to reopen.

These improvements occurred mainly because of new or upgraded sewage treatment plants, better treatment of industrial wastes, and bans on use of detergents, household cleaners, and water conditioners that contained phosphates.

Despite this important progress many problems remain. Each August a large zone severely depleted of dissolved oxygen is likely to stretch across the center of Lake Erie. The oxygen-poor water in this zone kills fish and microorganisms that support the lake's food web. During the last 10 years, the time that the zone lasts has increased from two weeks to a month and scientists do not know why. Possible causes include oxygen depletion by zebra mussels (Case Study, p. 267), undetected inputs of phosphates from fertilizers through storm runoff sewers, an unknown naturally occurring cycle, or climate change.

More bad news. According to a 2000 survey by the EPA, more than three-fourths of the shoreline of the Great Lakes is not clean enough for swimming or for supplying drinking water. The EPA and Environment Canada have identified 43 highly polluted shoreline areas. Nonpoint land runoff of pesticides and fertilizers from urban sprawl now surpasses industrial pollution as the greatest threat to the lakes. Sediments in 26 toxic hot spots (Figure 22-8) remain heavily polluted.

About half of the toxic compounds entering the lakes come from atmospheric deposition of pesticides, mercury from coal-burning plants, and other toxic chemicals from as far away as Mexico and Russia. Toxic chemicals such as PCBs have built up in food chains and webs (Figure 22-6), contaminating many types of sport fish and depleting populations of birds, river otters, and other animals feeding on contaminated fish. A recent survey by Wisconsin biologists found that one fish in four taken from the Great Lakes is unsafe for human consumption. Another problem

has been an 80% drop in EPA funding for cleanup of the Great Lakes since 1992.

Some environmentalists call for banning the use of toxic chlorine compounds such as bleach in the pulp and paper industry around the Great Lakes. They would also ban new incinerators (which can release toxic chemicals into the atmosphere) in the area, and they would stop the discharge into the lakes of 70 toxic chemicals that threaten human health and wildlife. Officials in the industries involved have successfully opposed such bans.

22-4 POLLUTION OF GROUNDWATER

Why Is Groundwater Pollution Such a Serious Problem? Not Easily Cleaned

Groundwater can become contaminated with a variety of chemicals because it cannot effectively cleanse itself and dilute and disperse pollutants.

According to many scientists, a serious threat to human health is the out-of-sight pollution of groundwater, a prime source of water for drinking and irrigation. Studies show that groundwater pollution comes from numerous sources (Figure 22-9, p. 502). People who dump or spill gasoline, oil, and paint thinners and other organic solvents onto the ground also contaminate groundwater.

Although experts rate groundwater pollution as a low-risk ecological problem, they consider pollutants in drinking water (much of it from groundwater) a high-risk health problem. Once a pollutant from a leaking underground tank or other source contaminates groundwater it permeates the nearby porous layers of sand, gravel, or bedrock in the aquifer like water saturating a sponge. This makes removal of the contaminant difficult and costly.

Then the contaminated water slowly flows through the aquifer and creates a widening *plume* of contaminated water. If this plume reaches a well used to extract groundwater, the polluted water can get into drinking water and into water used to irrigate crops.

When groundwater becomes contaminated, it cannot cleanse itself of *degradable wastes* as flowing surface water does (Figure 22-5). One reason is that groundwater flows so slowly—usually less than 0.3 meter or 1 foot per day—that contaminants are not diluted and dispersed effectively. Another problem is that groundwater usually has much lower concentrations of dissolved oxygen (which helps decompose many contaminants) and smaller populations of decomposing bacteria. Also, the usually cold temperatures of groundwater slow down chemical reactions that decompose wastes.

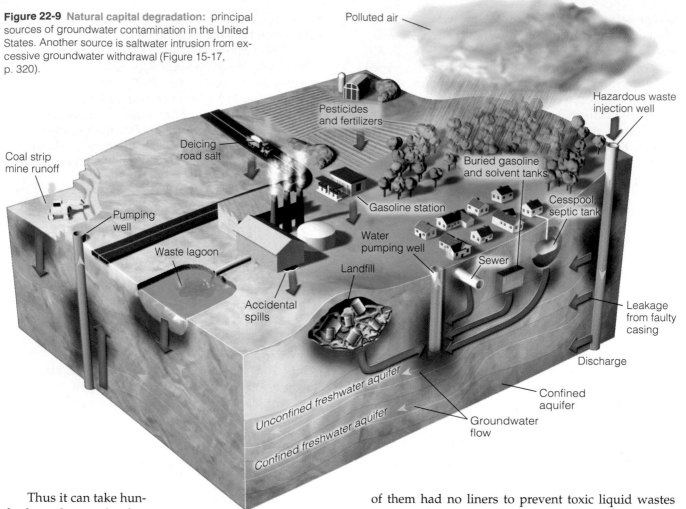

Figure 22-9 Natural capital degradation: principal sources of groundwater contamination in the United States. Another source is saltwater intrusion from excessive groundwater withdrawal (Figure 15-17, p. 320).

Polluted air

Hazardous waste injection well

Coal strip mine runoff

Deicing road salt

Pesticides and fertilizers

Pumping well

Buried gasoline and solvent tanks

Gasoline station

Cesspool, septic tank

Waste lagoon

Water pumping well

Sewer

Accidental spills

Landfill

Leakage from faulty casing

Discharge

Unconfined freshwater aquifer

Confined aquifer

Confined freshwater aquifer

Groundwater flow

Thus it can take hundreds to thousands of years for contaminated groundwater to cleanse itself of *degradable* wastes. On a human time scale, *nondegradable wastes* (such as toxic lead, arsenic, and fluoride) are there permanently.

What Is the Extent of Groundwater Pollution? Uncertain Overall but Serious in Some Areas

Leaks from chemical storage ponds, underground storage tanks and well piping, and seepage of agricultural fertilizers can contaminate groundwater.

On a global scale we do not know much about groundwater pollution because few countries go to the great expense of locating, tracking, and testing aquifers. But scientific studies in scattered parts of the world provide us with some *bad news*.

According to the EPA and the U.S. Geological Survey, one or more organic chemicals contaminate about 45% of *municipal* groundwater supplies in the United States. An EPA survey of 26,000 industrial waste ponds and lagoons in the United States found that one-third

of them had no liners to prevent toxic liquid wastes from seeping into aquifers. One-third of these sites are within 1.6 kilometers (1 mile) of a drinking water well.

The U.S. General Accounting Office estimated in 2002 that at least 76,000 underground tanks storing gasoline, diesel fuel, home heating oil, and toxic solvents were leaking their contents into groundwater in the United States (Figure 22-9). In California's Silicon Valley, where electronics industries use underground tanks to store a variety of organic solvents, local authorities found leaks in 85% of the tanks they inspected.

During this century, scientists expect many of the millions of underground tanks installed around the world in recent decades to corrode, leak, contaminate groundwater, and become a major global health problem. Determining the extent of a leak from a single underground tank can cost $25,000–250,000, and cleanup costs range from $10,000 to more than $250,000. If the chemical reaches an aquifer, effective cleanup is often not possible or is too costly. *Bottom line:* wastes we think we have thrown away or stored safely can escape and come back to haunt us.

According to the WHO, an estimated 70 million people in northern China and 30 million in northwestern India drink groundwater contaminated with high levels of naturally occurring *fluoride* (F⁻). This can cause crippling backbone and neck damage and a variety of dental problems.

Groundwater used as a source of drinking water can also be contaminated with *nitrate ions* (NO_3^-), especially in agricultural areas where nitrates in fertilizer can be leached into groundwater. Nitrite ions (NO_2^-) in the stomach, colon, and bladder can convert some of the nitrate ions in drinking water to organic compounds, which can cause cancer in various organs in more than 40 test animal species. The conversion of nitrates in tap water to nitrites in infants under 6 months old can cause a potentially fatal condition known as "blue baby syndrome," in which blood lacks the ability to carry sufficient oxygen to body cells.

Toxic *arsenic* (As) contaminates drinking water when a well is drilled into aquifers where soils and rock are naturally rich in arsenic. According to the WHO, more than 112 million people are drinking water with arsenic levels 5–100 times the WHO standard of 10 parts per billion (ppb). They include an estimated 30 million people in Bangladesh, 6 million in India's state of West Bengal, and 6 million in China. According to estimates by the WHO, long-term exposure to arsenic in drinking water is likely to cause 200,000–270,000 premature deaths from cancer of the skin, bladder, and lung in Bangladesh alone.

Good news. The United Nations Children's Fund (UNICEF) and several nongovernmental organizations in Bangladesh have started a program to evaluate wells serving several million people to identify those contaminated with arsenic and mark them with red paint.

Arsenic can also be released into the air and water by coal burning, copper and lead smelting, municipal trash incinerators, landfills containing arsenic-laden ash produced by coal-burning power plants, and use of certain arsenic-containing pesticides.

The international standard for arsenic in drinking water of 10 ppb was adopted in 1993 by the WHO and in 1998 by the European Union, and becomes the standard in the United States in 2006.

But according to the WHO and other scientists, even the 10-ppb standard is not safe. A 2001 study by the U.S. National Academy of Sciences found that routinely drinking water with arsenic levels of even 3 ppb poses a 1 in 1,000 risk of developing bladder or lung cancer. Many scientists call for lowering the standard to 3–5 ppb, but this would be very expensive.

Solutions: How Can We Protect Groundwater? Monitor and Say No

Prevention is the most effective and affordable way to protect groundwater from pollutants.

Figure 22-10 lists ways to prevent and clean up groundwater contamination. Treating a contaminated aquifer involves eliminating the source of pollution and drilling monitoring wells to determine how far, in what direction, and how fast the contaminated plume is moving. Then a computer model is used to project future dispersion of the contaminant in the aquifer. The final step is to develop and implement a strategy to clean up the contamination (Figure 22-10, right). This time-consuming and expensive process is somewhat like a blind surgeon trying to find and remove a cancer in your body before it grows too large.

Because of the difficulty and expense of cleaning up a contaminated aquifer, *preventing contamination is the most effective and cheapest way to protect groundwater resources* (Figure 22-10, left).

Underground tanks in the United States and a number of other developed countries are now strictly regulated. In the United States, thousands of old leaking tanks from gasoline stations and other facilities have been removed and the surrounding soil and groundwater have been treated to remove gasoline. This is expensive but there is little choice because the groundwater has already been contaminated.

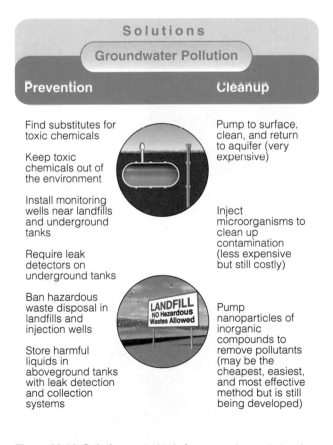

Figure 22-10 Solutions: methods for preventing and cleaning up contamination of groundwater. Which two of these solutions do you believe are the most important?

22-5 OCEAN POLLUTION

How Much Pollution Can the Oceans Tolerate? We Do Not Know

Oceans can disperse and break down large quantities of degradable pollutants if they are not overloaded.

The oceans can dilute, disperse, and degrade large amounts of raw sewage, sewage sludge, oil, and some types of degradable industrial waste, especially in deep-water areas. Also, some forms of marine life have been affected less by some pollutants than expected.

This has led some scientists to suggest it is safer to dump sewage sludge and most other harmful wastes into the deep ocean than to bury them on land or burn them in incinerators. Other scientists disagree, pointing out we know less about the deep ocean than we do

about the moon. They add that dumping harmful wastes in the ocean would delay urgently needed pollution prevention and promote further degradation of this vital part of the earth's life-support system.

How Do Pollutants Affect Coastal Areas? More People and Development Equal More Pollution

Pollution of coastal waters near heavily populated areas is a serious problem.

Coastal areas—especially wetlands and estuaries, coral reefs, and mangrove swamps—bear the brunt of our enormous inputs of pollutants and wastes into the ocean (Figure 22-11). This is not surprising because about 40% of the world's population lives on or within

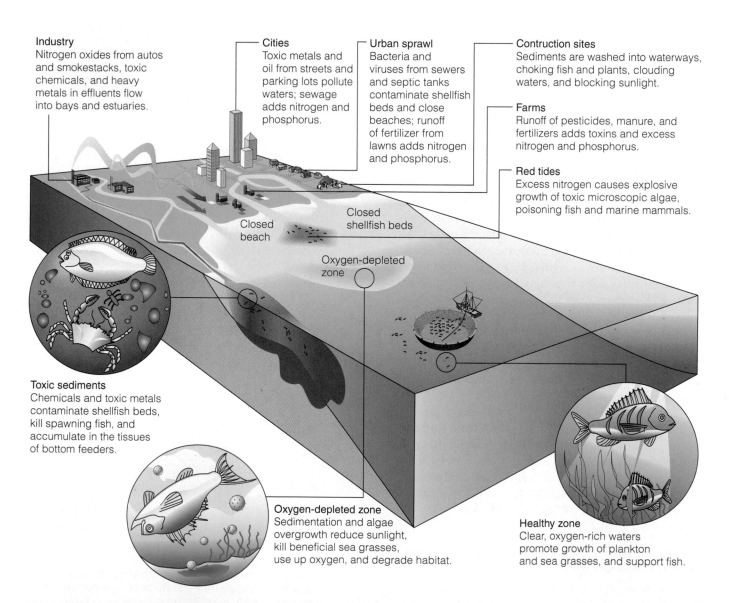

Industry
Nitrogen oxides from autos and smokestacks, toxic chemicals, and heavy metals in effluents flow into bays and estuaries.

Cities
Toxic metals and oil from streets and parking lots pollute waters; sewage adds nitrogen and phosphorus.

Urban sprawl
Bacteria and viruses from sewers and septic tanks contaminate shellfish beds and close beaches; runoff of fertilizer from lawns adds nitrogen and phosphorus.

Contruction sites
Sediments are washed into waterways, choking fish and plants, clouding waters, and blocking sunlight.

Farms
Runoff of pesticides, manure, and fertilizers adds toxins and excess nitrogen and phosphorus.

Red tides
Excess nitrogen causes explosive growth of toxic microscopic algae, poisoning fish and marine mammals.

Closed beach

Closed shellfish beds

Oxygen-depleted zone

Toxic sediments
Chemicals and toxic metals contaminate shellfish beds, kill spawning fish, and accumulate in the tissues of bottom feeders.

Oxygen-depleted zone
Sedimentation and algae overgrowth reduce sunlight, kill beneficial sea grasses, use up oxygen, and degrade habitat.

Healthy zone
Clear, oxygen-rich waters promote growth of plankton and sea grasses, and support fish.

Figure 22-11 Natural capital degradation: how residential areas, factories, and farms contribute to the pollution of coastal waters and bays. According to the UN Environment Programme, coastal water pollution costs the world $16 billion annually—$731,000 a minute—due to ill health and premature death.

100 kilometers (62 miles) of the coast and 14 of the world's 15 largest metropolitan areas (each with 10 million people or more) are near coastal waters.

In most coastal developing countries and in some coastal developed countries, municipal sewage and industrial wastes are dumped into the sea without treatment. For example, about 85% of the sewage from large cities along the Mediterranean Sea (with a coastal population of 200 million people during tourist season) is discharged into the sea untreated. This causes widespread beach pollution and shellfish contamination.

Recent studies of some U.S. coastal waters have found vast colonies of human viruses from raw sewage, effluents from sewage treatment plants (which do not remove viruses), and leaking septic tanks. According to one study, about one-fourth of the people using coastal beaches in the United States develop ear infections, sore throats, eye irritations, respiratory disease, or gastrointestinal disease.

Runoffs of sewage and agricultural wastes into coastal waters introduce large quantities of nitrate (NO_3^-) and phosphate (PO_4^{3-}) plant nutrients, which can cause explosive growth of harmful algae. These *harmful algal blooms* (HABs) are called red, brown, or green toxic tides, depending on their color. They can release waterborne and airborne toxins that damage fisheries, kill some fish-eating birds, reduce tourism, and poison seafood.

According to a 2004 report by the U.N. Environment Programme, each year some 150 large *oxygen-depleted zones* (sometimes inaccurately called dead zones) form mostly in temperate coastal waters and in landlocked seas such as the Baltic and Black Seas. These zones result from excessive nonpoint inputs of fertilizers and animal wastes from land runoff and deposition of nitrogen compounds from the atmosphere. This cultural eutrophication depletes dissolved oxygen. Without oxygen most of the aquatic life (except bacteria) dies or moves elsewhere. The biggest such zone in U.S. waters and the second largest in the world forms every summer in a narrow stretch of the Gulf of Mexico off the mouth of the Mississippi River (Figure 22-12).

Promising news. Scientists in Asia and elsewhere are experimenting with adding certain types of fine clay to the water to help control algal blooms. The idea is to find clay with particles fine and heavy enough to stick to the algae and remove them by weighing them down like microanchors. More tests are needed to determine the effectiveness of the method and to be sure that the clay particles do not harm other aquatic organisms.

The enclosed Baltic Sea is highly polluted because it receives runoff and air pollutants from a huge area with more than 70 million people and 15% of the world's industrial production. *Good news.* In 1980, the countries surrounding the Baltic Sea signed the *Helsinki Convention*—the world's first international agreement to reduce marine pollution. Despite some limitations,

Figure 22-12 Natural capital degradation: a large zone of oxygen-depleted water (less than 2 ppm dissolved oxygen) forms for half of the year in the Gulf of Mexico as a result of oxygen-depleting algal blooms. It is created mostly by huge inputs of nitrate (NO_3^-) and phosphate (PO_4^{3-}) plant nutrients from the massive Mississippi River basin. In 2002, the area of the zone (shown in green) was roughly equivalent to the area of the state of New Jersey. This problem is worsened by loss of wetlands, which help filter plant nutrients.

this agreement serves as an example of how countries can work together to help reduce common water pollution problems.

Preventive measures for reducing the number and size of such oxygen-depleted zones include *reducing nitrogen inputs from various sources* (Figure 22-7), *planting forest and grasslands to soak up excess nitrogen and keep it out of waterways, restoring coastal wetlands, improving sewage treatment to reduce discharge of nitrates into waterways, requiring further reduction of NO_x emissions from motor vehicles,* and *phasing in forms of renewable energy to replace the burning of fossil fuels.*

Case Study: The Chesapeake Bay: An Estuary in Trouble

Pollutants from six states contaminate the shallow Chesapeake Bay estuary, but cooperative efforts have reduced some of the pollution inputs.

Since 1960, the Chesapeake Bay—America's largest estuary—has been in serious trouble from water pollution, mostly because of human activities. One problem is that between 1940 and 2004, the number of people living in the Chesapeake Bay area grew from 3.7 million to 17 million, and may soon reach 18 million.

Another problem is that the estuary receives wastes from point and nonpoint sources scattered throughout a huge drainage basin that includes 9 large rivers and 141 smaller streams and creeks in parts of six states (Figure 22-13). Also, the bay is a huge pollution sink because it is quite shallow and only 1% of the waste entering it is flushed into the Atlantic Ocean.

Phosphate and nitrate levels have risen sharply in many parts of the bay, causing algal blooms and oxygen depletion. Commercial harvests of its once abundant oysters, crabs, and several important fish have fallen sharply since 1960 because of a combination of pollution, overfishing, and disease.

Studies show that point sources, primarily sewage treatment plants and industrial plants (often in violation of their discharge permits), account for about 60% by weight of the phosphates. Nonpoint sources—mostly runoff from urban, suburban, and agricultural land and deposition from the atmosphere—account for about 60% by weight of the nitrates.

In 1983, the United States implemented the Chesapeake Bay Program. It is the country's most ambitious attempt at *integrated coastal management* in which citizens' groups, communities, state legislatures, and the federal government have worked together to reduce pollution inputs into the bay. Strategies include establishing land-use regulations in the bay's six watershed states to reduce agricultural and urban runoff, banning phosphate detergents, upgrading sewage treatment plants, and better monitoring of industrial discharges. In addition, wetlands are being restored and large areas of the bay are being replanted with sea grasses to help filter out nutrients and other pollutants.

This hard work has paid off. Between 1985 and 2000, phosphorus levels declined 27%, nitrogen levels dropped 16%, and grasses growing on the bay's floor have made a comeback. This is a significant achievement given the increasing population in the watershed and the fact that nearly 40% of the nitrogen inputs come from the atmosphere.

However, there is still a long way to go, and a sharp drop in state and federal funding has slowed progress. According to a 2003 report prepared by the University of Maryland's School of Law, "After years of dialogue and billions in expenditures, the bay is no healthier than it was 10 years ago."

But despite some setbacks, the Chesapeake Bay Program shows what can be done when diverse groups work together to achieve goals that benefit both wildlife and people.

Solutions: Can Oysters Help Clean Up the Chesapeake Bay? Should We Bring Them Back?

Rebuilding the Chesapeake Bay's depleted oyster population with disease-resistant oysters could greatly reduce water pollution because oysters filter algae and silt from water.

Marine scientists are looking for ways to rebuild the Chesapeake Bay's once huge population of the eastern oyster as a way to help clean up the water. Oysters are filter feeders that vacuum up the algae and nutrient-laden suspended silt that cause many of the Chesapeake Bay's water pollution problems.

The bay's oyster population once served as a natural water purifier by filtering the bay's entire volume of water every 3 or 4 days. But overharvesting and two parasitic oyster diseases have reduced the oyster population to about 1% of its historic high. Consequently, today's oyster population needs about a year to filter the bay's water.

Computer models project that increasing the oyster population to only 10% of its historic high would improve water quality and spur the growth of underwater sea grass. Ways to do this include introducing disease-resistant Asian oysters, seeding beds with older oysters presumed to have some disease resistance, and setting aside 20–25% of the bay's oyster beds as sanctuaries to protect stocks from overharvesting.

In 2003, Virginia officials approved a plan to put 1 million Asian oysters in the bay. However, a 2003 study by the National Academy of Sciences recommended caution in introducing nonnative species into the Chesapeake Bay without carefully looking at the possible harmful ecological effects of such a project.

What Is Being Done to Control the Dumping of Pollutants into the Ocean? There Is No Away

Parts of the world's oceans are dump sites for a variety of toxic materials and sewage and garbage from ships.

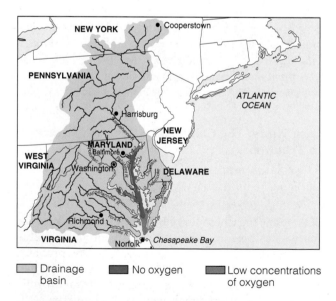

Figure 22-13 Natural capital degration: Chesapeake Bay, the largest estuary in the United States, is severely degraded as a result of water pollution from point and nonpoint sources in six states and from deposition of air pollutants.

Dumping industrial waste off U.S. coasts has stopped, although it still occurs in a number of other developed countries and some developing countries. But barges and ships still legally dump large quantities of **dredge spoils** (materials, often laden with toxic metals, scraped from the bottoms of harbors and rivers to maintain shipping channels) at 110 sites off the Atlantic, Pacific, and Gulf coasts of the United States.

Many countries also dump large quantities of *sewage sludge* into the ocean. It is a gooey mixture of toxic chemicals, infectious agents, and settled solids removed from wastewater at sewage treatment plants.

Good news. Since 1992, the United States has banned this practice. And 50 countries with at least 80% of the world's merchant fleet have agreed not to dump sewage and garbage at sea.

Bad news. This agreement is difficult to enforce and violations are common. Most ship owners save money by dumping wastes at sea and, if caught, get only small fines. Each year as many as 2 million seabirds and more than 100,000 marine mammals (including whales, seals, dolphins, and sea lions) die when they ingest or become entangled in fishing nets, ropes, and other debris dumped into the sea or discarded on beaches (see photo on p. viii).

Under the London Dumping Convention of 1972, 100 countries agreed not to dump highly toxic pollutants and high-level radioactive wastes in the open sea beyond the boundaries of their national jurisdictions. Since 1983, these same nations have observed a moratorium on the dumping of low-level radioactive wastes at sea, which in 1994 became a permanent ban.

But these agreements are hard to monitor and enforce. In 1992, it was learned that for decades, the former Soviet Union had been dumping large quantities of high- and low-level radioactive wastes into the Arctic Ocean and its tributaries.

What Are the Major Sources of Ocean Oil Pollution? Land-Based Activities, Not Tankers, Are the Major Culprits

Most ocean oil pollution comes from human activities on the land.

Crude petroleum (oil as it comes out of the ground) and *refined petroleum* (fuel oil, gasoline, and other processed petroleum products) reach the ocean from a number of sources. In 1989, the *Exxon Valdez* oil tanker went off course, hit rocks, and released large amounts of oil into Alaska's Prince William Sound in an accident that ended up costing about $7 billion (including cleanup costs and fines for damage against ExxonMobil). In 2002, the oil tanker *Prestige* sank off the coast of Spain and over two years leaked about twice as much oil as the *Exxon Valdez*. Such tanker accidents and blowouts at offshore drilling rigs (when oil escapes under high pressure from a borehole in the ocean floor) get most of the publicity because of their high visibility.

But much more oil is released from other smaller, day-to-day, and less visible activities. They include the normal operation of offshore wells, washing oil tankers and releasing the oily water, loading and unloading of oil tankers at ports, and leaks from oil pipelines, refineries, and storage tanks. Natural oil seeps also release large amounts of oil into the ocean at some sites.

Studies show that most ocean oil pollution comes from activities on land. According to a 2002 study by the Pew Oceans Commission, every 8 months an amount of oil equal to that spilled by the *Exxon Valdez* tanker drains from the land into the oceans. Almost half (some experts estimate 90%) of the oil reaching the oceans is waste oil dumped on the ground, poured down the drain, spilled, or leaked onto the land or into sewers by cities, industries, and people changing their own motor oil.

What Are the Effects of Oil Pollution on Ocean Ecosystems and Coastal Communities? Serious but Not Long-Lasting

Oil pollution can have a number of harmful ecological and economic effects, but most disappear within 3–15 years.

The effects of oil on ocean ecosystems depend on a number of factors: type of oil (crude or refined), type of aquatic system, amount released, distance of release from shore, time of year, weather conditions, average water temperature, and ocean currents.

Volatile organic hydrocarbons in oil immediately kill a number of aquatic organisms, especially in their vulnerable larval forms. Some other chemicals form tarlike globs that float on the surface and coat the feathers of birds (especially diving birds) and the fur of marine mammals. This oil coating destroys their natural insulation and buoyancy, causing many of them to drown or die of exposure from loss of body heat.

Heavy oil components that sink to the ocean floor or wash into estuaries can smother bottom-dwelling organisms such as crabs, oysters, mussels, and clams or make them unfit for human consumption. Some oil spills have killed reef corals.

Research shows that most (but not all) forms of marine life recover from exposure to large amounts of *crude oil* within about 3 years. But recovery from exposure to *refined oil*, especially in estuaries and salt marshes, can take 10–15 years. The effects of spills in cold waters and in shallow enclosed gulfs and bays generally last longer.

Oil slicks that wash onto beaches can have a serious economic impact on coastal residents, who lose income normally gained from fishing and tourist activities. Oil-polluted beaches washed by strong waves or currents become clean after about a year, but beaches in sheltered areas remain contaminated for several years. Despite the localized harmful effects, EPA experts rate oil spills as a low-risk ecological problem.

How Well Can We Clean Up Oil Spills? Not Very Well

Current methods can recover no more than about 15% of the oil from a major spill, explaining why prevention is the best strategy.

If they are not too large, oil spills can be partially cleaned up by mechanical, chemical, fire, and natural methods. *Mechanical methods* include using *floating booms* to contain the oil spill or keep it from reaching sensitive areas, *skimmer boats* to vacuum up some of the oil into collection barges, and *absorbent devices* such as large mesh pillows filled with feathers or hair to soak up oil on beaches or in waters too shallow for skimmer boats.

Chemical methods include using *coagulating agents* to cause floating oil to clump together for easier pickup or to sink to the bottom (where it usually does less harm) and *dispersing agents* to break up oil slicks. But these agents can damage some types of organisms. *Fire* can burn off floating oil, but crude oil is hard to ignite and burning it produces air pollution.

In time, the natural action of wind and waves mixes or emulsifies oil with water (like emulsified salad dressing), and bacteria biodegrade some of the oil. Scientists are developing *biological methods* in which "cocktails" of bacteria are sprayed on the oil to break it down into chemicals that the bacteria consume or that disperse harmlessly into the sea. Adding special nutrients required by the bacteria usually speeds up the decomposition process. This bioremediation cleanup by naturally occurring bacteria is cheaper and may be much more effective than other cleanup methods.

Scientists estimate that current methods can recover no more than 15% of the oil from a major spill. This explains why preventing oil pollution is the most effective and in the long run the least costly approach.

Good news. Because of concern over the 1989 *Exxon Valdez* oil spill, the Oil Pollution Act of 1990 set up a trust fund to provide up to $1 million per spill for cleanup. It also required that by 2015 all oil tankers operating in U.S. waters must be constructed with two hulls—one inside the other—to help protect against spills. Similar international laws have been established, and in 2002 the European Union voted to ban single-hull oil tankers from their waters by 2010 and by 2005 for the largest tankers. Some members of Congress have unsuccessfully proposed legislation to require double hulls for tankers in U.S. waters by 2007.

Bad news. In 2004, about half of the world's 10,000 oil tankers still had the older and more vulnerable single hulls. Cruise ships can also pollute coastal waters with oil and other waste—most of which is dumped at sea or in fragile coastal areas when the ships visit various ports. Scuba diving, anyone?

Solutions: How Can We Protect Coastal Waters? Think Prevention

Preventing or reducing the flow of pollution from the land and from streams emptying into the ocean is the key to protecting the oceans.

Figure 22-14 list ways analysts have suggested to prevent and reduce excessive pollution of coastal waters. Study this figure carefully.

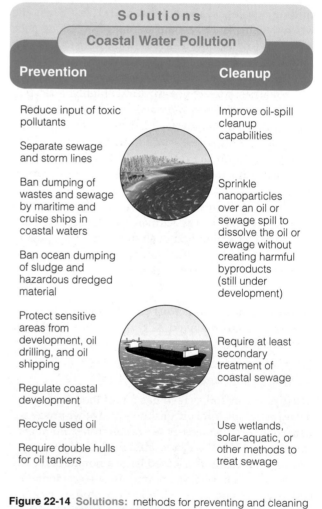

Solutions

Coastal Water Pollution

Prevention	Cleanup

Prevention:
- Reduce input of toxic pollutants
- Separate sewage and storm lines
- Ban dumping of wastes and sewage by maritime and cruise ships in coastal waters
- Ban ocean dumping of sludge and hazardous dredged material
- Protect sensitive areas from development, oil drilling, and oil shipping
- Regulate coastal development
- Recycle used oil
- Require double hulls for oil tankers

Cleanup:
- Improve oil-spill cleanup capabilities
- Sprinkle nanoparticles over an oil or sewage spill to dissolve the oil or sewage without creating harmful byproducts (still under development)
- Require at least secondary treatment of coastal sewage
- Use wetlands, solar-aquatic, or other methods to treat sewage

Figure 22-14 Solutions: methods for preventing and cleaning up excessive pollution of coastal waters. Which two of these solutions do you believe are the most important?

The key to protecting oceans is to reduce the flow of pollution from the land and from streams emptying into the ocean. Thus ocean pollution control must be linked with land-use and air pollution policies because about one-third of all pollutants entering the ocean worldwide come from air emissions from land-based sources.

22-6 PREVENTING AND REDUCING SURFACE WATER POLLUTION

How Can We Reduce Surface Water Pollution from Nonpoint Sources? Emphasize Prevention

The key to reducing nonpoint pollution, most of it from agriculture, is to prevent it from reaching bodies of surface water.

There are a number of ways to reduce nonpoint water pollution, most of it from agriculture. Farmers can reduce soil erosion, especially by keeping cropland covered with vegetation and by reforesting critical watersheds. They can also reduce the amount of fertilizer running off into surface waters and leaching into aquifers by using slow-release fertilizer, using none on steeply sloped land, and planting buffer zones of vegetation between cultivated fields and nearby surface water.

Applying pesticides only when needed and relying more on biological control of pests can reduce pesticide runoff. Farmers can control runoff and infiltration of manure from animal feedlots by planting buffers and locating feedlots and animal waste sites away from steeply sloped land, surface water, and flood zones.

Good news. In 2003 Smithfield Foods, a large pork producer, announced plans to build a facility in Utah to convert the wastes from 500,000 hogs—about half of its annual hog production in Utah—to make renewable biodiesel fuel for vehicles. In addition, researchers are experimenting with planting poplar trees to suck up waste from contaminated hog waste lagoons.

In 2002 a federal court forced the EPA to uphold the intent of the Clean Water Act and require about 15,500 of the nation's largest livestock feedlots or factory farms to apply for EPA runoff permits by 2006, develop plans to handle manure and wastewater, and file annual reports with the EPA. If this rule goes into effect, large livestock operations will have to obey the same pollution control regulations that have been applied to other industries since 1972.

Livestock producers who have successfully fought such regulation for over 30 years say the new rules will cost them too much, and they hope to persuade Congress to eliminate, delay, or weaken the new rules. Stay tuned for developments.

These tougher rules are spurring scientists to come up with better ways to deal with animal waste. They are exploring ways to burn it, convert it to natural gas, recycle undigested nutrients in manure back into animal feed, and extract valuable chemicals from manure to make plastics or even cosmetics.

Other scientists are looking at ways to rinse away many of the soluble and smelly ingredients in manure to leave tough, strawlike particles of fiber that can be pressed into fiberboard for making cabinets and furniture. The resulting fiber is called animal processed fiber, a formal name for processed cow and hog poop.

> **X** *HOW WOULD YOU VOTE?* Should we greatly increase efforts to reduce water pollution from nonpoint sources? Cast your vote online at http://biology.brookscole.com/miller14.

How Can We Reduce Water Pollution from Point Sources? Legal and Market Approaches

Most developed countries use laws to set water pollution standards, but in most developing countries such laws do not exist or are poorly enforced.

The Federal Water Pollution Control Act of 1972 (renamed the Clean Water Act when it was amended in 1977) and the 1987 Water Quality Act form the basis of U.S. efforts to control pollution of the country's surface waters. The Clean Water Act sets standards for allowed levels of key water pollutants and requires polluters to get permits specifying how much of various pollutants they can discharge into aquatic systems.

The EPA is also experimenting with a *discharge trading policy* that uses market forces to reduce water pollution (as has been done with sulfur dioxide for air pollution control, p. 454) in the United States. Under this program a water pollution source is allowed to pollute at a higher level than allowed in its permit by buying credits from permit holders with pollution levels below their allowed levels.

Some environmentalists support discharge trading. But they warn that such a system is no better than the caps set for total pollution levels in various areas, and call for careful scrutiny of the cap levels. They also warn that discharge trading could allow pollutants to build up to dangerous levels in areas where credits are bought. In addition, they call for gradually lowering the caps to encourage prevention of water pollution and development of better technology for controlling water pollution, neither of which is a part of the current EPA water pollution discharge trading system.

Bad news. According to Sandra Postel, director of the Global Water Policy Project, most cities in developing countries discharge 80–90% of their untreated sewage directly into rivers, streams, and lakes, which are used for drinking water, bathing, and washing clothes.

How Can We Reduce Water Pollution from Point Sources? The Technological Approach

Septic tanks and various levels of sewage treatment can reduce point-source water pollution.

In rural and suburban areas with suitable soils, sewage from each house can be discharged into a **septic tank** (Figure 22-15). About one-fourth of all homes in the United States are served by septic tanks.

In U.S. urban areas, most waterborne wastes from homes, businesses, factories, and storm runoff flow through a network of sewer pipes to *wastewater* or *sewage treatment plants*. Some cities have a separate network of pipes for carrying runoff of storm water from streets and parking lots. But 1,200 U.S. cities have combined the sewer lines for these two systems because it is cheaper.

Bad news. Heavy rains or too many users hooked up to the system can cause combined sewer lines to overflow and discharge untreated (raw) sewage directly into surface waters. According to the EPA, at least 40,000 such overflows occur each year in the United States. The EPA estimated that each year 1.8–3.5 million people get sick from swimming in waters contaminated by sewage overflows. The EPA estimates that it would cost $10 billion a year for a decade

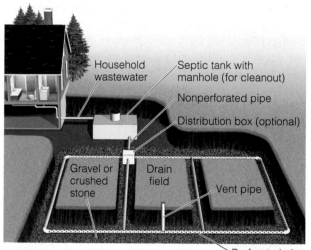

Figure 22-15 Solutions: *septic tank system* used for disposal of domestic sewage and wastewater in rural and suburban areas. This system separates solids from liquids, digests organic matter and large solids and discharges the liquid wastes in a network of buried pipes with holes over a large drainage or absorption field. As these wastes drain from the pipes and percolate downward, the soil filters out some potential pollutants, and soil bacteria decompose biodegradable materials. To be effective, septic tank systems must be properly installed in soils with adequate drainage, not placed too close together or too near well sites, and pumped out when the settling tank becomes full.

to install dual systems, add capacity, and repair the nation's $2 trillion aging sewer network. To help protect public health, environmentalists want Congress to change the Clean Water Act to require the EPA to monitor sewer leaks and overflows and report them to public health authorities.

Raw sewage reaching a treatment plant typically undergoes one or both of two levels of wastewater treatment. One is **primary sewage treatment.** It is a *physical* process that uses screens and a grit tank to remove large floating objects and solids such as sand and rock, and a settling tank that allows suspended solids to settle out as sludge (Figure 22-16). By itself, primary treatment removes about 60% of the suspended solids and 30–40% of the oxygen-demanding organic wastes from sewage but removes no phosphates, nitrates, salts, radioisotopes, or pesticides.

A second level is called **secondary sewage** treatment. It is a *biological* process in which aerobic bacteria remove up to 90% of dissolved and biodegradable, oxygen-demanding organic wastes. This is done by trickling wastewater through beds of gravel covered with films of aerobic bacteria or by passing it through an *aeration tank* where air is pumped through a slurry of aerobic bacteria before the wastewater is sent to a second settling tank.

A combination of primary and secondary treatment (Figure 22-16) removes 95–97% of the suspended solids and oxygen-demanding organic wastes, 70% of most toxic metal compounds and nonpersistent synthetic organic chemicals, 70% of the phosphorus (mostly as phosphates), 50% of the nitrogen (mostly as nitrates), and 5% of dissolved salts. But this process removes only a tiny fraction of long-lived radioactive isotopes and persistent organic substances such as some pesticides.

Because of the Clean Water Act, most U.S. cities have combined primary and secondary sewage treatment plants. According to the EPA, however, at least two-thirds of these plants have at times violated water pollution regulations. Also, 500 cities have failed to meet federal standards for sewage treatment plants, and 34 East Coast cities simply screen out large floating objects from their sewage before discharging it into coastal waters.

A third level of cleanup is **advanced** or **tertiary sewage treatment.** It is a series of specialized chemical and physical processes that remove specific pollutants left in the water after primary and secondary treatment. Its most widespread use is to remove phosphates and nitrates from wastewater before it is discharged into surface waters to help reduce nutrient overload. Advanced treatment is expensive and is used to treat only 5% of the wastewater in the United States.

Before discharge, water from primary, secondary, or advanced treatment undergoes *bleaching* to remove

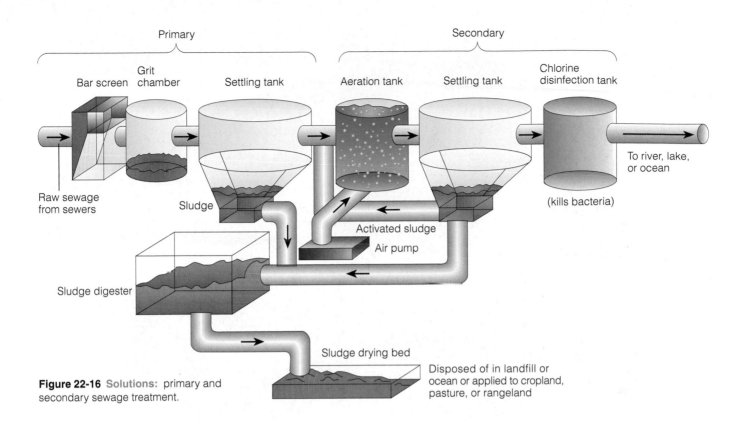

Figure 22-16 Solutions: primary and secondary sewage treatment.

Labels in figure:
Primary — Bar screen, Grit chamber, Settling tank
Secondary — Aeration tank, Settling tank, Chlorine disinfection tank
Raw sewage from sewers
Sludge
Activated sludge
Air pump
Sludge digester
Sludge drying bed
To river, lake, or ocean (kills bacteria)
Disposed of in landfill or ocean or applied to cropland, pasture, or rangeland

water coloration and *disinfection* to kill disease-carrying bacteria and some but not all viruses. The usual method for doing this is *chlorination.* But chlorine can react with organic materials in water to form small amounts of chlorinated hydrocarbons. Some of these chemicals can cause cancers in test animals and may damage the human nervous, immune, and endocrine systems (Case Study, p. 416).

Use of other disinfectants, such as *ozone* and *ultraviolet light,* is increasing. But they cost more and their effects do not last as long as chlorination.

What Should We Do with Sewage Sludge? An Unsettled Problem

Sewage sludge can be used as a soil conditioner, but this can cause health problems if it contains infectious bacteria and toxic chemicals.

Sewage treatment produces a gooey *sludge* containing a slimy mixture of bacteria-laden solids and often-toxic chemicals and metals when sewer systems mix industrial and household waste. In the United States, about 9% by weight of this sludge is placed in large circular digesters and kept warm for several weeks to allow anaerobic bacteria to decompose organic materials and produce compost for use as a soil conditioner. About 36% of the sludge, also known as *biosolids,* is used to fertilize farmland, forests, golf courses, cemeteries, parkland, highway medians, and degraded land. The remaining 55% is dumped in conventional landfills where it can contaminate groundwater, or is

incinerated. Such burning of waste can pollute the air with toxic chemicals, and it produces a toxic ash usually buried in landfills that the EPA says will eventually leak.

From an ecological standpoint, it is desirable to recycle plant nutrients in sewage sludge to the soil on land. But there are problems with using sewage sludge to fertilize crops (Figure 22-17, p. 512). As long as harmful bacteria and other pathogens and toxic chemicals are not present, sludge can fertilize land used for food crops or livestock. But removing bacteria (usually by heating), toxic metals, and organic chemicals is expensive and rarely done in the United States. According to a 2002 report by the National Academy of Sciences, the EPA is using outdated science to set standards for using sewage sludge as a fertilizer in the United States.

A growing number of alleged health problems and lawsuits have resulted from use of sludge to fertilize crops in the United States. To protect consumers and avoid lawsuits, some food packers such as Del Monte and Heinz have banned produce grown on farms using sludge as a fertilizer.

How Can We Improve Sewage Treatment? Eliminate Toxics

Preventing toxic chemicals from reaching sewage treatment plants would eliminate such chemicals from the sludge and water discharged from such plants.

Environmental scientist Peter Montague calls for redesigning sewage treatment systems to prevent toxic

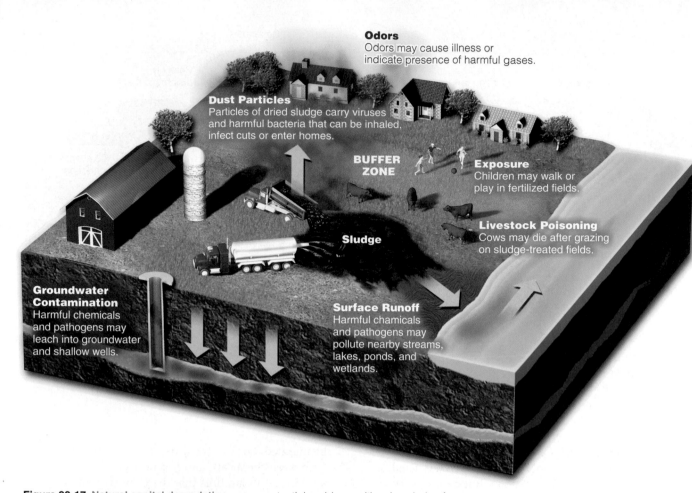

Figure 22-17 Natural capital degradation: some potential problems with using sludge from sewage treatment plants as a fertilizer on croplands. The EPA says that sludge is safe to use if applied following its guidelines. Scientists and people who have gotten sick from exposure to sludge fertilizer claim the guidelines are insufficient and not adequately enforced.

and hazardous chemicals from reaching sewage treatment plants and thus from getting into sludge and the water discharged from such plants.

He suggests several ways to do this. One is to require industries and businesses to remove toxic and hazardous wastes from water sent to municipal sewage treatment plants. Another is to encourage industries to reduce or eliminate toxic chemicals use and waste.

Another suggestion is to have more households, apartment buildings, and offices eliminate sewage outputs by switching to waterless *composting toilet systems* that are installed, maintained, and managed by professionals. Such systems would be cheaper to install and maintain than current sewage systems because they do not require vast systems of underground pipes connected to centralized sewage treatment plants. They also save large amounts of water. They work great. I used one for 15 years.

✗ HOW WOULD YOU VOTE? Should we ban the discharge of toxic chemicals into pipes leading to sewage treatment plants? Cast your vote online at http://biology.brookscole.com/miller14.

Solutions: How Can We Treat Sewage by Working with Nature? Ecological Purification

Natural and artificial wetlands and other ecological systems can be used to treat sewage.

John Todd has developed an ecological approach to treating sewage, which he calls *living machines* (Figure 22-1). More than 150 cities and towns in the United States use natural and artificial wetlands to treat sewage as a low-tech, low-cost alternative to expensive waste treatment plants.

For example, the coastal town of Arcata, California, created some 65 hectares (160 acres) of wetlands between the town and the adjacent Humboldt Bay. The marshes and ponds, developed on this land that was once a garbage dump, act as an inexpensive natural waste treatment plant. The project cost less than half the estimated cost of a conventional treatment plant.

Here is how it works. First, sewage goes to sedimentation tanks, where the solids settle out as sludge that is removed and processed for use as fertilizer. Next, the liquid is pumped into oxidation ponds, where bacteria break down remaining wastes. After a month or so, the water is released into the artificial marshes, where plants and bacteria carry out further filtration and cleansing. Then the purified water flows into the Humboldt Bay with its abundant marine life.

The marshes and ponds also serve as an Audubon Society bird sanctuary and provide habitats for thousands of otters, seabirds, and marine animals. The town celebrates its natural sewage treatment system with an annual "Flush with Pride" festival. However, some cities do not have the land available for this approach.

Mark Nelson has developed a small, low-tech, and inexpensive artificial wetland system to treat raw sewage from hotels, restaurants, and homes in developing countries (Figure 22-18). This *wastewater garden* system removes 99.9% of fecal coliform bacteria and more than 80% of the nitrates and phosphates from incoming sewage that in most developing countries is often dumped untreated into the ocean or into shallow holes in the ground. The water flowing out of such systems can be used to irrigate gardens or fields or to flush toilets and thus help save water.

Genetic engineering may also get into the act. The guts of some insects are resistant to pesticides. Researchers have isolated the gene that provides this resistance, and they have transferred it to easily cultured bacterial species. They envision passing contaminated water through a large vessel or *bioreactor* containing the genetically modified bacteria that consume the pesticides. Stay tuned about developments in this promising area of frontier science. This might be an interesting career choice.

Currently about 1.7 billion people do not have access to adequate sanitation. And the world's population is projected to add 2.9 billion more people between 2004 and 2050—an average of 172,000 new people per day needing access to safe drinking water and adequate sanitation. Without greatly increased investment in conventional and unconventional sewage treatment systems, the number of people with inadequate sanitation could reach 3 billion by 2050. Dealing with this important challenge will take scientific and engineering ingenuity and lots of money.

How Successful Has the United States Been in Reducing Water Pollution? Good and Bad News

Water pollution laws have significantly improved water quality in many U.S. streams and lakes, but there is a long way to go.

Great news. According to the EPA, the Clean Water Act of 1972 led to a number of improvements in U.S. water quality. Between 1992 and 2002, the number of Americans served by community water systems that met federal health standards increased from 79% to 94%.

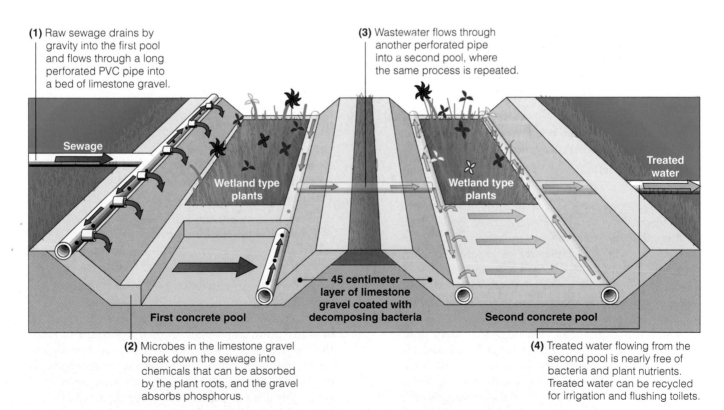

(1) Raw sewage drains by gravity into the first pool and flows through a long perforated PVC pipe into a bed of limestone gravel.

(3) Wastewater flows through another perforated pipe into a second pool, where the same process is repeated.

Sewage

Wetland type plants

Treated water

Wetland type plants

First concrete pool

45 centimeter layer of limestone gravel coated with decomposing bacteria

Second concrete pool

(2) Microbes in the limestone gravel break down the sewage into chemicals that can be absorbed by the plant roots, and the gravel absorbs phosphorus.

(4) Treated water flowing from the second pool is nearly free of bacteria and plant nutrients. Treated water can be recycled for irrigation and flushing toilets.

Figure 22-18 Solutions: *wastewater garden.* Hotels, restaurants, and homes in developing countries can use this small gravity-fed artificial wetland system to treat sewage. It uses only 1.9–3.8 square meters (20–30 square feet) of space per person.

Also, between 1972 and 2002, the percentage of U.S. stream lengths found to be fishable and swimmable increased from 36% to 60% of those tested. And the amount of topsoil lost through agricultural runoff was cut by about 1.1 billion metric tons (1 billion tons) annually. In addition, between 1972 and 2002, the proportion of the U.S. population served by sewage treatment plants increased from 32% to 74%. And between 1974 and 2002, annual wetland losses decreased by 80%. These are impressive achievements given the increases in the U.S. population and per capita consumption since 1972.

Bad news. In 2000, the EPA found that 45% of the country's lakes and 40% of the streams surveyed were too polluted for swimming or fishing. The number of polluted streams, lakes, and estuaries could be much higher because only 19% of the country's stream lengths, 43% of its lake and reservoir area, and 36% of its estuaries have been tested for water quality.

Runoff of animal wastes from hog, poultry, and cattle feedlots and meat processing facilities pollutes 7 of every 10 U.S. rivers. Most livestock wastes are not treated and are stored in lagoons that sometimes leak. They can also overflow or rupture as a result of excessive rainfall and spill their contents into nearby streams and rivers and sometimes into residential areas.

Fish caught in more than 1,400 different waterways and more than a fourth of the nation's lakes are unsafe to eat because of high levels of pesticides, mercury, and other toxic substances. A 2003 internal study by the EPA found that at least half of the country's 6,600 largest industrial facilities and municipal wastewater treatment plants have illegally discharged toxic or biological wastes into waterways for years without government enforcement actions or fines.

Should the U.S. Clean Water Act Be Strengthened or Weakened? A Raging Controversy

Some want to strengthen the Clean Water Act while others want to weaken it.

Some environmentalists and a 2001 report by the EPA's inspector general call for the Clean Water Act to be strengthened. Suggested improvements include increased funding and authority to control nonpoint sources of pollution, upgrading the computer system for monitoring compliance with the law, and strengthening programs to prevent and control toxic water pollution.

Other suggestions include providing more funding and authority for integrated watershed and airshed planning to protect groundwater and surface water from contamination, and expanding the rights of citizens to bring lawsuits to ensure that water pollution laws are enforced. The National Academy of Sciences also calls for halting the loss of wetlands, higher standards for wetland restoration, and creating new wetlands before filling any natural wetlands.

Many people oppose these proposals, contending that the Clean Water Act's regulations and government wetlands regulations are already too restrictive and costly. Farmers and developers see the law as a curb on their rights as property owners to fill in wetlands. They also believe they should be compensated for any property value losses resulting from federal regulations protecting wetlands.

State and local officials want more discretion in testing for and meeting water quality standards. They argue that in many communities it is unnecessary and too expensive to test for all the water pollutants required by federal law.

X *How Would You Vote?* Should the U.S. Clean Water Act be strengthened? Cast your vote online at http://biology .brookscole.com/miller14.

22-7 DRINKING WATER QUALITY

How Is Urban Drinking Water Purified? The High-Tech Centralized Approach

Centralized water treatment plants that operate much like wastewater treatment plants can provide safe drinking water for city dwellers.

Areas that depend on surface water for drinking usually store it in a reservoir for several days. This improves taste and clarity by increasing dissolved oxygen content and allowing suspended matter to settle. Next the water is pumped to a purification plant where it is filtered as needed and chlorinated to meet government drinking water standards. In areas with very pure groundwater sources, little treatment except disinfection is necessary.

How Can Modern Water-Purification Systems Be Protected from Terrorist Acts? A Difficult Problem

The United States is upgrading security on water purification and delivery systems, but it is difficult to protect such a vast and complex system.

In the United States there is increased concern over terrorists adding harmful chemicals or biological agents to reservoirs and other parts of the nation's vast network of water purification systems. Reservoirs are so huge that they are hard to poison with chemical or biological agents. Still, drinking water is hard to protect because of the large number of reservoirs, the vast network of water purification plants and water distribution systems, and accessibility of water systems on every street through fire hydrants and service connections.

Officials are working to find ways to make it harder to access or damage water purification plants and pipes. They are upgrading surveillance cameras and other security measures. They are also developing chemical tests and biological indicators that quickly indicate the presence of chemical or biological agents, and they are working on emergency response plans in case of contamination. A major problem is that protecting these systems will cost several billion dollars and so far Congress has not provided enough funds to get the job done.

How Can We Purify Rural Drinking Water in Developing Countries? The Low-Tech Decentralized Approach

Researchers have developed several simple and inexpensive ways for individuals and villages in developing countries to purify drinking water.

Ways to purify drinking water can be simple. In tropical countries without centralized water treatment systems, the WHO is urging people to purify drinking water by exposing a clear plastic bottle filled with contaminated water to intense sunlight. In the strong sunlight found in most tropical countries, heat and the sun's UV rays can kill infectious microbes in as little as 3 hours. Painting one side of the bottle black can improve heat absorption in this simple solar disinfection method. Where it has been used, incidences of dangerous childhood diarrhea have decreased by 30–40%.

In Bangladesh, households receive strips of cloth for filtering cholera-producing bacteria from drinking water. Villages where women use such strips to strain water have cut cholera cases in half.

Another simple method involves adding a small amount of a chlorine-disinfectant solution to plastic or clay water-storage vessels with a narrow mouth and cap and a spigot—similar to what U.S. campers frequently use. The storage vessel design helps protect the disinfected water from additional bacterial contamination. Trials in Zambia, Kenya, and India show that this approach can cut the rate of diarrheal disease in half. This highly publicized method is now in use in 15 developing countries.

How Well Is Drinking Water Quality Protected by Law? The Legal Approach

Most developed countries have laws establishing drinking water standards, but most developing countries do not have such laws or do not enforce them.

About 54 countries, most of them in North America and Europe, have standards for safe drinking water. The U.S. Safe Drinking Water Act of 1974 requires the EPA to establish national drinking water standards, called *maximum contaminant levels,* for any pollutants that may have adverse effects on human health. However, such laws do not exist or are not enforced in most developing countries.

Privately owned wells are not required to meet federal drinking water standards for two reasons. One is that it costs at least $1,000 to test each well and owners would need to retest their water every few years. The other is that some homeowners oppose mandatory testing and compliance.

Health scientists call for strengthening the U.S. Safe Drinking Water Act in several ways. One is to combine many of the drinking water treatment systems that serve fewer than 3,300 people with nearby larger systems. Another is to strengthen and enforce public notification requirements about violations of drinking water standards. They also call for banning all toxic lead in new plumbing pipes, faucets, and fixtures (current law allows fixtures with up to 10% lead to be sold as lead free). According to the Natural Resources Defense Council (NRDC), such improvements would cost about $30 a year per U.S. household.

However, water-polluting industries are pressuring elected officials to weaken the Safe Drinking Water Act. One proposal is to eliminate national tests of drinking water and public notification requirements about violations of drinking water standards.

A second proposal is to allow states to give drinking water systems a permanent right to violate the standard for a given contaminant if the provider claims it cannot afford to comply. Another suggestion is to eliminate the requirement that water systems use affordable, feasible technology to remove cancer-causing contaminants. Finally, there are suggestions to greatly reduce the EPA budget for enforcing the Clean Water Act.

> **X HOW WOULD YOU VOTE?** Should the Safe Drinking Water Act be strengthened? Cast your vote online at http://biology.brookscole.com/miller14.

Is Bottled Water the Answer? Solution or Expensive Rip-off?

Some bottled water is not as pure as tap water and costs much more.

Despite some problems, experts say the United States has some of the world's cleanest drinking water. Yet about half of all Americans worry about getting sick from tap water contaminants, and many drink bottled water or install expensive water purification systems.

Studies reveal that in the United States bottled water is 240 to 10,000 times more expensive than tap water. In addition, about one-fourth of it is tap water, bacteria contaminate about one-third of it, and various potentially harmful organic chemicals contaminate about one-fifth of it. On the other hand, some countries must rely on bottled water because some of their tap water is too polluted to drink.

Use of bottled water can also cause some environmental problems. For example, 1.4 million metric tons (1.5 million tons) of plastic bottles are thrown away globally each year, and toxic gases and liquids are released during the manufacture of plastic water bottles. In addition, greenhouse gases and other air pollutants are emitted by the fossil fuels burned to make plastic bottles and to deliver bottled water to suppliers.

Before drinking expensive bottled water and buying costly home water purifiers, health officials suggest that consumers have their water tested by local health authorities or private labs (not companies trying to sell water purification equipment). The goals are to identify what contaminants (if any) must be removed and to determine the type of purification needed to remove such contaminants. Independent experts contend that unless tests show otherwise, for most urban and suburban Americans served by large municipal drinking water systems, home water treatment systems are not worth the expense and maintenance hassles.

Buyers should check out companies selling water purification equipment and be wary of claims that the EPA has approved a treatment device. Although the EPA does *register* such devices, it neither tests nor approves them.

✗ *How Would You Vote?* Should pollution standards be established for bottled water? Cast your vote online at http://biology.brookscole.com/miller14.

How Can We Reduce Water Pollution? Individuals Matter

Shifting our priorities from controlling to preventing and reducing water pollution will require bottom-up political action by individuals and groups.

It is encouraging that since 1970 most of the world's developed countries have enacted laws and regulations that have significantly reduced point-source water pollution. Most of these improvements were the result of *bottom-up* political pressure on elected officials by individuals and organized groups. However, little has been done to reduce water pollution in most developing countries.

To health scientists and environmentalists the next step is to increase efforts to reduce and prevent water pollution in developed and developing countries by asking the question: *How can we avoid producing water pollutants in the first place?* Figure 22-19 lists ways to do this over the next several decades.

This shift to *preventing water pollution* will not take place in developed countries without *bottom-up* politi-

Solutions

Water Pollution

- Prevent groundwater contamination

- Greatly reduce nonpoint runoff

- Reuse treated wastewater for irrigation

- Find substitutes for toxic pollutants

- Work with nature to treat sewage

- Practice four R's of resource use (refuse, reduce, recycle, reuse)

- Reduce resource waste

- Reduce air pollution

- Reduce poverty

- Reduce birth rates

Figure 22-19 Solutions: methods for preventing and reducing water pollution. Which two of these solutions do you believe are the most important?

cal pressure on elected officials. It will not occur in developing countries without similar pressure from citizens as well financial and technical aid from developed countries. Figure 22-20 lists some actions you can take to help reduce water pollution.

What Can You Do?

Water Pollution

- Fertilize your garden and yard plants with manure or compost instead of commercial inorganic fertilizer.

- Minimize your use of pesticides.

- Never apply fertilizer or pesticides near a body of water.

- Grow or buy organic foods.

- Compost your food wastes.

- Do not use water fresheners in toilets.

- Do not flush unwanted medicines down the toilet.

- Do not pour pesticides, paints, solvents, oil, antifreeze, or other products containing harmful chemicals down the drain or onto the ground.

Figure 22-20 What can you do? Ways to help reduce water pollution.

It is a hard truth to swallow, but nature does not care if we live or die. We cannot survive without the oceans, for example, but they can do just fine without us.

ROGER ROSENBLATT

CRITICAL THINKING

1. Explain why dilution is not always the solution to water pollution.

2. For each of the eight categories of pollutants listed in Table 22-1, is it most likely to originate from (a) point sources or (b) nonpoint sources?

3. A large number of fish are found floating dead on a lake during the summer. You are asked to determine the cause of the fish kill. What reason would you suggest for the kill? What measurements would you make to verify your hypothesis?

4. Are you for or against banning injection of liquid hazardous wastes into deep wells below drinking water aquifers (Figure 22-9, p. 502)? Explain. What are the alternatives?

5. When you flush a toilet, where does the wastewater go? Trace the actual flow of this wastewater in your community from your toilet through sewers to a wastewater treatment plant and from there to the environment. Try to visit a local sewage treatment plant to see what it does with your wastewater. Compare the processes it uses with those shown in Figure 22-16 (p. 511). What happens to the sludge produced by this plant? What improvements, if any, would you suggest for this plant?

6. Congratulations! You are in charge of sharply reducing water pollution from nonpoint sources throughout the world. What are the three most important things you would do?

7. Congratulations! You are in charge of sharply reducing groundwater pollution throughout the world. What are the three most important things you would do?

8. Congratulations! You are in charge of providing safe drinking water for the poor and other people in developing countries. What are the three most important things you would do?

PROJECTS

1. In your community,
   a. What are the principal nonpoint sources of contamination of surface water and groundwater?
   b. What is the source of drinking water?
   c. How is drinking water treated?
   d. How many times during each of the past 5 years have levels of tested contaminants violated federal standards? Were violations reported to the public?
   e. Has pollution led to fishing bans or warnings not to eat fish from any lakes or rivers in your region?
   f. Is groundwater contamination a problem? If so, where, and what has been done about the problem?
   g. Is there a vulnerable aquifer or critical recharge zone that needs protection to ensure the quality of groundwater? Is your local government aware of this? What action (if any) has it taken?

2. Are storm drains and sanitary sewers combined or separate in your area? Are there plans to reduce pollution from runoff of storm water? If not, make an economic evaluation of the costs and benefits of developing separate storm drains and sanitary sewers, and present your findings to local officials.

3. Use library research, the Internet, and user interviews to evaluate the relative effectiveness and costs of home water purification devices. Determine the type or types of water pollutants each device removes and the effectiveness of this process.

4. Find out the price of tap water where you live. Then go to a grocery or other store and get prices per liter (or other volume unit) on all the available types of bottled water. Use these data to compare the price per liter of various brands of bottled water with the price of tap water.

5. Use the library or the Internet to find bibliographic information about *William Ruckelshaus* and *Roger Rosenblatt*, whose quotes appear at the beginning and end of this chapter.

6. Make a concept map of this chapter's major ideas, using the section heads, subheads, and key terms (in boldface). See material on the website for this book about how to prepare concept maps.

LEARNING ONLINE

The website for this book contains study aids and many ideas for further reading and research. They include a chapter summary, review questions for the entire chapter, flash cards for key terms and concepts, a multiple-choice practice quiz, interesting Internet sites, references, and a guide for accessing thousands of InfoTrac® College Edition articles. Log on to

http://biology.brookscole.com/miller14

Then click on the Chapter-by-Chapter area, choose Chapter 22, and select a learning resource.

23 Pest Management

Pest & Disease Control

CASE STUDY
Along Came a Spider— Biological Pest Control

Since agriculture began about 10,000 years ago, we have been competing with insect pests for the food we grow. Today we are not much closer to winning this competition than we were then, mostly because of the astounding abilities of insect pests to multiply and, through natural selection, rapidly develop genetic resistance to poisons we throw at them.

Some Chinese farmers use a biological strategy to help control insect pests. Instead of spraying their rice and cotton fields with poisons, they build little straw huts around the fields in the fall.

These farmers are encouraging insects' worst enemy, one that has hunted them for millions of years: *spiders* (Figure 23-1). The little huts are for hibernating spiders. Protected from the winter cold by the huts, far more of the hibernating spiders become active in the spring. Ravenous after their winter fast, they scuttle off into the fields to stalk their insect prey.

Even without human help, the world's 30,000 known species of spiders kill far more insects every year than insecticides do. A typical acre of meadow or woods contains an estimated 50,000–2 million spiders, each devouring hundreds of insects per year.

Entomologist Willard H. Whitcomb found that leaving strips of weeds around cotton and soybean fields provides the kind of undergrowth favored by insect-eating wolf spiders (Figure 23-1, left). He also sings the praises of a type of banana spider, which lives in warm climates and can keep a house clear of cockroaches.

In Maine, Daniel Jennings of the U.S. Forest Service uses spiders to help control the spruce budworm, which devastates spruce and fir forests in the Northeast. Spiders also attack the much-feared gypsy moth, which destroys tree foliage.

The idea of encouraging populations of spiders in fields, forests, and even houses scares some people because spiders have bad reputations. A few spider species, such as the black widow, the brown recluse, and eastern Australia's Sydney funnel web, are dangerous to people. But most spider species, including the ferocious-looking wolf spider, do not harm humans. Even the giant tarantula rarely bites people, and its venom is too weak to harm us or other large mammals unless we are allergic to it. As we seek new ways to coexist with the insect rulers of the planet, we would do well to be sure that spiders are on our side.

This chapter looks first at the advantages and disadvantages of the conventional chemical approach to pest control based on using synthetic chemical pesticides. Then it discusses the advantages and disadvantages of a variety of biological and ecological alternatives for controlling pest populations.

Figure 23-1 Natural capital: *working with nature.* Spiders are insects' worst enemies. Most spiders, such as the wolf spider (left) and the crab spider (right), found in many parts of the world, are harmless to humans.

A weed is a plant whose virtues have not yet been discovered.
RALPH WALDO EMERSON

This chapter addresses the following questions:

- What are pesticides, and what types are used?
- What are the advantages and disadvantages of using chemicals to kill insects and weeds?
- How well is pesticide use regulated in the United States?
- What are the alternatives to using conventional pesticides, and what are the advantages and disadvantages of each alternative?

23-1 PESTICIDES: TYPES AND USES

How Does Nature Keep Pest Populations under Control? Natural Enemies

Predators, parasites, and disease organisms found in nature control populations of most pest species as part of the earth's free ecological services.

A **pest** is any species that competes with us for food, invades lawns and gardens, destroys wood in houses, spreads disease, invades ecosystems, or is simply a nuisance. Worldwide, only about 100 species of plants (which we call weeds), animals (mostly insects), fungi, and microbes (which can infect crop plants and livestock animals) cause about 90% of the damage to the crops we grow.

In natural ecosystems and polyculture agroecosystems, *natural enemies* (predators, parasites, and disease organisms) control the populations of about 98% of the potential pest species as part of the earth's free ecological services and thus help keep any one species from taking over for very long.

When we clear forests and grasslands, plant monoculture crops (Figure 6-25, p. 118) and douse fields with pesticides, we upset many of these natural population checks and balances. Then we must devise ways to protect our monoculture crops, tree plantations, and lawns from insects and other pests that nature once controlled at no charge.

What Are Pesticides? Ways to Repel or Kill Pests

We use chemicals to repel or kill pest organisms as plants have done for millions of years to defend themselves against hungry herbivores.

To help control pest organisms, we have developed a variety of **pesticides** or **biocides**—chemicals to kill or control populations of organisms we consider undesirable. Common types of pesticides include *insecticides* (chemicals that kill insects by blocking reproduction, clogging their airways, or disrupting their nervous system), *herbicides* (chemicals that kill weeds by disrupting their metabolism and growth), *fungicides* (fungus killers), and *rodenticides* (rat and mouse killers). *Biocide is* a more accurate name for these chemicals because most pesticides kill other organisms as well as their pest targets.

We did not invent the use of chemicals to repel or kill other species; plants have been producing chemicals to ward off, deceive, or poison herbivores that feed on them for about 225 million years. This is a never-ending, ever-changing coevolutionary process: Herbivores overcome various plant defenses through natural selection; then new plant defenses are favored by natural selection in this ongoing cycle of evolutionary punch and counterpunch.

As the human population grew and agriculture spread, people began looking for ways to protect their crops, mostly by using chemicals to kill or repel insect pests. Sulfur was used as an insecticide well before 500 B.C.; by the 1400s, people were applying toxic compounds of arsenic, lead, and mercury to crops as insecticides. Farmers abandoned this approach in the late 1920s when the increasing number of human poisonings and fatalities prompted a search for less toxic substitutes. *Bad news.* Traces of these nondegradable toxic metal compounds are still found in soils dosed with them long ago.

In the 1600s, farmers used nicotine sulfate, extracted from tobacco leaves, as an insecticide. In the mid-1800s, two more natural pesticides were introduced: *pyrethrum* (obtained from the heads of chrysanthemum flowers) and *rotenone* (extracted from the roots of various tropical forest legumes). These *first-generation pesticides* were mainly natural chemicals or botanicals borrowed from plants that had been defending themselves against insects eating them and herbivores grazing on them. In other words, we learned to copy nature.

In addition to protecting crops, people have used chemicals produced by plants to repel or kill insects in their households, yards, and gardens. Compared with some commercial insecticides, these chemicals can be less expensive and less of a potential health hazard. See the website for this chapter to find out about natural chemicals and methods that can be used to control weeds and to repel or kill common pests such as ants, mosquitoes, cockroaches, flies, and fleas.

What Is the Second Generation of Pesticides? Chemistry and Natural Plants to the Rescue

Chemists have developed hundreds of chemicals that can kill or repel pests, and they have improved natural pesticides produced by plants.

A major pest control revolution began in 1939, when entomologist Paul Müller discovered that DDT (dichlorodiphenyltrichloroethane), a chemical known since 1874, was a potent insecticide. DDT was the first of the so-called *second-generation pesticides*. It soon became the world's most used pesticide, and Müller received the Nobel Prize in 1948 for his discovery. Since then, chemists have made hundreds of other pesticides by making slight modifications in the molecules in various classes of chemicals (Table 23-1).

Since 1970 chemists have returned to natural repellents and poisons produced by plants. They have copied nature by improving first-generation botanical pesticides and adding microbotanicals (Table 23-1). They also developed pesticides based on a variety of natural chemicals found in the leaves and seeds of the remarkably versatile neem tree (Figure 11-17, p. 211, and Solutions, p. 213).

In 2003, Colorado State University biologist Jorge Vivanco discovered that knapweed, an invasive weed that has taken over large areas of grazing land in the West, may hold the key to developing effective natural herbicides. He and his colleagues found that the roots of this plant secrete a toxic chemical compound (catechin) into the soil that can wipe out all other surrounding plants. They found that adding toxic catechin to the soil or spraying it on weeds killed undesirable plants within a week. This new natural herbicide, discovered by learning from nature, should soon be on the market. Scientific curiosity and observing nature pay off.

How Are Pesticides Used Today? Almost Everywhere

Since 1950 we have greatly increased our use of a variety of increasingly toxic synthetic pesticides on crops, lawns, golf courses, and in households.

Since 1950, pesticide use has risen more than 50-fold, and most of today's pesticides are more than 10 times as toxic as those used in the 1950s. About three-fourths of these second-generation pesticides are used in de-

Table 23-1 Major Types of Pesticides

Type	Examples	Persistence	Biologically Magnified?
Insecticides			
Chlorinated hydrocarbons	DDT, aldrin, dieldrin, toxaphene, lindane, chlordane, methoxychlor, mirex	High (2–15 years)	Yes
Organophosphates	Malathion, parathion, diazinon, TEPP, DDVP, mevinphos	Low to moderate (1–2 weeks), but some can last several years	No
Carbamates	Aldicarb, carbaryl (Sevin), propoxur, maneb, zineb	Low (days to weeks)	No
Botanicals	Rotenone, pyrethrum, and camphor extracted from plants, synthetic pyrethroids (variations of pyrethrum), rotenoids (variations of rotenone), and neonicotinoids (variations of nicotine)	Low (days to weeks)	No
Microbotanicals	Various bacteria, fungi, protozoa	Low (days to weeks)	No
Herbicides			
Contact chemicals	Atrazine, simazine, paraquat	Low (days to weeks)	No
Systemic chemicals	2,4-D, 2,4,5-T, Silvex, diuron, daminozide (Alar), alachlor (Lasso), glyphosate (Roundup)	Mostly low (days to weeks)	No
Soil sterilants	Tribulan, diphenamid, dalapon, butylate	Low (days)	No
Fungicides			
Various chemicals	Captan, pentachlorophenol, zeneb, methyl bromide, carbon bisulfide	Most low (days)	No
Fumigants			
Various chemicals	Carbon tetrachloride, ethylene dibromide, methyl bromide	Mostly high	Yes (for most)

veloped countries, but use in developing countries is soaring.

After growing rapidly, pesticide use on crops in the United States has leveled off since 1980, but nonagricultural uses have increased. About one-fourth of pesticide use in the United States is for ridding houses, gardens, lawns, parks, playing fields, swimming pools, and golf courses of pests.

According to the U.S. Environmental Protection Agency (EPA), the average lawn in the United States is doused with 10 times more synthetic pesticides per hectare than U.S. cropland. Golf courses get almost as much pesticide per hectare as lawns get. Children rolling around in the grass on treated lawns and in city parks can pick up dangerous levels of some of these chemicals. They are especially vulnerable because they are still developing and can absorb more of these chemicals in proportion to their body weight than adults do. Health scientists warn that exposures to these and other toxic chemicals early in life can increase the risk of developing learning disabilities, behavioral problems, some forms of cancer, and other chronic diseases in childhood and in adulthood. Pesticide company officials dispute such claims.

The EPA also estimates that 84% of U.S. homes use pesticide products such as bait boxes, pest strips, bug bombs, flea collars, pesticide pet shampoos, and weed killers. What pesticide products are used where you live?

Sending someone roses, carnations, or other cut flowers is a nice gesture. Most of the cut flowers sold in the United States are imported from countries such as Colombia and Ecuador. *Bad news for romance.* According to a 1995 report by the World Resources Institute, flowers from these countries are heavily dosed with fungicides, insecticides, and herbicides. This poses health threats to the tens of thousands of workers, most of them women, who work in flower farms and greenhouses for about $5 a day. Environmentalists urge us to buy organic flowers.*

Some pesticides, called *broad-spectrum agents,* are toxic to many species; others, called *selective* or *narrow-spectrum agents,* are effective against a narrowly defined group of organisms. Pesticides vary in their *persistence,* the length of time they remain deadly in the environment (Table 23-1). In 1962, biologist Rachel Carson warned against relying on synthetic organic chemicals to kill insects and other species we deem pests (Individuals Matter, p. 27).

---

*You can search for local farms, farmers' markets, and community groups that supply fresh and dried organic flowers at www.localharvest.org. There are not many organic flower growers in the United States, but this could change if the demand increased.

23-2 THE CASE FOR PESTICIDES

What Are the Advantages of Modern Synthetic Pesticides? Many Benefits

Modern pesticides save lives, increase food supplies, increase profits for farmers, work fast, and are safe if used properly.

Proponents of conventional chemical pesticides contend that their benefits outweigh their harmful effects. Conventional pesticides have a number of important benefits.

They save human lives. Since 1945, DDT and other chlorinated hydrocarbon and organophosphate insecticides probably have prevented the premature deaths of at least 7 million people (some say as many as 500 million) from insect-transmitted diseases such as malaria (carried by the *Anopheles* mosquito), bubonic plague (carried by rat fleas), and typhus (carried by body lice and fleas).

They increase food supplies. According to the UN Food and Agriculture Organization, about 55% of the world's potential human food supply is lost to pests—about two-thirds of that before harvest and the rest after. Pests before and after harvest destroy an estimated 37% of the potential U.S. food supply; insects cause 13% of these losses, plant pathogens 12%, and weeds 12%. Without pesticides, these losses would be worse, and food prices would rise. Figure 23-2 (p. 522) shows five of the most common insect pests in the United States and their ranges.

They increase profits for farmers. Pesticide companies estimate that every $1 spent on pesticides leads to an increase in U.S. crop yields worth approximately $4 (but studies have shown this benefit drops to about $2 if the harmful effects of pesticides are included).

They work faster and better than alternatives. Pesticides control most pests quickly at a reasonable cost, have a long shelf life, are easily shipped and applied, and are safe when handled properly by farm workers. When genetic resistance occurs, farmers can use stronger doses or switch to other pesticides.

When used properly, their health risks are very low compared with their benefits. According to Elizabeth Whelan, director of the American Council on Science and Health (ACSH), which presents the position of the pesticide industry, "The reality is that pesticides, when used in the approved regulatory manner, pose no risk to either farm workers or consumers." According to the EPA, the worst-case scenario is that synthetic pesticides in food cause 0.5–1% of all cancer-related deaths in the United States, or 3,000–6,000 premature deaths per year, far less than the estimated number of lives saved each year by pesticides.

Figure 23-2 Geographic range of five major pests in the lower 48 states of the United States. (Data from U.S. Department of Agriculture)

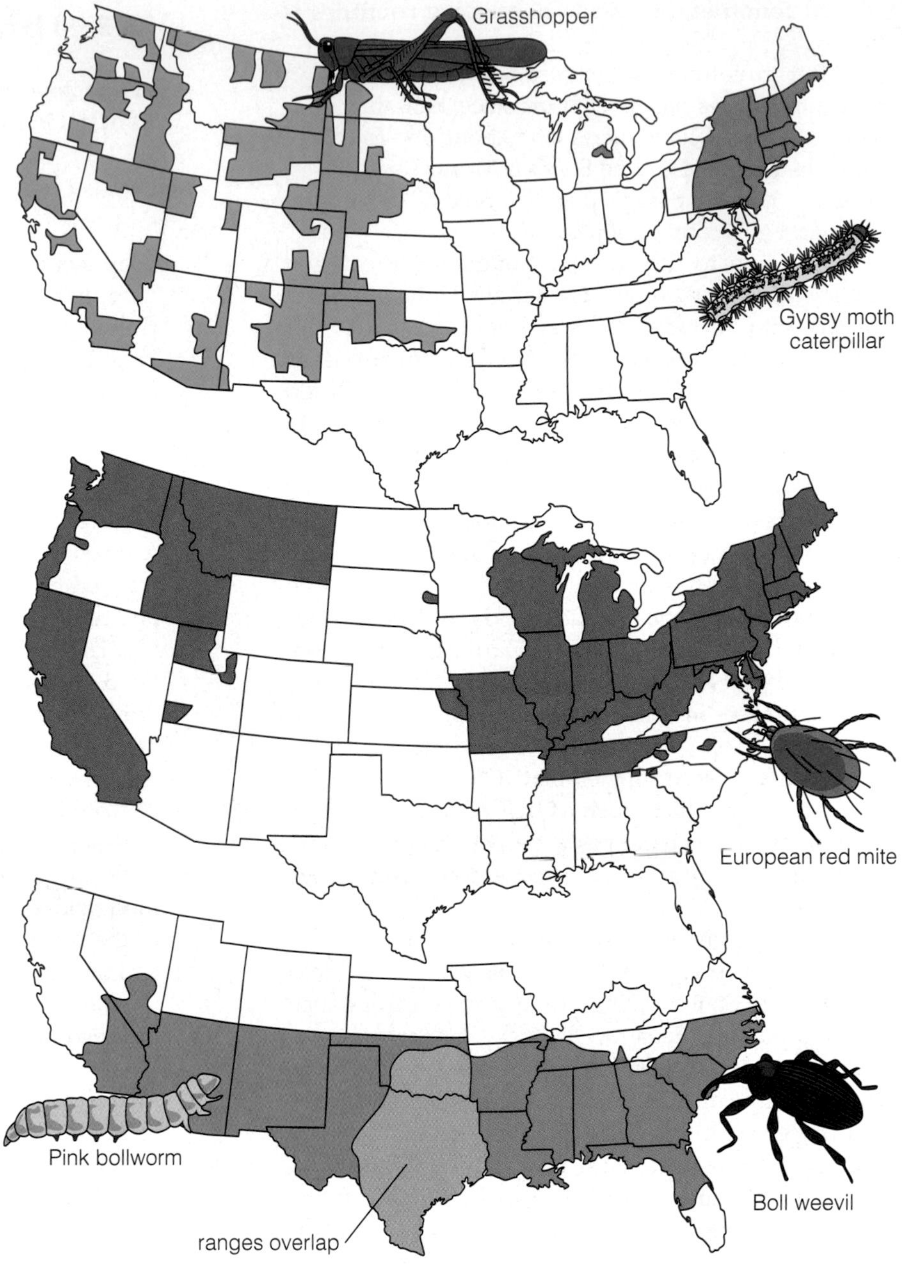

Grasshopper

Gypsy moth caterpillar

European red mite

Pink bollworm

Boll weevil

ranges overlap

According to studies by microbiologist Bruce Ames, we consume far more natural pesticides produced by plants than synthetic ones produced by humans. Ames also contends that exposure to natural pesticides in food causes more cancers than exposure to synthetic pesticides—although neither exposure poses much risk.

Newer pesticides are safer and more effective than many older pesticides. Greater use is being made of botanicals and microbotanicals (Table 23-1). Derived originally from plants, they are safer to users and less damaging to the environment than many older pesticides. Genetic engineering is also being used to develop pest-resistant crop strains and genetically altered crops that produce pesticides.

Many new pesticides are used at much lower rates per unit area than older products. For example, application amounts per hectare for many new herbicides are 1/100 the rates for older ones, and genetically engineered crops could reduce the use of toxic insecticides.

What Is the Ideal Pesticide? An Ongoing Search

Scientists work to develop more effective and safer pesticides, but through coevolution pests find ways combat the pesticides we throw at them.

Scientists continue to search for the ideal pest-killing chemical, which would have these qualities:

- Affect only the target organism
- Not cause genetic resistance in the target organism
- Disappear or break down into harmless chemicals after doing its job
- Be more cost effective than doing nothing

The search continues, but so far no known natural or synthetic pesticide chemical meets all or even most of these criteria.

23-3 THE CASE AGAINST PESTICIDES

What Is the Major Problem with Using Pesticides? Insects Have an Evolutionary Advantage

Insects can rapidly become genetically resistant to widely used pesticides.

Opponents of widespread pesticide use believe their harmful effects outweigh their benefits. They cite several serious problems with the use of conventional pesticides.

The major problem is that the widespread use of synthetic pesticides *accelerates the development of genetic* resistance to these chemicals by pest organisms. Insects breed rapidly (Figure 23-3), and within 5–10 years (much sooner in tropical areas) they can develop immunity to pesticides through natural selection and come back stronger than before (Spotlight, p. 524). Weeds and plant disease organisms also develop genetic resistance, but more slowly. Since 1945, about 520 species of insects (Figure 23-4), 280 plant disease organisms, and 150 weed species have developed genetic resistance to one or more pesticides.

Because of genetic resistance, many insecticides (such as DDT) no longer do a good job of protecting people from insect-transmitted diseases in some parts of the world. This has led to the resurgence of tropical diseases such as malaria. Genetic resistance can also put farmers on a *pesticide treadmill*, whereby they pay more and more for a pest control program that often becomes less and less effective.

What Are Other Problems with Using Pesticides? Some Serious Concerns

Pesticides can wipe out natural enemies of pest species, create new pest species, and end up in the environment, and some can harm wildlife and people.

Another problem is that *most pesticides kill beneficial species as well as the target pest species.* For example, most insecticides kill natural predators and parasites that help control the populations of insect pest species. Wiping out natural predators can unleash new pests,

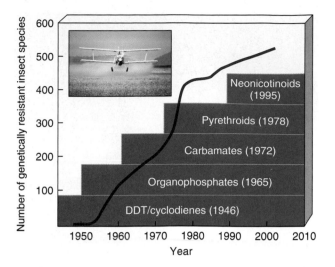

Figure 23-4 Between 1945 and 2000 about 520 insect species became genetically resistant to one or more widely used pesticides. Blue bars show time span over which types of pesticide groups have been used. Dates in parentheses indicate the year in which genetic resistance was first documented. (Data from U.S. Department of Agriculture and the Worldwatch Institute)

whose populations their predators had previously held in check, and can cause other unexpected effects (Connections, p. 173). Of the 300 most destructive insect pests in the United States, 100 were once minor pests that became major pests after widespread use of insecticides. With wolf spiders (Figure 23-1, left), wasps, predatory beetles, and other natural enemies out of the way, the population of a rapidly reproducing insect pest species can rebound and even get larger within days or weeks after initially being controlled.

Also, *pesticides do not stay put.* According to the U.S. Department of Agriculture (USDA), only 0.1–2% of the insecticide applied to crops by aerial (Figure 23-5, p. 524) or ground spraying reaches the target pests. Also, less than 5% of herbicide applied to crops reaches the target weeds. In other words, 98–99.9% of the pesticides and more than 95% of the herbicides we apply end up in the air, surface water, groundwater, bottom sediments, food, and nontarget organisms, including humans and wildlife (Figure 19-4, p. 411). Crops that have been genetically altered to release small amounts of pesticides directly to pests can help overcome this problem. But this can promote genetic resistance to such pesticides.

Some pesticides harm wildlife. According to the USDA and the U.S. Fish and Wildlife Service, each year pesticides applied to cropland in the United States wipe out about 20% of U.S. honeybee colonies and damage another 15%. This costs farmers at least $200 million per year from reduced pollination of vital crops. Pesticides also kill more than 67 million birds and 6–14 million fish, and menace about one of every five endangered and threatened species in the United States.

Figure 23-3 A boll weevil, just one example of an insect capable of rapid breeding. In the cotton fields of the southern United States, these insects lay thousands of eggs, producing a new generation every 21 days and as many as six generations in a single growing season. Attempts to control the cotton boll weevil account for at least one-fourth of insecticide use in the United States. For example, it typically takes about 114 grams (one-quarter pound) of pesticides to make one cotton T-shirt. Some farmers are increasing their use of natural predators and other biological methods to control this major pest.

Figure 23-5 A crop duster spraying an insecticide on grapevines south of Fresno, California. Aircraft apply about 25% of the pesticides used on U.S. cropland, but only 0.1–2% of these insecticides actually reach the target pests. To compensate for the drift of pesticides from target to nontarget areas, aircraft apply up to 30% more pesticide than ground-based application does.

Some pesticides can threaten human health. The World Health Organization (WHO) and the UN Environment Programme (UNEP) estimate that each year pesticides seriously poison at least 3 million agricultural workers in developing countries and at least 300,000 in the United States. This causes 20,000–40,000 deaths (about 25 in the United States) per year. Health officials believe the actual number of pesticide-related illnesses and deaths among the world's farm workers is greatly underestimated because of poor record-keeping, lack of doctors, inadequate reporting of illnesses, and faulty diagnoses.

Each year about 110,000 Americans, mostly children, get sick from misuse or unsafe storage of pesticides in the home, and about 20 die. According to studies by the National Academy of Sciences, exposure to pesticide residues in food causes 4,000–20,000 cases of cancer per year in the United States. Because roughly half of all people with cancer die prematurely, this amounts to about 2,000–10,000 premature deaths per year in the United States from exposure to legally allowed pesticide residues in foods. This is higher than the EPA estimate of 3,000–6,000 premature deaths per year. The pesticide industry disputes these claims.

Studies have linked exposure to some pesticides to childhood leukemia, Parkinson's disease, immune system disorders, and prostate and breast cancer.

Some scientists are becoming increasingly concerned about possible genetic mutations, birth defects, nervous system disorders (especially behavioral disorders), and effects on the immune and endocrine systems from long-term exposure to low levels of various pesticides (Case Study, p. 416). The pesticide industry disputes such claims.

Case Study: How Successful Have Pesticides Been in Reducing Crop Losses in the United States? Barely Holding the Line

A slightly higher percentage of the U.S. food supply is lost to pests today than in the 1940s.

Studies indicate that pesticides have not been as effective in reducing crop losses to pests in the United States as agricultural experts had hoped, mostly because of genetic resistance and reductions in natural predators.

David Pimentel, an expert in insect ecology, has evaluated data from more than 300 agricultural scientists and economists and come to three major conclusions.

SPOTLIGHT

A Superbug Called the Silverleaf Whitefly

The ideal insect pest would attack a variety of plants, be highly prolific and have a short generation time, have few natural predators, and be genetically resistant to a number of pesticides.

Bad news. The *silverleaf whitefly* has these characteristics, and farmers who have encountered it call it a *superbug*. This tiny white insect escaped from poinsettia greenhouses in Florida in 1986 and has become established in Florida, Arizona, California, and Texas.

It is known to eat at least 500 species of plants but does not like onions and asparagus and has no natural enemies. Dense swarms of these tiny insects attack plants, suck them dry, and leave them withered and dying.

U.S. crop losses from this insect are greater than $200 million a year and are growing. Scientists are scouring the world looking for natural enemies of this superbug. Stay tuned.

Critical Thinking

What is the ecological lesson to be learned from silverleaf whitefly?

What Goes Around Can Come Around

U.S. pesticide companies make and export to other countries pesticides that have been banned or severely restricted—or never even approved—for use in the United States. Other industrial countries also export banned and unapproved pesticides.

But what goes around can come around. In what environmentalists call a *circle of poison*, residues of some of these banned or unapproved chemicals exported to other countries can return to the exporting countries on imported food. Persistent pesticides such as DDT can also be carried by winds from other countries to the United States.

Environmentalists have urged the U.S. Congress—without success—to ban such exports. Supporters of pesticide exports argue that such sales increase economic growth and provide jobs and that banned pesticides are exported only with the consent of the importing countries. They also contend that if the United States did not export pesticides, other countries would.

In 1998, more than 50 countries met to finalize an international treaty that requires exporting countries to have informed consent from importing countries for exports of 22 pesticides and 5 industrial chemicals. In 2000, more than 100 countries developed an international agreement to ban or phase out the use of 12 especially hazardous persistent organic pollutants (POPs)—9 of them persistent chlorinated hydrocarbon pesticides such as DDT. In 2004, this treaty went into effect.

Critical Thinking

Should U.S. companies be allowed to export pesticides that have been banned, severely restricted, or not approved for use in the United States? Explain.

First, although the use of synthetic pesticides has increased 33-fold since 1942, about 37% of the U.S. food supply is lost to pests today compared to 31% in the 1940s. Since 1942 losses attributed to insects almost doubled from 7% to 13% despite a 10-fold increase in the use of synthetic insecticides.

Second, the estimated environmental, health, and social costs of pesticide use in the United States range from $4 billion to $10 billion per year. The International Food Policy Research Institute puts the estimate much higher, at $100–200 billion per year, or $5–10 in damages for every dollar spent on pesticides.

Third, alternative pest management practices could halve the use of chemical pesticides on 40 major U.S. crops without reducing crop yields.

Numerous studies and experience show that pesticide use can be reduced significantly without reducing yields, and in some cases, yields increase. Sweden has cut pesticide use in half with almost no decrease in crop yields. Campbell Soup uses no pesticides on tomatoes it grows in Mexico, and yields have not dropped. After a two-thirds cut in pesticide use on rice in Indonesia, yields increased by 15%.

> ☒ *HOW WOULD YOU VOTE?* Do the advantages of using synthetic chemical pesticides outweigh their disadvantages? Cast your vote online at http://biology.brookscole.com/miller14.

23-4 PESTICIDE REGULATION

How Are Pesticides Regulated in the United States? The Legal Approach

A federal law regulates pesticide use in the United States, but it can be improved.

The Federal Insecticide, Fungicide, and Rodenticide Act (FIFRA) was established by Congress in 1947 and amended in 1972. It requires EPA approval for use of all commercial pesticides. Pesticide companies must evaluate the biologically active ingredients in their products for toxicity to animals and, by extrapolation, to humans. EPA officials then review such data from pesticide companies to determine whether the pesticide can be registered for use. When a pesticide is approved for use on fruits or vegetables, the EPA sets a *tolerance level* specifying the amount of toxic pesticide residue that can legally remain on the crop when the consumer eats it.

The EPA banned or severely restricted the use of 56 active pesticide ingredients between 1972 and 2004. The banned chemicals include most chlorinated hydrocarbon insecticides, several carbamates and organophosphates, and the systemic herbicides 2,4,5-T and Silvex (Table 23-1). However, there is still controversy over the ban of DDT and other chlorine-containing pesticides (Case Study, p. 526).

Also, the 1996 Food Quality Protection Act (FQPA) increased public protection from pesticides. It requires manufacturers to demonstrate the safety of active ingredients in new pesticide products for infants and children. The EPA must also consider the effects of simultaneous exposures to more than one pesticide when setting pesticide tolerance levels.

However, banned or unregistered pesticides may be manufactured in the United States and exported to other countries (Connections, above left). Also, according to scientific literature reviewed by the EPA,

approximately 165 of the active ingredients approved for use in U.S. pesticide products are known or suspected human carcinogens. By 2004, only 43 of these pesticide chemicals had been banned by the EPA or discontinued voluntarily by manufacturers.

A study of Missouri children revealed a statistically significant correlation between childhood brain cancer and use of various pesticides in the home, including flea and tick collars, no-pest strips, and chemicals used to control pests such as roaches, ants, spiders, mosquitoes, and termites. Also, EPA scientists published a report in 2000 indicating that atrazine (widely used as a weed killer by farmers) could cause uterine, prostate, and breast cancer in humans and disrupt reproductive development.

Also, according to studies by the National Academy of Sciences, federal laws regulating pesticide use in the United States are inadequate and poorly enforced by the EPA, Food and Drug Administration (FDA), and USDA. Another study by the National Academy of Sciences found that up to 98% of the potential risk of developing cancer from pesticide residues on food grown in the United States would be eliminated if EPA standards were as strict for pre-1972 pesticides as they are for later ones.

The pesticide industry disputes these findings and says that eating food grown by using pesticides for the past 50 years has never harmed anyone in the United States. The industry also claims that the benefits of pesticides far outweigh their disadvantages.

Environmentalists and a number of health officials call for strengthening U.S. pesticide laws to help prevent contamination of groundwater by pesticides, improve the safety of farm workers who are exposed to high levels of pesticides, and allow citizens to sue the EPA for not enforcing the law. Pesticide manufacturers strongly oppose such changes and lobby elected officials to weaken FIFRA.

Pesticide control laws in the United States could be improved. But most other countries (especially developing countries) have not made nearly as much progress as the United States has in regulating pesticides.

Case Study: Revisiting DDT— from Riches to Rags

Since 1972 DDT has been banned in developed countries, and there is controversy over its continuing use in some developing countries to combat malaria.

After its discovery in 1939, DDT quickly became the world's most widely used pesticide. It was a cheap and effective weapon to kill crop-devouring insects and mosquitoes and other insects that transmitted infectious diseases such as malaria. There is little doubt that single-handedly this chemical has saved many millions of lives from infectious diseases.

DDT's role as a "chemical hero" began changing in 1962 when Rachel Carson published her book *Silent Spring*, which warned of the dangers of DDT and other broad-spectrum and persistent pesticides (Individuals Matter, p. 27). This led to much closer scrutiny of such pesticides and public pressure to ban DDT and its persistent chlorinated hydrocarbon chemical cousins that were also widely used as pesticides (Table 23-1).

In 1970, the U.S. Environmental Protection Agency was established. In 1972, the earlier Federal Insecticide, Fungicide, and Rodenticide Act (FIFRA) was amended to give the EPA control over the registration and regulation of pesticides in the United States.

In that same year, the EPA banned the use of DDT (and later the use of its similar chemical cousins) in the United States. The EPA banned DDT for several reasons. *First,* it is a broad-spectrum chemical that kills many beneficial insects along with its target species. *Second,* it is a persistent chemical that remains in one chemical form or another in the environment for up to 15 years and can be biologically magnified in food webs (Figure 19-4, p. 411). *Third,* it reduced populations of many birds and other species, especially those feeding at high trophic levels in food webs, such as eagles and peregrine falcons. *Fourth,* there was some preliminary but not conclusive evidence that it could cause cancer in humans. *Fifth,* it was becoming less effective because a growing number of insect pests that consume crops and transmit diseases had developed genetic resistance to DDT and other chlorinated hydrocarbon pesticides. Some contend political pressure from the public and a growing environmental movement also played a role in the ban of this chemical.

Pesticide manufacturers opposed the ban but were more than happy to supply more expensive alternatives. Debate over the DDT ban in the United States continues today. There was considerable evidence for its ecological harm and more evidence has accumulated. But pesticide industry scientists say there was not enough evidence then (and today) that DDT can cause cancer in humans—one of the key reasons used to ban the chemical under the FIFRA pesticide law.

Critics of the ban try to separate the possible harmful effects of DDT on humans from its effects on other species and ecosystems. They pose such questions as, Do we want to protect penguins or people?

Scientists say this is too simplistic because we cannot separate harm to the environment from harm to people. This is especially true for widely used and long-lived chemicals such as DDT that can build up in food webs and are now found in even the most remote parts of the world.

This was the heart of Rachel Carson's warning. Traces of these chemicals are everywhere, including

our bodies, and we should be concerned about their possible long-term effects on both the environment and human health. Critics of pesticides contend that the best way to reduce such risks is to prevent such chemicals from reaching the environment. This would spur us to look for safer, affordable alternative chemicals and for biological and ecological ways to control pests.

In addition, since 1975 there has been growing evidence that very low levels of chlorine-containing pesticides and a variety of other fat-soluble chemicals may disrupt the human immune, endocrine, and nervous systems by mimicking and disrupting the effects of natural hormones in our bodies (Case Study, p. 416). The scientific jury is still out on if or how these chemicals are harmful to humans.

Critics of the ban on DDT and other chlorinated hydrocarbon pesticides say the ban ended up increasing human deaths from exposure to pesticides. Why? Organophosphates that were less persistent and ecologically damaging replaced chlorinated hydrocarbon pesticides. But it turned out that these chemicals were hundreds and in some cases thousands of times more toxic to humans than DDT and its chemical cousins. As a result, these replacements killed a large number of farm workers and children playing in sprayed fields or otherwise coming into contact with organophosphate pesticides.

This led the EPA to ban the use of many organophosphates and then carbamates that followed them (Table 23-1). Since then new groups of pesticides such as botanicals and microbotanicals have been developed that are less harmful to humans and the environment.

Although DDT is banned in developed countries, it has not gone away. It is manufactured legally in several countries and is still used to treat crops and to kill disease-carrying insects in a number of developing countries.

In 2000, delegates from 122 countries agreed on a global pollution prevention treaty to control, reduce, phase out, and destroy stockpiles of 12 persistent organic pollutants (POPs). This list of chemicals, called the *dirty dozen,* includes DDT and eight other chlorine-containing persistent pesticides.

The treaty, which went into effect in 2004, allows 25 countries to continue using DDT to combat malaria until safer alternatives are available. This was allowed because the health benefits of using DDT to decrease malaria far outweigh the remote possibility of harm to people. Although DDT may prove to have some as-yet unknown harmful effects on humans, malaria kills about 1 million people a year—most of them children—and sickens and weakens several hundred million people.

In addition, spraying low levels of DDT indoors and on bed nets would not spread large amounts of the chemical into the environment compared to blanketing crop fields with DDT. This should slow the development of genetic resistance to DDT in malaria-carrying mosquitoes. Also, after the ban it was discovered that even when mosquitoes developed genetic resistance to DDT, it still acted as a repellent and irritant that drove nocturnal mosquitoes out of homes before they had a chance to bite. Despite this decision the World Bank and other international aid agencies do not provide loans or funds for malaria-control projects that involve the use of DDT.

Opponents argue that a complete ban on DDT will spur research efforts to find other cost-effective pesticides for killing malaria-causing mosquitoes and to find alternatives to using pesticides. They support WHO efforts to use a variety of methods to reduce the threat of malaria.

X *HOW WOULD YOU VOTE?* Should DDT and other persistent chlorine-containing pesticides still be used to control malaria throughout the world? Cast your vote online at http://biology .brookscole.com/miller14.

23-5 ALTERNATIVES TO CONVENTIONAL CHEMICAL PESTICIDES

What Should Be the Primary Goal of Pest Control? Pest Reduction Not Eradication

Reducing crop damage to an economically tolerable level should be the primary goal of pest control efforts.

In most cases, the primary goal of spraying with conventional pesticides is to eradicate pests in the area affected. However, critics say the primary goal of any pest control strategy should be to reduce crop damage to an economically tolerable level. The point at which the economic losses caused by pest damage outweigh the cost of applying a pesticide is called the *economic threshold.* Because of the risk of increased genetic resistance and other problems, continuing to spray beyond the economic threshold can make matters worse and can cost more than it is worth.

The problem is determining when the economic threshold has been reached. This involves careful monitoring of crop fields to assess crop damage and determine pest populations.

Many farmers do not want to bother doing this and instead are likely to use additional *insurance spraying* to be on the safe side. One method used to reduce unnecessary insurance spraying is the purchase of *pest-loss insurance.* It pays farmers for losses caused by pests and is usually cheaper than using excess pesticides.

Another source of increased pesticide use is *cosmetic spraying.* Extra pesticides are used because most consumers often buy only the best-looking fruits and vegetables even though there is nothing wrong with blemished ones. The only solution to this problem is consumer education. Would you buy blemished fruits and vegetables?

What Are Other Ways to Control Pests? Copy Nature

A mix of cultivation practices and biological and ecological alternatives to conventional chemical pesticides can help control pests.

Many scientists believe we should greatly increase the use of biological, ecological, and other alternative methods for controlling pests and diseases that affect crops and human health. A number of methods are available.

One is the use of various *cultivation practices* to fake out pest species. Examples are rotating the types of crops planted in a field each year, adjusting planting times so major insect pests either starve or get eaten by their natural predators, and growing crops in areas where their major pests do not exist. Also, farmers can increase the use of polyculture, which uses plant diversity to reduce losses to pests.

Homeowners can reduce weed invasions by cutting grass no lower than 8 centimeters (3 inches) high. This provides a dense enough cover to keep out crabgrass and many other undesirable weeds. Homeowners can also avoid growing plants such as roses that attract a number of insect pests and grow plants such as chrysanthemums and marigolds that repel many insect pests.

Genetic engineering can be used to speed up the development of pest- and disease-resistant crop strains (Figure 23-6). But there is controversy over whether the projected advantages of the increasing use of genetically modified plants and foods outweigh their projected disadvantages (Figure 14-19, p. 292).

We can increase the use of *biological pest control.* It involves importing natural predators (Figures 23-1 and 23-7), parasites, and disease-causing bacteria and viruses to help regulate pest populations. More than 1,000 species have been introduced to help control pest species in North America, with generally favorable results. For example, several species of European beetles are being used in the United States to help reduce the purple loosestrife plant that has invaded many U.S. wetlands (Figure 13-5, p. 256).

Biological control focuses on selected target species, is nontoxic to other species, and minimizes genetic resistance. Also, it can save large amounts of money—about $25 for every $1 invested in controlling 70 pests in the United States. However, biological agents cannot always be mass produced, are often slower acting and more difficult to apply than conventional pesticides, can sometimes multiply and become

Figure 23-6 The results of one example of using *genetic engineering* to reduce pest damage. Both tomato plants were exposed to destructive caterpillars. The normal plant's leaves are almost gone (left), whereas the genetically altered plant (right) shows little damage.

Figure 23-7 Natural capital: biological pest control. An adult convergent ladybug (right) is consuming an aphid (left).

pests themselves, and must be protected from pesticides sprayed in nearby fields.

Another strategy is *insect birth control.* This involves raising males of insect pest species in the laboratory and sterilizing them by exposure to radiation or chemicals. The sterile males are released into an infested area to mate with fertile wild females who then lay eggs that never hatch. This method has been used to control the screwworm fly, a major livestock pest from the southeastern United States (Figure 23-8), and the Mediterranean fruit fly (medfly) during a 1990 outbreak in California. However, problems include high costs, difficulties in knowing the mating time and behavior of each target insect, and the large number of sterile males needed. In addition, there are few species for which this strategy works, and sterile males must be released continually to prevent pest resurgence.

Sex attractants can also help control pests. Plants and animals have evolved a variety of natural attractants called *pheromones.* Scientists have identified many of these natural chemicals and use them to lure

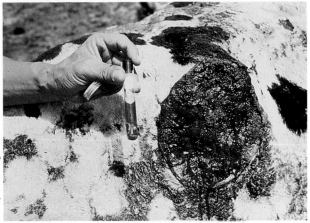

Figure 23-8 Infestation of a steer by screwworm fly larvae in Texas. An adult steer can be killed in 10 days by thousands of maggots feeding on a single wound.

Figure 23-9
Pheromones can help control populations of pests, such as the red scale mites that have infested this lemon grown in Florida.

pests such as Japanese beetles into traps or to attract their natural predators into crop fields (usually the more effective approach). Pheromones can also be released into the air to confuse insects and make it difficult for them to find mates. More than 50 companies worldwide sell about 250 pheromones to control pests (Figure 23-9).

These chemicals attract only one species, work in trace amounts, have little chance of causing genetic resistance, and are not harmful to nontarget species. However, it is costly and time consuming to identify, isolate, and produce the specific sex attractant for each pest or predator.

Another approach is to use *hormones that disrupt an insect's normal life cycle* (Figure 23-10) and prevent it from reaching maturity and reproducing (Figure 23-11, p. 530). Insect hormones are natural chemicals produced by insects to regulate their growth at various stages of their natural life cycle. By learning what hormones an insect needs at various stages in its life, scientists can use these chemicals to disrupt and kill the insect—another example of learning from nature.

Insect hormones have the same advantages as sex attractants. But they take weeks to kill an insect, often are ineffective with large infestations of insects, and sometimes break down before they can act. In addition, they must be applied at exactly the right time in the target insect's life cycle, can sometimes affect the target's predators and other nonpest species, and are difficult and costly to produce.

Some farmers have controlled some insect pests by *spraying them with hot water.* This has worked well on cotton, alfalfa, and potato fields and in citrus groves in Florida, and the cost is roughly equal to that of using chemical pesticides.

Another strategy is to *expose foods to high-energy gamma radiation.* Such *food irradiation* extends food shelf life and kills insects and parasitic worms (such as trichinae in pork). It also kills harmful bacteria such as salmonella, which infects at least 51,000 Americans and kills 2,000 each year, and *E. coli,* which infects more than 20,000 Americans and kills about 250 each year. According to the U.S. FDA and the WHO, more than 2,000 studies show that foods exposed to low doses of gamma radiation are safe for human consumption.

But critics of irradiating food argue that it forms trace amounts of certain chemicals that have caused cancer in laboratory animals. They also point out that the long-term health effects of eating irradiated food are unknown and consumers do not want old and possibly less nutritious food to be made to appear fresh and healthy by irradiation. They also support clear labeling of all irradiated foods so that consumers can

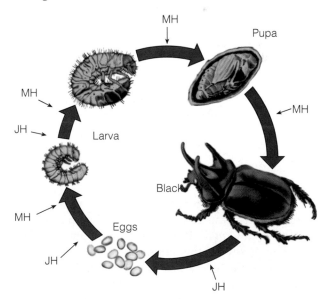

Figure 23-10 For normal insect growth, development, and reproduction to occur, certain juvenile hormones (JH) and molting hormones (MH) must be present at genetically determined stages in the insect's life cycle. If applied at the proper time, synthetic hormones disrupt the life cycles of insect pests and help control their populations.

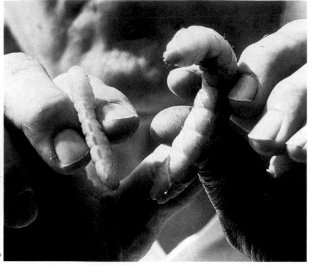

Figure 23-11 A use of hormones to prevent insects from maturing completely, making it impossible for them to reproduce. The stunted tobacco hornworm (left) was fed a hormone that prevents production of molting hormones. They eat but die when they cannot shed the skin off of their bulging bodies—somewhat like being trapped in a tight wet suit while you put on lots of weight. A normal hornworm is shown on the right.

make informed choices—a proposal that is opposed by sellers of such foods. Another problem: the poorly protected facilities for food irradiation contain radioactive isotopes that terrorists could steal and use to make dirty nuclear bombs.

Some consumers oppose irradiating food and refuse to eat it because they fear it is radioactive, but it is not. When food is irradiated it does not become radioactive any more than you do when you get a dental or chest X ray.

Is Integrated Pest Management the Answer? A Combined Ecological Approach

An ecological approach to pest control uses an integrated mix of cultivation and biological methods, and small amounts of selected chemical pesticides as a last resort.

An increasing number of pest control experts and farmers believe the best way to control crop pests is a carefully designed **integrated pest management (IPM)** program. In this approach, each crop and its pests are evaluated as parts of an ecological system. Then farmers develop a control program that includes cultivation, biological, and chemical methods applied in proper sequence and with the proper timing.

The overall aim of IPM is not to eradicate pest populations but to reduce crop damage to an economically tolerable level. Fields are monitored carefully to determine when an economically damaging level of pests has been reached. When this happens farmers first use biological methods (natural predators, parasites, and disease organisms) and cultivation controls, including vacuuming up harmful bugs. Small amounts of insecticides—mostly based on natural insecticides produced by plants—are applied only as a last resort. Also, different chemicals are used in order to slow the development of genetic resistance and to avoid killing predators of pest species.

In 1986, the Indonesian government banned 57 of the 66 pesticides used on rice and phased out pesticide subsidies over a 2-year period. It also launched a nationwide education program to help farmers switch to IPM. The results were dramatic. Between 1987 and 1992, pesticide use dropped by 65%, rice production rose by 15%, and more than 250,000 farmers were trained in IPM techniques. In Sri Lanka IPM increased rice yields 11–44% and increased farmer incomes 38–178%. Sweden and Denmark have used IPM to cut their pesticide use in half.

The experiences of these and other countries show that a well-designed IPM program can reduce pesticide use and pest control costs by at least half, cut preharvest losses from pests by half, and improve crop yields. It can also reduce inputs of fertilizer and irrigation water and slow the development of genetic resistance because pests are assaulted less often and with lower doses of pesticides.

Thus *IPM is an important form of pollution prevention* that reduces risks to wildlife and human health. Consumers Union estimates that if all U.S. farmers practiced IPM by 2020, public health risks from pesticides would drop by 75%.

Why Have More Farmers Not Switched to Integrated Pest Management? Politics in Action

Government subsidies for conventional pesticides, opposition by pesticide manufacturers, and a lack of experts to advise farmers hinder a widespread shift to integrated pest management.

Despite its promise, IPM, like any other form of pest control, has some disadvantages. It requires expert knowledge about each pest situation including a pest's life cycles, feeding habits, movements, and nesting habits. It is also slower acting than conventional pesticides, and methods developed for a crop in one area might not apply to areas with even slightly different growing conditions. Also, initial costs may be higher, although long-term costs typically are lower than those of using conventional pesticides.

Widespread use of IPM is hindered by government subsidies of conventional chemical pesticides and opposition from agricultural chemical companies, whose pesticide sales would drop sharply. There is also a lack of experts to help farmers shift to IPM.

A 1996 study by the National Academy of Sciences recommended that the United States shift from chemically based approaches to ecologically based pest management approaches. According to the study, within 5–10 years, such a shift could cut U.S. pesticide use in half, as it has in several other countries.

A growing number of scientists urge the USDA to use three strategies to promote IPM in the United States:

- Add a 2% sales tax on pesticides and use the revenue to fund IPM research and education

- Set up a federally supported IPM demonstration project on at least one farm in every county

- Train USDA field personnel and county farm agents in IPM so they can help farmers use this alternative

The pesticide industry has successfully opposed such measures.

Good news. Several UN agencies and the World Bank have joined together to establish an IPM facility. Its goal is to promote use of IPM by disseminating information and establishing networks among researchers, farmers, and agricultural extension agents involved in IPM.

X *HOW WOULD YOU VOTE?* Should governments heavily subsidize a switch to integrated pest management? Cast your vote online at http://biology.brookscole.com/miller14.

We need to recognize that pest control is basically an ecological, not a chemical, problem.

ROBERT L. RUDD

CRITICAL THINKING

1. Do you agree or disagree that because DDT and the other banned chlorinated hydrocarbon pesticides pose no demonstrable threat to human health and have saved millions of lives, they should again be approved for use on crops in the United States? Explain.

2. If increased mosquito populations threatened you with malaria or West Nile virus, would you spray DDT in your yard and inside your home to reduce the risk? Explain. What are the alternatives?

3. Explain how widespread use of a pesticide can (a) increase the damage done by a particular pest and (b) create new pest organisms.

4. Explain why biological pest control often is more successful on a small island than on a continent.

5. Should farmers be given government subsidies for switching to integrated pest management (IPM)? Explain your position.

6. Should certain types of foods be irradiated to help control disease organisms and increase shelf life? Explain. If so, should such foods be required to carry a clear label stating that they have been irradiated? Explain.

7. What changes, if any, do you believe should be made in the Federal Insecticide, Fungicide, and Rodenticide Act and the Food Quality Protection Act that regulate pesticide use in the United States?

8. Congratulations! You are in charge of pest control for the entire world. What are the three most important components of your global pest management strategy?

PROJECTS

1. How are bugs and weeds controlled in (a) your yard and garden, (b) the grounds of your school, and (c) public school grounds, parks, and playgrounds in your community?

2. List all pesticides used in or around your home. Compare the results for your entire class. Which ones could be eliminated?

3. Some research shows that although many people agree we need to make greater use of alternatives to conventional pesticides for controlling pests, when they are faced with an actual infestation from insects or rodents the first thing they do is spray with pesticides. Survey members of your class and other groups to help determine the validity of these research findings.

4. Use the library or the Internet to find bibliographic information about *Ralph Waldo Emerson* and *Robert L. Rudd*, whose quotes appear at the beginning and end of this chapter.

5. Make a concept map of this chapter's major ideas, using the section heads, subheads, and key terms (in boldface). Look at the website for this book for information about making concept maps.

LEARNING ONLINE

The website for this book contains study aids and many ideas for further reading and research. They include a chapter summary, review questions for the entire chapter, flash cards for key terms and concepts, a multiple-choice practice quiz, interesting Internet sites, references, and a guide for accessing thousands of InfoTrac® College Edition articles. Log on to

http://biology.brookscole.com/miller14

Then click on the Chapter-by-Chapter area, choose Chapter 23, and select a learning resource.

24 Solid and Hazardous Waste

Love Canal: There Is No "Away"

Between 1942 and 1953, Hooker Chemicals and Plastics (owned by OxyChem since 1968) sealed chemical wastes containing at least 200 different chemicals into steel drums and dumped them into an old canal excavation (called Love Canal after its builder, William Love) near Niagara Falls, New York.

In 1953, Hooker Chemicals filled the canal, covered it with clay and topsoil, and sold it to the Niagara Falls school board for $1. In 1957, Hooker warned the school board not to disturb the clay cap because of possible danger from the buried toxic wastes.

By 1959, an elementary school, playing fields, and 949 homes had been built in the 10-square-block Love Canal area (Figure 24-1). Some of the roads and sewer lines crisscrossing the dump site disrupted the clay cap covering the wastes. In the 1960s, an expressway was built at one end of the dump. It blocked groundwater from migrating to the Niagara River and allowed contaminated groundwater and rainwater to build up and overflow the disrupted cap.

Residents began complaining to city officials in 1976 about chemical smells and chemical burns their children received playing in the canal area, but their concerns were ignored. In 1977, chemicals began leaking from the badly corroded steel drums into storm sewers, gardens, basements of homes next to the canal, and the school playground.

In 1978, after media publicity and pressure from residents led by Lois Gibbs (a mother galvanized into action as she watched her children come down with one illness after another (see her Guest Essay on the website for this chapter), the state acted. It closed the school and arranged for the 239 homes closest to the dump to be evacuated, purchased, and destroyed.

Two years later, after protests from families still living fairly close to the landfill, President Jimmy Carter declared Love Canal a federal disaster area, had the remaining families relocated, and offered federal funds to buy 564 more homes. Because of the difficulty in linking exposure to a variety of chemicals to specific health effects (Section 19-2, p. 410), the long-term health effects of exposure to hazardous chemi-

Figure 24-1 The Love Canal housing development near Niagara Falls, New York, was built near a hazardous waste dump site. The photo shows the area when it was abandoned in 1980. In 1990, the EPA allowed people to buy some of the remaining houses and move back into the area.

cals for Love Canal residents remain unknown and controversial.

The dumpsite has been covered with a new clay cap and surrounded by a drainage system for pumping leaking wastes to a new treatment plant. In 1990, state officials began selling 260 of the remaining houses in the area—renamed Black Creek Village. Buyers must sign an agreement stating that New York state and the federal government make no guarantees or representations about the safety of living in these homes.

Love Canal sparked creation of the Superfund law, which forced polluters to pay for cleaning up abandoned toxic waste dumps and made them wary of producing new ones. In 1983 Love Canal became the first Superfund site. After spending close to $400 million in cleanup costs it was removed from the Superfund priority list in 2004.

The Love Canal incident is a vivid reminder of three lessons from nature: *We can never really throw anything away; Wastes often do not stay put;* and, *preventing pollution is much safer and cheaper than trying to clean it up.*

Solid wastes are only raw materials we're too stupid to use.
ARTHUR C. CLARKE

This chapter examines the solid and hazardous wastes we produce. It addresses the following questions:

- What is solid waste and how much do we produce?

- What can we do to reduce, reuse, and recycle solid waste?

- What are the advantages and disadvantages of burning or burying solid waste?

- What is hazardous waste and how can we deal with it?

- How can we detoxify hazardous waste?

- What are the advantages and disadvantages of burning or burying hazardous waste?

- What can we do to reduce exposure to lead, mercury, and dioxins?

- How is hazardous waste regulated in the United States?

- How can we make the transition to a more sustainable low-waste society?

24-1 WASTING RESOURCES

Why Should We Care about Solid Waste? Resource Waste and Pollution

Solid waste is a symptom of an unnecessary waste of resources whose production causes pollution and environmental degradation.

Solid waste is any unwanted or discarded material that is not a liquid or a gas. So what is the big deal? For most people, garbage trucks arrive and whisk away the solid waste they produce—out of sight, out of mind.

In nature there is essentially no solid waste because the wastes of one organism become nutrients for other organisms. But humans will always produce some solid waste. Indeed, we produce such wastes directly and indirectly in almost everything we do. The solid waste we produce directly is called *garbage*. But most people do not realize that mines, factories, food growers, and businesses supplying the goods and services they use are responsible for about 98% of the world's solid waste.

We need to be concerned about such waste production for two reasons. One is that much of it represents an unnecessary waste of the earth's precious resources. The other is that producing the solid products we use and often discard is responsible for huge amounts of air pollution (including greenhouse gases), water pollution, soil erosion, and land degradation (Figure 16-13, p. 343).

Some *good news* is that we could reduce our direct and indirect production of solid waste by 75–90%, as you will learn in this chapter.

Case Study: How Much Solid Waste Does the United States Produce? Affluenza in Action

The United States produces about a third of the world's solid waste and buries more than half of it in landfills.

What country produces the most solid waste? The answer is the United States that, with only 4.6% of the world's population, produces about one-third of the world's solid waste—a glaring symptom of affluenza (p. 14).

About 98.5% of the solid waste in the United States (and in most developed countries) comes from mining, oil and natural gas production, agriculture, sewage sludge, and industrial activities (Figure 24-2). This solid waste is produced *indirectly* to provide goods and services to meet the needs and growing wants of consumers.

Suppose you buy a desktop computer. You probably do not know that making it used 700 or more different materials obtained from mines, oil wells, and chemical factories all over the world. You may also be unaware that for every 0.5 kilogram (1 pound) of electronics it contains, approximately 3,600 kilograms (8,000 pounds) of solid and liquid waste was created somewhere in the world. Extracting these resources and converting them into your computer also required large amounts of energy produced mostly by burning fossil fuels, which emits pollutants and CO_2 into the air.

The remaining 1.5% of solid waste is **municipal solid waste (MSW)**—often called *garbage* or *trash*—generated mostly by homes and workplaces. This small

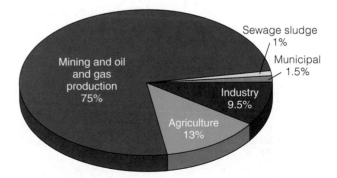

Figure 24-2 Natural capital degradation: sources of the estimated 11 billion metric tons (12 billion tons) of solid waste produced each year in the United States. Mining, oil and gas production, agricultural, and industrial activities produce 65 times as much solid waste as household activities. (Data from U.S. Environmental Protection Agency and U.S. Bureau of Mines)

part of the overall solid waste problem is still huge. *Bad news.* Between 1960 and 2001, the total amount of MSW in the United States each year increased 2.6-fold and is still rising. Each year the United States generates enough MSW to fill a bumper-to-bumper convoy of garbage trucks encircling the globe almost eight times! Between 1960 and 1990, the amount of MSW produced per person in the United States increased by 70%.

Canada is the world's second largest per capita producer of MSW. Japan and most developed countries in Europe produce about half as much MSW per person as the United States, and most developing countries produce about one-fourth to one-tenth as much.

What Is in U.S. Garbage? Paper Rules

Paper products make up the largest percentage of municipal solid waste in the United States, but electronic waste or e-waste is the fastest-growing type of solid waste.

Analysis of landfill content shows that paper makes up about 38% of the trash buried in U.S. landfills, followed by yard waste (12%), food waste (11%), and plastics (10%).

But *electronic waste* or *e-waste* consisting of discarded TV sets, cell phones, computers, and other electronic devices is the fastest-growing solid waste problem in the United States and the world. It is also a source of toxic and hazardous wastes such as polyvinylchloride (PVC) and compounds containing lead and mercury that can contaminate the air, surface water, groundwater, and soil. In the United States, only about 2% of such e-waste is recycled.

How do we know the composition of trash in landfills? Much of it comes from research by *garbologists* such as William Rathe who pioneered this field at the University of Arizona. These scientists are modern versions of archaeologists who examine people's trash and dig holes in garbage dumps and analyze what they find.

Many people think of landfills as huge compost piles where biodegradable wastes are decomposed within a few months. But garbologists looking at the contents of landfills found 50-year-old newspapers that were still readable and hot dogs and pork chops buried for decades that still looked edible. In landfills (as opposed to open dumps), trash can resist decomposition for perhaps centuries because it is tightly packed and protected from sunlight, water, and air.

What Does It Mean to Live in a High-Waste Society? A Throwaway Mentality

Most solid waste is a highly visible sign of how a society infected with affluenza wastes valuable resources.

According to architect and environmental designer William McDonough, the industrial revolution that has been taking place for about 275 years has a number of harmful consequences. It has put huge amounts of toxic material into the air, water, and soil. It has put hard-to-separate mixtures of potentially valuable resources in landfills or other holes all over the planet, where they are too difficult or expensive to retrieve and separate into resources. It has spurred thousands of complex government regulations, mainly designed to keep most people from being poisoned or harmed too quickly instead of keeping people and natural systems safe for the long term.

It has depleted and degraded the earth's natural capital (top half of back cover) and eroded biodiversity and human cultural diversity. Finally, it has counted these harmful consequences as economic progress because they raise the gross domestic product.

Here are a few of the solid wastes consumers throw away in the high-waste economy found in the United States:

- Enough aluminum to rebuild the country's entire commercial airline fleet every 3 months

- Enough tires each year to encircle the planet almost three times

- Enough disposable diapers each year that if they were linked end to end they would reach to the moon and back seven times

- About 2 billion disposable razors, 130 million cell phones, 50 million computers, and 8 million television sets each year

- Discarded carpet each year that would cover the state of Delaware

- About 2.5 million nonreturnable plastic bottles every hour

- About 670,000 metric tons (1.5 billion pounds) of edible food per year

- Enough office paper each year to build a wall 3.5 meters (11 feet) high across the country from New York City to San Francisco

- Some 186 billion pieces of junk mail (an average of 660 per American) each year, about 45% of which are thrown in the trash unopened

Strange things happen in a society infected with affluenza. For example, according to the United Nations Environment Programme, Americans spend more on trash bags each year than 90 other countries spend for everything. And American comedian Lily Tomlin observes, "We buy a wastebasket and take it home in a plastic bag. Then we take the wastebasket out of the bag, and put the bag in the wastebasket."

24-2 PRODUCING LESS WASTE

What Are Our Options? Management or Prevention

We can try to manage the solid wastes we produce or try to reduce or prevent their production.

We can deal with the solid wastes we create in two ways. One is *waste management*. This is a *high-waste approach* (Figure 3-18, p. 53) that views waste production as a largely unavoidable product of economic growth. It attempts to manage the resulting wastes in ways that reduce environmental harm, mostly by mixing and often crushing them together and then burying them, burning them, or shipping them off to another state or country. In effect, it mixes the wastes we produce together and then transfers them from one part of the environment to another.

The second approach is *waste reduction,* a *low-waste approach* that recognizes there is no "away." It views most solid waste as potential resources that we should be reusing, recycling, or composting. With this approach we should be taught to think of trash cans and garbage trucks as *resource containers*. Figure 24-3 lists ways to reduce waste. Study this figure carefully.

Waste reduction is based on the four R's for dealing with the wastes we produce: *refuse, reduce, reuse, or recycle (including composting).* It is the preferred solution because it tackles the problem of waste production at the front end—before it occurs—rather than at the back end after wastes have already been produced. It also saves matter and energy resources, reduces pollution (including emissions of greenhouse gases), helps protect biodiversity, and saves money.

Solutions: How Can We Reduce Solid Waste? The Sustainability Six

Reducing consumption and redesigning the products we produce are the best ways to cut waste production and promote sustainability.

Here are six ways to reduce resource use, waste, and pollution—what we might call the *sustainability six*. First, *consume less.* Before buying anything, ask questions such as: Do I really *need* this or do I just *want* it? Can I buy it secondhand (reuse)? Can I borrow or rent it (reuse)?

Second, *redesign manufacturing processes and products to use less material and energy.* A skyscraper built today includes about a third less steel than one the same size built in the 1960s because of the use of lighter-weight but higher-strength steel. The weight of cars has been reduced by about one-fourth by using such steel along with lightweight plastics and composite materials. Plastic milk jugs weigh 40% less that they did in the 1970s, and aluminum drink cans contain one-third less aluminum. All of these changes involve savings in energy use as well as materials.

Third, *redesign manufacturing processes to produce less waste and pollution.* Most toxic organic solvents can be recycled within factories or replaced with water-based or citrus-based solvents (Individuals Matter, p. 489). Hydrogen peroxide can be used instead of toxic chlorine to bleach paper and other materials. Three nontoxic ways to clean clothes are now available. One cleans clothes with water in computer-controlled machines and another uses a nontoxic silicone solvent in conventional dry-cleaning machines. A new method submerses clothes in liquid carbon dioxide. Check your local phone directory to locate dry cleaners that use these alternative methods.

Fourth, *develop products that are easy to repair, reuse, remanufacture, compost, or recycle.* A Xerox photocopier with

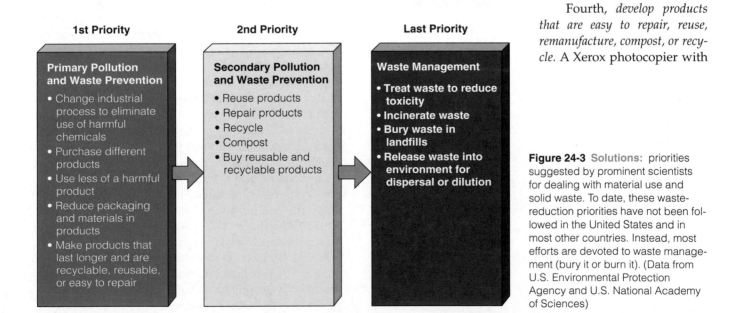

1st Priority

Primary Pollution and Waste Prevention

- Change industrial process to eliminate use of harmful chemicals
- Purchase different products
- Use less of a harmful product
- Reduce packaging and materials in products
- Make products that last longer and are recyclable, reusable, or easy to repair

2nd Priority

Secondary Pollution and Waste Prevention

- Reuse products
- Repair products
- Recycle
- Compost
- Buy reusable and recyclable products

Last Priority

Waste Management

- Treat waste to reduce toxicity
- Incinerate waste
- Bury waste in landfills
- Release waste into environment for dispersal or dilution

Figure 24-3 Solutions: priorities suggested by prominent scientists for dealing with material use and solid waste. To date, these waste-reduction priorities have not been followed in the United States and in most other countries. Instead, most efforts are devoted to waste management (bury it or burn it). (Data from U.S. Environmental Protection Agency and U.S. National Academy of Sciences)

every part reusable or recyclable for easy remanufacturing should eventually save the company $1 billion in manufacturing costs.

Fifth, *design products to last longer.* Today's tires have an average life of 97,000 kilometers (60,000 miles). Researchers believe this use could be extended to at least 160,000 kilometers (100,000 miles).

Sixth, *eliminate or reduce unnecessary packaging.* From an environmental standpoint, the preferred hierarchy for packaging is *no packaging* (nude products), *minimal packaging, reusable packaging,* and *recyclable packaging.* Canada has set a goal of using the first three of these packaging priorities to cut excess packaging in half. Here are some key questions for designers, manufacturers, and consumers to ask about packaging: Is it necessary? Can it use fewer materials? Can it be reused? Are the resources that went into it renewable? Does it contain the highest feasible amount of recycled material? Can it be biodegraded into harmless nutrients that are recycled in the earth's natural chemical cycles? Is it designed to be recycled easily? Can it be incinerated without producing harmful air pollutants or a toxic ash? Can it be buried and decomposed in a landfill without producing chemicals that can contaminate groundwater?

Figure 24-4 lists some ways you can reduce your output of solid waste.

Improvements in resource productivity and environmental design are very important. But we can do much better through a new *resource productivity*

What Can You Do?

Solid Waste

- Follow the four R's of resource use: Refuse, Reduce, Reuse, and Recycle.

- Ask yourself whether you relly need a particular item.

- Rent, borrow, or barter goods and services when you can.

- Buy things that are reusable, recyclable, or compostable, and be sure to reuse, recycle, and compost them.

- Do not use throwaway paper and plastic plates, cups, and eating utensils, and other disposable items when reusable or refillable versions are available.

- Use e-mail in place of conventional paper mail.

- Read newspapers and magazines online.

- Buy products in concentrated form whenever possible.

Figure 24-4 What can you do? Ways to reduce your output of solid waste.

revolution. In their 1999 book *Natural Capitalism,* Paul Hawken, Amory Lovins, and Hunter Lovins contend that we have the knowledge and technology to greatly increase resource productivity by getting 75–90% more work or service from each unit of material resources we use. To these analysts, the only major obstacles to such an economic and ecological revolution are laws, policies, taxes, and subsidies that continue to reward inefficient resource use and fail to reward efficient resource use. There are many fulfilling career choices for people wanting to become part of the resource productivity revolution.

24-3 THE ECOINDUSTRIAL REVOLUTION AND SELLING SERVICES INSTEAD OF THINGS

What Is the Ecoindustrial Revolution? Reducing Waste Production by Copying Nature

We can make industrial manufacturing processes more sustainable by redesigning them to mimic how nature deals with wastes.

There are growing signs that a new *ecoindustrial revolution* will take place over the next 50 years. The goal is to make industrial manufacturing processes cleaner and more sustainable by redesigning them to mimic how nature deals with wastes. Recall that in nature the waste outputs of one organism become the nutrient inputs of another organism, so all of the earth's nutrients are endlessly recycled.

One way we can mimic nature is to recycle and reuse most chemicals used in industries instead of dumping them into the environment. Another is to have industries interact in complex *resource exchange webs* where the wastes of one manufacturer become raw materials for another—similar to food webs in natural ecosystems (Figure 4-19, p. 69). This is happening in Kalundborg, Denmark, where an electric power plant and a number of nearby industries, farms, and homes work together to save money and reduce their outputs of waste and pollution. They do this by exchanging waste outputs and thus converting them into resources, as shown in Figure 24-5. Trace the connections in this diagram.

Today there are about 20 ecoindustrial parks similar to the one in Kalundborg in various parts of the world and more are being built or planned. Some are being developed on abandoned industrial sites, called *brownfields,* which are cleaned up and redeveloped.

In Europe at least one-third of all industrial wastes are sent to waste-material exchanges or clearinghouses where they are sold or given away as raw materials for other industries. About a tenth of the industrial waste

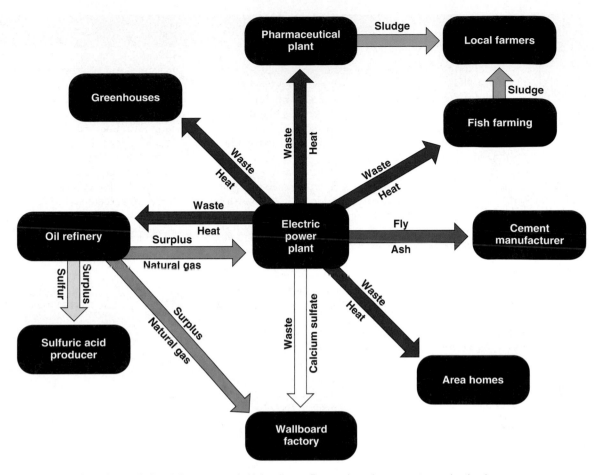

Figure 24-5 Solutions: *industrial ecosystem* in Kalundborg, Denmark, reduces waste production by mimicking a natural food web. The wastes of one business become the raw materials for another. Waste heat in the form of hot air or water can be piped from the power plant to several of the other sites.

in the United States is sent to such clearinghouses, a figure that could be greatly increased.

In addition to eliminating most waste and pollution, these industrial forms of *biomimicry* provide many economic benefits for businesses. They reduce the costs of controlling pollution and complying with pollution regulations. If a company does not add pollutants to the environment, it does not have to worry about government regulations or being sued because the wastes harm someone. The company also improves the health and safety of its workers by reducing their exposure to toxic and hazardous material and thus reduces company health-care insurance costs.

Biomimicry also stimulates companies to come up with new, environmentally beneficial chemicals, processes, and products that can be sold worldwide. Such companies have a better image among consumers based on results rather than public relations campaigns.

In 1975, the Minnesota Mining and Manufacturing Company (3M), which makes 60,000 different products in 100 manufacturing plants, began a Pollution Prevention Pays (3P) program. It redesigned equipment and processes, used fewer hazardous raw materials, identi-

fied hazardous chemical outputs (and recycled or sold them as raw materials to other companies), and began making more nonpolluting products.

By 1998, 3M's overall waste production was down by one-third, its air pollutant emissions per unit of production were 70% lower, and the company had saved more than $750 million in waste disposal and material costs. Since 1990, a growing number of companies have adopted similar programs. See the Guest Essay by Peter Montague on cleaner production on the website for this chapter.

What Is a Service-Flow Economy? Selling Services Instead of Things

Businesses can greatly decrease their pollution and waste by shifting from selling goods to selling services that the goods provide.

In the mid-1980s, German chemist Michael Braungart and Swiss industry analyst Walter Stahel independently proposed a new economic model that would provide profits while greatly reducing resource use and waste. Their idea for more sustainable economies

Ray Anderson

Ray Anderson (Figure 24-A) is CEO of Interface, a company based in Atlanta, Georgia, that makes carpet tiles. The company is the world's largest commercial carpet manufacturer, with 26 factories in six countries, customers in 110 countries, and more than $1 billion in annual sales.

Anderson changed the way he viewed the world and his business after reading Paul Hawken's book *The Ecology of Commerce.* In 1994, he announced plans to develop the nation's first totally sustainable green corporation.

He has implemented hundreds of projects with the goals of zero waste, greatly reduced energy use, and eventually zero use of fossil fuels by relying on renewable solar energy. By 1999, the company had reduced resource waste by almost 30% and reduced energy waste enough to save $100 million. One of Interface's factories in California runs mostly on solar cells to produce the world's first solar-made carpet.

To achieve the goal of zero waste, Anderson plans to stop selling carpet and lease it as a way to encourage recycling. For a monthly fee, the company will install, clean, and inspect the carpet on a monthly basis, repair worn carpet tiles overnight, and recycle worn-out tiles into new carpeting. As Anderson puts it, "We want to harvest yesterday's carpets and recycle them with zero scrap going to the landfill and zero emissions into the eco-system—and run the whole thing on sunlight."

Anderson is one of a growing number of business leaders committed to finding a more economically and ecologically sustainable way to do business while still making a profit for stockholders. Between 1993 and 1998, the company's revenues doubled and profits tripled, mostly because the company saved $130 million in material costs with an investment of less than $40 million. Anderson says he is having a blast.

Figure 24-A Ray Anderson

---

involves shifting from our current *material-flow economy* (Figure 3-18, p. 53) to a *service-flow economy* over the next few decades. Instead of buying most goods outright, customers would use eco-leasing, renting the *services* that such goods provide.

In a service-flow economy, a manufacturer makes more money on a product if it uses the minimum amount of materials, lasts as long as possible, and is easy to maintain, repair, remanufacture, reuse, or recycle.

There is evidence that such an economic shift based on *eco-leasing* is under way. Since 1992, the Xerox Corporation has been leasing most of its copy machines as part of its mission to provide *document services* instead of selling photocopiers. When the service contract expires, Xerox takes the machine back for reuse or remanufacture and has a goal of sending no material to landfills or incinerators. To save money, machines are designed to use recycled paper, have few parts, be energy efficient, and emit as little noise, heat, ozone, and copier chemical waste as possible. Canon in Japan and Fiat in Italy are taking similar measures.

Another example is Carrier, the world's leading maker of air conditioning equipment, which now leases *cooling services.* Carrier teams up with other service providers to install superefficient windows and more efficient lighting and to make other energy-efficiency upgrades that reduce the cooling needs of its customers. Carrier makes money by providing such services rather than installing equipment.

Dow and several other chemical companies are doing a booming business in leasing organic solvents (used mostly to remove grease from surfaces), photographic developing chemicals, and dyes and pigments. In this *chemical service business*, the company delivers the chemicals, helps the client set up a recovery system, takes away the recovered chemicals, and delivers new chemicals as needed.

Finally, Ray Anderson, CEO of a large carpet tile company, plans to lease rather than sell carpet (Individuals Matter, above). There are many entrepreneurial and career opportunities in the emerging service-flow economy.

24-4 REUSE

What are the Advantages and Disadvantages of Reuse? Improves Environmental Quality for Some, Can Create Hazards for Others

Reusing products is an important way to reduce resource use, waste, and pollution in developed countries but can create hazards for the poor in developing countries.

Reuse involves cleaning and using materials over and over and thus increasing the typical life span of a product. This form of waste reduction reduces use of matter and energy resources, cuts pollution and waste, creates local jobs, and saves money. Traditional forms of reuse include salvaging automobile parts from older cars in junkyards and salvaging bricks, doors, fine woodwork, and other items from old houses and buildings.

However, in today's high-throughput societies we have increasingly substituted throwaway tissues for reusable handkerchiefs, disposable paper towels and napkins for reusable cloth ones, throwaway paper plates and cups and plastic utensils for reusable plates, cups, and silverware, and throwaway beverage containers for refillable ones. We even have disposable cameras.

Reuse is alive and well in most developing countries but can be a health hazard for the poor. About 80% of the e-waste in the United States, including discarded TV sets, computers, and cell phones, is shipped to China, India, Pakistan, and other (mostly Asian) countries where labor is cheap and environmental regulations are weak or pooly enforced. Workers there, many of them children, dismantle the products to recover reusable parts and are thus exposed to toxic metals such as lead, mercury, and cadmium. The scrap left over is dumped in waterways and fields, or burned in open fires, which exposes the workers to toxic dioxins.

In cities such as Manila in the Philippines, Mexico City, and Cairo, Egypt, large numbers of people—many of them children—eke out a living by scavenging, sorting, and selling materials they get from open city dumps. This exposes them to toxins and infectious diseases.

Should We Use Refillable Containers?
Reviving Reuse

Refilling and reusing containers uses less resources and energy, produces less waste, saves money, and creates local jobs.

Two examples of reuse are refillable glass beverage bottles and refillable soft drink bottles made of polyethylene terephthalate (PET) plastic. Typically such bottles make 15 round-trips before they become too damaged for reuse and then are recycled. Reusing these containers saves energy (Figure 24-6) and reduces the pollution and wastes associated with using energy resources. Refilling beverage bottles also stimulates local economies by creating local jobs related to their collection and refilling. Moreover, studies by Coca-Cola and PepsiCo of Canada show that their soft drinks in 0.5-liter (16-ounce) bottles cost one-third less in refillable bottles than in throwaway bottles.

But big companies make more money by producing and shipping throwaway beverage and food containers at centralized facilities. This shift has put many small local bottling companies, breweries, and canneries out of business.

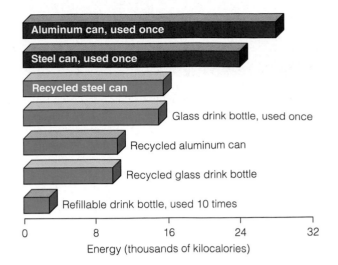

Figure 24-6 Energy consumption for different types of 350-milliliter (12-fluid-ounce) beverage containers. (Data from Argonne National Laboratory)

Denmark and Canada's Prince Edward Island have led the way by banning all beverage containers that cannot be reused. To encourage use of refillable glass bottles, Ecuador has a refundable beverage container deposit fee that is half of the cost of the drink. In Finland, 95% of the soft drink, beer, wine, and spirits containers are refillable, and in Germany, about three-fourths are refillable.

> ✗ **HOW WOULD YOU VOTE?** Do you support banning all beverage containers that cannot be reused, as Denmark has done? Cast your vote online at http://biology.brookscole.com/miller14.

What Are Other Ways to Reuse Things?
Reducing Throwaway Items

We can use reusable shopping bags, food containers, and shipping pallets, and borrow tools from tool libraries.

Cloth bags can be used to carry groceries and other items instead of paper or plastic bags. Both plastic and paper bags are environmentally harmful, and the question of which is more damaging has no clear-cut answer. To encourage people to bring reusable bags, stores in the Netherlands and Ireland charge for shopping bags. As a result, the use of plastic shopping bags dropped by 90–95% in both countries. In 2004, supermarkets in Shanghai, China's largest city, began charging shoppers for plastic bags in an attempt to reduce waste.

> ✗ **HOW WOULD YOU VOTE?** Should consumers have to pay for plastic or paper bags at grocery and other stores? Cast your vote online at http://biology.brookscole.com/miller14.

- Buy beverages in refillable glass containers instead of cans or throwaway bottles.

- Use reusable plastic or metal lunchboxes.

- Carry sandwiches and store food in the refrigerator in reusable containers instead of wrapping them in aluminum foil or plastic wrap.

- Use rechargeable batteries and recycle them when their usefull life is over.

- Carry groceries and other items in a reusable basket, a canvas or string bag, or a small cart.

- Use reusable sponges and washable cloth napkins, dishtowels, and handkerchiefs instead of throwaway paper ones.

Figure 24-7 What can you do? Ways to reuse some of the items you buy.

Other examples of reusable items are metal or plastic lunchboxes and plastic containers for storing lunchbox items and refrigerator leftovers, instead of using throwaway plastic wrap and aluminum foil.

Manufacturers can use shipping pallets made of recycled plastic waste instead of throwaway wood pallets. In 1991, Toyota shifted entirely to reusable shipping containers. A similar move by the Xerox Corporation saves the company more than $3 million per year.

Another example of reuse involves *tool libraries* (such as those in Berkeley, California, and Takoma Park, Maryland) where people can check out a variety of power and hand tools.

Figure 24-7 lists several ways for you to reuse some of the items you buy.

24-5 RECYCLING

What Is Recycling? An Environmental Success Story

Recycling is an important way to collect waste materials and turn them into useful products that can be sold in the marketplace.

Recycling involves reprocessing discarded solid materials into new, useful products. Recycling has a number of important benefits to people and the environment (Figure 24-8). Recycling also reduces unsightly and

costly litter. Picking up litter thrown along highways by thoughtless consumers costs the United States about $500 million a year. Households and workplaces produce five major types of materials that can be recycled: *paper products* (including newspaper, magazines, office paper, and cardboard), *glass, aluminum, steel,* and some types of *plastics.*

Materials collected for recycling can be reprocessed in two ways. *Primary* or *closed-loop* recycling occurs when waste is recycled into new products of the same type—turning used newspapers into new newspaper and used aluminum cans into new aluminum cans, for example.

Secondary recycling, also called *downcycling,* involves converting waste materials into different products. For example, used tires can be shredded and converted into rubberized road surfacing and newspapers can be converted to cellulose insulation.

Environmentalists distinguish between two types of wastes that can be recycled. One is *preconsumer* or *internal waste.* It consists of waste generated in a manufacturing process and recycled instead of being discarded. The other is *postconsumer* or *external waste* generated by consumer use of products. There is about 25 times more preconsumer than postconsumer waste. It is important to recycle both types.

In theory, just about anything is recyclable, but only two things count. *First,* will the item actually be recycled? Sometimes separated wastes collected for recycling are mixed with other wastes and sent to landfills or incinerated, mostly when prices for recycled raw materials fall sharply.

Second, will businesses and individuals complete the recycling loop by buying products that are made from recycled materials? If we do not buy those products, recycling does not work.

But we cannot close the loop and do our bit in creating a market for recycled materials unless we can easily identify whether a product is made entirely or partly from recycled material. This would be clear if governments require that all products made from recycled materials have an easily recognized logo and a label clearly showing the percentage of recycled material they contain, perhaps with a highly visible strip with a green bar extending from 0 to 100%.

Switzerland and Japan recycle about half of their MSW. The United States recycles about 30% of its MSW—up from 6.4% in 1960. This roughly 5-fold increase in recycling is an impressive achievement. But the country's total amount of solid waste has continued to increase although the MSW per person has leveled off since 1990. Studies indicate that with economic incentives and better design of waste management systems the United States and other developed countries could recycle 60–80% of their MSW.

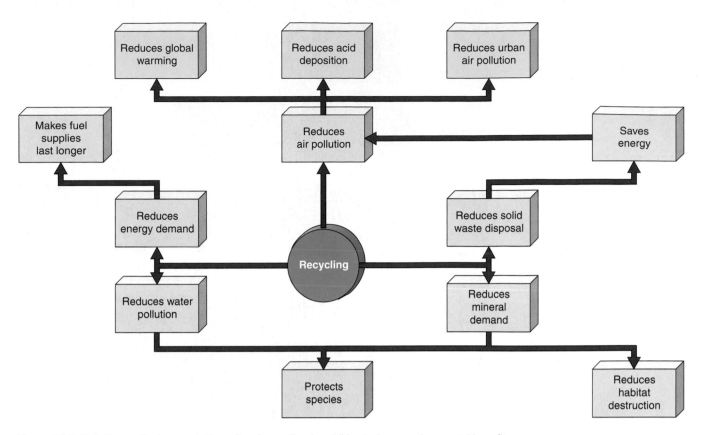

Figure 24-8 Solutions: *Environmental benefits of recycling.* In addition to these environmental benefits, recycling saves more money and creates far more jobs than burning wastes or disposing of them in landfills—assuming that all of these operations receive equal or no government subsidies or tax breaks. Which two of these benefits do you believe are the most important? Despite its many benefits, recycling is still an output approach that deals with wastes after they are produced instead of a way to reduce the overall flow of resources.

How Useful Is Composting? Recycling by Copying Nature

Composting biodegradable organic waste mimics nature by recycling plant nutrients to the soil.

Composting is a simple process in which we copy nature to recycle some of the biodegradable organic wastes we produce. The organic material produced by composting can be added to soil to supply plant nutrients, slow soil erosion, retain water, and improve crop yields.

Some cities in Austria, Belgium, Denmark, Germany, Luxembourg, and Switzerland recover and compost more than 85% of their biodegradable wastes. *Bad news.* Only about 5% of the paper, yard, and vegetable food waste in U.S. MSW is composted, but studies show that it could be raised to 35%.

Such wastes can be collected and composted in centralized community facilities, as is done in many European Union countries. The resulting compost can be used as an organic soil fertilizer, topsoil, or landfill cover. It can also be used to help restore eroded soil on hillsides and along highways, strip-mined land, overgrazed areas, and eroded cropland.

To be successful, a large-scale composting program must be located carefully and control odors, because people do not want to live near a giant compost pile or plant. Composting programs must also exclude toxic materials that can contaminate the compost and make it unsafe for fertilizing crops and lawns.

You can easily make your own compost by collecting organic wastes in a backyard bin. For details on composting, see the website for this chapter.

How Should We Recycle Solid Waste? To Separate or Not to Separate

There is disagreement over whether to send mixed urban wastes to centralized resource recovery plants or have individuals sort recyclables for collection and sale to manufacturers as raw materials.

One way to recycle is to send mixed urban wastes to a centralized *materials-recovery facility (MRF)* shown in Figure 24-9 (p. 542). There, machines or workers separate the mixed waste to recover valuable materials for

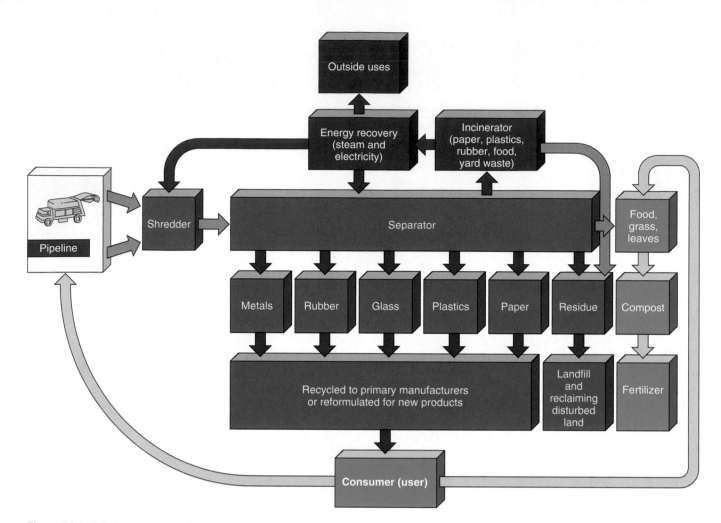

Figure 24-9 Solutions: a generalized *materials-recovery facility* (MRF) sorts mixed wastes for recycling and burning to produce energy. Because such plants need high volumes of trash to be economical, they discourage reuse and waste reduction.

sale to manufacturers as raw materials. The remaining paper, plastics, and other combustible wastes are recycled or burned to produce steam or electricity to run the recovery plant or to sell to nearby industries or homes. Ash from the incinerator is buried in a landfill. Trace the flow of materials through an MRF as diagrammed in Figure 24-9.

Such plants are expensive to build, operate, and maintain. They can emit toxic air pollutants, if not operated properly, and they produce a toxic ash that must be disposed of safely.

MRFs are hungry beasts that must have a large input of garbage to make them financially successful. Thus their owners have a vested interest in increasing *throughput* of matter and energy resources to produce more trash—the reverse of what prominent scientists believe we should be doing (Figure 24-3).

To many experts, it makes more sense economically and environmentally for households and busi-

nesses to separate their trash into recyclable categories such as glass, paper, metals, certain types of plastics, and compostable materials. Then these segregated wastes are collected and sold to scrap dealers, compost plants, and manufacturers.

The *source separation* approach has several advantages over the centralized approach. It produces much less air and water pollution and has lower start-up costs and operating costs than an MRF. It also saves more energy, provides more jobs per unit of material, and yields cleaner and usually more valuable recyclables. In addition, it educates people about the need for waste reduction, reuse, and recycling.

To promote separation of wastes for recycling, many communities use a *pay-as-you-throw (PAUT)* waste collection system. It charges households and businesses for the amount of mixed waste picked up but does not charge for pickup of materials separated for recycling.

Case Study: How Much Wastepaper Is Being Recycled? Encouraging News

Recycling paper has a number of environmental and economic benefits and is easy to do.

Paper (especially newspaper and cardboard) is easy to recycle. Recycling newspaper involves removing its ink, glue, and coating and then reconverting it to pulp that is pressed into new paper. A variety of affordable high-quality recycled papers are available to meet most printing demands (including this book).

About 42% of the world's industrial tree harvest is used to make paper. With 4.6% of the world's population, the United States consumes about 30% of the world's paper and buries or incinerates more than half of this paper. Currently the United States recycles about 49% of its wastepaper (up from 25% in 1989) and 70% of its corrugated cardboard containers. At least 10 other countries recycle 50–97% of their wastepaper and paperboard, with a global recycling rate of 43%. *Bad news.* Despite a 49% recycling rate, the amount of paper thrown away each year in the United States is more than all of the paper consumed in China. Also, about 95% of books and magazines produced in the United States are printed on virgin paper. Some individuals and groups have been letting magazine and book publishers know that they will no longer buy their products unless they greatly increase their use of recycled paper.

One problem associated with making paper is the chlorine (Cl_2) and chlorine compounds (such as chlorine dioxide, ClO_2), used to bleach about 40% of the world's pulp for making paper. These compounds are corrosive to processing equipment, hazardous for workers, hard to recover and reuse, and harmful when released to the environment. However, a growing number of paper mills (mostly in the European Union) are replacing chlorine-based bleaching chemicals with chemicals such as hydrogen peroxide (H_2O_2) or oxygen (O_2).

Environmentalists propose that governments require paper companies to use labels that list the recycled content of paper products and whether the paper was bleached with chlorine or a chlorine-free process.

In 2000, 90% of the copier paper purchased by the U.S. government (one of the country's largest buyers) had 30% recycled content. In 2002, Staples, a major office supply company, pledged to phase out purchases of paper products from endangered forests and achieve an average of 30% postconsumer recycled content in all paper products it sells. Manage-

ment will provide annual reports on progress toward these goals. The company adopted the goals in response to a 2-year grassroots effort called The Paper Campaign, a coalition of dozens of citizens' groups dedicated to protecting forests and increasing the use of recycled paper. Bottom-up political action and using the power of the pocketbook work.

Case Study: Is It Feasible to Recycle Plastics? Some Problems

Recycling many plastics is chemically and economically difficult.

Plastics are made of various types of large polymer or resin molecules made by chemically linking monomer molecules (petrochemicals) produced mostly from oil and natural gas (Figure 24-10).

Currently, only about 10% by weight of all plastic wastes in the United States are recycled, for three reasons. *First,* many plastics are difficult to isolate from other wastes because the many different resins used to make them are often difficult to identify and some plastics are composites of different resins. Most plastics also contain stabilizers and other chemicals that must be removed before recycling.

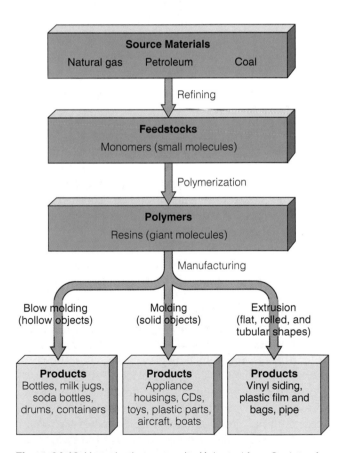

Figure 24-10 How plastics are made. (Adopted from Society of the Plastics Industry)

Second, recovering individual plastic resins does not yield much material because only small amounts of any given resin are used per product. *Third,* the price of oil used to produce petrochemicals for making plastic resins is so low that the cost of virgin plastic resins is much lower than that of recycled resins. An exception is PET (polyethylene terephthalate), used mostly in plastic drink bottles. It can be melted and remanufactured into products such as fleece, clothing, carpet, and nonfood packaging. However, the PET collected for recycling must not have other plastics mixed with it. For example, a single PVC (polyvinyl chloride) bottle in a truckload of PET can render it useless for recycling.

Thus, mandating that plastic products contain a certain amount of recycled plastic resins is unlikely to work. It could also hinder the use of recycled plastics in reducing the resource content and weight of many widely used items such as plastic bags and bottles.

Cargill Dow, a joint venture by a giant agricultural company (Cargill) and a chemical company (Dow), is manufacturing biodegradable and recyclable plastic containers made from a polymer called polyactide (ACT) made from the sugar in corn syrup. Instead of being sent to landfills, containers made from this *bio-plastic* could be composted to produce a soil conditioner.

Toyota, the world's No. 2 automaker is investing $38 billion in a process that makes plastics from plants. By 2020, it expects to control two-thirds of the world's supply of such bioplastics.

Does Recycling Make Economic Sense? Yes for Many Materials

Recycling materials such as paper and metals has important economic and environmental benefits.

Whether recycling makes monetary sense depends on how you look at the economic and environmental benefits and costs of recycling. Critics say recycling does not make sense if it costs more to recycle materials than to send them to a landfill or incinerator. They also point out that recycling is often not needed to save landfill space because many areas are not running out of space.

Critics concede that recycling may make economic sense for valuable and easy-to-recycle materials (such as aluminum, paper, and steel), but not for cheap or plentiful resources such as glass from silica and most plastics that are expensive to recycle.

Critics of recycling also argue that it should pay for itself. But proponents of recycling point out that conventional garbage disposal systems are paid for mostly by charges to households and businesses. So why should recycling be held to a different standard and forced to compete on an uneven playing field?

Proponents also point out that the primary benefit of recycling is not reducing the use of landfills and incinerators but the other important benefits it provides for people and the environment (Figure 24-8). They point to studies showing that the net economic, health, and environmental benefits of recycling far outweigh the costs. Also, they remind us that the recycling industry is an important part of the U.S. economy. It employs about 1.1 million people and its annual income is much larger than both the mining and the waste management industries together.

Cities that make money by recycling and have higher recycling rates tend to use a *single-pickup system* for materials to be recycled and garbage that cannot be recycled instead of a more expensive dual-pickup system. In single-pickup systems, dealing with recyclables costs about half as much per metric ton as disposing the same amount of waste in most modern landfills.

Successful systems also tend to use a *pay-as-you-throw system.* San Francisco, California, uses such a system to recycle almost half of its MSW.

> **✗ HOW WOULD YOU VOTE?** Do the advantages of recycling materials such as paper and metals outweigh the disadvantages? Cast your vote online at http://biology.brookscole.com/miller14.

Why Do We Not Have More Reuse and Recycling? Faulty Accounting and an Uneven Economic Playing Field

Prices of goods that do not tell the ecological truth, too few government subsidies and tax breaks, low landfill dumping costs, and price fluctuations hinder reuse and recycling.

Four factors hinder reuse and recycling. *First* is a faulty accounting system in which the market price of a product does not include the harmful environmental health costs associated with the product during its *life cycle* (Figure 24-11). Many scientists and economists believe that a life-cycle analysis should be made and published for products.

Second, there is an uneven economic playing field because in most countries resource-extracting industries receive more government tax breaks and subsidies than recycling and reuse industries. We get more of what we reward.

Third, charges for depositing wastes in landfills (called *tipping fees*) in the United States are lower than those in most of Europe. *Fourth,* the demand and thus the price paid for recycled materials fluctuate mostly because buying goods made with recycled materials is not a priority for most governments, businesses, and individuals.

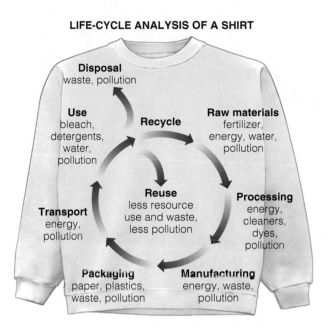

LIFE-CYCLE ANALYSIS OF A SHIRT

Disposal
waste, pollution

Use
bleach, detergents, water, pollution

Recycle

Raw materials
fertilizer, energy, water, pollution

Reuse
less resource use and waste, less pollution

Transport
energy, pollution

Processing
energy, cleaners, dyes, pollution

Packaging
paper, plastics, waste, pollution

Manufacturing
energy, waste, pollution

Figure 24-11 *Life-cycle analysis* of the resources used and pollutants produced over the lifetime of a product. Including all of these costs in the market prices of the products we buy would allow us to compare the harmful environmental costs of different products and thus make more informed choices about the products we buy.

How can we encourage reuse and recycling? Proponents say that leveling the economic playing field is the best way to start. Governments can *increase* subsidies and tax breaks for reusing and recycling materials (the carrot) and *decrease* subsidies and tax breaks for making items from virgin resources (the stick).

Another way to encourage recycling is to greatly increase use of the *pay-as-you-throw (PAUT)* system and encourage or require government purchases of recycled products to help increase demand and lower prices. Governments can also pass laws requiring companies to take back and recycle or reuse packaging discarded by consumers. Globally, at least 29 countries (most in the European Union) have such "take-back" laws. In the Netherlands, all packaging waste is banned from landfills. In 2002, the European Union adopted a ruling requiring its member countries to recycle 55–80% of all packaging waste by 2008.

Governments can also require manufacturers to take back and recycle or reuse appliances, computers and other electronic equipment, and motor vehicles at the end of their useful lives. Such *product stewardship* is required for car manufacturers and appliance makers in European Union (EU) countries and for major appliances in Japan.

The EU also requires companies to take back electronic products from consumers without charge and bans e-waste in municipal solid waste. Beginning in 2006, manufacturers must begin phasing out use of

toxic and hazardous materials in electronic products. These *product stewardship policies* create a strong economic inventive for companies to redesign products for safer and easier recycling, reuse, and remanufacturing.

The United States lags far behind the European Union in dealing with the problem of e-waste. However, in 2003, the office supplier Staples announced a program that allows store visitors to drop off their cell phones, PDAs, pagers, and rechargeable batteries for recycling. A portion of the proceeds from this recycling program will be donated to the Sierra Club to help support environmental education and conservation programs. In 2004, Staples, the Product Stewardship Institute, and the U.S. EPA announced a partnership to test a pilot program for recycling e-waste. Under this program several major electronics manufacturers will pay for recycling of their name brand products taken back to Staples.

✗ *HOW WOULD YOU VOTE?* Should governments pass laws requiring manufacturers to take back and reuse or recycle all packaging waste, appliances, electronic equipment, and motor vehicles at the end of their useful lives? Cast your vote online at http://biology.brookscole.com/miller14.

Finally, we can require labels on all products listing recycled content and the types and amounts of any hazardous materials they contain—similar to labels on food products that list ingredients and provide nutritional information. This can help consumers make more informed choices about the environmental consequences of buying certain products.

24-6 BURNING AND BURYING SOLID WASTE

What Are the Advantages and Disadvantages of Burning Solid Waste? A Faded Rose in Some Countries

Japan and a few European countries incinerate most of their municipal waste, but this is done less in the United States and in other European countries.

Globally, municipal solid waste is burned in over 1,000 large *waste-to-energy incinerators*, which boil water to make steam for heating water or space or for producing electricity. Trace the flow of materials through the process as diagrammed in Figure 24-12 (p. 246). Japan and Switzerland burn more than half of their MSW in incinerators compared to 16% in the United States and about 8% in Canada.

In some plants, called *mass-burn incinerators*, mixed trash is dumped into a huge furnace. This saves the expense and hazards of removing nonburnable material. But often there are air pollution and corrosion

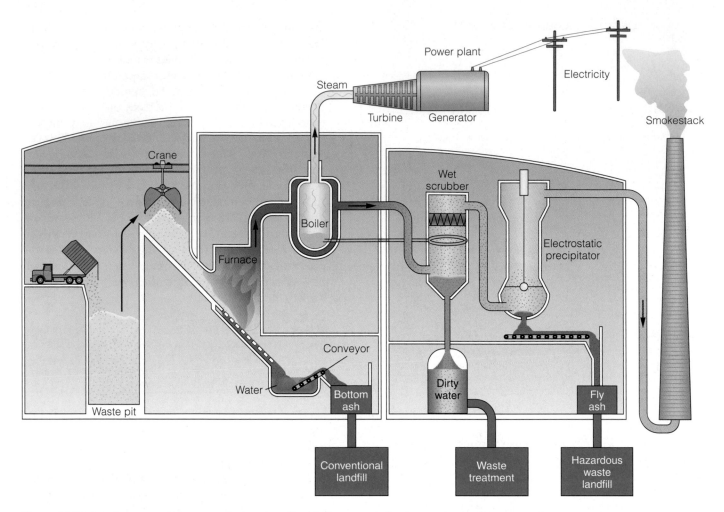

Figure 24-12 Solutions: *waste-to-energy incinerator* with pollution controls that burns mixed solid waste and recovers some of the energy to produce steam used for heating or producing electricity. (Adapted from EPA, *Let's Reduce and Recycle*)

problems because of the difficulty of having to constantly adjust combustion conditions for different mixes of trash.

In *refuse-derived fuel incinerators*, burnable waste is separated from unburnable and recyclable materials. Burning only combustible waste produces more energy and also leads to less air pollution.

Figure 24-13 lists the advantages and disadvantages of using incinerators to burn solid and hazardous waste. Study this figure carefully.

Since 1985, more than 280 new incinerator projects have been delayed or canceled in the United States because of high costs, concern over air pollution, and intense citizen opposition.

Figure 24-13 Trade-offs: advantages and disadvantages of incinerating solid waste. These trade-offs also apply to the incineration of hazardous waste. Pick the single advantage and disadvantage that you think are the most important.

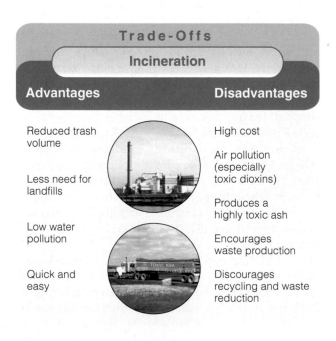

Trade-Offs

Incineration

Advantages	Disadvantages
Reduced trash volume	High cost
Less need for landfills	Air pollution (especially toxic dioxins)
Low water pollution	Produces a highly toxic ash
Quick and easy	Encourages waste production
	Discourages recycling and waste reduction

What Are the Advantages and Disadvantages of Burying Solid Waste? A Widely Used Last Resort

Most of the world's municipal solid waste is buried in landfills that will eventually leak toxic liquids into the soil and underlying aquifers.

About 54% by weight of the MSW in the United States is buried in sanitary landfills, compared to 90% in the United Kingdom, 80% in Canada, 15% in Japan, and 12% in Switzerland. There are two types of landfills. **Open dumps** are essentially fields or holes in the ground where garbage is deposited and sometimes covered with soil. They are rare in developed countries but widely used in many developing countries.

In newer landfills, called **sanitary landfills,** solid wastes are spread out in thin layers, compacted, and covered daily with a fresh layer of clay or plastic foam. Modern state-of-the-art landfills on geologically suitable sites and away from lakes, rivers, floodplains, and aquifer recharge zones are lined with clay and plastic before being filled with garbage, as seen in Figure 24-14. Note in the figure that the landfill bottom is covered with a second impermeable liner, usually made of several layers of clay, thick plastic, and sand. This liner collects *leachate* (rainwater contaminated as it percolates through the solid waste) and is intended to prevent its leakage into groundwater. Wells are drilled around the landfill to monitor any leakage.

Collected leachate is pumped from the bottom of the landfill, stored in tanks, and sent to a regular sewage treatment plant or an on-site treatment plant. When full, the landfill is covered with clay, sand, gravel, and topsoil to prevent water from seeping in.

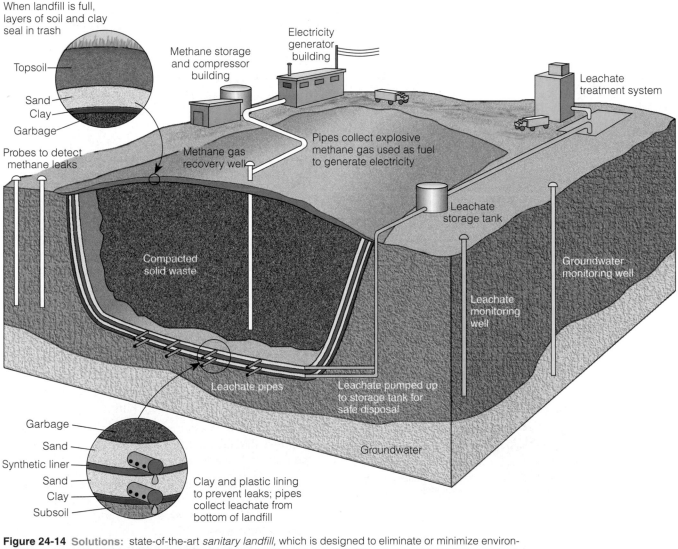

Figure 24-14 Solutions: state-of-the-art *sanitary landfill*, which is designed to eliminate or minimize environmental problems that plague older landfills. Even such state-of-the-art landfills are expected to leak eventually, passing both the effects of contamination and cleanup costs on to future generations. Since 1997, only this state-of-the art type of landfill can operate in the United States. As a result, many older and small landfills have been closed and replaced with larger local and regional modern landfills.

These new landfills are equipped with a connected network of vent pipes to collect landfill gas (consisting mostly of two greenhouse gases, methane and carbon dioxide) released by the underground decomposition of wastes. The methane is filtered out and burned in small gas turbines to produce steam or electricity for nearby facilities or sold to utilities for use as a fuel. In the United States Waste Management has decided that producing and selling power from methane produced by decomposing garbage in its many landfills is a major business opportunity. Figure 24-15 lists the advantages and disadvantages of using sanitary landfills to dispose of solid waste.

Thousands of older and abandoned landfills in the United States (and elsewhere) do not have gas collection systems and will emit methane and carbon dioxide, both potent greenhouse gases, for decades.

Contamination of groundwater and nearby surface water by leachate from unlined and lined older landfills is also a serious problem. Some 86% of older U.S. landfills studied have contaminated groundwater, and a fifth of all Superfund hazardous waste sites are former municipal landfills. In other words, most older landfills throughout the world are chemical time bombs that release greenhouse gases and can eventually leak hazardous chemicals.

X *How Would You Vote?* Do the advantages of burying solid waste in sanitary landfills outweigh the disadvantages? Cast your vote online at http://biology.brookscole.com/miller14.

24-7 HAZARDOUS WASTE

What Is Hazardous Waste? Toxic Threats

Developed countries produce about 80–90% of the world's solid and liquid wastes that can harm people, and most such wastes are not regulated.

Hazardous waste is any discarded solid or liquid material that is *toxic, ignitable, corrosive,* or *reactive* enough to explode or release toxic fumes. According to the UN Environment Programme, developed countries produce 80–90% of these wastes.

In the United States, about 5% of all hazardous waste is regulated under the Resource Conservation and Recovery Act (RCRA, pronounced "RICK-ra") and is often referred to as *RCRA hazardous waste*. RCRA does not regulate

- Radioactive wastes
- Hazardous and toxic materials discarded by households (Figure 24-16)
- Mining wastes
- Oil- and gas-drilling wastes (routinely discharged into surface waters or dumped into unlined pits and landfills)
- Liquid waste containing organic hydrocarbon compounds (80% of all liquid hazardous waste) cement kiln dust produced when liquid hazardous wastes are burned in a cement kiln
- Wastes from the thousands of small businesses and factories that generate less than 100 kilograms (220 pounds) of hazardous waste per month

About 72% of these hazardous wastes are produced by chemical and petroleum industries and another 22% are generated by mining and metal processing industries.

The amount of hazardous and toxic waste in the United States and other countries is likely to increase because of the projected 80% global increase in chemical

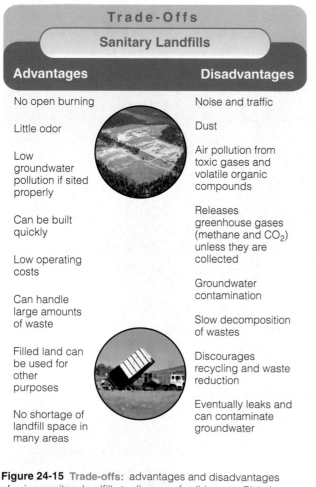

Trade-Offs

Sanitary Landfills

Advantages	Disadvantages
No open burning	Noise and traffic
Little odor	Dust
Low groundwater pollution if sited properly	Air pollution from toxic gases and volatile organic compounds
Can be built quickly	Releases greenhouse gases (methane and CO_2) unless they are collected
Low operating costs	Groundwater contamination
Can handle large amounts of waste	Slow decomposition of wastes
Filled land can be used for other purposes	Discourages recycling and waste reduction
No shortage of landfill space in many areas	Eventually leaks and can contaminate groundwater

Figure 24-15 Trade-offs: advantages and disadvantages of using sanitary landfills to dispose of solid waste. Pick the single advantage and disadvantage that you think are the most important.

What Harmful Chemicals Are in Your Home?

Cleaning

- Disinfectants
- Drain, toilet, and window cleaners
- Spot removers
- Septic tank cleaners

Paint

- Latex and oil-based paints
- Paint thinners, solvents, and strippers
- Stains, varnishes, and lacquers
- Wood preservatives
- Artist paints and inks

General

- Dry-cell batteries (mercury and cadmium)
- Glues and cements

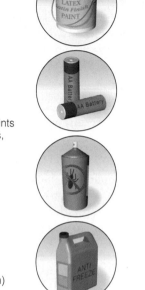

Gardening

- Pesticides
- Weed killers
- Ant and rodent killers
- Flea powders

Automotive

- Gasoline
- Used motor oil
- Antifreeze
- Battery acid
- Solvents
- Brake and transmission fluid
- Rust inhibitor and rust remover

Figure 24-16 Harmful chemicals found in many homes. Congress has exempted disposal of these household materials from government regulation. Make a survey to see which of these chemicals are in your home.

the EPA estimated that in a worst-case situation more than 1 million people could be killed or injured by a terrorist attack on any one of 123 major U.S. chemical plants.

Case Study: A Black Day in Bhopal, India

The world's worst industrial accident occurred in 1984 at a pesticide plant in Bhopal, India.

December 2, 1984, will long be a black day for India. On that date the world's worst industrial accident occurred at a Union Carbide pesticide plant in Bhopal, India.

An explosion in an underground storage tank released a large quantity of highly toxic methyl isocyanate (MIC) gas, used to produce carbamate pesticides. Investigation found that water leaking into the tank through faulty valves and corroded pipes caused an explosive chemical reaction.

Once in the atmosphere, some of the toxic MIC was converted to more deadly hydrogen cyanide gas. The toxic cloud of gas settled over about 78 square kilometers (30 square miles), exposing up to 600,000 people. Many were illegal squatters living near the plant because they had no other place to go. The deadly cloud spread through Bhopal without warning because the plant's warning sirens had been turned off to save money.

According to Indian officials, at least 8,000 people died within a few days after the accident. The International Campaign for Justice in Bhopal estimates that by 2003 the accident had killed 23,000 and climbing. It also puts the number of people suffering from chronic illnesses from the accident at 120,000–150,000. An international team of medical specialists estimated in 1996 that 50,000–60,000 people sustained permanent injuries such as blindness, lung damage, and neurological problems. With any of these estimates, this was the world's largest industrial tragedy.

production (including many of hazardous and toxic chemicals) between 1995 and 2020.

How Safe Are U.S. Chemical Plants from Terrorist Attacks? Toxic Terrorism

Large amounts of hazardous wastes could be released into the environment by terrorist attacks on major chemical plants in the United States.

Managers of industrial plants that manufacture and use chemicals work hard to prevent accidental release of chemicals that can harm workers or nearby residents. But accidents can happen, as thousands of people living near a pesticide manufacturing plant in Bhopal, India, learned in 1984 (Case Study, right).

The 2001 act of terrorism on New York City's World Trade Center Towers and the Pentagon has heightened concerns about terrorist acts against such plants. Roughly 20,000 industrial plants in the United States contain large quantities of hazardous chemicals. Analysts view such plants as easy targets for acts of sabotage by terrorists.

A 1999 government study found that chemical plant security ranged from "fair to very poor." In 2000,

Indian officials claim that Union Carbide probably could have prevented the tragedy by spending no more than $1 million to upgrade plant equipment and improve safety. According to an investigation by India's Central Bureau of Investigation (CBI), corporate managers of Union Carbide in the United States made a decision to save money by cutting back on maintenance and safety because the plant had proven to be a financial disappointment for the company. The CBI found that on the night of the disaster six safety measures designed to prevent a leak of toxic materials were inadequate, shut down, or malfunctioning.

After the accident, Union Carbide reduced the corporation's liability risks for compensating victims by selling off a portion of its assets and giving much of the profits to its shareholders in the form of special dividends. In 1994, Union Carbide sold its holdings in India and later was taken over by Dow Chemical.

In 1989, Union Carbide agreed to pay an out-of-court settlement of $470 million to compensate the victims (low estimate of 3,000 deaths) without admitting any guilt or negligence concerning the accident. In 1992, the Court of the Chief Judicial Magistrate for Bhopal charged Warren Anderson, CEO of Union Carbide at the time of the accident, with "culpable homicide" (the equivalent of manslaughter) and issued a warrant for his arrest. Since then he has refused to appear in court and the U.S. government has not responded to India's request to extradite him to India to stand trial. Every December since 1984, marchers in Bhopal have paraded an effigy of Warren Anderson through town and burned it. Dow, which bought up Union Carbide in 1999, refused to accept any of the company's alleged Bhopal liabilities.

What Can We Do with Hazardous Waste? Manage It or Produce Less

We can burn, bury, detoxify, reuse, recycle, or not produce hazardous wastes.

We can *manage* hazardous waste mostly by burning or burying it. This is an *output* approach that tries to figure out what to do with such wastes after we have produced them.

We can also use a *pollution prevention* or *waste reduction* approach. See the Guest Essay on this subject

by Lois Gibbs on the website for this chapter. This is an *input* approach that tries to reduce the production of such wastes and their release into the air, water, and soil. With this approach, scientists look for substitutes for toxic or hazardous materials, reuse or recycle them within industrial cycles, or use them as raw materials for new products (Figure 24-5) instead of burning or burying them.

Figure 24-17 lists the priorities that prominent scientists believe we should follow in dealing with hazardous waste. Study this figure carefully. Denmark is following these priorities but most countries are not.

How Can We Remove or Detoxify Hazardous Waste? Science to the Rescue

Chemical and biological methods can be used to remove hazardous wastes or to reduce their toxicity.

In Denmark, all hazardous and toxic waste from industries and households is delivered to 21 transfer stations throughout the country. The waste is then transferred to a large treatment facility. There, about three-fourths of the waste is detoxified by physical, chemical, and biological methods and the rest is buried in a carefully designed and monitored landfill.

Physical methods used to detoxify hazardous wastes include filtering out solids, distilling liquid mixtures to separate out harmful chemicals, and precipitating such chemicals from solution. Especially deadly wastes can be encapsulated in glass, cement, or ceramics and then isolated in storage sites.

Chemical reactions can also be used to convert hazardous chemicals to less harmful or harmless chem-

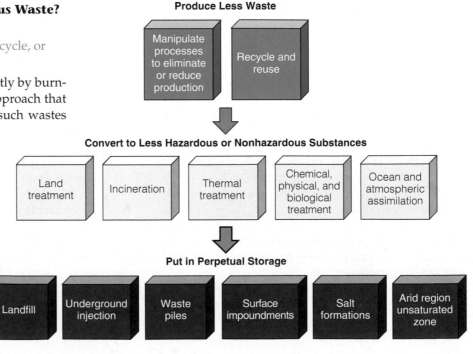

Figure 24-17 Solutions: priorities suggested by prominent scientists for dealing with hazardous waste. To date, these priorities have not been followed in the United States or most other countries. (Data from U.S. National Academy of Sciences)

icals. Scientists are testing the use of *cyclodextrin*—a type of sugar made from corn starch—to remove toxic materials such as solvents and pesticides from contaminated soil and groundwater. To clean up a site, a solution of cyclodextrin is injected. After this molecular-sponge material moves through the soil or groundwater and attracts various toxic chemicals, it is pumped out of the ground, stripped of its contaminants, and reused.

Some scientists and engineers consider biological treatment as the wave of the future for cleaning up some types of toxic and hazardous waste. One approach is *bioremediation*, in which bacteria and enzymes are used to help destroy toxic or hazardous substances or convert them to harmless compounds. See the Guest Essay by John Pichtel on this topic on the website for this chapter.

Another biological way to treat hazardous wastes is *phytoremediation*. It involves using natural or genetically engineered plants to absorb, filter, and remove contaminants from polluted soil and water (Figure 24-18). For example, researchers at the University of Georgia used genetic engineering to add extra detoxifying power to a cottonwood tree. They inserted a mercury-detoxifying gene of *E. coli* into the genome of a common soil bacterium, which was incorporated into day-old cottonwood trees. After reaching maturity such trees can extract more mercury ions (Hg^{2+}) from

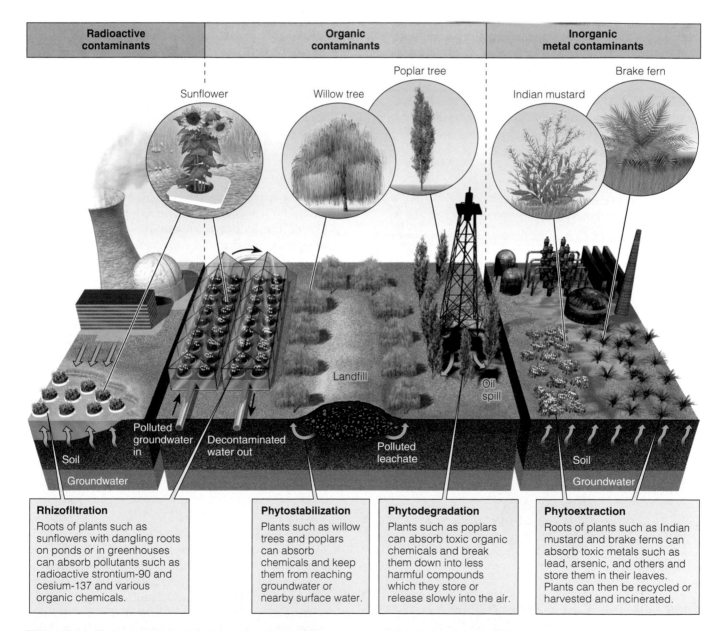

Radioactive contaminants	Organic contaminants	Inorganic metal contaminants

Sunflower

Poplar tree

Willow tree

Brake fern

Indian mustard

Polluted groundwater in

Decontaminated water out

Landfill

Oil spill

Polluted leachate

Soil

Groundwater

Soil

Groundwater

Rhizofiltration
Roots of plants such as sunflowers with dangling roots on ponds or in greenhouses can absorb pollutants such as radioactive strontium-90 and cesium-137 and various organic chemicals.

Phytostabilization
Plants such as willow trees and poplars can absorb chemicals and keep them from reaching groundwater or nearby surface water.

Phytodegradation
Plants such as poplars can absorb toxic organic chemicals and break them down into less harmful compounds which they store or release slowly into the air.

Phytoextraction
Roots of plants such as Indian mustard and brake ferns can absorb toxic metals such as lead, arsenic, and others and store them in their leaves. Plants can then be recycled or harvested and incinerated.

Figure 24-18 *Phytoremediation.* Ways that various types of plants can be used as pollution sponges to clean up soil and water and radioactive substances (left), organic compounds (center), and toxic metals (right). (Data from American Society of Plant Physiologists, U.S. Environmental Protection Agency, and Edenspace)

contaminated soil than conventional cottonwoods. Various plants have been identified as potential "pollution sponges" to help clean up soil and water contaminated with chemicals such as pesticides, organic solvents, radioactive metals, and toxic metals such as lead, mercury, and arsenic. Figure 24-19 lists advantages and disadvantages of phytoremediation. Study this figure carefully.

Another way to detoxify hazardous waste is with a plasma torch. **Plasma**—a fourth state of matter—is an ionized gas made up of electrically conductive ions and electrons. Passing electrical current through a gas to generate an electric arc and very high temperatures can create a plasma. This process can be carried out continuously in a *plasma torch* somewhat similar to a welding torch.

High temperatures from the torch can decompose liquid or solid hazardous organic material into ions and atoms that can be converted into simple molecules, cleaned up, and released as a gas. They can also convert hazardous inorganic matter into a molten glassy material that encapsulates toxic metals and keeps them from leaching into groundwater.

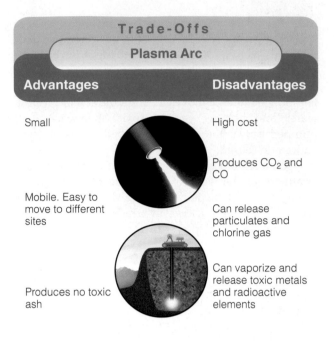

Figure 24-20 **Trade-offs:** advantages and disadvantages of using a *plasma arc torch* to detoxify hazardous wastes. Pick the single advantage and disadvantage that you think are the most important.

One approach is to feed solid or liquid waste material into a reaction chamber where the heat of the plasma breaks down organic molecules such as oil, solvents, and paint into their basic atoms, which then recombine into gases.

Another approach is to drill a small hole into a contaminated site such as an old landfill, insert the torch, and turn it on. The process is repeated over a grid pattern to decontaminate an entire area. Unlike incineration, the plasma process produces no toxic ash that must be disposed of safely.

Figure 24-20 lists the advantages and disadvantages of using a plasma arc torch to detoxify hazardous waste.

What Are the Advantages and Disadvantages of Burning and Burying Hazardous Waste? Last Resorts

Hazardous waste can be incinerated or disposed of on or underneath the earth's surface, but this can pollute the air and water.

Hazardous waste can be incinerated. This has the same mixture of advantages and disadvantages as burning solid wastes (Figure 24-13). Two major disadvantages of incinerating hazardous waste is that it releases air pollutants such as toxic dioxins and produces a highly toxic ash that must be safely and permanently stored.

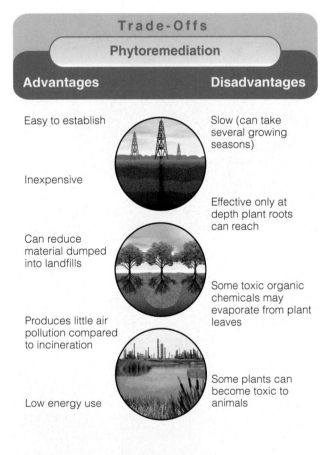

Figure 24-19 **Trade-offs:** advantages and disadvantages of using *phytoremediation* to remove or detoxify hazardous waste. Pick the single advantage and disadvantage that you think are the most important.

Most hazardous waste in the United States is disposed of on land in deep underground wells, surface impoundments such as ponds, pits, or lagoons, and state-of-the-art landfills.

In *deep-well disposal*, liquid hazardous wastes are pumped under pressure through a pipe into dry, porous geologic formations or zones of rock far beneath aquifers tapped for drinking and irrigation water. Theoretically, these liquids soak into the porous rock material and are isolated from overlying groundwater by essentially impermeable layers of rock.

Figure 24-21 lists the advantages and disadvantages of deep-well disposal of liquid hazardous wastes. Take time to study this figure. Many scientists believe current regulations for deep-well disposal in the United States are inadequate and should be improved.

✗ *How Would You Vote?* Do the advantages of deep-well disposal of hazardous waste outweigh their disadvantages? Cast your vote online at http://biology.brookscole .com/miller14.

Surface impoundments are excavated depressions such as ponds, pits, or lagoons into which liquid hazardous wastes are drained and stored (Figure 22-9, p. 502). As water evaporates, the waste settles and becomes more concentrated. Figure 24-22 lists the advantages and advantages of this method. Study this figure carefully. EPA studies found that 70% of these storage basins in the United States have no liners and as many as 90% may threaten groundwater. According to the

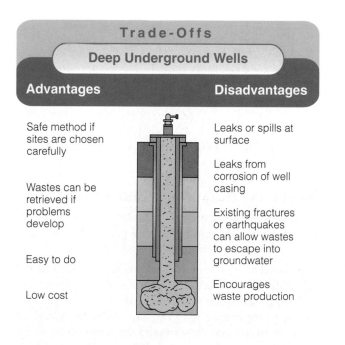

Figure 24-21 Trade-offs: advantages and disadvantages of injecting liquid hazardous wastes into deep underground wells. Pick the single advantage and disadvantage that you think are the most important.

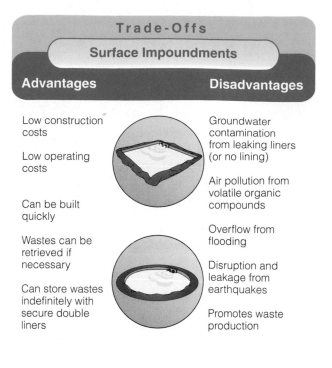

Figure 24-22 Trade-offs: advantages and disadvantages of storing liquid hazardous wastes in surface impoundments. Pick the single advantage and disadvantage that you think are the most important.

EPA, eventually all liners are likely to leak and can contaminate groundwater.

Sometimes liquid and solid hazardous wastes are put into drums or other containers and buried in carefully designed and monitored *secure hazardous waste landfills* (Figure 24-23, p. 554). Many large companies have secure landfills to treat their own hazardous waste but there are 23 commercial hazardous waste landfills in the United States.

Sweden goes further and buries its concentrated hazardous wastes in underground vaults made of reinforced concrete. By contrast, in the United Kingdom most hazardous wastes are mixed with household garbage and stored in hundreds of conventional landfills throughout the country.

Hazardous wastes can also be stored in carefully designed *aboveground buildings*. This is especially useful in areas where the water table is close to the surface or areas that are above aquifers used for drinking water. These structures are built to withstand storms and to prevent the release of toxic gases. Leaks are monitored and any leakage is collected and treated.

Each year there are more than 500,000 shipments of hazardous chemicals in the United States. Most go by trucks or trains to landfills and incinerators. Hazardous raw materials are also shipped to manufacturing plants. On average, these shipments result in about 13,000 accidents per year in the United States, involving about 100 deaths, than 10,000 injuries, and evacuations

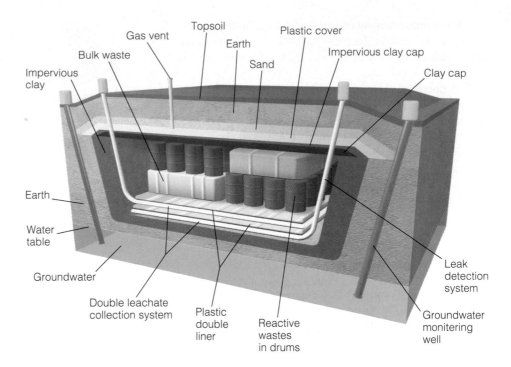

Figure 24-23 Solutions: secure hazardous waste landfill.

Labels: Gas vent · Topsoil · Bulk waste · Earth · Plastic cover · Impervious clay · Sand · Impervious clay cap · Clay cap · Earth · Water table · Groundwater · Double leachate collection system · Plastic double liner · Reactive wastes in drums · Leak detection system · Groundwater monitering well

of more than 500,000 people. Most communities do not have the equipment and trained personnel needed to deal with hazardous waste spills. How well is your community prepared to deal with such spills? In the future, some of the risks might be reduced by using solid nanoparticle materials to absorb the hazardous liquids.

Figure 24-24 lists ways you can reduce your environmental input of hazardous waste.

What Are Brownfields? Recycling Dangerous Sites

Many abandoned industrial and other hazardous waste sites are being cleaned up and put to use.

Brownfields are abandoned industrial and commercial sites usually contaminated with hazardous wastes. Examples include factories, junkyards, older landfills, and gas stations. Some 450,000–600,000 brownfield sites exist in the United States, often in economically distressed inner cities.

What Can You Do?

Hazardous Waste

- Use pesticides in the smallest amount possible.

- Use less harmful substances instead of commercial chemicals for most household cleaners. For example, use liquid ammonia to clean appliances and windows; vinegar to polish metals, clean surfaces, and remove stains and mildew; baking soda to clean household utensils, deodorize, and remove stains; borax to remove stains and mildew.

- Do not dispose of pesticides, paints, solvents, oil, antifreeze, or other products containing hazardous chemicals by flushing them down the toilet, pouring them down the drain, burying them, throwing them into the garbage, or dumping them down storm drains.

Figure 24-24 What can you do? Ways to reduce your input of hazardous waste into the environment.

Brownfields can be cleaned up and reborn as parks, nature reserves, athletic fields, ecoindustrial parks (p. 536), and neighborhoods. But first old oil and grease, industrial solvents, toxic metals, and other contaminants must be removed from their soil and groundwater. Such efforts have been hampered by concerns about legal liability for the contamination. However, Congress and almost half of the states have passed laws limiting the liability for developers and their lenders.

Brownfield redevelopment is now seen as a way to rebuild parts of cities, create jobs, and increase the tax base. By 2004, more than 40,000 former brownfield sites had been redeveloped in the United States and many other projects are under way.

24-8 CASE STUDIES: LEAD, MERCURY, AND DIOXINS

What Is the Threat from Lead? A Toxic Metal

Lead is especially harmful to children and is still used in leaded gasoline and household paints in about 100 countries.

Because it is a chemical element, lead (Pb) does not break down in the environment. Lead is a potent neurotoxin that can harm the nervous system, especially in young children. Each year, 12,000–16,000 American children under age 9 are treated for acute lead poisoning, and about 200 die. About 30% of the survivors suffer from palsy, partial paralysis, blindness, and mental retardation.

Research indicates that children under age 6 and unborn fetuses with even fairly low blood levels of lead are especially vulnerable to nervous system impairment, lowered IQ (by an average of 7.4 points), shortened attention span, hyperactivity, hearing damage, and various behavior disorders.

Good news. Between 1976 and 2000, the percentage of U.S. children ages 1 to 5 with blood lead levels above the current safety standard dropped from 85% to 2.2%, preventing at least 9 million childhood lead poisonings. The primary reason was that government regulations banned leaded gasoline in 1976 (with complete phaseout by 1986) and lead-based paints in 1970 (but illegal use continued until about 1978)—an excellent example of the power of pollution prevention.

Bad news. Even with the encouraging drop in average blood levels of lead, the U.S. Centers for Disease Control and Prevention estimates that at least 400,000 U.S. children still have unsafe blood levels of lead, caused by exposure from a number of sources. A major source is inhalation or ingestion of lead particles from peeling lead-based paint found in about 38 million houses built before 1960. Lead can also leach from water lines and pipes and faucets containing lead. In addition, a 1993 study by the U.S. National Academy of Sciences and numerous other studies indicate *there is no safe level of lead in children's blood.*

Health scientists have proposed a number of ways to help protect children from lead poisoning, as listed in Figure 24-25. Taking most of these actions will cost an estimated $50 billion in the United States. But health officials say the alternative is to keep poisoning and mentally handicapping millions of children.

Although the threat from lead has been reduced in the United States, this is not the case in many developing countries. About 80% of the gasoline sold in the world today is unleaded, but about 100 countries still use leaded gasoline. The World Health Organization (WHO) estimates that 130–200 million children around the world are at risk from lead poisoning, and 15–18 million children in developing countries have permanent brain damage because of lead poisoning—mostly

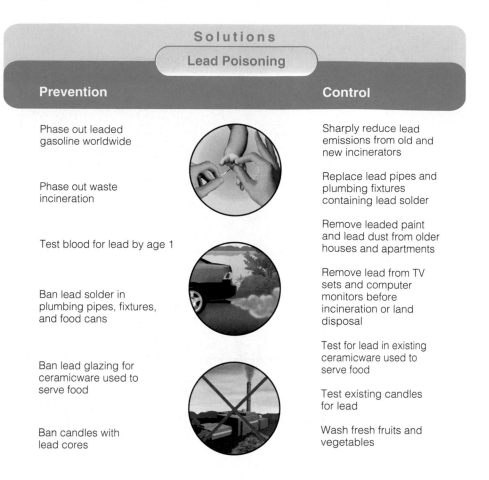

Figure 24-25 Solutions: ways to help protect children from lead poisoning. Which two of the solutions do you believe are the most important?

from use of leaded gasoline. *Good news.* China recently phased out leaded gasoline in less than three years.

What Is the Threat from Mercury? Some Fish May Come With a Side of Toxic Mercury

Mercury is released into the environment mostly by burning coal and incinerating wastes and can build to high levels in some types of fish consumed by humans.

Mercury—the only metal that is liquid at room temperature—is used in thermometers, dental fillings, fluorescent lights, mercury light switches, and other electrical equipment and is released into the atmosphere from burning coal and incinerating municipal and industrial wastes. Mercury compounds are also used as paint pigments, fungicides, insecticides, and in dry-cell batteries.

Once released into the atmosphere from natural or human sources, elemental mercury often is converted to more toxic inorganic and organic mercury compounds, as shown in Figure 24-26. Trace the paths in this diagram.

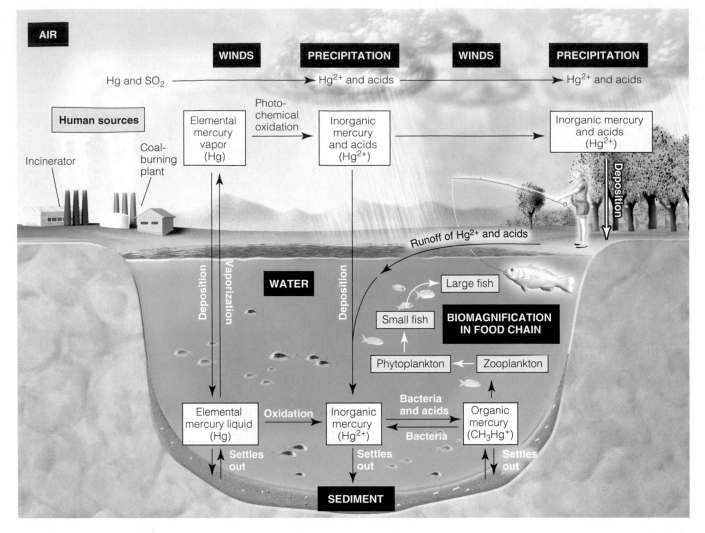

Figure 24-26 Cycling of mercury in aquatic environments, in which mercury is converted from one form to another. The most toxic form to humans is methylmercury (CH_3Hg^+), which can be biologically magnified in aquatic food chains. Some mercury is also released back into the atmosphere as mercury vapor.

Humans are exposed to mercury in two ways. One is by inhaling vaporized elemental mercury (Hg) or particulates of inorganic mercury (Hg^{2+}) salts (such as HgS and $HgCl_2$). The other is eating fish contaminated with methylmercury (CH_3Hg^+), which is very toxic and can be biologically magnified in food chains and webs. The greatest risk is brain damage from exposure to low levels of methylmercury (CH_3Hg^+) in fetuses and young children whose nervous systems are still developing.

Mercury is released naturally from rocks, soil, and volcanoes and by vaporization from the ocean. However, 50–75% of mercury emissions are believed to come from human activities, mostly coal burning and to a lesser degree waste incineration. Mercury is also released into the air when it is used to help extract gold and silver from ores, in some chlorine manufac-

turing processes, and when a fluorescent light bulb breaks.

These human-related sources emit elemental mercury (Hg) vapor and particles of inorganic mercury (Hg^{2+}) salts into the atmosphere. Within hours to days, these forms of mercury fall back to the earth's land and aquatic systems in rain, snow, or as dry particles. Some falls out locally and some in downwind areas.

Once moderately harmful inorganic mercury ions (Hg^{2+}) enter an aquatic system, bacteria often convert it to highly toxic methylmercury that can be biologically magnified in food chains and webs (Figure 24-26). This explains why high levels of methylmercury are often found in the tissues of sharks, swordfish, king mackerel, tilefish, and albacore (white) tuna feeding at high trophic levels in food chains and webs. In 2004, the Food and Drug Administration (FDA) and the U.S.

EPA advised women who may become pregnant, pregnant women, and nursing mothers not to eat shark, swordfish, king mackerel, or tilefish and to limit their consumption of albacore tuna to no more than 170 grams (6 ounces) per week. They also advised such individuals to check local advisories about the safety of fish caught in local lakes, rivers, and coastal areas.

Levels of methylmercury in lakes and lake organisms appear to be connected to acid deposition, because the conversion rate of inorganic mercury to methylmercury is higher in acidified lakes.

According to a 2001 study by the Centers for Disease Control and Prevention, 8% of American women of childbearing age risk having a baby born with irreversible neurological problems because of exposure of the fetus to mercury, mostly from the mother eating seafood contaminated with methylmercury. Problems include brain and nerve system damage that can result in cerebral palsy, delayed onset of walking and talking, learning disabilities, tremors, irritability, impaired coordination, and memory loss. In the United States, up to 300,000 babies born each year are at risk from such problems due to mercury exposure while in the womb.

Figure 24-27 lists ways to prevent or control human exposure to mercury. In its 2003 report on global mercury pollution, the UN Environment Programme recommended phasing out coal burning and waste incineration as rapidly as possible.

In 2003, environmental engineers David Mazyck and Chang Yu Wu at the University of Florida developed a way to remove much of the mercury from smokestack emissions. They inject tiny particles of silica and a chemical that, when exposed to UV light, produces highly reactive molecules that remove most of the mercury in smokestack gases. The mercury can be removed from the silica so that the silica can be reused. And the mercury can be sold for use in products such as fluorescent bulbs. The researchers have founded a company, Sol-gel Power Technologies, to develop the technology.

In 2000, the U.S. EPA officially determined that mercury is a hazardous substance as defined by the Clean Air Act, which requires that emissions of such substances be strictly controlled. However, in 2003 the Bush administration opposed international limits on mercury emissions and other mandatory measures aimed at reducing the risk of mercury exposure, mostly from burning coal and incinerating hazardous wastes. Also, government analysis shows that President Bush's proposed Clear Skies air pollution control program will reduce mercury pollution less than current regulations under the Clean Air Act and could allow emissions of mercury to triple by 2013. In 2003, by executive order President Bush exempted the country's more polluting coal-burning power plants from having to follow a rule in the Clean Air Act that would require them to upgrade their air pollution control equipment whenever they do a significant expansion (p. 156). This will further delay reducing mercury emissions by exempting the plants with the highest mercury emissions from the requirements of the Clean Air Act and result in *hot spots* near such plants with high mercury levels.

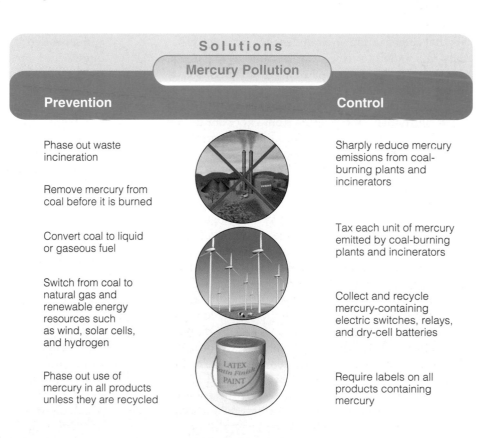

Solutions		
Mercury Pollution		
Prevention		**Control**
Phase out waste incineration		Sharply reduce mercury emissions from coal-burning plants and incinerators
Remove mercury from coal before it is burned		
Convert coal to liquid or gaseous fuel		Tax each unit of mercury emitted by coal-burning plants and incinerators
Switch from coal to natural gas and renewable energy resources such as wind, solar cells, and hydrogen		Collect and recycle mercury-containing electric switches, relays, and dry-cell batteries
Phase out use of mercury in all products unless they are recycled		Require labels on all products containing mercury

Figure 24-27 Solutions: ways to prevent or control inputs of mercury into the environment from human activities—mostly through coal-burning plants and incinerators. Which two of these solutions do you believe are the most important?

How Dangerous Are Dioxins? Controversy Over Unintended Culprits

Dioxins are potentially harmful chlorinated hydrocarbons produced as by-products of various industrial processes such as waste incineration and paper bleaching.

Dioxins (or polychlorinated dibenzodioxins) are a family of more than 75 different chlorinated hydrocarbon compounds. They form as by-products in high-temperature chemical reactions involving chlorine and hydrocarbons.

Dioxins (or polychlorinated dibenzodioxins) are mainly unwanted by-products or unintended consequences of a wide range of industrial processes. Natural processes such as forest fires and volcanic eruptions also produce them.

Worldwide, incineration of municipal and medical wastes accounts for about 70% of dioxin and furan releases to the atmosphere. Other sources include wood-burning fireplaces, coal-fired power plants, metal smelting and refining facilities, wood-pulp and paper mills, and sludge from municipal wastewater treatment plants.

Toxicology studies indicate that about 30 dioxin compounds have significant toxicity. One dioxin compound, TCDD, is the most toxic (Table 19-1, p. 414) and the most widely studied. Dioxins are persistent chemicals that linger in the environment for decades, especially in soil and human fat tissue. About 90% of the human exposure to trace levels of dioxins occurs through eating contaminated food.

There is concern about possible harmful health effects on humans and wildlife from exposure to low levels of dioxins. A 2001 draft report of an EPA-sponsored comprehensive review of the scientific literature by more than 100 scientists around the world came to three major conclusions.

First, TCDD is a human carcinogen, and other dioxin compounds are likely human carcinogens, especially for people who eat large amounts of fatty meats and dairy products. *Second,* the most powerful possible effects of exposure to low levels of dioxin on humans are disruption of the reproductive, endocrine, and immune systems (Case Study, p. 416) and harmful effects on developing fetuses. *Third,* very low levels of dioxin in the environment can cause serious damage to certain wildlife species. But industries producing dioxins say the dangers of long-term exposure of humans to low levels of dioxins are overestimated.

Because it will take decades to resolve these issues, some environmental and health scientists call for using a *precautionary strategy* to sharply reduce emissions of dioxins now. This would be done mostly by banning the use of chlorine for bleaching paper (as several European countries have done) and eliminating chlorinated hydrocarbon compounds that produce dioxins

from hazardous wastes burned in incinerators, iron ore sintering plants, and cement kilns.

A 2003 report by the National Academy of Sciences recommended that the United States take precautionary steps to reduce dioxin levels, especially in food. The panel of medical experts called for testing livestock forage and feed for dioxins, setting lower limits for dioxins in food products and dietary supplements, and testing the products for dioxin levels. They also said the government should encourage people to eat less fat and meat, which tend to have higher levels of dioxins. *Bad news.* These things are not being done.

24-9 HAZARDOUS WASTE REGULATION IN THE UNITED STATES

What Is the Resource Conservation and Recovery Act? Tracking Hazardous Waste from Cradle to Grave

U.S. firms producing fairly large amounts of hazardous waste must get a permit from the EPA and submit a record tracking these wastes from production to disposal.

In 1976, the U.S. Congress passed the *Resource Conservation and Recovery Act* (*RCRA,* mentioned previously) and amended it in 1984. This law has three major requirements. *First,* the EPA is to identify hazardous wastes and set standards for their management by states. *Second,* firms that store, treat, or dispose of more than 100 kilograms (220 pounds) of hazardous wastes per month must have a permit stating how such wastes are to be managed. *Third,* permit holders must use a *cradle-to-grave* system to keep track of waste they transfer from a point of generation (cradle) to an approved off-site disposal facility (grave) and submit proof of this to the EPA.

What Is the Superfund Act? Cleaning Up Abandoned Waste Sites

In the United States, the EPA identifies abandoned hazardous waste sites, cleans them up, and sends the bill to all responsible parties it can find.

In 1980, the U.S. Congress passed the *Comprehensive Environmental Response, Compensation, and Liability Act,* commonly known as the *CERCLA* or *Superfund* program. Through taxes on chemical raw materials, this law plus later amendments has provided a trust fund to achieve three goals.

One is to identify abandoned hazardous waste dump sites (Spotlight, p. 559), underground tanks leaking toxic chemicals, and other hazardous waste sites. Second is to protect and if necessary clean up groundwater near such sites and to clean up the sites.

When they can be found, responsible parties must pay for the cleanup. If no responsible parties can be found, the government cleans up the site using a fund financed by taxes on oil and chemical companies.

The third goal is to put the worst sites that represent an immediate and severe threat to human health on a *National Priorities List (NPL)* to be cleaned up via removal, treatment, incineration, bioremediation, or phytoremediation.

At the beginning of 2004, the EPA had identified about 11,300 sites in need of cleanup, including 1,250 on the priority list. Since the Superfund started in 1983, 300 sites have been cleaned up according to government standards and removed from the NPL. Cleanup has been substantially completed on about 72% of the remaining 1,250 priority sites at an average cost of $20 million per site.

The five states with the most sites are New Jersey (113), California (96), Pennsylvania (92), New York (90), and Michigan (67). California's Santa Clara County, the birthplace of the nation's semiconductor industry in what is called Silicon Valley, contains more Superfund priority sites than any other county in the nation. This is not surprising, because a single semiconductor plant uses 500 to 1,000 different chemicals, many of them toxic.

The former U.S. Office of Technology Assessment and the Waste Management Research Institute estimate that the Superfund list could eventually include at least 10,000 priority sites, with cleanup costs of up to $1 trillion, not counting legal fees—a glaring example of why preventing pollution is cheaper than cleaning it up.

In 1984, Congress amended the Superfund Act to give citizens the right to know what toxic chemicals are being stored or released in their communities. This included requiring 20,000 large manufacturing facilities to report their annual releases of 582 toxic chemicals into the environment. In order, the four industries emitting the largest amounts of toxic air pollutants in 2001 were mining (46%), electric utilities (17%), chemical production (9%), and primary metal production (8%). Between 1988 and 2001, toxic emissions in the United States dropped by 55%. However, this still means that 2.8 million metric tons (3.1 million tons) of toxic chemicals were legally released into the air, water, and ground during 2001—an average of 9.5 kilograms (21 pounds) for each American. You can look at this *Toxic Release Inventory* to find out what toxic chemicals are being stored and released in your neighborhood by going to the EPA website at www.epa.gov/tri/.

Who Should Pay for Cleaning Up Superfund Sites? Polluters or Taxpayers?

The Superfund law was designed to have polluters pay for cleaning up abandoned hazardous waste sites,

---

SPOTLIGHT
Using Honeybees to Detect Toxic Pollutants

Honeybees are being used to detect the presence of toxic and radioactive chemicals in concentrations as low as several parts per trillion. On their forays from a hive, bees pick up water, nectar, pollen, suspended particulate matter, volatile organic compounds, and radioactive material found in the air near the sites they visit.

The bees then bring these materials back to the hive. There they fan the air vigorously with their wings to regulate the hive's temperature. This action releases and circulates pollutants they picked up into the air inside the hive.

Scientists have put portable hives, each containing 7,000–15,000 bees, near known or suspected hazardous waste sites. A small copper tube attached to the side of each hive pumps air out. Portable equipment is used to analyze the air for toxic and radioactive materials.

To find out where the bees have gone to forage for food, a botanist uses a microscope to examine pollen grains to determine what kinds of plants they come from. These data can be correlated with the plants found in an area up to 1.6 kilometers (1 mile) from each hive. This allows scientists to develop maps of toxicity levels and hot spots on large tracts of land.

This approach is much cheaper than setting up a number of air pollution monitors around mines, hazardous waste dumps, and other sources of toxic and radioactive pollutants. These *biological indicator* species have helped locate and track toxic pollutants and radioactive material at more than 30 sites across the United States.

Critical Thinking

Because honeybees can pick up toxic pollutants anywhere they go, should honey from all beehives be tested for such pollutants before it is placed on the market? Explain.

---

but now taxpayers are footing the bill when the culprits cannot be found.

To keep taxpayers from footing most of the bill, cleanups of Superfund sites were based on the *polluter-pays principle.* The EPA is charged with finding the parties responsible for each site, requiring them to pay for the entire cleanup, and suing them if they do not. When the EPA can find no responsible party, it draws money out of the Superfund for cleanup. This fund was financed by taxes on oil and chemical companies. However, many of the remaining Superfund sites were created by mining and smelting industries that were not required to pay taxes into the fund.

Since the Superfund program began, polluters and their insurance companies have been working hard to do away with the *polluter-pays* principle at the heart of the program and make it mostly a *taxpayer-pays* approach. This strategy, which has been successful, has four components. *First,* deny responsibility (stonewall) to tie up the EPA in expensive legal suits for years. *Second,* sue local governments and small businesses to make them responsible for cleanup, both as a delaying tactic and to turn them into opponents of Superfund's strict liability requirements.

Third, mount a public relations campaign declaring that toxic dumps pose little threat, cleanup is too expensive compared to the risks involved, and the Superfund law is unfair, wasteful, and ineffective. *Fourth,* persuade Congress not to renew the tax on oil and chemical companies that financed the Superfund.

The EPA points out that the strict polluter-pays principle in the Superfund Act has been effective in making illegal dump sites virtually relics of the past—an important form of pollution prevention for the future. It has also forced waste producers, fearful of future liability claims, to reduce their production of such waste and to recycle or reuse much more of it.

Congress passed a law that went into effect in 2002 that generally eliminates financial liability for small business and residential property owners that contributed only small amounts of hazardous waste to Superfund sites. However, Congress has refused to renew the tax on oil and chemical companies that financed the Superfund after it expired in 1995. As a result, the Superfund is now broke and taxpayers, not polluters, are footing the bill for future cleanups when the responsible parties cannot be found. The addition of new sites has slowed down, and government funds available for cleanup have been reduced. This is an important issue for the one out of four Americans who lives within 6.4 kilometers (4 miles) of a superfund site. Are you or any members of your family one of these individuals?

X *HOW WOULD YOU VOTE?* Should the U.S. Congress reinstate the polluter-pays principle by using taxes from chemical, oil, mining, and smelting companies to reestablish a fund for cleaning up existing and new Superfund sites? Cast your vote online at http://biology.brookscole.com/miller14.

24-10 ACHIEVING A LOW-WASTE SOCIETY

What Is the Role of Grassroots Action? Making a Difference

In the United States, citizens have kept large numbers of incinerators, landfills, and hazardous waste treatment plants from being built in their local areas.

In the United States, individuals have worked together to prevent hundreds of incinerators, landfills, or treat-ment plants for hazardous and radioactive wastes from being built in or near their communities. Opposition has grown as numerous studies have shown that such facilities have traditionally been located in communities populated mostly by African Americans, Asian Americans, Latinos, and poor whites. This practice has been cited as an example of *environmental injustice.* See the Guest Essay on this subject by Robert Bullard on the website for this chapter.

Health risks from incinerators and landfills, when averaged over the entire country, are quite low, but the risks for people living near these facilities are much higher. They, not the rest of the population, are the ones whose health, lives, and property values are being threatened.

Manufacturers and waste industry officials point out that something must be done with the toxic and hazardous wastes produced to provide people with certain goods and services. They contend that if local citizens adopt a "not in my back yard" (NIMBY) approach, the waste still ends up in someone's back yard.

Many citizens do not accept this argument. To them, the best way to deal with most toxic or hazardous wastes is to produce much less of them, as suggested by the National Academy of Sciences (Figure 24-17). For such materials, their goal is "not in anyone's back yard" (NIABY) or "not on planet Earth" (NOPE) by emphasizing pollution prevention and use of the precautionary principle.

What Can Be Done at the International Level? The POPs Treaty

An international treaty calls for phasing out the use of harmful persistent organic pollutants (POPs).

In 2001, the UN Commission on Human Rights declared that being able to live free of pollution is a basic human right. There is a long way to go in converting this ideal into reality, but important progress has been made.

Between 1989 and 1994, an international treaty to limit transfer of hazardous waste from one country to another was developed. And in 2000, delegates from 122 countries completed a global treaty to control 12 *persistent organic pollutants (POPs).*

These widely used toxic chemicals are insoluble in water and soluble in fat. This means that in the fatty tissues of humans and other organisms feeding at high trophic levels in food webs, they can become concentrated at levels hundreds of thousand times higher than in the general environment (Figure 19-4, p. 411, and Figure 22-6, p. 498). These persistent pollutants can also be transported long distances by wind and water.

The list of 12 chemicals, sometimes called the *dirty dozen,* includes DDT, eight other chlorine-containing persistent pesticides (Table 23-1, p. 520), PCBs, dioxins, and furans. The goals of the treaty are to ban or phase out use of these chemicals and to detoxify or isolate stockpiles of them. About 25 countries can continue using DDT to combat malaria until safer alternatives are available. Developed nations will provide developing nations about $150 million per year to help them switch to safer alternatives to the 12 POPs.

Environmentalists consider the POPs treaty an important milestone in international environmental law because it uses the *precautionary principle* to manage and reduce the risks from toxic chemicals. This list is expected to grow as scientific studies uncover more evidence of toxic and environmental damage from some of the chemicals we use.

In 2000, the Swedish Parliament enacted a law that by 2020 would ban all chemicals that are persistent and can bioaccumulate in living tissue. This law also requires an industry to perform risk assessments on all old and new chemicals and show that these chemicals are safe to use, as opposed to requiring the government to show they are dangerous. In other words, chemicals are assumed guilty until their innocence can be established—the reverse of the current policy in the United States and most countries. There is strong opposition to this approach in the United States, especially by industries producing potentially dangerous chemicals.

How Can We Make the Transition to a Low-Waste Society? A New Vision

A number of the principles and programs discussed in this chapter can be used to make the transition to a low-waste society during this century.

According to physicist Albert Einstein, "A clever person solves a problem, a wise person avoids it." To prevent pollution and reduce waste production, many environmental scientists urge us to understand and live by four key principles:

- Everything is connected.
- There is no "away" for the wastes we produce.
- Dilution is not always the solution to pollution.
- The best and cheapest way to deal with waste and pollution is to produce less pollutants and to reuse and recycle most of the materials we use.

Currently, the order of priorities for dealing with solid waste in the United States and in most countries is the reverse of the order suggested by prominent scientists in Figure 24-3. It does not have to be that way. Some scientists and economists estimate that 60–80% of the solid waste we produce can be eliminated by a combination of *reducing waste production, reusing and recycling materials* (including composting), and *redesigning* manufacturing processes and buildings to produce less waste.

The governments of Norway, Austria, and the Netherlands have committed themselves to reduce their resource waste by 75%. Other countries are following their lead.

Likewise, there is growing interest in and use of increased *resource productivity, pollution prevention, ecoindustrial systems,* and *service-flow* businesses.

In addition, at least 24 countries have *eco-labeling programs* that certify a product or service as having met specified environmental standards. The first was Germany's *Blue Angel* program, begun in 1978. It now awards its seal of approval to more than 3,900 products and services. Other examples are India's *Ecomark,* Singapore's *Green Label,* and the *Green Seal* program in the United States.

Such revolutions start off slowly but can accelerate rapidly as their economic, ecological, and health advantages become more apparent to investors, business leaders, elected officials, and citizens.

The key to addressing the challenge of toxics use and wastes rests on a fairly straightforward principle: harness the innovation and technical ingenuity that has characterized the chemicals industry from its beginning and channel these qualities in a new direction that seeks to detoxify our economy.
ANNE PLATT MCGINN

CRITICAL THINKING

1. Use the second law of thermodynamics (p. 51) to explain why a *properly designed* source-separation recycling program takes less energy and produces less pollution than a centralized program that collects mixed waste over a large area and hauls it to a centralized facility where workers or machinery separate the wastes for recycling.

2. Are you for or against bringing about an *ecoindustrial revolution* in the country and community where you live? Explain. Do you believe it will be possible to phase in such a revolution over the next two to three decades? Explain.

3. Explain why some businesses participating in an exchange and chemical-cycling network (Figure 24-5, p. 537) might produce large amounts of waste for use as resources within the network rather than redesigning their manufacturing processes to reduce waste production. Is this acceptable? Explain.

4. Are you for or against shifting to a *service-flow economy* in the country and community where you live? Explain. Do you believe it will be possible to shift to such an economy over the next two to three decades? Explain.

5. In 2003, Changing World Technologies built a pilot plant to test a process it has developed that can convert a mixture of computers, old tires, turkey bones and feathers, and other wastes into oil by mimicking and speeding up the way that nature converts biomass into oil. If this recycling process turns out to be technologically and economically feasible, explain why it could increase waste production.

6. Would you oppose having a hazardous waste landfill, waste treatment plant, deep-injection well, or incinerator in your community? Explain. If you oppose these disposal facilities, how do you believe the hazardous waste generated in your community and your state should be managed?

7. Give your reasons for agreeing or disagreeing with each of the following proposals for dealing with hazardous waste:

a. Reduce the production of hazardous waste and encourage recycling and reuse of hazardous materials by charging producers a tax or fee for each unit of waste generated.

b. Ban all land disposal and incineration of hazardous waste to encourage recycling, reuse, and waste treatment and to protect air, water, and soil from contamination.

c. Provide low-interest loans, tax breaks, and other financial incentives to encourage industries producing hazardous waste to reduce, recycle, reuse, treat, and decompose such waste.

8. Congratulations! You are in charge of the world. List the three most important components of your strategy for dealing with **(a)** solid waste and **(b)** hazardous waste.

PROJECTS

1. Collect all of the trash (excluding food waste) that you generate in a typical week. Measure its total weight and volume. Sort it into major categories such as paper, plastic, metal, and glass. Then weigh each category and calculate the percentage in each category. What percentage of this waste consists of materials that could be recycled or reused? What percentage of the items could you have done without? Tally and compare the results for your entire class.

2. What percentage of the municipal solid waste in your community is **(a)** placed in a landfill, **(b)** incinerated, **(c)** composted, and **(d)** recycled? What technology is used in local landfills and incinerators? What leakage and pollution problems have local landfills or incinerators had? Does your community have a recycling program? Is it voluntary or mandatory? Does it have curbside collection? Drop-off centers? Buyback centers? A hazardous waste collection system? Devise a plan for improving the MSW system in your community and submit it to local officials.

3. Make a survey of the hazardous materials (Figure 24-16, p. 549) found in your house or apartment,

or in your family home if you live in a dorm. Which of these materials are actually used? Call city officials to find out how you can dispose of hazardous chemicals you do not need.

4. What hazardous wastes are produced at your school? What happens to these wastes?

5. Go to the EPA Superfund website at www.epa.gov /superfund/sites/npl/npl.htm. Click on your state to find out how many hazardous sites it has on the National Priority List. Find any sites close to where you live or go to school. Click on each site near you to learn about its history, what types of pollutants it contains, the sources of these pollutants, how it is being cleaned up, and progress toward this goal.

6. If possible, visit a local recycling center or materials-recovery facility to find out how it works, where the materials separated out for recycling go, how the prices of these separated materials have fluctuated in the last 3 years, and the major problems faced by the facility.

7. Go to a large office supply store and compare prices for comparable grades of copy paper made from virgin paper with those containing some recycled content. Make the same price comparison at a stationery store.

8. Use the library or the Internet to find bibliographic information about *Arthur C. Clarke* and *Anne Platt McGinn*, whose quotes appear at the beginning and end of this chapter.

9. Make a concept map of this chapter's major ideas, using the section heads, subheads, and key terms (in boldface). Look on the website for this book for information about making concept maps.

LEARNING ONLINE

The website for this book contains study aids and many ideas for further reading and research. They include a chapter summary, review questions for the entire chapter, flash cards for key terms and concepts, a multiple-choice practice quiz, interesting Internet sites, references, and a guide for accessing thousands of InfoTrac® College Edition articles. Log on to

http://biology.brookscole.com/miller14

Then click on the Chapter-by-Chapter area, choose Chapter 24, and select a learning resource.

25 Sustainable Cities

The Ecocity Concept in Curitiba, Brazil

Environmental and urban designers envision the development of more sustainable cities, called *ecocities* or *green cities*. The ecocity is not a futuristic dream. One of the world's most livable and sustainable major cities is Curitiba, Brazil, with more than 2.5 million people.

This city decided in 1969 to focus on mass transit. Curitiba probably has the world's best bus system, which each day carries more than three-fourths of its people throughout the city along express lanes dedicated to buses (Figure 25-1).

Bike paths run throughout most of the city. Cars are banned from 49 blocks of the city's downtown area, which has a network of pedestrian walkways connected to bus stations, parks, and bike paths. Because Curitiba relies less on automobiles, it uses less energy per person and has less air pollution, greenhouse gas emissions, and traffic congestion than most comparable cities.

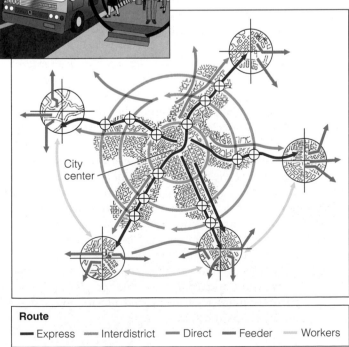

City center

Route

— Express — Interdistrict — Direct — Feeder — Workers

To reduce flood damage and add green space, the city transformed flood-prone areas along its rivers into a series of interconnected parks that are crisscrossed by bike paths. Volunteers have planted more than 1.5 million trees throughout the city. No tree in the city can be cut down without a permit, and two trees must be planted for each one cut.

The city recycles roughly 70% of its paper and 60% of its metal, glass, and plastic, which is sorted by households for collection three times a week. Recovered materials are sold mostly to the city's more than 500 major industries that must meet strict pollution standards. Most industries are in an industrial park outside the city limits. A major bus line runs to the park, but many of the workers live nearby and can walk or bike to work.

The city uses old buses as roving classrooms to give the poor basic skills needed for jobs. Other retired buses have become health clinics, soup kitchens, and some of the city's 200 day-care centers, which are open 11 hours a day and are free for low-income parents.

The poor receive free medical, dental, and child-care, and 40 feeding centers are available for street children. The city has a *build-it-yourself* system that gives each poor family a plot of land, building materials, two trees, and an hour's consultation with an architect.

In Curitiba, virtually all households have electricity, drinking water, and trash collection. About 95% of its citizens can read and write and 83% of the adults have at least a high school education. All schoolchildren study ecology. Polls show that 99% of the city's inhabitants would not want to live anywhere else.

This ecocity design is the brainchild of architect and former college teacher Jaime Lerner, who has served as the city's mayor three times since 1969. Under his leadership, the municipal government dedicated itself to two goals. *First,* find simple, innovative, fast, cheap, and fun solutions to problems. *Second,* establish a government that is honest, accountable, and open to public scrutiny.

An exciting challenge during this century will be to reshape existing cities and design new ones like Curitiba that are more livable and sustainable and that have a lower environmental impact.

Figure 25-1 Solutions: bus system in Curitiba, Brazil. This system moves large numbers of passengers around rapidly because each of the five major spokes has two express lanes used only by buses. Double- and triple-length bus sections are hooked together as needed, and boarding is speeded up by the use of extra-wide doors and raised tubes that allow passengers to pay before getting on the bus (top left).

The city is not an ecological monstrosity. It is rather the place where both the problems and the opportunities of modern technological civilization are most potent and visible.

PETER SELF

This chapter addresses the following questions:

- How is the world's population distributed between rural and urban areas, and what factors determine how urban areas develop?

- What are the major resource and environmental problems of urban areas?

- How do transportation systems shape urban areas and growth, and what are the advantages and disadvantages of various forms of transportation?

- What methods are used for planning and controlling urban growth?

- How can cities be made more sustainable and more desirable places to live?

25-1 URBANIZATION AND URBAN GROWTH

What Causes Urban Growth? Magnets for Business and Hope for the Poor

Many people move to cities because "push" factors force them out of rural areas and "pull" factors give them the hope of finding jobs and a better life in the city.

At the beginning of the industrial revolution about 275 years ago most people lived in rural areas and small villages and towns. Today almost half of the world's people live in densely populated urban areas, as rural people have migrated to cities with the hope of finding jobs and a better life.

Urban areas grow in two ways—by *natural increase* (more births than deaths) and by *immigration,* mostly from rural areas. This urban immigration is influenced by *push factors* such as poverty, lack of land to grow food, declining agricultural jobs, famine, and war that force people out of rural areas. Rural people are also *pulled* to urban areas in search of jobs, food, housing, a better life, entertainment, and freedom from religious, racial, and political conflicts.

People are also pushed and pulled to cities by government policies that favor urban over rural areas. For example, developing countries spend most of their budgets on economic development and job creation in urban areas, especially in capital cities where their leaders live. Some governments establish lower food prices in urban areas, which pleases city dwellers,

helps keep leaders in power, and attracts the rural poor.

What Are the Worldwide Patterns of Urbanization and Urban Growth? More People and More Poverty

Urban populations are growing rapidly throughout the world, and many cities in developing countries have become centers of poverty.

A country's **degree of urbanization** is the percentage of its population living in an urban area. **Urban growth** is the rate of increase of urban populations.

Five major trends are important in understanding the problems and challenges of urbanization and urban growth. First, *the proportion of the global population living in urban areas is increasing.* Between 1850 and 2004, the percentage of people living in urban areas increased from 2% to 48% (Figure 25-2). According to UN projections, by 2007 half of the world's people will live in urban areas and by 2030 about 60% will. Between 2004 and 2030 the world's urban population is projected to increase from 3.1 billion to 5 billion. Almost all of this growth will occur in already overcrowded cities in developing countries such as India, Brazil, China, and Mexico (Figure 25-2).

Second, *the number of large cities is mushrooming.* In 2004 more than 400 cities had a million or more people, and this is projected to increase to 564 by 2015. Today there are 18 *megacities* or *meagalopolises* (up from 8 in 1985) with 10 million or more people—most of them in developing countries (Figure 25-2).

A third trend is that *urbanization and urban populations are increasing rapidly in developing countries* (Figure 25-3). Between 2004 and 2030, the degree of urbanization in developing countries is expected to increase from 41% to 56%. In Latin American and Caribbean developing countries, 75% of the people are urban dwellers compared to only 35% in Africa and 39% in Asia.

Fourth, *urban growth is much slower in already heavily urbanized developed countries (with 76% urbanization) than in developing countries.* North America's urbanization is 79%, the highest in the world. By 2030 urbanization in developed countries is projected to increase to 84%.

Fifth, *poverty is becoming increasingly urbanized as more poor people migrate from rural to urban areas, mostly in developing countries.* The United Nations estimates that at least 1 billion poor people live in the urban areas of developing countries. If you visit poor areas of such a city your senses may be overwhelmed with a chaotic but vibrant crush of people, vehicles of all sorts, street vendors, traffic jams, noise, smells, smoke from wood and coal fires, and people sleeping on streets or living in crowed, unsanitary, and rickety and unsafe slums and shantytowns.

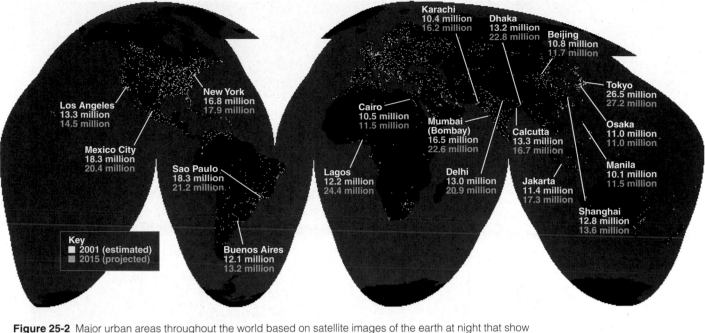

Figure 25-2 Major urban areas throughout the world based on satellite images of the earth at night that show city lights. Currently, the 48% of the world's people living in urban areas occupy about 2% of the earth's land area. Note that most of the world's urban areas are found along the coasts, and most of Africa and much of the interior of South America, Asia, and Australia are dark at night. This figure also shows the populations of the world's 18 *megacities* with 10 or more million people in 2001 and their projected populations in 2015. Note that all but four are located in developing countries. Use this figure to list in order the world's five most populous cities in 2001 and the five most populous ones projected for 2015. (National Geophysical Data Center/National Oceanic and Atmospheric Administration and United Nations)

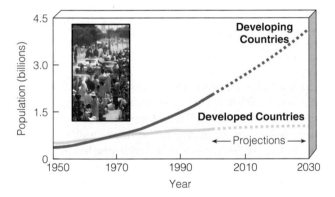

Figure 25-3 Urban population in developed and developing countries, 1950–2003, with projections to 2030. (Data from United Nations Population Division)

How Urbanized Is the United States? City-Dwellers Dominate

Almost eight of every ten Americans live in urban areas, about half of them in sprawling suburbs.

Between 1800 and 2004, the percentage of the U.S. population living in urban areas increased from 5% to 79%. The population has shifted in four phases. First, *people migrated from rural areas to large central cities.* Currently, three-fourths of Americans live in 271 *metropolitan areas* (cities with at least 50,000 people), and nearly half live in consolidated metropolitan areas containing 1 million or more residents (Figure 25-4, p. 566).

Second, many people *migrated from large central cities to suburbs and smaller cities.* Currently, about 51% live in the suburbs and 30% in central cities.

Third, many people *migrated from the North and East to the South and West.* Since 1980, about 80% of the U.S. population increase has occurred in the South and West, particularly near the coasts. California in the West, with 34.5 million people, is the most populous state, followed by Texas in the Southwest with 21.3 million. This shift is expected to continue.

Fourth, *some people have migrated from urban areas back to rural areas* since the 1970s, and especially since 1990.

How Has the Quality of Urban Life in the United States Changed? Progress and Challenges

The quality of urban life improved significantly for most Americans during the last century, but there is still a long way to go.

Since 1920, many of the worst urban environmental problems in the United States have been reduced significantly. Most people have better working and housing conditions, and air and water quality have improved (Figure 10-11, p. 181). Better sanitation, public water supplies, and medical care have slashed death rates and the prevalence of sickness from malnutrition and infectious diseases. And concentrating most of the

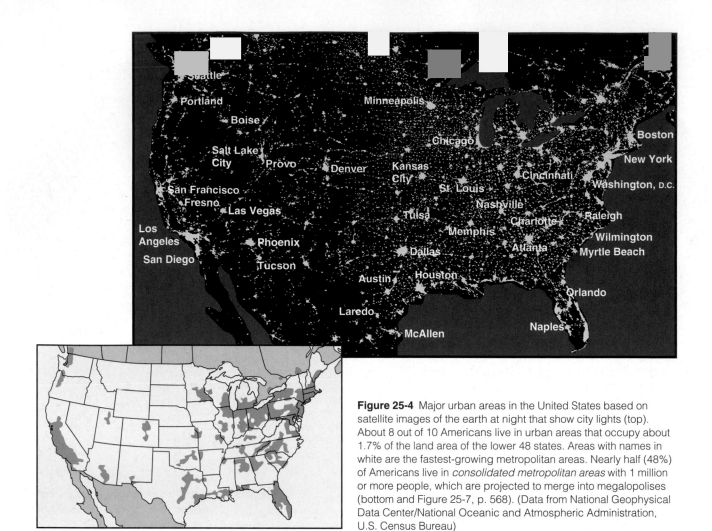

Figure 25-4 Major urban areas in the United States based on satellite images of the earth at night that show city lights (top). About 8 out of 10 Americans live in urban areas that occupy about 1.7% of the land area of the lower 48 states. Areas with names in white are the fastest-growing metropolitan areas. Nearly half (48%) of Americans live in *consolidated metropolitan areas* with 1 million or more people, which are projected to merge into megalopolises (bottom and Figure 25-7, p. 568). (Data from National Geophysical Data Center/National Oceanic and Atmospheric Administration, U.S. Census Bureau)

population in urban areas has helped protect the country's biodiversity by reducing the destruction and degradation of wildlife habitat.

However, a number of U.S. cities, especially older ones, have *deteriorating services* and *aging infrastructures* (streets, schools, bridges, housing, and sewers). Many also face *budget crunches* from rising costs as some businesses and people move to the suburbs or rural areas and reduce revenues from property taxes. And there is *rising poverty* in the centers of many older cities, where unemployment typically is 50% or higher.

What Is Urban Sprawl and What Are Its Effects? Paving Paradise and Driving to Get Anywhere

When there is ample and affordable land, urban areas tend to sprawl outward, swallowing up surrounding countryside.

Another major problem in the United States and other countries with lots of room for expansion is **urban sprawl.** Growth of low-density development on the edges of cities and towns gobbles up surrounding countryside—frequently prime farmland or forests—

and increases dependence on cars (Figure 25-5). The result is a far-flung hodgepodge of housing developments, shopping malls, parking lots, and office complexes—loosely connected by multilane highways.

Before sprawl, people in cities and small towns lived, shopped, and worked close to their homes and could meet most of their daily needs by walking. Every few blocks had a small grocery store, a pharmacy, professional offices, and other stores. Often shop owners lived above their stores or rented out such spaces. Most people could walk, bike, or take mass transit to neighborhood schools and parks without the need for a car.

Starting in 1945, most U.S. cities began spreading out and more people followed what was advertised as the "American Dream," living in their own house on their own piece of land away from the central city.

Six major factors promoted urban sprawl in the United States. *First,* ample land was available for most cities to spread outward. *Second,* federal government loan guarantees for new single-family homes for World War II veterans stimulated the development of suburbs. *Third,* low-cost gasoline and the federal and state funding of highways encouraged automobile use and

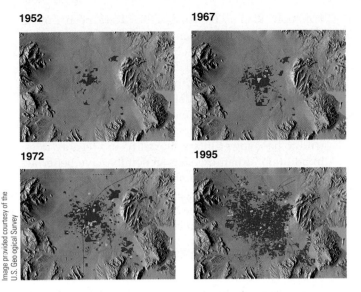

1952 **1967**

1972 **1995**

Image provided courtesy of the U.S. Geological Survey

Figure 25-5 *Urban sprawl* in and around Las Vegas, Nevada, 1952–1995—a process that has continued. Between 1970 and 2004, the population of water-short Clark County which includes Las Vegas more than quadrupled from 463,000 to 2 million. The growth is expected to continue but may be limited by lack of water (Spotlight, p. 326).

the development of once-inaccessible outlying tracts of land that were affordable for many Americans.

Fourth, tax laws encouraged home ownership by allowing deduction of interest on home loans from income taxes. *Fifth,* most state and local zoning laws required large residential lots and separation of residential and commercial use of land in new communities.

Sixth, most urban areas consist of numerous political jurisdictions, which rarely work together to develop an overall plan for managing and controlling urban growth and sprawl. *In a nutshell, urban sprawl is the product of affordable land, automobiles, cheap gasoline, and poor urban planning.*

Figure 25-6 shows some of the undesirable consequences of urban sprawl. Look carefully at this figure. Urban sprawl has increased travel time in automobiles, decreased energy efficiency, increased urban flooding problems, and destroyed prime cropland, forests, open space, and wetlands. It has also led to the economic decline of many central cities.

To pay for heavily mortgaged houses and cars, adults in a typical suburban family have to spend

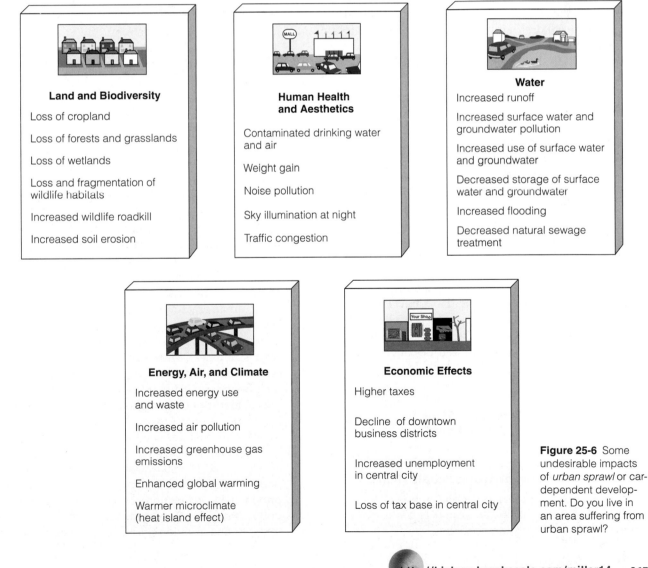

Land and Biodiversity

Loss of cropland

Loss of forests and grasslands

Loss of wetlands

Loss and fragmentation of wildlife habitats

Increased wildlife roadkill

Increased soil erosion

Human Health and Aesthetics

Contaminated drinking water and air

Weight gain

Noise pollution

Sky illumination at night

Traffic congestion

Water

Increased runoff

Increased surface water and groundwater pollution

Increased use of surface water and groundwater

Decreased storage of surface water and groundwater

Increased flooding

Decreased natural sewage treatment

Energy, Air, and Climate

Increased energy use and waste

Increased air pollution

Increased greenhouse gas emissions

Enhanced global warming

Warmer microclimate (heat island effect)

Economic Effects

Higher taxes

Decline of downtown business districts

Increased unemployment in central city

Loss of tax base in central city

Figure 25-6 Some undesirable impacts of *urban sprawl* or car-dependent development. Do you live in an area suffering from urban sprawl?

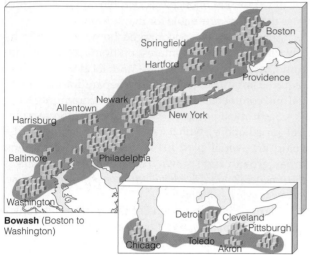

Figure 25-7 Two megalopolises: *Bowash*, consisting of urban sprawl and coalescence between Boston and Washington, D.C., and *Chipitts*, extending from Chicago to Pittsburgh.

many of their non-working hours driving to and from work, or running errands over a vast suburban landscape. This leaves many of them with little energy and time for their children and themselves, or getting to know their neighbors.

In 2003, Reid Ewing and other researchers discovered a connection between sprawling suburbs and spreading waistlines. They found that people living in suburbs, where it is hard to get anywhere on foot or by bicycle, are heavier than those in central cities and in pedestrian-friendly towns.

As they grow and sprawl outward, separate urban areas may merge to form a *megalopolis*. For example, the remaining open space between Boston, Massachusetts, and Washington, D.C., is rapidly urbanizing and coalescing. The result is an almost continuous 800-kilometer-long (500-mile-long) urban area that is sometimes called *Bowash* (Figure 25-7 and Figure 25-4, bottom).

Megalopolises developing all over the world include the area between Amsterdam and Paris in Europe, Japan's Tokyo–Yokohama–Osaka–Kobe corridor known as Tokohama, and the Brazilian industrial triangle made up of São Paulo, Rio de Janeiro, and Belo Horizonte.

25-2 URBAN RESOURCE AND ENVIRONMENTAL PROBLEMS

Case Study: What Are the Advantages of Urbanization? Concentrating People Helps

Urban areas can offer more job opportunities and better education and health, and can help protect biodiversity by concentrating people.

For more than 6,000 years, cities have been centers of economic development, education, jobs, technological developments, culture, social change, and political power. The high density of urban populations provides governments and businesses with significant cost advantages in delivering goods and services.

Urban residents in many parts of the world live longer than rural residents, and urban populations tend to have lower infant mortality and fertility rates. In addition, urban dwellers generally have better access to medical care, family planning, education, and social services than people in rural areas.

Urban areas also have some environmental advantages. For example, recycling is more economically feasible because of the large concentrations of recyclable materials, and per capita expenditures on environmental protection are higher in urban areas. Also, concentrating people in urban areas helps preserve biodiversity by reducing the stress on wildlife habitats.

Case Study: What Are the Disadvantages of Urbanization? Concentrating People Has Some Harmful Effects

Cities are rarely self-sustaining, and they threaten biodiversity, lack trees, grow little of their food, concentrate pollutants and noise, spread infectious diseases, and are centers of poverty, crime, and terrorism.

Although urban dwellers occupy only about 2% of the earth's land area, they consume about three-fourths of all resources. Because of this and their high waste output (Figure 25-8), most of the world's cities are not self-sustaining systems.

Large areas of land must be disturbed and degraded to provide urban dwellers with food, water, energy, minerals, and other resources. This decreases the earth's biodiversity. Also, as cities expand they destroy rural cropland, fertile soil, forests, wetlands, and wildlife habitats. At the same time, they provide little of the food they use. From an environmental standpoint, urban areas are somewhat like gigantic vacuum cleaners, sucking up much of the world's matter, energy, and living resources and spewing out pollution, wastes, and heat. Thus, urban areas have large *ecological footprints* (Figure 1-7, p. 10) that extend far beyond their boundaries. If you live in a city, calculate its ecological footprint by going to the website www.redefiningprogress.org/. Also see the Guest Essay on this topic by Michael Cain on this chapter's website.

In urban areas most trees, shrubs, and other plants are destroyed to make way for buildings, roads, and parking lots. Most cities thus largely lose the benefits provided by vegetation that would help absorb air pollutants, give off oxygen, help cool the air through transpiration, provide shade, reduce soil erosion, muf-

Inputs

Energy

Food

Water

Raw materials

Manufactured goods

Money

Information

Outputs

Solid wastes

Waste heat

Air pollutants

Water pollutants

Greenhouse gases

Manufactured goods

Noise

Wealth

Ideas

Figure 25-8 Natural capital degradation: Urban areas rarely are sustainable systems. The typical city depends on large nonurban areas of land and water for huge inputs of matter and energy resources and for large outputs of waste matter and heat. For example, according to an analysis by Mathis Wackernagel and William Rees, an area 58 times as large as that of London is needed to supply its residents with resources. They estimate that meeting the needs of all the world's people at the same rate of resource use as that of London would take at least three more earths.

fle noise, provide wildlife habitats, and give aesthetic pleasure. As one observer remarked, "Most cities are places where they cut down all or most of the trees and then name the streets after them."

As cities grow and water demands increase, expensive reservoirs and canals must be built and deeper wells drilled. This can deprive rural and wild areas of surface water and deplete groundwater faster than it is replenished.

Flooding also tends to be greater in cities, in some cases because they are built on floodplains or in low-lying coastal areas subject to natural flooding. Another reason is that covering land with buildings, asphalt, and concrete causes precipitation to run off quickly and overload storm drains. In addition, urban development often destroys or degrades wetlands that act as natural sponges to help absorb excess water.

Another threat is that many of the world's largest cities are in coastal areas (Figure 25-2) that could be flooded sometime in this century if sea levels rise due to projected global warming.

Because of their high population and resource consumption, urban dwellers produce most of the world's air pollution, water pollution, and solid and hazardous wastes. Also, pollutant levels are generally higher in urban areas because they are produced in a smaller area and cannot be as readily dispersed and diluted as are those produced in most rural areas.

Most urban dwellers are subjected to **noise pollution:** any unwanted, disturbing, or harmful sound that impairs or interferes with hearing, causes stress, hampers concentration and work efficiency, or causes accidents. Noise levels (Figure 25-9, p. 570) above 65 dBA are considered unacceptable, and prolonged exposure to levels above 85 dBA can cause permanent hearing damage.

In addition, high population densities in urban areas can increase the spread of infectious diseases, especially if adequate drinking water and sewage systems are not available.

Cities generally are warmer, rainier, foggier, and cloudier than suburbs and nearby rural areas mostly because of their buildings, pavement, and lack of green space. The enormous amounts of heat generated by cars, factories, furnaces, lights, air conditioners, and heat-absorbing dark roofs and roads in cities create an *urban heat island* surrounded by cooler suburban and rural areas.

Higher CO_2 and soot concentrations from fossil fuel–burning, cars, factories, and buildings intensify this heat-island effect. The higher CO_2 levels can increase plant growth, expecially opportunistic species such as ragweed, that can worsen asthma. Also, tiny soot and other particles help deliver pollen, mold, and other allergens deep into the lungs. This may help explain why childhood asthma rates have climbed

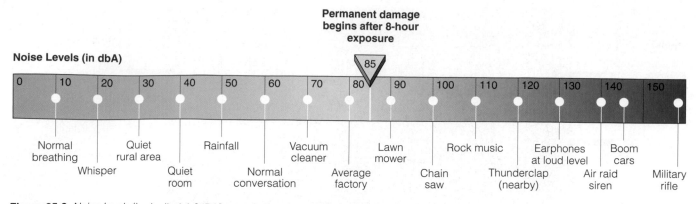

Figure 25-9 *Noise levels* (in decibel-A [dBA] sound pressure units) of some common sounds. Sound pressure becomes damaging at about 75 dBA and painful at around 120 dBA. At 180 dBA it can kill. Because the dB and dBA scales are logarithmic, sound pressure is multiplied 10-fold with each 10-decibel rise. Thus a rise from 30 dBA (quiet rural area) to 60 dBA (normal restaurant conversation) represents a 1,000-fold increase in sound pressure on the ear. You are being exposed to a sound level high enough to cause permanent hearing damage if you need to raise your voice to be heard above the racket, if a noise causes your ears to ring, or if nearby speech seems muffled. Ways to control noise include modifying noisy activities and devices, shielding noisy devices or processes, shielding workers from noise, moving noisy operations or machinery away from people, and using anti-noise (a new technology that cancels out one noise with another).

steadily in recent years in many U.S. and Canadian urban areas. As cities grow and merge, their heat islands also merge and can keep polluted air from being diluted and cleansed.

There are also problems with the artificial light created by urban areas (Figures 25-2 and 25-4). Lighting up the sky prevents people from seeing the glories of a starry night and hinders astronomers from conducting their research. There is also growing evidence that artificial light is affecting plants and animals in a variety of ecosystems. Species affected by such *light pollution* include turtles who lay their eggs on beaches at night and migrating birds that are lured off course by lights on high-rise office buildings and fatally collide with such structures. Chicago, Illinois, leads U.S. efforts to have major city-center buildings turn off their lights between 11 P.M. and dawn, and Toronto, Canada, has had a similar lights-out program since 1996.

Wesley College professor Marianne Moore and her colleagues have found evidence that artificial illumination can alter aquatic ecosystems and could ultimately decrease water quality. Minute zooplankton avoid predators by remaining well below the surface during the day and then rising to graze on algae at night. But artificial light from urban glows can discourage their nightly surface feeding. If their grazing is inhibited, algae populations could explode and these blooms would deplete dissolved oxygen needed by fish and decrease water quality.

Urban areas can intensify poverty and social problems. Crime rates also tend be higher in urban areas than in rural areas (Connections, p. 572). And urban areas are more likely and desirable targets for terrorist acts.

Case Study: How Do the Urban Poor in Developing Countries Live—Living on the Edge with Ingenuity and Hope

Most of the urban poor in developing countries live in crowded, unhealthy, and dangerous conditions, but many are better off than the rural poor.

Many of the world's poor live in crowded central city *slums*—multifamily tenements and rooming houses where three to six people live in a single room. Most of these dwellings have inadequate sanitation and ventilation and many are in unsafe structures. Others live in *squatter settlements* and *shantytowns* on the outskirts of most cities in developing countries, some perched precariously on steep hillsides subject to landslides. In these illegal settlements, people take over unoccupied land and build shacks from corrugated metal, plastic sheets, scrap wood, and other scavenged building materials. Still others live or sleep on the streets, having nowhere else to go.

Squatters living near the edge of survival in these areas usually lack clean water, sewers, electricity, and roads, and often are subject to severe air and water pollution and hazardous wastes from nearby factories (Case Study, p. 549). Their locations may be especially prone to landslides, flooding, earthquakes, or volcanic eruptions.

Most cities cannot afford to provide squatters with basic protections and services, and their officials fear that doing so will attract even more of the rural poor. Many city governments regularly bulldoze squatter shacks and send police to drive the illegal settlers out. The people later move back in or develop another shantytown somewhere else.

Despite joblessness, squalor, overcrowding, and environmental and health hazards, most squatters and slum residents are better off than the rural poor. With better access to family planning programs, they tend to have fewer children and better access to schools. They work, raise families, educate their children, and often have to care for their parents. Some work together to establish water supplies, sewer, and health care facilities, and schools. Many squatter settlements provide a sense of community and a vital safety net of neighbors, friends, and relatives.

Mexico City is an example of an urban area in crisis. About 18.3 million people—roughly one of every six Mexicans—live there (Figure 25-2). It is the world's second most populous city, and each year, about 210,000 new residents arrive.

Mexico City suffers from severe air pollution, close to 50% unemployment, deafening noise, overcrowding, traffic congestion, inadequate public transportation, and a soaring crime rate. More than one-third of its residents live in slums called *barrios* or in squatter settlements without running water or electricity.

At least 3 million people have no sewer facilities. This means huge amounts of human waste are deposited in gutters, vacant lots, and open sewers every day, attracting armies of rats and swarms of flies. When the winds pick up dried excrement, a *fecal snow* often falls on parts of the city. Open garbage dumps also contribute dust and bacteria to the atmosphere. This bacteria-laden fallout leads to widespread salmonella and hepatitis infections, especially among children.

Mexico City has one of the world's worst photochemical smog problems because of a combination of too many cars and polluting industries, a sunny climate, and topographical bad luck. The city lies in a high-elevation bowl-shaped valley surrounded on three sides by mountains—ideal conditions for thermal inversions that trap pollutants at ground level (Figure 20-7, top, p. 443). Since 1982, the amount of contamination in the city's air has more than tripled, and breathing that air is said to be roughly equivalent to smoking three packs of cigarettes a day.

The city's air and water pollution cause an estimated 100,000 premature deaths per year. Writer Carlos Fuentes has nicknamed this megacity "Makesicko City."

Water demands are pushing the city's aquifer beyond its limits. Energy costs to extract water have soared as wells have become much deeper. Withdrawal from aquifers caused parts of the city to subside by 9 meters (30 feet) during the twentieth century. Some areas now subside as much as 30 centimeters (1 foot) a year.

The city government has banned cars from a 50-block central zone, required catalytic converters on all cars made after 1991, phased out use of leaded gasoline, and replaced old buses, taxis, and delivery vehicles with cleaner vehicles running mostly on liquefied petroleum gas. The city also planted more than 25 million trees to help absorb pollutants and bought some land for use as green space.

Some progress has been made. The percentage of days each year in which air pollution standards are violated has fallen from 50% to 20%. But the city still has an inadequate mass transportation system and weak, poorly enforced air pollution standards for industries and motor vehicles. If you were in charge of Mexico City, what are the three most important things you would do?

> **X** *HOW WOULD YOU VOTE?* Should squatters around cities of developing countries be given title to land they live on? Cast your vote online at http://biology.brookscole.com/miller14.

25-3 TRANSPORTATION AND URBAN DEVELOPMENT

How Do Land Availability and Transportation Systems Affect Urban Development? Stack or Sprawl

Land availability determines whether a city grows vertically or spreads out horizontally and whether it relies mostly on mass transportation or the automobile.

The two main types of ground transportation are *individual* (such as cars, motor scooters, bicycles, and walking) and *mass* (mostly buses and rail systems). About 90% of all travel in the world is by foot, bicycle, motor scooter, or bus—mostly because only about 10% of the world's people can afford a car.

Land availability is a key factor determining the types of transportation people use. If a city cannot spread outward, it must grow vertically—upward and downward (below ground)—so it occupies a small land area with a high population density. Most people living in such *compact cities* like Hong Kong and Tokyo walk, ride bicycles, or use energy-efficient mass transit.

A combination of cheap gasoline, plentiful land, and a network of highways produces *dispersed cities*. They are found in countries such the United States, Canada, and Australia, where ample land often is available for outward expansion. Sprawling cities depend on the automobile; motor vehicles are increasing in both compact and dispersed cities. Today there are about 700 million cars, trucks, and buses in the world. By 2050, the number of motor vehicles is projected to increase sevenfold to 3.5 billion—2.5 billion of them in today's developing countries.

Is this sustainable? No one knows. Some analysts believe that phasing in motor vehicles with clean-burning hybrid and fuel cell engines would allow the

How Can Reducing Crime Help the Environment?

Most people do not realize that reducing crime can help improve environmental quality. Crimes such as robbery, assault, and shootings can have several harmful environmental effects.

It can drive people out of cities, which are our most energy-efficient living arrangements. Every brick in an abandoned urban building represents an energy waste equivalent to burning a 100-watt light bulb for 12 hours. Each new suburb means replacing farmland or reservoirs of natural biodiversity such as forests with dispersed, energy- and resource-wasting roads, houses, and shopping centers.

Crime can make people less willing to walk, bicycle, and use energy-efficient public transit systems. It also forces many people to use more energy to deter burglars. For example, trees and bushes near a house help save energy by reducing solar heat gain in the summer and providing windbreaks in the winter. But to help reduce break-ins many homeowners clear away trees and bushes near their houses, as well as installing yard lights and leaving indoor lights, TVs, and radios on to deter burglars.

The threat of crime also causes overpackaging of many items to deter shoplifting or poisoning of food or drug items.

Critical Thinking

Can you think of any environmental benefits of certain types of crimes?

world's motor vehicle fleet to double while emitting less air pollution than today's fleet.

Vehicle emissions are not the only problem. Producing motor vehicles and building roads, parking lots, and garages use huge amounts of energy and matter resources that produce pollution and environmental degradation. Also, motor vehicles take up space and thus cause congestion as their numbers multiply. More and more people could end up stuck in traffic jams in fuel-efficient and low-polluting cars going nowhere.

What Is the Role of Motor Vehicles in the United States? Cars Rule

Passenger vehicles account for almost all U.S. urban transportation and travel to work, and each year Americans drive as far as everyone else in the world combined.

America showcases the advantages and disadvantages of living in a society dominated by motor vehicles. With 4.6% of the world's people, the United States has almost a third of the world's motor vehicles. About two-thirds of the 225 million motor vehicles (excluding big trucks and buses) in the United States are cars and the remainder are sport utility vehicles (SUVs), pickup trucks, and vans.

Mostly because of urban sprawl and convenience, passenger vehicles are used for 98% of all urban transportation and 91% of travel to work in the United States. About three-fourths of all trips are less than 1.6 kilometers (1 mile) from home. About 75% of Americans drive alone to work, 5% commute on public transit, and 0.5% bicycle to work. Mostly because of urban sprawl and a network of highways, Americans drive about 4 trillion kilometers (2.5 trillion miles) each year, about the same distance driven by all other drivers in the world. Each year American vehicles consume about 43% of the world's gasoline. According to the American Public Transit System, if Americans doubled their use of mass transit from the current 5% to 10%, it would reduce U.S. dependence on oil by 40%.

Many governments in rapidly industrializing countries such as China want to develop an automobile-centered transportation system. Suppose China succeeds in having one or two cars in every garage and consumes oil at the U.S. rate. According to environmental leader Lester R. Brown, China would then need slightly more oil each year than the world now produces and would have to pave an area equal to half of the land it now uses to produce food.

What Are the Advantages and Disadvantages of Motor Vehicles? A Troubled Love Affair

Motor vehicles provide personal benefits and help run economies, but they also kill lots of people, pollute the air, promote urban sprawl, and lead to time- and gas-wasting traffic jams.

Motor vehicles have a number of important benefits. On a personal level, they provide mobility and are a convenient and comfortable way to get from one place to another. They also are symbols of power, sex, social status, and success for many people. For some they also provide escape from an increasingly hectic world.

From an economic standpoint, much of the world's economy is built on producing motor vehicles and supplying roads, services, and repairs for them. In the United States, for example, $1 of every $4 spent and one of every six nonfarm jobs is connected to the automobile. And five of the seven largest U.S. industrial firms produce motor vehicles or their fuel.

Despite their important benefits, motor vehicles have many harmful effects on people and the environment. They have killed almost 18 million people since 1885, when Karl Benz built the first automobile. Throughout the world they kill an estimated 1.2 million people each year—an average of 3,300 deaths per day—and injure another 15 million. Each year they also kill about 50 million wild animals and family pets.

In the United States, motor vehicle accidents kill more than 40,000 people a year and injure another 5 million, at least 300,000 of them severely. *Car accidents have killed more Americans than have all wars in the country's history.*

Motor vehicles are the world's largest source of air pollution. They emit six of the eight major air pollutants (Table 20-2, p. 438), which prematurely kill 30,000–60,000 people per year in the United States, according to the Environmental Protection Agency. Motor vehicles are also the fastest-growing source of carbon dioxide emissions—now producing almost one-fourth of them. In addition, they account for two-thirds of the oil used in the United States and one-third of the world's oil consumption.

Motor vehicles have helped create urban sprawl. At least a third of urban land worldwide and half in the United States is devoted to roads, parking lots, gasoline stations, and other automobile-related uses. This prompted urban expert Lewis Mumford to suggest that the U.S. national flower should be the concrete cloverleaf.

Another problem is congestion. If current trends continue, U.S. motorists will spend an average of 2 years of their lives in traffic jams, wasting about $60 billion a year in gasoline and lost time.

Building more roads may not be the answer. Many analysts agree with the idea, stated by economist Robert Samuelson, that "cars expand to fill available concrete."

Motor vehicles have harmful economic costs, mostly because of deaths and injuries, higher insurance rates, pollution, work time wasted in traffic jams, and decreased property values near noisy, congested roads. According to the International Center for Technology Assessment, such costs amount to $1,970–5,990 per American each year. Because these costs are not included in the prices of motor vehicles and gasoline, most people do not associate them with the motor vehicles they buy and use.

How Can We Reduce Automobile Use? Use Honest Accounting

We can reduce automobile use by having users pay for its harmful effects but this is politically unpopular.

Environmentalists and a number of economists suggest that one way to reduce the harmful effects of automobile use is to make drivers pay directly for most of the harmful costs of automobile use—a *user-pays* approach based on honest environmental accounting. One option is to include the estimated harmful costs of driving as a tax on gasoline. Such taxes would amount to about $1.30–2.10 per liter ($5–8 per gallon) of gasoline in the United States and would spur the use of more energy-efficient motor vehicles.

Proponents urge governments to use gasoline tax revenues to help finance mass transit systems, bike paths, and sidewalks. The government could reduce taxes on income and wages to offset the increased taxes on gasoline and thus make such a *tax shift* more politically acceptable.

Another way to reduce automobile use and congestion is to raise parking fees and charge tolls on roads, tunnels, and bridges—especially during peak traffic times. For example, densely populated Singapore is rarely congested because it taxes cars heavily and auctions the rights to buy a car. Also, anyone driving downtown pays a daily user fee of $3–6 that rises during rush hours. The government uses the revenue from taxes and fees to fund an excellent mass transit system. This approach is also being used in Oslo, Norway; Melbourne, Australia; and London, England. In London, charging $8 for any vehicle entering the central city during the workday has cut traffic congestion by a fourth and increased use of mass transit.

Scores of cities including Rome, Italy; Stockholm, Sweden; Copenhagen, Denmark; Prague, Czechoslovakia; Geneva, Switzerland; and Curitiba, Brazil, (p. 563) have established *car-free areas*.

More than 300 cities in Germany, Austria, Italy, Switzerland, and the Netherlands have a *car-sharing* network. Each member pays for a card that opens lockers containing keys to cars parked at designated spots around a city. Members reserve a car in advance or call the network and are directed to the closest locker and car. They are billed monthly for the time they use a car and the distance they travel. In Berlin, Germany, car sharing has cut car ownership by 75% and car commuting by nearly 90%.

Another way to reduce car use and accidents and save gasoline is the *electronic commute* in which people use computers and other telecommunication devices to do all or much of their work at home. Shopping online also reduces the need to travel to shopping malls and other stores, although this is offset partially by increased delivery truck trips.

Is It Feasible to Reduce Automobile Use in the United States? Kicking Auto Addiction Is Hard

Reducing car use in the United States is difficult because of political opposition from the public and powerful car-related industries, too little emphasis on establishing modern, efficient mass transit options, and addiction to cars.

Most analysts doubt that the approaches just discussed are feasible in the United States, for three reasons. *First*, it faces strong political opposition from two groups, one being the public, largely unaware of the huge hidden costs they are already paying. The other group is the politically powerful transportation-related industries such as oil and tire companies, road builders, carmakers, and many real estate developers. However, taxpayers might accept sharp increases in gasoline taxes if the extra costs were offset by decreases in taxes on wages and income.

Second, fast, efficient, reliable, and affordable mass transit options and bike paths are not widely available in most of the United States. In addition, the dispersed nature of most U.S. urban areas makes people dependent on cars.

Third, most people who can afford cars are virtually addicted to them, and many people in the U.S. and elsewhere who cannot afford a car hope to buy one someday.

What Are Alternatives to the Car? Use Your Muscles and Travel with Others

Alternatives include walking, bicycling, driving scooters, and taking subways, trolleys, trains, and buses.

There are a number of alternatives to cars, each with advantages and disadvantages. One widely used alternative is the *bicycle.* Because of their advantages (Figure 25-10), bicycles outsell cars by more than two to one.

Trade-Offs

Bicycles

Advantages	Disadvantages
Affordable	Little protection in an accident
Produce no pollution	
Quiet	Do not protect riders from bad weather
Require little parking space	
Easy to maneuver in traffic	Not practical for trips longer than 8 kilometers (5 miles)
Take few resources to make	Can be tiring (except for electric bicycles)
Very energy efficient	
Provide exercise	Lack of secure bike parking

Figure 25-10 Trade-offs: advantages and disadvantages of *bicycles.* Pick the single advantage and disadvantage that you think are the most important.

Trade-Offs

Motor Scooters

Advantages	Disadvantages
Affordable	Little protection in an accident
Produce less air pollution than cars	Does not protect drivers from bad weather
Require little parking space	
Easy to maneuver in traffic	Gasoline engines are noisy
Electric scooters are quiet and produce little pollution	Gasoline engines emit large quantities of air pollutants

Figure 25-11 Trade-offs: advantages and disadvantages of *motor scooters.* Pick the single advantage and disadvantage that you think are the most important.

Bicycles are widely used for urban trips in countries such as the Netherlands, China, and Japan. Bicycling and walking account for about 28% of the trips in the Netherlands, compared to only 6% in the United States. In Copenhagen, Denmark, 2,300 bicycles are available for public use at no charge. The system is financed by ads attached to the bicycle frames and wheels, and is so popular that each bicycle is used on average once every 8 minutes.

Only one of every 200 Americans bicycle to work. But one out of five say they would do so if safe bike lanes were available and if their employers provided secure bike storage and showers at work.

About 2 million *electric bicycles* are on the road worldwide, and about 400,000 more are added each year. Existing bikes can easily be converted to electric bikes, and new ones can be bought for $500–1,200. Bicycles powered by small fuel cells should be available within a few years.

Motor scooters have advantages and disadvantages (Figure 25-11) and are especially useful for people in developing countries who cannot afford a car. Electric scooters can reduce air pollution and noise.

Heavy-rail systems (subways, elevated railways, and metro trains) and *light-rail* systems (streetcars, trolley cars, and tramways) have their advantages and disadvantages (Figure 25-12). To be cost effective, rail systems must travel along densely populated corridors in urban areas. At one time the United States had an effective light-rail system, but it was dismantled to promote car and bus use (Case Study, p. 576).

The rail system in Hong Kong is one of the world's most successful for three reasons. *First,* the

Trade-Offs

Mass Transit Rail

Advantages	Disadvantages

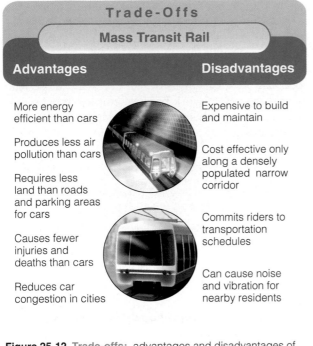

Advantages

More energy efficient than cars

Produces less air pollution than cars

Requires less land than roads and parking areas for cars

Causes fewer injuries and deaths than cars

Reduces car congestion in cities

Disadvantages

Expensive to build and maintain

Cost effective only along a densely populated narrow corridor

Commits riders to transportation schedules

Can cause noise and vibration for nearby residents

Figure 25-12 Trade-offs: advantages and disadvantages of *mass transit rail systems in urban areas.* Pick the single advantage and disadvantage that you think are the most important.

Trade-Offs

Buses

Advantages	Disadvantages

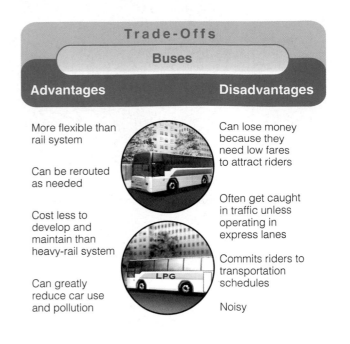

Advantages

More flexible than rail system

Can be rerouted as needed

Cost less to develop and maintain than heavy-rail system

Can greatly reduce car use and pollution

Disadvantages

Can lose money because they need low fares to attract riders

Often get caught in traffic unless operating in express lanes

Commits riders to transportation schedules

Noisy

Figure 25-13 Trade-offs: advantages and disadvantages of *bus systems in urban areas.* Pick the single advantage and disadvantage that you think are the most important.

city is densely populated, making it ideal for a rapid-rail system running through its corridor. *Second*, half the population can walk to a subway station in 5 minutes. *Third*, a car is an economic liability in this crowded city even for those who can afford one.

Buses are the most widely used form of mass transit within urban areas, mainly because they have more advantages than disadvantages (Figure 25-13). Curitiba, Brazil, has one of the world's best bus systems (Figure 25-1).

A *rapid-rail system between urban areas* is another option. In western Europe and Japan, high-speed bullet trains travel between cities at up to 330 kilometers (200 miles) per hour. Figure 25-14 lists the major ad-

vantages and disadvantages of such rapid rail systems. In 2004, Shanghai, China, began operating the world's first commercial high-speed magnetic levitation train between its airport and downtown. The train, suspended in air slightly above the track and propelled forward by strong repulsive and attractive magnetic forces, travels much faster than bullet trains.

In the United States, a high-speed bullet train network could replace airplanes, buses, and private cars for most medium-distance travel between major American cities (Figure 25-15, p. 576). Critics say such a system would cost too much in government subsidies. But this ignores the fact that motor vehicle trans-

Trade-Offs

Rapid Rail

Advantages	Disadvantages

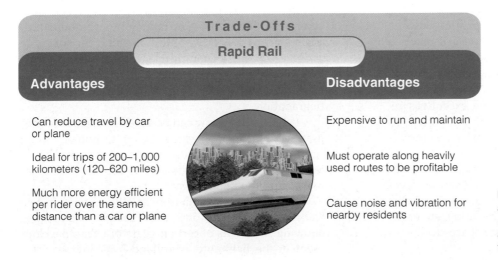

Advantages

Can reduce travel by car or plane

Ideal for trips of 200–1,000 kilometers (120–620 miles)

Much more energy efficient per rider over the same distance than a car or plane

Disadvantages

Expensive to run and maintain

Must operate along heavily used routes to be profitable

Cause noise and vibration for nearby residents

Figure 25-14 Trade-offs: advantages and disadvantages of *rapid-rail systems between urban areas.* Pick the single advantage and disadvantage that you think are the most important.

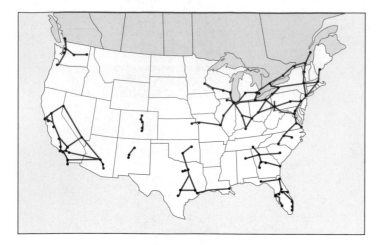

Figure 25-15 Solutions: potential routes for high-speed bullet trains in the United States and parts of Canada. Such a system could allow rapid, comfortable, safe, and affordable travel between major cities in a region. It would greatly reduce dependence on cars, buses, and airplanes for trips between these urban areas. (Data from High Speed Rail Association)

portation receives subsidies of $300–600 billion per year in the United States.

Case Study: Mass Transit in the United States— Destroying a Great System

In the early 1900s the United States had one of the world's best streetcar systems, but it was bought up and destroyed by several companies in order to sell cars and buses.

In 1917, all major U.S. cities had efficient electric trolley or streetcar (light-rail) systems. Many people think of Los Angeles as the original car-dominated city. But in the early 20th century Los Angeles had the largest electric-rail mass transit system in the United States.

That changed when General Motors, Firestone Tire, Standard Oil of California, Phillips Petroleum, and Mack Truck (which also made buses) formed a holding company called National City Lines. By 1950, the holding company had purchased privately owned streetcar systems in 83 major cities. It then dismantled these systems to increase sales of cars and buses.

The courts found the companies guilty of conspiracy to eliminate the country's light-rail system, but the damage had already been done. The executives responsible were fined $1 each, and each company paid a fine of $5,000, less than the profit returned by replacing a single streetcar with a bus.

During this same period, National City Lines worked to convert electric-powered commuter locomotives to much more expensive and less reliable diesel-powered locomotives. The resulting increased costs contributed significantly to the sharp decline of the nation's railroad system.

In the United States, 80% of federal gasoline tax revenue is used to build and maintain highways, and only 20% is used for mass transit, bike paths, and walkways. This encourages states and cities to invest in highways instead of mass transit. The federal tax code also penalizes mass transit users and those who bicycle or walk to work because employers who provide parking for their employees can deduct this expense from their taxes.

✗ HOW WOULD YOU VOTE? Should half of the gasoline tax be used to develop mass transit, bike lanes, and other alternatives to the car? Cast your vote online at http://biology.brookscole.com/miller14.

25-4 URBAN LAND-USE PLANNING AND CONTROL

What Is Conventional Land-Use Planning? Focusing on Growth

Most land-use planning in the United States leads to poorly controlled urban sprawl and environmental degradation and funds this often-destructive process with property taxes.

Most urban and some rural areas use some form of **land-use planning** to determine the best present and future use of each parcel of land.

Much land-use planning is based on the assumption that considerable future population growth and economic development should be encouraged, regardless of environmental and other consequences. Typically this leads to uncontrolled or poorly controlled urban growth and sprawl.

A major reason for this often destructive process in the United States and some other countries is that 90% of the revenue that local governments use to provide public services such as schools, police and fire protection, and water and sewer systems comes from *property taxes* levied on all buildings and property based on their economic value. Thus local governments often try to raise money by promoting economic growth because they usually cannot raise property tax rates enough to meet expanding needs. Typically the long-term result is poorly managed economic growth, leading to more environmental degradation.

Land-use planning can be aided by the use of geographic information system (GIS) technology (Figure 4-35, p. 84). Many cities and counties in the United States have used this technology to convert their planning maps into digital form.

In the 1990s, GIS data from satellite images, historical data, and census data were used to create maps showing snapshots of certain years of urban development in the Baltimore, Maryland–Washington, D.C., area (Figure 25-7) between 1862 and 1999. They were presented in an animated video showing how the cities merged into one gigantic urban area by the 1990s. The video helped a governor of Maryland win legislative approval for an antisprawl, smart growth program.

What Are the Advantages and Disadvantages of Using Zoning to Control Land Use? Useful but Improvable

Zoning is useful but can favor high-priced development over environmental protection and can discourage innovative solutions to urban problems.

Once a land-use plan is developed, governments control the uses of various parcels of land by legal and economic methods. The most widely used approach is **zoning,** in which various parcels of land are designated for certain uses.

Zoning can be used to control growth and protect areas from certain types of development. For example, cities such as Portland, Oregon, and Curitiba, Brazil, (p. 563) have used zoning to encourage high-density development along major mass transit corridors to reduce automobile use and air pollution.

Despite its usefulness, zoning has several drawbacks, one being that some developers can influence or modify zoning decisions in ways that cause destruction of wetlands, prime cropland, forested areas, and open space. Another problem is that zoning often favors high-priced housing, factories, hotels, and other businesses over protecting environmentally sensitive areas and providing low-cost housing. The reason is, again, that most local governments depend on property taxes for their revenue.

In addition, overly strict zoning can discourage innovative approaches to solving urban problems. For example, the pattern in the United States and in some other countries has been to prohibit businesses in residential areas, which increases suburban sprawl. Some urban planners want to return to *mixed-use zoning* to help reduce sprawl. For example, in the 1970s, Portland, Oregon, decided that it could cut driving and gasoline consumption by resurrecting the idea of neighborhood grocery stores. It worked.

How Is Smart Growth Used to Control Growth and Sprawl? Channeling Growth and Reining in the Car

Smart growth can control growth patterns, discourage urban sprawl, reduce car dependence, and protect ecologically sensitive areas.

There is growing use of the concept of **smart growth** or *new urbanism* to encourage more environmentally sustainable development that requires less dependence on cars, controls and directs sprawl, and reduces wasteful resource use. It recognizes that urban growth will occur. But it uses zoning laws and an array of other tools to channel growth to areas where it can cause less harm, discourage sprawl, protect ecologically sensitive and important lands and waterways, and develop more environmentally sustainable urban areas and neighborhoods that are more enjoyable places to live. Figure 25-16 (p. 578) lists smart growth tools used to prevent and control urban growth and sprawl. Which, if any, of these tools are being used in your community?

The most widely used ways to slow and control urban sprawl are to set growth boundaries around cities, preserve open space outside of urban areas, develop spaces within urban areas that have been left behind from urban sprawl, create new towns and villages within existing cities, and revitalize neighborhoods and downtown areas.

Portland, Oregon used some of these strategies to control sprawl and reduce dependence on the car, and it worked. Since 1975 Portland's population has grown by about 50% but its urban area has increased by only 2%. And abundant green space and natural beauty is just 20 minutes from downtown.

The city also encourages clustered, mixed-use neighborhood development consisting of stores, light industries, professional offices, high-density housing, and access to mass transit that allows most people to meet their daily needs without a car. Portland has further reduced car use by developing an excellent light-rail line and an extensive network of bus lines, bike lanes, and walkways. Employers are encouraged to give their employees bus passes instead of providing parking spaces. Downtown Portland is a vibrant and thriving community and, in 2000, *Money* magazine listed Portland as the most livable city in the United States. Curitiba, Brazil (p. 563) has also used a variety of such strategies to control sprawl and reduce dependence on the car. And car-free villages have been created in cities such as Munich, Germany; Vancouver, Canada; and Zurich, Switzerland. Several studies have shown that most forms of smart growth provide more jobs and spur more economic renewal than conventional economic growth.

China has taken the strongest stand of any country against sprawl. The government has designated 80% of the country's arable land as *fundamental land.* Building on such land requires approval from local and provincial governments and the State Council—somewhat like having to get congressional approval for a new subdivision in the United States. Developers violating these rules face the death penalty. National land-use planning also is used in Japan and much of western Europe.

Most European countries have been successful in discouraging urban sprawl and encouraging compact cities. They have controlled development at the national level and imposed high gasoline taxes to

Limits and Regulations

Limit building permits

Urban growth boundaries

Greenbelts around cities

Public review of new development

Zoning

Encourage mixed use

Concentrate development along mass transportation routes

Promote high-density cluster housing developments

Planning

Ecological land-use planning

Environmental impact analysis

Integrated regional planning

State and national planning

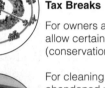

Protection

Preserve existing open space

Buy new open space

Buy development rights that prohibit certain types of development on land parcels

Taxes

Tax land, not buildings

Tax land on value of actual use (such as forest and agriculture) instead of highest value as developed land

Tax Breaks

For owners agreeing legally to not allow certain types of development (conservation easements)

For cleaning up and developing abandoned urban sites (brownfields)

Revitalization and New Growth

Revitalize existing towns and cities

Build well-planned new towns and villages within cities

Figure 25-16 Solutions: *smart growth* or *new urbanism tools* used to prevent and control urban growth and sprawl.

discourage car use and encourage people to live closer to workplaces and shops. High taxes on heating fuel also encourage people to live in apartments and small houses. Governments have used most of the resulting gasoline and heating fuel tax revenues to develop efficient train and other mass transit systems within and between cities.

Solutions: Land-Use Planning in Oregon— Control from the Top

Oregon has zoned rural land to prevent environmental degradation, controlled urban growth, and put land-use planning in the hands of state officials.

Since the mid-1970s, Oregon has had a comprehensive statewide land-use planning process based on three administrative decisions:

- To permanently zone all rural land in Oregon as forest, agricultural, or urban land

- To draw an urban growth line around each community in the state, with no urban development allowed outside the boundary

- To place control over land-use planning in state hands through the Land Conservation and Development Commission

Not surprisingly, the last action has been the most controversial. It is based on the idea that public good takes precedence over private property rights—a well-established principle in most European countries but generally opposed in the United States.

Oregon's plan has worked because it is not designed to "just say no" to development. Instead, it encourages certain kinds of development, such as dense urban development that helps prevent destruction of croplands, wetlands, and biodiversity in the surrounding area.

Because of the plan, most of the state's rural areas remain undeveloped. Before these land-use and planning laws, the state lost about 12,100 hectares (30,000 acres) of agricultural land a year; now it is only losing about 810 hectares (2,000 acres) a year.

How Can Urban Open Space Be Preserved and Used? Be Protective and Creative

Small and large parks, greenbelts, urban growth boundaries, cluster development, and greenways can be used to preserve open space.

One way to preserve open space outside a city is to employ Oregon's *urban growth boundary* model, used also in the states of Washington and Tennessee. A more traditional way is to preserve significant blocks of open space in the form of municipal parks. Central Park in New York City, Golden Gate Park in San Francisco, and Grant Park in Chicago are examples of large urban parks. Many European cities also have large- and medium-size parks.

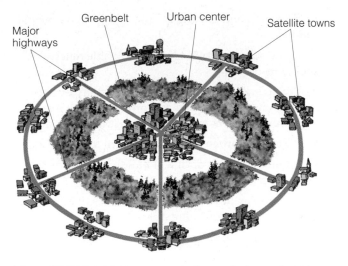

Figure 25-17 Establishing a *greenbelt* around a large city can control urban growth and provide open space for recreation and other nondestructive uses. Satellite towns sometimes are built outside the belt. Highways or rail systems can be used to transport people around the periphery or into the central city.

In 1883, Minneapolis, Minnesota, officials vowed to create "the finest and most beautiful system of public parks and boulevards of any city in America." In the eyes of many, this goal has been achieved. Today the city has 170 parks spaced so that most homes in Minneapolis are within six blocks of a green space.

Some cities provide open space and control urban growth by surrounding a large city with a *greenbelt* (Figure 25-17): an open area used for recreation, sustainable forestry, or other nondestructive uses. Satellite towns can be built outside the belt. Ideally, the outlying towns and the central city are linked by an extensive public transport system. Many cities in western Europe and Canadian cities such as Toronto and Vancouver have used this approach.

We can also let nature reclaim spaces we have developed as examples of *reconciliation ecology* (p. 247). On the west side of Manhattan, New York, for example, an abandoned elevated rail line now suppors abundant plant and animal life—an example of nature creating a self-seeding, self-sustaining urban landscape without human input.

Case Study: How Is New Urbanism Creating More Livable Spaces? Returning to Traditional Neighborhood Development

There is a growing movement to create mixed-use villages and neighborhoods within urban areas where people can live, work and shop close to their homes.

Since World War II, the typical approach to suburban housing development in the United States has been to bulldoze a tract of woods or farmland and build rows of houses on standard-size lots (Figure 25-18, middle). Many of these developments and their streets are named after the trees and wildlife they displaced such as Oak Lane, Cedar Drive, Pheasant Run, and Fox Fields.

In recent years, builders have increasingly used a pattern, known as *cluster development*, in which high-density housing units are concentrated on one portion of a parcel, with the rest of the land (often 30–50%) used for commonly shared open space (Figure 25-18, bottom). When done properly, high-density cluster developments are a win-win solution for residents, developers, and the environment. Residents get more open and recreational space, aesthetically pleasing surroundings, and lower heating and cooling costs because some walls are shared. Developers can cut their costs for site preparation, roads, utilities, and other forms of infrastructure.

Some communities are going further and using principles of new urbanism to develop entire villages and recreate mixed-use neighborhoods within existing cities. These principles include *walkability* with most things within a 10-minute walk of home and work by

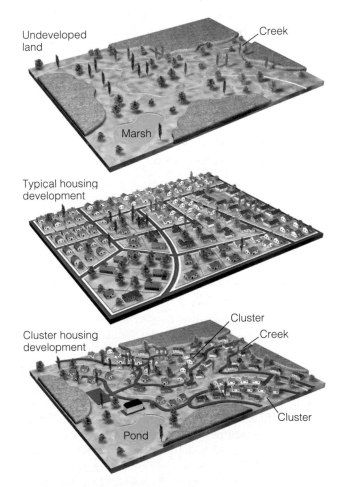

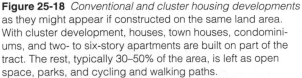

Figure 25-18 *Conventional and cluster housing developments* as they might appear if constructed on the same land area. With cluster development, houses, town houses, condominiums, and two- to six-story apartments are built on part of the tract. The rest, typically 30–50% of the area, is left as open space, parks, and cycling and walking paths.

recognizing that our bodies are biologically designed for walking; *mixed-use and diversity* where there is a mix of pedestrian-friendly shops, offices, apartments, and homes and people of different ages, classes, cultures, and races; *quality urban design* emphasizing beauty, aesthetics, and architectural diversity; *environmental sustainability* based on developemnt with minimal environmental impact; and *smart transportation* with high-quality trains connecting neighborhoods, towns, and cities. The goal is to create places that uplift, enrich, inspire the human spirit, and that are incubators of friendship, cooperation, and civic engagement.

One of the larger examples is the newly constructed mixed-use Mayfaire Village within the city of Wilmington, North Carolina. This 162-hectare (400-acre) village has clusters consisting of a town retail center with loft rental apartments above some stores, condominiums, apartments, rental houses, single-family homes, offices, hotels, and lots of green and recreational space. The town center is within easy walking or biking distance of the housing clusters. A major portion of the site is dedicated to open spaces such as soccer fields, parks, bike paths, and hiking trails, all within a few minutes of the housing and shopping clusters. About one-fourth of the site is preserved in its natural state. The village is located only a few minutes from Wrightsville Beach on the Atlantic Ocean. Other examples of such villages are Mizner Place in Boca Raton, Florida; Middleton Hills near Madison, Wisconsin; Phillips Place in Charlotte, North Carolina; Kentlands in Gaithersberg, Maryland; and Valencia, California (near Los Angeles).

25-5 MAKING URBAN AREAS MORE LIVABLE AND SUSTAINABLE

How Can We Make Cities More Sustainable, Desirable Places to Live? The Ecocity Concept

An ecocity allows people to walk, bike, or take mass transit for most of their travel, and recycles and reuses most of its wastes, grows much of its own food, and protects biodiversity by preserving surrounding land.

According to most environmentalists and urban planners, the primary problem is not urbanization but our failure to make cities more sustainable and livable. They call for us to make new and existing urban areas more self-reliant, sustainable, and enjoyable places to live through good ecological design. See the Guest Essay on this topic by David Orr on the website for this chapter.

A more environmentally sustainable city, called an *ecocity* or *green city*, emphasizes:

- Preventing pollution and reducing waste
- Using energy and matter resources efficiently

- Recycling and reusing at least 60% of all municipal solid waste
- Using solar and other locally available renewable energy resources
- Protecting and encouraging biodiversity by preserving surrounding land and protecting and restoring natural systems and wetlands within urban areas
- Promoting urban gardens and farm markets
- Promoting green design of buildings, including green roofs
- Using solar-powered living machines (Figure 22-1, p. 491) and wetlands to treat sewage (Solutions, p. 512)

An ecocity is a people-oriented city, not a car-oriented city. Its residents are able to walk, bike, or use low-polluting mass transit for most of their travel. An ecocity requires that all buildings, vehicles, and appliances meet high energy-efficiency standards. Trees and plants adapted to the local climate and soils are planted throughout to provide shade and beauty, supply wildlife habitats, and reduce pollution, noise, and soil erosion. Small organic gardens and a variety of plants adapted to local climate conditions often replace monoculture grass lawns.

Abandoned lots and industrial sites and polluted creeks and rivers are cleaned up and restored. Nearby forests, grasslands, wetlands, and farms are preserved. Much of an ecocity's food comes from nearby organic farms, solar greenhouses, and community gardens. There are also small gardens on rooftops and in yards, abandoned lots, and window boxes. People designing and living in ecocities take seriously the advice Lewis Mumford gave more than three decades ago: "Forget the damned motor car and build cities for lovers and friends."

The ecocity is not a futuristic dream, as you saw in the chapter opening case study about Curitiba, Brazil. Other more environmentally sustainable and livable cities include Waitakere City, New Zealand; Helsinki, Finland; Leicester, England; Portland, Oregon (p. 578); Davis, California; Olympia, Washington; and Chattanooga, Tennessee (Case Study, p. 581).

China is planning to develop 10 model environmental or ecocities. The first project focuses on transforming Suzhou, a rapidly expanding city of 2.2 million people just 64 kilometers (40 miles) from Shanghai. It is one of China's oldest cities that is internationally known for its combination of history, culture, and greenery. Green initiatives include relocating polluting industries outside of the city, a pilot project requiring local taxis to run on natural gas, building four light rail and subway lines, a seven-story height limit on buildings in the city's center, and landscaping the city's network of canals. It is promoting the use of solar water heaters, plans to phase out gasoline-powered motorcycles by 2007, and is planning a network of battery exchange and disposal centers to serve the rapidly increasing use of electric-powered bicycles and mopeds.

China has a long way to go in converting its urban sustainability goals into reality. But if successful, China could become model for the world in ecocity design.

Case Study: Chattanooga, Tennessee—From Brown to Green

Local officials and citizens have worked together to transform Chattanooga from a highly polluted city to one of the most sustainable and livable cities in the United States.

In the 1950s, Chattanooga was known as one of the dirtiest cities in the United States. Its air was so polluted by smoke from its coke ovens and steel mills that people sometimes had to turn on their headlights in the middle of the day. The Tennessee River flowing through the city's industrial wasteland bubbled with toxic waste.

People and industries fled the downtown area and left a wasteland of abandoned factories, boarded-up buildings, high unemployment, and crime.

Within two decades, Chattanooga transformed itself into one of the most livable cities in the United States. Efforts began in 1984 when civic leaders used a series of town meetings as part of a *Vision 2000* process—a 20-week series of community meetings brought together more than 1,700 citizens from all walks of life to build a consensus about what the city could be at the turn of the century. Citizens identified the city's main problems, set goals, and brainstormed thousands of ideas for solutions.

By 1995, Chattanooga had met most of its original goals, which included encouraging zero-emission industries to locate there and replacing its diesel buses with a fleet of quiet, zero-emission electric buses, made by a new local firm. The city reduced car use in the downtown by building satellite parking lots and providing free and rapid bus service to and from the city center. The city also launched an innovative recycling program after citizen activists and environmentalists blocked construction of a new garbage incinerator. Another project involved renovating much of the city's existing low-income housing and building new low-income rental units.

Chattanooga built the nation's largest freshwater aquarium, which became the centerpiece for downtown renewal. The city also developed a 35-kilometer-long (22-mile-long) riverfront park along both sides of the Tennessee River running through downtown. The park is filled with shade trees, flowers, fountains, and street musicians, and draws more than 1 million visitors per year.

As property values and living conditions have improved, people and businesses are moving back downtown. An abandoned place once filled with despair is now a vibrant community filled with hope. These accomplishments show what citizens, environmentalists, and business leaders can do when they work together to develop and achieve common goals.

In 1993, the community began the process again in *Revision 2000*. More than 2,600 participants identified additional goals and more than 120 recommendations for further improvements. One goal is to transform a blighted brownfield in South Chattanooga into an environmentally advanced, mixed community of residences, retail stores, and zero-emission industries where employees can live near their workplaces.

This new low-waste ecoindustrial park is modeled after the one in Kalundborg, Denmark (Figure 24-5, p. 537). Underground tunnels will link 30 industrial buildings to share heating, cooling, and water supplies and to use the waste matter and energy of some enterprises as resources for others. The new ecoindustrial area will also have an ecology center using a living machine (Figure 22-1, p. 491) to treat sewage, wastewater, and contaminated soils.

According to many environmentalists, urban planners, and economists, urban areas that fail to become more livable and ecologically sustainable over the next few decades are inviting economic depression and increased unemployment, pollution, and social tension. What is your community doing?

A sustainable world will be powered by the sun; constructed from materials that circulate repeatedly; made mobile by trains, buses, and bicycles; populated at sustainable levels; and centered around just, equitable, and tight-knit communities.

GARY GARDNER

CRITICAL THINKING

1. Do you prefer living in a rural, suburban, small-town, or urban environment? Describe the ideal environment in which you would like to live, and list the environmental advantages and disadvantages of living in such a place. Compare your answers with those of other members of your class.

2. Do you believe the United States or the country where you live should develop a comprehensive and integrated mass transit system over the next 20 years, including building an efficient rapid-rail network for travel within and between its major cities? How would you pay for such a system?

3. If you own a car or hope to own one, what conditions, if any, would encourage you to rely less on the automobile and to travel to school or work by bicycle, on foot, by mass transit, or by carpool or vanpool?

4. Do you believe Oregon's approach to land-use planning (Solutions, p. 578) should be used in the state or area where you live? Explain your position.

5. In June 1996, representatives from many countries met in Istanbul, Turkey, at the Second UN Conference on Human Settlements (nicknamed the City Summit). One issue was the question of whether housing is a universal *right* (a position supported by most developing countries) or just a *need* (supported by the United States and several other developed countries). What is your position on this issue? Defend your choice.

6. Some analysts suggest phasing out federal, state, and government subsidies that encourage sprawl by funding roads, single-family housing, and large malls and superstores. These would be replaced with subsidies that encourage sidewalks and bicycle paths, multifamily housing, high-density residential development, and a mix of housing, shops, and offices (mixed-use development). Do you support this approach? Explain.

7. Congratulations! You are in charge of the world. List the five most important features of your urban policy.

PROJECTS

1. Consult local officials to determine how land use is decided in your community. What roles do citizens play in this process?

2. For a class or group project, borrow one or more decibel meters from your school's physics or engineering department or from a local electronics repair shop. Make a survey of sound pressure levels at various times of day and at several locations. Plot the results on a map. Also, measure sound levels in a room with a sound system and from earphones at several different volume settings. If possible, measure sound levels at an indoor concert, a club, and inside and outside a boom car at various distances from the speakers. Correlate your findings with those in Figure 25-9, p. 570.

3. As a class project, **(a)** evaluate land use and land-use planning by your school, **(b)** draw up an improved plan based on ecological principles and the principles of sustainability listed in Figure 9-15, p. 174, and **(c)** submit the plan to school officials.

4. As a class project, use the following criteria to rate the community where you live or go to school on a green

index from 0 to 10. Rate the community for each of the following questions and average the results to get an overall score. Are existing trees protected and new ones planted throughout the city? Do you have parks to enjoy? Can you swim in any nearby lakes and rivers? What is the quality of your water and air? Is there an effective noise pollution reduction program? Does your city have a recycling program, a composting program, and a hazardous waste collection program, with the goal of reducing the current solid waste output by at least 60%? Is there an effective mass transit system? Are there bicycle paths? Are all buildings required to meet high energy-efficiency standards? How much of the energy is obtained from locally available renewable resources? Are environmental regulations for existing industry tough enough and enforced well enough to protect citizens? Do local officials look carefully at an industry's environmental record and plans before encouraging it to locate in your city or county? Is ecological planning used to make land-use decisions? Are city officials actively planning to improve the quality of life for all of its citizens? If so, what is the plan? Compare your answers with those by other members of your class.

5. Use the library or the Internet to find bibliographic information about *Peter Self* and *Gary Gardner*, whose quotes appear at the beginning and end of this chapter.

6. Make a concept map of this chapter's major ideas, using the section heads, subheads, and key terms (in boldface). Look on the website for this book for information about making concept maps.

LEARNING ONLINE

The website for this book contains study aids and many ideas for further reading and research. They include a chapter summary, review questions for the entire chapter, flash cards for key terms and concepts, a multiple-choice practice quiz, interesting Internet sites, references, and a guide for accessing thousands of InfoTrac® College Edition articles. Log on to

http://biology.brookscole.com/miller14

Then click on the Chapter-by-Chapter area, choose Chapter 25, and select a learning resource.

26 Economics, Environment, and Sustainability

Pollution Control

CASE STUDY

How Important Are Natural Resources?

Economics is the study of how individuals and societies choose to use limited or scarce resources to satisfy their unlimited wants. Recall that *economic growth* is an increase in a nation's capacity to provide people with goods and services, and *economic development* is the improvement of human living standards by economic growth. For more than 200 years there has been a debate on whether there are limits to economic growth.

Neoclassical economists such as Milton Friedman and Robert Samuelson view economic growth as both necessary and desirable. Neoclassical economists also view economic growth as unlimited because if we deplete a particular resource we should be able to use our ingenuity and technology to find a substitute. For example, if we run out of oil we should

be able to find a substitute such as hydrogen or nuclear power. Thus, natural capital is viewed as an important but not indispensable economic resource.

Ecological economists such as Herman Daly and Robert Costanza disagree. They point out that there are no substitutes for many natural resources, such as air, water, fertile soil, and biodiversity. They conclude that ultimately economic growth is limited because some forms of economic growth can deplete and degrade the quantity and quality of these irreplaceable forms of natural capital. They call for us to redesign our economic systems to encourage environmentally sustainable forms of economic development and to discourage environmentally harmful forms of economic growth.

In the middle of this debate are *environmental economists*. They generally agree with ecological economists that some forms of economic growth are not sustainable. But they believe we can modify the principles of neoclassical economics and reform current economic systems, rather than totally redesigning them. Figure 26-1 shows some of the forms of economic growth and development that most ecological and environmental economists want to encourage.

No-till cultivation

Forest conservation

Production of energy-efficient fuel-cell cars

Underground CO_2 storage using abandoned oil wells

High-speed trains

Deep-sea CO_2 storage

Solar-cell fields

Bicycling

Communities of passive solar homes

Water conservation

Recycling plant

Recycling, reuse, and composting

Landfill

Cluster housing development

Wind farms

Figure 26-1 Solutions: these are some components of more environmentally sustainable economic development favored by ecological and environmental economists. The goal is to have economic systems put more emphasis on conserving and sustaining the air, water, soil, biodiversity, and other natural resources that sustain all life and all economies.

This chapter discusses how we can use economics to promote environmental quality and sustainability. It addresses the following questions:

- What are economic systems and how do they work?

- How do economists differ in their views of economic systems, pollution control, and resource management?

- How can we monitor economic and environmental progress?

- What is full-cost pricing?

- What economic tools can we use to help us shift to full-cost pricing?

- How does poverty reduce environmental quality, and how can we reduce poverty?

- How can we shift to more environmentally sustainable economies over the next few decades?

26-1 ECONOMIC RESOURCES AND SYSTEMS

What Supports and Drives Economies? Three Types of Capital

An economic system produces and distributes goods and services by using natural, human, and physical resources

An **economic system** is the social institution through which goods and services are produced, distributed and consumed to satisfy people's unlimited wants in the most efficient possible way.

Recall that *capital* is any form of wealth used to sustain a business or produce more wealth. Three types of resources or capital are used to produce goods and services. One is **natural resources,** or **natural capital,** the materials produced by the earth's natural processes, which support all economies and all life (top half of back cover). See the Guest Essay on this topic by Paul Hawken on the website for this chapter.

A second resource type is **human resources** or **human capital,** people's physical and mental talents that provide skills and abilities, innovation, culture, and organization. A third type is **physical** or **manufactured resources,** items made from natural resources with the help of human resources, such as tools, machinery, equipment, factories, and shipping facilities.

What Is a Pure Free Market? A Theoretical Model of Supply, Demand, and Marginal Costs and Benefits

A purely free market is an ideal that does not match real world markets.

A **pure free-market economic system** is a theoretical ideal or model in which buyers (demanders) and sellers (suppliers) freely interact in *markets* without any government or other interference to make all economic decisions. No one can influence the price at which a good or service sells. All parties have full access to the market and enough information about the beneficial and harmful aspects of economic goods and services to make decisions.

According to these ideals, all economic decisions are governed solely by the competitive interactions of *demand* (the amount of a good or service that people want), *supply* (the amount of a good or service that is available), and *price* (the market cost of a good or service). *Supply* is represented on the graph in Figure 26-2 by the blue curve showing how much a producer of any good or service is willing to supply (measured on the horizontal *quantity* axis) for different prices (measured on the vertical *price* axis). *Demand* is shown on the orange curve showing how much consumers will pay for different quantities of the good or service.

The point at which the curves intersect is called the *market price equilibrium point*, where the supplier's

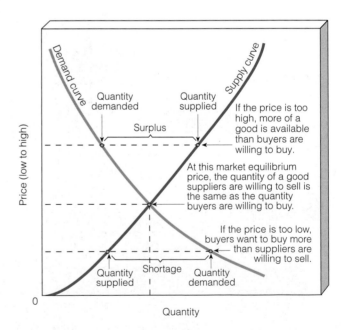

Figure 26-2 Supply, demand, and market equilibrium for a good or service in a pure market system. If all factors except price, supply, and demand are held fixed, market equilibrium occurs at the point at which the demand and supply curves intersect. This is the price at which sellers are willing to sell and buyers are willing to pay for the good or service provided.

price matches what buyers are willing to pay for the same quantity and a sale is made. In Figure 26-2 it is the point at which the supply and demand curves intersect. In a pure free market economy such competition between willing sellers and buyers is said to bring about the greatest efficiency of resource use. *Profit* or *loss* is the difference between the cost of producing something and the price buyers are willing to pay.

Changes in supply and demand can shift one or both curves back and forth, and thus change the equilibrium price. For example, when supply is increased (shifting the blue curve to the right) and demand remains the same the equilibrium price will go down. Similarly, when demand is increased (shifting the orange curve to the right) and supply remains the same, the equilibrium price will increase. Try moving the curves in Figure 26-2 in different ways that represent changes in supply and demand, and notice how the equilibrium price changes.

Two related economic concepts are those of *marginal costs* and *marginal benefits*. In economics, anything described as *marginal* usually refers to an increase in some measurement involving a certain number of units of a good or service and that number of units plus one. **Marginal cost** is the increase in total cost resulting from producing one more unit of a good or service. For example, a supplier (seller) might ask "How much would it cost and how much profit might I make if I produce one more unit of my product? Look at Figure 26-2 again, and note that if you start at any point on the quantity axis and move to the right, the supply curve takes you up on the price axis. The difference between a seller's starting price and the new price represents the seller's *marginal cost*—the cost of producing that one more unit. The seller's *marginal benefit* is the profit made by producing and selling one more unit.

Similarly, when a seller produces one more unit of a product or service, the increase in the benefit that it provides to a buyer is the **marginal benefit**. For example, as a buyer you might ask how much would I benefit if I buy one more shirt? You can think of marginal benefit of buying a new shirt as the *difference* between the benefit you gain from having ten shirts and the benefit you enjoyed from having nine. In this case, the shirtmaker's marginal cost is the *difference* between the cost of producing 1000 shirts and the cost of producing 1001. And your marginal cost is what it costs you to buy one more shirt. A sale occurs if both the seller and buyer find the marginal costs and benefits advantageous. In real world economics, marginal costs and benefits are what actually determine prices and benefits to consumers and costs and profits to producers.

In fact, the market economic systems found in the real world do not meet the theoretical conditions described above. *In practice truly free markets do not exist.*

Businesses strive to drive their competitors out of business and exert as much control as possible over the prices of the goods and services they provide. Companies lobby for government subsidies, tax breaks, or regulations that give their products a market advantage over their competitors. Some companies also try to withhold information from consumers about dangers posed by products unless the government requires them to provide such information.

Also, there are exceptions to the free market theory of supply and demand. Some consumers may buy a good or service regardless of its price. For example, raising the price of gasoline or cigarettes may not significantly reduce consumer demand because some buyers feel they have to have these products. Economists call this *price inelasticity*.

Why Have Governments Intervened in Market Economic Systems? Making Up for Market Deficiencies

Governments intervene in market systems to help provide economic stability, national security, and public services such as education, crime protection, and environmental protection.

Markets often work well in guiding the production of *private goods*, but experience shows that they cannot be relied upon to provide the adequate levels of *public services* such as national security and environmental protection. Thus governments intervene in market systems to help correct *market failures*. For example, a single seller or buyer (monopoly) or a single group of sellers or buyers (oligopoly) might come to dominate the market and thus control supply or demand and price for a good or service. Governments can prevent this and other market failures through laws and regulations. Governments can also promote economic stability by trying to control boom-and-bust cycles that occur in market systems.

Other reasons for government interventions are to

- Provide public services such as national security and education

- Provide an economic safety net for people who because of health, age, and other factors cannot meet their basic needs

- Protect people from fraud, trespass, theft, and bodily harm

- Establish and enforce civil rights and property rights

- Protect the health and safety of workers and consumers

- Prevent or reduce pollution and depletion of natural resources

- Manage public land resources such as national forests, parks, and wildlife reserves

26-2 ECONOMISTS' VIEWS OF POLLUTION CONTROL AND RESOURCE MANAGEMENT

How Do Economists Differ in Their View of Market-Based Economic Systems? The Importance of Natural Capital

Neoclassical economists see natural resources as a component of an economic system, and ecological economists see economic systems as a component of nature's economy.

Neoclassical economists view the earth's natural capital as a subset or part of a human economic system (Figure 26-3). Natural resources are seen as important but not vital because of our ability to find substitutes for scarce resources and ecosystem services.

To neoclassical economists, economic growth is necessary, desirable, and essentially unlimited. It is seen as the best way to provide jobs and distribute wealth. If the economy is growing, there is money available to provide jobs, develop new resources, and supply public services such as environmental protection, education, national security, and crime protection. In this prevailing view, the global economy is, and should be, hard-wired to accelerating growth based mostly on increasing throughputs of matter and energy resources. (Figure 3-18, p. 53).

Ecological and *environmental economists* disagree with this model. They view economic systems as subsystems of the environment that depend heavily on the earth's irreplaceable natural resources (Figure 26-4). According to ecological economist Herman Daly, the neoclassical model of an economy "ignores the origin of natural resources flowing into the system and the fate of wastes flowing out of the system. It is as if a biologist had a model of an animal that contained a circulatory system but had no digestive system that tied it firmly to the environment at both ends." Ecological economists also believe that conventional economic growth eventually is unsustainable because it can deplete or degrade the natural resources on which economic systems depend.

Ecological and environmental economists distinguish between unsustainable economic growth and environmentally sustainable economic development (Figure 26-5, p. 588). They call for making a shift from our current economy based on unlimited economic growth to a more *environmentally sustainable economy,* or *eco-economy.* See the Guest Essay on this topic by Herman Daly on the website for this chapter.

Various ecological and environmental economists have suggested eight strategies to help make the shift to an eco-economy over the next several decades.

- Use resources more efficiently

- Use indicators that monitor economic and environmental health.

- Have the market prices of goods and services include their harmful effects on the environment and human health (full-cost pricing).

- Phase out environmentally harmful government subsidies and tax breaks.

- Shift taxes by lowering taxes on income and wealth and increasing taxes on pollution and resource waste.

- Pass laws and regulations to prevent pollution and resource depletion in certain areas.

- Use tradable permits or rights to pollute or use resources within programs that limit overall pollution and resource use in given areas.

- Use *eco-labeling* to identify products produced by environmentally sound methods and thus help consumers make informed choices.

Let us look at these solutions in more detail.

Natural Resources   **Manufactured Resources**   **Human Resources**   **Goods and Services**

Figure 26-3 *Neoclassical economists* view the earth's natural capital, or natural resources, as a subset or part of a human economic system. They contend that economic growth is not limited by the earth's natural resources because we should be able to find substitutes for scarce resources and ecosystem services. Economic growth is seen as good and essentially unlimited.

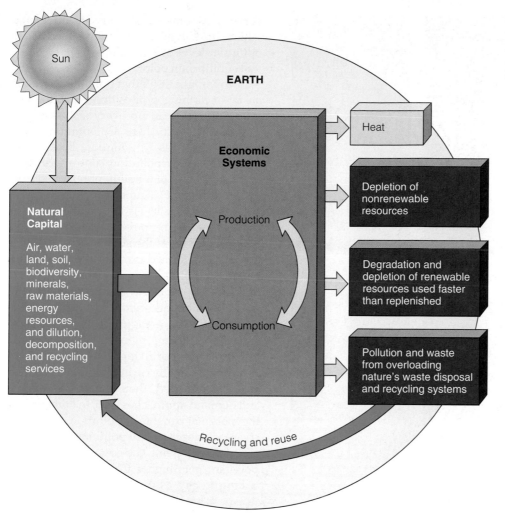

Figure 26-4 *Ecological economists* see all economies as human subsystems that depend on resources and services provided by the sun and the earth's natural resources.

Labels in figure:
Sun

EARTH

Economic Systems

Heat

Production

Consumption

Depletion of nonrenewable resources

Degradation and depletion of renewable resources used faster than replenished

Pollution and waste from overloading nature's waste disposal and recycling systems

Natural Capital

Air, water, land, soil, biodiversity, minerals, raw materials, energy resources, and dilution, decomposition, and recycling services

Recycling and reuse

How Can Technological Developments Improve Market Efficiency? Producing More with Less

Better technology and more efficient production systems can produce goods and services with fewer resources.

All economists agree that technological developments and more efficient production systems can mean that fewer resources are needed to produce the same amount of a good or service.

For example, between 1975 and 1995 the food output of the United States and most countries per unit of land increased while the inputs of labor and resources such as water and fertilizer needed to produce each unit of that output fell. However, ecological and environmental economists warn that such increases in the efficiency of food production might not be sustainable because of the degradation of soil, water, and other natural resources needed to produce food. These economists believe that this may be a factor in the leveling off of global grain production since 1985 (Figure 14-16, p. 287).

Economists agree that increases in technological and production efficiencies can cause significant changes to supply, demand, and prices for a good or service. For example, technological improvements can make it cheaper to produce a good or service and increase its supply at a particular price. This can cause its market equilibrium point (Figure 26-2) to shift to a lower price and thus benefit buyers. Trace such a shift in Figure 26-2.

Scarcity of a resource can stimulate research and development for new and more efficient technologies and production systems and a search for new reserves (Figure 16-10, p. 340) and for substitutes for such resources. It can also lead to increased recycling and reuse of a resource (Figure 16-16, p. 345).

For example, scarcity of a metal such as copper has led to improved technology for extracting and processing lower-grade ores, use of microorganisms that can be used to remove copper (and other metals) from its ore, increased recycling of copper, and use of aluminum as a substitute for copper in wiring. This can lead to lower prices and a rise in demand, which can eventually deplete reserves, raise prices, and stimulate a new search for improved technology and substitutes.

However, ecological and environmental economists warn that as demand keeps rising at some point the harmful environmental effects of extracting,

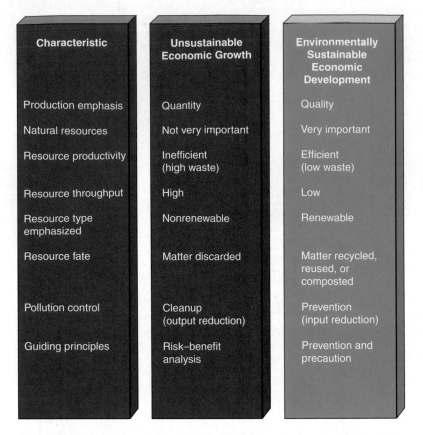

Characteristic	Unsustainable Economic Growth	Environmentally Sustainable Economic Development
Production emphasis	Quantity	Quality
Natural resources	Not very important	Very important
Resource productivity	Inefficient (high waste)	Efficient (low waste)
Resource throughput	High	Low
Resource type emphasized	Nonrenewable	Renewable
Resource fate	Matter discarded	Matter recycled, reused, or composted
Pollution control	Cleanup (output reduction)	Prevention (input reduction)
Guiding principles	Risk–benefit analysis	Prevention and precaution

Figure 26-5 Comparison of unsustainable economic growth and environmentally sustainable economic development according to ecological economists and many environmental economists.

processing, and using increasing amounts of various nonrenewable mineral and energy resources can limit their availability and that in some cases substitutes may not be available.

How Do Economists Value Pollution Control and Resource Management? Some Advocate Using Market Prices While Others Disagree

Economists think about optimum levels of pollution control and resource use, but there is considerable disagreement on how to arrive at those levels.

An important concept in environmental economics is that of *optimum levels* for pollution control and resource use. In the early days of a new coal mining operation, for example, the cost of removing coal is easy for developers to recover in sales of their product. However, after most of the more readily accessible coal has been removed, taking what is left can become too costly. In this case, the marginal cost of removal goes up with each unit of coal taken. Figure 26-6 shows this in terms of supply, demand, and equilibrium. Where the demand

curve crosses the supply curve is the point at which it no longer pays to remove the coal. The optimum level of coal mining is at or below that equilibrium point.

You might think that pollution control is an all-or-nothing proposition—that the best solution is to clean up every last bit of any pollutant. In fact, there are optimum levels for various kinds of pollution, because the marginal cost of pollution control also goes up for each unit of a pollutant removed from the environment. Figure 26-7 shows various possible optimal levels of pollution control. In this case, clean-up costs are shown on the blue curve. This is the supply curve, because the service supplied is removal of pollutants. Note that it slopes up more sharply (costing more) as we get closer to removing 100% of the pollutant in question.

The red line in Figure 26-7 represents the *demand* for cleanup by users of water or air. With their air or drinking water polluted, consumers are initially up in arms, but as the pollutant is removed, their concern is relieved and demand approaches zero. In other words, the marginal benefits of pollution control decrease with each unit of pollution removed.

At some point, the cost of removing the pollutant gets higher than what people are willing to pay, as their demand for clean-up lessens. That point is the equilibrium point, or the *optimum level* for clean-up.

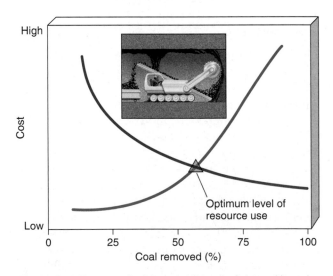

Figure 26-6 The cost of mining coal (blue line) rises with each additional unit removed. Mining a certain amount of coal is profitable, but at some point the cost of mining exceeds the monetary benefits (red line). That is, the marginal cost of mining increases while the marginal benefits decrease as more coal is removed. Where the two curves meet is theoretically the optimum level of resource use.

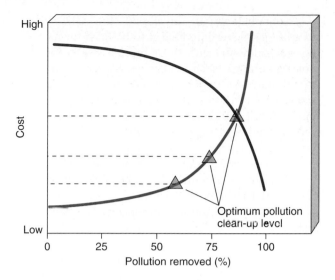

Figure 26-7 The cost of cleaning up pollution (blue line) rises with each additional unit removed. Cleaning up a certain amount of pollution is affordable, but at some point the cost of pollution control is greater than the harmful costs of the pollution to society. That is, the marginal cost of pollution clean-up increases (blue line) and the marginal benefits decrease (red line) as more pollution is removed. Where the blue curve intersects with any other curve is a point of optimum pollution control for the pollutant represented.

Another factor determining the shape and placement of the demand curve is how much people value their resources. If no one cares whether or not the water in a lake is clear or groundwater is pure, the optimum level of clean-up will be close to zero. But if they demand a clean lake and groundwater, the optimum level will rise. In other words, pollution control (or resource use) that is optimum for some will be high or low for others. The levels depend on human values as much as anything else.

Case Study: What is Cost-Benefit Analysis, and How Can It Be Improved? Weighing Costs and Benefits to Make Choices

Comparing likely costs and benefits of an environmental action can help with decision-making, but it is a limited tool.

Another widely used tool for making economic decisions about how to control pollution and manage resources is **cost-benefit analysis (CBA)**. It involves comparing estimated costs and benefits for actions such as implementing a pollution control regulation, building a dam on a river, filling in a wetland, or preserving an area of forest. It involves trying to estimate the optimum level of pollution clean up (Figure 26-7) or resource use (Figure 26-6).

CBA is one of the main tools economists and decision makers throughout the world use to help them make decisions about pollution control, biodiversity protection, and the construction of roads, airports, dams and other facilities.

Making a CBA involves determining who or what might be affected by a particular regulation or project, projecting potential outcomes, evaluating alternative actions, and determining who benefits and who is harmed. Then an attempt is made to assign monetary costs and benefits to each of the factors and components involved.

Direct costs involving land, labor, materials, and pollution-control technologies are often fairly easy to estimate. But indirect costs of things we value such as clean air and water that are not traded in the marketplace are difficult to make and are controversial. We can put estimated price tags on human life, good health, clean air and water, and various forms of natural capital such as an endangered species, a forest, a wetland, and other forms of natural capital. However, the monetary values that different people assign to such things vary widely because of different assumptions and value judgments. This can lead to a wide range of projected costs and benefits.

CBA is controversial because making accurate estimates of costs and benefits is difficult. They are also easy to manipulate by parties supporting or opposing a particular regulation or project.

Because of these drawbacks, CBA can lead to wide ranges of benefits and costs with a lot of room for error. For example, one U.S. industry-sponsored CBA estimated that compliance with a standard to protect U.S. workers from vinyl chloride would cost $65–90 billion. In the end, meeting the standard cost the industry less than $1 billion. A study by the Washington-based Economic Policy Institute found that the estimated costs made by industries for complying with proposed environmental regulations in the United States are almost always more (and often much more) than the actual costs of implementing the regulations.

Some environmental groups use CBA to help evaluate proposed environmental projects and regulations. But some environmentalists oppose putting too much emphasis on using this approach as a primary factor in decision making because the large uncertainties involved allow manipulation of the data and estimates to come up with a desired result.

If conducted fairly and accurately, CBA is a useful tool for helping making economic decisions. To minimize possible abuses and errors, environmentalists and economists advocate using the following guidelines for a CBA:

- Use uniform standards.
- Clearly state all assumptions used.
- Rate the reliability of data used.

- Estimate short- and long-term benefits and costs for all affected population groups.

- Compare the costs and benefits of alternative courses of action.

- Summarize the range of estimated costs and benefits.

Various economists make use of tools such as optimum levels and CBA in different ways. The neoclassical approach would tend to value resources strictly according to market information. They would look at the going prices for timber or coal, for example, or estimate the amount of money that tourists are likely to spend visiting a lake after it is cleaned up. They would calculate marginal costs and benefits of developing a resource or of cleaning up pollution somewhere, and then determine optimum levels of resource use or pollution control.

Environmental and ecological economists would tend to be less bound by market prices. They might assign higher-than-market values to resources to account for their importance to ecosystem health, biodiversity, and future generations. They would thus determine higher optimum levels of pollution control and lower optimum levels of resource use than would some neoclassical economists.

Have environmental regulations in the United States been worth the cost to industry, business, and consumers? There are different opinions on this issue depending on how CBA analyses are made. However, a 2003 joint study by the White House and the Office of Management and Budget (OMB) estimated that the total annual benefits from EPA regulations between 1992 and 2002 ranged from $121–193 billion compared to annual costs of $23.3–26.6 billion. Thus according to this CBA, the economic benefits of such regulations have exceeded their costs by a factor of 4.5 to 8.

26-3 MONITORING ENVIRONMENTAL PROGRESS

How Can We Evaluate Environmental Quality and Human Well-Being? Develop and Use New Indicators

We need new indicators to accurately reflect changing levels of environmental quality and human health.

Economists and environmental scientists want measurements that can indicate what is happening in an economy and in nature's economy. *Gross domestic product (GDP)*, and *per capita GDP* indicators provide a standardized and useful method for measuring and comparing the economic outputs of nations.

Economists who developed the GDP many decades ago never intended it to be used for measuring environmental quality or human well-being. But most governments and business leaders incorrectly use this narrowly defined indicator that way. The GDP is deliberately designed to measure the annual economic value of all goods and services produced within a country without attempting to make any distinction between goods and services that are environmentally or socially beneficial and those that are harmful.

Environmental and ecological economists and environmental scientists call for the development of new indicators to help monitor environmental quality and human well-being. One approach is to develop indicators that *add* to the GDP things not counted in the marketplace that enhance environmental quality and human well-being. They would also *subtract* from the GDP the costs of things that lead to a lower quality of life and depletion of natural resources.

One such indicator is the *genuine progress indicator (GPI)*, introduced in 1995 by Redefining Progress, a nonprofit organization that develops economics and policy tools to help evaluate and promote sustainability. (This group also developed the concept of ecological footprints, Figure 1-7, p. 10 and Figure 9-12, p. 172). Within the GPI, the estimated value of beneficial transactions that meet basic needs, but in which no money changes hands, are added to the GDP. Examples are unpaid volunteer work, healthcare for family members, childcare, and housework. Then the estimated harmful environmental costs (such as pollution and resource depletion and degradation) and social costs (such as crime) are subtracted from the GDP.

Genuine progress indicator	= GDP +	benefits not included in market transactions	−	harmful environmental and social costs

Figure 26-8 compares the per capita GDP and GPI for the United States between 1950 and 2000. Note that while the per capita GDP rose sharply, the per capita GPI stayed nearly flat and declined slightly between 1975 and 2000.

A similar indicator is the *Index of Sustainable Economic Welfare (ISEP)* developed by ecological economists Herman Daly and John Cobb. It combines estimates of income, natural resource depletion, environmental degradation, economic benefits from volunteer work, and distribution of income for different countries. The United Nations has developed an indicator of human well-being based on measurements of a country's standard of living, education, and life expectancy.

Others call for a much more detailed indicator based on carrying out *materials balance measurements*. It involves, for any good or service, measuring necessary inputs of matter and energy resources from the environment, their flow rates through the economy, and the outputs of pollution, waste, and heat into the environment (Figure 3-18, p. 53). This approach would give us detailed information on how conditions are changing, what problems are more serious than others, and what solutions work the best. However, it is difficult and costly to get such information, especially for nonmarket transactions, as discussed below.

The GPI and other environmental and social indicators are far from perfect and include many crude estimates. But without such indicators, we do not know much about what is happening to people, the environment, and the planet's natural resource base, and we have no effective way to measure what policies work. In effect, according to ecological and environmental economists, we are trying to guide national and global economies through treacherous economic and environmental waters at ever-increasing speeds without a good radar system.

The *good news* is that several of these indicators are available. The *bad news* is that they are not widely used and reported. Proponents call for us to get to the point that every time a change in GDP is reported we also get information on a change in one or more environmental and social indicators.

How Can We Assign Monetary Values to Resources Not Traded in the Marketplace? Ways to Represent Nature

Economists have developed several ways to estimate the nonmarket values of the earth's ecological services.

Environmental and ecological economists have developed various tools for estimating the values of the earth's ecological services, as discussed earlier on p. 204.

This involves estimating *nonuse values* not represented in market transactions. One is an *existence value* based on knowing that an old-growth forest or endangered species exists, even though we may never see them or use them. Another is *aesthetic value* based on putting a monetary value on a forest, species, or part of nature because of its beauty. A third type, called a *bequest* or *option value*, is based on the willingness of people to pay to protect some forms of natural capital for use by future generations.

Economists have developed several ways to estimate the monetary value of resources that cannot be priced by conventional means. One approach is to estimate a *mitigation cost* of how much it would take to offset an environmental damage. For example, how much would it cost to protect a forest from cutting, move an endangered species to a new habitat, or restore a statue damaged by air pollution?

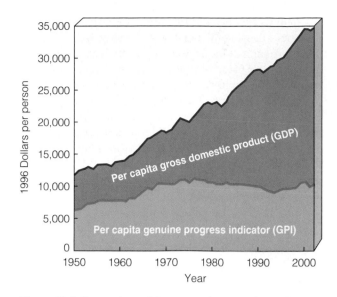

Figure 26-8 Comparison of the per capita gross domestic product (GDP) and per capita genuine progress indicator (GPI) in the United States between 1950 and 2000. (Data from Redefining Progress, 2002)

Another method is to estimate a *willingness to pay* by using a survey to determine how much people would be willing to pay to keep a particular species from becoming extinct, a particular forest from being cut down, or a specific river or beach from being polluted. This approach is controversial because people may inflate their estimates and not indicate the prices they would really pay to preserve various nonuse values.

How Can We Estimate the Future Value of a Resource? Assigning Discount Rates

Economists use discount rates to estimate the future value of a resource.

The **discount rate** is an estimate of a resource's future economic value compared to its present value. It is based on the idea that having something today may be worth more than it will be in the future. The size of the discount rate (usually given as a percentage) is a primary factor affecting how a resource such as a forest or fishery is used or managed.

At a zero discount rate, for example, a stand of redwood trees worth $1 million today will still be worth $1 million 50 years from now. However, most businesses, the U.S. Office of Management and Budget, and the World Bank typically use a 10% annual discount rate to evaluate how resources should be used. At this rate, the stand of redwood trees will be worth only $10,000 in 50 years. With this discount rate, it makes sense from an economic standpoint for an owner to cut these trees down as quickly as possible and invest the money in something else.

The value of discount rates are controversial. Proponents cite several reasons for using high (5–10%) discount rates. One is that inflation may reduce the value of their future earnings on a resource. Another is innovation or changes in consumer preferences could make a product or resource obsolete. For example, natural-looking composites of wood made from plastics may reduce the future use and market value of renewable redwood, and wind and hydrogen may greatly reduce the economic value of nonrenewable oil in the future. Also resource owners argue that without a high discount rate, they can make more money by investing their capital in some other venture.

Critics point out that high discount rates encourage such rapid exploitation of resources for immediate payoffs and that sustainable use of most renewable natural resources is virtually impossible. These critics believe that a 0% or even a negative discount rate should be used to protect unique and scarce resources. They also point out that moderate discount rates of 1–3% would make it profitable to use nonrenewable and renewable resources more sustainably or slowly. In addition, suppose that an acceptable substitute for a resource such as redwood or oil does not become available. Then these resources could become priceless. As you can see, there are no easy ways to make such decisions.

Economic return is not always the determining factor in how resources are used or managed. In some cases, farmers and owners of forests, wetlands, and other resources use *ethical concerns* in determining how they use and manage such resources. Their respect for the land and nature or what they believe to be their responsibility to future generations can override their desire for short-term profit at the expense of long-term resource and environmental sustainability.

26-4 HARMFUL EXTERNAL COSTS AND FULL-COST PRICING

What Are External or Indirect Costs? Hidden Costs

The direct price you pay for something does not include indirect environmental, health, and other harmful costs associated with its production and use.

All economic goods and services have *internal* or *direct costs* associated with producing them. For example, if you buy a car the direct price you pay includes the costs of raw materials, labor, and shipping, as well as a markup to allow the car company and its dealers some profits. Once you buy the car you must pay additional direct costs for gasoline, maintenance, and repair.

Making, distributing, and using any economic good or service also involve *indirect* or *external costs* or *benefits* not included in the market price and affecting people other than the buyer and seller. Economists call such costs and benefits **externalities.** A *positive externality* benefits someone not involved in an economic transaction. For example, if a car dealer volunteers to remove litter and debris from a stretch of highway you will benefit even if you have never bought a car from the dealer.

A *negative externality* is a harmful cost borne by someone not involved in an economic transaction. For example, extracting and processing raw materials to make a car uses nonrenewable energy and mineral resources, produces solid and hazardous wastes, disturbs land, and pollutes the air and water (Figure 16-13, p. 343). These external costs not included in the price of the car can have short- and long-term harmful effects on other people and on the earth's life-support systems.

Because these harmful external costs are not included in the market price of a car, most people do not connect them with car ownership. Still, the car buyer and other people in a society pay these hidden costs sooner or later, in the form of poorer health, higher costs of health care and insurance, higher taxes for pollution control, traffic congestion, and land used for highways and parking.

Similarly, in the United States, the price of gasoline was about $1.75 per gallon (46¢ per liter) in mid-2004. But this price did not include the external costs just listed along with lost work time while stalled in traffic jams and the harmful effects of urban sprawl (Figure 25-6, p. 567). According to a study by John Holtzclaw, when we include the harmful indirect costs, American consumers are really paying about $1.80–2.25 per liter ($3–8.60 per gallon)—depending mostly on whether the military costs of ensuring access to Middle Eastern oil are included. And these costs do not include the future harmful effects on climate change (Figure 21-13, p. 475) caused in part by CO_2 emissions from motor vehicles. In addition, the price of lumber does not include the value of the ecological services provided by intact forests (Figure 11-7, left, p. 200) or government subsidies to the timber industry.

What Is Full-Cost Pricing? Creating An Environmentally Honest Market

Including external costs in market prices informs consumers about the cost of their purchases on the earth's life-support systems and to human health.

For many economists, creating an *environmentally transparent or honest market system* is a way to deal with the harmful costs of goods and services. It requires including costs, as much as possible, in the market price of any good or service, such that its price would come as close as possible to its **full cost**—its actual internal costs plus its actual external costs.

This system would allow consumers to make more informed choices, because they would be aware of most or all the costs involved, which is one of the major goals of a truly free-market economy. It would likely cause consumers to give more thought to choosing fuel-efficient cars over much more expensive gas-guzzlers. Many people would probably conserve more water because its price would be much higher. They might also produce less trash, because the cost of collecting and disposing of nonrecyclable trash would go way up.

With full-cost pricing, some eco-friendly (or green) goods and services that now cost more would eventually cost less because internalizing external costs encourages producers to invent more resource-efficient and less-polluting methods of production, thereby cutting their production costs. Jobs would be lost in environmentally harmful businesses as consumers more often chose green products, but more jobs would be created in environmentally beneficial businesses. If a shift to full-cost pricing took place over several decades, most environmentally harmful businesses would have time to transform themselves into environmentally beneficial businesses. And consumers would have time to adjust their purchases and buying habits to more environmentally friendly products and services.

> ☒ *How Would You Vote?* Should full-cost pricing be used in setting the market prices of goods and services? Cast your vote online at http://biology.brookscole.com/miller14.

Full-cost pricing seems to make a lot of sense. So why is it not used more widely? There are various reasons. One is that many producers of harmful and wasteful goods would have to charge more and some would go out of business. Naturally, they oppose such pricing.

Also, it is difficult to put a price tag on many environmental and health costs. But to ecological and environmental economists, making the best possible estimates is far better than not including such costs in what we pay for most goods and services.

Phasing in such a system requires government action. Few if any companies will volunteer to reduce short-term profits by becoming more environmentally responsible. For example, assume you own an electronics company and you believe that we should all pay for the pollution resulting from production of electronics products. Assume also that your competitors do not believe this. Will you raise your prices to include the estimated costs of that pollution, while they

do not? If you do, your customers will probably buy their electronics from your competitors. Most consumers are looking for the best price, and by doing the right thing, you may eventually go bankrupt. So full-cost pricing has to be initiated across the market by an outside force, namely, the government.

Governments can use several strategies to encourage or force producers to work toward full-cost pricing including phasing out environmentally harmful subsidies, levying taxes on environmentally harmful goods and services, passing laws to regulate pollution and resource depletion, and using tradable permits for pollution or resource use. Let us look at these and other strategies in more detail.

26-5 WAYS TO IMPROVE ENVIRONMENTAL QUALITY AND SHIFT TO FULL-COST PRICING

How Can Ending Certain Subsidies and Tax Breaks Improve Environmental Quality and Reduce Resource Waste? Not Rewarding Environmentally Harmful Activities

We can improve environmental quality and help phase in full-cost pricing by removing environmentally harmful government subsidies and tax breaks.

Government subsidies and tax breaks can accelerate resource development, depletion, and degradation. We currently give depletion allowances and tax breaks to mining, oil, and coal companies for getting minerals and oil out of the ground. Taxpayers subsidize the cost of irrigation water for farmers and help to provide subsidies and low-cost loans to buy fishing boats.

One way to encourage a shift to full-cost pricing is to *phase out* environmentally harmful subsidies and tax breaks, which cost the world's governments about $1.9 trillion a year, according to studies by Norman Myers and other analysts. This is about 4.5% of the $42 trillion value of all of the goods and services produced throughout the world in 2004 and creates a huge economic incentive for environmental destruction and degradation.

On paper, phasing out such subsidies may seem like a great idea. But it involves political decisions that often are opposed successfully by powerful interests receiving the subsidies and tax breaks. They want to keep, and if possible increase, these benefits and often oppose subsidies and tax breaks for more environmentally beneficial competitors. For example, the fossil fuel and nuclear power industries in the United States (and in most developed countries) have gotten huge government subsidies compared to those for less harmful competing alternatives, such as conservation and wind power (Figure 18-34, p. 407). Removing these harmful subsidies and tax breaks would level the economic playing field and promote the use of the cheapest and least environmentally harmful energy alternatives.

Some countries are beginning to reduce environmentally harmful subsidies. Japan, France, and Belgium have phased out all coal subsidies. Germany has cut coal subsidies in half and plans to phase them out completely by 2010. China has cut coal subsidies by about 73% and has imposed a tax on high-sulfur coals. Between 1997 and 2001, these two actions helped reduce China's coal use by 5% at a time when its economy expanded by one-third. Thus shifts toward full-cost pricing can reduce resource use and pollution while encouraging more environmentally sustainable forms of economic growth and development.

How Can Green Taxes and Fees and Tax Shifting Improve Environmental Quality and Reduce Resource Waste? Making Polluters and Consumers Pay the Full Price

Taxes and fees on pollution and resource use can take us closer to full-cost pricing and shifting taxes from wages and profits to pollution and waste helps makes this feasible.

Another way to discourage pollution and resource waste is to *use green taxes or effluent fees* to help internalize many of the harmful environmental costs of production and consumption. Higher fees can also be charged for extracting lumber and minerals from public lands, using water provided by government-financed projects, and using public lands for livestock grazing.

Taxes can be levied on a per-unit basis on the release of pollution and hazardous or nuclear waste produced, and on the use fossil fuels, timber, and minerals. Figure 26-9 lists advantages and disadvantages of using green taxes and fees.

✗ *How Would You Vote?* Do the advantages of green taxes and fees outweigh the disadvantages? Cast your vote online at http://biology.brookscole.com/miller14.

To many analysts, the tax system in most countries is backwards. It can *discourage* what we want more of—jobs, income, and profit-driven innovation—and *encourage* what we want less of—pollution, resource waste, and environmental degradation. A more environmentally sustainable economic system would *lower* taxes on labor, income, and wealth and *raise* taxes on environmentally harmful activities.

With such a tax shift, for example, a tax on coal would include the increased health costs of breathing polluted air, damages from acid deposition, and estimated costs from climate change. Then taxes on wages and wealth could be reduced by the amount produced by the coal tax.

Shifting more of the tax burden from wages and profits to pollution and waste has a number of advantages (Figure 26-10). Some 2,500 economists, including eight Nobel Prize winners, have endorsed the concept of tax shifting. According to N. Gregory Mankiw, who chairs the President's Council of Economic Advisers: "Cutting income taxes while increasing gasoline taxes would lead to more rapid economic growth, less traffic congestion, safer roads, and reduced risk of global warming—all without jeopardizing long-term fiscal solvency. This may be the closest thing to a free lunch that economics has to offer." Such taxes would also stimulate the production and use of more fuel-efficient motor vehicles and reduce dependence on imported oil.

Economists also point out that successful implementation of green taxes would require such a tax

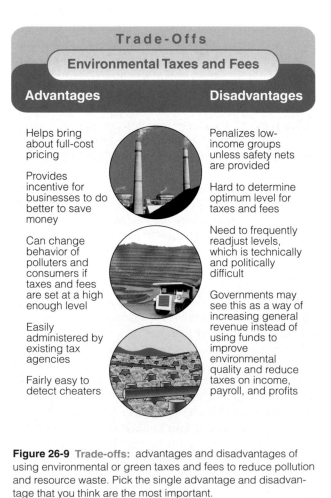

Trade-Offs	
Environmental Taxes and Fees	
Advantages	**Disadvantages**
Helps bring about full-cost pricing	Penalizes low-income groups unless safety nets are provided
Provides incentive for businesses to do better to save money	Hard to determine optimum level for taxes and fees
Can change behavior of polluters and consumers if taxes and fees are set at a high enough level	Need to frequently readjust levels, which is technically and politically difficult
Easily administered by existing tax agencies	Governments may see this as a way of increasing general revenue instead of using funds to improve environmental quality and reduce taxes on income, payroll, and profits
Fairly easy to detect cheaters	

Figure 26-9 Trade-offs: advantages and disadvantages of using environmental or green taxes and fees to reduce pollution and resource waste. Pick the single advantage and disadvantage that you think are the most important.

- **Decreases depletion and degradation of natural resources**

- **Improves environmental quality by full-cost pricing**

- **Encourages pollution prevention and waste reduction**

- **Stimulates creativity in solving environmental problems to avoid paying pollution taxes and thereby increases profits**

- **Rewards recycling and reuse**

- **Relies more on marketplace rather than regulation for environmental protection**

- **Provides jobs**

- **Can stimulate sustainable economic development**

- **Allows cuts in income, payroll, and sales taxes**

Figure 26-10 Solutions: Advantages of taxing wages and profits less and pollution and waste more. Pick the two advantages that you think are the most important.

shift. It would have to be phased in over 15 to 20 years to allow businesses to plan for the future and depreciate existing capital investments over their useful lives. And because consumption taxes place a larger burden on the poor and lower middle-class than do income taxes, governments would need to provide safety nets in the form of lifeline payments or credits for essentials such as food, fuel, and housing.

Nine western European countries have begun trial versions of such tax shifting, known as *environmental tax reform*. So far only a small amount of revenue has been shifted by taxes on emissions of CO_2 and toxic metals, garbage production, and vehicles entering congested cities. But such experience shows that this idea works.

✗ HOW WOULD YOU VOTE? Do you favor shifting taxes on wages and profits to pollution and waste? Cast your vote online at http://biology.brookscole.com/miller14.

How Can Environmental Laws and Regulations Improve Environmental Quality and Reduce Resource Waste? Encouraging Innovation

Environmental laws and regulations work best if they motivate companies to find innovative ways to control and prevent pollution and reduce resource waste.

Most economists agree that government intervention in the marketplace is needed to control or prevent pollution, reduce resource waste, and encourage full-cost pricing.

Regulation is a widely used form of government intervention. It involves enacting and enforcing laws that set pollution standards, regulate harmful activities such as releasing toxic chemicals into the environment, and require that certain irreplaceable or slowly replenished resources be protected from unsustainable use.

Many environmentalists and business leaders agree that *innovation-friendly regulations* can motivate companies to develop eco-friendly products and processes that can increase profits and competitiveness in national and international markets. But they also agree that some overly costly pollution control regulations discourage innovation. Some regulations are too prescriptive, for example, mandating specific technologies. Some set compliance deadlines that are too short to allow companies to find innovative solutions, and they discourage risk taking and experimentation.

Consider the difference between the United States and Sweden concerning their regulation of the pulp and paper industries. In the 1970s, strict U.S. regulations with short compliance deadlines forced companies to adopt the best available end-of-pipe water pollution treatment systems, which were costly.

By contrast, in Sweden the government started with slightly less strict standards and longer compliance deadlines but clearly indicated that tougher standards would follow. This more flexible and innovation-friendly approach gave companies time to focus on redesigning their production processes instead of relying mostly on waste treatment. It also spurred them to look for innovative ways to prevent pollution and improve resource productivity to meet stricter future standards. They developed processes for pulping and chlorine-free bleaching processes that met the emission standards, lowered operating costs, and gave them a competitive advantage in international markets.

Experience shows that an innovation-friendly regulatory process emphasizes pollution prevention and waste reduction and requires industry and environmental interests to work together in developing realistic standards and timetables. It sets goals, but frees industries to meet them in any way that works, and establishes standards strict enough to promote real innovation, allowing enough time for it. It also uses market incentives such as emissions and resource-use charges and tradable pollution and resource-use permits to encourage compliance and innovation.

Finally, pollution control regulations have to be designed to improve environmental quality while not being too costly. Recall that the marginal cost for removing a specific pollutant from gases or wastewater being discharged rises with each additional unit of that pollutant that is removed (Figure 26-7).

There are problems with the regulatory approach. One is that ecological economists, health scientists, and business leaders often disagree in their estimates of the harmful costs of pollution. Even scientists in the same field often have different estimates of such costs because they lack data and hold different assumptions.

Also, many regulations are geared toward achieving optimum levels of pollution over a large area such as a whole state. Some critics raise environmental justice questions about who benefits and who suffers from such regulations. Levels may be optimum for the state, but not for the people living near or downwind or downriver from a polluting power plant, incinerator, or factory. They are being exposed to much higher levels of pollution than are the majority of people in the state. See the Guest Essay on this topic by Robert Bullard on the website for this chapter.

In addition, assigning monetary values to lost lives, ecosystems, and ecological services is difficult and controversial, and varies widely because of lack of data and different assumptions and value judgments. But assigning little or no value to such things means they will not be counted at all in determining optimum pollution levels.

Figure 26-11 shows the evolution of several phases of environmental management through regulation. The period between 1970 and 1985 can be viewed as the *resistance-to-change management era*, during which many companies and government regulators developed an adversarial relationship, and companies resented and actively resisted environmental regulations (Figure 26-11, left). In addition, many government regulators thought they had to prescribe ways for reluctant companies to clean up their pollution emissions. Most companies responded by hiring outside environmental consultants (who usually favored end-of-pipe pollution control solutions) and by using lawyers to oppose or find legal loopholes in the regulations. They also lobbied elected officials to have environmental laws and regulations overthrown, weakened, or changed to allow for easy compliance.

By 1985, most company managers accepted environmental regulations and continued to rely mostly on pollution control. However, they placed little emphasis on trying to find innovative solutions to pollution and resource waste problems because the regulations were too strict and the market rewards too low.

In the 1990s, a growing number of company managers began to realize that environmental improvement is an economic and competitive opportunity instead of a cost to be resisted. This was the beginning of the *innovative management era*, which environmental and business visionaries project will go through several phases over the next 40–50 years (Figure 26-11).

During this time, many consumers began buying green products. Some firms also recognized that their shareholder value depends in part on having a good environmental record. A growing number of firms began looking for innovative and profitable ways to reduce resource use, pollution, and waste (Figure 24-5, p. 537 and Individuals Matter, p. 538). As a result, the environment is becoming an important component of business strategic planning. And many corporations now routinely issue environmental and sustainability reports to their stockholders.

This has been stimulated by the more than $2 trillion that exists in environmentally and socially screened investment funds and environmental and sustainability concerns of shareholders, insurance companies, bond-ranking agencies, and managers of state pension funds. In 2002, for example, institutional investors managing over $4.5 trillion in assets wrote the 500 largest global corporations asking them for full disclosure of the emissions of climate-changing gases and their policies on reducing the risks from climate change.

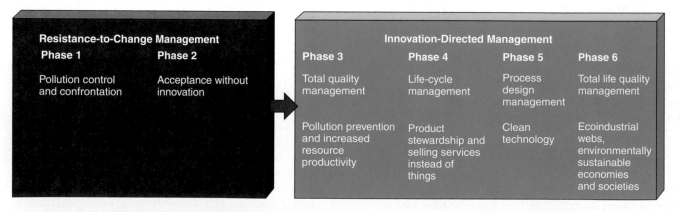

Resistance-to-Change Management		Innovation-Directed Management			
Phase 1	Phase 2	Phase 3	Phase 4	Phase 5	Phase 6
Pollution control and confrontation	Acceptance without innovation	Total quality management	Life-cycle management	Process design management	Total life quality management
		Pollution prevention and increased resource productivity	Product stewardship and selling services instead of things	Clean technology	Ecoindustrial webs, environmentally sustainable economies and societies

Figure 26-11 Solutions: evolution of environmental management.

Should We Rely More on Tradable Pollution and Resource-Use Permits? The Marketplace Can Work

The government can set a limit on pollution emissions or use of a resource, give pollution or resource use permits to users, and allow them to trade their permits in the marketplace.

A market-approach is for the government to *grant tradable pollution and resource-use permits*. The government sets a limit or cap on total emissions of a pollutant or use of a resource such as a fishery. Then it issues or auctions permits that allocate the total among manufacturers or users.

A permit holder not using its entire allocation can use it as a credit against future expansion, use it in another part of its operation, or sell it to other companies. In the United States, this approach has been used to reduce the emissions of sulfur dioxide and several other air pollutants, as discussed on p. 454. Tradable rights can also be established among countries to help preserve biodiversity and reduce emissions of greenhouse gases and other pollutants with harmful regional or global effects.

Figure 26-12 lists advantages and disadvantages of using tradable pollution and resource-use permits. The effectiveness of such programs depends on how high or low the initial cap is set and the rate at which the cap is reduced.

> **✗ HOW WOULD YOU VOTE?** Do the advantages of using tradable pollution and resource use permits to reduce pollution and resource waste outweigh the disadvantages? Cast your vote online at http://biology.brookscole.com/miller14.

How Can Eco-Labeling Improve Environmental Quality and Reduce Resource Waste? Informing Consumers

Labeling environmentally beneficial goods and resources extracted by more sustainable methods can help consumers decide what goods and services to buy.

We can use *product eco-labeling* to encourage companies to develop green products and services and to help consumers select more environmentally beneficial products and services. Eco-labeling programs have been developed in Europe, Japan, Canada, and the

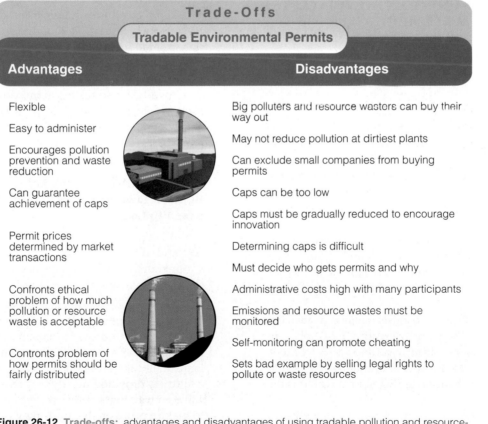

Trade-Offs

Tradable Environmental Permits

Advantages	Disadvantages
Flexible	Big polluters and resource wastors can buy their way out
Easy to administer	May not reduce pollution at dirtiest plants
Encourages pollution prevention and waste reduction	Can exclude small companies from buying permits
Can guarantee achievement of caps	Caps can be too low
Permit prices determined by market transactions	Caps must be gradually reduced to encourage innovation
	Determining caps is difficult
	Must decide who gets permits and why
	Administrative costs high with many participants
Confronts ethical problem of how much pollution or resource waste is acceptable	Emissions and resource wastes must be monitored
	Self-monitoring can promote cheating
Confronts problem of how permits should be fairly distributed	Sets bad example by selling legal rights to pollute or waste resources

Figure 26-12 Trade-offs: advantages and disadvantages of using tradable pollution and resource-use permits to reduce pollution and resource waste. Pick the single advantage and disadvantage that you think are the most important.

Germany:
Blue Angel (1978)

Canada:
Environmental
Choice (1988)

United States:
Green Seal (1989)

Nordic Council:
White Swan (1989)

European Union:
Eco-label (1992)

China:
Environmental
label (1993)

Figure 26-13 Solutions: symbols used in some of the *eco-labeling* programs that evaluate green or environmentally favorable products.

United States where, for example, the *Green Seal* labeling program has certified more than 300 products; see Figure 26-13.

Eco-labels are also being used to identify fish caught by sustainable methods (certified by the Marine Stewardship Council) and to certify timber produced and harvested by sustainable methods (evaluated by organizations such as the Forestry Stewardship Council; Solutions, p. 205).

26-6 REDUCING POVERTY TO IMPROVE ENVIRONMENTAL QUALITY AND HUMAN WELL-BEING

How Is the World's Wealth Distributed? Flowing up to the Rich

Since 1960, most of the financial benefits of global economic growth have flowed up to the rich rather than down to the poor.

Poverty is usually defined as the inability to meet one's basic economic needs. According to a 2000 World Bank study, half of humanity is trying to live on less than $3 (U.S.) a day and one of every five people on the planet is struggling to survive on an income of roughly $1 (U.S.) per day.

Poverty has numerous harmful health and environmental effects (Figure 1-11, p. 13, and Figure 19-18, p. 429) and has been identified as one of the five major causes of the environmental problems we face (Figure 1-10, p. 13).

Most neoclassical economists believe a growing economy can help the poor by creating more jobs, enabling more of the increased wealth to reach workers, and providing greater tax revenues that can be used to help the poor help themselves. Economists call this the *trickle-down* effect.

However, since 1960, most of the benefits of global economic growth as measured by income have flowed up to the rich rather than down to the poor (Figure 26-14). Since 1980, growth of this *wealth gap*

has increased. According to Ismail Serageldin, the planet's richest three people have more wealth than the combined GDP of the world's 47 poorest countries and their 600 million people. In Vandana Shiva's words, "Resources move from the poor to the rich, and pollution moves from the rich to the poor." South African President Thabo Mbeki told delegates at the 2003 Johannesburg World Summit on Sustainable Development, "A global human society based on poverty for many and prosperity for a few, characterized by islands of wealth, surrounded by a sea of poverty, is unsustainable."

These trends do not mean that economic growth causes poverty. Instead, they mean that for a variety of reasons rich nations and individuals have devoted only a small fraction of their wealth to helping reduce poverty and its harmful effects on the environment and human well-being.

Poverty is also sustained by corruption, absence of property rights, insufficient legal protection, and inability of many poor people to borrow money to grow crops or start a small business.

Case Study: What Is the Role of the World Bank in Economic Development? Controversy over Big Loans

The World Bank makes loans to developing countries for their economic development, but a number of these loans have had harmful environmental and social effects.

The World Bank is the major player in global economic development. It was formed in 1945 after World War II to provide loans for rebuilding Europe and Japan. In the 1950's the focus shifted to providing loans to aid the economic development of developing countries, provided mostly by private investors in the 150 countries that jointly own the bank. Investors hope to make a profit on the funds they put up, mostly from interest paid on the loans. The United States provides more of the investment capital than any other country and the presidents of the bank have all been Americans.

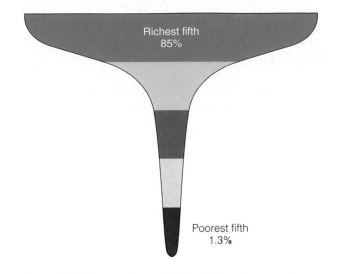

Figure 26-14 Data on the *global distribution of income* show that most of the world's income has flowed up; the richest 20% of the world's population receive more of the world's income than all of the remaining 80%. Each horizontal band in this diagram represents one-fifth of the world's population. This upward flow of global income has accelerated since 1960 and especially since 1980. This trend can increase environmental degradation by increasing average per capita consumption by the richest 20% of the population and causing the poorest 20% of the world's people to use renewable resources faster than they are replenished in order to survive. (Data from UN Development Programme and Ismail Serageldin, "World Poverty and Hunger—A Challenge for Science," *Science* 296 (2002): 54–58)

A number of World Bank loans for large-scale dams, roads into tropical forests, and mining operations have been environmentally destructive and controversial.

Critics accuse the bank of making loans for large-scale projects without evaluating the long-term environmental and social impacts and without requiring adequate safeguards to help reduce or eliminate harmful environmental and social impacts. Critics also call for the bank to focus more on making moderate and small-scale loans that benefit the poor directly.

Another problem is that to make enough money to pay the interest on their loans, many developing countries sell their mineral, timber, and other resources to developed counties at low prices. This depletes their natural capital and can eventually leave them without enough resources to support future economic development. Also, interest payments can deplete national budgets, leaving little for health, education, and other important programs.

In recent years, environmentalists and representatives of the poor have staged large-scale protests against such policies. In response, the bank has begun trying to carry out more detailed reviews of the environmental and social impacts of its loans. But it remains to be seen how such reviews will affect the bank's lending policies.

How Can We Reduce Poverty? Help the Poor Help Themselves

We can sharply cut poverty by forgiving the international debts of the poorest countries and greatly increasing international aid and small individual loans to help the poor help themselves.

Analysts point out that reducing poverty requires the governments of most developing countries to make policy changes. One is to shift more of the national budget to help the rural and urban poor work their way out of poverty. Another is to give villages and the urban poor title to common lands and to crops and trees they plant.

Encouraging sustainable forms of economic development can help reduce global poverty but analysts say that by itself this is not enough. Analysts suggest that one way to help reduce global poverty is to forgive at least 60% of the $2.4 trillion debt that developing countries owe to developed countries and international lending agencies and all of the $422 billion debt of the poorest and most heavily indebted countries on the condition that the money saved on the debt interest be spent on meeting basic human needs. Currently, developing countries pay almost $300 billion per year in interest to developed countries to service this debt. According to environmental economist John Peet, this inability to "service their debt assures perpetual poverty for the poor nations, and, effectively, perpetual servitude to the rich nations."

Critics say that many countries relieved of some debt will take on more debt, and they want assurances that most of the savings from debt relief are passed on to the poor in the form of titles to land, education, jobs, and better health care.

Developed countries can increase nonmilitary government and private aid to developing countries, with mechanisms to assure that most of the aid goes directly to the poor to help them become more self-reliant and to help provide social safety nets such as welfare, unemployment payments, and pension benefits that are available in most developed countries. Developed countries also need to mount a massive global effort to combat malnutrition and the infectious diseases that kill millions of people prematurely, helping perpetuate poverty. Another approach is for lending agencies to make small loans to poor people who want to increase their income (Solutions, p. 601). They should also make investments in small-scale infrastructure that help the poor such as solar cell power facilities in villages, small-scale irrigation projects, and farm-to-market roads. Lending agencies can also make investments in helping sustain and restore the resource bases of fisheries, forests, and small-scale agriculture that provide more than half of the world's jobs.

Developed countries can help developing countries create more environmentally sustainable economies, or eco-economies. According to Robert B. Shapiro, former CEO of Monsanto, "If emerging economies have to relive the entire industrial revolution with all its waste, its energy use, and its pollution, I think it's all over."

Finally, there is a need for both developed countries and developing countries to stabilize their populations. Many analysts also call for encouraging the political and social empowerment of the poor, especially women. This can be done by putting more decision making at the local level and integrating human rights with sustainable development.

According to the United Nations Development Programme (UNDP), it will cost about $50 billion a year to provide universal access to basic services such as education, health, nutrition, family planning, safe water, and sanitation. The UNDP notes that this is less than 0.1% of the world's annual income and is only a fraction of what the world devotes each year to military spending (Figure 26-15).

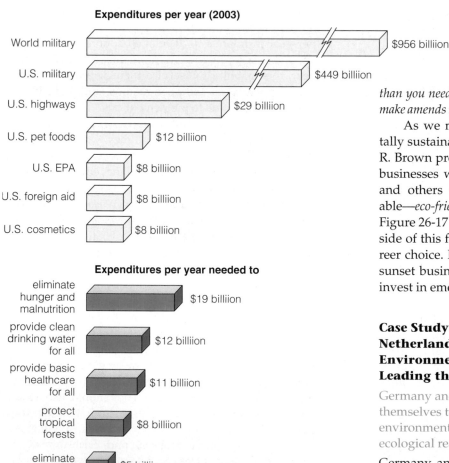

Expenditures per year (2003)

World military — $956 billiion
U.S. military — $449 billiion
U.S. highways — $29 billiion
U.S. pet foods — $12 billiion
U.S. EPA — $8 billiion
U.S. foreign aid — $8 billiion
U.S. cosmetics — $8 billiion

Expenditures per year needed to

eliminate hunger and malnutrition — $19 billiion
provide clean drinking water for all — $12 billiion
provide basic healthcare for all — $11 billiion
protect tropical forests — $8 billiion
eliminate illiteracy — $5 billiion

Figure 26-15 What should our priorities be? (Data from United Nations, World Health Organization, U.S. Department of Commerce, and U.S. Office of Management and Budget)

26-7 MAKING THE TRANSITION TO MORE ENVIRONMENTALLY SUSTAINABLE ECONOMIES

How Can We Make a Transition to an Eco-Economy? Copy Nature and Use Full-Cost Pricing

An eco-economy copies nature's four principles of sustainability and environmental economic strategies.

On page 586, I listed environmental economic strategies that have been discussed at length in this chapter. Eco-economies will increasingly employ these strategies—environmentally sensitive economic indicators, full-cost pricing, tax shifting, and eco-labeling, among many others.

An eco-economy mimics the processes that sustain the earth's natural systems (Figure 9-15, p. 174). Paul Hawken and several other business leaders and economists have suggested ways for using these guidelines and various economic tools for making the transition to more environmentally sustainable eco-economies over the next several decades. They echo what has been discussed in this chapter and are summarized in Figure 26-16 (p. 602). Hawken's simple golden rule for an eco-economy is this: *"Leave the world better than you found it, take no more than you need, try not to harm life or the environment, and make amends if you do."*

As we make the transition to more environmentally sustainable economies during this century, Lester R. Brown projects that some environmentally harmful businesses will decline and become *sunset businesses*, and others that are more environmentally sustainable—*eco-friendly, businesses* will grow in importance. Figure 26-17 (p. 603) lists some of both types. The right side of this figure might give you some ideas for a career choice. Forward-looking owners and investors in sunset businesses will use their profits and capital to invest in emerging eco-friendly businesses.

Case Study: How Are Germany and the Netherlands Working to Achieve More Environmentally Sustainable Economies? Leading the Way

Germany and the Netherlands have dedicated themselves to making their economies more environmentally sustainable for economic and ecological reasons.

Germany and the Netherlands are working to make their economies more environmentally sustainable. In Germany sales of environmental protection goods and services—already more than $600 billion per year—are projected to rise.

Microloans to the Poor

Most of the world's poor want to earn more, become more self-reliant, and have a better life. But they have no credit record. Also, they have few if any assets to use for collateral to secure a loan to buy seeds and fertilizer for farming or tools and materials for a small business.

For almost three decades an innovative tool called *microlending* or *microfinance* has helped deal with this problem. For example, since economist Muhammad Yunus started it in 1976, the Grameen (Village) Bank in Bangladesh has provided more than $4 billion in microloans (varying from $50 to $500) to several million mostly poor, rural, and landless women in 40,000 villages. About 94% of the loans are to women who start their own small businesses as sewers,

weavers, bookbinders, peanut fryers, or vendors.

To stimulate repayment and provide support, the Grameen Bank organizes microborrowers into five-member "solidarity" groups. If one member of the group misses a weekly payment or defaults on the loan, the other members of the group must make the payments.

The Grameen Bank's experience has shown that microlending is both successful and profitable. For example, less than 3% of microloan repayments to the Grameen Bank are late, and the repayment rate on its loans is 90–95%, much higher than the repayment rate for conventional loans by commercial banks throughout most of the world.

About half of Grameen's borrowers move above the poverty line within 5 years, and domestic vio-

lence, divorce, and birth rates are lower among most borrowers.

Microloans to the poor by the Grameen Bank are being used to develop day-care centers, health clinics, reforestation projects, drinking water supply projects, literacy programs, and group insurance programs. People also use them to bring small-scale solar and wind power systems to rural villages.

Grameen's model has inspired the development of microcredit projects in more than 58 countries that have reached 36 million people (including dependents), and the number is growing rapidly.

Critical Thinking

Why do you think there has been little use of microloans by international development and lending agencies such as the World Bank and the International Monetary Fund? How might this situation be changed?

Mostly because of stricter air pollution regulations, German companies have developed some of the world's cleanest and most efficient gas turbines and invented the world's first steel mill that uses no coal. Germany sells these and other improved environmental technologies globally. Germany is also one of the world's leading manufacturers of wind turbines.

In 1977, the German government started the Blue Angel *eco-labeling* program to inform consumers about products that cause the least environmental harm (Figure 26-13). Most international companies use the German market to test and evaluate green products.

German car companies, as part of a recycling revolution, are required to pick up and recycle all domestic cars they make, and such *take-back* requirements are being extended to almost all products to reduce use of energy and virgin raw materials. Germany is selling these recycling technologies to other countries.

Germany's government has supported research and development aimed at making it the world's leader in solar-cell and wind turbine technology and hydrogen fuel.

Finally, Germany provides about $1 billion per year in green foreign aid to developing countries, much of it is designed to stimulate demand for German technologies and products.

In 1989, the Netherlands—a tiny country with about 16 million people—began implementing a National Environmental Policy Plan, or Green Plan, as a result of widespread public alarm over declining environmental quality. The goal is to slash production of many types of pollution by 70–90% and achieve the world's first environmentally sustainable economy, ideally within a few decades.

The government began by identifying eight major areas for improvement: climate change, acid deposition, eutrophication, toxic chemicals, waste disposal, groundwater depletion, unsustainable use of renewable and nonrenewable resources, and local nuisances (mostly noise and odor pollution).

Then the government formed task forces consisting of people in industry, government, and citizens' groups for each of the eight areas, asking each task force to agree on targets and timetables for drastically reducing pollution. Each group was free to pursue whatever policies or technologies it wanted, but if a group could not agree, the government would impose its own targets and timetables and stiff penalties for industries not meeting certain pollution reduction goals.

Each task force focused on four general themes. (1) life-cycle management (Figure 26-11, right); (2) energy efficiency, with the government committing $385 million per year to energy conservation programs;

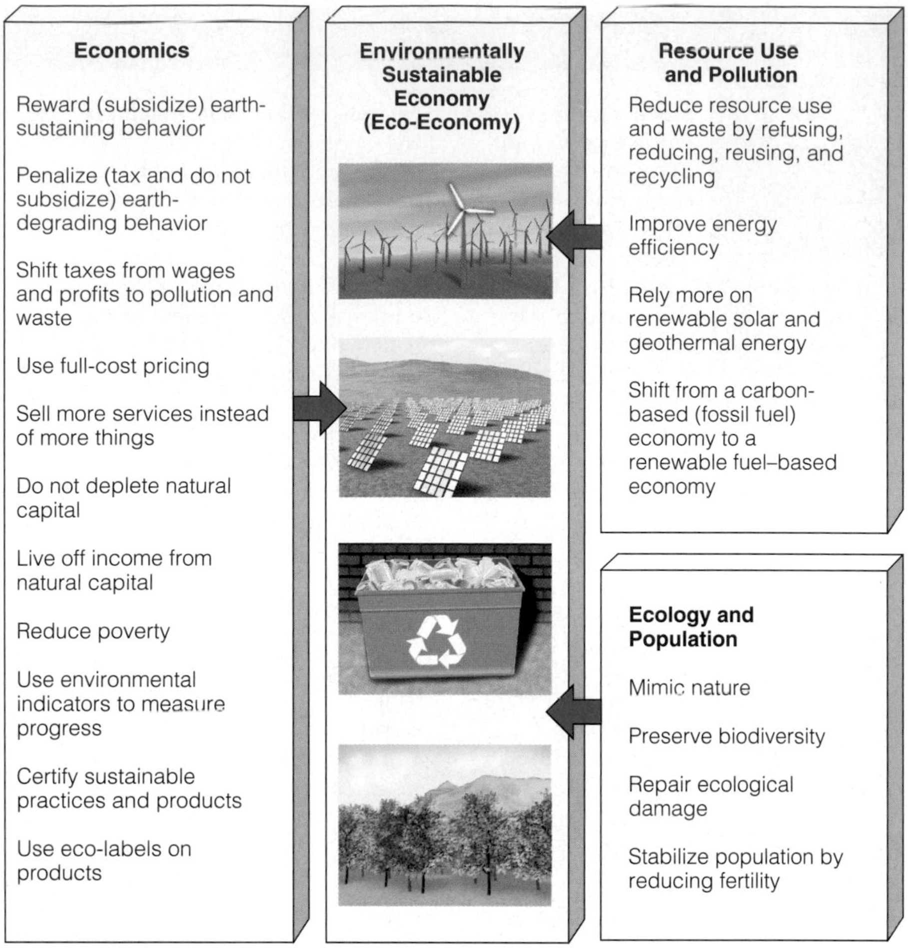

Figure 26-16 Solutions: principles for shifting to more environmentally sustainable economies or eco-economies during this century.

Economics	Environmentally Sustainable Economy (Eco-Economy)	Resource Use and Pollution
Reward (subsidize) earth-sustaining behavior		Reduce resource use and waste by refusing, reducing, reusing, and recycling
Penalize (tax and do not subsidize) earth-degrading behavior		Improve energy efficiency
Shift taxes from wages and profits to pollution and waste		Rely more on renewable solar and geothermal energy
Use full-cost pricing		Shift from a carbon-based (fossil fuel) economy to a renewable fuel–based economy
Sell more services instead of more things		
Do not deplete natural capital		
Live off income from natural capital		**Ecology and Population**
Reduce poverty		Mimic nature
Use environmental indicators to measure progress		Preserve biodiversity
Certify sustainable practices and products		Repair ecological damage
Use eco-labels on products		Stabilize population by reducing fertility

(3) environmentally sustainable technologies, also supported by a government program; and (4) improving public awareness through a massive government-sponsored public education program.

Many of the country's leading industrialists like the Green Plan becaue they can make investments in pollution prevention and pollution control with less financial risk and a high degree of certainty about long-term environmental policy. And they are free to deal with the problems in ways that make the most sense for their businesses. Industrial leaders have also learned that creating more efficient and environmentally sound products and processes often reduces costs and increases profits as they are sold at home and abroad.

The Netherlands plan is the first attempt by any country to foster a national debate on the issue of environmental sustainability and to encourage innovative solutions to environmental problems. Is the plan working? There is a long way to go, but the news is encouraging. Most of the target groups have met or exceeded their goals on schedule. A huge amount of environmental research by the government and private sector has taken place. This has led to an increase in or-

ganic agriculture, greater reliance on bicycles in some cities, and more ecologically sound new housing developments. But some of the more ambitious goals such as decreasing CO_2 levels may have to be revised downward or even abandoned.

Shifting to eco-economies over the next several decades this will require bold leadership by business leaders and elected officials and bottom-up political pressure from concerned citizens. Forward-looking investors, corporate executives, and political leaders are recognizing that *the environmental revolution is also an economic revolution.*

Converting the economy of the 21st century into one that is environmentally sustainable represents the greatest investment opportunity in history.

LESTER R. BROWN AND CHRISTOPHER FLAVIN

CRITICAL THINKING

1. Should we attempt to maximize economic growth by producing and consuming more and more economic goods and services? Explain. What are the alternatives?

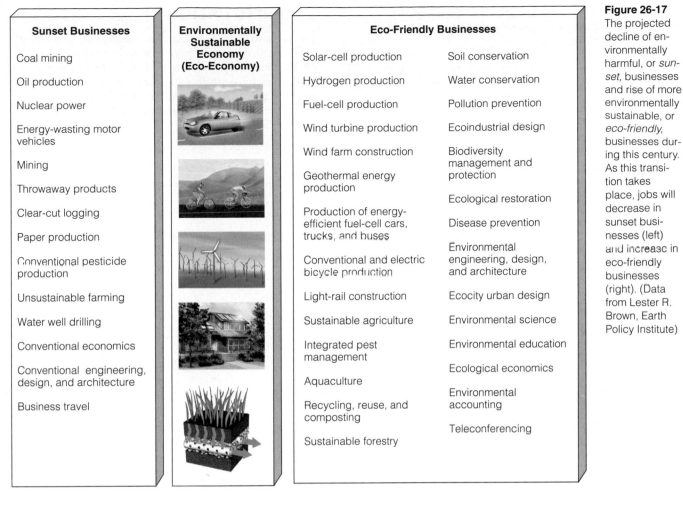

Sunset Businesses

Coal mining

Oil production

Nuclear power

Energy-wasting motor vehicles

Mining

Throwaway products

Clear-cut logging

Paper production

Conventional pesticide production

Unsustainable farming

Water well drilling

Conventional economics

Conventional engineering, design, and architecture

Business travel

Environmentally Sustainable Economy (Eco-Economy)

Eco-Friendly Businesses

Solar-cell production

Hydrogen production

Fuel-cell production

Wind turbine production

Wind farm construction

Geothermal energy production

Production of energy-efficient fuel-cell cars, trucks, and buses

Conventional and electric bicycle production

Light-rail construction

Sustainable agriculture

Integrated pest management

Aquaculture

Recycling, reuse, and composting

Sustainable forestry

Soil conservation

Water conservation

Pollution prevention

Ecoindustrial design

Biodiversity management and protection

Ecological restoration

Disease prevention

Environmental engineering, design, and architecture

Ecocity urban design

Environmental science

Environmental education

Ecological economics

Environmental accounting

Teleconferencing

Figure 26-17
The projected decline of environmentally harmful, or *sunset*, businesses and rise of more environmentally sustainable, or *eco-friendly*, businesses during this century. As this transition takes place, jobs will decrease in sunset businesses (left) and increase in eco-friendly businesses (right). (Data from Lester R. Brown, Earth Policy Institute)

2. According to one definition, *sustainable development* involves meeting the needs of the present human generation without compromising the ability of future generations to meet their needs, as discussed in Chapter 1. What do you believe are the needs referred to in this definition? Compare this definition with the definition of environmentally sustainable economic development given in Figure 26-5, p. 588.

3. Suppose that over the next 20 years the current harmful environmental and health costs of goods and services are internalized so that their market prices reflect their total costs. What harmful and beneficial effects might such full-cost pricing have on your lifestyle?

4. Explain why you agree or disagree with the proposals that various analysts have made for sharply reducing poverty, as discussed on pages 599–600, Which two of these proposals do you believe are the most important?

5. Explain why you agree or disagree with each of the major principles for shifting to a more environmentally sustainable economy listed in Figure 26-16, p. 602. Which three do you believe are the most important?

6. Congratulations! You are in charge of the world. List your five most important actions for shifting to eco-economies over the next 50 years.

PROJECTS

1. List all the economic goods you use, and then identify those that meet your basic needs and those that satisfy your wants. Identify any economic wants you **(a)** would be willing to give up, **(b)** you believe you should give up but are unwilling to, and **(c)** hope to give up in the future. Relate the results of this analysis to your personal impact on the environment. Compare your results with those of your classmates.

2. Pick one of the suggestions listed under "Eco-Friendly Business" in Figure 26-17 (above) and develop a business plan for a company that would provide the service you selected. For example, assume you'll go into the business of ecological restoration and describe: **(a)** your service, **(b)** your customers, **(c)** your mission statement, and **(d)** your strategy for promoting the business.

3. Pick a regulation in the state or country where you live (such as a water pollution law or regulation) and examine how it affects businesses and other organizations. Determine whether it is an innovation-friendly regulation and explain why or why not. If not, how could it be made more innovation-friendly?

4. Interview officials at a company in your town or region to get their views on one or more environmental regulations affecting their industry. Describe the effects of that regulation on the company and the company's approach toward dealing with it.

5. Use the library or the Internet to find bibliographic information about *Gaylord Nelson, Lester R. Brown,* and *Christopher Flavin,* whose quotes appear at the beginning and end of this chapter.

6. Make a concept map of this chapter's major ideas, using the section heads, subheads, and key terms (in boldface type). Look on the website for this book for information about making concept maps.

LEARNING ONLINE

The website for this book contains study aids and many ideas for further reading and research. They include a chapter summary, review questions for the entire chapter, flash cards for key terms and concepts, a multiple-choice practice quiz, interesting Internet sites, references, and a guide for accessing thousands of InfoTrac® College Edition articles. Log on to

http://biology.brookscole.com/miller14

Then click on the Chapter-by-Chapter area, choose Chapter 26, and select a learning resource.

27 Politics, Environment, and Sustainability

Rescuing a River

In the 1960s, Marion Stoddart (Figure 27-1) moved to Groton, Massachusetts, on the Nashua River, then considered one of the nation's filthiest rivers. For decades, industries and towns along the river had used it as a dump. Dead fish bobbed on its waves, and at times the water was red, green, or blue from pigments discharged by paper mills.

Instead of thinking nothing could be done, Stoddart committed herself to restoring the Nashua and establishing public parklands along its banks.

She did not start by filing lawsuits or organizing demonstrations. Instead she created a careful cleanup plan and approached state officials with her ideas. They laughed, but she was not discouraged and began practicing the most time-honored skill of politics: one-on-one persuasion. She identified power brokers in the riverside communities and began to educate them, win them over, and get them to cooperate in cleaning up the river.

She also got the state to ban open dumping in the river. When federal matching funds promised for building a treatment plant failed to materialize, Stoddart gathered 13,000 signatures on a petition sent to President Richard Nixon. The funds arrived in a hurry.

Stoddart's next success was getting a federal grant to beautify the river. She hired high school dropouts to clear away mounds of debris. When the river cleanup was completed, she persuaded communities along the river to create a riverside park and woodlands along both banks.

Now, four decades later, the Nashua is still clean. Several new water treatment plants have been built, and a citizens' group founded by Stoddart keeps watch on water quality. The river supports many kinds of fish and other wildlife, and its waters are used for canoeing and recreation. The project is testimony to what a committed individual can do to bring about change from the bottom up by getting people to work together. For her efforts, the UN Environment Programme named Stoddart an outstanding worldwide worker for the environment.

Politics is the process by which individuals and groups try to influence or control the policies and actions of governments at local, state, national, and international levels. Politics is concerned with who has power over the distribution of resources and who gets what, when, and how. Many people think of politics in national terms, but what directly affects most people is what happens in their local community.

© Seth Resnick

Figure 27-1 Individuals matter: Marion Stoddart canoeing on the Nashua River near Groton, Massachusetts. She spent more than two decades spearheading successful efforts to have this river cleaned up.

Politics is the art of making good decisions on insufficient evidence.

LORD KENNET

This chapter discusses how we can use politics to promote environmental quality and sustainability. It addresses the following questions:

- What major environmental and political challenges do we face in this century?

- How do democracies work, and what factors hinder the ability of democracies to deal with environmental problems?

- How do we influence, develop, and implement environmental policy?

- What is the role of environmental law in dealing with environmental problems?

- What are the major types and roles of environmental groups and their opponents?

- What types of global environmental policies and treaties exist, and how might they be improved?

27-1 ENVIRONMENTAL AND POLITICAL CHALLENGES FOR THIS CENTURY

What Changes in Environmental Awareness and Focus Have Taken Place: Some Major Shifts

There have been seven shifts in the way we view and deal with environmental problems.

This chapter examines the strengths and weaknesses of political systems in dealing environmental problems. Before doing this we need to understand the nature of the environmental problems we face in terms of nations and the international community of nations.

Since the 1970s there have been seven shifts in the types and focus of the environmental problems we face. One is increasing concern about *the harmful effects of human activities on biodiversity and other forms of natural capital* that support all life and economies. This is leading to increased emphasis on protecting and restoring entire ecosystems instead of focusing primarily on keeping individual species from becoming prematurely extinct. To be effective this will require international cooperative efforts.

A second is a *shift from local to regional and global concerns* about emissions of air and water pollutants that can be transported from one region or country to another. One example is a significant increase in emissions of sulfur and nitrogen compounds, especially in Asia, that can blanket large regions with smog and harmful and acid forming chemicals. Another example is rising levels of carbon dioxide in the atmosphere from burning fossil fuels and clearing forests that can affect regional and global climate patterns. A third example is rising levels of nitrogen compounds in the atmosphere and aquatic systems because of emissions of gaseous nitrogen compounds by power plants and motor vehicles and rapidly growing runoff of nitrogen fertilizers from cropland and urban land into rivers, lakes, and coastal waters.

A third shift involves *growing concern over the threat of climate change and its potential to disrupt ecological, economic, and political systems.* Many analysts consider this threat and the related problem of biodiversity loss to be the two most important environmental problems we face.

A four shift is a *growing awareness of the pollution problems of developing countries*—especially those in the heavily populated urban areas of China, India, Mexico, and Brazil—that are undergoing rapid industrialization and economic growth. Many people in these countries now have some of the world's highest levels of exposure to tiny particles in indoor and outdoor air, lead (mostly from burning leaded gasoline), and exposure to infectious organisms in drinking water. A fifth and related shift is a *growing awareness of the harmful effects of poverty on the environment and human health* (p. 13).

Sixth is *increasing concern about possible effects of trace amounts of some synthetic organic (carbon-based) chemicals on human health and wildlife.* Examples are pesticides, plastics, industrial chemicals, drugs, and food additives. We know little about the potentially harmful environmental and health effects of trace amounts of such chemicals.

So far our approach has been to assume that such chemicals are innocent until shown to be harmful. Now there is a shift, led by the European Union, to have nations and the international community assume that such chemicals are potentially harmful until shown to be harmless. This leads to increased emphasis on preventing pollutants from reaching the environment instead of trying to clean them up after they have been dispersed into the environment.

The seventh shift involves *relying more on the international community to deal with environmental problems in an increasingly globalized world and economy.* So far we have not been very effective in bringing about this important shift.

In the 1970s the United States led the world in recognizing and dealing with local and national environmental problems. However, since then it has given up a leadership position in dealing with the increasingly urgent regional and global environmental problems we face. Environmental leadership is now coming mostly from the nations of the European Union. However, major nations such as the United States, China, and India will have to assume leadership positions for global environmental efforts to succeed. This will require cooperative efforts among government, business, and environmental leaders and bottom-up political pressure from individual citizens and environmental organizations.

27-2 DEALING WITH ENVIRONMENTAL PROBLEMS IN DEMOCRACIES

What Is a Democracy, and How Do Democratic Governments Work? Government By and For the People

In a democracy people elect others to govern, and can freely express their opinions and beliefs.

Democracy is government by the people through elected officials and representatives. In a *constitutional democracy,* a constitution provides the basis of government authority, limits government power by mandating free elections, and guarantees free speech.

Political institutions in constitutional democracies are designed to allow gradual change to ensure economic and political stability. In the United States, for example, rapid and destabilizing change is curbed by a system of checks and balances that distributes power among the three branches of government—*legislative, executive,* and *judicial*—and among federal, state, and local governments.

In passing laws, developing budgets, and formulating regulations, elected and appointed government officials must deal with pressure from many competing *special-interest groups.* Each group advocates passing laws, providing subsidies or tax breaks, or establishing regulations favorable to its cause and weakening or repealing laws, subsidies, taxes, and regulations unfavorable to its position.

Some special-interest groups, such as corporations, are *profit-making organizations,* and others are nonprofit *nongovernmental organizations (NGOs).* Examples of NGOs are labor unions and major and grassroots environmental organizations.

What Factors Hinder the Ability of Democracies to Deal with Environmental Problems? A Short-Term Outlook

Democracies are designed to deal mostly with short-term, isolated problems.

The deliberate design of democracies to promote stability is highly desirable. But several related features of democratic governments hinder their ability to deal with environmental problems. One is a tendency to react to short-term, isolated environmental problems, instead of regarding them as parts of a whole and acting to prevent them from occurring. Many important environmental problems such as climate change, biodiver-

sity loss, and long-lived hazardous waste have long-range effects, are related to one another, and require integrated long-term solutions emphasizing prevention.

But because elections are held every few years, most politicians seeking re-election are compelled to focus on short-term, isolated problems rather than on complex, interrelated, time-consuming, and long-term problems.

Most politicians will no longer be in office when harmful long-term effects from environmental problems appear. And there is no powerful political constituency representing future generations or long-term environmental sustainability.

Another problem is that most elected officials must spend much of their time raising money to get reelected—leaving them too little time for dealing with real issues. Finally, too many political leaders do not understand how the earth's natural systems work and how they support all life, economies, and societies. This lack of ecological literacy is dangerous in these times when we are moving through treacherous ecological waters at an increasing speed.

27-3 DEVELOPING, INFLUENCING, AND IMPLEMENTING ENVIRONMENTAL POLICY

What Principles Can Guide Us in Making Environmental Policy Decisions? Principles Are Important

Several principles can guide us in making environmental decisions.

An **environmental policy** consists of laws, rules, and regulations related to an environmental problem that are developed, implemented, and enforced by a particular government agency. Analysts have suggested that legislators and individuals evaluating existing or proposed environmental policy should be guided by several principles:

- *The humility principle*: Our understanding of nature and of the consequences of our actions is quite limited.

- *The reversibility principle*: Try not to do something that cannot be reversed later if the decision turns out to be wrong. For example, most biologists believe the current large-scale destruction and degradation of forests, wetlands, wild species, and other components of the earth's biodiversity is irreversible on a human time scale.

- *The precautionary principle*: When much evidence indicates that an activity threatens human health or the environment, take measures to prevent or reduce harm, even if some of the cause-and-effect relationships are not fully established scientifically. In such cases, it is better to be safe than sorry.

- *The prevention principle*: Whenever possible, make decisions that help prevent a problem from occurring or becoming worse.

- *The polluter pays principle*: Develop regulations and use economic tools such as full-cost pricing to insure that polluters bear the cost of the pollutants and wastes they produce.

- *The integrative principle*: Make decisions that involve integrated solutions to environmental and other problems.

- *The public participation principle*: Citizens should have open access to environmental data and information and the right to participate in developing, criticizing, and modifying environmental policies.

- *The human rights principle*: All people have a right to an environment that does not harm their health and well-being.

- *The environmental justice principle:* Establish environmental policy so that no group of people bears an unfair share of the harmful environmental risks from industrial, municipal, and commercial operations or from the execution of environmental laws, regulations, and policies. Environmental justice, as discussed in Chapter 26, means that every person is entitled to protection from environmental hazards regardless of race, gender, age, national origin, income, social class, or any other factor. See the Guest Essay on this subject by Robert D. Bullard in the website for this chapter.

How Can Individuals Affect Environmental Policy? All Politics Is Local

Most improvements in environmental quality are the result of millions of citizens putting pressure on elected officials and of individuals developing innovative solutions to environmental problems.

A major theme of this book is that *individuals matter*. History shows that significant social change usually comes from the *bottom up* when individuals join with others to bring about change. Without grassroots political action by millions of individual citizens and organized groups, the air you breathe and the water you drink today would be much more polluted, and much more of the earth's biodiversity would have disappeared. Figure 27-2 lists ways you can influence and change government policies in constitutional democracies. Which, if any, of these things do you do?

In developed countries many people devote much of their lives to acquiring more things instead of participating in building more just and sustainable communities. In the United States, only a small percentage of people vote or participate in political campaigns and elections.

What Can You Do?

Influencing Environmental Policy

- Become informed on issues

- Run for office (especially at local level)

- Make your views known at public hearings

- Make your views known to elected representatives

- Contribute money and time to candidates for office

- Vote

- Form or join nongovernment organizations (NGOs) seeking change

- Support reform of election campaign financing

Figure 27-2 What can you do? Ways you can influence environmental policy.

In addition, an increasing number of Americans and citizens of some other countries are too busy, tired, distracted, or cynical to participate in helping make their communities better places to live. Instead of becoming involved in their communities and helping develop environmental policies, many believe they can perform their environmental duty by recycling, buying environmentally friendly products, planting a tree, composting, eating organic food, buying shade-grown coffee, driving a fuel-efficient vehicle, and perhaps mailing a check to an environmental organization.

These are important and responsible activities that help the environment But in order to influence environmental policy people need to actively work together to improve communities and neighborhoods, as the citizens of Chattanooga, Tennessee (p. 581) and Curitiba, Brazil (p. 563) have done These and other cases have demonstrated the validity of the insight of Aldo Leopold: "All ethics rest upon a single premise: that the individual is a member of a community of interdependent parts." Thus what we do at the local level has global implications—much like dropping a pebble in a lake and watching the resulting ripples spread outward. This is the meaning of the slogan, "Think globally and act locally."

Case Study: What Is Environmental Leadership? An Option for Each of Us

Environmental leaders provide vision, focus, resources, and other types of support to people who want to make changes to environmental policies, and each of us can play a leadership role.

Leaders are persons whom other people choose to follow because of their vision, credibility, courage, or charisma. Good leaders help people to focus their energy, to set goals, and to pursue those goals efficiently. And leaders often provide the resources and moral support that people need to keep going when their goals seem to be slipping away. So if a group wants to influence environmental policy, an energetic leader is indispensable.

History judges political leaders by whether and how they respond to the great issues of their day. Today's political leaders face climate change, loss of biodiversity, poverty, and other related environmental problems.

One such politician, Prime Minister Tony Blair of the United Kingdom, believes that environmental degradation is the key issue for this generation and that "climate change is unquestionably the most urgent environmental challenge." He calls for the United Kingdom and other nations of the world to reduce carbon dioxide emissions by 60% by 2050 and for a "new international consensus to protect the environment and combat the devastating impacts of climate change."

Solutions to global environmental problems will require leadership from the United States, the world's wealthiest and most powerful society. However, instead of leading us toward solutions, the majority of elected U.S. officials are leading a charge in the opposite direction. They are weakening environmental laws, withdrawing from international efforts to deal with climate change, weakening national and international biodiversity protection, lowering nonmilitary foreign aid for fighting poverty, and encouraging the use of fossil fuels, instead of increasing energy conservation and relying much more on renewable energy resources.

According to a 2004 article by environmental writer Bill McKibben, "It is odd for American environmentalists to . . . realize that we no longer play a leading role of any kind. If you spend much time at international conferences, you see that we are no more the center of gravity, the fount of new ideas. Long before President Bush ditched the Kyoto treaty, we were drifting toward the back of the pack."

Since 1980, the United States has been mired in the politics of confrontation, polarization, and deadlock on environmental and many other major issues. There are many reasons for this situation, but many analysts say that one is a lack of vision and leadership.

The United States is also losing out as a leader in the development and sales of key environmental technologies. According to a 2004 column by Thomas L. Friedman, the United States is beginning to lose its competitive and innovative edge in science and technology to Japan and several western European nations and possibly in the near future to the rapidly developing countries of China and India. He points out that "the percentage of Americans graduating with bachelor's degrees in science and engineering is less than half of the comparable percentage in China and Japan, and that U.S. government investments are lagging in physics, chemistry, and engineering."

Before 1980, the United States was on the cutting edge of developing promising environmental technologies such as wind turbines, solar cells, and fuel-efficient cars. This began changing when in 1980 Congress gutted research and development support for these and other promising environmental technologies.

Now Denmark has taken over as the leader in developing and selling wind turbines and Germany and Japan have become leaders in the development of solar cells. Japan, led by Toyota Motor Company, has become the global leader in the development of hybrid motor vehicles. In 2004, investment analysts warned that because of a lack of sufficient government R & D and tax breaks, the United States may loose the race to dominate the development and sales of hydrogen-powered fuel cells. Canada, Japan, and the European Union are each devoting more money on fuel-cell and hydrogen research than the United States. Go to parts of western Europe and you will see many examples of green architecture and increasingly green and livable cities and downtowns—ideas that are just beginning to receive some attention in the United States. As the saying goes, "you snooze, you lose."

According to some critics, much of this failure of vision and leadership on the part of the United States is the result of the immense political power of the country's coal, oil, nuclear power, mining, and automobile industries. For decades these mature and profitable industries have received huge taxpayer-financed government subsidies and tax breaks and have succeeded in preventing other more environmentally sustainable businesses from receiving similar levels of government support (Figure 18-34, p. 407). In economics and politics, you get more of what you reward.

Some scientists warn of a dangerous trend in the United States involving suppressing, discrediting, or altering scientific facts and falsely labeling ideas accepted as sound science as junk science. On February 18, 2004, more than 60 scientists, including several Nobel laureates, released a statement accusing the Bush administration of deliberately distorting scientific fact "for partisan political ends." These and other scientists warn that the integrity of the government scientific advisory process is being undermined by suppressing studies not favorable to political goals, replacing scientific advisory committees with members more partial to industry positions, selectively gutting research budgets for studies that might produce results undermining political goals, and firing or transferring government scientists who speak out or release scientific information unfavorable to the administration's political goals. Administrative officials deny such charges.

The way out of this quagmire, some analysts say, is for individuals to work together to elect ecologically literate officials who will provide environmental leadership for the United States and the world. Thus leadership at the grassroots level is perhaps the key to solving a leadership crisis at higher levels.

Each of us can provide leadership on environmental or other issues in three ways. One is to *lead by example*, using our own lifestyles and beliefs to demonstrate that change is possible and beneficial.

A second approach is to *work within existing economic and political systems to bring about environmental improvement*. We can influence political decisions by campaigning and voting for candidates and by communicating with elected officials. We can also send a message to companies making harmful environment products or policies by *voting with our wallets* and letting them know what we have done. You would be surprised at how few consumer complaints it takes for a company to change its ways because of a fear of having a bad public image. We can also work within the system by choosing environmental careers (Individuals Matter, below).

A third form of environmental leadership is to *run for a local office*. Look in the mirror. Maybe you are one who can make a difference as an office holder.

A fourth form involves *proposing and working for better solutions to environmental problems*. Leadership is more than being against something. It also involves coming up with alternatives and getting people to work together to achieve them, especially in today's often-hostile political climate.

Solutions: Election Finance Reform in the United States: Taking the Government Back

Drastic reform of the election financing system can reduce the influence of special-interest money in elections in the United States.

It should come as no surprise that in the U.S. political system (and those in most other democratic countries), people with money and power have greatest influence in deciding who gets elected, what causes they support after being elected, and whether they get reelected.

Most analysts and about 80% of citizens polled agree that the U.S. political system is based more on money than on citizens' votes and concerns. This is one reason that voter turnout for federal elections in the United States dropped from 74% in 1900 to 49% in 2000. And many people who do vote often feel they are simply choosing the lesser of two evils.

According to social analyst Paul H. Ray, survey upon survey shows that over 70% of U.S. voters are unhappy with the country's system for financing elections. This explains why a growing number of people of all political persuasions see *drastic election finance reform* as the single most important way to reduce the influence of special-interest money in local, state, and federal elections in the United States. They urge American citizens to focus their efforts on this crucial

Environmental Careers

INDIVIDUALS MATTER

In the United States and other developed countries, the *green job market* is one of the fastest-growing segments of the economy.

Many employers are actively seeking environmentally educated graduates. They are especially interested in people with scientific and engineering backgrounds and double majors (business and ecology, for example) or double minors. Other possibilities are majors in business administration and environmental law.

Throughout this book I have exposed you to various career possibilities in this exciting and challenging field. They include environmental engineering, sustainable forestry and range management, parks and recreation management, air and water quality control, solid waste and hazardous waste management, recycling, urban and rural land-use planning, computer modeling, ecological restoration, and soil, water, fishery, and wildlife conservation and management.

Environmental careers can also be found in education, environmental planning, environmental management, environmental health, toxicology, geology, ecology, conservation biology, chemistry, climatology, population dynamics and regulation (demography), law, risk analysis, risk management, accounting, environmental journalism, design and architecture, energy conservation and analysis, renewable-energy technologies, hydrology, consulting, public relations, activism and lobbying, economics, diplomacy, development and marketing, publishing (environmental magazines and books), and teaching and law enforcement (pollution detection and enforcement teams).

Critical Thinking

Have you considered an environmental career? Why or why not?

issue as the key to making government more responsive to ordinary people on environmental and other matters.

One suggestion for reducing the excessive influence of powerful special interests is to let the people (taxpayers) *alone* finance election campaigns, with low spending limits. Candidates could use their own money, but could not accept direct or indirect donations from any other individuals, groups, or parties.

With such a reform, elected officials could spend their time governing instead of raising money and catering to powerful special interests. Office seekers would not need to be wealthy. Special-interest groups would be heard because of the validity of their ideas, not the size of their pocketbooks.

Some steps have been taken to reform election financing. Thanks to the work of many citizens, candidates running for office in four states—Arizona, Maine, Massachusetts, and Vermont—have the option of rejecting all private campaign contributions and qualifying for full public financing of their campaigns within certain spending limits.

X *How Would You Vote?* Do you support financing federal, state, and local election campaigns with taxpayer funds only? Cast your vote online at http://biology.brookscole.com/miller14.

How Can Organizations Change to Foster Better Policy Making? Be Nimble, Flexible, and Adaptive in a Rapidly World

To achieve more sustainable environmental policies, most organizations will be more effective if they shift from hierarchical to network models.

Businesses, governments, and all organizations are entering a new era in which to thrive and survive, most must shift from rigid, slow-acting, top-down hierarchical organizations to something more flexible that can adapt quickly to changing conditions. This new organizational model is that of a *network* instead of that of a *hierarchy.*

In a hierarchy consisting of a pyramid of increasingly powerful layers of decision makers, information takes a long route to the top, and decisions take time to filter down. People in such organizations are used more as information transmitters than as innovators of new ideas. The result is a lack of good information at the top, a rigid set of controls and rules, and too little innovation. In today's rapidly changing and increasingly globalized and interconnected world, such hierarchical structures are rapidly going the way of dinosaurs.

Hierarchies are necessary for the stable functioning of some parts of governments and for some businesses. But many organizations are shifting to a flatter, leaner, and more adaptable network structure without as many middle- and senior managers. In such organizations, information flows rapidly to all members of the network, not unlike the way energy and matter flow through food webs in ecosystems. It is one way of achieving more organizational sustainability by copying nature. More open and democratic information flow makes it much easier to adapt to changing conditions, and it promotes cooperation and innovation. Some analysts say that the European Union is slowly emerging as an example of a new and more flexible network form of government.

In the network model, leaders still have a vital role. Their job is to develop vision, values, and objectives for their organizations. Then they must promote feedback from employees, encourage innovation and adaptation, and establish employee performance goals.

An important aspect of emerging network organizations is their use of *adaptive management* strategies (Figure 11-23, p. 218) to cope with new information and changing conditions, to learn from experience, and to modify plans quickly as needed. This approach uses the basic techniques of science (Figure 3-2, p. 33) and systems analysis (Figure 4-36, p. 85) to develop computer models for examining alternative plans and projecting possible outcomes or scenarios. The primary goal is to anticipate problems rather than simply react to them.

Plans should be flexible and easy to change in response to new information, unexpected developments, and changing conditions. Progress toward goals is regularly monitored and evaluated, and this information is used as feedback to adapt the plan as needed. Any group seeking to influence environmental or other policies will need to adopt such organizational structures and strategies.

Case Study: How Is Environmental Policy Made in the United States? A Complicated and Thorny Process

Formulating, legislating, and executing environmental policy in the United States is a complex, difficult, and controversial process.

The federal government consists of three separate but interconnected branches: legislative, executive, and judicial. The *legislative branch,* called the Congress and composed of the House of Representatives and the Senate, has two main duties. One is to approve and oversee government policy by passing laws that establish a government agency or instruct an existing agency to take on new tasks or programs. The other is to oversee the functioning and funding of various agencies of the executive branch concerned with carrying out government policies.

The *executive branch* consists of the president, and a staff that together oversee the various agencies authorized by Congress to carry out government policies.

Major agencies responsible for environmental policy are listed in Figure 27-3. The president also proposes annual budgets, legislation, and appointees for executive positions, which must be approved by Congress, and tries to persuade Congress and the public to support his or her policy proposals.

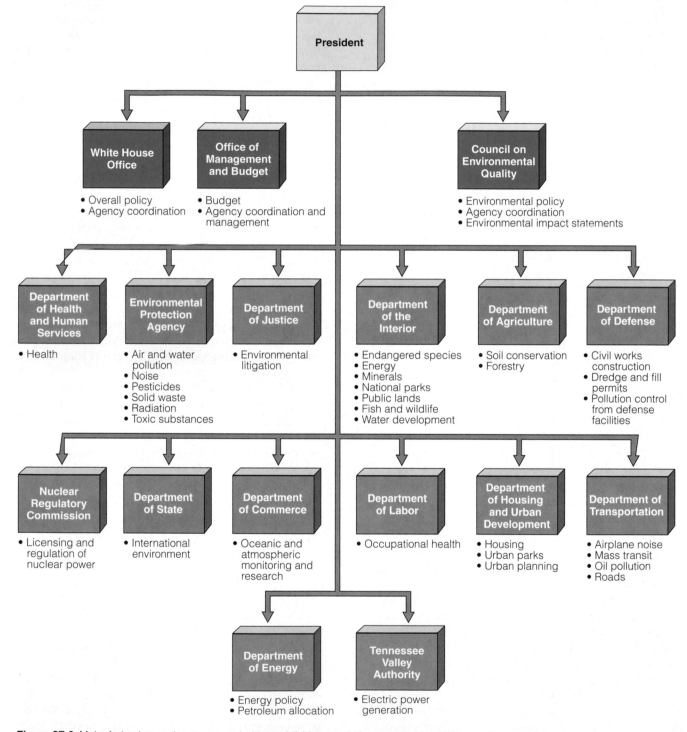

Figure 27-3 Major federal agencies concerned with establishing regulations and implementing environmental laws in the United States. Such agencies are established by Congress but are run by the president as part of the executive branch of government. This diagram shows only the environmental responsibilities of these agencies. Many have a broad range of other responsibilities.

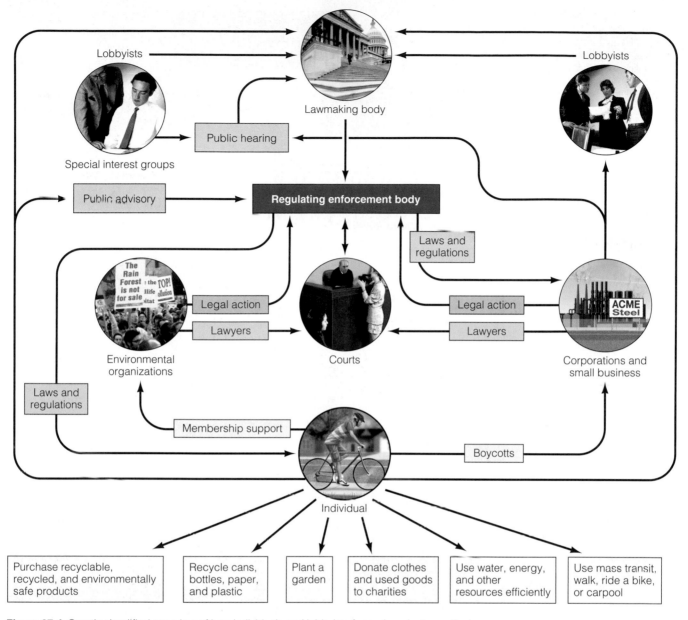

Figure 27-4 Greatly simplified overview of how individuals and lobbyists for and against a particular environmental law interact with the legislative, executive, and judicial branches of government in the United States. The bottom of this diagram also shows some ways in which individuals can bring about environmental change through their own lifestyles. See the website for this book for details on contacting elected representatives.

The *judicial branch* consists of a complex and layered series of courts at the local, state, and federal levels. These courts enforce and interpret different laws passed by legislative bodies.

The major function of the federal government is to develop and implement *policy* for dealing with various issues. Policy is typically composed of *laws* passed by the legislative branch, *regulations* instituted by the executive branch to put laws into effect, and *funding* to implement and enforce the laws and regulations. Figure 27-4 is a greatly simplified overview of how individuals and lobbyists for and against a particular en-

vironmental law interact with the three branches of government in the United States. Trace the flows of information and feedback in this diagram.

Several steps are involved in establishing federal environmental policy (or any other policy). First, lawmakers must acknowledge that an environmental problem exists and that the government has a responsibility to address it. Next, an interested party (such as a citizen, a group, a legislator, or the president) creates a bill hoping to pass it into law to deal with the problem. Converting a bill into a law is a complex process that you can trace in Figure 27-5 (p. 614). An important

Figure 27-5 How a bill introduced into the U.S. House of Representatives becomes a law. Individual citizens and lobbying groups (Figure 27-4) can influence how the bill is written before it is introduced and what happens to it at every stage of this complex process. Once a bill is signed into law, it goes to appropriations committees in both houses for agreement on how much funding it will receive. Without adequate funding, a law cannot be implemented. Continued intervention by individuals and lobbying groups can be very important at this stage.

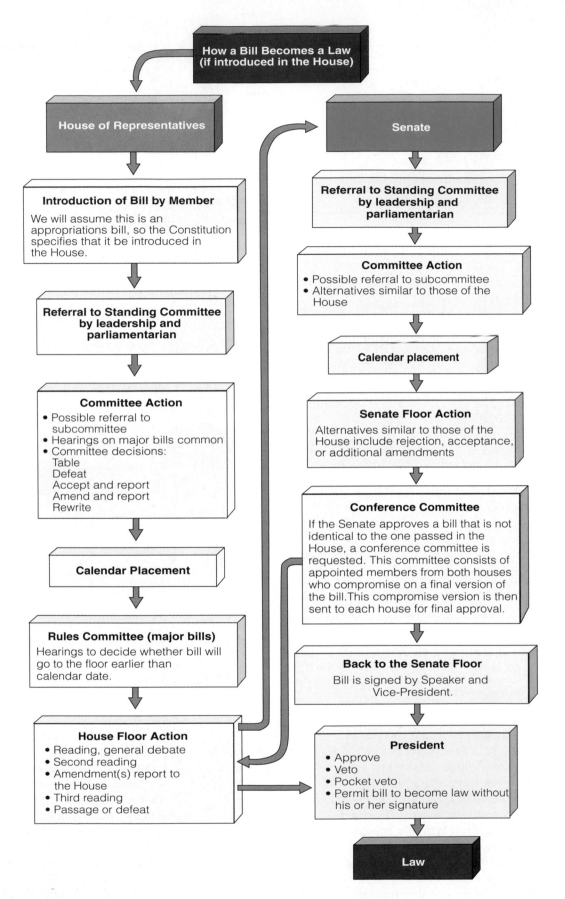

How a Bill Becomes a Law (if introduced in the House)

House of Representatives

Introduction of Bill by Member
We will assume this is an appropriations bill, so the Constitution specifies that it be introduced in the House.

Referral to Standing Committee by leadership and parliamentarian

Committee Action
• Possible referral to subcommittee
• Hearings on major bills common
• Committee decisions:
  Table
  Defeat
  Accept and report
  Amend and report
  Rewrite

Calendar Placement

Rules Committee (major bills)
Hearings to decide whether bill will go to the floor earlier than calendar date.

House Floor Action
• Reading, general debate
• Second reading
• Amendment(s) report to the House
• Third reading
• Passage or defeat

Senate

Referral to Standing Committee by leadership and parliamentarian

Committee Action
• Possible referral to subcommittee
• Alternatives similar to those of the House

Calendar placement

Senate Floor Action
Alternatives similar to those of the House include rejection, acceptance, or additional amendments

Conference Committee
If the Senate approves a bill that is not identical to the one passed in the House, a conference committee is requested. This committee consists of appointed members from both houses who compromise on a final version of the bill. This compromise version is then sent to each house for final approval.

Back to the Senate Floor
Bill is signed by Speaker and Vice-President.

President
• Approve
• Veto
• Pocket veto
• Permit bill to become law without his or her signature

Law

factor in this process is **lobbying** in which individuals or groups use public pressure, personal contacts, and political action to persuade legislators to vote or act in their favor. Some lobbyists are unpaid individuals who believe in a particular issue and others are paid professionals. The complex ballet of lobbyists competing for influence among legislators is a fascinating and important process.

Most environmental bills are evaluated by as many as 10 committees in the House of Representatives and the Senate. Effective proposals often are weakened by this fragmentation and by lobbying from groups opposing the law. Nonetheless, since the 1970s, a number of important environmental laws have been passed in the United States, as discussed throughout this text. Figure 27-6 lists some of the major environmental laws passed in the United States since 1969.

Passing a law is not enough to make policy. The next step involves trying to get enough funds appropriated to implement and enforce the law. Indeed, *developing and adopting a budget is the most important and controversial activity of the executive and legislative branches.*

Once a law has been passed and funded, the appropriate government department or agency must draw up regulations or rules for implementing it. Legislatures often give agencies considerable leeway for filling in the details on how a law will work. The resulting rules can put teeth into the law, or if drawn weakly, can render the law toothless.

An affected group may take the agency to court for failing to implement and enforce the regulations effectively, or for enforcing them too rigidly.

Politics plays an important role in the policies and staffing of environmental regulatory agencies—depending on what political party is in power and the prevailing environmental attitudes. Industries facing environmental regulations often put political pressure on regulatory agencies and lobby to have the president appoint people to high positions in such agencies who come from the industries being regulated. In other words, the regulated try to take over the agencies and become the regulators—described by some as "putting foxes in charge of the henhouse."

In addition, people in regulatory agencies work closely with and often develop friendships with officials in the industries they are regulating. Some industries offer regulatory agency employees high-paying jobs in an attempt to influence their regulatory decisions. This can lead to what is called a *revolving door,* as employees move back and forth between industry and government.

According to social scientists, the development of public policy in democracies often goes through a *pol-*

1969	National Environmental Policy Act (NEPA)
1970	Clean Air Act
1971	
1972	Clean Water Act; Coastal Zone Management Act; Federal Insecticide, Fungicide, and Rodenticide Act; Marine Mammal Protection Act
1973	Endangered Species Act
1974	Safe Drinking Water Act
1975	
1976	Resource Conservation and Recovery Act; Toxic Substances Control Act; National Forest Management Act
1977	Soil and Water Conservation Act; Clean Water Act; Clean Air Act Amendments
1978	National Energy Act
1979	
1980	Superfund (CERCLA); National Energy Act Amendments; Coastal Zone Management Act Amendments
1981	
1982	Endangered Species Act Amendments
1983	
1984	Hazardous and Solid Waste Amendment Act (SARA); Safe Drinking Water Act Amendments
1985	Endangered Species Act Amendments
1986	Superfund Amendments and Reauthorization
1987	Clean Water Act Amendments
1988	Federal Insecticide, Fungicide, and Rodenticide Act Amendments; Endangered Species Act Amendments
1989	
1990	Clean Air Act Amendments; Reauthorization of Superfund; Waste Reduction Act
1991	
1992	Energy Policy Act
1993	
1994	
1995	Endangered Species Act Amendments
1996	Safe Drinking Water Act Amendments

Figure 27-6 Some major environmental laws and their amended versions enacted in the United States since 1969. A more detailed list is found on the website for this chapter.

icy life cycle consisting of four stages: *recognition, formulation, implementation,* and *control.* Figure 27-7 (p. 616) illustrates this cycle and shows the general positions of several major environmental problems in the policy life cycle in the United States and most other developed countries. Carefully study this figure.

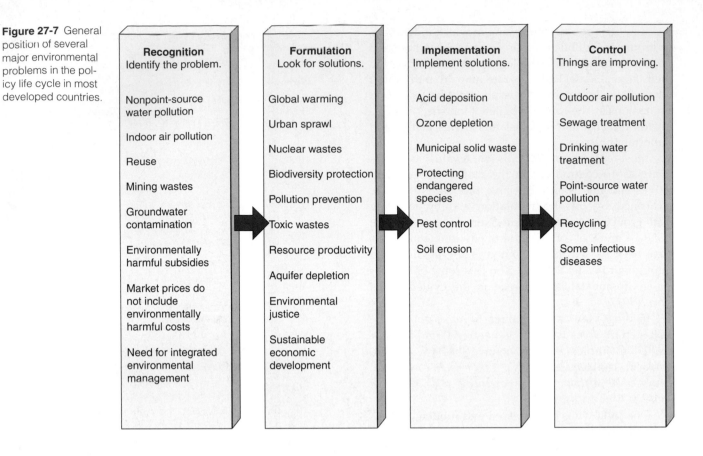

Figure 27-7 General position of several major environmental problems in the policy life cycle in most developed countries.

Recognition
Identify the problem.

Nonpoint-source water pollution

Indoor air pollution

Reuse

Mining wastes

Groundwater contamination

Environmentally harmful subsidies

Market prices do not include environmentally harmful costs

Need for integrated environmental management

Formulation
Look for solutions.

Global warming

Urban sprawl

Nuclear wastes

Biodiversity protection

Pollution prevention

Toxic wastes

Resource productivity

Aquifer depletion

Environmental justice

Sustainable economic development

Implementation
Implement solutions.

Acid deposition

Ozone depletion

Municipal solid waste

Protecting endangered species

Pest control

Soil erosion

Control
Things are improving.

Outdoor air pollution

Sewage treatment

Drinking water treatment

Point-source water pollution

Recycling

Some infectious diseases

27-4 ENVIRONMENTAL LAW

What Is Environmental Law and How Does It Evolve? A Mix of Legislation and Tradition

The body of environmental laws is constantly evolving through legislation and lawsuits.

Environmental law is a body of statements defining what is reasonable environmental behavior for individuals and groups, according to the larger community, and attempting to balance competing social and private interests. It includes statutory laws, administrative laws, and common laws.

Statutory laws are those developed and passed by legislative bodies such as federal and state governments. **Administrative laws** consist of administrative rules and regulations, executive orders, and enforcement decisions related to the implementation and interpretation of statutory laws. **Common law** is a body of unwritten rules and principles derived from thousands of past legal decisions along with commonly accepted practices, or *norms,* within a society. Most of it consists of *case law,* a body of legal opinions derived from past court decisions. The body of laws is continuously evolving, as almost every major environmental regulation is challenged in court.

Most environmental lawsuits are *civil suits*—those brought to settle disputes or damages between one party and another. Many common law cases are settled using the legal principle of *nuisance.* A nuisance occurs when people use their property in a way that causes annoyance or injury to others. For example, a homeowner may bring a nuisance suit against a nearby factory because of the noise it generates.

In such a civil suit, the **plaintiff,** the party bringing the charge, seeks to collect damages for injuries to health or for economic loss from the **defendant,** the party being charged. The plaintiff may also seek an *injunction,* by which the defendant would be required to stop whatever action is causing the harm. An individual or a clearly identified group may bring such a suit. A *class action suit* is a civil suit filed by a group, often a public interest or environmental group, on behalf of a larger number of citizens who allege similar damages but who need not be listed and represented individually.

Using the principles of common law, the court may side with the plaintiff if it finds that the loss of sleep, health problems, or other damage from the noise is greater than the cost of eliminating or reducing the noise. Short of closing the factory, often the court tries to find a reasonable or balanced solution to the problem. For example, it may order the factory to reduce noise to certain levels or to eliminate it during certain periods, such as at night.

Another principle used in common law cases is *negligence* in which a party causes damage by knowingly acting in an unlawful or unreasonable manner. For example, a company may be found negligent if it fails to handle hazardous waste in a way required by a statutory law. A court may also find a company negligent if it fails to do something a reasonable person would do, such as testing waste for certain harmful chemicals before dumping it into a sewer, landfill, or river. Generally, negligence is harder to prove than nuisance.

What Factors Hinder the Effectiveness of Environmental Lawsuits in the United States? Mostly Money and Time

Environmental lawsuits are expensive and difficult to win.

Several factors limit the effectiveness of environmental lawsuits. *First,* any person bringing the suit must establish that she or he has the legal right or *legal standing* to do so in a particular court. To have such a right, plaintiffs must show that they have personally suffered health or financial losses from some alleged environmental activity.

Second, bringing any lawsuit is expensive—too much so for most individuals. Large organizations, on the other hand, can often afford to defend themselves in court for months or years.

Third, public interest law firms cannot recover attorneys' fees unless Congress has specifically authorized it in the laws that those firms are seeking to have enforced. By contrast, corporations can reduce their taxes by deducting their legal expenses—in effect having the public pay for part of their legal fees. In other words, the legal playing field is uneven and in financial terms is stacked against individuals and groups of private citizens filing environmental lawsuits.

Fourth, to stop a nuisance or to collect damages from a nuisance or an act of negligence, plaintiffs must establish they have been harmed in some significant way and that the defendant caused the harm. Doing this can be difficult and costly. Suppose a company (the defendant) is charged with causing cancer in individuals by polluting a river. If hundreds of other industries and cities dump waste into that river, establishing that the defendant is the culprit is very difficult and requires expensive investigation, scientific research, and expert testimony. In addition, it is hard to establish that a particular chemical caused the plaintiffs' cancers.

Fifth, most states have *statues of limitations,* laws that limit how long a plaintiff can take to sue after a particular event occurs. These statutes often make it essentially impossible for victims of cancer, which may take 10–20 years to develop, to file or win a negligence suit.

Sixth, the court, or series of courts if the case is appealed, may take years to reach a decision. During that time a defendant may continue the allegedly damaging action unless the court issues a temporary injunction against it until the case is decided.

Finally, some corporations, developers, and government agencies file *strategic lawsuits against public participation (SLAPPs)* against citizens who publicly criticize a business for some activity such as polluting or a government agency for not performing its legal obligation to protect the public. SLAPPs range from $100,000 to $100 million but average $9 million per suit. For example, in Texas when a woman publicly called a nearby landfill a dump the landfill owners sued her husband for $5 million for failing to "control his wife." Judges who recognize them for what they are throw out about 90% of the SLAPPs that go to court. But individuals and groups hit with SLAPPs must hire lawyers, and typically spend 1–3 years defending themselves.

Most SLAPPs are not meant to be won, but are intended to intimidate individuals and activist groups—to keep them from exercising their democratic rights to criticize or oppose projects they believe are environmentally harmful. Once fear and rising defense costs shake the victim of a SLAPP, the defendant often drops the suit. Sometimes a company or government agency offers to drop the lawsuit if the defendants agree to stop their protest and never discuss the case or oppose the plaintiff again.

Some citizen activists have fought back with counter suits and have been awarded damages. For example, a Missouri woman who was sued for criticizing a medical waste incinerator won an $86.5 million judgment against the incinerator's owner.

Even after paying such awards, corporations and developers generally save money by filing such suits. Unlike the people they are suing, they can count legal and liability insurance costs as business expenses and write them off on their taxes. In other words, they get all taxpayers to pay much of the cost of lawsuits against a few taxpayers who are exercising their rights as citizens.

Because of the numerous difficulties just discussed, an increasing number of environmental lawsuits are being settled out of court. Some are settled privately and others by *mediation,* in which a neutral party tries to resolve the dispute in a way that is acceptable to both parties. Mediation is much less costly and time consuming and may provide a more satisfactory resolution of a dispute than going to court. But a settlement drawn up by mediation is not legally binding unless the terms of the agreement make it so. Thus months of mediation can result in an agreement that polluters may ignore.

Despite many obstacles, proponents of environmental law have accomplished a great deal since the 1960s. In the United States, more than 20,000 attorneys in 100 public interest law firms and groups specialize partly or entirely in environmental and consumer law. In addition, many other lawyers and scientific experts participate in environmental and consumer lawsuits as needed and sometimes without charge.

Analysts have suggested three major reforms to help level the legal playing field for citizens suffering environmental damage. *First,* pressure Congress to pass a law allowing juries and judges to award citizens their attorney fees, to be paid by the defendants, in successful lawsuits.

Second, establish rules and procedures for identifying frivolous SLAPP suits so that cases without factual or legal merit could be dismissed within a few weeks rather than years. *Third,* raise the fines for violators of environmental laws and punish more violators with jail sentences. Polls indicate that 84% of Americans consider damaging the environment to be a serious crime.

Case Study: What Are the Major Types of Environmental Laws in the United States? A Variety of Approaches

U.S. environmental laws set pollution standards, screen toxic substances, evaluate environmental impacts, encourage resource conservation, and protect various ecosystems and species from harm.

Concerned citizens have persuaded Congress to enact a number of important federal environmental and resource protection laws (Figure 27-6) that seek to protect environmental quality by using various approaches. One is to *set standards for pollution levels* (as in the Clean Air Acts and the Federal Water Pollution Control Act). Another is to *screen new substances for safety* (as in the Toxic Substances Control Act).

A third type of legislation *encourages resource conservation* (the Resource Conservation and Recovery Act and the National Energy Act). A fourth type *sets aside or protects various ecosystems, resources, and species* (the Endangered Species Act and the Wilderness Act).

A fifth approach is to *require evaluation of the environmental impact of an activity proposed by a federal agency,* as in the *National Environmental Policy Act* or *NEPA* passed in 1970. Under NEPA, an *environmental impact statement* *(EIS)* must be developed for every major federal project likely to have an important effect on environmental quality. The EIS must describe why the proposed project is needed, its short-term and long-term beneficial and harmful environmental impacts, ways to lessen harmful impacts, and an evaluation of alternatives. An EIS typically takes 6–9 months to develop and is often hundreds of pages long. The documents must be published and are open to public comment.

NEPA does not prohibit environmentally harmful government projects but it requires federal agencies to take environmental consequences into account in making decisions and exposes proposed projects and their likely harmful effects to public scrutiny. At least 36 U.S. states and a number of other countries—including Canada, Sweden, France, New Zealand, and Australia—have passed laws similar to NEPA.

Environmentalists have used EISs to block harmful projects or get them modified to reduce their environmental impacts. Many agree that NEPA has helped federal agencies to evaluate and reduce the harmful environmental impacts of their projects and activities.

Critics say EISs are costly and can unnecessarily delay projects by requiring too much analysis (paralysis by analysis). Proponents say analysis is needed to make government agencies think more seriously about the impact of proposed projects and to examine alternatives.

Opponents of environmentalists have targeted NEPA as a law that needs to be weakened or repealed. In 2003, the Bush administration began looking at ways to overhaul NEPA and asked Congress to exempt the Department of Defense (DOD) from having to file EISs in the interests of national security. Environmentalists oppose this because national security is involved in only a few of the wide range of potentially damaging projects undertaken by the DOD, and such projects can dealt with individually.

Some environmental laws and presidential executive orders contain glowing rhetoric about goals but little guidance on how to meet them, leaving this task to regulatory agencies and the courts. In other cases, these acts specify one or more of the following general principles for setting regulations. *First,* expose people to no unreasonable risk (food regulations in the Food, Drug, and Cosmetic Act). *Second,* expose people to little or no risk (the zero-discharge goals of the Safe Drinking Water and Clean Water Acts). *Third,* set standards based on best available technology (the Clean Air, Clean Water, and Safe Drinking Water Acts). *Fourth,* use cost–benefit analysis (the Toxic Substances Control Act). *Fifth,* make the polluter pay (the Superfund Law until recently)

27-5 ENVIRONMENTAL GROUPS AND THEIR OPPONENTS

What Are the Roles of Major Environmental Groups? Watchdogs and Agents of Change

Environmental groups monitor environmental activities, work to pass and strengthen environmental laws, and work with corporations to find solutions to environmental problems.

The spearhead of the global conservation and environmental movement consists of more than 100,000 nonprofit NGOs working at the international, national, state, and local levels—up from about 2,000 in 1970. The growing influence of these organizations is one of the most important changes influencing environmental decisions and policies.

NGOs range from grassroots groups with just a few members to global organizations like the 5-million-member World Wide Fund for Nature with offices in 48 countries. Other international groups with large memberships include Greenpeace, the World Wildlife Fund, the Nature Conservancy (Solutions, p. 216), Grameen Bank (Solutions, p. 601), and Conservation International.

Using e-mail and the Internet, environmental NGOs have organized themselves into an array of powerful international networks. Examples include the Pesticide Action, Climate Action, International Rivers, Women's Environment and Development, and Biodiversity Action Networks. They collaborate across borders, gathering environmental information, monitoring environmental change, and acting quickly. They monitor the environmental activities of governments, corporations, and international agencies such as the World Bank and the World Trade Organization (WTO). They expose corruption and violations of national and international environmental agreements, such as CITES, which prohibits international trade of endangered species. Groups such as Conservation International and the Nature Conservancy have brokered debt-for-nature swaps where developing countries agree to protect ecologically important areas in exchange for reduction of their international debt. NGOs are also watchdogs for environmental accountability at the local level.

In the United States, more than 8 million citizens belong to over 30,000 NGOs dealing with environmental issues. They range from small grassroots groups to large heavily funded groups, led by chief executive officers and staffed by expert lawyers, scientists, economists, lobbyists, and fund raisers. The 10 largest of these—sometimes called the "group of 10"—are the World Wildlife Fund, Sierra Club, National Wildlife Federation, Audubon Society, Greenpeace, Friends of the Earth, Natural Resources Defense Council, Wilderness Society, Ducks Unlimited, and Izaak Walton League. Many of these and other smaller environmental groups doubled and tripled their memberships between 1980 and 2000, mostly in response to increased fundraising activities and to attempts to weaken or repeal environmental laws and regulations. However, many members do not actively participate in the activities of such organizations and up to half do not renew their memberships.

The large groups have become powerful and important forces within the political system, by working individually and together in coalitions to help persuade Congress to pass and strengthen environmental laws and to fight off attempts to weaken or repeal them.

However, these large environmental groups must guard against being subverted by the political system they work to improve. This is a risk because these groups rely heavily on corporate donations, and many of them have corporate executives as board members, trustees, or council members.

The *good news* is that instead of acting as adversaries, some industries and environmental groups are working together to find solutions to environmental problems. For example, Environmental Defense has worked with McDonald's to redesign its packaging system to eliminate polyethylene foam clamshell hamburger containers. It has also worked with General Motors to help remove high-pollution cars from the road and with various multinational corporations to set targets for reducing their carbon dioxide emissions. The World Resources Institute is collaborating with leading businesses to build a market among corporations for electrical power produced from renewable energy resources.

Some environmental groups have shifted some resources from demonstrating and litigating to publicizing research on innovative solutions to environmental problems. For example, to promote the use of chlorine-free paper, Greenpeace Germany printed a magazine using such paper and encouraged readers to demand that magazine publishers switch to chlorine-free paper. Shortly thereafter, several major magazines made that shift.

What Are the Roles of Grassroots Environmental Groups? Citizen Action from the Bottom Up

Thousands of citizens' groups working to improve environmental quality form the base of the global environmental movement.

The base of the environmental movement in the United States and throughout the world consists of thousands of grassroots citizens' groups organized to improve environmental quality, often at the local level. According to political analyst Konrad von Moltke, "There isn't a government in the world that would have done anything for the environment if it weren't for the citizen groups."

These groups carry out a number of environmental roles. One is to work with individuals and communities to oppose harmful projects such as landfills, waste incinerators, nuclear waste dumps, clear-cutting of forests, and various development projects. They have also pressured government officials to take action when group members have been victims of environmental harm or of environmental injustice because of the unequal distribution of environmental risks. See the Guest Essay on environmental justice by Robert D. Bullard on the website for this chapter.

Grassroots groups have also formed land trusts and other local organizations to save wetlands, forests, farmland, and ranchland from development. They have helped restore degraded rivers and wetlands, and have converted abandoned urban lots into community gardens and parks. Some groups are coalitions of workers and environmentalists who aim to improve worker safety and health.

International examples of grassroots NGOs are Kenya's Green Belt Movement (Individuals Matter, p. 214) in which citizens plant trees on public and private land; India's long-standing Chipko movement where villagers protect trees by hugging them and thus placing themselves between the trees and axes and chainsaws; and Sri Lanka's Sarvodaya Shramadana movement, which has developed wells for drinking water, gardening, and other small-scale improvement projects in 12,000 villages.

Taken together, a loosely connected network of grassroots NGOs working for bottom-up political, social, economic, and environmental change can be viewed as an emerging citizen-based *global sustainability movement*. These millions of citizens are becoming informed and empowered by access to the World Wide Web, cell phones, e-mail, faxes, GIS mapping programs, and other components of the global communications web.

As Jeremy Rifkin puts it, "We are rapidly moving from geopolitics to biosphere politics." According to Rifkin, the Internet, coupled with our better understanding of how the earth sustains itself, will allow us, for the first time in human history, to really think globally and act locally.

The late John W. Gardner, former cabinet official and founder of Common Cause, suggested using the following basic rules for effective political action by grassroots organizations:

- Have a full-time continuing organization.

- Limit the number of targets and hit them hard. Groups dilute their effectiveness by taking on too many issues.

- Organize for action, not just for study, discussion, or education.

- Form alliances with other organizations on a particular issue.

- Communicate positions in an accurate, concise, and moving way.

- Persuade and use positive reinforcement.

- Concentrate efforts mostly at the state and local levels.

Some grassroots environmental groups use nonviolent and nondestructive tactics of protest marches, pickets, road blocks, tree sitting (Individuals Matter, p. 206), confronting illegal whaling ships, street theater, and other devices for generating publicity to help educate and sway members of the public to their causes. Many of these tactics are borrowed from Mahatma Gandhi's successful nonviolent civil disobedience strategy in helping win India's independence from Great Britain and the U.S. civil rights movement. Some find the tactics of these groups controversial while others admire them for standing up for their beliefs in nonviolent ways.

X *HOW WOULD YOU VOTE?* Do you support the use of nonviolent and nondestructive civil disobedience tactics by environmental groups and individuals? Cast your vote online at http://biology.brookscole.com/miller14.

Much more controversial are militant environmental groups that break into labs to free animals used to test drugs or that destroy property such as bulldozers and SUVs. Most environmentalists oppose such tactics because they involve illegal and destructive acts, give other environmentalists a bad name, and play into the hands of environmentalists' political opponents.

Case Study: Environmental Action by Students in the United States—Making a Difference

Many student environmental groups work to bring about environmental improvements in their schools and local communities.

Since 1988, there has been a boom in environmental awareness on a number of college campuses and public schools across the United States.* Most student environmental groups work with members of the faculty and administration to bring about environmental improvements in their schools and local communities.

Many of these groups make environmental audits of their campuses or schools.** Then they use the data gathered to propose changes that will make their campus or school more ecologically sustainable, usually saving money in the process.

---

*See *Ecodemia: Campus Environmental Stewardship at the Turn of the 21st Century* (Washington, D.C.: National Wildlife Federation, 1995) and the *Campus Environmental Yearbook*, published annually by the National Wildlife Federation.

**Details for conducting such audits are found in April Smith and the Student Environmental Action Coalition, *Campus Ecology: A Guide to Assessing Environmental Quality and Creating Strategies for Change* (Los Angeles: Living Planet Press, 1993), and Jane Heinze-Fry, *Green Lives, Green Campuses*, available free on the website for this textbook.

Such audits have resulted in numerous improvements. For example, Morris A. Pierce, a graduate student at the University of Rochester in New York, developed an energy management plan adopted by that school's board of trustees. Under this plan, a capital investment of $33 million is projected to save the university $60 million over 20 years. Students have also helped convince almost 80% of universities and colleges in the United States to develop recycling programs.

At Bowdoin College in Maine, chemistry professor Dana Mayo and student Caroline Foote developed the concept of *microscale experiments*, in which smaller amounts of chemicals are used. This has reduced toxic wastes. Today more than half of all undergraduates in chemistry in the United States use such microscale techniques, as do universities in a growing number of other countries. At Carnegie-Mellon University, students in introductory chemistry courses can carry out a number of experiments using a virtual or "no-spill" laboratory.

Students at Oberlin College in Ohio helped design a more sustainable environmental studies building. At Northland College in Wisconsin, students helped design a "green" dorm that features a large wind generator, panels of solar cells, recycled furniture, and waterless (composting) toilets.

At Minnesota's St. Olaf College, students have carried out sustainable agriculture and ecological restoration projects. Students at Brown University studied the impacts of lead and other toxic pollutants in low-income neighborhoods in nearby Providence, Rhode Island.

A 1997 report by the National Wildlife Federation's Campus Ecology Program found that 23 student-researched and student-motivated projects had saved the participating universities and colleges $16.3 million. According to this study, implementing similar programs in the nation's 3,700 universities and colleges could help improve environmental quality and environmental education and save more than $15 billion. Such student-spurred environmental activities and research studies are spreading to universities in at least 42 other countries. See Noel Perrin's Guest Essay on this topic on this chapter's website.

How Successful Have Environmental Groups and Their Opponents Been? Achievements and Setbacks

Environmental groups have helped educate the public and business and political leaders about environmental issues and pass environmental laws, but an organized movement has undermined many of these efforts.

Since 1970, a variety of environmental groups in the United States and other countries have helped increase understanding of environmental issues by the general public and some business and government leaders. They have also gained public support for an array of environmental and resource-use laws in the United States and other countries. In addition, they have helped individuals deal with a number of local environmental problems.

Polls show that more than 80% of the U.S. public strongly support environmental laws and regulations and do not want them weakened. But polls also show that less than 10% of the U.S. public views the environment as one of the nation's most pressing problems. As a result, environmental concerns often do not get transferred to the ballot box. As one political scientist put it, "Environmental concerns are like the Florida Everglades, a mile wide but only a few inches deep." And since 1980 a well-organized and well-funded movement has undermined much of the improvement in environmental understanding and support for environmental concerns in the United States.

One problem is that the focus of environmental issues has shifted from easy-to-see dirty smokestacks and burning rivers to more complex and controversial environmental problems that are less visible, harder to understand and solve, and that have long-range harmful effects. Examples are climate change, wasted energy, ozone depletion, biodiversity loss, nonpoint-source water pollution, and unseen groundwater pollution. Explaining such complex issues to the public and mobilizing support for often controversial, long-range solutions to such problems is difficult. See the Guest Essay on environmental reporting by Andrew C. Revkin on the website for this chapter.

Another problem is that many environmentalists have brought mostly bad news. History shows that bearers of bad news are not received well, and opponents of environmentalists have used this to undermine environmental concerns.

History also shows that people are moved to bring about change mostly by an inspiring, positive vision of what the world could be like, one that provides hope for the future. So far, environmentalists with a variety of beliefs and goals have not worked together to develop broad, compelling, and positive visions that can be used as road maps for a more sustainable future for humans and other species.

Instead of using confrontation, some people are working to mediate environmental disputes by getting each side to listen to one another's concerns, to try to find areas of agreement, and to work together to find solutions (Solutions, p. 622). This approach is being used successfully in places such as the Netherlands (p. 602), the Chesapeake Bay area (p. 505), the Great

How Can We Improve Environmental Laws and Regulations? Time for a Checkup

SOLUTIONS

Environmentalists agree that some government laws and regulations go too far and that bureaucrats sometimes develop and impose unfair and excessively costly regulations. They argue that the solution is to stop regulatory abuse, not to throw out or seriously weaken the body of laws and regulations that help protect the public good.

According to environmental economist William Ashworth,

Government regulation did not fall out of the sky; it was erected, piece-by-piece, as an attempt to deal with the damage caused by unrestrained property rights and the unregulated free-market system. . . . We do not need to deconstruct regulation, but to reconstruct it.

To accomplish this, a growing number of analysts urge environmentalists to take a hard look at existing environmental laws and regulations. Which laws or parts of laws have worked, and why? Which have failed, and why? Which government bureaucracies concerned with developing and enforcing environmental and resource regulations have abused their power or have not been responsive enough to the needs of ordinary people? How can such abuses be corrected? What existing environmental laws (or parts of such laws) and regulations should be repealed or modified?

What environmental problems lend themselves to market-based approaches (free-market environmentalism), and which ones do not? What roles should pollution prevention, waste reduction, and the precautionary principle play in environmental legislation and regulation?

These are important issues that environmentalists, business leaders, elected officials, and government regulators need to address with a cooperative, problem-solving spirit.

Critical Thinking

Identify an environmental law in the United States (or in the country where you live) that you believe needs to be improved. How would you improve it?

Lakes (p. 500), Chattanooga, Tennessee (p. 581), and Curitiba, Brazil (p. 563).

What Are the Goals of the Environmentalists' Opponents in the United States? Undermine, Weaken, and Crush

Since 1980, a political movement has attempted to discredit, weaken, and destroy the environmental movement in the United States.

Despite general public approval, there is strong opposition to many environmental proposals, laws, and regulations by three major groups. *First*, some corporate leaders, some corporations, and other powerful people see environmental laws and regulations as threats to their wealth and power. *Second*, some citizens see environmental laws and regulations as threats to their private property rights (p. 242) and jobs. *Third*, some state and local government officials resent having to implement federal environmental laws and regulations without federal funding (unfunded mandates) or disagree with certain regulations.

Since 1980, businesses, individuals, and some elected officials have mounted a strong campaign to weaken or repeal existing environmental laws and regulations, change the way in which public lands are used (p. 199), and destroy the reputation and effectiveness of the U.S. environmental movement.

Some groups seeking to weaken or do away with environmental laws and regulations have many members and large budgets. Examples are the National Farm Bureau and the Cattleman's Association.

Some of these groups are genuine grassroots organizations, while others are primarily lobbying groups for various industries. People for the West, for example, is an organization claiming to represent ordinary rural people. But almost all of its budget and 12 of its 13 board members come from mining and timber companies. Sometimes it takes careful research to determine who is behind a group with an environmentally friendly name.

Because of the efforts of their opponents, major environmental groups in the United States have spent most of their time and money since 1980 trying to prevent existing environmental laws and regulations from being weakened or repealed.

27-6 GLOBAL ENVIRONMENTAL POLICY

Solutions: Should We Expand the Concept of National and Global Security? The Big Three

Many analysts believe that environmental security is as important as military and economic security.

Countries are legitimately concerned with *military security* and *economic security*. However, ecologists point out that all economies are supported by the earth's natural capital. According to environmental expert Norman Myers,

If a nation's environmental foundations are degraded or depleted, its economy may well decline, its social fabric deteriorate, and its political structure become destabilized as growing numbers of people seek to sustain themselves from declining resource stocks. Thus, national security is no longer about fighting forces and weaponry alone. It relates increasingly to watersheds, croplands, forests, genetic resources, climate, and other factors that, taken together, are as crucial to a nation's security as are military factors.

Proponents of this view call for all countries to make environmental security a major focus of diplomacy and government policy at all levels. This would be implemented by having a council of advisers made up of highly qualified experts in environmental, economic, and military security who integrate all three security concerns in making major decisions.

X *How Would You Vote?* Do you believe that environmental security is just as important as economic and military security? Cast your vote online at http://biology.brookscole.com/miller14.

What Is the Role of International Environmental Organizations? Major Players

International environmental organizations gather and evaluate environmental data, help develop environmental treaties, and provide funds and loans for sustainable economic development.

A symphony of international environmental organizations helps shape and set environmental policy. Perhaps the most influential is the United Nations with a large family of organizations such as the UN Environment Programme (UNEP), the World Health Organization (WHO), United Nations Development Programme (UNDP), and the Food and Agriculture Organization (FAO).

Other organizations that make or influence environmental decisions are the World Bank, the Global Environment Facility (GEF), and the World Conservation Union (IUCN). The website for this chapter has a list of major international environmental organizations.

These organizations have played important roles in

- Expanding understanding of environmental issues

- Gathering and evaluating environmental data

- Helping develop and monitor international environmental treaties

- Providing funds and loans for sustainable economic development and reducing poverty

- Helping more than 100 nations develop environmental laws and institutions

Despite their often limited funding, these diverse organizations have made important contributions to global and national environmental progress since 1970.

Since the 1972 UN Conference on the Human Environment in Stockholm, Sweden, progress has been made in addressing environmental issues at the global level. Political scientist Keith Caldwell credits the Stockholm conference with forcing many governments to develop domestic environmental programs and legitimizing not harming the biosphere as an object of national and international concern and policy. It also created the United Nations Environment Programme (UNEP) to help develop the global environmental agenda. But the UNEP is a small and underfunded agency. Figure 27-8 (p. 624) lists some of the good and bad news about international efforts to deal with global environmental problems such as poverty, climate change, biodiversity loss, and ocean pollution. Carefully study this figure.

The primary focus of the international community on environmental problems has been the development of various international environmental laws and nonbinding policy declarations called *conventions*. Over 500 international environmental treaties and agreements—known as *multilateral environmental agreements* (MEAs)—have been developed, two-thirds of them signed in recent decades. The website for this chapter lists some of the major MEAs.

To date, the Montreal and Copenhagen Protocols for protecting the ozone layer are the most successful examples of how the global community can work together to deal with a serious global environmental challenge (p. 488). Figure 27-9 (p. 624) lists some major problems with MEAs and solutions to these problems.

During the past 30 years James Gustave (Gus) Speth has either created or run many of the world's most important environmental organizations, including the Natural Resources Defense Council, the White House Council on Environmental Quality, and the United Nations Development Program, and is now dean of Yale University's School of Forestry and Environmental Studies. In 1994, he wrote the book *Red Sky at Morning*, summarizing his insider view of global environmental efforts. He gives a failing grade to almost every effort.

Here are some of his observations. The United States "led the fight for national-level [environmental] policies in the 1970s. But with the exception of the Montreal and Copenhagen Protocols for protecting the ozone layer, the United States has largely failed to give international leadership on the global environmental agenda. . . . The world needs a United States that leads by example and diplomacy, with generosity and compassion."

"The current system of international efforts to help the environment simply isn't working. . . . The climate

Trade-Offs

Global Efforts on Environmental Problems

Good News	Bad News

Good News

Environmental protection agencies in 115 nations

Over 500 international environmental treaties and agreements

UN Environment Programme (UNEP) created in 1972 to negotiate and monitor international environmental treaties

1992 Rio Earth Summit adopted key principles for dealing with global environmental problems

2002 Johannesburg Earth Summit attempted to implement policies and goals of 1992 Rio summit and find ways to reduce poverty

Bad News

Most international environmental treaties lack criteria for monitoring and evaluating their effectiveness

1992 Rio Earth Summit led to nonbinding agreements without enough funding to implement them

By 2003 there was little improvement in the major environmental problems discussed at the 1992 Rio Earth Summit

2002 Johannesburg Earth Summit failed to provide adequate goals, deadlines, and funding for dealing with global environmental problems such as climate change, biodiversity loss, and poverty

Figure 27-8 Trade-offs: good and bad news about international efforts to deal with global environmental problems. Pick the single piece of good news and the single piece of bad news that you think are the most important.

Solutions

International Environmental Treaties

Problems	Solutions

Problems

Take a long time to develop and are weakened by requiring full consensus

Poorly monitored and enforced

Lack of funding for monitoring and enforcement

Treaties are not integrated with one another

Solutions

Do not require full consensus among regulating parties

Establish procedures for monitoring and enforcement

Increase funding for monitoring and enforcement

Harmonize or integrate existing agreements

Figure 27-9 Major problems with global environmental treaties and agreements and solutions to these problems.

convention is not protecting climate, the biodiversity convention is not protecting biodiversity, the desertification convention is not preventing desertification" and . . . "the Law of the Sea is not protecting fisheries. Nor are they poised to do so in the immediate future." And protection of the world's forest has not even "reached the point of a convention."

According to Speth, "Global environmental problems have gone from bad to worse and governments are not yet prepared to deal with them." However, he does cite some success in the protocols on protecting the ozone layer, the Convention on the Trade in Endangered Species (CITES), and the Ocean Dumping Convention.

Speth says that "In general, the issue with major treaties is not weak enforcement or weak compliance; the issue is weak treaties. These agreements are easy for governments to slight because their impressive goals are not followed by clear requirements, targets, and timetables . . . and an unwillingness to commit financial resources for real incentives." He says that "international law is still far too dominated by the outmoded concept that only governments get to play" instead of following a principle in the 1972 Rio convention that "environmental issues are best handled with the participation of all concerned citizens."

Speth and several other national and international environmental leaders call for establishing a World Environment Organization (WEO) to help develop and oversee global environmental policies. He says it is strange that we have World Trade Organization (WTO), World Health Organization (WHO), and a number of other similar organizations but no World Environmental Organization (WEO). He asks us to imagine what might have happened with global environmental policies if the developed nations "had put as much energy into a WEO as they have put into the WTO."

Case Study: Is Encouraging Global Free Trade Environmentally Helpful or Harmful? A Difficult Issue

There is concern that current rules for reducing global trade barriers may weaken national laws that protect the environment, consumers, and workers.

Like it or not, we are in an age of rapid economic, political, and social globalization. Expanding international trade is seen as a way to stimulate economies around the world, help distribute wealth, and decrease poverty.

According to the economic *comparative advantage* hypothesis, each country or place in a country often has an advantage over other areas in producing or selling one or more types of goods or services. These advantages can result from various factors such as a pool of cheap or technically skilled labor or availability of resources such as minerals, good soil, ample water, solar energy, or wind.

International free trade allows companies providing goods and services to take advantage of the cheapest labor and resources anywhere in the world. In theory, this should also lower the market prices of goods and services for consumers. For example, recently some major American companies have been using Indian workers as computer programmers because their hourly wage is about one-fifth the wage of programmers in the United States. This has alarmed American workers who have lost well-paying jobs in this and other fields where jobs are being *outsourced* to other countries.

On April 15, 1994, representatives of 120 nations signed the Uruguay Round of the General Agreement on Tariffs and Trade (GATT). This is a revised version of the 1948 GATT convention, which attempted to lower tariff barriers to world trade between member nations. The new GATT established a World Trade Organization (WTO) and gave it the status of a major international organization similar in stature and power to the United Nations and the World Bank. The WTO's role is to enforce the new GATT rules of world trade and settle disputes about these rules between nations.

Currently, the WTO has 141 member countries. Representatives of the Quad Countries—the United States, Canada, Japan, and the European Union—determine most WTO policy. GATT and the WTO are the result of negotiations among the world's major industrial nations, which regulate 90% of all international trade.

GATT has removed some trade barriers that restrict international trade by developing nations. But critics complain that the system is designed to ensure that the primary role of most developing nations is to supply raw materials, such as minerals, timber, and agricultural commodities, to developed nations, which turn these imported resources into high-priced goods traded in international markets. As a result, critics say that such developing nations get little of the income generated by international trade.

Most WTO officials are trade experts and corporate lawyers, representing primarily the interests of transnational corporations. The WTO has no elected representatives and is not subject to freedom of information laws or public review of its proceedings and decisions. Any member country can charge any other WTO member country with violating one of the WTO's complex international trade rules.

When this occurs, several things happen. *First*, the case is decided in secret by a tribunal of three anonymous WTO judges, usually corporate lawyers with no particular expertise in the issues being decided. There are no conflict-of-interest restraints on tribunal members, and no information about their possible conflicts of interest is available to the public.

Second, all documents, transcripts, and details of the proceedings are kept secret and only the results are announced.

Third, only official government representatives of the countries involved can submit documents or appear before the tribunal.

Fourth, decisions are binding worldwide and can be appealed only to another tribunal of judges within the WTO. A final panel ruling can be appealed to the entire WTO but can be overturned only by agreement of all WTO members. This is virtually impossible because the winning country is unlikely to vote to overturn a ruling in its favor.

Any country (or part of a country) that violates a ruling of WTO panels has four choices: amend its laws to comply with WTO rules, pay annual compensation to the winning country, pay high tariffs imposed on the disputed goods by the WTO, or find itself shunned and locked out of global commerce.

Proponents argue that this significant transfer of power from nations to the WTO is necessary and beneficial for several reasons. *First*, globalization of trade is inevitable and we have to take part in guiding it. *Second*, reducing global trade barriers will benefit developing countries, whose products often are at a competitive disadvantage in the global marketplace because of trade barriers erected by developed countries.

Third, reducing global trade barriers can stimulate economic growth in all countries by allowing consumers to buy more things at lower prices. *Fourth,* globalization of trade will raise the environmental and health standards of developing countries.

However, opponents contend that current WTO rules will have several harmful consequences. *First,* they will increase the economic and political power of transnational corporations and decrease the power of small businesses, citizens, and democratically elected governments. *Second,* they will eliminate many jobs and lower wages in developed countries and eventually in developing countries as transnational companies move their operations throughout the world in search of cheap labor, natural resources, and lower environmental standards. *Third,* current WTO rules will weaken environmental and health and safety standards in developed countries.

According to critics, some WTO rules and omissions of principles affect the ability of national, state, and local governments to protect the environment, the health of citizens, and worker health and safety. Here are some examples:

- Governments cannot set standards for how imported products are produced or harvested. This means, for example, that government purchasing policies cannot discriminate against materials produced by child labor or slave labor. Also, they cannot require that items be manufactured from recycled materials or that fish-harvesting methods be protective of dolphins or turtles, for example.

- Current WTO rules do not recognize the rights of countries to take action to protect the atmosphere, the oceans, and other parts of the global commons. There is concern that some provisions of international treaties to protect biodiversity and the ozone layer and to reduce the threats of global warming might be ruled illegal under WTO rules.

- All national, state, or local environmental, health, and safety laws and regulations must be based on globally accepted scientific evidence and risk analysis showing there is a worldwide scientific consensus on the danger. Otherwise, they are considered to be trade barriers that exceed WTO international standards. Because of the inherent scientific and other uncertainties in determining health risks and carrying out risk analysis, this requirement is almost impossible to meet. In other words, the burden of proof falls on those trying to prevent pollution rather than on polluters (Figure 27-10). This means that chemicals, products, and technologies cannot be banned on the basis

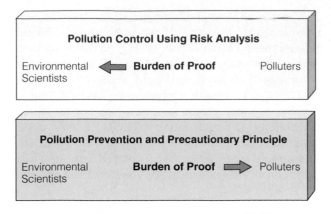

Figure 27-10 The environmental burden of proof falls on different parties, depending on the system used.

of the pollution prevention and precautionary principles—two of the foundation stones of modern environmental protection.

- National laws covering packaging, recycling, and eco-labeling of items involved in international markets are illegal barriers to trade. Enforcing this rule can effectively cancel the third mainstay of modern environmental protection: *consumers' right to know* about the safety and content of products through labeling.

If allowed to stand, environmentalists contend that the four WTO rules just described will force us back to an earlier and less effective era of using end-of-pipe pollution cleanup based on uncertain and easily manipulated risk assessment as the primary ways for dealing with pollution.

On a more positive note, nations may be able to use the WTO to reduce environmentally harmful subsidies that distort the economic playing field. But this would also prevent using subsidies to reward companies producing environmentally beneficial goods and services—an important part of the proposed strategy for developing eco-economies (Figure 26-16, p. 602).

Solutions: Improving Trade Agreements: Environmentally Sustainable Trade

Critics call for changing trade agreements to make WTO proceedings more open and to establish global standards for protecting the environment, consumers, and workers.

Critics of current trade agreements would rewrite and correct what they believe are serious weaknesses in GATT and turn it into GAST: the *General Agreement for Sustainable Trade.*

They offer several suggestions for doing this.

- Set minimum environmental, consumer protection, and worker health and safety standards for all participating countries.

- Open all discussions and findings of any GATT panel or other WTO body to global public scrutiny and inputs from experts on the issues involved.

- Incorporate the precautionary and pollution prevention principles into WTO rules.

- Protect the rights of consumers to know about the health and environmental impact of imported products they purchase by allowing eco-labeling programs.

- Recognize the right of countries to use trade measures to protect the global commons.

- Allow countries to require that imported or exported items be manufactured totally or partially from recycled materials.

- Allow national, state, or local governments to restrict imports of products from countries shown by international investigation to use child or slave labor or violate universally recognized human rights.

- Allow international environmental agreements and treaties to prevail when they conflict with WTO rules or the rules of any other trade agreement.

- Set up and fund a World Environmental Organization (WEO) as a counterbalance to the power and influence of the WTO over global and national environmental policy.

Unless citizens and NGOs exert intense pressure on legislators in the Quad countries governing the WTO, critics warn that such safeguards will not be incorporated into GATT and other international trade agreements.

X *How Would You Vote?* Do you favor changing trade agreements to make WTO proceedings more open and to establish global standards for protecting the environment, consumers, and workers? Cast your vote online at http://biology.brookscole.com/miller14.

Severn Cullis-Suzuki

INDIVIDUALS MATTER

At the age of 12 Severn Cullis-Suzuki and three of her Vancouver, Canada, schoolmates raised money to go to the 1992 Rio Earth Summit. She was invited to address the delegates. In her speech, which had a great impact and received a standing ovation, she said: "In school you teach us not to fight with others, to work things out, to respect others, to clean up our mess, not to hurt other creatures, to share, not be greedy. Then why do you go out and do these things you tell us not to do? You grownups say you love us, but I challenge you, please, to make your actions reflect your words."

She went on to get a B.S. degree in biology from Yale University and was invited to attend the 2002 Johannesburg Earth Summit as a member of UN Secretary-General Kofi Annan's World Summit advisory panel.

As a member of today's younger generation, what advice does she have? "When I was little the world was simple. But as a young adult, I am learning that as we have to make choices—education, career, lifestyle—life gets more and more complicated. She reports that much of what she wanted for the world's future when she was 12 "was idealistic and naïve."

"Today I am no longer a child, but I'm worried about what kind of environment my children will grow up in. . . . We are not facing up to our mess. We are not facing up to our lifestyles. . . . Real environmental change depends on us. We can't wait for our leaders. . . . The challenges are great, but if we accept individual responsibility and make sustainable choices, we will rise to the challenges."

Can We Develop More Environmentally Sustainable Political and Economic Systems in the Next Few Decades? Establishing a New Partnership

We need to work together to find and implement innovative solutions to local, national, and global environmental, economic, and social problems.

Environmentalists call for people from all political persuasions and walks of life to work together to develop a positive vision for a transition to more environmentally sustainable societies and economies throughout the world (Individuals Matter, above right).

A major goal would be to promote the development of creative experiments at local levels—such as the one in Curitiba, Brazil (p. 563)—that could be spread to other areas over the next few decades. A second goal is to get citizens, business leaders, and elected officials to cooperate in trying to find and implement innovative solutions to local, national, and global environmental, economic, and social problems.

Some progress is being made in developing more environmentally sustainable societies, especially in several European Union countries and in Canada. The World Economic Forum, The Yale Center for Environmental Law and Policy, and the Center for International Earth Science Information Network (CIESIN) at Columbia University have developed an *Environmental Sustainability Index (ESI)*. It measures the overall environmental progress of 142 countries using 20 core indicators, including urban air quality, resource use, and environmental regulation.

In 2002, Finland, Norway, Sweden, Canada, and Switzerland had in order the highest ESI scores. Kuwait, the United Arab Emigrates, North Korea, Iraq, and Saudi Arabia had the lowest score. The United States ranked 45th out of the 142 countries evaluated. It received a high rating for decreasing water pollution and conventional air pollutants and active discussion of environmental policy. But it received a low rating on reducing greenhouse gas emissions and reducing resource waste.

Proponents recognize that making a cultural shift to more environmentally sustainable societies over the next few decades will be controversial, and like all significant change it will not be predictable, orderly, or painless.

According to business leader Paul Hawken, making this change means

> *thinking big and long into the future. It also means doing something now. It means electing people who really want to make things work, and who can imagine a better world. It means writing to companies and telling them what you think. It means never forgetting that the cash register is the daily voting booth in democratic capitalism.*

Several guidelines have been suggested for fostering cooperation instead of confrontation as we deal with important environmental problems. *First,* recognize that business is not the enemy. Businesses are here to make money for their investors and stockholders. So why not reward them and encourage new environmental innovations by shifting government subsidies from earth-degrading activities to earth-sustaining activities and by shifting taxes from income and wealth to pollution and resource waste? Environmentalists and leaders of corporations could thus become partners in a joint quest for environmental and economic sustainability.

Second, shift the emphasis for dealing with environmental problems to preventing or minimizing them. *Third,* use well-designed and carefully monitored marketplace solutions to prevent most environmental problems instead of relying primarily on laws, regulations, and litigation in dealing with environmental problems.

Fourth, cooperate and innovate to find win-win solutions to environmental problems instead of using confrontational tactics to come up with less effective I-win-you-lose solutions in which the earth always ends up losing.

Fifth, stop exaggerating. People on both sides of thorny environmental issues should take a vow not to exaggerate or distort their positions in attempts to play win-lose environmental games. They should recognize that there are trade-offs in any environmental

decision—as presented throughout this book—and work together to find balanced win-win solutions that are implemented in a flexible and adaptive manner.

In working to make the earth a better place to live, we should be guided by historian Arnold Toynbee's observation, "If you make the world ever so little better, you will have done splendidly, and your life will have been worthwhile," and by George Bernard Shaw's reminder that "indifference is the essence of inhumanity."

In the end, it all comes back to each of us taking responsibility. We all have to decide whether we want to be part of the problem or part of the solution to the environmental challenges we face.

As the wagon driver said when they came to a long, hard hill, "Them that's going on with us, get out and push. Them that ain't, get out of the way."
ROBERT FULGHUM

CRITICAL THINKING

1. What are the greatest strengths and weaknesses of the system of government in your country with respect to **(a)** protecting the environment and **(b)** ensuring environmental justice for all? What three major changes, if any, would you make in this system?

2. Explain why you agree or disagree with the nine principles recommended by some analysts for use in making environmental policy decisions listed on pp. 607–608.

3. Rate the last four presidents of the United States (or leaders of the country where you live) on a scale of 1–10 in terms of their ability to act as environmental leaders.

4. Suppose a presidential candidate ran on a platform calling for the federal government to phase in a tax on gasoline so that, over 5–10 years, the price of gasoline would rise to $5–8 a gallon (as is the case in Japan and most western European nations). The candidate argues that this tax increase is necessary to encourage oil and gasoline conservation, reduce air pollution, slow global warming, and enhance future economic, environmental, and military security. The candidate also says the tax revenue would be used to reduce taxes on wages and profits and to provide an economic safety net for the poor and lower middle class. Would you vote for this candidate who wants to greatly increase the price of gasoline? Explain.

5. What are the advantages and disadvantages of using only public funds to finance all election campaigns? Explain why you support or oppose such an idea. Why might the major environmental groups in the United States not support such a reform? Should they?

6. Explain why you agree or disagree with each of three solutions given on p. 618 for leveling the legal playing field for citizens who have suffered environmental harm. Try to interview an environmental lawyer and a corporate lawyer to get their views on this.

7. Explain why you agree or disagree with each of the suggestions listed on pp. 626–627 for improving or revising the current rules of the World Trade Organization. Do you favor establishing a World Environmental Organization? Explain.

8. Some people say: "Most people will not become involved in making the world a better place so why should I?" "Individuals cannot make a difference." "People will act only when there is a crisis, and by then it will be too late." Do you have any of these attitudes? Compare your response with those of other members of your class. Some analysts say that these are merely excuses or rationalizations some people use to avoid getting involved, and as such, are forms of intellectual dishonesty and civic and ethical laziness. Do you agree with this assessment?

9. Congratulations! You are in charge of formulating environmental policy in the country where you live. List the three most important components of your policy. Compare your views with those of other members of your class and see if you can agree on a consensus policy.

PROJECTS

1. Polls have identified five categories of U.S. citizens in terms of their concern over environmental quality: **(1)** those involved in a wide range of environmental activities, **(2)** those who do not want to get involved but are willing to pay more for a cleaner environment, **(3)** those who are not involved because they disagree with many environmental laws and regulations, **(4)** those who are concerned but do not believe individual action will make much difference, and **(5)** those who strongly oppose the environmental movement. To which group do you belong? Compare the results of members in your class and determine the percentage in each category. As a class, conduct a similar poll on your campus.

2. Have each member of your class select a particular environmental law, and evaluate it in terms of **(a)** its use of or failure to use the principles listed on pp. 607–608 and **(b)** the role that environmental organizations and citizen actions played in its development. Compare the results of these analyses.

3. Pick a particular environmental law in the United States or in the country where you live. Use the library or the Internet to evaluate the law's major strengths and weaknesses. Decide whether the law should be weakened, strengthened, or abolished, and explain why. List the three most important ways you believe the law should be improved.

4. Try to interview **(a)** a lobbyist for an industry seeking to weaken a specific environmental law, **(b)** a lobbyist for

an environmental group seeking to strengthen the law, **(c)** an EPA official supporting strengthening the law, and **(d)** an elected representative who must make a decision about the law. Compare their views and perspectives, and come to a conclusion about what should be done. If you cannot get interviews, set up a mock discussion panel with class members taking each of these roles.

5. What student environmental groups, if any, are active at your school? How many people actively participate in them? Pick one of these groups and find out what environmentally beneficial things have they done. What actions taken by this group, if any, do you disagree with? Why?

6. Use the library or the Internet to learn about and evaluate the effectiveness of a particular national or international environmental group. What is your chosen group's mission? How successful has the group been in fulfilling its mission? To what do you attribute its success or failure?

7. Use the Internet and the local phonebook to identify environmental or conservation groups in your area. Contact them for information on their activities and their achievements and report this information to your class.

8. Use the library or the Internet to learn about and evaluate a specific environmentally related decision made by the World Trade Organization in determining whether or not a member nation violated one of its international trade rules. Explain why you agree or disagree with this decision.

9. Use the library or the Internet to find bibliographic information about *Lord Kennet* and *Robert Fulghum,* whose quotes appear at the beginning and end of this chapter.

10. Make a concept map of this chapter's major ideas, using the section heads, subheads, and key terms (in boldface type). Look at the website for this book for information about making concept maps.

LEARNING ONLINE

The website for this book contains study aids and many ideas for further reading and research. They include a chapter summary, review questions for the entire chapter, flash cards for key terms and concepts, a multiple-choice practice quiz, interesting Internet sites, references, and a guide for accessing thousands of InfoTrac® College Edition articles. Log on to

http://biology.brookscole.com/miller14

Then click on the Chapter-by-Chapter area, choose Chapter 27, and select a learning resource.

28 Environmental Worldviews, Ethics, and Sustainability

CASE STUDY
Biosphere 2: A Lesson in Humility

In 1991, eight scientists (four men and four women) were sealed into Biosphere 2, a $200 million facility designed to be a self-sustaining life-support system (Figure 28-1) and to increase our understanding of the earth's life-support system: Biosphere 1.

The 1.3-hectare (3.2-acre) sealed system of interconnected domes was built in the desert near Tucson, Arizona. It had a tropical rain forest, savanna, desert, lakes, streams, freshwater and saltwater wetlands, and a mini-ocean with a coral reef.

The facility was stocked with more than 4,000 species of organisms. Sunlight and external natural gas–powered generators provided energy. The Biospherians were to be isolated for 2 years and to raise their own food using intensive organic agriculture, breathe air recirculated by plants, and drink water cleansed by natural recycling processes.

From the beginning there were many unexpected problems. The life-support system began unraveling. Large amounts of oxygen disappeared when soil organisms converted it to carbon dioxide. Additional oxygen had to be pumped in from the outside to keep the Biospherians from suffocating.

The nitrogen and carbon recycling systems also failed to function properly. Levels of nitrous oxide rose high enough to threaten the occupants with brain damage and had to be controlled by outside intervention. Carbon dioxide skyrocketed to levels that threatened to poison the humans and spurred the growth of weedy vines that choked out food crops. Plant nutrients leached from the soil and polluted the water systems.

Tropical birds died after the first freeze. An Arizona ant species got into the enclosure, proliferated, and killed off most of the system's introduced insect species. After the majority of the introduced insect species became extinct, the facility was overrun with cockroaches and katydids. All together, 19 of the Biosphere's 25 small animal species became extinct. Before the 2-year period was up, all plant-pollinating insects became extinct, thereby dooming to extinction most of the plant species.

Despite many problems, the facility's waste and wastewater were recycled, and the Biospherians were able to produce 80% of their food supply.

Scientists Joel Cohen and David Tilman, who evaluated the project, concluded, "No one yet knows how to engineer systems that provide humans with life-supporting services that natural ecosystems provide for free."

Columbia University took over Biosphere 2 as a research facility for a few years, but abandoned it in 2003.

Figure 28-1 Biosphere 2, constructed near Tucson, Arizona, was designed to be a self-sustaining life-support system for eight people sealed into the facility in 1991. The experiment failed because of a breakdown in its nutrient cycling systems.

The main ingredients of an environmental ethic are caring about the planet and all of its inhabitants, allowing unselfishness to control the immediate self-interest that harms others, and living each day so as to leave the lightest possible footprints on the planet.

ROBERT CAHN

This chapter involves discussions of values, beliefs and ethics—subjects on which people disagree. Yet solving environmental problems requires grappling with our values and beliefs, implementing them in our lives, discussing them with one another, and working to agree on courses of action for solving problems.

This chapter addresses the following questions:

- What human-centered environmental worldviews guide most industrial societies?

- What are some life-centered and earth-centered environmental worldviews?

- What ethical guidelines might be used to help us work with the earth?

- How can we live more sustainably?

28-1 ENVIRONMENTAL WORLDVIEWS IN INDUSTRIAL SOCIETIES

What Is an Environmental Worldview? Beliefs and Values

Your environmental worldview is how you think the world works, what you believe your role in it should be, and what you believe is right and wrong in terms of environmental behavior.

People disagree about how serious our environmental problems are and what we should do about them. The conflicts arise mostly out of differing **environmental worldviews**: how people think the world works, what they believe their role in it should be, and what they believe is right and wrong in terms of environmental behavior (**environmental ethics**).

People with widely differing environmental worldviews can take the same data, be logically consistent, and arrive at quite different conclusions because they start with different assumptions, beliefs, and values.

Figure 28-2 (p. 632) shows some different types of environmental worldviews. Most can be characterized as *human-centered, earth-centered*, or some combination of both. As you move from the center of Figure 28-2 to the outside rings, the worldviews become less human-centered and more earth-centered and devoted to sustaining the earth's natural systems (ecosystems), life-forms (biodiversity), and life-support system (bio-

sphere) for the benefit of humans and other forms of life.

How Do We Decide What Is Right and Wrong Environmental Behavior? A Search for Ethical Principles

Several philosophies can help us decide what is right and wrong environmental behavior.

There are several major moral philosophical systems that deal with the concepts of what constitutes right and wrong environmental (or other) behavior.

One is *universalism*, developed by philosophers such as Plato (427–347 B.C.E.) and Immanuel Kant (1724–1804). According to this view, there are basic principles of ethics, or rules of right and wrong, that are universal and unchanging. Some believe that God or other sources of wisdom have provided ethical guidelines. Others believe that these rules can be discovered through reason, experience, and knowledge.

Another philosophy is *utilitarianism*, developed by Jeremy Bentham (1748–1832) and John Stuart Mill (1806–1873). It says that an action is right if it produces the greatest satisfaction or pleasure for the greatest number of people. Critics contend that this assumes that we can somehow quantitatively measure happiness and compare it among different people.

A related moral philosophy is *consequentialism*, which proposes that whether an act is morally right or wrong depends on its consequences. In effect, we determine correct moral conduct *solely* by conducting a cost-benefit analysis of the consequences of our actions. Thus an action is morally right if its consequences as a whole are more favorable than unfavorable.

Another philosophy of right and wrong is *relativism*, promoted by ancient Greek teachers called Sophists who disagreed with Plato's universalist ethics. It asserts that moral values of right and wrong are relative to cultures, eras, or situations and there are no absolute principles of right and wrong.

According to the philosophy of *rationalism*, we can develop principles of right and wrong by using logic to analyze ideas and arguments. This philosophy based on the power of reason was developed by philosophers such as René Descartes (1596–1650), Benedict De Spinoza (1632–1677), and Gottfried W. Leibniz (1646–1716).

According to *nihilism*, often associated with Friedrich Nietzsche (1844–1900), the concepts of values and moral beliefs are useless because nothing can be known or communicated. There is no purpose or meaning to life except the struggle to survive in which "might is right."

Because this is not a philosophy textbook, the brief summaries of several moral philosophies just given

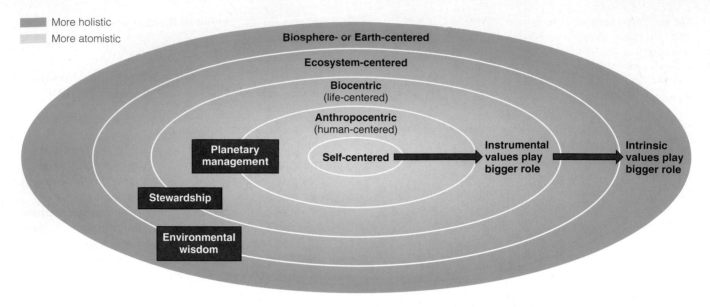

More holistic
More atomistic

Biosphere- or Earth-centered
Ecosystem-centered
Biocentric (life-centered)
Anthropocentric (human-centered)

Planetary management
Self-centered → Instrumental values play bigger role → Intrinsic values play bigger role

Stewardship

Environmental wisdom

Figure 28-2 Environmental worldviews lie on a continuum running from more self- and human-centered (center) to biosphere- or earth-centered (outer rings). (Diagram developed by Scott Spoolman)

are necessarily superficial and incomplete. There are many complex arguments involved in evaluating these and other moral philosophies that you can explore by studying philosophy.

What Is the Difference between Instrumental and Intrinsic Values? Usefulness versus Existence

Life forms have value because of their usefulness to us or to the biosphere or merely because they exist.

What we value largely determines how we act. Environmental philosophers normally divide values into two types. One type, called **instrumental** or **utilitarian,** values life-forms because of their usefulness to us or to the biosphere. The concept of preserving natural capital and biodiversity because they sustain life and support economies is an instrumental value based mostly on the usefulness of these natural goods and services to us.

A second type, called **intrinsic** or **inherent,** values a form of life just because it exists, regardless of whether it has any usefulness to us.

As worldviews go from being human-cenetered to more earth-centered (moving out from the center of Figure 28-2) instrumental values play a lesser role, and intrinsic values become more important. The view that a wild species, an ecosystem, biodiversity, or the biosphere has value only because of its usefulness to us is called an **anthropocentric** (human-centered) instrumental value.

In contrast, the view that these forms of life are valuable simply because they exist, independently of

their use to human beings, is called a **biocentric** (life-centered) intrinsic value. According to a *biocentric worldview,* all species and ecosystems and the biosphere have both intrinsic and instrumental value, we are just one of many species, and we have an ethical responsibility not to impair the long-term sustainability and adaptability of the earth's natural systems for all life.

What Are the Major Human-Centered Environmental Worldviews? Managers and Stewards

In the human-centered view, we are the planet's most important species and should become managers or stewards of the earth.

Some people in today's industrial consumer societies have a **planetary management worldview**. According to this human-centered environmental worldview, we are the planet's most important and dominant species and we can and should manage the earth mostly for our own benefit. Other species and parts of nature are seen as having only *instrumental value* based on how useful they are to us. This is a form of utilitarianism.

Figure 28-3 (left) summarizes the four major beliefs or assumptions of this environmental worldview. Here are three variations of this environmental worldview:

- The *no-problem school.* We can solve any environmental, population, or resource problems with more economic growth and development, better management, and better technology.

- The *free-market school.* The best way to manage the planet for human benefit is through a free-market global economy with minimal government interference and regulations. All public property resources should be converted to private property resources and the global marketplace, governed by pure free-market competition, should decide essentially everything.

- The *spaceship-earth school.* The earth is like a spaceship: a complex machine that we can understand, dominate, change, and manage to prevent environmental overload and provide a good life for everyone. This view developed as a result of photographs taken from space showing the earth as a finite planet, or an island in space (Figure 5-1, p. 88). This led many people to see that the earth is our only home and we had better treat it with care.

Another anthropocentric environmental worldview is the stewardship worldview. According to this increasingly popular view, we have an ethical responsibility to be caring and responsible managers, or *stewards,* of the earth. We can and should make the world a better place for our species and others through love, care, knowledge, and technology. Figure 28-3 (center) summarizes the major beliefs of this worldview.

According to the stewardship view, as we use the earth's natural capital, we are borrowing from the earth and from future generations, and we have an ethical responsibility to pay the debt by leaving the earth in at least as good a condition as we enjoy.

In thinking about our responsibility toward future generations, some analysts believe we should consider the wisdom of the 18th century Iroquois Confederation of Native Americans: *In our every deliberation, we must consider the impact of our decisions on the next seven generations.*

28-2 LIFE-CENTERED AND EARTH-CENTERED ENVIRONMENTAL WORLDVIEWS

Can We Manage the Planet? Look at Biosphere 2

Some analysts doubt that we can effectively manage the planet.

Some people believe any human-centered worldview will eventually fail because it wrongly assumes we now have or can gain enough knowledge to become effective managers or stewards of the earth.

According to these analysts, the unregulated global free-market approach will not work because it is based on increased losses or degradation of the earth's natural

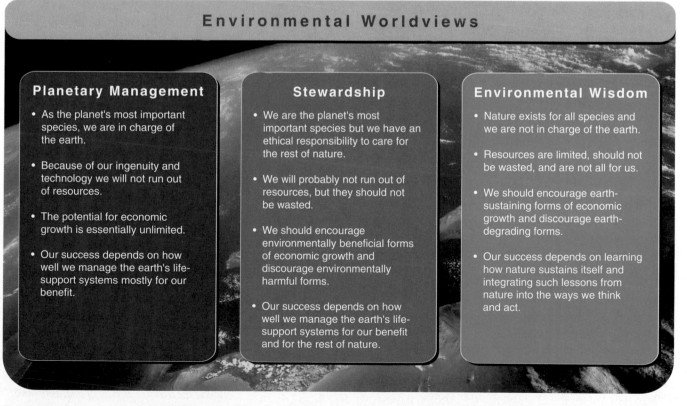

Environmental Worldviews

Planetary Management
- As the planet's most important species, we are in charge of the earth.
- Because of our ingenuity and technology we will not run out of resources.
- The potential for economic growth is essentially unlimited.
- Our success depends on how well we manage the earth's life-support systems mostly for our benefit.

Stewardship
- We are the planet's most important species but we have an ethical responsibility to care for the rest of nature.
- We will probably not run out of resources, but they should not be wasted.
- We should encourage environmentally beneficial forms of economic growth and discourage environmentally harmful forms.
- Our success depends on how well we manage the earth's life-support systems for our benefit and for the rest of nature.

Environmental Wisdom
- Nature exists for all species and we are not in charge of the earth.
- Resources are limited, should not be wasted, and are not all for us.
- We should encourage earth-sustaining forms of economic growth and discourage earth-degrading forms.
- Our success depends on learning how nature sustains itself and integrating such lessons from nature into the ways we think and act.

Figure 28-3 Comparison of two opposing environmental worldviews. Which of these environmental worldviews comes closer to your own environmental worldview?

capital and focuses on short-term economic benefits regardless of the harmful long-term environmental and social consequences.

The image of the earth as an island or ship in space has played an important role in raising global environmental awareness. But critics argue that thinking of the earth as a spaceship that we can manage is an oversimplified and misleading way to view an incredibly complex and ever-changing planet. This view was supported by the failure of Biosphere 2 (p. 630).

These critics point out that we do not even know how many species live on the earth, much less what their roles are and how they interact with one another and their nonliving environment. We have only an inkling of what goes on in a handful of soil, a meadow, a pond, or any other part of the earth.

As biologist David Ehrenfeld puts it, "In no important instance have we been able to demonstrate comprehensive successful management of the world, nor do we understand it well enough to manage it even in theory." According to environmental educator David Orr: "On balance, I think, we are becoming more ignorant because we are losing cultural knowledge about how to inhabit our places on the planet sustainably, while impoverishing the genetic knowledge accumulated through millions of years of evolution."

What Are Life-Centered and Eco-Centered Worldviews? Deciding What to Protect and Sustain

Some believe we should focus on protecting species from premature extinction and others on not destroying or degrading ecosystems, biodiversity, and the earth's life support systems.

People disagree over how far we should extend our ethical concerns for various forms or levels of life (Figure 28-4). Some critics believe that human-centered environmental worldviews should be expanded to recognize the *inherent* or *intrinsic value* of all forms of life regardless of their potential or actual use to us.

Most people with a life-centered worldview believe we have an ethical responsibility to avoid causing the premature extinction of species through our activities, for three reasons. *First,* each species is a unique storehouse of genetic information that should be

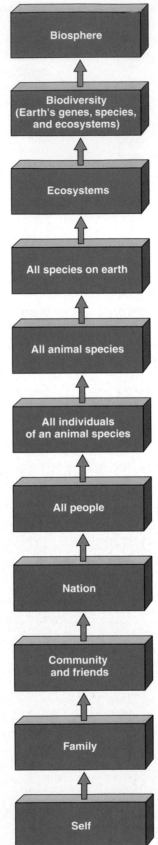

Figure 28-4 Levels of ethical concern. People disagree over how far we should extend our ethical concerns on this scale. How far up this scale would you extend your own ethical concerns?

respected and protected because it exists (intrinsic value). *Second,* each species is a potential economic good for human use (instrumental value). *Third,* populations of species are capable through evolution and speciation of adapting to changing environmental conditions.

Trying to decide whether all or only some species should be protected from premature extinction resulting from human activities is a difficult and controversial ethical problem. It is hard to know where to draw the line and be ethically consistent.

Here are three of the issues involved: *First,* should all species be protected from premature extinction because of their intrinsic value, or should only certain ones be preserved because of their known or potential instrumental value to us or to their ecosystems?

Second, should all insect and bacterial species be protected, or should we attempt to exterminate those that eat our crops, harm us, or transmit disease organisms?

Third, should we emphasize protecting keystone and foundation species over other species that play lesser roles in ecosystems?

Some people believe we must go beyond focusing on species. They believe we have an ethical responsibility not to degrade the earth's ecosystems, biodiversity, and biosphere for this and future generations of humans and other species. In other words, they have an *earth-centered,* or *ecocentric,* environmental worldview, devoted to preserving the earth's biodiversity and the functioning of its life-support systems for all forms of life.

Why should we care about the earth's biodiversity? According to the late enviromentalist and systems expert Donella Meadows,

Biodiversity contains the accumulated wisdom of nature and the key to its future. If you wanted to destroy a society, you would burn its libraries and kill its intellectuals. You would destroy its knowledge. Nature's knowledge is contained in the DNA within living cells. The variety of genetic information is the driving engine of evolution and the source of adaptability.

What Is the Environmental Wisdom Worldview? Learning to Work with the Earth

According to this view we are not in charge, should not waste the earth's finite resources, and should live more sustainably by mimicking the ways in which the earth has sustained itself for billions of years.

As we move out from the center of Figure 28-2, we see life-centered and then earth-centered worldviews. One earth-centered worldview is called the **environmental wisdom worldview.** Its major beliefs are summarized in Figure 28-3 (right). In many respects, it is the opposite of the planetary management worldview (Figure 28-3, left).

According to this worldview, because we are not in charge of the biosphere, we should learn how it has maintained itself over billions of years and use what we learn to sustain ourselves and the biosphere into the future (Figure 9-15, p. 174).

This worldview includes the belief that the earth does not need us managing it in order to go on, whereas we do need the earth in order to survive. According to this view we cannot save the earth because it does not need saving. What we need to save is the existence of our own species and other species that may become extinct because of our activities.

Sustainability expert Lester W. Milbrath asks us to try this thought experiment: "Imagine that, suddenly, all the humans disappeared, but all the buildings, roads, shopping malls, factories, automobiles, and other artifacts of modern civilization were left behind. What then? After three or four centuries, buildings would have crumbled, vehicles would have rusted and fallen apart, and plants would have recolonized fields, roads, parking lots, even buildings. Water, air, and soil would gradually clear up; some endangered species would flourish. Nature would thrive splendidly without us." See Lester Milbrath's Guest Essay on this topic on the website for this chapter.

Others say we do not need to be biocentrists or ecocentrists to value life and the earth. They point out that the human-centered stewardship environmental worldview also calls for us to value individuals, species, and the earth's life-support systems as part of our responsibility as the earth's caretakers.

What Is the Deep Ecology Environmental Worldview? Thinking Deeply about Our Responsibilities

The beliefs of deep ecology call for us to think more deeply about our obligations toward both human and nonhuman life.

A related ecocentric environmental worldview is the *deep ecology worldview,* which consists of eight premises developed in 1972 by Norwegian philosopher Arne Naess, in conjunction with philosopher George Sessions and sociologist Bill Devall. *First,* each nonhuman form of life on the earth has inherent value, independent of its value to humans. *Second,* the fundamental interdependence and diversity of life-forms contribute to the flourishing of human and nonhuman life on earth.

Third, humans have no right to reduce this interdependence and diversity except to satisfy vital needs. *Fourth,* present human interference with the nonhuman world is excessive, and the situation is worsening rapidly. *Fifth,* because of the damage caused by this interference, it would be better for humans, and much better for nonhumans, if there were a substantial decrease in the human population.

Sixth, basic economic and technological policies must therefore be changed. *Seventh,* the predominant ideology must change such that measurements of the *quality of life* focus on the overall health of the environment and all living things, rather than on the material wealth of individuals and societies. *Eighth,* those who subscribe to these points have an obligation directly or indirectly to try to implement the necessary changes.

Naess also described some lifestyle guidelines compatible with the basic beliefs of deep ecology. They include appreciating all forms of life, consuming less, emphasizing satisfaction of vital needs rather than wants, working to improve the standard of living for the world's poor, working to eliminate injustice toward fellow humans or other species, and acting nonviolently.

Deep ecology is not an ecoreligion, nor is it antireligious or antihuman, as some of its critics have claimed. It is a set of beliefs that would have us think more deeply about the inherent value of all life on the earth and about our obligations toward all life.

What Is the Ecofeminist Environmental Worldview? Give Women a Fair Chance

Women should be given the same rights as men and treated as equal partners in our joint quest to develop more environmentally sustainable and just societies.

French writer Françoise d'Eaubonne coined the term *ecofeminism* in 1974. It includes a spectrum of views on the relationships of women to the earth and to male-dominated societies (patriarchies). Most ecofeminists agree that we need a life-centered or earth-centered environmental worldview. However, they believe a main cause of our environmental problems is not just human centeredness, but specifically male centeredness (*androcentrism*).

Many ecofeminists argue that the rise of male-dominated societies and environmental worldviews since the advent of agriculture is primarily responsible for our violence against nature and for the oppression of women and minorities as well. To such ecofeminists, our shift from hunter–gatherer to agricultural and industrial societies changed our view of nature from that of a nurturing mother to that of a foe to be conquered.

Ecofeminists note that women earn less than 10% of all wages, own less than 1% of all property, and in most societies have far fewer rights than men. These analysts also argue that to become primary players in the male power-and-domination game, most women are forced to emphasize the characteristics deemed masculine and become "honorary men."

Some ecofeminists suggest that oppression by men has driven women closer to nature and made them more compassionate and nurturing. As oppressed members of society, they argue, many women have more experience in dealing with interpersonal conflicts, bringing people together, acting as caregivers, and identifying emotionally with injustice, pain, and suffering. Thus women with such qualities are in a better position to help lead as we struggle to develop more environmentally sustainable and just societies. Such societies would be based on cooperation, rather than confrontation and domination, and on finding *win-win* solutions to our environmental and human problems instead of *win-lose* solutions often associated with our current male-dominated societies.

Ecofeminists argue that women should be treated as equal partners with men. They do not want just a fair share of the patriarchal pie or to be given token roles or co-opted into the male power game. They want to work with men to bake an entirely new pie. They hope to heal the rift between humans and nature and end oppression based on sex, race, class, and cultural and religious beliefs.

Ecofeminists are not alone in calling for us to encourage the rise of *life-centered people* who emphasize the best human characteristics: gentleness, caring, compassion, nonviolence, cooperation, and love.

X **How Would You Vote?** Which one of the following comes closest to your environmental worldview: planetary management, stewardship, environmental wisdom, deep ecology, ecofeminist? Cast your vote online at http://biology .brookscole.com/miller14.

Which Worldview Is More Likely to Prove Correct? Catastrophe vs. Continuing Growth

Using images of economic or ecological collapse can deter us from preventing or slowing the spread of environmental degradation.

The planetary management, stewardship, and environmental wisdom worldviews differ over whether there are physical and biological limits to economic growth, beyond which ecological and economic collapse are likely to occur. This argument has been going on since Thomas Malthus published his book *The Principles of Political Economy* in 1836.

In 2000, conservation biologist Carlos Davidson proposed a way to bridge the gap between differing worldviews and to help motivate the political changes needed to halt or slow the spread of environmental degradation.

Davidson disagrees with the view of some economists that technology will allow continuing economic growth without causing serious environmental damage. However, he also disagrees with the view that continuing economic growth based on consuming and degrading natural capital will lead to ecological and economic crashes.

Instead of crashes, he suggests the metaphor of a gradually unraveling tapestry to describe the effects of environmental degradation. He asks us to think of nature as a diverse tapestry—an incredible variety of biomes, aquatic systems, and ecosystems.

The tapestry is losing threads, more in some areas than in others, and is even torn in some places, but it is unlikely to simply fall apart. However, he believes that degradation in parts of the earth's ecological tapestry is occurring; is likely to increase because of problems such as climate change and biodiversity; and must be prevented and slowed by using the pollution prevention and precautionary principles.

However, Davidson believes that using catastrophe metaphors such as "ecological collapse" and "going over a cliff" can hinder these efforts. He argues that repeated predictions of catastrophe, like the boy who cried wolf, will be heard at first, but later ignored. As a result, people will lose motivation to prevent more tears in nature's tapestry or to look more deeply at the causes of the damage.

X **How Would You Vote?** Do you believe there are physical and biological limits to human economic growth? Cast your vote online at http://biology.brookscole.com/miller14.

28-3 LIVING MORE SUSTAINABLY

What Are the Main Components of Environmental Literacy? Understanding and Caring about How the Earth Works and Sustains Itself

Environmentally literate citizens and leaders are needed to build more environmentally sustainable and just societies.

Most environmentalists believe that learning how to live more sustainably requires a foundation of environmental education. They cite some of the key goals of environmental education or ecological literacy.

- *Develop respect or reverence for all life.*

- *Understand as much as we can about how the earth works and sustains itself, and use such knowledge to guide our lives, communities, and societies.*

- *Look for connections within the biosphere and between our actions and the biosphere.*

- *Use critical thinking skills to become seekers of environmental wisdom instead of overfilled vessels of environmental information.*

- *Understand and evaluate our environmental worldview and see this as a lifelong process.*

- *Learn how to evaluate the beneficial and harmful environmental consequences of our lifestyle and professional choices, today and in the future.*

- *Foster a desire to make the world a better place and act on this desire.*

Specifically, an ecologically literate person should have a basic comprehension of:

- Concepts such as environmental sustainability, natural capital, exponential growth, carrying capacity, and risks and risk analysis

- Environmental history (to help keep us from repeating past mistakes)

- The laws of thermodynamics and the law of conservation of matter

- Basic principles of ecology

- Ways to sustain biodiversity

- Sustainable agriculture and forestry

- Sustainable cities

- Sustainable water use

- Nonrenewable and renewable energy resources

- Soil and mineral resources

- Pollution prevention and waste reduction

- Environmentally sustainable economic and political systems

- Environmental ethics

According to environmental educator Mitchell Thomashow, four basic questions should be at the heart of environmental literacy.

First, where do the things I consume come from? *Second,* what do I know about the place where I live? *Third,* how am I connected to the earth and other living things? *Fourth,* what is my purpose and responsibility as a human being?

How we answer these questions determines our *ecological identity.* What are your answers to these four questions?

Figure 28-5 summarizes guidelines and strategies for achieving more sustainable societies that have been discussed throughout this book.

How Can We Learn from the Earth? Seeking Environmental Wisdom

In addition to formal learning, we need to learn by experiencing nature directly.

Formal environmental education is important, but is it enough? Many analysts say no and urge us to take the

Figure 28-5 Solutions: guidelines and strategies for achieving more sustainable societies.

time to escape the cultural and technological body armor we use to insulate ourselves from nature and to experience nature directly.

They suggest we kindle a sense of awe, wonder, mystery, and humility by standing under the stars, sitting in a forest or taking in the majesty and power of an ocean.

We might pick up a handful of soil and try to sense the teeming microscopic life within it that keeps us alive. We might look at a tree, mountain, rock, or bee and try to sense how it is a part of us and we a part of it as interdependent participants in the earth's life-sustaining recycling processes.

Many psychologists believe that consciously or unconsciously we spend much of our lives in a search for roots: something to anchor us in a bewildering and frightening sea of change. As philosopher Simone Weil observed, "To be rooted is perhaps the most important and least recognized need of the human soul."

Earth-focused philosophers say that to be rooted, each of us needs to find a *sense of place:* a stream, a mountain, a yard, or any piece of the earth we feel at one with—a place we know, experience emotionally, and love. When we become part of that place, it becomes a part of us. Then we are driven to defend it from harm and to help heal its wounds.

This might lead us to recognize that the healing of the earth and the healing of the human spirit are one and the same. We might discover and tap into what Aldo Leopold calls "the green fire that burns in our hearts" and use this as a force for respecting and working with the earth and with one another.

How Can We Live More Simply? Escaping from Affluenza

Some people are voluntarily adopting lifestyles in which they enjoy life more by consuming less.

Many analysts urge us to *learn how to live more simply.* Seeking happiness through the pursuit of material things is considered folly by almost every major religion and philosophy. Yet it is preached incessantly by modern advertising that encourages us to buy more and more things. Some affluent people in developed countries are adopting a lifestyle of *voluntary simplicity,* doing and enjoying more with less by learning to live more simply. Voluntary simplicity is based on Mahatma Gandhi's *principle of enoughness:* "The earth provides enough to satisfy every person's need but not every person's greed. . . . When we take more than we need, we are simply taking from each other, borrowing from the future, or destroying the environment and other species."

Most of the world's major religions have similar teachings. For example, "Why do you spend your money for that which is not bread, and your labor for that which does not satisfy?" (Christianity: Old Testament, Isaiah 55:2). "Eat and drink, but waste not by excess" (Islam: Koran 7.31). "One should abstain from acquisitiveness" (Hinduism: Acarangastura 2.119). "He who knows he has enough is rich" (Taoism: Tao Te Ching, Chapter 33).

Implementing these principles means asking ourselves, "How much is enough?" The answer is not easy because people in affluent societies are conditioned to want more and more, and they often think of such wants as vital needs (Spotlight, p. 639).

Voluntary simplicity is a form of *environmentally ethical consumption.* It begins by asking a series of questions before buying anything: Do I really need this, or do I merely want it? Can I buy it secondhand (reuse)? Can I borrow, rent, lease, or share it? Can I build it myself?

The decision to buy something triggers another set of questions: Is the product produced in an environmentally sustainable manner? Did the workers who produced it get fair wages for their work, and did they have safe and healthful working conditions? Is it designed to last as long as possible? Is it easy to repair, upgrade, reuse, and recycle?

Figure 28-6 summarizes some ethical guidelines proposed by various ethicists and philosophers for living more sustainably or simply on the earth. In the words of biologist David Suzuki, "Family, friends,

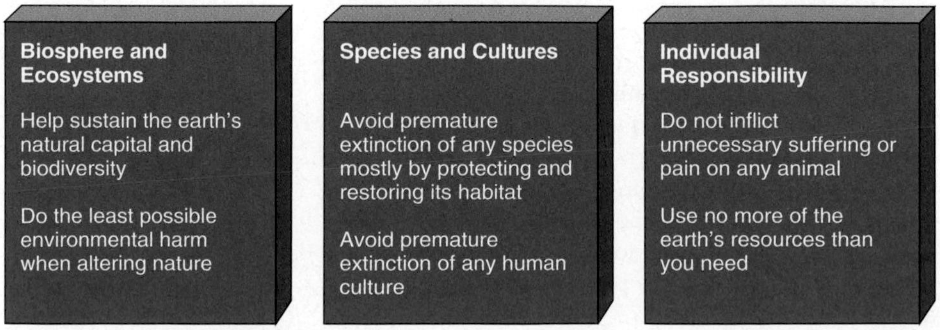

Biosphere and Ecosystems	Species and Cultures	Individual Responsibility
Help sustain the earth's natural capital and biodiversity	Avoid premature extinction of any species mostly by protecting and restoring its habitat	Do not inflict unnecessary suffering or pain on any animal
Do the least possible environmental harm when altering nature	Avoid premature extinction of any human culture	Use no more of the earth's resources than you need

Figure 28-6 Solutions: some ethical guidelines for living more sustainably.

What Are Our Basic Needs?

Obviously, each of us has a basic need for enough food, clean air, clean water, shelter, and clothing to keep us alive and in good health. According to various psychologists and other social scientists, each of us also has other basic needs:

■ A secure and meaningful livelihood to provide our basic material needs

■ Good physical and mental health

■ The opportunity to learn and give expression to our intellectual, mechanical, and artistic talents

■ A nurturing family and friends and a peaceful and secure community that help us develop our capacity for caring and loving relationships while giving us the freedom to make personal choices

■ A clean and healthy environment that is vibrant with biological and cultural diversity

■ A sense of belonging to and caring for a particular place and community

■ An assurance that our children and grandchildren will be able to meet these same basic needs

A difficult but fundamental question is asking how much of the stuff we are all urged to buy helps us meet these basic needs. Indeed, psychologists point out that many people buy things in the hope or belief they will make up for not meeting some of the basic needs listed here.

Critical Thinking

1. What basic needs, if any, would you add to or remove from the list given here?

2. Which of the basic needs listed here (or additional ones you would add) do you feel are being met for you? What are your plans for trying to fulfill any of your unfulfilled needs? Relate these plans to your environmental worldview.

---

community—these are the sources of the greatest love and joy we experience as humans. . . . None of these pleasures requires us to consume things from the Earth, yet each is deeply fulfilling."

How Can We Be More Effective as Environmental Citizens? Avoid Mental Traps and Despair, Be Adaptable, and Enjoy Life

We can help make the world a better place by not falling into mental traps that lead to denial and inaction and by keeping our empowering feelings of hope slightly ahead of our immobilizing feelings of despair.

When we first encounter an environmental problem, our initial response often is to find someone or something to blame, such as greedy industrialists or uncaring politicians. It is the fault of such villains, and we are the victims. This response can lead to despair, denial, apathy, and inaction because we feel powerless to stop or influence these forces.

Upon closer examination we may realize that we all make some direct or indirect contributions to the environmental problems we face. Yet, we do not want to feel guilty or bad about the environmental harm our lifestyles may be inflicting. Thus we try not to think about it much—another path to denial and inaction.

According to primatologist Jane Goodall, "The greatest danger to our future is apathy. . . . Can we overcome apathy? Yes, but only if we have hope. . . .

Technology alone is not enough. We must engage with our hearts also."

Analysts suggest that we move beyond blame, guilt, fear, denial, and apathy by recognizing and avoiding common mental traps that lead to denial, indifference, and inaction. These traps include *gloom-and-doom pessimism* (it is hopeless), *blind technological optimism* (science and technofixes will save us), *fatalism* (we have no control over our actions and the future), *extrapolation to infinity* (if I cannot change the entire world quickly, I will not try to change any of it), *paralysis by analysis* (searching for the perfect worldview, philosophy, solutions, and scientific information before doing anything), and *faith in simple, easy answers.*

We will all accomplish more if we keep our empowering feelings of hope slightly ahead of our immobilizing feelings of despair. It is also important not to use guilt and fear to try to motivate other people.

Recognizing that there is no single correct or best solution to the environmental problems we face is also important. Indeed, one of nature's most important lessons is that preserving diversity—in this case, being flexible and adaptable in trying a variety of solutions to our problems—is the best way to adapt to the earth's and life's largely unpredictable, ever-changing conditions.

Finally, we should have fun and take time to enjoy life. Laugh every day and enjoy nature, beauty, friendship, and love. This empowers us to become good earth citizens who practice *good earthkeeping.*

What Are the Major Components of the Environmental Revolution? A Call for Greatness

The message of environmentalism is not gloom and doom, fear, and catastrophe but hope, a positive vision of the future, and a call for greatness in dealing with the environmental challenges we face.

The *environmental revolution* that many environmentalists call for us to bring about during this century would have several components:

■ A *biodiversity protection revolution* devoted to protecting and sustaining the genes, species, natural systems, and chemical and biological processes that make up the earth's biodiversity.

■ An *efficiency revolution,* that minimizes the wasting of matter and energy resources.

■ A *solar–hydrogen revolution* based on decreasing our dependence on carbon-based nonrenewable fossil fuels and increasing our dependence on forms of renewable solar energy that can be used to produce hydrogen fuel from water.

■ A *pollution prevention revolution* that reduces pollution and environmental degradation from harmful chemicals, preventing their release into the environment by recycling or reusing them, and learning to live without them.

■ A *sufficiency revolution,* dedicated to meeting the basic needs of all people on the planet while affluent societies learn to live more sustainably by living with less.

■ A *demographic revolution* based on reducing fertility to bring the size and growth rate of the human population into balance with the earth's ability to support humans and other species more sustainably.

■ An *economic and political revolution* in which we use economic systems to reward environmentally beneficial behavior and to discourage environmentally harmful behavior.

Opponents of such a cultural change like to paint environmentalists as messengers of gloom, doom, and hopelessness. But *the message of environmentalism is not gloom and doom, fear, and catastrophe but hope and a positive vision of the future.*

We should rejoice in our environmental accomplishments, but the real question is, *Where do we go from here?* How can past environmental advances be transferred to developing countries? How can humans make the cultural transition to more environmentally sustainable societies based on learning from and working with nature?

As you have read in this book, we have an incredible array of technological and economic solutions to the environmental problems we face. The challenge for all of us is to implement such solutions by converting environmental wisdom and beliefs into political action.

This requires becoming involved in making the world a better place. As Gandhi said many years ago, "We must become the change we want to see." This requires understanding that *individuals matter.* Virtually all of the environmental progress we have made during the last few decades occurred because individuals banded together to insist that we can do better.

This journey begins in your own community because in the final analysis *all sustainability is local.* We help make the world more sustainable by working to make our local communities more sustainable. This begins with your own lifestyle. This is the meaning of the motto, "Think globally, act locally."

Throughout this book I have used various figures to list things that you can do to act as a responsible environmental citizen. I suggest that you review such actions by looking at Figures 11-25 (p. 222), 12-16 (p. 249), 14-30 (p. 303), 15-27 (p. 330), 18-36 (p. 408), 20-22 (p. 460), 21-19 (p. 483), 21-26 (p. 488), 22-20 (p. 516), 24-4 (p. 536), 24-7 (p. 540), 24-24 (p. 554), 27-2 (p. 608) and the more general list in Figure 28-5 (p. 637).

As you review these ideas, I suggest you mark off the things you are doing. Then try to pick out two or three of the items in each list that you believe are the most important things to do and combine them into a master list. Each week try to carry out at least one of these actions until you have worked through your entire list.

What Is the Earth Charter? Four Ethical Guidelines

Respect and care for life and biodiversity and build more sustainable, just, democratic, and peaceful societies for present and future generations.

In March 2000, the Earth Charter was finalized. More than 100,000 people in 51 countries and 25 global leaders in environment, business, politics, religion, and education took part in creating this charter. It is a document creating an ethical and moral framework to guide the conduct of people and nations toward each other and the earth. Here are its four guiding principles:

■ Respect earth and life in all its diversity.

■ Care for life with understanding, love, and compassion.

■ Build societies that are free, just, participatory, sustainable, and peaceful.

■ Secure earth's bounty and beauty for present and future generations.

It is an incredibly exciting time to be alive as we struggle to implement such ideals by entering into a new relationship with the earth that keeps us all alive and supports our economies. The transition to a more sustainable world will not be easy but it can be done if enough of us care.

Envision the earth's life-sustaining processes as a beautiful and diverse web of interrelationships—a kaleidoscope of patterns, rhythms, and connections whose very complexity and multitude of possibilities remind us that cooperation, sharing, honesty, humility, and love should be the guidelines for our behavior toward one another and the earth.

When there is no dream, the people perish.
PROVERBS 29:18

CRITICAL THINKING

1. Some analysts argue that the problems with Biosphere 2 resulted mostly from inadequate design and that a better team of scientists and engineers could make it work. Explain why you agree or disagree with this view.

2. This chapter has summarized a number of different environmental worldviews. Analyze them and find the beliefs you agree with, as a description your own environmental worldview. Which of your beliefs were added or modified as a result of taking this course? Compare your answer with those of your classmates.

3. Explain why you agree or disagree with the following ideas: **(a)** Everyone has the right to have as many children as he or she wants. **(b)** Each member of the human species has a right to use as many resources as he or she wants. **(c)** Individuals should have the right to do anything they want with land they own. **(d)** Nature should be used, not preserved. **(e)** Species exist to be used by humans. **(f)** All forms of life have an intrinsic value and therefore have a right to exist. **(g)** All organisms are interconnected and interdependent. **(h)** We have no right to harm and kill animals by using them for furs and to test for toxic chemicals, pharmaceutical drugs, or cosmetics. Are your answers consistent with the beliefs of your environmental worldview that you described in question 2?

4. Theologian Thomas Berry calls the industrial consumer society built on the human-centered, planetary management environmental worldview the "supreme pathology of all history." He says, "We can break the mountains apart; we can drain the rivers and flood the valleys. We can turn the most luxuriant forests into throwaway paper products. We can tear apart the great grass cover of the western plains, and pour toxic chemicals into the soil and pesticides onto the fields, until the soil is dead and blows away in the wind. We can pollute the air with acids, the rivers with sewage, the seas with oil. . . . We can invent computers capable of processing ten million calculations per second. And why? To increase the volume and speed with which we move natural resources through the consumer economy to the junk pile or the waste heap. . . . If, in these activities, the topography of the planet is damaged, if the environment

is made inhospitable for a multitude of living species, then so be it. We are, supposedly, creating a technological wonderworld. . . . But our supposed progress . . . is bringing us to a wasteworld instead of a wonderworld." Explain why you agree or disagree with this assessment.

5. Some analysts believe learning environmental wisdom by experiencing the earth and forming an emotional bond with its life-forms and processes is unscientific, mystical nonsense based on a romanticized view of nature. They believe better scientific understanding of how the earth works and improved technology are the best ways to achieve sustainability. Do you agree or disagree? Explain.

6. Try to answer the following fundamental ecological questions about the corner of the world you inhabit: Where does your water come from? Where does the energy you use come from? What kinds of soils are under your feet? What types of wildlife are your neighbors? Where does your food come from? Where does your waste go?

7. How do you feel about **(a)** carving huge faces of people in mountains, **(b)** driving an off-road motorized vehicle in a desert, grassland, or forest, **(c)** using throwaway paper towels, tissues, napkins, and plates, **(d)** wearing furs, and **(e)** having tropical fish, birds, snakes, or other wild animals as pets? Are your answers consistent with the beliefs of your environmental worldview that you described in question 2?

8. Review your experience with the mental traps described on p. 639. Which of these traps have you fallen into? Were you aware you had been ensnared by any of them? Do you plan to free yourself from these traps? How?

PROJECTS

1. Use a combination of the major existing societal, economic, and environmental trends, possible new trends, and your imagination to construct three different scenarios of what the world might be like in 2060. Identify the scenario you favor, and outline a strategy for achieving this alternative future. Compare your scenarios and strategies with those of your classmates.

2. Make an environmental audit of your school. Rate each of the following items on a scale of 1 through 10, with 10 being the highest rating. What proportion of each of the major types of matter resources used are recycled, reused, or composted? What priority does your school give to buying recycled materials? How much does your school emphasize energy efficiency, use of renewable forms of solar energy, and environmental design in developing new buildings and renovating existing ones? Does it use ecologically sound planning in deciding how its grounds and buildings are managed and used? Does your school limit the use of toxic chemicals in its buildings and on its grounds? What proportion of its food purchases comes from nearby farmers? What proportion of the food it purchases is grown by sustainable or organic agriculture? Average the number assigned to each category to come up with an overall environmental rating of your school. Compare your ratings with those of other members of your class and come up with a list of the five most important things that need to be done to improve the environmental rating of your school. Share your findings with school officials.

3. Does your school's curriculum provide *all* graduates with the basic elements of environmental literacy? To what extent are the funds in its financial endowments invested in enterprises that are working to develop or encourage environmental sustainability? Over the past 20 years, what important roles have its graduates played in making the world a better and more sustainable place to live? Using such information, rate your school on a 1–10 scale in terms of its contributions to environmental awareness and sustainability. Develop a detailed plan illustrating how your school could become better at achieving such goals, and present this information to school officials, alumni, parents, and financial backers.

4. If you knew you were going to die and had an opportunity to address everyone in the world for 5 minutes, what would you say? Write out your 5-minute speech and compare it with those of other members of your class.

5. Write an essay in which you identify key environmental experiences that have influenced your life and thus helped form your current ecological identity. Examples may include **(a)** fond childhood memories of special places where you connected with the earth through emo-tional experiences, **(b)** places you knew and cherished that have been polluted, developed, or destroyed, **(c)** key events that forced you to think about environmental values or worldviews, **(d)** people or educational experiences that influenced your understanding of and concern about environmental problems and challenges, and **(e)** direct experience and contemplation of wild places. Share your experiences with other members of your class.

6. Use the library or the Internet to find bibliographic information about *Robert Cahn,* whose quote is found at the beginning of this chapter.

7. Make a concept map of this chapter's major ideas, using the section heads, subheads, and key terms (in boldface type). Look on the website for this book for information about making concept maps.

LEARNING ONLINE

The website for this book contains study aids and many ideas for further reading and research. They include a chapter summary, review questions for the entire chapter, flash cards for key terms and concepts, a multiple-choice practice quiz, interesting Internet sites, references, and a guide for accessing thousands of InfoTrac® College Edition articles. Log on to

http://biology.brookscole.com/miller14

Then click on the Chapter-by-Chapter area, choose Chapter 28, and select a learning resource.

A P P E N D I X   1

UNITS OF MEASURE

LENGTH

Metric

1 kilometer (km) = 1,000 meters (m)
1 meter (m) = 100 centimeters (cm)
1 meter (m) = 1,000 millimeters (mm)
1 centimeter (cm) = 0.01 meter (m)
1 millimeter (mm) = 0.001 meter (m)

English

1 foot (ft) = 12 inches (in)
1 yard (yd) = 3 feet (ft)
1 mile (mi) = 5,280 feet (ft)
1 nautical mile = 1.15 miles

Metric–English

1 kilometer (km) = 0.621 mile (mi)
1 meter (m) = 39.4 inches (in)
1 inch (in) = 2.54 centimeters (cm)
1 foot (ft) = 0.305 meter (m)
1 yard (yd) = 0.914 meter (m)
1 nautical mile = 1.85 kilometers (km)

AREA

Metric

1 square kilometer (km^2) = 1,000,000 square meters (m^2)
1 square meter (m^2) = 1,000,000 square millimeters (mm^2)
1 hectare (ha) = 10,000 square meters (m^2)
1 hectare (ha) = 0.01 square kilometer (km^2)

English

1 square foot (ft^2) = 144 square inches (in^2)
1 square yard (yd^2) = 9 square feet (ft^2)
1 square mile (mi^2) = 27,880,000 square feet (ft^2)
1 acre (ac) = 43,560 square feet (ft^2)

Metric–English

1 hectare (ha) = 2.471 acres (ac)
1 square kilometer (km^2) = 0.386 square mile (mi^2)
1 square meter (m^2) = 1.196 square yards (yd^2)
1 square meter (m^2) = 10.76 square feet (ft^2)
1 square centimeter (cm^2) = 0.155 square inch (in^2)

VOLUME

Metric

1 cubic kilometer (km^3) = 1,000,000,000 cubic meters (m^3)
1 cubic meter (m^3) = 1,000,000 cubic centimeters (cm^3)
1 liter (L) = 1,000 milliliters (mL) = 1,000 cubic centimeters (cm^3)
1 milliliter (mL) = 0.001 liter (L)
1 milliliter (mL) = 1 cubic centimeter (cm^3)

English

1 gallon (gal) = 4 quarts (qt)
1 quart (qt) = 2 pints (pt)

Metric–English

1 liter (L) = 0.265 gallon (gal)
1 liter (L) = 1.06 quarts (qt)
1 liter (L) = 0.0353 cubic foot (ft^3)
1 cubic meter (m^3) = 35.3 cubic feet (ft^3)
1 cubic meter (m^3) = 1.30 cubic yards (yd^3)
1 cubic kilometer (km^3) = 0.24 cubic mile (mi^3)
1 barrel (bbl) = 159 liters (L)
1 barrel (bbl) = 42 U.S. gallons (gal)

MASS

Metric

1 kilogram (kg) = 1,000 grams (g)
1 gram (g) = 1,000 milligrams (mg)
1 gram (g) = 1,000,000 micrograms (μg)
1 milligram (mg) = 0.001 gram (g)
1 microgram (μg) = 0.000001 gram (g)
1 metric ton (mt) = 1,000 kilograms (kg)

English

1 ton (t) = 2,000 pounds (lb)
1 pound (lb) = 16 ounces (oz)

Metric–English

1 metric ton (mt) = 2,200 pounds (lb) = 1.1 tons (t)
1 kilogram (kg) = 2.20 pounds (lb)
1 pound (lb) = 454 grams (g)
1 gram (g) = 0.035 ounce (oz)

ENERGY AND POWER

Metric

1 kilojoule (kJ) = 1,000 joules (J)
1 kilocalorie (kcal) = 1,000 calories (cal)
1 calorie (cal) = 4,184 joules (J)

Metric–English

1 kilojoule (kJ) = 0.949 British thermal unit (Btu)
1 kilojoule (kJ) = 0.000278 kilowatt-hour (kW-h)
1 kilocalorie (kcal) = 3.97 British thermal units (Btu)
1 kilocalorie (kcal) = 0.00116 kilowatt-hour (kW-h)
1 kilowatt-hour (kW-h) = 860 kilocalories (kcal)
1 kilowatt-hour (kW-h) = 3,400 British thermal units (Btu)
1 quad (Q) = 1,050,000,000,000,000 kilojoules (kJ)
1 quad (Q) = 2,930,000,000,000 kilowatt-hours (kW-h)

TEMPERATURE CONVERSIONS

Fahrenheit(°F) to Celsius(°C):
  $°C = (°F - 32.0) \div 1.80$
Celsius(°C) to Fahrenheit(°F):
  $°F = (°C \times 1.80) + 32.0$

APPENDIX 2

MAJOR EVENTS IN U.S. ENVIRONMENTAL HISTORY

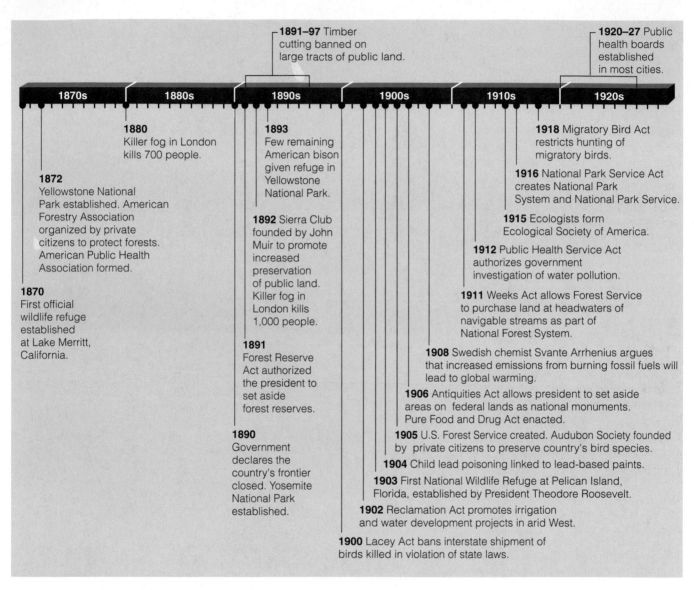

1891–97 Timber cutting banned on large tracts of public land.

1920–27 Public health boards established in most cities.

| 1870s | 1880s | 1890s | 1900s | 1910s | 1920s |

1880 Killer fog in London kills 700 people.

1893 Few remaining American bison given refuge in Yellowstone National Park.

1918 Migratory Bird Act restricts hunting of migratory birds.

1872 Yellowstone National Park established. American Forestry Association organized by private citizens to protect forests. American Public Health Association formed.

1892 Sierra Club founded by John Muir to promote increased preservation of public land. Killer fog in London kills 1,000 people.

1916 National Park Service Act creates National Park System and National Park Service.

1915 Ecologists form Ecological Society of America.

1870 First official wildlife refuge established at Lake Merritt, California.

1891 Forest Reserve Act authorized the president to set aside forest reserves.

1912 Public Health Service Act authorizes government investigation of water pollution.

1911 Weeks Act allows Forest Service to purchase land at headwaters of navigable streams as part of National Forest System.

1908 Swedish chemist Svante Arrhenius argues that increased emissions from burning fossil fuels will lead to global warming.

1906 Antiquities Act allows president to set aside areas on federal lands as national monuments. Pure Food and Drug Act enacted.

1890 Government declares the country's frontier closed. Yosemite National Park established.

1905 U.S. Forest Service created. Audubon Society founded by private citizens to preserve country's bird species.

1904 Child lead poisoning linked to lead-based paints.

1903 First National Wildlife Refuge at Pelican Island, Florida, established by President Theodore Roosevelt.

1902 Reclamation Act promotes irrigation and water development projects in arid West.

1900 Lacey Act bans interstate shipment of birds killed in violation of state laws.

1870–1930

Figure 1 Examples of the increased role of the federal government in resource conservation and public health and establishment of key private environmental groups, 1870–1930.

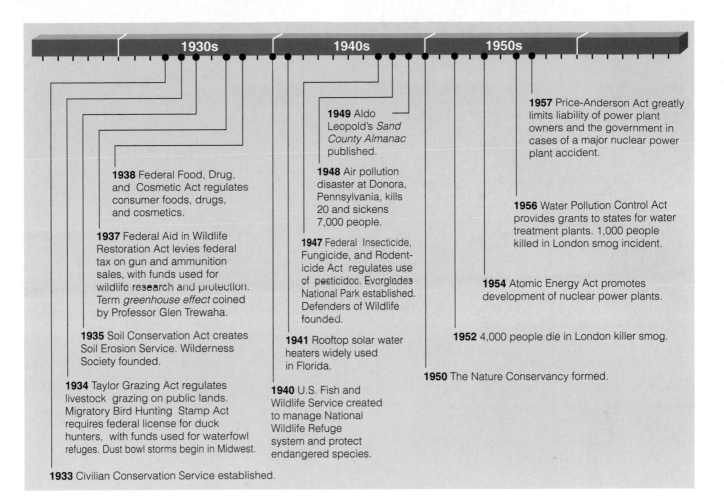

1930–1960

Figure 2 Some important conservation and environmental events, 1930–1960.

The timeline shows the following events:

1930s

1938 Federal Food, Drug, and Cosmetic Act regulates consumer foods, drugs, and cosmetics.

1937 Federal Aid in Wildlife Restoration Act levies federal tax on gun and ammunition sales, with funds used for wildlife research and protection. Term *greenhouse effect* coined by Professor Glen Trewaha.

1935 Soil Conservation Act creates Soil Erosion Service. Wilderness Society founded.

1934 Taylor Grazing Act regulates livestock grazing on public lands. Migratory Bird Hunting Stamp Act requires federal license for duck hunters, with funds used for waterfowl refuges. Dust bowl storms begin in Midwest.

1933 Civilian Conservation Service established.

1940s

1949 Aldo Leopold's *Sand County Almanac* published.

1948 Air pollution disaster at Donora, Pennsylvania, kills 20 and sickens 7,000 people.

1947 Federal Insecticide, Fungicide, and Rodenticide Act regulates use of pesticides. Everglades National Park established. Defenders of Wildlife founded.

1941 Rooftop solar water heaters widely used in Florida.

1940 U.S. Fish and Wildlife Service created to manage National Wildlife Refuge system and protect endangered species.

1950s

1957 Price-Anderson Act greatly limits liability of power plant owners and the government in cases of a major nuclear power plant accident.

1956 Water Pollution Control Act provides grants to states for water treatment plants. 1,000 people killed in London smog incident.

1954 Atomic Energy Act promotes development of nuclear power plants.

1952 4,000 people die in London killer smog.

1950 The Nature Conservancy formed.

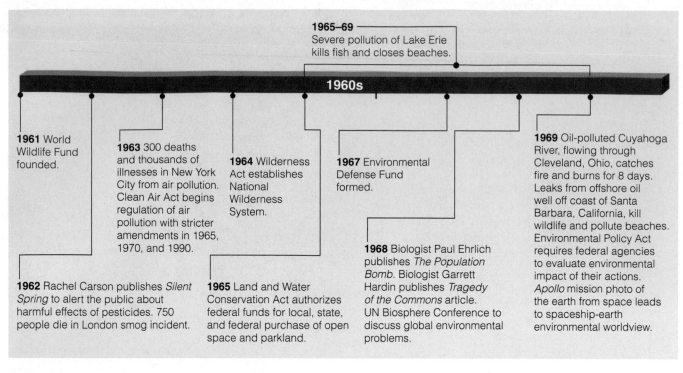

1965–69
Severe pollution of Lake Erie kills fish and closes beaches.

1960s

1961 World Wildlife Fund founded.

1963 300 deaths and thousands of illnesses in New York City from air pollution. Clean Air Act begins regulation of air pollution with stricter amendments in 1965, 1970, and 1990.

1964 Wilderness Act establishes National Wilderness System.

1967 Environmental Defense Fund formed.

1969 Oil-polluted Cuyahoga River, flowing through Cleveland, Ohio, catches fire and burns for 8 days. Leaks from offshore oil well off coast of Santa Barbara, California, kill wildlife and pollute beaches. Environmental Policy Act requires federal agencies to evaluate environmental impact of their actions. *Apollo* mission photo of the earth from space leads to spaceship-earth environmental worldview.

1962 Rachel Carson publishes *Silent Spring* to alert the public about harmful effects of pesticides. 750 people die in London smog incident.

1965 Land and Water Conservation Act authorizes federal funds for local, state, and federal purchase of open space and parkland.

1968 Biologist Paul Ehrlich publishes *The Population Bomb*. Biologist Garrett Hardin publishes *Tragedy of the Commons* article. UN Biosphere Conference to discuss global environmental problems.

1960s

Figure 3 Some important environmental events during the 1960s.

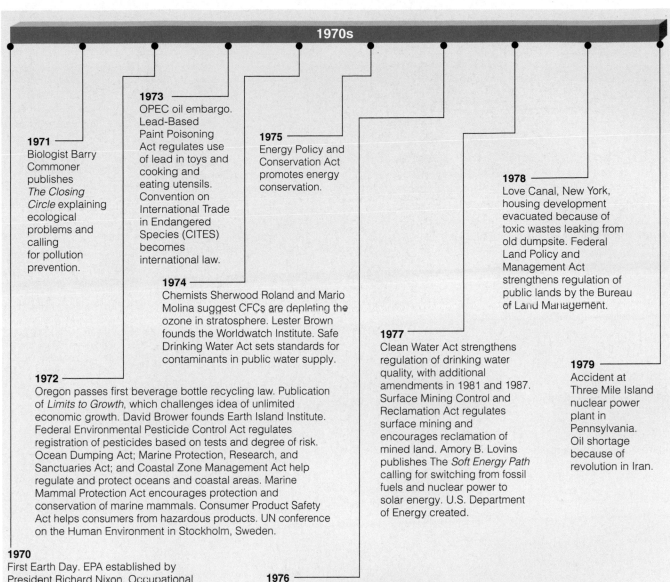

1970s

1970
First Earth Day. EPA established by President Richard Nixon. Occupational Health and Safety Act promotes safe working conditions. Resources Recovery Act regulates waste disposal and encourages recycling and waste reduction. National Environmental Policy Act passed. Clean Air Act passed. Natural Resources Defense Council created.

1971
Biologist Barry Commoner publishes *The Closing Circle* explaining ecological problems and calling for pollution prevention.

1972
Oregon passes first beverage bottle recycling law. Publication of *Limits to Growth*, which challenges idea of unlimited economic growth. David Brower founds Earth Island Institute. Federal Environmental Pesticide Control Act regulates registration of pesticides based on tests and degree of risk. Ocean Dumping Act; Marine Protection, Research, and Sanctuaries Act; and Coastal Zone Management Act help regulate and protect oceans and coastal areas. Marine Mammal Protection Act encourages protection and conservation of marine mammals. Consumer Product Safety Act helps consumers from hazardous products. UN conference on the Human Environment in Stockholm, Sweden.

1973
OPEC oil embargo. Lead-Based Paint Poisoning Act regulates use of lead in toys and cooking and eating utensils. Convention on International Trade in Endangered Species (CITES) becomes international law.

1974
Chemists Sherwood Roland and Mario Molina suggest CFCs are depleting the ozone in stratosphere. Lester Brown founds the Worldwatch Institute. Safe Drinking Water Act sets standards for contaminants in public water supply.

1975
Energy Policy and Conservation Act promotes energy conservation.

1976
National Forest Management Act establishes guidelines for managing national forests. Toxic Control Substances Act regulates many toxic substances not regulated under other laws. Resource Conservation and Recovery Act requires tracking of hazardous waste and encourages recycling, resource recovery, and waste reduction. Noise Control Act regulates harmful noise levels. UN Conference on Human Settlements.

1977
Clean Water Act strengthens regulation of drinking water quality, with additional amendments in 1981 and 1987. Surface Mining Control and Reclamation Act regulates surface mining and encourages reclamation of mined land. Amory B. Lovins publishes The *Soft Energy Path* calling for switching from fossil fuels and nuclear power to solar energy. U.S. Department of Energy created.

1978
Love Canal, New York, housing development evacuated because of toxic wastes leaking from old dumpsite. Federal Land Policy and Management Act strengthens regulation of public lands by the Bureau of Land Management.

1979
Accident at Three Mile Island nuclear power plant in Pennsylvania. Oil shortage because of revolution in Iran.

1970s

Figure 4 Some important environmental events during the 1970s, sometimes called the *environmental decade*.

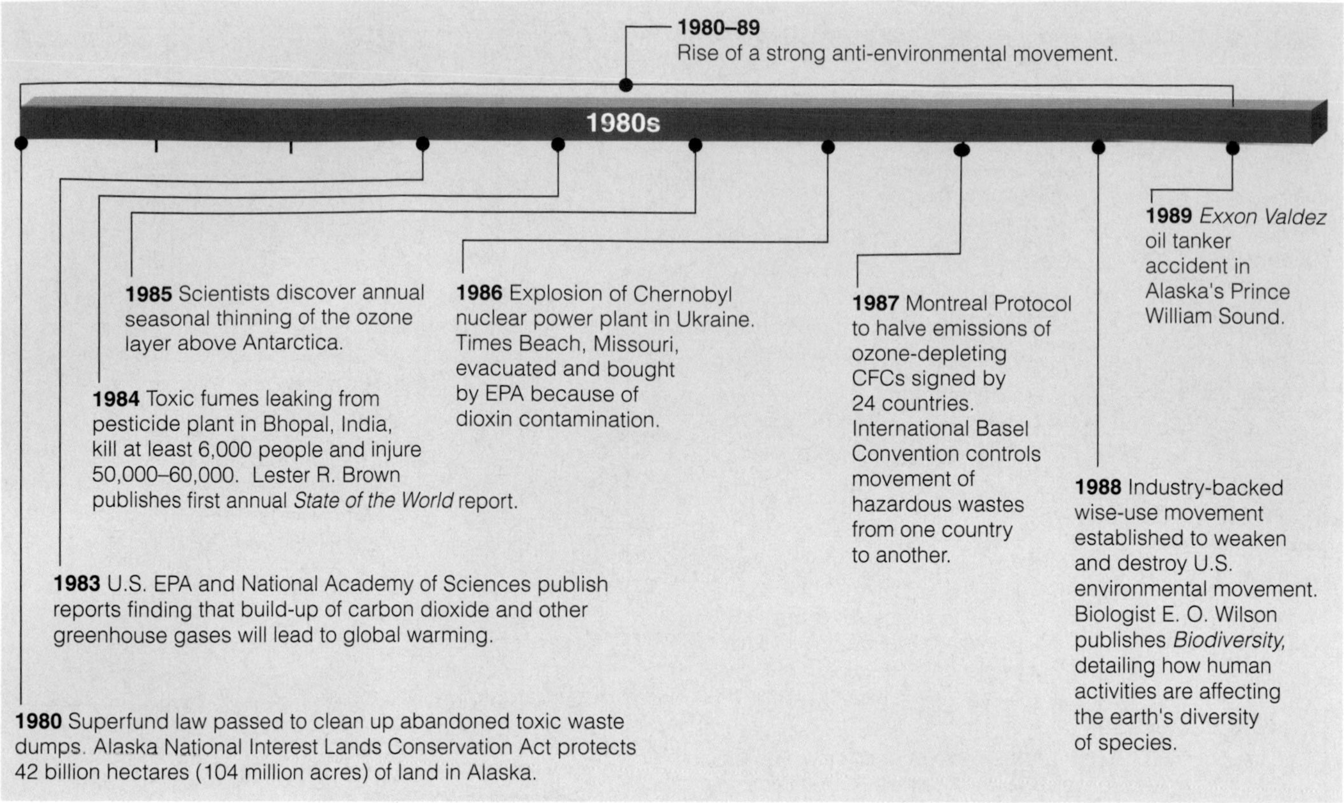

1980–89
Rise of a strong anti-environmental movement.

1980s

1985 Scientists discover annual seasonal thinning of the ozone layer above Antarctica.

1984 Toxic fumes leaking from pesticide plant in Bhopal, India, kill at least 6,000 people and injure 50,000–60,000. Lester R. Brown publishes first annual *State of the World* report.

1986 Explosion of Chernobyl nuclear power plant in Ukraine. Times Beach, Missouri, evacuated and bought by EPA because of dioxin contamination.

1987 Montreal Protocol to halve emissions of ozone-depleting CFCs signed by 24 countries. International Basel Convention controls movement of hazardous wastes from one country to another.

1989 *Exxon Valdez* oil tanker accident in Alaska's Prince William Sound.

1988 Industry-backed wise-use movement established to weaken and destroy U.S. environmental movement. Biologist E. O. Wilson publishes *Biodiversity*, detailing how human activities are affecting the earth's diversity of species.

1983 U.S. EPA and National Academy of Sciences publish reports finding that build-up of carbon dioxide and other greenhouse gases will lead to global warming.

1980 Superfund law passed to clean up abandoned toxic waste dumps. Alaska National Interest Lands Conservation Act protects 42 billion hectares (104 million acres) of land in Alaska.

1980s

Figure 5 Some important environmental events during the 1980s.

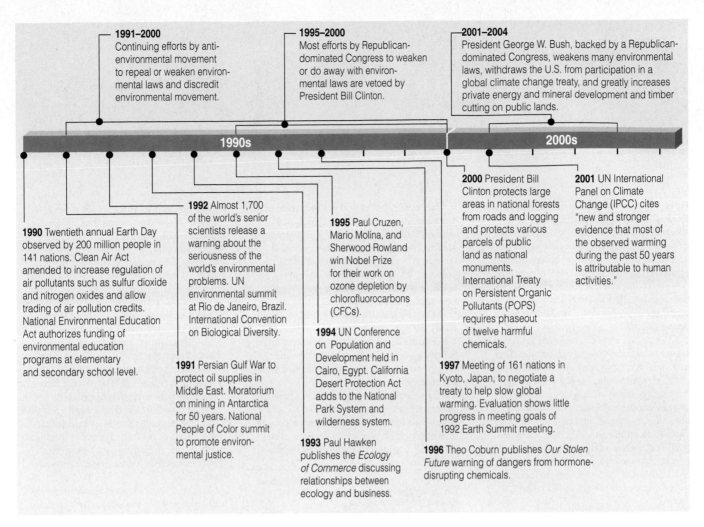

1991–2000 Continuing efforts by anti-environmental movement to repeal or weaken environmental laws and discredit environmental movement.

1995–2000 Most efforts by Republican-dominated Congress to weaken or do away with environmental laws are vetoed by President Bill Clinton.

2001–2004 President George W. Bush, backed by a Republican-dominated Congress, weakens many environmental laws, withdraws the U.S. from participation in a global climate change treaty, and greatly increases private energy and mineral development and timber cutting on public lands.

1990s

2000s

1990 Twentieth annual Earth Day observed by 200 million people in 141 nations. Clean Air Act amended to increase regulation of air pollutants such as sulfur dioxide and nitrogen oxides and allow trading of air pollution credits. National Environmental Education Act authorizes funding of environmental education programs at elementary and secondary school level.

1992 Almost 1,700 of the world's senior scientists release a warning about the seriousness of the world's environmental problems. UN environmental summit at Rio de Janeiro, Brazil. International Convention on Biological Diversity.

1991 Persian Gulf War to protect oil supplies in Middle East. Moratorium on mining in Antarctica for 50 years. National People of Color summit to promote environmental justice.

1995 Paul Cruzen, Mario Molina, and Sherwood Rowland win Nobel Prize for their work on ozone depletion by chlorofluorocarbons (CFCs).

1994 UN Conference on Population and Development held in Cairo, Egypt. California Desert Protection Act adds to the National Park System and wilderness system.

1993 Paul Hawken publishes the *Ecology of Commerce* discussing relationships between ecology and business.

2000 President Bill Clinton protects large areas in national forests from roads and logging and protects various parcels of public land as national monuments. International Treaty on Persistent Organic Pollutants (POPS) requires phaseout of twelve harmful chemicals.

1997 Meeting of 161 nations in Kyoto, Japan, to negotiate a treaty to help slow global warming. Evaluation shows little progress in meeting goals of 1992 Earth Summit meeting.

1996 Theo Coburn publishes *Our Stolen Future* warning of dangers from hormone-disrupting chemicals.

2001 UN International Panel on Climate Change (IPCC) cites "new and stronger evidence that most of the observed warming during the past 50 years is attributable to human activities."

1990–2004

Figure 6 Some important environmental events, 1990–2004.

APPENDIX 3

SOME BASIC CHEMISTRY

How Can Elements Be Arranged in the Periodic Table According to Their Chemical Properties? Chemists have developed a way to classify the elements according to their chemical behavior, in what is called the *periodic table of elements* (Figure 1). Each of the horizontal rows in the table is called a *period*. Each vertical column lists elements with similar chemical properties and is called a *group*.

The partial periodic table in Figure 1 shows how the elements can be classified as *metals*, *nonmetals*, and *metalloids*. Most of the elements found to the left and at the bottom of the table are *metals*, which usually conduct electricity and heat and are shiny. Examples are sodium (Na), calcium (Ca), aluminum (Al), iron (Fe), lead (Pb), and mercury (Hg). Atoms of such metals achieve a more stable state by losing one or more of their electrons to form positively charged ions such as Na^+, Ca^{2+}, and Al^{3+}.

Nonmetals, found in the upper right of the table, do not conduct electricity very well and usually are not shiny. Examples are hydrogen (H),* carbon (C), nitrogen (N), oxygen (O), phosphorus (P), sulfur (S), chlorine (Cl), and fluorine (F). Atoms of some nonmetals such as chlorine, oxygen, and sulfur tend to gain one or more electrons lost by metallic atoms to form negatively charged ions such as O^{2-}, S^{2-}, and Cl^-. Atoms of nonmetals can also combine with one another to form molecules in which they share one or more pairs of their electrons.

The elements arranged in a diagonal staircase pattern between the metals and nonmetals have a mixture of metallic and nonmetallic properties and are called *metalloids*. Figure 1 also identifies the ele-

---

*Hydrogen, a nonmetal, is placed by itself above the center of the table because it does not fit very well into any of the groups.

ments required as *nutrients* for all or some forms of life and elements that are moderately or highly toxic to all or most forms of life. Six nonmetallic elements—carbon (C) oxygen (O), hydrogen (H), nitrogen (N), sulfur (S), and phosphorus (P)—make up about 99% of the atoms of all living things.

What Are Ionic and Covalent Bonds? Sodium chloride (NaCl) consists of a three-dimensional network of oppositely charged *ions* (Na^+ and Cl^-) held together by the forces of attraction between opposite charges (Figure 2). The strong forces of attraction between such oppositely charged ions are called *ionic bonds*. Because ionic compounds consist of ions formed from atoms of metallic (positive ions) and nonmetallic (negative ions) elements, they can be described as *metal-nonmetal compounds*.

Figure 3 shows the chemical formulas and shapes of the molecules for several

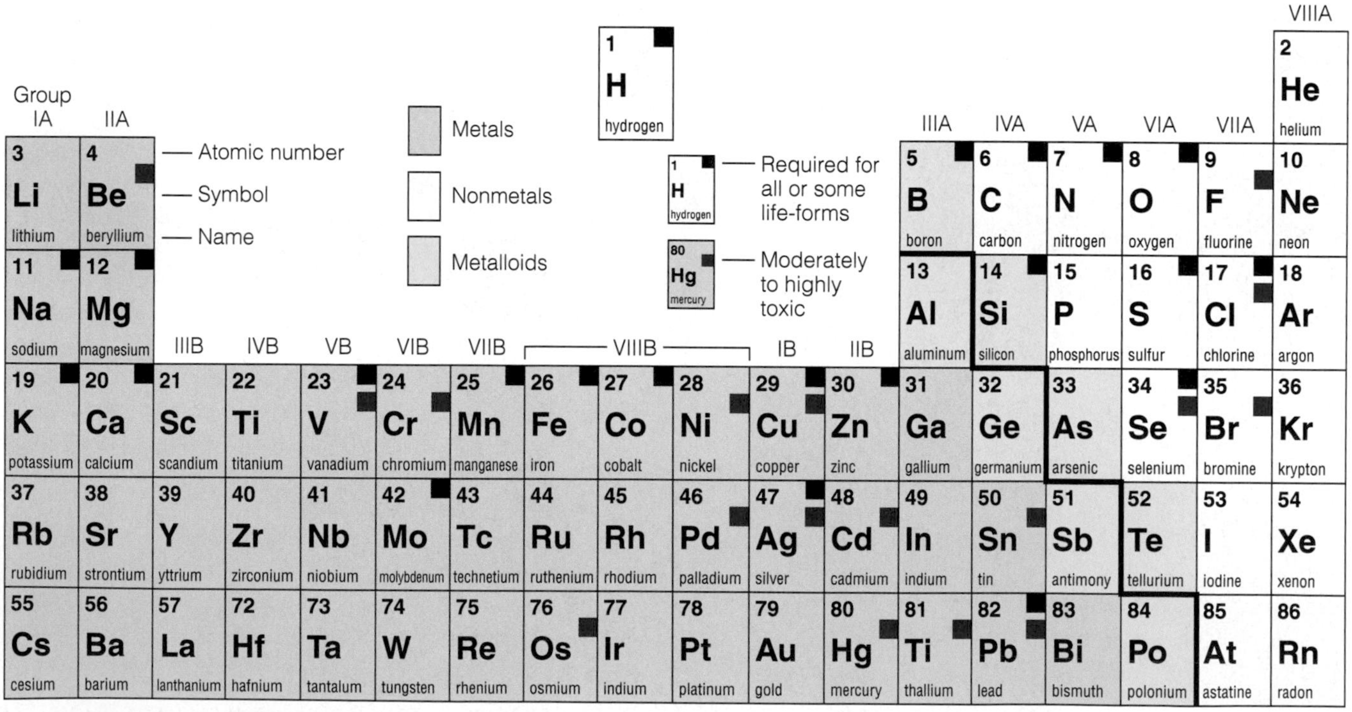

Figure 1 Abbreviated *periodic table of elements*. Elements in the same vertical column, called a *group*, have similar chemical properties. To simplify matters at this introductory level, only 72 of the 115 known elements are shown.

common covalent compounds, formed when atoms of one or more nonmetallic elements (Figure 1) combine with one another. The bonds between the atoms in such molecules are called *covalent bonds* and form when the atoms in the molecule share one or more pairs of their electrons. Because they are formed from atoms of nonmetallic elements, molecular or covalent compounds can be described as *nonmetal-nonmetal compounds*.

What Are Hydrogen Bonds? Ionic and covalent bonds form between the ions or atoms *within* a compound. There are also weaker forces of attraction *between* the molecules of covalent compounds (such as water) resulting from an unequal sharing of electrons by two atoms.

For example, an oxygen atom has a much greater attraction for electrons than does a hydrogen atom. Thus, in a water molecule, the electrons shared between the oxygen atom and its two hydrogen atoms are pulled closer to the oxygen atom, but not actually transferred to the oxygen atom. As a result, the oxygen atom in a water molecule has a slightly negative partial charge, and its two hydrogen atoms have a slightly positive partial charge (Figure 4, p. A-4, top).

The slightly positive hydrogen atoms in one water molecule are then attracted to the slightly negative oxygen atoms in another water molecule. These forces of attraction *between* water molecules are called *hydrogen bonds* (Figure 4, p. A-4, top). Hydrogen bonds also form between other covalent molecules or portions of such molecules containing hydrogen and nonmetallic atoms with a strong ability to attract electrons.

What Are Nucleic Acids? *Nucleic acids* are made by linking hundreds to thousands of four different types of monomers, called *nucleotides*. Each nucleotide consists of (1) a phosphate group, (2) a sugar molecule containing five carbon atoms (deoxyribose in DNA molecules and ribose in RNA molecules), and (3) one of four different nucleotide bases (represented by A, G, C, and T, the first letter in each of their names) (Figure 5, p. A-4).

In the cells of living organisms, these nucleotide units combine in different numbers and sequences to form *nucleic acids* such as various types of DNA and RNA. Hydrogen bonds formed between parts of the four nucleotides in DNA hold two DNA strands together like a spiral staircase, forming a double helix (Figure 6, p. A-4). DNA molecules can unwind and replicate themselves.

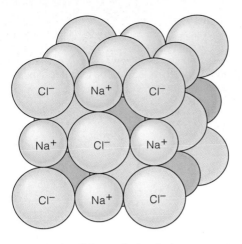

Figure 2 A solid crystal of an *ionic compound* such as sodium chloride consists of a three-dimensional array of opposite charged ions held together by *ionic bonds* resulting from the strong forces of attraction between opposite electrical charges. They are formed when an electron is transferred from a metallic atom such as sodium (Na) to a nonmetallic element such as chlorine (Cl). Such compounds tend to exist as solids at normal room temperature and atmospheric pressure.

H_2
hydrogen

O_2
oxygen

N$_2$
nitrogen

Cl_2
chlorine

NO
nitrogen oxide

CO
carbon monoxide

HCl
hydrogen chloride

H_2O
water

NO_2
nitrogen dioxide

CO_2
carbon dioxide

SO_2
sulfur dioxide

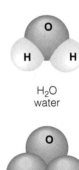
O_3
ozone

CH_4
methane

NH_3
ammonia

SO_3
sulfur trioxide

H_2S
hydrogen sulfide

Figure 3 Chemical formulas and shapes for some molecular compounds formed when atoms of one or more nonmetallic elements combine with one another. The bonds between the atoms in such molecules are called *covalent bonds*. Molecular compounds tend to exist as gases or liquids at normal room temperature and atmospheric pressure.

Figure 4 *Hydrogen bonds.* Slightly unequal sharing of electrons in the water molecule creates a molecule with a slightly negatively charged end and a slightly positively charged end. Because of this electrical polarity, hydrogen atoms of one water molecule are attracted to oxygen atoms of another water molecule. These forces of attraction *between* water molecules are called *hydrogen bonds.*

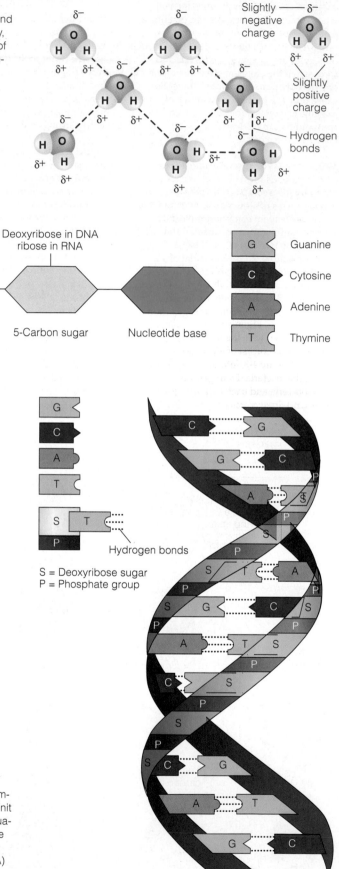

Figure 5 Generalized structure of nucleotide molecules linked in various numbers and sequences to form large nucleic acid molecules such as various types of DNA (deoxyribose nucleic acid) and RNA (ribose nucleic acid). In DNA the 5-carbon sugar in each nucleotide is deoxyribose; in RNA it is ribose. The four basic nucleotides used to make various forms of DNA molecules differ in the types of nucleotide bases they contain—guanine (G), cytosine (C), adenine (A), and thymine (T).

Figure 6 Portion of the double helix of a DNA molecule. The helix is composed of two spiral (helical) strands of nucleotides, each containing a unit of phosphate (P), deoxyribose (S), and one of four nucleotide bases: guanine (G), cytosine (C), adenine (A), and thymine (T). The two strands are held together by hydrogen bonds formed between various pairs of the nucleotide bases. Guanine (G) bonds with cytosine (C), and adenine (A) with thymine (T).

APPENDIX 4

CLASSIFYING AND NAMING SPECIES

How Can Species Be Classified? Biologists classify species into different *kingdoms,* on the basis of similarities and differences in characteristics such as their **(1)** modes of nutrition, **(2)** cell structure, **(3)** appearance, and **(4)** developmental features.

In this book, the earth's organisms are classified into six kingdoms: *eubacteria, archaebacteria, protists, fungi, plants,* and *animals.* Most bacteria, fungi, and protists are *microorganisms:* organisms so small they cannot be seen with the naked eye.

Eubacteria consist of all single-celled prokaryotic (Figure 4-3, left, p. 58) bacteria except archaebacteria. Examples are various cyanobacteria and bacteria such as *Staphylococcus* and *Streptococcus.*

Archaebacteria are single-celled bacteria that are evolutionarily closer to eukaryotic cells than to eubacteria. Examples are **(1)** methanogens that live in anaerobic sediments of lakes and swamps and in animal guts, **(2)** halophiles that live in extremely salty water, and **(3)** thermophiles that live in hot springs, hydrothermal vents, and acidic soil.

Protists (Protista) are mostly single-celled eukaryotic organisms such as diatoms, dinoflagellates, amoebas, golden brown and yellow-green algae, and protozoans. Some protists cause human diseases such as malaria and sleeping sickness.

Fungi are mostly many-celled, sometimes microscopic, eukaryotic organisms such as mushrooms, molds, mildews, and yeasts. Many fungi are decomposers. Other fungi kill various plants and cause huge losses of crops and valuable trees.

Plants (Plantae) are mostly many-celled eukaryotic organisms such as red, brown, and green algae and mosses, ferns, and flowering plants (whose flowers produce seeds that perpetuate the species). Some plants such as corn and marigolds are *annuals,* which complete their life cycles in one growing season; others are *perennials,* which can live for more than 2 years, such as roses, grapes, elms, and magnolias.

Animals (Animalia) are also many-celled eukaryotic organisms. Most, called *invertebrates,* have no backbones. They include sponges, jellyfish, worms, arthropods (insects, shrimp, and spiders), mollusks (snails, clams, and octopuses), and echinoderms (sea urchins and sea stars). Insects play roles that are vital to our existence (p. 64). *Vertebrates* (animals with backbones and a brain protected by skull bones) include fishes (sharks and tuna), amphibians (frogs and salamanders), reptiles (crocodiles and snakes), birds (eagles and robins), and mammals (bats, elephants, whales, and humans).

How Are Species Named? Within each kingdom, biologists have created subcategories based on anatomical, physiological, and behavioral characteristics. Kingdoms are divided into *phyla,* which are divided into subgroups called *classes.* Classes are subdivided into *orders,* which are further divided into *families.* Families consist of *genera* (singular, *genus*), and each genus contains one or more *species.* Note that the word *species* is both singular and plural. Figure 1 shows this detailed taxonomic classification for the current human species.

Most people call a species by its common name, such as robin or grizzly bear. Biologists use scientific names (derived from Latin) consisting of two parts (printed in italics or underlined) to describe a species. The first word is the capitalized name (or abbreviation) for the genus to which the organism belongs. This is followed by a lowercase name that distinguishes the species from other members of the same genus. For example, the scientific name of the robin is *Turdus migratorius* (Latin for "migratory thrush"), and the grizzly bear goes by the scientific name *Ursus horribilis* (Latin for "horrible bear").

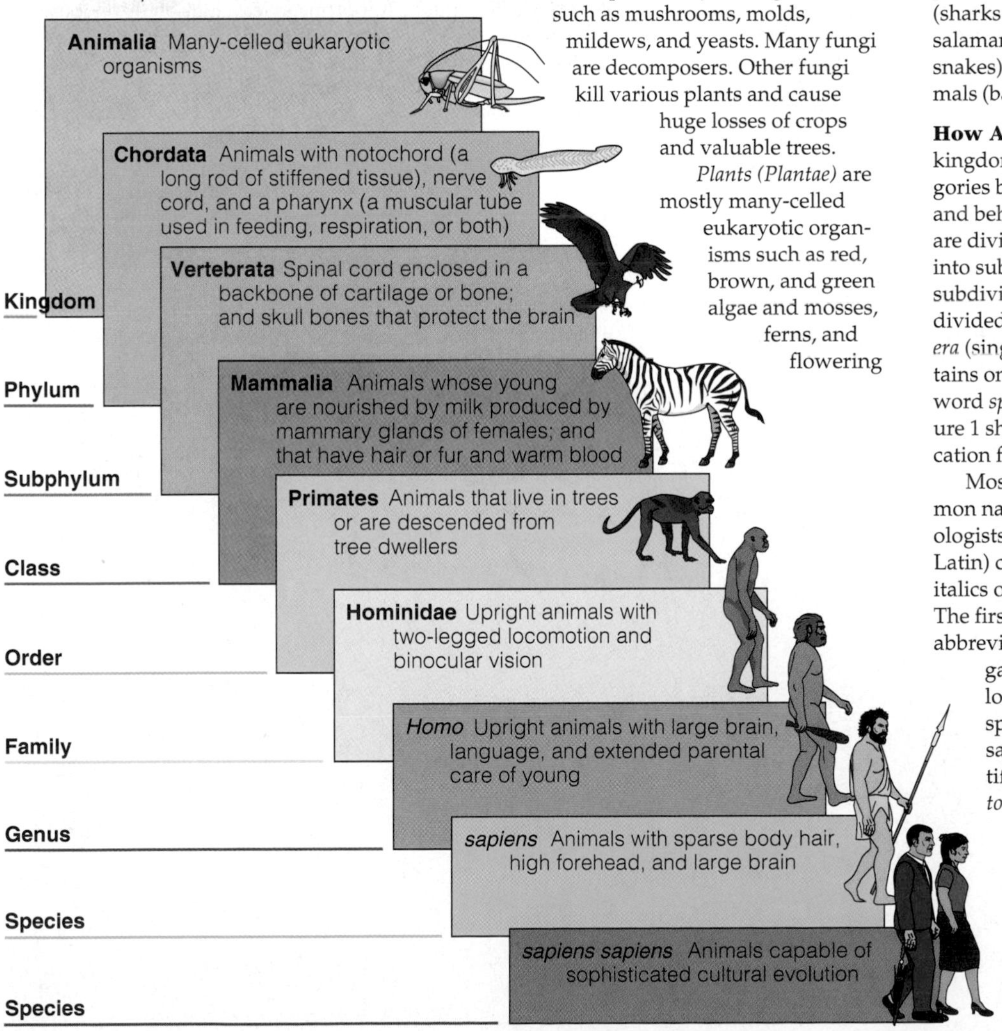

Kingdom
Animalia Many-celled eukaryotic organisms

Phylum
Chordata Animals with notochord (a long rod of stiffened tissue), nerve cord, and a pharynx (a muscular tube used in feeding, respiration, or both)

Subphylum
Vertebrata Spinal cord enclosed in a backbone of cartilage or bone; and skull bones that protect the brain

Class
Mammalia Animals whose young are nourished by milk produced by mammary glands of females; and that have hair or fur and warm blood

Order
Primates Animals that live in trees or are descended from tree dwellers

Family
Hominidae Upright animals with two-legged locomotion and binocular vision

Genus
Homo Upright animals with large brain, language, and extended parental care of young

Species
sapiens Animals with sparse body hair, high forehead, and large brain

Species
sapiens sapiens Animals capable of sophisticated cultural evolution

Figure 1 Taxonomic classification of the latest human species, *Homo sapiens sapiens.*

APPENDIX 5

BRIEF HISTORY OF THE AGE OF OIL

Some milestones in the Age of Oil:

- **1905:** Oil supplies 10% of U.S. energy.

- **1925:** United States produces 71% of the world's oil.

- **1930:** Because of an oil glut, oil sells for 10¢ a barrel.

- **1953:** U.S. oil companies account for about half of the world's oil production and the United States is the world's leading oil exporter.

- **1955:** United States has 20% of the world's estimated oil reserves.

- **1960:** OPEC formed so developing countries, with most of the world's known oil and projected oil reserves, can get a higher price for their oil.

- **1973:** United States uses 30% of the world's oil, imports 36% of this oil, and has only 5% of the world's proven oil reserves.

- **1973–1974:** OPEC reduces oil imports to the West and bans oil exports to the U.S. because of its support for Israel in the 18-day Yom Kippur War with Egypt and Syria. World oil prices rise sharply (Figure 17-11, p. 358) and lead to double-digit inflation in the United States and many other countries and a global economic recession.

- **1975:** Production of estimated U.S. oil reserves peaks.

- **1979:** Iran's Islamic Revolution shuts down most of Iran's oil production and reduces world oil production.

- **1981:** Iran–Iraq war pushes global oil prices to a historic high (Figure 17-13, p. 359).

- **1983:** Facing an oil glut, OPEC cuts its oil prices.

- **1985:** U.S. domestic oil production begins to decline and is not expected to increase enough to affect the global price of oil or to reduce U.S. dependence on oil imports (Figure 17-12, p. 358).

- **August 1990–June 1991:** United States and its allies fight the Persian Gulf War to oust Iraqi invaders of Kuwait and to protect Western access to Saudi Arabian and Kuwaiti oil supplies.

- **2004:** OPEC has 67% of world oil reserves and produces 40% of the world's oil. U.S. has only 2.9% of oil reserves, uses 26% of the world's oil production, and imports 55% of its oil.

- **2010:** U.S. could be importing at least 61% of the oil it uses as consumption continues to exceed production (Figure 17-12, p. 358).

- **2010–2030:** Production of oil from the world's estimated oil reserves is expected to peak. Oil prices expected to increase gradually as the demand for oil increasingly exceeds the supply—unless the world decreases demand by wasting less energy and shifting to other sources of energy.

- **2010–2048:** Domestic U.S. oil reserves projected to be 80% depleted.

- **2042–2083:** Gradual decline in dependence on oil.

GLOSSARY

abiotic Nonliving. Compare *biotic*.

acid See *acid solution*.

acid deposition The falling of acids and acid-forming compounds from the atmosphere to the earth's surface. Acid deposition is commonly known as *acid rain*, a term that refers only to wet deposition of droplets of acids and acid-forming compounds.

acid rain See *acid deposition*.

acid solution Any water solution that has more hydrogen ions (H^+) than hydroxide ions (OH^-); any water solution with a pH less than 7. Compare *basic solution, neutral solution*.

active solar heating system System that uses solar collectors to capture energy from the sun and store it as heat for space heating and water heating. Liquid or air pumped through the collectors transfers the captured heat to a storage system such as an insulated water tank or rock bed. Pumps or fans then distribute the stored heat or hot water throughout a dwelling as needed. Compare *passive solar heating system*.

adaptation Any genetically controlled structural, physiological, or behavioral characteristic that helps an organism survive and reproduce under a given set of environmental conditions. It usually results from a beneficial mutation. See *biological evolution, differential reproduction, mutation, natural selection*.

adaptive management Flexible management that views attempts to solve problems as experiments, analyzes failures to see what went wrong, and tries to modify and improve an approach before abandoning it. Because of the inherent unpredictability of complex systems, it often uses the precautionary principle as a management tool. See *precautionary principle*.

adaptive radiation Process in which numerous new species evolve to fill vacant and new ecological niches in changed environments, usually after a mass extinction. Typically, this takes millions of years.

adaptive trait See *adaptation*.

administrative laws Administrative rules and regulations, executive orders, and enforcement decisions related to the implementation and interpretation of statutory laws.

advanced sewage treatment Specialized chemical and physical processes that reduce the amount of specific pollutants left in wastewater after primary and secondary sewage treatment. This type of treatment usually is expensive. See also *primary sewage treatment, secondary sewage treatment*.

aerobic respiration Complex process that occurs in the cells of most living organisms, in which nutrient organic molecules such as glucose ($C_6H_{12}O_6$) combine with oxygen (O_2) and produce carbon dioxide (CO_2), water (H_2O), and energy. Compare *photosynthesis*.

affluenza Unsustainable addiction to overconsumption and materialism exhibited in the lifestyles of affluent consumers in the United States and other developed countries.

age structure Percentage of the population (or number of people of each sex) at each age level in a population.

agricultural revolution Gradual shift from small, mobile hunting and gathering bands to settled agricultural communities in which people survived by learning how to breed and raise wild animals and to cultivate wild plants near where they lived. It began 10,000–12,000 years ago. Compare *environmental revolution, hunter–gatherers, industrial–medical revolution, information and globalization revolution*.

agroecology See *sustainable agriculture, low-input agriculture, and organic farming*.

agroforestry Planting trees and crops together.

air pollution One or more chemicals in high enough concentrations in the air to harm humans, other animals, vegetation, or materials. Excess heat and noise are also considered forms of air pollution. Such chemicals or physical conditions are called air pollutants. See *primary pollutant, secondary pollutant*.

albedo Ability of a surface to reflect light.

alien species See *nonnative species*.

allele Slightly different molecular form found in a particular gene.

alley cropping Planting of crops in strips with rows of trees or shrubs on each side.

alpha particle Positively charged matter, consisting of two neutrons and two protons, that is emitted as a form of radioactivity from the nuclei of some radioisotopes. See also *beta particle, gamma rays*.

altitude Height above sea level. Compare *latitude*.

anaerobic respiration Form of cellular respiration in which some decomposers get the energy they need through the breakdown of glucose (or other nutrients) in the absence of oxygen. Compare *aerobic respiration*.

ancient forest See *old-growth forest*.

animal manure Dung and urine of animals used as a form of organic fertilizer. Compare *green manure*.

annual Plant that grows, sets seed, and dies in one growing season. Compare *perennial*.

anthropocentric Human-centered. Compare *biocentric*.

aquaculture Growing and harvesting of fish and shellfish for human use in freshwater ponds, irrigation ditches, and lakes, or in cages or fenced-in areas of coastal lagoons and estuaries. See *fish farming, fish ranching*.

aquatic Pertaining to water. Compare *terrestrial*.

aquatic life zone Marine and freshwater portions of the biosphere. Examples include freshwater life zones (such as lakes and streams) and ocean or marine life zones (such as estuaries, coastlines, coral reefs, and the deep ocean).

aquifer Porous, water-saturated layers of sand, gravel, or bedrock that can yield an economically significant amount of water.

arable land Land that can be cultivated to grow crops.

area strip mining Type of surface mining used where the terrain is flat. An earthmover strips away the overburden, and a power shovel digs a cut to remove the mineral deposit. After removal of the mineral, the trench is filled with overburden, and a new cut is made parallel to the previous one. The process is repeated over the entire site. Compare *dredging, mountaintop removal, open-pit mining, subsurface mining*.

arid Dry. A desert or other area with an arid climate has little precipitation.

artificial selection Process by which humans select one or more desirable genetic traits in the population of a plant or animal species and then use *selective breeding* to produce populations containing many individuals with the desired traits. Compare *genetic engineering, natural selection*.

asexual reproduction Reproduction in which a mother cell divides to produce two identical daughter cells that are clones of the mother cell. This type of reproduction is common in single-celled organisms. Compare *sexual reproduction*.

atmosphere The whole mass of air surrounding the earth. See *stratosphere, troposphere*.

atmospheric pressure A measure of the mass per unit area of air.

atom Minute unit made of subatomic particles that is the basic building block of all chemical elements and thus all matter; the smallest unit of an element that can exist and still have the unique characteristics of that element. Compare *ion, molecule*.

atomic number Number of protons in the nucleus of an atom. Compare *mass number*.

autotroph See *producer*.

background extinction Normal extinction of various species as a result of changes in local environmental conditions. Compare *mass depletion, mass extinction*.

bacteria Prokaryotic, one-celled organisms. Some transmit diseases. Most act as decomposers and get the nutrients they need by breaking down complex organic compounds in the tissues of living or dead organisms into simpler inorganic nutrient compounds.

barrier islands Long, thin, low offshore islands of sediment that generally run parallel to the shore along some coasts.

basic solution Water solution with more hydroxide ions (OH^-) than hydrogen ions (H^+); water solution with a pH greater than 7. Compare *acid solution, neutral solution*.

benthos Bottom-dwelling organisms. Compare *decomposer, nekton, plankton*.

beta particle Swiftly moving electron emitted by the nucleus of a radioactive isotope. See also *alpha particle, gamma rays*.

bioaccumulation An increase in the concentration of a chemical in specific organs or tissues at a level higher than would normally be expected. Compare *biomagnification*.

biocentric Life-centered. Compare *anthropocentric*.

biocides See *pesticides*.

biodegradable Capable of being broken down by decomposers.

biodegradable pollutant Material that can be broken down into simpler substances (elements and compounds) by bacteria or other decomposers. Paper and most organic wastes such as animal manure are biodegradable but can take decades to biodegrade in modern landfills. Compare *degradable pollutant, nondegradable pollutant, slowly degradable pollutant*.

biodiversity Variety of different species (*species diversity*), genetic variability among individuals within each species (*genetic diversity*), variety of ecosystems (*ecological diversity*), and functions such as energy flow and matter cycling needed for the survival of species and biological communities (*functional diversity*).

biofuel Gas or liquid fuel (such as ethyl alcohol) made from plant material (biomass).

biogeochemical cycle Natural processes that recycle nutrients in various chemical forms from the nonliving environment to living organisms and then back to the nonliving environment. Examples are the carbon, oxygen, nitrogen, phosphorus, sulfur, and hydrologic cycles.

bioinformatics Applied science of managing, analyzing, and communicating biological information.

biological amplification See *biomagnification*.

biological community See *community*.

biological diversity See *biodiversity*.

biological evolution Change in the genetic makeup of a population of a species in successive generations. If continued long enough, it can lead to the formation of a new species. Note that populations—not individuals—evolve. See also *adaptation, differential reproduction, natural selection, theory of evolution*.

biological oxygen demand (BOD) Amount of dissolved oxygen needed by aerobic decomposers to break down the organic materials in a given volume of water at a certain temperature over a specified time period.

biological pest control Control of pest populations by natural predators, parasites, or disease-causing bacteria and viruses (pathogens).

biomagnification Increase in concentration of DDT, PCBs, and other slowly degradable, fat-soluble chemicals in organisms at successively higher trophic levels of a food chain or web. Compare *bioaccumulation*.

biomass Organic matter produced by plants and other photosynthetic producers; total dry weight of all living organisms that can be supported at each trophic level in a food chain or web; dry weight of all organic matter in plants and animals in an ecosystem; plant materials and animal wastes used as fuel.

biome Terrestrial regions inhabited by certain types of life, especially vegetation. Examples are various types of deserts, grasslands, and forests.

biopharming Use of genetically engineered animals to act as biofactories for producing drugs, vaccines, antibodies, hormones, industrial chemicals such as plastics and detergents, and human body organs.

biosphere Zone of earth where life is found. It consists of parts of the atmosphere (the troposphere), hydrosphere (mostly surface water and groundwater), and lithosphere (mostly soil and surface rocks and sediments on the bottoms of oceans and other bodies of water) where life is found. Sometimes called the *ecosphere*.

biotic Living organisms. Compare *abiotic*.

biotic potential Maximum rate at which the population of a given species can increase when there are no limits on its rate of growth. See *environmental resistance*.

birth rate See *crude birth rate*.

bitumen Gooey, black, high-sulfur, heavy oil extracted from tar sand and then upgraded to synthetic fuel oil. See *tar sand*.

breeder nuclear fission reactor Nuclear fission reactor that produces more nuclear fuel than it consumes by converting nonfissionable uranium-238 into fissionable plutonium-239.

broadleaf deciduous plants Plants such as oak and maple trees that survive drought and cold by shedding their leaves and becoming dormant. Compare *broadleaf evergreen plants, coniferous evergreen plants*.

broadleaf evergreen plants Plants that keep most of their broad leaves year-round. Examples are the trees found in the canopies of tropical rain forests. Compare *broadleaf deciduous plants, coniferous evergreen plants*.

buffer Substance that can react with hydrogen ions in a solution and thus hold the acidity or pH of a solution fairly constant. See *pH*.

calorie Unit of energy; amount of energy needed to raise the temperature of 1 gram of water 1°C (unit on Celsius temperature scale). See also *kilocalorie*.

cancer Group of more than 120 different diseases, one for each type of cell in the human body. Each type of cancer produces a tumor in which cells multiply uncontrollably and invade surrounding tissue.

carbon cycle Cyclic movement of carbon in different chemical forms from the environment to organisms and then back to the environment.

carcinogen Chemicals, ionizing radiation, and viruses that cause or promote the development of cancer. See *cancer*. Compare *mutagen, teratogen*.

carnivore Animal that feeds on other animals. Compare *herbivore, omnivore*.

carrying capacity (K) Maximum population of a particular species that a given habitat can support over a given period.

cell Smallest living unit of an organism. Each cell is encased in an outer membrane or wall and contains genetic material (DNA) and other parts to perform its life function. Organisms such as bacteria consist of only one cell, but most of the organisms we are familiar with contain many cells. See *eukaryotic cell, prokaryotic cell*.

CFCs   See *chlorofluorocarbons*.

chain reaction   Multiple nuclear fissions, taking place within a certain mass of a fissionable isotope, that release an enormous amount of energy in a short time.

chemical   One of the millions of different elements and compounds found naturally and synthesized by humans. See *compound, element*.

chemical change   Interaction between chemicals in which there is a change in the chemical composition of the elements or compounds involved. Compare *nuclear change, physical change*.

chemical evolution   Formation of the earth and its early crust and atmosphere, evolution of the biological molecules necessary for life, and evolution of systems of chemical reactions needed to produce the first living cells. These processes are believed to have occurred about 1 billion years before biological evolution. Compare *biological evolution*.

chemical formula   Shorthand way to show the number of atoms (or ions) in the basic structural unit of a compound. Examples are H_2O, $NaCl$, and $C_6H_{12}O_6$.

chemical reaction   See *chemical change*.

chemosynthesis   Process in which certain organisms (mostly specialized bacteria) extract inorganic compounds from their environment and convert them into organic nutrient compounds without the presence of sunlight. Compare *photosynthesis*.

chlorinated hydrocarbon   Organic compound made up of atoms of carbon, hydrogen, and chlorine. Examples are DDT and PCBs.

chlorofluorocarbons (CFCs)   Organic compounds made up of atoms of carbon, chlorine, and fluorine. An example is Freon-12 (CCl_2F_2), used as a refrigerant in refrigerators and air conditioners and in making plastics such as Styrofoam. Gaseous CFCs can deplete the ozone layer when they slowly rise into the stratosphere and their chlorine atoms react with ozone molecules.

chromosome   A grouping of various genes and associated proteins in plant and animal cells that carry certain types of genetic information. See *genes*.

chronic undernutrition   An ongoing condition suffered by people who cannot grow or buy enough food to meet their basic energy need. Compare *malnutrition, overnutrition*.

civil suit   Lawsuit in which a plaintiff seeks to collect damages for injuries or for economic loss or have the court issue a permanent injunction against further wrongful action. Compare *class action suit*.

class action suit   Civil lawsuit in which a group files a suit on behalf of a larger number of citizens who allege similar damages but who need not be listed and represented individually. Compare *civil suit*.

clear-cutting   Method of timber harvesting in which all trees in a forested area are removed in a single cutting. Compare *seed-tree cutting, selective cutting, shelterwood cutting, strip cutting*.

climate   Physical properties of the troposphere of an area based on analysis of its weather records over a long period (at least 30 years). The two main factors determining an area's climate are *temperature*, with its seasonal variations, and the amount and distribution of *precipitation*. Compare *weather*.

climax community   See *mature community*.

coal   Solid, combustible mixture of organic compounds with 30–98% carbon by weight, mixed with various amounts of water and small amounts of sulfur and nitrogen compounds. It forms in several stages as the remains of plants are subjected to heat and pressure over millions of years.

coal gasification   Conversion of solid coal to synthetic natural gas (SNG).

coal liquefaction   Conversion of solid coal to a liquid hydrocarbon fuel such as synthetic gasoline or methanol.

coastal wetland   Land along a coastline, extending inland from an estuary, that is covered with salt water all or part of the year. Examples are marshes, bays, lagoons, tidal flats, and mangrove swamps. Compare *inland wetland*.

coastal zone   Warm, nutrient-rich, shallow part of the ocean that extends from the high-tide mark on land to the edge of a shelflike extension of continental land masses known as the continental shelf. Compare *open sea*.

coevolution   Evolution in which two or more species interact and exert selective pressures on each other that can lead each species to undergo various adaptations. See *evolution, natural selection*.

cogeneration   Production of two useful forms of energy, such as high-temperature heat or steam and electricity, from the same fuel source.

cold front   Leading edge of an advancing mass of cold air. Compare *warm front*.

commensalism   An interaction between organisms of different species in which one type of organism benefits and the other type is neither helped nor harmed to any great degree. Compare *mutualism*.

commercial extinction   Depletion of the population of a wild species used as a resource to a level at which it is no longer profitable to harvest the species.

commercial inorganic fertilizer   Commercially prepared mixture of plant nutrients such as nitrates, phosphates, and potassium applied to the soil to restore fertility and increase crop yields. Compare *organic fertilizer*.

common law   Body of unwritten rules and principles derived from thousands of past legal decisions. It is based on evaluation of what is reasonable behavior in attempting to balance competing social interests. Compare *statutory law*.

common-property resource   Resource that people normally are free to use; each user can deplete or degrade the available supply. Most are renewable and owned by no one. Examples are clean air, fish in parts of the ocean not under the control of a coastal country, migratory birds, gases of the lower atmosphere, and the ozone content of the upper atmosphere (stratosphere). See *tragedy of the commons*.

community   Populations of all species living and interacting in an area at a particular time.

competition   Two or more individual organisms of a single species (*intraspecific competition*) or two or more individuals of different species (*interspecific competition*) attempting to use the same scarce resources in the same ecosystem.

competitive exclusion principle   No two species can occupy exactly the same fundamental niche indefinitely in a habitat where there is not enough of a particular resource to meet the needs of both species. See *ecological niche, fundamental niche, realized niche*.

complexity   In ecological terms, refers to the number of species in a community at each trophic level and the number of trophic levels in a community.

compost   Partially decomposed organic plant and animal matter used as a soil conditioner or fertilizer.

compound   Combination of atoms, or oppositely charged ions, of two or more different elements held together by attractive forces called chemical bonds. Compare *element*.

concentration   Amount of a chemical in a particular volume or weight of air, water, soil, or other medium.

condensation nuclei   Tiny particles on which droplets of water vapor can collect.

coniferous evergreen plants   Cone-bearing plants (such as spruces, pines, and firs) that keep some of their narrow, pointed leaves (needles) all year. Compare *broadleaf deciduous plants, broadleaf evergreen plants*.

coniferous trees   Cone-bearing trees, mostly evergreens, that have needle-shaped or scalelike leaves. They produce wood known commercially as softwood. Compare *deciduous plants*.

consensus science   See *sound science*.

conservation   Sensible and careful use of natural resources by humans. People with this view are called *conservationists*.

conservation biologist   Biologist who investigates human impacts on the diversity of life found on the earth (biodiversity) and develops practical plans for preserving such biodiversity. Compare *conservationist, ecologist, environmentalist, environmental scientist, preservationist, restorationist*.

conservation biology Multidisciplinary science created to deal with the crisis of maintaining the genes, species, communities, and ecosystems that make up earth's biological diversity. Its goals are to investigate human impacts on biodiversity and to develop practical approaches to preserving biodiversity.

conservationist Person concerned with using natural areas and wildlife in ways that sustain them for current and future generations of humans and other forms of life. Compare *conservation biologist, ecologist, environmentalist, environmental scientist, preservationist, restorationist.*

conservation-tillage farming Crop cultivation in which the soil is disturbed little (minimum-tillage farming) or not at all (no-till farming) to reduce soil erosion, lower labor costs, and save energy. Compare *conventional-tillage farming.*

constancy Ability of a living system, such as a population, to maintain a certain size. Compare *inertia, resilience.* See *homeostasis.*

consumer Organism that cannot synthesize the organic nutrients it needs and gets its organic nutrients by feeding on the tissues of producers or of other consumers; generally divided into *primary consumers* (herbivores), *secondary consumers* (carnivores), *tertiary (higher-level) consumers, omnivores,* and *detritivores* (decomposers and detritus feeders). In economics, one who uses economic goods.

contour farming Plowing and planting across the changing slope of land, rather than in straight lines, to help retain water and reduce soil erosion.

contour strip mining Form of surface mining used on hilly or mountainous terrain. A power shovel cuts a series of terraces into the side of a hill. An earthmover removes the overburden, and a power shovel extracts the coal, with the overburden from each new terrace dumped onto the one below. Compare *area strip mining, dredging, mountaintop removal, open-pit mining, subsurface mining.*

controlled burning Deliberately set, carefully controlled surface fires that reduce flammable litter and decrease the chances of damaging crown fires. See *ground fire, surface fire.*

conventional-tillage farming Crop cultivation method in which a planting surface is made by plowing land, breaking up the exposed soil, and then smoothing the surface. Compare *conservation-tillage farming.*

convergent plate boundary Area where earth's lithospheric plates are pushed together. See *subduction zone.* Compare *divergent plate boundary, transform fault.*

coral reef Formation produced by massive colonies containing billions of tiny coral animals, called polyps, that secrete a stony substance (calcium carbonate) around themselves for protection. When the corals die, their empty outer skeletons form layers and cause the reef to grow. They are found in the coastal zones of warm tropical and subtropical oceans.

core Inner zone of the earth. It consists of a solid inner core and a liquid outer core. Compare *crust, mantle.*

corrective feedback loop See *negative feedback loop.*

corridors Long areas of land connecting habitat that would otherwise become fragmented.

cost–benefit analysis (CBA) Estimates and comparison of short-term and long-term benefits (gains) and costs (losses) from an economic decision.

cover crops The planting of crops such as alfalfa, clover, or rye immediately after harvest to help protect and hold the soil.

critical mass Amount of fissionable nuclei needed to sustain a nuclear fission chain reaction.

crop rotation Planting a field, or an area of a field, with different crops from year to year to reduce soil nutrient depletion. A plant such as corn, tobacco, or cotton, which removes large amounts of nitrogen from the soil, is planted one year. The next year a legume such as soybeans, which adds nitrogen to the soil, is planted.

crown fire Extremely hot forest fire that burns ground vegetation and treetops. Compare *controlled burning, ground fire, surface fire.*

crude birth rate Annual number of live births per 1,000 people in the population of a geographic area at the midpoint of a given year. Compare *crude death rate.*

crude death rate Annual number of deaths per 1,000 people in the population of a geographic area at the midpoint of a given year. Compare *crude birth rate.*

crude oil Gooey liquid consisting mostly of hydrocarbon compounds and small amounts of compounds containing oxygen, sulfur, and nitrogen. Extracted from underground accumulations, it is sent to oil refineries, where it is converted to heating oil, diesel fuel, gasoline, tar, and other materials.

crust Solid outer zone of the earth. It consists of oceanic crust and continental crust. Compare *core, mantle.*

cultural eutrophication Overnourishment of aquatic ecosystems with plant nutrients (mostly nitrates and phosphates) because of human activities such as agriculture, urbanization, and discharges from industrial plants and sewage treatment plants. See *eutrophication.*

cyanobacteria Single-celled, prokaryotic, microscopic organisms. Before being reclassified as monera, they were called blue-green algae.

DDT Dichlorodiphenyltrichloroethane, a chlorinated hydrocarbon that has been widely used as an insecticide but is now banned in some countries.

death rate See *crude death rate.*

debt-for-nature swap Agreement in which a certain amount of foreign debt is canceled in exchange for local currency

investments that will improve natural resource management or protect certain areas in the debtor country from harmful development.

deciduous plants Trees, such as oaks and maples, and other plants that survive during dry seasons or cold seasons by shedding their leaves. Compare *coniferous trees, succulent plants.*

decomposer Organism that digests parts of dead organisms and cast-off fragments and wastes of living organisms by breaking down the complex organic molecules in those materials into simpler inorganic compounds and then absorbing the soluble nutrients. Producers return most of these chemicals to the soil and water for reuse. Decomposers consist of various bacteria and fungi. Compare *consumer, detritivore, producer.*

deductive reasoning Using logic to arrive at a specific conclusion based on a generalization or premise. It goes from the general to the specific. Compare *inductive reasoning.*

defendant The individual, group of individuals, corporation, or government agency being charged in a lawsuit. Compare *plaintiff.*

deforestation Removal of trees from a forested area without adequate replanting.

degradable pollutant Potentially polluting chemical that is broken down completely or reduced to acceptable levels by natural physical, chemical, and biological processes. Compare *biodegradable pollutant, nondegradable pollutant, slowly degradable pollutant.*

degree of urbanization Percentage of the population in the world, or a country, living in areas with a population of more than 2,500 people (higher in some countries). Compare *urban growth.*

democracy Government by the people through their elected officials and appointed representatives. In a *constitutional democracy,* a constitution provides the basis of government authority and puts restraints on government power through free elections and freely expressed public opinion.

demographic transition Hypothesis that countries, as they become industrialized, have declines in death rates followed by declines in birth rates.

demography The study of the size, composition, and distribution of human populations and the causes and consequences of changes in these characteristics.

depletion time The time it takes to use a certain fraction, usually 80%, of the known or estimated supply of a nonrenewable resource at an assumed rate of use. Finding and extracting the remaining 20% usually costs more than it is worth.

desalination Purification of salt water or brackish (slightly salty) water by removal of dissolved salts.

desert Biome in which evaporation exceeds precipitation and the average amount of precipitation is less than 25 centimeters (10 inches) a year. Such areas have little vegetation or have widely spaced, mostly low vegetation. Compare *forest, grassland.*

desertification   Conversion of rangeland, rain-fed cropland, or irrigated cropland to desertlike land, with a drop in agricultural productivity of 10% or more. It usually is caused by a combination of overgrazing, soil erosion, prolonged drought, and climate change.

detritivore   Consumer organism that feeds on detritus, parts of dead organisms, and cast-off fragments and wastes of living organisms. The two principal types are *detritus feeders* and *decomposers.*

detritus   Parts of dead organisms and cast-off fragments and wastes of living organisms.

detritus feeder   Organism that extracts nutrients from fragments of dead organisms and their cast-off parts and organic wastes. Examples are earthworms, termites, and crabs. Compare *decomposer.*

deuterium (D; hydrogen-2)   Isotope of the element hydrogen, with a nucleus containing one proton and one neutron and a mass number of 2.

developed country   Country that is highly industrialized and has a high per capita GNP. Compare *developing country.*

developing country   Country that has low to moderate industrialization and low to moderate per capita GNP. Most are located in Africa, Asia, and Latin America. Compare *developed country.*

dieback   Sharp reduction in the population of a species when its numbers exceed the carrying capacity of its habitat. See *carrying capacity.*

differential reproduction   Phenomenon in which individuals with adaptive genetic traits produce more living offspring than do individuals without such traits. See *natural selection.*

dioxins   Family of 75 different chlorinated hydrocarbon compounds formed as unwanted by-products in chemical reactions involving chlorine and hydrocarbons, usually at high temperatures.

discount rate   The economic value a resource will have in the future compared with its present value.

dissolved oxygen (DO) content   Amount of oxygen gas (O_2) dissolved in a given volume of water at a particular temperature and pressure, often expressed as a concentration in parts of oxygen per million parts of water.

distribution   Area over which we can find a species. See *range.*

disturbance   A discrete event that disrupts an ecosystem or community. Examples of *natural disturbances* include fires, hurricanes, tornadoes, droughts, and floods. Examples of *human-caused disturbances* include deforestation, overgrazing, and plowing.

divergent plate boundary   Area where earth's lithospheric plates move apart in opposite directions. Compare *convergent plate boundary, transform fault.*

DNA (deoxyribonucleic acid)   Large molecules in the cells of organisms that carry genetic information in living organisms.

domesticated species   Wild species tamed or genetically altered by crossbreeding for use by humans for food (cattle, sheep, and food crops), pets (dogs and cats), or enjoyment (animals in zoos and plants in gardens). Compare *wild species.*

dose   The amount of a potentially harmful substance an individual ingests, inhales, or absorbs through the skin. Compare *response.* See *dose-response curve, median lethal dose.*

dose-response curve   Plot of data showing effects of various doses of a toxic agent on a group of test organisms. See *dose, median lethal dose, response.*

doubling time   The time it takes (usually in years) for the quantity of something growing exponentially to double. It can be calculated by dividing the annual percentage growth rate into 70.

drainage basin   See *watershed.*

dredging   Type of surface mining in which chain buckets and draglines scrape up sand, gravel, and other surface deposits covered with water. It is also used to remove sediment from streams and harbors to maintain shipping channels. See *dredge spoils.* Compare *area strip mining, contour strip mining, mountaintop removal, open-pit mining, subsurface mining.*

drift-net fishing   Catching fish in huge nets that drift in the water.

drought   Condition in which an area does not get enough water because of lower-than-normal precipitation or higher-than-normal temperatures that increase evaporation.

early successional plant species   Plant species found in the early stages of succession that grow close to the ground, can establish large populations quickly under harsh conditions, and have short lives. Compare *late successional plant species, mid-successional plant species.*

earthquake   Shaking of the ground resulting from the fracturing and displacement of rock, which produces a fault, or from subsequent movement along the fault.

ecological diversity   The variety of forests, deserts, grasslands, oceans, streams, lakes, and other biological communities interacting with one another and with their nonliving environment. See *biodiversity.* Compare *functional diversity, genetic diversity, species diversity.*

ecological efficiency   Percentage of energy transferred from one trophic level to another in a food chain or web.

ecological footprint   Amount of biologically productive land and water needed to supply each person or population with the renewable resources they use and to absorb or dispose of the wastes from such resource use. It measures the average environmental impact of individuals or populations in different countries and areas.

ecological niche   Total way of life or role of a species in an ecosystem. It includes all physical, chemical, and biological conditions a species needs to live and reproduce in an ecosystem. See *fundamental niche, realized niche.*

ecological restoration   Deliberate alteration of a degraded habitat or ecosystem to restore as much of its ecological structure and function as possible.

ecological succession   Process in which communities of plant and animal species in a particular area are replaced over time by a series of different and often more complex communities. See *primary succession, secondary succession.*

ecologist   Biological scientist who studies relationships between living organisms and their environment. Compare *conservation biologist, conservationist, environmentalist, environmental scientist, preservationist, restorationist.*

ecology   Study of the interactions of living organisms with one another and with their nonliving environment of matter and energy; study of the structure and functions of nature.

economic decision   Deciding what goods and services to produce, how to produce them, how much to produce, and how to distribute them to people.

economic depletion   Exhaustion of 80% of the estimated supply of a nonrenewable resource. Finding, extracting, and processing the remaining 20% usually costs more than it is worth. May also apply to the depletion of a renewable resource, such as a fish or tree species.

economic development   Improvement of living standards by economic growth. Compare *economic growth, environmentally sustainable economic development.*

economic growth   Increase in the capacity to provide people with goods and services produced by an economy; an increase in gross domestic product (GDP)). Compare *economic development, environmentally sustainable economic development, sustainable economic development.* See *gross domestic product.*

economic resources   Natural resources, capital goods, and labor used in an economy to produce material goods and services. See *natural resources.*

economics   The study of how individuals and societies choose to use limited or scarce resources to satisfy their unlimited wants.

economic system   Method that a group of people uses to choose what goods and services to produce, how to produce them, how much to produce, and how to distribute them to people. See *pure free-market economic system.*

economy   System of production, distribution, and consumption of economic goods.

ecosphere   See *biosphere.*

ecosystem Community of different species interacting with one another and with the chemical and physical factors making up its nonliving environment.

ecosystem services Natural services or natural capital that support life on the earth and are essential to the quality of human life and the functioning of the world's economies. See *natural resources.*

ecotone Transitional zone in which one type of ecosystem tends to merge with another ecosystem. See *edge effect.*

edge effect The existence of a greater number of species and a higher population density in a transition zone (ecotone) between two ecosystems than in either adjacent ecosystem. See *ecotone.*

electromagnetic radiation Forms of kinetic energy traveling as electromagnetic waves. Examples are radio waves, TV waves, microwaves, infrared radiation, visible light, ultraviolet radiation, X rays, and gamma rays. Compare *ionizing radiation, nonionizing radiation.*

electron (e) Tiny particle moving around outside the nucleus of an atom. Each electron has one unit of negative charge and almost no mass. Compare *neutron, proton.*

element Chemical, such as hydrogen (H), iron (Fe), sodium (Na), carbon (C), nitrogen (N), or oxygen (O), whose distinctly different atoms serve as the basic building blocks of all matter. Two or more elements combine to form compounds that make up most of the world's matter. Compare *compound.*

endangered species A wild species with so few individual survivors that the species could soon become extinct in all or most of its natural range. Compare *threatened species.*

endemic species Species that is found in only one area. Such species are especially vulnerable to extinction.

energy Capacity to do work by performing mechanical, physical, chemical, or electrical tasks or to cause a heat transfer between two objects at different temperatures.

energy efficiency Percentage of the total energy input that does useful work and is not converted into low-quality, usually useless heat in an energy conversion system or process. See *energy quality, net energy.* Compare *material efficiency.*

energy productivity See *energy efficiency.*

energy quality Ability of a form of energy to do useful work. High-temperature heat and the chemical energy in fossil fuels and nuclear fuels are concentrated high-quality energy. Low-quality energy such as low-temperature heat is dispersed or diluted and cannot do much useful work. See *high-quality energy, low-quality energy.*

environment All external conditions and factors, living and nonliving (chemicals and energy), that affect an organism or other specified system during its lifetime.

environmental degradation Depletion or destruction of a potentially renewable resource such as soil, grassland, forest, or

wildlife that is used faster than it is naturally replenished. If such use continues, the resource becomes nonrenewable (on a human time scale) or nonexistent (extinct). See also *sustainable yield.*

environmental ethics Human beliefs about what is right or wrong environmental behavior.

environmentalism A social movement dedicated to protecting the earth's life support systems for us and other species.

environmentalist Person who is concerned about the impact of people on environmental quality and believe that some human actions are degrading parts of the earth's life-support systems for humans and many other forms of life. Compare *conservation biologist, conservationist, ecologist, environmental scientist, preservationist, restorationist.*

environmental justice Fair treatment and meaningful involvement of all people regardless of race, color, sex, national origin, or income with respect to the development, implementation, and enforcement of environmental laws, regulations, and policies.

environmentally sustainable economic development Development that *encourages* forms of economic growth that meet the basic needs of the current generations of humans and other species without preventing future generations of humans and other species from meeting their basic needs and *discourages* environmentally harmful and unsustainable forms of economic growth. It is the economic component of an *environmentally sustainable society.* Compare *economic development, economic growth.*

environmentally sustainable society Society that satisfies the basic needs of its people without depleting or degrading its natural resources and thereby preventing current and future generations of humans and other species from meeting their basic needs.

environmental movement Efforts by citizens at the grassroots level to demand that political leaders enact laws and develop policies to curtail pollution, clean up polluted environments, and protect pristine areas and species from environmental degradation.

environmental policy Laws, rules, and regulations related to an environmental problem that are developed, implemented, and enforced by a particular government agency.

environmental resistance All the limiting factors that act together to limit the growth of a population. See *biotic potential, limiting factor.*

environmental revolution Cultural change involving halting population growth and altering lifestyles, political and economic systems, and the way we treat the environment so that we can help sustain the earth for ourselves and other species. This involves working with the rest of nature by learning more about how nature sustains itself. See *environmental wisdom worldview.* Compare *agricultural revolution, hunter-*

gatherers, industrial-medical hunter–gatherers, industrial–medical revolution, information and globalization revolution.

environmental science An interdisciplinary study that uses information from the physical sciences and social sciences tolerant of how the earth works, how we interact with the earth, and how to deal with environmental problems.

environmental scientist Scientist who uses information from the physical sciences and social sciences to understand how the earth works, learn how humans interact with the earth, and develop solutions to environmental problems. Compare *conservation biologist, conservationist, ecologist, preservationist, restorationist.*

environmental wisdom worldview Beliefs that **(1)** nature exists for all the earth's species, not just for us, and we are not in charge of the rest of nature; **(2)** resources are limited, should not be wasted, and are not all for us; **(3)** we should encourage earth-sustaining forms of economic growth and discourage earth-degrading forms of economic growth; and **(4)** our success depends on learning to cooperate with one another and with the rest of nature instead of trying to dominate and manage earth's life-support systems primarily for our own use. Compare *frontier environmental worldview, planetary management worldview, spaceship-earth worldview, stewardship worldview.*

environmental worldview How people think the world works, what they think their role in the world should be, and what they believe is right and wrong environmental behavior (environmental ethics).

EPA U.S. Environmental Protection Agency; responsible for managing federal efforts to control air and water pollution, radiation and pesticide hazards, environmental research, hazardous waste, and solid waste disposal.

epidemiology Study of the patterns of disease or other harmful effects from toxic exposure within defined groups of people to find out why some people get sick and some do not.

epiphyte Plant that uses its roots to attach itself to branches high in trees, especially in tropical forests.

erosion Process or group of processes by which loose or consolidated earth materials are dissolved, loosened, or worn away and removed from one place and deposited in another. See *weathering.*

estuary Partially enclosed coastal area at the mouth of a river where its freshwater, carrying fertile silt and runoff from the land, mixes with salty seawater.

eukaryotic cell Cell containing a *nucleus,* a region of genetic material surrounded by a membrane. Membranes also enclose several of the other internal parts found in a eukaryotic cell. Compare *prokaryotic cell.*

euphotic zone Upper layer of a body of water through which sunlight can penetrate and support photosynthesis.

eutrophication Physical, chemical, and biological changes that take place after a lake, estuary, or slow-flowing stream receives inputs of plant nutrients—mostly nitrates and phosphates—from natural erosion and runoff from the surrounding land basin. See *cultural eutrophication*.

eutrophic lake Lake with a large or excessive supply of plant nutrients, mostly nitrates and phosphates. Compare *mesotrophic lake, oligotrophic lake*.

evaporation Conversion of a liquid into a gas.

even-aged management Method of forest management in which trees, sometimes of a single species in a given stand, are maintained at about the same age and size and are harvested all at once. Compare *uneven-aged management*.

evergreen plants Plants that keep some of their leaves or needles throughout the year. Examples are ferns and cone-bearing trees (conifers) such as firs, spruces, pines, redwoods, and sequoias. Compare *deciduous plants, succulent plants*.

evolution See *biological evolution*.

exhaustible resource See *nonrenewable resource*.

existence value See *intrinsic value*.

exotic species See *nonnative species*.

experiment Procedure a scientist uses to study some phenomenon under known conditions. Scientists conduct some experiments in the laboratory and others in nature. The resulting scientific data or facts must be verified or confirmed by repeated observations and measurements, ideally by several different investigators.

exploitation competition Situation in which two competing species have equal access to a specific resource but differ in how quickly or efficiently they exploit it. See *interference competition, interspecific competition*.

exponential growth Growth in which some quantity, such as population size or economic output, increases at a constant rate per unit of time (such as 2% per year). An example is the growth sequence 2, 4, 8, 16, 32, 64 and so on; when the increase in quantity over time is plotted, this type of growth yields a curve shaped like the letter J. Compare *linear growth*.

external benefit Beneficial social effect of producing and using an economic good that is not included in the market price of the good. Compare *external cost, full cost*.

external cost Harmful social effect of producing and using an economic good that is not included in the market price of the good. Compare *external benefit, full cost, internal cost*.

externalities Social benefits ("goods") and social costs ("bads") not included in the market price of an economic good. See *external benefit, external cost*. Compare *full cost, internal cost*.

extinction Complete disappearance of a species from the earth. This happens when a species cannot adapt and successfully reproduce under new environmental conditions or when it evolves into one or more new species. Compare *speciation*. See also *endangered species, mass depletion, mass extinction, threatened species*.

family planning Providing information, clinical services, and contraceptives to help people choose the number and spacing of children they want to have.

famine Widespread malnutrition and starvation in a particular area because of a shortage of food, usually caused by drought, war, flood, earthquake, or other catastrophic events that disrupt food production and distribution.

feedback loop Circuit of sensing, evaluating, and reacting to changes in environmental conditions as a result of information fed back into a system; it occurs when one change leads to some other change, which eventually reinforces or slows the original change. See *negative feedback loop, positive feedback loop*.

feedlot Confined outdoor or indoor space used to raise hundreds to thousands of domesticated livestock. Compare *rangeland*.

fermentation See *anaerobic respiration*.

fertility The number of births that occur to an individual woman or in a population.

fertilizer Substance that adds inorganic or organic plant nutrients to soil and improves its ability to grow crops, trees, or other vegetation. See *commercial inorganic fertilizer, organic fertilizer*.

first law of thermodynamics In any physical or chemical change, no detectable amount of energy is created or destroyed, but in these processes energy can be changed from one form to another; you cannot get more energy out of something than you put in; in terms of energy quantity, you cannot get something for nothing (there is no free lunch). This law does not apply to nuclear changes, in which energy can be produced from small amounts of matter. See *second law of thermodynamics*.

fishery Concentrations of particular aquatic species suitable for commercial harvesting in a given ocean area or inland body of water.

fish farming Form of aquaculture in which fish are cultivated in a controlled pond or other environment and harvested when they reach the desired size. See also *fish ranching*.

fish ranching Form of aquaculture in which members of a fish species such as salmon are held in captivity for the first few years of their lives, released, and then harvested as adults when they return from the ocean to their freshwater birthplace to spawn. See also *fish farming*.

fissionable isotope Isotope that can split apart when hit by a neutron at the right speed and thus undergo nuclear fission. Examples are uranium-235 and plutonium-239.

floodplain Flat valley floor next to a stream channel. For legal purposes, the term often applies to any low area that has the potential for flooding, including certain coastal areas.

flows See *throughputs*.

flyway Generally fixed route along which waterfowl migrate from one area to another at certain seasons of the year.

food chain Series of organisms in which each eats or decomposes the preceding one. Compare *food web*.

food web Complex network of many interconnected food chains and feeding relationships. Compare *food chain*.

forest Biome with enough average annual precipitation (at least 76 centimeters, or 30 inches) to support growth of various tree species and smaller forms of vegetation. Compare *desert, grassland*.

fossil fuel Products of partial or complete decomposition of plants and animals that occur as crude oil, coal, natural gas, or heavy oils as a result of exposure to heat and pressure in the earth's crust over millions of years. See *coal, crude oil, natural gas*.

fossils Skeletons, bones, shells, body parts, leaves, seeds, or impressions of such items that provide recognizable evidence of organisms that lived long ago.

free-access resource See *common-property resource*.

freons See *chlorofluorocarbons*.

freshwater life zones Aquatic systems where water with a dissolved salt concentration of less than 1% by volume accumulates on or flows through the surfaces of terrestrial biomes. Examples are *standing* (lentic) bodies of freshwater such as lakes, ponds, and inland wetlands and *flowing* (lotic) systems such as streams and rivers. Compare *biome*.

front The boundary between two air masses with different temperatures and densities. See *cold front, warm front*.

frontier worldview Viewing undeveloped land as a hostile wilderness to be conquered (cleared, planted) and exploited for its resources as quickly as possible. Compare *environmental wisdom worldview, planetary management worldview, spaceship-earth worldview, stewardship worldview*.

frontier forest See *old-growth forest*.

frontier science Preliminary scientific data, hypotheses, and models that have not been widely tested and accepted. Compare *junk science, sound science*.

full cost Cost of a good when its internal costs and its estimated short- and long-term external costs are included in its market price. Compare *external cost, internal cost*.

functional diversity Biological and chemical processes or functions such as energy flow and matter cycling needed for the survival of species and biological communities. See *biodiversity, ecological diversity, genetic diversity, species diversity*.

fundamental niche The full potential range of the physical, chemical, and biological factors a species can use if there is no competition from other species. See *ecological niche*. Compare *realized niche*.

fungicide Chemical that kills fungi.

Gaia hypothesis Hypothesis that the earth is alive and can be considered a system that operates and changes by feedback of information between its living and nonliving components.

gamma rays A form of ionizing electromagnetic radiation with a high energy content emitted by some radioisotopes. They readily penetrate body tissues. See also *alpha particle, beta particle*.

gangue Waste or undesired material in an ore. See *ore*.

GDP See *gross domestic product*.

gene flow Movement of genes between populations, which can lead to changes in the genetic composition of local populations.

gene mutation See *mutation*.

gene pool The sum total of all genes found in the individuals of the population of a particular species.

generalist species Species with a broad ecological niche. They can live in many different places, eat a variety of foods, and tolerate a wide range of environmental conditions. Examples are flies, cockroaches, mice, rats, and human beings. Compare *specialist species*.

genes Coded units of information about specific traits that are passed on from parents to offspring during reproduction. They consist of segments of DNA molecules found in chromosomes.

gene splicing See *genetic engineering*.

genetic adaptation Changes in the genetic makeup of organisms of a species that allow the species to reproduce and gain a competitive advantage under changed environmental conditions. See *differential reproduction, evolution, mutation, natural selection*.

genetically modified organism (GMO) Organism whose genetic makeup has been modified by genetic engineering.

genetic diversity Variability in the genetic makeup among individuals within a single species. See *biodiversity*. Compare *ecological diversity, functional diversity, species diversity*.

genetic drift Change in the genetic composition of a population by chance. It is especially important for small populations.

genetic engineering Insertion of an alien gene into an organism to give it a beneficial genetic trait. Compare *artificial selection, natural selection*.

genome Complete set of genetic information for an organism.

geographic isolation Separation of populations of a species for long times into different areas.

geology Study of the earth's dynamic history. Geologists study and analyze rocks and the features and processes of the earth's interior and surface.

geothermal energy Heat transferred from the earth's underground concentrations of dry steam (steam with no water droplets), wet steam (a mixture of steam and water droplets), or hot water trapped in fractured or porous rock.

global climate change A broad term that refers to changes in the earth's climate mostly as a result of changes in temperature and precipitation.

globalization Broad process of global social, economic, and environmental change that leads to an increasingly integrated world. See *information and globalization revolution*.

global warming Warming of the earth's atmosphere because of increases in the concentrations of one or more greenhouse gases primarily as a result of human activities. See *greenhouse effect, greenhouse gases*.

grassland Biome found in regions where moderate annual average precipitation (25–76 centimeters, or 10–30 inches) is enough to support the growth of grass and small plants but not enough to support large stands of trees. Compare *desert, forest*.

greenhouse effect A natural effect that releases heat in the atmosphere (troposphere) near the earth's surface. Water vapor, carbon dioxide, ozone, and several other gases in the lower atmosphere (troposphere) absorb some of the infrared radiation (heat) radiated by the earth's surface. This causes their molecules to vibrate and transform the absorbed energy into longer-wavelength infrared radiation (heat) in the troposphere. If the atmospheric concentrations of these greenhouse gases rise and they are not removed by other natural processes, the average temperature of the lower atmosphere will increase gradually. Compare *global warming*. See also *natural greenhouse effect*.

greenhouse gases Gases in the earth's lower atmosphere (troposphere) that cause the greenhouse effect. Examples are carbon dioxide, chlorofluorocarbons, ozone, methane, water vapor, and nitrous oxide.

green manure Freshly cut or still-growing green vegetation that is plowed into the soil to increase the organic matter and humus available to support crop growth. Compare *animal manure*.

green revolution Popular term for introduction of scientifically bred or selected varieties of grain (rice, wheat, maize) that, with high enough inputs of fertilizer and water, can greatly increase crop yields.

gross domestic product (GDP) Annual market value of all goods and services produced by all firms and organizations, foreign and domestic, operating within a country.

gross primary productivity (GPP) The rate at which an ecosystem's producers capture and store a given amount of chemical energy as biomass in a given length of time. Compare *net primary productivity*.

ground fire Fire that burns decayed leaves or peat deep below the ground surface. Compare *crown fire, surface fire*.

groundwater Water that sinks into the soil and is stored in slowly flowing and slowly renewed underground reservoirs called aquifers; underground water in the zone of saturation, below the water table. Compare *runoff, surface water*.

habitat Place or type of place where an organism or population of organisms lives. Compare *ecological niche*.

habitat fragmentation Breakup of a habitat into smaller pieces, usually as a result of human activities.

half-life Time needed for one-half of the nuclei in a radioisotope to emit its radiation. Each radioisotope has a characteristic half-life, which may range from a few millionths of a second to several billion years. See *radioisotope*.

hazard Something that can cause injury, disease, economic loss, or environmental damage. See also *risk*.

hazardous chemical Chemical that can cause harm because it is flammable or explosive, can irritate or damage the skin or lungs (such as strong acidic or alkaline substances), or can cause allergic reactions of the immune system (allergens). See also *toxic chemical*.

hazardous waste Any solid, liquid, or containerized gas that can catch fire easily, is corrosive to skin tissue or metals, is unstable and can explode or release toxic fumes, or has harmful concentrations of one or more toxic materials that can leach out. See also *toxic waste*.

heat Total kinetic energy of all the randomly moving atoms, ions, or molecules within a given substance, excluding the overall motion of the whole object. Heat always flows spontaneously from a hot sample of matter to a colder sample of matter. This is one way to state the second law of thermodynamics. Compare *temperature*.

herbicide Chemical that kills a plant or inhibits its growth.

herbivore Plant-eating organism. Examples are deer, sheep, grasshoppers, and zooplankton. Compare *carnivore, omnivore*.

heterotroph See *consumer*.

high An air mass with a high pressure. Compare *low*.

high-input agriculture See *industrialized agriculture*.

high-quality energy Energy that is concentrated and has great ability to perform useful work. Examples are high-temperature heat and the energy in electricity, coal, oil, gasoline, sunlight, and nuclei of uranium-235. Compare *low-quality energy*.

high-quality matter Matter that is concentrated and contains a high concentration of a useful resource. Compare *low-quality matter*.

high-throughput economy The situation in most advanced industrialized countries, in which ever-increasing economic growth is sustained by maximizing the rate at which matter and energy resources are used, with little emphasis on pollution prevention, recycling, reuse, reduction of unnecessary waste, and other forms of resource conservation. Compare *low-throughput economy, matter-recycling economy*.

HIPPO Acronym for habitat destruction and fragmentation, invasive species, population growth, pollution, and over-harvesting.

homeostasis Maintenance of favorable internal conditions in a system despite fluctuations in external conditions. See *constancy, inertia, resilience*.

host Plant or animal on which a parasite feeds.

human capital See *human resources*.

human resources Physical and mental talents of people used to produce, distribute, and sell an economic good. Compare *manufactured resources, natural resources*.

humus Slightly soluble residue of undigested or partially decomposed organic material in topsoil. This material helps retain water and water-soluble nutrients, which can be taken up by plant roots.

hunter–gatherers People who get their food by gathering edible wild plants and other materials and by hunting wild animals and fish. Compare *agricultural revolution, environmental revolution, industrial–medical revolution, information and globalization revolution*.

hydrocarbon Organic compound of hydrogen and carbon atoms. The simplest hydrocarbon is methane (CH_4), the major component of natural gas.

hydroelectric power plant Structure in which the energy of falling or flowing water spins a turbine generator to produce electricity.

hydrologic cycle Biogeochemical cycle that collects, purifies, and distributes the earth's fixed supply of water from the environment to living organisms and then back to the environment.

hydropower Electrical energy produced by falling or flowing water. See *hydroelectric power plant*.

hydrosphere The earth's *liquid water* (oceans, lakes, other bodies of surface water, and underground water), *frozen water* (polar ice caps, floating ice caps, and ice in soil, known as permafrost), and *water vapor* in the atmosphere. See also *hydrologic cycle*.

identified resources Deposits of a particular mineral-bearing material of which the location, quantity, and quality are known or have been estimated from direct geological evidence and measurements. Compare *undiscovered resources*.

igneous rock Rock formed when molten rock material (magma) wells up from the earth's interior, cools, and solidifies into rock masses. Compare *metamorphic rock, sedimentary rock*. See *rock cycle*.

immature community Community at an early stage of ecological succession. It usually has a low number of species and ecological niches and cannot capture and use energy and cycle critical nutrients as efficiently as more complex, mature communities. Compare *mature community*.

immigrant species See *nonnative species*.

immigration Migration of people into a country or area to take up permanent residence.

indicator species Species that serve as early warnings that a community or ecosystem is being degraded. Compare *keystone species, native species, nonnative species*.

inductive reasoning Using observations and facts to arrive at generalizations or hypotheses. It goes from the specific to the general and is widely used in science. Compare *deductive reasoning*.

industrialized agriculture Using large inputs of energy from fossil fuels (especially oil and natural gas), water, fertilizer, and pesticides to produce large quantities of crops and livestock for domestic and foreign sale. Compare *subsistence farming*.

industrial–medical revolution Use of new sources of energy from fossil fuels and later from nuclear fuels, and use of new technologies, to grow food and manufacture products. Compare *agricultural revolution, environmental revolution, hunter–gatherers, information and globalization revolution*.

industrial smog Type of air pollution consisting mostly of a mixture of sulfur dioxide, suspended droplets of sulfuric acid formed from some of the sulfur dioxide, and a variety of suspended solid particles. Compare *photochemical smog*.

inertia Ability of a living system to resist being disturbed or altered. Compare *constancy, resilience*.

infant mortality rate Number of babies out of every 1,000 born each year that die before their first birthday.

infiltration Downward movement of water through soil.

information and globalization revolution Use of new technologies such as the telephone, radio, television, computers, the Internet, automated databases, and remote sensing satellites to enable people to have increasingly rapid access to much more information on a global scale. Compare *agricultural revolution, environmental revolution, hunter–gatherers, industrial–medical revolution*.

inherent value See *intrinsic value*.

inland wetland Land away from the coast, such as a swamp, marsh, or bog, that is covered all or part of the time with freshwater. Compare *coastal wetland*.

inorganic compounds All compounds not classified as organic compounds. See *organic compounds*.

inorganic fertilizer See *commercial inorganic fertilizer*.

input Matter, energy, or information entering a system. Compare *output, throughput*.

input pollution control See *pollution prevention*.

insecticide Chemical that kills insects.

instrumental value Value of an organism, species, ecosystem, or the earth's biodiversity based on its usefulness to us. Compare *intrinsic value*.

integrated pest management (IPM) Combined use of biological, chemical, and cultivation methods in proper sequence and timing to keep the size of a pest population below the size that causes economically unacceptable loss of a crop or livestock animal.

intercropping Growing two or more different crops at the same time on a plot. For example, a carbohydrate-rich grain that depletes soil nitrogen and a protein-rich legume that adds nitrogen to the soil may be intercropped. Compare *monoculture, polyculture, polyvarietal cultivation*.

interference competition Situation in which one species limits access of another species to a resource, regardless of whether the resource is abundant or scarce. See *exploitation competition, interspecific competition*.

internal cost Direct cost paid by the producer and the buyer of an economic good. Compare *external benefit, external cost, full cost*.

interplanting Simultaneously growing a variety of crops on the same plot. See *agroforestry, intercropping, polyculture, polyvarietal cultivation*.

interspecific competition Attempts by members of two or more species to use the same limited resources in an ecosystem. See *competition, competitive exclusion principle, intraspecific competition*.

intertidal zone The area of shoreline between low and high tides.

intraspecific competition Attempts by two or more organisms of a single species to use the same limited resources in an ecosystem. See *competition, interspecific competition*.

intrinsic rate of increase (r) Rate at which a population could grow if it had unlimited resources. Compare *environmental resistance*.

intrinsic value Value of an organism, species, ecosystem, or the earth's biodiversity based on its existence, regardless of whether it has any usefulness to us. Compare *instrumental value*.

invertebrates Animals that have no backbones. Compare *vertebrates*.

ion Atom or group of atoms with one or more positive (+) or negative (−) electrical charges. Compare *atom, molecule*.

ionizing radiation Fast-moving alpha or beta particles or high-energy radiation (gamma rays) emitted by radioisotopes. They have enough energy to dislodge one or more electrons from atoms they hit, forming charged ions in tissue that can react with and damage living tissue. Compare *nonionizing radiation*.

isotopes Two or more forms of a chemical element that have the same number of protons but different mass numbers because they have different numbers of neutrons in their nuclei.

J-shaped curve Curve with a shape similar to that of the letter J; can represent prolonged exponential growth. See *exponential growth*.

junk science Scientific results or hypotheses presented as sound science but not having undergone the rigors of the peer review process. Compare *frontier science, sound science*.

kerogen Solid, waxy mixture of hydrocarbons found in oil shale rock. Heating the rock to high temperatures causes the kerogen to vaporize. The vapor is condensed, purified, and then sent to a refinery to produce gasoline, heating oil, and other products. See also *oil shale, shale oil*.

keystone species Species that play roles affecting many other organisms in an ecosystem. Compare *indicator species, native species, nonnative species*.

kilocalorie (kcal) Unit of energy equal to 1,000 calories. See *calorie*.

kilowatt (kW) Unit of electrical power equal to 1,000 watts. See *watt*.

kinetic energy Energy that matter has because of its mass and speed or velocity. Compare *potential energy*.

K-selected species Species that produce a few, often fairly large offspring but invest a great deal of time and energy to ensure that most of those offspring reach reproductive age. Compare *r-selected species*.

K-strategists See *K-selected species*.

kwashiorkor Type of malnutrition that occurs in infants and very young children when they are weaned from mother's milk to a starchy diet low in protein. See *marasmus*.

lake Large natural body of standing freshwater formed when water from precipitation, land runoff, or groundwater flow fills a depression in the earth created by glaciation, earth movement, volcanic activity, or a giant meteorite. See *eutrophic lake, mesotrophic lake, oligotrophic lake*.

land degradation Occurs when natural or human-induced processes decrease the future ability of land to support crops, livestock, or wild species.

landfill See *sanitary landfill*.

land-use planning Process for deciding the best present and future use of each parcel of land in an area.

late successional plant species Mostly trees that can tolerate shade and form a fairly stable complex forest community. Compare *early successional plant species, mid-successional plant species*.

latitude Distance from the equator. Compare *altitude*.

law of conservation of energy See *first law of thermodynamics*.

law of conservation of matter In any physical or chemical change, matter is neither created nor destroyed but merely changed from one form to another; in physical and chemical changes, existing atoms are rearranged into different spatial patterns (physical changes) or different combinations (chemical changes).

law of tolerance The existence, abundance, and distribution of a species in an ecosystem are determined by whether the levels of one or more physical or chemical factors fall within the range tolerated by the species. See *threshold effect*.

LD50 See *median lethal dose*.

LDC See *developing country*.

leaching Process in which various chemicals in upper layers of soil are dissolved and carried to lower layers and, in some cases, to groundwater.

less developed country (LDC) See *developing country*.

life-cycle cost Initial cost plus lifetime operating costs of an economic good. Compare *full cost*.

life expectancy Average number of years a newborn infant can be expected to live.

limiting factor Single factor that limits the growth, abundance, or distribution of the population of a species in an ecosystem. See *limiting factor principle*.

limiting factor principle Too much or too little of any abiotic factor can limit or prevent growth of a population of a species in an ecosystem, even if all other factors are at or near the optimum range of tolerance for the species.

linear growth Growth in which a quantity increases by some fixed amount during each unit of time. An example is growth that increases by units of two in the sequence 2, 4, 6, 8, 10, and so on. Compare *exponential growth*.

liquefied natural gas (LNG) Natural gas converted to liquid form by cooling to a very low temperature.

liquefied petroleum gas (LPG) Mixture of liquefied propane (C_3H_8) and butane (C_4H_{10}) gas removed from natural gas and used as a fuel.

lithosphere Outer shell of the earth, composed of the crust and the rigid, outermost part of the mantle outside the asthenosphere; material found in earth's plates. See *crust, mantle*.

loams Soils containing a mixture of clay, sand, silt, and humus. Good for growing most crops.

lobbying The process by which individuals or groups use public pressure, personal contacts, and political action to persuade legislators to vote or act in their favor.

logistic growth Pattern in which exponential population growth occurs when the population is small, and population growth decreases steadily with time as the population approaches the carrying capacity. See *S-shaped curve*.

low An air mass with a low pressure. Compare *high*.

low-input agriculture See *sustainable agriculture*.

low-quality energy Energy that is dispersed and has little ability to do useful work. An example is low-temperature heat. Compare *high-quality energy*.

low-quality matter Matter that is dilute or dispersed or contains a low concentration of a useful resource. Compare *high-quality matter*.

low-throughput economy Economy based on working with nature by recycling and reusing discarded matter, preventing pollution, conserving matter and energy resources by reducing unnecessary waste and use, not degrading renewable resources, building things that are easy to recycle, reuse, and repair, not allowing population size to exceed the carrying capacity of the environment, and preserving biodiversity and ecological integrity. See *environmental worldview*. Compare *high-throughput economy, matter-recycling economy*.

low-waste society See *low-throughput economy*.

LPG See *liquefied petroleum gas*.

macroevolution Long-term, large-scale evolutionary changes among groups of species. Compare *microevolution*.

macronutrients Chemical elements that organisms need in large amounts to live, grow, or reproduce. Examples are carbon, oxygen, hydrogen, nitrogen, phosphorus, sulfur, potassium, calcium, magnesium, and iron. Compare *micronutrients*.

magma Molten rock below the earth's surface.

malnutrition Faulty nutrition, caused by a diet that does not supply an individual with enough protein, essential fats, vitamins, minerals, and other nutrients needed for good health. Compare *overnutrition, undernutrition*.

mangrove swamps Swamps found on the coastlines in warm tropical climates. They are dominated by mangrove trees, any of about 55 species of trees and shrubs that can live partly submerged in the salty environment of coastal swamps.

mantle Zone of the earth's interior between its core and its crust. Compare *core, crust*. See *lithosphere*.

manufactured resources Manufactured items made from natural resources and used to produce and distribute economic goods and services bought by consumers. These include tools, machinery, equipment, factory buildings, and transportation and

distribution facilities. Compare *human resources, natural resources.*

manure See *animal manure, green manure.*

marasmus Nutritional deficiency disease caused by a diet that does not have enough calories and protein to maintain good health. See *kwashiorkor, malnutrition.*

marginal benefit Increase in benefit when a seller produces one more unit of a product or service. Compare *marginal cost.*

marginal cost Increase in total cost resulting from producing one more unit of a good or service. Compare *marginal benefit.*

mass The amount of material in an object.

mass depletion Widespread, often global period during which extinction rates are higher than normal but not high enough to classify as a mass extinction. Compare *background extinction, mass extinction.*

mass extinction A catastrophic, widespread, often global event in which major groups of species are wiped out over a short time compared with normal (background) extinctions. Compare *background extinction, mass depletion.*

mass number Sum of the number of neutrons (n) and the number of protons (p) in the nucleus of an atom. It gives the approximate mass of that atom. Compare *atomic number.*

mass transit Buses, trains, trolleys, and other forms of transportation that carry large numbers of people.

material efficiency Total amount of material needed to produce each unit of goods or services. Also called *resource productivity.* Compare *energy efficiency.*

matter Anything that has mass (the amount of material in an object) and takes up space. On the earth, where gravity is present, we weigh an object to determine its mass.

matter quality Measure of how useful a matter resource is, based on its availability and concentration. See *high-quality matter, low-quality matter.*

matter-recycling economy Economy that emphasizes recycling the maximum amount of all resources that can be recycled. The goal is to allow economic growth to continue without depleting matter resources and without producing excessive pollution and environmental degradation. Compare *high-throughput economy, low-throughput economy.*

mature community Fairly stable, self-sustaining community in an advanced stage of ecological succession; usually has a diverse array of species and ecological niches; captures and uses energy and cycles critical chemicals more efficiently than simpler, immature communities. Compare *immature community.*

maximum sustainable yield See *sustainable yield.*

MDC See *developed country.*

median lethal dose (LD50) Amount of a toxic material per unit of body weight of test animals that kills half the test population in a certain time.

megacity City with 10 million or more people.

meltdown The melting of the core of a nuclear reactor.

mesosphere Third layer of the atmosphere; found above the stratosphere. Compare *stratosphere, troposphere.*

mesotrophic lake Lake with a moderate supply of plant nutrients. Compare *eutrophic lake, oligotrophic lake.*

metabolism Ability of a living cell or organism to capture and transform matter and energy from its environment to supply its needs for survival, growth, and reproduction.

metamorphic rock Rock produced when a preexisting rock is subjected to high temperatures (which may cause it to melt partially), high pressures, chemically active fluids, or a combination of these agents. Compare *igneous rock, sedimentary rock.* See *rock cycle.*

metastasis Spread of malignant (cancerous) cells from a tumor to other parts of the body.

metropolitan area See *urban area.*

microclimates Local climatic conditions that differ from the general climate of a region. Various topographic features of the earth's surface such as mountains and cities typically create them.

microevolution The small genetic changes a population undergoes. Compare *macroevolution.*

micronutrients Chemical elements that organisms need in small or even trace amounts to live, grow, or reproduce. Examples are sodium, zinc, copper, chlorine, and iodine. Compare *macronutrients.*

microorganisms Organisms such as bacteria that are so small that they can be seen only by using a microscope.

micropower systems Systems of small-scale decentralized units that generate 1–10,000 kilowatts of electricity. Examples include microturbines, fuel cells, and household solar panels and solar roofs.

midsuccessional plant species Grasses and low shrubs that are less hardy than early successional plant species. Compare *early successional plant species, late successional plant species.*

mineral Any naturally occurring inorganic substance found in the earth's crust as a crystalline solid. See *mineral resource.*

mineral resource Concentration of naturally occurring solid, liquid, or gaseous material in or on the earth's crust in a form and amount such that extracting and converting it into useful materials or items is currently or potentially profitable. Mineral resources are classified as *metallic* (such as iron and tin ores) or *nonmetallic* (such as fossil fuels, sand, and salt).

minimum-tillage farming See *conservation-tillage farming.*

minimum viable population (MVP) Estimate of the smallest number of individuals necessary to ensure the survival of a population in a region for a specified time period, typically ranging from decades to 100 years.

mixture Combination of one or more elements and compounds.

model An approximate representation or simulation of a system being studied.

molecule Combination of two or more atoms of the same chemical element (such as O_2) or different chemical elements (such as H_2O) held together by chemical bonds. Compare *atom, ion.*

monoculture Cultivation of a single crop, usually on a large area of land. Compare *polyculture, polyvarietal cultivation.*

more developed country (MDC) See *developed country.*

mountaintop removal Type of surface mining that uses explosives, massive shovels, and even larger machinery called draglines to remove the top of a mountain to expose seams of coal underneath a mountain. Compare *area strip mining, contour strip mining.*

multiple use Use of an ecosystem such as a forest for a variety of purposes such as timber harvesting, wildlife habitat, watershed protection, and recreation. Compare *sustainable yield.*

municipal solid waste Solid materials discarded by homes and businesses in or near urban areas. See *solid waste.*

mutagen Chemical or form of radiation that causes inheritable changes (mutations) in the DNA molecules in the genes found in chromosomes. See *carcinogen, mutation, teratogen.*

mutation A random change in DNA molecules making up genes that can yield changes in anatomy, physiology, or behavior in offspring. See *mutagen.*

mutualism Type of species interaction in which both participating species generally benefit. Compare *commensalism.*

native species Species that normally live and thrive in a particular ecosystem. Compare *indicator species, keystone species, non-native species.*

natural capital See *natural resources.*

natural gas Underground deposits of gases consisting of 50–90% by weight methane gas (CH_4) and small amounts of heavier gaseous hydrocarbon compounds such as propane (C_3H_8) and butane (C_4H_{10}).

natural greenhouse effect Heat buildup in the troposphere because of the presence of certain gases, called greenhouse gases. Without this effect, the earth would be nearly as cold as Mars, and life as we know it could not exist. Compare *global warming.*

natural ionizing radiation Ionizing radiation in the environment from natural sources.

natural law See *scientific law.*

natural radioactive decay Nuclear change in which unstable nuclei of atoms spontaneously shoot out particles (usually alpha or beta particles) or energy (gamma rays) at a fixed rate.

natural rate of extinction See *background extinction.*

natural recharge Natural replenishment of an aquifer by precipitation, which percolates downward through soil and rock. See diagram on top half of back cover, *recharge area.*

natural resources The earth's natural materials and processes that sustain life on the earth and our economies. Compare *human resources, manufactured resources.*

natural selection Process by which a particular beneficial gene (or set of genes) is reproduced in succeeding generations more than other genes. The result of natural selection is a population that contains a greater proportion of organisms better adapted to certain environmental conditions. See *adaptation, biological evolution, differential reproduction, mutation.*

negative feedback loop Situation in which a change in a certain direction provides information that causes a system to change less in that direction. Compare *positive feedback loop.*

nekton Strongly swimming organisms found in aquatic systems. Compare *benthos, plankton.*

net energy Total amount of useful energy available from an energy resource or energy system over its lifetime, minus the amount of energy *used* (the first energy law), *automatically wasted* (the second energy law), and *unnecessarily wasted* in finding, processing, concentrating, and transporting it to users.

net primary productivity (NPP) Rate at which all the plants in an ecosystem produce net useful chemical energy; equal to the difference between the rate at which the plants in an ecosystem produce useful chemical energy (gross primary productivity) and the rate at which they use some of that energy through cellular respiration. Compare *gross primary productivity.*

neutral solution Water solution containing an equal number of hydrogen ions (H^+) and hydroxide ions (OH^-); water solution with a pH of 7. Compare *acid solution, basic solution.*

neutron (n) Elementary particle in the nuclei of all atoms (except hydrogen-1). It has a relative mass of 1 and no electric charge. Compare *electron, proton.*

niche See *ecological niche.*

nitrogen cycle Cyclic movement of nitrogen in different chemical forms from the environment to organisms and then back to the environment.

nitrogen fixation Conversion of atmospheric nitrogen gas into forms useful to plants by lightning, bacteria, and cyanobacteria; it is part of the nitrogen cycle.

noise pollution Any unwanted, disturbing, or harmful sound that impairs or interferes with hearing, causes stress, hampers concentration and work efficiency, or causes accidents.

nondegradable pollutant Material that is not broken down by natural processes. Examples are the toxic elements lead and mercury. Compare *biodegradable pollutant, degradable pollutant, slowly degradable pollutant.*

nonionizing radiation Forms of radiant energy such as radio waves, microwaves, infrared light, and ordinary light that do not have enough energy to cause ionization of atoms in living tissue. Compare *ionizing radiation.*

nonnative species Species that migrate into an ecosystem or are deliberately or accidentally introduced into an ecosystem by humans. Compare *native species.*

nonpersistent pollutant See *degradable pollutant.*

nonpoint source Large or dispersed land areas such as crop fields, streets, and lawns that discharge pollutants into the environment over a large area. Compare *point source.*

nonrenewable mineral resource A concentration of naturally occurring nonrenewable material in or on the earth's crust that can be extracted and processed into useful materials at an affordable cost.

nonrenewable resource Resource that exists in a fixed amount (stock) in various places in the earth's crust and has the potential for renewal by geological, physical, and chemical processes taking place over hundreds of millions to billions of years. Examples are copper, aluminum, coal, and oil. We classify these resources as exhaustible because we are extracting and using them at a much faster rate than they were formed. Compare *renewable resource.*

nontransmissible disease A disease that is not caused by living organisms and does not spread from one person to another. Examples are most cancers, diabetes, cardiovascular disease, and malnutrition. Compare *transmissible disease.*

no-till farming See *conservation-tillage farming.*

nuclear change Process in which nuclei of certain isotopes spontaneously change, or are forced to change, into one or more different isotopes. The three principal types of nuclear change are natural radioactivity, nuclear fission, and nuclear fusion. Compare *chemical change, physical change.*

nuclear energy Energy released when atomic nuclei undergo a nuclear reaction such as the spontaneous emission of radioactivity, nuclear fission, or nuclear fusion.

nuclear fission Nuclear change in which the nuclei of certain isotopes with large mass numbers (such as uranium-235 and plutonium-239) are split apart into lighter nuclei when struck by a neutron. This process releases more neutrons and a large amount of energy. Compare *nuclear fusion.*

nuclear fusion Nuclear change in which two nuclei of isotopes of elements with a low mass number (such as hydrogen-2 and hydrogen-3) are forced together at extremely high temperatures until they fuse to form a heavier nucleus (such as helium-4). This process releases a large amount of energy. Compare *nuclear fission.*

nucleus Extremely tiny center of an atom, making up most of the atom's mass. It contains one or more positively charged protons and one or more neutrons with no electrical charge (except for a hydrogen-1 atom, which has one proton and no neutrons in its nucleus).

nutrient Any food or element an organism must take in to live, grow, or reproduce.

nutrient cycle See *biogeochemical cycle.*

oil sand Deposit of a mixture of clay, sand, water, and varying amounts of a tarlike heavy oil known as bitumen. Bitumen can be extracted from tar sand by heating. It is then purified and upgraded to synthetic crude oil. See *bitumen.*

oil shale Fine-grained rock containing various amounts of kerogen, a solid, waxy mixture of hydrocarbon compounds. Heating the rock to high temperatures converts the kerogen into a vapor that can be condensed to form a slow-flowing heavy oil called shale oil. See *kerogen, shale oil.*

old-growth forest Virgin and old, second-growth forests containing trees that are often hundreds, sometimes thousands of years old. Examples include forests of Douglas fir, western hemlock, giant sequoia, and coastal redwoods in the western United States. Compare *second-growth forest, tree plantation.*

oligotrophic lake Lake with a low supply of plant nutrients. Compare *eutrophic lake, mesotrophic lake.*

omnivore Animal that can use both plants and other animals as food sources. Examples are pigs, rats, cockroaches, and people. Compare *carnivore, herbivore.*

open dumps Fields or holes in the ground where garbage is placed and sometimes covered with soil. They are rare in developed countries, but widely used in many developing countries. Compare *santitary landfill.*

open-pit mining Removing minerals such as gravel, sand, and metal ores by digging them out of the earth's surface and leaving an open pit. Compare *area strip mining, contour strip mining, dredging, mountaintop removal, subsurface mining.*

open sea The part of an ocean that is beyond the continental shelf. Compare *coastal zone.*

ore Part of a metal-yielding material that can be economically and legally extracted at a given time. An ore typically contains two parts: the ore mineral, which contains the desired metal, and waste mineral material (gangue).

organic compounds Compounds containing carbon atoms combined with each other and with atoms of one or more other elements such as hydrogen, oxygen, nitrogen, sulfur, phosphorus, chlorine, and fluorine. All other compounds are called *inorganic compounds*.

organic farming Producing crops and livestock naturally by using organic fertilizer (manure, legumes, compost) and natural pest control (bugs that eat harmful bugs, plants that repel bugs, and environmental controls such as crop rotation) instead of using commercial inorganic fertilizers and synthetic pesticides and herbicides. See *sustainable agriculture*.

organic fertilizer Organic material such as animal manure, green manure, and compost, applied to cropland as a source of plant nutrients. Compare *commercial inorganic fertilizer*.

organism Any form of life.

other resources Identified and undiscovered resources not classified as reserves. Compare *identified resources, reserves, undiscovered resources*.

output Matter, energy, or information leaving a system. Compare *input, throughput*.

output pollution control See *pollution cleanup*.

overburden Layer of soil and rock overlying a mineral deposit. Surface mining removes this layer.

overfishing Harvesting so many fish of a species, especially immature fish, that not enough breeding stock is left to replenish the species and it becomes unprofitable to harvest them.

overgrazing Destruction of vegetation when too many grazing animals feed too long and exceed the carrying capacity of a rangeland or pasture area.

overnutrition Diet so high in calories, saturated (animal) fats, salt, sugar, and processed foods and so low in vegetables and fruits that the consumer runs high risks of diabetes, hypertension, heart disease, and other health hazards. Compare *malnutrition, undernutrition*.

oxygen-demanding wastes Organic materials that are usually biodegraded by aerobic (oxygen-consuming) bacteria if there is enough dissolved oxygen in the water. See also *biological oxygen demand*.

ozone depletion Decrease in concentration of ozone (O_3) in the stratosphere. See *ozone layer*.

ozone layer Layer of gaseous ozone (O_3) in the stratosphere that protects life on earth by filtering out most harmful ultraviolet radiation from the sun.

PANs Peroxyacyl nitrates. Group of chemicals found in photochemical smog.

parasite Consumer organism that lives on or in and feeds on a living plant or animal, known as the host, over an extended period of time. The parasite draws nourishment from and gradually weakens its host; it may or may not kill the host. See *parasitism*.

parasitism Interaction between species in which one organism, called the parasite, preys on another organism, called the host, by living on or in the host. See *host, parasite*.

parts per billion (ppb) Number of parts of a chemical found in 1 billion parts of a particular gas, liquid, or solid.

parts per million (ppm) Number of parts of a chemical found in 1 million parts of a particular gas, liquid, or solid.

parts per trillion (ppt) Number of parts of a chemical found in 1 trillion parts of a particular gas, liquid, or solid.

passive solar heating system System that captures sunlight directly within a structure and converts it into low-temperature heat for space heating or for heating water for domestic use without the use of mechanical devices. Compare *active solar heating system*.

pasture Managed grassland or enclosed meadow that usually is planted with domesticated grasses or other forage to be grazed by livestock. Compare *feedlot, rangeland*.

pathogen Organism that produces disease.

PCBs See *polychlorinated biphenyls*.

per capita GDP Annual gross domestic product (GDP) of a country divided by its total population at midyear. It gives the average slice of the economic pie per person. Used to be called per capita GNP. See *gross domestic product*.

percolation Passage of a liquid through the spaces of a porous material such as soil.

perennial Plant that can live for more than 2 years. Compare *annual*.

permafrost Perennially frozen layer of the soil that forms when the water there freezes. It is found in arctic tundra.

permeability The degree to which underground rock and soil pores are interconnected and thus a measure of the degree to which water can flow freely from one pore to another. Compare *porosity*.

perpetual resource An essentially inexhaustible resource on a human time scale. Solar energy is an example. Compare *nonrenewable resource, renewable resource*.

persistence How long a pollutant stays in the air, water, soil, or body. See also *inertia*.

persistent pollutant See *slowly degradable pollutant*.

pest Unwanted organism that directly or indirectly interferes with human activities.

pesticide Any chemical designed to kill or inhibit the growth of an organism that people consider undesirable. See *fungicide, herbicide, insecticide*.

petrochemicals Chemicals obtained by refining (distilling) crude oil. They are used as raw materials in manufacturing most industrial chemicals, fertilizers, pesticides, plastics, synthetic fibers, paints, medicines, and many other products.

petroleum See *crude oil*.

pH Numeric value that indicates the relative acidity or alkalinity of a substance on a scale of 0 to 14, with the neutral point at 7. Acid solutions have pH values lower than 7, and basic or alkaline solutions have pH values greater than 7.

phosphorus cycle Cyclic movement of phosphorus in different chemical forms from the environment to organisms and then back to the environment.

photochemical smog Complex mixture of air pollutants produced in the lower atmosphere by the reaction of hydrocarbons and nitrogen oxides under the influence of sunlight. Especially harmful components include ozone, peroxyacyl nitrates (PANs), and various aldehydes. Compare *industrial smog*.

photosynthesis Complex process that takes place in cells of green plants. Radiant energy from the sun is used to combine carbon dioxide (CO_2) and water (H_2O) to produce oxygen (O_2) and carbohydrates (such as glucose, $C_6H_{12}O_6$) and other nutrient molecules. Compare *aerobic respiration, chemosynthesis*.

photovoltaic cell (solar cell) Device that converts radiant (solar) energy directly into electrical energy.

physical change Process that alters one or more physical properties of an element or a compound without altering its chemical composition. Examples are changing the size and shape of a sample of matter (crushing ice and cutting aluminum foil) and changing a sample of matter from one physical state to another (boiling and freezing water). Compare *chemical change, nuclear change*.

physical resources See *manufactured resources*.

phytoplankton Small, drifting plants, mostly algae and bacteria, found in aquatic ecosystems. Compare *plankton, zooplankton*.

pioneer community First integrated set of plants, animals, and decomposers found in an area undergoing primary ecological succession. See *immature community, mature community*.

pioneer species First hardy species, often microbes, mosses, and lichens, that begin colonizing a site as the first stage of ecological succession. See *ecological succession, pioneer community*.

plaintiff The individual, group of individuals, corporation, or government agency bringing the charges in a lawsuit. Compare *defendant*.

planetary management worldview Beliefs that (1) we are the planet's most important species; (2) we will not run out of resources because of our ingenuity in developing and finding new ones; (3) the potential for economic growth is essentially limitless; and (4) our success depends on how well we can understand, control, and manage the earth's life-support systems mostly for our own benefit. See *spaceship-earth worldview*. Compare *environmental wisdom worldview, frontier worldview, stewardship worldview*.

plankton Small plant organisms (phytoplankton) and animal organisms (zooplankton) that float in aquatic ecosystems.

plantation agriculture Growing specialized crops such as bananas, coffee, and cacao in tropical developing countries, primarily for sale to developed countries.

plasma An ionized gas consisting of electrically conductive ions and electrons. It is known as a fourth state of matter.

plates See *tectonic plates*. Various-sized areas of the earth's lithosphere that move slowly around with the mantle's flowing asthenosphere. Most earthquakes and volcanoes occur around the boundaries of these plates. See *lithosphere, plate tectonics*.

plate tectonics Theory of geophysical processes that explains the movements of lithospheric plates and the processes that occur at their boundaries. See *lithosphere, tectonic plates*.

point source Single identifiable source that discharges pollutants into the environment. Examples are the smokestack of a power plant or an industrial plant, drainpipe of a meatpacking plant, chimney of a house, or exhaust pipe of an automobile. Compare *nonpoint source*.

poison A chemical that adversely affects the health of a living human or animal by causing injury, illness, or death.

politics Process through which individuals and groups try to influence or control government policies and actions that affect the local, state, national, and international communities.

pollutant A particular chemical or form of energy that can adversely affect the health, survival, or activities of humans or other living organisms. See *pollution*.

pollution An undesirable change in the physical, chemical, or biological characteristics of air, water, soil, or food that can adversely affect the health, survival, or activities of humans or other living organisms.

pollution cleanup Device or process that removes or reduces the level of a pollutant after it has been produced or has entered the environment. Examples are automobile emission control devices and sewage treatment plants. Compare *pollution prevention*.

pollution prevention Device or process that prevents a potential pollutant from forming or entering the environment or sharply reduces the amount entering the environment. Compare *pollution cleanup*.

polychlorinated biphenyls (PCBs) Group of 209 different toxic, oily, synthetic chlorinated hydrocarbon compounds that can be biologically amplified in food chains and webs.

polyculture Complex form of intercropping in which a large number of different plants maturing at different times are planted together. See also *intercropping*. Compare *monoculture, polyvarietal cultivation*.

polyvarietal cultivation Planting a plot of land with several varieties of the same crop.

Compare *intercropping, monoculture, polyculture*.

population Group of individual organisms of the same species living in a particular area.

population change An increase or decrease in the size of a population. It is equal to (Births + Immigration) − (Deaths + Emigration).

population density Number of organisms in a particular population found in a specified area or volume.

population dispersion General pattern in which the members of a population are arranged throughout its habitat.

population distribution Variation of population density over a particular geographic area. For example, a country has a high population density in its urban areas and a much lower population density in rural areas.

population dynamics Major abiotic and biotic factors that tend to increase or decrease the population size and age and sex composition of a species.

population size Number of individuals making up a population's gene pool.

population viability analysis (PVA) Use of mathematical models to estimate a population's risk of extinction. See *minimum viable population*.

porosity Percentage of space in rock or soil occupied by voids, whether the voids are isolated or connected. Compare *permeability*.

positive feedback loop Situation in which a change in a certain direction provides information that causes a system to change further in the same direction. Compare *negative feedback loop*.

potential energy Energy stored in an object because of its position or the position of its parts. Compare *kinetic energy*.

poverty Inability to meet basic needs for food, clothing, and shelter.

ppb See *parts per billion*.

ppm See *parts per million*.

ppt See *parts per trillion*.

prairies See *grasslands*.

precautionary principle When there is scientific uncertainty about potentially serious harm from chemicals or technologies, decision makers should act to prevent harm to humans and the environment. See *pollution prevention*.

precipitation Water in the form of rain, sleet, hail, and snow that falls from the atmosphere onto the land and bodies of water.

predation Situation in which an organism of one species (the predator) captures and feeds on parts or all of an organism of another species (the prey).

predator Organism that captures and feeds on parts or all of an organism of another species (the prey).

predator–prey relationship Interaction between two organisms of different species in which one organism, called the *predator*, captures and feeds on parts or all of another organism, called the *prey*.

preservationist Person concerned primarily with setting aside or protecting undisturbed natural areas from harmful human activities. Compare *conservation biologist, conservationist, ecologist, environmentalist, environmental scientist, restorationist*.

prey Organism that is captured and serves as a source of food for an organism of another species (the predator).

primary consumer Organism that feeds on all or part of plants (herbivore) or on other producers. Compare *detritivore, omnivore, secondary consumer*.

primary pollutant Chemical that has been added directly to the air by natural events or human activities and occurs in a harmful concentration. Compare *secondary pollutant*.

primary productivity See *gross primary productivity, net primary productivity*.

primary sewage treatment Mechanical sewage treatment in which large solids are filtered out by screens and suspended solids settle out as sludge in a sedimentation tank. Compare *advanced sewage treatment, secondary sewage treatment*.

primary succession Ecological succession in a bare area that has never been occupied by a community of organisms. See *ecological succession*. Compare *secondary succession*.

probability A mathematical statement about how likely it is that something will happen.

producer Organism that uses solar energy (green plant) or chemical energy (some bacteria) to manufacture the organic compounds it needs as nutrients from simple inorganic compounds obtained from its environment. Compare *consumer, decomposer*.

prokaryotic cell Cell that does not have a distinct nucleus. Other internal parts are also not enclosed by membranes. Compare *eukaryotic cell*.

proton (p) Positively charged particle in the nuclei of all atoms. Each proton has a relative mass of 1 and a single positive charge. Compare *electron, neutron*.

pure free-market economic system System in which all economic decisions are made in the market, where buyers and sellers of economic goods interact freely, with no government or other interference. Compare *capitalist market economic system*.

pyramid of energy flow Diagram representing the flow of energy through each trophic level in a food chain or food web. With each energy transfer, only a small part (typically 10%) of the usable energy entering one trophic level is transferred to the organisms at the next trophic level.

radiation Fast-moving particles (particulate radiation) or waves of energy (electromagnetic radiation). See *alpha particle, beta particle, gamma rays*.

radioactive decay Change of a radioisotope to a different isotope by the emission of radioactivity.

radioactive isotope See *radioisotope*.

radioactive waste Waste products of nuclear power plants, research, medicine, weapon production, or other processes involving nuclear reactions. See *radioactivity*.

radioactivity Nuclear change in which unstable nuclei of atoms spontaneously shoot out "chunks" of mass, energy, or both at a fixed rate. The three principal types of radioactivity are gamma rays and fast-moving alpha particles and beta particles.

radioisotope Isotope of an atom that spontaneously emits one or more types of radioactivity (alpha particles, beta particles, gamma rays).

rain shadow effect Low precipitation on the far side (leeward side) of a mountain when prevailing winds flow up and over a high mountain or range of high mountains. This creates semiarid and arid conditions on the leeward side of a high mountain range.

range See *distribution*.

rangeland Land that supplies forage or vegetation (grasses, grasslike plants, and shrubs) for grazing and browsing animals and is not intensively managed. Compare *feedlot, pasture*.

range of tolerance Range of chemical and physical conditions that must be maintained for populations of a particular species to stay alive and grow, develop, and function normally. See *law of tolerance*.

rare species A species that has naturally small numbers of individuals (often because of limited geographic ranges or low population densities) or has been locally depleted by human activities.

realized niche Parts of the fundamental niche of a species that are actually used by that species. See *ecological niche, fundamental niche*.

recharge area Any area of land allowing water to pass through it and into an aquifer. See *aquifer, natural recharge*.

reconciliation ecology The science of inventing, establishing, and maintaining new habitats to conserve species diversity in places where people live, work, or play.

recycling Collecting and reprocessing a resource so that it can be made into new products. An example is collecting aluminum cans, melting them down, and using the aluminum to make new cans or other aluminum products. Compare *reuse*.

reforestation Renewal of trees and other types of vegetation on land where trees have been removed; can be done naturally by seeds from nearby trees or artificially by planting seeds or seedlings.

reliable runoff Surface runoff of water that generally can be counted on as a stable source of water from year to year. See *runoff*.

renewable resource Resource that can be replenished rapidly (hours to several decades) through natural processes. Examples are trees in forests, grasses in grasslands, wild animals, fresh surface water in lakes and streams, most groundwater, fresh air, and fertile soil. If such a resource is used faster than it is replenished, it can be depleted and converted into a nonrenewable resource. Compare *nonrenewable resource* and *perpetual resource*. See also *environmental degradation*.

replacement-level fertility Number of children a couple must have to replace them. The average for a country or the world usually is slightly higher than 2 children per couple (2.1 in the United States and 2.5 in some developing countries) because some children die before reaching their reproductive years. See also *total fertility rate*.

reproduction Production of offspring by one or more parents.

reproductive isolation Long-term geographic separation of members of a particular sexually reproducing species.

reproductive potential See *biotic potential*.

reserves Resources that have been identified and from which a usable mineral can be extracted profitably at present prices with current mining technology. See *identified resources, undiscovered resources*.

resilience Ability of a living system to restore itself to original condition after being exposed to an outside disturbance that is not too drastic. See *constancy, inertia*.

resource Anything obtained from the living and nonliving environment to meet human needs and wants. It can also be applied to other species.

resource partitioning Process of dividing up resources in an ecosystem so that species with similar needs (overlapping ecological niches) use the same scarce resources at different times, in different ways, or in different places. See *ecological niche, fundamental niche, realized niche*.

resource productivity See *material efficiency*.

respiration See *aerobic respiration*.

response The amount of health damage caused by exposure to a certain dose of a harmful substance or form of radiation. See *dose, dose-response curve, median lethal dose*.

restoration ecology Research and scientific study devoted to restoring, repairing, and reconstructing damaged ecosystems.

restorationist Scientist or other person devoted to the partial or complete restoration of natural areas that have been degraded by human activities. Compare *conservation biologist, conservationist, ecologist, environmental scientist, preservationist*.

reuse Using a product over and over again in the same form. An example is collecting, washing, and refilling glass beverage bottles. Compare *recycling*.

Richter scale A measurement used by scientists to determine the magnitude of earthquakes.

riparian zones Thin strips and patches of vegetation that surround streams. They are very important habitats and resources for wildlife.

risk The probability that something undesirable will result from deliberate or accidental exposure to a hazard. See *risk analysis, risk assessment, risk–benefit analysis, risk management*.

risk analysis Identifying hazards, evaluating the nature and severity of risks (*risk assessment*), using this and other information to determine options and make decisions about reducing or eliminating risks (*risk management*), and communicating information about risks to decision makers and the public (*risk communication*).

risk assessment Process of gathering data and making assumptions to estimate short- and long-term harmful effects on human health or the environment from exposure to hazards associated with the use of a particular product or technology. See *risk–benefit analysis*.

risk–benefit analysis Estimate of the short- and long-term risks and benefits of using a particular product or technology. See *risk assessment*.

risk communication Communicating information about risks to decision makers and the public. See *risk, risk analysis, risk–benefit analysis*.

risk management Using risk assessment and other information to determine options and make decisions about reducing or eliminating risks. See *risk, risk analysis, risk–benefit analysis, risk communication*.

rock Any material that makes up a large, natural, continuous part of the earth's crust. See *mineral*.

rock cycle Largest and slowest of the earth's cycles, consisting of geologic, physical, and chemical processes that form and modify rocks and soil in the earth's crust over millions of years.

r-selected species Species that reproduce early in their life span and produce large numbers of usually small and short-lived offspring in a short period. Compare *K-selected species*.

r-strategists See *r-selected species*.

rule of 70 Doubling time (in years) = 70/(percentage growth rate). See *doubling time, exponential growth*.

ruminants Grazing animals with complex digestive systems that enable them to convert grass and other roughage into meat and milk.

runoff Freshwater from precipitation and melting ice that flows on the earth's surface into nearby streams, lakes, wetlands, and reservoirs. See *reliable runoff, surface runoff, surface water*. Compare *groundwater*.

rural area Geographic area in the United States with a population of less than 2,500. The number of people used in this definition may vary in different countries. Compare *urban area*.

salinity Amount of various salts dissolved in a given volume of water.

salinization Accumulation of salts in soil that can eventually make the soil unable to support plant growth.

saltwater intrusion Movement of salt water into freshwater aquifers in coastal and inland areas as groundwater is withdrawn faster than it is recharged by precipitation.

sanitary landfill Waste disposal site on land in which waste is spread in thin layers, compacted, and covered with a fresh layer of clay or plastic foam each day.

scavenger Organism that feeds on dead organisms that were killed by other organisms or died naturally. Examples are vultures, flies, and crows. Compare *detritivore*.

science Attempts to discover order in nature and use that knowledge to make predictions about what should happen in nature. See *frontier science, scientific data, scientific hypothesis, scientific law, scientific methods, scientific model, scientific theory, sound science*.

scientific data Facts obtained by making observations and measurements. Compare *scientific hypothesis, scientific law, scientific methods, scientific model, scientific theory*.

scientific hypothesis An educated guess that attempts to explain a scientific law or certain scientific observations. Compare *scientific data, scientific law, scientific methods, scientific model, scientific theory*.

scientific law Description of what scientists find happening in nature repeatedly in the same way, without known exception. See *first law of thermodynamics, law of conservation of matter, second law of thermodynamics*. Compare *scientific data, scientific hypothesis, scientific methods, scientific model, scientific theory*.

scientific methods The ways scientists gather data and formulate and test scientific hypotheses, models, theories, and laws. See *scientific data, scientific hypothesis, scientific law, scientific model, scientific theory*.

scientific model A simulation of complex processes and systems. Many are mathematical models that are run and tested using computers.

scientific theory A well-tested and widely accepted scientific hypothesis. Compare *scientific data, scientific hypothesis, scientific law, scientific methods, scientific model*.

secondary consumer Organism that feeds only on primary consumers. Compare *detritivore, omnivore, primary consumer*.

secondary pollutant Harmful chemical formed in the atmosphere when a primary air pollutant reacts with normal air components or other air pollutants. Compare *primary pollutant*.

secondary sewage treatment Second step in most waste treatment systems in which aerobic bacteria decompose up to 90% of degradable, oxygen-demanding organic wastes in wastewater. This usually involves bringing sewage and bacteria together in trickling filters or in the activated sludge process. Compare *advanced sewage treatment, primary sewage treatment*.

secondary succession Ecological succession in an area in which natural vegetation has been removed or destroyed but the soil is not destroyed. See *ecological succession*. Compare *primary succession*.

second-growth forest Stands of trees resulting from secondary ecological succession. Compare *old-growth forest, tree farm*.

second law of energy See *second law of thermodynamics*.

second law of thermodynamics In any conversion of heat energy to useful work, some of the initial energy input is always degraded to a lower-quality, more dispersed, less useful energy, usually low-temperature heat that flows into the environment; you cannot break even in terms of energy quality. See *first law of thermodynamics*.

sedimentary rock Rock that forms from the accumulated products of erosion and in some cases from the compacted shells, skeletons, and other remains of dead organisms. Compare *igneous rock, metamorphic rock*. See *rock cycle*.

seed-tree cutting Removal of nearly all trees on a site in one cutting, with a few seed-producing trees left uniformly distributed to regenerate the forest. Compare *clear-cutting, selective cutting, shelterwood cutting, strip cutting*.

selective cutting Cutting of intermediate-aged, mature, or diseased trees in an uneven-aged forest stand, either singly or in small groups. This encourages the growth of younger trees and maintains an uneven-aged stand. Compare *clear-cutting, seed-tree cutting, shelterwood cutting, strip cutting*.

septic tank Underground tank for treating wastewater from a home in rural and suburban areas. Bacteria in the tank decompose organic wastes, and the sludge settles to the bottom of the tank. The effluent flows out of the tank into the ground through a field of drainpipes.

sexual reproduction Reproduction in organisms that produce offspring by combining sex cells or *gametes* (such as ovum and sperm) from both parents. This produces offspring that have combinations of traits from their parents. Compare *asexual reproduction*.

shale oil Slow-flowing, dark brown, heavy oil obtained when kerogen in oil shale is vaporized at high temperatures and then condensed. Shale oil can be refined to yield gasoline, heating oil, and other petroleum products. See *kerogen, oil shale*.

shelterbelt See *windbreak*.

shelterwood cutting Removal of mature, marketable trees in an area in a series of partial cuttings to allow regeneration of a new stand under the partial shade of older trees, which are later removed. Typically, this is done by making two or three cuts over a decade. Compare *clear-cutting, seed-tree cutting, selective cutting, strip cutting*.

shifting cultivation Clearing a plot of ground in a forest, especially in tropical areas, and planting crops on it for a few years (typically 2–5 years) until the soil is depleted of nutrients or the plot has been invaded by a dense growth of vegetation from the surrounding forest. Then a new plot is cleared and the process is repeated. The abandoned plot cannot successfully grow crops for 10–30 years. See also *slash-and-burn cultivation*.

slash-and-burn cultivation Cutting down trees and other vegetation in a patch of forest, leaving the cut vegetation on the ground to dry, and then burning it. The ashes that are left add nutrients to the nutrient-poor soils found in most tropical forest areas. Crops are planted between tree stumps. Plots must be abandoned after a few years (typically 2–5 years) because of loss of soil fertility or invasion of vegetation from the surrounding forest. See also *shifting cultivation*.

slowly degradable pollutant Material that is slowly broken down into simpler chemicals or reduced to acceptable levels by natural physical, chemical, and biological processes. Compare *biodegradable pollutant, degradable pollutant, nondegradable pollutant*.

sludge Gooey mixture of toxic chemicals, infectious agents, and settled solids removed from wastewater at a sewage treatment plant.

smart growth Form of urban planning which recognizes that urban growth will occur but uses zoning laws and an array of other tools to prevent sprawl, direct growth to certain areas, protect ecologically sensitive and important lands and waterways, and develop urban areas that are more environmentally sustainable and more enjoyable places to live.

smelting Process in which a desired metal is separated from the other elements in an ore mineral.

smog Originally a combination of smoke and fog but now used to describe other mixtures of pollutants in the atmosphere. See *industrial smog, photochemical smog*.

soil Complex mixture of inorganic minerals (clay, silt, pebbles, and sand), decaying organic matter, water, air, and living organisms.

soil conservation Methods used to reduce soil erosion, prevent depletion of soil nutrients, and restore nutrients already lost by erosion, leaching, and excessive crop harvesting.

soil erosion Movement of soil components, especially topsoil, from one place to another, usually by wind, flowing water, or both. This natural process can be greatly accelerated by human activities that remove vegetation from soil.

soil horizons Horizontal zones that make up a particular mature soil. Each horizon has a distinct texture and composition that vary with different types of soils. See *soil profile*.

soil permeability Rate at which water and air move from upper to lower soil layers. Compare *porosity*.

soil porosity See *porosity*.

soil profile Cross-sectional view of the horizons in a soil. See *soil horizon*.

soil structure How the particles that make up a soil are organized and clumped together. See also *soil permeability, soil texture*.

soil texture Relative amounts of the different types and sizes of mineral particles in a sample of soil.

solar capital Solar energy from the sun reaching the earth. Compare *natural resources*.

solar cell See *photovoltaic cell*.

solar collector Device for collecting radiant energy from the sun and converting it into heat. See *active solar heating system, passive solar heating system*.

solar energy Direct radiant energy from the sun and a number of indirect forms of energy produced by the direct input. Principal indirect forms of solar energy include wind, falling and flowing water (hydropower), and biomass (solar energy converted into chemical energy stored in the chemical bonds of organic compounds in trees and other plants).

solid waste Any unwanted or discarded material that is not a liquid or a gas. See *municipal solid waste*.

sound science Scientific data, models, theories, and laws that are widely accepted by scientists considered experts in the area of study. These results of science are very reliable. Compare *frontier science, junk science*.

spaceship-earth worldview View of the earth as a spaceship: a machine that we can understand, control, and change at will by using advanced technology. See *planetary management worldview*. Compare *environmental wisdom worldview, stewardship worldview*.

specialist species Species with a narrow ecological niche. They may be able to live in only one type of habitat, tolerate only a narrow range of climatic and other environmental conditions, or use only one type or a few types of food. Compare *generalist species*.

speciation Formation of two species from one species because of divergent natural selection in response to changes in environmental conditions; usually takes thousands of years. Compare *extinction*.

species Group of organisms that resemble one another in appearance, behavior, chemical makeup and processes, and genetic structure. Organisms that reproduce sexually are classified as members of the same species only if they can actually or potentially interbreed with one another and produce fertile offspring.

species diversity Number of different species (species richness) and their relative abundances (species evenness) in a given area or community. See *biodiversity*. Compare *ecological diversity, genetic diversity*.

species equilibrium model See *theory of island biography*.

species evenness Number of individuals within a species in a community. See *species diversity, species richness*.

species richness Number of different species in a community.

spoils Unwanted rock and other waste materials produced when a material is removed from the earth's surface or subsurface by mining, dredging, quarrying, and excavation.

S-shaped curve Leveling off of an exponential, J-shaped curve when a rapidly growing population exceeds the carrying capacity of its environment and ceases to grow.

stability Ability of a living system to withstand or recover from externally imposed changes or stresses. See *constancy, inertia, resilience*.

statutory law Law developed and passed by legislative bodies such as federal and state governments. Compare *common law*.

stewardship worldview **(1)** we are the planet's most important species but we have an ethical responsibility to care for the rest of nature; **(2)** we will probably not run out of resources but they should not be wasted; **(3)** we should encourage environmentally beneficial forms of economic growth and discourage environmentally harmful forms of economic growth; and **(4)** our success depends on how well we can understand, control, and manage and care for the earth's life-support systems for our benefit and for the rest of nature. Compare *environmental wisdom worldview, planetary management worldview, spaceship earth worldview*.

stratosphere Second layer of the atmosphere, extending about 17–48 kilometers (11–30 miles) above the earth's surface. It contains small amounts of gaseous ozone (O_3), which filters out about 95% of the incoming harmful ultraviolet (UV) radiation emitted by the sun. Compare *troposphere*.

stream Flowing body of surface water. Examples are creeks and rivers.

strip cropping Planting regular crops and close-growing plants, such as hay or nitrogen-fixing legumes, in alternating rows or bands to help reduce depletion of soil nutrients.

strip cutting A variation of clear-cutting in which a strip of trees is clear-cut along the contour of the land, with the corridor narrow enough to allow natural regeneration within a few years. After regeneration, another strip is cut above the first, and so on. Compare *clear-cutting, seed-tree cutting, selective cutting, shelterwood cutting*.

strip mining Form of surface mining in which bulldozers, power shovels, or stripping wheels remove large chunks of the earth's surface in strips. See *area strip mining, contour strip mining, surface mining*. Compare *subsurface mining*.

subatomic particles Extremely small particles—electrons, protons, and neutrons—that make up the internal structure of atoms.

subduction zone Area in which oceanic lithosphere is carried downward (subducted) under the island arc or continent at a convergent plate boundary. A trench ordinarily forms at the boundary between the two converging plates. See *convergent plate boundary*.

subsidence Slow or rapid sinking of part of the earth's crust that is not slope-related.

subsistence farming Supplementing solar energy with energy from human labor and draft animals to produce enough food to feed oneself and family members; in good years enough food may be left over to sell or put aside for hard times. Compare *industrialized agriculture*.

subsurface mining Extraction of a metal ore or fuel resource such as coal from a deep underground deposit. Compare *surface mining*.

succession See *ecological succession, primary succession, secondary succession*.

succulent plants Plants, such as desert cacti, that survive in dry climates by having no leaves, thus reducing the loss of scarce water. They store water and use sunlight to produce the food they need in the thick, fleshy tissue of their green stems and branches. Compare *deciduous plants, evergreen plants*.

sulfur cycle Cyclic movement of sulfur in different chemical forms from the environment to organisms and then back to the environment.

superinsulated house House that is heavily insulated and extremely airtight. Typically, active or passive solar collectors are used to heat water, and an air-to-air heat exchanger is used to prevent buildup of excessive moisture and indoor air pollutants.

surface fire Forest fire that burns only undergrowth and leaf litter on the forest floor. Compare *crown fire, ground fire*. See *controlled burning*.

surface mining Removing soil, subsoil, and other strata and then extracting a mineral deposit found fairly close to the earth's surface. See *area strip mining, contour strip mining, dredging, mountaintop removal, open-pit mining*. Compare *subsurface mining*.

surface runoff Water flowing off the land into bodies of surface water. See *reliable runoff*.

surface water Precipitation that does not infiltrate the ground or return to the atmosphere by evaporation or transpiration. See *runoff*. Compare *groundwater*.

survivorship curve Graph showing the number of survivors in different age groups for a particular species.

sustainability Ability of a system to survive for some specified (finite) time.

sustainable agriculture Method of growing crops and raising livestock based on organic fertilizers, soil conservation, water conservation, biological pest control, and minimal use of nonrenewable fossil-fuel energy.

sustainable development See *environmentally sustainable economic development*.

sustainable living Taking no more potentially renewable resources from the natural world than can be replenished naturally and not overloading the capacity of the environment to cleanse and renew itself by natural processes.

sustainable society A society that manages its economy and population size without doing irreparable environmental harm by overloading the planet's ability to absorb environmental insults, replenish its resources, and sustain human and other forms of life over a specified period, usually hundreds to thousands of years. During this period, it satisfies the needs of its people without depleting natural resources and thereby jeopardizing the prospects of current and future generations of humans and other species.

sustainable yield (sustained yield) Highest rate at which a potentially renewable resource can be used without reducing its available supply throughout the world or in a particular area. See also *environmental degradation*.

symbiosis Any intimate relationship or association between members of two or more species. See *symbiotic relationship*.

symbiotic relationship Species interaction in which two kinds of organisms live together in an intimate association. Members of the participating species may be harmed by, benefit from, or be unaffected by the interaction. See *commensalism, interspecific competition, mutualism, parasitism, predation*.

synergistic interaction Interaction of two or more factors or processes so that the combined effect is greater than the sum of their separate effects.

synergy See *synergistic interaction*.

synfuels Synthetic gaseous and liquid fuels produced from solid coal or sources other than natural gas or crude oil.

synthetic natural gas (SNG) Gaseous fuel containing mostly methane produced from solid coal.

system A set of components that function and interact in some regular and theoretically predictable manner.

tailings Rock and other waste materials removed as impurities when waste mineral material is separated from the metal in an ore.

tar sand See *oil sand*.

tectonic plates Various-sized areas of the earth's lithosphere that move slowly around with the mantle's flowing asthenosphere. Most earthquakes and volcanoes

occur around the boundaries of these plates. See *lithosphere, plate tectonics*.

temperature Measure of the average speed of motion of the atoms, ions, or molecules in a substance or combination of substances at a given moment. Compare *heat*.

temperature inversion Layer of dense, cool air trapped under a layer of less dense, warm air. This prevents upward-flowing air currents from developing. In a prolonged inversion, air pollution in the trapped layer may build up to harmful levels. See *radiation temperature inversion, subsidence temperature inversion*.

teratogen Chemical, ionizing agent, or virus that causes birth defects. Compare *carcinogen, mutagen*.

terracing Planting crops on a long, steep slope that has been converted into a series of broad, nearly level terraces with short vertical drops from one to another that run along the contour of the land to retain water and reduce soil erosion.

terrestrial Pertaining to land. Compare *aquatic*.

territoriality Process in which organisms patrol or mark an area around their home, nesting, or major feeding site and defend it against members of their own species.

tertiary (higher-level) consumers Animals that feed on animal-eating animals. They feed at high trophic levels in food chains and webs. Examples are hawks, lions, bass, and sharks. Compare *detritivore, primary consumer, secondary consumer*.

tertiary sewage treatment See *advanced sewage treatment*.

theory of evolution Widely accepted scientific idea that all life forms developed from earlier life forms. Although this theory conflicts with the creation stories of many religions, it is the way biologists explain how life has changed over the past 3.6–3.8 billion years and why it is so diverse today.

theory of island biogeography The number of species found on an island is determined by a balance between two factors: the *immigration rate* (of species new to the island) from other inhabited areas and the *extinction rate* (of species established on the island). The model predicts that at some point the rates of immigration and extinction will reach an equilibrium point that determines the island's average number of different species (species diversity).

thermal inversion See *temperature inversion*.

thermocline Zone of gradual temperature decrease between warm surface water and colder deep water in a lake, reservoir, or ocean.

threatened species A wild species that is still abundant in its natural range but is likely to become endangered because of a decline in numbers. Compare *endangered species*.

threshold effect The harmful or fatal effect of a small change in environmental conditions that exceeds the limit of tolerance of an organism or population of a species. See *law of tolerance*.

throughput Rate of flow of matter, energy, or information through a system. Compare *input, output*.

throwaway society See *high-throughput economy*.

time delay Time lag between the input of a stimulus into a system and the response to the stimulus.

tolerance limits Minimum and maximum limits for physical conditions (such as temperature) and concentrations of chemical substances beyond which no members of a particular species can survive. See *law of tolerance*.

total fertility rate (TFR) Estimate of the average number of children who will be born alive to a woman during her lifetime if she passes through all her childbearing years (ages 15–44) conforming to age-specific fertility rates of a given year. In simpler terms, it is an estimate of the average number of children a woman will have during her childbearing years.

toxic chemical See *poison*. See *carcinogen, hazardous chemical, mutagen, teratogen*.

toxicity Measure of how harmful a substance is.

toxicology Study of the adverse effects of chemicals on health.

toxic waste Form of hazardous waste that causes death or serious injury (such as burns, respiratory diseases, cancers, or genetic mutations). See *hazardous waste*.

toxin See *poison*.

traditional intensive agriculture Producing enough food for a farm family's survival and perhaps a surplus that can be sold. This type of agriculture uses higher inputs of labor, fertilizer, and water than traditional subsistence agriculture. See *traditional subsistence agriculture*. Compare *industrialized agriculture*.

traditional subsistence agriculture Production of enough crops or livestock for a farm family's survival and, in good years, a surplus to sell or put aside for hard times. Compare *industrialized agriculture, traditional intensive agriculture*.

tragedy of the commons Depletion or degradation of a potentially renewable resource to which people have free and unmanaged access. An example is the depletion of commercially desirable fish species in the open ocean beyond areas controlled by coastal countries. See *common-property resource*.

transform fault Area where the earth's lithospheric plates move in opposite but parallel directions along a fracture (fault) in the lithosphere. Compare *convergent plate boundary, divergent plate boundary*.

transgenic organisms See *genetically modified organisms (GMOs)*.

transmissible disease A disease that is caused by living organisms (such as bacteria, viruses, and parasitic worms) and can spread from one person to another by air, water, food, or body fluids (or in some cases by insects or other organisms). Compare *nontransmissible disease.*

transpiration Process in which water is absorbed by the root systems of plants, moves up through the plants, passes through pores (stomata) in their leaves or other parts, and evaporates into the atmosphere as water vapor.

tree farm See *tree plantation.*

tree plantation Site planted with one or only a few tree species in an even-aged stand. When the stand matures it is usually harvested by clear-cutting and then replanted. These farms normally are used to grow rapidly growing tree species for fuelwood, timber, or pulpwood. See *even-aged management.* Compare *old-growth forest, second-growth forest, uneven-aged management.*

trophic level All organisms that are the same number of energy transfers away from the original source of energy (for example, sunlight) that enters an ecosystem. For example, all producers belong to the first trophic level, and all herbivores belong to the second trophic level in a food chain or a food web.

troposphere Innermost layer of the atmosphere. It contains about 75% of the mass of earth's air and extends about 17 kilometers (11 miles) above sea level. Compare *stratosphere.*

true cost See *full cost.*

ultraplankton Photosynthetic bacteria no more than 2 micrometers wide.

undergrazing Reduction of the net primary productivity of grassland vegetation and grass cover from absence of grazing for long periods (at least 5 years). Compare *overgrazing.*

undernutrition Consuming insufficient food to meet one's minimum daily energy needs for a long enough time to cause harmful effects. Compare *malnutrition, overnutrition.*

undiscovered resources Potential supplies of a particular mineral resource, believed to exist because of geologic knowledge and theory, although specific locations, quality, and amounts are unknown. Compare *identified resources, reserves.*

uneven-aged management Method of forest management in which trees of different species in a given stand are maintained at many ages and sizes to permit continuous natural regeneration. Compare *even-aged management.*

upwelling Movement of nutrient-rich bottom water to the ocean's surface. This can occur far from shore but usually occurs along certain steep coastal areas where the surface layer of ocean water is pushed away from shore and replaced by cold, nutrient-rich bottom water.

urban area Geographic area with a population of 2,500 or more. The number of people used in this definition may vary, with some countries setting the minimum number of people at 10,000–50,000.

urban growth Rate of growth of an urban population. Compare *degree of urbanization.*

urbanization See *degree of urbanization.*

urban sprawl Growth of low-density development on the edges of cities and towns. See *smart growth.*

utilitarian value See *instrumental value.*

vertebrates Animals that have backbones. Compare *invertebrates.*

volcano Vent or fissure in the earth's surface through which magma, liquid lava, and gases are released into the environment.

warm front The boundary between an advancing warm air mass and the cooler one it is replacing. Because warm air is less dense than cool air, an advancing warm front rises over a mass of cool air. Compare *cold front.*

water cycle See *hydrologic cycle.*

waterlogging Saturation of soil with irrigation water or excessive precipitation so that the water table rises close to the surface.

water pollution Any physical or chemical change in surface water or groundwater that can harm living organisms or make water unfit for certain uses.

watershed Land area that delivers water, sediment, and dissolved substances via small streams to a major stream (river).

water table Upper surface of the zone of saturation, in which all available pores in the soil and rock in the earth's crust are filled with water.

watt Unit of power, or rate at which electrical work is done. See *kilowatt.*

weather Short-term changes in the temperature, barometric pressure, humidity, precipitation, sunshine, cloud cover, wind direction and speed, and other conditions in the troposphere at a given place and time. Compare *climate.*

weathering Physical and chemical processes in which solid rock exposed at earth's surface is changed to separate solid particles and dissolved material, which can then be moved to another place as sediment. See *erosion.*

wetland Land that is covered all or part of the time with salt water or freshwater, excluding streams, lakes, and the open ocean. See *coastal wetland, inland wetland.*

wilderness Area where the earth and its community of life have not been seriously disturbed by humans and where humans are only temporary visitors.

wildlife management Manipulation of populations of wild species (especially game species) and their habitats for human benefit, the welfare of other species, and the preservation of threatened and endangered wildlife species.

wildlife resources Wildlife species that have actual or potential economic value to people.

wild species Species found in the natural environment. Compare *domesticated species.*

windbreak Row of trees or hedges planted to partially block wind flow and reduce soil erosion on cultivated land.

wind farm Cluster of small to medium-sized wind turbines in a windy area to capture wind energy and convert it into electrical energy.

worldview How people think the world works and what they think their role in the world should be. See *environmental wisdom worldview, planetary management worldview, spaceship-earth worldview, stewardship worldview.*

zero population growth (ZPG) State in which the birth rate (plus immigration) equals the death rate (plus emigration) so that the population of a geographic area is no longer increasing.

zone of aeration Zone in soil that is not saturated with water and that lies above the water table. See *water table, zone of saturation.*

zone of saturation Area where all available pores in soil and rock in the earth's crust are filled by water. See *water table, zone of aeration.*

zoning Regulating how various parcels of land can be used.

zooplankton Animal plankton. Small floating herbivores that feed on plant plankton (phytoplankton). Compare *phytoplankton.*

INDEX

Note: Page numbers of **boldface** type indicate definitions of key terms. Page numbers followed by italicized *f, t,* or *b* indicate figures, tables, and boxes.

Abiotic components in ecosystems, **61,** 63*f*
Abortion rate, family planning and reduction of, 189–90
Aboveground buildings, hazardous waste storage in, 553
Abyssal zone, open sea, 137
Accidentally-introduced nonnative species, 235*f,* 236–37
Acid(s), 40, 41*f*
Acid deposition, 81, 83, **444–49**
  causes and formation of, 444–45
  effects of, on aquatic ecosystems, 446, 447*f*
  effects of, on human health and materials, 445–46
  effects of, on plants and soils, 446–48
  global regions affected by, 446*f*
  prevention versus cleanup of, 448*f*
  strategies for reducing, 448–49
  in United States, 445*f,* 448
Acid mine drainage, 343, 344*f*
Acid rain. *See* Acid deposition
Acid shock, 446
Active solar heating system, 389*f,* **392***f*
Acute effects of toxics, 412
Adaptation, **90**
  ecological niches and, 91–93
  limits on, 93
Adaptive ecosystem management process, 218*f*
Adaptive management strategies, 611
Adaptive radiations, **96**
Adaptive trait, **90.** *See also* Adaptation
Administrative law, **616**
Advanced (tertiary) sewage treatment, **510**
Aerobic respiration, **66,** 78
Aesthetic value, 591
  of biodiversity, 196
Affluence, environmental quality and, 14. *See also* Affluenza
Affluenza, **14,** 52
  escaping from, 638–39
  solid waste production and, 533–34
Aflatoxin, 412
Aftershocks, earthquake, 337
Age of Oil, history of, A12
Age structure, population, 164
  in developed and developing countries, 185*f*
  effect of, on population growth, 184, 185*f,* 186*f*
  effect of AIDS on, in Botswana, 424*f*
  global aging and, 187*f*

making population and economic projections based on, 184–86
  reduced fertility and change in, 186–87
  rise in death rates from AIDS and change in, 187
  of select countries, 184*f*
Aging, global trends in, 187*f*
Agribusiness, 276
Agricultural revolution, **21–22,** 23*f*
  green revolution and, 276–78
  trade-offs (advantages/disadvantages) of, 23*f*
Agriculture
  controlling pests through cultivation methods of, 528
  effects of global warming on, 475*f,* 476–77
  government policy on, 301–2
  green revolution in, 276–79
  human nutrition and, 287–90
  impact of soil erosion and degradation on, 279
  increasing crop yields in, 291–94
  increasing meat production in, 294–97
  industrialized (high-input), 274–75, 276–78*f*
  irrigation and (*see* Irrigation)
  low-tech, in Kenya, 279*b*
  monoculture, 118*f*
  most important crops, 274
  pesticide use in (*see* Pesticides)
  plantation, 123, 275
  polyculture, 273, 303
  soil conservation and methods of, 284–86
  sustainable, 279*b,* 302–3
  systems of, 274, 275*f*
  traditional (low-input), 275–76, 278–79
  in United States, 276–78
  water pollution and, 495, 509, 511, 512*f*
  water waste and, 323
  yield rates, 286–87
Agroforestry, **278, 284,** 285*f*
AIDS (acquired immune deficiency syndrome), 423–24
  effect of, on population age structure, 424*f*
  HIV and, 422
  population decline due to, 187
Air circulation, climate and global, 105, 106*f,* 107*f*
Air pollutants
  formaldehyde as, 450
  hazardous, 453
  indoor, 449–51
  major classes of, 436*t*
  major outdoor, 438*t*
  mercury as, 557
  primary, 435, 436*f*
  radon gas, 450–51
  secondary, 436*f*
  sources and types of outdoor, 436*f*

Air pollution, 433–60
  acid deposition as cause of outdoor regional, 444–49
  coal burning as source of, 365
  defined, **435**
  disasters involving, 437*b*
  Earth's atmosphere and, 434–35
  effects of, on living organisms and materials, 451–53
  food production and, 290*f*
  indoor, 449–51
  as key environmental problem, 12*f*
  lichen as indicator of, 433
  motor vehicles as largest source of, 438*t,* 573
  outdoor, major types and sources of, 435–39
  photochemical and industrial smog as, 437*b,* 439–43
  preventing and reducing, 453–60
  in urban areas, 176*f,* 437
Air pressure, 102–3
Albedo, **467,** 468*f*
Alcock, James, 210
Algae, production of hydrogen from green, 404*b*
Algal blooms, 498, 505
Alien (nonnative) species, **146.** *See also* Nonnative species
Alleles, **90**
Alley cropping, **278, 284,** 285*f*
Alligators as keystone species, 149*b*
Allopatric speciation, 94
Alpha particles, **48**
Alpine tundra, 118–19
Altitude, climate, biomes, and effects of, 112*f,* **113**
Ammonification, nitrogen fixation and, 81
Amphibians as indicator species, 147*f*
Anaerobic respiration (fermentation), **66**
Anderson, Ray, development of sustainable green corporation by, 538*b*
Anemia, 288
Animal(s)
  accumulation of DDT in tissues of, 411*f*
  desert, 113–14
  grassland, 115, 116*f*
  keystone species, 148–49 (*see also* Keystone species)
  lab testing of, to determine toxicity of chemicals, 413–15
  tropical rain forest, 121, 122*f*
  zoos and aquariums for protection of endangered, 246–47
Animal manure, **285,** 398
Animal plankton, 129
Antarctic
  food web in, 69*f*
  stratospheric ozone depletion in, 485, 486*f*

Anthracite, 364. *See also* Coal
Anthropocentric (human-centered) instrumental value, **632**–33
Antibiotics, resistance of pathogens due to overuse of, 419–21
Anti-environmental movement in U.S., 28
Antiquities Act of 1906, 25
Appliances, energy efficient, 390
Aquaculture, 297, **300**–301
  advantages and disadvantages of, 301*f*
Aquariums, protecting endangered species in, 246–47
Aquatic biodiversity, 251–72
  human impacts on, 253–57
  in Lake Victoria, Africa, 251
  managing/sustaining global marine fisheries, 263–65
  overview of, 252–53
  protecting/sustaining marine, 257–63
  protecting/sustaining/restoring lakes and rivers, 267–71
  protecting/sustaining/restoring wetlands, 265–67
Aquatic life zones, **61**, 127–41
  biodiversity in (*see* Aquatic biodiversity)
  biological magnification of PCBs in, 498*f*
  coral reefs, 127, 135–36, 137*f*, 252, 263
  distribution of global, 128*f*
  effects of acid deposition on, 446, 447*f*
  energy flow in, 70*f*
  food web in, 69*f*, 497, 498*f*
  freshwater, 61, 137–42 (*see also* Freshwater life zones)
  layers of, 129–30
  limiting factors in, 62–64, 129–30
  marine (ocean saltwater), 61 (*see also* Marine life zones)
  mercury contamination in, 556*f*, 557
  range of tolerance for temperature in, 64*f*
  saltwater, 130–37
  terrestrial life zones versus, 129
  trade-offs (advantages/disadvantages) of life in, 129*f*
  two types of, 128
  types of organisms living in, 128–29
Aquifers, **308**–9. *See also* Groundwater
  confined, and unconfined, 308*f*
  depletion of, 309, 318–19, 320*f*, 321
  fossil, 309
  in Mexico City, 571
  natural recharge of, 308
  saltwater intrusions in, 320*f*
Aral Sea water diversion project, and ecological disaster, 316*f*, 317
Arboretum, 246
Arctic National Wildlife Refuge (ANWR), oil and gas development in, 359, 360*f*
Arctic sea ice, shrinking of, 467*f*
Arctic tundra, 118, 119*f*, 141
  drilling for oil in, 359, 360*f*
Area strip mining, **341***f*, 364
Argentine fire ant, 236*f*
Arsenic as water pollutant, 503
Artificial selection, **97**
  of crops, 291
Asexual reproduction, **169**
El-Ashry, Mohamed, 323, 407
Asian Brown Cloud, 442
Asian swamp eel, 256
Assimilation, nitrogen fixation and, 81
Asthenosphere, 332, 333*f*
Asthma, 452

Atmosphere, **59**, 60*f*, 434–35
  chemical makeup of, climate, and greenhouse effect, 109, 110*f*, 434*f*, 438
  climate and (*see* Climate; Climate change)
  comparison of measured changes in average temperature of, 462, 463*f*
  energy transfer by convection in, 107*f*
  key characteristics of, 434
  layers of, 434*f*
  ozone in, 435*f* (*see also* Ozone (O_3); Ozone depletion)
  pollution in (*see* Air pollution)
  stratosphere of, 59, 435 (*see also* Stratosphere)
  temperature of, 463*f*, 466, 467t, 469–74
  troposphere of, 434–35 (*see also* Troposphere)
Atmospheric pressure, **434**
Atom(s), **39**
  chemical compounds and bonds between, 41, A9
  in chemical equations, 47–48
  structure of, 39, 40*f*
Atomic number, **39**
Australia
  Great Barrier Reef, 262, 264
  solar energy in, 393
Automobiles. *See* Motor vehicles
Autotrophs, **64**. *See also* Producers

Baby-boom generation, United States, 184
Baby-bust generation, United States, 185
Background extinction, **95**
Bacon, Francis, 175
Bacteria
  antibiotic resistant, 419–20
  cleaning up oil spills using, 508
  fecal coliform, in water, 493*f*
  gut inhabitant mutualism, 155
  nitrogen cycle and role of, 80–81
  as pathogen, 419, 492t
  prokaryotic cell, 56, 58*f*
  in water, 492t, 493*f*
Bacterial source tracking (BST), 493
Balanced chemical equations, 47–48
Balance of nature, 159
Baleen whales, 259, 260*f*
Baltic Sea, water pollution in, 505
Bangkok, Thailand, air pollution in, 176*f*
Bangladesh
  flooding in, 327–28
  microlending to reduce poverty in, 601*b*
Barrier beaches, 133, 134*f*, 135*f*
Barrier islands, **133**, 134, 135*f*
Basal cell skin cancers, 487*f*
Bases, 40, 41*f*
Bathyal zone, open sea, 136–37
Bats, 91
  ecological role of, 230
  as keystone species, 143
Beal, John, stream restoration by, 270*b*
Bedrock, 74
Behavioral strategies for avoiding predation, 154
Bennett, Hugh, 281
Bentham, Jeremy, 631
Benthic zones, lake, 138, 139*f*
Benthos, **129**
Bequest (option) value, 591
  of biodiversity, 196
Beta particles, 48
Beverage containers, energy consumption for types of, 539*f*
Bhopal, India, toxic chemical release at, 549–50

Bicycles as transportation, 574*f*
Bioaccumulation of toxic chemicals, **411**
Biocentric (life-centered) values and worldview, **632**, 633–36
Biocides, **519**. *See also* Pesticides
Biocultural restoration, 221
Biodegradable pollutants, **48**
Biodegradable waste, composting of, 541
Biodiversity, 66, **67**
  aquatic (*see* Aquatic biodiversity)
  bioinfomatics and applied information about, 197
  changes in, over geological time, 96*f*
  in communities, 144, 145*f*, 146–50
  ecosystems and preservation of (*see* Ecosystem approach to preserving biodiversity)
  extinction of, (*see* Extinction of species)
  factors tending to decrease or increase, 195*f*
  global warming and, 475*f*
  impact of human activities on, 96, 195–96
  D. Meadows on value of, 634
  in mountain biomes, 125
  priorities for protecting, E.O. Wilson's, 221
  projected status of Earth's, in years 1998–2018, 196*f*
  reasons for protecting, 196
  in saltwater and freshwater life zones, 129, 130, 131, 132*f*, 135, 136*f*, 139*f*
  of species (*see* Species; Species diversity)
  species approach to preserving (*see* Species approach to preserving biodiversity)
  terrestrial (*see* Terrestrial biodiversity)
Biodiversity hot spots, 218
  conservation biology and identification of, 197
  extinction of species in, 229
  global, 219*f*
  in United States, 242*f*
Biodiversity loss, 5. *See also* Extinction of species
  deforestation as cause of, 203
  food production and, 290*f*
  as key environmental problem, 12*f*
Biofuels, **398**–400
Biogas production, 398
Biogeochemical (nutrient) cycles, **76**–84
Bioinformatics, **197**
Biological community, **59**. *See also* Community(ies)
Biological diversity, **67**. *See also* Biodiversity
Biological evolution, 87, **88***f*, 89*f*. *See also* Evolution
Biological extinction of species, 225. *See also* Extinction of species
Biological hazards, 410, 419–28. *See also* Disease
  bioterrorism as, 426–28
  HIV and AIDS, 422, 423–24
  malaria, 424–25
  nontransmissible and transmissible disease as, 419, 420*f*
  overuse of antibiotics and struggle against, 419–21
  pathogenic organisms, 419*f*
  reducing incidence of infectious disease, 425–26
  tuberculosis as case study, 421–22
  viruses and viral disease, 419, 422–23
Biological indicators, 433
Biological magnification of PCBs in aquatic ecosystem, 498*f*

Biological pest control, 518, 528
Biological weathering, 336
Biomagnification of toxic chemicals, **411**f
  in aquatic food chain, 498f
Biomass, 6, **68**
  burning of, as fuel, 397–98
  energy produced from (*see* Biomass energy)
  productivity rates of, 70–72
Biomass energy, 352, 397–400
  burning carbon compounds as, 397–98
  ethanol and methanol as, 399–400
  gaseous fuels produced as, 398–99
  solid biomass, 399f
  types of biomass fuel, 398f
Biomass plantations, 398
Biome(s), **61**
  along 39th parallel in United States, 62f
  climate and, 110–13
  desert, 113–15
  Earth's major, 111f
  forests, 120–25
  global air circulation and, 107f
  grasslands, 115–20
  mountains, 125–26
Biomimicry, 537
Biomining, 347
Biomonitors, 494
Biopharming, **98**
Biophilia, 231b
Biophobia, 231b
Bioprospectors, 154
Bioreactor, 513
Bioremediation, 551
Biosensors, 494
Biosolids (sludge), 511
Biosphere, 6, 57f, **59,** 60f
  solar energy, gravity and, 60–61
Biosphere reserve, 217, 262
Biosphere 2, Tucson, Arizona, 630, 633–34
Bioterrorism, 426–28
  common agents associated with, 427f
Biotic change, 160
Biotic components in ecosystems, **61,** 63f
Biotic potential of populations, **165**
Birds, 123
  effect of human activities on, 233f, 234
  effect of pesticides on embryos of, 35b
  evolutionary divergence of honeycreepers, 93f
  extinction of passenger pigeon, 224
  flyways of migratory, 245, 246f
  mutualistic relationships of, 155
  red-cockaded woodpeckers, 91–92
  resource partitioning among warblers, 150, 151f
  seabirds, 234
  songbirds, 233f
  specialized feeding niches of shorebirds, 92f
Birth control methods, effectiveness of, 182f
Birth rate (crude birth rate), **177**
  advantages/disadvantages of reducing, 187–88
  average, for select countries, 177f
  economic development and reduction of, 188–89
  as factor in human population size, 177
  factors affecting, 180–82
  global population growth and, 177–78
  reducing, by empowering women, 190–91
  reducing, with family planning programs, 189–90
  in U.S., years 1910-2004, 180f

Births as variable in population size, 164.
  See also Birth rate (crude birth rate)
Bison, 118
  near-extinction of American, 20
Bitumen, 360–61
Blair, Tony, 609
Bleaching, sewage treatment and, 510–11
Blind technological optimism, trap of, 639
Blue whale, near extinction of, 260–62
Boll weevil, 522f, 523f
Boreal forests, 123–25
Botanical gardens, 246
Botswana, AIDS and population age structure, 424f
Bottled water, 515–16
Bourlag, Norman, 276
Bowash megalopolis (Boston-New York-Washington DC), 568f
Brazil
  Curitiba, as sustainable city in, 563
  key demographic factors, 186f
  tropical deforestation in, 211, 212
Breeder nuclear fission reactors, **376–77**
Broadleaf evergreen plants, **121**
Brown, Lester R., 193, 289, 303, 602
Brown-air smog, 439
Brownfields, 220, 536
  recycling, 554
Bubonic plague, 167
Buffer, 444
Buffer zone, biosphere reserve, 217f
Buildings
  energy efficiency in design of, 379, 386–90
  energy loss from inefficient, 388f
  formaldehyde in, 450
  hazardous waste storage in aboveground, 553
  naturally cooled, 393
  net energy efficiency in heating, 382f
  radon gas in, 450–51
  saving energy in, 386–88
  sick-building syndrome, 449–50
  solar heated, 389f, 390, 391f
Bullet trains, 575, 576f
Bureau of Land Management (BLM), U.S., 28, 197
Burning
  of biomass as fuel, 397–98
  of hazardous wastes, 552–54
  of solid wastes, 545, 546f
Burying
  hazardous waste, 552–54
  solid waste, 547–48
Buses as transportation
  advantages and disadvantages of, 575f
  in Curitiba, Brazil, 563f
Bush, George H. W., 29
Bush, George W., presidential administration of
  air pollution policies, 438, 455, 557
  climate-change policies, 481–82
  public lands policy, 346
Bushmeat, 212, 239
Business
  developing sustainable, service flow, 537–38
  reducing water use by, 325–26
  urban growth linked to opportunities in, 564
Butterflies, population of monarch, 59f
Bycatch, 255, 263–64, 298

Cahn, Robert, 631
Calabrese, Edward, 415
California, 209

California Water Transfer Project, 317–18
Callender, Jim, wetlands restoration by, 266b
Camouflage coloration, 153f
Canada
  demographic data for, 181f
  expected damage from, in U.S. and, 337f
  fossil fuel deposits in, 357f
  James Bay Water Transfer Project in, 318
Canadian lynx-snowshoe hare predator-prey relationship, 168f
Cancer, **416**
  lung, 452
  skin, 487f, 488
Cap-and-trade emissions program, 455
Capital, 6
  human, 584
  natural, 6 (*see also* Natural capital)
  solar, 6, 351, 391–95
  types of, as driving force in economies, 584
Capitalist market economic systems, 584–85
  government intervention into, 585–86
Captive breeding, 246
Carbohydrates, simple, and complex, 42
Carbon
  organic compounds and, 41–42
Carbon cycle, 60f, **77,** 78–79f
  biospheric cycling of carbon, 77–78
  human impact on, 79f, 80, 465–66
Carbon dioxide (CO_2), 78
  as air pollutant, 437–39
  coal burning and emissions of, 365, 455–56, 464
  concentration of, in troposphere (years 1860-2003), 465f
  deforestation and, 203
  emissions of, per units of energy by various fuels, 360, 361f
  global temperature and levels of, 464, 465f
  photosynthesis and levels of, 474
  slowing global warming by removing and storing, 478–79, 480f
  storage of, in oceans, 472f, 473
Carbon monoxide as air pollutant, 438t
Carbon tetrachloride, 485
Carcinogens, **416**
Careers, environmental, 610b
Car-free areas, 573
Carnivores, 152–53
Carrying capacity (K), **165**
  diebacks as result of exceeded, 166–67
  logistic growth of population and, 166f
Carson, Rachel, 26, 27b, 142, 521, 526
Carter, Jimmy, 28
Case law, 616
Case reports, chemical toxicity, 413
Cash crops, 275
Cattle. *See* Livestock production
Cell(s), **56**
  eukaryotic, 56, 58f
  prokaryotic, 56, 58f
Center-pivot low-pressure sprinkler, 323, 324f
Centers for Disease Control (CDC), U.S., 409, 555
Central Arizona Project (CAP), 317f
Ceramics, 349
Cetaceans, 259, 260f, 261f
Chain reaction, nuclear, **49,** 50f
Channelization of streams, 327f, 328
Chaparral, 119–20
Chateaubriand, François-Auguste-René de, 195

Chattanooga, Tennessee, 581, 608
Chemical(s)
  defensive, 153
  inorganic and organic, as water pollutants, 492*t*
  inorganic compounds, 492
  ozone depleting, 484, 485*f*
Chemical analysis
  location of minerals by, 340
  of water, 494
Chemical bonds, 39, 41, A8–A10
Chemical change in matter (chemical reaction), 47
Chemical equations, balanced, 47–48
Chemical evolution, 88
Chemical formula, **41**, A9*f*
Chemical hazards, 410–18. *See also* Hazardous wastes; Toxic chemicals
  in air pollution, 453–54
  arsenic as, 503
  case studies on lead, mercury, chlorine, and dioxins, 554–58
  determining toxicity of, 413–16
  in homes, 549*f*
  hormesis controversy and, 415
  hormone disrupters as, 416, 417*f*
  indoor air pollutants as, 449–51, 458, 459*f*
  international agreements to reduce, 560–61
  lack of sufficient knowledge about, 417–18
  laws to protect against, 28, 532, 548, 557, 558–60, 618
  persistent organic pollutants (POPs), 560
  precautionary principle applied to, 418
  shift in awareness toward problems of, 606
  toxic and hazardous forms of, 416
  toxicology and assessment of, 411*f*, 412
  trace levels as, 412
  in water, 492*t*, 495, 496, 497, 498*f*, 499, 500–501, 502*f*, 504*f*
Chemical interactions, 412
Chemical nature of pollutants, 48
Chemical reaction, **47**
Chemical weathering, 336
Chemistry, basic principles of, A8–A10
  elements and periodic table, A8*f*
  hydrogen bonds and nucleic acids, A9, A10*f*
Chemosynthesis, **65**
Chernobyl nuclear power accident, 350, 433
Chesapeake Bay, pollution in, 505–6
Children
  effects of pesticides on, 521, 524
  malnutrition and disease in, 13–14, 287–88
  protecting, from lead poisoning, 555*f*
China, 11
  acid deposition in, 445
  air pollution in, 441, 442–43
  biogas production in, 398
  biological pest control in, 518
  carbon dioxide emissions by, 482
  coal as energy source in, 364, 365
  demographic data for, 191*f*
  desertification in, 282
  food production in, 289–90
  fuel efficiency standards in, 383
  land-use planning in urban areas, 577, 580
  light-rail and rapid-rail system in, 575–76
  motor vehicle use in, 572
  paper production in, 210
  population regulation in, 191–92

Three Gorges Dam in, 315*f*–16
water withdrawal and use in, 309*f*
wind, air pollution, and, 101–2
Chipko movement in India, 620
Chiras, Daniel D., 4
Chlorinated hydrocarbons, 42
Chlorination of water, 511
Chlorofluorocarbons (CFCs), 484
  projected concentrations of, in stratosphere, 489*f*
  reducing use of, 488–89
  stratospheric ozone depletion and role of, 484, 485*f*
  substitutes for, 489*b*
Chromosomes, **42**
Chronic effects of toxics, 412
Chronic undernutrition, **287**
Cichlids, 251
Circle of poison, 525*b*
Cities. *See* Urban areas; Urban growth; Urbanization
Citizens, transformation into environmental, 639
Civilian Conservation Corps (CCC), 26
Civil suits, 616
Clarke, Arthur C., 533
Class action lawsuits, 616
Clean Air Acts, U.S., 437*b*, 448, 557, 618
  New Source Review rule in, 456
  reduction of air pollution credited to, 453–54
Cleanup approach
  to acid deposition, 448*f*
  to air pollution, 454, 456*f*, 458*f*, 459*f*
  to coastal water pollution, 508*f*
  costs of, 589*f*
  to global warming, 479*f*
  to groundwater pollution, 503*f*
Clean Water Act, U.S., 509, 510, 513–14, 618
Clear-cutting forests, **201**, 202*f*
  advantages and disadvantages of, 203*f*
"Clear Skies" initiative, G. Bush administration, 455, 557
Climate, **105**–10
  Aral Sea eco-disaster and changes in, 317
  atmosphere, greenhouse effect, and, 109, 438
  biomes and, 110–13 (*see also* Biome(s))
  change in (*see* Climate change)
  Earth's axis and, 105, 106*f*
  effects of, 105*f*
  El Niño-Southern Oscillation (ENSO), 108, 109*f*
  global air circulation and, 105–6
  global circulation model of, 469, 471*f*
  global zones of, 106*f*
  La Niña, 108–9
  ocean currents, upwellings and, 106*f*, 107, 108*f*
  temperature and precipitation as determiners of, 462
  topography, and local, 109–10
  troposphere and, 434–35 (*see also* Troposphere)
  water shortages and, 311–12
  weather and, 102–4
  wind and, 107, 108*f*
Climate change, 461–84
  carbon dioxide and, 437–39
  characteristics of global warming, 467*t*
  factors affecting Earth's temperature and, 472–74
  global warming and, 466–69 (*see also* Global warming)
  greenhouse gases and, 464*t*, 465–66

human activities leading to, 465–69
methods of studying, 462, 463*f*
natural greenhouse effect and, 109, 434*f*, 464–65
past, 462–63
possible effects of, 474–77
projecting future, 469–72
reducing greenhouse gases, 481–84
shift in awareness to concern about, 606
solutions for dealing with threat of, 477–81
water quality and, 495
Climate change legacy, 482
Climax community, 159, 160
Clinton, Bill, 29
Clone, genetic, 97
Closed-loop recycling, 540
Closed nuclear fuel cycle, 367, 368*f*
Cloud cover, climate and changes in, 473
Cloud seeding, 322
Clownfish, mutualism and, 155*f*
Clumping, population dispersion as, 164*f*
  effects of, on population density, 167
Cluster housing developments in urban areas, 579*f*
Coal, **364**–66
  advantages and disadvantages of, as fuel, 365*f*
  converting, into gaseous and liquid fuel, 365–66
  costs of mining, 588*f*
  environmental impact of human use of, 365, 464–65
  extraction and processing of, 341–44, 364
  formation of, 364*f*
  nuclear energy versus, 371*f*
  uses and supplies of, 364–65
Coal-burning facilities
  energy waste in, 381–82
  reducing air pollution from, 455–56
Coal gasification, **365**, 456
Coal liquefaction, **365**
Coastal coniferous forests, 125
Coastal wetlands, **131**, 132*f*
Coastal zones, **130**–31
  global warming and, 475*f*
  integrated coastal management, 262, 506
  as marine life zone, 130–35
  pollution in, 504*f*, 505, 506*f*
  protecting, 508–9
  urbanization in, 569
Cobb, John, 590
Cockroaches, 92*b*
Coevolution, **90**–91, 161
Cogeneration, **382**–83
Cohen, Michael J., 434
Cold desert, 113
Cold front, **102**
Colinvaux, Paul A., 462
Colorado River Basin, 314*f*–15
Columbia River basin, managing, 268–70
Comanagement, 264
Combined heat and power (CHP) systems, 382–83
Command-and-control regulatory approach, 455
Commensalism, **156**
Commercial energy, 351
  fossil-fuel based, 351–52
  reducing waste of, 380*f*
  U.S. economy and flow of, 381*f*
  use of, globally and in United States, 352*f*
Commercial extinction, 255, 259
Commercial inorganic fertilizers, **285**, 286*f*
Commercial whaling, 262

Common Cause (organization), 620
Commoner, Barry, 27
Common law, **616**
Common-property resources, **10**
Community(ies), 57f, **59,** 143–62. *See also* Ecosystem(s)
  biodiversity in, 144, 145f, 146–50
  climax (mature), 159, 160
  competition and predation in, 150–54
  ecological stability and sustainability of, 160–61
  ecological succession in, 156–60
  keystone species in, 143, 148, 149b
  niche structure of, 144–45
  parasitism, mutualism, and commensalisms in, 154–56
  physical appearance and population distribution in, 144f
  species diversity, 144, 145–46
  species types in, 146–50
  structure of, 144–45
Comparative risk analysis, 428
Competition among species, 150
  avoidance of, 150–51
  as limiting factor on population size, 165
Competitor (K-selected) species, 169, 170f
Complexity, ecological, **160–61**
Compost, **286**
  of biodegradable organic waste, 541
Composting toilet systems, 512
Compounds, **39,** 41
  inorganic, 43
  organic, 41–42
Comprehensive Environmental Response, Compensation, and Liability Act (Superfund program), U.S., 28, 532, 558–60, 618
Concentration, **40**
  of pollutants, 48
Condensation, 77
Condensation nuclei, **77,** 103
Conduction, 45f
Coniferous evergreen trees, 123–25
Consensus (sound) science, **36**
Consequentialism, 631
Conservation biology, **197,** 247
Conservation era of U.S. environmental history, 24–26
Conservation of matter, law of, 46, **47,** 48
Conservation Reserve Program (CRP), 280
Conservation-tillage farming, **284**
  advantages and disadvantages of, 284f
Constancy, **160**
Constant loss population, 170, 171f
Constitutional democracy, 607
Consumers, **65**
  excessive consumption by (*see* Affluenza)
  feeding relationships among producers, decomposers, and, 65f
Consumers' Right to Know, 626
Consumption of resources
  environmentally ethical, 638
  excessive (*see* Affluenza)
  relationship to environmental problems, 14
Consumptive water use, 309
Containment vessels, nuclear-fission reactors, 366
Continental crust, 332, 333f
Continental drift, 95f, 333
Continental shelf, 131
Contour farming, **284,** 285f
Contour strip mining, **341**f, 342, 364
Contraceptive methods, effectiveness of, 182f

Control group, 414
Controlled experiment, 34, 414
  example of, 35b
Control rods, nuclear fission reactors, 366
Convection, 45f
  energy transfer by, in atmosphere, 107f
Convection cells, Earth's crust, 333, 334f
Conventional-tillage farming, **284**
Convention on International Trade in Endangered Species (CITES), 257, 259, 624
Conventions (international agreements), 623
Convergent plate boundaries, **334–35,** 336f
Cooling, natural, 393
Copenhagen Protocol, 488, 489f, 623
Coral reefs, **127,** 252
  development of, 263
  as ecosystem, 135, 136f
  human impact and major threats to, 135, 137f
Core, Earth's, **332,** 333f
Core area, biosphere reserves, 217f
Corporate Average Fuel Economy (CAFE) standards, 383, 404
Cosmetic spraying of pesticides, 527–28
Costanza, Robert, 583
Costa Rica
  ecological restoration in, 221
  nature reserves in, 216
Cost(s)
  of cleaning up pollution, 589f
  of controlling pollution, 588f
  internal, and external, 592–93
  mitigation, 591
  pricing based on full, 592–93
Cost-benefit analysis (CBA), **589**–90
Coupled global circulation models, 469, 470f, 471f
Covalent bonds, 41, A8–A9
Covalent (molecular) compounds, 41
Cover crops, **284**
Cowie, Dean, 1
Crime in urban areas, 570
  improving environment by reducing, 572b
Critical mass, **49,** 50f
Critical thinking, improving, 3–4
Crocodiles, American, 263
Crop(s)
  cash, 275
  cover, 284
  energy production from (*see* Biomass energy)
  genetically modified, 99, 291, 292f
  grain crops, most important, 274
  green revolution and increased yields in, 276, 277f
  increasing world yields of, 291–94
  monoculture, 118f, 123
  pesticide use on, 521, 524–25
  residues from, as biofuel, 398
  water used to irrigate, 309, 310f, 314, 323
Croplands, 274, 275
  ecological and economic services provided by, 276f
  increasing amount of, 293–94
  irrigated, 282–84, 293
Crop rotation, **286**
Crossbreeding crops, 291
Crown fire in forests, 207, 208f
  preventing, 208–9
Crude birth rate, **177.** *See also* Birth rate
Crude death rate, **177.** *See also* Death rate

Crude oil, **355,** 507. *See also* Oil
  refining, 355, 356f
Crust, Earth's, **332,** 333f, 334f
Crustal displacement, lake formation and, 138
Cullis-Suzuki, Severn, 627b
Cultural eutrophication of lakes, 140, **498,** 499f
  Great Lakes as case study of, 500f, 501
  Lake Washington recovery, 499–500
Cultural hazards, 410, 428, 430f. *See also specific types of cultural hazards, e.g.* Poverty
Cultural world (culturesphere), 6
Culture, environment and epochs of human evolution
  Agricultural Revolution, 21–22, 23f
  change and, 21
  environmental history in U.S., 24–30, A2–A7
  hunting and gathering societies, 21
  industrial-medical revolution and, 22, 23f
  information and globalization revolution, 23–24
  sustainability (*see* Sustainability; Sustainable living)
Curitiba, Brazil, as sustainable ecocity, 563
Cuyahoga River, cleanup of, 496
Cyanide gas release in Bhopal, India, 549–50
Cyanide heap leaching, 344
Cyclic fluctuations of population size, 167, 168f
Cyclodextrin, 551
Cyclone, tropical, 104f

Daly, Herman, 583, 591
Dams and reservoirs, 313–16. *See also* Lake(s)
  advantages and disadvantages of, 313–14
  along Colorado River basin, 314–15
  environmental impact of, 395
  for flood control, 315, 328–29
  freshwater supplies and, 313–14
  hydropower at, 315, 318
  Three Gorges Dam in China, 315–16
  water pollution in, 497–501
Darwin, Charles, 88, 164
Davidson, Carlos, 636
DDT, 27b, 47
  bioaccumulation and biomagnification of, 411f
  as pesticide, 519–20, 526–27
  unintended consequences of using, 173b
Death
  causes of human, globally, 429f
  population size and, 164
  premature, from air pollution, 453
Death rate (crude death rate)
  average, for select countries, 177f
  factors affecting, 182–83
  global population growth and, 177–78
  population decline due to AIDS and rise in, 187
D'Eaubonne, Françoise, 635
Debt-for-nature swaps, 213
Deceptive looks and behavior as defensive strategy, 153f, 154
Deciduous trees, 122–23
Decomposers, **65,** 66f, **129**
  feeding relationships among producers, consumers, and, 65f
Deductive reasoning, **35**
Deep-ecology environmental worldview, 635
Deep-ocean floor, 252

Deep underground storage of high-level radioactive wastes, 373f
Deep-well disposal of hazardous wastes, 553
Deer populations in United States, 238
Defendant in lawsuits, **616**
Deforestation, **203**
  environmental degradation caused by, 203f
  flooding caused by, 327, 328f
  reforestation projects, 213, 214b
  release of greenhouse gases by, 466
  tropical, 210–14
Degradable (nonpersistent) pollutants, **48**
Degree of urbanization, **564**
Demand (economics), 584–85
Democracy, **607**
  dealing with environmental problems in, 607
Democratization of learning and communication, 9
Demographic bottleneck, 171
Demographic transition, **188**, 189f
Demography, **177**
Dengue fever, 420f
Denmark, 609
Density-dependent population controls, 167
Density-independent population controls, 167
Department of Energy (DOE), U.S., 370, 373, 380
Depletion time of mineral resources, **345**–46
Deposit feeders, 137
Desalination, water, 321, **322**
Desertification, **281**–82
  causes and consequences of, 283f
  global sites of, 282f
Deserts, **113**–15
  human impacts on, 115f
  major types of, 113
  plants and animals of, 113–14
  temperature and precipitation in, 113f
Desiccation, 311
Detoxifying hazardous wastes, 550–52
Detritus, **65**
Detritus feeders, 66f
Detrivores, **65**, 66f
Deuterium-tritium and deuterium-deuterium nuclear fusion reactions, 50f, 377
Devall, Bill, 635
Developed countries, **8.** *See also names of individual countries, e.g.* Canada
  assistance of, in reducing global poverty, 599–600
  biological hazards in, 419–28
  connections among environmental problems in, 15
  environmental policy life cycle in, 615, 616f
  freshwater supplies in, 312
  oil consumption in, 358f, 359f
  population and population growth in, 8f
  population structure by age and sex, 185f
  reducing stream and river pollution in, 496–97
  urbanization in, 564, 565f
Developing countries, **8.** *See also names of individual countries, e.g.* Brazil
  age structure of select,
  biological hazards in, 419–28
  connections among environmental problems in, 15
  deaths from air pollution in, 453
  drinking water supply in, 515

freshwater supplies in, 312
gray-air smog in, 440–41
growing demand for oil in, 359f
population and population growth in, 8f
population structure by age and sex, 185f
reducing poverty in, 599–600
reducing stream and river pollution in, 497
reuse of materials and resources in, 538–39
shift in awareness toward environmental problems in, 606
urbanization in, 564, 565f
Dew point, 103
Dieback (crash), population, 166–67f
Differential reproduction, **90**
Dioxins, health threat of, 558
Dirty bomb, 375–76
Discharge trading policy, 509
Discount value of resources, **591**–92
Disease, 419–28
  asthma, 452
  bioterrorism, 426–28
  defenses of organisms against, 154
  emphysema, chronic bronchitis, and lung, 452f
  in forests, 206, 207f
  HIV and AIDS, 187, 422, 423–24
  infectious (*see* Infectious disease)
  malaria, 420f, 424–25
  mosquito as vector for three, 419, 420f
  overuse of antibiotics and, 419–21
  pathogens as cause of, 419f
  population declines due to, 187
  reducing incidence of, 425–26
  relationship of, to poverty and malnutrition, 13, 287
  tobacco, smoking, and, 409
  transmissible and nontransmissible, 419
  tuberculosis, 419, 421–22
  viral, 422–23
  water pollution as cause of human, 493t
Disinfection of water, 511
Dispersed cities, 571
Dissolved oxygen (DO) content, **64**
  water quality and, 494f
Distillation of water, 322
Distribution (range) of species, **58**
Disturbance, **159**
  affects of, on succession and species diversity, 159
  opportunists species and environmental, 169
Divergent evolution, 93f, 94
Divergent plate boundaries, **334**–35, 336f
DNA (deoxyribonucleic acid), 42–43
  effects of ionizing radiation on, 49
  structure of, A10f
Dorr, Ann, 349
Dose, toxic chemical, **410**, 412
  median, and average lethal, 413, 414t
  sensitivity to different, 411f
Dose-response curve, 413f, **414**
  no threshold, and threshold types of, 414f, 415
Doubling time of population growth, **178**
Downcycling, 540
Drainage basin, **140**, **307**
Dredge spoils, **507**
Dredging, 341f
Drift-net fishing, 298, 299f
Drinking water, 13, 14–15. *See also* Freshwater
  access to safe, 495
  arsenic in, 503

bottled, 515–16
groundwater pollution and, 501–3
purification of, 491, 511, 512–13, 514–15
reducing disease by disinfecting, 511
Drip irrigation systems, 323, 324f
Drought, 311
  threats of, to amphibians, 148
Dry climate, 311
Dry deposition, 444
Dry steam geothermal energy, 400
Dubos, René, 6
Dunes, primary and secondary, 135f
Durant, Will, 332
Durian fruit, 143
Dust Bowl, United States, 281f

Early loss population, 170, 171f
Early successional plant species, **156,** 157f
Earth
  albedo (reflectivity) of, 467, 468f
  atmosphere (*see* Atmosphere)
  biomes of, 111f (*see also* Biome(s))
  climate and tilted axis of, 105, 106f (*see also* Climate)
  climate change (*see* Climate change)
  evolution and adaptation of life on, 88–91 (*see also* Adaptation; Evolution)
  geological processes of, external, 335–36
  geologic processes of, internal, 332–33, 334f, 335f
  gross primary productivity of, 70, 71f
  just right conditions for life on, 87
  life-support systems (*see* Life-support systems, Earth's)
  natural greenhouse effect on, 61, 79, 109, 110f, 434f, 438, 464–65
  as ocean planet, 130f
  origins of life on, 88
  poles of, ozone depletion and, 485–86
  projected status of biodiversity on, 196f
  structure of, 60f, 332, 333f
  summary of chemical and biological evolution on, 87, 88f, 89f
  temperature of, 463f, 471f, 472–74
  topography of, and local climate, 109–10
  urban areas at night on, 565f
  as water planet, 306
  wind patterns and rotation of, 107f
Earth Charter, 640–41
Earthquakes, **336**–37
  expected damage from, in Canada and U.S., 337f
  major features and effects of, 337f
Earth-sheltered house, 461
Earth tubes, 393
Easter Island, environmental degradation and population diebacks on, 32, 166
Ebola virus, 422
Echo-boom generation, United States, 186
Ecocities, 580–81
  Chattanooga, Tennessee, 581, 608
  Curitiba, Brazil, as, 563, 608
Eco-economy. *See* Environmentally-sustainable economy
Ecofeminist environmental worldview, 635–36
Ecoindustrial revolution, 536–38
  reducing waste by copying nature, 536, 537f
Eco-labeling programs, 586, 597–98f, 601
Eco-leasing, 538
Ecological diversity, **67**
Ecological economists, 583, 586, 587f
Ecological efficiency, **68**
Ecological extinction of species, 225

Ecological footprint, 10–11
  estimating your personal, 11
  per capita, 10
  of three countries, 10f
  of urban areas, 568–70
Ecological literacy, goals of, 637
Ecological niche(s), **91–93**
  of alligators, 149b
  of bird species in coastal wetlands, 92f
  broad, and narrow, 91–93
  in communities, 144–45
  evolutionary divergence and specialized, 93f
  fundamental, and realized, 91
  resource partitioning and specialization in, 150f
  of rocky and sandy shores, 133, 134f
  in tropical rain forests, 122f
Ecological restoration, **220–21**
  in Costa Rica, 221
  of streams and rivers, 270b, 329
Ecological services, 6
  provided by croplands, 276f
  provided by forests, 200f, 204
  provided by freshwater systems, 138f, 253
  provided by marine systems, 130f, 252–53
  provided by rivers, 268f
Ecological succession, **156–60**
  disturbances affecting, 159
  ecosystem characteristics at immature and mature stages of, 158t
  factors affecting rate of, 159
  predictability of, 159–60
  primary, 156–57
  secondary, 157–58
  shifting community composition as, 156
Ecology, 56–59
  of communities (see Community(ies))
  connections in nature and, 56, 57f
  defined, **6**, 55, **56**
  Earth's life-support systems and, 59–61
  of ecosystems (see Ecosystem(s))
  living more sustainably by learning from, 174–75
  organisms and, (see Organism(s))
  of populations (see Population ecology)
  reconciliation, 247–49
  research in, 84–86
  species and (see Species)
  wind and, 101–2
Economically-depleted resources, 11
Economic development, **7**
  demographic transition and, 188, 189f
  economic growth versus, 7–8
  environmentally sustainable, 17–18, 583f
  reduced birth rates linked to, 188–89
  trade-offs (advantages/disadvantages) of, 8f
  unsustainable, versus environmentally-sustainable, 588f
  World Bank and global, 598–99
Economic growth, 7
  comparison of unsustainable, and environmentally sustainable, 588f
  economic development versus, 7–8
  reducing poverty by fostering, 599–600
Economic incentives and disincentives
  government subsidies and tax breaks, 593–94
  green taxes and effluent fees as, 594–95
Economic projections, age structure of human population and, 184–86
Economic resources, 585–86

Economics, **583–604**
  costs and pricing issues related to, 592–93
  growth and development (see Economic development; Economic growth)
  improving environmental quality using methods of, 593–98
  market forces (see Market economic systems and forces)
  of mineral resource supplies, 346
  monitoring progress of environment in context of, 590–92
  neoclassical, ecological, and environmental schools of, 583, 586, 587f
  pollution control and resource management from perspective of, 586–90
  of recycling, 544–45
  reducing air pollution through methods of, 454–55
  reducing poverty, improving environment and quality of life using methods of, 598–600
  resources and systems related to, 584–86
  shift to service flow, 537–38
  transition to environmentally sustainable, 586, 600–602 (see also Environmentally-sustainable economy)
Economic security, 622–23
Economic services
  provided by croplands, 276f
  provided by forests, 200f, 204
  provided by freshwater systems, 138f, 253
  provided by marine systems, 130f, 252–53
Economic system(s), **584**
  developing sustainable, 627–28
  government intervention into market, 585–86
  pure free-market, 584–85
Economic threshold of pesticide use, 527
Economists
  ecological, 583, 586, 587f
  environmental, 583, 586
  neoclassical, 583, 586
Economy(ies). See also Economics
  capital as driving force in, 584
  high-throughput (high-waste), 52, 53f, 535
  low-throughput (low-waste), **53f**–54
  matter-recycling-and-reuse, 52–53
  service-flow, 537–38
  transition to environmentally sustainable, 586–87, 600–2 (see also Environmentally-sustainable economy)
Ecosystem(s), 55–86. See also Community(ies)
  abiotic and biotic components of, 61, 63f
  biodiversity in, 66–67 (see also Biodiversity)
  biological components of, 64–66
  biomes and aquatic life zones, 61, 62f (see also Aquatic life zones; Biome(s))
  defined, **59**
  Earth's life-support systems and, 59–61
  ecological studies of, 84–86
  ecological succession at different stages of, 158t
  ecology concepts applied to, 56–59
  energy flow in, 67–69
  factors limiting population growth in, 62–64
  feeding (trophic) levels in, 67, 68f
  food chains and food webs in, 67, 68f, 69f
  global warming effects on, 476
  industrial, 536, 537f
  matter cycling in, 60f, 67f, 76–84

  primary productivity of, 70–72
  priorities for protecting, E.O. Wilson's, 221
  role of insects in, 55–56
  soils in, 72–76
  structure of, as natural capital, 67f
  tolerance limits in, 61–62
Ecosystem approach to preserving biodiversity, 197f
  conservation biology and, 197
  ecological restoration, 220–21
  forest resources, 199–210
  human impacts on biodiversity and, 195–96
  national parks, 214–15
  nature reserves, 215–20
  public lands in U.S. and, 197–99
  reintroduced species, 194–95
  tropical deforestation and, 210–14
Eco-tourism, 230, 240
Eddington, Arthur S., 54
Education, key goals and questions of environmental, 637
Edward, Michael, 420
Effluent fees, 594–95
Egg pulling, 246
Egypt, water resources in, 305
Ehrenfeld, David, 634
Ehrlich, Paul, 27
Eiler, John, 403
Einstein, Albert, 561
Eisley, Loren, 128
Election finance reform and environmental policy, 610–11
Electrical grid, vulnerability of, 406
Electricity, generation of
  from biomass, 397–400
  heat stored in water (geothermal) and, 400–401
  hydropower plants and, 315, 318, 395, 396f
  solar cells and, 394–95
  solar generation of high-temperature, 393–94
  wind and, 396–97
Electric motors, 383
Electromagnetic radiation, **44–45**
Electromagnetic spectrum, 44f
Electron(s)(e), **39**
Electron givers, 40
Electronic commute, 573
Electronic waste (e-waste), 534
Electron receivers, 40
Elements, **39**
  periodic table of, A8f
Elephants, reduction in ranges/habitat for, 232f
El Niño-Southern Oscillation (ENSO), 108f, 109f
Emerson, Ralph Waldo, 519
Eminent domain, 242
Emissions cap, 455
Emissions trading policy, 454–55, 480–81
Endangered species, 149b, **225**
  aquatic, 255, 256, 257–63
  blue whale, 260–62
  examples of, 226–27f
  private landowners and protection of, 242–43
  protecting with international treaties, 240–41
  reintroduced gray wolf, 194
  species prone to becoming, 228f
  threat of habitat loss and degradation to, 231–34

threat of nonnative species to, 234–38

threat of poaching and hunting to, 238–39

threat of predator control, markets, climate change, and pollution to, 240

Endangered Species Act (ESA), U.S., 194, 199, 241, 259, 618

accomplishments of, 244b

arguments for strengthening, 244–45

arguments for weakening, 243–44

Endocrine system, chemical hazards to, 416–17

Energy, 11, 44–46. *See also* Energy resources

consumption of, for types of beverage containers, 539f

defined, **44**

electromagnetic radiation as, 44f–45

environmental problems connected to matter and, 52–54

flow (*see* Energy flow)

global consumption of, 352f

kinetic, and potential, 44

laws governing changes in, 51–52

net, 354–55

one-way flow of high-quality, and life on Earth, 60f

quality and usefulness of, 46

solar, 351

work, transfer of heat, and, 44, 45

Energy conversion devices, energy efficiency of, 381f

Energy efficiency, **51, 380**

of common devices, 381f

in industry, 382–83

in homes and office buildings, 379, 386–88

net, 382

reducing energy waste, 388–91

as strategy, 407f

in transportation, 383–86

ways to improve, 382–91

Energy flow

annual, in aquatic ecosystem, 70f

in ecosystems, 67–69

pyramid of, 67–68, 69f

to and from sun, 61f

Energy policies, U.S. priorities in, 407f

Energy prices, strategy of artificially-high, or artificially-low, 406

Energy productivity, **51**. *See also* Energy efficiency

Energy quality, **46**

Energy resources, 11, 340, 350–78

alternative 353–54

biomass as renewable, 397–400

coal as nonrenewable, 364–66

commercial (*see* Commercial energy)

consumption of, globally, 352f

consumption of by fuel, in U.S., 353f

economics, politics, and, 405–6

efficiency as (*see* Energy efficiency)

evaluating, 351–55

geothermal as renewable, 400–401f

net energy and, 354–55

hydrogen as, 401–4

hydropower as renewable, 395, 396f

micropower as renewable, 404–5

natural gas as nonrenewable, 362–63

nuclear accidents, 350, 369, 376

nuclear energy as nonrenewable, 366–77

oil (fossil fuel) as nonrenewable, 351, 355–62

policy priorities for, in U.S., 407f

responding to global warming by conserving, 478, 479f

solar as renewable, 351, 391–95

strategies for developing sustainable, 405–8

water cycle as renewable, 395

wind as renewable, 396–97

Environment

defined, **6**

degradation of (*see* Environmental degradation; Environmental problems)

effects of wind on global, 101–2, 106f, 107f

history of (*see* Environmental history)

human impact on (*see* Human impact on environment)

Environmental audits, 620–21

Environmental awareness, recent shifts in, 606–7

Environmental careers, 610b

Environmental citizens, becoming, 639

Environmental degradation, **9.** *See also* Environmental problems

T. Blair on, 609

on Easter Island, 32

of natural capital, 10f (*see also* Natural capital degradation)

Environmental economists, 583, 586

Environmental education, key goals and questions of, 637

Environmental era of environmental history in United States (years 1960-2004), 26–30

Environmental ethics, **16**, 592, **631**

four guidelines of, 640

global warming and, 483

levels of ethical concern, 634f

A. Leopold's land ethic, 30, 199

philosophical principles and, 631–32

sustainable living and, 638f

Environmental groups, 618–22

achievements and setbacks of, 621–22

global, 623–25

goals of opponents of, 622

grassroot, 560, 608, 619–20

mainstream, 618–19

student, 620–21

Environmental history, 20–31, A2–A7

agricultural revolution and, 21–22, 23f

case study: Aldo Leopold's land ethic, 30

conservation era in United States (years 1832-1960), 24–26

environmental era in United States (years 1960-2004), 26–30

events in U.S., years 1870-1930, A2f

events in U.S., years 1930-1960, A3f

events in U.S., years 1960s, A4f

events in U.S., years 1970s, A5f

events in U.S., years 1980s, A6f

events in U.S., years 1990s, A7f

human cultural changes and, 21

hunting and gathering societies and, 21

industrial-medical revolution, 22, 23f

information and globalization revolution, 23, 24f

near-extinction of American bison, 20–21

tribal and frontier eras in United States, 24

Environmental Impact Statement (EIS), 618

Environmental indicators, 234

Environmentalism, **6**

Environmental justice, environmental policy and, 608

Environmental law, 616–18. *See also* Environmental law, United States; International agreements: Treaties

evolution of, 616–17

improving environmental quality and reducing waste with, 595–96

improving regulations and, 622b

Environmental law, United States, 616–18, A2–A7

air pollution, 437b, 448, 453–54, 456, 557, 618

antiquities, 25

endangered species, 28, 194, 241, 259

Environmental Protection Agency, 28

factors hindering lawsuits related to, 617–18

federal agencies concerned with implementing, 612f

forests, 25, 208–9

hazardous and toxic wastes, 28, 532, 548, 557, 558–60, 618

improving, 622b

individuals' and lobbyists' interactions with government regarding, 613f

introduction of bills into House of Representatives, 614f

land management, 28

lobbying and, 615

major historical, 615f

major types of, 618

mining-related, 331, 342

national parks, 26

nuclear energy, 368

pesticides, 525–26

soil conservation and food security, 280, 281

water pollution and water quality, 496, 509, 510, 513, 514, 515, 618

wilderness preservation, 27, 218

Environmental leadership, 608–10

failure of United States to provide, 606, 609, 623–24

Environmental literacy, main components of, 636–37

Environmentally-sustainable economic development, **17–18f**, 583f. *See also* Economic development

Environmentally-sustainable economy, 600–2

copying nature to make transition to, 600

cost-benefit analysis and, 589–90

determining harmful costs in, 592

developing new indicators of environmental quality and human quality of life, 590–91

eco-labeling in, 597, 598f, 601

environmental laws and regulations in, 595–96

environmental permits in, 597

estimating future value of resources in, 591–92

full-cost pricing in, 592–98, 600

government subsidies and taxes in, 593–95

market prices on pollution control and resource use, 588–89

monetary values of non-trade resources, 591

pollution control and resource management in, 588–89

principles for shifting to, 602f

progress toward, in Germany and Netherlands, 600–2

reducing poverty in, 598–600

strategies for transition to, 586

technology, reduced resource use, and improved market efficiency for, 587–88

trade agreements and, 625–27

Environmentally-sustainable political systems, 627–28
Environmentally-sustainable society, **6**–7
  components of environmental revolution and building, 640
  environmental citizenship and, 639
  guidelines and strategies for developing, 637f
  voluntary simplicity and, 638–39
Environmentally-sustainable trade, 626–27
Environmental management, changes in, 596f
Environmental movement, **26**–27
  grassroot groups in, 560, 608, 619–20
  mainstream groups in, 618–19
Environmental permits, advantages and disadvantages of tradable, 597f
Environmental pessimists, 16
Environmental policy, **607**–15, 622–28
  democratic political systems and making of, 607
  developing more sustainable political and economic systems as, 627–28
  election finance reform and, 610–11
  environmental leadership and, 608–10
  free trade and, 625–26
  global, 622–28
  improving trade agreements and, 626–27
  individual's influence on, 608 (*see also* Individuals, influence and role of)
  international agreements and (*see* International agreements)
  life cycle of, 615, 616f
  organizational change and better, 611
  principles guiding formation of, 607–8
  role of international environmental organizations in, 623–25
  security issues tied to, 622–23
  in United States, 611–16 (*see also* Environmental law, United States)
Environmental problems, 5–19
  author's introduction to, 1
  biodiversity loss, 5, 12f, 203 (*see* Biodiversity loss; Endangered species; Extinction of species)
  connections among, 14–15
  economic issues and, 7–9 (*see also* Economics; Economy(ies))
  exponential growth as factor in, 5–6
  food and (*see* Food production; Food resources)
  globalization and, 8–9
  greatest current and future, 17
  human activities and unintended, 38f, 173b, 251, 266
  human population growth (*see* Human population growth)
  key, and causes of, 12–13f
  matter and energy laws, and, 52–54
  pollution as, 11–12 (*see also* Air pollution; Water pollution)
  population growth and, 7–8
  poverty associated with, 13–14, 238, 239 (*see also* Poverty)
  resource problems and, 12–15
  resources and, 9–11
  stratospheric ozone depletion leading to, 486f
  sustainability of present course and, 16–18
  sustainable living and, 6–7(*see also* Sustainability; Sustainable living)
  of urban areas, 566, 567f, 568
  waste and, 12f (*see also* Waste)

Environmental protection
  establishing policy of, in U.S., 607–15
  nonviolent civil disobedience and, 206b
  precautionary principle and, 161, 229
  U.S. history and eras of, 24–50
  U.S. law and (*see* Environmental law, United States)
Environmental Protection Agency (EPA), U.S., 28, 428, 437, 439, 449, 458, 498, 510, 514, 521, 560
  establishment of, 526
Environmental quality
  affluence and, 14
  improving, by establishing green taxes and fees, 594–95
  improving, by reducing poverty, 598–600
  improving, by removing subsidies and tax breaks, 593–94
  improving, with eco-labeling, 597–98, 601
  improving, with environmental laws and regulations, 595–96
  improving, with marketplace methods, 597
  making transition to environmentally-sustainable economy and improved, 600–2
  new indicators for evaluating, 590
Environmental resistance, **165**
Environmental revolution, components of, 640
Environmental risks to humans, 17f, 428f. *See also* Risk(s)
Environmental science
  critical thinking and study of, 3–4
  defined, **6**
  importance of studying, 1
  "junk," 609
  learning skills for study of, 1–3
Environmental Sustainability Index (ESI), 627–28
Environmental wisdom, 4
Environmental wisdom worldview, **16**–17, 30, 632f, **635**
  planetary management and stewardship views versus, 633f
  sustainable society and, 637–38
Environmental worldview(s), **16**–17, 630–42
  balance-of-nature, 159
  deep-ecology, 635
  defined, **631**
  ecofeminist, 635–36
  environmental wisdom, 16–17, 30, 632f, 633f, 635, 637–38
  frontier, 24
  human-centered, 631, 632–33
  in industrial societies, 631–33
  life-centered, and Earth-centered, 633–36
  planetary management, and stewardship, 16, 632, 633f
  potential results of, 636
  spaceship-Earth, 27
  sustainable living and, 636–41
  wisdom, 16–17, 30, 632f
Epidemiological studies on toxic chemicals, 413
Epidemiological transition, 419
Epiphytes, 121, 156
Equatorial upwellings, 108f
Erosion, 335–36. *See also* Soil erosion
Essential Study Skills for Science Students (Chiras), 4
Estuary(ies), **131**, 132f, 252
  pollution in, 504, 505, 506f
Ethanol as fuel, 399
Ethics. *See* Environmental ethics

Ethiopia, water resources in, 305
Eukaryotic cells, **56**, 58f
Euphotic zone, **129**, 136
European Union, environmental leadership by, 606, 609
Eutrophication, lake, 139–40, **498**–501
Eutrophic lake, **139**f
Evaporation, 77
Even-age forest management, **200**, 201f
Everglades, restoration of Florida, 266, 267f
Evergreen coniferous forests, 123–25
Evergreen trees, 124
Evolution, 87–100
  adaptation and, 88–93
  artificial selection, genetic engineering, and future of, 97–100
  chemical, and biological, 87, 88f
  coevolution, 90–91, 161
  defined, **88**–90
  ecological niches and, 91–93
  macroevolution, 90
  microevolution and, 90
  misconceptions about, 93
  natural selection and, 90
  origins of life and, 88
  speciation, extinction, biodiversity, and, 94–96
  theory of, 90
Exclusive Economic Zones (EEZ), ocean, 262, 264
Executive branch of government, 607
  making of environmental policy and, 612
Existence (intrinsic) value, 591, 632
  of species biodiversity, **196**, 230
Exotic (nonnative) species. See Nonnative species
Experiments, **33**
  determining chemical toxicity using laboratory, 413–15
Exponential growth, **5**
  of human population, 178
  of populations, 166, 167f
External costs, 592
External geologic processes, 335–36
Externalities, negative and positive, **592**
Extinction of species, **94**–95, 225–29. *See also* Endangered species; Threatened (vulnerable) species
  background, 95
  basic and secondary causes of, 231f
  characteristics of vulnerability to extinction, 228f
  estimating rate of, 227–28, 229
  habitat loss, degradation, and fragmentation linked to, 231–34
  human activities and, 96, 224, 228–29
  on islands, 145, 146f
  mass, 95
  mass depletion and, 96
  nonnative species as cause of, 146, 234–38
  passenger pigeon, 224
  poaching and hunting as cause of, 238–39
  potential of, 225, 226–27f, 228f, 256–57
  reconciliation ecology and, 247–49
  research and legal approach to preventing, 240–45
  role of climate change and pollution in, 240
  role of market for exotic pets and plants in, 240
  role of predator control in, 239–40
  sanctuary approach for preventing, 245–47, 262
  threat of, to amphibians, 147–48
  three types of, 225

Extinction spasm, 227
Extrapolation to infinity, trap of, 639
Exxon Valdez oil spill, 507

Facilitation, rate of ecological succession and, 159
Faith in simple and easy answers, trap of, 639
Falkenmark, Malin, 311
Family planning, **189**
  in China and India, 191–92
  empowering women for better, 190–91
  in Iran, 190
  reducing birth and abortion rates by using, 189–90
  in Thailand, 176
Farm Act of 1985, U.S., 280–81
Farming. *See* Agriculture
Farms, protecting endangered species on, 246
Fatalism, trap of, 639
Fecal coliform bacteria in water, 493f
Federal Insecticide, Fungicide, and Rodenticide Act (FIFRA) of 1947, U.S., 525–26
Federal Land Policy and Management Act, U.S., 28
Federal Water Pollution Control Act of 1972, U.S., 509, 618
Feedback loops in systems, **37**
Feeding modes in open-sea organisms, 137
Feeding niches, specialized, 92f
Feeding relationships among producers, consumers, and decomposers, 65f
Feedlots, 275, 295
Fermentation (anaerobic respiration), **66**
Fertility, human, **178**–79
  changes in U.S., 179–80
  factors affecting, 180–83
  reduced, and population declines, 186–87
  replacement level, 179
  total rate of (TFR), 179
Fertilizers, 285–86
  commercial inorganic, 285, 286f
  organic, 285–86
  sewage sludge as, 511, 512f
Filter feeders, 137
Fire in forests, 207–9
  global warming and, 476
  reducing damage from, 208–9
  types of fires, 207, 208f
  using goats to manage, 209b
First-generation pesticides, 519
First law of thermodynamics, **51**
Fish
  aquaculture of, 297
  biodiversity of marine, 253–54f
  effect of human activities on marine, 255–56
  effect of nonnative species on, 251, 256–57
  government-subsidized fleets for harvesting, 300
  harvesting of, 297–98, 299f
  harvest yields, 299f, 300f
  major types of commercially harvested, 297, 298f
  mercury in, 555, 556f
  mutualism and, 155
  salmon, 268, 269f, 270
  threats to biodiversity of, 251, 314, 316
  world catch of, 299f
Fish and Wildlife Service, U.S., 198, 240

Fisheries, **297**
  aquaculture, 300–1
  freshwater, 271
Fisheries, marine, 255–56, 274
  government-controlled access to, 264–65
  managing for biodiversity and sustainability in, 263–64
  market-controlled access to, 265
  solutions for managing, 263f
Fish farming, **300**–1
Fish ranching, **300**–1
Fitness, 93
Flavin, Christopher, 602
Flooding, 327–29
  in Bangladesh, 327–28
  causes of, 327
  reducing risk of, 328–29
  in urban areas, 569
Flood irrigation, 323, 324f
Floodplains, 141, **327**
  as stream zones, 140f, 141
Florida Everglades, restoration of, 266, 267f
Flowing (lotic) bodies of water, 137
Flows (throughputs), system, **36**
Fluidized-bed combustion, 456
Flying foxes (bats), 143
Flyways, migratory birds, 245, 246f
Food(s). *See also* Food production; Food resources
  bushmeat as, 212, 239
  genetically modified, 99, 291–92
  new types of, 293
  pesticides and increased supply of, 521
  prices of, 301–2
  waste of, 294
Food, Drug, and Cosmetic Act, U.S., 618
Food and Agriculture Organization (FAO), 623
Food chains, **67**
  bioaccumulation and biomagnification of toxins in, 411f, 498f
  model of, 68f
Food production, 273–304
  in China, 289–90
  environmental impact of, 289, 290f, 295b
  fish, 297–301
  global warming and, 475f, 476–77
  government agricultural policy and, 301–2
  green revolution, and traditional techniques of, 276–79
  human nutrition and, 287–89
  increasing crop yields, 291–94
  increasing fish and shellfish production, 297–301
  increasing meat production, 294–97
  location of world's principal types of, 275f
  perennial polyculture and, 273
  rates of, 286–87
  soil conservation, effects on, 284–86
  soil erosion and degradation, effects on, 279–84
  sustainable agriculture and, 302–3
  systems and types of, 274–76
  in urban areas, 294
Food Quality Protection Act (FQPA) of 1996, U.S., 525
Food resources
  production of (*see also* Food production)
  protecting, with alternative methods, 527–31

protecting, with chemical pesticides, 519–27
  supply problems as key environmental problem, 12f
Food Security Act (Farm Act) of 1985, U.S., 280–81
Food web, **67**
  in Antarctic, 69f
  industrial ecosystem modeled as, 536, 537f
  in lakes, and effects of water pollution, 497, 498f
  of living organisms in soil, 74f
Foreshocks, 337
Forest(s), **120**–25, 199–205
  building roads into, 200, 201f, 239
  certification of sustainably-grown timber from, 205
  deforestation, environmental impact of, 203
  deforestation of tropical, 210–14
  disease and insect pests in, 206, 207f
  ecological and economic services provided by, 200f, 204
  effects of acid deposition on, 448
  effects of global warming on, 476f
  environmental effects of deforestation, 203
  evergreen coniferous, 123–25
  fires in, 207–9, 476
  global warming and, 475f
  harvesting methods in, 200–1, 202f, 203
  human impact on, 125 (*see also* Deforestation)
  major types of, 199–200
  management methods, 200, 209, 210f
  mangrove swamps, 133
  as natural capital, 200f
  protection of, through nonviolent civil disobedience, 206b
  status of U.S., 205–6
  status of world's, 203–4
  sustainable management of, 204–5, 209, 210
  temperate deciduous, 122–23
  temperate rain forests, 125
  tropical rain forest, 120–22, 210–14
  in United States, 205–10
Forest Reserve Act of 1891, U.S., 25
Forest Service, U.S., 197, 207, 208
Formaldehyde, 450
Formosan termite, 237b
Fossil aquifer, 309
Fossil fuels, 78
  air pollution and, 365, 435–36, 437b, 439
  as commercial energy source, 351–52 (*see also* Coal; Natural gas; Oil)
  food production and, 278
  global warming and burning of, 464, 465f, 466
  government subsidies for 593–94
  raising taxes on, 406
  strategies for reducing dependence on, 403–4, 405–6
  U.S. and Canadian deposits of, 357f
Fossils, **88**
Foundation species, 149–50
Founder effect, 171
Fox, speciation of Arctic, and gray, 94f
Free-access resources, **10**
Free-market school of planetary management, 632–33
Free trade agreements, 625–27

Freshwater, 307–13
  availability of, 307
  dams and reservoirs to increase supplies of, 313–16
  groundwater as source of, 308–9, 501–3
  increasing supplies of, 312
  resources, in United States, 309, 310f, 311
  shortages of, 311–13
  surface water as source of, 307
  uses of, 309, 310f
  water pollution in lakes, 497–501
  water pollution in streams and rivers, 495–97
  withdrawal rate of, 309
Freshwater life zones, 61, 128, **137–42**
  biodiversity in, 253–54
  distribution of global, 128f
  ecological and economic services provided by, 138f, 253
  ecosystem components, 63f
  effect of plant nutrients on, 138–40
  human impact on, 141–42
  inland wetlands, 141
  lakes, 138, 139f, 267–69
  lakes, water pollution in, 497–501
  life zones in, 138, 139f
  protecting, sustaining, and restoring, 267–71
  streams and rivers, 140, 141f, 268–71
  streams and rivers, pollution in, 495–97
Friedman, Milton, 583
Frogs as indicator species, 147f
Front (weather), **102**
Frontier environmental worldview, **24**
Frontier science, 35, **36**
Frost wedging, 336
Frozen zoos, 247
Fuel(s)
  biofuels, 398–400
  carbon dioxide emissions, by select, 361f
  coal as, 364–66
  ethanol and methanol as, 399–400
  gasoline, cost of, 384b, 592
  hydrogen as, 401–4
  natural gas as, 362, 363f
  oil as, 355–62
  uranium oxide, 366–68
Fuel cell(s), 379
  cars powered by, 384, 385f, 386f
  hydrogen and, 386
  research on, in Europe, 609
Fuel cell vehicles, 385–86
Fuel economy, motor vehicle, 383f
Fuelwood, shortage of, 352. See also Biomass
Fulghum, Robert, 628
Full-cost pricing, 592, **593**
  ecological services provided by forests and, 204
  improving environmental quality to shifting to, 593–95
  mineral mining and, 345
Fumigants, 520t
Functional diversity, **67**
Fundamental niche, **91**
Fungi, mycorrhizae, 155f
Fungicides, 519, 520t

Gamma rays, **48**
Gandhi, Mahatma, 206b, 638
Ganges River, pollution in, 497
Gangue, 344
Garbage. See Solid wastes
Garbologists, 534
Gardner, Gary, 581
Gasohol, 399

Gasoline, cost of, 384b, 592
Gender benders (hormone disrupters), 417
Gene(s), **42**
  alleles as alternative forms of, 90
  diversity in (see Genetic diversity)
  transferring, between species, 97–98
Gene banks, 245–46
Gene pool, **90**
General Agreement for Sustainable Trade (GAST), 626–27
General Agreement on Tariffs and Trade (GATT), 625
Generalist species, **91**
General Mining Law of 1872, U.S., 199, 331
Gene splicing, **97**
Genetically modified food, 99, 291–92
Genetically modified organisms (GMOs), **97**, 98f
Genetic damage, 49
Genetic diversity, **58**, 59f, 67
  effects of, on population size, 171
  preserving, 229
Genetic drift, 171
Genetic engineering, **97–98**, 274
  concerns about, 98–99
  increasing crop yields with, 291, 292f, 293
  protecting food plants against pests using, 528f
  transferring genes between species as, 97–98
  water purification and, 513
Genetic information, preserving species', 229
Genetic resistance to pesticides, 522, 523f
Genetic variability, 90. See also Genetic diversity
  response to toxic chemicals and, 411
Genome, **43**
Genuine progress indicator (GPI), 590, 591f
Geographic information systems (GIS), 84f, 576
Geographic isolation, speciation and, **94**
Geology, 332
  earthquakes and volcanic eruptions, 336–38
  Earth structure, 332
  geologic processes, Earth's interior, 332–33, 334f
  geologic processes, Earth's surface, 335–36
  mineral resources, 340–49 (see also Mineral resources; Mining)
  minerals, rocks, and rock cycle, 338–40
  plate tectonics and, 333, 334f, 335f
  soil (see Soil; Soil conservation; Soil erosion)
Geothermal energy, **400–1**
  advantages/disadvantages of, 401f
Geothermal heat pumps, 400
Germany, transition to environmentally-sustainable economy in, 600–1
Giant pandas, 92
Gibbs, Lois Marie, 532
Glacial periods, 462, 463f
Glaciation, lake formation and, 138
Glaciers
  melting, 469
  studying climate change by testing, 462
Gleck, Peter H., 314
Global circulation model, 469, 471f
Global climate change, **466.** See also Climate change
Global cooling, 462, 463f
Global dimming, 474
Global Environment Facility (GEF), 623

Globalization, **8–9**
  affluenza and, 14
  information-globalization revolution, 23, 24f
Global security and environmental policy, 622–23
Global sustainability movement, 620
Global Treaty on Migratory Species (1979), 259
Global warming, 79–80, 109, 110f, 438, **466.** See also Climate change
  computer models of projected, 469–71
  defined, **466**
  deforestation as cause of, 203, 466
  difficulties in responding to threat of, 483–84
  effects of, on trees and forests, 476f
  human activities and, 465–69
  major characteristics of, 467t
  ocean currents and, 472f
  past climate change and, 462, 463f
  possible effects of, 474–77
  preparing for, 482, 483f
  prevention versus cleanup approach to, 479f
  schools of thought on how to respond to, 477–78
  solutions for dealing with, 477–84
  speed of, 471–72
  water quality and, 495
  water resources and, 318
Global Water Policy Project, 509
Gloom-and-doom pessimism, trap of, 639
Goats, forest management using, 209b
Goodall, Jane, 639
Goods and services, 586f
Gorilla, hunting and poaching threat to, 238, 239f
Government(s). See also Politics
  agricultural policies of, 301–2
  control of access to fisheries by, 264–65
  democracies, 607
  election finance reform, 610–11
  environmental law and policy in U.S. and, (see also Environmental law, United States)
  environmentally harmful subsidies given by, 593–94
  environmental problems and, 607
  intervention of, into market economic systems, 585–86
  management of freshwater supplies by, 312–13
  pesticide regulations, in U.S. 525–26
  reduction of global warming threat by, 479–80
  regulations (see Regulations)
  tax policies (see Taxes)
Grain, 274
  conversion of, into animal protein, 297f
  global production of, 287f
Grain crops, most important, 274
Grameen Bank, Bangladesh, 601b
Grasshopper effect, air pollution and, 441
Grasslands, **115–20**
  chaparral, 119–20
  human impact on, 120f
  major types of, 115
  polar, 118–19
  savannas and tropical, 115–17
  wind and soil erosion in, 281
Grassroots environmental groups, 608, 619–20
  achieving low-waste society by actions of, 560

Gravimeter, 340
Gravity, 60
Gray-air smog, 441
Gray water, 326
Great Barrier Reef, Australia, 262, 264
Great Depression of 1930s, 26
Great Lakes
   alien species in, 267–68
   biological magnification of PCBs in, 498f
   case study on pollution and cleanup in, 500f, 501
Greenbelt around urban areas, 579f
Green Belt Movement, Kenya, 214, 620
Green Building Council, 387
Greenhouse effect, **109,** 464–65. *See also* Climate change
   chemical makeup of atmosphere and, 109, 110f, 434f, 438
   increased carbon dioxide levels and, 438
   natural, 61, 79, 110f, 434f
Greenhouse gases, 61, **109**
   carbon dioxide as, 437–39, 464–65f (*see also* Carbon dioxide (CO$_2$))
   climate change caused by (*see* Climate change; Global warming)
   intercepting, 478, 479f
   international efforts to reduce, 481–82
   major, from human activities, 464t, 465, 466
   natural greenhouse effect and, 464–65
   reducing emissions of, 481–82, 483f
   U.S. role in emissions of, 466
Green job market, 610b
Green manure, **286**
Green revolution(s), **276**–79
   expanding, 292–93
   genetic, 291–92
   increased food production from, 276, 277f
Green Seal, 598f
Green taxes (effluent fees), 594–95
Gross domestic product (GDP), **7,** 590, 591f
Gross national income (GNI),
Gross primary productivity (GPP), **70**
   of Earth's regions, 71f
   net primary productivity versus, 71f
Ground fires in forests, 208
Groundwater, **308**–9
   aquifers, 308–9, 319–21
   arsenic concentrations in U.S., 503
   pollution of, 501–3
   preventing depletion of, 320f
   solutions for protecting, 503f
   sources of contamination in, 502f
   system, 308f
   withdrawal of, 309, 318, 319f, 320–21
Gulf of Mexico, oxygen-depleted zone, and pollution in, 505f
Gut inhabitant mutualism, 155

Habitat, **58, 91**
   designation of critical, 241–42
   effects of climate change on wildlife, 240
   loss, degradation, and fragmentation of, as extinction threat, 148, 231–33
   population size and, 170
   threats to marine biodiversity and loss/degradation of, 253–55
Habitat conservation plans (HCPs), 243
Habitat corridors, 217
Habitat islands, 146
Hadley cells, 107f
Half-life of radioisotopes, **48,** 49f
Halons, 485
Hardin, Garrett, 10, 27
Harmful algal blooms (HABs), 505

Hawken, Paul G., 44, 177
   golden rule on eco-economy, 600
   on sustainable political and economic systems, 628
Hazard(s)
   biological, 410, 419–28
   chemical, 410–18
   cultural, 410 (*see also specific hazards, e.g.* Poverty)
   physical, 410 (*see also specific hazards, e.g.* Earthquakes; Flooding)
   risk analysis of, 428–32
Hazardous air pollutants, (HAPs), 453–54
Hazardous chemicals, **416.** See also Chemical hazards; Hazardous wastes
Hazardous wastes, **548**–54
   brownfields, 554
   burning and burying, 552–54
   detoxifying, 550–52
   electronic (e-waste), 534
   Great Lakes basin and, 500f
   individual's role in reducing, 554f
   industrial disasters involving (Bhopal, India), 549–50
   international agreements on, 560–61
   lead, mercury, and dioxins as, 554–58
   at Love Canal, New York, 28, 532
   regulation of, in United States, 548, 558–60
   strategies for handling, 550
   terrorism and chemical plants with, 549
   toxic threat of, 548–49
   water pollution and, 502f
Hazards
   biological, 419–28
   chemical, 410–18
   cultural, 410, 428, 430f
   disease as (*see* Disease)
   major types of, 410
   natural (earthquakes, volcanoes), 336–38
   risk analysis of, 428–32
Health. *See* Human health
Healthy Forests Initiative (2003), 208–9
Heat, **45**
   high temperature, 393–94
   means of transferring, 45f
   ocean storage of, 472f, 473
   as water pollutant, 492t
Heating
   solar, 389f, 391–92
   space, 382f, 388, 389f, 391, 392f
   water, 389–90
Heavy crude oil, 355
Heavy-rail systems, 574
Heliostats, 393
Helsinki Convention, 505
Hemingway, Ernest, 21
Hepatitis B virus (HBV), 422
Herbicides, 519, 520t
Herbivores, 152
Heterotrophs, **65.** *See also* Consumers
Hexachlorobutadiene, 485
High (weather), **103**
High-input (industrialized) agriculture, 274, **275**
   in United States, 276, 279f
High-level radioactive wastes, 367
   options for storage/disposal of, 372–73
   storage of, at nuclear power plants, 371–72
   transport of, in U.S., 374f
   Yucca Mountain, Nevada, storage of, 373–75
High-quality energy, **46f,** 354
   one-way flow of, and life on Earth, 60f

High-quality matter, **43**–44
High seas, 264
High-speed trains, 575f, 576f
High-throughput (high-waste) economy and society, throwaway mentality in, 534
Hill, Julia (Butterfly), 206b
HIPPO (habitat destruction and fragmentation), **231**
Hirschorn, Joel, 432
HIV (human immunodeficiency virus), AIDS and, 187, 422, 423–24
Homes
   air pollution in, 449–51
   chemical hazards in, 549f
   designing for energy efficiency, 379, 386–88
   earth-sheltered, 461
   ecoroofs on, 388
   energy loss from, 388f
   naturally-cooled, 393
   net energy efficiency in heating, 382f, 388, 389f
   pesticide use in and near, 521
   reducing water use in, 325–26
   solar-heated, 389f, 391, 392f
   straw bale, 387f
   superinsulated, 387f
Honeybees, detection of toxic pollutants by, 559b
Hormesis, 415
Hormonally-active agents (HAAs), 417f
Hormone(s)
   chemical disruptors of human, 416–417f
   controlling insects by disrupting, 529f, 530f
Hormone blockers, 417f
Hormone mimics, 417f
Hot dry-rock zones, 400
Hot water geothermal energy, 400
Human(s)
   basic needs of, 639b
   environmental worldviews centered on, 631, 632–33
   evolution of, 99–100
   homes of (*see* Homes)
   impact of (*see* Human impact on environment)
   population (*see* Human population; Human population growth)
   quality of life, 565–66, 590–91
   respiratory system of, 451, 452f
Human body. *See also* Human health
   effects of ionizing radiation on, 49
   temperature regulation in, 37
Human capital (human resources), **584,** 586f
Human culture
   diversity in, 67
   Earth's life-support systems and, 15f
   environmental history and changes in, 21–24
Human-dominated systems, natural systems versus, 172f
Human health
   biological hazards (disease) to, 410, 419–28 (*see also* Disease)
   chemical hazards to, 410–18
   effects of acid deposition on, 445
   effects of air pollution on, 442, 445, 451–53, 458
   effects of global warming on, 475f
   effects of ionizing radiation of, 49
   effects of ozone depletion and ultraviolet radiation on, 487–89

Human health (cont'd)
  effects of water pollution on, 492, 493t, 505
  greatest risks to, 17f, 428f, 429f
  impact of coal burning on, 365
  impact of food production on, 290f
  infectious disease and, 13 (see also Infectious disease)
  nutrition and (see Human nutrition)
  pesticides and, 521, 524
  poisons and (see Poison)
  risk, probability, and hazards to, 410
  risk analysis of hazards to, 428–32
  threat of lead, mercury, and dioxins to, 554–58
  tobacco, smoking, and damage to, 409
  toxicology and assessment of threats to, 410–16
Human impact on environment, 171–75
  air pollution, 12f, 176f (see also Air pollution)
  on aquatic biodiversity, 253–57
  on biodiversity, 195–96
  on bird species, 233–34
  on carbon cycle, 79f, 80
  climate change, greenhouse gases, and 109, 437–39, 465–69
  on deserts, 115f
  on extinction of species, 96, 224, 228–29
  extracting, processing of mineral resources as, 331f, 340–49, 364
  in fisheries, 255–56
  food production and, 289, 290f, 294, 295b, 296
  on freshwater systems, 140, 141–42, 256, 309, 312–29
  on grasslands, 120f
  greenhouse gases and, 464t
  on hydrologic cycle, 77
  on marine/saltwater systems, 133, 135, 253–56
  on mountains, 126f
  on nitrogen cycle, 81
  pesticide use and (see Pesticides)
  on phosphorus cycle, 82–83
  shift in awareness toward concern about, 606
  species extinction and, 228–29
  on sulfur cycle, 84
  sustainability principles and, 174f
  types of natural-system alteration by, 171–74
  unintended consequences of, 38f, 173b, 251, 266
  urbanization as, 564–76
  water pollution and (see Water pollution)
Human nutrition
  food supplies and, 287–89
  global caloric intake, average, 288f
  macronutrients, 287
  malnutrition, 13, 287–88, 302–3
  micronutrients and, 287, 288
  overnutrition, 288–89
  undernutrition, 287
Human population, 176–93
  age structure of, 184–87
  birth and death rates, and size of, 177
  case studies, India and China, 191–92
  demography as study of, 177
  dieback of, 166
  factors affecting size of, 177–83
  global aging of, 187f
  global warming and, 475f
  growth of (see Human population growth)

influencing size of, 187–91
  land-use planning for improved habitat for, 576–80
  projections, 179f, 184–86
  reduced fertility and declines in, 186–87
  rise in death rates and decline in, 187
  in urban areas (see Urban areas; Urbanization)
  variation in sensitivity to toxic chemicals in, 411f
  world's most populous countries, 178f
Human population growth, 7
  case studies on slowing, 176, 190, 191–92
  in developed and developing countries, 8f
  doubling time, 178
  effect of age structure on, 184, 185f, 186f
  exponential growth of, 5f, 177–78
  factors affecting birth and fertility rates, 180–82
  factors affecting death rates, 182–83
  fertility rates and, 178–79
  new vision for reducing global, 192–93
  in United States, and fertility rates, 179–80
  in United States, and role of immigration, 183
Human resources (human capital), 584, 586f
Human rights principle and environmental policy, 608
Humility principle, and environmental policy, 607
Humus, 73
Hunter-gatherer cultures, 21
Hunting
  species extinction caused by, 238–39
  of whales, 259–62
Hurricanes, 104
Hutchinson, G. Evelyn, 56
Hybrid gas-electric internal combustion engine, 384, 385f, 457
Hydrobromoflurocarbon (HBFC), 485
Hydrocarbons, 42
Hydrogen, isotopes of, 40f
Hydrogen as fuel, 379, 401–4
  advantages/disadvantages of, 402f
  immediate energy strategies and potential of, 403–4
  impact of, on Earth's atmosphere, 403
  production of, from green algae, 404b
  replacing oil with, 401–3
  role of fuel cells, 386
  storage of, 402–3
Hydrogen bonds, 306, A9, A10f
Hydrogen chloride, 485
Hydrogen energy system, 403f
Hydrologic (water) cycle, 60f, 76f–77, 128
  availability of freshwater and, 307
  groundwater, surface water, and, 309
  human impact on, 77
  producing electricity from, 395
Hydropower, 6, 395
Hydrosphere, 59, 60f
Hydrothermal ore deposits, 348f

Iceberg, as freshwater source, 322
Ice caps
  global warming and melting of, 466, 467f, 468
  studying climate change by testing, 462, 463
Ice cores, 462, 463f

Identified resources of nonrenewable minerals, 340
Igneous rock, 338
Immigration and emigration
  effects of, on population size, 164
  island species diversity and, 145
  policy on, in United States, 183
  urban growth and, 564
Immune system, chemical hazards to human, 416
Inbreeding, 171
Incandescent light bulbs, 381, 390
Incineration
  of biomass, 397–98
  of hazardous waste, 552–54
  of solid wastes, 545–46
Income, global distribution of, 599f
Index of Sustainable Economic Welfare (ISEP), 590
India
  air pollution in, 442–43
  Chipko movement in, 620
  demographic data for, 191f
  ecological footprint of, 10f
  population regulation in, 191–92
  toxic chemical release in Bhopal, 549–50
  water pollution in, 497
Indicator species, 147–48
  for air pollution, 433
  amphibians (frogs) as, 147f–48
  for water pollution, 494
Individuals, influence and role of
  author's story, 1
  cleaning up rivers, 605
  creating green sustainable corporations, 538b
  on delaying global warming and reducing carbon dioxide, 482, 483f
  on energy use and waste, 408f
  environmental policy and, 608f
  protecting species, 249f
  reducing CO₂ emissions, 483f
  reducing and preventing air pollution, 450f, 459
  reducing exposure to indoor air pollution, 460f
  reducing exposure to UV radiation, 488f
  reducing hazardous waste, 554f
  reducing solid waste, 536f
  reducing water pollution, 516f
  reducing water use and waste, 330f
  reuse of materials, 540f
  in stream restoration, 270b
  sustainable agriculture and food production, and, 303f
  sustaining terrestrial biodiversity, 222f
Individual transfer quotas (ITQs), 265
Individual transportation in urban areas, 571, 572–74
Indoor air pollution, 449–51
  formaldehyde as, 450
  radon gas as, 450–51
  strategies for reducing, 458, 459f
  types and sources of, 449, 450f
Inductive reasoning, 34
Industrialized (high-input) agriculture, 274, 275
  in United States, 276, 279f
Industrial-medical revolution, 22–23
  trade-offs (advantages/disadvantages) of, 23f
Industrial Revolution, air pollution and, 437b

Industrial smog, 437b, **440**
  in China and India, 441, 442–43
  in developing countries, 440–41
  factors influencing development of,
      441–42
  temperature inversions and, 442, 443f
Industrial societies, environmental world-
      views in, 631–33
Industry
  brownfields and abandoned, 220,
      536
  reducing water waste by, 325–26
  release of toxic chemicals by, 549–50
  terrorism and chemical hazards
      from, 549
  waste produced by, 326
  water needs for, 310f
  water pollution caused by, 495
Inertia (persistence), **160**
Infant mortality rate, **182**
Infectious agents. See Pathogens
Infectious disease, 13, 419
  bioterrorism and, 426–28
  deadliest, 421f
  dengue fever, 420f
  as density-dependent population
      control, 167
  HIV and AIDS, 187, 422, 423–24
  influenza, 422
  malaria, 420f
  pathogens causing, 419f
  reducing incidence of, 425, 426f
  tuberculosis, 419, 421–22
  viral, 419, 422–23
  water pollution as cause of, 492, 493t
  yellow fever, 420f
Infiltration, water, **74,** 77
Influenza (flu), 422
Information and globalization revolution,
      23–**24**
  trade-offs (advantages/disadvantages)
      of, 24f
Infrared radiation, 109
Inherent value, **632,** 634
Inhibition, rate of ecological succession and,
      159
Injunctions, legal, 616
Inland wetlands, freshwater, **141**
Innovation-directed environmental man-
      agement era, 596
Inorganic compounds, **43**
Inorganic fertilizers, **285,** 286f
Input pollution control, **12.** See also
      Prevention approach
Inputs, system, **36**
Insect(s)
  alternative methods of controlling pest,
      527–31
  Argentine fire ant, 236f
  birth control for, 528, 529f
  boll weevil, 522f, 523f
  butterflies, 59f
  chemical pesticides for control of,
      519–25
  cockroaches, 92b
  ecological role of, 55–56
  in forests, 206, 207f
  Formosan termite, 237b
  hormone disruption of life cycle of, 529f,
      530f
  as pests (see Pest(s))
  praying mantis, 55f
  sex attractants (pheromones) for, 528,
      529f
  silverleaf whitefly as, 524b

  threats from nonnative, 146–47
  as vectors of disease, 419, 420f
Insecticides, 519, 520t
In situ mining, 347
Instrumental value, **632**
  of biodiversity and species, **196,** 229
Insurance spraying, pesticide use and, 527
Integrated coastal management, 262, 506
Integrated pest management (IPM), **530**–31
Integrative principle and environmental
      policy, 608
Intercropping, **278**
Interglacial periods, 462
Intergovernmental Panel on Climate
      Change (IPCC), 463
  reports of, on global warming, 469–72,
      476
Intermediate disturbance hypothesis, 159
Internal combustion engine
  energy waste by, 381
  hybrid gas-electric vehicles, 384, 385f
Internal (direct) costs, 592
Internal geologic processes, 332–35
International agreements, 623–25
  on biodiversity and endangered species,
      257, 259, 624
  conventions, 623
  multilateral environmental agreements
      (MEAs), 623
  on ocean dumping, 507, 624
  on protecting stratospheric ozone,
      488–89
  on reducing greenhouse gas emissions,
      481–82
  on reducing hazardous and toxic wastes,
      560–61
  on reducing marine pollution, 505
  shift in awareness toward need for more,
      606
  on trade, 625–27
International Convention on Biological
      Diversity (1995), 259
International Environmental organizations,
      623–25
International Whaling Commission (IWC),
      259
Internet
  critical thinking and, 3
  non-governmental environmental groups
      and use of, 619
Interplanting, **278**
Interspecific competition, **150**
Intertidal zones, **133,** 134f
Intrinsic rate of increase (r), population
      growth, **165,** 166f
Intrinsic (existence) value, **632,** 634
  of biodiversity and species, **196,** 230
Introduced species. See also Nonnative
      species
  Formosan termite, 237b
  Nile perch, in Lake Victoria, 251
  reducing threats from, 237–38
  role of accidentally introduced, 235f,
      236–37
  role of deliberately introduced, 146–47,
      234, 235f, 236
Ion(s), **39**
  charges and concentration of, 40, 41f
  compounds and chemical bonding of, 41
Ionic bonds, 41, A8–A9
Ionic compounds, 41
Ionizing radiation, 44f, **45**
Iraq
  oil in, 356
  water resources in, 305

Ireland, 166–67
Irregular behavior in population-size
      change, 167, 168f
Irrigation
  drip, 323, 324f
  global warming and reduced water for,
      476–77
  increasing amount of land under, 293
  major systems of, 323, 324f
  reducing water waste in, 323–25
  soil salinization and waterlogging caused
      by, 282–84
  water withdrawal for, 309, 310f, 314
Irruptive population-size changes, 167, 168f
Island(s)
  barrier, 133, 134, 135
  number of species on, 145–46
Island biogeography, theory of, **145,** 146f,
      233
Islands of biodiversity, mountains as,
      125–26
Isotopes, **40**
Israel, 322
  Red Sea coral reef development in, 263

Jackson, Wes, 273
James Bay Watger Transfer Project, 318f
Janzen, Daniel, 221
Jet streams, 102
Jordan, water resources in, 305
Jordan River basin, 305
J-shaped curve of human population
      growth, 5f
Judicial branch of government, 607
  making of environmental policy and, 613
Junk science, 36

Kant, Immanuel, 631
Kenaf, 210
Kenya
  Green Belt Movement in, 214, 620
  low-tech, sustainable agriculture
      in, 279b
Kerogen, 361
Keystone species, **148**–49
  alligators as, 149b
  bats, 230
  flying foxes (bats) in tropical forests, 143
  gray wolves, 194
  sea otters, 163
Kinetic energy, **44**
Krebs, Charles J., 168
K-selected species, **169,** 170f
Kudzu vine, 234, 236f
Kwashiorkor, 287
Kyoto Protocol, 29, 481–82

Lacandon Maya Indians, Mexico, 212
Lacey Act of 1900, 241
Lake(s), **138**–40. See also Dams and
      reservoirs
  Aral Sea water disaster, 316f–17
  biodiversity in, 251
  effect of plant nutrients on, 138, 139f
  eutrophication of, 139–40, 498–99
  protecting, sustaining, and restoring,
      267–71
  water pollution in, 497–501
  zones of, 138, 139f
Lake Okeechobee, Florida, 266, 267f
Lake Victoria, Africa, threats to biodiversity
      in, 251
Lake Washington, Washington, recovery
      from cultural eutrophication in,
      499–500

Land
    availability of, and urbanization,
        571–72
    classification of, 285
    conservation of private, 216
    increasing amount of cultivated,
        293–94
    increasing amount of irrigated, 293
    species protection and privately-owned,
        242–43
Land degradation, **279**
Land disposal of solid and hazardous
    wastes, 552–54
Land ethic, Aldo Leopold's, 30, 199
Landfills
    sanitary, 547f, 548f
    secure hazardous waste, 553, 554f
Land Institute, Salina, Kansas, 273
Land-use planning, **576**–80
    advantages and disadvantages of zoning
        as, 577
    conventional, 576–77
    in Oregon, 577, 578–79
    smart-growth concept as, 577, 578f
    traditional neighborhood development
        as, 579–80
    urban growth boundaries, 578, 579f
La Niña climate shift, 108–9
Large marine systems, 264
Large-scale hydropower, 395
    advantages and disadvantages of, 396f
Las Vegas, Nevada
    urban sprawl around, 567f
    water consumption in, 326b
Late loss population, 170, 171f
Lateral recharge, 308
Late successional plant species, 157f
Latitude, **113**
    climate, biomes, and effects of, 112f, 113
    species diversity at different, 145f
Law. See Environmental law; Environmental
    law, United States
Law of conservation of energy, **51**
Law of conservation of matter, 46, **47**, 48
Law of progressive simplification, 14
Law of thermodynamics, 51–52
Law of the Sea, United Nations, 262
Law of tolerance, **62**
LD50 (median lethal dose), **413**, 414t
Leachate, landfills and, 547
Leaching, **74**
Lead as human health threat of, 554–55
Leadership, environmental, 608–10
Learning skills
    critical thinking and, 3–4
    improving, 1–3
Legislative branch of government, 607
    creation of environmental law in U.S.,
        614f
    making of environmental policy and,
        611–12, 613f, 614f, 615
Leibniz, Gottfried W., 631
Leopold, Aldo, 30f, 197, 199, 222, 225, 274
Levees to control flooding, 327f, 328
Lichens, 336
    as indicators of air pollution, 433
Life
    levels of organization in, 57f
    origins of, 88
Life-centered worldviews, 633–36
Life-cycle analysis of products, 545f
Life-cycle cost, energy efficiency and, **381**
Life expectancy, **182**
Lifestyle, estimating risks in one's,
    431–32

Life-support systems, Earth's, 59–61
    Biosphere 2 as attempt to create self-
        sustaining, 630, 633–34
    human culture and, 15f
    major components of, 59, 60f
    solar energy, 60–61
    sun, cycling of matter, gravity, and, 60
Lighting, energy efficient, 381, 390
Light pollution in urban areas, 570
Light-rail systems, 574, 575f
Light-water reactors (LWR), nuclear, 366,
    367f
Limiting factor principle, **62**
Limiting factors in ecosystems, **62**–64
    adaptation and, 93
    in aquatic life zones, 129–30
    phosphorus as, in soil, 82
    on population growth, 165
    precipitation and temperature as, 111,
        112f, 120f
    in terrestrial life zones, 110–13f, 115f
Limnetic zones, lakes, 138, 139f
Liquified natural gas (LNG), **362**
Liquified petroleum gas (LPG), **362**
Lithosphere, **59**, 60f, 332, **333**f
Littoral zones, lakes, 138, 139f
Livestock production, 274. See also Range-
    lands
    environmental problems associated with,
        116–17, 212, 294, 295b, 296
    feedlots, 275, 295
    overuse of antibiotics in, 420–21
    rangelands and, 294–97
    water pollution and, 509, 511, 512f
Living machines, ecological wastewater
    treatment and, 491, 512–13
Living sustainably, 6–7. See also Sustainable
    living
Living systems. See also Systems
    characteristics of, versus human-
        dominated systems, 172f
    human impacts on, 171–75 (see also
        Human impact on environment)
    precautionary principle for protecting,
        161, 264
    second law of thermodynamics in, 52f
    stability and sustainability of, 160–61
Lobbying, **615**
Local extinction of species, 225
Logistic growth, **166**f
London Dumping Convention of 1972, 507
Longlining, 298, 299f
Longwall mining, 342f
Los Angeles, California
    air pollution in, 442, 443f
    destruction of electric-rail system in, 576
Love Canal, New York, hazardous waste
    dump in, 28, 532
Lovins, Amory, 44, 379, 380
Low (weather), **103**
Low-energy precision application (LEPA)
    sprinklers, 323
Low-input agriculture
    sustainable, 302–3
    traditional, 275–76, 278–79
Low-level radioactive wastes, 372
Low-quality energy, **46**f
Low-quality matter, 43f, **44**
Low-throughput (low-waste) economy and
    society, **53**f–54
    achieving, 560–61
    reducing waste in, 535
Lungs
    effect of ultrafine particles in air on, 458
    emphysema in human, 452f

Maathai, Wangari, reforestation work of,
    214b
MacArthur, Robert H., 102, 145, 150, 169
McGinn, Anne Platt, 271, 561
MacKibben, Bill, 206
Macroevolution, **90**
Macronutrients, human nutrition and need
    for, 287
Madagascar, 132f
Magnetometer, 340
Magnitude, earthquake, 336
Mainstream environmental groups, 618–19
Malaria, 424–25
    life cycle of, 425f
    spraying pesticides to control insects car-
        rying, 173b
    worldwide distribution of, 420f
Malignant melanoma, 487f, 488
Malnutrition, **287**–88
    reducing, 302–3
    relationship of, to poverty and disease, 13
Malthus, Thomas, 636
Man and Biosphere (MAB) Programme, 217
Mangrove forest swamps, 133
Mantle, Earth's, **332**, 333f, 334f
Mantle plumes, 333
Manufactured (physical) resources, **584**,
    586f
Manure
    animal, 285, 398
    green, 286
Marginal benefit, **585**
Marginal cost, **585**
Marine animals, protection of, 241
Marine biodiversity, 129, 130, 131, 132f, 135,
    136f, 252
    commercial whaling and, 259–62
    difficulties of protecting, 257
    integrated coastal management for pro-
        tection of, 262
    international agreements and marine
        sanctuaries for protection of, 262
    legal agreements and awareness for pro-
        tecting, 257–59
    reconciliation ecology and protection of,
        263
Marine life zones, 61, 130–37, 257–63. See
    also Ocean(s)
    barrier islands, 133, 134f, 135
    coastal zones, 130–35
    coral reefs, 127, 135–36, 252, 263
    dunes, 135f
    ecological and economic services pro-
        vided by, 130f, 252–53
    ecological importance of oceans, 130
    ecological niches on rocky and sandy
        shores, 133, 134f
    estuaries, coastal wetlands, and man-
        grove swamps of, 130–33
    food web in Antarctic, 69f
    human impact on, and degradation of,
        133, 135, 137f
    open sea, biological zones in, 136–37
    salt marsh ecosystem, 132f
    water pollution in, 504–9
Marine Mammal Protection Act (1972), U.S.,
    241, 259
Marine protected areas (MPAs), 262
Marine reserves, 262
Market economic systems and forces. See
    also Economics
    free-market systems, 584–85
    government intervention into, 585–86
    market control of access to fishers, 265
    mineral resources and, 346

pollution control, resource use, and, 588–89
reducing air pollution with, 454–55
reducing/preventing greenhouse gas emissions with, 480–81
reducing water pollution with, 509–10
technology and improvement of market efficiency, 587–88
transparent, and full-cost pricing, 592–93
Market failures, 585
Market price equilibrium point, 584, 585*f*
Marsh
freshwater, 141
saltwater, 132*f*
Marsh, George Perkins, 25
Mass-burn incinerators, 545–46
Mass depletion, **96**
Mass extinctions, **95,** 96
Mass number, **40**
Mass transit, 571
rail systems, 574–76
in United States, 576
Material(s)
effects of acid deposition on, 445–46
as substitutes for mineral resources, 349
Material efficiency, **44**
Materials-recovery facilities (MRFs), 541, 542*f*
Mathematical models, 36
Matter, 38–44
atoms and, 39–40
chemical bonds and, 41, A8–A9
defined, **39**
energy and, 44–46
energy laws and, 51–52
environmental problems connected to energy and, 52–54
four states of, 43
genes, chromosomes, and DNA molecules as, 42–43
inorganic compounds as, 43
ions and, 40, 41*f*
law of conservation of, 46–48
levels of organization in, 57*f*
as nature's building block, 39
nuclear changes in, 48–51
organic compounds (carbon-based) and, 41–42
physical and chemical changes in, 46–47
quality and usefulness of, 43–44
types of, 38–39
Matter cycling in ecosystems, 60*f*, 67*f*, 76–84
biogeochemical (nutrient) cycles and, 76
carbon cycle, 77, 78*f*, 79, 80
hydrologic (water) cycle, 76*f*, 77
nitrogen cycle, 80*f*–81
phosphorus cycle, 81, 82*f*, 83
sulfur cycle, 83*f*–84
Matter quality, 43*f*, 44
Matter-recycling-and-reuse economy, **52–53**
Mature community, 160
Maximum sustained yield (MSY), fisheries, 264
Mead, Margaret, 17–18, 489
Meadows, Donella, 634
Measurement, units of, A1
Meat
bushmeat, 212, 239
conversion of grain into, 297*f*
fish, 297–301
Meat production, 294–97
environmental impact of, 294–96
rangelands and pasture for, 294
sustainable, 296–97
Mechanical weathering, 336

Median lethal dose, **413**
Mediation, legal conflict and, 617
Medical revolution. *See* Industrial-medical revolution
Megacities and megalopolis, 564, 565*f*, 568*f*
Megareserves, 216*f*
Melanoma, 487*f*, 488
Melis, Tasios, 404*b*
Mercury
cycling of, in aquatic environments, 556*f*, 557
health threat of, 558
preventing and controlling inputs of, 557*f*
Mesotrophic lakes, **140**
Metallic mineral resources, 11, 340. *See also* Mineral resources
Metals, 40
Metamorphic rock, **339**
Metapopulations, 171
Metastasis, **416**
Methane (CH_4), 362, 464*t*, 466
concentration of, in troposphere (years 1860–2003), 465*f*
global warming and emissions of, 474
Methane hydrates, 474
as source of natural gas, 362
Methanol economy, 400
Methanol as fuel, 400*f*
Methyl bromide, 485
Methyl chloroform, 485
Methylmercury, 556–57
Metropolitan areas. *See* Urban areas
Mexico
demographic data for, 181*f*
solutions for tropical deforestation in, 212–13
Mexico City, environmental problems of, 320, 571
Microbes, 56–57
Microclimates, topography and production of, 109–10
Microevolution, **90**
Microirrigation systems, 323–24
Micronutrients, human nutrition and, 287, 288
Microorganisms (microbes), ecological role of, 56–57, 230f. *See also* Bacteria; Viruses
Micropower systems, 404, **405**
advantages of, 406*f*
decentralized power system using, 405*f*
Microscale experiments, 621
Microturbines, 388
Middle East
energy resources (oil) in, 356
water resources and conflicts in, 305, 322
Midgley, Thomas, Jr., 484
Midsuccessional plant species, **156,** 157*f*
Milbrath, Lester W., 635
Military security, 622–23
Mill, John Stuart, 631
Mimicry, 153*f*, 154
Mineral, **338**
hardrock, 331
Mineral resources, 11
categories of nonrenewable, 340*f*
environmental effects of using, 343
finding, removing, and processing nonrenewable, 340–42
life cycle of, 343–44, 345*f*
limits on extraction and use of, 344–45
substitutes for, 349
supplies of, 345–49
Minimum-tillage farming, 284

Mining
costs of coal, 588*f*
environmental impact of, 331*f*, 364
of lower-grade ores, 344, 346–47
major methods of, 341–42
in United States on public lands, 331
water pollution caused by, 495
Minneapolis, Minnesota, land-use planning in, 579*f*
Minnesota Mining and Manufacturing Company (3M), 537
Mitigation banking, 265
Mitigation cost, 591
Mixed-use zoning, 577
Model(s), **33**
of complex systems, 36–37
coupled global circulation, 469, 470*f*
Molecular economy, 347
Molecule(s), **39**
in chemical equations, 47–48
of DNA, 42–43
Molina, Mario, 484
Molten rock geothermal energy, 400
Monoculture(s), 293
in grasslands, 118*f*
green revolutions in high-input, 276
tree plantations as, 123, 200
Monomers, 42
Montague, Peter, 511
Montreal Protocol, 488, 489*f*, 623
Mosquito as vector for diseases, 420*f*, 424, 425*f*
Motor scooters as transportation, 574*f*
Motor vehicles, 383–86
advantages and disadvantages of, 572–73
alternatives to, 574–76
average fuel economy in, 383*f*
cost of gasoline for, 384*b*, 592
fuel cell, 384–86
ethanol and methanol as fuel for, 399, 400*f*
hybrid gas-electric, 384, 385*f*, 457
internal and external economic costs of, 592
motor scooters, 574*f*
reducing air pollution from, 456–57, 458*f*
saving energy in use of, 383
in United States, 572–74
in urban areas, 571–76
Mountains, 125–26
ecological importance of, 125
human impact on, 126*f*, 342
rain shadow effect and, 110*f*
Mountaintop removal, **342**
Moveable marine reserves, 262
Muir, John, 25*f*, 126, 218
Müller, Paul, 519
Multilateral environmental agreements (MEAs), 623
Multispecies management concept for fisheries, 264
Municipal solid waste (MSW), **533–34**
burning, 545–48
burying (landfills), 547*f*, 548
recycling, 540–45
Mushroom spores, 286
Muskegs (acidic bogs), 124
Mutagens, 90, **416**
Mutations, **90,** 416
pesticides and, 524
Mutualism, 143, **154–56**
examples of, 155*f*
Mycorrhizae fungi, 155*f*
Myers, Norman, 96, 289, 623

Naess, Arne, 635
Nano solar cells, 394
Nanotechnology, 347–48
Nashua River, Massachusetts, 605
National Acid Precipitation Assessment
    Program (NAPAP), 448
National ambient air quality standards
    (NAAAQS), 453
National Environmental Policy Act (NEPA),
    618
National Forest System, U.S., 197, 198f
    logging in, 210f
    management of, 209–10
National Marine Fisheries Service (NMFS),
    241
National park(s), 214–15
    in United States, 198, 199f, 214–15, 266
National Park Service (NPS), U.S., 198
National Park Service Act of 1916, 26
National Park System, U.S., 26, 198f–99
National Priorities List (NPL), U.S., 559
National Resource Lands, U.S., 197, 198f
National security, expanding concept of,
    622–23
National Wild and Scenic Rivers Act (1968),
    271
National Wilderness Preservation System,
    U.S., 27, 199
National Wildlife Refuges, U.S., 198f, 245
Native species, 146
Natural capital, 6, 584, 586f
    assigning monetary value to untraded,
        591
    atmosphere as, 434f, 435f
    biodiversity, 96f
    biodiversity hotspots, 219f
    biogeochemical cycles as, 76–84
    biological evolution as, 89f
    in biomes, 114f, 116f, 117f, 119f, 121f, 123f,
        124f
    in communities, 144f, 150f, 157f, 158f
    croplands as, 276f
    degradation of (see Natural capital
        degradation)
    detrivores as, 66
    Earth's, 7f, 87f
    Earth's climate and ocean currents as,
        106f, 470f, 472f
    Earth's crust and mantle as, 333f, 334f
    Earth's surface and atmosphere as, 59, 60f
    economists' perspectives on, 586, 587f
    ecosystem structure as, 67f
    endangered and threatened species,
        226–27f, 228f
    energy resources, nonrenewable, 351f,
        357f, 364f
    in forests, 200f, 211f
    in freshwater ecosystems, 138f
    future (discount) value of, 591–92
    groundwater system as, 308f
    human activities and degradation of, 38f
        (see also Human impact on environ-
        ment)
    levels of organization of matter as, 57f
    lost, in extinct species, 224, 225f
    marine biodiversity as, 253–54f, 260f
    material resources as, 9f (see also
        Resource(s))
    mineral resources, 339f, 340f
    net primary productivity of ecosystems
        as, 72f
    pharmaceutical plants in tropical forests,
        211f
    precipitation and temperature as biome,
        112f
    in saltwater ecosystems, 130f, 137f, 138f
    soil as, 73f, 74f, 75f
    songbirds, 233f
    species as, 58f
    sustainability and preservation of, 6–7,
        174f
    threats to, 238f, 242f
    use and degradation of, 10f
    water budget as, 307f
    water resources, 138f, 268, 308f, 496f
Natural capital degradation, 12f, 13f
    of agricultural land, 280f, 282f, 283f
    in aquatic ecosystems, 137f, 256f, 258f,
        268f
    of atmosphere, 440f, 444f, 445f, 446f, 447f
    in biogeochemical cycles, 77, 79f–80, 81f,
        82–83, 84
    in deserts, 115f
    food production as cause of, 290f
    in forests, 125f, 201f, 203f
    in grasslands, 120f
    habitat reduction of wild species as, 232f
    human activities and, 38f, 172f, 196f (see
        also Human impact on environment)
    introduced species and, 236f
    projected global biodiversity (years
        1998–2018), 196f
    of water resources, 311f, 316f, 320f, 328f,
        494f, 499f, 500f, 502f, 504f, 505f, 506f,
        512f
Natural gas, 362–63
    advantages and disadvantages
        of, 363f
    forms of, 362
    as fuel, 362–63
    future of, in the United States, 363
    supplies of, 363
Natural greenhouse effect, 61, 79, 109, 110f,
    434f, 438
    Earth's temperature and climate linked
        to, 464
Natural (scientific) law, 34
Natural radioactive decay, 48
Natural recharge, 308
Natural resources, 6, 583, 586f. See also Nat-
    ural capital; Resource(s)
Natural Resources Conservation Service,
    U.S., 281, 285
Natural selection, 90
Natural systems, characteristics of, versus
    human-dominated systems, 172f
Natural world. See Biosphere
Nature Conservancy, work of, 216
Nature reserves, 215–20
    adaptive ecosystem management as,
        217–18
    biosphere reserves, 217
    in Costa Rica, 216
    Nature Conservancy and private lands
        as, 216
    need for greater formation of, 215–16
    size of, 216f, 217
    top priority areas for (biodiversity
        hotspots), 218, 219f
    wilderness as, 218–19
    wilderness protection in the U.S., 219–20
Needs, basic human, 639b
Neem tree, 213–14
Negative (corrective) feedback loop, 37
Negligence lawsuits, 617
Neighborhood development in urban areas,
    579f, 580
Nekton, 129
Nelson, Mark, 513, 584
Neoclassical economists, 583, 586

Nervous system, chemical hazards to
    human, 416
Net energy, 354–55, 382
Net energy efficiency, 382
    for two types of space heating, 382f
Net energy ratios for select energy systems,
    354f
Netherlands
    ecological footprint of, 10f
    progress toward environmentally-
        sustainable economy in, 600–602
Net primary productivity (NPP), 70, 274
    in aquatic life zones, 130
    estimated average, of select ecosystems,
        72f
    gross primary productivity versus, 71f
    human use of biomass production, 72
    limits on population species imposed by,
        70–72
    in terrestrial and aquatic ecosystems, 72f
Neurotoxins, 416
Neutral solution, 40, 41f
Neutrons (n), 39
New urbanism, 578–80
Niche, 91. See also Ecological niche(s)
Nicotine, 409
Nietzsche, Friedrich, 631
Nigeria, key demographic factors in, 186f
Nihilism, 631
Nile perch, 251
Nitrate ions, 80
    as water pollutant, 503
Nitrification, 80–81
Nitrite ions, 80
Nitrogen, 39
Nitrogen cycle, 60f, 80–81
Nitrogen fixation, 80–81
Nitrogen oxide (NO_2) as air pollutant, 438t,
    455
Nitrous oxide
    concentration of, in troposphere (years
        1860–2003), 465f
    as greenhouse gas, 464t
Nixon, Richard, 28
Noise pollution, 569, 570f
Nondegradable pollutants, 48
Nongovernmental organizations (NGOs),
    607, 618–19
Nonionizing radiation, 44f, 45
Nonmetallic mineral resources, 11, 340
Nonmetals, 40
Nonnative species, 146, 234–38
    accidentally introduced, 235f, 236–37
    deer population as example of, in U.S.,
        238
    deliberately introduced, 234, 235f, 236,
        251
    as extinction threat in aquatic systems,
        256–57
    in Great Lakes, 267–68
    killing, in ship ballast, 257b
    reducing threat of, 237–38
Nonpersistent (degradable) pollutants, 48
Nonpoint sources of pollution, 12
    water pollution, 494f, 495, 505–6, 509
Nonrenewable mineral resources, 340–49
    environmental effects of using, 343–45
    finding, removing, and processing
        mineral, 340–42
    supplies of, 345–49
Nonrenewable resources, 9, 11
    energy, 350–78 (see also Energy
        resources)
    minerals, 11, 340–49
    production and exhaustion cycle of, 11f

Nonreplenishable aquifers, 309
Nonthreshold dose-response model, 414f, 415
Nontransmissible disease, **419**
Nonuse value, 591
  of biodiversity, 196
Nonviolent civil disobedience, Julia Butterfly Hill and, 206b
No-problem school of planetary management, 632
Norms, environmental law and social, 616
No-till farming, 284, 479
Nuclear chain reaction, 50f
Nuclear changes in matter, 48–51
  defined, **48**
  fission, 49–50
  fusion, 50–51
Nuclear energy, 366–77
  advantages/disadvantages of conventional, 370–71
  aging nuclear plants, 375
  breeder reactors, 376–77
  brief history of, 368–70
  Chernobyl nuclear disaster, 350, 433
  coal energy versus, 371f
  costs of, 376
  fission reactors for production of, 366, 367f
  fuel cycle, 366–67, 368f
  fusion as alternative form of, 377
  future of, in United States, 377
  government subsidies for, 594–95
  net energy ratio of, 355
  reactors in the United States, 369f
  threat of "dirty" bombs and, 375–76
  vulnerability of, to terrorist attacks, 371–72, 374–76
  waste produced by, 371–75
Nuclear fission, **49,** 50f
Nuclear fission reactors, 366, 367f
  advanced light-water reactors (ALWRs), 376
  breeder, 376–77
  costs and safety of, 376
  decommissioning aging, 367, 375
  energy waste at, 381
  location of, in the United States, 369f
  net energy efficiency of, 382f
  operation of, 366
  radioactive waste storage at, 371–72
  vulnerability of, 371
  wastes produced by, 371–75
Nuclear fuel cycle, 366–67, 368f
  advantages and disadvantages of, 370f
Nuclear fusion, **50–51**
  as alternative to nuclear fission, 377
  solar energy as, 60–61
Nuclear Regulatory Commission (NRC), 370
Nucleic acids, 42, A9–A10. See also DNA (deoxyribonucleic acid)
Nucleotides, general structure of, A10f
Nucleus (atom), **39**
Nucleus (cell), 56, 58f
Nuisance lawsuits, 616
Nutrient cycles (biogeochemical cycles), **76–84**
Nutrition. See Human nutrition

Ocean(s). See also Marine life zones
  biodiversity in, 252, 253–54f
  as carbon-storage and heat reservoir, 472–73
  coastal zones of, 130, 131f
  ecological importance of, 130

effect of currents in, on climate, 106f, 107–8, 472f
life zones of, 131f
mineral extraction from seawater, 348
open sea zone of, 130, 131f, 136–37
pollution in (see Ocean pollution)
protecting biodiversity of, 257–63
sea levels, 468, 477
upwellings in, 108f
Ocean fisheries, 274, 255–56, 263–65, 274
Oceanic crust, 332, 333f
Ocean pollution, 504–9
  Chesapeake Bay as case study, 505, 506f
  coastal areas affected by, 504–5
  dumping of pollutants in, 506–7
  oil as pollutant in, 507–8
  protection of coastal waters, 508–9
  sources of, 504f, 505
  tolerance levels for, 504
Off-shore drilling, 357f
Ogallala Aquifer, depletion of, 319, 320, 321f
Oil, 355–62
  advantages/disadvantages of conventional, 360f
  carbon dioxide emissions from burning, 360, 361f
  cost of (years 1950–2003), 358f
  crude (petroleum), 355–56
  development of, in Arctic National Wildlife Refuge, 359, 360f
  food production and, 278
  growing demand for, 358, 359f
  history of Age of Oil and future of, A12
  off-shore drilling for, 357f
  oil shale and tar sands as sources of, 360–62
  refining crude, 355, 356f
  reserves of, 358
  U.S. supplies of, 356, 357f, 358
  as water pollutant, 507–8
  world supplies of, 356
Oil sand, **360**–61
Oil shale, 361, 362f
Oil spills, 507
  effects of, 507–8
  methods of cleaning, 508
Old-growth forests, **199**
Oligotrophic lake, **138**, 139f
Omnivores, **65**
Open dumps, **547**
Open nuclear fuel cycle, 368f
Open-pit mining, **341**f
Open sea, 130, 131f
  organisms of, 137
  pollution in, 504–8
  zones of, 136–37
Open space, urban, 578–79
Opportunist (r-selected) species, **169,** 170f
Optimal levels of pollution, 588–89
Optimum sustained yield (OSY) of fisheries, 264
Oral rehydration therapy, 426
Ore, 343–44
  biomining, 347
  grade of, 344
  mining lower-grade, 344, 346–47
Ore mineral, 343
Organelles, 56, 58f
Organic chemicals as water pollutants, 492t
Organic compounds, 41–**42**
Organic fertilizers, **285**–86
Organic solar cells, 394

Organism(s), **56,** 57f. See also names of specific organisms, e.g. Animal(s)
  in aquatic life zones, 128–29 (see also Aquatic life zones)
  effects of air pollution on living, 442, 445, 451–53
  effects of global warming on, 476
  eukaryotic, and prokaryotic, 56, 58f
  fossils of, 88
  genetically modified, 97–99
  tolerance limits of, 61–62
Organization of Petroleum Exporting Countries (OPEC), 356
Orr, David W., 634
Other resources, mineral, **340**
Outdoor air pollution, 435–39
  carbon dioxide as, 437–39
  disasters involving, 437b
  effects of, on troposphere's temperature, 473–74
  major types of, 435, 436t, 437, 438t
  photochemical and industrial smog as, 437b, 439–43
  reducing, from coal-burning facilities, 455, 456f
  regional, from acid deposition, 444–49
  strategies for reducing, 459f
Output pollution control, **12.** See also Cleanup approach
Outputs, system, **36**
Overburden, **341**
Overconsumption, 638
Overfishing, 255–56, 263–64, 298, 300
Overgrazing, **295,** 296f
Overhunting, 148
Overnutrition, **288**–89
Overshoot of carrying capacity, 166, 167f
Overturn in lakes, 138
Oxygen, 39
  dissolved, and water quality, 494f
Oxygen cycle, 60f
Oxygen-demanding wastes as water pollutant, 492t, 496f
Oxygen-depleted coastal zones, 505
Oxygen sag curve, 495, 496f
Oysters, water purification by, 506
Ozone (O_3), 39
  as air pollutant, 438t
  as disinfectant, 511
  photochemical, 435f
  stratospheric layer of, 435f (see also Ozone depletion)
Ozone depletion, 484–88
  causes of, 484, 485f
  at Earth's poles, 485–86
  health threats of, 486–88
  hydrogen as fuel and potential of, 403
  major characteristics of, 467t
  reducing, 488–89
  threat of, 484
Ozone hole/ozone thinning, 485

Paine, Robert, 148, 149
Paper
  alternative plants for production of, 210
  recycling of waste, 540, 543
  solid waste, including, 534
Paracelsus, 410
Paralysis by analysis, trap of, 639
Parasites, 419
Parasitism, **154**
  threat of, to amphibians, 148
Parent material (C horizon), soil, 73f, 74
Partial zero-emission vehicles (PZEVs), 457
Passenger pigeon, extinction of, 224

Passive solar heating system, 389f, **391**, 392f
Pastures, **294**
Patented land, 331
Pathogens, 419f. *See also* Disease
　antibiotic resistance in, 419–21
　of trees and forests, 206, 207f
　in water, 492t
Pay-as-you-throw (PAUT) waste collection
　system, 542, 545
Peanut butter, aflatoxins in, 412
PCB (polychlorinated biphenyls), 413
　biological magnification of, 498f
Peer review, 33–34
Per capita ecological footprint, **10**
Per capita GDP (gross domestic product), **7**,
　590, 591f
Percolation (water), 77
Perennial polyculture, 273
Periodic table, A8f
Permafrost, **118**
Perpetual resource, **9**
Persistence, **160**
　of pesticides, 521
　of pollutants, **48**
　of toxins, 411
Persistent organic pollutants (POPs), inter-
　national treaty on control of, 560
Persistent (slowly degradable) pollutants,
　48
Pest(s), 55, **519**
　alternative methods of controlling, 518
　chemical control of (*see* Pesticides)
　in forests, 206, 207f
　natural enemies of, 519
　primary goal of, 527–28
　range of five major, in U.S., 522f
　silverleaf whitefly as, 524b
Pesticides, **519–21**
　alternatives to, 518, 527–31
　bioaccumulation and biomagnification of
　　DDT, 411f
　R. Carson on, 26, 27b, 526
　case against, 522–25
　case for, 521–22
　circle of poison and, 525b
　contemporary use of, 520–21
　experiment on effects of, on bird
　　embryos, 35b
　first-generation, 519
　genetic resistance to, 522, 523f
　ideal, 522
　major types of, 520t
　regulation of, in U.S., 525–27
　second-generation, 519–20
　threat of, to wild species, 240
Pesticide treadmill, 523
Pest-lost insurance, 527
Petrochemicals, **356**
Petroleum, **355**. *See also* Oil
Pets, market for exotic, 240
pH, **40**
　acid deposition and, 444, 445f
　scale, 40, 41f
Pharmaceutical products, 148
　from tropical forests, 211f
Pheromones, controlling insects using, 528,
　529f
Phosphorus cycle, 60f, 81–82f
　human impact on, 82–83
Photochemical oxidants, 439
Photochemical smog, 437b, **439–43**
　in China and India, 442–43
　factors influencing formation of, 441–42
　formation of, 440f
　industrial smog and, 440–41

in Mexico City, 571
　temperature inversions and, 442, 443f
　trees and, 439–40
Photosynthesis, 60, **64**, 78
　CO_2 levels and, 474
Photovoltaic (PV) cells, **394**
Physical change in matter, **46–47**
Physical components of ecosystems, 61–64
Physical hazards, 410. *See also* specific
　physical hazards, e.g. Flooding
Physical (manufactured) resources, **584**, 586f
Physical weathering, 336
Phytodegradation, 551f
Phytoextraction, 551f
Phytoplankton, 64, **129**
Phytoremediation, 551
　advantages and disadvantages, 552f
Phytostabilization, 551f
Pimentel, David, 234
Pinchot, Gifford, 25
Pioneer species, **156**
Plaintiff in lawsuits, **616**, 617
Planetary management worldview, **16, 632**,
　633f
Plankton, **129**, 152
Plant(s). *See also* Tree(s); Vegetation
　desert, 113–14
　early successional, midsuccessional, and
　　late successional, 156, 157f
　effects of acid deposition on, 446–48
　in evergreen coniferous forests, 124f
　first-generation pesticides produced by,
　　519, 520
　genetically modified, 97, 98f
　market for exotic, 240
　most important food, 274 (*see also*
　　Crop(s))
　pharmaceuticals derived from tropical
　　forests, 211f
　roots of, and mycorrhizal fungi, 155f
　in terrestrial communities, 144f
　in tropical rain forests, 121
Plantation agriculture, 123, **275**
Plant nutrients
　effects of, on lakes, 138, 139f, 140
　as water pollutant, 492t
Plasma, **43, 552**
Plasma arc torch, detoxifying hazardous
　wastes with, 552f
Plastics
　production of, 543f
　recycling, 543–44
　reusing items made of, 539
Plate tectonics, **333**, 334f, 335f
Plato, 631
Plume, groundwater pollution, 501
Plutonium, 376
Poaching as threat to wild species, 238–39
Point sources of pollution, **12**
　of water pollution, **494f**, 495, 505–6,
　　509–11
Poison, **413**
　arsenic, 503
　defensive, 153
　dioxins as, 558
　lead as, 554–55
　mercury as, 555–57
　toxicity ratings and lethal doses, 412, 413,
　　414t
Polar grasslands, 118, 119f
Polar ice
　global warming and melting of, 466, 467f,
　　468
　studying climate change by testing, 462,
　　463

Polar vortex, 485
Poles, ozone thinning over Earth's, 467t,
　485–86
Politics, **605–29**
　developing, influencing, and imple-
　　menting environmental policy,
　　607–15
　developing more sustainable political
　　systems, 627–28
　environmental groups and opponents,
　　618–22
　environmental law, 616–18 (*see also* Envi-
　　ronmental law; Environmental law,
　　United States)
　environmental policy, global, 622–28
　environmental problems in democracies,
　　607
　political and environmental challenges of
　　twenty-first century, 606–7
　sustainable energy resources and role of,
　　405–6, 407
Pollutants. *See also* Chemical hazards
　air (*see* Air pollutants)
　categories of, 48
　chemical nature, concentration, and per-
　　sistence of, 48
　shift in concerns about transport
　　of, 606
　water (*see* Water pollutants)
Polluter pays principle, 559–60
　environmental policy and, 608
Pollution, **11–12**
　air, 12f, 176f (*see also* Air pollution)
　causes and threats of, 11–12
　costs of controlling, versus costs of
　　cleaning up, 588f, 589f
　economic solutions for (*see* Economics)
　laws to control (*see* Environmental
　　law)
　noise, 569, 570f
　optimum levels for control of, 588–89
　point, and nonpoint sources of, 12 (*see
　　also* Nonpoint sources of pollution;
　　Point sources of pollution)
　shift in awareness toward concerns
　　about, 606
　solutions to problem of, 12
　threat of, to wild species, 240
　tradable permits for, 597
　water (*see* Water pollution)
Pollution cleanup, **12**. *See also* Cleanup
　approach
Pollution prevention, **12**. *See also* Prevention
　approach
Polyculture, 273, **278**, 293, 303
Polymers, 42
Polyvarietal cultivation, **278**
Population(s), **57f–58**. *See also* Population
　dynamics
　age distribution, 164
　biotic potential of, 165
　dispersion patterns, 164f
　growth (*see* Population growth)
　human (*see* Human population; Human
　　population growth)
　metapopulations, 171
　size (*see* Population size)
　variation in sensitivity to toxic chemicals
　　in, 411f
Population, human. *See* Human population;
　Human population growth
Population change, calculation of, **177**
Population crash (die-back), 166, 167f
Population density, 164
　effect of, on population growth, **167**

Population dynamics, 164–68
  carrying capacity (K) and, 165, 166f
  factors governing population-size changes, 164–65
  limits on population growth, 165
  major characteristics of populations, 164
  population change curves in natures, 167–68
  population density, population growth, and, 164, 167
  population diebacks, 166–67
  population growth, logistic and exponential, 166
  predation and, 168
  reproductive patterns and survival, 168–70
  of southern sea otters, 163
Population ecology, 163–75
  case study of sea otters and, 163
  effects of genetic variations on population size and, 171
  human impacts on natural systems and, 171–75
  population dynamics, carrying capacity and, 164–68
  reproductive patterns, survival, and, 168–70, 171f
Population growth, 7, 8f
  doubling time, 178
  effect of population density on, 167
  exponential and logistic, 166
  human (see Human population growth)
  intrinsic rate of increase (r), 165, 166f
  limits on, 62–64
  reducing hunger and malnutrition by reducing, 302
  resources and competitors as limits on, 165
  in urban areas, 564, 565f
Population size, 164–65. See also Population growth
  carrying capacity and, 165, 166f
  growth and decrease factors in, 165f
  habitat and, 170
  limits on, imposed by net rate of biomass production, 70–72
  role of genetic variation in, 171
  role of predation in, 168
  types of fluctuations in, 167, 168f
  variables in, 164
Portland, Oregon, land-use planning in, 577, 578–79
Positive feedback loop, 37
Possibility versus probability, 410
Postel, Sandra, 323, 509
Potential energy, 44
Poverty, 598–600
  economic development and, 598–99
  global distribution of wealth/income, and, 598, 599f
  illegal smuggling of wild species linked to, 238, 239
  as human health risk, 428, 429f
  microlending as solution for, 601b
  reducing, by helping poor to help themselves, 599–600
  relationship of environmental problems to, 13–14
  relationship of malnutrition, and disease to, 13, 287–88
  shift in awareness toward problem of, 606
  urbanization and, 564, 566, 570–71
  water shortages and, 312
Prairie dogs, 239–40

Prairie potholes, 141
Prairies, 115. See also Grasslands
Precautionary principle, 161, 418, 626f
  chemical hazards and, 418f, 558, 561
  climate change, global warming, and, 478
  in fisheries, 264
  protecting natural systems and, 161
  species extinction and, 229
Precipitation, 62, 77
  acid deposition/acid rain, 81
  as biome limiting factor, 111, 112f, 120f
  climate and seasonal changes in, 105
  effects on outdoor air pollution, 441
  freshwater availability and, 310f
  rain shadow effect, 110f
Preconsumer waste, recycling, 540
Predation, 151. See also Predators; Prey
  defenses against, 153–54
Predator-prey relationship, 151
  snowshoe hare and Canadian lynx, 168f
Predators, 151
  defenses against, 153–54
  as keystone species, 194
  population size and role of, 168
  relationship of, to prey, 151
  sharks as, 151–52
  species extinction threats due to control of, 239–40
  strategies of, 152–53
Prescribed burns, 208
Preservation movement, 25
Prevention approach
  acid deposition and, 448–49
  to air pollution, 454, 456f, 458f, 459f
  chemical hazards, 418
  to coastal water pollution, 508f, 509
  to forest fires, 208
  global warming and, 479f
  groundwater contamination, 503f
  integrated pest management as, 530–31
  reducing threats from nonnative species, 237–38
  waste reduction, 535–36
Prevention principle and environmental policy, 608
Prey, 151
  avoidance or defense against predators among, 153–54
Price (economics), 584
Price-Anderson Act, U.S., 368
Price inelasticity, 585
Primary air pollutants, 435, 436f
Primary (closed-loop) recycling, 540
Primary sewage treatment, 510, 511f
Primary succession, 156, 157f
Primm, Stuart, 96, 229
Principle of enoughness, 638
Private goods, 585
Private ownership, 10
  land conservation through, 216
  of water supplies, 312–13
Probability, risk and, 410
Producers, 64
  feeding relationships among consumers, decomposers, and, 65f
  rate of biomass production by, 70
Product(s)
  eco-labeling of, 586, 597–98f
  life-cycle analysis of, 545f
Product stewardship, 545
Profit-making organizations, 607
Profundal zones, lake, 138, 139f
Prokaryotic cell, 56, 58f. See also Bacteria
Property rights, protection of endangered species and, 242–43

n-Propyl bromide, 486
Proteins, 42
  bushmeat as source of, 239f
  conversion of, grain into animal, 297f
Protons (p), 39
Protozoans, 419
Public lands, U.S., 197, 198f, 199
  controversy over management of, 199
  major types of, 197–99
  mineral resources and mining on, 331, 346
Public participation principle and environmental policy, 608
Public services, 585
Pumped-storage hydropower, 395
Pure free-market economic system, 584–85
Purple loosestrife, 256f
Purse-seine fishing, 298, 299f
Pyramid of energy flow, 67, 68, 69f

Quagga mussel, 268
Quality of life, human
  new indicators for evaluating, 590–91
  in urban areas, 565–66

Radiation
  electromagnetic, 44–45
  heat transfer through, 45f
  infrared, 109
  ionizing, 44f, 45
  nonionizing, 44f, 45
  ultraviolet, 109
Radiation-measuring equipment, 340
Radioactive "dirty" bomb, 375–76
Radioactive isotopes (radioisotopes), 48
  half-life of, 48, 49f
Radioactive materials
  sites contaminated by, 350
  as water pollutant, 492t
Radioactive wastes, 367, 371–75
  high-level, 367, 371–75
  low-level, 372
Radon gas, 450, 451f
Rain Forest Alliance, 205
Rain shadow effect, 110f
Random dispersion of populations, 164f
Rangelands, 274, 294. See also Livestock production
  overgrazing of, 295, 296f
  riparian zones on, 296f
  sustainable management of, 296–97
Range (distribution) of species, 58
Range of tolerance, 61, 62
  for temperature in aquatic ecosystem, 64f
Rapid-rail transportation system, 575f, 576f
Rationalism, 631
Ray, G. Carleton, 252
Ray, Paul H., 610
Reagan, Ronald, 28, 439
Realized niche, 91
Reasoning used in science, 34–35
Reconciliation ecology, 247–49
  applied to urban areas, 579
  implementation of, 248–49
  as new conservation strategy, 247–48
  protecting aquatic marine systems with, 263
  protecting bluebirds with, 248b
Recreational rivers, 271
Recreation services, 229–30
Recycling, 11, 540–45
  benefits of, 541f
  composting as, 541
  economics of, 544
  factors hindering, 544–45

Recycling (cont'd)
　materials-recovery facilities, 541, 542f
　of plastics, 543–44
　separating solid wastes for, 541–42
　two types of, 540
　of wastepaper, 540, 543
Red-cockaded woodpeckers, 91–92
Red tides, 101
Rees, William, 11
Refinery, oil, 355, 356f
Reforestation projects, 213, 214b
Refuse-derived fuel incinerators, 546
Regulations. *See also* Environmental law;
　　Environmental law, United States
　improving environmental law and gov-
　　ernment, 622b
　improving environmental quality and
　　reducing waste with government,
　　595–96
　U.S. federal agencies concerned with
　　environmental, 612f
Reimichen, Thomas, 269
Relativism, 631
Reliable runoff, **307**
Remote sensing, 84
Renewable resources, **9**–10
Replacement-level fertility, **179**
Reproduction
　asexual, 169
　differential, 90
　sexual, 90, 169
Reproductive isolation, speciation
　　and, **94**
Reproductive patterns, 168–70
　asexual, and sexual, 90, 169
　competitor (K-selected), 169, 170f
　opportunist (r-selected), 169, 170f
Reproductive time lag, 166
Research frontier, 130, 252
Reserve(s), mineral, **340**
Reservoirs. *See* Dams and reservoirs
Resilience, **160**
Resistance-to-change environmental man-
　　agement era, 596
Resource(s), 6, 7f, **9**–11. *See also* Natural
　　capital
　assigning monetary value to, 591
　common-property (free-access), 10
　consumption of, and environmental
　　problems, 14
　degradation of free renewable, 9–10
　economics and (*see* Economics)
　future (discount) value of, 591–92
　as limiting factor on population growth,
　　165
　natural, **6**
　nonrenewable (*see* Nonrenewable
　　resources)
　perpetual, and renewable, 9–10
　urban areas and use of, 568, 569f, 570
Resource Conservation and Recovery Act
　　(RCRA), U.S., 548, 558, 618
Resource exchange webs, 536, 537f
Resource partitioning, **150**f
　among birds, 150, 151f
Resource productivity (material efficiency),
　　44
Respiratory system, human, 451, 452f
Response to toxic substances, **412**
Restoration
　of streams, 270b
　of wetlands, 266–67
Reuse, **11,** 538–40
　advantages and disadvantages of, 538–39
　factors hindering, 544–45

reducing throwaway items and encour-
　　aging, 539–40
　of refillable containers, 539
Reverse osmosis, water and, 322
Reversibility principle, and environmental
　　policy, 607
Rhinoceros, reduction in habitat for black,
　　232f
Rhizofiltration, 551f
Rice, 274, 277f, 287f
Richter scale, **336**
Risk(s), **410**
　analysis of (*see* Risk analysis)
　biological hazards as, 410, 419–28
　chemical hazards as, 410–18
　cultural hazards, 410, 428, 430f
　greatest, to humans, 17f, 428, 429f
　physical hazards, 410
　probability and, 410
　of tobacco use, 409
Risk analysis, **428**–32
　becoming better at, 431–32
　comparative, 428f
　estimating risks of technology, 428–29
　international trade agreements
　　and, 626f
　risk management and, 430–31
　risk perception and, 431
　usefulness of, 429–30
Risk assessment, **410,** 428
Risk management, **410,** 428, 430–31
Risk perception, 431
　factors distorting, 431
Rivers and streams, 140–41
　channelization of, 327f, 328
　characteristics and zones of, 140f–41
　cleaning up polluted, 605
　Colorado River, 314–15
　global, 311f
　human impact on, 141–42
　individual efforts to clean up, 605
　pollution in, 145f
　protecting, sustaining, and restoring,
　　267–71
　restoring using trees, 270b
　in U.S., 310f
　water pollution in, 495–97
　wild and scenic, 271
　world systems of, 128f
Roads, impact of building, in forests, 200,
　　201f, 212, 239
Robins, effect of pesticides on embryos of,
　　35b
Rock, **338**
　as nonrenewable resource, 340
　types and recycling of, 338–39
Rock cycle, **339**f
Rocky Mountain Institute, 379
Rocky shores, 133, 134f
Rodenticides, 519
Rodricks, Joseph V., 417–18
Rogers, Will, 14
Room-and-pillar mining, 342f
Roosevelt, Franklin D., 26
Roosevelt, Theodore, 25, 26f, 245
Rosenblatt, Roger, 517
Rosenzweig, Michael, 204, 247, 249, 263
Rowland, Sherwood, 484
R-selected species, 169, 170f
Ruckelshaus, William, 492
Rudd, Robert L., 531
Rule of 70, **178**
Runoff, 77
　reliable, 307, 309
　surface, 77, **140,** 307

Safe Drinking Water Act, U.S., 515, 618
Safe harbor agreements, 243
Sagebrush Rebellion, 28
Salinity, **64,** 128
Salinization, **283**
Salmon
　life cycle, of, 269f
　managing Columbia River basin and
　　needs of, 268–70
　rebuilding populations of, 271f
Salt marsh ecosystem, 132f
Saltwater aquifers, storing CO_2 in,
Saltwater desalination, 321, **322**
Saltwater intrusions, 320f
Saltwater life zones, 128, 130–37. *See also*
　　Marine biodiversity; Marine life zones
Samuelson, Robert, 583
Sanchez, Pedro, 279b
Sanctuary approach to protecting wild
　　species, 245–47, 262
Sandy shores, 133, 134f, 135f
Sanitary landfill, **547**
　advantages/disadvantages of, 548f
　state-of-the-art, 547f
SARS (severe acute respiratory syndrome),
　　422
Sarvodaya Shramadana movement, Sri
　　Lanka, 620
Saudi Arabia, 319, 322, 356, 359
Savanna, 115–17
Scenic rivers, 271
Science, **33**–36
　environmental, 6
　frontier science versus sound, 35–36
　hypothesis, models, and theories of,
　　33–34
　junk, 36
　methods of, 34
　reasoning used in, 34–35
　validity of findings of, 35
Scientific Certification Systems (SCS), 205
Scientific data, **33**
　baseline ecological, 85–86
Scientific hypotheses, **33,** 34
Scientific (natural) law, **34**
Scientific methods, **34**
Scientific theory, **34**
Scientists, opinions of, on climate change
　　and global warming, 469–71
Sea anemone, 155f
Sea lamprey, 268
Sea levels, climate change, global warming,
　　and rising, 468, 475f, 477
Sea otters, population dynamics of, 163
Seasonal wetlands, 141
Sea star (*Piaser orchaceus*), 148, 149–50
Seawater, mineral extraction from, 348
Secondary air pollutants, **436**f
Secondary recycling (downcycling), 540
Secondary sewage treatment, **510,** 511f
Secondary succession, **156,** 157, 158f
Second-generation pesticides, 519–20
Second-growth forests, **200**
Second law of thermodynamics, **51**–52
　in living systems, 52f
Secure hazardous waste landfills, 553, 554f
Security
　environmental policy and, 622–23
　terrorism and, 371–72, 374–76, 406,
　　426–28, 514–15, 549
Sediment as water pollutant, 492t
Sedimentary rock, **338**–39
Seed banks, 245–46
Seed-tree cutting, **201,** 202f
Seismic surveys, 340

Selective breeding, 97
Selective cutting (tree harvesting), **201,** 202*f*
Self, Peter, 564
Semidesert, 113
Sense of place, 638
Septic tank, **510**
Service flow economy, 537–38
Sessions, George, 635
Sewage
    ecological purification of, 491
    reducing water usage in handling, 326
    treatment of, 491, 510, 511–13*f*
    as water pollutant, 509, 510, 511, 512*f*
Sewage sludge, 507
Sexually transmitted disease (STD), 423–24
Sexual reproduction, **170**
    genetic variability and, 90
Shale oil, **361,** 361*f*
Shanty towns, 570
Sharks, 151–52
Shellfish, major types of commercially harvested, 297, 298*f*
Shelterbelts, **285**
Shelterwood cutting, **201,** 202*f*
Shifting cultivation, **22**
Ships, killing invader species in ballast of, 257*b*
Shiva, Vandana, 293
Shop-till-you-drop virus, 14
Shorebirds, specialized feeding niches of, 92*f*
Shore upwelling, 108*f*
Short-grass prairies, 117
Silent Spring (Carson), 26, 27, 526
Silverleaf whitefly as, 524*b*
Simple carbohydrates, 42
Simple living, 638–39
Sinkholes, 320
Skin, human, threat of UV radiation and cancer to, 487*f*, 488
Slash-and-burn cultivation, **22**
Slobodkin, Lawrence B., 161
Slowly degradable (persistent) pollutants, **48**
Small-scale hydropower, 395
Smart-growth concept in urban areas, **577**
    tools for, 578*f*
Smelting, **344**
Smog
    industrial, 437*b*
    origins of term, 437*b*
    photochemical, 439–43
Smoking, health hazards of, 409
Snowshoe hare and Canadian lynx predator-prey relationship, 168*f*
Sodium chloride (NaCl), 339
Soil(s), **72**–76
    effects of acid deposition on, 446–48
    erosion of (*see* Soil erosion)
    food web of living organisms in, 74*f*
    importance of, 72–73
    loss of fertility in, 279
    maintaining fertility of, 285–86
    nutrients in, as limiting factor, 62
    profile (layers) of mature, 73*f*–74, 75*f*
    salinization and waterlogging of, 282–84
    texture, porosity, and permeability of, 75–76
    volcanic, 338
Soil conservation, **284**–86
    conservation tillage and, 284
    inorganic commercial fertilizers and, 285, 286*f*
    methods of, 284, 285*f*

organic fertilizers and, 285–86
    tillage methods for, 284, 285*f*
    in United States, 280–81
Soil erosion, **279**–84, 335–36
    causes of, 279
    desertification and, 281, 282*f*, 283*f*
    Dust Bowl and, 281
    extent of global, 279, 280*f*
    food production and, 281, 290*f*, 295*b*, 296*f*
    reducing (*see* Soil conservation)
    salinization, and waterlogging as, 282–84
    tillage methods for reduction of, 284, 285*f*
    in United States, 280–81
Soil Erosion Act of 1935, U.S., 281
Soil horizons, **73**
Soil moisture detectors, 323
Soil permeability, **76**
Soil porosity, **76**
Soil profile, **73***f*
    of principal soil types, 75*f*
Soil sequestration, 479, 480*f*
Soil texture, **75**
Solar capital, **6,** 7*f*
    as renewable energy resource, 391–95
Solar cells, 379, **394,** 395*f*
Solar cookers, 393
Solar dimming, 474
Solar energy, **6,** 7*f*, 351
    flow of energy to/from Earth and, 60*f*, 61*f*
    generating high-temperature heat and electricity with, 393–94
    heating houses with, 382*f*, 389*f*
    net energy efficiency of passive, 382*f*
    passive and active, 389*f*, 391–92
    solar cells for production of, 394–95
Solar thermal systems, 393
Solid biomass fuels, 398*f*, 399*f*
Solid wastes, **533**–48
    burning and burying, 545–48
    ecoindustrial revolution and reduction of, 536–37*f*
    individual's role in reducing output of, 536*f*
    paper as component of, 534, 543
    priorities for dealing with, 535*f*
    producing less, 535–36
    production of, in U.S., 533–34
    recycling, 540–45
    reuse and reduction of, 538–40
    service flow economy and, 537–38
Somatic damage, 49
Songbirds, threatened species of, 233*f*
Sophists, 631
Sound (consensus) science, **36**
Source zone, streams, 140*f*
Space heating, net energy efficiency in, 382*f*
Spaceship-earth school environmental worldview, 27, 633
Special-interest groups, 607
Specialist species, 91*f*, 92–93
    bird, 92*f*, 93*f*
Specialized ecological niches, 91–93
Speciation, **94**
    allopatric, and sympatric, 94
Speciation crisis, 229
Species, 56–57
    classifying and naming, A11
    competition and predation among, 150–54
    crossbreeding of, 97*f*
    ecological niches of, 91–93
    endangered, 149*b*
    distribution (range) of, 58, 64*f*

diversity of, in communities, 144–46 (*see also* Biodiversity)
    diversity of, at different latitudes, 145*f*
    diversity of, in polluted and non-polluted streams, 145*f*
    effect of pesticides on beneficial, 523
    evolution of, 94
    extinction of (*see* Extinction of species)
    foundation, 149–50
    generalist, 91
    indicator, 147–48
    keystone (*see* Keystone species)
    native, 146
    nonnative (invasive, alien), 146, 267–68
    number of, on islands, 145–46
    parasitism, mutualism, and commensalisms among, 154–56
    pioneer, 156
    preservation of ecosystems and, 221–22
    protection of diversity in (*see* Species approach to protecting biodiversity)
    reasons for preserving wild, 229–30
    reproductive patterns, 169
    specialist, 91–93
    survivorship curves, 170, 171*f*
    threatened, 149*b*
    transfer of genes between (genetic engineering), 97–98
Species approach to protecting biodiversity, 197*f*, 224–50
    biophilia, 231*b*
    Endangered Species Act and, 241–43, 244*b*, 259
    extinction threats from habitat loss, degradation, and fragmentation, 231–34
    extinction threats from nonnative species, 234–38
    extinction threats from poaching and hunting, 238–39
    extinction threats from predator control, markets, climate change, and pollution, 239–40
    extinct species and, 224, 225–29
    importance of wild species and reasons for, 229–30
    individual contribution to, 249*f*
    marine species, 257–65
    reconciliation ecology and, 247–49
    research and legal approach, 240–45
    sanctuary approach, 245–47, 262
Species area relationship, 228
Species diversity, **67.** *See also* Biodiversity; Species approach to protecting biodiversity
    in communities, 144–46
    ecosystem stability and, 160–61
    effects of disturbances on succession and, 159
Species equilibrium model, **145**
Speth, James Gustave (Gus), 623–24
Spiders, biological pest control using, 518
Spinoza, Benedict De, 631
Sri Lanka, 620
Squamous cell carcinoma, 487*f*
Squatter settlements, 570
Stable population-size changes, 167, 168*f*
Standing (lentic) bodies of water, 137
Statues of limitations, 617
Statutory law, **616**
Stegner, Wallace, 218
Stewardship worldview, **16,** 632*f*, 633*f*
Stoddart, Marion, 605
Strategic lawsuits against public participation (SLAPPs), 617, 618

Stratosphere, **59,** 435
  ozone in, beneficial versus harmful, 435*f*
  ozone depletion in, 484–88 (*see also*
    Ozone depletion)
  protecting ozone in, 488–89
Streams. *See* Rivers and streams
Strip cropping, **284,** 285*f*
Strip cutting (tree harvesting), **201,** 202*f,* 203
Strong, Maurice, 351
Student environmental movements, 620–21
Subduction process, 335
Subduction zone, 336
Subsidence, 320, 337
Subsistence agriculture, traditional, 275,
    278–79
Subsoil (B horizon), 73*f,* 74
Subsurface mining, **341,** 342*f*
Succulent plants, 114
Sudan, water resources in, 305
Sulfur cycle, **83***f*
  human impact on, 84
Sulfur dioxide (SO_2) and sulfur dioxide
    (SO_3), 83, 440–41, 454–55
Sun. *See* Solar energy
Superfund Act, U.S., 28, 532, 558–60, 618
Superinsulated house, 387*f*
Supply (economics), 584–85
Surface fire in forests, 207, 208*f*
Surface impoundments of hazardous
    wastes, 553*f*
Surface litter layer (O horizon), soil, 73*f*
Surface mining, **341**
Surface Mining Control and Reclamation
    Act of 1977, 342
Surface runoff, 77, **140, 307**
Surface water, **140,** 307, 509–14
  protecting, in United States, 513–14
  reducing pollution in, from nonpoint
    sources, 509
  reducing pollution in, from point sources,
    509–11
  runoff and, 77, 140, 307
  sewage sludge and pollution of, 511, 512*f*
  sewage treatment and protection of,
    511–13
Surge valves, 323
Survivorship curves, **170,** 171*f*
Suspended particulate matter (SPM) as air
    pollutant, 438*t*
  components of, 441*f*
  methods for reducing, 457*f*
  reducing fine and superfine particles,
    441*f*
Sustainability, 6–7
  of agriculture, 302–3
  ecocity concept and, 563, 580–81
  ecological stability and, 160–61
  in global trade, 626–27
  forest management for increased, 204,
    205*f*
  low-waste society and, 535–36, 560–61
  of present course, 16–18
  principles of, 174*f*
Sustainable (low-input) agriculture, **302–3**
  making transition to, 303
Sustainable cultivation, 22
Sustainable living, 6–7, 636–41. *See also*
    Environmentally sustainable society
  basic terms for, 6
  becoming environmental citizens, 639
  components of environmental revolution
    and, 640
  environmental literacy and, 636–37
  environmental wisdom and, 637–38
  ethical guidelines for, 638*f,* 640–41

future scenario of, 461
  guidelines and strategies for developing,
    637*f*
  meaning of, 6–7
  natural capital and, 6 (*see also* Natural
    capital)
  reducing waste and, 535–36
  simple living and escape from affluenza,
    638–39
  in urban areas, 563, 580–81
Sustainable trade, 626–27
Sustainable yield, **9**
Suzuki, David, 330, 435
Swamps, 141
  mangrove, 133*f*
Sweden, innovation-friendly environmental
    regulations in, 595
Synergistic interaction, **37**
  acid deposition and, 447
  chemical, 412
Synergy, **37–38**
Synfuels, 365–66
Syria, water resources in, 305
Systems, 36–38
  analysis of, 85*f*
  components of, 36*f*
  defined, **36**
  Earth's life support (*see* Life-support sys-
    tems, Earth's)
  environmental surprises and, 38
  feedback loops in, 37
  living (*see* Living systems)
  models of, 36–37
  synergy and, 37–38
  time delays in, 37

Taigas, 123–25
Tailings, 344
Tall-grass prairie, 117
Tar sand, **360**–61
Taxes
  ending environmentally-harmful policies
    on, 593–94
  green, 594–95
  on motor vehicles, 573
  property, 576
  reducing global-warming threats with
    energy and carbon, 479
  shifting from wages and profits to pollu-
    tion and waste, 595*f*
  toxic waste cleanup and, 559–60
Technological optimists, 16
Technology
  environmental impact of, 15
  estimating risks associated with, 428–29
  reducing resource use and improving
    market efficiency with, 587–88
  reducing water pollution using, 510–11
Technology transfer, reducing global-warm-
    ing threat with, 479–80
Tectonic plates, **333,** 334*f,* 335*f*
  boundaries between, 334–35, 336*f*
Temperate deciduous forests, 122, 123*f*
Temperate desert, 113, 114*f*
Temperate grasslands, 117
Temperate rain forests, 125
Temperate shrublands, 119
Temperature, **45**
  as biome limiting factor, 111, 112*f,* 120*f*
  climate and Earth's average, 463*f,* 471*f*
  factors affecting Earth's, 472–74
  projected, global warming and Earth's,
    471*f*
  range of tolerance for, 64*f*
  water and moderation of, 306

Temperature inversions, **442,** 443*f*
Temperature regulation, 37
Teratogens, **416**
Terracing, **284,** 285*f*
Terrestrial biodiversity, 194–223
  conservation biology and bioinfomatics
    for protection of, 197
  ecological restoration and, 220–21
  human impacts on, 195–96
  keystone predators and, 194
  managing and sustaining forests for pro-
    tection of, 199–210
  national parks for protection of, 214–15
  nature reserves for protection of, 215–20
  priorities for protecting, 221–22
  species and ecosystem approach to pre-
    serving biodiversity in, 196, 197*f*
  strategies for sustaining, 222*f*
  tropical deforestation and loss of,
    210–14
  U.S. public lands and protection of,
    197–99
Terrestrial ecosystems. *See also* Biome(s)
  components of, 63*f*
  differences between aquatic life zones
    and, 129–30
  human impacts on biodiversity of,
    195–96
  limiting factors in, 62–64
  plant species in communities, 144*f*
Terrorism
  bioterrorism, 426–28
  electrical grid and threat of, 406
  hazards chemicals and threat of, 549
  nuclear energy and threat of, 371–72,
    374–76
  September 11, 2001 events in U.S., 371
  water purification systems and threat of,
    514–15
Thailand, programs for slowing population
    growth in, 176
Thames River, reducing pollution in, 496
Theory of evolution, **90**
Theory of island biogeography, **145,** 146*f*
Thermal pollution, 309
Thermodynamics, laws of, 51–52
Thomashow, Mitchell, 637
Thoreau, Henry David, 18, 25*f*
Threatened (vulnerable) species, 149*b,* **225**
  aquatic, 255–63
  examples, 226–27*f*
  species prone to becoming, 228*f*
Three Gorges Dam, China, 315*f,* 316
Three Mile Island nuclear accident,
    369, 376
Threshold dose-response model, 414*f,* 415
Throughputs, system, **36**
Thunderheads, 102
Thyroid disrupters, 417
Tiger
  poaching of, 238–39
  reduced habitat of Indian, 232*f*
Tiger salamanders, 91
Timber
  certifying sustainable grown, 205
  harvesting, from U.S. national forests,
    209, 210*f*
  harvesting methods, 200–203
  reducing need for, 210
Time delays, systems and effects of, **37**
Tobacco, health hazards of, 409
Todd, John, 306, 491*f,* 512–13
Tolerance, rate of ecological succession and,
    159
Tool libraries, 540

Toothed whales, 259, 260*f*
Topography and microclimates, 109–10
Topsoil
   A horizon, 73*f*
   erosion of (*see* Soil erosion)
   protection of, 281
Tornadoes, 103*f*
Total fertility rate (TFR), **179**
   decline in, for select countries, 179*f*
   population projections based on, 179*f*
   reducing with family planning programs,
      176
   for United States, years 1917–2004, 179*f*
Toxic chemicals, **416**
   in air pollution, 453–54
   animal testing of, 413–15
   biological magnification of, 498*f*
   detection of, by honeybees, 559*b*
   determining toxicity, 413–15
   dose of, 410, 412, 413*f*, 414*f*
   estimates of toxicity of, 415–16
   factors affecting harm caused by, 411–12
   hazardous chemicals versus, 416
   in homes, 549*f*
   as human health hazard, 410–12, 554–58
   laws to control, 618
   persistent organic pollutants (POPs), 560
   as poison (*see* Poison)
   precautionary principle applied to, 418
   reducing from sewage, 511–12
   release of, by industrial plants, 549–50
   terrorism and release of, 549
   trace levels of, 412
   variations in sensitivity to, 411*f*
   water pollution and, 498*f*, 500*f*, 504*f*
Toxicity, **410**–12
   ratings of, and average lethal dose for
      humans, 414*t*
   studies to determine, 413–15
   validity of estimates on, 415–16
Toxicology, 410–16
   assessment of chemical hazards, 410–12
   assessment of dose hazard, 412, 413, 414
   assessment of trace levels, 412
   case reports and studies on, 413
   estimating toxicity of chemicals, 412–15
   hormesis and, 415
   poisons, 412
   validity of toxicity estimates, 415–16
Toxic Release Inventory (TRI), 453
Toxin, **413**. *See also* Poison
Toxin-laced mining wastes, 343
Toxic Substances Control Act, 618
Toynbee, Arnold, 14
Trace levels of toxic chemicals, 412
Tradable pollution, 597
Trade
   global free, 625–26
   improving agreements related to, 626–27
Trade-offs (advantages/disadvantages), 4
   of agricultural revolution, 23*f*
   of aquaculture, 301*f*
   of bicycles and motor scooters, 574*f*
   of burying and storing hazardous wastes,
      553*f*
   of China's Three Gorges Dam, 315*f*
   of clear-cutting forests, 203*f*
   of coal and synthetic fuels, 365*f*
   of coal versus nuclear energy, 371*f*
   of conservation tillage, 284*f*
   of conventional natural gas, 363*f*
   of conventional nuclear fuel cycle, 370*f*
   of dams and reservoirs, 313*f*
   of economic development, 8*f*
   of ethanol fuel, 399*f*

   of genetically-modified food and crops,
      292*f*
   of geothermal energy, 401*f*
   of global efforts to solve environmental
      problems, 624*f*
   of green taxes and fees, 594*f*
   of groundwater withdrawal, 319*f*
   of heavy oils from oil shale and oil sand,
      362*f*
   of hydrogen as energy source, 402*f*
   of industrial-medical revolution, 23*f*
   of information-globalization revolution,
      24*f*
   of inorganic commercial fertilizers, 286*f*
   of large-scale hydropower, 396*f*
   of life in aquatic environments, 129*f*
   of logging in U.S. national forests, 210*f*
   of methanol fuel, 400*f*
   of oil/gas drilling in Arctic National
      Wildlife Refuge, 360*f*
   of passive versus active solar heating,
      391*f*, 392*f*
   of phytoremediation, 552*f*
   of solar cells, 395*f*
   of solar generation of high-temperature
      heat and electricity, 393*f*
   of solid biomass energy production, 3
      99*f*
   of solid biomass, ethanol and methanol
      fuel, 399*f*, 400*f*
   of solid waste incineration, 546*f*
   of tradable environmental permits, 597*f*
   of wind-generated energy, 397*f*
   of wind power, 397*f*
Trade winds, 108*f*
Traditional intensive agriculture, **275,**
   278–79
Traditional subsistence agriculture, **275,**
   278–79
Tragedy of the commons, 9–**10**
   Easter Island, 32, 166
Transform fault, **335,** 336*f*
Transgenic organisms, **97,** 98*f*
Transition zone
   in biosphere reserves, 217*f*
   in streams, 140*f*, 141
Transmissible diseases, **419**. *See also* Infectious disease
Transpiration, 77
Transportation
   alternative, 574–76
   motor vehicle, 383–86
   saving energy in, 383–86
Trawler fishing, 297, 299*f*
Treaties. *See also* International agreements
   problems and solutions of
      environmental, 624*f*
   protecting endangered species with
      international, 240–41, 257–59, 262
   protecting stratospheric ozone with,
      488–89
   reducing greenhouse gas emissions with,
      481–82
Tree(s). *See also* Forest(s)
   alternatives to, for paper production,
      210
   coniferous evergreen, 124
   deciduous, 122–23
   effects of acid deposition on, 448
   effects of global warming on, 476*f*
   environmental conditions and distribu-
      tion of maple, 64*f*
   fire and, 207–9
   harvesting methods, 200–3
   neem, 213–14

   pathogens and insects affecting,
      206, 207*f*
   photochemical smog and, 439–40
   shelterbelts of, 285
Tree farms, **200,** 201*f*
Tree plantations, 123, **200,** 201
Trench, at converging tectonic plates, 335,
   336*f*
Tribal era of U.S. environmental history, 24
Trickle-down effect, 598
Tromp, Tracey, 403
Trophic levels in ecosystems, **67,** 68*f*
Tropical cyclones, 103, 104*f*
Tropical desert, 113
Tropical forests, 210–14. *See also* Tropical
   rain forests
   causes of deforestation in, 211–12
   neem tree for reforestation of, 213–14
   pharmaceutical plants from, 211*f*
   plant and animal niches in, 122*f*
   rate of deforestation in, and reasons for
      caring about, 210–11
   reducing deforestation and degradation
      of, 212–13*f*
   strategies for sustaining, 213*f*
Tropical grasslands, 115–17
Tropical rain forests, 120–22
   ecosystem of, 121*f*
   flying foxes (bats) as keystone species in,
      143
   slash-and-burn, and shifting cultivation
      techniques in, 22*f*
   specialized plant and animal niches in,
      122*f*
Troposphere, **59,** 434–35
   air pollutants in, 435–36
   average temperature of, 462, 463*f*
   greenhouse effect in, 61, 109, 110*f*, 438,
      465–66
   warming of, 466–77 (*see also* Global
      warming)
   weather and climate influenced by,
      434–35
Tuberculosis, 419, 421–22
   global distribution of, 422*f*
Tucson, Arizona, water consumption in,
   325, 326*b*
Tundra, 118, 119, 359
Turkey, water resources in, 305
Turner, Ray, 489*b*
Turtles, protection of marine, 257, 258*f*
Typhoons, 104

Ultrafine particles in air pollution, 441*f*,
   458
Ultraplankton, **129**
Ultraviolet (UV) radiation, 109, 306
   as disinfectant, 511
   ozone depletion, and human health prob-
      lems caused by, 484, 486–88
   reducing exposure to, 488*f*
   threat of, to amphibians, 148
Undernutrition, **287**
Undiscovered resources, mineral, **340**
Uneven-aged forest management, **200**
Uniform dispersion of populations, 164*f*
Union Carbide Company, 549–50
United Nations, environmental programs
   of, 623
United Nations Conference on the Human
   Environment, 1972, 623
United Nations Development Programme
   (UNDP), 623
United Nations Environment Programme
   (UNEP), 623

United States
  acid deposition in, 445f, 448
  agriculture and food production in, 276–78
  air pollution in, 437, 438t, 449–51
  air pollution disasters in, 437b, 443f
  baby-boom generation in, 179, 184
  baby-bust generation in, 185
  biodiversity hot spots in, 242f
  biomas energy in, 398
  biomes along 39th parallel of, 62f
  birth rates in, years 1910-2004, 180f
  Chesapeake Bay, pollution in, 506, 507f
  coal reserves in, 364
  coastal water and ocean pollution in, 505f, 506f
  commercial energy use in, 352f, 353f, 381f
  ecocity land-use planning in, 577, 578, 579–80, 581
  deaths from tobacco use in, 409f
  deer populations in, 238
  demographic changes in, years 1900–2000, 181f
  demographic data for Mexico, Canada, and, 181f
  Dust Bowl, 281f
  earthquake damage zones in Canada and, 337f
  echo-boom generation in, 186
  ecological footprint of, 10f
  effectiveness of birth control methods in, 182f
  election finance reform in, 610–11
  energy future of (finding fossil-fuel substitutes), 352–53
  energy use, and gross domestic product in, 381f
  environmental law in (see Environmental law, United States)
  environmental policy in, 607–15
  eras of environmental history in, 24–30
  failure of, to provide environmental leadership, 606, 623–24
  fertility and birth rates in, 179–80
  forests and forest management in, 205–10
  fossil fuel deposits in, 357f
  freshwater resources in, 309, 310f, 311
  Great Lakes, pollution in, 500f, 501
  goals of opponents of environmental movement in, 622
  greenhouse gas emissions by, 466
  groundwater contamination in, 502f, 503
  groundwater withdrawal in, 319, 320–21
  immigration and population growth in, 183f
  key demographic indicators in Brazil, Nigeria, and, 186
  land-use planning in, 576–80
  major events in environmental history of, A2–A7
  mass transit in, 576
  motor vehicle use in, 572–74
  natural gas as fuel used in, 363
  nonnative species in, 234, 236f, 237
  nuclear energy in, 366–67
  oil supplies, consumption, and imports, 356, 357f, 358
  pesticide regulation in, 525–26
  pesticide use in, 524–25
  pests in, five major, 522f
  population by age and sex, 185f
  population growth in, 1900–2004, 180f
  public lands in (see Public lands, U.S.)
  real cost of gasoline in, 384b, 592

  reducing stream and river pollution in, 496–97
  risks faced by people living in, 430f
  soil conservation in, 280–81
  soil erosion in, 280–81
  solid wastes produced in, 533f, 534
  strategies for reducing fossil-fuel dependency, 403–4
  strategies for sustainable energy, 405–8
  support for environmental movement in, 621
  total fertility rates for years, 1917–2004, 179f
  urbanization in, 565, 566f
  urban problems of, 565–68
  water withdrawal in, 309f
  wetlands protection in, 265–66
  wind energy, good locations for, 397
U.S. Global Change Research Program (USGCRP), 463
Universalism, 631
Universities and colleges, environmental action at, 620–21
Upwelling, ocean, 108f
Uranium
  fission of, 50f
  isotopes of, 40f
Urban areas, 563–82. See also Urbanization
  air pollution in, 176f, 437, 438f, 441, 442, 443f
  availability of land in, 571–72
  crime in, 570, 572b
  ecocity concept in Curitiba, Brazil, 563
  environmental problems of, 567f
  fostering sustainability of, 580–81
  global, 565f
  greenbelts and growth boundaries around, 578, 579f
  growing food in, 294
  growth of (see Urban growth)
  megacity and megalopolis, 564, 565f, 568f
  microclimates produced by, 110
  neighborhood development in, 579f, 580
  quality of life in, 565–66
  resource and environmental problems in, 568–71
  smart-growth concept applied to, 577–78f
  transportation alternatives in, 572–76
  in United States, 565, 566f
Urban growth, 564
  causes of, 564
  land-use planning supporting, 571–72, 576–77
  patterns of, 564, 565f
  transportation and, 571–76
Urban growth boundary, 578, 579f
Urban heat island, 569
Urbanization, 564–68
  advantages of, 568
  causes of, 564
  degree of, 564
  environmental disadvantages of, 568–70
  global patterns of, 564–65
  land availability and, 571–72
  in Mexico City, 571
  patterns of, 564, 565f
  poverty and, 564, 566, 570–71
  sprawl and, 566–68
  transportation and, 572–76
  in United States, 565, 566f
Urban sprawl, 566–68, 573
  land-use planning to avoid, 577, 578f
User-pays approach, motor vehicles and, 573

Utilitarianism, 631
Utilitarian value, 632

Vaccines, 423
Van Doren, Mark, 1
Van Leeuwenhoek, Antoni, 100
Vector, disease, 419, 420f
Vegetation
  flooding caused by removal of, 327, 328f
  removal of, in urban areas, 568–69
Viravidaiya, Mechai, population-growth reduction in Thailand and role of, 176f
Viruses, 419, 422–23
  Ebola, 422
  hepatitis B (HBV), 422
  HIV, 187, 422
  influenza, 422
  as pathogens, 419f
  reproduction in, 421f
  SARS, 422
  West Nile, 422–23
Volcanoes, 101, 337, 338f
  lake formation and, 138
Voluntary candidate conservation agreements, 243
Voluntary simplicity, 638–39
Vulnerable species. See Threatened (vulnerable) species

Wackernagel, Mathis, 11
Wait-and-see strategy on climate change, 478
Ward, Barbara, 6
Warm front, 102
Warm-rock reservoir deposits, 400
Warning coloration, 153f, 154
Waste, 532–62
  achieving low-waste society, 560–61
  burning and burying, 545–48, 552–54
  case studies: lead, mercury, chlorine, and dioxins as, 554–58
  detoxifying, 550–52
  ecoindustrial revolution as solution to, 536–38
  of energy, 380f
  groundwater pollution caused by degradable, 501–2
  hazardous, 532, 548–58
  hazardous waste regulations in U.S., 558–60
  mining, 343
  oxygen-demanding, as water pollutant, 492t, 496f
  producing less, 535–36
  production of, as key environmental problem, 12f
  radioactive, 367, 371–75
  recycling as solution to, 540–45
  reducing water, 322–26
  of resources, 533–34
  reuse as solution to, 538–40
  solid, 533–48
  taxing, 595f
Waste management, 535
Wastepaper, recycling, 540, 543
Waste-to-energy incinerator, 545, 546f
Wastewater garden system, 513
Wastewater treatment, 510
  ecological, 491
Water, 306. See also Water pollution; Water resources
  drinking, 514–16 (see also Drinking water)
  ecological significance of, 306
  fresh (see Freshwater)
  groundwater, (see also Groundwater)

in hydrosphere, 59, 60f
measuring quality of, 493–94
reducing waste of, 322–26
scarcity of, 311–12
solubility of toxins in, 411
sustainable use of, 329f
unique properties of, 306–7
Water cycle, 60f. *See also* Hydrologic (water)
  cycle
  producing electricity from, 395
Water heaters, energy efficient, 389
Water hot spots in United States, 310f
Water hyacinth, 251
Waterlogging, **283**–84
Water mining, 309
Water pollutants
  arsenic as, 503
  dumping of, 506–7
  major categories of, 492t, 493
  nitrate ions as, 503
  oil as, 507–8
Water pollution, 491–517
  defined, **492**
  drinking water quality and, 493, 495, 503,
    514–16
  food production and, 290f
  of freshwater lakes, 497–501
  in freshwater streams and rivers, 495–97
  in groundwater, 501–3
  individual role in reducing, 516f
  as key environmental problem, 12f
  mineral extraction/processing as cause
    of, 331f
  in oceans and coastal waters, 504–9
  preventing and reducing surface, 509–14
  solutions for reducing, 516f
  species diversity and abundance, and,
    145f
  threat of, to amphibians, 148
  types, effects, and sources of, 492–95
Water Quality Act of 1987, U.S., 509
Water resources, 305–30
  cloud seeding and iceberg towing to
    increase, 322
  conflict over, 305, 310
  dams and reservoirs for greater supplies
    of, 313–16
  desalination for greater supplies of,
    321–22
  excessive amounts of (flooding),
    327–29
  food production and degradation of, 290f
  global warming and, 475f
  groundwater withdrawal and reduced,
    318–21
  impact of food production on, 317,
    323–26
  insufficient supplies of, 311–13
  in Middle East, 305, 322
  polluted (*see* Water pollution)
  reducing waste of, 322–26
  supply, renewal, and use of, 307–11
  sustainable use of, 329f, 330

transferring, from one location to
    another, 316–18
  unique properties of water, 306–7
Water scarcity, global, 311–12
Watershed, **140, 307**
Water stress, 311
Water table, **308**
Water vapor as greenhouse gas, 464
Water withdrawal, 309, 318–22
  rate of, 319–21
  reducing, 312, 330f
Wave(s), 44
Wavelength, 44
Wealth gap, global, 598, 599f
Weather, **102**–4. *See also* Climate
  air masses and changes in, 102
  global warming and possible extremes
    of, 475f
  pressure changes and, 102–3
  tornadoes, tropical cyclones and, 103f,
    104f
  troposphere and formation of,
    434–35
Weather extremes, 103, 109
Weathering, **336**
Wenzel, Richard, 420
West Nile virus, 422–23
Wet deposition, 444
Wetlands
  coastal, 131, 132f, 504–5
  flood control and use of, 329
  inland, 141
  restoring, 266b, 267
  protection of, in United states, 265–66
  solutions for protecting, 265f
Wetland sewage treatment system, 513
Wet steam geothermal energy, 400
Whale(s), 260f, 261f
  commercial hunting of, 259
  near extinction of blue, 260–62
Whale Conservation and Protection Act
  (1976), U.S., 259
Wilderness, 27
  protection of, 218–19
  protection of, in United States, 219–20
Wilderness Act of 1964, U.S., 27, 218, 618
Wilderness recovery areas, 220
Wilderness Society, 218
Wildlands Project, The (TWP), 220
Wildlife
  effects of climate change and pollution
    on, 240
  effects of habitat loss, degradation, and
    fragmentation on, 231–34
  effects of nonnative species on, 234–38
  effects of pesticides on, 523
  effects of predator control on, 239–40
  extinction of (*see* Extinction of species)
  importance of, 229–30
  poaching and hunting of, 238–39
  reasons for protecting, 229–30
  reconciliation ecology and conservation
    of, 247–49

research and legal approach to protect-
    ing, 240–45
  sanctuary approach to protecting, 245–47
  U.S. refuge system for, 245
Wildlife refuges, 245
  Arctic, 359–60
Wild rivers, 271
Willingness to pay, resource value and, 591
Wilson, Edward O., 67, 96, 145, 154, 169,
    195, 225, 229, 249
  on biophilia, 231b
  priorities for protecting biodiversity,
    221–22
Wind
  ecological effects of, 101–2
  effects on outdoor air pollution, 441
  generation of electricity with, 396f, 397f
  global air circulation, climate, and, 106f,
    107f, 108
  soil erosion caused by, 281
  trade winds and El Niño-Southern
    Oscillation, 108f, 109f
Windbreaks, 285f
Wind energy, 6, 396, 397f
Wind turbines, 396
Wise-use movement, 28
Wolf, reintroduced populations of gray, 194
Women
  reducing birth rates by empowering,
    190–91
  typical workday for rural African, 190f
World Bank, 623
  global economic development and role
    of, 598–99
World Conservation Union (IUCN), 227,
    262, 623
World Environment Organization (WEO),
    proposed, 625, 627
World Health Organization (WHO), 291,
    409, 421, 423–24, 425, 426, 437, 453, 475,
    492, 555, 623
World Trade Organization (WTO), 313
  environment, human health, and impact
    of, 625–26
Worldviews. *See* Environmental world-
    view(s)
World Wildlife Fund, 205

Xeriscaping, 325
Xerox Corporation, 538

Yellow fever, 420f
Yellowstone National Park, 194
Yucca Mountain, Nevada, nuclear waste
  storage facilitiy, 373–75

Zebra mussel, 268, 501
Zone of aeration, **308**
Zone of saturation, **308**
Zoning, land-use planning using, **577**
Zooplankton, **129**
Zoos, protecting endangered species in,
  246–47